UNIVERSAL DESIGN HANDBOOK

Wolfgang F. E. Preiser Editor in Chief

University of Cincinnati
Cincinnati, Ohio

Elaine Ostroff Senior Editor

Adaptive Environments Center
Boston, Massachusetts

Foreword by Robert Ivy, FAIA
Editor in Chief, *Architectural Record*

McGRAW-HILL

New York   Chicago   San Francisco   Lisbon
London   Madrid   Mexico City   Milan
New Delhi   San Juan   Seoul
Singapore   Sydney   Toronto

McGraw-Hill

A Division of The McGraw·Hill Companies

Copyright © 2001 by The McGraw-Hill Companies, Inc. All rights reserved. Printed in the United States of America. Except as permitted under the United States Copyright Act of 1976, no part of this publication may be reproduced or distributed in any form or by any means, or stored in a data base or retrieval system, without the prior written permission of the publisher.

1 2 3 4 5 6 7 8 9 0   DOC/DOC   0 7 6 5 4 3 2 1

P/N 0-07-135957-5
Part of
ISBN 0-07-137605-4   REF   720·87 PRE

The sponsoring editor for this book was Wendy Lochner, the editing supervisor was Tom Laughman, and the production supervisor was Pamela Pelton. It was set in Times Roman by North Market Street Graphics.

Printed and bound by R. R. Donnelley & Sons Company.

 This book was printed on recycled, acid-free paper containing a minimum of 50% recycled, de-inked fiber.

McGraw-Hill books are available at special quantity discounts to use as premiums and sales promotions, or for use in corporate training programs. For more information, please write to the Director of Special Sales, Professional Publishing, McGraw-Hill, Two Penn Plaza, New York, NY 10121-2298. Or contact your local bookstore.

The authors of Chapter 4 (c)—Roger Coleman; 33 (c)—C. J. Walsh; 66 (c)—Judy Brewer, including *WCAG 1.0, ATAG 1.0,* and *UAAG 1.0* excerpts (c) World Wide Web Consortium (Massachusetts Institute of Technology, Institut National de Recherche en Informatique et en Automatique, Keio University) all rights reserved, http://www.w3.org/consortium/legal; and 69 (c)—Leslie Kanes Weisman, have licensed the use of the material contained in their chapters to McGraw-Hill for the *Universal Design Handbook.*

The CD-ROM included with this book contains the following material in the public domain, provided by the U.S. Access Board:

• Proposed Rule, Americans with Disabilities Act (ADA) Accessibility Guidelines for Building & Facilities and Architectural Barriers Act (ABA) Accessibility Guidelines
• Accessible Rights-of-Way: A Design Guide
• Detectable Warnings: Synthesis of U.S. and International Practice
• Amendments to the ADAAG

DEDICATION

Ronald L. Mace, FAIA, was an internationally recognized architect, product designer, educator, and mentor. He coined and passionately promoted the concept of universal design, a design philosophy that challenged convention and provided a design foundation for a more usable world. Ron died unexpectedly on June 29, 1998. This handbook is dedicated to him and his visionary dream to make universal design common design practice.

Ron's pioneering work in accessible design, which provided a foundation for universal design, was instrumental in the passage of national legislation prohibiting discrimination against people with disabilities, particularly the Fair Housing Amendments Act of 1988 and The Americans with Disabilities Act of 1990. In 1989, Ron established the federally funded Center for Accessible Housing, later renamed the Center for Universal Design, at the College of Design at North Carolina State University in Raleigh. Ron was also president of Barrier Free Environments, Inc. (BFE), and a principal of BFE Architecture, P.A., in Raleigh, North Carolina. In 1988, he was appointed to the College of Fellows of the American Institute of Architects.

In celebrating Ron's life at a memorial service, there were many intense recollections. One colleague noted, "Ron had an incredible standard of excellence. He attended to design detail in an unequaled way because of his deep understanding and appreciation of how people work and how they use places, spaces and products."

Ron's last public appearance was at the first Designing for the 21st Century International Conference on Universal Design, held in June 1998. Beaming with pleasure, he tooled around from place to place followed by people wanting to talk with him. He was thrilled with the outpouring of energy and commitment, and with people from all over the world talking about how they were trying to, and often succeeding in, creating and/or teaching universal design. He was surrounded by hundreds of people who had chosen to make universal design their life's work. He was so happy.

So it is with love, admiration, and gratitude that we dedicate this *Universal Design Handbook* to Ron Mace. He would have appreciated this book with its representation of the diverse, international effort to create human-centered environments, products, and information. Without Ron's vision and exceptional communication skills, this book and the work it documents would not exist.

CONTENTS

Part 4   Public Policies, Systems, and Issues

Part 5   Residential Environments

Part 6   Universal Design Practices

Part 7   Education and Research

Part 8 Case Studies

Part 9 Information Technology

Part 10 The Future of Universal Design

CONTRIBUTORS

EDITOR IN CHIEF

Wolfgang F. E. Preiser, Ph.D., Dipl.-Ing. *University of Cincinnati, Cincinnati, Ohio* (CHAP. 9)
Wolfgang F. E. Preiser is a professor of architecture at the University of Cincinnati and an international consultant in building performance evaluation. He is published widely, with 13 books on topics ranging from post-occupancy evaluation to facility programming and design research in general. He has lectured worldwide and has won numerous awards and honors, including the Fulbright Fellowship, the Progressive Architecture Annual Award for Applied Research, the Rieveschl Award for Scholarly and Creative Works at the University of Cincinnati, the EDRA Career Achievement Award, and others.

SENIOR EDITOR

Elaine Ostroff, Ed.M. *Adaptive Environments Center, Boston, Massachusetts* (CHAPS. 1 and 43)
Elaine Ostroff is the founding director of the Adaptive Environments Center and an international consultant on universal design education. She directs the Global Universal Design Education Project and the Access to Design Professions Project, lectures on universal design worldwide, and has authored numerous publications on universal design as well as on the Americans with Disabilities Act. She has been honored by the Environmental Design Research Association, Brandeis University, the Council for Exceptional Children, the ADA National Technical Assistance Network, and the Radcliffe Institute.

CONTRIBUTING AUTHORS

Susan Balandin *University of Sydney, Lidcombe, New South Wales, Australia* (CHAP. 38)
Susan Balandin, Ph.D., FSPA, is trained as a speech pathologist and is a senior research fellow at the Centre for Developmental Disability Studies in Sydney, Australia.

Singanapalli Balaram *National Institute of Design, Ahmedabad, India* (CHAP. 5)
Singanapalli Balaram is head of the Design Foundation Studies, National Institute of Design. He is vice president of the Society of Industrial Designers of India and author of related books and articles. He serves on several advisory boards, including Design Issues (U.S.A.), the Governing Council of the Centre for Environmental Planning and Technology, and the Governing Council of the National Institute of Design.

Kim Allen Beasley *Beasley Architectural Group, Alexandria, Virginia* (CHAP. 47)
Kim Allen Beasley was the architectural director for the Paralyzed Veterans of America. He led the ADA consulting effort for the 1996 Olympic Games. He invented an adaptable stadium seat, and he holds numerous awards, including the 1992 Medal of Excellence from *Engineering News Record*. He is coauthor of *Design for Hospitality, Accessible Home Design*, and *Fair Housing Design Guide*.

Richard Best *Rowntree Foundation, York, United Kingdom* (CHAP. 14)
Richard Best has been the director of the Joseph Rowntree Foundation and of its housing arm, the Joseph Rowntree Housing Trust, since 1988. He has acted as adviser to the parliamentary committee covering housing and the environment, and he is currently one of the advisers to the minister for housing and planning.

Judy Brewer *World Wide Web Consortium: MIT, INRIA, Keio* (CHAP. 66)
Judy Brewer directs the Web Accessibility Initiative at the World Wide Web Consortium (W3C). Prior to joining W3C, she worked on initiatives to increase access to mainstream technology for people with disabil-

ities and to improve the dialogue between industry and the disability community. Judy has a background in management, technical writing, education, applied linguistics, and disability advocacy.

Keith T. Bright   *University of Reading, Reading, United Kingdom* (CHAP. 44)
Keith T. Bright is director of Access Design and Management, and director of the Research Group for Inclusive Environments at the University of Reading. He is a member of the new British standards sub-committee preparing the standard, "Access to Buildings for Disabled People," advising the government on access to the built environment.

Olav Rand Bringa   *Norwegian Council on Disability, Oslo, Norway* (CHAP. 29)
Olav Rand Bringa is a member of the Norwegian Society of Chartered Engineers and has been employed in public and private enterprises in the fields of planning, land development, construction, and accessibility design. Bringa has written a number of books and articles, mainly on the subjects of accessibility and universal design.

Margaret Calkins   *IDEAS, Kirtland, Ohio* (CHAP. 22)
Margaret Calkins, president of IDEAS, Inc., a consultation, education, and research firm dedicated to exploring the therapeutic potential of the environment—social and organizational as well as physical—as it relates to frail and impaired adults, is author of *Design for Dementia: Planning Environments for the Elderly and the Confused.*

Robyn Chapman   *Spastic Centre of New South Wales, Sydney, New South Wales, Australia* (CHAP. 38)
Robyn Chapman, B.App.Sc.Phty., is trained as a physiotherapist and is currently the manager of the Rural Outreach Assessment and Development Service at the Spastic Centre of New South Wales in Sydney, Australia.

Jon Christophersen   *Norwegian Building Research Institute, Oslo, Norway* (CHAP. 13)
Jon Christophersen, an architect, is the author of numerous publications on housing quality, accessible housing, special needs housing, and the Norwegian State Housing Bank's requirements for life span dwellings. His recent work includes recommendations for accessibility in the European Union's railways and methods for assessing accessibility in existing settings, including transportation.

Roger Coleman   *Royal College of Art, London, United Kingdom* (CHAP. 4)
Roger Coleman is a senior research fellow at the Royal College of Art (RCA) and, since 1999, director of the RCA's Helen Hamlyn Research Centre. He established both the DesignAge program and the European Design for Ageing Network. He is director of the R&D company London Innovation, is a jury member for the RSA Student Design Awards, and has authored numerous publications.

Shauna Corry   *North Dakota State University, Fargo, North Dakota* (CHAP. 56)
Shauna Corry is an associate professor and facility management program coordinator at North Dakota State University. Her research focus is universal design and accessibility issues. She has conducted ADA audits of more than 60 buildings and has participated in the Universal Design Education Project I in 1994. She completed an interdisciplinary Ph.D. from Washington State University in the fall of 2000.

Thomas D. Davies, Jr.   *Paralyzed Veterans of America, Washington, D.C.* (CHAP. 47)
Thomas D. Davies, Jr., is the senior associate architecture director for the Paralyzed Veterans of America. He coauthored *Handbook for Design on Specially Adapted Housing, Design for Hospitality, Accessible Home Design,* and the *Fair Housing Design Guide.* He also helped lead the ADA consulting effort for the 1996 Olympic Games.

Brian R. M. Everton   *ProductABILITY Consulting, Winnipeg, Manitoba, Canada* (CHAP. 16)
Brian R. M. Everton is a professional interior designer who has also completed graduate studies in industrial design and business management. After more than 10 years in health care product design and manufacturing, he is now the managing consultant for the not-for-profit ProductABILITY Consulting in Winnipeg, Manitoba, Canada.

Valerie Fletcher   *Adaptive Environments Center, Boston, Massachusetts* (CHAP. 60)
Valerie Fletcher is executive director of Adaptive Environments in Boston and was cochair of Designing for the 21st Century II in June 2000. She spent much of her career in public mental health and is the former deputy commissioner of the Massachusetts Department of Mental Health. She has a masters in ethics and public policy from Harvard University.

Gilbert Geis   *University of California-Irvine, Irvine, California* (CHAP. 18)
Gilbert Geis is Professor Emeritus, Department of Criminology, Law, and Society, School of Social Ecology, University of California-Irvine. He is a former president of the American Society of Criminology and is the recipient of its Edwin H. Sutherland Award for outstanding research. His Ph.D. in sociology is from the University of Wisconsin.

Larry Goldberg *CPB/WGBH, National Center for Accessible Media, Boston, Massachusetts* (CHAP. 67)
Larry Goldberg is director of the Media Access Group at the WGBH Educational Foundation in Boston. He is the founding chairperson of the Working Group on Advanced Television Closed Captioning and an inventor and patent holder for Rear Window™, a closed-captioning system for movie theaters, amusement parks, and other public attractions.

Selwyn Goldsmith *London, United Kingdom* (CHAP. 25)
Selwyn Goldsmith, a retired architect, is the author of *Designing for the Disabled*, published in three editions (1963, 1967, and 1976) by the Royal Institute of British Architects. His more recent books, *Designing for the Disabled: The New Paradigm* (1997) and *Universal Design* (2000), were published by Butterworth-Heinemann. He has a severe physical disability—the effect of polio in 1956 at age 23.

Susan Goltsman *Moore Iacofano Goltsman, Berkeley, California* (CHAPS. 19 AND 64)
Susan Goltsman, FASLA, is a founding principal of Moore Iacofano Goltsman with over 20 years' experience in environmental design, planning, and education with a specialty in children and youth. She holds degrees in environmental psychology, landscape architecture, and environmental design, and she serves on the board of the Landscape Architecture Foundation.

Louis-Pierre Grosbois *École d'Architecture, La Villette, Paris, France* (CHAP. 27)
Louis-Pierre Grosbois is an architect, consultant, and urban planner; he is also a professor at the École d'Architecture de Paris in La Villette, with an emphasis on design for all. His major publications include: *Handicap Physique et Construction* (1983), *Improving Our Living Conditions* (1986), and *Living in an Accessible City: From Uses to Designing* (1998).

Marcelo Pinto Guimarães *Federal University of Minas Gerais, Belo Horizonte, Brazil* (CHAP. 57)
Marcelo Pinto Guimarães, a professor of design at the Universidade Federal de Minas Gerais (UFMG) in Belo Horizonte, Brazil, is director of ADAPTSE, a research center on accessibility at UFMG and a director of the Independent Living Center of Belo Horizonte (CVI-BH). His degree is from the State University of New York at Buffalo.

Gloria Gutman *Gerontology Research Centre, Simon Fraser University, Vancouver, British Columbia, Canada* (CHAP. 36)
Gloria Gutman directs the Gerontology Research Centre and the Gerontology Diploma and Masters programs at Simon Fraser University in Vancouver, British Columbia. She has authored/edited 19 books and more than 100 articles, reports, and papers on seniors housing, long-term care, shelter and care of persons with dementia, and health promotion/aging.

Hans Haenlein *University of Reading, Reading, United Kingdom* (CHAP. 44)
Hans Haenlein joined the University of Reading as professor of architecture with over 20 years' experience in the design of education and community buildings and research into briefing and inclusive environments. He is currently designing a concept building for the International Centre for Inclusive Environments at the University of Reading, Reading, United Kingdom.

Brigitte Halbich *HEWI Heinrich Wilke GmbH, Bad Arolsen, Germany* (CHAP. 39)
Brigitte Halbich is a bathroom design consultant with HEWI, a German manufacturer of barrier-free sanitary fittings. In cooperation with Bruce A. Corson, M.E., M.A. (AIA), she is developing a seminar concept on universal design for HEWI, Inc., in Lancaster, Pennsylvania, as a service for designers in the United States.

James D. Harrison *National University of Singapore, Singapore* (CHAP. 40)
James D. Harrison is an associate professor in the Department of Architecture at the National University of Singapore. For more than 15 years he has been teaching and researching accessibility and universal design and is a U.K.-registered architect. He has been actively engaged in promoting access issues in Singapore and the Asia-Pacific region, including working as an expert for U.N. ESCAP and other key authorities.

Matthias Hürlimann *archi-netz, Zürich, Switzerland* (CHAP. 37)
Mattias Hürlimann, a graduate of the Technical University in Zürich, Switzerland, is an architect focusing on universally designed housing for persons who are elderly and/or disabled, both in new construction and adapted, existing housing stock.

Chitose Ikeda *NEC Design, Ltd., Tokyo, Japan* (CHAP. 55)
Chitose Ikeda became involved in corporate communication research and overseas public relations strategy development after having worked in research and development of information communication devices. She is currently engaged in universal design. She is also a member of the Design Committee, Universal Design Working Group, in the Communications Industry Association of Japan (CIAJ).

Assunta D'Innocenzo   *National Research Council (CNR), Rome, Italy* (CHAP. 15)
Assunta D'Innocenzo has a degree in architecture. She is director of research and a consultant for local bodies, housing companies, the CNR, Ministry of Public Works, and the national associations of older people on housing. She is an author and an expert for the Ministry of Public Works and a member of the CIB Task Group 19.

Louise Jones   *Eastern Michigan University, Ypsilanti, Michigan* (CHAP. 52)
Louise Jones, a professor of interior design, Eastern Michigan University, is a noted speaker at national and international conferences. She has authored more than 45 publications related to interior design education, design research, and universal design. Recent design work integrates instructional technology, sustainable design, and universal design into academic exemplars for the twenty-first century.

Stanton Jones   *University of Oregon, Eugene, Oregon* (CHAP. 51)
Stanton Jones, a professor of landscape architecture at the University of Oregon, has spent most of his career focusing upon issues of equality and justice as they pertain to the creation of places that simultaneously support human use, celebrate human diversity, and preserve or enhance the natural and cultural processes that are inherent in any given site.

Satoshi Kose   *Building Research Institute, Tsukuba, Japan* (CHAP. 17)
Satoshi Kose is the director of the Housing and Building Economy Department, Building Research Institute, in Japan. His major area of research is safety, human factors, and universal design. He has developed dwelling design guidelines toward the aging society, is on the editorial boards of *Building Research & Information* and *Journal of Architectural and Planning Research*.

M. Powell Lawton (*Deceased*)   *Philadelphia Geriatric Center, Philadelphia, Pennsylvania* (CHAP. 7)
M. Powell Lawton was director of research at the Philadelphia Geriatric Center for 30 years and was senior research scientist at the time of his death. He was also adjunct professor of human development at the Pennsylvania State University and professor of psychiatry at Temple University School of Medicine. His doctorate was in clinical psychology from Teachers' College, Columbia University. His research on how older people use their environment has influenced the practice of a generation of architects, planners, and designers.

Alan Leibert   *Card Europe Limited, Hertfordshire, United Kingdom* (CHAP. 68)
Alan Leibert is managing director of Card Europe, which was a partner in the DISTINCT project.

Danise R. Levine   *RERC on Universal Design, University at Buffalo, Buffalo, New York* (CHAP. 41)
Danise R. Levine is an architectural research associate at the IDEA Center and has worked on ADA compliance studies of university and public school systems, home modifications, development of fair housing education and outreach activities, and World Wide Web site development. She is the coauthor of four technical reports and multimedia presentations on universal design.

Rachael Luck   *University of Reading, Reading, United Kingdom* (CHAP. 44)
Rachael Luck is a registered architect working within the Research Group for Inclusive Environments at the Department of Construction Management and Engineering, the University of Reading. She is honorary secretary serving on the Council for the Design Research Society (DRS) and a member of the Architect's Registration Board.

Shauna Mallory-Hill   *Technical University of Eindhoven, Eindhoven, Netherlands* (CHAP. 16)
Shauna Mallory-Hill is an architect experienced in barrier-free and universal design consulting, education, and research. She was the editor of *ACCESS*, 2nd edition; director of the CIBFD (now the Universal Design Institute) from 1993 through 1996; and the recipient of NAAW and DJ Memorial awards for education. Shauna now lives in the Netherlands where she is completing a Ph.D.

Sandra Manley   *University of the West of England, United Kingdom* (CHAP. 58)
Sandra Manley is associate head of the School of Planning and Architecture at the University of the West of England in the United Kingdom. She has developed, in collaboration with colleagues, a degree in architecture and planning that has people-centered design as one of its key themes.

Mamoru Matsumoto   *City and Regional Development, Ministry of Land, Infrastructure, and Transport, Tokyo, Japan* (CHAP. 21)
Mamoru Matsumoto, J.C.E.A., has a B.S. in agriculture from Hokkaido University. He is the director of the Parks and Greens Division, City Bureau, Ministry of Construction in Japan. He is engaged in a number of research projects on outdoor settings with universal design, including "Barrier-Free Green Space Planning," conducted by the Japanese Institute of Landscape Architecture.

John Mathiason *Associates for International Management Services, Mt. Tremper, New York* (CHAP. 11)
John Mathiason is the managing director of Associates for International Management Services and was formerly the deputy director of the Division for the Advancement of Women of the United Nations Secretariat. He has participated in the training of trainers on the Rules in Latin America and Asia. He has a Ph.D. from the Massachusetts Institute of Technology.

Sanjoy Mazumdar *University of California-Irvine, Irvine, California* (CHAP. 18)
Sanjoy Mazumdar teaches at the University of California-Irvine. He is former chair of the Environmental DesignResearch Association, the Cultural Aspects of Design Network, and is editorial board member of the *Journal of Architectural and Planning Research.*

Yoshisuke Miyake *SEN, Inc., Osaka, Japan* (CHAP. 48)
Yoshisuke Miyake, J.C.E.A., J.I.A., has an M.L.A. from Harvard University. He is president of SEN, Inc., in Osaka, Japan, Also, he is chairman of the Barrier-Free Park Design Joint Research Committee, of the Organization for Landscape and Urban Greenery Technology Development.

Patricia A. Moore *Arizona State University, Tempe, Arizona* (CHAP. 2)
Patricia A. Moore is president of Moore Design, Inc., and a visiting professor of design at Arizona State University. As an internationally renowned gerontologist and designer, she is a leading authority on the requirements and behaviors of elders and all people as they progress throughout their lifespan.

Annalisa Morini *National Research Council (CNR), Rome, Italy* (CHAP. 15)
Annalisa Morini has a degree in civil engineering. She is a researcher at the CNR, working in the Central Institute for Industrialization and Building Technology (ICITE). She is a member of CIB Task Group 19, an evaluator of European Union research, and the author of books and papers in technical journals and conference proceedings.

Ruth Morrow *University of Sheffield, Sheffield, South Yorkshire, United Kingdom* (CHAP. 54)
Ruth Morrow, having practiced architecture in Great Britain and Germany, taught at several schools of architecture throughout Great Britain and Ireland before coordinating the DraWare project at the University College in Dublin. She is now a lecturer at the University of Sheffield, where her work continues to focus on design for all people.

James L. Mueller *J. L. Mueller, Inc., Chantilly, Virginia* (CHAPS. 45 AND 49)
James L. Mueller is an industrial designer who has worked in the field of design for people with disabilities since 1974. His consulting firm serves individuals, designers, employers, manufacturers, and public organizations on the subjects of design and accommodation of people with disabilities.

Abir Mullick *RERC on Universal Design, University at Buffalo, Buffalo, New York* (CHAPS. 41 AND 42)
Abir Mullick is an associate professor in the Department of Architecture at the University at Buffalo. He has served as a universal design expert to the National Kitchen and Bath Association and has published extensively in design, architecture, planning, rehabilitation, and human factors journals.

Sylvia Nowlan *Gerontology Research Centre, Simon Fraser University, Vancouver, British Columbia, Canada* (CHAP. 36)
Sylvia Nowlan is a graduate student at the University of Calgary School of Environmental Design, specializing in universal design. She has a degree in psychology.

Roxane Offner *Lighthouse International, New York, New York* (CHAP. 61)
Roxane Offner is an ADA consultant to Lighthouse International. She was deputy advocate in the New York State Office of Advocate for the Disabled, focusing on issues of aging, health, building codes, and universal design. She has been involved in the disability rights movement for over 30 years and was one of the founders of the first Independent Living Centers in New York.

Rex J. Pace *Center for Universal Design, North Carolina State University, Raleigh, North Carolina* (CHAP. 34)
Rex J. Pace is an architectural designer with 12 years' experience in universal and accessible design. He is presently coordinator of technical assistance and a lead designer at the Center for Universal Design. He previously was an integral part of the nationally recognized consulting firm of Barrier Free Environments, Inc.

Kenneth J. Parker *National University of Singapore, Singapore* (CHAP. 32)
Kenneth J. Parker is a senior lecturer in the Department of Building and has been researching at the National University of Singapore from 1991. He is a Fellow of the Chartered Institution of Building Service Engineers (CIBSE) in the United Kingdom, and Secretary (Hon.) of the International Commission on Technology and Accessibility (ICTA—Asia Pacific).

Jake Pauls   *Consulting Services in Building Use and Safety, Silver Spring, Maryland* (CHAP. 23)
After 20 years of research with the National Research Council of Canada, Jake Pauls moved to the United States in 1987. He now operates an independent consultancy in building use and safety. A certified professional ergonomist (CPE), he is known for bridging among ergonomics, public health, and development of codes and standards addressing life safety and building usability.

Annette Pedersen   *Theory Department, Western Australian School of Visual Arts, Edith Cowan University, Perth, Western Australia, Australia* (CHAP. 53)
Annette Pedersen is a sociolegal scholar with a feminist art historical background. She has extensive experience in coordinating equity-inclusivity projects at the University of Western Australia, where she was teaching contemporary cultural studies in the English department. Currently, she is based in the theory department of the Western Australian School of Visual Arts at Edith Cowan University.

John P. Petronis   *Architectural Research Consultants, Inc., Albuquerque, New Mexico* (CHAP. 62)
John P. Petronis is the founder and president of Architectural Research Consultants (ARC), Inc., located in Albuquerque, New Mexico. He holds master's degrees in architecture and business administration and is a certified planner and registered architect.

Karen D. Piltner   *KOOB Agentur für Public Relations, Mülheim a.d. Ruhr, Germany* (CHAP. 39)
Karen D. Piltner is a marketing consultant specializing in the target group of seniors, and she is the author of numerous articles on home modification and design. She holds a philological degree in literature, philosophy, and fine arts from the University of Bochum in Germany, and completed a postdegree in business administration and marketing.

Mark A. Proffitt   *Dorsky, Hodgson & Partners, Cleveland, Ohio* (CHAP. 22)
Mark A. Proffitt is an architectural researcher with Dorsky, Hodgson & Partners, an architectural firm that specializes in design for aging. His primary responsibilities include post-occupancy evaluations and programming for the elderly design studio. He has coauthored a book on the creation and evaluation of an innovative health center.

Avi Ramot   *Israel Center for Accessibility, Jerusalem, Israel* (CHAP. 31)
Avi Ramot is an accredited social worker with a degree in social work from Hebrew University in Jerusalem, and a Ph.D. in sociology from the State University of New York in Albany. He has worked for 27 years in rehabilitation, and is the director of SHEKEL, the Israel Center for Accessibility.

Laurie Ringaert   *Universal Design Institute and University of Manitoba, Winnipeg, Manitoba, Canada* (CHAP. 6)
Laurie Ringaert is an occupational therapist and holds a master's degree in community health sciences. She is director of the Universal Design Institute and a faculty member of the School of Architecture at the University of Manitoba in Winnipeg. She is also the chair of the Canadian Centre on Disability Studies.

Gary M. Robb   *National Center on Accessibility, Indiana University, Bloomington, Indiana* (CHAP. 20)
Gary M. Robb is executive director of the National Center on Accessibility at the Indiana University Bradford Woods Outdoor Center. He is also an associate professor of recreation and park administration at Indiana University.

Robert Robie   *Architectural Research Consultants, Inc., Albuquerque, New Mexico* (CHAP. 62)
Robert Robie is vice president of ARC, Inc. He is a registered architect with 20 years' experience in the planning and design of public facilities. Prior to joining ARC, he was a staff architect with the Albuquerque Public Schools, where he managed multiple small- and large-scale school construction projects.

John P. S. Salmen   *Universal Designers and Consultants, Inc., Takoma Park, Maryland* (CHAP. 12)
John P. S. Salmen, A.I.A., is a licensed architect, the president of Universal Designers and Consultants, Inc., and the publisher of *Universal Design Newsletter*. He is a recognized expert and consultant on the Americans with Disabilities Act and its Standards for Accessible Design. He is a leader in the emerging field of universal design.

Jim Singh Sandhu   *Inclusive Design Research Associates, Newcastle upon Tyne, United Kingdom* (CHAPS. 3 AND 68)
Jim Singh Sandhu has worked in the field for more than 30 years in 25 countries and has published 200 papers, reviews, book chapters, and books. A founder member of the European Institute for Design and Disability and past president, he has also been a policy adviser to the British government and the European Commission.

Jon A. Sanford *Atlanta VAMC, Decatur, Georgia* (CHAP. 22)
Jon A. Sanford is a research architect at the Atlanta VA Rehab R&D Center on Geriatric Rehabilitation and director of research at Extended Home Living Services in Wheeling, Illinois. He previously served as the coordinator for evaluation studies at the Center for Universal Design, North Carolina State University, and helped develop the Principles of Universal Design.

Maria Cristina Sará-Serrano *Associates for International Management Services, Mt. Tremper, New York* (CHAP. 11)
Maria Cristina Sará-Serrano is president of Associates for International Management Services and was the United Nations representative for Disabled Peoples' International for 13 years. She participated on behalf of DPI in the negotiations of the Standard Rules and has directed training of trainers on the Rules in Latin America and Asia. She has a diploma from Santiago College, Santiago, Chile.

Katsushi Sato *Department of Housing and Architecture, Japan Women's University, Tokyo, Japan* (CHAP. 59)
Katsushi Sato is an associate professor in the department of housing and architecture at Japan Women's University in Tokyo. He is an expert on accessible environments with the Social Development Division of the United Nations Economic and Social Commission for Asia and the Pacific (UNESCAP).

Mary Ann Clarke Scott *Gerontology Research Centre, Simon Fraser University, Vancouver, British Columbia, Canada* (CHAP. 36)
Mary Ann Clarke Scott has 12 years' experience in architecture and planning and is a registered architect in British Columbia. She is principal of Generations: Architecture Planning Research, which specializes in environments for elderly people and people who are disabled, as well as in universal design, consulting to development, design, health care, and government clients.

Scott M. Shea *Scheuber & Darden Architects, Aurora, Colorado* (CHAP. 35)
Scott M. Shea is a practicing architect. As a research associate at the IDEA Center from 1993 to 1996, his responsibilities included research and education incentives focusing on fair housing and design for seniors and people with disabilities. He recently completed *Fair Housing in Colorado*, a handbook for complying with federal and state fair housing laws.

Aaron Steinfeld *California Partners for Advanced Transit and Highways, University of California, Berkeley, California* (CHAP. 50)
Aaron Steinfeld received his Ph.D. from the University of Michigan. His research includes work on driver interfaces, automobile head-up displays, augmented reality, perception, short-term memory, and rehabilitation. He is a research specialist at California PATH, University of California at Berkeley, where he works on developing advanced transportation technologies.

Edward Steinfeld *Rehabilitation Engineering Research Center (RERC) on Universal Design, University at Buffalo, Buffalo, New York* (CHAPS. 24, 35, AND 50)
Edward Steinfeld is a registered architect, educator, and design researcher. His recent interests include research on anthropometrics of wheelchair users, technical studies on visitability in housing, universal design of transportation systems, methods for measuring the usability of products and environments, and computerized access audits and design of universal products for residential buildings.

Molly Follette Story *Center for Universal Design, North Carolina State University, Raleigh, North Carolina* (CHAPS. 10 AND 49)
Molly Follette Story is a researcher and designer specializing in products for diverse populations, including individuals with disabilities and older adults. She has taught industrial design since 1984 at the Georgia Institute of Technology and North Carolina State University. She has been on the staff of the Center for Universal Design since 1994.

Cynthia Stuen *Lighthouse International, New York, New York* (CHAP. 61)
Cynthia Stuen is senior vice president for education at Lighthouse International, and the director of its Center for Education. Among her more recent publications are *Family Involvement: Maximizing Rehabilitation Outcomes for People with Impaired Vision* and *The Aging Eye and Low Vision*. She has more than 50 publications for consumer and professional audiences.

Gihei Takahashi *Toyo University, Kawagoe City, Japan* (CHAP. 30)
Gihei Takahashi is an associate professor in the Department of Architecture, Faculty of Engineering, Toyo University. He is chairman of the Normalization Environment Subcommittee of the Architectural Institute of Japan (AIJ), chairman of the Universal Design Committee of the Japan Institute of Architecture (JIA), and a member of the planning committee of the Japan Council on Disability (JCD).

Noriko Takayanagi   *NEC Corporation, Tokyo, Japan* (CHAP. 55)
Noriko Takayanagi is currently engaged in universal design research and development in the NEC Corporate Design Division, after having been involved in the design development of audiovisual and information communication devices. For the last 4 years, she has led the collaboration with Tama Art University on universal design.

Riadh Tappuni   *UN-ESCWA, Beirut, Lebanon* (CHAP. 63)
Riadh Tappuni is United Nations First Human Settlements officer at the Economic and Social Commission for Western Asia (ESCWA), and he is the regional housing reconstruction coordinator for Pec/Peje at the United Nations Interim Administration Mission in Kosovo.

Fred Tepfer   *University of Oregon, Eugene, Oregon* (CHAP. 46)
Fred Tepfer, a licensed architect in Oregon, is planning associate in the University of Oregon Planning Office and ADA coordinator for physical barriers. He teaches at the University of Oregon in education (educational facilities) and in architecture (architectural programming). He is also in private practice, focusing on schools, public and private housing, and nonprofit organizations.

Gregg Vanderheiden   *Trace R&D Center, University of Wisconsin-Madison, Madison, Wisconsin* (CHAP. 65)
Gregg Vanderheiden is a professor of industrial engineering and founder/director of the Trace R&D Center at the University of Wisconsin-Madison. His research currently focuses on universal design of information technology and telecommunications products, and he recently completed a study of what motivates the industry to adopt universal design of their products.

Fabrizio Vescovo   *Architect, Rome, Italy* (CHAP. 26)
Architect Fabrizio Vescovo is an Executive Member of the Latium Council, Department for Town Planning and Housing. He is adviser to the Order of Architects of Rome and a member of the Interministerial Committee for Public Works and Social Affairs on accessibility. He has authored many books and publications on urban planning and environmental quality.

C. J. Walsh   *Sustainable Design, Dublin, Ireland* (CHAP. 33)
C. J. Walsh is an architect, fire safety engineer, and technical controller. He is chief technical consultant at Sustainable Design International, a multidisciplinary consultancy. He is a member of the International Council for Research and Innovation in Building and Construction (CIB) and the Expert Group on the European Concept for Accessibility (ECA).

Leslie Kanes Weisman   *New Jersey Institute of Technology, Newark, New Jersey* (CHAP. 69)
Leslie Kanes Weisman, Assoc. AIA, is professor and former associate dean of the School of Architecture at the New Jersey Institute of Technology. She is the author of the award-winning book, *Discrimination by Design: A Feminist Critique of the Man-Made Environment*, and the coeditor of *The Sex of Architecture*.

Polly Welch   *University of Oregon, Eugene, Oregon* (CHAP. 51)
Polly Welch, an architect and professor at the University of Oregon, has been involved in accessibility and diversity issues for 20 years as a practitioner, as a consultant to public and private organizations on user accommodation, and as a public administrator in a housing agency. She is the editor of *Strategies for Teaching Universal Design* (1995).

Maarten Wijk   *Delft University of Technology, Delft, Netherlands* (CHAP. 28)
Maarten Wijk is a professor in the Department of Architecture at the Delft University of Technology. His interests include ergonomics of the built environment. He is a contract researcher and adviser at EGM onderzoek bv, which is the R&D Department of EGM Architects, Dordrecht, the Netherlands. His focus of work is quality management, accessibility, and labor conditions.

Leslie C. Young   *Center For Universal Design, North Carolina State University, Raleigh, North Carolina* (CHAP. 34)
Leslie C. Young, a professional designer, author, and specialist with more than 25 years' experience in design for people with disabilities, is director of design at the Center for Universal Design, North Carolina State University. Formerly, she was key staff of Barrier Free Environments, Inc., a nationally recognized consulting firm.

John Zeisel   *Hearthstone Alzheimer Care, Lexington, Massachusetts* (CHAP. 8)
John Zeisel is president of Hearthstone Alzheimer Care. He has pioneered the use of nonpharmacological treatments for Alzheimer's disease, including the use of the physical environment as treatment. Author of *Inquiry by Design*, he has taught at Harvard University, the University of Minnesota, McGill University, and the University of Quebec at Montreal. He is presently carrying out research on the brain's "environmental system."

FOREWORD

Have you ever broken your arm and tried to open a door? Been late for a plane and had to negotiate a crowded airport laden down with luggage? Picked up your 98-year-old aunt from synagogue or church? Struggled to understand directional signage in the street of an unfamiliar city? Felt left out at a neighborhood meeting because you couldn't see or hear the speaker? Reached to settle your belongings on a high shelf that lies just beyond your reach? If so, you have faced the consequences of design, an activity that has the power to affect our daily lives, for good or ill. Yet, for much of our history, designers and builders have ignored large segments of the population.

As human beings, we do not exist in the abstract. Throughout the continuum of our lives, we grow and shrink in stature, strengthen and weaken in musculature, quicken and diminish in visual and aural acuity—our physical state, including our senses and our strength, varies radically from age to age. In addition to more obvious physical challenges, our abilities and our infirmities can ebb with the barometric pressure, the pollen count, and our psychological sense of well-being.

The diverse portrait of the human condition thus avoids moralizing: We are not "good" one year and "bad" the next. Instead, for designers, it demands that we ask questions. Do the objects, spaces, places, and cities that we design enhance our experience? Enable us to perform tasks at peak efficiency? Encourage our passage or our rest? Ennoble or degrade us? Welcome our collaboration or exclude us? Simplify or improve our lives?

In its highest sense, universal design, which this book addresses, suggests a change of focus. Rather than codifying rigid rules for the abstract, perfect person (as earlier handbooks did), it asks us to remove our rose-colored glasses, see the world as it really is, and adjust our designs accordingly. The operative point of view for designers (whether architects, landscape architects, interior designers, engineers, industrial designers, Web designers, or wayfinders), becomes one of empathy for the human condition; in universal design, solutions reflect the diversity of human abilities—throughout the range of life.

Although codes may help ensure compliance where the society has proved intransigent, the ultimate answer to universal design lies in employing our full imaginative and aesthetic gifts in a new way of seeing. Digital technology may prove an ally in this effort, allowing us to "see" in ways we may not have imagined. But the solution to a more sympathetic physical environment lies in the human consciousness—refocused, enlightened, and responsive to the fullness of who we are, and what we can accomplish. Read the following pages, and learn.

Robert Ivy, FAIA
Editor in Chief
Architectural Record Magazine
New York, NY

PREFACE

This book is inspired by the unprecedented and growing worldwide interest in the design of environments and products that respect the diversity of human beings. The attention comes from many sectors: It includes market-oriented interests that see the spending power of aging baby boomers with a significantly expanded customer base; aging advocates speaking for themselves; families and clients who want to live productive lives and wish to remain in their own homes; disability rights policy makers; activists who assert their right to equality and inclusive environments for people with disabilities. The engagement with universal design is evidenced by an increasing number of international conferences with participants from every continent. Numerous study tours to the United States by international groups of designers, government officials, and businesses wanting to understand the relationship between the Americans with Disabilities Act and universal design further confirm the need to create this publication in an attempt to communicate the remarkable range of effort in the arena of universal design in the United States, as well as around the world.

The primary purpose of this book is to help encourage the practice of good design by making universal design information available to designers, design educators, decision makers in business and government, as well as advocates. The editors have combined their collective expertise in universal design and environment/behavior studies to shape a publication that illustrates the range of worldwide policies and practices.

Why should one be concerned with universal design? In a period of rapidly changing demographics, with an aging population that will have a longer life expectancy, it is of paramount concern to anticipate a range of disabilities that will affect an increasing number of the population. For instance, in the United States, more than 50 million people have varying degrees of disabilities, including mental and physical. This number is expected to increase, and thus, by providing this segment of the population with accessibility, safety, convenience and satisfaction, universal design will be practiced in the spirit of its "inventors." Perhaps the most vocal and effective of these was the late Ronald L. Mace, FAIA, to whom this book is dedicated.

The question arises how lasting changes can be effected in the design professions regarding the awareness and understanding of the concept of universal design and its implications for the planning and design of facilities. These changes must occur in the design world, as well as with the agencies that are concerned with aging and people with disabilities. It appears that the one major realistic hope for such change lies in education and in teaching of future professionals in schools of design, including architecture, engineering, planning, graphic, industrial, interior design, and information technology. This is why the editors of this volume have put a major emphasis on design education, both within the university as well as within community settings. This has included the curriculum and teaching of required introductory courses for the past 15 years. They have reinforced this through exercises and field studies of real buildings in upper-level courses, including thesis programs and design projects and 30 years of community and continuing education programs.

From the outset, it was clear that this book would reflect a global perspective. Contributors from 18 primarily industrialized countries were invited to contribute chapters on the experience with universal design and with accessibility standards over the last 20-plus years, and to project into the future needed directions and changes. Clearly, some contributors were not in agreement with the legislative developments in this field in the United States, Canada, and the United Kingdom, and it is hoped that their critical stance will encourage discourse among the users of this volume.

Language and translation issues had to be addressed, as a number of contributors reside in non-English-speaking countries. This presented a challenge of word-by-word editing and of retaining the proper meaning of words and terminology in the authors' home countries, as opposed to the United States. Measurement systems had to be dealt with, in terms of differences between metric and nonmetric, and it was decided to retain the original system in a given author's country, and to provide conversions where deemed feasible and useful.

Personal, life-changing events occurred to several authors: First and most tragic, Powell Lawton (author of Chap. 7), passed away. He will be missed and remembered for his humanity and great contributions to both gerontology and the environment and behavior field. One author served the United Nations in Kosovo while preparing his chapter, a true challenge; two author groups had their prime authors become parents during the year-long process of writing their manuscripts; and finally, several authors changed their professional affiliations.

An editorial process was evolved for this book, which consisted of three rounds of reviews for each manuscript, building upon each other. Sixteen review categories were established to assure consistency of content, structure, and format. They included such issues as: the sequence in which items were to be addressed in a given chapter; the introduction linking a given chapter topic to universal design; the background and historical evolution of the topic, building type, or issue being addressed in a given chapter; the use of illustration materials; cross-referencing among chapters in the book; and the balancing of topics in a given section.

Finally, human aspects of shepherding 69 author groups through the arduous editing process can only be understood by someone who has experienced this him- or herself. Extracting manuscripts from busy authors in a timely manner is quite a challenge, indeed. Electronic communications were both an asset that made the book possible and added some challenges as well, in accommodating the range of technological expertise and computer capacity.

As a result, this handbook reflects an in-depth review of the state of the art in the progression to more universal designing. The introduction to each chapter provides a brief summary that will help orient the reader.

ACKNOWLEDGMENTS

The editors wish to acknowledge the crucial role of the National Endowment for the Arts (NEA) and the National Institute on Disability and Rehabilitation Research (NIDRR) for their support for universal design through their policies as well as grants for education, research, product development, technical assistance, and promotion. The impact of both agencies is reflected in the work of a number of the authors from several countries. In addition to their research and program support, both agencies have used their position to inform other government agencies and the public about universal design. Three national Rehabilitation Engineering Research Centers on universal design are now in place, with NIDRR support. Two on the built environment are the Center for Universal Design at North Carolina State University and the RERC on Universal Design, University at Buffalo. The RERC on Information Technology Access is at the Trace Research and Development Center, University of Wisconsin, Madison.

We are most appreciative of the 69 author teams who worked diligently, often rewriting their chapters three and four times to reach the clarity that would reflect the excellence of their work. The language barriers were daunting, and we are humbled by those authors who produced work in "U.S. English"—fully aware of our own limits in their native languages.

Thanks are owed to the persons who got the editors interested in the field of accessible and universal design in the first place, and the agencies, which helped sustain that interest through grants and other support.

In the case of Editor in Chief Preiser, the National Science Foundation (NSF) assisted research leading to the publication of *Public Building Accessibility: A Self-Evaluation Guide*[*] in the late 1970s. This guide would allow businesses, hospitals, or any organization for that matter, to check out their facilities for accessibility without necessarily hiring consultants. Later, the NEA provided several grants to help develop an electronic "Guidance System for Visually Impaired Persons"[†] for buildings and the campus of the University of New Mexico. Ray Marshall, coordinator for special services at that university, is thanked for his excellent cooperation and for providing subjects for the research that lead to a prototype of that guidance system. We also thank William Davis at the New Mexico School for the Visually Handicapped in Alamogordo, New Mexico, for assisting in the testing and research, which was awarded the 1984 *Progressive Architecture* Annual Award in Applied Research.

Preiser was inspired by the late Ronald L. Mace, who coined the term *universal design* and who shared his vision in the book *Design Intervention: Toward a More Humane Architecture.*[‡]

Senior Editor Ostroff is extremely appreciative of the significant role played by the many individuals who actively participated on the board of directors of Adaptive Environments since 1978. Their sustaining support, commitment, and fiduciary risk taking allowed her and the staff of a small nonprofit organization to create participatory, user-centered design education programs—first in accessible, then in universal, design. Other significant support has come from NEC Foundation of America, who, along with the NEA, supported ongoing work in universal design education.

[*] Gray, K., et al., *Public Building Accessibility: A Self-Evaluation Guide,* University of New Mexico, School of Architecture and Planning, Albuquerque, NM, 1978.

[†] Preiser, W. F. E., "A Combined Tactile/Electronic Guidance System for Visually Impaired Persons in Indoor and Outdoor Spaces," in E. Chigier (ed.), *Design for Disabled Persons,* Freund Publishing House, London, UK, 1988.

[‡] Preiser, W. F. E., J. C. Vischer, and E. T. White (eds.), *Design Intervention: Toward a More Humane Architecture,* Van Nostrand Reinhold, New York, 1991.

The editor in chief acknowledges the role that Senior Editor Elaine Ostroff, played in recruiting more than 45 authors into this book project. Her national and international network reflects her work of more than 30 years in the field of accessibility and in universal design, and her knowledge of this emerging field was invaluable in the editorial process. Ostroff, the founding director of Adaptive Environments in Boston, developed the Universal Design Education project noted throughout the handbook and was instrumental in organizing the international conferences "Designing for the 21st Century," held in New York City in 1998, and Providence, Rhode Island, in 2000. The growing number of presenters and attendees is evidence of the fact that universal design is sweeping the world as a movement, apparently unstoppable and particularly significant in aging societies.

We wish to thank the many individuals and organizations that helped shape this book, as well as refine the emphasis of the contents. Working in the global arena, a number of difficulties had to be overcome, ranging from communications in general, to an equitable editorial process, which recognizes regional and cultural differences in universal design, as well as accessibility for persons with disabilities in different countries.

The CD-ROM that accompanies this handbook was made possible by the cooperation of the U.S. Access Board, who has provided the electronic files of the most current ADA Accessibility Guidelines and related technical assistance materials.

Perhaps the most important source of inspiration and continuing involvement in the field of universal design have been our students and colleagues over the course of the past 25 years. Without their help and genuine care, we would not have been able to carry out this and numerous other projects.

We thank Wendy Lochner, senior editor for architecture at McGraw-Hill, for shepherding us through the daunting process of creating this innovative handbook on universal design, without any precedent to emulate. Tom Laughman of North Market Street Graphics was a true facilitator in managing the book production, including copyediting, as well as the page composition and proofing process for McGraw-Hill.

Deirdre Price was an excellent project assistant in charge of word processing and handling three rounds of manuscript reviews, as well as multiple communications with our authors. Last, but not least, Gary DeVoe played an invaluable role in the final submission by compiling an electronic record of 69 chapter drafts with over 400 images on a set of three CDs.

Finally, our spouses, Cecilia Fenoglio-Preiser and Earl Ostroff, provided strong support throughout this long and challenging project.

P · A · R · T · 1

INTRODUCTION

CHAPTER 1

UNIVERSAL DESIGN: THE NEW PARADIGM

Elaine Ostroff, Ed.M.
Adaptive Environments Center, Boston, Massachusetts

1.1 INTRODUCTION

The *Universal Design Handbook* is a unique compilation that brings together the rich international experience of what we call universal design. Universal design is not a trend but an enduring design approach that assumes that the range of human ability is ordinary, not special. This book describes and illustrates the extraordinary growth in the international movement to create environments and products for all people. During the past 15 years, the approach to design that accommodates people with functional limitations has been changing from narrow code compliance to meet the specialized needs of a few to a more inclusive design process for everybody.

Contrary to the negative assumption that attention to the needs of diverse users limits good design, the experience of imaginative designers around the world reveals the range of applications that delight the senses and lift the human spirit when universal design is integral to the overall concept. Universal design is assuming growing importance as a new paradigm that aims at a holistic and integrated approach to design, ranging in scale from product design to architecture, and urban design on one hand, and systems controlling the ambient environment and information technology, on the other. The terminology differs from one country to another; there are significant cultural differences in how the movement has evolved in each country, but the similarities are more apparent than the differences as they transcend national laws, policies, and practices.

There is a confluence of factors generating the need for more universally designed products, environments, and amenities—it includes the competitive, global nature of business today, the flourishing communications technology industry, the international disability movement, and the rapidly growing aging and disabled populations all over the world. Everyone is likely at some time to experience the misfit between themselves and their environment. Ambient conditions or stress may create problems with using products or buildings. Aging increases the potential for vulnerability in the environment. People worldwide are living longer, the aging population will double in the next 20 years (McNeil, 1997), and a child born today has a 50 percent chance of living to be 100 years old.

In this introductory chapter, we note the multiple influences affecting universal design. We also discuss the different terminology used throughout the book to label the design process that respectfully includes all people, and in so doing, we illuminate the differences between meeting legislated national accessibility requirements and the practice of universal design.

We explain the structure of the book and then identify key international themes that are discussed by authors from different countries. Late-breaking developments are included and future challenges are highlighted.

1.2   INTERNATIONAL INFLUENCES LEADING TO UNIVERSAL DESIGN

What has inspired this worldwide appetite for universal design efforts around the world? The U.S. Supreme Court Decision in 1954, *Brown vs. the Board of Education,* established the precedent that "separate is not equal." Ostroff, in Chap. 43, "Universal Design Practice in the United States," considers this precedent of equal opportunity in education as the milestone that marks the beginning of an approach to design that respects all users. Accessibility features that are a thoughtless add-on after the basic design of a place or a product have a stigmatizing quality not unlike the segregated "back of the bus" practices that were once the norm in the United States. There are two major, distinctive threads that can be traced historically—the legislative measures that included specialized requirements to accommodate people with disabilities, primarily affecting the larger-scale built environment and the non-regulated market-driven responses to an aging society, primarily relating to products. Both of these threads are well covered within the handbook, by several authors. Sandhu's Chap. 3, "An Integrated Approach to Universal Design: Toward the Inclusion of All Ages, Cultures, and Diversity," provides a broad historic view of the legislative and civil rights measures around the world. Others detail the legislative developments in several countries—the United States, the United Kingdom, Italy, Canada, Japan, and Israel. Chapter 4, "Designing for Our Future Selves," by Coleman, offers a rich perspective from the United Kingdom on ways that the changing demographics are leading to new business opportunities and a more inclusive design consciousness. In Chap. 50, "Universal Design in Automobile Design," Steinfeld and Steinfeld discuss the innovative design changes in the highly competitive automotive industry in order to respond to the safety and comfort needs of an older society. Ikeda and Takayanagi, in Chap. 55, "Universal Design Research Collaboration Between Industry and a University in Japan," detail the mutually beneficial experience in which the attention to older and disabled consumers fuses two separate objectives: (1) commercialization and profit for industry and (2) the education of product designers.

In addition to the nondiscriminatory accommodations for people with disabilities and the attention to the aging demographics that were the initial influences for universal design, there is an increasing awareness of conditions that have not traditionally been thought of as universal design issues. Attention to these issues is expanding the scope of universal design to encompass a broad range of environmental concerns. They include the prevalence of "sick buildings" and their concomitant health problems related to poor indoor air quality that have been a major source of disablement for growing numbers of people. Diseases of the respiratory system are the highest cause of disability in children under age 18, with asthma the most common disabling condition (Wenger et al., 1996). The U.S. Access Board is directing a quarter of its research money, beginning in 2000, for work on indoor environmental air quality (Lamielle, 1999). Acoustics and lighting are also assuming importance in the creation of optimum, nondisabling environments. Weisman, in Chap. 69, "Creating the Universally Designed City: Prospects for the New Century," further stretches the potential for universal design.

1.3   TERMINOLOGY

The terms used to describe environments that promote human functioning differ in many countries, as will be seen throughout this book. There has also been a developmental change in the language used in some countries, reflecting not only the evolution from initial efforts to remove barriers that exclude people to a more inclusive design approach, but changing social policies as well. What follows are the most common terms and a discussion of what they mean in different contexts.

Universal Design.   *Universal design* is a term that was first used in the United States by Ron Mace (1985), but the concepts were also expressed in other countries. *Universal design* and *inclusive design* are terms often used interchangeably in the United States to label a design approach that implies equity and social justice by design. Although there are other terms that are frequently used, such as *life span design* and *transgenerational design,* Mullick and Steinfeld (1997) explain that what separates universal design from these terms is universal design's focus on social inclusion. This relates to the "separate is not equal" precedent of equal opportunity.

Unfortunately, the term *universal design* has inappropriately been adopted by some people, especially in the United States, as a trendy synonym for compliance with Americans with Disabilities Act Standards for Accessible Design. We see the poor design and the problems created by this confusion, especially in thoughtless new designs that end up looking like retrofits. Ramps added in new construction are a good example, where none would have been needed if the architects had considered the needs of all users as fundamental in the earliest stages of the programming process, rather than a technical requirement to be added at the end of the design process. Perhaps this misunderstanding is the result of good intentions—the use of what may be thought of as a politically correct term, but it inhibits the creative process invited by universal design. Welch and Jones (1999) note, "This indicates that significant systemic and attitudinal barriers stand in the way of real change."

Ron Mace noted that minimum standards are an important part, but not the definition of universal design. His 1988 definition of universal design is quoted in several chapters: "Universal design is an approach to design that incorporates products as well as building features which, to the greatest extent possible, can be used by everyone."

Barrier-Free Design.   The initial term used around the world was *barrier-free design* and related to the efforts that began in the late 1950s to remove barriers for disabled people from the built environment. An international conference held in Sweden in 1961 cited extensive efforts throughout Europe, Japan, and the United States, primarily by rehabilitation organizations, to "reduce the barriers to the disabled" (ISRD, 1961). Around that same time, a related international effort began, shifting care for people with disabilities who had been institutionalized and removed from mainstream society back to community-oriented programs and facilities. Christophersen and Gulbrandsen (2000) highlight the Norwegian policy shift from "institutional care to special needs housing to equality and inclusion" as parallel to international trends. Lusher, in her detailed article on the development of access laws and codes in the United States (Lusher, 1989) reported on the efforts by the President's Committee on the Employment of the Handicapped and the Veterans Administration to study ways that the federal government might increase accessibility.

More recently, in the United States, the term *barrier-free* has been perceived negatively, as a feature prescribed only for use by disabled people. In Europe and Japan, the term has been used more broadly to describe universal design. However, in Europe, the term *design for all* has been increasingly used since 1967, and now the more popular term in Japan is *universal design.*

Accessibility.   This term has a very different meaning for some European experts involved in the European Concept for Accessibility (Wijk, 1996). For them, accessibility is the umbrella issue for all parameters that influence human functioning in the environment. They define accessibility as an environmental quantity. In the United States, *accessible design* became more widely used in the 1970s as a more positive term than *barrier-free design,* but it was and is still very much linked to legislated requirements.

1.4   *STRUCTURE OF THE* UNIVERSAL DESIGN HANDBOOK

The 69 chapters of the handbook are organized in 10 parts, moving from the most comprehensive to more specific topics. Every part has an international mix of authors. Each chapter has a similar structure, beginning with an introduction that is essentially an overview that summarizes the contents of the chapter for the reader. A synopsis of each part follows.

Part 1: Introduction.   This part introduces the reader to the key themes of the book and also provides a more experiential reminder of the environmental interactions faced by users on a daily basis. It highlights the challenges and opportunities to enable people, by design.

Part 2: Premises and Perspectives in Universal Design.   The chapters in this part reflect upon broad themes in universal design—the genealogy of universal design; responses to demographic changes and market needs; the implications for universal design in industrializing nations where the socioeconomic conditions sharply divide the haves and have-nots; the importance of individualized assessment within an environment/behavior framework; and the role that the human brain plays in environmental perception and universal design. The critical involvement of the user is highlighted and universal design evaluation as a continuous effort to improve the quality of design through appropriate feedback mechanisms is described.

Part 3: Universal Design Guidelines and Accessibility Standards.   This part encompasses the development of performance-based criteria and guidelines for universal design in varied settings including outdoor play for children and natural environments as well as international policy-making efforts. National governmental efforts in several countries show the development and application of accessibility standards. Included are issues relating to prescriptive standards and challenges they present to universal design. The Principles of Universal Design are critiqued in relation to a complex user group—people with Alzheimer's disease. This section also has comparisons of accessible housing built to national standards in Europe, and problems when accessibility by legislation is interpreted by the courts, along with issues of safety standards in residential stair design.

Part 4: Public Policy Systems and Issues.   This part includes the analysis of universal design issues in mass transportation and also identifies complex issues in the development of national policies for accessibility and universal design. Historic developments are detailed, including struggles and successes—both in industrialized countries and in developing economies. It presents approaches to universal design based on high-level planning as well as sustainable human and social development.

Part 5: Residential Environments.   This part begins with a well-illustrated description of the next-generation universal single-family home and moves on to multifamily housing and group living arrangements, addressing both the design as well as government policies that impact design. Housing policies from the United States, Canada, Switzerland, Australia, Germany, and Southeast Asia are reviewed. Two chapters deal specifically with housing elements: one on universally designed kitchens and appliances and another on bathrooms.

Part 6: Universal Design Practices.   With an emphasis on the creation of places and products, this part begins with the civil rights history leading to universal design practice in the United States. It includes a briefing process for universal design from the United Kingdom and illustrates universal design primarily in settings and building types in the United States. These include offices and workplaces, educational environments, sports and entertainment facilities, landscape design, as well as the universal design of products and automobiles. The practice of landscape architecture is illustrated through the aesthetic spirituality of universally designed outdoor gardens and parks in Japan.

Part 7: Education and Research.   The chapters in this part describe teaching experiences in professional schools of design in the United States, Australia, and the United Kingdom. It also includes a partnership between a college of design and a corporation in Japan. Advances and obstacles are detailed in each setting. Examples of universal design research are presented, including post-occupancy evaluation and the use of rating scales for inclusive design at the building and the urban design scale.

Part 8: Case Studies.   This case study part of the book examines specific projects that include either accessibility or universal design, and from which practical lessons can be derived. Examples include urban design projects in the United Kingdom and the United States; a United Nations regional project in Southeast Asia; an award-winning building serving people with vision impairments in New York City; capital planning in a local public school district in Arizona; the systematic rebuilding of the war-ravaged central district in Beirut, Lebanon; and a transportation-oriented multibuilding campus in Oakland, California.

Part 9: Information Technology.   This part details the methods and public policy by which the emerging communication and information technology is being made more universally usable. The incentives in new U.S. legislation are described and strategies for designers of telecommunications equipment are introduced. The international guidelines for the Internet and sophisticated innovations in every aspect of mass media are presented and European developments in Smart Card technology are explained.

Part 10: The Future of Universal Design.   The book concludes with a visionary chapter that builds a bridge to the future. It underscores both the growth as well as the massive problems in cities around the world, challenging planners and designers of cities to apply principles such as diversity, inclusivity, equity, and environmental health in the rethinking of our cities.

1.5   INTERNATIONAL THEMES

There are several themes that appear frequently throughout the book, and they surface in different parts, reflecting experiences from different countries. The following sections analyze and relate some of the authors who address similar themes.

Design-for-All

Grosbois, in Chap. 27, "The Evolution of Design for All in Public Buildings and Transportation in France," provides both historic architecturally based theory as well as illuminating case examples of how design for all was achieved in transportation systems and in public buildings. He cites Vitruvius, the often-quoted Roman architect, who said that architecture should be based on three qualities: (1) *Firmitas* (solidity of construction), (2) *Voluptas* (the aesthetic experience), and (3) *Commoditas* (adaptation to use). Vitruvius recommends, "Laying the building out so ingeniously that nothing could hinder its use." Grosbois also references architect Alvar Aalto's critique of formalism and its inhuman nature. He uses a door handle that Aalto designed and the remarkable Fiskars scissors, produced by a Finnish industrialist, as examples of design for all.

There are welcome connections between Grosbois' analysis of design for all in the Lille metro system and the Grenoble Tram with Chap. 24, "Universal Design in Mass Transportation," by Steinfeld, in which he emphasizes the importance of mass transportation in urban environments, in promoting social and economic viability. He introduces the concept of the universally accessible *travel chain*. The travel chain includes the information needed to understand and gain access to transportation, the stations, and the vehicles. Steinfeld includes guidance for universally designed elements of transportation-related design from information strategies and wayfinding, to level changes and security systems in terminals, to passenger loading and seating.

Aging as a Stimulus for Universal Design

Numerous authors emphasize the unprecedented change in demographics, with an aging population that will have a longer life expectancy than ever before, and the implications for design. Only a few are noted here. Moore, in Chap. 2, "Experiencing Universal Design," reminds the reader of the more affluent and style-conscious baby boomers who want function, convenience, and good design—all at once. Sandhu's bell-curve diagram in Chap. 3 illustrates the huge increase in users from when designers go beyond the old norms based on young, able-bodied users, to a more inclusive, universal design approach. He also cites the humane recommendations to the United Nations from the International Federation for Ageing that point to universal design. Coleman, in Chap. 4, also reframes the thinking about the recipients of the design process in this developing context of social and demographic change. He explains the concept of "designing for our future selves" to help designers and policy makers move beyond some statistical "other." Kose, in Chap. 17, "The Impact of Aging on Japanese Accessibility Design Standards," charts the growth of aging populations in industrializing nations. He notes that Japan has the fastest-growing aging population. By 2015, more than 25 percent of the country will be over 65 years of age, and the Japanese government has developed design guidelines to accommodate these changes. Harrison, in Chap. 40, "Housing for Older Persons in Southeast Asia: Evolving Policy and Design," describes the pressure of demographics and changing cultural patterns that lead to older people living apart from their families, in new urban housing developments where they have access to services and public transportation.

Information Technology

This exploding field is both a source of worldwide economic development and, currently, the most productive arena for universal design practice. Several authors illuminate aspects of information technology that increasingly impact the daily experience of every individual and the work of every corporation and government agency. Vanderheiden, in Chap. 65, "Fundamentals and Priorities for Design of Information and Telecommunication Technologies," explains the congruence between the technology needs of a mobile workforce and consumers with physical and sensory limitations. Brewer, in Chap. 66, "Access to the World Wide Web: Technical and Policy Perspectives," details the international effort to create standards and tools that assure access to this fast-growing medium. Goldberg, in Chap. 67, "Universal Design in Film and Media," highlights the expanded access to film and television now possible through digital technology. Sandhu and Leibert describe the European work to provide wide usability in another emerging field in Chap. 68, "User-Centered Deployment and Integration of Smart Card Technology: The DISTINCT Project." In a related development with far-reaching commercial implications, the U.S. Access Board (2000) has recently published the regulations and technology standards to be used by the U.S. government in the purchase and use of accessible electronic and information technology.

These unique standards encompass the types of technology covered and lay out the minimum level of access required. They provide both technical criteria specific to various types of technologies and performance-based requirements, which focus on the functional capabilities of covered technologies.

Government Policies and Universal Design

The book intentionally reflects divergent views of government policies and laws that relate to universal design. While many authors, as seen throughout the book, describe architectural and transportation designs that are not limited by accessibility standards, a few authors are critical of legislated approaches and believe that they thwart the practice of universal design. Salmen, in Chap. 12, "U.S. Accessibility Codes and Standards: Challenges for Universal Design," cites court challenges as proof of the problems with legislated requirements. He proposes performance criteria instead of prescriptive design standards. Mazumdar and Geis, in Chap. 18, "Interpreting Accessibility Standards: Experiences in the U.S. Courts," have related concerns about legislated means to assure accessible design.

Wijk, in Chap. 28, "The Dutch Struggle for Accessibility Awareness," recounts the failed results in the Netherlands of separate accessibility requirements in influencing designers and builders. Subsequent efforts produced some improvements, but he questions any requirements and labels that give designers the excuse to marginalize the needs of diverse users. He illustrates the continuing gap between what is built and what all people need, and he proposes changes in the professional educational process as one approach to good design for everyone.

Goldsmith, in Chap. 25, "The Bottom-Up Methodology of Universal Design," has a similar view. He asserts that the top-down prescriptive legislated requirements for accessibility in both Great Britain and the United States do not make buildings more convenient for everyone. Although he acknowledges that these statutory controls for making public buildings accessible have massively extended their general use, to achieve universal design he recommends a bottom-up approach that would begin with the needs of all users. He suggests, "For architects who are keen to take it on, informative design guidance would help."

Bringa, in Chap. 29, "Norway's Planning Approach to Implement Universal Design," explains the new approach of the Ministry of the Environment to develop a high-level strategy that would introduce accessibility as a goal in planning, to overcome fragmented, poorly coordinated technical approaches. In addition to national policy and research, a series of pilot projects as well as educational programs for planners are underway and being evaluated.

Urban Design

Universal design is comparatively new at the urban scale, and several authors bring important experience to this emerging area. D'Innocenzo and Morini, in Chap. 15, "Accessible Design in Italy," provide a thorough review of the development of accessibility in Italy, with a focus on public buildings and urban design. They emphasize the renewal of existing buildings and areas rather than new construction. Many of their case studies illustrate the way accessibility is regulated and applied. Noting that the focus in Italy in the coming decades will be on improving the quality of existing neighborhoods at the urban scale, they introduce a remarkable process of Neighborhood Agreements. Municipalities can apply for funding of this experimental tool for urban renewal that includes local planning and priority setting. They point to the "city for all" as the goal for urban and building processes.

Fletcher, in Chap. 60, "A Neighborhood Fit for People: Universal Design on the South Boston Waterfront," introduces another neighborhood-oriented urban design project in her Boston-based experience. Unlike the Italian emphasis on renewal of urban areas, Fletcher's focus is on the biggest new development in Boston's recent history. Acres of industrial waste, parking lots as well as prime waterfront property are part of the highly political project that will include massive building projects. She presents the strategies involving multiple partnerships, education for user/experts, and technical assistance to introduce universal design principles at the urban scale in the overall planning of the South Boston Waterfront. Manley also illustrates the application of universal design at the urban scale as she, with her university students, address the challenge in Chap. 58, "Creating an Accessible Public Realm." Her focus is to remove barriers in the disabling urban environment by using a mainstream, universal design approach that will enable all people to walk safely on city streets and sidewalks, minimizing reliance on the use of private automobiles. She introduces one method—collaborative street audits—which have proved useful in towns in the west of England. Involving local citizens with and without disabilities, audits were conducted and the results entered into a geographic information system (GIS) for capital planning. Manley explains the advantages as well as the disadvantages of the highly participatory and complex audit process. She provides the readers with extensive process information that can be applied elsewhere.

Tappuni, in Chap. 63, "Access in Rebuilding Beirut's Center," describes the process of creating a socially inclusive urban environment in the context of a developing economy. In Lebanon, the policies, criteria, and methodologies were established as part of the major reconstruction of a city center leveled by war.

User Involvement

User involvement in the teaching, learning, and practice of universal design is addressed by a number of authors. Ringaert, in Chap. 6, "User/Expert Involvement in Universal Design," introduces the primacy of the consumer expressed in the independent living paradigm as the basis for the involvement and contribution of people who are disabled in accessibility consulting work in Winnipeg and other parts of Canada. The role of the user/expert involves extensive education to enable people to offer more than their own personal experience as consultants, however valuable that may be. Ringaert describes the training programs in which user/experts acquired the technical expertise and cross-disability awareness to become consultants. She illustrates the application of that knowledge in a city-wide access consultation that involved multiple users, including design students and street kids, in Winnipeg. Noting that this is only the beginning, she emphasizes the need for further development and research in the work with user/experts.

Pedersen's experience in Western Australia also involves the users, but in a very different context. In Chap. 53, "Designing Cultural Futures at the University of Western Australia," Pedersen describes two teaching/research project experiences with architecture students in experimental design studios. Both projects explored an inclusive pedagogy that would lead to the students' appreciation of collaboration, diversity, and cultural equity. One studio involved the collaborative design of housing with a remote indigenous community; in the second studio, the students worked with disability consultants to experiment with universal design within the university. The teaching context that Pedersen discusses is intensely complex, and lightly sketched here to identify the range of issues surrounding user involvement.

Sustainability

Sustainability is mentioned by several authors as a component of their universal design process. Weisman and Walsh both discuss sustainability in depth. Walsh, in Chap. 33, "Sustainable Human and Social Development: An Examination of Contextual Factors," introduces a broad person-centered view of sustainable development that he believes can be a more appropriate and powerful context for universal design. He provides a detailed background through United Nations and European reports and treaties that promote sustainable design. He also criticizes the lack of success in many European design-for-all efforts to gain mainstream support for accessibility. Walsh proposes changes to the European Disability Agenda that would broaden research and development along with extensive monitoring and higher-level coordination by the European Union.

Weisman, on the other hand, in her postscript to the book (Chap. 69) sees universal design as providing the context and a base for environmentally healthy communities. She documents the connections between social and environmental health, linking environmental risk to poverty and racism. Her discussion of massive urbanization and the impact of modern building and in-adequate public transport emphasizes the challenge we face in sustainable urban planning. Her vision for the universally designed city not only defines the environmental imperatives ahead, but also offers great examples that illustrate solutions that work. The editors are mindful that sustainability issues and solutions in industrializing nations may be of different scale but share many characteristics of the industrialized world.

1.6 NEW DEVELOPMENTS IN UNIVERSAL DESIGN

Although the book is comprehensive and voluminous, there are still other remarkable developments in universal design that could not be captured within these pages, as we faced the constraints of what could be included in one volume, as well as the familiar phenomenon of those who produce great work but do not write about it. Fortunately, there are several relatively new periodicals, Web sites, and a new CD-ROM that document many of these efforts.

To highlight some recent European work, we note several designs that are featured in *Crisp and Clear,* a new glossy, well-illustrated periodical of the European Institute of Design and Disability (2000). Recent issues examine the Copenhagen Metro in Denmark, which is currently in design by the Italian firm, Giugiaro. The interviews with the designers reveal their enthusiasm in combining high style and ergonomics to create design for all. Also featured is the famous London taxi TX1, known as the "Black Cab," and the world's most universally designed taxi. Jevon Thorpe designed it in response both to his father's needs as a wheelchair user and to the requirements of the Disability Discrimination Act. Selected as an example of brilliant British creativity by the Design Council's Millennium Products Initiative, TX1 includes a recharging unit for mobile phones as well as a swing-out seat, an integral child seat, and a simple ramp that extends easily from the low-floor cab.

The *Universal Design* magazine from Japan is another highly visual periodical. It is relatively new, and began in 1997 primarily as a Japanese-language quarterly (Universal Design Consortium, 2000). It initially had a strong focus on health care settings but now includes schools, corporate, and residential facilities and services. The *Universal Design Newsletter,* a U.S. quarterly periodical that includes international developments, has published on universal design and accessibility since 1993. The vast range of articles have included golf course design, theater design, playgrounds, supermarkets, concert halls, historic Williamsburg, and talking signs. Back issues are available in bound volumes (Universal Designers and Consultants, 2000). The *Global Universal Design Educator's Online News* is the timeliest periodical, e-mailed monthly with international news on universal design. Back issues are archived and searchable online (Ostroff, 2000).

The *Proceedings of "Designing for the 21st Century II"* is readily available online (Ostroff, 2000) and includes papers from 15 countries that were presented at the international conference held in June 2000 in Providence, Rhode Island.

The *Exemplars of Universal Design* is a new CD-ROM that illustrates universal design excellence in 32 well-photographed projects selected by a jury from international submissions. They represent the work of many design disciplines—architecture, landscape architecture, industrial design, interior design, and graphic design. Supported by the National Endowment for the Arts and NEC Foundation of America, the CD-ROM was produced by and is available from the Center for Universal Design (2000). A superb and easy-to-use universally usable teaching tool, it relates the applicable Principles of Universal Design to each project. The descriptive text inserted to explain each image is a model of accessible visual media for people who use screen readers.

1.7 FUTURE CHALLENGES

The book illuminates many of the problems facing our society and the opportunities for a more broadly conceived universal design approach to mitigate these challenges. However, the practice of universal design is relatively young and has just begun to grapple with issues that affect its long-term development. The book introduces significant but limited efforts in each of the following three areas that will address the future success of universal design:

1. Education of design professionals
2. Evaluation of the impact of universally designed products and environments
3. Communication about universal design to the general public

Education of Design Professionals. Until universal design is infused in preprofessional and continuing professional education, the attitudes of designers will limit their understanding and appreciation of diversity. They will continue to shape their designs for a mythical average norm, creating barriers that exclude the contributions and participation of millions of people all over the world.

Evaluation of the Impact of Universally Designed Products and Environments. Universal design evaluation has barely begun. Although there is a rich history of environmental behavior research to draw from, there has been little research in universal design. Until we understand the impact of universal design on the usability and quality of buildings and products, we cannot establish meaningful criteria for future design.

Communication About Universal Design to the General Public. Mainstream communication is essential. Until the general public—which includes decision makers and clients—appreciates the value of universal design, it will not assume importance to designers. This publication will reach the professionals who want to integrate universal design into their mainstream design practice. To significantly influence the process of design, we need to reach the rest of society.

1.8 BIBLIOGRAPHY

Access Board, *Electronic and Information Technology Accessibility Standards,* Federal Register, Washington, DC, 21 December 2000.

Center for Universal Design, *Exemplars of Universal Design,* Center for Universal Design, North Carolina State University, Raleigh, NC, 2000.

Christophersen, J., and O. Gulbrandsen, *Studentboliger for funksjonshemmede* (Housing for disabled students), Project report no. 293, Norwegian Building Research Institute, Oslo, Norway, 2000.

European Institute for Design and Disability, *Crisp & Clear,* European Institute for Design and Disability, Aarhus, Denmark, 2000; www.design-for-all.org.

International Society for Rehabilitation of the Disabled (ISRD), *Proceedings of the Physically Disabled and Their Environment,* ISRD, Stockholm, Sweden, 12–18 October 1961.

Lamielle, M., Press release on environmental air quality research, National Center for Environmental Health Strategies, Vorhees, NJ, 28 September 1999.

Lusher, R. H., "Handicapped Access Laws and Codes," *Encyclopedia of Architecture,* John Wiley & Sons, New York, 1989.

Mace, R., *Universal Design, Barrier Free Environments for Everyone,* Designers West, Los Angeles, 1985.

McNeil, J., *Americans with Disabilities: 1994–95,* U.S. Bureau of the Census, Current Population Reports, U.S. Government Printing Office, Washington, DC, 1997.

Mullick, A., and E. Steinfeld, "Universal Design: What It Is and What It Isn't," *Innovation, the Quarterly Journal of the Industrial Designers Society of America,* 16:1, 1997.

Ostroff, E. (ed.), *Global Universal Design Educator's Online News,* Elaine Ostroff, Westport, MA, 2000; www.universaldesign.net.

Ostroff, E. (ed.), *Proceedings of "Designing for the 21st Century II, An International Conference on Universal Design,"* Adaptive Environments Center, Providence, RI, 14–18 June 2000; www.adaptenv.org/21century/.

Universal Design Consortium, *Universal Design,* Universal Design Consortium, Tokyo, Japan, 2000.

Universal Designers & Consultants, *Universal Design Newsletter,* Universal Designers & Consultants, Takoma Park, MD, 2000.

Welch, P., and S. Jones, "The Power of Imagination," in T. Mann (ed.), *Proceedings of the 30th Annual Conference of the Environmental Design Research Association,* Orlando, FL, 1999.

Wenger, B. L., S. Kaye, and M. P. LaPlante, *Disabilities Among Children,* Disability Statistics Abstract (15), National Institute for Disability and Rehabilitation Research, Washington, DC, 1996; http://dsc.ucsf.edu.

Wijk, M., *European Concept for Accessibility,* CCPT, Rijswijk, Netherlands, 1996.

CHAPTER 2
EXPERIENCING UNIVERSAL DESIGN

Patricia A. Moore, M.A., M.Ed.
Arizona State University, Tempe, Arizona

2.1 INTRODUCTION

This chapter begins with a personal scenario that illustrates the environmental interactions that many people face every day. It explains the empathic research through which the author gained in-depth experience about hostility of the everyday social and built environment. Critical parameters for universal design are identified, ranging from the need for self-determination to changing demographics, the reorientation of health care and the increasing technology in homes. The chapter identifies both the opportunities as well as the challenges for designers to address human needs, by design.

2.2 A DAY IN THE LIFE OF MANY PEOPLE IN THE INDUSTRIALIZED WORLD

Each day begins with a familiar routine. It may be the first light rousing us from sleep, or an annoying beep from that infernal alarm clock interrupting our favorite dream of the touch-down pass, or the sounds of traffic, or the ring of the telephone startling us awake in the strange bed of a hotel on yet another business trip. Instinctively we pat the nightstand in search of our glasses, or blink into consciousness, confirming that there is no time for a few more winks.

Free of the covers, wiggling toes locate slippers, or bravely manage barefoot to the bathroom, in a sleepy shuffle or an urgent run to the toilet. Soon we emerge refreshed from a hot shower, or the quick splash of water on our face. There's a shave or a makeup regimen, combing and styling of ample or precious locks, perhaps some medication or vitamins and a search for the uniform of the day.

After fumbling with buttons, reaching for hooks, and struggling with zippers, there are school lunches to pack, a gulp of instant coffee, or the pleasure of a sumptuous breakfast and the perusal of the morning paper. A mad dash for the bus or train, or a resigned wait in bumper-to-bumper traffic, or the repositioning of the garden hose, introduces us to yet another day of work, leisure, or play.

By the time we have joined the other members of our families, our neighbors, our coworkers, or the throngs of our communities, we have already encountered enumerable challenges and tasks, each activity defined by how we accomplish it, manage it, or avoid it.

Over the course of our lives, the quality of each day is determined by how well our individual capacities match the requirements with which we are presented. Our successes and our failures are the direct result of these interactions between the environment and our abilities, whether physical, psychological, social, or financial.

Perhaps the most significant variable in this delicate blend and precarious balance is the feature of design and the role it plays in recognizing needs and providing solutions. By its very nature, design assumes that individuals are capable of utilizing the places and products that are conceived and created.

Designers, regardless of their expertise and focus, should be chartered with the mandate that accepts all users as equal and all designs as enabling their use. But as anyone who has not been able to appropriately manage a chore and achieve a desired result can attest, everyday design is anything but forgiving, supportive, or universal in scope.

Throughout the world, regardless of the distinctions of culture, ethnicity, and geography, people are very much the same in their desire to cope with the challenges of daily life, to be accepted, appreciated, and of value to others. We may wear different clothes, eat different foods, speak different languages, and live in different ways, but we are essentially alike in our needs for shelter, sustenance, and purpose. That we are all individuals is what makes us all the same, yet just as it is with snowflakes, no two of us are alike. Our distinctions define our uniqueness. Our uniqueness defines our commonality.

The Lifespan Challenge

Our shared experience is the life span, that which is our beginning and our end. From the moment we are born, until the time of our death, we are consumers, users of goods and services, residents of domiciles, participants in communities. We are a tribe. We are a team. We are in need of support, and we are providers of care.

Throughout the span of our lives, we learn new skills, fail to accomplish goals, succeed at tasks. When we rose to our feet and took our first steps, it was the start of a personal journey of change and growth. With the love of our parents, the support of our families, and the camaraderie of our friends, we proceed through the course of our lives.

We recover from illness, injury, and surgery. We bear children. We break a leg on a ski vacation or in a fall in our home. We twist our ankle on a staircase. We slip in the tub and fracture a hip. We sustain a back injury in a car accident. The debilities we sustain can be temporary or permanent. We experience normative changes to our abilities as we age, with reductions in our capacity for hearing, seeing, movement, and strength.

With a lifetime of challenge, we rely on the care of people and the support of places and products that make us more able than we are alone. This is the universal mandate. Architects and designers become our pathfinders when they view the life span as a whole and when they recognize that each individual, regardless of his or her situation, is deserving of an appropriate design response. This is universality, by design.

The Universal Factor: Self-Determination

For the infant, the toddler, and the child, parents and extended families provide a home, nourishment, and nurturing. We anticipate the innocents' needs and desires, and we help shape their goals and their dreams with our selections on their behalf. We do so with love and pleasure because this is our duty and the nature of civilized, social behavior.

This is our role as primary purchasers, those capable of making choices and providing for life's requirements on behalf of individuals who, by the nature of their youth, vulnerability, and their financial status, are dependent on their parents and family members. Those who benefit from our selections and support as decision makers are primary users, the beneficiaries of our duty, devotion, and concern.

Without our applied means, our young would be made unable. As time passes, our children mature into adults with the same charges and responsibilities for their offspring, assuming the role of the primary purchaser, maintaining the primary user.

And so it is with other relationships—the grandparent who benefits from the gift of a new gizmo or gadget that might not attract their attention, but is selected by a loving grandchild who is aware of the item and confident that it will provide happiness and enhanced quality of life.

The realm of purchases made on behalf of others presents the advertising community with boundless positioning opportunities for potential consumers that might otherwise not recognize the impact of an offering. The tradition of the early "adopter," that consumer who always buys the latest technology and product, impacting the purchase decisions of family and friends, affords a valuable messenger to those end users with the greatest unmet needs.

The Question of Age

With the achievement, in the past century, of enhanced health and subsequent longevity, a new dimension of the cycle of provision has presented a heightened challenge for people of all ages. The existence of an ever-growing population of elders throughout the world has identified another range of generational support.

Our elders, dealing with changed capacity, reduced ability, and increased need, require the same accommodations and compensations in late life that they found necessary at life's start. Made increasingly more dependent and less autonomous by situations of poor health, chronic physical conditions, debilitating illness, fixed incomes, and changed social status, elders find themselves struggling to manage independent lifestyles in their homes and communities.

Uncomfortable and unwilling to simply have their daily needs met for them with the loving support of family and friends, our elders desire the means by which they can live independently, secure in the knowledge that features of architecture and design augment their abilities and enhance their capabilities for their entire life span.

This added challenge is one that the design community best addresses with a universal approach to the creation and introduction of specific technologies that address the plethora of activities presented to each of us, each and every day.

Acceptance and utilization of new offerings in architecture and design will require a user-centered awareness and understanding, focusing on the coexistence of multiple generations, representing a broad range of abilities and capacities. Our homes, workplaces, and communities must provide for daily living solutions that simultaneously address the distinctions of generations, their level of health, and amount of wealth.

Assumption that any one stage of life, or individual, is somehow more worthy of attention and provision than another is not only morally intolerable, it is economically infeasible.

As the baby boom cohort approaches late life, subsequent generations, their numbers minuscule in comparison, will find that hands-on caregiving will be difficult if not impossible. The only reasonable approach for planning and provision will be thoughtful design responses that augment personal care and overcome environmental barriers.

If people of all ages and abilities are to find quality for the entirety of their life span, it will be through the universal accomplishments of architects and designers.

2.3 THE DESIGN DILEMMA

With the end of World War II, and the first attempts to demonstrate the importance of building barrier-free environments, in large part, on behalf of our injured veterans, the advocacy community began promoting a sensibility of barrier removal. The further maiming of soldiers in the Korean War, and especially with the insistence of the injured Vietnam veterans, accessibility became a central mandate for the design of usable spaces.

Unwilling to simply return to a sedentary life, in the comfort and confines of their homes, cared for by loving spouses and families, these veterans demanded accessible homes and workplaces; improvements in the design of personal mobility devices, automobiles, and mass transit; and, perhaps, the most important advancement, a recognition by other members of their communities that they had a right to these accommodations. Their demands were part of the movement that led to the passage of the first federal legislation requiring architectural accessibility in federally financed facilities.

When architect Ron Mace, himself a wheelchair user as a result of having contracted polio as a child, first highlighted the need for universal design, a new movement in architecture was born.

Coincidentally, the industrial design community found itself being called upon to become more responsive to meeting the growing needs of a diverse consumer population with the creation of universally usable products for home life and the workplace. Guilty of perpetuating the myth that only the youngest of consumers were interested in new approaches and developing technologies, industrial designers did not have the regulatory requirements that prompted their colleagues in architecture to enable all consumers, by design.

The traditional course of design and development implied that consumers could be classified into two standards, or types. There were the average people ("normal" individuals), and there were "special" populations (persons with disabilities and elderly persons).

This insistence that somehow it was reasonable to identify and isolate people according to their age and physical and mental capacities resulted in the creation of environments and the production of products that, by their very nature, imposed a range of impending failures for people who were viewed as too old or unable.

The ultimate result of this segregation is the prejudice of *them* and *us,* a less-than-subtle prejudice that formulates good, real, and serious design as those solutions that meet the needs of the privileged and growing minority of the youngest, healthiest, and most able-bodied consumers over the needs and requirements of older people and individuals who manage life with physical and cognitive concerns.

The Empathic Model

In the mid-1970s, the disparity created by reinforcing stereotypes of inability with disabling products presented the author with the challenge of merging industrial design practices and methodologies with those of the social sciences and medical community. The focus was the creation of a better balance and opportunities for consumers with the presentation of products that were developed with the additional support of psychological and sociological considerations.

One of the most far-reaching research methodologies for enhanced understanding of user needs is that of immersion exercises, which result in a greater empathy with individuals and their unique, personal requirements (Pastalan, 1973).

In an attempt to better understand the conditions and situations created by a world designed and built for the most able members of society, the author undertook this approach with a unique empathic technique, in an endeavor to better understand a full range of consumer needs.

Utilizing prosthetic devices to reduce the physical capacities of a woman in her mid-20s, as seen in Fig. 2.1, in conjunction with theatrical makeup and molds, the author was able to transform into various women of approximately 80 years of age, as seen in Fig. 2.2.

A range of characterizations from impoverished to wealthy, ambulatory to reliant on wheelchair use for mobility, coupled with distinctions of reduced endurance, hearing, strength, and vision, allowed the author to travel to 116 cities throughout the United States and Canada for a period of more than 3 years (1979–1982) and experience life as an elderly woman.

The empathic portrayals allowed the author to experience the daily frustrations of living in a world that had been conceived and built with the most able and well users as its focus. Surrounded by a myriad of tasks, which comprise the completion of the activities each of us must accomplish in order to live an autonomous and independent lifestyle, inability was constant. It was reinforced by the negative reactions of people more able and capable.

Within weeks of beginning the experience, depression, self-doubt, and reduced self-esteem resulted. While the initial premise for the research was to understand and measure the challenges older people face interfacing with the built environment and the manufactured world,

FIGURE 2.1   Photo of author at the time of the empathic experience.

perhaps the more important result was the first-hand experience that individuals face when they are made unable, by design.

The pairing of age and ability in the determination of criteria for environmental and product design promotes a careful balance of the normative changes that occur in our bodies and capacities over the course of our life spans, as well as those distinctions that exist from illness and injury. The design community must address all consumers, all needs, and all situations, as equal in the weighting of mandates for design resolutions.

While it is not usually possible for one individual to fully experience and comprehend the life circumstance of another person, every person can consider others in making choices and determinations that affect all people equally.

FIGURE 2.2   Photo of author in character as a woman, age 85.

Offerings are not universally designed if only some consumers can open the package, read the instructions, use the product, and manage it within the setting they call home and community. The design imperative is to provide the necessary means for every person they can possibly serve.

2.4   REDEFINING DISABILITY

The role of language in the determination of cultural bias is often a subject of considerable debate and occasional ridicule, but regardless of individual beliefs, the labels that are utilized to describe individuals and their circumstances cannot be dismissed in terms of their significance.

The Western World distinguishes itself in terms of discrimination on the basis of age and ability. Our advertisements depict the superior qualities of youth and fitness. Our heroes are often those from the realms of entertainment, fashion, and sports. The average person is left to evaluate himself or herself on the standard of the fashion magazine cover girl, or the leading man in the latest action thriller.

For those of us who are of a certain age or ability level, comparisons to the most beautiful and the fittest present disappointing results. In an effort to determine when prejudice is acquired, images of adults, representing an age range from 20 to 70 years, each of whom displayed a visible distinction of physical capacity, were shown to preschoolers. The study, in which pictures were shown to about 1000 children and teens, was conducted in the New York area in 1989, with participants from Connecticut, New Jersey, and New York (Moore, 2000).

The children were asked to describe what they saw. Without exception, the children reported that they saw men and women performing specific tasks, such as cooking a meal, exercising their pet, or doing their job. The same images were then displayed to teenagers. Again, each participant reported similarly. This time, the teens saw "handicapped" people "trying" to perform the task depicted, or "disabled" men and women "attempting" to accomplish a chore. In a few, short years, these children developed a negative bias to individuals whom they viewed as less than normal and not average.

Specifically, the study's findings showed that the prejudices were formed after the age of 5 and by the age of 12. It speaks to the fact that adults, media, and multiple other, unknown sources are influencing the attitudes and, thereby, behaviors of young people. In a word, we are "taught" to be against and, therefore, afraid to grow older because of the mistaken beliefs and mythology that surround the unknown that is the future. Images in the media and advertising reinforce the fear of aging and perpetuate the stigma of being viewed as an elderly person. Inability to do a task, or "being handicapped," is linked and associated with advanced age.

The understanding that every person has a specific and distinct capacity underscores the fact that we all have different abilities and characteristics when compared with one another. When technologies and assistive devices, products, and services provide us an opportunity to enhance our physical and cognitive abilities, we are enabled by design.

2.5   DESIGN OR ART

The challenge for the combined specialists in the design community, the architects, engineers, graphic, interior, and product designers is not only to provide places and things that address the concerns of today and tomorrow, but also to provide for adaptations and modifications of existing environments and products that fail to embrace all people as equal. There can be no tolerance of any setting or item that defines, distinguishes, or segregates individuals on the basis of their capacity and ability.

When a person encounters a physical setting, way-finding signage, food packaging, or a home appliance that they find uncomfortable, unsafe, or unusable, they are experiencing an interface with art, not design. There will always be a place for things of beauty, artifacts that define culture and personality, but within the realm of the built environment and the manufactured world, consumers have a right to expect their unique needs will be met by design. This universal requirement for a quality life is the ultimate challenge for planners, architects, contractors, engineers, researchers, designers, and manufacturers involved in all areas of consumer provision.

Quality Is Designed

Thus, quality is not an accident, it is designed. Regardless of the industry, the one axiom all companies purport to follow is that of the provision of *quality* goods and services. Sensibly, would a company deliberately go to market selling low- or poor-quality products? And yet, throughout the industrialized world, consumers are left to contend with items that do not address their unique and individual needs. Consumers are left to morph themselves to the things and spaces that surround them. Instead of being supported by manufactured goods and built environments, people throughout the world are reminded that they must fit the quality of the product or place.

It can be argued, therefore, that those architects, designers, and engineers who are mindful and responsive in their offerings are creating quality by design. Further, if design is to be the great equalizer, quality design is mindful of each individual, and thereby must be universally usable.

The Universal Opportunity

Environments and products of all types are of benefit to each of us throughout the course of the life span. Every manufactured element can be enhanced and improved by a universal approach and appeal.

From the features of the transportation that delivers us to our jobs, to the quality of the lighting in the workplace that allows us to do our work, and the placement and design of the tools that we use to achieve our goals, we are dependent on design.

Every manufactured object and every built environment is the result of design. The extent to which all products and places function on behalf of every user is the effect of universal design.

OXO Good Grips: The Creation of a Universal Icon

The completion of the simplest of household chores can become an impossible feat when conditions such as arthritis limit grip, grasp, and necessary strength. For housewares guru, Sam Farber, the failure to manage cooking tasks had a particularly personal frustration. But the subject of the concern wasn't Farber; it was his beloved wife, Betsey, who struggled with ordinary kitchen tools. Together they gave birth to what has arguably become the icon of universal design, the OXO Good Grips products.

In the mid-1980s, Farber presented the challenge of creating a new approach to the design of traditional cooking gadgets to Smart Design, a New York City–based consultancy for product development. The Smart Design staff teamed with the author, and an array of medical practitioners to understand the impact of arthritis and reduced capacity. An intensive study of form and function, for a full range of users managing a myriad of health conditions, resulted in an aesthetically pleasing and highly utilitarian complement of food preparation tools, as seen in Fig. 2.3.

The market success of the Good Grips gadgets has been due to their emphasis on consumer comfort and usability in concert with engaging design. These personal prosthetics address the unique functional concerns of each user as equal to any other, a true universal response.

And with the addition of garden tools, office products, and a rethinking of every type of hand-held product, the universal design mandate has demonstrated that by design, a better mousetrap can be built!

Creating the Toolbox

From the moment we are able to hold a rattle in our hand, we live with tools. As we mature, our tools expand from playthings for our personal amusement to the objects we require to perform much loftier tasks, the *activities of daily living* (ADLs).

From a health care perspective, ADLs define our level of functioning, or our ability for independent living. From a consumer perspective, the activities we undertake throughout the course of each day determine our quality of life and capacity for autonomy.

FIGURE 2.3   Photo of OXO Good Grips.

An assistive device, like stainless-steel grab bars, might afford us the opportunity to safely and successfully position our bodies for toileting, but it does little to reinforce our dignity, maintain our self-concept, or define our home as aesthetically pleasing. A more effective solution, such as a toilet designed with a sculptural handle feature, or a seat that can be raised or lowered to a height appropriate to personal requirements, addresses universal needs for efficiency, security, usability, and style.

Consider the plethora of opportunities for universal creativity that exist in our homes. From bath and kitchen, dining and recreation, home office and bedroom, the place where we live, our "nest," defines the quality of daily life by the products and places, in which we live. Designs and features fill our individual toolboxes with the means by which we engage each day.

When our toolbox fails to provide a place or product we can safely, efficiently, and happily engage, the result can be frustration, injury, or displacement. Universal solutions have never been so necessary as when a person cannot function independently without them.

2.6   CHANGING HOUSEHOLD PROFILES

The definition of the traditional family has broadened dramatically since the birth of the baby boom generation. The images of the earliest television clans, with a stay-at-home mother, vacuuming in heels and pearls, Dad dining each evening in jacket and tie, bemused by the tales of precocious, yet perfect children, have evolved with the cultural press of women in the workforce, divorce, single-parenting, and so-called nontraditional couples. The mobility of the population has further reduced the familial supports that augmented parenting and care of elder parents.

The impact of this dramatic social change has yet to be fully realized, but strained finances, shared resources, and limited contact with parents and children have already defined the American family as a unit under stress.

With the aging of its populace, the United States faces the challenge of providing for the unmet needs of elders in isolation, as well as those who find themselves balancing their lives with those of their dependent grandchildren, or elders too sick and frail to manage independently.

The relative absence of younger people to assist in caring for the dramatic rise in the number of elders serves as a mandate for other means that will provide for the activities of daily living. That provision will be accomplished by design.

The Universal House

For more than 3 decades, architects, designers, and engineers have engaged in the exercise of creating a home that would anticipate and know the individual needs of its residents and provide them accordingly. Such smart homes would utilize an array of technologies to provide for comfort and safety, as well as performing more efficiently, and economically as less brilliant abodes.

From the earliest features of creature comfort and personal safety, such as automatic garage door openers, light timers, and thermostats, residents have craved for the benefits of higher IQs for their homes. This quest for technological improvements can be attributed to a growing awareness that universal features and solutions support the broadest range of consumer use and need.

Whether a homeowner ambulates with the assistance of a conveyance such as a walker or wheelchair, or if a member of their family hears with the augmentation of an aid, or the "lady of the house" stands 5 feet, on tiptoe, the complexion of residents demands an eclectic response.

When the design of a home responds to the full array of all residents, their families, and friends, then it is a universal house. Anything less is limited decorating.

Tomorrow's Hospital: The Home

At the end of the last century, the acute care health center had been dramatically redefined from the place where it began. The traditional hospital had been reoriented from its role as the primary source for health and healing, to one that essentially provides major diagnostics and surgeries. Birthing, physical therapy, elective surgeries, radiologic diagnoses, and enumerable health procedures have been designated to the status of outpatient services, or subacute care.

This transition, initially prompted by the network of insurance providers and reimbursement agencies as a means to reduce rising costs, soon proved itself to be a popular choice for patients who responded favorably to the opportunity to be able to return to their own beds and homes after treatment, in lieu of the confines of frightening and sterile hospital environments.

The promotion of a holistic approach to health has also heightened patient awareness of the differences in quality offered by various care options and has encouraged an aggressive response by both government- and private-sector providers. Given the choice, patients typically prefer home care to institutional care, and the corresponding improvement in outcomes is demonstrable.

For this approach to health maintenance and associated preventative care to be fully effective and successful, home modifications and necessary technological interfaces must be accomplished. The underlying essential element for this realization will be appropriate design solutions for the products and environments defined as the nests we call home.

2.7   SOFT TECH/HARD TECH

As technological interventions increase in range and frequency, consumers will confront a potentially negative dilemma. Can there be too much technology in our lives? Is technology always the answer? How do we manage emerging technologies as solutions and not substitutions?

Headlines and research remind us daily that Americans are woefully overweight and underactive. Suburbia is absent of sidewalks and overrun with highways. Processed foods and convenience distribution have made the dining room a place used only occasionally, and typically at holidays. Cell phones ring incessantly and e-mail has taken the place of snail mail. Convenience argues with tradition with every detail of daily living.

The risk of machines taking the place of people is one that becomes a critical concern when the result is isolation for the user. While Internet shopping might provide those with a scarcity of time an opportunity to acquire necessary goods and services, strolling through the aisles of the local grocery store or mall might be a source of precious social contact. When we no longer possess the capacity to drive a car, or ride as a passenger, we are at risk of isolation. Universal solutions for personal transportation provide for the potential of choice in our lives.

While the next accomplishment of science might bring the potential for increased life span, we should also consider, will the years be enhanced in turn? Care must be taken that we do not create homes and communities commanded by impersonal devices in the place of family and friends. Perhaps it will be best to always remember how a hug took away the fears of childhood, and a whisper of reassurance can be more influential than anything.

2.8   CONCLUSIONS

If the goal of equality for all is to be achieved, then equitable opportunities are the key. When we divide people into groups (i.e., the haves and the have-nots), we risk the objective of achievement of quality of life for all people. But when we deny the possibility for attainment of life's quality because our segregation creates and sustains a bias of a hierarchy by ability (i.e., the abled and the disabled), then we have deemed failure as reasonable and, worse, inevitable.

There can be no tolerance for an attitude of *them* and *us* in any circumstance. This arrogance that speaks to a claim that a situation is someone else's problem or concern is a potential condemnation of everyone.

The next time that you look into a mirror, open a door, or, a can of tuna, think of doing this under different circumstances. The next time you use an ATM, make a long distance telephone call, or record a movie on your VCR, think of doing this under different circumstances. See yourself, in any circumstance, and imagine the universal possibilities of inclusion by design.

And when you are incensed, inconvenienced, or injured because of the inadequacies of a product or a place, imagine what intervention might have occurred with a universal design.

2.9 BIBLIOGRAPHY

Moore, P., Personal Communication with the editors of this volume. December 2000. The study was conducted by Moore Design Associates.

Pastalan, Leon, R. K. Mautz, and J. Merrill, "The Simulation of Age Related Sensory Losses: A New Approach to the Study of Environmental Barriers," in W. F. E. Preiser (ed.), *Environmental Design Research,* Dowden, Hutchinson and Ross, Stroudsburg, PA, 1973.

P · A · R · T · 2

PREMISES AND PERSPECTIVES IN UNIVERSAL DESIGN

CHAPTER 3

AN INTEGRATED APPROACH TO UNIVERSAL DESIGN: TOWARD THE INCLUSION OF ALL AGES, CULTURES, AND DIVERSITY

Jim S. Sandhu, Dip.A.D.(Hons.), M.Des.R.C.A., F.C.S.D., F.R.S.A., Churchill Fellow
Inclusive Design Research Associates,
Newcastle upon Tyne, United Kingdom

3.1 INTRODUCTION

This chapter describes the genealogy of universal design, including conceptual and morphological issues. Historically, these are embedded in the struggles for equality and citizenship. The role of legislation in preserving civic rights, as well as in bringing about environmental changes, is discussed. Examples are culled from the author's extensive experiences in various circumstances. Attitudinal barriers are by far the biggest problem and are critical to the elimination of design barriers. This chapter advocates an inclusive approach to designing for all ages, including people from developing countries, rather than a compartmentalized approach. Characteristics of universal design with particular reference to information technology (IT) and its impact on the various design professions are outlined. The key message is that a multi-disciplinary approach in which integration and convergence are central is important. This chapter also covers sustainability, obstacles to progress in universal design, and its impact on older people and children.

3.2 TOWARD UNIVERSAL DESIGN

Some of you may remember the poem by Oliver Wendell Holmes about the wonderful one-hoss shay. This was a horse-drawn carriage that was going along the road when suddenly it stopped and collapsed, completely: The axles broke, the shaft broke, the wheels came off, the seat shattered, and the spokes of the wheels splintered. The driver found himself sitting on what was, in effect, a pile of firewood.

Holmes held this up to be an example of very good design. He argued that with most things, bits and pieces progressively go wrong until some vital, irreplaceable component gives way, and that is that—The whole thing has to be thrown away. What we have is a bad design, one suggesting that a weak component should have been made as strong as every other part. In the same way, there is no point in overdesigning other components so that they last forever, because they will only be thrown away with everything else. Holmes argued that the very best design is the one whose components all have exactly the same lifetime, which was the case with the one-hoss shay. All very logical and rational—just like universal design, as some would have us believe. However, the one-hoss shay was, of course, a metaphor for the whole of life and its various stages, not just death, dying and old age.

So what is *universal design?* The phrase covers a range of interpretations, which are often used interchangeably and overlappingly with design for all, barrier-free design, transgenerational design, *design for the broader average* or *design for the "non-average,"* as coined by the author for a new course in 1973 at the Polytechnic of Central London. Typically, in line with the whole process of new-concept generation, adherents of some of these approaches claim exclusive rights and insights to their own brand and phraseology of terms.

Universal design is a concept that extends beyond the issues of compliance with accessibility standards for people with disabilities and offers a powerful rationale for responding to the broad diversity of users who have to interact with the built environment.

Universal design is one aspect of a larger trend in the design fields described by Weisman as the "politics of inclusion and wholeness" in place of the patronizing "politics of tolerance and competing interests." Fundamentally, it is an approach that values and celebrates human diversity. Above all, it highlights a major paradigm shift—from treating people as part of the medical model, as dependent, passive recipients of care and services, to a model in which everyone is treated as an equal citizen and disability is seen merely as a social construct.

3.3   HISTORY

The genealogy of universal design has been hinted at previously in terms of the evolution of the various generic names. Conceptually, these started to emerge in slipshod fashion from the fuzzy mists of history. No doubt Thomas Paine's *The Rights of Man* had an early ideological impact, followed by his influence on the French Revolution with its short-lived secular emphasis on equality for all.

Although the parallel Industrial Revolution in the United Kingdom brought the benefits of making many more products available to a greater proportion of the population, the necessary standardization involved in the production process led to an even bigger gap between the product and the user. *Fordism* was a prime example with its, "You can have your car in any color as long as it's black." Historically, mass production can be seen as a setback for universal design.

An even more superficial influence was the Russian Revolution with its glossy focus on a brave new world and a golden age for all comrades. Closer to our times, in the 1960s, were two parallel movements that had far-reaching consequences: the evolution of the normalization principle in Sweden and the civil rights movement in the United States. The latter, in particular, influenced disabled activists to take the center stage to introduce a new conceptual framework—that disability is a condition imposed by society, that attitudes and inadequacies of the built environment were the main culprits. (The author has close first-hand experience of the historical developments in both Sweden and the United States, albeit, in the early 1970s— first, as a Churchill Fellow in disability to Sweden and then as a consultant to the New York State Department for Mental Hygiene—the most horrendous title for a public service imaginable, but also one that neatly encapsulates professional attitudes of a certain time in the context of this chapter.)

Prior to the Americans with Disabilities Act (ADA), there had been several legislative measures undertaken in the United States focusing on accessibility (e.g., the Architectural Barriers Act of 1968, Section 504 of the 1973 Rehabilitation Act, Section 508 of the 1986 Rehabilitation Amendments Act, etc.). However, it was the United Kingdom that had led the field covering most aspects of accessibility such as in housing, transport, employment, education, and so forth, starting in 1944. At the European level, the European Council of Ministers passed a binding Resolution A.P.(72)5 in 1972:

> On the planning and equipment of buildings with a view to making them more accessible to the physically handicapped.

Unlike the present ADA, no one in authority took the slightest notice of this resolution because there were no compliance measures built into it. With this in mind, we have to remember that design is no accident, but a mirror of its age.

The interregnum prior to the emergence of universal design was composed of key actors like Gunnar Dybwad, a senior Geneva-based World Health Organization (WHO) official who talked about *integrative architecture* in the early 1960s; Kenneth Bayes of the Design Research Unit in the United Kingdom, who set up the Centre on Accessible Environments in 1969; and Karl Grunewald of the Swedish Social Services Department and his team who started to translate the *normalization principle* into the built environment supported by Selwyn Goldsmith in the United Kingdom in the early 1970s. Then came a new set of actors who, unhappy with one-off conferences and seminars, set to move the emerging concept into the crucial educational arena. Ron Mace (who first used the term *universal design*), Elaine Ostroff, and James Pirkl were key players in the United States. The author set up his "Design for the Non-Average" course at the School of Architecture, Polytechnic of Central London in 1973, which not only entailed lectures by users and eminent thinkers in the field but also visits to doss houses, old people's homes, mental hospitals, the House of Commons, and centers of good practice.

By far, the biggest breakthrough came about with the ADA and several similar enactments in Canada, Australia, Spain, Holland, and with the United Kingdom's recent Disability Discrimination Act. At the international level, we have the UN Standard Rules for the Equalization of Opportunity—especially Rule 5 on Accessibility or Article 13 of the Amsterdam Treaty of the European Union, which focuses on equal opportunities and came into force on May 1, 1999. These developments have posed some fundamental doubts. How do you overcome pervasive attitudinal barriers when physical barriers can be neatly addressed with a few code-compliant measures? The inherent limitations of the emerging design standards in general have produced yet another reason for the concept of universal design.

A more urgent reason for supporting the concept was a realization that the very tools used to bring about greater accessibility were providing potent symbols of separateness as well. The official terminology of the legislative and clinical worlds tended to emphasize people's differences and suggested that their identity was their disability. Instead of accessible toilets, we have clearly marked "disabled toilets," very often reinforced by the outdated international symbol with its wheelchair user—just in case you forget who you are. Attitudinally and perversely, the good intention has turned into a handicapping environment, just like the black and white benches of apartheid. It is difficult to imagine special toilets for specific age groups—with their own distinctive symbols.

So, what about older people in all these developments? Paradoxically and strangely, in the context of design, the old are a very young phenomenon. In the past, they were simply subsumed under some category or other of disability. They have come to the fore by default. It is based on the fact and recent realization that the rise in the numbers of the very old will bear heavily on the resources in the future. No doubt in some sectors universal design is seen purely as a pragmatic solution to the resource question. In fact, the real problem, once again, lies elsewhere—in the imagination and in designers' attitudes. To paraphrase Bishop Berkley from the last century, "The old are a loveable and a delightful lot—and most welcome to Heaven, but of no earthly use." So what purpose do the old serve then? Who are they, and do the traditional divisions of human life into childhood, youth, middle age, and old age still fit our experience in the twenty-first century?

In the context of universal design, of course, none of these divisions matter—by definition. Age is a construct, a pattern woven by society, which changes out of recognition as times change. Even the traditional image of old age as a time of superior wisdom is passing away. In the developed world, we no longer have elders whose counsel is precious and who must be respected. Old people, who have become veiled in outworn stereotypes and new prejudices, are now the great excluded. Universal design, precisely because it does not focus on separate groups but on a broad spectrum of enabling environments for as many people as possible, offers great hope.

3.4   CHARACTERISTICS

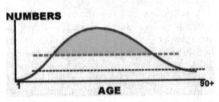

FIGURE 3.1   Graph illustrating principles of universal design.

Some of the earliest baseline efforts toward universal design focused on the functional fit of products and environments to people, resulting in anthropometric and human factors research. Sadly, much of the data that reached designers was based on the average, young, white, and able-bodied male as represented by the shaded part of the bell curve (see Fig. 3.1). The present inclusive approach, using the best principles of universal design, has shown that it is possible to increase the range and number of people to include all those above the bottom dotted line. And it makes excellent marketing sense.

Only 10 years ago, designers were still largely ignorant about using new universal design data that were beginning to emerge from various research efforts. What was considered to fall under the special or disability umbrella became another word for separate development. They confused *data* with *information* and *information* with *knowledge*. This attitude can be largely attributed to the development, use, pervasiveness, and seduction of *telematics*—the repository for quick answers rather than deep knowledge. *Telematics* is a largely European term to cover the integration and convergence across information-processing, communication, and media technologies.

The fact that the design professions are fast changing as never before, largely due to the impact of new technology, will have a major effect on the concept, evolution, and practice of universal design. We can be sure that this change will continue at an even more accelerated pace. The design professions are already aware of the following irreversible changes:

- That largely due to developments in information technology, designers are increasingly working in the service sector—much closer to the coalface—people.
- That designers can no longer be stereotyped as people who draw pretty pictures. They now use the awesome power of telematics and can call on local or international databases for wide-ranging information and images.
- That due to these developments, the dividing line between the various design professions is becoming hazy.
- That telematics is enabling untrained people to have easy access to specific design skills and information—giving universal design a new twist.
- That designers are becoming versatile at integrating software and hardware interfaces into a wide variety of products (e.g., smart equipment, smart homes, smart toilets, etc.).
- That generalist designers who have a wide range of interdisciplinary interests to complement their specialized skills are the keenest observers, the ablest communicators, and best able to work in the context of universal design.

These developments mean that designers are no longer simply focusing on products and the built environment but also on the design and provision of public services in the context of a barrier-free information society (see Fig. 3.2). For a totally inclusive approach to universal

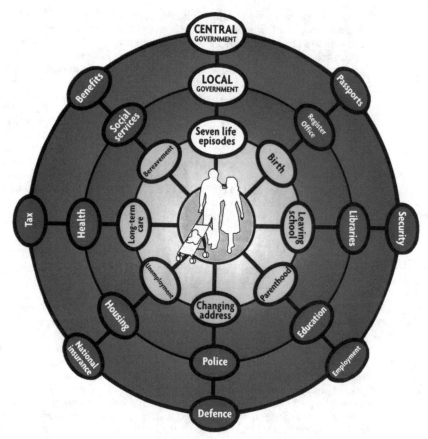

FIGURE 3.2   Public services and life cycles.

design, it is crucial to add transport, infrastructures, assistive technology, information technology, and more to this list. It is critical, therefore, that designers understand the various networks, bureaucracies, and processes that impact people during their lifetimes and sometimes enmesh them.

The approach to universal design, as adumbrated by the Center for Universal Design, the European Institute for Design and Disability, and others recognizes that accessible systems, reliable information sources, and enabling environments can maximize choice and enhance the ability of the individual to live independently and to exercise citizenship proactively. Underlying this is the fact that although not everyone needs assistive technology or specialized products like wheelchairs, everyone does need good design—whatever the context. Good design enables, and bad design disables, irrespective of the user's abilities.

Universal design is all about integrating the continuum of the micro- and macroperspectives of the surrounding world. Central to these perspectives is the designer's ability to carry out the type of detailed multilevel analysis that is required. At the core of this context-specific process is the need to understand the application of seamlessness to the user/task/environment/product interrelationship. It necessarily implies a multidisciplinary team approach with input from a broader range of specialities and areas of expertise (e.g., transport, demography, sociology, psychology, human factors, etc.). Increasingly, this synergy will be crucial to the further evolution of universal design.

Typical examples of multidisciplinary approaches in the European context involving the author are the European Commission–funded TURTLE, NEWT, and DISTINCT projects, which can be accessed at www.tag.co.uk/turtle, www.prosoma.com.lu, and distinct@newcastle.ac.uk. The first two centered on transport in northeastern England and Berlin, covering such topics as usage, accessibility, quality and mode of information provision, preferences, public information booths, infrastructure data, and the development of a real-time networked public transport information system. The large European consortium consisted of designers, engineers, ergonomists, computer experts, psychologists, a sociologist, various companies, service providers, and volunteers from a panel of 1000 users. The DISTINCT project (see Chap. 68, "User-Centered Deployment and Integration of Smart Card Technology: The DISTINCT Project," by Sandhu and Leibert) focused entirely on the use of smart cards and related interfaces to enable all citizens to access a wide range of public services across seven sites in Europe. The services that were covered included health, leisure, transport, education, housing, banking, parking, and so forth, all using the DISTINCT card as a single conduit. Although the consortium was different from the one described previously, it comprised a similar range of disciplines but also included local authorities and their elected representatives.

3.5   SUSTAINABILITY

Another essential element in the evolution of universal design will be the development of a more systematic infrastructure to enable the cyclical flow of resources and energy within the built environment and services domain. Universal design means that we have to pay greater attention to sustainability and the quality of the natural environment. It imposes a whole new moral tone and synergy on our activities. No nation is an island anymore due to globalization and the death of physical distances, and we have to take this into account. Attitudinal distances remain, however, and they are a major stumbling block.

In terms of these developments, it is not a lack of information that is the problem; rather, the problem is a lack of commitment. It could not be otherwise, given that we are assailed daily by near instantaneous newsflashes of disasters and disputes: Chernobyl and its consequences; limbs being hacked off by rebels in Sierra Leone; the mind-numbing plight of some of the 23 million refugees from around the world; the hopeless sight of some of the 120,000 children abandoned to misery in 600 orphanages in Romania alone; the unbelievable medieval treatment of women in some countries. Can politics and universal design sit side-by-side in these circumstances?

Right now, the fastest urbanization in history is underway. How does this figure into the overall theme of this book? How can universal design be used to make sense of cities that are already cauldrons of chaotic living conditions, communities of teeming crowds, and environments from Dante's inferno? How can universal design grapple with the ever-widening cycle of poverty and creeping paralysis in the world's bustees, barriados, favelas, kampongs, gecekondos, and the like? The phenomenon appears to be constant, but the treatment is not. Perhaps, it can never be, until world governments make it a priority to control pollution; deterioration of air, water, and sanitation; increased accidents; and conflict. Universal design has a powerful role in breaking this cycle, which results in long-term illness, misery, and poverty. A UNESCO study has shown that, worldwide, the most common form of employment for individuals with a disability resulting from these conditions is begging (Helander, 1993).

Still within the context of sustainability, but at the other end of poverty, there is evidence that suggests that denser cities use less energy for travel. This is crucial to minimizing pollution and its consequences. The question is, How do we add and incorporate the energy factor into universal design?

What new design metrics, standards, codes, morphologies, structures, and taxonomies do we need to develop to speed up the processes of amelioration and to enhance the quality of life? And how is the quality of life affected by the fact that the built environment is getting smarter by the day—with chips in walls, floors, chairs, ceilings, cars, wheelchairs, shoes, and so on? How can we harness the power of the chip to make it more accessible to older people and other marginalized groups? In what ways can the chip emulate the example of the wind-up radio to directly benefit citizens of developing countries?

Adherents of universal design should be aware of one additional problem: Governments everywhere, faced with hard monetary choices, are cutting back on social security. Developed countries that, earlier this century, proudly pronounced themselves to be welfare states, promising their citizens care from the "cradle to the grave," recognize that the state simply cannot afford such blanket provision in the new millennium. And this is happening at a time when the number of potential beneficiaries is rising steadily.

Developing countries, struggling to get out of the poverty trap, never had the resources to set up such services in the first place. Theoretically, they appear to be at a stage where things could only get better. The potential impact of universal design in these circumstances could be immense.

3.6 STUMBLING BLOCKS

Lessons from developing countries corroborate that social inclusion, equal opportunities, and mainstreaming are integral to universal design. In 1998, the Danish government, along with local activists, initiated the Advocacy and Action (AAA) Programme in six African countries: Ethiopia, Kenya, Lesotho, Uganda, Zambia, and Zimbabwe. The main objective was to bring together representatives of disability organizations and external experts to discuss action on the UN Standard Rules. They then concentrated on specific aspects, in each country, such as inclusive education, employment, or habitat.

The AAA workshops identified six major barriers to the enjoyment of full human rights by marginalized groups in these countries, which are supported by the author's extensive experience of developing countries around the world:

1. Hostile attitudes by officials and "ignorant" members of the public.
2. Lack of resources and institutional capacity.
3. Lack of enforcement. Although there are laws and regulations, there is often neither money nor the will to ensure that they are enforced.
4. Urban emphasis in activities, leaving large tracts of the countryside uncovered.
5. Environmental barriers that make it extremely difficult for citizens, especially those with a disability, to access services in both urban and rural areas.
6. Inadequate services in health care, education, training, and rehabilitation.

If universal design is to be increasingly propagated in developing countries, as it should be, then its proponents need to understand these local conditions and to work alongside local people. Anything otherwise and they would have Gabriel Garcia Marquez (1983) flung at them:

> The interpretation of our reality through patterns not our own serves only to make us ever more unknown, ever less free, ever more solitary.

3.7 AGING

We need to examine the very foundations of the Principles of Universal Design, which have been developed largely by American colleagues in order to realize that these are not set in stone but will continue to evolve. It should be noted that other basic principles, which are also simply but cogently spelled out in the International Federation on Ageing's (IFA) *Declara-*

tion on the Rights and Responsibilities of Older Persons, have a more human face because they are not formalistic. The face, however, is ageless—as it should be. In fact, seen through the mirror of superimposed reality, the image reflects back the timeline of a person in one image. The ultimate goals of universal design are encapsulated in the five Rights of the IFA, with minor modifications, which were based on the UN's "Principles for Older Persons." Interestingly, they all apply to children as well, thereby supporting the maxim, "Design for the old and you include the young—design for the young and you exclude the old":

The built environment should maximize *independence.*

It should enable full *participation* in society.

It should enhance the provision and process of *care.*

It should provide a platform for *self-fulfillment.*

It should enhance individual *dignity.*

A number of current trends in society and the global market clearly justify the need for action regarding older people. Society is aging, and because women tend to live longer than men, this group is becoming increasingly female. The statistics on European demography are by now well known (Sandhu and Wood, 1990). At the end of 1994, the European Union (EU) comprised 70 million people over the age of 60. This represents almost 20 percent of the total population, and projections indicate a situation whereby one-quarter of the population will be over 60 by the year 2020. There is evidence that this is paralleled in most other developed countries, including the United States and Canada.

Not only are older people making up an increasing percentage of the overall population, the number of older people over 80—particularly at risk of frailty and dependence—will increase even more. In this context, a 300 percent increase in the period from 1960 to 2020 is forecast. This has major consequences for the financing and organization of health and care services. In addition, the family composition, the context in which most caring occurs, is changing: An increasing number of women are entering the labor market, and family members are becoming widely dispersed. As the prevalence of both temporary and permanent disability and functional problems is increasing due to aging, accident rates, and modern health technology provisions (enabling people to live longer, thereby increasing the risk of disability and ill health), the burden on social protection and care systems is increasing dramatically.

Care services are often fragmented with many different professionals involved, with little cooperation and coordination between services. Universal design, in combination with information and communication technologies, could make a practical and useful contribution in enhancing the integration of these services. Choice and independence are fundamental criteria for quality services for this group of users. The question is, How can universal design incorporate these issues into its framework in order to begin providing for real-life solutions?

3.8   CHILDREN

Childhood is the very basis for the type of life, quality of knowledge, and experience found in later years. In the context of concepts such as the quality of life and sustainable development, which are economically feasible, ecologically prudent, and socially just, childhood takes on a whole new meaning. In the civic sense, a *Society for All* has been widely recognized, especially in the developed world when speaking of the *Information Society.* Best practice in this area has clearly demonstrated that it is possible to bridge the gaps between old and young, urban and rural, men and women, educated and illiterate, disabled and able-bodied. Universal design has made it possible to establish these links and make these developments real and touchable in our everyday lives.

Pipe dreams? In the developed world, not at all—Technology has kept its promise, overcome many hurdles, and is capable of delivering the goods, even though it has not yet percolated to every nook and cranny. The vast range of alternative means of access to computer-based systems is a prime example. There are vast segments of the population, however, to which these developments are denied for one reason or another—the millions of children worldwide who are orphans, refugees, incarcerated in institutions that are not fit for animals, as beggars all over, aimlessly fleeing in the thousands from war and conflict as in southern Sudan, unsheltered, disabled, and deprived in every way possible.

What is it that a child misses when all he has seen is blood and savagery? What does she miss by not being able to handle things, by not walking, by not being able to ask questions or voice fury or fantasy? Why does it matter that he should do these things? Does she in fact learn to compensate or not? How can we help him to make up for his deprivation and lack of experience or, in some way, not to miss what he has missed? What, if any, are the permanent effects of these missed experiences?

If universal design is to fulfill its role, we need to find urgent answers to these conundrums. In terms of specific deprivations, how do we provide solutions for those who are:

- Sensorily deprived through not having normal sensory experiences
- Culturally deprived through poor environmental conditions and pattern of life
- Maternally deprived through not having a mother substitute provided for them
- Recreationally deprived through not having any play facilities
- Linguistically deprived through little contact with others
- Depersonalized through traumatizing experiences, being treated as one group, one crowd, having no personal possessions, and undergoing mass routines

These aspects of deprivation may contribute additional problems for children who already suffer from physical and mental impairments: emotional insecurity, further intellectual impairment, and social incompetence.

The author's first half of 30 years' working in the field was entirely devoted to designing and researching solutions to some of these issues in several countries, which mainly focused on children and the built environment (Sandhu and Hendricks-Jansen, 1976) and how to use it to compensate for sensory-motor problems. In line with universal design, the practical outcomes were based on an in-depth understanding of child development generally and the role of play and playthings in particular (Sandhu, 1982). Play is a powerful tool for social inclusion.

Despite the title, it was with social inclusion in mind that the author helped to organize a seminal exhibition called *Playthings for the Handicapped Child* at the Royal College of Art in 1971. It was the springboard for the Toy Libraries movement in the United Kingdom, and for the first time, catalogs began to appear with new sections for "special children." As catalogs became glossier, so did playthings. Designers focused largely, however, on either restyling previously successful products to meet the changing needs, or the introduction of gimmicks calculated to match the prevailing imagination of children. There was a noticeable increase in television advertising and television character merchandising (e.g., Barbie, Action Man, the Hulk, etc.). This artificial creation of demand was based mainly on the adult world expressing its childhood psyche, dreams, and fantasy and had very little to do with good design. Needless to say, this practice continues to this day.

The *Playthings* exhibition was an incredible success. It was featured several times in all of the major British media, including the American *Time* magazine. It traveled for a year and a half around Britain. It is to the great credit of the Royal College of Art that from then on, disability and aging became integral to its overall remit. Historically, the key people who brought this about were the Rector Sir Robin Darwin, Sir Hugh Casson, Sir Misha Black, Professors L. Bruce Archer and Chris Cornford—all household names in the United Kingdom. The latter two nonknights were key to the change.

It is difficult to convey the general rigidity in which the value of play is expressed and passed over in favor of the real business of education. In the context of playthings, perhaps, the most overused word these days is *educational*. Advertisers splash *developmental* or *instructional* across their catalogs as the irresistible inducement to parents and teachers. The fact is that most playthings are merely props, that most children learn to play by instinct, and experiment is buried in glossy hyperboles. Playthings still tend to be thought of like convenience food, which, no doubt, has its uses to some parents (like TV, which is a perfect time filler). However, there is a danger of producing not only toy-trained children but parents as well—people who know how to fit a jigsaw or manipulate electronic consoles but have little idea of fun.

Most people have become so conditioned that they forget that a pile of empty boxes or wooden blocks may, at the right moment in development, have more educational potential than, say, a number game. The crucial goal of universal design is to bring fun back to playthings and to clarify the interregnum between the empty shoe boxes and Action Man.

3.9 CURRENT OBSTACLES TO PROGRESS

As mentioned earlier, a great number of largely attitudinal problems are a stumbling block to progress in realizing universal design. Eight of these problems are briefly itemized:

1. In some non-design-related sectors, universal design is presented as a *panacea* for all the ills of civilization. It gives false hope, especially to marginalized groups who understand the concept even less than the so-called expert. It also implies that design is a facile discipline, instead of being one of the most complex and demanding disciplines—even more so when trying to achieve universal design objectives.

2. Design professionals have *vested interests* in propagating their own profession to the exclusion of others, which detracts from the synergy that is required. On one hand, telematic developments are breaking down professional barriers by making various skills and information readily accessible to all; on the other, professional bureaucracies, which are often made up of business, administrative, or management personnel, seem to be battening down the hatches.

3. Partly linked with the preceding is *clientelism* and the need to make a fast buck—which means developing countries and marginalized groups tend to get ignored. Just as capital tends to chase cheap labor and quick returns, the design professions largely tend to chase prestigious and well-paying clients.

4. There is a strong *resistance* to innovation not just by professional bodies but also by public bureaucracies of civic authorities. Traditionally, established bureaucracies have seen change as a threat to their status quo. This impacts on the various design professions who have to interface with them.

5. Professionals rarely like to admit *ignorance* about multidisciplinary and multilevel issues (e.g., the relevance of service considerations, information access, etc.). It is a paradox that at one level the design professions pride themselves on their ability to access key information on the Internet and use this as a leverage with users faced with mundane, day-to-day problems, but they are unable to admit that it requires a multilevel approach to solve these.

6. Public service innovation can be stymied by *financial constraints* due to economic circumstances. This, again, largely impacts on developing countries.

7. There is a *communication gap* between the various actors: architects, designers, users, researchers, developers, administrators, suppliers, manufacturers, bureaucrats, venture capitalists, and the like.

8. Increasingly, there is an *information gap* due to technology, between the information rich and the information poor, which exacerbates the problems outlined in item 7. This means that technology itself is beginning to create additional problems for marginalized groups who do not have access to information technology for whatever reason.

3.10 OTHER CHALLENGES TO UNIVERSAL DESIGN AND SOCIAL ACCESSIBILITY

At a broad social level, some of the major challenges facing universal design can be summarized as follows:

- To create a legal framework that enhances and protects existing democratic rights (e.g., privacy protection, democratic structures, etc.)
- To establish practical rules of engagement, which will encourage people to use new technologies, especially in the context of quality content, easy access, and reasonable tariffs
- To promote awareness of the real opportunities that are available for the citizen in the areas of work, education, health, environment, new services, and the like
- To ensure that products and markets meet the highest standards to satisfy consumer needs, in the context of intellectual property, common technical standards, choice of products and services, and so on.

3.11 CONCLUSION

Universal design is not yet a coherent and systematic approach to designing for people. It has many missing pieces in its complex jigsaw puzzle. The recent focus on convergence, vigorously advocated by the UN Standard Rules and the European Commission holds great promise for the propagation, practice, and evolution of universal design. The Commission's Fifth Framework probably means little to non-European readers, but it is going to be a powerful engine for new developments and concepts concerning the built environment and the information society. At the hub of these developments is a truly multidisciplinary approach to resolving social and technical issues.

Companies need to see designers as pathfinders for identifying the lifestyles that are likely to occur and the products and services that are relevant to changing situations. This means that, more than ever, they have to identify design as a core competence, support it, and rely on it.

As we struggle for a world that is seamlessly accessible, seamlessly supportive, and seamlessly caring, we must never forget that no machine will ever be able to replace the superior wisdom that comes only with age and experience. Not now, nor in 100 years' time. It is wisdom and not intelligence, per se, that makes us unique and human. It should once again take the center stage. However, neither should we forget that it is childhood that provides the foundation for that superior wisdom. Universal design is an important tool in this continuous and stupendous process of living. It not only spans the entire life cycle, but it is also inclusive of cultural differences, and it encapsulates the developed and the developing worlds. It implies that designers can learn and benefit from understanding certain lifestyles and practices of the latter. At its best, it affirms life at every level—human, animal, or vegetable.

Finally, it is incumbent on all those concerned with the topic not only to have a mission but also a vision, to see the world as it can be, not merely as it is. Sustainability is no longer merely a concept—It is on the top of the world agenda. At the core of this concept is the redefining of wealth to include natural capital: clean air, fresh water, an effective ozone layer, fertile land, clean sea, and the abundant diversity of species. This is best expressed in the words of Wendell Berry (1995):

> We have lived by the assumption that what was good for us would be good for the world. We have been wrong. We must change our lives, so that it will be possible to live by the contrary assumption that what is good for the world will be good for us. And that requires that we make an effort to know the world and to learn what is good for it. We must learn to cooperate in its processes, and to yield to its limits.

3.12   BIBLIOGRAPHY

Berry, W., *Private Property and the Common Wealth: Another Turn of the Crank,* Counterpoint, Washington, D.C., 1995.

Helander, E., *Prejudice and Dignity: An Introduction to Community-Based Rehabilitation,* United Nations Development Programme, New York, 1993.

International Disability Foundation, *The World Disability Report,* IDF, Geneva, Switzerland, 1999.

International Federation on Ageing, *Declaration on the Rights and Responsibilities of Older Persons,* IFA, Washington, D.C., undated.

Marquez, G. G., *The Fragrance of Guava: In Conversation with Plinco Apulayo,* Ann Wright, (transl.), Verso, London, UK, 1983.

Sandhu, J. S., *Playaids Catalogue,* Newcastle Polytechnic, Newcastle upon Tyne, UK, 1982.

———, *Demography and Market Sector Analysis of People with Special Needs in Thirteen European Countries: A Report on Telecommunication Usability Issues,* European Commission's RACE TUDOR Project, Newcastle Polytechnic, Newcastle upon Tyne, UK, 1990.

——— (Ed.), *Usability Issues for People with Special Needs. A report for the European Commission,* RACE-TUDOR R1088 Project, University of Northumbria, Newcastle upon Tyne, UK, 1992.

———, "A Holistic Approach to the Design-for-All Concept," *European Institute for Design and Disability Newsletter,* no. 2, 1995.

Sandhu, J. S., and H. Hendricks-Jansen, *Environmental Design for Handicapped Children,* Saxon House, Farnsborough, UK, 1976.

Weisman, L. K., "Towards an Architecture of Inclusion," *Outreach ACSA Women's Issues Newsletter,* October 1994.

Welch, P. (Ed), *Strategies for Teaching Universal Design,* Adaptive Environments Center, Boston, and MIG Communications, Berkeley, CA, 1995.

3.13   RESOURCES

Web Sites

Access to the Internet—Bobby.   www.cast.org/bobby

ACT Centre: Accessible Page Design Home Page.   www.ott.igs.net/~starling/acc/index.htm

Americans with Disabilities Act.   http://janweb.icdi.wvu.edu/kinder

The Center for Universal Design.   www.design.ncsu.edu/cud/index.html

COST219—Access to telecommunications.   www.stakes.fi/cost219

Include.   www.stakes.fi/include

Physical Disability.   www.eskimo.com/~jlubin/disabled.html

Royal National Institute for the Blind.   www.rnib.org.uk

Trace Research and Development Center.   www.trace.wisc.edu

The Web Accessibility Initiative.   www.w3.org/pub/WWW/Disabilities

CHAPTER 4
DESIGNING FOR OUR FUTURE SELVES

Roger Coleman, M.A. (Hons.) Edin., F.R.C.A., F.R.S.A.
Royal College of Art, London, United Kingdom

4.1 INTRODUCTION

There is an emerging international trend toward developing and adopting universal design and associated strategies in response to social and demographic change, to accessibility legislation and related building codes, and to consumer pressure and new market opportunities. This trend is not uniform. There are marked differences between developments in different countries, resulting from cultural and historic factors, and in particular from the contributions made by individuals and organizations. By tracing the history of local strands, understanding what makes them distinct, and identifying elements of convergence and difference, the concept of universal design is enriched through diversity. In this chapter a specific U.K. history is described, linking concepts of "design for need" and "socially useful production" that originated in the 1970s, with a growing understanding of the facts and implications of population aging, and later initiatives intended to raise awareness and change practice within the design profession, education, and industry. Population aging raises important issues that cannot readily be addressed by legislation; consequently there is a need to gather data, construct a cogent and convincing business case for "inclusive design," and provide exemplars of designs capable of stimulating an older marketplace by offering improved quality of life and responding to the rapidly changing lifestyles of older people. The Royal College of Art (RCA) in London has played a significant role in these developments in the United Kingdom, and in Europe, as described in this chapter. However, there are many other histories that need to be gathered together to give a full picture of the origins and diversity of what is emerging as a new, international movement in design.

Background: Living Longer

One day, cycling through his village, the author came upon a large notice reading, "Happy Birthday Roger—40 today!" On closer inspection a more sinister subtext was revealed, in the form of a second notice that declared "No more squash and tennis, carpet bowls and dominoes from now on. Roger, park your invalid carriage here." Not a celebration but a public demonstration of just how deeply ingrained is the fear of growing old, and the concomitant and largely erroneous association of age with dependency and disability. The truth is rather different. Since 1900, U.K. life expectancy has increased on average by some 2.5 years per decade, one of the great achievements of the twentieth century (Kirkwood, 1999). As people age, they become more diverse as individual life-courses give rise to divergent experiences, interests, activities, and capabilities. In the context of aging populations, such diversity will increase, in particular as people explore the new possibilities opened up by the 25 or more years of life expectancy that have been added in the course of the twentieth century (Laslett, 1996). (See Fig. 4.1.)

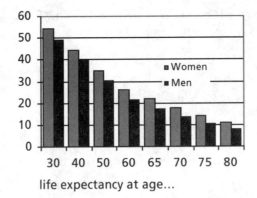

FIGURE 4.1 U.K. life expectancy by age. *(Continuous Mortality Investigation Bureau, U.K. Institute of Actuaries)*

However, as people age, they change physically, mentally, and psychologically (Haigh, 1993). The gradual accumulation of multiple minor impairments—of reductions in eyesight, hearing, dexterity, mobility, and memory—impacts on their ability to maintain an active and independent lifestyle (Laslett, 1998). Although not disabled in the conventional sense, older people are increasingly disabled by a mismatch between the environment they live in—buildings, products, and services—and their own changing capabilities. Faced with a dramatic and unprecedented growth of older age groups, it will be necessary to challenge common assumptions about the nature and spread of disability within the whole population (Coleman, 1993b). For example, although the wheelchair has come to symbolize disability, in reality only a minority of disabled people use wheelchairs; consequently a wheelchair-friendly world is not necessarily a disability-friendly world. Furthermore, the majority of wheelchair users are old, as are the majority of disabled people, while the great majority of older people are not disabled (Gill, 1997). Until assumptions about the nature and prevalence of disabilities more accurately reflect the facts, the needs of the disabled population as a whole are as unlikely to be addressed as are the needs and aspirations of aging populations (Smith, 1990; Steenbekkers and van Beijsterveldt, 1998).

Disabled by Design

In America *people with disabilities* is currently the accepted term, whereas in Europe many people object to the implication that disability is something they are born to, or acquire by accident, or by aging. Here, one preferred term is *disabled people,* which suggests that disability is something people have thrust upon them by an inadequate environment. The author endorses this view. In his opinion, people are disabled by design, rather than their particular capabilities, and everyone is likely at some point in their life to experience disability as a mismatch between themselves and their environment (Laslett, 1998). Since the human habitat, in its urban manifestation in particular, is constructed and designed almost entirely by human beings, if that world does not work for an individual at any particular time, then that person is disabled by design, if not by intention (see Fig. 4.2). The reality of aging populations will give high profile to this issue, particularly in Europe with its heritage of old buildings and cities, which are frequently, by their very design, disabling.

FIGURE 4.2 Disabled by design.

Enabled by Design

However, just as design can *disable,* it can also *enable* (Benkzton, 1993; Goldsmith, 1984). Mismatches can be eliminated through appropriate user-aware design (Coleman, 1994b; Fisk, 1993). Unfortunately, examples of good enabling design, particularly of products, are few and far between. Changing this reality requires not just legislation, which at the moment impacts more on the built environment than on goods and services; it requires a change in attitudes among designers, manufacturers, and service providers. That change will not come about without a clear and convincing market case being made, without readily accessible information and design guidance, and without good-practice examples. Older designers may find it hard to relinquish elitist attitudes of the past, whereas younger designers can be sensitized to the opportunities for innovation that come from a user-aware design practice and methodology.

With the worldwide aging of populations and the considerable social and economic impact this will have in the twenty-first century, there is now an opportunity to place the enabling potential of design center-stage in creating an open, accessible, and integrated world for the future—an opportunity on which the universal design movement in the United States, and the growing focus on design-for-all and inclusive design in Europe can deliver. Design for Our Future Selves is a related concept developed as part of the DesignAge program at the RCA under the author's leadership (Coleman, 1993a, 1994a and b). In this chapter the origins of this thinking are described, as well as how and why that particular concept was developed and put into practice, and how the idea is being taken forward within the recently established Helen Hamlyn Research Centre at the RCA.

4.2   EARLY BEGINNINGS

Socially Useful Production

DesignAge was established in 1991, under the author's continuing direction, and with the support of the Helen Hamlyn Foundation. His interest in the project stemmed from having worked on the development of products and services for older and disabled people as part of an initiative launched by the Greater London Council (GLC) in 1984. During a downturn in the U.K. arms industry, the workforce at Lucas Aerospace proposed an alternative corporate plan based on 150 "socially useful" product ideas (Wainwright and Elliot, 1982). Although never taken up by the company, this people-centered approach to economic development excited interest in many countries. In an attempt to put theory into practice, the GLC took on the leader of the Lucas initiative, Mike Cooley, as director of technology at the Greater London Enterprise Board, where his major achievement was the establishment of a Londonwide group of Technology Networks. By acting as catalysts for collaborations between designers, researchers, engineers, community-based organizations, and skilled workers, the Technology Networks set out to close the gap between producer and consumer. What that vision urgently required was concrete examples to make it understandable and convincing. As coordinator of one of these Technology Networks, the author had the job of creating such examples. Unfortunately, Margaret Thatcher's government disbanded the GLC before the initiative had time to prove itself. However, the Technology Network the author was running managed to continue as an independent research and development organization, London Innovation Ltd., and bring some of these socially useful products and services to the marketplace.

Example Projects

Some of these socially useful projects included the Cloudesley Chair, a component-based seating system for disabled children aged from 6 months to 16 years; the Mini-Gym, an exercise machine for people in wheelchairs; the Neater Eater, which allows people with intention tremor to feed themselves independently; Mobile, a household repair service for disabled and elderly tenants in two London boroughs; and the Lynch Motor, an innovative electric traction motor, ideal for environmentally-friendly electric vehicles—mopeds, city-cars, etc.—and outboard and inboard marine drives. Concepts taken to working prototype stage included a road-rail bus for rural areas, capable of linking existing roads and railways; and a low-cost, energy-efficient self-build housing system. Research activities encompassed the ergonomic and user evaluation of a low-floor bus for London Transport, and research for British Telecom into problems associated with telephone dialing.

As a consequence, when asked what *socially-friendly* meant, it was possible to point to real products and services developed with a high level of input from end users, and designed in an attractive, user-friendly way.

From Design for Disability . . .

The author's interest in these issues began when a friend developed multiple sclerosis. After a period of decline, she found it difficult to care for herself, and a decision was made that she should be placed in institutional care—standard practice 20 years ago. However, it was clear to the author that the friend's apartment, not her condition, was the source of the problem, and that with some design input this could be rectified. The case was put to her local authority, and it was eventually agreed that if her kitchen could be suitably improved she could stay at home. With another of her friends the author designed, built, and installed a new wheelchair-friendly kitchen for her, and she lived at home for many years (see Fig. 4.3).

What this woman wanted above all was not a "disabled" kitchen but a stylish cooking space. Personal surroundings communicate strong messages about identity, social position, and values, which makes meeting people's aspirations as important as functionality and problem solving, if not more so. This thinking was carried on into the Technology Networks initiative, where the author's experience convinced him there was a fundamental problem with concepts like design for disability and special needs design, which were current at that time. Not only was there little recognition of the aspirational and identity issues involved (the main emphasis being on problem solving), but the end users were treated as a marginal and dependent group, and the dominant aesthetic was medical.

The consequence was and continues to be, products that people do not like using, a mar-

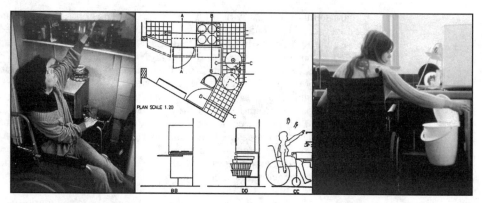

FIGURE 4.3   Wheelchair kitchen, before and after.

ket for them dominated by the National Health Service, and cost constraints that reinforce a medical aesthetic and make it uneconomic to invest in good design (Gardner, Powell, and Page, 1993; Mapstone, 2000). In Europe, markets are fragmented by country and by culture in a way that is quite unlike that in the United States, and it is not surprising that the benchmark "universal" product, Good Grips™ made its way from there to Europe, not vice versa (Coleman, 1997b, 1999b). Legal issues of entitlement and obligation to supply, which determine whether people will be provided with equipment by the state, have predominated in Europe, and this has made it difficult for manufacturers, retailers, and statutory services to improve the quality and provision of suitable products (Mandelstam, 1997). What was needed was a reorientation of mainstream design and production—one that would design out disability by making products and services as user-friendly as possible. But the problem, in the United Kingdom and in Europe, was that there was neither a willingness among mass-market manufacturers to embrace an inclusive or design-for-all approach, nor was there a willingness among specialist manufacturers to move outside the safety of markets dominated by statutory services.

What would drive this change was not clear at that time, but if it was to be consumer-led, as the author firmly believed, then the way to convince hardheaded business people was to identify new market opportunities. At London Innovation, research into telephone dialing problems articulated a clear message from the users consulted—a broad range of people with specific disabilities and age-related multiple minor impairments. The message was, "Why a special phone for us; why not a better phone for everyone?" As a consequence, the researchers set out to attach numbers to these problems as a way of arguing the market case for a better phone for all, not an easy task prior to 1991, as until then the U.K. National Census did not record disability-related statistics. Unfortunately, that message was ahead of its time and was not acted on by British Telecom, but one thing the figures revealed was the rapid growth in the number of older people, and the author saw in this the market opportunity to drive change in the United Kingdom and Europe, where the population-aging trend was firmly established.

. . . to New Design for Old

In 1986 a seminal exhibition, sponsored by the Helen Hamlyn Foundation, was held in the Boilerhouse Gallery of the Victoria and Albert Museum in London (Manley, 1986). Its intention was to demonstrate that with thoughtful design it is possible to ensure that older people have the opportunity to live life to the full and maintain a natural dignity and essential independence. Some of the finest design talent in Europe was brought together in this exhibition. The brief was to develop a collection of new designs for older people; forward-looking concepts that would challenge the commercial world to recognize an increasingly significant market for age-friendly products with a universal appeal. The contributors included: from the United Kingdom, architect and designer Alan Tye; Nick Butler (director of BIB Design Consultants); Kenneth Grange (director of Pentagram); and cutlery designer and manufacturer David Mellor; from Finland, Antti and Vuokko Nurmesniemi; from Italy, Vico Magistretti and Massimo and Lella Vignelli; and from the United States and the Netherlands, Robert Blaich (director of design at the consumer electronics giant, Philips).

This heavyweight team produced a mold-shattering exhibition. Furniture, clothing, consumer durables, bathroom and kitchen fittings, door furniture, and personal items showed that products for older people could combine functionality with the appearance and material qualities that made them a pleasure to use and own (see Fig. 4.4). *New Design for Old* demonstrated that addressing the needs of older people in the design of everyday items was a route to innovation, where the lessons learned from one sector of society could be applied to design as a whole. What emerged was the beginning of a new design approach offering both human and commercial benefits.

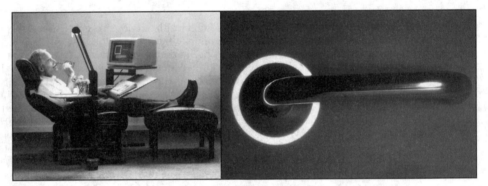

FIGURE 4.4   "New Design for Old" exhibition: the "Jefferson Chair" by Niels Diffrient; luminescent door furniture by Alan Tye.

Student Design Awards

On display alongside the professional exhibits were the results of a competition organized as part of the Student Design Awards of The Royal Society for the Encouragement of Arts, Manufactures & Commerce (RSA). This prestigious annual event was established in 1924, and is now the premier European student design competition. Since 1986 the *New Design for Old* section has become the flagship of the competition, dealing as it does with an important social issue in keeping with the RSA's mission: to work to create a civilized society based on a sustainable economy (Coleman, 1996). Sponsored by the Helen Hamlyn Foundation and other bodies, including the European Union, British Gas, IDEO, the Wellcome Foundation, and the Mercers Company, this section of the competition has gathered strength over the past 13 years, with over 1000 student entries, while some 5000 students have worked to the annual brief as part of their degree courses. Winning entries range from consumer products to interior designs and intergenerational games (see Fig. 4.5).

In 1994, Gavin Pryke's innovative redesign of the glass jar, developed at the Royal College of Art, was not only the winning entry in the *New Design for Old* section, but winner of the Master's Medal, a personal commendation awarded by the late Jean Muir, fashion designer and master of the faculty of Royal Designers for Industry (RDI), who admired the seamless combination of functionality, appearance, and performance that had come from an in-depth study of user and industry priorities.

FIGURE 4.5   RSA Student Design Award winners: 1994, "A jar we can open" by RCA Ceramics and Glass Student Gavin Pryke; 1995, door security by University of East London Industrial Design student Sally Muddel; 1992, flexible handled cutlery by RCA Industrial Design Engineering student Susanna Steele.

Design for Need

Having stimulated interest on a national basis, the Helen Hamlyn Foundation next entered into an extended collaboration with the Royal College of Art (RCA), leading to a dedicated research center exploring the social dimension of design. The RCA has a long-standing interest in social issues. In 1976 it staged the Design for Need conference, which brought practitioners from around the world to London (Bicknell and McQuiston, 1977). Subjects ranged from ecology, environmental policy, and the recycling of materials to self-build housing, workplace design, designing out disability, equipment for emergencies and disasters, and beyond to design education in developing countries. Two of the closing papers brought the overall thrust of the conference into sharp focus. Mike Cooley talked about "Design for Social Use," and Victor Papanek's subject was "Because People Count: 12 Methodologies for Action." Both argued for a shift in emphasis from the object to the user, from the producer to the consumer.

In retrospect, this then-radical philosophy is remarkably close to contemporary concerns in the United States to extend the relevance of universal design beyond a narrow association with the Americans with Disabilities Act (ADA) and other disability legislation. The designers gathering at the RCA in 1976 shared a confidence in the power and effectiveness of design as a tool for improving the quality of people's lives and a belief in the ethical obligations that come with power. They were sure they could change the world for the better, and do so by design—a vision and sense of purpose in common with that of some sections of the universal design movement now.

4.3   DESIGNAGE

The collaboration between the Helen Hamlyn Foundation and the Royal College of Art began with the establishment of DesignAge, an action-research program exploring the implications for design of aging populations. This brief presented the author with a unique opportunity to take some of the ideas previously worked on at London Innovation from the margins to the mainstream of design thinking and practice. Because DesignAge was not a teaching program, it could develop as a cross-disciplinary activity engaging with postgraduate students from fashion and textiles to vehicle design, and from photography to industrial design and engineering (Coleman, 1994a). Though unique in being an entirely postgraduate university, the RCA is tiny and highly specialized, and so it was also essential to collaborate with other institutions; with manufacturing, retailing, and design companies, and with older people's organizations.

The results have included exhibitions, conferences, competitions, new designs and products, publications, and a growing body of information about the subject gathered together in a special collection at the RCA. All of this has helped to raise awareness and give design students and professionals alike access to information they would otherwise not be able to source in the often short time frame of a college project or commercial development program (Coleman and Pullinger, 1993; Coleman, 1997a; Myerson, 1999).

Population Aging

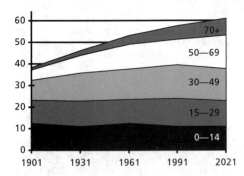

FIGURE 4.6   Aging in the United Kingdom 1900–2021: millions by age-group. *(Coleman from U.K. census, 1991)*

One of the things to emerge from the survey of demographic, ergonomic, and social data carried out at the RCA was the fact that preconceptions about aging are decidedly out of phase with reality (Coleman, 1993b). For the whole of recorded history, until as recently as the late seventeenth century, no more than 1 in 10 of the U.K. population was aged over 60. Currently one in three is over 50, and by 2020 almost half the adult population will be aged 50 or over (Laslett, 1996). This trend will continue and probably be replicated worldwide. In addition, due to plummeting birthrates, the world population will not rise exponentially, as many environmentalists fear, but in all likelihood stabilize within the next 100 years. The result will be a radical and probably irreversible change in the shape of human populations. In the past, populations were predominantly young, but over the last 150 years this historical trend has been reversed. In the United Kingdom, between 1900 to 2020, almost the entire growth of the population has been, and will continue to be, in the 50-plus age group. (For related statistics, see Figs. 4.6 and 4.7.)

Conventional wisdom has it that the youth market is the economic motor for the future, and much of the consumer culture of the late twentieth century has been built on that assumption. However, what is clear from the facts is that every additional consumer added to the marketplace over the past century has been an older person. An opportunity has been missed to develop products and services that more closely match the needs and aspirations of the

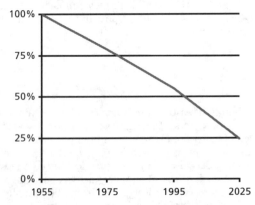

FIGURE 4.7   Global fertility rates: percentage of world population reproducing at above replacement rate. *[After World Health Organization (WHO) Report, 1998.]*

only consumer sector with real growth potential (Buck, 1990). In the future, mainstream consumer markets will have to concern themselves with the substantial rise in the number of people who are less than able-bodied yet wish to enjoy an active and independent lifestyle—or these markets will face a reduction in growth, turnover, and profitability (Coleman, 1994b, 1999c).

There is an interesting convergence here between environmental concerns and universal design. The environmental damage due to human activity is increasingly recognized, and a stable or falling global population is seen as an essential component in a sustainable future. But stable or reducing populations will be old, not young. As a consequence, design in the twenty-first century will have to concern itself with both sustainability and inclusivity—not from the worthy and somewhat patronizing motives of the past, but with a new realism that acknowledges the convergence of social and economic interests in creating a stable and equitable world in which people of all ages and abilities can live independently, contribute to society, and fulfill their personal aspirations (Coleman, 1998).

Age and Disability

In this context it is important to understand the relationship between aging and disability. In many research and development programs there is an implicit assumption that the needs of older people and disabled people are the same—a failing that has held back progress. More recently the European Union has recognized the importance of adopting an inclusive or design-for-all approach (Ballabio, 1998), and efforts are now being concentrated on developing the design and management tools and strategies that can deliver a meaningful response to issues of population aging and disability (e.g., Norris and Peebles, 2000). However, there remains a widespread ignorance of the underlying facts, and therefore an inability to interpret them and respond accordingly.

Dr John Gill, of the U.K. Royal National Institute for the Blind (RNIB), has constructed figures for the incidence of a range of impairments across the continent of Europe (Gill, 1997) (see Fig. 4.8). Seen alongside the data on population aging, these figures lead the author to believe that in developing ideas about barrier-free access and universal design, the needs of some groups have been advanced very successfully, while those of other groups have been

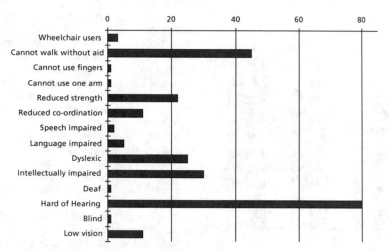

FIGURE 4.8 Disability (millions) in geographic Europe (800M). *[After John Gill, U.K. Royal National Institute for the Blind (RNIB).]*

overlooked or ignored. Contrary to common assumption, most disabled people are old, not young. The number of people with age-related hearing loss far outweighs those with profound deafness; just as the number of people with low vision far outweighs those who are registered blind. Add to this the number of people wearing glasses due to age-related presbyopia, and conclusions about the composition of the "disabled" community that differ radically from prevailing assumptions are inescapable.

There are issues of definition here. First, the number of people registered as "disabled" does not necessarily give a true indication of the number whose lives are affected. Official measures vary from country to country, and are used to determine who qualifies for welfare and other benefits. Second, while the great majority of older people are not in any officially recognized sense disabled, many suffer the consequences of multiple minor impairments which though individually below disability thresholds, can accumulate and significantly challenge their ability to live an active and independent life (Haigh, 1993; Fisk and Roberts, 1997). Adding their number to the "official" figures would force a significant revision of assumptions about the nature, extent, and impact of impairment among the whole community.

What is needed is a new way of understanding and quantifying disability that more accurately reflects the number of people likely to be disabled if their needs are not taken into account in the design process: a shift in the perception of disability away from physical and mental condition toward inadequacies in design, and a shift in professional attitudes away from the stigma of what have for too long been referred to as "aids and adaptations" toward an ethos and aesthetic of enabling design. This will strengthen the case for universal or inclusive design methods, while applying this new understanding will present important challenges and opportunities to industry and the business sector.

The subject is complex, and guidelines will not necessarily deliver the improvements they are meant to produce, especially if they are not updated in line with evolving knowledge and understanding. The numbers of what might be called *critical users* will vary with the particular subject under consideration, be it a telephone or a bus. In each case a key task will be to identify the whole range of potential users, and among them those vulnerable to exclusion by inappropriate design. Taking the needs and capabilities of these critical or vulnerable users into account, both in the design process and in trials and postproduction evaluation, will be essential in ensuring that good intentions lead to real-life benefits (Freudenthal, 1999; Nayak, 1998). Even here, given the rapid evolution of lifestyles and aspirations among older age groups, it is difficult to determine precisely by whom and under what conditions products and services will be used.

It also is necessary to factor in the 15 percent of the current European population aged 65 or over, the 1 percent per annum projected increase in the retired population over the next two decades, and the fact that the 85-plus age group is the fastest-growing population segment in Europe. What this means is that there will be many more retired people and more frail old people. However, there is some indication that in the United States, the rate of increase in age-related disability, cardiovascular disease, osteoarthritis, and so forth, is falling relative to the growth of the older age groups (see Fig. 4.9). This is encouraging news, and offers the possibility that successful medical and design interventions, in particular those that encourage and support independent living, could eventually deliver a zero growth rate for age-related disability in the developed world, despite population aging.

In developing and third-world countries the impact of population aging will be significantly different, placing higher stress on the social and economic fabric. In particular, the movement of young people away from rural areas to fast-growing conurbations is likely to leave older people without the support of the extended families they previously relied on. This, coupled with

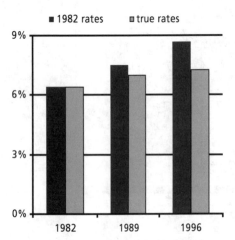

FIGURE 4.9 Age-related disability in the United States; 1982 rates against actual rates (1996). *(WHO Report 1998)*

limited health and welfare provision, will make it even harder for them to sustain life quality into old age.

Challenges

The overriding challenge is to create an open, accessible, and integrated world for the future: a supportive environment of buildings, products, and services that make it possible for everyone to live independent and fulfilling lives, for as long as possible. The more effectively this can be achieved, the less strain will be placed on social and welfare systems, and the more older people will be encouraged to spend the now considerable wealth they control on the goods and services that deliver independence and quality of life. There is considerable market potential here that could stimulate economic growth. Unfortunately, there is no real consumer offer for older people, no rationale for spending, and a lack of understanding of their wants and needs.

The second challenge, therefore, is to develop a consumer offer that closely matches the aspirations of older people. At the RCA the DesignAge program has sought to build an understanding of this issue by working directly with older people, through a collaboration with the University of the Third Age (U3A), a fast-growing and self-organized association of retired people. U3A members attend regular user forums at the RCA, where they meet students, participate in focus groups and other research activities, and discuss consumer issues with professional designers and industry managers (Coleman, 1997d) (see Fig. 4.10). Bringing older users into the college gives young students an opportunity to talk through ideas, develop concepts, and later test prototypes and research specific issues of styling, aesthetics, and usability. The RCA students and U3A members build up a high degree of mutual trust, and all find the experience interesting and enjoyable. This interaction with older consumers gives students a rapid insight into how to develop appropriate products and services, and into the pitfalls that await them if their approach can be seen to be in the least patronizing.

FIGURE 4.10   User forums with members of the U.K. University of the Third Age (U3A).

A Fresh Map of Life

Peter Laslett, one of the first to identify the population aging trend, and author of a key work on its social and economic implications, has pointed out that because this trend is so new, older people are moving into uncharted territory where there are no precedents and no role models (Laslett, 1996). To understand this change, and to add substance to the essentially empirical insights coming from the collaboration with U3A, DesignAge has also developed a relationship with marketing and trends research companies like TMS Partnership, the leading U.K. fashion and clothing industry analyst, and the London-based Henley Centre for Forecasting. As a consequence, both organizations have revisited existing data from large surveys (Coleman, 1993a).

By reexamining purchasing activity between 1976 and 1992, TMS has established that while there was a small but persistent decline in the 20- to 24-year-old share of the U.K. clothing market, over the same period the 65-plus share doubled. The conclusion drawn by John Harrison, Director of TMS, is that the clothing industry as a whole can only contract and lose high street presence if it fails to stimulate older people to respond more to fashion. To do that, it must offer greater choice in both ranges of clothes and types of outlets, and importantly, a better understanding of how body shapes change with age, and how cutting, styling, and sizing need to change to reflect this. Similarly, in reviewing data from its 1996 Leisure Tracking Survey, the Henley Centre unconventionally segmented the over-50s rather than the under-50s, to arrive at a comparison of older people's leisure preferences by age with those of younger people as a whole. The results show a marked preference for restaurants as opposed to pubs; the importance of walking as a leisure activity well into one's 70s; a rising reliance on the car among the over-65s; and a continuing interest in do-it-yourself home improvements. Understanding these trends helps to identify new opportunities for products and services that more precisely meet the requirements of older people.

Building on this and other data, the Henley Centre has developed a matrix of wants, needs, and aspirations that is beginning to effectively track older people's behavior. Increasingly, they are looking on later life as an opportunity for self-realization, a time to do many of the things that were not possible while they had responsibilities in the shape of young families and careers. In order to achieve that goal certain things become important: managing time to make the most of these new possibilities; managing the aging process itself, by staying fit and healthy and later by finding ways to remain independent for as long as possible; living with the fragmented family structures that are the consequence of increased mobility, lower birthrates, and the rise in divorce and single-person households; understanding how to manage money and ensure financial security over many years; and maintaining the social networks of family and friends that give a sense of identity and belonging.

. . . and Opportunities

In a recent book (Coleman, 1999c), the author has described what he sees as the three broad areas of opportunity arising from these changes.

First, the mainstream provision of goods and services in both public and private sectors, and importantly, in the leisure and retail sectors, along with the built environments that surround them, can benefit from becoming age-friendly. Here the concept of "barrier-free" architecture has a role to play in removing obstacles and encouraging the participation of people of all ages and abilities. As does universal or considerate design in everyday objects: packaging that can be opened without taking a knife to it, chairs that are easy to get in and out of, clothing that is light, warm, easy to put on and to clean, but still elegant and good looking; signage and information graphics that are clear and easy to read, controls that are simple and easy to use. These things will broaden the appeal of goods and make life better for all that use them, not just older people.

Second, niche products and services targeted at older people in ways that are positive and nonpatronizing, like 50-plus insurance policies, activity holidays, well designed, adaptable housing and age-friendly IT products. There are other opportunities waiting to be developed. For instance, the bulk of the do-it-yourself and interior decor market is targeted at younger householders, while there is a need for older people to make changes to their homes that will help them live independently. Unfortunately, people are not aware of what products are available, how they could benefit from them, or how they could improve their homes with good, age-friendly interior design.

Third, there is a growing market for specialist products and services—from personalized interfaces and other assistive products, through home-based services and on to institutional care—that address the needs of the minority of older people with disabilities, and those who require regular care. At present, this market sector is poorly serviced. Health care and welfare services have been reluctant to spend money on research and development, and as a consequence most technical aids have been designed with functionality and economy uppermost and are depressing to look at and stigmatizing to use.

Manufacturers that invest in good design and seek to break the prevailing institutional mold will find their products and services have a marketability outside the institutional sector. To understand how this approach can be put into practice the author has brought together some brief case studies, ranging from design methods to innovations in product development and marketing.

4.4   CASE STUDIES

The first example, from the international design company IDEO, is a scenario developed in response to a product challenge issued jointly by DesignAge and the U.K. Design Business Association in November 1993. The scenario explores a range of proposed improvements to the London bus system, telling the stories of "critical" users of the system. The designers at IDEO use this and related techniques to help them quickly get under the skin of different users, and think through the product or service they are working on from perspectives other than their own (Moggridge, 1993). By describing three individuals and meeting their specific needs, the designers have identified many improvements which they have integrated into the system in ways that make it work better for everyone (Coleman, 1997b).

The Urban Bus: A Scenario

The first character is Agnes; she is 73 years old and would like to obtain shelter from the weather while she waits for the bus. She would also like to know how long she has to wait, and not have to climb steps to board the bus. The key to meeting her needs is an enclosed shelter with a raised floor and ramped access, allowing both ambulant and wheelchair users a level entry to the bus (see Fig. 4.11). The shelter incorporates an information system and a ticketing machine. The bus driver's seat turns around so that he can face her and make eye contact with her, without having to twist his neck.

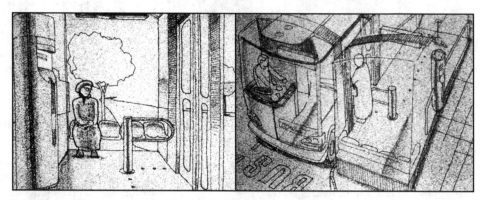

FIGURE 4.11   Agnes's story. *(IDEO)*

The second character, Sarah, has two younger children. Her priority is getting help managing children, shopping, and money all at the same time. One way is to offer her a prepayment system. At the supermarket, she can charge up a prepayment card for using with the bus service, and so avoid having to search for change when on the bus. She offers the driver the card. He slips it into the reader and returns it to her. This makes life simple for Sarah and also gives the transport company useful information about who uses the bus and when. Once on the bus, Sarah discovers there is plenty of room for her bags. (See Fig. 4.12.)

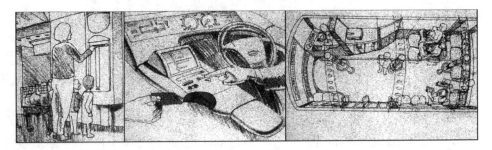

FIGURE 4.12   Sarah's story. *(IDEO)*

The third character, Sam, is from Malaysia and is visiting London on business. His priorities are payment options, an easy-to-use system, and some help with finding his way around. If one is unfamiliar with a transport system it can be very difficult to use. It can be difficult to locate the bus stop, and in some cities tickets have to be bought in advance and are not available on the bus. Here the stop is clearly marked, with a structural element that includes a ticketing machine. The instructions are simple and logical, with a choice of languages. The machine will take coins or credit cards, a great boon for anyone who does not have the right currency or coins. As a final touch, the ticket incorporates a route map with icons marking key buildings and monuments. (See Fig. 4.13.)

FIGURE 4.13   Sam's story. *(IDEO)*

User Consultation

The next example is that of a German company, Heinrich Wilke GmbH (HEWI), that has specialized in architectural ironmongery and associated products for barrier-free environments. At a practical level this has meant ensuring that their entire range functions for people of all ages and abilities. However, an otherwise barrier-free environment can be marred by one inappropriate detail, and such mistakes are often made simply because builders, tradesmen, and designers do not have the right information at hand. Consequently, HEWI publishes illustrated brochures for architects, builders, and planners on subjects like barrier-free living and adaptable housing, which have been produced in close cooperation with potential users, caregivers, and professional specialists. In conjunction with a program of educational seminars and workshops for tradesmen and other installers, this makes the necessary understanding and expertise available to people in the industry.

The company has also embarked on a new communication policy built around an open-ended dialogue with its immediate customers—architects and building managers—and the end users of these products and environments. This strategy was launched in September 1996, with an international conference bringing together architects, planners, and product developers, along with experts from the fields of sociology, design theory, and gerontology, and most important, potential end users and representatives from the areas of health care insurance, social services, and housing management. The company's aim is to not manipulate the market, but to understand and respond to the needs and aspirations of the people who use its products—which in turn implies that product development should be the end result of a process of consultation, which is now the case at HEWI (Coleman, 1997b).

Age-Friendly Interiors

The German company Bisterfeld und Weiss (B+W), a well-known manufacturer of quality contract furniture, identified a market for furniture for institutional settings. The company first developed an easy chair with a footrest, arms that fold down to make it easy to transfer to and from a wheelchair, and a side pocket for personal items. The product was successful, but designer Arno Votteler was aware of aesthetic shortcomings in this sort of product. As a consequence, the company decided to develop a range of furniture that would do away with the stigma associated with existing institutional products. The result was a collection of tables, chairs, cupboards—in fact all the items required to create an attractive personal environment—especially for people living in a small space. Such furniture could bring functionality to the domestic environment, without compromising on appearance, and bring the warmth and taste of the home into institutional settings and sheltered housing (see Fig. 4.14).

FIGURE 4.14   New range of age-friendly furniture by *B+W.*

A specification was developed, based on market research and consultation with end users and care professionals. Design and development was carried out over two years, and all items were given ergonomic trials in institutional settings, and user-tested for performance and acceptability. The appearance of the furniture is restful, with attention given to the hygiene issues that are important for care managers. The overall intention was to communicate safety, stability, comfort, and quality through a domestic styling that would break the institutional mold and offer a mix of durability, functionality, appearance, and value that could form the core of a new care sector offer that puts feeling and well-being first, and bridges the gap between home and institution (Coleman, 1997b, 1999b).

4.5 DESIGN FOR OUR FUTURE SELVES

It was not enough to research these important social changes in an academic way. If the DesignAge program were to be successful it would also have to engage the hearts and minds of RCA students. RCA graduates reach a very high standard during their two-year practice-based MA course, and the level of hiring by industry is well over 90 percent in the design disciplines. This offered a way to influence the design culture in major companies around the world, but that influence would not be felt if the student body did not engage with the research carried out by DesignAge. The average RCA student is 27 years old, and though many arrive with a high degree of social concern and some professional experience, it is not easy for them to grasp the significance of designing for people whose wants, needs, and aspirations are not the same as their own.

The theme of the program Design for Our Future Selves was chosen with this in mind: to offer young designers a personal way into the issues; and to encourage them to think in terms of products and services that engage with aging as a natural part of the life-course. If young designers conceive of products or buildings they cannot use in their old age, then they have failed the challenge. The Design for Our Future Selves concept focused student project work not on older people per se, but on themselves in the future. The program has concentrated on the convergence between social and commercial imperatives, on arguing the case for age-friendly design and encouraging industry and the design profession to recognize the opportunities offered by an older consumer market. It has done this by bringing together factual and trends information, along with design-relevant tools and guidance (Coleman, 1999a and d).

Working with Older Users

Rather than concentrate on what is wrong with current products, the emphasis has been on future generations of products and services that work well for everyone, not simply in functional terms, but by responding to changing wants, needs, and aspirations in an holistic way. To add flesh to the bones of this concept, exemplars have come from an annual competition at the RCA, and from collaborations with industry and professional designers. A good example of collaboration, and of the benefits of end-user involvement in the design process, is the 1996 Sitting Pretty initiative taken by the U.K. South-Eastern Museums Service. Respected designers were invited to submit proposals for new seating for two of London's museums, and then to work with older and disabled people to ensure the resulting designs would not only meet the needs of these groups, but would appeal to all museum visitors.

Leighton House attracts many visitors, including a high proportion of older people. Located in South Kensington, not far from the Albert Hall and the RCA, it was the home of Lord Leighton, a famous Victorian painter. The original decor is well preserved, including an internal courtyard with columns, fountain, and mosaics inspired by Arabian architecture. The challenge here was to design a range of seating that would complement a unique museum, and at the same time cater for a wide range of visitors.

DesignAge worked with furniture designer David Colwell, providing two groups of older users for him to interview and test prototypes on. An important concern to emerge from this experience was the height range the seating should cater for, ranging from older women, who could be less than 5 ft tall, to men measuring well over 6 ft. The final design has many features that are age-friendly. A broad base and cross-shaped undercarriage allow the front edge of the seat to curve gently inwards, and so accommodate different thigh lengths without restricting circulation in shorter people. Generous seats allowed for larger posteriors, and long arms with a good handgrip make sitting down and standing up easier. The coloring, upholstery, and detailing were chosen to match the very special setting, and the range includes a chair with arms, a stool, and a corner seat, again with arms, all of which can be arranged in "conversational" groups.

David Colwell went on to incorporate key elements of his design into a general-purpose chair which is now produced by his company, Trannon Furniture. Manufactured from sustainable coppiced ash, the chair is lightweight (a benefit for anyone who has to move it), modern in its styling, yet domestic in the use of natural materials. The result is a thoughtful combination of people-friendly and sustainable elements and a fitting design for the future (Coleman, 1997b).

Shopping in the Future

Beginning in 1992, the design team of Safeway Stores (U.K.) worked closely with RCA tutors and students over a 12-month period, to develop a range of innovations offering benefits to customers of all ages and abilities by extrapolating from the particular needs of older consumers. In parallel with this, a young manager action group investigated older consumer issues from a company perspective. The combined results were presented to the Safeway board of directors, and many of the lessons learned were transformed into improvements in store design, packaging, customer service, and other aspects of the company's business.

The first stage of the collaboration identified many features that create problems for older customers, from bending and stretching associated with high and low shelves to information design, signage, labeling, lighting, and glare (see Fig. 4.15). This initial audit resulted in design guidelines and a store checklist highlighting the needs of older and less able users. However, since shopping is a social experience as well as a practical necessity, a group of RCA industrial design engineering students undertook a further, more general study of the sensory quality of the environment as it might affect the older shopper. Information was gathered on shop organization, layout, and user behavior by a variety of methods. These included photography and video recording, discussions with managers and customers, and a work-study analysis of the supermarket environment in action.

The team identified a range of sensory factors of particular importance to older people. Since older people experience changes in the way they see, hear, and move, the students argued that an environment that set out to enhance sensory feedback and pleasure would be

FIGURE 4.15 Bending and peering, Safeway Stores, United Kingdom.

especially attractive to older people. By integrating changes in layout with new and emerging technology they developed scenarios of shopping in the future to demonstrate how considering the needs of older people could lead to new concepts in store design. An important consideration was how to make the change from a retailing space where efficiencies of stocking, turnover, throughput of customers, and minimal staffing take precedence, to something more akin to a social space, where people gather and meet out of choice rather than necessity.

These scenarios were illustrated with collages and picture stories, in which human-centered technology was combined with spatial and organizational changes. The question addressed in each case was: How can superstores be brought back into the town center where most older people live, and still offer choice within a smaller space, while making it attractive and convenient for everybody? (Coleman, 1994a)

The Sensual Supermarket: A Scenario

The year is 2005. Alice and her young grandson Henry are going on a shopping expedition—it is more fun than staying in and shopping by TV, and the exercise will do them good—but it looks as if it might rain, so Alice calls up a local "European Rickshaw." Soon the doorbell rings and the driver is waiting for them. The Rickshaw is a small, lightweight taxi that will take a wheelchair or two passengers plus luggage or store purchases. It costs a quarter of the price of a London black cab and runs on environment-friendly electricity. Because of its low capital cost and fuel economy the fares are cheap. Alice uses it frequently as it saves having a car and as she no longer enjoys driving.

The Rickshaw avoids the worst of the traffic in the bus lane, and is allowed into the pollution-free central area. Soon they arrive at Alice and Henry's favorite shop, which looks more like a street-market or arcade than an old-fashioned supermarket. The designers have discovered ways of breaking out of the gridlock of shelving and checkouts that made shopping such a bore in the last century, and have opened up the shop frontage with an "active facade" of moveable panels incorporating information, advertising, and large-scale visual elements (see Fig. 4.16). They have also used the latest scanning and electronic warehousing technology to do away with checkouts and high-density shelving.

FIGURE 4.16   RCA Industrial Design Engineering students scenario for Safeway stores—Shopping in the Future, an active facade (1993).

Changes like these have made it possible to humanize the environment by reorganizing it around social focus points like the café, newsstand, and bakery. Alice likes it because it is busy and sociable, like shops used to be! She can choose her vegetables and cheese personally while the boring staple items can be ordered and paid for with a hand-held "ticker" and collected separately or delivered. In a clever way this has brought back the intimacy and conviviality of the traditional market, while combining it with the modern benefits of convenience, choice, and value.

As they enter the shop Alice takes a "ticker," an electronic shopping list (see Fig. 4.17), which she uses to order all the items she wants parceled up for her. The list tells her what she has ordered and what it all costs. She can "tick" individual items by scanning the bar codes on the product, or at the shelf edge, or from a catalog in the café area, and add or subtract goods until she has ordered everything she wants and can afford. There are also in-store displays that will read the bar code on a product and tell her all about what's in it, where it comes from, give her recipes and tips and more. Henry loves finding out where everything comes from and seeing the people who produce it and how they live.

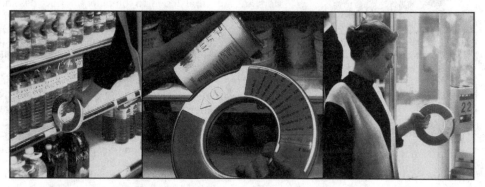

FIGURE 4.17   Shopping in the Future, scanning technologies.

The shelving is curved so that Alice can easily reach a large area from one point, and there is none of the bending and stretching that there used to be. There is also far less shelving, with goods displayed in small quantities, and less confusion, all of which makes the routine side of shopping quick, simple, and convenient. Once she is sure she has everything she wants Alice pays off her "ticker" with her switch card. While her goods are being collected and packed in the electronic warehouse downstairs, she and Henry buy fresh pasta in the delicatessen, a cake for tea from the bakery, and pick up Alice's prescription refill at the pharmacy, which has a direct link to her doctor's office.

This human-centered approach is carried through into the detail of fixtures, fittings, and products, with fiber-optic display lighting to reduce glare and improve presentation, glass jars and lids that are easy to handle and open, and attractive chairs that are surprisingly easy to get in and out of. Even the crockery in the café works well for older people like Alice, who often find cups, plates, and saucers difficult to grip and carry. There are shopping trolleys that combine functionality with convenience and ergonomic fit, and even offer a perch seat for a short rest.

After visiting the newsstand for Henry's favorite comic book, they decide they are ready to go back. Alice stops to check the electronic classifieds on her way out—she is looking for a garden shed—and, seeing two that might do, enters her phone number so she can be called later by the sellers. By now the sun is shining and Alice and her grandson decide to walk back home via the park. They take the cake and comic book with them, leaving all the bulky goods to be delivered later.

In reality, shopping is not like that at all, but why? The problem is, very few people can imagine such things are possible, and this is where creative designers can help, by offering new visions of a people-friendly future. As an example, the previous scenario was written 8 years ago. All of the elements of Alice's shopping trip were designed then, and many of them were prototyped or modeled at life size and could have gone into production almost immediately. Since then, Safeway Stores U.K. has introduced many changes to their shopping environments, and is now a world leader in customer self-scanning.

Student Competition

In addition to working with companies, an important strand of the DesignAge program has been a regular competition open to final-year MA students in all the design disciplines taught at the RCA. Entries are exhibited and judged at the annual summer show. This gives a high profile to the DesignAge agenda, both within the RCA, and outside too, as the show has thousands of visitors. The competition began in 1994, and over the years the many entries have served to establish a body of work that gives substance to the concept of Design for Our Future Selves (Myerson, 1999). The entries extend from fashion, furniture, and transport design to the metal and jewelry crafts, and from architecture, interiors, and industrial design, to computer interface design, ceramics, and glass. The rigor and talent of staff and students in all the RCA studios have been key to achieving one of the central aims of DesignAge—to demonstrate that by including the needs of older people it is possible to design better solutions for all ages and abilities. The diversity of the competition projects supports this concept by demonstrating how traditional objects can be redesigned, and new concepts developed. Some of the designs represent a sense of freedom and escape, such as yachts or cars or flotation platforms to view the underwater world from. But most home in on the practicalities of restricted lives, with ideas to support household chores, reading, sitting, standing, and staying warm, and extending into humanizing hospitals and residential care. (For examples of winning designs, see Figs. 4.18 and 4.19.)

FIGURE 4.18 DesignAge competition winner 1999—"Pull the plug" by Martin Bloomfiled *(RCA Industrial Design Engineering).*

Personal alarms, steps, and purses that cannot be snatched are small advances that could make life worth living for many older people. These designs offer greater self-determination, mobility, andchoice for a section of the community that has been overlooked until recently. By no means are all of the projects aimed directly at the mature market, but in tackling such issues as living alone, they bring the age issue into focus.

FIGURE 4.19 DesignAge competition winner 1998—The "Tate Stool" a lightweight, portable seat for museum visitors by Olof Kolte, RCA Furniture Design.

The Next Step

The DesignAge program is now a core theme of the Helen Hamlyn Research Centre. Launched in January 1999, the Centre has a broad social mission to explore design for all ages and abilities in a period of rapid social and technological change. This Design for Our Future

Selves agenda embraces four research themes: age (the continuation of the DesignAge program); work (supporting changing patterns of working and communicating, in the office, in the home, and on the move); mobility (enabling greater mobility, freedom, and access for people travelling within and between cities); and care (health care and rehabilitation to support independent living). The Research Centre investigates these themes through think tanks, externally funded research projects, and a research associate program, offering some of the best RCA graduates the opportunity to work for an additional year in collaboration with industry partners.

Projects are set which advance the research center's agenda, exploit the particular talents of individual students, and in some cases take their graduation or competition projects through to production. And so, Design for Our Future Selves has become the inspiration for a broader program of research around social change issues.

4.6   THE BROADER CONTEXT

Another DesignAge initiative is the European Design for Ageing Network (DAN), linking educators, researchers, designers, marketers, and manufacturers from across Europe, sharing information and best practice, and stimulating collaboration between them (e.g., Hewer et al., 1995; Juul-Andersen and Jensen, 1997). With its own Web site, database of contacts and information, and access to the RCA's special collection, the network acts as a primary source of information with links to over 300 partners around the world. Local and national networks add to the effectiveness of this loose association of researchers and practitioners, and a number of substantial design-led research projects have resulted from the work of the DAN network. The DAN Web site, previously maintained by the Netherlands Design Institute, has now been integrated into the Helen Hamlyn Research Centre Web site at www.hhrc.rca.ac.uk/DesignAge.

DAN is one of a series of initiatives, many of which have been supported by the European Union. These include the International Gerontechnology Society, spearheaded by the Finnish health and welfare research organization STAKES, and the European Institute for Design and Disability, in which the Irish and Spanish groups have taken a lead, with the U.K. arm being the U.K. Institute for Inclusive Design. In the Netherlands, a not-for-profit organization has been established to promote design for all, while in Germany there is a very active DAN group.

4.7   GENERAL LESSONS LEARNED FOR UNIVERSAL DESIGN

Universal design has primarily focused on disability, with a strong emphasis on accessibility and the built environment. What an examination of population aging reveals is that the large majority of people with disabilities is old, not young; that age-related impairments differ from those stemming from genetic and traumatic causes; and that older people are affected by progressive, multiple minor impairments that can seriously affect life quality without being recognized as disabling. An understanding of the causes and consequences of population aging makes it clear that "disability" is part of the modern, extended life span and is something that nearly everyone will experience. Enabling design can significantly reduce the link between impairment and disability (Goldsmith, 1997). The challenge for universal design is to fully understand these factors and give adequate attention to them by developing environments products and services that work for people across the whole life span (Pirkl and Babbic, 1998; Pirkl, 1993). A failure to do so could lead to older people suffering from a loss of independence, being excluded from access to products and services (particularly where information technology is concerned), and thus being effectively discriminated against (Brewer-Janse et al., 1997; Fozard, 1997).

The opportunity presented by population aging lies in the strength of the trend from youthful to mature populations. As a result of this radical and probably irreversible shift in age balance, the numbers of people likely to be affected by inappropriate or disabling designs will be very large indeed. Given that the age-related factors are understood and incorporated into the body of universal design, it can enhance life quality and offer real value to older people, creating market demand on a scale sufficient to stimulate industry and commerce to respond in ways that legislation cannot. The rapidly changing lifestyles of older people offer an opportunity to develop new products and services that further enhance life quality and, in particular, enable active independence—the primary desire of older people—thereby impacting on the incidence of age-related disability and the social and economic costs arising from it.

In the United Kingdom, in particular, as a result of the Technology Foresight exercise carried out by the Department of Trade and Industry, the aging population has been identified as a key priority for the future. As a result, research initiatives have been launched that seek to enhance the quality of life of older people and to involve older users in research programs. Industrial interest in such matters is growing, and the design management community is keen to develop a better understanding of the issues and have access to appropriate design strategies, tools, and guidance. In response to this, the author, along with colleagues from Cambridge University Engineering Design Centre, from the recently established Cambridge University Interdisciplinary Research Centre on Ageing (CIRCA), the Design for Ability Unit at Central St. Martins School of Art and Design, and the U.K. Design Council, are working on assembling the necessary data and information and preparing it for dissemination to the design community. The focus of this research is inclusive design—strategies and methodologies that industry and design professionals can adopt in order to achieve progressive improvements in product and service design by better understanding and including the needs of the whole population.

4.8 CONCLUSIONS

In Europe, many organizations and individuals have advanced the theory and practice of a design approach that places the user on center-stage and seeks to create an open, accessible, and integrated world for the future through design that is age-friendly, barrier-free, and inclusive. Only some are mentioned in this chapter, and the history traced is a particular one and not representative of the diversity of initiatives in Europe. What matters is not the terminology used, but the fact that a new international movement in design is emerging. This movement is not a single initiative, but one with parallel, yet different, manifestations in the many European countries involved, that complements rather than replicates the development of universal design in America, and is enriched by this. It is a vision of a people-friendly world that recognizes and celebrates diversity in ways that are sustainable and people-centered. And this movement is the logical outcome of a period in which people in general have moved closer to understanding that they constitute one human race, living on a small planet, whose collective future security and well-being depend on learning to live together.

4.9 BIBLIOGRAPHY

Ballabio, E., "The European Commission R&D Initiatives to Promote Design for All," *Proceedings of International Workshop on Universal Design,* Building Research Institute, Ministry of Construction and Japan International Science and Technology Exchange Centre, Tokyo, Japan, 1998.

Benkzton, M., "Designing for Our Future Selves: The Swedish Experience," *Applied Ergonomics,* vol. 24, no. 1, 1993.

Bicknell, J., and L. McQuiston (eds.), "Design for Need: The Social Contribution of Design," *Conference Proceedings,* Pergamon Press & Royal College of Art, London, UK, 1977.

Brouwer-Janse, M. D., R. Coleman, J. L. Fulton-Suri, J. Fozard, G. de Vries, and M. Yawitz, "User Interfaces for Young and Old," *Interactions,* vol. 4, no. 2, 1997.

Buck, S. (ed.), *The 55-Plus Market: Exploring a Golden Business Opportunity,* McGraw-Hill, New York, 1990.

Coleman, R. (ed.), *Designing for Our Future Selves,* Royal College of Art, London, UK, 1993a.

———, "A Demographic Overview of the Ageing of First World Populations," *Applied Ergo-nomics,* vol. 24, no. 1, 1993b.

———, "Design Research for Our Future Selves," *Royal College of Art Research Papers,* vol. 1, no. 2, 1994a.

———, "Age: The Challenge for Design," In Myerson, J. (ed.), *Design Renaissance: Selected Papers from the International Design Congress,* Open Eye and the Chartered Society of Designers, Horsham, UK, 1994b.

———, "The Case for Inclusive Design—An Overview," *Proceedings of the 12 Triennial Congress,* International Ergonomics Association and Human Factors Association of Canada, Toronto, Canada, 1994c.

——— (ed.), *10 years of RSA New Design for Old,* The Royal Society for the Encouragement of Arts Manufactures and Commerce, London, UK, 1996.

———, "Mit alter Menschen arbeiten: ein kollaborative Designprozeß," and "Verbesserung der Lebens-qualität alteren Menschen durch Design," *Handbuch der Gerontechnik,* second edition, Ecomed-Verlag, Landsberg, Bavaria, Germany, 1997a.

——— (ed.), *Design für die Zukunft,* DuMont Buchverlag, Cologne, Germany, 1997b.

———, "Breaking the Age Barrier," *RSA Journal,* Nov./Dec. 1997c.

——— (ed.), *Working Together: A New Approach to Design,* Royal College of Art, London, UK, 1997d.

———, "Improving the Quality of Life of Older People by Design," In Graafmans, J., V. Taipale, and N. Charness (eds.), *Gerontechnology: A Sustainable Investment in the Future,* Studies in Health Technology and Informatics, vol. 48, IOS Press, Amsterdam, Netherlands, 1998.

———, "Designing for Our Future Selves." (The 1998 Bernard Isaacs Memorial Lecture), *Gerontology* (Special Edition on Ageing and Technology), vol. 26, nos. 1–2, 1999a.

———, "What Design Can Do," In Peto, J. (ed.), *Design Process Progress Practice,* The Design Museum, London, UK, 1999b.

———, "Inclusive Design—Design for All," In Jordan, P. W., and W. S. Green (eds.), *Human Factors in Product Design: Current Practice and Future Trends,* Taylor & Francis, London, UK, 1999c.

———, "Design for the Future," In Japanese, in *Design for the 21st Century: An International Conference on Universal Design.* Toshi Bunka Sha, Tokyo, Japan, 1999d.

Coleman, R., and D. J. Pullinger (eds.), *Designing for Our Future Selves.* Special edition of *Applied Ergonomics,* vol. 24, no. 1, 1993.

Fisk, D., and W. Rogers, *Human Factors and the Older Adult,* Academic Press, 1997.

Fisk, J., "Design for the Elderly: A Biological Perspective," *Applied Ergonomics,* vol. 24, no. 1, 1993.

Fozard, J. L., "Ageing and Technology: A Developmental View," In Rogers, W. A. (ed.), *Designing for an Aging Population: Ten Years of Human Factors and Ergonomics Research,* Human Factors and Ergonomics Society, 1997.

Freudenthal, A., *The Design of Home Appliances for Young and Old Consumers,* Ageing and Ergonomics Series, no. 2, Delft University Press, Delft, Netherlands, 1999.

Gardner, L., L. Powell, and M. Page, "An Appraisal of a Selection of Products Currently Available to Older Consumers," *Applied Ergonomics,* vol. 24, no. 1, 1993.

Gill, J., *Access Prohibited? Information for Designers of Public Access Terminals,* Royal National Institute for the Blind, London, UK, 1997.

Goldsmith, S., *Designing for the Disabled,* third edition, Royal Institute of British Architects, London, UK, 1984.

———, *Designing for the Disabled: A New Paradigm,* Architectural Press, London, UK, 1997.

Haigh, R., "The Ageing Process: A Challenge for Design," *Applied Ergonomics,* vol. 24, no. 1, 1993.

Hewer, S., et al., *The DAN Teaching Pack: Incorporating Age-Related Issues into Design Courses,* Royal Society for the Encouragement of Arts Manufactures and Commerce, London, UK, 1995.

Juul-Andersen, K., and E. Jensen, *Design Guidelines for Elderly and Disabled Persons' Housing,* Danish Centre for Technical Aids for Rehabilitation and Education, Taastrup, Denmark, 1997.

Kirkwood, T., *Time of Our Lives: The Science of Human Ageing,* Weidenfeld & Nicolson, London, UK, 1999.

Laslett, P., *A Fresh Map of Life: The Emergence of the Third Age,* second edition, Weidenfeld & Nicolson, London, UK, 1996.

———, "Design Slippage Over the Life-Course," In Graafmans, J., V. Taipale, and N. Charness (eds.), *Gerontechnology: A Sustainable Investment in the Future,* Studies in Health Technology and Informatics, vol. 48, IOS Press, Amsterdam, Netherlands, 1998.

Mandelstam, M., *Equipment of Older and Disabled People and the Law,* Jessica Kingsley Publishers, London, UK, 1997.

Manley, D. (ed.), *New Design for Old,* Victoria and Albert Museum and the Helen Hamlyn Foundation, London, UK, 1986.

Mapstone, N., *Fully Equipped: The Provision of Equipment to Older or Disabled People by the NHS and Social Services in England and Wales,* Audit Commission Publications, London, UK, 2000.

Moggridge, W., "Design by Story-Telling," *Applied Ergonomics,* vol. 24, no. 1, 1993.

Myerson, J. (ed.), *Design for Our Future Selves: DesignAge Competitions at the Royal College of Art 1994–98,* Helen Hamlyn Research Centre, London, UK, 1999.

Nayak, L., "Design Participation of the Thousand Elders," In Graafmans, J., V. Taipale, and N. Charness (eds.), *Gerontechnology: A Sustainable Investment in the Future,* Studies in Health Technology and Informatics, vol. 48, IOS Press, Amsterdam, Netherlands, 1998.

Norris, B., and L. Peebles, *Older Adult Data: The Handbook of Measurements and Capabilities of Older Adult,* Data for Design Safety, Department of Trade and Industry, London, UK, 2000.

Pirkl, J. J., *Transgenerational Design: Products for an Aging Population,* Van Nostrand Reinhold, New York, 1993.

———, and A. Babic, *Guidelines and Strategies for Designing Transgenerational Products,* Copely, 1988.

Sixsmith, A., and J. Sixsmith, "Older People, Driving and New Technology," *Applied Ergonomics,* vol. 24, no. 1, 1993.

Smith, D. B. D., "Human Factors and Aging: An Overview of Research Needs and Application Opportunities," *Human Factors,* vol. 32, no. 5, 1990.

Steenbekkers, L., and C. van Beijsterveldt (eds.), *Design-Relevant Characteristics of Ageing Users: Backgrounds and Guidelines for Product Innovation,* Delft University Press, Delft, Netherlands, 1998.

Wainwright, H., and D. Elliot, *The Lucas Plan: A New Trade Unionism in the Making?* Allison & Busby, London, UK, 1982.

4.10 RESOURCES

Age Concern England.   Astral House, 1268 London Road South, London SW16 4ER, tel. +44 20 8679 8000. The leading U.K. age-related charity. Age concern also coordinates a European network through its sister organization, Eurolink Age. www.ace.org.uk/ and www.eurolinkage.org/.

ARJO Ltd.   St Catherine Street, Gloucester, GL1 2SL, UK, tel. +44 080702 430 430, fax. +44 1452 525207. Manufactures the "Stedy" and other care products.

Design Council of England.   34 Bow Street, London WC2E 7AT, UK, tel. +44 20 7420 5200, fax. +44 020 7420 5300. Promotes and champions design in the United Kingdom. It has taken a keen interest in population aging as it affects design, through its "Living Longer" program, and is actively engaged in research into inclusive design. www.design-council.org.uk/.

Design for Ability.   Central Saint Martins, Southampton Row, London WC1B 4AP, tel. +44 20 7514 7015, fax. +44 20 7514 7016. Has produced a CD-ROM—*Design Aid, a Design Resource about Disabled People as Consumers*—essential information for designers and manufacturers of products and anyone interested in understanding the lifestyles of disabled people. Copies can be obtained direct from Design for Ability.

Engineering Design Centre at Cambridge University.   Is actively researching usability and rehabilitation and inclusive design; details can be found at www.edc.eng.cam.ac.uk/ and http://rehab-www.eng.cam.ac.uk/projects/include/.

European Commission DG13. Rue de la Loi/Weststraat 200, B-1049 Brussels, Belgium, tel. +32 2 299 1802, fax. +32 2299 0248. Is the information technology branch of the European Commission. DG13 is active in promoting access to technology for people of all ages and abilities through the application and development of design-for-all methodologies. The Fifth Framework Program (FP5) defines the European Union's strategic priorities for research, technological development, and demonstration activities for the period 1998–2002. FP5 has been conceived to help solve problems and to respond to major socioeconomic challenges such as increasing Europe's industrial competitiveness, job creation, and improving the quality of life for European citizens of all ages and abilities. Emphasis is placed throughout on the process of innovation to ensure the output of EU research is translated into tangible benefits for all. www.cordis.lu/fp5/

Helen Hamlyn Research Centre (HHRC), The Royal College of Art. Kensington Gore, London SW7 2EU, UK, tel. + 44 20 7590 4242, fax. +44 20 7590 4244. Maintains a special collection of c.1300 items on design and aging at the Royal College of Art (RCA) in London. The collection can be searched by keyword, author, and title, and abstracts can be consulted via the World Wide Web. The HHRC Web site also hosts the European Design for Ageing Network (DAN) members' database, and has new items and extensive links to related programs. The HHRC Web site can be found at www.hhrc.rca.ac.uk/, and a special section of the site is devoted to the DesignAge program at www.hhrc.rca.ac.uk/DesignAge.

HEWI Heinrich Wilke GmbH. Prof.-Bier-Straße 1-5, 34454 Bad Arolsen, tel. +49 56 91 82 0, fax. +49 56 91 82 319, e-mail info@hewi.de, www.hewi.com/.

IDEO Europe. White Bear Yard, 144a Clerkenwell Road, London EC1R 5DF, www.ideo.com/.

Royal National Institute for the Blind (RNIB). 224 Great Portland Street, London W1A 6AA, UK, tel. +44 20 7388 1266. The RNIB is a leading U.K. charity and conducts research into design-related subjects, in particular in the field of telecommunications and new technology, and develops design-for-all guidance. www.rnib.org.uk/.

Royal National Institute for the Deaf (RNID). 105 Gower Street, London WC1E 6AH, UK, tel. +44 20 7387 8033, fax. +44 20 7388 2346. www.rnid.org.uk/.

Royal Society for the Encouragement of Arts, Manufactures and Commerce (RSA). 8 John Adam Street, London WC2N 6EZ, UK, tel. +44 20 7930 5115, fax. +44 20 7839 5805. The RSA has a long-standing interest in design matters, honors senior designers with membership of its 100-strong Faculty of Royal Designers for Industry (RDIs), and organizes the premier European student design competition, the RSA Student Design Awards (SDAs). New Design for Old has been a section of the SDAs for 12 years. www.rsa.org.uk/sda/.

United Kingdom Institute for Inclusive Design (UKIID). tel. +44 20 8682 0518, fax. +44 20 8682 3027. The UKIID is the U.K. branch of the European Institute for Design and Disability.

CHAPTER 5
UNIVERSAL DESIGN AND THE MAJORITY WORLD

Singanapalli Balaram, Professor, M.Des. (NID), Fellow (SIDI)
National Institute of Design, Ahmedabad, India

5.1 INTRODUCTION

Today, in many parts of the world, there are indications that universal design is going to stay. It may change its name but not the content. The presence of universal design, however, is not even. While in some countries it is very strongly pulsating, in some other countries it is hardly existing. In the economically sharply divided world of haves and have-nots, the situation in the world of have-nots often goes unnoticed. But the world cannot go forward with a major part of its body missing. It is too important to be left out. The attempt of this chapter is to discuss that part of the world of the have-nots to assess the situation of universal design and related issues. Its focus will be on India whose situation maybe considered as similar to other have-not countries. In the absence of reliable scientific data, the arguments have been necessarily based on personal experience and indirect information.

5.2 BACKGROUND

At the outset, it ought to be recognized that the problems, perceptions, and, therefore, the possible solutions of the have world are different from those of the have-not world. The reasons are many: The influence of thousands of years' traditions, the colonial or other forms of dominance by foreign powers, economic insufficiency, pressures of the commercial world, and overpopulation are some of them.

Design is an activity that is based on socioeconomic criteria and that will, therefore, naturally follow suit in being different in these countries. It is too simplistic to treat design as an isolated aesthetic activity existing outside the mainstream forces of economics, culture, and society. Design in the economically developing world is beyond an activity that is concerned with formal qualities and superficial aspects aimed at boosting sales. It is a serious activity that is concerned with playing a key role in economic and social development of the people. Moreover, centuries-old traditions are bound to have a profound influence on people's behavior and patterns of choice, which go beyond physical rationale. It is in this context that universal design needs to be viewed differently from the way it is viewed in the Western countries.

5.3   *TERMINOLOGY*

Ways of saying are ways of seeing. As one says, so one sees. Saying influences, positively or negatively, people's attitudes toward a subject. Hence, before one proceeds further, it is necessary to clarify the importance of terms used in this chapter, as terms are crucial in forming or removing an intellectual or emotional bias.

There are four words that are used most frequently, which need to be released from an associative bias. The first term among these is *third world,* which refers to the relatively poorer countries in the world. But, in fact, the generalized usage of *third* and *first* discriminates the two, projecting the former as deficient in more ways than one. A better way of looking at this reality follows:

> If one considers the number of people belonging to a country as important as the money possessed by that country, one would see that the majority of the world's population lives in the so-called third world. Therefore, the author feels that it is appropriate to use the term *majority world* in place of *third world* and to refer to the few rich countries of the world as the *minority world.*

Second, the term *development,* used in the context of developing countries, is very general and loaded with negative connotations such as something inferior in every way. Originally, this term was coined to mean industrial development in some parts of the world. To avoid this negative bias, therefore, it is appropriate to use a precise word and replace *development* with *industrialization.* In this chapter, the developed and developing countries will be called, appropriately, *industrialized* and *industrializing* countries.

Third, *advanced countries* discriminates, implying unfairly that the other countries are backward and shows them in poor light. Hence, the term *advanced countries* will not be used in this chapter.

Finally, one must recognize the importance of culture in human life more than the wealth and technical prowess. Many countries of the world are either rich in cultural diversity or rich with centuries of cultural traditions that continue to influence the people's present lives. Few countries are blessed with both. Such countries will be referred to positively as *tradition-bound countries* in this chapter.

For all other purposes of general reference, it is perhaps better to distinguish between various regions with geographical location, such as Eastern, Western, European, Occidental, Oriental, Asian, and so on. In this chapter, the term *nonaverage people* is occasionally substituted to cover generally all categories of people—elderly, disabled, anomalous, and so on.

5.4   *DEFINING UNIVERSAL DESIGN IN THE MAJORITY-WORLD CONTEXT*

Universal design argues for the importance of wholesomeness and the importance of making, through design, the so-called weak component in the society as strong as every other part. In the industrialized and relatively wealthy countries, this weakness is merely physical or sensory. It is concerned with people's diminished abilities, either temporary or permanent, such as impairment, retardation, advanced age, pregnancy, and so on. In the nonindustrialized majority-world countries, the weakness is beyond physical disabilities. The weakness is essentially social construct, which severely inhibits the equal participation of certain sections of society in public and private life.

The economic, social, and cultural realities of nonindustrialized countries affect universal design in three ways:

1. The prevalence of oppressing realities such as poverty, population pressures, illiteracy, and lack of infrastructures calls for universal design solutions vastly different from those elsewhere.

2. These realities also make implementation of universal design extremely difficult.

3. Since success breeds success and vice versa, the very weaknesses mentioned earlier will create more problems of the disability, old age, and the like, due to malnutrition, inadequate prenatal and postnatal care, inadequate sanitation and hygiene, ignorance of child medication, unsafe and accident-prone situations due to bad road conditions, and poor industrial environments.

Universal design, therefore, needs to be viewed from this angle differently from the perspective of the industrialized countries. Universal design for majority-world countries can then be defined as a concept that extends not only beyond issues of accessibility of the built environment, but that also covers the social, cultural, and economic issues, which are major influences in uniting normal or average people and people with different physical, mental, or psychological abilities.

If one accepts universal design as an "approach that values and celebrates human diversity," majority-world countries such as India and China have overwhelming and perhaps unmanageable cases. For instance, India's cultural, religious, linguistic, and ethnic diversities are almost impossible to comprehend.

5.5 *DEMOGRAPHIC SITUATION: FIGURES SPEAK*

According to the United Nations' estimates, the world elderly population constituted 14.8 percent of the total world population in 1985. While the elderly population in industrialized countries is expected to increase by 77 percent over 1985 figures, the comparative percentage increase in majority-world countries is approximately 207. China and India will then be the most populous countries, accounting for 48 percent of the world's elderly population (UN report, 1998). The aging scenario in India is alarming with 70 million elderly people above the age of 60 years at present. This current figure is projected to reach 177 million by the year 2025.

In the case of disability, India already has one-third of the world's disabled population and one-half of the world's blind population, and these figures are increasing. While an accurate census is difficult due to huge and illiterate village populations, an approximate 63 million people in India suffer from impairment in physical or mental form.

From 1985 to 2000, the population aged 60 and above and 70 and above was expected to increase by 93 percent and 10.8 percent, respectively, while the total population increased only by 40 percent during the same period. By the year 2025, it is estimated that one out of every seven elderly persons in the world would be from India. In terms of gender distinction, it must be noted that the female ratio in the elderly population is increasing. Similar trends could be observed in other Asian countries such as China, Sri Lanka, South Korea, Bangladesh, and Pakistan.

5.6 *TRADITIONS OF UNIVERSAL DESIGN*

It is important to note that the concept of universal design is not new, especially in some tradition-rich countries. *Universal design* is a term coined by the late Ronald Mace 15 years ago. But it is an ancient concept that has been extant in cultures such as those in India for ages. For instance, India has the marvellous tradition of wearing unstitched, uncut garments, which continues even today in the new millennium and is certain to continue in the future. These garments have reached perfection in design and have become timeless classics. Since space does not permit a detailed discussion, let us just look at two examples. Figure 5.1 shows a dhoti, the lower-body garment worn by men; Fig. 5.2 shows a saree, which covers the upper and lower portions of the body as well as the head, and is worn by women.

FIGURE 5.1   Example of universal design in the Indian tradition—the *dhoti,* a lower garment worn by men.

Both of these garments are extremely elegant, highly functional, and can be worn in infinite ways to suit the occasion, whether it is work on the farm, a religious ceremony, or a social function such as marriage. There are codes of wearing. Judged by the way it is worn, the same cloth creates a strong identity as to the wearer's caste, community, social status, and even marital status. Besides, these garments are also highly symbolic. While being worn, they can be put to other functions, such as carrying a child, carrying objects, and so on. When they are not being worn, these garments can be put to other uses, such as a baby swing, packing material, and the like. The same cloth can be worn by everybody, irrespective of the age, physical dimension, or temporary or permanent disability. They are extremely adaptable solutions and excellent examples of universal design in practice.

5.7   *MYTHOLOGICAL INFLUENCE*

In tradition-bound countries such as India, culture plays a powerful role. In most Eastern countries, people who are elderly, even if they have impairments, are traditionally respected. In the joint family system, they command authority and continue to make major decisions as the wise seniors of the family.

FIGURE 5.2   Example of universal design in the Indian tradition—the *saree,* an upper and lower garment worn by women.

The Indian mind is considered to be preoccupied with the mythical domain, and even in the twenty-first century, Indian mythologies keep influencing people with certain societal and familial attitudes. The great Indian epic, *Mahabhaarata,* has it that Dhritarashtra, a totally blind and very old king, ruled India (then called Bhaarat) as well as the royal family until the end of the Kurukshetra War as mentioned in the epic. He had a family of 100 sons who were married and all lived as part of the joint family structure.

There is also a very strong ethical code in every Indian mind toward the elders and those with disabilities in the family. The son or the daughter has the unavoidable duty to look after those who are old and disabled, particularly if they are the parents. In the other great Indian epic, *Ramayana,* there is a crucial episode of Shravana, a youth who carried his old and severely disabled parents on his shoulder and took them on a pilgrimage. King Dasharatha accidently killed Shravana, and the orphaned disabled parents cursed the king. The whole story of *Ramayana* is the realization of this curse. The Shravana story is repeated to young children even today as the role model of a good dutiful son. In fact, it is more than that: It is a warning to the ruling class!

5.8   *SOCIOCULTURAL PROMINENCE*

Although the joint family system is on the decline due to urbanization, it is still largely present in the villages, which constitute 80 percent of India. In Indian cities, there are homes for the aged and for the disabled, but these are occupied primarily either by the orphaned or the poor of society. A majority of the population would consider it an extremely unsocial act to send their disabled or elderly relative to institutional care. Society would consider it an affront to their status to do so. As a chain effect, the quality of institutional care has also remained very poor.

It is traditional in India and other Asian countries that the elders are neither discriminated against, nor are they separated from others in the various processes of living. On the other hand, they are given a place of pride and respect. In the religious rituals, the blessings of the elders are sought before commencing the ritual and after its completion.

There are traditional festivals in India that encourage full participation of people with diverse abilities and ages. For instance, at Garba, a community festival dance in Gujarat, people of all ages—from a 3-year-old child to an 80-year-old person—irrespective of gender and age, dance before the Goddess. The dance is in circular form, which is infinite and nondiscriminating. This dance can be truly called a universal dance.

Another festival called Sankranti, a kite-flying festival, is also universal, where families celebrate by jointly flying colorful kites. People of all ages, genders, and abilities participate, dividing the tasks as they like and are capable of. The elderly and the disabled relatives take on the easier task of holding the reel of thread while the relatively tougher task of flying the kite and running after the kite is done by the energetic youth and children. Thus, everyone feels important and contributes according to his or her own specific, individual ability.

Similarly, in the villages, the old or disabled people take on the lighter yet important tasks of watching the drying grain or taking care of the toddlers. Elderly people are considered invaluable in swinging the infant or telling stories and lores to young children as they could spare enough time, which the parents cannot. Such social customs are extremely important as they make the people who are elderly or disabled and the others mutually dependent. Furthermore, such participation gives the people who are elderly and/or disabled a heightened purpose of living.

In India, when a person crosses the age of 60, it is celebrated with an important traditional ceremony called *shashti purti* (literally meaning *the completion of 60*). It is an indication of cultural importance given to the elderly in Indian society.

There is also a very important microlevel, traditional self-governing system in villages that continues even today. This is called the *Panchayati* system. It is a local governance where it becomes the prerogative of the elders of the community to participate and lead the lifestyle of the people.

In Chinese thought, the family is the fundamental social unit and not the individual. The family is hierarchically organized, where elders are placed high. Within such a close-knit family system, individuals have a duty toward each other, whether they are elderly, average, nonaverage, or socially disadvantaged.

5.9 MAJOR CONCERNS

The contemporary situation in India is startling, and India can be taken as an example of a highly populated, tradition-bound country in the process of industrialization whose situation could be seen as similar in many other majority-world countries. Following are some of its concerns.

Problems of Integration

More important than access to and interaction with the built environment, what the people who are elderly and people who are disabled confront is the segregation from the rest. True, in countries like India, the joint family system exists. But this is threatened by urbanization and attitudes of the younger generation, which are becoming influenced by the industrialization and lifestyles of the industrialized countries. More important than discrimination is the lack of purpose and participation in the family, which can lead to great psychological trauma and barriers for most people.

Problems of Communication

Even in the new millennium, India can still be called a country with an oral tradition since nearly two-thirds of its total population is illiterate. Therefore, written communication and print media have very little effect on the majority of the people. This itself is a barrier to cross. Visual communication is obviously the answer, but, like everything else, visual communication requires visual literacy on the part of the people with whom it is communicating. This cannot be expected to happen easily in such a vast society. People prefer to ask directions at a nearby shop even if there is a good map and signage. It is also a more interactive method!

A country with a tradition such as this can be addressed in only one way. The designers have to learn the language of the masses to be able to communicate with the signs they understand and the methods they accept. If the designers use oral communication instead of visual communication, perhaps then, it could evolve to be a universal design that caters to both the literate as well as the illiterate people. One fact that proves this argument is the high rate of success of songs and music in popular Indian films. This factor is so important that no popular Indian film can be found without songs.

Problems of Living

The wheelchair is accepted universally as an aid to mobilize people with lower-limb impairment. In countries like India, the wheelchair is unusable. This is because in rural areas of India, which hold the majority of the population, people cannot use wheelchairs due to uneven terrain. More important, the villagers do not sit on a chair at home. They either squat on the ground or sit on a raised wooden plank. Rural women cook, store, and do all the household work at ground level. They need a different kind of mobility device that operates at ground level. One of the innovations that has been tried is to take the traditional Indian wooden seat (called a *patla* in the local language, Gujarati), and put wheels, brakes, and other necessary conveniences to it. This could be a universal design because impaired, as well as unimpaired, women could use the patla as they have done in the past.

Problems of Function

Most of the artificial feet in the world are not actually feet but boots that are partially covered by the trousers. This gadget is unusable by 80 percent of the Indian people who are farmers.

The farmers do not wear trousers; instead, they wear a dhoti, which is lifted up for convenience of work in the fields, exposing the leg until above the knee. Besides this, the farmers often have to work in mud with their bare feet, they have to climb trees with their bare feet, and they have to perform many actions with their bare feet. As a solution to this, and to avoid discrimination, an Indian doctor, P. K. Sethi, developed a foot, popularly known as the *Jaipur foot*, that looks very much like the real foot in terms of colors and shades. At the same time, it is flexible enough to allow the user to climb a tree or to work in mud and water.

Problems of Production

While the country is on its way to industrialization, the vast community of craftspeople in villages remain without work. The village artisans and craftspeople, who suddenly found themselves unwanted by society, need to be made useful to society and to participate meaningfully in community activities as equals. This could be done if these artisans' skills were used in producing devices for the people who are disabled and elderly and for those people whose needs they could comprehend better. Such products could be of common use for average people, as well as the special-needs people. The advantage of village artisans producing universal design products is that these common products can be maintained on a sustainable basis without depending on the city-based workshops.

Problems of Acceptance

In tradition-bound countries like India, people do not readily accept new devices. There are strong hindrances to acceptance based on the semantic notions associated with the new product. There are strong taboos with certain forms and colors. Social distinctions get added to such taboos, reinforcing them further. For instance, antipollution masks introduced by a corporation were rejected by the people because they resembled a pig face and the pig is a cultural taboo. The pig is considered extremely low by both of the dominant religions of India—Hinduism and Islam.

This issue can be addressed by making the designers and craftspeople aware of culture-specific semiotics and semantics so that the design is not rejected outright, and at least it provokes the user to think about its viability. The product must also cater to a cross section of people across religious affinities. In a country like India where, more often than not, religion dominates the people's minds, a design like the preceding one not only alienates the design profession there but also has lasting effects. First, it reaffirms the belief that "old is gold and new is trash." Second, it might prevent any further acceptance of any redesigned product in the future.

Problems of Marketing

More often than not, market forces cater to the urban haves rather than the rural have-nots. They are quite insensitive to the fact that although the urban populace may be cosmopolitan enough to take religio-cultural slurr in its stride, the rural populace may not be so giving. Hence, any marketing venture has to be not only appealing but also acceptable. This is possible only after conducting extensive market research to find out and determine what is acceptable to the majority of the population.

5.10 POLITICS OF INCLUSION

L. K. Weisman had pleaded for the politics of inclusion and wholeness in place of patronizing politics of tolerance and competing interest. While agreeing to Weisman's (2000) argument, one would wonder about how to define *wholeness* and *inclusion* in the context of tradition-bound, culturally complex, and vastly diverse countries like India. If universal design wishes to eliminate discrimination in India, then it stands to consider other socially governed inhuman forms of discrimination as well. The people of the dalit community, for example, are untouchables even today. The caste system is prevalent throughout Indian society, though more apparent in villages. Access is denied for certain communities of people to several important places such as worship, festivals, public functions, and even common water sources. Regardless of their physical impairment, these groups of people are excluded. Caste-based exclusion is only one of the many forms of social attitudes existing in India. The community of eunuchs (*hijdas* in local language) is avoided by the society as much as it avoids leprosy-cured people. People who are black-skinned or fat or short, or even anomalous, face similar problems of exclusion.

5.11   AREAS OF DESIGN INTERVENTION

A very important role can be played by the designers in the majority-world countries, as they have the advantage of not committing the same mistakes that the industrialized countries have committed. On the contrary, by seizing the opportunity the majority-world designers can be role models for others to correct the mistakes. The following areas could be considered for major thrust for designers in these countries:

1. Societal attitudes
2. Educating for the future
3. Positive thinking by the user groups
4. Networking
5. Increasing usability range

Societal Attitude

Umberto Eco, in his *Serendipities: Language and Lunacy* (1999), says:

> The members of a culture A cannot recognise the members of the culture B as normal human beings (and vice versa) and define them as "barbarians," that is, etymologically stuttering and non-speaking beings and therefore non-human or sub-human beings.

This observation is applicable to other differences in human beings, too. It is applicable not only among the people who are literate and those who are not, but also among white-skinned and black-skinned, among urban and rural, among people who are living in plains and those living in hills; among rich and poor—The list is endless. This barbarian attitude forms the basis of opposition to universal design.

The advantaged group, deep down, would not like the other disadvantaged group to join them and be treated equally. This is perhaps the biggest obstacle for the proponents and designers of universal design to address immediately. As long as this is not addressed, the other efforts will not have their full effect. Therefore, the first major task facing the designers is to promote public opinion toward a positive *universal* attitude. Such awareness communication should cover the whole range of issues, from the language and terms of addressing the people who are nonaverage, to the social behavior in the company of the people who are nonaverage. In societies where illiteracy is dominant, oral, visual, and other forms of nonverbal communication can be employed effectively. The approach must be wholistic and not limited merely to people with physical and/or mental impairments.

Educating for the Future

Change in societal attitude will not happen overnight. It is a slow process and requires a long time for the new concepts to be accepted. The establishment has already set notions and definite views about several aspects of life and do not accept change readily, no matter how persuasive the communication is. It is thus advisable to start inculcating the right attitude toward people who are aged, disabled, disadvantaged, and so forth, from a very young age to the child as part of his or her regular education in schools, colleges, and universities. Children are the future society. It is less difficult and more effective to mold a young mind, and a better generation will soon make up the world. In majority-world countries where most children do not go to school, nonformal methods and nonformal educational devices (e.g., toys, games, and dolls) could be specially designed and employed.

Positive Thinking by the User Groups

Communication is interpersonal, and it works both ways. When there is a change in the attitude of one group that is at the receiving end, a corresponding change in the attitude is necessary for the other group that is at the originating end. While the average public needs to change their attitude toward the people who are nonaverage (e.g., the impaired, aged, leprosy cured, less abled, temporarily disabled, etc.), others also need to change their own attitude about the others and about themselves. A poor and low image of oneself by the nonaverage person would make him or her suspect even a genuine concern and care. Such doubts will hamper her or his joining society as an equal member and not being pitied or treated specially. To be treated equally and avail equal opportunities, the fear of acceptance has to be dispelled by the individual.

Equal opportunities should not be mistaken for *same* opportunities. A visually impaired person cannot compete with an able-bodied person in a typing job, and a hearing-impaired person cannot compete with an able-bodied person in a telephone operator job. The visually imparied, however, can compete equally with able-bodied people in a telephone operator job, and the hearing-impaired person can compete equally with able-bodied people in the typing job. Nonaverage people should be seen as people with different capabilities rather than people with lesser capabilities.

In a documentary film entitled *Listening to Shadows* (Sarkar, 1999), Ranchodbhai Soni, an Indian teacher who is visually impaired, talks with undaunted confidence that he does not regret being congenitally blind. He asks the society to put away its pity because he has no desire for sight. "What you have never known you cannot miss," he says and quoting Alexander Pope concludes "Whatever is, is right." Such positive attitudes in the nonaverage people are essential to be promoted through good design.

Networking

In many majority-world countries, efforts are being made by concerned people from all walks of life and professions toward the welfare of people who are elderly and such. Such groups and individuals include designers, journalists, doctors, teachers, voluntary organizations, governmental and nongovernmental organizations (NGOs), social workers, religious organizations, and parents and relatives of the nonaverage people. Many of them are doing excellent work with a high rate of success. The missing aspect is that these efforts remain in isolated pockets, and they are thus weak in spread and collective strength. This problem is particularly acute in countries with dense population and inadequate communications.

If the individuals and organizations network and develop links, they can pool their resources, benefit by sharing each others' experiences, complement each others' efforts, and generate solidarity as well as the symbiotic effect. We all know that the symbiotic effect will be much more than the sum of the individual efforts. The new information technology can play a major role in enabling such networking, in giving access to information, and in doing so with speed.

In nonindustrialized countries, where most of the population lives in rural areas, this may work only in urban areas. For rural needs, a different strategy of networking could be worked out. Oral networking, radio networking, and postal networking could be thought of, which could be used by village people.

Increasing Usability Range

There are many ways in which universally designed products and environments can bring equality among the average and the nonaverage people. Since universal design should be viewed as more than access, this chapter avoids the term *barrier-free access*.

Some of the general principles involved in universal design are necessary to reiterate. Segregation in any form means discrimination. This, in turn, means nonequality and nondemocracy. Therefore, for the good of universal design, it is essential to minimize the specialized products for special groups. Instead of designing separately for different groups, add value to products that are meant for average people so that they can also take care of other groups of people, such as those who are aged, disabled, or somehow separated and have special needs. Adding value in usual products by adding universal features would increase the product's *usability range*. This feature could be a virtue in the product for the consumer. By sharing interests, a closeness also develops between the average and nonaverage groups, which is the basis for a healthy society.

5.12   UNIVERSAL DESIGN: STRATEGIES OF APPLICATION

Victor Papanek called *Design for the Real World* (1972) a kind of design that is more meaningful. Universal design follows the same ideology. It should be viewed as a value and not as design style or a design fashion. Being so, its applications differ from the conventional applications of design. A quick look at some of the applications follows.

Addition: Extension of Use

Many existing products and systems can be extended with simple additions in design without much change in the overall product. Yet, such small change can include and greatly benefit people with special needs. Currency notes need not have Braille markings but raised or cut distinctions to help people who are visually impaired, as well as illiterate people. Medicine bottles can have similarly designed labels and large-print dosage instructions to aid people who are elderly and people with poor vision. Besides large print, dosage instructions could be visual for people who are illiterate. Road signage, in addition to language, could have auditory and three-dimensional signs. Bathrooms and toilets can have grab bars and textured floors. All alarms should be multisensory—auditory, visual, and olfactory.

Hindsight: Adaption of Existing Places and Products

Existing buildings, spaces, and products cannot be completely changed, but they could be made universal by a small renovation or adaption, taking into consideration the strong unchangeable cultural habits prevailing in a given country, especially in tradition-bound countries. Sometimes, it could be the introduction of an intermediate product. A project by a design student in India is a good example of this.

In India, the toilet is done by people in a squatting position. It is a cultural practice that is as strong as eating with one's hand without cutlery. But this causes serious problems for people who are aged, for those suffering from arthritis, and for those using crutches, and so on, as they cannot squat. A new toilet stool has been designed that could be pushed on top of the Indian squatting-type toilet. The new adaption is simple, foldable, and light enough to be lifted and hung on the wall to save space, an important criterion in highly populated countries. This is discussed in detail in the latter part of this chapter as a case study. Similar renovations are possible by providing tiny wheels to the low, ground-level seats (*patias* in Hindi) used by rural women in India who perform most functions at ground level.

Foresight: Prediction of Future Needs and Situations

If hindsight is necessary for buildings and products already in production, the well-designed products to be produced in the future should have foresight and future needs well considered in their conception. The products, systems, communications, and buildings should be designed so as to grow with the user, and to take into account the projected pressures of population, economy, and social change. Products must have modular provisions, and the like, which may become very useful aids when needed. In India, people build small houses initially, but keep provisions to expand them later when their needs grow and when they can afford it with their savings. This is universal design in attitude. It needs only to be turned toward old age, disability, and so on, and the related needs of inclusion. A trend is already emerging in India where people in urban localities make provisions for an elevator while building the apartment houses. In rural and semiurban localities, many people make provisions for a second entry, which will avoid the staircase and allow easy access for people who have difficulty walking, as well as for people using carriages and other personal wheeled items.

Bridging the Gap Between People

There is a need for products that act as a bridge between different people and their needs, whether that difference be cultural or physical. In India, the colonial times saw a number of such products, which have built bridges between the Indian and European cultures. The Anglo-Indian toilet is one of the well-known examples. This toilet accommodates the Indian squatting position as well as the European chair-level seating. It allows not only people from different cultures but different physical needs to be able to use the same toilet with no need for special provision.

A project by a student at the National Institute of Design bridges another gap. It is a hand printer that can be used both by a person who is visually impaired and by one who is not. It is a Braille-cum-English printer that allows written communication between people with different abilities. Development of more such products would eliminate product discrimination. Another good universal concept is the design of embossed and scented greeting cards, which would bring joy to both nonaverage people and average people.

The Right for Beauty

A thing of beauty is a joy forever. This is an old proverb. In the present context, it needs modification. A thing of beauty is a right for everybody. The reality, however, is different. Most of the socially conscious products or designs for the real world reveal the stepchild treatment in comparison with the commercial products. Even the award-winning products in the special-needs category, while functioning well, look ugly. Such products send out signals of pity or compromise in form, color, and style. It is humiliating to use such products, as they indignify the person using them. It is crucial that designers wake up to this reality and prove professionally that all people, irrespective of abilities, age, and other differences, have equal right to aesthetics.

There is no reason why a walking stick of the visually impaired should not be as elegant as the fancy walking stick of the rich man. Going by the evidence of international awards, some of the world's most elegant designs and some of world's most beautiful ladies are from the majority-world countries. The world designers would do well to recognize aesthetics as an integral part of function—one without the other is still a bad design.

5.13 GOVERNMENTAL POLICIES

The governments in the majority-world countries are preoccupied with the basic issues facing people, such as poverty alleviation, literacy, better infrastructure, roads, drinking water, sanitation, and so on. Most of these efforts concern average people, and there is hardly any time and resources left for the needs of the people who are elderly, disabled, and other people who are not average and who are presently a voiceless minority in comparison. Besides, people prefer family care to state care. There are further problems that hinder the process of the few welfare measures taken by these governments. Due to the economic backwardness, corruptive practices are rampant; whatever state allocations are made for the social welfare programs do not reach the deserving people.

The parliaments in the majority-world countries, however, make policies and laws in the interest of the people who are disabled or elderly. The government of India was party to the convention held by the economic and social commission for Asia and the Pacific held in 1992 at Beijing. As a consequence, it passed an act by the parliament in 1995 called the Persons with Disabilities (equal opportunities, protection of rights, and full participation) Act. It also passed a national policy on older persons in 1989. Similar efforts are afoot by the respective governments in other countries as well.

The government of the Philippines enacted the accessibility law in 1983 for the purpose of enhancing the mobility of persons with disabilities. This was followed up in 1992 by the Magna Carta for the Disabled Persons, which was passed by the House and Senate and signed by the president. These laws implement rules and regulations, especially for the public services. The Philippines also established the National Council for the Welfare of Disabled Persons (NCWDP).

The government of India has realized the key importance of meaningful occupation, as well as income generation for people with disabilities. Therefore, in 1995, it established an organization called the National Centre for the Promotion of Employment for the Disabled People (NCPEDP). The main regulation in India is the mandatory reservation of at least 3 percent of the total jobs in any company. The government, however, realizes its limitations and depends on the NGO sector as an important instrument for implementing the endeavors of the state and for catering to the needs of society.

The most important gap in many majority-world countries is the nonexistence of data related to people with disabilities, people who are aged, and other people who need help. Although there are many institutes and voluntary organizations working for the cause in different ways, there is neither coordination nor connection between these groups. Thus, most of these efforts are duplicated or diluted. With the exception of very few, the mainstream architects and designers in these countries are not practicing universal design. As a result, most of the people with special needs are left to take care of their problems in their own way. It is important to notice that there are hardly any efforts to bring together the requirements of average and nonaverage people.

The governments in majority-world countries are presently preoccupied with political instabilities and essential needs of people such as food, shelter, employment, education, health, and so on. But the growing number of people who are aged, disabled, and otherwise disadvantaged is not only a great consumer force, but also a great vote force, and democratic governments have to be mindful of this democratic power.

5.14 SERVICE ORIENTATION

For obvious self-interest, the commercial world today is not in favor of people buying fewer products. More products constantly promote growing discrimination, not only among the abled and disabled, old and young, but also among men and women, children and adults, and so forth. The more they discriminate, the more different products they can produce and sell.

Not only are there different products for people of one kind or another, but there are also products for yesterday and products for today, products for the morning and products for the evening. Many times, designers are unwilling partners in this overproliferation of products that promote discrimination rather than universality.

In the nonindustrialized countries such as India, the population is larger, which means that human services will be more economical than the products that offer similar services. Human services provide much needed employment to people. Houses are still built because they can be serviced inexpensively by hand, rather than produced as an industrial product. The laundry-person's services are cheaper than the cost of a washing machine combined with the cost of running it. The nonindustrialized countries can exploit this human wealth, which is ideal for universal design. Human beings can cater both to the average and nonaverage people equally well. The expenses of equipment that reads a book for people who are visually impaired can be easily replaced by the human being who costs less. The person, moreover, not only reads but also assists the impaired persons in other ways, too.

5.15 ECONOMICS OF UNIVERSAL DESIGN

All over the world, special products and services for people with special needs are prohibitively expensive. Years ago, in his book *Design for the Real World* (1972), Victor Papanek gave the example of a transistor radio, which costs $10 in the United States, while a hearing aid using the same components costs an astronomical $110. The reason given often is the low production numbers. This reason will not be valid with the majority-world countries, such as India and China, where a vast population lives. An increasing percentage of this population is aged, disabled, and disadvantaged in many other ways. There is thus the existence of a huge market. Besides, in countries like India, some of the people who are disabled and aged have been employed in producing devices for the population that is disabled or aged. One would not see why this enormous workforce and intellectual capital cannot be employed to produce universal design products. In fact, like elsewhere in the world, in this part of the globe, too, there will be a shift in target markets in the future. There are already indications. The wheeled patla, mentioned earlier in this chapter, is commercially produced by the local Blind People's Association (BPA) in Ahmedabad.

As Roger Coleman says in his *Designing for Our Future Selves: Meeting the Needs of the Older Consumer Through Universal and Age-Friendly Design Strategies* (1999):

> Unless the products and services on offer work well for and appeal across a wider range of age and capability—unless they are "universal" and "inclusive" in conception design and promotion, they will not fully capture the market that already exists and are unlikely to stimulate future growth in the market place.

In tradition-bound countries, this argument will find additional support since the people who are aged not only have accrued earnings but also have the authority and influence in family buying. This is people's tradition.

5.16 THE ROLE OF RELIGION

For many people, it is human nature to believe in God, particularly in times of distress. Jean Paul Sartre rightly observed that if there is no God, there is a need to invent one. In old age and in disability, people tend to become more religious. Religion provides psychological solace. In the case of some religions, people are provided practical support, too, as a social service. Besides, religion is able to bring people together, regardless of their differences and abilities. The only exception to this rule is the caste system, which is observed by the Hindu religion.

In tradition-bound countries, particularly like India, religion plays a greater role in the lives of people. As mentioned earlier, Indian mythology supports and promotes the importance of people who are aged or disabled. To the people who are aged or disabled, religious rituals and practices also provide occupations that are highly respectable by society and light enough to be performed easily. Religion, when consciously realized and positively applied, regardless of the faith one follows, could help greatly the cause of universality in majority-world countries.

5.17 CONSUMER FORUMS

In India and other majority-world countries, the consumer movement is getting stronger, and consumer forums are increasingly playing a very constructive role in empowering people. In India, they have often made the manufacturers be more responsible. Even the governments and nonformal sectors offering services have been held responsible by these forums. On one hand, the consumer forums are fighting for the rights of the people as consumers of products and services; on the other hand, the forums are educating the people about their rights. Some of these forums have well-equipped testing laboratories and market research infrastructure. One such forum in India is the Consumer Education and Research Centre (CERC) in Ahmedabad, which is well equipped, active, and runs a monthly magazine to spread consumer awareness to the public. So far, most of these forums have been concentrating on average people as consumers. It is time that the increasing nonaverage population's needs should also be considered by these forums, and universal design should be a component in the evaluation of a product or services. The term *product* is used here in a very broad, generic way and includes spaces, landscapes, and buildings.

5.18 CASE STUDY

Project.   Toilet seat attachment

Designer.   Makarand Kulkarni, Ahmedabad

One of the important problems being forced by people in India is using toilets. Indian toilets are located at ground level. Like all other work that is done at ground level, the Indian cultural practice is to excrete in a squatting position at ground level and to use water for cleaning that part of the body. The squatting position requires bending the knees totally (i.e., 360 degrees and pressing the calves and lower part of the thighs together tightly). For most people who are elderly, this tight bending is painful. It is likewise painful for people who suffer from arthritis and other joint problems. Moreover, people who use crutches and those who use wheelchairs find it impossible to use the ground-level toilets.

The problem was taken up at National Institute of Design by a senior student as part of the system design course. An initial survey conducted by the team indicated the following issues: Most of the contemporary Indian homes prefer the use of Western toilets or the Western-cum-Indian toilets, as mentioned elsewhere in this chapter. But this is not easy for most of the people who are in the middle-class income group. A typical middle-class house is usually an apartment with an already-existing ground-level Indian toilet situated in a narrow room, approximately 5.5 ft × 3 ft. The cost involved in adapting the existing toilet to fit a Western-type commode, along with plumbing and reconstruction, is very high and often not affordable.

The other members in the family are used to the Indian toilet pot and want to continue using it. Even the Western toilet is not much help to people with special needs, as it also has no armrests and no grab bars to assist them.

After a number of interactions with people with different needs, a brief was developed to design a device that could be adapted to the existing Indian ground-level toilet while keeping the cultural habits of the people in mind as well.

After various concept explorations in the form of drawings and models, an appropriate design solution was reached, and a life-size working model was made.

The new device is a foldable toilet seat. Indian people do not like to keep the objects of toilets outside. It is not uncommon that, after touching the objects in the toilet, people wash their hands well. The foldable seat, therefore, could be stored in the toilet room itself. It is light enough to be lifted by people who are elderly as well as weak, and it can be hung on the wall so that the floor space is saved for moving better. The new device has armrests, which also function as grab bars to assist the user in getting up or sitting down and in balancing oneself. It occupies much less space in the folded, as well as in the open, position and can easily be accommodated even in the smallest of toilet rooms. (See Figs. 5.3 through 5.5.)

FIGURE 5.3 The new toilet seat placed on top of the coventional Indian toilet.

FIGURE 5.4   The new toilet seat in a tucked-away position so that others can use the conventional toilet.

FIGURE 5.5 The new toilet is light enough to hang on the door to save space and to enable easy cleaning of the floor.

Its body is made of polypropylene, which has good mechanical strength and is, at the same time, light enough to make the device portable. There are two variations in the commonest Indian toilet pots, and the new device can fit on any type existing in the house, requiring no modification. It can also fit on top of the Western-cum-Indian toilet as well. It is comfortable for different body sizes because of its geometrical form and adequate thigh clearance.

As the height of this toilet seat matches the height of a standard wheelchair, it facilitates easy transfer of a person from a wheelchair to this toilet seat.

People in the Indian culture use a lot of water in the toilet, often by cleaning the whole floor. The new device takes this phenomenon into consideration. Not only can the new device be hung on the wall, but it is also rustproof. It uses standard parts, such as toilet covers, which makes the maintenance easier and possible, even in small towns and villages. Its cost has been brought down, so that a maximum number of people can afford the device.

A prototype is presently being built that will be field-tested thoroughly, and the feedback will be incorporated into the design before it is given to a manufacturer.

5.19   *GENERAL LESSONS LEARNED FOR UNIVERSAL DESIGN*

In a summary fashion, the general lessons learned are as follows. They are discussed in detail elsewhere in this chapter.

- Designers must promote public opinion toward a positive universal attitude. In illiteracy-dominant countries, this needs to be done through verbal and nonverbal methods of communication.
- It is very important to make children sensitive to the issues of universal design through nonformal methods of teaching and nonformal aids such as toys and games.
- Interpersonal communication between people who are nonaverage would not only bring solidarity but would promote a healthy, positive attitude in life. This needs encouragement through design.
- A better networking between various groups working for the cause of nonaverage people is essential for synergetic advantage. This needs to be done through rural communication.
- By consciously adding universal design features in any product, built environment, or communication, designers add value by increasing usability range.
- Universal design is possible even in poor countries.
- Universal design should not be looked upon as a magic formula that can give answers to all the ills of society. It has its limitations.

5.20   *CONCLUSION*

The nonindustrialized, labor-intensive countries, which are usually referred to as *developing countries,* have a disadvantage in terms of lack of capital and infrastructures. At the same time, they could be better suited to implement universal design and even to export universal design products and services to other industrialized countries because of their human capital and flexibility to change. The existence of a vast craft production sector and an enormous small-scale production sector makes it ideal for making universal designs with different choices for different cultures.

It was mentioned in the beginning of this chapter that universal design for majority-world countries extends not only beyond issues of a built environment's accessibility, but it also covers the social, cultural, and economic issues, which are major influences in uniting average people and people with different abilities. The question is what the prospects of change are for these social, cultural, and economic features. Can they survive under the impact of industrialization, new means of communication, and globalization? Based on past experience, one could conclude that in the tradition-rich countries where social and cultural features are deeply entrenched, major change will take a long time to come by. Take two examples from India.

EXAMPLE 1   *On the positive side, there is the existence of the joint family system. It is still very strong in rural areas, of which 80 percent of India is composed. In most of the Indian films, which reflect the popular trend, the strongest familial bonds can be seen. The government of India encourages joint families by giving special tax provisions under the Hindu Undivided Family (HUF) Act. Of course, there is increasing urbanization and breaking up of this system in urban areas. But it is comparatively slow.*

EXAMPLE 2   *The other example, a negative one this time, is the caste system. In spite of the great strides India made in terms of democratization, technological developments, and scientific progress, the caste system and the inequalities created by it still persist. There are child adop-*

tions, sperm banks, test-tube babies, and increasing opportunities for different communities to come together due to modern communications. Yet, a look at the matrimonial columns in major Indian dailies are indisputable evidence of the caste system's prevalence.

This mind-set will continue for a long time and will add to the gulf that already exists between the rich and the poor, between the educated and the uneducated, and between the privileged and the marginalized. Accessibility is not a problem for the rich, the educated, nor the privileged. The abundance of labor available in India will take care of that part of the population. It is the vast majority of the rest of the population that faces the problem of accessibility acutely. Universal design in the majority world has to address these culture-specific needs and influence society in bridging the gulf in a sustainable manner. For the majority-world designers, it is a challenging situation where design must play an active social role. This condition is quite different from the other economically well-to-do part of the world.

It is necessary for all concerned to recognize the contextual differences existing between different parts of the world and the fact that the design solutions also need to be different in order to work there. This chapter has attempted to articulate some of the differences prevailing in many forms in the highly populated and economically developing countries. Though a major part of the world's population lives there, their voices are often unheard. In Indian culture there is a wise saying, "Vasudhaiva Kutumbakam," which translates to, "The world is one family." It is indeed so, and it is imperative that the unheard voices must be heard and acted upon. When all products, buildings, and services are designed for all, keeping in mind all strata of human needs, it is universal design and it is better design.

The interesting fact about India is that despite the fact that it is 80 percent villages and that two-thirds of its more than 1 billion people are illiterate and poor, it is one of the most scientifically progressive countries in the world, bustling with some of the most cutting-edge technologies anywhere on the globe. Craftspeople are using electronic marketing; citizens are using electronic voting; and some state ministries are using daily teleconferencing. Indian cities like Bangalore and Hyderabad have become information technology centers that match the best, and Indian software expertise is eagerly sought after by the rich countries such as America and Germany. The situation is no different in China. What does this indicate? This indicates that money, education, and urbanization may facilitate, but these are no match to human potential and human progress. What is achieved by India in the field of information technology can be achieved in the field of universal design in any majority-world countries. To say that universal design in these countries must be the kind that suits the people's needs there—physical, social, cultural, and psychological—is not a contradiction of terms, indeed.

5.21 BIBLIOGRAPHY

Balaram, S., "Barrier-Free Architecture," *Indian Architect and Builder,* November 1999.

Balaram, S., *Design for Special Needs,* Encyclopedia Brittanica, Asian ed., New Delhi, India, 1999.

Balaram, S., *Thinking Design,* National Institute of Design, Ahmedabad, India, 1998.

Coleman, R., *Designing for Our Future Selves: Meeting the Needs of the Older Consumer Through Universal Age-Friendly Design Strategies,* Helen Hamlyn Research Centre, London, UK, 1999.

Eco, U., *Serendipities: Language and Lunacy,* Orion Books Ltd., London, UK, 1999.

Kose, S. (ed)., *Universal Design* (conference papers), Building Research Institute, Tsukuba, Japan, 1999.

Papanek, V., *Design for the Real World,* Paladin, Frogmore, UK, 1972.

Russel, P., *The Brain Book,* Routledge and Kegan Paul Ltd., London, UK, 1982.

Sarkar, K., *Listening to Shadows,* (film), National Institute of Design, Ahmedabad, India, 1999.

Whiteley, N., *Design for Society,* Reaktion Books Ltd., London, UK, 1993.

5.22 RESOURCES

Blind Peoples Association
Vastrapur, Ahmedabad—380 016, India
Contact: Bhushan Punani
Telephone: 079-440082

National Centre for Promotion of Employment for Disabled People
25, Green Park Extension
Yusuf Sarai, New Delhi—110 016, India
Contact: Javed Abidi
E-mail: ncped@nde.vsnl.net.in

National Institute of Design
Paldi, Ahmedabad—380 007, India
Contact: S. Balaram
E-mail: nid@vsnl.com

CHAPTER 6
USER/EXPERT INVOLVEMENT IN UNIVERSAL DESIGN

Laurie Ringaert, B.Sc., B.M.R.-O.T., M.Sc.
Universal Design Institute and University of Manitoba,
Winnipeg, Canada

6.1 INTRODUCTION

The purpose of this chapter is to describe and recognize user/expert involvement in universal design. The user/expert is recognized as an important contributor in both the universal design and the independent living paradigms. The Universal Design Institute in Winnipeg, Canada, developed and delivered national universal design access consulting introductory workshops for persons with disabilities and participated in subsequent projects, including an urban access audit project that included trained consultants as well as design students, street kids, and the participation of the community. The universal design access consultant workshops and subsequent projects have demonstrated how user/experts can make an important contribution to the universal design process. However, this project is only a beginning, and there is need for further development in this area.

6.2 BACKGROUND

As the baby boomer population ages, there is a growing need to make our urban environments more caring, more user friendly. Reinventing city centers according to universal design principles involves:

1. Studying the needs of a broad variety of users (including seniors and those with disabilities)
2. Actively involving these and other users in needs assessment and policy making for future improvements

The universal design movement is committed to ensuring that all spaces, products, and communications meet the needs of people of all ages and various levels of ability and that design in general contributes to quality of life. Universal design has been previously defined throughout this book. It strives to achieve safety, comfort, and convenience for all citizens in the community. Part of the universal design paradigm is user involvement in the process. Traditionally, any users have been minimally involved in design projects. The designer for a client has carried out designs with little input from the prospective user groups. This deficit has included persons with disabilities and seniors, along with nontraditional users such as maintenance personnel. The Universal Design Institute, formerly the Canadian Institute for Barrier-Free Design (CIBFD), sought to rectify this situation by developing and delivering a universal design access consultant training program for user/experts, in this case, persons with disabilities, and by subsequently involving them in various projects. The institute's recognition of consumers with disabilities as user/experts arises from the independent living paradigm.

The Independent Living Paradigm

The independent living (IL) movement began in the 1970s as the result of grassroots efforts to influence disability policies. The movement has been described as a product of a number of contemporary social movements, including the rise of consumerism, civil rights, and the self-help focus (Zukas, 1975; DeJong, 1979). Within the independent living paradigm, the person with a disability is defined as a consumer rather than as a patient or client. This paradigm provides an alternative to the medical and rehabilitation models, which focus on the limitations of the individual and on his/her inadequate performance of daily living tasks (DeJong, 1979; Crewe and Zola, 1984; Enns, 1986). Within the independent living paradigm, problems are defined in terms of barriers in the environment, including economic, architectural, or support systems, rather than in terms of the consumer's physical and/or mental disabilities (DeJong, 1979). As well, the paradigm emphasizes that pathology can be found in unprotected rights and in overdependency on relatives and professionals (Dunn, 1994).

Harlan Hahn, a well-known American disability rights advocate, has described the "minority-group" model (1988). This model emphasizes that environmental barriers have more impact than biological or psychological forces in shaping major life experiences of persons with disabilities. He argues that having to live with the shared constraints of inaccessible physical, social, and communicative environments and being denied equal access to education, employment, transportation, and housing while contending with negative stereotypes and minimal political power has constructed a distinct minority-group experience which includes the segregation, discrimination, and exploitation of people with disabilities (Hahn, 1988). Hahn's minority perspective is congruent with the principles of the independent living movement, which place emphasis on the environment rather than on the individual.

The independent living movement defines "independent living" as

> The process of translating into reality the theory that, given appropriate supportive services, accessible environments, and pertinent information and skills, severely disabled individuals may actively participate in all aspects of society (Crewe and Zola, 1983, p. 25).

The movement recognizes "independence" through dependence upon social and technological support. Such an interpretation presents a vision of "independence" as *mutual dependence,* also referred to as *interdependence* (Townsend and Ryan, 1991).

"Independence" according to the independent living movement, has been described by the following principles:

> To take part in all aspects of society as nondisabled people do (Derksen, 1983)
> To live outside of an institution (Kibele, 1989)
> To have control over one's life (Rock, 1988; Frieden and Cole, 1985)
> To include decision making (Rock, 1988; Kibele, 1989)
> To include freedom of choice (Lord and Osborne-Way, 1987)
> To be able to engage in risk taking (DeJong 1979, Rock 1988)

The primary meaning of independence to the independent living movement is the ability to have control over one's life in the community (Crewe and Zola, 1984). One way that persons with disabilities gain control of their lives in the community is through their day-to-day experience with facilitators and barriers in the built environment. They learn which situations have to be avoided; what works; and what has to be advocated for future betterment. Through this day-to-day interaction over time, they become experts in maneuvering through the built environment. This expertise provides a sense of control. Recognizing persons with disabilities as user/experts in the process of universal design rather than as "patients" in a medical model with minimal input, is in congruence with the independent living paradigm.

Recognition of Consumers with Disabilities as Experts

In recent years there has been more recognition of user/experts providing input to design projects; however, very little has actually been documented in the literature about this process. Lifchez (1987) discusses the importance of having persons with disabilities in the design classroom. Ostroff (1997) describes the importance of user/experts in the design process and defines them as

> ... anyone who has developed natural experience in dealing with the challenges of our built environment. User/experts include parents managing with toddlers, older people with changing vision or stamina, people of short stature, limited grasp or who use wheelchairs. These diverse people have developed strategies for coping with the barriers and hazards they encounter everyday. The experience of the user/expert is usually in strong contrast to the life experience of most designers and is invaluable in evaluating both existing products and places as well as new designs in development.

Anecdotal evidence of user/expert involvement includes calls for advice from designers and building contractors to disability-specific groups such as the local association for people with paraplegia, and disability generic groups such as independent living organizations. Calls are often made to individuals known personally to the designers or contractors. The problem with this approach is the premise that anyone with a disability has expertise in all access or universal design issues, which is an incorrect assumption. The specific study of universal design is a complex subject that develops over a period of years and involves learning at a variety of levels. Another issue is that the person seeking advice may often have to call a number of disability organizations to put together a complete picture. It is recognized that a diverse group of user/experts offer a wider range of perspectives as a means of understanding the responses of possible user populations (Ostroff, 1997).

Another approach to recognition of disability consumers to the process is through committees. Committees that include representatives from a variety of user groups have been assembled at municipal and at provincial and state levels. Representatives also have been asked to provide input to national building codes, standards, and guidelines. For instance, in Canada, the Canadian Standards Association includes a variety of consumer representatives on its standards development committees.

There are two main drawbacks to the previously discussed approaches. In most cases these representatives represent a unilateral dimension to universal design. For instance, the representative from a paraplegic association is an expert on wheelchair mobility needs only. Importantly, most representatives are not paid for their consultation. There is little monetary recognition for the contribution the experts make to the design.

The universal access consultant introductory workshops project recognized the wealth of knowledge and experience already available in these user/experts and built upon their skills. The intent was to provide broad-base training in universal design to persons with disabilities who already had experience in their own disability-specific access issues. They would then be able to provide consultation to designers, contractors, and others on broader universal design issues. The expectation was that they would also initiate their own consulting business and charge a fee for their services. This would then provide recognition for their work and provide them with an income. The hope was that this expertise would guide future development projects including those in urban areas.

Winnipeg: A Hub of Consumer Activity

The development of this unique course in Winnipeg, Canada, was not unusual if one considers the developmental background of the disability movement in Canada. Winnipeg is recognized as a hub of disability consumer activity and the birthplace of several disability organizations. Several of the strongest leaders in the movement came from Winnipeg, and as a result, many of the organizations were formed in Winnipeg. In 1980, Rehabilitation International held a conference in Winnipeg, and were perceived by people with disabilities as shutting them out from equal participation. As a result, Disabled Peoples' International was formed when a group of consumers held a meeting on their own in a room outside of the conference (Dreidger, 1989). The headquarters for this organization was housed in Winnipeg for almost two decades thereafter. The Council of Canadians with Disabilities also has its head office in Winnipeg. The Universal Design Institute—the only center addressing universal design research in Canada—was born in Winnipeg as the Canadian Institute for Barrier-Free Design, and recently the Canadian Center on Disability Studies was formed in the same city. The two centers now jointly carry out research for the World Health Organization's International Classification of Impairment Disability and Handicap, because again Winnipeg is recognized as this disability consumer hub of activity.

6.3 NATIONAL ACCESS CONSULTANT INTRODUCTORY WORKSHOPS

The National Access Consultant Introductory Workshops began as a pilot project through the Manitoba League of Persons with Disabilities. That organization contracted the CIBFD/Universal Design Institute to develop and deliver a course to five individuals in Winnipeg (see Fig. 6.1). The goal of the course was to train persons with disabilities as universal access consultants so that public and private sectors of their communities could employ them. This initial pilot course was carried out the fall of 1997.

Because of the success of the pilot course, the CIBFD/Universal Design Institute applied for funding to carry out a similar program across the country. The Institute received funding from Human Resources Development Canada Opportunities Fund for Persons with Disabilities to carry out the two-year project in 1998–2000. The funder dictated that the only eligible individuals for the program were those with disabilities who were unemployed and ineligible for employment insurance, as the mandate of the funder was to enhance employment opportunities.

The first two months of the project included developing a curriculum, deciding on teaching materials, ordering and compiling teaching materials, preparing information packages for disability organizations at the prospective sites, developing advertising for local newspapers, and sending out information packages. Tanis Woodland, who had been trained in the Winnipeg pilot, was hired as the administrative assistant to make all arrangements for the workshops including advertising, compilation of materials, arrangements for speakers, meeting facilities, and any other requirements that arose.

Workshop Development

The sites.   The sites where the workshops were held included St. Johns, Newfoundland; Thunder Bay, Ontario; Regina, Saskatchewan; Victoria, British Columbia; Edmonton,

FIGURE 6.1 Graduating class of the Universal Access Consulting Workshop pilot program in Winnipeg, Canada. Participants include (left to right) Tanis Woodland, David Tweed, Teresa Swedick, Gail Finkel (instructor), Ross Eadie, and Richard Friesen.

Alberta; and Halifax, Nova Scotia. There was also one individual who attended the Edmonton site from Whitehorse, Northwest Territories. Thus the workshops were held in seven sites including the Winnipeg pilot.

Advertising to attract participants. Initially a wide variety of disability organizations were contacted, provided with an information package, and asked if they knew of people who would qualify for the workshops. This was the sole method of advertising for the first three workshops. However, the Institute found this method not to be reliable in providing enough participants. The next workshops were advertised using a combination of contacts to disability organizations and newspaper advertisements. Requests were received from many nonqualifying (employed and nondisabled) individuals as well as from qualifying applicants.

The participants. A total of 36 people took the course (including Winnipeg) while a total of 24 people successfully completed all requirements, including attendance during the two weeks and successful completion of two assignments. Participants had a variety of disabilities including varying levels of hearing, vision, mobility, upper extremity agility, and mental health functioning.

Curriculum

The workshop curriculum was developed in consultation with the instructors, Betty Dion and Gail Finkel, two well-known universal design consultants in Canada, and the coordinator, Laurie Ringaert, director of the CIBFD/Universal Design Institute. It was decided that the format for the course would be two five-day sessions, with one month in between. One instructor taught the first five-day session, then there was one month of study time when participants could complete the first assignment. This was followed by a five-day session taught by the second instructor and completion of a second assignment one month after that week. If participants successfully passed both assignments and had a good attendance record, they then received a certificate and were eligible for one year of mentoring with either instructor.

The curriculum included the following topics: introduction to universal design concepts; the interaction of the range of human functioning and the built environment; human rights legislation; introduction to the building process; introduction to commonly used building codes, standards, and guidelines; how to provide an access audit proposal; how to conduct an access audit; how to prepare an access audit report; introduction to the role of the access consultant; introduction to how to start your own consulting business.

During each week-long session, participants had the opportunity to visit an actual audit site and practice auditing methods. Two guest speakers were also brought in. During week one, a building code official from the participating city spoke for one hour, and during week two a person from the Business Development Bank spoke on how to develop a small business. A sample of the curriculum is shown in Table 6.1.

A wide variety of Canadian and American accessibility/universal design materials were required for the workshops. Both Canadian and American materials were used for several reasons. Many essential documents have not yet been produced in Canada. Some American documents, such as the Americans with Disabilities Act (ADA), are very important for any North American access consultants to be aware of. As many documents as were available in alternate formats were ordered so that persons with visual or agility disabilities could access them. In many cases it was difficult to retrieve documents in alternate formats.

Participants were involved in simulation activities where they were to experience, for example, what it was like to move about using a wheelchair or while blindfolded. Unlike traditional simulation activities, however, these were persons with disabilities who were required to try out other "disabilities." For instance, a wheelchair user would have to move about while blindfolded; a person who was blind would have to use a wheelchair. This was the first experience of this kind for many of the participants, and they commented on how enlightening it was for them.

Preliminary Outcomes

Television episode. The Canadian Broadcasting Corporation (CBC) program *On the Move* produced a television episode of the project. This is a weekly program of the Independent Living Foundation that highlights activities related to disability across Canada. The video documented the individuals trained in Winnipeg who were carrying out a project and also interviewed the director of the CIBFD/Universal Design Institute regarding the program. This program subsequently won a Human Rights award.

Round Table for Participants. The Universal Design Institute hosted the first universal design conference in Canada (Universal Design in the City: Beyond 2000) in the fall of 1999. Individuals who participated in the program were invited to attend the conference and had their expenses paid. Approximately 15 individuals attended the conference. A round table luncheon was held where individuals discussed their experience with the course and the kind of work they were now involved with. All individuals felt that the course had been a positive experience for them. Some individuals had participated in paid work opportunities; others were involved on a voluntary basis. Some were just contemplating how to get their consulting businesses started, while others were well underway in their consulting business development. It seemed the individuals from Winnipeg had gained the most experience, probably because the Universal Design Institute as well as the Manitoba League of Persons with Disabilities were both in the city and could refer opportunities directly to them.

Workshop. A workshop was also held at the conference where the Universal Design Institute, the University of Winnipeg's Institute of Urban Studies, and the City of Win-

TABLE 6.1 National Access Consultant Workshop Curriculum Excerpt
Lesson Plan—Week Two

Day 3
9:30–10:00 Accessibility Evaluation Guide—Audit method based on CSA
10:00–10:45 Audit process
11:00–12:00 Writing an audit report
1:00–4:00 Do an audit

Materials
1. Handout: Evaluation guide (handed out in week one)

2. Various reports

3. Tape measurers

4. Clip boards

5. Light meter

6. Tape recorder, batteries, tape

Day 4
9:30–10:00 Discussion of audit
10:00–10:45 Writing the report and making recommendations
11:00–12:00 Source book—audit method
1:00–2:00 Maintaining seniors independence—audit method
2:00–2:45 The Workplace Workbook
3:00–4:00 Reading plans: MS Society, universally designed office space

Materials
1. Report outline

2. Handout: Source Book

3. Handout: Maintaining Seniors Independence

4. Handout: Workplace Workbook

5. MS plans and scales

Day 5
9:30–10:00 Codes and guidelines review
10:00–12:00 Speaker: Business Development Bank—How to develop a business plan
1:00–2:00 Marketing your skills
2:00–2:45 Assignment 2, case management
3:00–3:45 Topics of students choice
3:45–4:00 Student evaluation

Materials
1. Handout: Assignment 2 (given first day)

2. Handout: Case Management information sheet and timecard

3. Handout: Instructor evaluation forms, envelopes

nipeg Access Advisory Committee brought together the workshop participants as well as representatives from other city, provincial, and state access committees. They discussed projects and issues they were dealing with. It was one of the first meetings of its kind. Participants expressed the need to continue the communication between cities and to meet again. The Institute plans to facilitate this communication.

Project involvement.   The City of Winnipeg hired two of the workshop-trained individuals to provide consultation on a streetscaping redevelopment project. The city was redoing its main street. The city requested that the landscape firm hire two of the access consultants to provide consultation to the project. This was the first time that it had been done by the city.

The Manitoba League of Persons with Disabilities received a grant to study employment offices across the province. Again persons who had been trained in the project were hired. In another project, funded by the Active Living Alliance, two of the consultants were hired to carry out universal design audits of recreational facilities across the province. A booklet will be developed upon completion of this project. Yet another project involved two user/expert universal design access consultants who conducted a miniaudit of the University of Manitoba through the Universal Design Institute. One user/expert was blind, and the other had a mobility disability.

Universal design booklet.   The Universal Design Institute produced a booklet entitled *Is Your Business Open to All?* (Ringaert, Knutson, Rapson, 2000) which employed some of the workshop participants as models for the book and used their expertise in focus groups. This booklet describes to businesses the importance of universal design and that universal access consultants were trained across Canada. It suggests they contact the Institute for further information. In this way, the Institute will assist in directing future contact jobs to the individuals.

Audit.   One large project that the workshop participants in Winnipeg were involved with was an audit of downtown Winnipeg. The following pages provide information on this specific project.

6.4   USER/EXPERT INVOLVEMENT IN A UNIVERSAL DESIGN AUDIT OF A CITY

In 1998/99 Winnipeg, Manitoba, conducted a city accessibility audit that could provide a model for other cities. The City of Winnipeg initiated steps toward universal design by contracting (through the Winnipeg Development Agreement 5C funding) the CIBFD/Universal Design Institute to conduct a universal design audit of its downtown area. The goal of the City of Winnipeg Access Audit was to establish a baseline audit of accessibility and to provide recommendations. The City of Winnipeg Access Advisory Committee (AAC) was responsible for overseeing the project. (The AAC is the mayor's designated committee mandated to provide advice to the city council on issues related to access to information, services, and properties within the City of Winnipeg.) There are similarities between this project and the urban design project described by Sandra Manley (Chap. 58, "Creating an Accessible Public Realm").

Key Elements of the Audit

Participatory action paradigm in the process.   Because of the involvement of community stakeholders in all phases of the project—from initiation to research design, research process, and receipt of the final draft report—this project is an example of participatory action research.

User/expert involvement.   People with disabilities played an active role in designing the research, conducting the audit, and analyzing the data collected. The audit team leaders were persons with a variety of disabilities—mobility-, agility-, hearing-, or visual—who previously had taken a course in universal design access consultation from the Universal Design Institute. A range of age groups and background was represented by these individuals as well as by street kids, design students, and others who were also involved in the three-person audit teams (see Fig. 6.2).

FIGURE 6.2   City of Winnipeg audit: user/experts take measurements in the field.

Community consultation.   A technical steering committee (TSC) comprised of city department representatives, including designers, engineers, and members of the business community, provided an important perspective during monthly consultation on the project. In addition, a Delphi survey was distributed to this group and other community organizations to ensure broad stakeholder involvement in the process. A Delphi survey provided a community prioritization process. It established that safety and ease of access were the most urgent issues to address, beginning with sidewalks and curb cut design, then building entrances, and moving on to other building issues and then parks. It is important to note that by addressing the accessibility issues, safety would also be met. For example, a pothole in the middle of the sidewalk is both an accessibility and a safety issue, as is inadequate lighting. Community members also set priorities for the order in which specific sites within the scope of the city audit should be upgraded. A sample excerpt of the Delphi survey is shown in Table 6.2.

Development of the Universal Design Audit Process

Audit preparation, instrument development, team member training.   Accessibility audits of cities and communities, as well as accessibility codes, standards, guidelines, and checklists were reviewed to determine the types of data that should be considered in this accessibility audit. The Institute could not find examples of public rights-of-way audit checklists, methodologies, or guidelines that covered the broad spectrum required for a more universal design audit. For instance, none of the examples located covered all areas that the team deemed necessary for a universal design audit of a city that would incorporate the needs of people with visual, hearing, mobility, agility, and cognitive disabilities. The Institute therefore developed checklists and an accompanying database to suit its requirements through

TABLE 6.2   Delphi Survey Sample Excerpt

INSTRUCTIONS

1. The following second round of the Delphi survey is divided into three basic sections:

- *Section A* asks you to rank order the eight major elements that were audited.
- *Section B* asks you to rank order the problems/issues found in each of the major elements.
- *Section C* asks you to rank order the identified sub areas of the Centre Plan Area.

2. *For Section A.*   Please review the rank order of the eight (8) major Centre Plan Area elements, in terms of pedestrian accessibility issues, that the first survey respondents gave to the major elements. (1 is the highest importance and 8 is the least important.) Please review the prioritizations given and ensure that you agree with these prioritizations. If you do not agree please reprioritize.

3. *For Section B.*   Please review the rank order of the identified sub areas of the downtown City of Winnipeg Centre Plan Area and the problems/issues that were identified for each of the above mentioned major elements, in terms of pedestrian accessibility issues. Each of the major elements has associated with it a varying number of problems/issues, (1 is the highest importance, 2 is the next highest importance and so forth) for each of the problems/issues identified. At the end of each section there is an "other" item listed, if you feel that any other problem(s) should be listed, please indicate it (them) in the space provided and what you think it's (their) ranking should be. (If you do not think there are any other problems/issues, leave this part of the question blank.) Please review the prioritizations given by survey respondents and ensure that you agree with these prioritizations. If you do not agree please reprioritize.

4. *For Section C.*   Please rank order the eight (8) identified sub areas of the downtown City of Winnipeg Centre Plan area in terms of pedestrian accessibility issues. (1 is the highest importance and 8 is the least important.) Please review the prioritizations given by the survey respondents and ensure that you agree with these prioritizations. If you do not agree please reprioritize.

Section B

AUDIT AREA PROBLEM/ISSUES

Within each of the major elements listed, there are various problems/issues resulting in terms of pedestrian accessibility (to, from, and within these major elements).

NOTE: The definitions for the various elements, as noted in the first survey, have been relocated to the back of this survey.

1. Walkways/Sidewalks

The following rankings indicate the first survey respondent's prioritizations. Please review these prioritizations (1 being the most important and 12 being the least important), and ensure that you agree with these prioritizations. If you do not agree, please reprioritize in the space provided. If you feel that any other problems/issues are important to consider, please specify in the space provided

First Survey	Second Survey
Ranking	Ranking
1	___ —Clear path of travel along walkway and/or width for clear path of travel along walkway
2	___ —Condition of walkway
3	___ —Missing walkways
4	___ —Ramps or appropriate entrances to buildings or structures
5	___ —Cross slopes along the path of travel
6	___ —Running slopes along the path of travel

TABLE 6.2 Delphi Survey Sample Excerpt (Continued)

_7_	___ —Placement of street furniture along path of travel
_8_	___ —Lighting along the path of travel
_9_	___ —Placement of street fixtures along path of travel
_10_	___ —Color or texture contrasts that act as directional clues or warning surfaces
_11_	___ —Bushes, trees, and/or tree branches hanging over walkway
_12_	___ —Walkway surface texture
___	___ —Other (please specify)

consultation with the user/experts and by combining accessibility criteria from a number of primary sources. A two-day training session was held for all team members.

Actual audit.   The audit took place from September to mid-November 1998. This audit area constituted approximately 314 hectares or about 777 acres of Winnipeg, which is a prairie city with a population of approximately 600,000 people. In all, 847 blocks, 29 city-owned or leased buildings/offices, and 35 parks, parkettes, river walkways, and/or green areas were audited. Three main elements were considered: streets/sidewalks, civic buildings (leased, rented or owned), and parks.

Data input and site revisiting for information confirmation.   Data was entered into a database. Sites were revisited to clarify or elaborate on audit form notes. The project coordinator reviewed and compared all of the original data collected with that which was entered into the database, to attempt to ensure data consistency and database entry accuracy.

Data trends and analysis.   The project coordinator and the director of the Universal Design Institute examined the data and determined areas of increased frequency to determine the main problem areas. The combination of the Delphi survey results along with universal design theory assisted in making determination of priority areas to be addressed.

Recommendations and action plans.   A combination of the data analysis, review of existing city policies, stakeholder feedback, and universal design and independent living theory (participation of persons with disabilities in decision making) formed the basis for the recommendations and action plans. A draft of these recommendations and action plans was given to the access advisory committee and technical steering committee for feedback before the final report.

Ownership-Facilitated Change

The key to the success of this audit was undoubtedly the combined involvement of critical user groups and influential city representatives and business people. Local government and business constituencies acquired greater understanding and appreciation for the issues because they were directly involved in the auditing process. In fact, many of the safety problems that the Institute reported at the monthly meetings were rectified immediately.

6.5 *LESSONS LEARNED FOR UNIVERSAL DESIGN*

The training workshops and the subsequent projects demonstrated the valuable contribution of user/experts to the design process. Information that only they could have gained through their life experiences was contributed to the projects. It is important to note that many of the employment opportunities that have occurred have taken place in Winnipeg because the Universal Design Institute and the Manitoba League of Persons with Disabilities, which initiated the pilot, are located there. Even though the original intent was that they would work independently, in many of these projects, the universal design access consultants most often work in teams of two representing differing abilities. It seems that they appreciate the user expertise offered by each of their disabilities as well as teaming the functional abilities to assist in the audit. For instance, a consultant who is blind often works with a consultant with a mobility disability. The individual who is blind is able to bend down and hold the tape measure, while the other individual reads it. A blind consultant also worked with a deaf consultant, each contributing their abilities and knowledge. Part of the teaming process may be related to this work being relatively new for them, and they are continuously learning from each other.

Very few of the individuals from outside Winnipeg participated in the mentoring opportunity. They were reminded several times but only a few actually took the opportunity. It is unclear why this opportunity was not acted upon. The relatively slow start in employment for those living in cities other than Winnipeg can be accounted for by a number of reasons: (1) they were previously unemployed, many for several years, and may have had little employment experience; (2) the area of universal design consultation is relatively new, and it still requires development from a number of angles;* (3) the course was new and this was the first time it was taught; and (4) these individuals may need more support than was able to be provided in this project. An independent evaluator is currently carrying out an external evaluation of the project.

Regarding the city audit, the methodology for the process had to be established as the work progressed. The user/expert involvement in the process was key in determining recommendations. The concept of ownership was very important to the process. Because the participants in the project included people with disabilities, and also designers, policy makers, seniors, building code officials, and even street kids, they all had their opinions heard and felt empowered to bring about change. The novel concept of establishing three-person teams comprised of various kinds of experts worked well. The design students had firsthand learning from the expertise of the consultants with disabilities, and the consultants valued the involvement of the student designers as experts of another kind. Even the street kids brought their own real-world experience to the process. The Universal Design Institute is now refining the procedures for data collection, data entry, and data analysis in order to streamline the process and construct a final model that can be easily adopted by other cities committed to establishing a more caring, adaptable urban environment. The institute is pursuing developing research to create an electronic access map based upon the database. The city will now have to consider the recommendations and determine which to adopt and put into practice. It is important that universal design audits (including large-scale ones such as cities, communities, and universities) include a universal design/participatory action approach to a physical and policy accessibility audit. Changes in policies will direct the future. An overriding policy and commitment to universal design will in turn direct all departments in their work. Increased annual budget commitments to accessibility and changes to the construction standards will ensure proper universal access designs that are up-to-date with trends in universal access. Consultation with universal design access consultants on new construction is essential. Training of employees on disability issues and provision of accessible services is also important. There must also be a commitment to ongoing community consultation on access issues, which can be facilitated through an access advisory committee. Most important, there must be a commitment to ensure that an access plan is implemented and evaluated and to ongoing maintenance of an access database as repairs, renovations, and new construction occur.

---

* For instance, in the United States, where Americans with Disabilities Act (ADA) consultants are common, these individuals would have an easier time finding work.

6.6 CONCLUSIONS

The universal design access consultant workshops and the subsequent projects have demonstrated how user/experts can make an important contribution to the universal design process. As a result of this project, these individuals have additional universal design training added to their many years as user/experts. The fact that many have been hired in a number of projects also demonstrates that their expertise is being recognized. However, most of the activity has occurred through work done in Winnipeg, where the Universal Design Institute is located. The institute is actively pursuing national projects that will encourage employment of those trained across the country. There is a great deal of work ahead to ensure that urban designers recognize the important contribution of user/experts. Perhaps this project of the Universal Design Institute can foster further work in this area elsewhere in the world.

6.7 NOTE

The Universal Design Institute acknowledges the work of David Rapson (project manager, universal design consultant), Tanis Woodland (database consultant, universal design consultant), all those involved on the project team, as well as the City of Winnipeg Access Advisory Committee (Judy Redmond), Winnipeg Development Agreement (Jackie Halliburton), Technical Steering Committee, and all community participants on the City of Winnipeg project. The Institute also acknowledges the work of Betty Dion, Gail Finkel, and Tanis Woodland for their work on the National Access Consultant Introductory Workshops.

6.8 BIBLIOGRAPHY

Crewe, N. and I. Zola, *Independent Living For Physically Disabled People,* Jossey-Bass Inc., San Francisco, 1984.

DeJong, G., "Independent Living: From Social Movement To Analytic Paradigm," *Arch Phys Med Rehabil,* 60, 1979.

Derksen, J., *The Disabled Consumer Movement: Policy Implications for Rehabilitation Service Provision,* Coalition of Provincial Organizations of the Handicapped, Winnipeg, MB, Canada, 1983.

Dreidger, D., *The Last Civil Rights Movement: Disabled Peoples' International,* St. Martin's Press, New York, 1989.

Dunn, P., "Government Policy Innovations In Barrier-Free Housing, Accessible Transportation and Personal Supports, Paper presented at the National Independent Living Conference, Winnipeg, MB, Canada, 1994.

Enns, H., An excerpt from the historical development of attitudes towards the handicapped: a framework for change. In D'Aubin, A. (ed.) *Defining the Parameters of Independent Living,* COPOH, Winnipeg, MB, Canada, 1986.

Frieden, L., and J. Cole, "Independence: The Ultimate Goal of Rehabilitation of Spinal Cord-Injured Persons," *American Journal of Occupational Therapy,* vol. 39, no. 6, 1985.

Hahn, H., "The Politics Of Physical Differences: Disability and Discrimination," *J. Soc. Issues,* 44, 1988.

Kibele, A., "Occupational Therapy's Role In Improving the Quality of Life for Persons with Cerebral Palsy," *American Journal of Occupational Therapy,* vol. 43, no. 6, 1989.

Lifchez, R., *Rethinking Architecture: Design Students and Physically Disabled People,* University of California Press, Berkeley, CA, 1987.

Lord, J., and L. Osborne-Way, *Toward Independence and Community: A Qualitative Study Of Independent Living Centres In Canada,* Secretary of State, Ottawa, ON, Canada, 1987.

Ostroff, E., "Mining Our Natural Resources: The User as Expert," *Innovation, the Quarterly Journal of the Industrial Designers Society of America,* vol. 16, no. 1, 1997.

Ringaert, L., B. Knudtson, and D. Rapson, *Is Your Business Open to All?* Universal Design Institute, Winnipeg, MB, Canada, 2000.

Rock, B., Independence: "What It Means to Six Disabled People Living in The Community," *Disability, Handicap & Society,* vol. 3, no. 8, 1988.

Townsend, E., and B. Ryan, "Assessing Independence in Community Living," *Canadian Journal of Public Health,* 82, Jan./Feb. 1991.

Zola, I., "Involving the Consumer in the Rehabilitation Process: Easier Said than Done," In *Technology for Independent Living II,* American Association for the Advancement of Science, Washington, DC, 1982.

Zukas, H., CIL history, In report on the state of the art conference, Center for Independent Living, Berkeley, CA, 1975.

6.9 RESOURCES

Universal Design Institute.   E-mail: universal_design@umanitoba.ca. Web site: www.arch.umanitoba.ca/UofM/CIBFD.

Finkel, G. (ed.) *Access: A Guide to Accessible Design for Designers, Builders, Facility Owners and Managers,* Universal Design Institute, Winnipeg, Manitoba, Canada, 2000.

Frye, J., K. Frye, and R. Sandilands. *Accex: Universal Design Expert Software System,* Universal Design Institute, Winnipeg, Manitoba, Canada, 2000.

Inter-Organization Access Committee, *Supplement to Universal Design Guidelines: Focusing on the Needs of People with Visual Impairments,* Manitoba League of Persons with Disabilities, Winnipeg, Manitoba, Canada, 1997.

Jones, M., *People with Disabilities and Entrepreneurs,* The Canadian Centre on Disability Studies, Winnipeg, Manitoba, Canada, 1997.

Universal Design Institute, *Universal Design Resource 2000,* CD-ROM, 2000, available from AHEAD, www.ahead.org.

CHAPTER 7

DESIGNING BY DEGREE: ASSESSING AND INCORPORATING INDIVIDUAL ACCESSIBILITY NEEDS

M. Powell Lawton, Ph.D.
Philadelphia Geriatric Center, Philadelphia, Pennsylvania

7.1 INTRODUCTION

Universal design is the central topic of this chapter. For the present purpose, *universal design* may be defined as the best approximation of an environmental facet to the needs of the maximum possible number of users. This definition implies an effort at inclusiveness. A circular staircase may meet the aesthetic need of a particular consumer, but as a standard feature in a new housing tract, it would clearly limit access for some, and unnecessarily so, given the possibility of a custom order. There is also an opposite-directional aspect of universal design. It is possible that a feature explicitly created for a special user group may, in fact, turn out to be useful for all (e.g., a front porch stair rail). Thus, expanding usership capability and importing creative ideas from the specific to the universal are equally important design strategies.

7.2 BACKGROUND

The most universal aspect of the design process is the assumption that all people share a basic set of common needs and that all design must spring from the accommodation of such needs. Perhaps the best-known conceptualization of needs has been the Maslow (1964) hierarchy, which arranges needs from the most basic, or visceral, level through needs to know and to manipulate the external world, proceeding up to self-actualization, a level that can be attained only as lower-level needs become satisfied. In this conception, universality is greatest at the most basic, life-maintaining level, and it decreases as limitations on individual competence and on environmental resources increase. Maslow made an important distinction between *deficiency motivation* (i.e., stimulated by a lack of need satisfaction at the most basic levels) and *growth motivation* (stimulated by active search for new experience). Preiser (1988, 1991; see also Chap. 9) has used a similar *habitability framework* as the theoretical beginning of the design process, featuring three *habitability levels*.

The thoughts pursued in this chapter speak to the theme of this section, "Transcending the ADA Legacy," by suggesting that the Americans With Disabilities Act (ADA) has made major steps in reducing deficiency, but has further to go to enhance growth. In the process, an attempt will be made to reconcile the dialectic of universality and individuality by searching for variation in design solutions in the context of individual needs, as exhibited within a common need structure. To perform this task, a number of theoretical concepts that link person and environment will be utilized. In turn, these concepts will lead to design solutions that reflect the priorities of users of different environmental types. One of the first conceptual issues that requires attention is the question of the relationship between characteristics of the person and those of the environment.

Human Needs Versus Environmental Affordances

Needs, as construed for the present purpose, are clearly a property of the person. Much has been written to explore the alternative view that person and environment are one, versus their duality (Altman and Rogoff, 1987; Ittelson, 1976). It is convenient to deal with them as separate entities because the assessment task is more easily accomplished in that fashion than in more global concepts like the *behavior setting* (Barker, 1968), even though this latter concept may do better justice to the philosophic indivisibility of person and environment.

Personal needs are seen as differentially capable of being satisfied as a function of the ability of the environment to support those needs. Gibson's (1979) concept of *affordance* describes an environmental structure present in the physical configuration that allows a behavior to occur if the behaving individual discovers the structure and emits the behavior that uses the structure. An affordance thus is not the cause of any behavior. The individual must use the environment actively. The author's view is that personal need provides the motivation to search for affordances and that individual characteristics of the person determine the content of the behavior that effectively uses the affordance.

This theoretical view implies a parallel set of personal needs and environmental affordances whose behavioral outcomes and subjective states are capable of being evaluated in terms of quality of life. An early short list of environmental affordances was provided by Brill and Krauss (1970) as a set of continua (left-hand column of Table 7.1).

TABLE 7.1   Environmental Affordances and Personal Need Analogues

Environment*	Personal need
Communality—privacy	Privacy
Sociopetality—sociofugality	Affiliation
Informality—formality	Order
Familiarity—remoteness	Novelty
Accessibility—inaccessibility	Function
Ambiguity—legibility	Cognition
Diversity—homogeneity	Stimulation
Adaptability—fixity	Autonomy
Comfort—discomfort	Comfort

* *Source:*   From Brill and Krauss (1970, p. 50).

It is not difficult to find terms that refer to propensities of the person to relate to these environmental properties. The right-hand column of Table 7.1 shows the author's matching of commonly used personal needs from sources such as Murray (1938), Edwards (1953), and others, with the environmental dimensions suggested by Brill and Krauss. A definitive single set of universal needs/affordances is unlikely to be attainable, primarily because of the idiosyncrasies of nomenclature and the likelihood that in choosing category names, some will prefer broad categories (e.g., autonomy) and others will prefer more numerous but narrower categories (e.g., detachment, power, dominance, etc.). It thus would seem that a set of 8 to 12 needs could roughly define the universe. Although they cannot be reviewed here in detail, a continuous search for *dimensions* of environment, such as those shown in Table 7.1, has gone on, some of it reviewed in Lawton et al. (1997). One result of this work was the Professional Environmental Assessment Protocol, developed to guide experts in the assessment of the ecological qualities of nursing home environments (Lawton et al., 2000; Norris-Baker et al., 1999; Weisman et al., 1994). This approach, in turn, is presently being adapted to develop a set of quality-of-life indicators for the nursing home, with parallel measures for the personal, social, and physical environments (Kane, Kane, & Lawton, 1999). Later in this chapter, the question of the completeness of the needs/affordances shown in Table 7.1 will be discussed further.

The core of a model for optimizing environmental design is, therefore, a set of human needs, which, in generic form, are likely to be universal, and a parallel set of environmental affordances. The next requirement is a means to account for individually tailored content on both the person and environment sides. That is, there are multiple routes toward the fulfillment of any higher-order need and, therefore, multiple variations in any affordance that may be consistent with the need. Recognition of such personal idiosyncrasies forces recognition of the generic issue of the study of person and environment, the relationship between the individual and the aggregate. If the ideal is universality, what happens to the individual?

The ADA attempted to resolve this dilemma, first, by extending the boundaries of design for the majority so as to include a simply broader range of personal mobility and access affordance—*expanded design.* Many solutions are seen as possible at little extra cost by simply thoughtful, planned design. Beyond the accessibility requirements of ADA, repeated instances of discovery that universal design can be both better and cheaper have occurred in the recent history of design development. The tub grab bar, differentiation of wall and floor color/texture, and task lighting are only a few such examples. The second approach was to design solutions tailored to the special needs of users for whom expanded universal design was not helpful. The many technologies for special signaling, enhanced motor behaviors, and communication fall into this category. The universal design ideal has begun the process of influencing special-user design to result in *mainstreaming* design's replacing *handicapped* design. That is, user acceptance and the social message conveyed by the design are greatly enhanced when the aesthetics and the commonplace look of products are given the same attention as their functional quality. The combination of expanded design and special-user design form an improved approximation to universal design. How is it possible to move beyond this first-generation achievement? The next section will make general suggestions for this task.

7.3   *MOVING BEYOND ADA*

The psychological and social sciences have a potential contribution to the goal of universal design from three related vantage points. First is the extension of the concept of universal design from a deficiency-reduction purpose into the concept of a growth-promotion purpose. Second is to amplify the lexicon of needs and affordances. Third is to suggest strategies and techniques by which needs and affordances may be assessed and applied to design problems.

Growth Promotion

In the narrowest view, physical access represents a small proportion of all human activity. Access is, of course, a precondition for the satisfaction of a large number of other needs. Its instrumental value is thus high. Nonetheless, the focus on access has a tendency to obscure one's view of the larger and more diverse set of goals that access serves.

The history of research in mental health and in the psychology of need satisfaction was slow in catching up with Maslow's distinction between deficiency reduction and growth promotion. By now, however, there is general acceptance of the duality of mental health, in the sense that the many forms of psychopathology characterize deviations below "normality," while a separate set of personal characteristics denotes positive mental health. Examples of negative indicators are depression, anxiety, antisocial behavior, and the many other symptoms of mental illness. Examples of positive dimensions of mental health may be illustrated from Carol Ryff's (1989) work: autonomy, environmental mastery, personal growth, positive relationships, purpose, and self-acceptance. Although people who are mentally ill are likely to have fewer of the markers of positive mental health, the positive and negative aspects of mental health are far from being two ends of a single continuum.

The same duality has been established when the focus is narrowed to emotional states. When the frequency and duration of people's affect states are measured, positive affect is inversely correlated with negative affect, but the strength of their relationship is only moderate, in the neighborhood of a correlation coefficient ranging from 0.30 to 0.60, depending on circumstances. More concretely, this means that some people who are frequently sad may also be frequently happy and vice versa, while others may be infrequently either happy or sad.

Yet, another established relationship is seen in the tendency for externally engaging events, for example, social interaction, diverting activities, novel environments, to be relatively strongly associated with positive-emotion states but less so with negative states. Health and internalized cognitions such as self-concept, in turn, are inversely related to negative states but less to positive states (Bradburn, 1969; Lawton, 1983b; Lawton, DeVoe, and Parmelee, 1993). In particular, a positive evaluation of their homes and neighborhoods by older people was related to the frequency of positive feelings but not to negative feelings (Lawton, 1983a).

Growth Promotion and Deficiency Reduction

Extrapolating such findings into the environmental area, one would thus look toward stimulating and interesting environmental enhancements to fortify the positive aspects of mental health, but would expect them to be less effective in counteracting negative symptoms such as anxiety and depression. In contrasting fashion, removing environmental features that make health care more difficult or that elicit feelings of personal inadequacy may counteract some of the negative aspects of mental health. There is thus good reason to search further for ways that environmental design, human services, and, especially, self-motivated effort, all may be enhanced to pay better attention to the growth-promoting side of life. The next section will begin by making a further theoretical link between needs/affordances and outcomes, and then revisit the task of suggesting a structure with which to begin a lexicon for universal needs and affordances.

7.4   NEEDS, AFFORDANCES, OUTCOMES, AND QUALITY OF LIFE

The previous section was devoted to two aspects of mental health. Positive and negative aspects of mental health are best seen as outcomes of the transactions between people and their environments. People with particular needs who live in environments capable of satisfying those needs and who either proactively or reactively utilize those environmental affordances are more likely to experience positive behaviors and emotional states, and they are less likely to experience negative ones. The bottom line by which a fulfilling life is judged is based on the mix of behaviors and affects judged in such evaluational terms.

The antecedents of mental health in these terms fall into the realm of *quality of life* (QOL). Quality of life is composed of a set of value judgments, made on the basis of both subjective and social-normative standards, of how well various domains of the person's everyday life function in leading to their overall mental health. The number of possible domains may be many, but their number and relative importance vary with the individual. Examples are domains of family, friends, marriage, home, spirituality, and leisure time. Each may be seen as a suboutcome, and any of them may sometimes become a major life goal of its own. For example, some people specialize in their work, others in a hobby, others in their homes or their families. These domains of QOL represent the idiosyncratic aspects of the person-environment transaction where the larger categories of needs and affordances represent the universals. It will be argued that the needs and affordances are a template that should be used to map the specific aspects of QOL for individuals.

The Basic Set of Needs and Affordances

The parallel set of needs and affordances shown in Table 7.1 was based on the single array of environmental features offered a number of years ago by Brill and Krauss (1970). Not being driven by the concept of need, they made no particular effort (a) to represent all needs or (b) to arrange the environmental aspects in any hierarchy. In fact, introducing these two principles does not change the array a great deal. The set to be described next is suggested as basic for the present purpose. It revises the set suggested by Brill and Krauss primarily by introducing a rough hierarchical ordering and by augmenting the needs at the upper and lower levels of the hierarchy, and it extends the content somewhat beyond the user group of the Professional Environmental Assessment Protocol (PEAP) on which it is based (Weisman et al., 1994; i.e., physically or mentally frail older people living in a nursing home). The model is thus meant to apply to any designated user group in any context.

The hierarchy represented in Table 7.2 is based roughly on Maslow's hierarchy, which is a continuum of physiological needs to self-actualizing needs. The low end, missing from Table 7.1, may be designated as "security and safety" {i.e., the needs that must be satisfied to maintain life—physiological functioning such as nourishment, health, and safety from external threats such as weather or bodily harm ["health and safety" level in Preiser's (1983) formulation]}. The basis on which a rough order is suggested here is that as one moves upward from safety and security, the content of needs varies increasingly across individuals, the person's dependence upon fixed aspects of the environment decreases, and the ultimate outcome becomes mixed with proportionately more positive affective elements. In the case of safety and security, such outcomes as accidents, starvation, or victimization by criminals are potentially negative for all people. Environmental solutions that address these problems, such as stair rails, home-delivered meals, and target hardening are relatively powerful forces in reducing negative affect. Although one may feel warmly secure (positive affect) when a security threat is removed, the more typical outcome is, nonetheless, the reduction of negative outcomes as security increases up to the point of full adequacy. If an environment cannot satisfy basic physical needs, there is no opportunity to pursue other needs higher in the hierarchy. On the other hand, a person's willingness to pursue needs higher in the hierarchy may clearly involve taking the risk of loss of complete security.

Similarly, near the basic level is Brill and Krauss's accessibility, the broader term *function* being preferred, termed *functional and task performance* level, according to Preiser (1983). This is the heart of ADA. Assuming physical health and security, being mobile is clearly the first stepping-stone toward satisfaction of many other needs. The environmental antecedents for functionality are well known, but there is more to functionality than body access. Included in the affordances for functional behavior are a variety of conditions such as lighting, auditory features, signaling mechanisms, response modes, many of them contributed to by diverse disciplines such as engineering and ergonomics, in addition to the usual design professions. See Fisk and Rogers (1996) for a summary of age-relevant research in human factors. On the affective side, functional need frustration is commonly a sense of distress, deprivation, anxiety, and often

TABLE 7.2   Need-Affordance Analysis Model for Older People in Nursing Homes

Need satisfaction behavior	Possible measure	Affordance	Environmental indicator examples
Security/Safety Resident feels secure about personal safety, availability of health care, accident minimization, provision of basic life-maintenance services.	1. Preference 2. Consumer satisfaction 3. Personal competence	Protective, safe, nurturing environment.	Front desk with receptionist Door alarms in dementia units Floors with nonslip surface 2-min response time for resident signal
Function Residents move about and perform basic self-care.	1. Preference 2. Consumer satisfaction 3. Scales for independence in activities of daily living	Structure, furnishings, and staff practices encourage independence in self-care.	Storage space accessible Toilets near activity spaces Roll-in shower stall Wall-floor color contrast
Cognition People comprehend the meaning and signals communicated by their environment.	1. Preference 2. Consumer satisfaction 3. Cognitive screening	Effort to provide orienting features to foster comprehension of environment.	Resident handbook in large print Building structure comprehensible Sufficient light without glare Thoughtful orientational signage
Comfort Residents experience physical well-being and staff are alerted to identifying and ameliorating discomfort.	1. Preference 2. Consumer satisfaction 3. Mental health 4. Nonverbal signs of pleasure or distress	Responsive HVAC regulatory system, furniture, staff training in recognizing pain.	Vestibule draft buffer Room HVAC control Outdoor seating in sun and shade Low ambient noise level
Order Residents know what to expect, what is expected of them, and what behaviors should occur where.	1. Preference 2. Consumer satisfaction	Spaces and objects have self-communicating designations for appropriate behavior	"Props" visible that denote denote activity (e.g., religious symbol for worship area) Posted schedules for the week "Preview" capability of seeing what is going on in an area prior to entering it
Autonomy Residents take initiative and make choices for their lives and care.	1. Preference 2. Consumer satisfaction 3. Perceived control	Fixed and semi-fixed environmental features emphasize alternatives and encouragement of choice.	Movable seating Large, small, and private social spaces Menu alternatives Sitting areas with and without TV
Privacy Physical, visual, auditory, and communication privacy available.	1. Preference 2. Consumer satisfaction	Physical privacy and staff behavior supporting privacy.	High percentage of single rooms Toe-to-toe bed configuration in shared rooms Door signal, used by staff Private staff consultation spaces Staff training on confidentiality
Stimulation Sensory, cognitive, and behavioral stimulation at optimal level.	1. Preference 2. Consumer satisfaction 3. Varied activity program participation 4. Openness to experience	Environmental and behavioral variety.	Homelike décor Colors, textures, vistas, furnishings varied within moderate range Low-stimulation areas available

TABLE 7.2   Need-Affordance Analysis Model for Older People in Nursing Homes (*Continued*)

Need satisfaction behavior	Possible measure	Affordance	Environmental indicator examples
			Rich menu of possible activities Familiar-looking objects and spaces
Affiliation Social interaction or distance at desired level.	1. Preference 2. Consumer satisfaction 3. Family and friends quality of life 4. Extraversion-introversion	Physical and behavioral social opportunity.	Seating in high-traffic areas Seating with outdoor views Adequate 2-, 3-, and right-angled seating Small spaces for small social groups Scheduled social, intergenerational activities
Individuality (self-actualization) Maintain personal identity, ties with past, and pursuit of personal goals.	1. Preferences 2. Consumer satisfaction 3. Self-esteem scale	Social and physical environment encourage preservation of resident's past and affirmation of present and future self.	Individuality of room decor Self-markers from past (hobby, props) Staff training module on linking with past Personalization of room entrance Staff concern for end-of-life quality
Spirituality Opportunity for religious observance, meditation, feeling of unity with being or cultural guide beyond the human level.	1. Preference 2. Consumer satisfaction 3. Attitudes toward religion, spirituality, and cultural identity		Physical symbols of cultural, national, or religious background evident but not obtrusive Dedicated religious observance space Provision for diversity of symbols and observance when appropriate

anger at the barriers imposed by both personal disability and poor environmental design. There is an equally strong opportunity, however, for positive feelings to emerge, associated with successful environmental coping. Self-efficacy is a powerful motivator partly because of the frequent sense of elation that comes from mastering a new task and managing one's environment.

Cognition.   This is a need almost as basic as security and function. A primary task of life maintenance is to comprehend the external world and apply its meaning to all other phases of life. The pathologies of cognition are well known and include the many forms of biological disorders of the nervous system as well as psychologically based misreading of the meaning of environment. Within the normal range of cognitive functioning, environmental legibility and cognition have been some of the most studied of the environmental affordances. Much of this research has dealt with comprehending aspects of the environment that stimulate growth and new experience. On the affective level, it is clear how failure to read the environment is associated with deprivation and anxiety, but it is just as clear how excitement and new skills emerge from successful comprehension of environmental affordances and ability to accept new challenges and enrichment.

Comfort. This is a much more difficult need to define than the first two, but the sense of comfort bridges the life maintenance and the life-enriching ranges almost equally. Clearly, if safety and function are not satisfied, there can be no comfort. There is a very broad range, however, over which comfort can be positive rather than simply indicating the absence of distress. For example, Kolcaba (1991) has distinguished three forms of comfort: (1) relief from a previous state of distress, (2) ease as an ongoing state, and (3) transcendence, the most active form. Personal preference comes into play over some ranges of environmental variation, such as temperature, lighting, and ventilation, to say nothing of cognitive and aesthetic preferences. Beginning at this level of need, a primary affordance is the flexibility of the environment, in terms either of its capacity to be manipulated or its ranges of choices available to meeting individual needs. Comfort and the remaining needs to be described were called *psychological comfort and satisfaction* by Preiser (1983). He characterized them at that time as "not codified." Their complex nature has made them resistant to formalized assessment. Part of the purpose of this chapter is to suggest that it is time to undertake that difficult task.

Order. Knowing and being comfortable in one's world are possible when *order* is present. The skills of planning and looking into the future are proactive ways of satisfying the need for order. A world that is unpredictable or chaotic can be the source of intense negative emotion. By contrast, predictability and familiarity are well-known sources of emotional contentment. Socially and culturally legible environmental messages are a source of familiarity and predictability. As noted in relation to security, most people learn to take risks by judiciously choosing the unpredictable and unfamiliar at the right time.

Autonomy. The need for *autonomy* is a very complex need, one that is present at every stage of human development. Regardless of stage, people differ in the strength and pervasiveness of such a need. Its many faces may be seen in independence, dominance, assertiveness, uniqueness, choice, the wish for control, and other traits. Lack of ability to satisfy the need for autonomy is associated less with psychopathology than with passive dependence and lack of self-confidence. Major theories of behavior have been built around personal control as the central feature of human motivation (Schulz and Heckhausen, 1996). Various manifestations of this need appear as people construct their own environments in ways that afford privacy, territorial marking, and control over their uses. Anxiety and depression may be associated with lack of autonomy and exhilaration with independence and the successful exercise of power or feeling of freedom.

Affiliation. Where autonomy implies some status separation or distance from others, the need for *affiliation* distinguishes those who thrive on relationships and interaction with others from those who are more comfortable at a social distance from others (i.e., isolates). Environmental psychology has devoted considerable attention to those conditions that foster social interaction, called *sociopetality,* and those that discourage interaction, termed *sociofugality.* Positive interchange with others has typically been associated with positive affective states and involuntary isolation with negative states. Again, as in the case of most higher-order needs, personal preference is the critical determinant of the mental health outcome of socially rich or deprived environments; some people distance themselves from others by choice and do not find isolation stressful. Cultural variations and social norms condition all need-satisfying behaviors, but their influence on affiliation is particularly notable. The choice between affiliation and isolation is important to provide in all environmental design.

Stimulation. Other people are one source of stimulation, but people do appear to have differing need levels for stimulation in much broader ways, whether sensory, cognitive, or affective (Berlyne and Madsen, 1973; Csikszentmihaly and Graef, 1980). There is such a thing as a deprived level of stimulation, with pathological behavioral and emotional outcomes. In contrast, an anxious and behaviorally disorganized response is possible to overstimulation, represented in the extreme by some people with dementia who may require a stimulation-dampened environment (Hall, Kirschling, and Todd, 1986). Environmental diversity and novelty are more commonly associated with enjoyment and other positive affects, the most intense being the experience of *flow* (Csikszentmihaly and Graef, 1980). In unusual instances, and for unusual people, flow is a peak experience with complete focus on the experience to the exclusion of awareness of anything else.

Individuality.   The need for *individuality*, although universal, has very different manifestations for different people. Need fulfillment may be so difficult for the cognitively impaired or very limited person as to strip him or her of full awareness of personal identity. Such unusually low awareness is unfortunately often interpreted by both caregiving professionals and designers as a lack of all sensitivity to individual needs in such people. Such ascribed "nonpersonhood" may be counteracted by active attempts to help mark their individuality. For most people, such fulfillment begins with a sense of dignity and moves further through past memories and on to a subjective schema of the self against which all contemporary experience is given meaning. In Maslow's (1964) terms, self-actualization is a developmental level achieved only by the few who have developed need-satisfying systems for all lower needs, and who are able to be creative in whatever domain of life their gifts lead them. Typically, people construct or choose environments that reflect their individuality, whether through housing type, automobile, yard, household decor, hobby, or manner of dress. Successful identity marking is likely to be associated with positive feelings ranging from contentment to the peak of flowlike self-actualization. It should also be noted, however, that the peaks of creativity or selfless contribution to others are by no means attained in the context of positive affect alone. Conflict and anxiety are often mixed with creative behavior.

Spirituality.   Does *spirituality* qualify as a universal need? It seems unlikely, yet those for whom the conception of merging with some power greater than that of any human being is highly important, would place this need at the pinnacle of any need hierarchy. One cannot automatically assign the absence of spiritual need fulfillment to the deficiency realm, but the spiritual realm in those who value it clearly belongs in the growth motivation and the positive affect zone. One might also think of cultural tradition as a related source of need fulfillment, in the sense that merging with an organized symbolic system is a source of personal strength and affective experience.

7.5   *UNIVERSALITY AND INDIVIDUALITY REVISITED*

The list of needs and their implied environmental affordances is suggested as the universal component of the design process. Is this assertion anything more than an academic exercise?

There is a next step capable of translating this theoretical array into environmental design in a way that maximizes universality but that allows for diversity. The translation is based on formal assessment, divided into four steps, the most important of which is the first.

Step 1.   The systematic consideration of each need, for every prospective user. The critical question is, How will this need be best accommodated? In a way, the need list is a set of reminders, a checklist that needs to be completed before takeoff. Sometimes, a need is neglected by planners because they simply did not think to consider it. Working from a need-affordance matrix at the very least reminds the planner to take each into account systematically. In other instances, a relevant environmental affordance is not obvious. Perhaps the most malignant type of omission is the assumption that some need is irrelevant or of low priority because of some socially stereotypic presumed property of the user group. For example, under the assumption that the only really important need of a person with a motor disability is an attribute that enhances functional performance, that person's need for positively toned stimulation and mainstreaming (e.g., aesthetic quality) may be ignored.

Step 2.   Systematic assessment of individuals on the characteristics of representative members of the user group. The social and behavioral sciences have developed a technology for such formal assessment. Earlier practice has often been to identify the average ability or range of ability of the user group to satisfy a particular need and to use this average as the basis for the design process. In the procedure advocated here, assessment should be applied in a way that defines the limits of individual competence and leads to a broadening of the affordances of the environment—stretching the fit. *Competence* is a continuum along which the quality of a person's behavior as judged by social norms may be evaluated (White, 1959). In fact, the term should be used in the plural. The author, for example, has defined *competences* in biological health, functional health, cognition, time use, and social behavior as basic dimensions that characterize individual behaviors (Lawton, 1982). Each of these dimensions may be assessed by a variety of standardized measures, resulting in a profile of competencies and incompetencies unique to that individual. This is the point at which the artificial distinction between person and environment, alluded to earlier, becomes a major barrier to the task of behavioral design. Is there really such a person as an *incompetent* person? A *disabled* person? In contrast, the transactional view clearly represents the reality. Competent behavior is the favorable outcome of any person acting in an environmental context. Incompetence and disability are thus not personal terms but transactional. A key aspect of the individual assessment step is to think in terms of choice and alternatives, as contrasted with a design that is so closely tailored to the average as to allow little variation in the need-satisfying abilities of the planned user group or little future change in the users' abilities.

Step 3.   Assessment of an existing or prospective environment in terms of its affordances. The technology for assessing environments is less well developed than that for assessing people. In addition, the goal of assessing affordances is not likely to be attained in its strictest sense. An affordance is not always evident to either user or scientist. Instead, attributes of the environment about which there is some hypothesis regarding its potential usefulness are the focus of assessment. In analogous fashion to the discussion of competence, environmental affordances cannot be studied in the abstract, but always must be assessed in light of the transaction with the user.

Step 4.   The design process itself, at this point, is still an art rather than a science. Empirical science has led to the designation of needs and recognition of their universality by demanding that every user and every environment be assessed in terms of the way each need might be satisfied. Assessment of people and assessment of environments, while directed toward broadening the area of possible fit between user and environment, cannot be expected to lead literally to universal design. Although post-occupancy evaluation (POE) has long been an essential component of person-environment research (Moore, 1982; Preiser, Rabinowitz, and White, 1988; Reizenstein and Zimring, 1980), its technology needs to grow in concert with growing understanding of the complexities of the transaction. The science-informed art of design must begin with such knowledge, and then proceed through the usual processes of iterations and successive approximation. To address the issues in universal design, this process continues to apply the principles of stretching the fit, embedding choice, and anticipating change. As has already happened in the mainstream of universal design, many previously narrowly designed features will be found possible in stretch-fitting style that does not diminish the potential satisfaction of needs other than function. This art-of-design phase utilizes the material from the three previous steps but now depends on fitting patterns of expertise to the task. Weisman (1998) has written compellingly of the process by which experienced designers apply their knowledge in dynamically shifting fashion, when each new design task modifies the patterns in the designer's mind for use in the next process [see Alexander, Ishikawa, and Silverstein's classic development of the concept of pattern language (1977)].

7.6 APPLYING THE NEED STRUCTURE

Table 7.2 portrays examples of how the first three steps might be identified, beginning with safety and security. The example of a nursing home is used, but the method is generic. The process begins with the universal aspect of design, that is, a brief definition of each need, in the first column. The second column notes what personal attributes might be assessed formally in order to define the full range and central tendencies of user characteristics. Design may cease to be totally universal as a result of such individual assessment, but the first purpose of such assessment is to remind the designer of the limits of individual ability and, at that point, to start the process of stretching the fit. The details of assessment are just beginning to become specified. Three measurement approaches follow.

Determining Preferences.   The technology of assessment is probably less necessary for the majority of the population because they can verbalize their desires and act on them relatively easily. For physically and especially mentally impaired people, as well as socially low-powered people, there is value in the attempt to survey formally some of the varieties of ways people prefer to satisfy their needs. Few such survey instruments exist. For frail older people in long-term care, Van Haitsma et al. (1999) designed *The Preferences for Everyday Living Inventory,* which is available in a preliminary version. Preferences are so varied that the length of such instruments tends to become unwieldy. In addition, preferences are often context-specific, so that the design specifics may be less than universal. In another sense, the preference survey aids universal design because of its ability to display the broad range of choice that would be ideal to achieve universality. Nonetheless, some method for matching preferences to environmental affordance should be a standard component of the design process.

Consumer Satisfaction.   Consumer satisfaction is a postdesign component of the process. That is, asking the consumer for feedback regarding an existing design solution for use in the next design solution is an important aspect of post-occupancy evaluation. The environment behavior literature is full of such devices (Carp, 1975; Lawton, 1979). Since congressional action in 1987 mandated that input from nursing home residents be included in their program of care, a number of surveys designed for this purpose have been constructed (Applebaum, Straker, and Geron, 2000; Cohen-Mansfield and Ejaz, 1999). Because of wide variation in user competence (see next paragraph), the measures themselves are group-specific.

User Competence and Well-Being.   For some, but not all, needs, variations in the abilities or competencies of the user and in the well-being of the user may be determined by existing measures. Such assessment is most important when there are deficits in the abilities required to meet needs at the lower levels of the hierarchy. Limitations in cognitive ability put severe constraints on the use of universal design, for example, for people with dementia, developmental disabilities of intellectual function, or impairments in judgment. Thus, screening tests are important and readily available in many treatment situations, for intellectual impairment, sensory function, or difficulty in performing activities of daily living (ADL). Outcomes of some need-affordance transactions may be assessed with overall measures of mental health such as self-esteem or depression, or with domain-specific instruments such as quality of life among friends, family, housing, use of time. Finally, the possibility of using personality measures such as introversion-extraversion, openness to experience, locus of control, or attitudes toward spirituality may be helpful. There is a large literature in assessment for each of many user groups [for adults in general, see Robinson, Shaver, and Wrightsman (1991); for older adults, see Lawton and Teresi (1994)].

7.7   *LESSONS LEARNED FOR UNIVERSAL DESIGN: A SKELETON PROCEDURAL OUTLINE*

There is a growing demand for the services of a consultant in behavioral design for new residential facilities for elders and for renovation and reuse design. Sometimes, the major mission of either the design-oriented or the behavior-oriented consultant is to sensitize the sponsors, board, administration, and prospective staff to the principles of universal design merged with user-specific needs when appropriate. These constituencies usually find gratification in becoming involved in design discussions, but they usually also require instruction and expert input as the process unfolds. A capsule outline of principles to be conveyed follows.

- Define the user group.
- Specify the ranges of user characteristics.
  - Needs
  - Competences
  - Preferences
  - Cultural and social patterns
- Define desired outcomes.
  - Deficiency reduction
  - Growth promotion
- Specify environmental affordances likely to enhance outcomes.
  - Needs—meeting environmental features
  - User-group-specific compensatory design features
  - Consumer survey, focus group, and other approaches to elicit environmental preference
- Feedback cycle where universality is the goal.
  - Redesign in expanded-design mode, according to the principles of universal design
  - Redesign in special-user mode, preserving universality to the maximum degree possible while accommodating special needs
  - Reshape both expanded and special-user trial designs in the direction of mainstreaming
- Post-occupancy evaluation.
  - Observe environment in use
  - Identify misuses, errors, and ambiguities of place designation
  - Identify magnet areas or features
  - Group discussion and instruction on optimal use of the environment

The end result of a clear planning process with interchange between client subgroups and the design team/consultants is very likely to result not only in a more effective design but in a continuing process over time to optimize the use of the new environment.

7.8   CONCLUSION

It is not realistic to think that every design decision will result in optimum usability for every person. The argument has been advanced that human needs are, however, universal and that it is possible to pursue a universally relevant procedure as the design process proceeds. That procedure involves asking questions systematically about every need for every user. Specification of the user group and assessment of the preferences and the competences of the user group as to the central tendency and range, as its members pursue satisfaction of needs, follows. One result from this process will be that growth-inducing needs will be far better represented than has been achieved by traditional prosthetic design planning. As needs and competences are assessed, environmental design affordances are matched to the personal characteristics. At this stage, some design decisions will result in stretching the fit to reduce the gap between universal design and special-user design. The major result, however, will be achieving the universal goal of representing every common need in the design, if not by the same solution in every case.

Bidirectionality is an ideal of universal design. The search for universal principles to guide design should lead to maximum inclusiveness; special attention would be directed toward any design feature that excluded a potential user. Not every user can be accommodated in this way, however. Approaching the problem from the opposite direction (i.e., recognizing that needs, competences, preferences, and cultural tradition may not all be universal), can motivate creativity in design. With an open mind, the designer can then linger longer to wonder whether what appeared to be group-specific can actually lead to an innovation that is useful beyond that group.

7.9 BIBLIOGRAPHY

Alexander, C., S. Ishikawa, and M. Silverstein, *A Pattern Language,* Oxford University Press, New York, 1977.

Altman, I., and B. Rogoff, "World Views in Psychology: Trait, Interactional, Organismic, and Transactional Perspectives," in D. Stokols and I. Altman (eds.), *Handbook of Environmental Psychology,* vol. 1, John Wiley, New York, 1987, pp. 7–40.

Applebaum, R. A., J. K. Straker, and S. M. Geron, *Assessing Satisfaction in Health and Long-Term Care,* Springer, New York, 2000.

Barker, R. G., *Ecological Psychology,* Stanford University Press, Stanford, CA, 1968.

Berlyne, D. E., and K. B. Madsen (eds.), *Pleasure, Reward, Preference,* Academic Press, New York, 1973.

Bradburn, N., *The Structure of Psychological Well-Being,* Aldine, Chicago, 1969.

Brill, M., and R. Krauss, "Planning for Community Mental Health Centers," in H. Sanoff and S. Cohn (eds.), *Proceedings of the Environmental Design Research Association,* vol. 1, EDRA, Edmond, OK, 1970.

Carp, F. M., "Long-Range Satisfaction with Housing," *Gerontologist,* 15: 27–34(a), 1975.

Cohen-Mansfield, J., and F. Ejaz, *Consumer Surveys in Long-Term Care,* Springer, New York, 1999.

Csikszentmihaly, M., and R. Graef, "The Experience of Freedom in Daily Life," *American Journal of Community Psychology,* 8: 401–414, 1980.

Edwards, A. L., *Manual for the Personal Preference Schedule,* Psychological Corporation, New York, 1953.

Fisk, A. D., and W. A. Rogers (eds.), *Handbook of Human Factors and the Older Adult,* Academic Press, San Diego, CA, 1996.

Gibson, J. J., *The Ecological Approach to Visual Perception,* Houghton-Mifflin, Boston, 1979.

Hall, G., M. V. Kirschling, and S. Todd, "Sheltered Freedom—An Alzheimers Unit in an ICF," *Geriatric Nursing,* 7 (May/June): 132–136, 1986.

Ittelson, W. H., "Some Issues Facing a Theory of Environment and Behavior," in H. M. Proshansky, W. H. Ittelson, and L. G. Rivlin (eds.), *Environmental Psychology,* 2nd ed., Holt, Rinehart, & Winston, New York, 1976, pp. 51–59.

Kane, R. A., R. L. Kane, and M. P. Lawton, *Measurement, Indicators, and Improvement of the Quality of Life in Nursing Homes,* contract with Health Care Financing Administration, Division of Health Care Services Research and Policy, University of Minnesota, Minneapolis, MN, 1999.

Kolcaba, K. Y., "A Taxonomic Structure for the Concept Comfort," *IMAGE: Journal of Nursing Scholarship,* 23: 237–240, 1991.

Lawton, M. P., "Social Science Methods for Evaluating Housing Quality for Older People," *Journal of Architectural Research,* 7: 5–11, 1979.

———, "Competence, Environmental Press, and the Adaptation of the Elderly," in M. P. Lawton, P. G. Windley, and T. O. Byerts (eds.), *Environmental Theory and Aging,* Springer, New York, 1982, pp. 33–59.

————, "Environment and Other Determinants of Well-Being in Older People," *The Gerontologist,* 23: 349–357, 1983a.

————, "The Dimensions of Well-Being," *Experimental Aging Research,* 9: 65–72, 1983b.

————, M. R. DeVoe, and P. Parmelee, "The Relationship of Events and Affect in the Daily Lives of an Elderly Population," *Psychology and Aging,* 19: 469–477, 1993.

————, and J. Teresi (eds.), *Annual Review of Gerontology and Geriatrics,* vol. 14: *Assessing Older People,* Springer Publishing, New York, 1994.

————, G. D. Weisman, P. Sloane, and M. Calkins, "Assessing Environments for Older People with Chronic Illness," *Journal of Mental Health and Aging,* 3: 83–100, 1997.

————, ————, ————, C. Norris-Baker, M. Calkins, and S. I. Zimmerman, "Professional Environmental Assessment Procedure for Special Care Units for Elders with Dementing Illness," *Alzheimer Disease and Associated Disorders,* 14: 28–38, 2000.

Maslow, A. H., *Motivation and Personality,* Harper, New York, 1964.

Moore, G. T. (ed.), "Applied Architectural Research: Post-Occupancy Evaluation of Buildings," *Environment and Behavior,* 14 (6): 643–724, 1982.

Murray, H. A., *Explorations in Personality,* Oxford, New York, 1938.

Norris-Baker, L., J. Weisman, M. P. Lawton, and P. Sloane, "Assessing Special Care Units for Dementia: The Professional Environmental Assessment Protocol," in E. A. Steinfeld and G. S. Danford (eds.), *Measuring Enabling Environments,* Plenum, New York, 1999.

Preiser, W. F. E., "The Habitability Framework: A Conceptual Approach Towards Linking Human Behavior and Physical Environment," *Design Studies,* 4: 84–91, 1983.

————, "Design Innovation and the Challenge of Change," in W. F. E. Preiser, J. C. Vischer, and E. T. White (eds.), *Design Intervention: Toward a More Humane Architecture,* Van Nostrand Reinhold, New York, 1991, pp. 335–351.

————, H. Z. Rabinowitz, and E. T. White, *Post-Occupancy Evaluation,* Van Nostrand Reinhold, New York, 1988.

Reizenstein, J. E., and C. M. Zimring, "Evaluating Occupied Environments," *Environment and Behavior,* 12: 427–558, 1980.

Robinson, J. P., P. R. Shaver, and L. S. Wrightsman (eds.), *Measures of Personality and Social Psychological Attitudes,* Academic Press, San Diego, CA, 1991.

Ryff, C. D., "Happiness Is Everything, or Is It? Explorations on the Meaning of Psychological Well-Being," *Journal of Personality and Social Psychology,* 57: 1069–1081, 1989.

Schulz, R., and J. Heckhausen, "A Life-Span Model of Successful Aging," *American Psychologist,* 51: 702–714, 1996.

Van Haitsma, K., K. Ruckdeschel, B. Carpenter, and M. P. Lawton, *The Preference for Everyday Living Inventory,* Philadelphia Geriatric Center, Philadelphia, 1999.

Weisman, G., *Toward a Model for Architectural Research and Design: Pragmatism, Place, and Patterns,* Architectural Institute of Japan, Osaka, Japan, 1998.

————, M. P. Lawton, P. D. Sloane, L. Norris-Baker, and M. Calkins, *Professional Environmental Assessment Protocol (PEAP): A standardized method of expert evaluation of dementia special care units,* School of Architecture, University of Wisconsin at Milwaukee, Milwaukee, WI, 1994.

White, R. W., "Motivation Reconsidered: The Concept of Competence," *Psychological Review,* 66: 297–333, 1959.

CHAPTER 8
UNIVERSAL DESIGN TO SUPPORT THE BRAIN AND ITS DEVELOPMENT

John Zeisel, Ph.D.
Hearthstone Alzheimer Care, Lexington, Massachusetts

8.1 INTRODUCTION

This chapter describes how universal design can be approached using what we know about the brain and the way it handles information about the environment. Essentially, everyone has similar brain structure and design. No matter what one's background or physical abilities, parts of everyone's brain help them find their way, remember important landmarks, and use the environment to accomplish their goals. Environments can either support people in using their "environmental brains" or undermine these brain functions, diverting mental energy that might be usefully employed elsewhere. This chapter proposes that eight fundamental environmental characteristics, and their corresponding mind-brain functional areas, make up the universal environmental brain functions that healthy and healing design must support.

Research on environments for people with Alzheimer's disease uncovered these needs. Environments are especially important for this group because the disease causes particular damage to the parts of their brains that perceive, understand, interpret, and act on the environment. After describing the eight characteristics, a case study demonstrates how these principles apply to assisted living treatment residences for people with Alzheimer's disease.

8.2 BACKGROUND ON DESIGN AND ENVIRONMENT FOR PEOPLE WITH ALZHEIMER'S DISEASE

Because people with Alzheimer's get lost in unfamiliar and sometimes even in familiar settings, because they sometimes mistake even familiar objects, and because they often do not recognize familiar friends and family, we can summarize one part of their illness as being unable to manage environment. Calkins (1988) was one of the earliest environmental designers to codify design approaches to ameliorating these difficulties (see Chap. 22, "Design for Dementia: Challenges and Lessons for Universal Design," by Calkins et al.). In her groundbreaking book, *Designing for Dementia,* Calkins identified the interactions between such social phenomena as territoriality, familiarity, and residentiality and specific physical supports for those with Alzheimer's disease. Other gerontologists, social scientists, and designers have not only recognized that the physical environment can support the needs of those with Alzheimer's when designed correctly, but that it can also be harmful. Because people with dementia are anxious in unfamiliar, new, and strange surroundings, such as institutional settings, these can cause anxiety and agitation, which in turn can lead to striking out and other forms of aggression. Incorrectly planned environments can lead to increased symptoms; correctly designed environments can lead to the reduction of symptoms.

But what has this to do with designing for everyone—universal design? Cognitively intact people react the same way to environments as do those with Alzheimer's and other dementias. Everyone tends to be more relaxed, productive, and effective in comfortable and familiar settings. Most people are anxious in strange, new, and unfamiliar settings. There is, however, a major difference: The part of the brain that soothes anxiety in these situations, the hippocampus, functions in intact brains and is damaged in the brains of people with Alzheimer's. If one wants to understand what people generally feel in different environments, what better way to do this but to observe people with Alzheimer's in those settings—their feelings are the same as others, just with less self-control.

People generally experience things the way those with Alzheimer's experience them when they are "emotionally hijacked" (Goleman, 1995). Whenever a fight-or-flight situation arises, such as when someone is suddenly frightened, confronted with something startling, or surprised without warning, the brain short-circuits the hippocampus. People tend to act with less reflection and more gut instinct when under stress than when they have time to think. For example, when late for a train, someone is more likely to get lost on the way to the platform. When someone is running out of gas, he or she is more likely to make a wrong turn and miss the next gas station. And when people with guns hear sounds they fear are burglars, they are more likely to shoot (often another family member up getting a late-night sandwich) rather than calmly ask who might be there.

Alzheimer's Design: A Chance to Study Universal Design

People who suffer from Alzheimer's disease provide a unique opportunity to study the way the brain has developed to cope with environment. Design for this group provides a chance to see how design that respects the brain can support function and thought. Why is this the case? Alzheimer's disease is manifested in fibrillary plaques and tangles in the brain that impair, in addition to memory activities, a person's ability to cope with most normally complex environments. Designers of such environments and researchers who study both the environmental needs of this group as well as the effect of environmental design on them through post-occupancy evaluation have learned a great deal that is applicable to design for everyone—universal design.

The Brain: A Universal Design Organ

Over millions of years, the brains of Homo sapiens have developed to cope with the physical world around them in remarkably efficient ways. Certain parts of the brain have developed to make sure people historically and today find their way to food sources, to mates, and to get home. Parts of the brain have developed to help people know where they are and others to make sure they know what time it is. Altogether, people have developed to fit into their environment, to read it, and to act on it to help survival and procreation.

Design has a choice. Pay attention to these developments in order to support functioning and survival, or consider the environment independent of people and how they cope with it. Choosing to pay attention makes designers responsible for the effects of their professional actions and encourages collaboration between designers and the social and neural sciences. Choosing to see design as independent from people frees designers to act as they have since the beginning of design modernism, when breaking with tradition was seen as a requirement for design creativity (Brolin, 1976). Choosing to pay attention, however, does not mean giving up on creativity. Rather, it means to develop new traditions that are both creative and respectful of the human brain and body.

8.3 THREE LEVELS OF ENVIRONMENTAL INFLUENCE: PASSIVE, FUNCTIONAL, AND PROACTIVE

During the end of the twentieth century, environmental psychologists, designers, and hygienists have defined the relationship between environment and people's health in many ways—among them are sick buildings and functional buildings. A new paradigm is emerging in the beginning of the twenty-first century—healing environments.

Passive. *Sick building syndrome* is a term coined for large office buildings with a high incidence of upper respiratory illness among its users. Sick buildings are most directly the result of a conflict between energy conservation and sealed buildings: reducing planned fresh air intake to keep energy costs low and limiting outside air infiltration through tight construction to keep toxic chemicals from offgassing materials in the building. Identifying sick buildings is one way to frame the relationship between environments and health; namely, that a healthy building is one that does not make you sick. The building's sickness becomes an issue only if the quality of air and other indoor characteristics falls below an acceptable quality threshold.

Functional. A second, more *active* way to design for health rests in the work of functional designers who systematically gather information about and incorporate the needs of users in their design. For example, hospital architects and planners are generally required to develop extensive functional programs before designing and laying out spaces. Considerations that might come up include making sure that the distances staff members have to walk to perform their jobs is kept to a minimum; and that it is as easy as possible to see into patients' rooms from corridors, independent of the impact on patients' sense of privacy.

Another group of practitioners who contribute extensively to functional building technology are teachers, researchers, and design and planning professionals involved in facility programming. Such professionals in their practice demonstrate that not only can environments contribute to the "health" of communities, organizations, social institutions and people, but also that such contributions can be predicted, planned, measured, and evaluated.

Proactive. Traditionally—in planning, architecture, interior and landscape design—practitioners have claimed that the environments they design have beneficial, even healing, powers. Spiritual buildings and gardens soothe the soul and raise the spirits; office interiors increase productivity; and hospital buildings, interiors, and gardens are therapeutic to patients' recovery. Such claims present environments as tools for proactive healing.

This progression from *not sick* to *healthy* to *healing*—reflecting in part Maslow's hierarchy of needs (1954)—can be a useful taxonomy to distinguish how different environments relate to health.

- Building is not sick, it doesn't make people sick (no discomfort).
- Building is healthy, it helps people do their job (satisfaction).
- Building heals, it contributes actively to people's well-being (quality of life).

Parallel Movements

The movement toward healing buildings can be seen in varied design disciplines and problems, some of which are reviewed here.

In organizational management, there is now a strong and growing movement away from traditional hierarchical structures, toward more egalitarian, empowering, team organizations. And this movement is not gaining force because of misspent idealism; companies are finding that it contributes to their bottom line.

As Vischer points out in *Workspace Strategies* (1996), companies such as IBM Canada and Anderson Consulting have developed new ways to locate and house employees that reflect dynamic organizational structures. These include telecommuting, where a communications network is established to enable certain groups to work on the road and from a home office as effectively as if they were all housed in the same building; and such innovations as hotelling, where employees call in advance of coming into the office to reserve a workstation, phone, and file.

Newspapers, like the *Minneapolis Star Tribune,* are not only redesigning their logo and their front pages but are also replanning their newsroom to support ergonomics for employees and improved team storytelling. Newly organized workplaces house teams of reporters, editors, photographers, and graphic artists who all contribute equally to the most reader-responsive stories possible. The 350-person newsroom is now organized on a clearly legible "urban plan" with parks, coffee shops, street lights, and newsstands where newspaper staff can meet and plan stories together more easily. Such design contributes to the long-term health of both employees and the newspaper (Zeisel, Anderson, and Lockwood, 1996).

Exploring Healing Environments for People with Alzheimer's Disease

Designing residences for people with Alzheimer's disease and related dementias provides designers with special challenges and opportunities to explore the question of healing environments. These diseases are not mental illnesses, they are diseases of the brain—brain matter is lost over time. It is difficult for persons with these diseases to lay down new memories, like remembering a message just taken over the phone or the name of a person just met, but it is easy for them to draw on deep memories of the past, of their long life. It is difficult for persons with these diseases to carry out complex tasks, like organizing a multicourse meal or balancing a checkbook, but it is easy for them to understand environments that are presented clearly and legibly.

Environments that heal—healing environments—for people with Alzheimer's disease clearly have to represent and reflect deep memories. What are these deep memories? They may be memories of workplaces, of traditional houses, of streets they have lived on, or they may be even more profound environmental memories.

What are profound memories? Fireplaces represent warmth, safety, and food. In traditional house settings, the Inglenook (the covered hearth with built-in benches on both sides) evokes such profound feelings. The kitchen is one place that strongly evokes food, family, and friendly communication. One's own personal objects, such as a familiar photograph or an old housecoat, bring back memories of friendship and caring to demented residents. Music evokes great joy and profound sadness among Alzheimer's residents, letting them know that they are still alive. And just outside the front door—like in the clearing in front of a cave—people know they can feel the elements but are still close enough to home to be safe.

Deep Memories Translated into Healing Environments

How do such deep and profound memories translate into healing environments—homes and gardens—for people with dementia? Certain key areas and issues help translate these principles into physical form:

- *Exit control* provides residents with enough actual security to enable them to feel and be safe in the entire planned environment, and thus to be able to be free and independent in their actions.

- *Social places* planned to appear different and unique, and not too numerous to be confusing, help cue residents to appropriate behaviors—helping them to draw on deep memories and be competent in their actions.

- *Private away places* provide residents with the chance to collect themselves and get away from the pressures of spending all day and night with a community of the same people.
- *Walking paths* with clear destinations and visual diversions to stop and look at along the way create environments where residents' need to wander becomes purposeful activity—taking a walk.
- *Healing gardens* accessible to residents all day in all seasons, sometimes to plant and sometimes to shovel snow, give residents contact with another deep memory—the outdoors, the seasons, the weather, flowers, and trees.
- *Residentiality*—scale, furniture, and decor that convey the feeling of home and homeyness—relaxes residents by tapping into their deep understanding that at home everything they do is okay. They can relax because it's safe and friendly.
- *Independence* is something that residential living provides residents by focusing environmental design on supporting what residents can do instead of what they cannot.
- *Comprehensiblity* of the ambient and spatial environment—colors, sounds, sights, smells, space—is important for people who are confused by disordered and strange complexity. In understandable settings, demented residents can handle themselves well and are much less upset (Zeisel, Hyde, and Levkoff, 1994).

Detailed Criteria for Healing Environments

Exits.   The brains of Alzheimer's residents cannot hold cognitive maps and they frequently forget how to return home. Therefore, people with dementia should only leave their homes when accompanied by someone else. Doorways from a residence that open to the larger public community, therefore, need to be controlled. Residents who spend so much time indoors become agitated by doors with mixed messages: On the one hand, windows and hardware on the door attract residents and seem to invite them to go out; on the other hand, locks and keypads prevent their use. Exit doors that are less visible—more unobtrusive—with no attracting hardware reduce agitation. Increasing the visibility and making more inviting any safe door to a secure healing garden further diverts attention from doors that exit to dangerous areas.

In this century other signals can be chosen so that they do not disturb the ambiance of the residential setting—such as chimes rather than alarms. The less obvious the door, the signal, the hardware, and the other side of the door, the greater independence will the resident have in his or her safe environment.

Walking Paths.   One of the symptoms of Alzheimer's disease for certain people is the desire to walk, perhaps looking for something without knowing precisely what. While aimless wandering can be a problem for staff in a facility that has no place for this activity, a well-designed pathway can transform wandering into walking. A pathway can achieve this goal if it is interesting and does not dead-end. Such a pathway need not be a specially designed circular track, but rather can be the thoughtful connection of corridors that pass through common areas, and connect up again to corridors going in another direction. Interest along the path is important so that those walking always have some goal in sight—the next interesting picture, view, or plant. And interest at the end of the path, a social space or a fireplace, provides a place to walk to, a destination that gives purpose to each trip.

Common Space.   Residents in Alzheimer's assisted living facilities spend almost all their time in the facility and together. To satisfy their need for diversity and to reduce boredom and agitation, it is essential to have at least two if not three different common spaces—dining room, kitchen, living room, foyer. The more the settings of these rooms are different and interesting, the easier it is for staff to manage smaller family-like activities there and for residents to feel stimulated by the differences in ambiance they can sense.

Private Areas.   Because residents spend so much time together, they also need places to be alone, to avoid the pressures of social interaction. Just because someone is demented does not mean that they can stand being together with others 24 hours a day. Individual spaces that residents can use to get away by themselves can include private bedrooms or small, out-of-the-way corner sitting areas in a living room or garden. Residents with visiting family members who just need to sit together quietly can also use places like this. Private areas are also places residents and their families use to decorate and furnish personally, thus creating a soothing mood that triggers positive memories.

Healing Gardens.   Not every residence is able to provide its residents a safe and secure outdoor area immediately adjacent to the residential area. Yet, this ideal gives residents a sense of nature, weather, and plants. If nothing else, Alzheimer's residents enjoy being outdoors and are relaxed by being able to get out of the confinement they feel inside. Yet, a healing garden is even more than a place to get out, it is a sanctuary where a basic drive to have contact with normal forces can be met.

   If such an amenity is not easily provided (e.g., if the residential area is on an upper floor of an urban building), designers and operators need to arrange alternatives. One possibility is to have an outdoor area nearby that residents can use regularly accompanied by staff—on grade or on a roof patio. When such an arrangement is to be made, the path there and back must be thought through carefully to avoid creating anxiety and a breakdown in safety.

Residentiality.   Home, fireplace, front porch, and garden are residential environmental design elements that create positive mood in residents by touching deep-seated memories. The familiarity of residential furniture, spaces, decorations, and lighting fixtures relaxes everyone in Alzheimer's facilities—residents, their families, and staff members. Managing the size of features to be residential—a scale people can relate to and grasp easily—can be soothing itself. A refrigerator, a window, a small room, a small group gathering are all familiar, understandable, and manageable elements for demented residents and everyone with whom they interact.

Independence.   Details in the environment such as handrails and floors that prevent slips and falls contribute to the independence and autonomy of residents, because they support each person's ability to do things on their own. It may seem obvious that a toilet so low that it prevents an older person from standing up alone limits independence, but it does. Any nonprosthetic or unsafe design element has this effect. The safer the environment, the more likely staff are to permit residents to move about by themselves and make independent choices.

Comprehensibility.   Alzheimer's residents are not confused by everything around them. When the sounds, sights, and smells they experience are familiar, they can cope with them and enjoy them. A common myth and mistake in design for dementia care is that if everything is sedate and bland, residents will be soothed. This is not the case. Soothing can be anxiety producing if taken to an extreme. What is needed is to create enough activity to keep residents interested, and to make sure that the activity provided is understandable to them. Colors are fine, and traditional patterns for wallpaper are better than abstract patterns. A television is fine, and recorded films that have fewer rapid changes for advertisements are more satisfying than random violence and loud noises from the television. Comprehensibility comes from sensible, common-sense management.

8.4 CASE STUDY—AN ASSISTED LIVING RESIDENCE

How do these principles apply to actual design of treatment settings for people with Alzheimer's disease? The principles have been applied to six assisted living residences for people with Alzheimer's disease in Massachusetts and New York, and perhaps to others based on designers and programmers having read earlier published articles describing the eight characteristics.

One of the consciously designed projects is located in a renovated hospital building in Woburn, Massachusetts, about 20 minutes north of Boston. (See Fig. 8.1.) The assisted living residence for people with Alzheimer's disease houses 26 people cared for by a 24-hour staff, with several people who come to participate in the program just for the day. The treatment residence is located entirely on one floor and is *secured* for residents with magnetic doorlocks controlled by a coded push pad limiting egress and access to those who know—and can remember—the code. This code can be changed if a higher-functioning resident learns the code. The doors have no windows to the outside, and if they do, these are painted over. A tall, decorative fence surrounds the garden. The fence provides safety, and also potentially reduces agitation. It prevents views out to activities that might, by their interest, present attractive nuisances that might encourage residents to leave.

The main central pathway in the residence is essentially a straight line from one end to the other. But this straight corridor provides an interesting walking path with its many wall decorations—photographs chosen by residents and thus understandable, reminiscence shadow boxes with mementos of residents' lives, decorated boards announcing events, staff members' names, and residents' faces. The walking path ends at one end with a fireplace/hearth and living room where resident meetings are held, small group activities are organized, and there is a television set. Almost at the other end is a room with a tile floor in which painting and other messier activities can be run, and in the middle, one passes a large dining room and residential kitchen.

The common spaces are each different from one another and, thus, can stimulate different moods in residents' minds. The living room is carpeted, has a unique decorative border near the ceiling, and has white flowing curtains. The kitchen/dining room has windows along one side, dining chairs and tables, a faux-wood-tile floor, and a residential kitchen with wooden cabinets and a breakfast counter at one end. Another common room is adjacent to the porch and garden and has less light and more active furniture. While residents may not remember the precise attributes of each room, they are likely to remember the feel of each.

The bedrooms generally provide the opportunity for residents to have *privacy* and be surrounded by their personal furniture and mementos. Every bedroom has a door and is therefore private. Most have a dedicated bathroom for each bedroom, while three sets of bedrooms each share a common bathroom—behind the entry door to the apartment. Residents all have their own furniture, wall hangings, and other decorations—all memory cues that are intended to reduce agitation and improve memory.

There is a lovely therapeutic healing garden accessible from a wide porch adjacent to the common room. The outdoor porch is covered and wide enough to sit on—although it is cold in winter in New England—and provides a view over the garden. A gentle ramp leads down to the completely enclosed and safe garden—a half level below—designed with both landscaped care and residential and wayfinding principles (Zeisel and Tyson, 1999). There is a clear walking path, planting boxes, benches, and landmarks to help orient residents.

Each of the rooms in the assisted living residence is scaled like a residential space—perhaps with the exception of the dining room that can seat all 26 residents. The ceilings are low, the furniture residential in style, and unique decorative borders reflecting the use of the room grace the walls near the ceiling. And the residence—with 26 people—provides the opportunity for everyone who lives, works, and visits there to get to know each other. While the number is larger than a nuclear family, it is about the size of an extended family unit or a small residential community.

Because the entire residence is safe, both inside by virtue of the finishes and fixtures, and outside by virtue of secure doors and fences, residents are generally free to be as independent as their physical capacity enables them to be. Staff members, secure in their knowledge that residents will not wander away, do not feel they have to constantly follow and hold residents up. The "lean rail" along the walls in each hallway even enables residents who might otherwise be unsteady on their feet to make their way to where they are going by themselves.

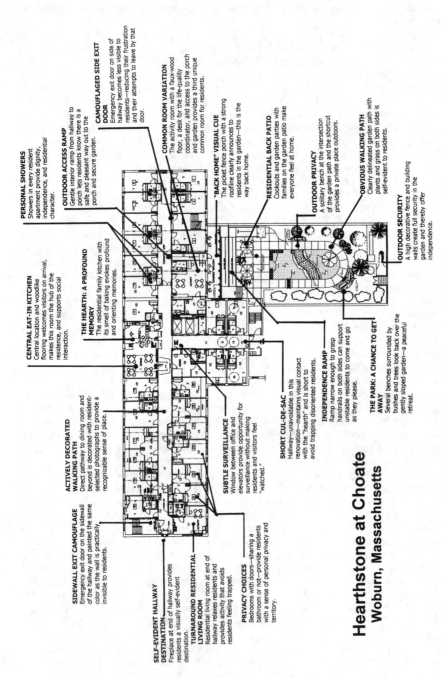

Hearthstone at Choate
Woburn, Massachusetts

PERSONAL SHOWERS
Showers in every resident apartment provide dignity, independence, and residential character.

OUTDOOR ACCESS RAMP
Gentle interior ramp from hallway to porch lets residents know there is a safe and pleasant way out to the porch and secure garden.

CAMOUFLAGED SIDE EXIT DOOR
Emergency exit door on side of hallway becomes less visible to residents—reducing their frustration and their attempts to leave by that door.

COMMON ROOM VARIATION
The activity room with a faux-wood floor, a desk for the life-quality coordinator, and access to the porch and garden provides a third unique common room for residents.

"BACK HOME" VISUAL CUE
The picket fence porch with a strong roofline clearly announces to residents in the garden—this is the way back home.

RESIDENTIAL BACK PATIO
Cookouts and garden parties with families on the garden patio make everyone feel at home.

OUTDOOR PRIVACY
A solitary bench at the intersection of the garden path and the shortcut provides a private place outdoors.

OBVIOUS WALKING PATH
Clearly delineated garden path with plants and grass on both sides is self-evident to residents.

OUTDOOR SECURITY
A high decorative fence and building walls create full security in the garden and thereby offer independence.

CENTRAL EAT-IN KITCHEN
Central location and woodlike flooring welcomes visitors on arrival, makes this room the hub of the residence, and supports social interaction.

THE HEARTH: A PROFOUND MEMORY
The residential family kitchen with its smell of baking evokes profound and orienting memories.

SHORT CUL-DE-SAC
Hallway—unavoidable in this renovation—maintains visual contact with the "hearth" and is short to avoid trapping disoriented residents.

INDEPENDENCE RAMP
Ramp narrow enough to grasp handrails on both sides can support unstable residents to come and go as they please.

THE PARK: A CHANCE TO GET AWAY
Several benches surrounded by bushes and trees look back over the gently sloped garden—a peaceful retreat.

SUBTLE SURVEILLANCE
Window between office and elevators provide opportunity for surveillance without making residents and visitors feel "watched."

ACTIVELY DECORATED WALKING PATH
Direct pathway to dining room and beyond is decorated with resident-selected photographs to provide a recognizable sense of place.

SIDEWALL EXIT CAMOUFLAGE
Emergency exit door on the sidewall of the hallway and painted the same color as the wall is practically invisible to residents.

SELF-EVIDENT HALLWAY DESTINATION
Fireplace at end of hallway provides residents a visually self-evident destination.

TURNAROUND RESIDENTIAL LIVING ROOM
Residential living room at end of hallway relaxes residents and provides activity that avoids residents feeling trapped.

PRIVACY CHOICES
Bedrooms with doors—sharing a bathroom or not—provide residents with a sense of personal privacy and territory.

FIGURE 8.1 Annotated plan of Hearthstone at Choate, Woburn, Massachusetts.

There are no strange sounds, views, or other sensory stimuli in the residence. Furniture is familiar; the arbor in the garden is the same as in many residential yards, the photos on the wall present comforting and interesting sights. There is no overhead public announcement system, and no strange and shiny floors waxed to meet regulations for cleanliness, as might be found in long-term-care institutions. And the radio and television are not left on all day—programs are chosen and videotapes and audiotapes used that present familiar tunes and shows.

In sum, the entire design and layout of this assisted living residence for people with Alzheimer's—its architecture, landscape architecture, and interiors—are planned to augment residents' memories and ability to function on their own. By taxing the parts of residents' brains that are still working well, and relieving the parts that are damaged, the whole person is supported. He or she feels at home, in control of himself or herself—as much as his or her age allows—and competent.

8.5   HEALING DESIGN PERFORMANCE CRITERIA

The eight categories defined in Table 8.1 appear to reflect specific brain abilities that have developed to help individuals cope with the environment, and that need to be supported by the environments if people are to be healed by them. Table 8.1 highlights specific areas of the brain, their capabilities, identifies performance criteria for the environment, and suggests design approaches. What people with Alzheimer's disease lose because of their disease is what all people need to have supported in order to be nurtured and have one's health supported.

The Goal: Quality of Life

Abraham Maslow (1954) included among the components of life quality for cognitively intact people half a century ago survival, maintenance, and personal actualization. For those with Alzheimer's disease, a similar model can be constructed. Three levels of needs must be met for these people to experience life quality, with the highest level representing the highest quality.

Figure 8.2 illustrates these levels. The first two levels of needs that caregivers must meet to provide quality of life require only brief explanation. The most basic set of needs are physiological: the need for safety, health, nourishment, and shelter. Meeting these needs forms the basis for meeting the other sets of needs. The next set are behavioral needs for appropriate functioning and use of the environment. Help is required to meet these needs because damage to the brain of people with dementia specifically affects those areas that control social behaviors, impulse control, and environmental cognition.

The highest life quality level, self-actualization, deserves explanation because popular misconception holds that people with dementias do not sense themselves the way cognitively intact people do. However, the part of the brain that controls mood and emotion, the amygdala, is one of the last to be affected. A person's moods and emotions are, therefore, readily accessible to themselves and to caregivers until very late in the disease. It is difficult to maintain positive mood and emotion for this population over time, because people with Alzheimer's disease are so very sensitive. They feel negative feelings as readily as positive ones; they cannot control their anger, sadness, and anxiety. The highest goal of life quality treatment is not only to maintain positive mood, but also to maintain emotional stability around social norms—recognizing and dealing with other people and personal norms—maintaining a sense of self.

Two treatment modalities are available to maintain quality of life in assisted living residences specially built for those with Alzheimer's disease and related dementias: *environmental design* and *management operations*. Design and management can be analyzed separately; in life they are integrally connected.

TABLE 8.1   Healing Design Performance Criteria Responsive to Cognitive Neuroscience

Brain location*	Brain capability	Design performance criteria	Possible design approaches
1. Parietal and occipital lobes	Ability to hold a cognitive map; ability to be in the present	Naturally mapped environments in which all the information needed to find one's way around is embedded in the setting rather than needing to be kept in mind	Clear destinations; clear landmarks, nodes, edges, districts, and pathways; all of these with a unique character that defines and distinguishes them
2. Anterior occipital lobe and hippocampus	Ability to hold onto the memory of objects and places	Environments that are safe even when a person is lost; clarity of surrounding environments; safety from external forces	Clear boundaries between home territory and public areas; clarity of territory; ability to have privacy when needed
3. Frontal lobe	Having a sense of self	Personal environments that provide cues for one's own identity and autobiography	Movable elements such as furniture and decorations that evoke memories of a person's culture, personal history, family, and achievements
4. Hippocampus and amygdala	Ability to remember places visited in the past; ability to hold on to moods, feelings, and emotions	Spaces that evoke different and strong moods and emotions so that people feel—and thus know deeply—where they have been	Varied spaces, each evoking a different mood and emotion
5. Hippocampus	Ability to perceive and process new places; ability to retrieve hard-wired memories	Significant places that focus on hard-wired memories such as food, warmth, social support, and nature	Environments that include strongly evocative elements such as a fireplace, kitchen, view out over garden
6. Frontal lobe and motor cortex of the parietal lobe	Awareness of physical and other limitations and disabilities; natural sense of self-control and independence	Environments that are prosthetic in that they naturally make up for losses in mobility and limb strength and are safe	Environments with rails to lean on in hallways, toilets high enough to get on and off with little arm strength, soft materials on the floor to cushion falls
7. Anterior and medial temporal lobe, and parietal lobe losses; sensory cortex strengths	Receptive and expressive language centers; senses of smell, touch, and hearing	Environmental messages and cues in nonverbal form that take advantage of multiple sensory modalities at the same time	Environments in which sensory stimuli such as smells of food, sounds of music, comforting soft materials are all orchestrated to augment verbal and written messages
8. Supra chiasmatic nuclei (SCN)	Sense of time and circadian rhythms; ability to sense nature, the passage of time, and the seasons out of doors	Outdoor environments that provide contact with nature, weather, time of day, and plants—natural cues to the passage of time	Gardens with clear pathways, lively planting areas, hard surfaces to walk on, benches to sit down on, shady areas, trees, and plants

*Note: Areas of the brain are presented as if these were the only areas being used for each set of functions. This oversimplification is presented for illustrative and descriptive purposes. Generally, although one area may be more involved than others, more than one associated area often influences behaviors and brain responses.

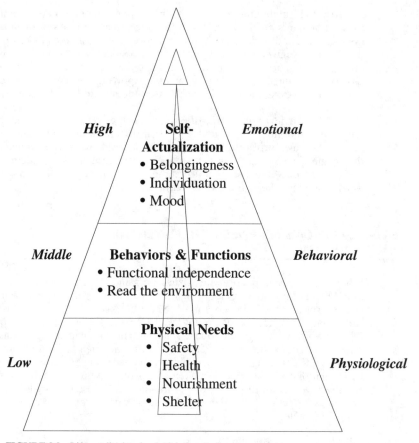

FIGURE 8.2 Life-quality levels of Alzheimer's disease and care.

Before describing these two treatment approaches in assisted living, however, it may be prudent to ask if assisted living is even an acceptable setting to care for people with dementia. Some might ask, how can those with dementia benefit from the autonomy, personal dignity, and privacy that assisted living provides? Aren't people with dementia violent, dependent, and confused? Don't people with dementia require constant surveillance?

As discussed earlier, even those with dementia can experience quality of life. It is our lack of understanding of the disease that limits the ability of people with dementia to enjoy these liberties, not their incapacities. Practice and research bear out that many of the negative behaviors commonly associated with the disease are reactions to inadequate care and treatment, rather than symptoms that are integral to the disease.

Autonomy.   If people with Alzheimer's are in a completely safe setting, and if their families are willing to negotiate the risks of freedom with caregivers, people with Alzheimer's can enjoy autonomy—as long as they do not injure themselves and others. Every person with dementia—except for the very end stages—has remaining skills: gardening, cooking, laughing, listening, walking, chatting, and drinking coffee or tea. They can be independent and autonomous in employing their skills, as long as they are protected from the dangers created by their trying to employ abilities they have lost.

Dignity.   The ability to feel, express, and respond to emotion is an ability that most people with Alzheimer's maintain until well into the disease. Emotions include happiness, satisfaction, sadness, and pride, among others. People feeling dignity, linked to pride, is something that every caregiver sees daily whenever a resident successfully completes a task: getting dressed, eating, or mixing a bowl of cookie dough. If one applies a narrow definition of dignity to this group, they may have very little. Applying their own expressed definition of pride to their actions, one can perceive and better understand how dignified they feel.

Privacy.   Whenever one enters the room of a person with dementia in assisted living, one sees a person at home. Surrounded by their own furniture and photographs of family members in their own private space, many of these residents seem well and satisfied. Only when they begin to answer questions and make conversation does their dementia strike the visitor. It is as if, shrouded in a shell of private territory, they are protected from the questions and words that make their illness so evident.

8.6   GENERAL LESSONS LEARNED FOR UNIVERSAL HEALING DESIGN

When taking care of people with Alzheimer's and other dementias, one learns to listen carefully to what they say so that one can understand what they mean even when the words are not quite clear. One learns not to say "No," but to gently divert a confused person from difficult situations. When words do not seem to make sense to the other person, one learns to replace words with hugs and touch. These practical management lessons can be usefully carried over to daily life with colleagues, family, and friends. Using them in "normal" society helps people get along with others better, and generally be better people.

Design that touches people's deep and profound understanding of their surroundings is equally powerful for those without dementia as for those with such a disease. Healing design principles for people with Alzheimer's disease can be translated into archetypal deep healing design principles.

- Feeling safe, secure, and free
- Understanding what is expected in one's community
- Being able to get away by oneself and unwind
- Knowing where one is going and having fun getting there
- Enjoying the outdoors and the changes of seasons and weather
- Knowing that anything one does is okay because the place is safe and familiar
- Celebrating what one can do, not what one fails at
- Not having to struggle to understand one's surroundings

Environments for living, work, and play that provide their users with these profound opportunities are likely to be healing environments.

8.7   CONCLUSIONS

Although the study of the brain and mind have been the subject of intensive study for several centuries, the link to environmental design has only recently been made (Changeux, 1985). The next steps in this quest for the brain's environmental system lie in research to test and elaborate these linkages between neuroscience, design, and behavior, in the practical application of what is already known, and in the development of further theory to point out new directions.

Research needs to progress on two fronts—in neuroscience and in environment-behavior (E-B) studies. Employing neuroscientific methods, including magnetic resonance imaging (MRI) and positron-emission tomography (PET) scans, will further understanding of how the brain processes environmental information. Zeisel, in the second edition of *Inquiry by Design* (Zeisel, 2001) describes the present state of knowledge linking neuroscience, the brain, and how the brain processes and acts on environmental information. But there is a great deal more that needs to be discovered. Employing E-B methods, including natural observation, experimental designs, and even pen-and-pencil questionnaires and tests will help systematically test and thus refine notions of how environments contribute to everyone's quality of life and well-being by incorporating healing design principles.

A particularly significant direction to pursue would be to elaborate more and different user groups to whom these principles apply and to determine the ways present environments do or do not support these needs. Looking at the actual environmental lives of people with different physical, cultural, mental, occupational, and other characteristics can lead to further understanding of the principles' universality.

Practical application through policy, design, construction, and facilities management will not only improve the quality of life of those in new settings, but will provide further research sites. Regulations of all sorts explicitly and implicitly influence the brain-supportive character of most settings—among these are requirements for fire safety, locks, signage, privacy, and physical accessibility, to name a few—and are applied to almost all settings that are designed and constructed. Such policies can be reviewed for their brain compatibility. Designers can begin to apply a mind-brain perspective in addition to their present aesthetic and functional perspective. Asking, "How does it work, look, and affect the brain of users?" will make their buildings, landscapes, cities, and interiors richer and friendlier.

Theory is important because, without further theory elaborating on the initial concepts presented here, the field of mind-brain-behavior-environment will stagnate. There is only so much that blind empiricism can uncover, and so little is known at the turn of the millennium. Theory development based on present knowledge and on new data as they are uncovered is needed to know where to look for the most fruitful new research directions. Theory is needed to determine what combinations of disciplines will be most fruitful in integrating neuroscience, environmental design, and the social and psychological sciences. And theory is needed to fruitfully select elements from present knowledge in each of these fields, and to combine them in ways that shed light on the present state of the art and its intellectual future.

8.8 BIBLIOGRAPHY

Brolin, B., *The Failure of Modern Architecture,* Van Nostrand Reinhold, New York, 1976.

Calkins, M. P., *Design for Dementia: Planning Environments for the Elderly and the Confused,* National Health Publishing, Owings Mills, MD, 1988.

Changeux, J.-P., *Neuronal Man,* Princeton University Press, Princeton, NJ, 1985.

Goleman, D., *Emotional Intelligence: Why It Can Matter More Than IQ.* Bantam Books, New York, 1995.

Maslow, A., *Motivation and Personality,* Harper and Row, New York, 1954.

Vischer, J., *Workspace Strategies: Environment as a Tool for Work,* Chapman & Hall, New York, 1996.

Zeisel, J., "Healing Environments: Healthful Buildings," in Loeb Fellowship Forum, Harvard University, Cambridge, MA, 1996.

———, "Life-Quality Alzheimer Care in Assisted Living" in Benjamin Schwartz and Ruth Brent (eds.), *Aging, Autonomy, and Architecture: Advances in Assisted Living,* Johns Hopkins University Press, Baltimore, MD, 1999.

———, *Inquiry by Design,* 2nd ed., Cambridge University Press, Cambridge, UK, 2001.

————, and M. Tyson, "Alzheimer's Treatment Gardens," in Clare Cooper Marcus and Marni Barnes (eds.), *Healing Gardens: Therapeutic Benefits and Design Recommendations,* John Wiley & Sons, New York, 1999.

————, J. Anderson, and R. Lockwood, "Newsroom Team Design: A Case Study of the *Minneapolis Star Tribune,*" in J. Zeisel (ed.), *Inquiry by Design,* 2nd ed., Cambridge University Press, Cambridge, UK, 2001.

————, J. Hyde, and S. Levkoff, "Best Practices: An Environment-Behavior (E-B) Model for Alzheimer Special Care Units," in *American Journal of Alzheimer's Care & Research,* 1994.

CHAPTER 9

TOWARD UNIVERSAL DESIGN EVALUATION

Wolfgang F. E. Preiser, Ph.D., Dipl.-Ing.
University of Cincinnati, Cincinnati, Ohio

9.1 INTRODUCTION

This chapter presents the concept and framework of universal design evaluation (UDE), which is based on consumer-feedback-driven, preexisting evolutionary evaluation process models developed by the author [i.e., post-occupancy evaluation (POE) and building performance evaluation (BPE)]. The intent of UDE is to evaluate the impact on the user of universally designed environments. Working with Mace's definition of universal design, "an approach to creating environments and products that are usable by all people to the greatest extent possible" (Mace, 1991 in Welch, 1995), protocols are needed to evaluate the outcomes of this approach. Possible strategies for evaluation in the global context are presented, along with examples of case study evaluations that are presently being carried out. Initiatives to introduce universal design evaluation techniques in education and training programs are outlined. Exposure of students in the design disciplines to philosophical, conceptual, methodological, and practical considerations of universal design is advocated as the new paradigm for "design of the future."

9.2 FEEDBACK AND EVALUATION

In a world that is consumer-feedback driven, new paradigms for the business world, management, and the design of products, buildings, and systems are emerging (Petzinger, 1999). Companies following the old economy and paradigms, such as Procter & Gamble, suddenly find themselves outmaneuvered and outpaced by companies following the new Internet-driven paradigms. Whereas in the past consumer products were designed and marketed by producers, often without regard to the consumers' needs and wants, the companies of the new economy seek continuous feedback on the quality and desirability of their products, and are thus consumer-demand driven, not supply driven. Similarly, whereas in the past, economic constraints were available capital, the new constraints revolve around creativity and all the design activities that are fed by it, such as software development, the movie industry and entertainment, and so forth. Universal design should be no exception.

Thus, we can see that while past mechanical models were top-down, hierarchical, and largely based on extraction of land, energy, and materials, today's "currency" and source of value is information, knowledge, and creativity.

Designers' clients are usually not the users of the environments or products they design. To bridge the gap of not understanding and of possible misinterpretation in the interface between designer and user, more direct channels of communication must be established. Gaining relevant information from users themselves about their particular needs and behaviors will help ensure satisfactory functioning of the new designs. The case of large housing developments, in which the "anonymous" customers or "average" citizens of a certain income bracket are the prospective users, shows how important this kind of input into the decision-making process at the programming stage is. The damage or behavioral cost caused by erroneous design decisions cannot be expressed in dollars, but certainly it is running high, just considering accidents caused by buildings which result in injury and death. Today mass production proliferates mistakes and increases the danger of false assumptions and irreversible decisions in planning and design. Ironically, with today's technology, it is possible to create individualized mass-produced products, such as cars with hundreds of options, following consumer choices.

Existing environments and products in use provide the best simulation models momentarily available. Too little use is being made of these invaluable sources of information. Feedback and evaluation, knowledge-based development, programming, and design are linked activities drawing information from systematic scrutiny of actual performance (National Academy of Sciences, 1986 and 1987).

Criteria for universally designed products and environments should be based on the evaluation of existing ones and modified if necessary. The crucial point in every evaluation is the problem of what to compare the findings to, the question of what matters to users. An evaluation should be administered in such a way as to detect fits and misfits, degrees of user satisfaction, and levels of building performance.

Product evaluation, in terms of performance and customer satisfaction, is an accepted procedure in industry. Product improvement is routine and forms an integral part of the price calculation. It would seem natural and more economical in the long run to thoroughly investigate the possibilities of user-oriented "product" evaluations in a complex field, such as universal design.

While the constructed environment cannot be called a "product" in its strict meaning, because it is dynamic and changing over time, feedback and evaluation costs should be included in the life cycle costing of buildings as an "extra" that might be written off against taxes, since they represent an extraordinary service for future times. Built-in evaluation costs for the financing of large-scale evaluative environmental research should become the starting point for an ongoing national program for universally designed housing, especially in the large retirement communities of the "Sunbelt" states in the United States.

Statistical surveys document the allocations of research funds in industry. Huge sums are spent every year to tailor different (not necessarily better or safer) cars, and the development of a pill can cost millions. One cannot understand, however, why only a negligible amount of money is spent on research related to a matter so crucially important as housing, transportation, and the built environment in general.

A rational design process with feedback through ongoing evaluation efforts, if graphically represented, resembles a loop, whereby the information feedback through continuous evaluation would lead to better-informed design assumptions and, finally, to better solutions. Due to the exclusive character of information gathered in evaluative research and stored and updated in an information system, the designer should be enabled to make more informed and user-oriented decisions.

Feedback/Feedforward-Based Performance Evaluation

The field of building performance evaluation (BPE) is developing in the United States and other industrialized nations. Why do the so-called "industrializing" nations not engage in inquiry about how built and natural environments affect people? Clearly, priorities in some parts of the globe are different, and while some societies appear sane, whole, and in consonance with nature, others pollute and destroy the environment at an unprecedented rate and with disastrous short- and long-term results. The existence of the field of BPE can be rationalized as a means of reducing these costs and bringing environments and their occupants into a state of dynamic equilibrium.

Adapted from the interdisciplinary field of cybernetics—"the theoretical study of control and communication in machines and physiological systems . . ." (*Encyclopaedia Britannica,* 1976)—the purpose of BPE and the universal design performance framework presented here is to introduce new concepts and approaches to the applied fields of architecture, environmental design, and planning by considering the perceived quality of products and built environments and the effect on their users. Figure 9.1 depicts an ourobol, the symbol for a cybernetic concept.

The systems approach appears appropriate for this field: It holistically links diverse phenomena that influence relationships between people, products, and their surroundings, including the physical and social environments. Like any other living species, humans are thought of as organisms seeking equilibrium within a dynamic, ever-changing environment.

The interactive nature of relationships between people and their surroundings is also recognized in the line of research that studies the impact of human actions on the physical environment, both built and natural. BPE is multidisciplinary, and its manifestations are disparate pieces of research that are largely applied in nature. Perhaps this is due to the fact that a coherent theoretical framework has been lacking in the past.

FIGURE 9.1   Feedback/feedforward.

Universal Design Performance

The goal of universal design is to achieve universal design performance (i.e., the characteristics of products or occupied facilities or buildings that are perceived to support or impede human activities, in terms of universal design criteria to support individual, communal, or organizational goals).

A philosophical base and set of objectives are the universal design principles:

1. They define the degree of fit between individuals or groups and their environment, both natural and built.

2. They refer to the attributes of products or environments that are perceived to support or impede human activity.

3. They imply the objective of minimizing adverse effects of products, environments, and their users, such as discomfort, stress, distraction, inefficiency, and sickness, as well as injury and death through accidents, radiation, toxic substances, etc.

4. They constitute not an absolute, but a relative concept, subject to different interpretations in different cultures and economies, as well as temporal and social contexts. Thus, they may be perceived differently over time by those who interact with the same facility or building, such as occupants, management, maintenance personnel, and visitors.

The nature of basic feedback systems was discussed by von Foerster (1985): The evaluator makes comparisons between the outcomes (O) which are actually sensed or experienced, and the expressed goals (G) and expected performance criteria (C), which are usually documented in the functional program and made explicit through performance specifications. Von Foerster observed that "even the most elementary models of the signal flow in cybernetic systems require a (motor) interpretation of a (sensory) signal," and further, "the intellectual revolution brought about by cybernetics was simply to add to a 'machine,' which was essentially a motoric power system or a sensor that can 'see' what the machine or organism is doing, and, if necessary, initiate corrections of its actions when going astray." The evolutionary feedback process in building delivery in the future is shown in Fig. 9.2. The motor in such a system is the programmer, designer, or evaluator who is charged with the responsibility of ensuring that buildings meet state-of-the-art performance criteria.

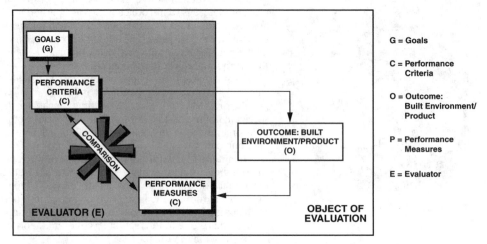

FIGURE 9.2   Basic feedback system.

The environmental design or delivery process is goal oriented. It can be represented by a basic system model with the goal of achieving universal design performance criteria (Fig. 9.2).

1. The universal design performance framework conceptually links the overall client goals (G), namely those of achieving habitability or environmental quality, with the elements in the system that are described in the following items.

2. Performance evaluation criteria (C) are derived from the client's goals (G), standards, and state-of-the-art criteria for a building type. Universal design performance is tested or evaluated against these criteria by comparing them with the actual performance (P) (see item 5).

3. The effector (E) moves the system and refers to such activities as planning, programming, designing, constructing, activating, occupying, and evaluating an environment or building.

4. The outcome (O) represents the objective, physically measurable characteristics of the product, environment, or building under evaluation (e.g., its physical dimensions, lighting levels, and thermal performance).

5. The actual performance (P) refers to the performance as observed, measured, and perceived by those using a product, and occupying or assessing an environment, including the subjective responses of occupants and objective measures of the environment.

Any number of subgoals (Gs) for achieving environmental quality can be related to the basic system (Preiser, 1991) through modified effectors (Es), outcomes (Os), and performance (Ps). Thereby, the outcome becomes the subgoal (Gs) of the subsystem with respective criteria (Cs), effectors (Es), and performance of the subsystem (Ps). The total outcome of the combined basic and subsystems is then perceived (P) and assessed (C) as in the basic system (see Fig. 9.3).

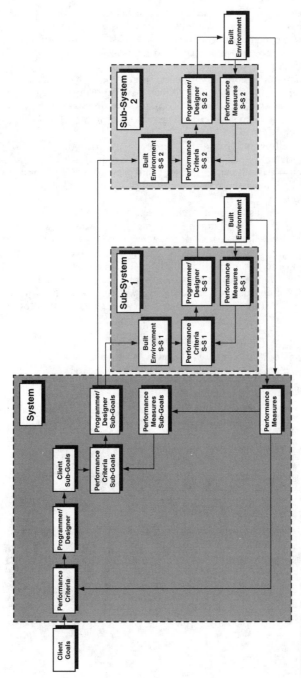

FIGURE 9.3 Feedback system with subsystems.

Performance Levels

Subgoals of building performance may be structured into three performance levels pertaining to user needs: (1) the health/safety/security level, (2) the function and efficiency level, and (3) the psychological comfort and satisfaction level. With reference to these levels, a subgoal might include safety, adequate space and spatial relationships of functionally related areas, privacy, sensory stimulation, or aesthetic appeal. For a number of subgoals, performance levels interact and may also conflict with each other, requiring resolution.

Framework elements include products/buildings/settings, users, and user needs. The physical environment is dealt with on a setting-by-setting basis. Framework elements are considered in groupings from smaller to larger scales or numbers, or from lower to higher levels of abstraction, respectively.

For each setting and user group, respective performance levels of pertinent sensory environments and quality performance criteria are required (e.g., for the acoustic, luminous, gustatory, olfactory, visual, tactile, thermal, and gravitational environments). Also relevant is the effect of radiation on the health and well-being of people, from both short-term and long-term perspectives.

User needs versus the built environment or products are construed as performance levels. Grossly analogous to the human needs hierarchy (Maslow, 1948) of self-actualization, love, esteem, safety, and physiological needs, a three-level breakdown of performance levels reflects occupant needs in the physical environment. This breakdown also parallels three basic levels of performance requirements for buildings (i.e., firmness, commodity, and delight), which the Roman architect Vitruvius (translated in 1960) had pronounced.

These historic constructs ordering user needs were transformed and synthesized into the "habitability framework" (Preiser, 1983) by devising three levels of priority depicted in Fig. 9.4:

1. Health, safety, and security performance
2. Functional, efficiency, and workflow performance
3. Psychological, social, cultural, and aesthetic performance

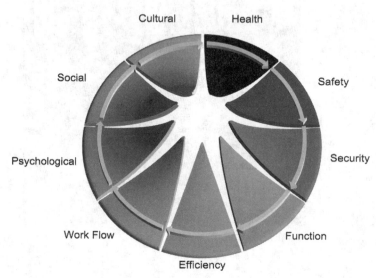

FIGURE 9.4   Universal design performance criteria. (*Graphics courtesy of MABorger Design.*)

These three categories parallel the levels of standards and guidance designers should or can avail themselves of: Level 1 pertains to building codes and life safety standards that projects must comply with. Level 2 refers to the state-of-the-art knowledge about products, building types, and so forth, exemplified by agency-specific design guides or reference works like *Time-Saver Standards: Architectural Design Data* (Watson, Crosbie, and Callender, 1997). Level 3 pertains to research-based design guidelines, which are less codified, but nevertheless of importance for designers. Examples can be found in Chap. 22 of this volume, by Calkins, Sanford, and Proffitt.

Shown in Fig. 9.5 are the relationships and correspondences between the habitability framework and the Principles of Universal Design devised by the Center for Universal Design (1997). In summary, the framework presented here systematically relates products, buildings, and settings to users and their respective needs vis à vis the product or environment. It represents a conceptual, process-oriented approach that accommodates relational concepts to applications in any type of product, building, or environment. This framework can be transformed to permit stepwise handling of information concerning person-environment relationships (e.g., in the programming specification, design and hardware selection for acoustic privacy).

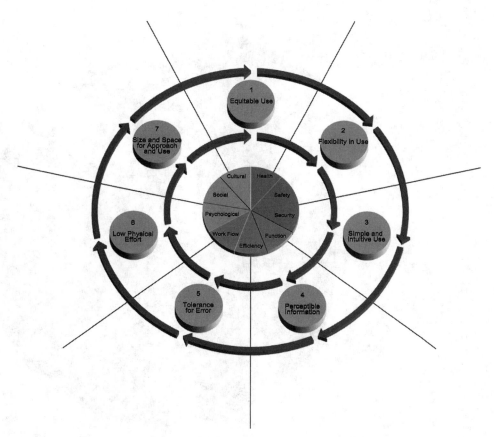

FIGURE 9.5   The Principles of Universal Design and performance criteria. (*Graphics courtesy of MABorger Design.*)

9.3 EVOLVING EVALUATION PROCESS MODELS: FROM POE TO BPE AND UDE

As previously outlined, universal design evaluation is the process of systematically comparing the actual performance of universally designed products, buildings, places, and systems with explicitly documented criteria for their expected performance. It is based on the post-occupancy evaluation (POE) process model (Fig. 9.6) developed by Preiser, Rabinowitz, and White (1988). While POE is primarily focused on the built environment, it can be beneficially applied to universal design in all its manifestations. For a synopsis, see Preiser (1999).

Post-Occupancy Evaluation: An Overview

Many parties participate in the use of buildings, including investors, owners, operators, maintenance staff, and perhaps most important of all, the end users (i.e., the actual persons occupying the building). The focus in this chapter is on occupants and their needs as they are affected by building performance, and on occupant evaluations of buildings. The term *evalu-*

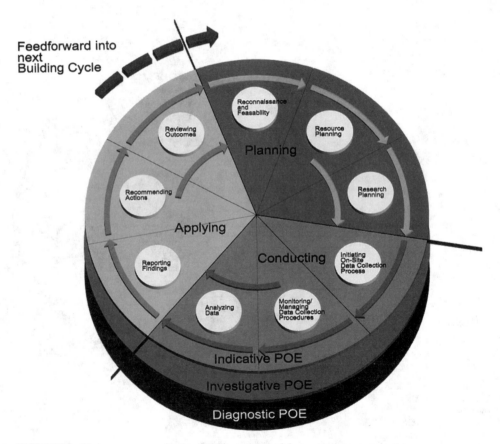

FIGURE 9.6   Post-occupancy evaluation (POE) process model.

ation contains the world "value," and thus occupant evaluations must state explicitly whose values are referred to in a given case. An evaluation must also state whose values are used as the context within which performance will be tested. A meaningful evaluation focuses on the values behind the goals and objectives of those who wish their buildings to be evaluated, or those who carry out the evaluation.

There are differences between the quantitative and qualitative aspects of building per-form-ance and the respective performance measures. Many aspects of building performance are in fact quantifiable, such as lighting, acoustics, temperature and humidity, durability of materials, amount and distribution of space, and so on. Qualitative aspects of building per-formance pertain to the ambiance of a space (i.e., the appeal to the sensory modes of touch-ing, hearing, smelling, kinesthetic and visual perception, including color). Furthermore, the evaluation of qualitative aspects of building performance, such as aesthetic beauty or visual compatibility with a building's surroundings, is somewhat more difficult and less reliable. In other cases, the expert evaluator will pass judgment. Examples are the expert ratings of scenic and architectural beauty awarded chateaux along the Loire River in France, as listed in travel guides. The higher the apparent architectural quality and interest of a building, the more stars it will receive. Recent advances in the assessment methodology for visual aes-thetic quality of scenic attractiveness are encouraging. It is hoped that someday it will be possible to treat even this elusive domain in a more objective and quantifiable manner (Nasar, 1988).

In the following text, the topic of post-occupancy evaluation and its theoretical base are presented.

The POE Context

The post-occupancy evaluation (POE) is not the end phase of a building project, but rather, it is an integral part of the entire building delivery process. It is also part of a process in which a POE expert draws on available knowledge, techniques, and instruments in order to predict a building's likely performance over a period of time.

At the most fundamental level, the purpose of a building is to provide shelter for activities that could not be carried out as effectively, if at all, in the natural environment. A building's performance is its ability to accomplish this. POE is the process of the actual evaluation of a building's performance once in use by human occupants.

A POE necessarily takes into account the owners', operators', and occupants' needs, per-ceptions, and expectations. From this perspective, a building's performance indicates how well it works to satisfy the client organization's goals and objectives, as well as the needs of the individuals in that organization. A POE can answer, among others, these questions: Does the facility support or inhibit the ability of the institution to carry out its mission? Are the mate-rials selected safe (at least from a short-term perspective) and appropriate to the use of the building? In the case of a new facility, does the building achieve the intent of the program that guided its design?

The Process of POE

Several types of evaluations are made during the planning, programming, design, construction, and occupancy phases of a building project. They are often technical evaluations related to questions about the materials, engineering, or construction of a facility. Examples of these evaluations include structural tests, reviews of load-bearing elements, soil testing, and mechan-ical systems performance checks, as well as postconstruction evaluation (physical inspection) prior to building occupancy.

Technical tests usually evaluate some physical system against relevant engineering or performance criteria. While technical tests indirectly address such criteria by providing a better and safer building, they do not evaluate it from the point of view of occupant needs and goals, or performance and functionality as it relates to occupancy. The client may have a technologically superior building, but it may provide a dysfunctional environment for people.

Other types of evaluations are conducted that address issues related to operations and management of a facility. Examples are energy audits, maintenance and operation reviews, security inspections, and programs which have been developed by professional building managers. While not POEs, these evaluations are relevant to questions similar to those previously described.

The process of POE differs from these and technical evaluations in several ways:

1. A POE addresses questions related to the needs, activities, and goals of the people and organization using a facility, including maintenance, building operations, and design-related questions. Other tests assess the building and its operation, regardless of its occupants.

2. The performance criteria established for POEs are based on the stated design intent and criteria contained in or inferred from a functional program. POE evaluation criteria may include, and are not solely based on, technical performance specifications.

3. Measures used in POEs include indices related to organizational and occupant performance, such as worker satisfaction and productivity, as well as measures of building performance previously referred to (e.g., acoustic and lighting levels, adequacy of space and spatial relationships, etc.).

4. POEs are usually "softer" than most technical evaluations. POEs often involve assessing psychological needs, attitudes, organizational goals and changes, and human perceptions.

5. POEs measure both successes and failures inherent in building performance.

Purposes and Types of POEs

A POE can serve several purposes, depending on a client organization's goals and objectives. POEs can provide the necessary data for the following:

1. To measure the functionality and appropriateness of design and to establish conformance with performance requirements as stated in the functional program. A facility represents policies, actions, and expenditures that call for evaluation. When POE is used to evaluate design, the evaluation must be based on explicit and comprehensive performance requirements contained in the functional program statement previously referred to.

2. To fine-tune a facility. Some facilities incorporate the concept of "adaptability," such as in the case of office buildings, where changes are frequently necessary. In that case, routinely recurring evaluations contribute to an ongoing process of adapting the facility to changing organizational needs.

3. To adjust programs for repetitive facilities. Some organizations build what is essentially the identical facility on a recurring basis. POE identifies evolutionary improvements in programming and design criteria, and it also tests the validity of underlying premises that justify a repetitive design solution.

4. To research effects of buildings on their occupants. Architects, designers, environment-behavior researchers, and facility managers can benefit from a better understanding of building-occupant interactions. This requires more rigorous scientific methods than design practitioners are normally able to use. POE research in this case involves thorough and precise measures, and more sophisticated levels of data analysis, including factor analysis and cross-sectional studies for greater generalizability of findings.

5. To test the application of new concepts. Innovation involves risk. Tried-and-true concepts and ideas can lead to good practice, and new ideas are necessary to make advances. POE can help determine how well a new concept works once applied.

6. To justify actions and expenditures. Organizations have greater demands for accountability and POE helps generate the information to accomplish this objective.

POE Process Model

The POE process model (Fig. 9.6) shows three levels of effort that can be part of a typical POE, as well as the three phases and nine steps that are involved in the process of conducting POEs:

1. Indicative POEs give an indication of major strengths and weaknesses of a particular building's performance. They usually consist of selected interviews with knowledgeable informants, as well as a subsequent walk-through of the facility.

2. Investigative POEs go into more depth. Objective evaluation criteria are explicitly stated either in the functional program of a facility, or they have to be compiled from guidelines, performance standards, and published literature on a given building type.

3. Diagnostic POEs correlate physical environmental measures with subjective occupant response measures. Case study examples of POEs at these three levels of effort can be found in Preiser, Rabinowitz, and White (1988).

Benefits and Limitations of Current POE Practice

Each of the just-noted types of POEs can result in several benefits and uses. Recommendations can be brought back to the client, and remodeling can be done to correct problems. Lessons learned can influence design criteria for future buildings, as well as provide information about buildings in use to the building industry. This is especially relevant to the public sector, which designs buildings for its own use on a repetitive basis.

The many uses and benefits (i.e., short-, medium- and long-term) which result from conducting POEs appear in the following lists. They refer respectively to: immediate action; the three- to five-year intermediate time frame that is necessary for the development of new construction projects; and the long-term time frame that is necessary for strategic planning, budgeting, and master planning of facilities (i.e., ranging from 10 to 25 years). These benefits provide the motivation and rationale for committing to POE as a concept and for developing POE programs.

Short-term benefits of POEs include the following:

- Identification and solutions to problems in facilities
- Proactive facility management responsive to building user values
- Improved space utilization and feedback on building performance
- Improved attitude of building occupants through active involvement in the evaluation process
- Understanding of the performance implications of changes dictated by budget cuts
- Better informed design decision making and understanding of the consequences of design

Medium-term benefits include the following:

- Built-in capacity for facility adaptation to organizational change and growth over time, including recycling of facilities into new uses

- Significant cost savings in the building process and throughout the life cycle of a building
- Accountability for building performance by design professionals and owners

    Long-term benefits include the following:

- Long-term improvements in building performance
- Improvement of design databases, standards, criteria, and guidance literature
- Improved measurement of building performance through quantification

The most important benefit of a POE is its positive influence upon the delivery of humane and appropriate environments for people through improvements of the programming and planning of buildings. POE is a form of product research that helps designers develop a better design in order to support changing requirements of individuals and organizations alike.

POE provides the means to monitor and maintain a good fit between facilities and organizations, and people and activities that they support. POE can also be used as an integral part of a proactive facilities management program.

9.4   AN INTEGRATIVE FRAMEWORK FOR BPE

In 1997, the POE process model was developed into an integrative framework for building performance evaluation (Preiser and Schramm, 1997), involving the six major phases of the building delivery and life cycle (i.e., planning, programming, design, construction, occupancy, and recycling of facilities). In the following, the integrative framework for building performance evaluation is outlined. The time dimension was the major added feature, plus internal testing/review/troubleshooting cycles in each of the six phases.

The integrative framework attempts to respect the complex nature of performance evaluation in the building delivery cycle, as well as the life cycle of buildings. This framework defines the building delivery cycle from an architect's perspective, showing its cyclic evolution and refinement toward a moving target, achieving better building performance overall and better quality as perceived by the building occupants (see Fig. 9.7).

At the center of the model is actual building performance, both measured quantitatively and experienced qualitatively. It represents the outcome of the building delivery cycle, as well as building performance during its life cycle. It also shows the six subphases previously referred to: strategic planning, programming, construction, design, occupancy, and recycling. Each of these phases has internal reviews and feedback loops. Furthermore, each phase is connected with its respective state-of-the-art knowledge contained in building type-specific databases, as well as global knowledge and the literature in general. The phases and feedback loops of the framework can be characterized as follows:

Phase 1—Planning.   The beginning of the building delivery cycle is the strategic plan which establishes medium- and long-term needs of an organization through market/needs analysis, which in turn is based on mission and goals, as well as facility audits. Audits match needed items, including space, with existing resources, in order to establish actual demand.

Loop 1—Effectiveness Review.   Outcomes of strategic planning are reviewed in relation to big issue categories, such as corporate symbolism and image, visibility in the context surrounding the site, innovative technology, flexibility and adaptive reuse, initial capital cost, operating and maintenance cost, and costs of replacement and recycling at the end of the useful life of a building.

Phase 2B—Programming.   Once effectiveness review, cost estimating, and budgeting have occurred, a project has become a reality and programming can begin.

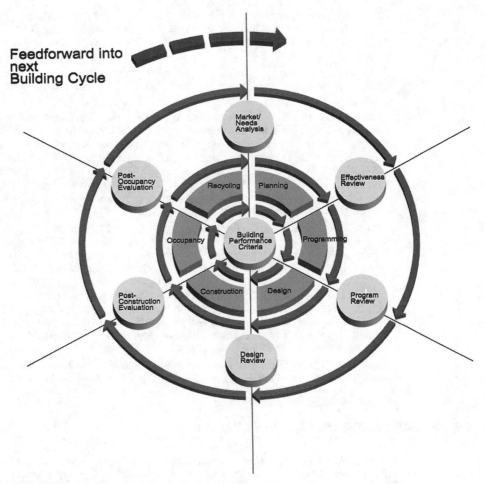

FIGURE 9.7   Building performance evaluation (BPE) integrative framework.

Loop 2B—Program Review.   The outcome of this phase is marked by a comprehensive documentation of the program review involving the client, the programmer, and representatives of the actual occupant groups.

Phase 3B—Design.   This phase contains the steps of schematic design, design development, and working drawings/construction documents.

Loop 3B—Design Review.   The design phase has evaluative loops in the form of design review or troubleshooting involving the architect, the programmer, and representatives of the client organization. The development of knowledge-based and computer-aided design (CAD) techniques make it possible to apply evaluations during the earliest design phases. This allows designers to consider the effects of design decisions from various perspectives, while it is still not too late to make modifications in the design.

Phase 4B—Construction.   In this phase construction managers and architects share in construction administration and quality control to assure contractual compliance.

Loop 4B—Postconstruction Evaluation.   The end of the construction phase is marked by postconstruction evaluation, an inspection that results in "punch lists," that is, items that need to be completed prior to commissioning and acceptance of the building by the client.

Phase 5B—Occupancy.   During this phase, move-in and start-up of the facility occur, as well as fine-tuning by adjusting the facility and its occupants to achieve optimal functioning.

Loop 5B—POE: Building performance evaluation during this phase occurs in the form of POEs carried out 6 to 12 months after occupancy, thereby providing feedback on what works in the facility and what does not. Post-occupancy evaluations will assist in testing hypotheses made in prototype programs and designs for new building types, for which no precedents exist. Alternatively, they can be used to identify issues and problems in the performance of occupied buildings, and further suggest ways to solve these. Further, POEs are ideally carried out in regular intervals; that is, in two- to five-year cycles, especially in organizations with recurring building programs.

Phase 6B—Recycling.   On the one hand, recycling of buildings to similar or different uses has become quite common. Lofts have been converted into artists' studios and apartments; railway stations have been transformed into museums of various kinds; office buildings have been turned into hotels; and factory space has been remodeled into offices or educational facilities. On the other hand, this phase might constitute the end of the useful life of a building when the building is decommissioned and removed from the site. In cases where construction and demolition waste reduction practices are in place, building materials with potential for reuse will be sorted and recycled into new products. At this point, hazardous materials, such as chemicals and radioactive waste, are removed in order to reconstitute the site for new purposes.

Loop 6B—Feedforward into the Next Building Cycle.   The lessons learned are fed into the predesign phases of the next project, thus benefitting the quality of outcome.

9.5   TOWARD UNIVERSAL DESIGN EVALUATION

The book *Building Evaluation Techniques* (Baird, et al., 1996) showcased a variety of building evaluation techniques, many of which would lend themselves to adaptation for purposes of UDE. In the same volume, this author (Preiser, 1996) presented a chapter on a three-day POE training workshop and prototype testing module, which involved the facility planners and designers and (after one year of occupancy) the end users, a formula that has proven to be very effective in generating useful performance feedback data. A proposed UDE process model is shown in Fig. 9.8.

Major benefits and uses of post-occupancy evaluations or performance feedback data are well-known and they include, when applied to UDE:

- Identify problems and develop universal design solutions
- Learn about the impact of practice on universal design and on building users in general
- Develop guidelines for enhanced universal design concepts and features in products, buildings, urban settings, and systems
- Create greater awareness in the public of successes and failures in universal design

It is critical to formalize and document, in the form of qualitative criteria and quantitative guidelines and standards, the expected performance of facilities in terms of universal design.

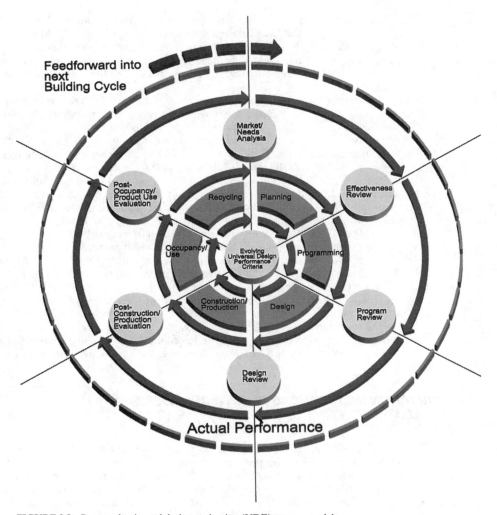

FIGURE 9.8 Proposed universal design evaluation (UDE) process model.

9.6 POSSIBLE STRATEGIES FOR UNIVERSAL DESIGN EVALUATION

In the previously referenced models, it is customary to include ADA standards for accessible design as part of a routine evaluation of facilities. The ADA standards only provide information on compliance with prescriptive technical standards, and nothing about performance—how the building or setting actually works for a range of users. The Principles of Universal Design (Center for Universal Design, 1997) constitute a user-oriented set of performance criteria to be considered. There is the need to identify and consider data gathering methods that include interviews, surveys, direct observation, photography, and the in-depth case study approach, among others.

Other authors in this book address assessment tools for universal design at the building (Corry, Chap. 56) and urban design scales (Guimaraes, Chap. 57, Manley, Chap. 58, and Ringaert, Chap. 6). In addition, the International Building Performance Evaluation (IBPE) project and consortium created by the author has attempted to develop a universal data collection toolkit that can be applied to any context and culture, while respecting cultural differences.

The author proposes to advance the state of the art through a collection of case study examples of different building types, with a focus on universal design, including living and working environments, public places, transportation systems, recreational and tourist sites, and so forth. These case studies will be structured in a standardized way, including videotaped walkthroughs of different facility types, and with various user types. The universal design critiques would focus on the three levels of performance previously referred to [i.e., (1) health, safety, security; (2) function, efficiency, work flow; and (3) social, psychological, cultural, and aesthetic performance] (Preiser, 1983). Other POE examples are currently under development through the Rehabilitation and Engineering Research Center (RERC) at Buffalo. One study focuses on wheelchair users, another on existing buildings throughout the United States. The RERC Web site explains that research in more detail.

Furthermore, methodologically appropriate ways of gathering data from populations with different levels of literacy and education (Preiser and Schramm, 2002) are expected to be devised. It is hypothesized that through these methodologies, culturally and contextually relevant universal design criteria will be developed over time. This argument was eloquently presented in Chap. 5 by Balaram when discussing universal design in the context of an industrializing nation like India.

The role of the user as "user/expert" (Ostroff, 1997) should also be carefully analyzed. The process of user involvement is often cited as central to successful universal design, but has not been systematically evaluated. Ringaert discusses the key involvement of the user in Chap. 6.

9.7 EDUCATION AND TRAINING IN UNIVERSAL DESIGN EVALUATION TECHNIQUES

Welch (1995) presented strategies for teaching universal design developed in a national pilot project involving 21 design programs throughout the United States. The initial learning from that project can be used in curricula in all schools of architecture, industrial design, interior design, landscape architecture, and urban design, when they adopt a new approach to embracing universal design as a paradigm for design in the future. That way, students will be familiarized with the values, concept, and philosophy of universal design at an early stage, and furthermore, through field exercises and case study evaluations, they will be exposed to real-life situations. As noted in Welch, it is important to have multiple learning experiences. Later on in the curriculum, these first exposures to universal design should be reinforced through in-depth treatment of the subject matter, and by integrating universal design into the studio courses as well as evaluation and programming projects.

A number of authors in this book, including Jones (Chap. 52), Morrow (Chap. 54), Pedersen (Chap. 53), and Welch and Jones (Chap. 51) offer current experiences and future directions in universal design education and training.

9.8 CONCLUSIONS

For universal design to become viable and truly integrated into the building delivery cycle of mainstream architecture and the construction industry, it will be critical to have all future students in these fields familiarized with universal design on one hand, and to demonstrate to practicing professionals the viability of the concept through a range of POE-based UDEs, including exemplary case study examples, on the other.

The "performance concept" and "performance criteria," made explicit and scrutinized through POEs, have now become an accepted part of good design, by moving from primarily subjective, experience-based evaluations to more objective evaluations based on explicitly stated performance requirements in buildings.

Critical in the notion of performance criteria is the focus on the quality of the built environment as perceived by its users/occupants. In other words, building performance is seen to be critical beyond aspects of energy conservation, life-cycle costing, and the functionality of buildings, but it focuses on users' perceptions of buildings.

For data gathering techniques for POE-based UDEs to be valid and standardized, the results need to become replicable.

Such evaluations have become most cost effective due to the fact that shortcut methods have been devised that allow the researcher/evaluator to obtain valid and useful information in a much shorter time frame than was previously possible. Thus, the cost of staffing and other expenses have been considerably reduced, making them affordable, especially at the "indicative" level previously described.

9.9   ACKNOWLEDGMENTS

The graphics for this chapter were designed by Martin Borger of the School of Architecture and Interior Design, University of Cincinnati, OH.

9.10   BIBLIOGRAPHY

Baird, G., et al. (eds.), *Building Evaluation Techniques,* McGraw-Hill, London, UK, 1996.

Center for Universal Design, *The Principles of Universal Design* (Version 2.0), North Carolina State University, Raleigh, NC, 1997.

Maslow, H., "A Theory of Motivation," *Psychological Review* 50: 370–398, 1948.

Nasar, J. L. (ed.), *Environmental Aesthetics: Theory, Methods and Applications,* MIT Press, Cambridge, MA, 1988.

National Academy of Sciences, Building Research Board (Wolfgang F. E. Preiser, Committee Chairman), *Programming Practices in the Building Process: Opportunities for Improvement,* National Academy Press, Washington, DC., 1986.

———, *Post-Occupancy Evaluation Practices in the Building Process: Opportunities for Improvement,* National Academy Press, Washington, DC, 1987.

Ostroff, E., "Mining Our Natural Resources: The User as Expert," *Innovation, The Quarterly Journal of the Industrial Designers Society of America,* vol. 16, no. 1, 1997.

Petzinger, Thomas, "A New Model for the Nature of Business: It's Alive!" *The Wall Street Journal,* February 26, 1999.

Preiser, W. F. E., "The Habitability Framework: A Conceptual Approach Toward Linking Human Behavior and Physical Environment," *Design Studies,* vol. 4, no. 2, April 1983.

———, "Design Intervention and the Challenge of Change," in W. F. E. Preiser, J. C. Vischer, and E. T. White, *Design Intervention: Toward a More Humane Architecture,* Van Nostrand Reinhold, New York, 1991.

———, "POE Training Workshop and Prototype Testing at the Kaiser-Permanente Medical Office Building in Mission Viejo, California, USA," in G. Baird, et al. (eds.), *Building Evaluation Techniques,* McGraw-Hill, London, UK, 1996.

————, "Post-Occupancy Evaluation: Conceptual Basis, Benefits and Uses," in J. M. Stein and K. F. Spreckelmeyer (eds.), *Classical Readings in Architecture,* McGraw-Hill, New York, 1999.

————, H. Z. Rabinowitz, and E. T. White, *Post-Occupancy Evaluation,* Van Nostrand Reinhold, New York, 1988.

———— and U. Schramm, "Building Performance Evaluation," in D. Watson et al. (eds.), *Time-Saver Standards: Architectural Design Data,* McGraw-Hill, New York, 1997.

———— and U. Schramm, "Intelligent Office Building Performance Evaluation in the Cross-Cultural Context: A Methodological Outline," *Intelligent Building,* vol. I, no. 1, 2002 (forthcoming).

Vitruvius, *The Ten Books on Architecture* (translated by M. H. Morgan), Dover Publications, New York, 1960.

von Foerster, H., *"Epistemology and Cybernetics: Review and Preview,"* Lecture at Casa della Cultura, Milan, Italy, 18 February, 1985.

Watson, D., M. J. Crosbie, and J. H. Callender (eds.), *Time-Saver Standards: Architectural Design Data* (seventh edition), McGraw-Hill, New York, 1997.

Welch, P. (ed.), *Strategies for Teaching Universal Design,* Adaptive Environments Center, Boston, MA, 1995.

P · A · R · T · 3

UNIVERSAL DESIGN GUIDELINES AND ACCESSIBILITY STANDARDS

CHAPTER 10

PRINCIPLES OF UNIVERSAL DESIGN

Molly Follette Story, M.S., IDSA

*Center for Universal Design, North Carolina State University,
Raleigh, North Carolina*

10.1 INTRODUCTION

Across the international community of research professionals, there are multiple definitions of universal design. Some definitions are broader; some are more narrow; some emphasize certain aspects over others. None is perfect; consensus is unnecessary. Differing definitions are a sign of healthy engagement with the concept, of practitioners seeking descriptions that are useful for various specific purposes.

The Center for Universal Design at North Carolina State University defines *universal design* as "the design of all products and environments to be usable by people of all ages and abilities, to the greatest extent possible."

This chapter describes the process of developing the Principles of Universal Design, which was coordinated by the Center for Universal Design, and presents design examples that satisfy each principle. The chapter also discusses uses for and cites recent applications of the principles.

10.2 WHY CREATE PRINCIPLES OF UNIVERSAL DESIGN?

Throughout most of its history, the universal design concept has suffered from a lack of established criteria that define what makes a design most usable. Instead, universal design has most often been communicated through presentation of good examples that embody certain aspects of the concept rather than provide concrete descriptions of its characteristics.

Because universal design can be manifested in any design to varying degrees, the late Ronald L. Mace liked to present a hierarchy of universal design examples, from designs requiring the smallest amount of interaction with users to ones requiring the greatest. The ones requiring the least interaction are the most universally usable. He used doors as an example. In order, Mace's hierarchy of interaction is as follows:

1. No door—just an opening in a wall
2. An "air door"—an opening in an outside wall where the heat or air conditioning stops
3. A powered door with a motion detector
4. A powered door with a pressure sensor mat

5. A powered door with a remote button that is pressed to open the door

6. A nonpowered door with no latch, which swings open when pushed

7. A door with a lever door handle

8. A door with a doorknob

9. A door with an automatic closing mechanism and no latch

10. A door with an automatic closing mechanism and a lever door handle

11. A door with an automatic closing mechanism and a doorknob

12. A revolving door—requiring the most interaction—which is unusable for people who use wheelchairs, push strollers, or pull wheeled luggage

While presentation of these examples is helpful, it requires audience members to interpret and internalize the approach for themselves. It demands substantial active involvement of listeners, and requires the presenter to offer a very large number and a very wide range of examples to ensure that all aspects of the concept have been conveyed.

So how do designers know when they have achieved universal design? If maximum usability for the widest diversity of users is the goal, how usable is usable enough? How can a designer evaluate the shortcomings of a given design and determine what could be done to improve it?

Until recently, the only usability criteria available were found in codes and standards. Some criteria are provided by accessibility building codes, such as those contained in the U.S. Americans with Disabilities Act Standards for Accessible Design (ADA Standards). Other sets of usability criteria are available in some of the American National Standards Institute (ANSI) and International Standards Organization (ISO) standards. The concept of usability is addressed in a few American and international standards, but their scope is limited to various information technologies such as interactive systems and software. For example, ISO 13407 addresses human-centered design processes for interactive systems. The 17 parts of the ISO 9241 series address ergonomic requirements for office work with video display terminals (VDTs), including dialog principles (Part 10) and guidance on usability (Part 11). The forthcoming American standard, ANSI 200, will be an extension of the ISO 9241 human computer interaction standard and will contain accessibility criteria not included in the ISO standard.

Typically, however, existing standards provide only minimum requirements to accommodate people with disabilities and fall substantially short of ideal conditions. John Salmen describes the limitations of such prescriptive standards in Chap. 12, "U.S. Accessibility Codes and Standards: Challenges for Universal Design." These standards also apply only to specific products and environments. Guiding principles are needed that articulate the full range of criteria for achieving universal design for all types of designs, as well as clarify how the concept of universal design may pertain to specific designs under development and suggest how usability of those designs could be maximized.

10.3   *HISTORY OF THE PRINCIPLES OF UNIVERSAL DESIGN*

From 1994 to 1997, the Center for Universal Design conducted a research and demonstration project funded by the U.S. Department of Education's National Institute on Disability and Rehabilitation Research (NIDRR). The project was titled Studies to Further the Development of Universal Design. One of the activities of the project was to develop a set of universal design guidelines.

On April 28 and 29, 1995, the project team convened a meeting of 10 experts on universal design at the offices of the Center for Universal Design at North Carolina State University in Raleigh, North Carolina. The group comprised 10 professionals active in the field of universal design, from seven institutions across the United States; it included architects, product designers, engineers, and environmental design researchers. The group members spent 2 days together, amassing their substantial collective knowledge of universal design and listing all the maxims, guidelines, and concepts they could articulate to describe the concept. They explicitly agreed that rather than claiming to be experts on all aspects of good design, the group would develop universal design guidelines to address only issues of design usability for the widest diversity of individuals.

Project staff members subsequently analyzed the results, grouped and merged the bits of data, added to them, shaped them, and, in collaboration with the other authors, developed a list they called the *Principles of Universal Design*. The draft principles were mailed to a dozen colleagues in universal design research and practice around the country for their review and comment, and their suggestions were incorporated into the final document.

Several versions of the Principles of Universal Design preceded the most current one. The earliest draft, dated May 22, 1995, included 10 principles:

- *Simple Operation.*   The design should be easy to understand and operate with no redundant or unnecessary steps and a minimum of complexity.
- *Intuitive Operation.*   The design should work the way that would be expected in the native culture and should not require substantial learning to operate.
- *Redundant Feedback.*   The design should provide feedback in multiple sensory modes.
- *Gradual Level Changes.*   The design should avoid the need to ascend or descend. When necessary, level changes should have gradual transitions and should be achieved in the same manner by all users.
- *Space for Approach and Movement.*   The design should provide ample space and clearances for close approach and maneuvering, even by people who use wheelchairs or scooters.
- *Low Physical Demand.*   Use of the design should require the expenditure of a minimum of energy and cause a minimum of fatigue.
- *Comfortable Reach Range.*   Use of the design should require a minimum range of reach.
- *Minimization of and Tolerance for Error.*   The design should not be damaged nor cause injury by accidental improper actions.
- *Alternate Methods of Use.*   The design should allow for a wide variety of human operation styles.
- *Perceptible Information.*   Information provided by the design should be detectable and discriminable.

Each of these principles was accompanied by a set of questions. For example, for "Simple Operation," the questions were as follows:

- Does each control perform only one function?
- Can the design be operated without simultaneous operations—particularly, with one hand?
- Has the number of steps involved in operating the design been minimized?
- Is immediate feedback given for each input?
- Is there no limit on reaction time or can reaction time be varied to suit the user?

By the second version, dated July 26, 1995, the number of principles had been reduced from 10 to 6, and the tone of the document had changed from a list of attributes to commands. Its six principles were as follows:

- *Make It Easy to Understand.*   The design should be easy to understand, regardless of the user's knowledge or language skills.
- *Make It Easy to Operate.*   The design should require minimum user effort and cause minimum fatigue.

- *Communicate with the User.*   The design should communicate effectively, regardless of the user's sensory abilities.
- *Design for User Error.*   The design should minimize the risk of damage or injury caused by user mistakes.
- *Accommodate a Range of Methods of Use.*   The design should allow methods of use according to the user's preferences and abilities.
- *Allow Space for Access.*   The design should provide ample space for approach and maneuvering regardless of the user's body size and position.

Each of these principles was accompanied by a list of guidelines. For example, for "Make It Easy to Understand," the guidelines were as follows:

- The design should be usable without training.
- Displays and controls should be placed in expected locations.
- Each control should perform only one function.
- The design should offer only as much information as is necessary for its operation.
- The design should be usable even if attention is divided.

The authors of the Principles of Universal Design struggled with the concept of *equitable use,* which first appeared in the third draft, dated August 31, 1995. Its initial definition was "The design does not disadvantage or stigmatize any group of users." Some on the team thought that because this issue was so fundamental to the concept, it should not be a principle but, rather, part of the definition of universal design. It was the only principle that did not address usability but, rather, egalitarianism. In the end, though, because it was so essential, the team decided that this aspect of universal design needed to be articulated just as much as and, in some ways, maybe even more than the others. From this version on, "Equitable Use" was included among the Principles of Universal Design, prominently placed first on the list.

10.4   PRINCIPLES OF UNIVERSAL DESIGN

In December, 1995, the Center for Universal Design published Version 1.1 of the Principles of Universal Design. Version 2.0 followed in April, 1997. The seven principles are as follows:

    Principle 1: Equitable Use

    Principle 2: Flexibility in Use

    Principle 3: Simple and Intuitive Use

    Principle 4: Perceptible Information

    Principle 5: Tolerance for Error

    Principle 6: Low Physical Effort

    Principle 7: Size and Space for Approach and Use

Each of these principles is defined and is followed by a set of guidelines that describe the key elements that should be present in a design that adheres to the principle (see Fig. 10.1).

The purpose of the Principles of Universal Design and their associated guidelines is to articulate the concept of universal design in a comprehensive way. The principles reflect the authors' belief that the basic universal design principles apply to all design disciplines, including environments, products and communications. The principles are intended to guide the design process, allow the systematic evaluation of designs, and assist in educating both designers and consumers about the characteristics of more usable design solutions.

FIGURE 10.1   The Principles of Universal Design, Version 2.0. *(Copyright © 1997 by North Carolina State University. Major funding provided by the National Institute on Disability and Rehabilitation Research.)*

The Principles of Universal Design

Version 2.0—April 1, 1997

The Center for Universal Design

North Carolina State University

Compiled by advocates of universal design, listed in alphabetical order: Bettye Rose Connell, Mike Jones, Ron Mace, Jim Mueller, Abir Mullick, Elaine Ostroff, Jon Sanford, Ed Steinfeld, Molly Story, and Gregg Vanderheiden.

Universal Design

The design of products and environments to be usable by all people, to the greatest extent possible, without adaptation or specialized design.

Principle 1: Equitable Use

The design is useful and marketable to people with diverse abilities.

Guidelines:

1a. Provide the same means of use for all users—identical whenever possible; equivalent when not.

1b. Avoid segregating or stigmatizing any users.

1c. Make provisions for privacy, security, and safety equally available to all users.

1d. Make the design appealing to all users.

Principle 2: Flexibility in Use

The design accommodates a wide range of individual preferences and abilities.

Guidelines:

2a. Provide choice in methods of use.

2b. Accommodate right- or left-handed access and use.

2c. Facilitate the user's accuracy and precision.

2d. Provide adaptability to the user's pace.

Principle 3: Simple and Intuitive Use

Use of the design is easy to understand, regardless of the user's experience, knowledge, language skills, or current concentration level.

Guidelines:

3a. Eliminate unnecessary complexity.

3b. Be consistent with user expectations and intuition.

3c. Accommodate a wide range of literacy and language skills.

3d. Arrange information consistent with its importance.

3e. Provide effective prompting and feedback during and after task completion.

Principle 4: Perceptible Information

The design communicates necessary information effectively to the user, regardless of ambient conditions or the user's sensory abilities.

Guidelines:

4a. Use different modes (pictorial, verbal, tactile) for redundant presentation of essential information.

4b. Maximize "legibility" of essential information.

4c. Differentiate elements in ways that can be described (i.e., make it easy to give instructions or directions).

(Continued)

4d. Provide compatibility with a variety of techniques or devices used by people with sensory limitations.

Principle 5: Tolerance for Error

The design minimizes hazards and the adverse consequences of accidental or unintended actions.

Guidelines:

5a. Arrange elements to minimize hazards and errors—most used elements, most accessible; hazardous elements eliminated, isolated, or shielded.

5b. Provide warnings of hazards and errors.

5c. Provide fail safe features.

5d. Discourage unconscious action in tasks that require vigilance.

Principle 6: Low Physical Effort

The design can be used efficiently and comfortably and with a minimum of fatigue.

Guidelines:

6a. Allow user to maintain a neutral body position.

6b. Use reasonable operating forces.

6c. Minimize repetitive actions.

6d. Minimize sustained physical effort.

Principle 7: Size and Space for Approach and Use

Appropriate size and space is provided for approach, reach, manipulation, and use regardless of user's body size, posture, or mobility.

Guidelines:

7a. Provide a clear line of sight to important elements for any seated or standing user.

7b. Make reach to all components comfortable for any seated or standing user.

7c. Accommodate variations in hand and grip size.

7d. Provide adequate space for the use of assistive devices or personal assistance.

The authors of the Principles of Universal Design envisioned that beyond the principles and guidelines, there would eventually be two additional levels of detail to the work. If Level 1 was the principles and Level 2 was the guidelines, Level 3 would be tests and Level 4 would be strategies. The tests in Level 3 would be similar to the questions proposed in earlier versions, and would serve to allow designers to query a design for universal usability. At Level 4, which would contain strategies for meeting the guidelines and passing the tests, the document would become discipline-specific. For example, for Principle 3, Simple and Intuitive Use, the strategies might describe the following:

- For architecture—methods of creating clear wayfinding
- For products—issues of correspondence and cognitive mapping
- For software—features of programs that follow well-known standards

10.5   EXAMPLES OF THE PRINCIPLES OF UNIVERSAL DESIGN

It is useful to illustrate the Principles of Universal Design with examples. Each of the designs presented here demonstrates a good application of one of the guidelines associated with the principles. The design solutions included here are not necessarily universal in every respect, but each is a good example of a specific guideline and helps to illustrate its intent.

Principle 1: Equitable Use.   The design is useful and marketable to people with diverse abilities. Designs should appeal to diverse populations and offer everyone a comparable and nonstigmatizing way to participate.

- A public telephone with a keyboard and a light-emitting diode (LED) display as well as a volume control can be used by people with hearing impairments to make calls, too. Multiple phones should be offered at various heights, each with a shelf on which to rest belongings or an elbow.
- The water play area in a children's museum shown in Fig. 10.2 simulates a meandering brook and invites enjoyment for everyone in and around the water. It is appealing to and usable by people who are short or tall, young or older.

FIGURE 10.2   "Meandering book" in children's museum invites participation.

Principle 2: Flexibility in Use.   The design accommodates a wide range of individual preferences and abilities. Designs should provide for multiple ways of doing things. Adaptability is one way to make designs universally usable.

- A children's playground (Fig. 10.3), if creatively designed to accommodate children with and without disabilities, can offer multiple play modes. Children may play alone or with others; they may have different abilities or different play styles; and they may choose to act out a different adventure each time they visit.

FIGURE 10.3   Creative children's playground supports a range of play scenarios.

- Optional computer hardware, such as a trackball or Braille printer, and software, such as Easy Access, can offer choices of input and output options.

Principle 3: Simple and Intuitive Use.   Use of the design is easy to understand, regardless of the user's experience, knowledge, language skills, or current concentration level. Make things work in an expected way.

- Placing cross-culturally meaningful icon labels next to controls on equipment can communicate the function of each control without the need for language.
- The prototype electronic thermostat designed at the Center for Universal Design provides information in visual, audible, and tactile formats (Fig. 10.4). The functions are clearly laid out and labeled, readouts are provided in both digital (visible) and analog (visible and tactile) formats, and the thermostat's voice output (audible) helps the user know what is happening when he or she pushes the buttons. For example, when the user presses one of the keys that has a raised tactile arrow, the thermostat announces "72 degrees." If the user keeps the down arrow depressed, the thermostat will count down: "71, 70, 69, 68. . . ." When the user lets go, the thermostat repeats "68 degrees." The other control buttons also cause the thermostat to speak.

Principle 4: Perceptible Information.   The design communicates necessary information to the user, regardless of ambient conditions or the user's sensory abilities. Designs should provide for multiple modes of input and output.

- The subway fare machine shown in Fig. 10.5 offers a push button for selecting instructions to be presented audibly. The fare machine also provides tactile lettering in all-capital letters, which is easier to feel with the fingertips, and high-contrast printed lettering in capital and lowercase letters, which is easier to see with low vision. Redundant audible feedback is also helpful for cognitive processing.
- Stereo and computer equipment cable terminals that are different sizes, shapes, and colors are easier to connect correctly because each cable fits in only one place. The distinctive cables are also easier to describe when using telephone or online technical assistance.

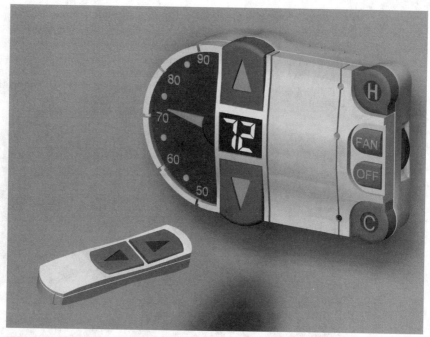

FIGURE 10.4 Prototype thermostat design by the Center for Universal Design.

FIGURE 10.5 Subway fare machines with high-contrast and tactile lettering.

Principle 5: Tolerance for Error.   The design minimizes hazards and the adverse consequences of accidental or unintended actions. Designs should make it difficult to make a mistake, but if someone does, it should not result in injury to the person or the product.

- The deadman switch activated by a secondary bar that runs across the handle of some power lawn mowers (Fig. 10.6) requires the user to squeeze the bar and the handle together to make the mower blade spin. If the two are not held together, the blade stops turning.

FIGURE 10.6   Deadman switch on lawn mower handle requires conscious use.

- The Undo option in computer software allows the user to correct mistakes without penalty.

Principle 6: Low Physical Effort.   The design can be used efficiently and comfortably and with a minimum of fatigue. Designs should minimize strain and overexertion.

- A microphone and voice recognition software on a computer (Fig. 10.7) eliminate the need for highly repetitive keystrokes or manual actions of any kind. This feature prevents as well as accommodates disabilities of the hand.
- Whether used out of necessity or for convenience, remote controls—for example, for televisions, stereo equipment, lights, fans, and electric garage door openers—eliminate most of the physical effort required to operate many devices.

Principle 7: Size and Space for Approach and Use.   Appropriate size and space is provided for approach, reach, manipulation, and use regardless of the user's body size, posture, or mobility. Designs should accommodate variety in people's body sizes and ranges of motion.

- A full-length sidelight window next to the front door of a home (Fig. 10.8) provides a view of the outside for persons of any height.
- Kitchens with counters available at multiple heights—for example, 28 in high for sitting and 36 in high for standing—accommodate people of varying heights, postures, positions, and preferences.

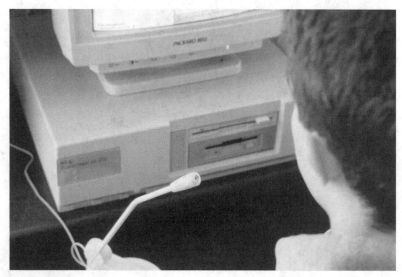

FIGURE 10.7   Computer hardware with microphone works with voice-recognition software.

10.6   APPLICATIONS OF THE PRINCIPLES OF UNIVERSAL DESIGN

The Principles of Universal Design were published in a book, *The Universal Design File: Designing for People of All Ages and Abilities* (Story, Mueller, and Mace, 1998). The book contains an introduction to the range of human abilities and a total of 96 images representing 2 to 5 photographic examples of each of the guidelines associated with the principles. It also includes a set of seven case studies of companies that have successfully practiced universal design. The full set of 14 case studies is available in another publication, *Case Studies on Universal Design* (Mueller, 1997).

The National Endowment for the Arts and the National Building Museum sponsored Search for Excellence in Universal Design, the first national search for and collection of images showing excellent examples of universal design in practice. Completed in 1996, the project was managed by Universal Designers and Consultants of Rockville, Maryland, in cooperation with the Adaptive Environments Center and the Center for Universal Design. The Principles of Universal Design were used as the basis for judging the entries in the fields of architecture, graphic design, industrial design, interior design, and landscape architecture. The results were published as a collection of photographic slide images of the 38 selected designs with accompanying text, titled *Images of Excellence* (Universal Designers and Consultants, 1996). Now available on CD-ROM (Universal Designers and Consultants, 2000), the Images collection was intended to support universal design education and practice.

The Center for Universal Design developed a second collection of universal design exemplars on CD-ROM in December 2000. *Universal Design Exemplars* (Center for Universal Design, 2000a) was funded by the National Endowment for the Arts and NEC Foundation of America. The Principles of Universal Design again served as judging and organizing criteria for the collection of 32 projects. Project descriptions include the design's background and multiple images show the object or space in use or in the context for which it was designed. The interactive CD format allows the user to explore the selected projects in depth through images and text, and to cross-reference projects by design discipline and the project's relationship to one or more of the Principles of Universal Design.

FIGURE 10.8   Sidelight next to front door of home provides view for residents of all heights.

Practitioners all over the world have adopted the Principles of Universal Design, translated them into several languages, and applied them to a variety of purposes. They are being used not only in the United States but also, for example, in Japan and in Norway, Finland, and several other European countries. Sandra Manley adapted the principles for use in her urban design work with students and the community in Bristol, England (see Chap. 58, "Creating an Accessible Public Realm"). Margaret Calkins, Jon Sanford, and Mark A. Proffitt recommend additional guidelines that would make the principles more appropriate in designing facilities for people with Alzheimer's disease (see Chap. 22, "Design for Dementia: Challenges and Lessons for Universal Design").

10.7 FUTURE DEVELOPMENT OF THE PRINCIPLES OF UNIVERSAL DESIGN

The Principles of Universal Design continue to be a work in progress and will be fine-tuned and reissued as need be and time permits.

The quality that most distinguishes *accessible design* from *universal design* is the visual and functional integration of the accessibility features, which needs to be designed into products and environments from the outset. This design integration removes any stigma associated with use of the design and results in the social inclusion of the broadest diversity of users. The concept of design integration must be captured in the Principles of Universal Design in some way, most likely as part of Principle 1.

In addition, the needs of individuals with multiple chemical sensitivities are not addressed in the current version of the Principles of Universal Design, as highlighted by Valerie Fletcher in Chap. 60, "A Neighborhood Fit for People: Universal Design on the South Boston Waterfront." This omission must be corrected in future versions.

Future work on the principles will also focus on the development of the tests and strategies previously referred to in this chapter. As mentioned earlier, at the level of strategies, at least, and possibly also tests, the document will need to be discipline-specific. The fields of telecommunications and information technology are leading the way in this area. One version of a set of tests for related hardware, software, and services is embodied in Sec. 255 of the Telecommunications Act of 1996 and Sec. 508 of the 1998 Amendments to the Rehabilitation Act. In response, a collection of strategies for compliance has been developed by the Trace Research and Development Center at the University of Wisconsin at Madison and can be found on the center's Web site (www.trace.wisc.edu/world).

By trying to be so general, the Principles of Universal Design suit nothing perfectly. They may need to be filtered depending on application to identify those issues that are relevant to specific designs. For example, for consumer product design, the tests may relate to phases of use, as follows:

Packaging

- Can you carry the package?
- Can you read what is printed on the package?
- Can you understand what is printed on the package?
- Can you understand how to open the package?
- Can you easily open the package?

Instructions

- Can you read the instructions? If not, is an alternative format offered?
- Can you understand the instructions? If not, can you contact the manufacturer's customer service department for help? If so, are the service representatives helpful?

Similar lists of questions could be developed for product installation, use, storage, maintenance, repair, and disposal. Subtasks associated with using various products will also vary widely. The ultimate document containing principles, guidelines, tests, and strategies may be electronic, which would greatly simplify its customization by industry and phase of use.

Researchers at the Center for Universal Design, in consultation with practicing design and marketing professionals and consumers from across the United States, have developed a set of tests for products that they call the Universal Design Performance Measures (Center for Universal Design, 2000b). The measures reflect the Principles of Universal Design and are intended to be easier to apply in practice. The Universal Design Performance Measures are designed to be a tool for both professional product designers and consumers. Design professionals can use the designer's version of the measures to evaluate products during the development process. Individuals can use the consumer's version of the measures to evaluate products in the store before buying them or to evaluate products they already own.

TABLE 10.1   Comparison Between Principle 2 of the Principles of Universal Design and the Universal Design Performance Measures for Products

Principles of Universal Design	Consumers' product evaluation survey	Designers' product evaluation survey
2a. Provide choice in methods of use.	2a. I can use this product in whatever way(s) are safe and effective for me.	2a. The product offers any potential user at least one way to use it safely and effectively.
2b. Accommodate right- or left-handed access and use.	2b. I can use this product with either my right or left side (hand or foot) alone.	2b. This product can be used by either right- or left-dominant users, including amputees with or without prostheses.
2c. Facilitate the user's accuracy and precision.	2c. I can use this product precisely and accurately.	2c. This product facilitates (or does not require) the user's accuracy and precision.
2d. Provide adaptability to the user's pace.	2d. I can use this product as quickly or as slowly as I want.	2d. This product can be used as quickly or as slowly as the user wants.

The consumer's and designer's versions of the Universal Design Performance Measures each comprise a set of 29 statements corresponding to the 29 guidelines in the Principles of Universal Design. The principles and the two versions of the performance measures all address the same issues, but each takes a different approach. A sample comparison of the second section of the Principles of Universal Design and the designer's and consumer's versions of the Universal Design Performance Measures is shown in Table 10.1.

Testing results indicate that the designer's version of the Universal Design Performance Measures provides a good relative assessment of universal usability, but the measures are not an absolute tool for achieving universal design. The only way to be certain that designs work well for a wide range of people is to ask people directly. Elaine Ostroff promotes the invaluable role of users in both education and practice, and has created the term *user/expert* to distinguish their participation (Ostroff, 1997).

The consumer's version of the Universal Design Performance Measures is expected to help guide personal purchasing decisions. The measures are also being employed by individuals participating in a design evaluation service being established by the Center for Universal Design. As a service to industry, products are mailed to and reviewed for usability by a diverse group of consumers from across the United States, and the results are reported back to manufacturers.

10.8   INTERNATIONAL CHALLENGES FOR UNIVERSAL DESIGN

The globe is shrinking and, increasingly, everyone must deal with cultural diversity. For example, how do individuals who speak Japanese use computers with keyboards that bear Roman characters? They use a system called *romaji-kanji keyboard conversion,* in which pressing a sequence of keystrokes on a standard QWERTY keyboard causes kanji or kana characters to appear on the monitor. Beyond the differences in written character representation, there are other basic differences in the ways the English and Japanese languages are written. English starts from the left side of the page and, at least traditionally, Japanese starts on the right; one language moves across the page first, and the other moves down the page first.

Other cultural conventions that help to structure various built environments also differ. American light switches move up and down; in other countries, switches sometimes move left and right. In the United States, the faucet lever is raised to start the water; in some countries, the lever is pushed down. Drivers in different countries even drive on opposite sides of the road.

How will universal design proponents reconcile these fundamental differences in convention? Cognitive maps, which set up user expectations of the way the world will be organized, are powerfully important to usability. They determine what a user's intuition will be when confronted with a new design.

Cultural diversity is one of the most challenging aspects of individual variance that must be addressed to achieve true global universal design—design for people of all ages and abilities, all sizes and shapes, both genders, and all nationalities. Singanapalli Balaram writes convincingly of the need for culturally appropriate design in Chap. 5, "Universal Design and the Majority World". Researchers around the world will have to work together to achieve this goal. Will it be necessary to homogenize all standards and conventions, or can a way be found to honor differing ones?

10.9   CONCLUSION

Articulating and describing attributes that make design universally usable is not a trivial challenge. Developing tests that can guide a designer to create flawless universal design without user input is an impossibility. The efforts described in this chapter are ambitious attempts to put into words a concept that embraces human diversity and applies to all designs.

By its nature, any design problem can be successfully addressed through multiple solutions. Choosing the most appropriate single design solution requires an understanding of and selection between tradeoffs in usability. This demands a high level of knowledge and a commitment to soliciting user input throughout the design process.

The Principles of Universal Design help articulate and describe all the different aspects of universal design. The principles' purpose is to guide others, and, in spite of their generic nature, they have proven to be useful in shaping projects of various types all over the world. It is the author's hope that they will support and inspire continuing advancement in the field of universal design.

10.10   BIBLIOGRAPHY

Center for Universal Design, *Universal Design Exemplars* (CD-ROM), Center for Universal Design, North Carolina State University, Raleigh, NC, 2000a.

———, *Universal Design Performance Measures for Products,* Center for Universal Design, North Carolina State University, Raleigh, NC, 2000b.

Mueller, J. L., *Case Studies on Universal Design,* Center for Universal Design, North Carolina State University, Raleigh, NC, 1997.

Ostroff, E., "Mining Our Natural Resources: The User as Expert," *Innovation, the Quarterly Journal of the Industrial Designers Society of America,* 16(1), 1997.

Story, M. F., J. L. Mueller, and R. L. Mace, *The Universal Design File: Designing for People of All Ages and Abilities.* Center for Universal Design, North Carolina State University, Raleigh, NC, 1998.

Universal Designers and Consultants, *Images of Universal Design Excellence,* Universal Designers and Consultants, Takoma Park, MD, 1996.

———, *Images of Universal Design Excellence,* Universal Designers and Consultants, Takoma Park, MD, 2000.

10.11 RESOURCES

Adaptive Environments Center, Inc.
Valerie Fletcher, Executive Director
374 Congress Street, Suite 301
Boston, MA 02210
(617) 695-1225
TTY: (617) 695-1225
Internet: www.adaptenv.org
E-mail: adaptive@adaptenv.org

Center for Inclusive Design and Environmental Access (IDEA)
Edward Steinfeld, Director
School of Architecture and Planning
State University of New York at Buffalo
112 Hayes Hall
Buffalo, NY 14214-3087
(716) 829-3483, extension 327
Internet: www.arch.buffalo.edu/~idea
E-mail: idea@arch.buffalo.edu

Center for Universal Design
Molly Follette Story, Interim Director
North Carolina State University
School of Design, Box 8613
Raleigh, NC 27695-8613
(800) 647-6777
TTY: (800) 647-6777
Internet: www.design.ncsu.edu/cud
E-mail: cud@ncsu.edu

Disability Statistics Center
University of California at San Francisco
3333 California Street, Suite 340
Campus Mail Box 0646
San Francisco, CA 94118
(415) 502-5210
TDD: (415) 502-5205
Fax: (415) 502-5208
Internet: dsc.ucsf.edu/UCSF
E-mail: distats@itsa.ucsf.edu

National Center for Health Statistics
U.S. Department of Health and Human Services
Centers for Disease Control and Prevention
Division of Data Services
Hyattsville, MD 20782-2003
(301) 458-4636
Internet: www.cdc.gov/nchs

Trace Research and Development Center
Gregg C. Vanderheiden, Director
University of Wisconsin at Madison

5901 Research Park Boulevard
Madison, WI 53719-1252
(608) 262-6966
TTY: (608) 263-5406
Internet: www.trace.wisc.edu
E-mail: info@trace.wisc.edu

Universal Design Newsletter
Universal Designers and Consultants, Inc.
6 Grant Avenue
Takoma Park, MD 20912
(301) 270-2470
Fax: (301) 270-8199
Internet: www.universaldesign.com
E-mail: UDandC@erols.com

U.S. Architectural Barriers and Compliance Board
Access Board
1331 F Street, NW, Suite 1000
Washington, DC 20004-1111
(202) 272-5434; (800) 872-2253
TTY: (202) 272-5449; (800) 993-2822
Fax: (202) 272-5447
Internet: www.access-board.gov
E-mail: info@access-board.gov

U.S. Department of Justice
950 Pennsylvania Avenue, SW
Washington, DC 20530-0001
Internet: www.usdoj.gov/crt/ada/adahom1.htm
E-mail: web@usdoj.gov

CHAPTER 11
UNITED NATIONS STANDARDS AND RULES

Maria Cristina Sará-Serrano, Dipl.
with John Mathiason, Ph.D.
Associates for International Management Services
Mt. Tremper, New York

11.1 INTRODUCTION

The United Nations Standard Rules for the Equalization of Opportunities for Persons with Disabilities were adopted to provide a normative framework for public policies and programs to enable persons with disabilities to participate fully and equally in all aspects of life. They have particular relevance for universal design in that they provide a framework within which designs can be placed and an incentive for governments, who have accepted the Rules by voting for them in United Nations bodies, to implement universal design principles. The imperatives can be seen in an analysis of the content of the Rules as they apply to universal design issues. The effectiveness of the Rules depends on the extent to which they are cited and used by advocates for persons with disabilities, and for this, new efforts are underway to equip advocates to use them at national and international levels.

11.2 BACKGROUND

Over time, each country in the world has dealt with accessibility and universal design in some way. Some countries elaborated detailed laws, regulations, and guidelines, but most dealt with accessibility by ignoring it. In a globalizing world, unevenness in accessibility is always ethically wrong and is increasingly economically unacceptable. The experience of any person with a disability flying from an accessible airport in one country to an inaccessible one in another is poignant testimony to the truth of this proposition.

The need for global standards was recognized for many years, first by persons with disabilities themselves, and then by policymakers in all countries. The United Nations Standard Rules for the Equalization of Opportunity for Persons with Disabilities, adopted by the United Nations General Assembly in 1993, are a response to this need.

While some have said that the Standard Rules are neither standard nor are they rules, they are the only universally accepted normative statements on what accessibility should mean. They constitute a point of reference, a normative floor, for elaborating national standards and guidelines and ensuring that persons with disabilities, wherever they live and wherever they may travel, have a minimum level of accommodation to their special needs and circumstances. They provide a politically agreed consensus on the basis of which universal design can be implemented.

The applicability of the Standard Rules to universal design can be understood in terms of the origins, limitations, content, and monitoring of the Rules. This chapter provides an overview of these elements.

11.3   ORIGINS OF THE STANDARD RULES

The issue of disability had been on the United Nations social development agenda since the beginning of the organization. Its original focus, like that of most countries, was on disabled veterans, soldiers who had been wounded in World War II. The first two heads of the United Nations Disability Unit were disabled veterans. This was politically logical. Disabled veterans were almost exclusively men, had been disabled as adults, were felt by society to merit attention because of their service, and had special governmental programs. Nonveteran, nonmale persons with disabilities were largely invisible.

It was assumed that the rights of persons with disabilities were protected by the human rights standards expressed in the Universal Declaration of Human Rights adopted by the United Nations in 1948.

As the century progressed, more and more human rights advocates recognized that more detailed norms would have to be agreed upon if the rights of particular segments of the population were to be adequately protected. Perhaps the first to recognize this were advocates for women's human rights, who sensed that the mere prohibition of discrimination on the basis of sex was not sufficient to promote enjoyment of other human rights. In 1969, at the time of the first United Nations Conference on Human Rights, a Declaration on the Elimination of Discrimination against Women was adopted. From the early 1970s there was a movement to convert these norms into a United Nations Convention on the Elimination of All Forms of Discrimination against Women, which culminated in the adoption of the Convention by the General Assembly in 1979.

Advocates for the rights of persons with disabilities took note of these developments. They pressed successfully to have 1982 designated as International Year for Disabled Persons. The Year provided an opportunity to reflect on the nature of disability and the public policies necessary to address it. The end result of the Year was the drafting of a World Plan of Action Concerning Disabled Persons and its adoption by the General Assembly at its 37th session in 1982. The World Plan mentioned accessibility in 8 of its 201 paragraphs, mostly in terms of stating that human settlements, transportation, and information should be accessible, but not elaborating further. In United Nations usage, a plan of action is a broad menu for possible action by governments, but does not constitute a normative statement as such.

By 1988 there was a concern among the members of the community concerned with disability that the World Plan of Action was not functioning as well as it should and that member states were not taking it seriously. At a meeting of the Commission for Social Development two delegates, one from Sweden with a disability, and the other from Italy, with a child with a disability, were particularly concerned. They worked together so that the delegations of Sweden and Italy respectively argued that an international human rights convention for persons with disability should be drafted. While the idea of a convention inspired some interest on the part of other delegations, it was not strongly supported. There were a number of reasons for this. Many delegations were concerned that there might be a spread in the number of human rights conventions and that this would dilute the human rights regime. Others were concerned that a convention might propose binding obligations for states that they could not meet given the level of development or given the amount of public resources available.

In the United Nations, whenever an idea has been advanced that receives some support there is an effort to find a compromise that will at least meet part of the concerns. In this case, the Commission for Social Development decided to begin the process of drafting an interme-diate type of human rights document. This would be in the form of a declaration that would try to specify what states should do but would not have the status of a treaty that would make compliance obligatory.

As a result, the Commission proposed to the Economic and Social Council that a process begin to draft what were called Standard Rules for the Equalization of Opportunities for Per-sons with Disabilities. The Council authorized the Commission to establish a Special Commit-tee to draft the Rules. The Special Committee met on three occasions. Using the World Plan of Action as its basis, it prepared a draft that could be adopted by the General Assembly.

The process of negotiating the text was unusual in that the Special Committee made provi-sion to have full participation of representatives of international nongovernmental organiza-tions (NGOs) concerned with disability. As a result, organizations such as Disabled Peoples' International, the World Blind Union, and the World Federation of the Deaf participated actively. In some cases they participated as nongovernmental organizations. However, many interested governments made sure that representative organizations of persons with disability were included on national delegations. As a result, the draft text reflected much of the views held by organizations representing persons with disabilities.

When the time came for the Special Committee to meet, Disabled Peoples' International had a representative to the United Nations Office in Vienna, Austria. This meant that this representative (one of the authors of this chapter) was present at all of the negotiations that took place between 1991 and 1993. As an NGO representative, she tried to influence govern-ment delegations to adopt the disability community's point of view. One matter of immediate concern to her was that there were very few women involved in the negotiations, either from the government or NGO sides. She saw one of her tasks as being to keep a focus on gender in the discussions.

NGO representatives were surprised that the content of the Rules could be negotiated rather quickly. Partly this was because many of the Rules were vague and subject to interpre-tation. Partly it was because most of the negotiators were genuinely committed to the cause of disability. Although there were some intense discussions, the main content was agreed upon early in the process.

Instead of being concerned with the content of the rules, the main debates in the Commit-tee were focused on the monitoring mechanism. The persons with disability wanted a moni-toring committee to be funded from the regular budget of the organization. Governments, especially those that were major contributors to the UN budget, were opposed to additional expense. The end result was a compromise that would not cost as much. But the representa-tives of the disability community wondered if it would work as well.

11.4 *WHAT THE RULES ARE AND WHAT THEY AREN'T*

The Standard Rules were adopted by consensus. This means that the governments accepted the content of the package as a whole even though they might have had reservations with regard to some of the elements. The Standard Rules do not have the same status as a conven-tion. In international treaty law, a convention is mandatory for all states that become party to it. The process of drafting and ratifying an international convention is lengthy. For example, it took over 15 years for the International Covenant on Civil and Political Rights to come into force because of the difficulty of many governments in ratifying it. In most countries, an inter-national convention takes on the same status as domestic law adopted by Parliament, and in most countries acceptance of an international convention means that all national laws, regu-lations, and procedures have to be brought into conformity with the convention. The Stan-dard Rules do not have this status. Governments are not legally obligated to implement the Rules' provisions. Rather, they accept a moral obligation to implement as many of the provi-sions as they can.

The Rules constitute a standard against which national policies and procedures can be measured. States that are in compliance with the Rules have "bragging rights" internationally. For states not in compliance, the Rules provide a tool for advocacy groups to embarrass governments and encourage them to bring policies, laws, and procedures into conformity with the Rules. Because the Rules are global in nature they can be used for comparison across countries. They can also be used as a basis for establishing agreements on specific items such as accessibility. It is this characteristic of the Rules that makes them particularly applicable to the issue of universal design.

The Standard Rules have a simple structure. They have five parts: (1) definitions, (2) preconditions, (3) target areas, (4) implementation measures, and (5) monitoring. Each of these has relevance to universal design.

11.5   BASIC DEFINITIONS AND PRECONDITIONS

A fundamental question for policy is the definition of *disability*. It is the basis for determining for what "reasonable accommodation" (in the sense used in the Americans with Disabilities Act) must be made. It sets the conditions around which accessibility must be designed and built.

The answers are not obvious. Does a disability have to be permanent or can it be temporary? What impediments lead to disability?

The Rules reflect a consensus that could be reached in the early 1990s. The Rules[1] define *disability* as:

> I. The term "disability" summarizes a great number of different functional limitations occurring in any population in any country of the world. People may be disabled by physical, intellectual or sensory impairment, medical conditions or mental illness. Such impairments, conditions or illnesses may be permanent or transitory in nature.

The Rules also define the term *handicap* in the following terms:

> 18. The term "handicap" means the loss or limitation of opportunities to take part in the life of the community on an equal level with others. It describes the encounter between the person with a disability and the environment. The purpose of this term is to emphasize the focus on the shortcomings in the environment and in many organized activities in society, for example, information, communication and education, which prevent persons with disabilities from participating on equal terms.

The Rules draw on the World Health Organization's International Classification of Impairments, Disabilities and Handicaps (ICIDH), which made a distinction between individual characteristics (impairments that could lead to disability) and social contexts (the consequences of these in a society). The terms used have become somewhat "politically incorrect," since they have a negative connotation and have been found, particularly by advocates for persons with disabilities, to underplay the role of the social, economic, and physical environment in guaranteeing equalization of opportunities. As a result, the WHO is in the process of revising the ICIDH.[2] The current draft uses a concept called *disablement*.[3]

Disablements are:

- Losses or abnormalities of bodily function and structure (impairments)
- Limitations of activities (disabilities)
- Restrictions in participation (formerly called handicaps)

In accordance with the social model, the ICIDH-2 also includes a list of contextual or environmental factors, including physical conditions such as climate and terrain, as well as aspects of the social and human built environment, including social attitudes, laws, policies, and social and political institutions.

The new terminology, while improving the correctness and clarity of language and making it clear that disability is more a social than a medical question, does not change the underlying definitions used in the Standard Rules, although when the Rules are revised, their definitional language will doubtless be updated.

Within both the older and newer definitions, the purpose of universal design could be seen as reducing restrictions on participation and overcoming limitations of activities.

For the Rules to be implemented, certain preconditions must exist. Four are specified: (1) awareness-raising, (2) medical care, (3) rehabilitation, and (4) support services. The logic is that unless there is awareness that dealing with disability is important, there will be little incentive for action. There must be adequate medical care to address the source of impairments and a minimum amount of rehabilitation to address limitations. There should be support services primarily in the form of assistive devices. Rule 4 states about support services that:

> States should ensure the development and supply of support services, including assistive devices for persons with disabilities, to assist them to increase their level of independence in their daily living and to exercise their rights.

11.6   ACCESSIBILITY IN THE RULES

The main substantive section of the Rules is called *Target Areas* and includes eight rules that deal, respectively, with all aspects of life in society. The first of these, Rule 5, is on accessibility. This highlights the central role of accessibility in achieving equalization of opportunities.

The basic goal of the Rules on accessibility is that:

> States should recognize the overall importance of accessibility in the process of the equalization of opportunities in all spheres of society. For persons with disabilities of any kind, States should (a) introduce programmes of action to make the physical environment accessible; and (b) undertake measures to provide access to information and communication.

Because the Rules are intergovernmentally adopted, their focus is on the role of the state rather than on society generally. They try to specify what states can do, within the limitations of state roles. Government responsibility is clearer in terms of physical environment than in terms of information and communication. Both aspects, however, are relevant to universal design. If principles of universal design are included in all governmental regulations of the physical environment in all countries, clearly accessibility will be enhanced. Similarly, if principles of universal design are included when information and communication systems are developed, accessibility will be ensured.

The Rules are precise in terms of physical accessibility, where government responsibility is clear in all countries. A sense that there is also a responsibility for accessible information is also beginning to be reflected in laws in some countries. The first mandate refers to the responsibility of governments to legislate accessibility standards. The Rule states:

> States should initiate measures to remove the obstacles to participation in the physical environment. Such measures should be to develop standards and guidelines and to consider enacting legislation to ensure accessibility to various areas in society, such as housing, buildings, public transport services and other means of transportation, streets and other outdoor environments.

When the Rules were agreed upon, not all states felt that they were in a position to agree to the use of mandatory language even on the adjustment of public areas to accessibility. Thus, the term *to consider enacting* was used. The Rules did not specify whether the standards and guidelines should be applied only to new construction or should involve retrofitting old constructions. Persons with disabilities were clear in their belief that the standards should apply to all areas within the purview of governments.

In drafting the Rule, the negotiators were aware that standards can only be implemented and enforced if those involved in design and construction were aware of them. Even where the building industry is primarily in the private sector, the government can influence compliance through communication, education, and training. The Rule specifies:

> 2. States should ensure that architects, construction engineers and others who are professionally involved in the design and construction of the physical environment have access to adequate information on disability policy and measures to achieve accessibility.

How states ensure this is left open. However, it is clear that a number of means are available, including codifying, assembling, and distributing information; ensuring that accessible design is included in the curricula of architectural and engineering schools; and including knowledge about accessibility in certifying examinations for professionals.

The Rules recognize the critical role of design in ensuring accessibility by stating that:

> 3. Accessibility requirements should be included in the design and construction of the physical environment from the beginning of the designing process.

The adoption of universal design criteria is entirely consistent with this Rule.

Finally, the Rules call for providing for participation of persons with disabilities in the design process. This language was strongly advocated by the organizations of persons with disabilities who were involved in the negotiations and its adoption reflects the resonance that the disability community's position had in government delegations. The Rule states:

> 4. Organizations of persons with disabilities should be consulted when standards and norms for accessibility are being developed. They should also be involved locally from the initial planning stage when public construction projects are being designed, thus ensuring maximum accessibility.

This Rule presupposes that there are organizations of persons with disabilities, that they can be mobilized to participate, and that a structure of openness can be achieved.

However, without this participation, designs will not necessarily be adequate. Anyone who has stayed in an "accessible" hotel room in different parts of the world can question whether persons with disabilities were consulted when the designs were made, or whether a wide range of persons with disabilities was surveyed about their needs.

The accessibility norms regarding communication are focused on ensuring that information presentation designs accommodate various disabilities. The Rule states that:

> 6. States should develop strategies to make information services and documentation accessible for different groups of persons with disabilities. Braille, tape services, large print and other appropriate technologies should be used to provide access to written information and documentation for persons with visual impairments. Similarly, appropriate technologies should be used to provide access to spoken information for persons with auditory impairments or comprehension difficulties.
>
> 9. States should encourage the media, especially television, radio and newspapers, to make their services accessible.

10. States should ensure that new computerized information and service systems offered to the general public are either made initially accessible or are adapted to be made accessible to persons with disabilities.

Implied in the Rule is that whenever a design involves communication or infrastructure, it should include accessibility criteria. The case of elevators that include Braille control markings and audible indicators of floors, as well as controls that are at the level of persons using wheelchairs, is an example.

In other target areas, specific accessibility norms that are relevant for universal design are included. Rule 7, on employment, states that government action programs should include:

(a) Measures to design and adapt workplaces and work premises in such a way that they become accessible to persons with different disabilities;
(b) Support for the use of new technologies and the development and production of assistive devices, tools and equipment and measures to facilitate access to such devices and equipment for persons with disabilities to enable them to gain and maintain employment;

Similarly, Rule 10, on culture, says that:

2. States should promote the accessibility to and availability of places for cultural performances and services, such as theatres, museums, cinemas and libraries, to persons with disabilities.
3. States should initiate the development and use of special technical arrangements to make literature, films and theatre accessible to persons with disabilities.

And Rule 11, on recreation and sports, specifies that:

States should initiate measures to make places for recreation and sports, hotels, beaches, sports arenas, gym halls, etc., accessible to persons with disabilities. Such measures should encompass support for staff in recreation and sports programmes, including projects to develop methods of accessibility, and participation, information and training programmes.

Reflecting the nature of agreements that could be reached a decade ago, the accessibility norms in the Rules emphasize accessibility to public spaces but are silent on norms and standards that can be applied in, for example, private housing or nonpublic aspects of the private sector. These are areas where implementation of universal design would be encouraged only indirectly by the Rules.

11.7   IMPLEMENTATION MEASURES

The part of the Rules dealing with implementation measures includes a number of norms that have relevance for universal design and its implementation. They provide a normative incentive for everyone to take the situation of persons with disabilities into account when making policy or decisions. They also emphasize the importance of action at the local level, and the need to support local programs by preparing and disseminating material on accessibility, among other things.

Rule 14, on policymaking and planning, states:

4. The ultimate responsibility of States for the situation of persons with disabilities does not relieve others of their responsibility. Anyone in charge of services, activities or the provision of

information in society should be encouraged to accept responsibility for making such programmes available to persons with disabilities:

5. States should facilitate the development by local communities of programmes and measures for persons with disabilities. One way of doing this could be to develop manuals or checklists and provide training programmes for local staff.

Another implementation measure suggests the utility of assembling public funds or using public policies to increase the incentives for using accessible design. Rule 16, on economic policies, says that:

3. States should consider the use of economic measures (loans, tax exemptions, earmarked grants, special funds, and so on) to stimulate and support equal participation by persons with disabilities in society.

4. In many States it may be advisable to establish a disability development fund, which could support various pilot projects and self-help programmes at the grass-roots level.

Implementing the international norms in the context of universal design involves more than knowing the Standard Rules. It involves using them to obtain leverage to influence public policy and private attitudes in favor of actions that will improve accessibility. This means taking advantage of two elements in the Rules: (1) participation of persons with disabilities, and (2) the monitoring mechanism.

Participation of Persons with Disabilities

The strongest proponents of the Rules have been organizations of persons with disabilities. They were actively involved in the drafting process and have been among those most concerned with implementation. They have the greatest stake in seeing to it that universal design principles are adopted.

The Rules throughout call for participation of persons with disabilities in all aspects of implementation. For this to happen, however, efforts must be made to strengthen and support organizations of persons with disabilities both nationally and internationally. Generally speaking, organizations of persons with disabilities are not well-funded. Ensuring that such organizations are given a voice in design issues, where their added value in terms of improved design is clear, is one way of increasing their support and importance.

Monitoring Mechanism

International human rights treaties inevitably have monitoring mechanisms that are instruments for implementation. Usually this is in the form of an expert committee that reviews reports of states that are parties to a given convention as an inducement to comply with its provisions. In this, the potential for international embarrassment when there is noncompliance is one of the strongest means of inducing states to implement their obligations.

Because the Standard Rules are not a treaty, no provision was made for a standing expert committee to monitor states for compliance. As noted earlier, the kind of monitoring mechanism that would be created was the part of the Rules that took the longest to negotiate. In the end, a compromise was reached. The Commission for Social Development would be the formal monitoring body, but a Special Rapporteur would assist it. The Rules state:

2. The Rules shall be monitored within the framework of the sessions of the Commission for Social Development. A Special Rapporteur with relevant and extensive experience in disability issues and international organizations shall be appointed, if necessary, funded by extrabudgetary resources, for three years to monitor the implementation of the Rules.

The institution of special rapporteur is an old one in United Nations human rights practice. It is based on the notion that an individual expert can undertake fact-finding to provide an independent basis for intergovernmental bodies, especially the Commission on Human Rights, to make determinations about human rights situations. It has been applied from time to time in other specialized subsidiary bodies and its application to the Standard Rules was a logical compromise.

While the Special Rapporteur was initially only mandated for three years, his term has already been extended once by the Economic and Social Council to August 2000.[4] The Special Rapporteur's responsibility is:

> 5. The Special Rapporteur shall send a set of questions to States, entities within the United Nations system, and intergovernmental and non-governmental organizations, including organizations of persons with disabilities. The set of questions should address implementation plans for the Rules in States. The questions should be selective in nature and cover a number of specific rules for in-depth evaluation. In preparing the questions the Special Rapporteur should consult with the panel of experts and the Secretariat.
>
> 6. The Special Rapporteur shall seek to establish a direct dialogue not only with States but also with local non-governmental organizations, seeking their views and comments on any information intended to be included in the reports. The Special Rapporteur shall provide advisory services on the implementation and monitoring of the Rules and assistance in the preparation of replies to the sets of questions.

The Special Rapporteur has reported annually to sessions of the Commission and has made recommendations on a number of aspects of the Rules. He has not focused specifically on issues of universal design.

One difficulty is that the Special Rapporteur is funded from extrabudgetary sources, which has made the office somewhat uncertain. He has a mandate to organize an advisory group made up of representatives designated by the main international organizations of persons with disabilities, but recently this group had difficulty meeting because of a lack of resources. Its most recent meeting took place before the 2000 session of the Commission for Social Development, in February 2000.

11.8   CONCLUSIONS

There is a growing consensus among organizations of persons with disabilities that the Standard Rules, while providing a good starting point for directing public policy toward equalization of opportunities, are not sufficient to provide the necessary universal guidance. Two alternatives have been suggested. The first is to revive the idea of drafting a human rights convention on the rights of persons with disabilities that would contain, in an obligatory manner, the provisions of the Standard Rules. An effort to include this possibility in the resolution on disability adopted by the Commission for Social Development in February 2000 was not successful. The second is to revise the Rules in the light of experience to incorporate those elements that are missing. For example, technological changes in the information field, particularly those relating to the Internet, require more detailed guidance than the rules currently provide. Further guidance would be useful also if universal design criteria were to be applied to all countries. The next review of the Long-term Strategy to Implement the World Programme of Action concerning Disabled Persons to the Year 2000 and Beyond, which is scheduled for 2002, should begin to answer some of these questions and determine future directions, including possible revision of the Standard Rules.

The monitoring mechanism that was extended for a further two years in 2000 will also be reviewed during 2002 and either the existing method, based on a Special Rapporteur, or a new approach will be decided upon. In either case, it will be important to ensure that this mechanism is used actively to promote implementation of the accessibility elements of the Standard Rules, including promotion of universal design approaches.

For the moment, the Standard Rules will have to be used as an advocacy tool and largely self-monitored. The main actors in this will be persons with disabilities through their organizations, as well as government offices concerned with equalization of opportunities. To support this effort, the United Nations has organized or supported a number of seminars and workshops. These have included training in the use of the Standard Rules, examination of the meaning of the Rules and advocacy strategies. Recent examples include the Seminar on Internet Accessibility and Persons with Disabilities: an ASEAN Perspective, held in Bangkok in July 1999; the Training Seminar for Central America and the Spanish-Speaking Caribbean on the Standard Rules, organized by Disabled Peoples' International in Santo Domingo in April 1998; and the Interregional Seminar and Symposium on International Norms and Standards Relating to Disability, organized by the Equal Opportunities Commission, Hong Kong SAR, in cooperation with Hong Kong University, Faculty of Law, Centre for Comparative and Public Law, in December 1999. In addition, a number of nongovernmental organizations, like the World Blind Union, have organized their own training programs on the Rules. Both the United Nations and a private initiative, WorldEnable, post information about the Rules on the Internet.

Persons interested in both universal design and the implementation of international norms for equalization of opportunities for persons with disabilities should make an effort to understand the Standard Rules and find ways to apply them to their particular circumstances.

11.9   NOTES

1. The Rules are found in General Assembly resolution A/RES/48/96 and are reproduced in the United Nations Website at www.un.org/esa/socdev/dissre00.htm. United Nations documents are organized in terms of numbered paragraphs, and references to the specific Rules are made in those terms.

2. The revision of a WHO classification is a long and laborious process. The current state of the ICIDH-2 can be found at www.who.int/icidh/brochure/content.htm.

3. www.who.int/icidh/brochure/concepts.htm

4. Since the adoption of the Rules in 1993 there has only been one Special Rapporteur, Bengt Lindquist of Sweden, a former parliamentarian and leader of organizations of persons with disability. His term of office was recently extended for an additional two years, to 2002.

11.10   RESOURCES

Disabled Peoples' International.   www.escape.ca/~dpi/

Persons with Disabilities, United Nations.   www.un.org/esa/socdev/enable/

CHAPTER 12

U.S. ACCESSIBILITY CODES AND STANDARDS: CHALLENGES FOR UNIVERSAL DESIGN

John P. S. Salmen, A.I.A.
Universal Designers and Consultants, Inc., Takoma Park, Maryland

12.1 INTRODUCTION

The United States has been a crucible for the initiation of the universal design movement. The Americans with Disabilities Act has brought great attention to the concept of accessibility. Some people would like to see a similar regulatory approach applied to universal design; however, the lessons learned in the United States indicate that effective implementation of accessibility codes and standards is dependent upon the education of and the availability of appropriately formatted information to those who must create or implement the design. This chapter highlights the regulatory processes that result in prescriptive standards, contrasts these with performance criteria, and identifies the challenges facing designers and developers in an increasingly global economy.

12.2 BACKGROUND: ACCESSIBILITY VERSUS UNIVERSAL DESIGN

There is a profound difference between universal design and accessibility. *Accessibility* is a function of compliance with regulations or criteria that establish a minimum level of design necessary to accommodate people with disabilities. *Universal design,* however, is the art and practice of design to accommodate the widest variety and number of people throughout their life spans. As more is learned about human needs and abilities, and as technologies develop, the practice of universal design improves, evolves, and changes. In truth, it might be better to think of this field as "universal designing," so as to focus on the decision-making process, rather than some end product that may be improved in the future. The term *universal designing* was coined in 1997, by Dr. Edward Steinfeld, AIA, during a meeting of faculty and advisors to the Universal Design Education Project, in Boston, Massachusetts. The more designers know about users, the better they can design. But designers will probably never be able to know it all. This reality demands collaborative efforts between designers, environmental decision makers, and users as the diversity and complexity of our global society increases. In this fluid context, it is undesirable to establish fixed criteria for what constitutes a universal design. Instead, universal design should be considered a receding horizon line toward which good design will forever advance.

12.3   *HISTORY OF ACCESSIBILITY CODES IN THE UNITED STATES*

Accessibility codes are a relatively new phenomenon, appearing only in the last 40 years of the twentieth century. The first accessibility technical standard in the United States was ANSI A117.1, *Specifications for Making Buildings and Facilities Accessible to, and Usable by the Physically Handicapped,* published in 1961 (see Chap. 25, "The Bottom-Up Methodology of Universal Design," by Goldsmith). While ANSI A117.1 (1961) was not immediately adopted, it became the model for nearly all the accessibility codes and regulations that followed.

Accessibility standards were a direct outgrowth of the changing demographics that followed advances in medical technology. Medical technology allowed many people to live who otherwise would have died in World War II, the Korean War, and the polio epidemic of the early 1950s. For the first time in history, there was a large population of young Americans using wheelchairs. Public and private efforts for veterans and other survivors provided rehabilitation and educational opportunities, resulting in a large, well-educated population of young people with disabilities who would not accept discriminatory practices that limited their participation in society. They took advantage of the changing landscape of American politics and opinion, inspired by the civil rights movement of the 1960s that culminated in the passage of the landmark U.S. Civil Rights Act of 1964. That was the first legislation prohibiting discrimination based on race; it established the foundation for future civil rights–based laws. With technical standards available and the nondiscrimination concept affirmed, accessibility regulations development began.

Laws, Regulations, and Standards

In the United States, there are three formal levels of guidance that affect the accessibility of the built environment—laws, regulations, and standards.

Laws are promulgated by a legislative body such as the U.S. Congress or a state legislature to address a public concern. The Americans with Disabilities Act (ADA) was a law passed by the U.S. Congress to address discrimination against people with disabilities. Laws are general rules that require more specific regulations to be developed, implemented, and enforced. The law usually identifies an agency or agencies responsible for the development and/or enforcement of the regulations. The ADA directed the U.S. Architectural and Transportation Barriers Compliance Board (Access Board) to develop technical criteria for accessibility compliance and the U.S. Department of Justice (DOJ) to write the implementing regulations and adopt specific standards for enforcement.

Regulations are usually created by an enforcement agency of a government entity. Regulatory agencies are required to consider public input in their processes, and are encouraged to adopt in part or in whole existing recognized standards from the public sector. For example, the regulations to implement Title III of the ADA were published by the DOJ (28 CFR Part 36, 1991), detailing the procedures to prohibit discriminatory practices. The regulation also requires that certain places "be designed, constructed and altered in compliance with accessibility standards established by this part."

Standards are technical criteria surrounding an issue area. Standards that have received broad input during their development are considered to be more dependable and authoritative. One balanced process that ensures consideration of all input is called *consensus-based standards making,* in which all affected parties have input and a vote on the standard. The technical regulations that the DOJ adopted were heavily based on standards that were developed in a consensus process approved by a private organization, the American National Standards

Institute (ANSI). It is important to note, however, that a standard is not enforceable until it is referenced or adopted into a regulation. For example, the Access Board developed the ADA Accessibility Guidelines (ADAAG) based on the ANSI A117.1 consensus-based standard. The ADAAG was not enforceable until it was adopted by the DOJ as the appendix to the ADA Title III regulations and became the ADA Standards for Accessible Design (ADA Standards).

ANSI A117.1—Where It All Started

After the development of the first ANSI Standard in 1961, technical criteria of accessibility were incrementally refined and expanded over time, through a back-and-forth process between private and federal agencies. On the private side, the ANSI A117 Standard was referenced or adopted for most of the construction regulations known as *model building codes*. On the federal side, the ANSI Standard was also used as a basis for regulations, including the Minimum Guidelines and Requirements for Accessible Design (MGRAD) in 1981; the Uniform Federal Accessibility Standard (UFAS) in 1985; and the ADA Accessibility Guidelines (ADAAG) in 1991.

The Architectural Barriers Act of 1968 was the first federal law to reference the ANSI Standard and lay the foundation for accessibility criteria in a series of other federal laws, including Sec. 504 of the Rehabilitation Act of 1973, the Fair Housing Amendments Act of 1988 (see Chap. 35, "Fair Housing: Toward Universal Design in Multifamily Housing," by Steinfeld and Shea), and eventually the Americans with Disabilities Act (ADA) of 1990, which collectively require accessibility in most newly constructed public, private, and commercial buildings.

ANSI Versus ADAAG

A major effort was undertaken from 1994 to 1996 by the ANSI A117 Committee and the ADAAG Review Federal Advisory Committee established by the Access Board, which were simultaneously revising both the ANSI A117.1 Standard and the ADAAG. They worked together to substantially harmonize the format, organization, and technical requirements of both the ADAAG and ANSI A117.1. This resulted in a 1998 version of the ANSI Standard and a corresponding set of proposals for revisions to the ADAAG, which were ultimately used only in part for the version of ADAAG 2000 proposed by the Access Board in late 1999. (The Access Board guidelines are included on the CD accompanying this handbook.)

State and Local Building Codes

Disability and accessibility advocates have not limited their efforts solely to the national level. Throughout the United States, state and local code officials have grappled with the needs and demands of citizens with disabilities to ensure access through state and local building codes and statutes. During the 1970s, state and municipal building authorities adopted their own versions of what they thought was necessary to make buildings accessible. Unfortunately, almost no two versions were alike, and the plethora of regulations led to a design and construction nightmare in the 1970s and 1980s when developers, designers, and manufacturers were unable to standardize any design details that would remain the same across state borders and sometimes between localities within the same state.

In the early 1990s, the differences between accessibility codes were somewhat reduced by the further development of later versions of ANSI A117.1 and the introduction of the ADAAG in 1991. Most states and municipalities have adopted one of the two, often through adoption of a model building code which uses ANSI A117.1. However, at the turn of the millennium, it is still impossible for national developers to cost-effectively design buildings to one single standard, because in almost every instance, states and municipalities have added local amendments and interpretations to "enhance" their access requirements. The unfortunate result of these enhancements is the creation of some ambiguity and conflict. For instance, Florida requires a lavatory in the accessible toilet stall, which conflicts with the clear floor area beside the toilet. Although there is a process to determine the appropriate standard to use when standards conflict, the resolution is not simple since it is often difficult to determine which criteria is more stringent or more beneficial to people with disabilities. For example, the state of Minnesota requires a vertical grab bar above the horizontal side bar beside accessible toilets. Does this bar obstruct the useful area above the horizontal grab bar, or does it assist semiambulatory people in getting up and down from a seated position?

How Minimums Become Maximums

Accessibility regulations typically establish minimum criteria which are often perceived as or become absolutes, or maximums, in the building industry. Often, when facing compliance with a regulation that is not part of the intended development program, the tendency on the part of building owners or developers is to provide no more space, material, or equipment than is needed. Designers are subsequently directed to be as efficient as possible and to design no more than the regulations require. This forces the designers to design the facility to the minimum requirements possible, resulting in the minimums becoming the maximums.

12.4   THE ADA

The ADA has two sections directly affecting the construction of buildings: Title II—State and Local Governments, and Title III—Public Accommodations and Commercial Facilities. Neither of these sections uses the term *universal design,* but both try to encourage designs that meet the needs of all people by setting a starting point with minimum criteria for accessible design.

Title II sets out regulations that assure nondiscrimination in the provision of services. These include requirements for new construction and alterations of state and municipally owned facilities. The regulatory agency, DOJ, allows designers to use either the UFAS or the ADA Standards until such time that Title II ADA Standards are promulgated. The final version of the ADAAG 2000 is likely to include the Title II ADA Standards.

Title III sums up the accessibility requirements for buildings that are open for commercial transactions with the public in the following statement: "Prohibition of discrimination. No individual shall be discriminated against on the basis of disability in the full and equal enjoyment of the goods, services, facilities, privileges, advantages or accommodations of any place of public accommodation by any private entity who owns, leases, (or leases to) or operates a place of public accommodation" [28 CFR Part 36.201 (a)].

Title III requires that any new construction, alteration, or addition to a building that affects commerce be accessible as defined by the ADA Standards. It also states that businesses that provide services to the public—places of public accommodation—remove barriers to the extent that doing so is "readily achievable,"—that is, "easily accomplishable and able to be carried out without much difficulty or expense" [28 CFR Part 36.304 (a)]—or, in other words, cheap and easy.

Alterations trigger one of the more commonly overlooked and confusing aspects of the ADA, known as the *path of travel rule.* This rule requires that up to 20 percent of the cost of an alteration be expended to ensure that there is an accessible path of travel from the property line to the altered area. Above the 20 percent, this cost would be considered a *disproportionate* expense, and thus the regulation is sometimes called the *disproportionality rule.* The rule requires that the rest rooms, drinking fountains, and telephones that serve the affected path of travel be made accessible, to the maximum extent feasible. This can produce dramatic and potentially costly additions to alteration projects. For example, if a hotel plans to spend $100,000 to renovate a ballroom, it would be required to spend up to an additional $20,000 to upgrade the rest rooms, drinking fountains, and telephones serving the ballroom, as well as those serving the path of travel from the property line to the ballroom (including parking, curb ramps, entry doors, and elevators) if those elements do not presently meet all the criteria for new construction found in the ADA Standards. The prioritization of barrier removal within the 20 percent additional cost is left up to the building owner, with recommendations from DOJ, that entry be addressed first, followed by primary function areas, bathrooms, and other features in that order. Exceptions that are typically not considered to be part of an alteration and therefore do not trigger the path of travel requirement include HVAC system modifications, lighting, fire suppression systems, roofing, and finish materials.

Civil Rights Versus Building Codes

The ADA is a civil rights law. The civil rights terminology of the ADA states that all people, regardless of their abilities, should have access to goods and services provided by businesses. The regulations promulgated by the U.S. DOJ and the technical assistance for voluntary compliance are intended to provide guidance to building designers and owners on how to meet the broad civil rights requirements of the law. This places a difficult responsibility on designers and building owners, who, for the most part, have never studied law or civil rights interpretations.

Designers and building owners are responsible for providing facilities and spaces that are safe for the public to use, and they rely on the building codes to guide them in this mandate. The civil rights nature of the ADA and the resemblance of its specific criteria to building codes lead many to believe that compliance with the letter of the law will ensure compliance with the law. Sometimes, unfortunately, this is not the case. When faced with ADA Title III lawsuits, U.S. courts are often turning to the spirit rather than the letter of the law (see Chap. 18, "Interpreting Accessibility Standards: Experiences in the U.S. Courts," by Mazumdar and Geis).

The problem is exacerbated by the fact that the ADA enforcement agency, the U.S. DOJ, does not provide definitive interpretations unless legal action is underway. Even if a building code has been certified by the DOJ as meeting or exceeding the minimum requirements of the ADA Standards, local building officials are not empowered to allow variances from the design criteria. Furthermore, these officials' interpretations can be challenged if the DOJ, in responding to a complaint, believes that the ADA civil rights requirements have not been met. However, if a DOJ-certified building code is followed, that fact can be used as rebuttable evidence if a complaint is filed.

This situation makes the practice of architecture and the operation of buildings increasingly difficult and costly. There is little certainty of compliance, outside of literal adherence to the prescriptive criteria, and great risk of legal action by anyone who feels that the design or operation of the facility creates an environment that limits their "equal enjoyment" and thus violates their civil rights. In the cases of *PVA v. Ellerbe Beckett* and *Lara et al. v. Cinemark USA, Inc.*, designers and developers who had followed the letter of the law in the design of arenas and stadium theaters (and, in the Cinemark case, had complied with a DOJ-certified building code) were sued by disability advocates who maintained that the location of the seats did not provide a "comparable line of sight" to that provided to standing spectators in the arena and most of the other attendees in the cinemas.

12.5 PERFORMANCE VERSUS PRESCRIPTIVE CRITERIA

For many years, a debate has been ongoing in the U.S. design and construction community regarding the best format for building codes—performance or prescriptive. Simply stated, *per-*

formance codes identify the ultimate operation or function of an element or space, whereas *prescriptive codes* explain to a greater or lesser degree how a space or element must be designed to satisfy the codes. For example, a performance code might say that a wall assembly must be able to withstand a fire for a specific period of time. A prescriptive code would say that the wall must be constructed of specific materials organized in a specific design, such as two layers of ½-in gypsum wallboard over a 4-in metal stud wall.

It is generally easier to write performance criteria because the concept of how an element or space should operate is fairly easily described. However, it is generally more difficult to evaluate and enforce performance criteria because the designer or building official must understand how all the individual elements work and how they work with each other. As building technology becomes increasingly sophisticated, materials and assemblies of materials may have characteristics that are not readily apparent or easily understood.

Prescriptive codes, on the other hand, are very tedious to develop and write because the language must be exact and yet applicable to a broad variety of situations. Prescriptive codes are easier to follow and enforce because they require less understanding of the function or purpose of the design and rely instead on evaluation of whether the design complies with the written letter of the law.

U.S. accessibility codes have tended to be prescriptive in nature. Disability advocates have argued that there is a large information gap between people with disabilities and the design and construction industry, regarding the needs and abilities of people with disabilities. This gap makes it difficult, they argue, for designers to produce designs that meet the intent of performance-based codes. Disability advocates also prefer prescriptive standards because they give clear evidence of compliance or discrimination. This argument is also presented by building officials who desire easily interpreted and therefore easily enforceable standards.

Unfortunately, prescriptive criteria can stifle design creativity. Where the design of a facility or element is rigidly specified, there is little opportunity for investigation of alternative or innovative designs, or use of new materials or products, without risking misinterpretation and noncompliance. Although there are sections in most accessibility codes that allow equivalent designs, there is no generally accepted methodology to determine what is equivalent. Section 2.2 of the ADA Standards accepts equivalent facilitation, which is stated as follows: *"Departures from particular technical and scoping requirements of this guideline by the use of other designs and technologies are permitted where the alternative designs and technologies used will provide substantially equivalent or greater access to and usability of the facility."* Few building owners or design firms have the knowledge base to be confident in attempting an alternate design.

12.6   LACK OF A STATISTICAL ERGONOMIC DATABASE

The specificity of the level of detail in the technical criteria of U.S. accessibility standards might be justified if there was a substantial body of ergonomic data upon which they were based. However, this is not the case. The criteria in the standards are largely based on studies of relatively small population samples. The research conducted at Syracuse University in the late 1970s for the ANSI A117.1 standard used a sample of approximately 200 individuals that included 58 wheelchair users (Steinfeld, 1979). Most of the criteria in the present ANSI and the ADA Standards are direct descendants of that small study group.

Although the body of statistical data is limited, the attempt to develop accessibility standards has gone on unhindered. The process has attempted to overcome this limitation by using the best judgments of groups of experienced and interested individuals. In the absence of more recent definitive statistical information, regulators have often based their decisions on personal experiences influenced by anecdotal evidence presented at regulatory committee meetings.

The development of dependable statistical ergonomic data should be the single most important goal for those involved in the development of accessibility criteria. The Access Board and the National Institute on Disability and Rehabilitation Research (NIDRR) were able to fund and initiate efforts in this direction in the late 1990s. With this data it will be less difficult to establish more reliable standards and to justify alternative solutions that meet the intent of equivalent facilitation and performance-based language.

12.7 THE FUTURE OF ACCESSIBILITY STANDARDS

There are many problems with existing accessibility standards, and it is questionable whether they will ever encourage designers to practice universal design in their present form. However, if one thing is clear, it is that because of the growing population of aging persons, many of whom have some type of disability, combined with the need for access in all areas, accessibility regulations will, in all likelihood, increase in the coming years. It is likely that there will be a continued proliferation of specific criteria aimed at specialized areas of the environment, extending beyond the scope of new residential and commercial buildings to include recreation and leisure activities, travel and tourism venues, and more historic and cultural sites. The Access Board is in the process of developing guidelines for cruise ships, wilderness trails, sports fields, playgrounds, swimming pools, and other similar facilities. There is no reason to think that this direction of their efforts will change.

International Consistency

As international communication and travel increase, it will become increasingly difficult to justify national or regional differences in accessibility criteria. U.S. travelers with disabilities already expect that multinational hotel chains will provide the same level and quality of accessibility in their properties in foreign countries as they do in the United States, whether it is required by building codes in those countries or not. The marketplace is demanding accessibility at a rate that is outstripping accessibility standards. Businesses that wish to capture this growing market of aging consumers recognize that they must provide a consistent level of quality across their chains from Bangor to Bangkok.

It is inevitable that regional differences in accessibility criteria will start to vanish as medical technology and commercial products become more uniformly distributed. The sooner that accessibility regulators around the world recognize this, and begin to pool their efforts and resources to establish researched and ergonomically-based performance criteria, the sooner accessibility will proliferate around the globe.

Performance standards can transcend national boundaries, whereas prescriptive criteria fail to allow for regional, if not local, differences. And, most important, performance standards can act as an intermediate step to the development of an understanding of how to create universal designs.

12.8 CONCLUSION

It is the author's opinion that most building designers, developers, and construction professionals lack an understanding of the changing needs and abilities of our society and thus of how to develop appropriate universal design solutions. The present generation of designers and facility owners and operators has had little involvement with and, consequently, little understanding of the needs of the growing populations of aging people or of people with disabilities. The problem would be easy if only one or two groups had responsibility for environmental decisions that impact users. With the civil rights nature of the issue, however, the actual population of environmental decision makers expands dramatically, as does legal liability. Maintenance workers can replace a moveable trash receptacle to a position where it blocks an accessible route, thus negating the carefully designed plans of the architect. A new sales clerk can inadvertently violate the civil rights of a person with a disability by placing a promotional display on the lowered portion of a cashier counter required for use by people of short stature or customers who use wheelchairs.

Further regulation will never resolve these types of problems. The solution is appropriately targeted educational materials and programs. Architects and sales clerks need very different kinds of information and education, yet they both need education. Designers of amusement parks need information that is different from that needed by designers of multifamily housing. Naval architects attempting to design more accessible cruise vessels have concerns and constraints that are vastly different from those of landscape architects who are designing interpretive trails in a wilderness forest.

While basic principles, such as the turning radius of an average wheelchair, may be the same everywhere, the information needed by individuals who are expected to apply that information is very different. Anyone who has studied the ADA Standards will recognize that parts of them are very difficult to understand, even for people who are very familiar with the document. Finding all the information that is appropriate for the specific situation with which one is concerned can sometimes be very difficult and frustrating, especially when an obscure though important section is pointed out after the design is constructed. With our increasingly sophisticated means of communications, it should soon be possible to format and deliver appropriate types and levels of information for each person along the spectrum of environmental decision making. When designers needs technical criteria or examples, they should be cross referenced and easily available, or when maintenance workers are replacing a lavatory, they should be able to easily find the correct mounting height without having to search through volumes of technical criteria. Unfortunately, here at the beginning of the twenty-first century, such interactive and customized formats are still limited in availability and not widely disseminated. Although there are some accessible design CAD programs used in architectural offices, and some pocket guides for use in the field, many more targeted, technologically appropriate, and user-friendly tools need to be developed. Once society begins to develop appropriate informational and educational tools, there will be more widespread application of accessibility, and eventually a demand by consumers for the superior performance of universal design will overtake the need for prescriptive standards.

12.9   BIBLIOGRAPHY

Council of American Building Officials, and American National Standards Institute, CABO/ANSI A117.1-1992, *American National Standard: Accessible and Usable Buildings and Facilities,* ANSI, New York, 1992.

Lara et al. v. Cinemark USA, Inc., District of Texas EP-97-CA-502-H.

Steinfeld, E., "Accessible Buildings for People with Walking and Reaching Limitations," U.S. Department of Housing and Urban Development, Office of Policy Development and Research, Washington, DC, 1979.

U.S. Department of Justice, "Nondiscrimination on the Basis of Disability by Public Accommodations and in Commercial Facilities; Final Rule," *Federal Register,* 56(144): 35544–35961, 1991.

12.10   RESOURCES

Universal Design Newsletter.   Published quarterly by Universal Designers & Consultants, Inc., Takoma Park, MD 20912.

Universal Design Online.   www.UniversalDesign.com.

CHAPTER 13

ACCESSIBLE HOUSING IN FIVE EUROPEAN COUNTRIES: STANDARDS AND BUILT RESULTS

Jon Christophersen, B.Arch.(Hons.)
Norwegian Building Research Institute, Oslo, Norway

13.1   INTRODUCTION

This chapter presents some results of a study of barrier-free dwellings in Norway, Scotland, England, Germany, and Italy (Christophersen, 1997). As a background, the study looked into housing policy aims and goals, particularly with regard to people who are disabled and elderly, as well as the existence, content, and ways of implementing standards, statutory requirements, and guidelines. The main conclusion of the study is that, broadly speaking, the built results conform to the ideals of universal design, not as a stated aim but as a consequence of creating housing that conforms to long-established housing traditions.

13.2   BACKGROUND

Past Studies

A number of studies have listed and compared European standards, statutory requirements, and guidelines, which in many cases have been in existence for a number of years, (e.g., see Ambrose, 1997; le Franc, 1992; Pickles, 1996; Sheridan, 1999). In summing up the studies, it appears that interest has been directed mainly toward the technical contents and the applicability of standards, rather than the actual built results. Any attempts to identify or evaluate the design of the buildings, whether in general terms or in the context of universal design, seem to be lacking. This may be characterized as a major problem, since it is the buildings that provide the lasting, most easily perceived and commonly experienced result of the efforts to create accessible environments.

Problems of Comparing Diverse Countries

There are a number of obstacles in such a comparative study. The legal aspects of building and the ways and means of dwelling procurement vary considerably from country to country.

There are also considerable and obvious differences in climatic conditions, in the way of life, in prevailing construction methods, and in traditional types of housing and dwelling plans. For example, the harsh Norwegian winters and the wet, windy conditions in Scotland contrast with the hot Italian summer. As for ways of building, the Italian multistory blocks of flats, which have dominated the public housing sector since World War II, have little in common with the single- and two-story timber-frame houses that prevail in Norway. Furthermore, Norway has hardly any public housing of the type found in the other four countries.

Nevertheless, enough similarities exist to be able to make meaningful comparisons. The basic human requirements for dwelling functions (e.g., living, eating, sleeping, etc.) are accommodated in the same ways in all five countries. Furthermore, the strength and techniques required of wheelchair users in order to negotiate the built environment are dependent on a combination of human conditions and wheelchair technology, which hardly differ among the five countries in this study.

13.3 MAIN ISSUES—SIMILARITIES AND DIFFERENCES

Four main points are listed in Table 13.1. The following paragraphs deal with each in more detail.

TABLE 13.1 Differences and Similarities of Legal Systems and Practice

Differences	Similarities
Legal systems and the ways and means of implementing barrier-free design have little in common.	Almost identical political policies, intentions, theories, and main aims of barrier-free planning.
Dwelling types and layouts differ greatly (but conform to traditional house and dwelling plans in each country).	The specifications for barrier-free solutions are similarly structured and similar in content; only the details differ.

Almost Identical Theoretical Basis and Political Policy Statements

The dominating theoretical approach is the same everywhere. Building on the commonly used gap model and the general aims of universal design, it stresses the relationship between individual capabilities and the built environment. Lowering or doing away with architectural barriers is seen as a way to make the built environment more functional for everybody. [For an illustration of this principle, see Aslaksen, Bergh, Bringa, and Heggem (1997).] Politically, the main aims of the central authorities are normalization, integration, equality, and high quality of life regardless of age or capability. Thus, there are practically no differences of political rhetoric, stated aims of policy, or the governing principles in any of the five countries. Their governments seem equally aware of the need for nonhandicapping environments in order to minimize the social costs and preserve the quality of life for an aging population. The proportion of elderly people in the population is expected to rise dramatically everywhere. That elderly and disabled people should have the choice to live in their own and ordinary homes, as well as in ordinary dwelling areas, is generally accepted as the best solution.

A Diversity of Organizational Structures and Systems

Legal systems and the ways in which housing is provided make up a framework for the implementation of barrier-free design. The variations, regarding both legal systems and housing provision, are considerable. It is, however, true to say that where statutory requirements exist, they are usually given in the form of performance criteria with reference to national standards or other technical specifications. The Italians, in particular, have produced a large number of guidebooks with detailed technical specifications to explain their statutory requirements [see Prestinenza, 1992; and Chap. 26 (by Fabrizio Vescovo) in this handbook]. In most guidelines and schemes for incentives, clearly stated technical requirements explain and exemplify how performance criteria may be satisfied.

- The Italian requirements for barrier-free design are laid down by the central authorities and apply to the whole of the country. The most rigorous of the accessibility requirements apply only to public housing—5 percent of all new, public-sector housing must be fully accessible, whereas private developments are treated more leniently. However, Italy has no comprehensive building code or building regulations. Instead, the legal requirements are contained in a succession of different laws and ministerial decrees. The latter have the same status as building regulations in other countries. Fundamental to the Italian thinking is a system of three levels of accessibility, given in law 13/1989 and decree 236/98:

  1. *Accessible.*   A wheelchair user should be able to enter, move about, and use all functions of a building unaided.
  2. *Visitable.*   A wheelchair user should be able to enter the building and the individual apartments, and be able to access the bathrooms and the living room.
  3. *Adaptable.*   No part of the building has to be accessible, but the planner must show that the building can be made accessible at a later date.

  Chapter 15, "Accessible Design in Italy" (by d'Innocenzo and Morini), describes the Italian approach in more detail.

- The German system, being designed to work in a federation of states, is almost totally decentralized, with the responsibility for upholding standards and setting demands being left to the individual German states (Länder). Technical specifications for accessibility are, however, contained in two German standards documents: the Deutsche Industrie Norm (DIN) 18024 and 18025. These norms are applicable to all German states that elect to use them. With the possible exception of Berlin, the state of Hesse seems to be the one that has the most comprehensive sets of legal requirements for barrier-free housing. The German case in this study is, therefore, limited to Hesse. The Hesse building code (Bauordnung) requires barrier-free access and elevators in buildings that have more than four stories. In addition, a set of legally binding technical guidelines, "Technische Wohnungsbau-Richtlinien," sets rigorous standards for public housing, including barrier-free design according to the DIN norms in at least a part of all public housing. In response to the expected aging of the population, the Germans have carried out a number of large research and evaluation studies.

- In Norway, economic incentives for barrier-free dwellings are provided by the Norwegian State Housing Bank. The economic incentives for accessibility are conditional upon compliance with the Bank's requirements, which are firmly based on recommendations from the Norwegian Society for the Handicapped and the Norwegian Building Research Institute. The economic incentives have proved a successful way of promoting the Norwegian life span standard over the last 15 years, but seem to have had less success recently. This system of state financing is peculiar to Norway—the Bank has financed some 80 percent of all new dwellings in Norway since the World War II. A central building code and regulations exist, but the statutory requirements for accessibility have marginal effect on housing design—at least for the dominating types of construction, since the bulk of the statutory requirements apply to multi-story housing. Although relevant in urban settings—particularly the city centers—most new construction in Norway has been in the form of low-rise, timber-frame buildings for a number of years. It may also be noted that public housing as found elsewhere in Europe hardly exists in Norway (Christophersen, 1995).

- The British legislative system for buildings is basically the same as the Norwegian (i.e., the building code and building regulations being the central documents). Scotland and England do not share the same set of regulations, but at the time this study was done, neither country had statutory requirements for barrier-free housing, whether private or public. The

TABLE 13.2 Comparison of Legislation and Guidelines for Accessibility in the Five Countries

	Norway	Scotland	England	Germany*	Italy
Code/regulations	Requirements for multistory developments	None	None	Federal level, none, legislation in some Länder	Statutory requirements for barrier-free design
Legislation for barrier-free design	Life span standard required for state funding of special needs housing	Requirements for housing association financing	Local authorities may demand barrier-free design	Extra legislation for special needs/public housing	Percentage of barrier-free designs required in public housing
Incentives	Grants and extra loans for life span standard	Some incentives through building association financing	Housing corporation requirements[†] guidelines	Public financing: grants/loans	Grants for public housing

* Varies between the individual German states.
[†] Two sets of requirements: one for general-needs housing, one for wheelchair-accessible houses.

statutory requirements for accessibility specifically excluded dwellings in both countries. On the one hand, England did not even require elevators in multistory developments; on the other, the elevator requirements applied only to buildings where they were installed. Neither were there incentives, economic or otherwise, for barrier-free construction. Building barrier-free housing in both Scotland and England was thus wholly dependent on information and the idealism of local housing providers such as builders, housing associations, or voluntary organizations. Guidelines do, however, exist, notably the lifetime homes standards—one set for England, developed by the Joseph Rowntree Foundation [see Table 14.1, Chap. 14 ("Lifetime Homes: Achieving Accessibility for All"), by Richard Best], and another for Scotland (see Scottish Homes Research and Innovation Services, 1995). Both sets of guidelines are based on recommendations made by the Access Committee for England. For English public housing, there are also requirements in the Housing Corporation's Scheme Development Standards. The English building regulations have recently been amended to include some accessibility features, as outlined in Chap. 14, Table 14.2.

Table 13.2 gives a summary of requirements in the five countries at the time the study was undertaken.

Similar Technical Requirements

The specifications used in all five countries—whether in the form of standards, statutory requirements, or guidelines—are drawn up in much the same way, detailing the necessary (i.e., minimum) free floor space, maximum gradients, and changes of level relating to passage by a person using a manually powered wheelchair. In all cases, the requirements aim to allow wheelchair users to enter a dwelling and move around inside it without assistance. The space requirements vary somewhat from country to country.

The two basic requirements concern free floor space for turning a wheelchair and the width of accessible path. The turning space is specified as a circle, a square, or both (Fig. 13.1). All countries give two dimensions for accessible paths: continuous and at a point (Fig. 13.2). Of all specifications, this is the one that varies most, from 0.75 to 0.9 m at a point and 0.9 to 1.2 m for a continuous, accessible path.

Requirements for the widths of doors and the height of thresholds are important supplements, the latter being also the maximum change of grade. It ranges from 15 mm in the

Norway
England/
Scotland
Germany
Italy
1.4 m
1.4 m
1.5 m
1.5 m
1.5 m
1.7 m
0.3
1.4 m

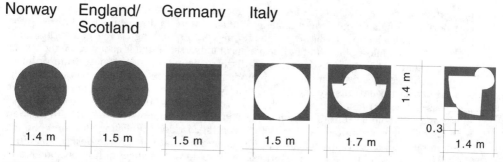

FIGURE 13.1 Comparison of requirements for turning space.

English criteria to 20 mm in Scotland and to 25 mm in Germany, Italy, and Norway. The wording of the requirements for doorwidths varies, but the basic dimension (the clearance) is remarkably consistent: 0.8 m in Norway, Germany, and Italy. Only the British require less: 750 mm clear space. The Germans have an additional requirement for 0.9m clearance of external doors to afford passage with the large wheelchairs that many use outside. In all countries a consequence of the requirement for free passage through a door is that the doorframes in external doors have to be wider than internal doors.

Interestingly, mobility impairments, quantified as maximum barriers for wheelchair users in manually powered wheelchairs, make up the majority of requirements for accessibility in housing. This clearly reflects the fact that this type of disability is the one that is best known and the one most commonly regarded as relevant to the problems of building construction and layout. Requirements for other types of disabilities are obviously underdeveloped. Where they exist, their chief concern is to avoid obstacles that may injure the users, such as preventing people from falling or colliding with cantilevers. The use of guide paths and tactile and color contrast could well be developed, as could requirements for the use of nonpoisonous materials and finishes, issues of noise control in buildings, and talking signs in outdoor areas. Presently, these issues play little part in the statutory requirements. For a further discussion, see Chap. 61, "Adding Vision to Universal Design," by Stuen and Offner.

13.4 BARRIER-FREE DWELLINGS CONFORM TO TRADITIONAL HOUSING

Of particular interest in the context of universal design is that barrier-free dwellings do not differ significantly from traditional housing in any of the five countries. The only distinguish-

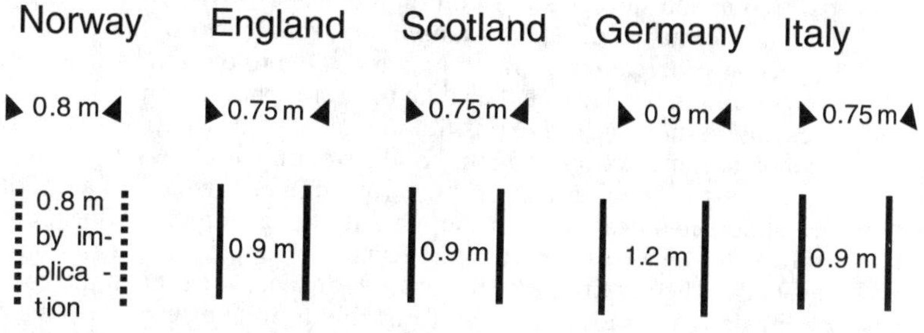

Norway
England
Scotland
Germany
Italy
0.8 m
0.75 m
0.75 m
0.9 m
0.75 m
0.8 m by im- plica - tion
0.9 m
0.9 m
1.2 m
0.9 m

FIGURE 13.2 Comparison of requirements for widths of passages. Bottom, continuous passage; top, at a point.

ing features of barrier-free layouts are slightly larger bathrooms and somewhat wider halls or corridors. For England and Scotland, an additional feature (a downstairs toilet in two-story houses) must also be noted. The rest of the rooms conform to long-established norms. Thus, one of the main philosophical and political aims of universal design is achieved on the whole: dwellings that can be used by elderly and disabled people do not deviate from ordinary dwellings. Barrier-free dwellings are accommodated everywhere in ordinary types of housing, and the barrier-free layouts are achieved with only a minimum of alterations to standard plan and house types. German designers do, however, have some problems with the space requirements and room dimensions that are necessary in order to comply with the DIN. In Norway, a popular single-story house/plan type with a central corridor connecting all rooms and the main entrance practically disappeared at the same time that barrier-free design was introduced. It is, however, likely that this shift had more to do with changes of fashion and greater attention to the cost of building than with the requirements for barrier-free design.

Layouts Differ

When designing housing that conforms to national traditions, the built results will vary considerably from country to country. This has to do with differences of tradition, not with accessibility, as the barrier-free plans exhibit the same characteristics as non-barrier-free plans.

In Italy and Germany, a central corridor runs like a spine through the entire plan, giving access to all or most rooms. The German corridor usually connects to the entrance door. In Italy, the entrance door often opens directly into the living room. The rooms, both in the Italian and in the German plans, are all of approximately equal size, though Italian living rooms are particularly small. In contrast, Norwegian plans have a small entry or hallway, and some rooms cannot be reached from a central circulation area. Bedrooms are comparatively small and the living rooms large, often connecting to a large kitchen. German kitchens are similar to the Norwegian (i.e., larger than those found in Italian) plans. British plans have particularly small rooms. The tradition of the narrow frontage-terraced house continues. England and Scotland are the only countries in this study where all main dwelling functions do not have to be situated at the entrance level. Bedrooms are commonly found on the upper floor and, in the case of a motion disability, may be reached by means of an internal elevator, for which space must be provided and structural preparations must be made.

Floor Plans

The plans in Fig. 13.3 illustrate in more detail the points made in Table 13.3. The examples also illustrate differences between the dominant types of housing: The Norwegian example shows the ground-floor plan of a two-story, detached house, which has three bedrooms, a bathroom, and storage spaces on the floor above. The English example is a terraced house and the German and Italian examples are from blocks of flats. The German case illustrates a particular problem for German planners. Due to the DIN's space requirements for barrier-free solutions, barrier-free flats take up more space than non-barrier-free solutions. This creates difficulties with the positioning of load-bearing walls, of achieving uniform and economic construction grids, and of distributing dwellings of varying sizes within a building. In contrast to the German problem, the Italian barrier-free and non-barrier-free solutions are almost identical. The reason is partly the generous size of the bathrooms in most new Italian dwellings, and partly the technical content of the requirements for barrier-free dwellings. The requirements for open floor space are seen in the context of a complex series of circulation patterns, making it possible to minimize space consumption through careful planning.

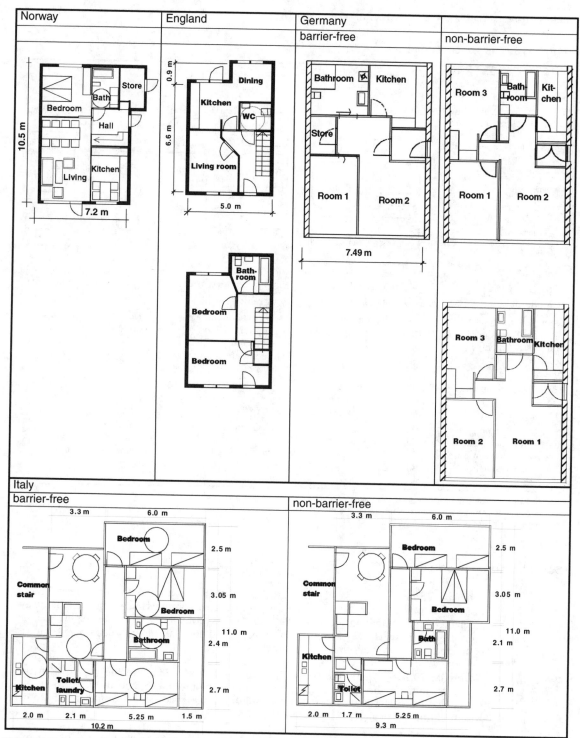

FIGURE 13.3 Comparison of floor plans.

TABLE 13.3 Comparison of Plan Types*

Norway	England and Scotland	Germany	Italy
Entry through hall.	Entry through hall/corridor.	Entry to central corridor.	Entry through living room.
Central hall.	Narrow hallway off center.	Long central corridor.	Long central corridor.
Doors to some rooms off the central corridor. Several rooms can only be accessed from other rooms.	Doors to all rooms off the hallway.	Doors to all rooms are off the central corridor. No direct access between rooms. All rooms are approximately equal size.	Doors to most rooms are off the central corridor.
Large living room, small bedrooms.	Small rooms. Toilet is on the ground floor†; bathroom is upstairs.	Living room and bedrooms are approximately equal size.	Large bedrooms, small living room.
Area of kitchen approximately as main bedroom. Kitchen and living room closely connected.	Kitchen and living are combined in one room.	Area of kitchen and bedrooms approximately equal in barrier-free solutions; small kitchen in non-barrier-free flats.	Small kitchen. Direct access from kitchen to living room common.

* The Norwegian and, particularly, the British types refer to the ground floor of a two-story house.

† The downstairs toilet is a particular feature of the lifetime criteria; the Housing Corporation only requires a downstairs toilet in houses for five people or more.

Some Important Rooms

Entry or Hallway. The sketches in Figs. 13.4 through 13.6 bring out the differences mentioned previously. On the one hand, there are the compact solutions that dominate in Norway, England, and Scotland; on the other, the central and rather space-consuming spine used in Germany and Italy. The Norwegian and the British solutions take up between 4.5 and 6 m², as compared with more than 10 m² in the other two.

Bathroom/Toilet. Again, the Norwegian and British solutions are the smallest. In the German solutions, the DIN's requirements are commonly interpreted in the most space-consuming manner, whereas in the Italian solutions, the generous space has mainly to do with tradition, particularly the standard use of the bidet.

Bedrooms. Bedroom size is partly determined by the choice of a standard size of bed and partly by the requirements for open floor space. In addition, the German and Italian require-

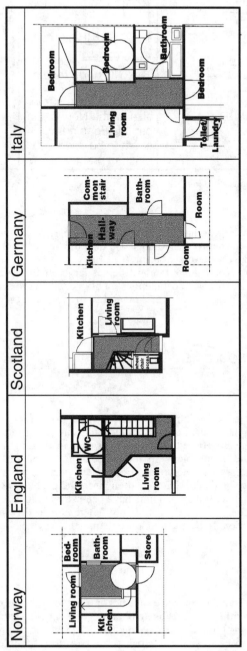

FIGURE 13.4 Comparison of solutions for entry/hall.

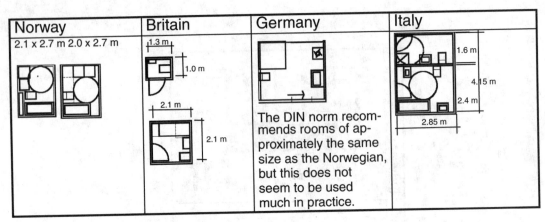

FIGURE 13.5 Comparison of bathrooms and toilets.

ments for public-sector housing specify minimum space standards. Interestingly, English space standards for bedrooms do not exist, only the Scottish homes have a space standard as shown in Fig. 13.6 for lifetime housing.

Although the Norwegian standard bed is larger than the Scottish, the minimum area required for bedrooms is almost identical in both countries: just over 10 m². In contrast, German and Italian requirements are 14 m² and 16 m², respectively.

13.5 SOME COMMON PROBLEMS

External Access

The impossibility of negotiating steep slopes when using a manual wheelchair has particular architectural implications: Long unsightly ramps are often the simplest way of overcoming the level differences between entrance doors and car parking facilities. Topographic characteristics aggravate the problem. So, too, do some forms of construction, as it is far easier to achieve level access to a concrete or brick building than to a timber house. English and Scottish builders have given considerable thought to the problem and generally seem to achieve good results, whereas Italian planners have been struggling and the Norwegians have had to make compromises. In the comparatively flat region in the German state of Hesse, level access does not seem to be much of a problem.

A Lack of Statistics

None of the five countries produce statistics or reliable estimates on the distribution of barrier-free dwellings. Neither is it possible to obtain figures showing what numbers of new, accessible dwellings are being built. Of the five countries, the statistics from the Norwegian State Housing Bank seem to be the most reliable, partly because of the detailed scrutiny to which the Bank subjects all proposals; however, the statistics cover only the housing projects financed by the bank. Projects financed through other sources, presently some 50 percent of all new construction, are not included in the Bank's statistics.

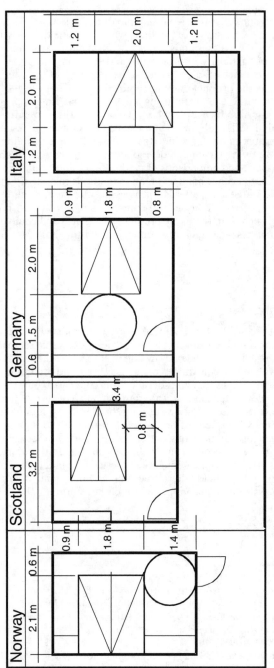

FIGURE 13.6 Comparison of bedrooms.

13.6 *FURTHER STUDIES/LESSONS LEARNED*

This study only looked into a small sample of designs from a limited selection of European countries. A logical extension of the study would be to include other countries, such as the rest of Scandinavia, the Netherlands, France, and Spain—particularly the first two, where the development of barrier-free house types and solutions has gone on for a number of years. There is also a distinct need for better statistics: None of the countries in the study had reliable statistics on the production of accessible housing, nor are good statistics on the standard of accessibility in the existing housing stock available. Lately, probably due to the expected rise of the proportion of people who are elderly in the population, central authorities have turned attention both to the ways and means of producing new, barrier-free housing and the accessibility status of the existing stock. Methods for evaluating the level of accessibility based on field surveys have been developed in Scandinavia and are being worked out in the United Kingdom, and comparisons of methodology and results could produce some interesting findings. On the other hand, evaluation methods relating to universal design are lacking, as are reliable statistical methods for assessing the number and severity of barriers in the existing housing stock. Analysis and comparisons of historical housing models could yield considerable insight into the likely presence of barriers and could lead to the development of methods for universal design improvements to existing housing and residential areas.

13.7 *CONCLUSIONS*

The design of accessible or barrier-free housing in Norway, Scotland, England, Germany, and Italy follows the broad principles of universal design. It would seem that guidelines and requirements for accessibility have been drawn up in such a way that accessible rooms and spaces can be created within the constraints of an accepted and traditional framework for housing, although these traditions vary considerably from country to country. The examples of new housing that were selected for this study are representative of current housing trends in the five participating countries, and they show that new, barrier-free solutions conform to long-established national traditions. This holds true for most aspects of the designs, including the types of housing, layout, plan form, dimensions, and the use of space. Interestingly, universal design does not seem to have been a precondition for the designs. It may, therefore, be surmised that the universal designs that have come about are the results of a pragmatic approach. The planners have been working within a field where tradition dominates and where solutions to new requirements are sought within the confines of established norms.

13.8 *BIBLIOGRAPHY*

Ambrose, Ivor, *Lifetime Homes in Europe and the UK,* Hoersholm: Joseph Rowntree Foundation/ Danish Building Research Institute, Hoersholm, Denmark, 1997.

Aslaksen, Finn, Steinar Bergh, Olav Rand Bringa, and Edel Heggem, *Universal Design—Planning and Design for All,* The Norwegian State Council on Disability, Oslo, Norway, 1997, http://home .sol.no/~obringa/universal.htm.

Christophersen, Jon, *The Growth of Good Housing,* The Norwegian State Housing Bank, HB-3061, Oslo, Norway, 1995.

———, *Varieties of Barrier Free Design—Accessible Housing in Five European Countries, a Comparative Study,* Project Report 211, The Norwegian Building Research Institute, Oslo, Norway, 1997.

Le Franc, Caroline, *Synthetic Board of Results,* Sophia Antipolis: Sigma Consultants/Ministère de l'Équipement et du Logement, Paris, France, 1992.

Pickles, Judith, *Guidance on the Design of Housing for Elderly and Disabled People—a Report on Practice throughout Scotland,* Scottish Homes, Edinburgh, UK, 1996.

Prestinenza, Luigi, *La Legislazione Italiana sulle barriere architettoniche,* Ordine degli architetti della provincia di Catania, Catania, Italy, 1994.

Joseph Rowntree Foundation, Lifetime Homes, York, UK, undated.

The Scottish Homes Research and Innovation Services, *Design of Barrier Free Housing,* The Scottish Homes Research and Innovation Services, Edinburgh, UK, 1995.

Sheridan, Linda, *A Comparative Study of the Control and Promotion of Quality of Housing in Europe,* Department of the Environment, Transport and the Regions, London, UK, 1999.

13.9   RESOURCES

General

Access Committee for England, 12 City Forum, 250 City Road, London EC1V 8AF, England. 1992. *Building Homes for Successive Generations.*

Bundesministerium für Raumordnung, Bauwesen und Städtebau, Deichmannsaue. 53179 Bonn. 1993. *Ältere Menschen und ihr Wohnquartier.*

Bundesministerium für Familie, Senioren, Frauen und Jugend, Rochusstrasse 8-10, Bonn, Germany. 1994. *Betreutes Wohnen—Lebensqualität sichern.*

Christophersen, Jon, 1990, *Livsløpsboliger, fungerer eller feiler,* Project Report no. 70. Norwegian Building Research Institute, Forskningsveien 3B, 0314 Oslo, Norway.

Flade, Antje, 1995, *Wohnen im Alter aus Psychologischer Sicht.* Institut Wohnen und Umwelt, Annastrasse, 64285 Darmstadt, Germany.

Fontana, Gaetano, 1993, *Sperimentare la norma.* Laboratorio tipologico nationale, Rome, Italy.

d'Innocenzo, Assunta, 1995, *La residenza degli anziani.* The Italian Association of Housing Co-operatives, FrancoAngeli, Viale Monza 106, 20127 Milano, Italy.

Prestinenza, Luigi, *Manuale per una progettazione senza barriere architettoniche.* Edil Stampa.

Prestinenza, Luigi, 1992, *Progettare la Sicurezza.* La Nuova Italia Scientifica, Via Sardegna 50, 00187, Roma, Italy.

Schnieder, Bernd, and Weiss, Christine, 1996, *Richtlinien für die Wohnversorgung.* Justus Liebig Universität Giessen. Bismarckstrase 37, 35390 Giessen, Germany.

Wüstenrot Stiftung, 1993, *Selbständigkeit durch Betreutes Wohnen im Alter.* Deutscher Eigenheimverein, Hohenzollernstrasse 46, 71638 Ludwigsburg, Germany.

Bayerisches Staatsministerium des Innern, Oberste Baubehörde, Franz-Josef-Strauss-Ring 4, 80539 München, Germany. 1995. *Wohnen ohne Barrieren.*

Laws, Regulations, Standards, and Guidelines

Norway

Byggeforskrift (Building regulations). Oslo, Norway. 1985.

Krav til livsløpsboliger (Minimum requirements for state loans and grants for special needs housing). The Norwegian State Housing Bank, Oslo, Norway. 1996.

United Kingdom

Building Regulations (England), Section M. Department of Environment, Transport and the Regions, London, UK, 1992.

Building Regulations (Scotland), Parts A, Q, and T. The Scottish Office Development, Construction and Building Control Group, Edinburgh, UK, 1994.

British Standard 5619, *Mobility Housing,* 1978.

British Standard 5810, *Access for the Disabled to Buildings,* 1979.

Scheme Development Standards, 1993, revised 1995. The Housing Corporation, 149 Tottenham Court Rd, London W1P 0BN, UK. Minimum requirements for housing corporation grants.

Germany

Hessische Bauordnung (Building regulations, Hesse), 1993. Hessisches Ministerium für Landesentwicklung, Wohnen, Forsten und Naturschutz, Hölderlindstrasse 1–3, 65187 Wiesbaden, Germany.

Technische Wohnungsbau Richtlinien 17, August 1992. Staatsanzeiger für das Land Hessen, p. 2153, changed 11 October 1993, p. 2771.

Deutsche Industrie Norm (DIN-Norm) 18024 and 18025, parts 1 and 2.

Italy

Law 118/1971, decree 384/1978.

Law 13/1989, decree 236/1989.

Law 104/1992.

Quarantelli, Paolo, *Assessorato opere, reti di servizi e mobilità.* Regione Lazio, undated.

CHAPTER 14
LIFETIME HOMES: ACHIEVING ACCESSIBILITY FOR ALL

Richard Best, B.A. Hons., O.B.E.
Rowntree Foundation, York, United Kingdom

14.1 INTRODUCTION

In the United Kingdom, many organizations representing people who are disabled have lobbied for several years for measures to persuade or compel those building new homes to incorporate features which, in the United States, would add up to universal design. Despite fine words from politicians, progress was slow until the end of the 1990s, when—picking up on unfulfilled promises from their predecessors—a new government decided to introduce compulsory measures for all future home building.

On October 25, 1999, new regulations came into force in the United Kingdom which require all new housing to be accessible to everyone, including elderly people and people with disabilities. This date goes down as a momentous day in the history of the struggle for a less disabling environment.

14.2 ACCESSIBILITY IN THE UNITED KINGDOM

The United Kingdom's Building Regulations, for some years, have required new public buildings to be accessible; the local planning authorities in every area enforce these requirements. Now the appropriate part of the Building Regulations—Part M—has been extended to residential buildings to make all new homes more accessible and convenient. But the measure has not been without controversy.

There was a small flurry of antagonism toward the proposals when the government went out to consultation a year before the new regulations became mandatory. An abortive "Save Our Doorsteps" campaign, based on a nostalgia for the great British doorstep, fizzled out when architects produced numerous pictures of fine buildings from the Victorian, Georgian, and earlier periods which had no doorsteps!

Within the construction industry, there were other concerns. House builders were anxious that their costs would rise, sales would be more difficult, and profits would fall. They were concerned that rain might be blown under the door, leading to claims against the guarantees they give for the first 10 years of a new home's life. And they thought some sites might have to be left undeveloped because it would prove impossible not to approach the home up or down a flight of steep steps.

In the event—as expected by those who have been implementing higher standards of accessibility for many years—these fears are proving unfounded. It is true that, in settling the price they pay for land, house builders must take into account any additional expense which may be involved in achieving the new standards. However, the extra cost for the developer is likely to be passed back to landowners who may receive a fraction less for their sites. For most homes, no extra expenditure should be needed. But in order to achieve full accessibility—in particular, for sufficient space to incorporate a downstairs toilet in a small two-story family house—a slightly bigger space standard will be needed than has been the case for the smallest new homes of the 1980s and 1990s. Most homes can be made more accessible and can achieve the government's new requirements simply by good design and within the same footprint.

Nevertheless, the construction industry's workforce—notoriously untrained in the United Kingdom—may need to learn some new techniques for ensuring that level thresholds create no new problems. However, the technology now exists and there are no particular obstacles to builders' meeting the required specification.

Government is supporting the National House Building Council in programs of publicity and training to spread the word about the changed practices that will help implement the new measures smoothly. It is not expected that the building control officers will be tolerant toward those who are reluctant to comply with the new regulations. The fact is that it is possible to incorporate accessibility even in circumstances that look difficult at first. For example, in relation to steeply sloping sites, developers have discovered that they were already in a position to achieve a level access to each home because they were bringing the car alongside the house to meet the demands of their customers. In the tiny fraction of cases where implementing the new standards is not possible, a special exemption can be secured.

The end result is that Britain's housing will be different in the future than it was in the past: not only will all new homes be easy to visit by those with mobility problems, but each new home will be suitable for life, for all but those with the most severe disabilities. It will not be necessary to segregate wheelchair users in special accommodation, separate from the rest of society. Furthermore, it will not be necessary for people to give up their homes later in life if, through illness or accident, they face an impairment. Life will be easier for parents pushing babies in baby carriages, for grandparents who come to stay, for teenagers who break a leg playing football, and for all those families who encounter the many hazards of a lifetime.

The remainder of this chapter describes both the mandatory requirements laid down in the Building Regulations for all house building and the rather higher standards established by the Joseph Rowntree Foundation for Lifetime Homes which are not only accessible but adaptable. With very little extra expense, but a bit more thought, Lifetime Homes can be created which go further than the government's new requirements. It is hoped that, in the years ahead, those house builders who want to have a marketing edge over their competitors will go the extra mile and adopt the superior standards which meet these criteria.

But even if the position remains as now, with changes confined to the new minimum regulatory standards, there is widespread satisfaction that—with special thanks to the construction minister—all those moving into new homes, or visiting them, will benefit from the government's foresight in insisting upon new standards of accessibility for the new century.

14.3   LIFETIME HOMES STANDARDS

Over a 10-year period, a working group organized by the Joseph Rowntree Foundation—comprising a number of representatives from disability organizations, architects, and housing providers, many of them people with disabilities—devised 16 standards which would make all new homes more accessible from the outset, but would also make them easy to adapt, if required, at a later date. These 16 standards together achieve the goal of Lifetime Homes, which will never force their occupants to relocate when they do not wish to leave, despite increased frailty or impairment. The criteria have been widely publicized and promoted by the foundation and by organizations of and for people with disabilities. Prior to the mandatory requirements, some local government planning authorities had already insisted that housing providers and house builders should satisfy these criteria. In addition, for publicly funded new homes, almost all these standards have been required for about 2 years.

The cost of incorporating Lifetime Homes standards varies according to the site, specification, and size of the dwelling. The smallest homes are the most problematic, since they may need to be bigger to include the required downstairs toilet, but three-bedroom, five-person homes are unlikely to present any cost problems. Figures show extra capital costs of about £1500 (on top of total costs of around £50,000) for the small homes but only about £250 for the larger properties with total costs of around £75,000.

The Joseph Rowntree Housing Trust, which undertakes innovative operational projects with support from the Joseph Rowntree Foundation, has built 600 homes, of differing sizes, in accordance with the Lifetime Homes standards. Five conclusions have been reached:

- With careful attention to detail, the standards can be achieved within the same footprint as virtually all new house types, even where high densities or narrow frontages are required.

- Costs are not material for average family homes, bungalows, or, of course, apartments in developments with elevators. Extra expenditure is incurred for the very smallest two-bedroom, two-story properties and for maisonettes—i.e., two-story buildings with one apartment above another. However, the very small, cramped two-story buildings common in the 1980s have proved unpopular on other grounds, and the house-building industry is moving away from starter homes of this kind.

- The Lifetime Homes standards are not suitable for people with the most severe mobility difficulties, particularly those using larger varieties of wheelchairs which require bigger turning circles. But they will accommodate most wheelchair users, allowing not just visitability but permanent occupation of the property.

- Older people are able to stay in their own homes for longer, saving themselves—or the state—expensive charges for living in residential accommodation. Just a few weeks of such savings will repay the modest initial investment in achieving the Lifetime Homes standards. Similarly, the much lower costs for adaptations at a later date repay several times over the initial extra costs of meeting Lifetime Homes standards.

- People with disabilities—including those with progressive mobility impairment—are greatly attracted by the opportunity to live in nonspecialist housing that is not stigmatized and separated from the rest of the community.

Table 14.1 sets out the details of the Lifetime Homes standards. These have not been made compulsory by the central government, although the construction minister, Nick Raynsford, M.P., extolled the "benefit of a little extra thought and care that needs to go into housing design in order to add the flexibility and adaptability found in Lifetime Homes." He contin-

TABLE 14.1 The Lifetime Homes Standards

Recommended, not required, by government

1. Where there is a car parking adjacent to the home, it should be, or should be capable of enlargement to attain, 3300 mm in width.

Comment: The general provision for a car parking space is 240 mm in width. If an additional 900 mm in width is not provided at the outset, there must be provision (e.g., a path or grass verge) to make possible enlargement of the overall width to 3300 mm if required at a later date.

2. The distance from the car parking space to the home should be kept to a minimum and should be level or gently sloping.

3. The approach to all entrances should be level or gently sloping.

Comment: Where the topography prevents a level approach, a maximum gradient of 1:12 is permissible on an individual slope of less than 5 m or 1:15 if between 5 and 10 m, and 1:20 where it is more than 10 m (providing there are top, bottom, and intermediate landings of not less than 1.2 m, excluding the swing of doors and gates). Paths should be a minimum of 900 mm.

4. Entrances should:
   a. Be illuminated.
   b. Have level access over the threshold.
   c. Have a covered main entrance.

The threshold upstand should not exceed 15 mm.

5. a. Communal stairs should provide easy access.
   b. Where homes are reached by an elevator, it should be fully wheelchair accessible.

Minimum dimensions for communal stairs:
   Uniform rise not more than 170 mm
   Uniform tread not less than 250 mm
   Handrails extend 300 mm beyond top and bottom step
   Handrail height 900 mm from each nosing

Minimum dimensions for elevators:
   Clear landing entrances 1500×1500 mm
   Internal 1100×1400 mm
   Elevator controls between 900 and 1200 mm from the floor and 400 mm from the elevator's internal front wall

6. The widths of the doorways and hallways should conform to the specifications in the next column.

Doorway clear opening width (mm):	Corridor/passageway width (mm):
750 or wider	900 (when approach is head-on)
750	1200 (when approach is not head-on)
775	1050 (when approach is not head-on)
900	900 (when approach is not head-on)

The clear opening width of the front door should be 800 mm.
There should be 300 mm to the side of the leading edge of doors on the entrance level.

7. There should be space for turning a wheelchair in dining areas and living rooms and adequate circulation space for wheelchair users elsewhere.

A turning circle of 1500 mm in diameter or a 1700×1400 mm ellipse is required.

8. The living room should be at entrance level.

9. In houses of two or more stories, there should be space on the entrance level that could be used as a convenient bed space.

10. There should be:
    a. A wheelchair-accessible entrance-level toilet room.
    b. Drainage provision enabling a shower to be fitted in the future.

The drainage provision for a future shower should be provided in all dwellings.

Dwellings of three or more bedrooms: For dwellings with three or more bedrooms, or on one level, the toilet room must be fully accessible. A wheelchair user should be able to close the door from within the closet and achieve side transfer from a wheelchair to at least one side of the toilet. There must be at least 1100 mm of clear space from the front of the toilet bowl.

TABLE 14.1   The Lifetime Homes Standards (*Continued*)

Recommended, not required, by government

	The shower provision must be within the closet or adjacent to the closet (the toilet could be an integral part of the bathroom in a flat or bungalow).
	Dwellings of two or fewer bedrooms: In small two-bedroom dwellings where the design has failed to achieve this fully accessible toilet, the Part M standard toilet will meet this standard.
11. Walls in bathrooms and toilets should be capable of taking adaptations such as handrails.	Wall reinforcements should be located between 300 and 1500 mm from the floor.
12. The design should incorporate: 　a.　Provision for a future stair lift. 　b.　A suitably identified space for a through-the-floor lift from the ground to the first floor for example, to a bedroom next to a bathroom.	There must be a minimum of 900 mm of clear distance between the stair wall (on which the lift would normally be located) and the edge of the opposite handrail/balustrade. Unobstructed landings are needed at top and bottom of stairs.
13. The design should provide for a reasonable route for a potential hoist from a main bedroom to the bathroom.	Most timber trusses today are capable of taking a hoist and tracking. Technological advances in hoist design mean that a straight run is no longer a requirement.
14. The bathroom should be designed to incorporate ease of access to the bath, toilet, and wash basin.	Although there is not a requirement for a turning circle in bathrooms, sufficient space should be provided so that a wheelchair user could use the bathroom.
15. Living room window glazing should begin at 800 mm or lower and windows should be easy to open/operate.	People should be able to see out of the window while seated. Wheelchair users should be able to operate at least one window in each room.
16. Switches, sockets, ventilation, and service controls should be at a height usable by all (i.e., between 450 and 1200 mm from the floor).	This applies to all rooms, including the kitchen and bathroom.

ued, "A number of local authorities encourage use of Lifetime Homes standards and house builders may want to go beyond the government's new Building Regulations, as a minimum regulatory standard, as they strive for excellence." He further said, "Lifetime Homes will help to make homes even more adaptable to long term needs." And he added, "Lifetime Homes standards in particular offer clear lifestyle benefits that homes built in earlier decades will not have."

Most of the Lifetime Homes standards have now been adopted by the Housing Corporation, the government's agency for supporting the registered social landlords (housing associations) who build subsidized accommodation.

14.4 GOVERNMENT REGULATIONS

The government drew up its new regulations for accessibility after extensive consultation. Revised Building Regulations (Part M) were published in October 1998 to allow time for the industry to adjust to the new system. As of October 1999, all new homes must meet these requirements. The construction minister explained, "The implementation of the Part M Regulations will improve the accessibility and convenience of new housing for everyone." And further, "Homes built to these standards can help to reduce future costs either for adaptations or for residential care." (See Table 14.2.)

TABLE 14.2   Key Elements of the Building Regulations (Part M)

General: The approach to the house should be wide enough for wheelchair users, even when there is a parked car on the right-hand side.	*Specific:* An additional 900 mm in width must be added to the space for a car.
General: The approach should not be too steep; ideally, it should be level.	*Specific:* If the plot gradient is less than 1:20, then no part of the approach must be steeper than 1:20. If the plot gradient is between 1:15 and 1:20, then individual slopes of 5 m or less may have gradients up to 1:12 and individual slopes 5 to 10 m in length may have gradients up to 1:15 (there must be top, bottom, and intermediate flat landings of not less than 1.2 m, excluding the swing of doors and gates. Part M makes provision for a stepped access if plot gradients are steeper than 1:15). Paths should be at least 900 mm wide.
General: An accessible threshold at entrance level should be provided.	*Specific:* Where unavoidable, a maximum 15 mm of upstand is permitted. Detailed information in a separate industry guide is available.
General: Doorways and corridors should be wide enough to allow wheelchair users to maneuver into and out of rooms.	*Specific:* The entrance door must always be at least 775 mm in width. Internal doors and corridors should conform to the following:

Doorway clear opening width (mm):   *Corridor/passageway width (mm):*

Doorway clear opening width (mm):	Corridor/passageway width (mm):
750 or wider	900 (when approach is head-on)
750	1200 (when approach is not head-on)
775	1050 (when approach is not head-on)
900	900 (when approach is not head-on)

The clear opening width of the front door should be 800 mm. There should be 300 mm to the side of the leading edge of doors on the entrance level.

General: Communal stairs in blocks of flats should provide ease of access to people with disabilities who are ambulant.	*Specific:* Uniform rise not more than 170 mm Uniform tread not less than 250 mm Handrails extend 300 mm beyond top and bottom step Handrail height 900 mm from nosing
General: A stepped change of level within an entrance story should allow ease of access to people with disabilities than who are ambulant.	*Specific:* Flight clear width 900 mm Suitable continuous handrail on each side where there are more than three rises Rises and goings in accordance with other parts of the Building Regulations
General: Wheelchair users should be able to use any elevator provided in a block of flats.	*Specific:* Clear landing entrances 1500 × 1500 mm Minimum internal elevator car dimensions 900 × 1250 mm Elevator controls between 900 and 1200 mm from the floor and 400 mm from the elevator's internal front wall. There should be tactile control buttons, and visual and audible story indicators for blocks of more than three stories.
General: Switches and sockets should be at a convenient height for all.	*Specific:* Switches and socket outlets for lighting and other equipment in habitable rooms should be between 450 and 1200 mm from finished floor level.
General: All homes should have an entrance-level toilet room which is accessible to someone using a wheelchair.	*Specific:* Clear usable space between front of toilet bowl and opposite wall/door should be 750 mm minimum. The distance from the central line of the cistern and the adjoining wall should be 450 mm minimum Where oblique access is provided, there should be 250 mm minimum to the side of the door.

14.5 *SOME COMMENTS FROM EXPERIENCE*

Since the Joseph Rowntree Housing Trust started building to Lifetime Homes standards in the mid-1990s, a number of lessons have been learned from the development process and from continuous feedback received from residents and property managers. The following are some of the key points covered in the guide, *Meeting Part M of the Building Regulations and Designing Lifetime Homes,* published by the Joseph Rowntree Foundation, in addition to some comments from consumers.

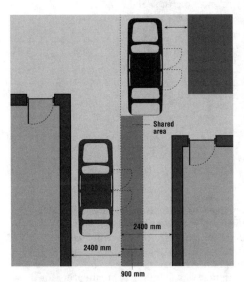

FIGURE 14.1 Flexible space between houses.

Outside the Home

When building detached or semidetached houses side by side, it is not necessary for there to be a width of twice the required maximum (3300 mm) between them. As long as two cars can pass each other, the capability of widening the space by taking in a path or garden strip, can involve a different position for each vehicle. Figure 14.1 illustrates one such solution.

In achieving a gentle approach to entrances, some sites are so steep that steps are unavoidable, but level access can still be achieved from the car parking space, as Fig. 14.2 illustrates.

While a significant proportion of those interviewed about the level thresholds at the entries of their homes had not noticed this feature, but expressed satisfaction when it was pointed out, the great majority appreciated the absence of steps. One user commented, "My granny slipped on her icy doorstep and broke her hip. I have always been nervous of steps in Winter." Another said, "I can visit everyone living in this area, in my wheelchair, because there are no doorsteps to negotiate. But there is no way I can go to my brother's house: he has two big steps and cannot lift me and my chair."

In the early days after adoption of these standards, some difficulties were encountered in ensuring that there would be no rain penetration under the front door. Although all new shops and offices now comply with requirements for level thresholds, and their construction ensures that no rain will come in under the doors, those working in the house-building industry sometimes need extra training. A simple solution to part of the problem is to use a prefabricated door within its frame, ensuring a perfect seal.

The government has published industry guidance documents on accessible thresholds. There are now numerous examples of different approaches—all of which can be successful—often using the proprietary upstand/threshold, which must be no more than 15 mm high. As one person observed, "It is good that there are no draughts under the door. This saves on our fuel bills." Another added, "Our carpet, on its underlay, comes to the same height as the threshold so the wheelchair can travel smoothly over this little obstacle." Figure 14.3 illustrates one solution.

Inside the Home

The width of corridors and doors can make or break the accessibility of a home. Internal doorways and hallways need to conform to the dimensions set out previously. A useful mechanism for achieving these is a 300-mm unobstructed space to the side of the door on the entrance level, making the passages wider than the door and its frame. Figure 14.4 gives details.

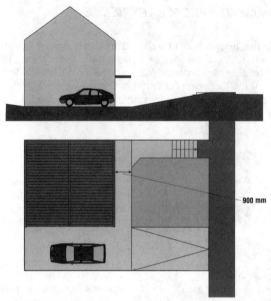

FIGURE 14.2   Level access from car to entrance.

Stair lifts can be expensive items, should they need to be fitted. Landings halfway up the stairs add to complexity and cost—although they can provide a convenient opportunity to pause for those with respiratory or heart problems—and turns in the stairs are similarly problematic in adding cost for stair lifts. But Lifetime Homes standards also provide space for through-the-floor lifts that can travel between the main living room and a bedroom above, carrying the full wheelchair, which is not possible with many chairlifts. The engineering for these is extremely simple, and costs are predicted to fall. One client noted, "All the houses we were

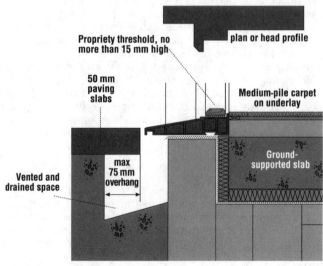

FIGURE 14.3   Level access—one example.

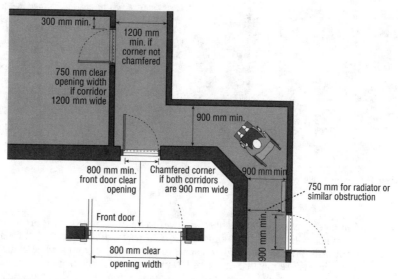

FIGURE 14.4 Widths for corridors and doors.

thinking of buying were unsuitable for a through-the-floor lift because, although a cupboard could contain the lift on the ground floor, the space above was always blocked by a wall, or the bath, or other obstacles. Now I can travel straight up, in my chair, to the main bedroom which is en-suite to the bathroom."

The government regulations require an accessible toilet room on the ground floor for all houses, and Fig. 14.5 illustrates typical dimensions. However, Lifetime Homes incorporate a

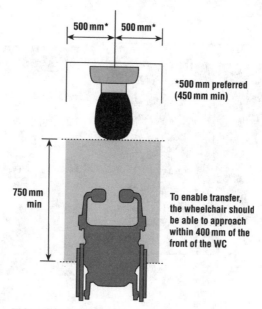

FIGURE 14.5 Access to ground-floor toilet.

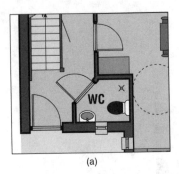

(a)

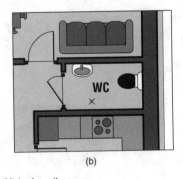

(b)

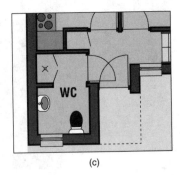

(c)

FIGURE 14.6   (a–c): Including a shower (X) in the toilet room.

drainage provision that enables a shower to be fitted in a downstairs toilet room at a later date. In homes with three bedrooms or more, a wheelchair user should be able to close the door from within the toilet room and achieve side transfer from a wheelchair to at least one side of the toilet. There must be at least 1100 mm of clear space from the front of the toilet bowl, and the door of the toilet room will often need to open outward to allow accessibility. Figure 14.6, *a* through *c,* illustrates possibilities; the shower drainage is shown with an "X."

Lifetime Homes standards, again going beyond the government's regulations, require provision for a reasonable route for a potential hoist from a main bedroom to the bathroom. Well-thought-out design—as in Fig. 14.7—provides for a future track and hoist through a removable floor-to-ceiling panel. Technological advances mean that tracks no longer have to go in a straight line, and a second-best solution would be a route for a hoist via the landing.

One attractive feature of the Lifetime Homes is that windows in living rooms should begin at 800 mm or lower and should be easy to open and operate. A home owner commented, "I can sit and enjoy the view while in my chair." Another said, "The children can see what is happening outside without climbing up." See Fig. 14.8 for details.

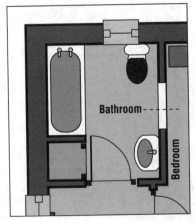

FIGURE 14.7   Bathroom with route for hoist from bedroom.

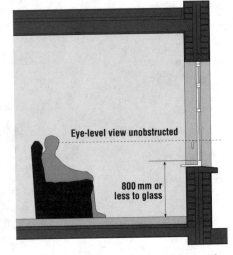

Eye-level view unobstructed

800 mm or less to glass

FIGURE 14.8   Enjoying the view while seated.

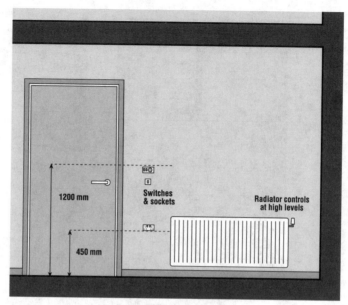

FIGURE 14.9   Heights for switches, outlets, and controls.

Most wheelchair users, and many people who find it difficult to bend and stretch, particularly appreciate the convenience of switches and socket outlets that are easy to reach. Figure 14.9 illustrates the government's regulations, which require the same standards as for Lifetime Homes. One client said, "I have never been able to change a fuse before because the fuse box always seems to be situated about six feet above the ground." Added another, "With my bad back, I used to find it difficult to reach the power points at floor level to plug in any electrical appliances and even to change the heat with the radiator control. Now I can reach all the sockets and switches."

14.6   TWO EXAMPLES

Figures 14.10 and 14.11 illustrate how new accommodation can best be designed to incorporate the U.K. government's new Building Regulations and to go that small step further to achieve full Lifetime Homes standards. These examples are drawn from the guide published by the Joseph Rowntree Foundation and edited by Caitriona Carroll, Julie Cowans, and David Darton (1999).

This two-bedroom home of 71.8 m² (768 ft²) has an internal frontage of 4.9 m. Although it is a small property, the thoughtful design makes the most of the available space; the plan shows how it meets all the standards. Note the shower provision in the toilet room. The simple layout, coupled with the lobby provision, gives a spacious feeling to the main rooms. The large "X" shows where a through-the-floor lift could travel from the living room to the main bedroom—if this is required later.

Figure 14.11 shows a three-bedroom home of 82 m² (882 ft²), this time with an internal frontage of 6 m. It shows the space for a through-the-floor lift which has little or no impact on the overall design or use of the house initially. Of course, if the lift was installed later, the

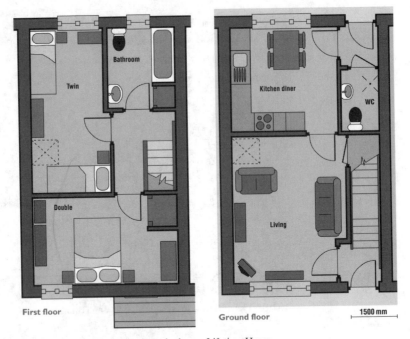

Bathroom

Twin

Double

First floor

Kitchen diner

WC

Living

Ground floor

1500 mm

FIGURE 14.10   Example of two-bedroom Lifetime Home.

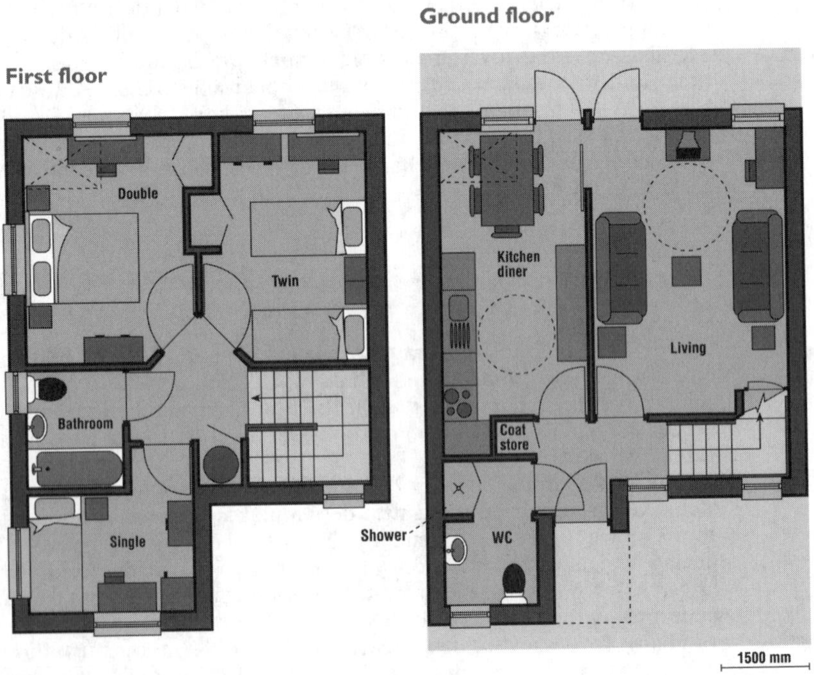

Ground floor

First floor

Double

Twin

Bathroom

Single

Kitchen diner

Living

Coat store

Shower

WC

1500 mm

FIGURE 14.11   Example of three-bedroom Lifetime Home.

household would have to adapt by repositioning furniture (or by doing without some furniture). But similar flexibility is required in all housing when an occupant acquires a disability. A sliding door between the dining area and the living room, coupled with French doors leading to the garden, makes this a very successful semi-open design.

There will be a huge variety of ways in which, for homes of all kinds, the new requirements can be met. The essential starting point for achieving change is simply the acceptance by designers that these criteria for greater accessibility are essential—not some kind of add-on.

14.7 CONCLUSIONS

This chapter has spelled out the arrangements which have emerged for achieving accessibility for all, by law, in the United Kingdom. Throughout the 1990s those active in the various organizations representing people with disabilities argued for a removal of the barriers which prevented so many people from entering newly built homes. Work in devising a set of accessibility standards created the Lifetime Homes standards, which have found favor with a number of housing providers. But the real breakthrough has resulted from a determination by government to regulate universal change for accessibility, affecting all those building new homes.

No increase in house prices is expected as a result of this measure, although occasionally those selling land for development might get marginally less where house builders had intended to provide particularly small homes, which must now achieve slightly higher space standards. The benefits accrue not only to those who move in and enjoy the extra spaciousness and attention to design detail, and not only to those who can visit everyone in these new homes. There is also the benefit to individuals and to the state in the long term through savings in the costs of adaptations and/or time spent in residential care or special accommodation, which are no longer necessary.

Of course, the great majority of homes are already constructed, and the new measures do not apply to existing property. It will take many years for truly accessible housing to represent the majority of all accommodation. But the expectation is that the approach now being applied universally to new homes will have an impact on the aspirations and attitudes of existing property owners; gradually, as properties are improved, modernized, and extended, it can be hoped that the owners will incorporate many accessibility features. Although overall space standards may be fixed, and the overall design may make accessibility more difficult, there are always modifications—such as raising and lowering power points and switches when a property is rewired—which can be installed with little or no extra expense. Furthermore, publicly funded landlords—housing associations and local authorities—are increasingly ensuring that wherever wholesale modernization takes place, accessibility is a priority in the remodeled home.

Concentrating on better access to all new homes should increase consumer understanding of the issues and heighten demand for better accessibility in old homes.

And are the new occupiers well satisfied with their accessible homes? The Joseph Rowntree Foundation has commissioned the Consumers' Association to undertake a satisfaction survey of 500 occupants of Lifetime Homes. It is the foundation's hope—and expectation—that these residents will provide a firm endorsement for universal design.

14.8 BIBLIOGRAPHY

Bonnett, David, and Nigel King, *Shared Ownership: Lifetime Homes: Making It Work,* Shared Ownership Lifetime Homes Group, London, UK, 1998.

Brewerton, Julie, and David Darton, *Designing Lifetime Homes,* Joseph Rowntree Foundation, York, UK, 1997.

Carroll, Caitriona, Julie Cowans, and David Darton, *Meeting Part M of the Building Regulations and Designing Lifetime Homes,* Joseph Rowntree Foundation, York, UK, 1999.

Cobbold, Christopher, *A Cost Benefit Analysis of Lifetime Homes,* York Publishing Services/Joseph Rowntree Foundation, York, UK, 1997.

Department of the Environment, Transport and the Regions, *Approved Document M: Access and Facilities for Disabled People; 1999 Edition* (Building Regulations), The Stationery Office, London, UK, 1998.

———, *Accessible Thresholds in New Housing: Guidance for House-Builders and Designers,* The Stationery Office, London, UK, 1999.

Drury, Andrew, *Standards and Quality in Development: A Good Practice Guide,* National Housing Federation, London, UK, 1998.

Housing Corporation, *Scheme Development Standards, 1998 Edition,* Housing Corporation, London, UK, 1998.

Sangster, Kim, *Costing Lifetime Homes,* Joseph Rowntree Foundation, York, UK, 1997.

CHAPTER 15
ACCESSIBLE DESIGN IN ITALY

Assunta D'Innocenzo and Annalisa Morini
National Research Council (CNR), Rome, Italy

15.1 INTRODUCTION

The principal aim of this chapter is to provide a general idea of how accessibility is studied, regulated, and applied in Italy, with an emphasis on public buildings and urban design. The chapter has three sections:

1. The accessibility concept, its evolution, and its rules and technical prescriptions
2. The application of accessibility in the planning, design, and building phases of different types of construction and environments
3. Accessibility as one of the key tools for improving the quality of any type of environment

This chapter will focus more on accessibility in renewal than on accessibility in new construction. Examples of different solutions are shown in order to evaluate the effects and to test guidelines.

15.2 BACKGROUND

The notion of accessibility has changed substantially in Italy, as in many other industrialized countries, due to the evolution of the concept of integration—that is, true inclusion, not segregation. Rather than tackling the problem in a comprehensive way in Italy, details and technical prescriptions were added to existing regulations over the years, resulting in professionals having to identify solutions while keeping in mind the principles of accessible environments. There is a lack of uniformity of definition even at the international level. The Italian word for "accessibility" nowadays means *universal design,* even if there is a difference in application. In fact, because of the emphasis on renewing and improving existing areas rather than building *ex-novo,* accessibility is applied more in the built environments and in difficult design situations such as historical centers, postwar peripheral neighborhoods, and industrial zones undergoing transformation of use.

Accessibility is, then, one of the design elements used to create a new equilibrium in cities and in the countryside, in downtowns and in peripheral suburbs, in residential buildings and in social and cultural buildings. With the evolution of the concept, application of accessibility was extended to each type of built environment, even including green, archaeological, or tourist areas. Of course, there is a gap (and in some cases, a huge one) between the concept and its application. But the change in societal attitude is significant. Now everybody knows that it is the right of every person to use as independently as possible not only his/her own house, but also the external environment.

15.3 DEVELOPMENT OF ACCESSIBILITY LAWS AND STANDARDS IN ITALY

In Italy, the concept of accessibility has changed its meaning over the years, starting with the aim of reducing difficulties for people with impairments, and currently attempting to make environments universally accessible by including (or at least trying to include) everybody as users of environments. This means not designing specific solutions for particular users, but developing solutions that can be used by the majority of the people. From an initial focus on public buildings, there has been a shift to applying the concept to every type of built environment.

History

Italian laws on accessibility were first initiated at the end of the 1960s in a partial, rather than a comprehensive way, with rules established for specific structures.

Principal measures were established in the following years[1]

1978 (DPR 384, 1978). To remove the architectural barriers from structures of public interest (social buildings, schools, services such as transport and related stations, etc.).

1986 (Law 41, 1986). To put into practice plans for the accessibility of public buildings and external urban spaces. This law was modified and integrated in 1992, with the Framework Law on Disability (Legge 104, 1992).

1989 (Law 13, 1989). To regulate the accessibility and adaptability of private buildings and residential public buildings with their related external areas, in both new construction and rehabilitation.

1996 (DPR 503, 1996). To revise all the previous regulations on the removal of architectural barriers in buildings, spaces, and public services.

In addition to national regulations, regions also passed laws which integrate regulations adopted at the central level during those years.[2]

National and regional laws adopted after 1989 exhibit the transformation in the concept of accessibility, having evolved from the limited precincts of accommodating disability toward including transversally the whole of society. Terms such as "invalid, mutilated, cripple" constituted the principal categories of users addressed by the initial rules.

A concrete example of this trend is the law that was also issued in 1989 by Regione Lombardia in northern Italy, which established the requirement for interventions to ensure each citizen, regardless of age, sex, physical and psychological characteristics, or temporary or permanent changes in ability, has the greatest autonomy in all activities conducted within the built environment (Legge Regione Lombardia, 1989). Its passage launched the phrase "the accessible city,"[3] reflecting a way of thinking that involves diversity and tolerance with respect to the entire built environment.

Even if these are goals not completely met, the emphasis of the main social and professional actors is on the wider theme of accessibility within urban spaces. Accessibility, therefore, orients thinking toward modifying daily environments in the city to meet the needs of *all* residents. In this sense the accessibility requirements in Italy are analogous to the term *universal design* as used in the United States.

Evolution in the Concept of Barriers

An interesting factor in the evolution of regulations in Italy is the changing meaning of the word "barrier" over the years. The earliest regulation, issued in 1967 (Circolare Ministero Lavori Pubblici 425, 1967), defined architectural barriers as obstacles that people with physical disabilities encounter in moving within urban areas and buildings, consisting essentially of changes in level in paths of travel (steps, differences in level, steep slopes, etc.) and passages too narrow for some users (elevators, doors, corridors, etc.).

The term *barrier* is principally used in criteria governing regulations for new construction. In fact, the law specifically "aims to remove these difficulties as much as possible, by avoiding changes of level in the path of travel or by providing an alternate route which can facilitate the mobility of people with disability problems."

Physical obstacles were narrowly described in the earliest regulations. Related issues of design and use of existing environments—such as the lack of adequate or appropriately designed parking according to regulations; unprotected pedestrian crossings; paths of travel by obstacles (signs, street furniture, etc.); lenient enforcement allowing illegal parking; lack of service equipment; or lack of barrier-free transportation—were not mentioned. When legislation was later revised in the law of 1989 (Legge 13, 1989), which is still in force, the words *architectural barriers,* correctly defined, gained more precision. According to Article 2 of this law, architectural barriers are:

- Physical obstacles that are a source of problems for the unhindered mobility of all users (but particularly for people who temporarily or permanently, for any reason, have limited mobility or loss of mobility)

- Obstacles that limit or prevent the comfortable and safe use of space, equipment, or components

- The lack of symbols and signals which, when present, allow wayfinding and orientation in places as well as the identification of sources of danger for all users (but particularly for people with visual/hearing impairment or loss)

The concept of accessibility started from a restricted application and became more comprehensive, due to a better understanding of barriers.

First of all, a *barrier* is a physical obstacle restricting mobility, which exists and will remain in the urban environment, but which can be avoided by different solutions suitable to each specific user need.

Second, a *barrier* is anything that does not allow the comfortable and safe use of space, equipment, and/or components. This term includes the disorder of social behavior and inadequacy of urban policies, for example, the lack of addresses (often rules related to the occupancy of streets are not adequate or not applied); the lack of coordination (telephone boxes are decided in Italy by Telecom, bus stops by the municipal transport agencies, etc.), and irresponsible behavior of car drivers (e.g., failure to respect pedestrian crossings, parking illegally in no-parking zones). As is reported by Sandra Manley in Chapter 58, the lack of coordination and responsibility is not solely an Italian problem. The extension of its meaning is measurable in the results of management of the city by local authorities. Where there is strong urban policy and enforcement, the occasions of disorder and transgression are lower and the social acceptance of rules by citizens is higher.

Third, a *barrier* is anything that does not allow users to orient themselves and recognize places or sources of danger. This is the most innovative aspect of accessibility, strongly linked to the process of improving the quality of the existing environment. *To orient* means to identify places in their geographical and directional features and to retain spatial characteristics in the memory. If people can orient themselves and recognize the different parts of the city, they acquire a sense of security and familiarity by reducing the normal tensions created by encountering new situations. Everyone needs orientation cues, but these are particularly important for children and older people, who experience anxiety in new environments. This is more evident in the modern parts of Italian cities, which developed without planning or regulation over the last 30 years under the pressure of quickly producing a large quantity of housing. Intervention for renewal of urban areas that had become strange and hostile, aimed for a new equilibrium by the use of small-scale and demarcated neighborhoods, which revive the familiar and unitary aspect of places. An example of that process is described in Sec. 15.8, which is partially devoted to *Contratti di quartiere* (neighborhood agreements).

Accessibility, Visitability, Adaptability

The second important innovation was introduced in 1989 and confirmed in 1996. This second accessibility legislation phase applied to all existing buildings and renewal areas, both public and private, as well as new construction. This law introduced the concept of three different levels of accessibility.

1. *Total accessibility.*   Where people with reduced motor or sensory capacities can travel to a building, easily enter and gain access to all of its parts, and use spaces and equipment in conditions of independence and safety. Total accessibility is applied differently according to building types and must ensure external access, and guarantee that every common area of the building is accessible (i.e., at least one route of travel must be usable by persons with disabilities).

2. *Partial accessibility, or visitability.*   Where people with reduced capacities can have access to the principal spaces in a building where activities are carried out, and where there is at least one accessible toilet in the building;

3. *Deferred accessibility, or adaptability.*   Where the built environment can be modified with limited costs in order to allow use by people with reduced capacities.

The Italian concepts of accessibility are addressed and compared with four other European countries (Norway, Scotland, England, and Germany) in Chap. 13, by Jon Christophersen.

The different modes of applying the accessibility requirements for specific sectors are described in the following sections.

15.4 ACCESSIBILITY LEGISLATION AND TECHNICAL REQUIREMENTS

The application of accessibility requirements necessitates the careful evaluation of the design for barriers, including the layout, dimensions of spaces and circulation routes, choice of finishes, and level of artificial lighting. Every technical aspect during the process from concept to construction must be designed, not for the average user, but for the possible extreme capacities of users, whether they be hearing impairments or mobility difficulties. Further, these requirements are applied to every part of the built environment, not only within buildings, but also in the external areas and in the urban environment. It is easier to adopt appropriate design decisions during new construction, and it is more difficult to intervene in or make modifications to existing buildings. Any plan for rehabilitating a neighborhood or a part of the city must take into account the new requirements. All people have the right to move around and use the city as much as possible.

Accessibility regulations describe design criteria, starting with general criteria for the whole building, followed by specific criteria for the accessibility, visitability, and adaptability of components (ranging from kitchens and bathrooms to doors, floors, and elevators) including technical solutions.

Specifications are provided for:

- The general layout and dimensions of spaces
- Technical components
- Technical fittings and materials

Layout and Dimensions of Spaces

The general layout of buildings in many cases is modified to facilitate the use of the whole built environment by people of varying abilities. Entrances, corridors, elevators, and circulation routes inside buildings are enlarged or widened. Traffic flow and orientation needs are specified. Changes in level and narrow or restricted passages are to be avoided. Ramps are required as alternatives to stairs. To provide an idea of the level of detail in the requirements, height is taken as an example. The appropriate height is described for every element, starting from those in a dwelling and extending to those in a building and finally to those in the external environment:

- *Dwellings.* Height specifications apply to the glass part of a door; handles of doors and windows, plugs, switches, buzzers, windowsills, balcony parapets, and thresholds between the internal dwelling and the terrace or balcony.
- *Common spaces of the building.* Height specifications apply to elevator buttons, intercom systems, mailboxes, and stairs (which must have the same riser height for each step).
- *Buildings.* Height specifications apply to public services. Benches, for example, must be installed with a seat at the height of a person using a wheelchair.
- *External environment.* Height specifications apply to telephone booths, buttons for pedestrian crossings, vending machines, and wastebaskets.

Technical Components

Specifications for technical components are precise to not only allow accessibility but also to facilitate the usability of the environment. Explanatory notes are provided to eliminate confusion and misinterpretation. For instance, specifications for doors include not only the appropriate width (i.e., no more than 120 cm for one door, no less than 80 cm for the entrance door, 75 cm for the others), but also:

- The allowable distance between one door and another along corridors, or the access to a building.
- The type of door permitted (e.g., revolving doors are not accessible).
- Safety requirements, such as specifying that the lower part of the door must not be easily breakable, to protect users and the door from impact with a wheelchair, and, the glass part, if any, must be inserted above the height of 40 cm.
- Ease-of-operation requirements, which, for example, specify that the pressure required to open the door should not exceed 8 kg; the height of handle installation should be between 85 and 95 cm, and the recommended width is 90 cm.

Requirements for furniture provide another example. In public spaces, waiting areas must be provided with an appropriate number of seats and be designed for user comfort. Where tables and desks are provided, free space must also be provided to allow for the circulation and use of persons in wheelchairs. Automatic machines of any kind used by the public, located inside or outside a building, must be usable by a person using a wheelchair as well as by those with reduced arm or hand mobility and must meet specific requirements in terms of position, height, and type of operating mechanisms.

Fittings and Materials

Appropriate fittings are very important to ensure functional efficiency and the easy use of the environment. For instance, to ensure the functionality of toilets, accessibility requirements specify the following:

- The correct position of grab bars, and of an emergency bell located close to the toilet and the bathtub.
- The shower must be at floor level, and be equipped with a folding seat and mobile shower head.
- In public spaces the position of the grab bar is fixed.
- In case of renovation, bidets may be omitted and a tub with the shower may be substituted, in order to have more space for a wheelchair.

The appropriate use of materials can facilitate mobility, support the use of the environment, and improve orientation cues. Materials must be suitable as well as durable over their lifetime of use, maintaining the required strength and tolerance characteristics.

Some of the requirements for the important element of flooring follow: Flooring must be slip-resistant, with a choice of suitable material or appropriate finish of the material. It can be designed to orient people with way-finding cues or to distinguish different spaces by the change of color or type of material. Differences in level must be avoided, and are restricted to a maximum of 2.5 cm. In pathways the material must be homogeneous in order to avoid mobility restrictions due to varying characteristics of flooring material. Since flooring must be smooth and level, materials utilized must not crack, splinter, heave, or bulge.

15.5   DESIGN GUIDELINES FOR PUBLIC BUILDINGS

Regulations for public buildings have evolved over the years. Public buildings were the first to be subject to accessibility regulations. This new direction toward integration was considered urgent enough for public decision makers to decide that state and public agencies must set an example by modifying their properties. The earliest buildings to be modified were public institutions, particularly schools, because the need to integrate children with disabilities into mainstream environments had both government and public support.

The most recent accessibility law, passed in 1996, also directed special attention to schools. An article was specifically devoted to school buildings to ensure that furniture and educational materials have the appropriate characteristics for use by children with varying disabilities.

After schools, spaces for cultural events were covered next by accessibility laws, with specific requirements applying to entrances, pedestrian pathways, and parking areas, as well as to public transport services and parking for cars used by people with motor difficulties.

The case studies described in the following paragraphs range from buildings used by the most frail users (e.g., the elderly and children) on one hand, to a market and theater that draw nearby residents to their activities, on the other.

The Intermediate School of Baselga di Piné

This public renewal project concerns a building constructed in the early 1970s. Located in the province of Trento, in northeastern Italy (ITEA, 1999), it stands in a wide, flat area occupied by buildings of social and cultural interest, such as the municipal library, the public health center, the fire station, and tennis courts. The design for renewal was intended to provide a new organization of primary and intermediate school facilities, along with more efficient use of space. The gym was enlarged. The design was developed by ITEA, the agency for social housing, which also implemented the renewal project in 1996.

Work included not only enlargement of the complex, but also improvement of the facilities and accessibility of the existing building. The new building included, in addition to the gymnasium, another smaller gymnasium, a swimming pool, a multifunctional room, and an auditorium (see Fig. 15.1). The external area consisted of parking, green spaces, and areas for sports and recreational activities, all of which were made accessible.

FIGURE 15.1   Renovated school near Trento.

Nursing Home in Bagnacavallo

This public rehabilitation case study shows a recent design of a new type of nursing home and a day center located at Bagnacavallo, in the province of Ravenna (Galli, 1995). In this historical center, dwellings for older people were linked with a day center for social activities (see Fig. 15.2). The accessibility law was a fundamental driver for the design to achieve three goals:

1. To improve the quality of life for elderly users, through functional design that guarantees the maximum environmental comfort, health care, and stimulation of social interaction in the local community
2. New uses for ancient buildings that are deteriorating and losing their original function
3. Generation of urban architecture that would also attract other inhabitants to live in the center

The residential area is composed of 16 dwellings, 5 at ground level and 11 on the first floor. Four of them are double and 12 are single units, providing shelter for 20 residents. The dwellings are flexible, so they can be modified with changing needs. Moreover, tenants can choose to spend their time at home or in common spaces.

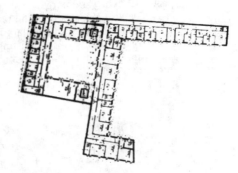

FIGURE 15.2   Nursing home and day center near Ravenna.

The Market Nomentano, Rome

This is one of the oldest daily markets in Rome, located in the square Alessandria in the central area close to via Veneto (CNR ICITE, Comune di Roma, 1991). There are still many daily markets in Italy, generally used for the selling of fresh vegetables and fruits. The public rehabilitation project, completed in 1987, modified the spaces, totaling around 1820 m^2 (see Fig. 15.3). There were 75 merchants, principally of food. To make the building usable by people with physical disabilities, the work was carried out by the Municipality of Rome's Department of Public Works.

Because the building entrances had some steps, a new ramp was built at one of the principal entrances from Alessandria street, and the toilets were made barrier-free. The selling area was all on one floor, with underground warehouse space exclusively for merchants and a flat on the upper story for the manager, accessible only by a staircase.[4]

15.6   *DESIGN GUIDELINES FOR THE URBAN ENVIRONMENT*

In the early 1990s, Italy addressed the renewal of the existing building stock. Simultaneously a phase of experimentation began for the improvement of outdoor areas, for their own intrinsic value rather than as adjuncts to buildings.

Pedestrian pathways, squares, green areas, parking, urban parks, and natural and archaeological areas were progressively improved in terms of usability, comfort, and safety, while also providing access to all population groups, especially those who were frail or disabled. The latest regulation of 1996 (DPR, 1996) requires three levels of accessibility in each public space. Public spaces, including new construction, rehabilitated areas, and those intended for temporary use must all conform to these new rules. The regulations cover the following elements:

- Pathways and pedestrian areas
- Pedestrian crossings
- Bus stops
- Public parking
- Green and sports areas
- Street furniture (including signals and facilities for resting)
- Streets with limited circulation

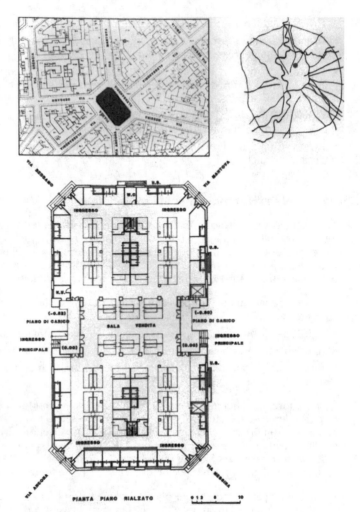

FIGURE 15.3   The rehabilitated market Nomentano in Rome.

Because the requirements are so all-inclusive, there is a risk that they will not be applied. Comprehensive renewal is difficult when different institutions and planning levels are involved in spaces that include traffic control, parks, and residential neighborhoods. Interventions must maintain the existing urban fabric, especially in historical areas. Responsive design should result in the application of regulations that are sensitive to the morphology of the area while avoiding solutions that result in new social exclusion.

Meeting the challenge of achieving accessibility for all, requires transversal design criteria that link every level of the urban scale, from one building to the entire city. Accessibility must be one of the requirements that guide normal design decisions and should not be considered only after these decisions have been made.

These are the *main guidelines* for universal design in the urban environment:

- To improve accessibility through different programs of urban and building upgrading.

- To apply accessibility as a transverse discipline that cuts across and connects sectors.

- To house older people in residential buildings, integrated with other users.
- To encourage older people to use existing and new facilities and services located in the community, through accessible routes and technological supports.
- To avoid creating "pedestrian islands." Circulation routes must allow a choice of transportation modes both on foot or by vehicle, for a diversity of people with their own personal preferences.
- To link morphological and functional elements by circulation routes, while permitting the basic goals of circulation such as walking, traversing, entering, using, and resting.
- To ensure adequate levels of comfort, suitability, and safety of pedestrian routes during the day and the night.

The Accessibility Plan of the Historic Center in Pistoia

This plan analyzed the accessibility levels of the historic center of Pistoia, a Tuscan municipality of 92,000 inhabitants. The study was carried out in early 1998, and it has now entered the design phase (Comune di Pistoia, 1999). With T. Alderighi (coordinator), A. D'Innocenzo, and C. Salvini as architects, the objectives of the plan are to facilitate and sustain the daily activities of people in the historic zone, regardless of their physical conditions. These include:

- To walk or "ambulate" for pleasure, or to visit tourist places or enjoy events in the historical area
- To traverse open and closed spaces to reach desired destinations
- To be able to enter buildings and spaces where communal activities take place
- To use services and facilities located in the historical area in conditions of comfort and security

Therefore, the plan must examine both the physical and functional characteristics of the historical area. A specific evaluation rating of accessible levels was selected for each morphological and functional element, by sampling. The selection was based on the intensity of use of circulation routes at various hours. The results were synthesised into a thematic map to describe the total accessibility of the historical center, which was then used to define the scope of the rehabilitation efforts. Specific studies, such as the following, aided the process:

- The public parks and garden plans; the intensity of use over the hours of the day; the pedestrian route plan; and the traffic plan, both private and public, including bicycles and vehicles
- The results of discussions between professionals and urban user associations, including older people, people with disabilities, and so forth

The main design guidelines of the *Accessibility Plan* focused on:

- Improving the existing pedestrian network, including strengthening its structure, security, and usability
- Improving the quality of life conditions in the historical center
- Preserving the morphological character of the existing urban fabric
- Assigning priority to rehabilitation operations

The historical center of Pistoia is not an isolated pedestrian island but is well-linked to different attractive destinations, encouraging the flow of people. This enhances the safety of the pedestrian pathways. Multiple uses and different modes of access reduce the dangers of "unsafe places," both during the day and the night.

The Accessibility Plan chose to first renovate those pedestrian routes where the main urban functions are located, and where it is possible to highlight a visual axis to foreshorten the urban vista. The main pedestrian routes have conveniently sheltered furniture for sitting, resting, phoning, refreshing, and informing. Additional services and facilities were planned. The existing garden areas were reorganized and refurbished to increase usability for people with disabilities.

Urban Pedestrian Pathway, Rome

An urban pedestrian pathway crossed the via del Corso in Rome, linking two famous ancient squares, Trevi and Pantheon (see Fig. 15.4). The intention of design was not only to focus on architectural barriers, but also to upgrade the environment with new paving and different types of lights (Progetto Roma, 1998). Attention was paid to all of the technical elements of surfacing, such as manhole covers, drainage, and entrances to the existing and restructured underground service ducts; to the main pathways for people with motor disabilities; and to facilities for tourist information, as well as acoustic and visual signage.

The design raised the road level up to the sidewalk level, in order to obtain a single-level walkway. The new paving included three large longitudinal fields of rough cobblestones, resembling the traditional surface of streets. The central pathways consisted of smooth polished cobblestones, which are easier for people with motor disabilities to use. The pedestrian route runs directly above that of the central underground services, incorporating the manhole covers in the design.

The two lateral walkways had a different design composed of closely fitting cut cobblestones with a darker and rougher surface. The longitudinal design of the paving was interfaced with transverse banding marking the different technical elements. The bar sections were placed in pairs, with a 60-cm distance between them, which is the dimension of the technical element they enclose.

Accessible Venice

The Agency for Promoting Tourism in Venice distributed a map of the city showing the most accessible pathways for persons with physical disabilities (Informa Handicap, 1999). Venice was chosen for its unique characteristics (and therefore, difficulties in renovating for accessibility). It also shared the design challenges of the historical centers of many other Italian cities. Nevertheless, it is important to note that accessibility maps are available for many Italian cities. Some are also posted on related Internet sites.

Map areas colored yellow are reachable by persons with disabilities by crossing only bridges with platform stair lifts, the keys for which can be obtained in any agency office in Venice. The different areas of the city can be visited by using accessible boats. Accessible stops are located on the map by black dots, while bridges with stair-lifts are identified by small black squares. Accessible toilets are marked by the legend "WC," the European symbol for toilet.

Information on how to reach Venice by air, rail, and car, including information on accessible parking (some available free of charge) is available in the information pamphlet. Though water taxis are still not available for people with disabilities, the Administration Council has already resolved to buy accessible transport. Water taxis serving the historical center are not accessible. The stops for ship lines are accessible, even though the ramps are too steep in some cases.

Park and Sport Facilities in Trento

This huge urban improvement project was located in the neighborhood of Mattarello in the city of Trento. It was provincially funded and built on land provided by the municipality. It

FIGURE 15.4 A new pathway in the center of Rome.

was begun in 1994 and finished in 1997 by the local Building Agency ITEA (ITEA, 1999). The park included areas for sports, leisure, and a children's playground (see Fig. 15.5). The area totaled about 25,000 m², adjoining a football field. The park included:

- The visitor center in the central area of the park, with a bar, a warehouse, toilets, and service rooms for the bowling alley.
- The football field, with a new structure for 400 spectators, including a first-aid room, toilets, and dressing rooms.

FIGURE 15.5   Park and sport installation in Trento.

- A central multifunctional area, with a volleyball court, a roller-skating rink, and space for traditional festivals.
- Two tennis courts, located in one corner.
- Green space at the center, reserved for informal football and running games, children's play, and water play for toddlers.
- Two new accessible parking spaces, in addition to the already existing one. Due to careful design, full accessibility (including pedestrian pathways) was made possible for all parts.

15.7   DESIGN GUIDELINES FOR INFRASTRUCTURE AND TRANSPORT

This section will deal with public and local transport. The assistance provided by the Italian railway system for passengers with disabilities, and the signage system for the orientation of people with visual impairment or loss, are all intended to achieve more comfortable pedestrian accessibility in a neighborhood of Rome. The railway system is selected as an example rather than airports and the air network, because in Italy trains are used more frequently than airlines for domestic travel.

Public Local Transport

An inquiry carried out by the Federtrasporti Association, which represents local public transport agencies in Italian cities, showed that only 32 agencies of a total of 148 are equipped with accessible transport vehicles (Ministero dei Trasporti, 1998). Of the 10,239 public buses in use, only 886 were designed to be also used by people with problems of mobility. Though these results are not encouraging, there are new signs of change in the supply of public transport.

During a congress in 1998 entitled "Mobility: A Right for Everybody," some critical factors were pointed out (Ministero dei Trasporti, 1998):

- The quantitative and qualitative aspects of demand by users with physical disabilities are not well known.
- The majority of the transport system is still not accessible.
- The purchase of appropriate accessible buses must be accompanied by a technical review of the entire system of routes and bus stops, because of barriers arising from the height of the pavement and of improper parking blocking bus stops. Also, there is a lack of integration, coordination, and cooperation among the different parts of the transport system.
- Information on the type of services supplied is not widely disseminated.

In many Italian cities new flexible transport services are growing, allowing better mobility of users with special needs. For example, Florence has operated a transport service for persons with disabilities for the last five years. The service called "Personal Bus" operates on demand using telematic and information technologies. This experience has shown that it is possible to plan and manage a transport system that is accessible and also economically efficient.

The Railway System: Assistance and Support

In 152 railway stations across Italy, information and assistance for people with disabilities is provided by Customer Assistance Centers (FS, 1999). The services include: train information; seat booking; wheelchairs; escort from the station entrance to the train or to other railway connections, and vice versa; lift truck with elevating platform and porter service on request, and only for one piece of luggage. By calling at least 24 hours before a journey within Italy, or three days before international travel, the disabled traveler can telephone the center directly to arrange for services.

Passengers using wheelchairs are asked to telephone ahead to reserve wheelchair places on the train, as well as the use of special elevating platforms, where available. At that time they can also specify if they will travel with their own wheelchair, or if they need to borrow a wheelchair from the Italian Railway Company. The Customer Assistance Center has one or two wheelchairs available upon request at the station of departure, a connecting train station, or the station of arrival, and also for transfer from the train to other modes of transport such as a taxi or private car. Passengers using a wheelchair, either rigid or collapsible, are entitled to travel in railroad cars bearing the international symbol of disability. Access to the car is by lift truck, operated by railway staff. Passengers using collapsible wheelchairs may also travel in ordinary cars, but they must inform the Customer Assistance Center in advance of their requirements for assistance.

Extra services are provided for a fee in the following cities: Milan, Bologna, Florence, Rome, Naples, Padua, and Venice. Passengers can be transported between their homes and the station if traveling on the Eurostar train service. On a certain number of Intercity/Eurocity and Eurostar trains there are cars with facilities for two passengers using wheelchairs, plus two accompanying persons. These cars are equipped with an accessible toilet facility and a travel area with a large window, a table, a grab bar, and a call button. Cars may be accessed by lift truck, but electric wheelchairs are not allowed. Since not all stations are equipped with the lift truck, the traveler must check specifically for this service in the information booklet on available assistance. At the present time (1999) 82 out of a total of 152 principal stations have this service, while 70 are not yet equipped.

Though not yet widely adopted, there are some experimental projects to support the mobility of people with visual problems. Pilot lights were installed in three railway stations—Arezzo, Florence, and Ferrara—in order to orient persons with low vision. Because of the high costs associated with the installation and maintenance of the system, it is unlikely that this solution will be extended to other cities. However, other orientation devices, such as tactile markings in corridors, grab bars with contrasting colors, and Braille instructions, have been tested and will probably be implemented when railway stations are upgraded. One of them is the tactile pathway for persons with visual or hearing impairments. The DPR 503/96 regulation makes the accessibility of urban environments for people with visual and hearing limitations obligatory. Therefore, agencies for the subway and railway systems are progressively modifying stations to improve access. Professionals with technical expertise, in cooperation with the Association of Citizens with Disabilities, are also developing some guidelines for the safety of disabled travelers.

The most popular accessibility feature is the LOGES system which is a guided pathway for the visually impaired, providing direction and safety information by four different types of cues. These are, according to Italferr, the Italian Railway Association for design and engineering, tactile cues sensed by foot; tactile cues for cane users; sound cues; and, color contrasts. The LOGES system is now being implemented in renovated stations in Rome: Figure 15.6 shows Aurelia station. The tactile pathway starts at the drop-off area, or the public transport stop at the station, and leads to the principal railway services—waiting and ticket areas, vending machines, toilets, telephone booths, drinking fountains, elevators, platforms, escalators, and stairs. In the main stations, the tactile pathway also links other public spaces such as shops, police stations, and luggage storage. The map of the tactile pathway is available in every railway station, installed on suitable reading desks or on the walls at appropriate points (e.g., at the beginning of a pathway, at the end, at each different level, etc.).

FIGURE 15.6   Pathway supporting people with visual impairment; detail of Aurelia Station, Rome.

Testing Flexible Mobility in the Pietra Papa Neighborhood, Rome

As part of the European research program COST, research on urban rehabilitation and pedestrian mobility is being undertaken in Pietra Papa, a semisuburban area of Rome. The objective was to identify design solutions and techniques of intervention that result in the improvement of the quality of the urban environment, as well as pedestrian safety (Marticingh, 1998). Experiments tested the feasibility of functional and environmental transformation, especially the pedestrian spaces of the city. The entire project was intended to be exemplary for the quality of life that is possible in the city.

The experimental application allowed verifying the feasibility of a functional and environmental transformation of the city sites in general, specifying the problems related to the use of pedestrian spaces. Further, it encouraged the study of design solutions used to slow traffic and to improve pedestrian spaces in other European countries, which meet the standards for the purpose of evaluating their compliance with Italian regulations. On the one hand, the challenge was to control vehicular traffic for pedestrian safety without hampering the flow, while on the other, to increase the quality of the pedestrian experience by improving comfort and attractiveness. Pedestrians must be treated as the principal users of such areas and their needs must be carefully studied, while design solutions for mobility were integrated with overall urban renewal.

Potential strategies included:

- The use of islands, where pedestrians have priority and traffic circulation is reduced
- New devices to slow down car traffic
- The coexistence of pedestrian mobility with vehicular traffic

Five types of pathways were defined: (1) exclusively for pedestrians, except for emergency and service vehicles; (2) principally for pedestrians, with cars allocated only 50 percent of the street; (3) pedestrian areas from which A and B zones are reachable and where cars have between 25 and 50 percent of the street; (4) raised sidewalks for pedestrians, occupying 25 percent of the street width, for walking to shopping and services; and (5) service sidewalks with a minimum width of 1.5 m, to guarantee safe pedestrian mobility. Each pathway and pedestrian crossing can also be used by people with different needs, according to their preference.

Design criteria for the square retained the concept of green space, which permits appreciation of the change of seasons. The space was paved with traditional stone materials, "sampietrini" and travertine, and was enhanced by the installation of a fountain, since Romans enjoy sitting on the edges of fountains in the summer. Public lighting was used to enhance safety from assaults or car accidents.

15.8   UNIVERSAL DESIGN: A TOOL FOR URBAN RENEWAL

In the coming decades, design research in Italy will focus on improving the quality of existing neighborhoods at the urban scale. Most of the new construction envisioned in town plans has been constructed in spaces allocated for them. The criteria to be used in renewing the existing urban fabric are being experimentally tested at the local and governmental levels. A key topic in design research is the improvement of the functional aesthetic and social integration of badly degraded existing neighborhoods located in the historic centers, or at the periphery of cities. A common rule guiding the most recent experiments is to achieve multiple and diverse positive effects by the renewal, ranging from employment to health and social benefits. The improvement of the existing environment can require interventions in the following three areas:

1. *Social.*   Favoring a mix of income levels

2. *Functional.*   Supplying neighborhoods with an integrated system of facilities, urban spaces, and pathways accessible to the majority of users

3. *Aesthetic.*   Improving environmental quality and morphological identity

Neighborhood Agreements

Neighborhood agreements are a key tool for the program of urban rehabilitation recently launched by the Ministry of Public Works (1998). This program, funded experimentally, is accessed by municipalities through a national competition for the rehabilitation of public residential neighborhoods. The winning municipalities carry out an urban improvement program on residential buildings and public open spaces, including pedestrian routes, squares, and neighborhood facilities, as well as on residential services for special groups, including public spaces for the elderly, young people, and so forth.

There are four experimental objectives for housing improvements under this program:

1. *Morphological quality.*   Meaning the conservation and the modification of the urban fabric

2. *Ecosystem quality.*   Focusing on bioarchitecture and urban ecology

3. *Level of inclusion.*   Meaning accessibility for disabled people, flexibility, and increased value to new users.

4. *Overall quality.*   Emphasizing quality control and warranty of the product

The program receives interministerial financial support. In addition to funds for construction from the Housing Ministry, the program receives contributions from the Labor Ministry for the hiring of unemployed people or public utility workers. These workers can be employed in building rehabilitation enterprises, in social/cultural activities, or in activities supporting special groups, such as older citizens, young at risk, and people with disabilities. Through activities such as consultations, cultural events, planning competitions, and evaluations of proposed designs, municipalities must have a process of user participation in the implementation of the program. Two case studies follow.

Neighborhood Agreement for the Foro Boario area in Pinerolo.   The first case study is the social and environmental rehabilitation program of a dilapidated area located in the urban center of Pinerolo, a municipality of 35,000 inhabitants about 36 km from Torino (Comune di Pinerolo, 1999). The Foro Boario area, covered by the Neighborhood Agreement of Pinerolo, included a complex of highly degraded residential buildings built in the late 1950s and some unused buildings that used to be a slaughterhouse. This area was lacking in primary services. Therefore, users gravitated to the adjacent neighborhood, which was difficult to reach because a canal separated these two areas. The pedestrian routes were disjointed, lacking both aesthetic and functional identity. Furthermore, the area lacked a gathering place for inhabitants, public transportation, or community social activity. The existing four-story public residential buildings, occupied almost entirely by older people, were inadequate (see Fig. 15.7). They lacked heating and elevators and were full of architectural barriers inside and outside the buildings. The pedestrian routes and open spaces were neither usable nor secure, nor were they adequately equipped. Therefore, the older residents were isolated and suffered from loneliness and depression.

The experimental objectives of the Neighborhood Agreement in Pinerolo were:

• To improve the quality of the existing residential buildings in order to meet the requirements for accessibility, comfort, and security of the older residents

• To insert new dwellings into the existing residential context, for the purpose of improved social integration

FIGURE 15.7   Pinerolo building before rehabilitation.

- To provide new services and facilities, particularly for the elderly people
- To introduce morphological characteristics to create spatial identity and reduce neighborhood isolation
- To encourage intergenerational integration to help ameliorate the aging of the neighborhood and to enhance interaction with other parts of the city

The agreement resulted in major renewal and improvement[5] through the environmental and technological rehabilitation of 68 existing dwellings, including the removal of architectural barriers; the addition of elevators, ventilation, and heating; and improvement of the facades of the buildings. A new residential area consisting of two buildings around a central square was constructed. The square was for year-round use, and it had diagonal pedestrian routes that crossed in the center, and also a winter garden heated by solar energy.

The residential building located on the south side of the square contains 30 dwellings, with floor areas of 33 and 48 m^2 respectively. It has units on the ground floor for the elderly, and dwellings on the first floor for young couples. The building located on the north side was designed for mixed use, with a communal space for the elderly, commercial and artisan spaces, and 25 dwellings. The square is the new neighborhood center, where inhabitants get together to use services and from which residents enter their houses. The elderly people who live in the old as well as in the new buildings are attracted to the garden square, where they can find facilities to use, to rest and to enjoy. Finally, the environmental quality of the pedestrian routes, gardens, and linkages of the neighborhood to the city was improved through modifications for accessibility, usability, and security.

Neighborhood Agreement for the Pietralata Vecchia area in Rome.   The municipality of Rome collaborated with IACP, the Agency for Social Housing, to implement the Neighborhood Agreement for Pietralata Vecchia, a particularly degraded area at the periphery of Rome. It was a dense and large settlement, disadvantaged by social exclusion. Above all, children had no safe spaces where they could gather or play (Di Michele, 1999). The program's objectives were to renovate the public buildings, which had been built in the 1950s, and to develop an integrated system of protected pathways safe for children and older residents. These pathways linked existing buildings with a variety of existing sport centers, churches, and so forth, and new facilities such as the new day center for older people.

15.9 FUTURE TRENDS

In the future the city "designed for all" will be a widely accepted goal for urban and building policies. This is not only due to the increasing number of older people[6], but also because of citizen demand for more appealing living environments. The need for a more accessible, safe, comfortable, attractive, and functional city is growing. Urban spaces for work and leisure must be open 24 hours of the day, in line with social changes requiring flexibility in the use of the city.

Design attention has shifted from merely the home to its relationship to the city. The quality of life has risen in the last century, permitting better social models, including an increase of cultural activities, free time, and recreational travel. The older people of today (and more so those of tomorrow) are the products of a period of economic growth very different from that following the last World War. Citizens have higher expectations and they are cosmopolitan in the cultural and technological products they consume. Persons with disabilities are no different. Today, people with visual or motor impairments can participate in sports and activities once believed beyond reach, such as skiing and horseback riding. It is their right to travel independently and to enjoy full independence in their preferred cultural or leisure activities—like any other citizen. The societal responsibility is to meet the challenge in the urban environment by providing quality of life for every resident, regardless of cultural, ethnic, physical, age, or sexual characteristics. Therefore, at the start of the new millennium Italian policies are in line with the notion of universal design, which seeks to include everyone. This approach still requires harmonization at the European and also the global levels. The shift to universal design will require investments that may strain the economies of Europe, which are growing at different rates. Each country also has its own cultural and social models for dwellings and cities. Thus, it is essential to learn from international experience with different approaches to regulation and implementation, and to share knowledge about potential design choices for accessibility.

Italians believe that they must foster a design culture that seeks to enable an independent, active, and integrated life for all users. The principal tendencies for the future are the following:

- Renewing the most deteriorated neighborhoods; modifying and equipping them for accessibility, functionality and urban structure
- Improving safety and security in cities; intervening to reduce risk and to promote a culture of solidarity, participation, and social control (Cardia, 1999)
- Developing flexible and compatible transport systems with integrated planning for the various modes of transport such as a pedestrian, bicycle, public and private transport
- Promoting citizen participation and urban consensus for urban renewal policies in order to meet the real needs of users

According to de Luzenberger (1999), the experimental Neighborhood Agreements are among the most advanced European processes for citizen participation.

15.10 CONCLUSIONS

Italy is taking concrete steps toward the Principles of Universal Design through its approach to the design of accessibility in urban areas. Despite many constraints, progress is being made—particularly the preservation of historic heritage architecture, which tended to be neglected during earlier urban transformations and renewal periods.

The new approach is based on a policy of differences, which recognize different users, as opposed to the policy of standardization which is designed for the "average man." The differences among people have become a positive value, which directs urban strategies and prevents social exclusion. With the free flow of capital, goods, and services through globalization, the city must cater to a more complex and evolving society. The city for all will be a necessity driven by deep social changes that are forecast for the future.

Within this framework, universal design will result from participation of society in its different representative forms, including volunteers, associations, and so forth, in urban renewal policy, while the various technical professionals will need to improve their strategies for wider diffusion of universal design.

The improvement of accessibility in the city should be the first criterion of design. The Italian policy of requiring different levels of accessibility shows the commitment to the principle, while developing an integrated approach to urban and residential renovation. The system of priorities ensures an increasing level of accessibility from total accessibility to visitability, thereby improving the urban environment for all users.

Flexible design solutions will ensure that urban environments improve the quality of life of all citizens over time. Coordination between planning, design, services, social assistance, and mobility will become a tool of critical importance.[7]

15.11 NOTES

1. From the application point of view, there is no difference between Legge and other decrees (DLgs, DM, DPR, DPCM, circolari): they must be obligatorily respected as norm at the same level. The origins of the initiative and the approval are different: Law is established by parliament, while the other decrees are delegated by parliament to government or a ministry, generally for a specific topic.

2. A collection of national and regional laws on the concepts of accessibility and architectural barriers is provided in a book by F. Vescovo (Vescovo, 1990). See also Chap. 26 by F. Vescovo in this book.

3. For more information on the notion of accessibility in cities, see also the book by D'Innocenzo et al. (D'Innocenzo et al., 1991).

4. Another example of social and public spaces is the renewal of the theater La Fenice in Senigallia, in the province of Ancona, which is particularly interesting because it had to preserve the archaeological ruins of the Roman period that were underneath (Righetti, 1999).

5. The experimental program is coordinated by A. D'Innocenzo and G. Pavoni.

6. Italy will lead all other European countries in the percentage of persons aged 65 and over (26.5 percent), according to the European Union population projections for the year 2030. (The Finnish Environment, 1999).

7. The authors are indebted to Satya Brinks, a wonderful friend and a great colleague, for her review of the manuscript for this chapter. They are also indebted to Annalisa Marinelli for her technical support when visiting the Aurelia railway station, and to ITALFERR, the Italian railway association for design and engineering, for information on the LOGES system.

15.12 BIBLIOGRAPHY

Cardia, C. "Sicurezza urbana," *Costruire,* n. 193, dossier n. 44, pp. 83–90, 1999.

Circolare Ministero Lavori Pubblici, *Standards residenziali,* 20 January, n. 425, 1967.

Comune di Pinerolo, *Contratto di Quartiere "Area ex Foro Boario,"* 1999.

Comune di Pistoia, "Piano/Progetto guida per l'accessibilità urbana del Centro storico di Pistoia," 1999.

CNR ICITE, Comune di Roma, *L'Edilizia Annonaria in Italia,* parte prima, Palombi, Roma, pp. 80–81, 1991.

Decreto del Ministero dei Lavori Pubblici, *Prescrizioni tecniche necessarie a garantire l'accessibilità, l'adattabilità e la visitabilità negli edifici privati e di edilizia residenziale pubblica sovvenzionata e agevolata, ai fini del superamento e dell'eliminazione delle barriere architettoniche,* 14 June, n. 236, 1989.

Decreto del Presidente della Repubblica, *Regolamento concernente norme di attuazione dell'art. 27 della Legge 30 marzo 1971, n. 118 a favore degli invalidi civili in materia di barriere architettoniche e di trasporti pubblici,* 27 April, n. 384, 1978.

Decreto del Presidente della Repubblica, *Regolamento recante norme per l'eliminazione delle barriere architettoniche negli edifici, spazi e servizi pubblici,* 24 July, n. 503, 1996.

De Luzenberger, G., *EASW, European Awareness Scenario Workshop,* CUEN-Ministero dell'Ambiente, 1999.

Di Michele, A., Roma, "Contratto di Quartiere 'Pietralata Vecchia', A misura di bambino," *Edilizia Popolare,* n. 261–262, pp. 80–87, 1999.

D'Innocenzo, A. et al., *La città accessibile,* Gangemi, Napoli, 1991.

Edilizia Popolare, *Numero speciale,* 255, 1998.

Ferrovie dello Stato, *Services for disabled passengers,* 1999.

Galli, C., "Progetto sperimentale di RSA con alloggi autonomi e centro diurno, Bagnacavallo, Ravenna," *SAIEDUE, 'Anziani e ambiente costruito,'* pp. 72–73, 1995.

Informahandicap, 1999. sito web: www.comune.venezia.handicap.it.

ITEA, "Baselga di Piné, Scuola media," *Abitare in Trentino,* pp. 127–129, 1999.

ITEA, "Trento (Mattarello), Parco attrezzato e impianti sportivi," *Abitare in Trentino,* pp. 61–64, 1999.

Legge, *Disposizioni per la formazione del bilancio annuale e pluriennale dello Stato,* 28 February, n 4, 1986.

Legge, *Disposizioni per favorire il superamento e l'eliminazione delle barriere architettoniche negli edifici privati,* 9 January, n. 13, 1989.

Legge, *Legge quadro per l'assistenza, l'integrazione sociale e i diritti delle persone handicappate,* 5 February, n. 104, 1989.

Legge Regione Lombardia, 20 February, n. 6, 1989.

Marticingh, L., "Una proposta di intervento a Pietra Papa, Roma: aspetti caratterizzanti ed apporti innovativi," *Paesaggio Urbano,* n. 2, pp. 33–42, 1998.

Ministero Affari Esteri, Comune di Roma, Università degli studi di Roma La Sapienza, "Percorso pedonale protetto Trevi-Pantheon," *Progetto Roma, La città del 2000,* pp. 56–57, 1992.

Ministero dei Trasporti, *Conto Nazionale Trasporti, Supplemento,* pp. 235–244, 1998.

Modulo, *Numero speciale "Case per Anziani,"* 211, 1995.

Righetti, P., "Teatro sospeso," *Modulo,* 248, February, pp. 24–28, 1999.

The Finnish Environment, *Housing for Older People in the EU Countries,* Ministry of the Environment, Helsinki, 1999.

15.13 RESOURCES

Alzheimer Italia
via T. Marino, 7
20121 Milano
Italy

**Association for the Development of Information Technology
Systems for People with Disabilities (ASPHI)**
via Arienti, 6/8
40124 Bologna
Italy
Telephone: +39 051277811; Fax: +39 051224116
Web site: www.asphi.it

Italian Association for Muscular Dystrophy (UILDM)
via P. Santacroce, 5
00167 Roma
Italy
Telephone: +39 066638149; Fax: +39 066638148
Web site: www.uildm.org

Italian Association for People with Motorial Disabilities (AIP)
via C. Cerbara, 20
00147 Roma
Italy

Italian Railway Company
via di Castel Bolognese, 30
00153 Roma
Italy
Disabled People Service: Telephone: +39 064881726
Fax: +39 0647307617
Web site: www.fs-on-line.com/

Ministry of Public Health
Lungotevere Ripa, 1
00153 Roma
Italy
Telephone: +39 0655945526
Web site: www.sanità.interbusiness.it

Ministry of Public Works, Department of Urban Areas and Residential Buildings
via Nomentana, 2
00161 Roma
Italy
Telephone: +39 0644125115; Fax: +39 0644124122
Web site: www.llpp.it

Ministry of Social Affairs
via V. Veneto, 56
00187 Roma
Italy
Disabled People Service: Telephone: +39 0648161350
Fax: +39 0648161331
Older People Service: Telephone: +39 0648161341
Fax: +39 0648161339

Ministry of Transportation
piazza della Croce Rossa, 1
00161 Roma
Italy
Telephone: +39 0644101
Web site: www.trasportinavigazione.it/Principale.htm

National Association of Municipalities (ANCI)
via dei Prefetti, 46
00186 Roma
Italy
Telephone: +39 06688091; Fax: +39 06621313
Web site: www.anci.it

European Guide for Same Opportunities
HELIOS II, Guida Europea di Buona Prassi, Commissione
Europea, DGV (the appendix contains a list of other guides
published by member states)

Tourist Guides on Accessibility
http://andi.casaccia.enea.it/Andi/COIN/TOUR/italiano/guide.
htm

CHAPTER 16
ACCESSIBILITY STANDARDS AND UNIVERSAL DESIGN DEVELOPMENTS IN CANADA

Shauna Mallory-Hill, M.Arch.
Technical University of Eindhoven, The Netherlands

Brian Everton, B.I.D., P.I.D.I.M.
ProductABILITY Consulting, Winnipeg, Manitoba, Canada

16.1   INTRODUCTION

This chapter explores how universal design has evolved in Canada in two ways: as a new design philosophy and as a legal requirement. Usability as primary design goal is a fundamental change in the way most designers think about and practice design. Only 50 years ago, no one in the building industry would have thought that making something usable meant anything more than taking into account standards and measures based on a standing male form. Today's definition of usability means taking into account a much broader set of design requirements and, in some cases, some extensive lateral thinking. Emerging out of changing market demands and improving awareness, universal design philosophy is gaining popularity among Canadian designers.

The other aspect of universal design that is discussed here is its role as a legal requirement. Usability is mandated in Canada in two forms of legal requirements: building codes and human rights legislation. The evolution of these two types of legislation reflects the evolution of the universal design movement in Canada over the past 30 years. However, many issues have arisen from trying to achieve universal design through legal means, particularly with regard to application and enforcement.

16.2   BACKGROUND

Universal design is the design of products and environments to be usable by all people, to the greatest extent possible, without the need for adaptation or specialized design. The intent of universal design is to simplify life for everyone by making products, communications, and the built environment more usable by as many people as possible at little or no extra cost. Universal design benefits people of all ages and abilities. (Center for Universal Design, 1999)

Today's concept of universal design—to make designs that are usable by the widest variety of persons possible—is a relatively new definition. Universal design as an architectural principle used to mean something quite different. The ancient Roman architect, Vitruvius, described the universal harmony and perfection of the human body in terms of how a "well-built man fits with extended hands and feet exactly into the most perfect geometrical figures, circle and square" (Wittkower, 1988). This single observation has influenced the thinking and, ultimately, the designs of generations of great masters since, particularly those of the High Renaissance such as Bramante and Leonardo da Vinci. Cesariano was even so bold as to assert that with the Vitruvian figure "one can define the proportions of everything in the world" (Wittkower, 1988). In more recent examples, one sees the outstretched hand of a male figure define Corbusier's universal measuring device "*Le Modulor*" (Corbusier, 1961) and the detailed anthropomorphic measures for a human (male) in *Architectural Graphic Standards* (Ramsey and Sleeper, 1988), a widely used reference book. For a very long time now, great design has been for and about the celebration of the optimal male form or "universal man" (Fig. 16.1).

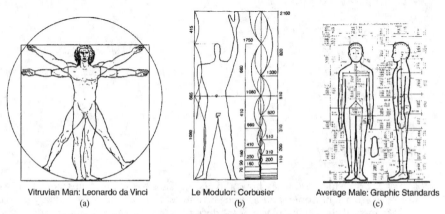

Vitruvian Man: Leonardo da Vinci   Le Modulor: Corbusier   Average Male: Graphic Standards
(a)   (b)   (c)

FIGURE 16.1 "Universal man": (*a*) "Vitruvian Man" (Leonardo da Vinci), (*b*) "*Le Modulor*" (Corbusier, 1961), and "Average Male" (Ramsey and Wittkower, 1994).

The concept of a universal man is based on a high ideal of human form that most people simply do not fit. Yet, surprisingly, it was not until the 1970s that design reference books of anthropometrics and measures, such as *Architectural Graphic Standards,* started to include dimensions for females and children. Dimensions for persons with disabilities, such as a person seated in a wheelchair, have only started to be included in the past 10 years. Even with the inclusion of a broader scope of human dimensions, measures of human form are still based on averages. Consequently, people who do not fall into the definition of average are forced to adapt or change their behavior in order to be able to function in environments based on these measures.

Given the long-standing history of design standards and guidelines being based on the universal man and average dimensions, the new concept of universal design is quite revolutionary. It represents a substantial and fundamental shift in perspective—from considering people in environments to be disabled to considering environments and products to be disabling.

Origins of Universal Design in Canada

It is not easy to identify precisely what motivated this change of design philosophy in North America, and, in particular, Canada. Although the formal definition of universal design is attributed to the American architect Ronald Mace in 1985, the origins of universal design thinking in Canada appear before this time, particularly from a cultural perspective.

It could be said that Canadians have always been sensitive to the recognition and accommodation of a variety of people. Canada has a long history of government policy not only to promote the immigration of people from foreign lands, but also to encourage cultural diversity after they arrive, the so-called Canadian mosaic.

This cultural diversity is so strongly defended that it has nearly driven the country apart several times. Canada's founding nations, or "two solitudes"—the English and the French—appear to be particularly at odds with one another. The province of Quebec, home of the majority of Canada's French-speaking population, has threatened to separate from Canada several times. One of Canada's most contentious culturally based universal design problem is the accommodation of both official languages. In 1977 the government of Quebec passed *Le Charte de la langue francaise* dictating that, among other things, all signage in the province must be in French only. However, over the years Canada has, through a commitment to consensus and compromise, managed to remain whole. Indeed, Canada has an internationally recognized reputation as a peacekeeping nation.

Although cultural diversity has had its place in the Canadian consciousness for a long time, it was not until the postwar era that the most significant changes in awareness about physical diversity seem to have begun. In the 1950s, the return of veterans with disabilities, the polio epidemic, and advances in medical technology were factors that combined to result in greater independent living of persons with disabilities in Canada. The eventual deinstitutionalization of persons with disabilities by the end of the decade led to the further growth of advocacy and support organizations willing to fight for and encourage research in the reduction of barriers in the physical environment. These organizations eventually brought about the first barrier-free design building code, as discussed in Sec. 16.3.

During the 1960s and 1970s, examples of buildings with radical new features began to appear. For example, Ten Ten Sinclair, an apartment block built in Winnipeg in 1975, is a building with many universal design features. Ten Ten was designed on the basis that the suites had to work for persons with and without disabilities. Although Ten Ten has been updated several times since, the original designers were thinking usability well beyond the norms of the period (see Fig. 16.2). Most similar apartments of the time had high windows (above 48 in), yet the window sills in the bedroom and living areas of Ten Ten are low, providing a much more enjoyable space in which to sit. The carefully laid out washrooms provide needed storage as well as turning space. The lever handles on the passage doors work for people with snow-covered mitts and arthritic hands. A low shelf for groceries and purses next to apartment mailboxes is convenient for all tenants.

(a)   (b)

FIGURE 16.2   Comparison of (*a*) standard and (*b*) Ten Ten's universal design floor plans.

Independent Living Movement in Canada

By the 1980s the Independent Living Movement had formally taken hold, developing out of the earlier consumer movement of the 1970s, in particular the Coalition of Provincial Organizations of the Handicapped, and the philosophies of the American disability theorist Gerben DeJong. DeJong observed that current society had focused on the limitations of the person, but that the future held opportunities for disciplines outside the rehabilitation and medical fields to have an impact upon "a wide variety of environmental variables" (DeJong, 1981). The mission of the Canadian Independent Living Movement (also discussed in Chap. 6, "User/Expert Involvement in Universal Design," by Ringaert) is: "To promote and enable the progressive process of citizens with disabilities taking the responsibility for the development and management of personal and community resources" (CAILC, 1999). In 1982 the United Nations declared International Year for Disabled Persons, and the Canadian government released the Obstacles report, lending further legitimacy to disability issues.

Following the introduction of the Americans with Disabilities Act (ADA) in 1990 in the United States, barrier-free design gained even more attention in Canada. Learning from the American experience with the ADA and Canada's own barrier-free legislation over the past decade, designers have become aware that a more user-conscious approach benefits more people than just those with disabilities. Today, barrier-free design is gradually evolving into universal design as a new focus for researchers, educators, designers, and ultimately legislators in this area.

16.3   EVOLUTION OF ACCESS LEGISLATION IN CANADA

The evolution of access legislation began in Canada with barrier-free design. It was here, in the development of barrier-free design codes and human rights legislation, that the concept of designing something to make it usable for someone other than the so-called average person was first introduced into law.

Building Codes

The publication of *Building Standards for the Handicapped 1965, Supplement No. 7 to the National Building Code of Canada* (NRC, 1965) marked the first attempt to make buildings accessible and usable by persons with physical disabilities in Canada. The requirements outlined in Supplement 7 were not mandatory, but, because of the pressure from local organizations, several cities in Canada adopted the supplement as part of their building codes. Supplement 7 was eventually incorporated into Part 3 of the National Building Code of Canada (NBCC) in 1985 (NRC, 1985).

The accessibility requirements of Section 3.8—Barrier-Free Design, covered in NBCC95 Part 3—Fire Protection, Occupant Safety and Accessibility of the NBCC, apply to all new buildings that, regardless of area or height, contain assembly occupancies, care or detention occupancies, or industrial occupancies and to all buildings of all other categories that exceed 600 m² in area or three stories in height. The only total exemptions for the accessibility requirements are for agricultural buildings or buildings that are not intended to be occupied on a daily or full-time basis, such as pumphouses, automatic telephone exchanges, or sub-stations. Notably, Part 3 requirements do not apply to residential dwellings and small shops.

Barrier-free requirements do, however, apply to residential dwellings and many small shops and businesses under NBCC Part 9—Housing and Small Buildings. The most common application of these requirements is to new apartment buildings. However, apartment buildings that are three stories or less without an elevator do not need to provide barrier-free environments above the entrance level. Access to individual suites is also not required if they are higher than 600 mm above the entrance level.

With the exception of fire codes, building codes are not normally retroactive. However, building codes do apply to new extensions or alterations, or when an existing building has a substantial change of occupancy. In response to a growing movement toward the renovation of existing building stock in Canada, the Canadian Commission on Building and Fire Codes (CCBFC) is starting to develop special knowledge and criteria in this area (Canadian Codes Centre, 1997).

Access Criteria

Limited in scope and number of provisions, early barrier-free code requirements caused problems in meeting the intent of the regulations. Architects, builders, and building inspectors encountered constant problems in the practical application of the standards. Even when the letter of the law was followed, unsatisfactory conditions could still result, such as 700-mm door widths that allowed a person using a wheelchair to pass only straight on, and not at a 90° angle off of a narrow hallway. For many persons with disabilities, such as persons with visual or hearing impairments, the manual-wheelchair perspective of the building code regulations fell also well short of addressing their needs.

Subsequent changes to the NBCC have improved the barrier-free design criteria, but building codes are inherently limited. Building codes generally concentrate on ensuring life and health safety in buildings and, consequently, tend to focus on where the lack of usability potentially could result in injury or loss of life. The minimum requirements are provided without any explanation of alternatives that may be better in a particular situation. As Salmen so aptly observes in Chap. 12 of this handbook, "U.S. Accessibility Codes and Standards: Challenges for Universal Design," by Salmen, when it comes to satisfying code requirements, minimums have a way of becoming maximums, leading to compliance but not necessarily full accessibility.

Application and Enforcement

Despite the inclusion of barrier-free regulations in the NBCC, the application and enforcement of barrier-free design standards have not been homogeneous across Canada. Originally, under the British North America Act and its later successor, the Constitution Act, the responsibility for building regulation in Canada was given to the provinces and territories. This responsibility was then generally delegated to municipalities. The result of this arrangement was the development of multiple provincial and municipal building codes, each based on particular regional needs. Not surprisingly, this led to much confusion in the design and building industry and made doing business in more than one region in Canada difficult (Canadian Codes Centre, 1997).

The first National Building Code of Canada was created in 1941 to overcome the problem of multiple building codes and act as model code for building regulations that could be adopted across Canada. However, many provinces and municipalities continued to use modified versions of the NBCC and its subsequent revised editions. This is particularly true of the Barrier-Free section of the NBCC, which was replaced by local supplements in provinces such as British Columbia and Manitoba.

The large differences in accessible facilities between provinces that exist today is perhaps a reflection of the degree to which local authorities adopted, and sometimes extended, the barrier-free requirements from the NBCC. However, it is more likely this difference has more to do with enforcement or lack of enforcement, particularly when comparing facilities in rural and major urban centers. In 1998 local building authorities from all provinces in Canada signed a memorandum of understanding agreeing to adopt the NBCC 1995 with as few changes as possible, including Section 3.8—Barrier-Free Design (NRC-CNRC 1995). Compliance evaluation and enforcement of building codes still remain the task of the "local authority having jurisdiction," which usually means inspectors provided by a provincial or municipal office.

Building Code Development

Although the building codes in Canada are primarily concerned with fire safety and health, accessibility has continued to be an essential part. Over the years, accessible building code requirements have been changed and updated thanks to the open and consensus-based process used to develop and maintain the National Building Code of Canada.

Published by the National Research Council of Canada (NRC), the NBCC is updated approximately every 5 years. Although the NRC's Institute for Research in Construction (IRC) provides technical advice, the content of the code is ultimately decided by standing committees and task groups of volunteers. These volunteers come from across Canada and are drawn from various segments of the construction industry, ranging from contractors to building users.

The Canadian Commission on Building and Fire Codes (CCBFC) leads the process of code development. The Commission's decision making is supplemented by input from the standing committees and task groups. The development of the 1995 code included input from the Standing Committee on Barrier-Free Design. Members of the public can also submit comments and suggestions, even if they are not on a committee. Finally, all of the proposed code undergoes a public review process before being approved and published. Over the years, this very open and participatory process has proven to be a very useful vehicle for including and improving accessibility and usability requirements in the code.

Input from Research and Standards

As described previously, the National Research Council's Institute for Research in Construction is mandated to provide research support for the National Building Code of Canada. The IRC can provide technical advice as well as undertake studies when data is missing. However, not all of the research input to the code is necessarily done by the IRC but by other groups as well, as noted in Sec. 16.5.

In addition to the requirements in the code, the NBCC references almost 200 documents directly and many more indirectly. Such supplements and reference works are particularly important for accessibility requirements because they provide an explanation of the intent and reasoning behind the requirements. Not all solutions that simply meet the letter of the law are necessarily the most optimal.

The National Research Council publishes an additional user's guide specifically for Part 3. Other referenced documents are prepared by standards-writing associations in Canada [e.g., the Canadian Standards Association (CSA)] and the United States [e.g., the American National Standards Association (ANSI)], or by organizations that are interested in promoting the proper application of barrier-free design codes, such as the Canadian Paraplegic Association (CPA). Supplements such as CPA's ACCESS Manual (Lane, de Forest, and Hill, 1989; Finkel, 2000) were created to help to explain the ergonomic and functional requirements of persons with disabilities, as well as provide much needed illustrations of mandatory code requirements and desirable alternatives. Supplements can also be found in digital form. ACCEX is an interactive software program containing various Canadian accessibility codes, standards, and guidelines as well as the American standard, ADAAG. ACCEX also features an expert system. This is a special computer program that can tell architects how the NBCC and British Columbia accessibility codes apply to a particular design (Frye, Frye, and Sandilands, 2000).

Future Development

The current plan for the NBCC is to change from a prescriptive code to a performance-based code. Prescriptive codes tend to concentrate on specifying solutions, such as the type and quality of materials, method of construction, and workmanship required. The key feature of performance-based codes is that instead of prescribing desired solutions, the codes concentrate on specifying a desired behavior or effect. For example, an existing prescriptive requirement, "a clear opening for a door must be 900 mm minimum," might be changed to read, "doors shall be passable by a person using a wheelchair." By concentrating on specifying the required effect, it is felt that performance-based codes give designers and builders the opportunity to use their creativity and new technologies to develop unique solutions more freely. Furthermore, performance-based codes are thought to save time and money by allowing contractors to choose from locally available systems and components, as long as performance targets are met. However, to ensure that solutions are code compliant, a greater onus will be on designers, builders, and inspectors to define what the desired performance is (i.e., requirements analysis) and on the ability to judge whether it is being achieved (i.e., performance evaluation). For this reason, it is anticipated that in the future building industry professionals are going to have to know much more about human function and accessibility than they did when they simply had to follow a prescriptive code.

Although prescriptive codes are easier to specify, they can stand in the way of cheaper and more efficient solutions. A performance-based approach allows for a large number of solutions, but it requires a high degree of problem definition. A performance specification includes the following features:

- *Functional performance requirements*—declarative statements of human needs
- *Performance criteria*—quantification of the performance statement into measurable indices
- *Performance evaluations*—tests for determining if solutions meet the criteria (Ware, 1972)

In preparation for this performance-based approach, the Canadian Codes Centre has been working on an objective-based code that includes intent statements to accompany the present code. This process began in 1996 and is expected to be included in the next NBCC, anticipated in 2004.

Critics of the performance-based approach to building design argue that designers and regulators lack sufficient knowledge to be able to describe what a building, or a part of a building, is to do and how to measure performance before and after it is built. It is not yet apparent whether research and educational institutions in Canada are prepared to provide this needed expertise. Because of this, performance-based codes will likely have a heavy impact on the level of liability and risk allocation for building authorities, contractors, and design professionals. In countries that have already implemented performance-based codes, like New Zealand, there has been a noticeable increase in lawsuits, filed against local government officers.

The performance-based reforms will likely provide increased flexibility and room for innovation, but with this freedom will be a high level of responsibility. Satisfying performance-based codes requires a much more holistic approach, through legislative regulation that is complemented by expertise, accountability, and responsible allocation of risk. For an additional discussion of performance and prescriptive-based codes, see Chap. 12, "U.S. Accessibility Codes and Standards: Challenges for Universal Design," by Salmen.

16.4   HUMAN RIGHTS LEGISLATION

The second type of legislation that promotes universal design in Canada is human rights legislation. As compared to building codes, human rights legislation is more general in nature. The requirements are concerned with providing equality for individuals. They apply not only to the design of the built environment, but to the provision of goods and services as well as employment.

The Canadian Human Rights Act (CHRA) was proclaimed by the Canadian Parliament in 1976 to 1977. The purpose of the CHRA is to extend the laws so that every individual has an equal opportunity without being hindered by discriminatory practices based on race, national or ethnic origin, color, religion, age, sex, marital status, family status, disability, or conviction of an offense for which a pardon has been granted. This act also established the Canadian Human Rights Commission (CHRC) in 1978. The CHRC has three main objectives:

- To promote knowledge of human rights in Canada and to encourage people to follow principles of equality
- To provide effective and timely means for resolving individual complaints
- To help reduce barriers to equality in employment and access to services (CHRC, 1998a)

In 1998, the CHRA was amended to demand that employers and organizations recognize that people with disabilities are full members of society (CHRC, 1998a). The CHRA specifically identifies that accommodation must be provided for persons with disabilities to participate equally and that accommodation is central to full social and economic integration into mainstream society. The concept of accommodating the particular needs of individuals unless doing so causes undue hardship is not new, and case law has already upheld such accommodation as a right, not a privilege. However, the CHRC made the amendment because it codifies the requirements for accommodation and specifies that undue hardship is to be measured against health, safety, and cost.

Application and Scope

The CHRC is responsible for human rights issues and their application at the federal level in Canada. Separate provincial and territorial human rights commissions are responsible for carrying out the provisions of the Human Rights Code within each province and municipality.

The requirements of the CHRA principally describe what constitutes a discriminatory practice, such as the denial of services to particular customers or failure to accommodate reasonable occupational requirements of an employee. Because the act is stated in performance-based terms, the commissioners are largely responsible for interpreting the specific applicability of the act for each individual case. Under the CHRA, the governor in council may also make regulations prescribing standards of accessibility to services, facilities, or premises.

Enforcement

Human rights legislation in Canada is enforced through a complaint-driven process. This means that individuals who feel they have been discriminated against are responsible for contacting the commission and initiating any legal action. Final judgments can be enforced under an order of the Federal Court of Canada. Punishment under the CHRA can include punitive fines up to $50,000. However, the complaint process includes a review and hearings that can often be lengthy. This can have the effect of deterring some people, such as seniors and those with disabilities, from pursuing cases.

Future Development

The current development of the CHRA is based on the results of various federal task forces that recommend changes relating to disability issues in Canadian society, the first dating back to the *Obstacles Report* of 1981 (Special Committee on the Disabled and Handicapped, 1981). Six years after the 1990 introduction of the ADA in the United States, the report of the Canadian Federal Task Force on Disability Issues, *Equal Citizenship for Canadians with Disabilities: The Will to Act* (Scott Task Force, 1996), recommended that the government should introduce a Canadians with Disabilities Act (CDA) with enforcement and monitoring mechanisms. However, instead of creating a CDA, the federal government has opted to collaborate with the provincial and territorial governments on a plan called "In Unison," to "improve the efficiency, effectiveness and coordination of the existing provincial and territorial programs and delivery system" (CHRC 1998b). As such, the initiative seeks to deal with the duplication and patchwork of services and encourage a process of accountability in consultation with the stakeholder groups.

The current In Unison approach, however, is causing concern that it will erode the rights of Canadians with disabilities through the further devolution of the power of the federal government. This shift in focus will likely see a weakening of the federal rights of equality by the off-loading of the enforcement responsibility to the various provincial and territorial governments. This may result in various structures of accountability and standards of performance across Canada, rather than a single approach based on clear national standards.

16.5   *THE CURRENT STATUS OF UNIVERSAL DESIGN IN CANADA*

The previous sections examine the nature of the laws that are meant to enforce the design of accessible and usable facilities and services in Canada. The following sections highlight how the concept of universal design has been embraced and developed further in the areas of education, research, and design practice in Canada.

Universal Design Education

Universal design is taught in Canada as part of university courses on design theory or barrier-free building code compliance evaluation. For example, the Department of Interior Design at Mount Royal College, Calgary, Alberta, offers a course titled Building Code—Universal Accessibility. The Faculty of Architecture at the University of Manitoba offers an interdisciplinary elective course, Introduction to Universal Design. This course is taught by the staff of the Universal Design Institute (UDI), formerly known as the Canadian Institute for Barrier-Free Design (CIBFD). The UDI also provides general educational support to all five departments of the Faculty of Architecture.

Professional design organizations have begun to further legitimize and promote universal design by accrediting academic programs. For example, the Mount Royal College and the University of Manitoba programs are accredited by the Foundation for Interior Design Education Research (FIDER), the accreditation body for interior design programs throughout North America. FIDER includes within its definition of professional interior design the concept of "ensuring the compliance to universal accessibility while meeting the client's needs, goals and life safety requirements" (FIDER, 1999).

An expanding vision of the end user among professionals is helping to move the topic of usability in design education beyond barrier-free design. At first, building code requirements encouraged designers and design educators to include users of manual wheelchairs as end users. Even so, persons using wheelchairs were still considered as only potential customers or visitors to the built environment. More often than not, provisions for them were added on rather than integrated into the design. As awareness grew in the 1980s and 1990s, the concept of the end user was expanded, and designers needed to consider a variety of mobility and sensory issues. Moreover, students needed to be taught that end users with disabilities could not only be potential customers or visitors, they could be employees or permanent residents as well.

In October 1999, a group of international design educators and professionals convened at the Universal Design in the City—Beyond 2000 conference, hosted by the City of Winnipeg Access Advisory Committee, the Universal Design Institute, and the University of Winnipeg Institute of Urban Studies. Delegates compared definitions and actions for the entrenchment of the ideals of universal design. During this discussion it became apparent that the vision of the end user had expanded even further to include cross-disability issues, ethnic and language requirements, gender issues, and the infinite variety of ages, shapes, and sizes of the human body. This is not to suggest that all design educators or practitioners agree upon or incorporate this vision of broadly based accommodation. However, it is suggested that defining and accommodating the needs of end users is becoming more important as a design issue.

Other Educational Programs.   Not all universal design education in Canada is directed at the level of formal academic programs in design. For example, the UDI developed and delivered a community education program that taught many of the basic concepts of universal design. This program, delivered in several centers across Canada and described in detail in Chap. 6, "User/Expert Involvement in Universal Design," by Ringaert, was a training program for persons with disabilities. The goal was to use workshops to train several individuals with disabilities, who had no previous training in any of the design disciplines, to become access consultants.

Another source of universal design education is the Design Exchange (DX). Located in Toronto, DX is an independent nonprofit, educational organization committed to promoting an understanding of design by business and the public. DX recognizes universal design as a field of specialization for designers (DX, 1999) and has played an educational role through the sponsorship of various conferences and presentations on universal design issues. For example, DX cosponsored the Humane Village Congress with the International Council of Societies of Industrial Design and facilitated presentations by persons such as Alexander Manu, who is a board member of Design for the World, a nonprofit association of international organizations and individuals that brings together the resources of various disciplines to promote solutions beyond the scope of any one single field of practice. DX has also sponsored various international exhibitions that promote the concepts of universal design. In the spring of 2000, DX showcased the remounted exhibition *Unlimited by Design,* which originally had been curated by the Cooper-Hewitt National Design Museum in New York. This museum exhibit addressed the importance of universal design as a means of making life easier for everyone.

Design Competitions.   Various design competitions act as a stimulant to the universal design education and research process in Canada. For example, in 1999 Du Pont Canada and the Design Exchange invited Canadian design students to participate in the Du Pont Third Annual Student Design Competition. Included in the criteria used to judge competition entries was the following:

> Inclusive design and direct benefit and applicability to people's lives—a design which is universally accessible and doesn't limit its use to able-bodied persons. A design which serves specific needs in common everyday life. (Du Pont, 1999)

Competitions in Canada not only encourage students to do universal design, they also encourage professionals. For example, the National Post Design Effectiveness Awards are given for commercial Canadian innovations in the fields of industrial design, consumer products, transportation, furniture, built environment, urban environment, visual communication, new media, and corporate strategy. Among the requirements for entries is that they must be "placed into distribution," or be manufactured or built into their final form and be available to the public. The judging criteria in the 1999 competition included user benefits such as "comfort, safety, ease of use, durability, universal function and improved quality of life" (National Post Design Effectiveness Awards, 1999).

Universal Design Research

A number of governmental, university-based organizations and individuals in Canada undertake research related to universal design. Examples of some of these follow. Often these organizations work with each other or together with community groups on various projects.

Government Organizations. Two of the largest government-based organizations which conduct or sponsor research in the area of universal design are the Canadian Mortgage and Housing Corporation (CMHC) and the National Research Council of Canada (NRC).

The CMHC is the federal government's housing agency responsible for administering the National Housing Act. This legislation is designed to aid in the improvement of housing and living conditions in Canada. Under Part IX of this act, the government of Canada provides funds to CMHC to "conduct research into the social, economic and technical aspects of housing and related fields, and to undertake the publishing and distribution of the results of this research." Over the years, the CMHC has consulted with seniors and people with disabilities across the country. It published several books, including design guidelines for housing elderly people, persons with disabilities, and persons with dementia, as well as texts describing new, innovative approaches for meeting long-term housing needs, such as Flexhousing. *Flexhousing,* also described in Chap. 36 of this handbook, "Progressive Housing Design and Home Technologies in Canada," by Scott et al., involves innovative design programming to allow the built environment to easily evolve over time to meet the changing needs of the residents.

The National Research Council, as mentioned earlier in this chapter, provides secretarial and technical support to the Canadian Commission on Building and Fire Codes (CCBFC) and its related committee operations. The CCBFC, the Institute for Research in Construction (IRC) and the Canadian Codes Centre (CCC) work together to develop and maintain all of the national model building codes of Canada. Although the IRC does very little applied research in the area of universal design, it can sponsor research by outside researchers when needed and play an important technical advisory role regarding universal design requirements considered for inclusion in the building codes.

University-Based Research Centers. Several centers that contribute to universal design research are housed within universities in Canada. These include the Universal Design Institute (UDI) and the Canadian Center of Disability Studies (CCDS) at the University of Manitoba, as well as the Gerontology Research Center (GRC) at Simon Fraser University in British Columbia.

In addition to its previously mentioned role in education, the UDI is also committed to research in universal design. Established in 1990, the Canadian Institute for Barrier-Free Design became the Universal Design Institute in May 2000. The UDI undertakes research activities to develop design solutions and standards for the built environment and products, as well as explanatory documents for design professionals relating to barrier-free building codes and guidelines (CIBFD, 1999). For example, in 1999, the UDI completed research on the dimensional requirements for the built environment based on power wheelchairs and scooters for both the National Building Code and the Canadian Standards Association (Rapson, Ringaert, and Qiu, 1999).

The Canadian Centre on Disability Studies (CCDS) was started in 1995. It developed from the involvement of various Canadian disability organizations, and community groups and representatives from different faculties at the University of Manitoba. One of the research projects of the CCDS was *Actualizing Universal Design* (Finkel and Gold, 1999). This research investigated all aspects and characteristics that comprise a universally acceptable environment and the development of a process-oriented instrument based on accepted universal values. The goal of the second phase of this study was to create a useful tool by identifying the categories of functionality to be addressed and defining all the elements and characteristics of universal values, regional principles, and local features that impact on the physical environment at each of the micro-, meso-, and macroenvironmental scales.

The Gerontology Research Center (GRC) at Simon Fraser University was established in 1982 (see also Chap. 36). GRC provides education and undertakes research on issues relating to adult development and aging (GRC, 1999). Housed within the GRC is the Dr. Tong Louie Living Laboratory. The laboratory opened in 1997 and is a state-of-the-art facility devoted to environmental-behavior research (GRC/BCIT, 1999). The living laboratory contains a full-scale simulated residential space for the study of different environmental and product designs for older persons and persons with disabilities. Cameras, remote sensors, and computers make up a sophisticated data acquisition and monitoring system for analyzing various designs in use by volunteers who are older or have disabilities. This provides a unique opportunity for designers, manufacturers, and consumers to come together to improve product design and person-environment fit.

Other Research.   Individual researchers in Canada also have contributed to the ideals of universal design. An example is Romedi Passini, who published his groundbreaking book *Wayfinding in Architecture* in 1984 (Passini, 1984). Based on studies of persons with visual impairments, Passini challenged conventional thinking on what had previously been thought of as a purely visual task. He argued that the stimulation of other senses as well—for example, by using floor textures or acoustical cues—is an important aspect of wayfinding in buildings. Along with Paul Arthur, Passini has since published two other books on wayfinding (Arthur and Passini, 1990; 1992), including an evaluation and design guide for wayfinding in public buildings published by Public Works Canada.

Universal Design in Professional Practice

The current marketplace in Canada has promoted some interesting examples of what can be considered as universal design. Most of these have been the result of the involvement of broader populations in the design process. Designers are driven to a more participatory approach to design for many reasons, not the least of which is a post–Ralph Nader educated consumer who has come to expect that systems designers should be accountable for how their products will impact everyone.

In Canada, the needs of consumers are changing, opening attractive new markets for designers and their products, while at the same time challenging designers to identify what these new needs are. Improved building codes and human rights legislation have enhanced the financial and political status of persons with disabilities by ensuring access to employment and facilities, yet many individuals are still faced with architectural barriers left over from a legacy of existing building stock and infrastructure. The baby boom of the postwar era and universal health care have created an active yet increasingly aging population that is likely to find itself short of suitable housing in the future. Continuing immigration of people into Canada from numerous foreign lands is constantly reshaping cultural expectations and needs. Increasing globalization has industry looking beyond the domestic consumer to develop products that are able to compete at a global level.

The following sections highlight some examples from practice in Canada that illustrate how designers are starting to use an inclusive approach to the process of design to achieve higher usability and acceptance by consumers and building users (see also the discussion of briefing and universal design in Chap. 44, "Project Briefing for an Inclusive Universal Design Process," by Luck et al.). It should be noted that, examples of universal design housing in Canada are not included here as these are described in Chap. 36, "Progressive Housing Design and Home Technologies in Canada," by Scott et al.

Placemaking.   Manitoba architect Michael Boreskie believes that achieving universal design begins in the programming (briefing) and design process. During design, Boreskie engages the client and user groups in a process called *placemaking* (Boreskie, 1999a). According to Boreskie, successful placemaking tries to be responsive to the needs, tangible and intangible, of the whole person over time. In order to achieve this end, the design process seeks to be inclusive and participatory by engaging the full life experience of as many users as is practical in town hall–style meetings and seminars.

The placemaking approach was used to design St. Timothy, a 21,000-ft^2 facility in Winnipeg, Manitoba, including a worship space, social spaces, offices, classrooms, ecological sanctuary, and sustainable garden (see Fig. 16.3). Placemaking resulted in the inclusion of

FIGURE 16.3 Placemaking table group engaged in designing the St. Timothy facility, Winnipeg, Manitoba.

qualitative dimensions of life, wide consensus on what needed to be accomplished and on the prioritization of issues, high user buy-in, improved fund raising, and a significantly enhanced appropriateness of the built environment to users' needs. For example, the St. Timothy users group selected chairs, instead of pews, for the worship area in order to make the space very flexible and allow for the easy inclusion of persons using wheelchairs in locations that are best suited to their desires and comfort (Boreskie, 1999b).

A similar participatory approach was used by members of a grassroots community organization in the West Broadway area of Winnipeg who came together for the purpose of redeveloping their neighborhood. Through the establishment of various social and political stakeholder groups, the community members wrestled the control of the design process away from the formal system. Now the community has control and works collaboratively with the designers, developers, and builders. By being directly involved in the decision-making process, the community has developed a very strong voice in the establishment of the goals and objectives for any redevelopment in the neighborhood.

Product Design.   ProductABILITY Consulting in Winnipeg is a nonprofit design service. This group undertakes product and interior design using a multidisciplinary team of experts who work in collaboration with occupational therapists, social workers, and home economists to create innovative solutions for a wide range of human functional needs. ProductABILITY works extensively in the field of custom housing design for people with disabilities. However, because most of the projects involve other family members who have a variety of abilities, the approach taken is to strive for universal solutions. ProductABILITY draws on seven years of research in cataloging over 45,000 products that have some aspect of compensatory or broad-range functional use. One of the most interesting aspects of ProductABILITY's design of residential suites is the way in which everyday products and good design improve the usability of their residential solutions.

The now-defunct Canadian Aging and Rehabilitation Product Development Corporation

(ARCOR) also used a highly participatory process in the development of products. These products were aimed at a broad market and had features that were adaptable to functional changes presented by aging or disability. To develop the products, ARCOR adopted a multidisciplinary approach drawing on the fields of engineering, medical rehabilitation, and marketing and on the expertise of existing manufacturers as well as consumers. Focus-group sessions were used to combine the knowledge of a broad spectrum of users. By listening to learn what were the unsatisfied needs of the consumers, ARCOR developed new product prototypes. Two successfully commercialized products were eventually realized.

Canadian product designer Paul Arthur, mentioned earlier in this chapter, has worked for over 25 years on the development of a universal language for wayfinding. His pictographic systems increase the ability to communicate to a greater population through the use of logical icons (Arthur, 1998). Arthur's work has been applied in numerous installations in Canada, including Toronto's Hospital for Sick Children, George Brown College, and the Memorial University in Newfoundland. More recently, Arthur's work is being utilized within electronic media for corporations such as Canada Post.

16.6 CONCLUSIONS—THE FUTURE OF UNIVERSAL DESIGN IN CANADA

This chapter provides a profile of how universal design has evolved and manifested itself in design legislation as well as in education, research, and practice in Canada. Whereas legislation forms a foundation of accessible and usable design, the growth of barrier-free design and its successor, universal design, as a design philosophy is providing the real incentive for development and innovation.

One of the essential limitations of designing with building codes is that they will always represent a minimum standard. Canada does benefit by using a broad-based consensus process that allows people from all segments of the construction industry, including consumers, to develop and improve its building codes. Consequently, substantial efforts have been made to improve the scope and content of the access-related requirements in the building codes over the past years. However, concerns remain over application and enforcement of access legislation in Canada because of the lack of a uniform approach nationally.

In terms of application, building codes are not retroactive, nor do they apply to all facilities in all regions in the same way, potentially leaving many disabling environments in place. The change from prescriptive-based to performance-based building codes may allow designers the chance to create more innovative solutions for equity within the built environment. However, the design professions are just beginning to understand the meaning of *equity* and *access* under the compulsory requirements of the prescriptive code. Designers lack the information and education they need to truly understand the functional requirements of universal design in a performance-based regulatory environment. Application of the performance-based human rights legislation to building practice and products has already proven to be difficult. This is because the intent of the law is explained (i.e., "do not discriminate on the basis of ability"), but the guidelines on how to achieve the intent (for example, "all doors in a building must be 920 mm") are not provided.

In terms of enforcement, building code compliance evaluation and control remains heavily reliant on the ability of local authorities within each region in Canada. The existing system for enforcement of the prescriptive-based code includes many inconsistencies. In the future, much more judgment will be necessary to establish whether a requirement of performance has truly been met. This suggests that local authorities will need to reevaluate their current procedures and training in order to effectively and uniformly enforce performance-based codes.

In the case of human rights legislation in Canada, individual members of the public, not trained officials, are required to enforce the legislation through a potentially lengthy complaint process. One might expect that this would potentially deter people from coming forward. However, the Human Rights Commission statistics show dramatic increases in complaints and actions taken against designers, builders, and development companies, as well as private and public buildings, based on discriminatory designs that are barriers to equal participation. The expectation for equality by today's educated consumer appears to be now well-rooted in the Canadian mindset, which no doubt will lead to even greater calls for accountability within the design and construction industry.

Universal design in Canada continues to be primarily driven forward by the forces of social and economic change. Canadian society is based on the recognition of a diverse cultural base. An ongoing commitment to immigration inevitably will diversify the population and be a catalyst for change. Other, previously marginalized, voices from within Canada will not be silenced either, as the continually growing strength of the lobbies for persons with disabilities and the aboriginal community forces all Canadians to be more responsive to their needs. The largest market force in Canada, however, will be the graying population of postwar baby boomers. The relative affluence of this group presents many opportunities for designers; however, some fear that not all Canadians will necessarily benefit as these more affluent older consumers, worried over personal and property security, turn to solutions such as gated retirement communities, heralding an unwelcome return to segregated and exclusionary environments.

Despite the examples of how education and research in universal design have started to take hold in some centers in Canada, it is obvious that much more can be done. More research is needed to help define user requirements and develop standards for evaluating design performance. Canadian design schools, with encouragement and promotion by professional organizations and industry, need to move quickly to develop more programs to help students learn how to address the diversified makeup of the Canadian society. Even more pressing is the need to provide continuing education programs and new design methods for practicing professionals who are immediately facing the impacts of the shifts in the regulatory and economic environments of the marketplace.

16.7 ACKNOWLEDGMENTS

The authors wish to thank Dr. John Frye for his expert input on Canadian barrier-free building design codes and standards referred to throughout this chapter.

16.8 BIBLIOGRAPHY

Arthur, P. *Pictographic Systems,* www.inforamp.net/~zlam/papictos.html, 1998.

———— and R. Passini, *1-2-3 Evaluation and Design Guide to Wayfinding: Helping Visitors Find Their Way Around Public Buildings,* Public Works Canada, Ottawa, 1990.

———— and ————, *Wayfinding: People, Signs, and Architecture,* McGraw-Hill, London, 1992.

Boreskie, M., *The Human Art of Placemaking,* www.boreskie.mb.ca/boreskie, 1999a.

————, personal communication, 10 November, 1999b.

————, personal communication, 23 November, 1999c.

CAILC, *What Is the Canadian Association of Independent Living Centres?,* www.cailc.ca/bkgrnd.htm, 1999.

Canadian Codes Centre, *How National Codes Are Developed and Updated,* www.nrc.ca/codes/about_E.shtml, 1997.

CCDS, *CCDS Projects Update—Spring 1999,* www.escape.ca/~ccds/resproj.html, 1999.

Center for Universal Design, *Definition of Universal Design,* www.design.ncsu.edu/cud/univ_design/ud.htm, 1999.

CHRC, *Canadian Human Rights Act,* chap. H-6, www.chrc-ccdp.ca/about/about.asp, 1998a.

———, *Canadian Human Rights Commission—1998 Annual Report,* www.chrc-ccdp.ca/ar-ra98/disab-defic.asp?1=e, 1998b.

CIBFD, *The Canadian Institute for Barrier-Free Design Inc.,* www.arch.umanitoba.ca/UofM/CIBFD/index.htm, 1999.

Corbusier, Le., *The Modulor: A Harmonious Measure to the Human Scale, Universally Applicable to Architecture and Mechanics,* 2d ed., Faber and Faber, London, 1961.

DeJong, G., "The Influence of Environmental Barriers on Independent Living Outcomes," *Rehab Brief: Bringing Research into Effective Focus,* 4(5), 1981.

Du Pont, *Du Pont Student Competition—Life by Design,* www.designexchange.org/life.html, 1999.

DX, *Experts Bureau,* www.designexchange.org/bus2.html, 1999.

FIDER, *Definition of Interior Design,* www.fider.org/defin.htm, 1999.

Finkel, Gail (ed.), *ACCESS: A Guide to Accessible Design,* 3d ed., Universal Design Institute, Winnipeg, Manitoba, Canada, 2000.

Finkel, G., and Y. Gold, *Actualizing Universal Design: P4 = Power: People, Places, Participation, and Process,* F * G Consortium, Winnipeg, Manitoba, Canada, 1999.

Frye, J., K. Frye, and R. Sandilands, *ACCEX: Universal Design Expert Software System,* Universal Design Institute, Winnipeg, Manitoba, Canada, 2000.

GRC, *The Gerontology Research Centre,* www.harbour.sfu.ca/gero, 1999.

GRC/BCIT, *The Tong Louie Living Laboratory,* www.harbour.sfu.ca/gero/livinlab.html, 1999.

Lane, John, Claude de Forest, and Shauna Hill (eds.), *ACCESS—A Guide for Designers,* 2nd ed., Canadian Paraplegic Association and Canadian Institute for Barrier-Free Design, Winnipeg, Manitoba, 1989.

National Post Design Effectiveness Awards, Call for Entries, www.designexchange.org/deacal21.html, 1999.

NRC-CNRC, *Building Standards for the Handicapped 1965, Supplement No. 7 to the National Building Code of Canada,* National Research Council of Canada, Ottawa, 1965.

———, *National Building Code of Canada,* Canadian Codes Center, National Research Council of Canada, Ottawa, 1985.

———, *National Building Code of Canada,* Canadian Codes Center, National Research Council of Canada, Ottawa, 1995.

Passini, R., *Wayfinding in Architecture,* Van Nostrand Reinhold, London, 1984.

ProductABILITY, *Ten Ten Sinclair Housing Inc.—State of the Art Suite 223,* www.tenten.mb.ca/apt223/apt223.html, 1999.

Ramsey, C. G., and H. R. Sleeper, *Architectural Graphic Standards,* 8th ed., John Ray Hoke, Jr. (ed.), Wiley, Chichester, UK, 1988.

Rapson, D., L. Ringaert, and J. Qiu, *New Dimensions for Universal Design Codes and Standards with Considerations of Power Wheelchairs and Scooters,* Canadian Institute for Barrier-Free Design, Winnipeg, Manitoba, 1999.

Scott Task Force (Canadian Federal Task Force on Disability Issues), *Equal Citizenship for Canadians with Disabilities: The Will to Act,* Government of Canada, Ottawa, 1996.

Special Committee on the Disabled and Handicapped, *Obstacles Report,* House of Commons Report, Government of Canada, Ottawa, 1981.

Ware, T. E., "Performance Specifications for Office Space Interiors," in B. E. Foster (ed.), *Performance Concept in Buildings Joint RILEM-ASTM-CIB Symposium Proceedings,* Philadelphia, 1–5 May 1972, U.S. National Bureau of Standards, Washington, D.C., 1972, pp. 357–354.

Wittkower, R., *Architectural Principles in the Age of Humanism,* St. Martin's Press, New York, 1988.

16.9 RESOURCES

CMHC, "Housing for Elderly People—Design Guidelines," *Maintaining Seniors' Independence Through Home Adaptations,* Canadian Mortgage and Housing Corporation, Ottawa, 1993.

CMHC/SCHL, *New Made-to-Convert Housing,* Canadian Mortgage and Housing Corporation, Ottawa, 1994.

CMHC/SCHL, *Housing Choices for Canadians with Disabilities,* Canadian Mortgage and Housing Corporation, Ottawa, 1995.

CMHC/SCHL, *Housing Persons with Disabilities,* Canadian Mortgage and Housing Corporation, Ottawa, 1996.

CMHC/SCHL, *FlexHousing: Homes that Adapt,* Canadian Mortgage and Housing Corporation, Ottawa, 1998.

CMHC/SCHL, *FlexHousing PowerKit: A Guide to Building Homes That Adapt to Life's Changes,* Canadian Mortgage and Housing Corporation, Ottawa, 1998.

CMHC/SCHL, *Housing Options for Persons with Dementia,* Canadian Mortgage and Housing Corporation, Ottawa, 1999.

CMHC/SCHL, *Planning Housing and Support Services for Seniors,* Canadian Mortgage and Housing Corporation, Ottawa, 1999. Includes community and residential design.

CSA, *Barrier-Free Design,* B651, Canadian Standards Association, Rexdale, Ontario, 1990.

CHAPTER 17
THE IMPACT OF AGING ON JAPANESE ACCESSIBILITY DESIGN STANDARDS

Satoshi Kose, Eng.D.
Building Research Institute, Tsukaba, Japan

17.1 INTRODUCTION

This chapter discusses the implications of rapid aging in Japan for building accessibility. It describes the introduction of the new, special Accessible and Usable Buildings Law (1994) and the Dwelling Design Guidelines for the Aging Society (1995). Since the contents of the law will be explained more fully in another chapter, the emphasis here is on the universal design features of the dwelling design guidelines. Japan is among perhaps the first countries to succeed in persuading the general public to accept design-for-all concepts for dwellings, and maybe the only one that has used economic incentives at the national level rather than using legal enforcement. The lessons learned and possible future directions are also discussed from the perspective of life in cities and their environs. It is, after all, the quality of life that determines the satisfaction of residents, and accessibility is one of the most crucial factors. Another factor that will determine satisfaction level is usability, and the Dwelling Design Guidelines exactly respond to these two factors with the concept of universal design.

17.2 AGING AND URBANISM IN JAPAN

The notion of the emergence of an aging society in Japan is relatively new, and its consequences have rarely been taken seriously, even by the government, until fairly recently. However, Japan is becoming a society with a rapidly growing population of elderly people (National Institute of Population Problems, 1986; National Institute of Population and Social Security Research, 1997). In 1970, just 7 percent of the population was age 65 or older. In 1994, the figure had reached 14 percent. It took only 24 years—that is, less than a generation—for Japan to double this percentage, a rate that has never been paralleled. The Japanese population will continue to grow older, and in the year 2015, more than 25 percent will be age 65 or older. This situation has never before been experienced in any developed nation; thus, Japan will be the first to reach this astonishing level. With such a high rate of aging, Japan must prepare for its future.

Another important fact is that more and more people in Japan live in urban areas. The official statistics show that in 1988 more than 76 percent of the Japanese population lived in cities and suburban areas. This is a fairly new phenomenon, but it requires difficult choices for those who have lived in rural areas for most of their lives. They must choose whether they will continue to live there with substandard social services or move into urban or suburban areas where their children's families live in order to receive a more acceptable level of services in addition to informal in-family care. In a sense, it is a choice between aging in place and relocation for better community care.

It must be pointed out that in-family care has been implicitly integrated into the system of social welfare services in Japan, at least until now. Moreover, the social assumption was that the eldest son's family would take care of the aging parents, preferably as an extended family. Unfortunately, the social context that supported this is becoming less prevalent, partly as a result of social change and the decreasing numbers of children.

Design requirements in dwellings for the aging population have never been properly met by the designers, although the aging population is increasing (Kose and Nakaohji, 1991; Kose, Ohta, Tanaka, and Watanabe, 1992). The situation is similar in the larger community environment. It is true that barrier-free design has long been tried in larger buildings with the assistance of design guidelines, but this has only been on a voluntary negotiated basis because virtually nothing is required by the Building Standard Law of Japan.

At the beginning of the 1990s, some local governments had begun to introduce barrier-free requirements in their local building ordinances. The Building Standard Law of Japan allows local governments some freedom to issue such ordinances. These requirements are becoming a new trend toward the recognition of the right of access to buildings, as opposed to the past assumption that barrier-free design is a form of welfare or a charity measure.

17.3   THE INFLUENCE OF INTERNATIONAL ACTIONS ON JAPANESE ACCESSIBILITY LEGISLATION

The United Nations International Year of Disabled Persons was 1982. Many activities took place around the world, including in Japan, where several design guidelines for buildings and facilities were issued by both the central and local governments (Kose, 1992b). The International Decade of Disabled Persons followed from 1983 to 1992, but there was no comprehensive summing up of the accomplishments for accessibility in the built environment in 1992. Most people seemed to be content with the fact that nonmandatory guidelines had worked well enough for the provision of accessibility.

The most remarkable move on the international scene during that period came from the United States, with the enactment of the Americans with Disabilities Act (ADA) in July 1990. It is a great milestone in the sense that the civil rights movement in the United States reached a stage where the rights of persons with disabilities were recognized as equal to those of all other citizens. Although the government agency responsible for the enforcement of accessibility to businesses and government services is the Department of Justice, it certainly meant that the United States reached a new, heightened level of accessibility. Many had assumed that the United States was a nation where the public sector would not usually intervene in the activities of private business, but the ADA showed that the United States took a step beyond that in regard to civil rights and accessibility. If businesses or even state and local governments do not abide by the ADA, they run the risk of complaints or possibly of being sued.

17.4   EARLY ACCESSIBILITY LEGISLATION AND THE BUILDING STANDARD LAW

In the past in Japan, the accessibility issue was mostly under the control of the Welfare Ministry and its departments. As early as 1974, Machida City, one of the local municipalities, issued Design Guidelines of Buildings and Facilities Toward Realizing a Welfare City, which aimed at creating an accessible environment for wheelchair users from the welfare perspective. Although it was not mandatory, the local government tried to influence building developers as much as they could to include barrier-free design for wheelchair users in larger buildings and facilities. Although they succeeded to some extent and other local governments followed, voluntary compliance was not as effective as desired and the move toward realizing barrier-free environments was slow. Part of the problem could be attributed to insufficient coordination in the local governments between the Buildings Control Department and the Welfare Department.

To overcome this limitation, some local governments went one step further and introduced local building ordinances that incorporated mandatory accessibility requirements. Kanagawa Prefecture started the discussion in 1988, and after careful preparation the revised building ordinance was enforced in 1990.

Overview of the Building Standard Law

The most important point is that the Japanese Building Standard Law was established in 1950 and no revision has ever been made regarding the basic philosophy. It became evident, therefore, that the law is far removed from the current social context. At the time when the law was enacted, the population that was age 65 or older constituted only 5 percent, and persons with disabilities were almost invisible. They stayed in their homes and were cared for primarily by family members. The law therefore assumed healthy adults as representative, major users of buildings. The only reference to other age groups was for pupils of primary schools, pertaining to dimensions of stairs, possibly because it was the only situation where children move around unaccompanied by parents and other adults. No system for revisions was incorporated into the law, and while there were several occasions when revisions were made, they related to technical issues such as mitigation measures for earthquakes or large fires, and not to social or contextual matters.

The Building Standard Law of Japan, however, does include some freedom for local governments to issue ordinances applicable to their localities. The law defines only the nationwide minimum level, and allows local specific needs to be met through local decisions. The original idea behind this freedom to make additions was to address specific local conditions, such as geographic or climatic factors, and, in particular, wind forces from typhoons, but not social issues. The additions must relate to structural and fire safety and sanitation, and they cover only large-scale public buildings. Smaller buildings can never be covered, nor can functional and use requirements be the focus of regulation.

However, and as noted earlier, the Kanagawa Prefecture attempted to include in the ordinance usability requirements for people with wheelchairs. It succeeded in persuading the local council chamber to pass the ordinance. The requirements that could not be included were left with the existing recommendations, Design Guidelines for Buildings and Facilities: Toward Realizing a Welfare City, which were under the control of the Welfare Department.

Following the lead of the Kanagawa Prefecture, the Osaka Prefectural Government took another step. As local municipality government law allows the local governments to establish ordinances to fulfill the goal of their policy measures, Osaka gave Design Guidelines for Buildings and Facilities: Toward Realizing a Welfare City the status of an ordinance rather than being just recommendations. This meant that the requirements were enforceable and not voluntary. It is arguable whether the general ordinance is as strictly enforceable as the local building ordinance, because the former can be overrestrictive as far as the rights of building clients are concerned. The reasoning of the local governments was that the local council chamber passed the ordinance, which meant that the voice of the public was in favor of the introduction of accessibility requirements.

Some other local governments were ready to follow these prefectures and introduced similar ordinances as part of their regular building ordinances. The recognition of the right of access to buildings has become a new trend, in contrast to the past assumption that barrier-free design is a form of welfare or charity.

Problems with Local Government Ordinances

It is helpful to compare the requirements in different local governments in order to gain some insight into problems that must be solved. Unlike countries where national standards on accessibility exist that can override local governments, such as the United Kingdom or the United States, Japan has never had anything of the kind. That meant that local governments established their own guidelines in the past, and they tended to incorporate the existing requirements of the welfare guidelines into the new ordinances. Such existing requirements were, unfortunately, inconsistent between different local governments because they often did not have any logical basis for decision making, either technical or social. The discrepancies between different local governments included the type of buildings to be covered, with large-scale residential buildings mostly excluded, except in some; the sizes of buildings, with minimum floor area varying from one local government to another, for example, from 1000 m^2 to 500 m^2; elevator requirements that could be either obligatory or voluntary; and the dimensions and numbers of respective features, such as clear width of doors and passages for various spaces. No one can argue that the requirements in one local government are absolutely superior to those in others, because such distinctions are not based on research. Designers and clients may encounter problems when they request building permits from one local government for a building design that has already been accepted in another. This problem is caused by the lack of performance-based standards (Kose, 1993).

17.5   A NEW APPROACH: ACCESSIBLE AND USABLE BUILDINGS LAW

In the early 1990s, the central government attempted to find ways to integrate new trends toward accessibility into their own policy initiatives. As a result, the new Japanese with Disabilities Act, which is handled mainly by the Ministry of Health and Welfare, was enacted in 1993. A completely updated version of the old Law on Measures for the Disabled, this complete revision appeared to have been accelerated by the ADA. In effect, it paved the way to establish a special law by the Ministry of Construction, the Accessible and Usable Buildings Law. This was the launch of a new era to ensure that environments would accommodate the aging society. Proposed in the spring of 1994, the new law passed the Diet on June 28, 1994 (Law 44, 1994). Government officials had feared there would be arguments against enacting the law, but the time was ripe and it passed without major opposition.

Overview of the Accessible and Usable Buildings Law

The law did not replace the Building Standard Law but provided an alternative procedure to obtaining a building permit if the building incorporated accessible and usable design features. Design guidelines were issued at two levels, basic and recommended, which described what constitutes accessible and usable buildings. Public buildings and facilities beyond a certain size were requested to abide by these guidelines. Through the use of local building ordinances, the law allocated some authority to the governors of prefectures, which enabled them to require owners and managers of public buildings in their localities to be accessible or usable.

New Policies and Incentives

The government also introduced preferential interest rate schemes and subsidies linked to the law at the same time, which are expected to work as incentives. This was previously done by some local governments for the installation of elevators in railway stations, for example. New measures include tax exemption for unavoidable floor area increase, as well as other provisions. The difficulty with that is that the effort necessary to apply for such benefits does not always justify the extra work in terms of time and cost.

Policy measures by the Ministry of Construction were also issued in June 1994 as a move toward a barrier-free built environment, ranging from housing to urban infrastructure (Ministry of Construction, 1994). The inclusion of housing issues was an attempt to integrate the whole problem of living environments. Previously, dwelling design solutions relied completely on efforts by the private sector and funding by individual residents, unless they were part of public rental housing schemes.

The attempt to solve the housing problems of the aging society was originally raised when the population forecast was announced by the National Institute of Population Problems in 1986 (Kose, 1987). The Ministry of Construction asked for research and development funds to tackle the emerging issue. Being responsible for promoting the five-year R&D project, the researchers with the Building Research Institute recognized the housing problems to be more urgent than building design accessibility, and they continued efforts to make dwellings more suitable to the increasingly aging society.

17.6 MANUFACTURED HOUSING: THE BEGINNINGS OF UNIVERSAL DESIGN

As early as the mid-1980s, several housing manufacturers, and prefabricated housing manufacturers in particular, foresaw the aging of their clients as a business opportunity and began to showcase their design concepts for the aging society as strengths. During the research processes, the Building Research Institute collaborated with some of these housing manufacturers in order to integrate the revolutionary yet feasible housing design guidelines. Since most Japanese housing consists of owner-occupied, detached houses, which are usually privately financed, it was crucial to obtain the support of prospective buyers.

Through various surveys and experiments, these five-year research efforts attempted to bring to fruition the Design Guidelines of Dwellings for the Aging Society (Kose, 1991; Kose, 1992a; Kose, Ohta, Tanaka, and Watanabe, 1992). The draft design guidelines were released in the spring of 1992 (Kose, 1994) and were finally issued in June 1995 by the Ministry of Construction. They required that three basic criteria be met: (1) floors without level differences, (2) handrail installation where keeping the body balance is crucial, and (3) corridors and door widths that permit assisted wheelchair passage (Kose, 1997). These were the problems commonly encountered in modifying existing dwellings (Nomura, Kose, et al., 1990), and it was predicted that these requirements would change the expectations for the future.

The guidelines were quickly incorporated into the government housing loan schemes by the Housing Loan Corporation of Japan (Kose, 1996) because they were considered essential for the realization of policy measures for housing for the aging society. They were not mandatory. However, coupled with the preferential loan schemes, they provided strong incentives. Design standards of major housing manufacturers were changed almost overnight when the Housing Loan Corporation decided to drastically change the system—in other words, to require either energy-conscious design, highly durable design, or design for the aging society.

Figures 17.1 and 17.2 are an example of a floor plan of a house, and Figs. 17.3 through 17.8 show how some of the design features of dwellings changed for the better in these 10 years. Around 10 years ago, even housing for senior citizens ignored the decreasing capabilities of the aging residents, and small floor-level differences constituted tripping hazards. Nowadays, most of these problems are nonexistent, not just in manufactured housing, but in dwellings made by local builders. The most remarkable change is the introduction of the accessible bathroom unit that eliminated level differences between the dressing area and the wet area. Since this was realized without any cost increase, one can call this a true example of universal design (Kose, 1998).

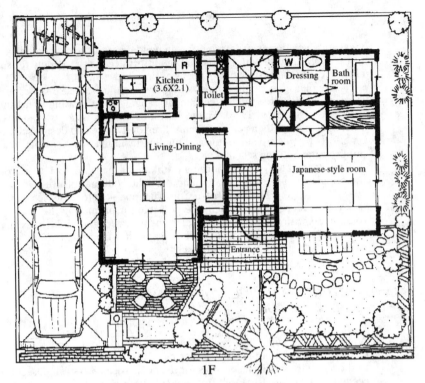

FIGURE 17.1 Plan of the ground floor of a detached house designed for the aging society.

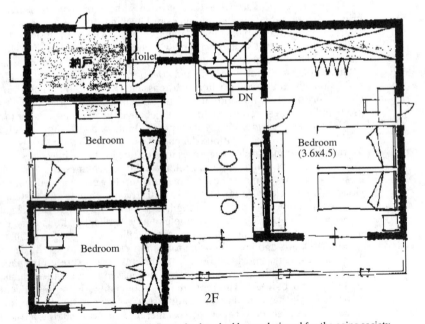

FIGURE 17.2 Plan of the upper floor of a detached house designed for the aging society.

FIGURE 17.3 Entrance gate of a detached house with entry at grade level.

17.7 THE NEED TO REVISE THE BUILDING STANDARD LAW TO INCORPORATE ACCESSIBILITY REQUIREMENTS

The problems designers and clients are faced with due to building control are probably not unique to Japan. However, it is true that the limitations of the present Building Standard Law are several in the Japanese system. The following issues need urgent solution:

1. The assumptions about building users, as far as their capabilities are concerned, are inadequate, as pointed out in part by the Accessible and Usable Building Law.

FIGURE 17.4 Flush (nonelevated) floor finish to prevent accidental falls in dwellings.

FIGURE 17.5   Another approach: the tatami floor is higher than wooden floor, enabling senior residents to get out of futon bed onto tatami.

FIGURE 17.6   Toilet with flush (nonelevated) floor and handrail installation.

FIGURE 17.7   Bathroom with the wet area is lowered to prevent water from coming out.

FIGURE 17.8   No threshold between dressing area and wet area, with careful design of bathroom fixtures.

2. There is no explicitly stated established system for routine revisions.

3. There are no considerations for compromises when alterations to existing buildings are at stake.

4. While there is too much emphasis on specification requirements, the performance standards are too vague.

5. The delegation of authority to local ordinances is too restrictive.

The Growing Need to Adapt Existing Buildings

Among the aforementioned problems, the lack of distinction between new and existing buildings seems to cause the most serious difficulties in real-world situations, followed by the lack of a logical basis for specified requirements. Clearly, the two are closely correlated. The basic idea behind the Japanese Building Standard Law appears to assume a 30- to 60-year life span for buildings, depending on the type of construction. The Japanese assumed that timber structures would be replaced in 30 years, and reinforced concrete or steel structures in 60 years. The thinking was that if change in society was slow, the country could perhaps wait until every building was replaced by a new one. However, as noted earlier, the growth of the aging population is much faster than had been anticipated. Japanese society cannot wait that long, nor can physical barriers be allowed to remain. It is preferable to improve existing buildings, even if the more demanding new building accessibility requirements cannot be met. To attain this intermediate goal, it is necessary to introduce another, lower, level of requirements that are applicable only to alterations of existing buildings. The idea of two levels of standards is common in countries where the expected life of buildings is long.

Problems That Need Solving

Presently, however, it is still not easy to revise the Building Standard Law. The government has insisted that the law define just the minimum level of requirements. In reality, however, not only clients but also architects and designers tend to think that current requirements are at an optimum level. A previous survey revealed that architects and designers preferred mandatory accessibility requirements because clients were not willing to accept voluntary ones.

Without support from society, the government officials argue, it is virtually impossible to introduce new accessibility concepts into the Building Standard Law. However, as society is aging fast, the public has not yet reached the stage where it will accept that barrier-free standards should be included in the requirements. There was one opportunity in 1998 when a complete revision of the Building Standard Law into a performance-based standard was conducted as part of the government's move toward deregulation. Unfortunately, the introduction of accessibility concepts into the law was denied. It will be some time before the proposal for the next revision will become a reality. The earliest occasion might be when the proposed Accessible Transportation Law takes effect. Arguments are sure to arise concerning the issue that if public transportation must be accessible and usable, why not public buildings?

Adaptation of Dwellings

The need for adaptation is common to dwellings (Lanspery and Hyde, 1996). The difficulty is more pronounced because of the mismatch between the capabilities of the residents and the design features of the dwellings. A solution that has worked in one house will not necessarily be effective in another. There have been attempts by local governments to assist in modifying dwellings, but their effectiveness has been limited. Even the new Care for the Aged Insurance System does not appear to support home modification; the emphasis is more on providing services through human resources rather than on the initial improvement of the physical environment. The completely outdated and shortsighted assumption prevailed that giving money for home modification is not a fair policy measure in the sense that it benefits only a small minority of the population, especially those who own their own homes.

In the meantime, the Housing Quality Assurance Law was enacted in 1999, and design for the aging was one of the performance requirements to be addressed. Since a recent survey on the effectiveness of the Dwelling Design Guidelines for the Aging Society suggested that they are accepted by the residents with satisfaction (Kose and Tanaka, 1998; Kose, Kumano, and Matsuzaki, 1999), their inclusion has been highly appreciated.

The law covers only new dwelling units, and it is not obligatory for them to be part of the scheme. However, the nonobligatory nature is likely to lead to a situation where better-quality housing providers would apply for the performance rating, and the expectation is that the general quality of dwellings will be upgraded in due course.

17.8 GENERAL LESSONS LEARNED FOR UNIVERSAL DESIGN

The most difficult part of the introduction of universal design concepts was how to persuade the general public to accept them. Examination of the Japanese accomplishments reveals that it is best when they are introduced with economic incentives into the market. Perhaps the best policy is to mix them with legal enforcement.

In a chapter of a book on human factors, Kose (2000) stressed that good design is what is needed by the users, and that there are six essential requirements—safety, accessibility, usability, affordability, sustainability, and aesthetics—to be eligible to be called "good design." Some of the requirements are enforceable, but not all. Kose stressed that the first four requirements must be met to qualify for being called "universal design," but perhaps affordability (i.e., economic concern) is most crucial.

17.9 CONCLUSIONS

The Asia-Pacific Decade of Disabled Persons began in 1993 with the initiative of the United Nations Economic and Social Commission for Asia and the Pacific (UN-ESCAP). It inevitably made Japan a forerunner on the issue of accessibility legislation in the Asia-Pacific region. Cooperation with the ESCAP officials, extensive discussions with specialists from different countries, both from governments and from nongovernmental organizations, have revealed the problems and prospects with which Japan is faced.

As noted earlier, with the end of the decade approaching, Japan is going to have a law on accessible transportation. It is sure to heighten the awareness of society, and it will lead to the revision of the Building Standard Law. After all, an inclusive society will be possible only when dwellings, buildings, transportation, and the built environment in general are designed for everybody.

17.10 BIBLIOGRAPHY

Kose, S., "Aging Population and the Impact on Buildings," *Kenchiku-Gijutsu* 427: 155–165, 1987 (in Japanese).
———, "Capability of the Elderly and Their Accident Experiences: Implication to Design of Safer, Easier-to-Use Dwellings," *Saigai-no-Kenkyu (Research on Disasters)* 22: 128–145, 1991 (in Japanese).

Kose, S., "Daily Living Capabilities of the Elderly and Their Accident Experiences: Implication to Design of Safer, Easier-to-Use Dwellings," *Equitable and Sustainable Environments, EDRA 23 Proceedings,* pp. 158–166, EDRA, Edmond, OK, 1992a.

———, *Barrier-Free Design in Japan, Building Design for Handicapped and Aged Persons, Council on Tall Buildings and Urban Habitat,* McGraw-Hill, Tokyo, Japan, 1992b, pp. 15–26.

———, "Recent Trends of Barrier-Free Building Ordinances and the Japanese Building Standard Law," *Fukushi-Rodo* 59: 12–23, 1993 (in Japanese).

———, "Ageing vs. Housing: Potential of Information Technology in the Japanese Context," *Potential of Information Technology for Solving Housing Problems of Aged People, Proceedings of an International Workshop,* pp. 1–5, Building Research Institute, Tsukuba, Japan, 1994.

———, "Possibilities for Change Toward Universal Design: Japanese Housing Policy for Seniors at the Crossroads," *Journal of Aging and Social Policy* 8(2&3): 161–176, 1996.

———, "Dwelling Design Guidelines for Accessibility in the Aging Society: A New Era in Japan?" in *Handbook of Japan-U.S. Environment Behavior Research: Toward a Transactional Approach,* Plenum, New York, 1997, pp. 25–42.

———, "From Barrier-Free to Universal Design: An International Perspective," *Assistive Technology* 10(1): 44–50, 1998.

——— (ed.), *Universal Design: An International Workshop,* Building Research Institute, Tsukuba, Japan, 1999.

———, "Universal Design for the Aging," to appear in *Encyclopedia of Ergonomics and Human Factors,* Taylor & Francis, London, UK, 2000.

———, and M. Nakaohji, "Housing the Aged: Past, Present, and Future; Policy Development by the Ministry of Construction of Japan," *The Journal of Architectural and Planning Research* 8(4): 296–306, 1991.

———, A. Ohta, Y. Tanaka, and K. Watanabe, "Examination of Design Effectiveness of Special Housing for the Aged: Is Japanese 'Silver-Housing' a Success?" *IAPS12 Proceedings,* vol. 3, pp. 161–166, University of Thessaloniki, Thessaloniki, Greece, 1992.

———, and Y. Tanaka, "The New Design Guidelines for Dwellings for the Ageing Society: How Are They Accepted by the Residents?" *People, Places and Public Policy—EDRA29 Proceedings,* pp. 53–56, EDRA, Edmond, OK, 1998.

———, I. Kumano, and A. Matsuzaki, "How Far Has the Design of Dwellings Improved in These Ten Years? Comparison of Two Groups of Houses from the Viewpoint of Design for the Aging Society." *Power of Imagination—EDRA30 Proceedings,* pp. 127–132, EDRA, Edmond, OK, 1999.

Lanspery, S., and J. Hyde (eds.), *Staying Put: Adapting Places Instead of the People,* Baywood Publishers, Amityville, NY, 1996.

Ministry of Construction, *Housing and Infrastructure Improvement Aimed at the Promotion of Healthy and Fulfilling Lifestyles: Guidelines for the Creation of Living Space Based on Citizen Welfare,* S. Kose, Tokyo, Japan, 1994.

National Institute of Population Problems, *Population Estimate in August 1986,* S. Kose, Tokyo, Japan, 1986.

National Institute of Population and Social Security Research. *Population Estimate in January 1997,* S. Kose, Tokyo, Japan, 1997.

Nomura, K., S. Kose, et al., *Housing Rehabilitation Manual for the Ageing Residents,* Japan Housing Reform Center, Tokyo, Japan, 1990 (in Japanese; revised in 1998).

17.11   RESOURCES

Government of Japan, Building Standard Law, 1950.

Government of Japan, Accessible and Usable Buildings Law, 1994.

Government of Japan, Housing Quality Assurance Law, 1999.

Government of Japan, Accessible and Usable Transportation Law, 2000.

U.S. Government, Fair Housing Amendment Act, 1988.

U.S. Government, Americans with Disabilities Act, 1990.

CHAPTER 18

INTERPRETING ACCESSIBILITY STANDARDS: EXPERIENCES IN THE U.S. COURTS[1]

Sanjoy Mazumdar, Ph.D., and Gilbert Geis, Ph.D.
University of California-Irvine, Irvine, California

18.1   INTRODUCTION

This chapter addresses a number of complex issues related to the provision of appropriate environments using legal means, such as the Americans with Disabilities Act (ADA). For persons with disabilities using wheelchairs, how is building accessibility being addressed in the United States? What does the law mandate? How have architects been responding to the need for more universally usable environments, given legal and regulatory requirements for accessibility?

A number of recent lawsuits based on the ADA adjudicated requirements for the design of sports and entertainment arenas in regard to spectators who use wheelchairs. The vagueness of the language of the law, the absence of proper action on the part of government enforcement agencies entrusted with administration of the provisions of the law, the persistence on the part of architects to make satisfactory accessibility a legal rather than a moral imperative, the various and contradictory rulings by judges, and the adversarial nature of the American legal system all contributed to the planning and construction of arenas that failed to meet the needs of patrons using wheelchairs.

18.2   BACKGROUND

The issue at first appears to be rather simple: Should not buildings be designed to provide satisfactory access and enjoyment to all who might reasonably need or desire to be involved in what goes on within them so that no one will feel discriminated or left out? Even for those who are able-bodied and without impairments, buildings are increasingly designed to improve comfort and pleasure and reduce stress. Buildings are routinely air conditioned and outfitted with elevators and conveniently located restrooms, for example. However, comfort and usability are important not only for the able-bodied, but also for persons with disabilities.

Accessibility and *universal design* mean more than only being readily able to enter a building and spaces inside. It also includes the ability to utilize and enjoy all the facilities contained in any building. Those who might be shut out if an equal-accessibility doctrine is not implemented include persons with mobility challenges or impairments, such as those using wheelchairs, persons with vision and hearing problems who might be unable to access signs and information or negotiate the environment satisfactorily, and those who are unable to operate the fixtures and furnishings in buildings, such as doors and door handles.

In the United States, how is accessibility being dealt with? A common technique for implementing public policy agendas is to pass laws. Designers need to follow applicable laws. But what does the law require? What is the state of the law with respect to accessibility? What are the architectural requirements for the construction of sports and entertainment arenas? How are the laws working? And how have persons using wheelchairs fared? These questions are taken up here.

This chapter will consider Title III of the ADA, which deals with public accommodations and commercial facilities, and includes provisions for access to buildings. This inquiry involves an examination of a number of federal court cases that dealt primarily with sports and entertainment arenas. Numerous cases have been filed and ruled on. It is impossible to review them all even when the scope is limited to arenas. The intent here is the provision of a sense of the American approach and how it is working, not comprehensiveness in mentioning every case.

Sports arenas and stadia are useful to study because they provide not only the usual concerns related to access and usability but also include consideration of lines of sight, dispersal, integration, and companion seating. Even though they deal primarily with persons using wheelchairs and do not address other forms of disability, they make valuable examples.

The legal cases involving access, as this research indicates, were a good deal more complicated than the opening paragraph suggests. They demonstrate how diverse judicial decisions muddied rather than clarified contradictory or ambiguous issues. The cases reviewed also highlighted the uneasy relationship between the architectural profession and the laws that bear or, arguably, do not bear upon it.

This chapter begins with a brief description of the American legal system, then takes up the ADA, the role of the courts, followed by a brief review of a few cases, before ending with a conclusion.

18.3 THE AMERICAN LEGAL SYSTEM

In the United States, Congress has the sole power to enact laws that apply to the entire nation. The introduction of a bill in Congress may be the outcome of various prods: a campaign promise, constituency pressure, personal interest, or lobbying by special interest groups, among others. The drafting of a law may reflect the involvement of those whom it is supposed to regulate and/or those who expect it to advance causes in which they believe or from which they will benefit. These negotiations are generally conducted away from public view.

Typically, the proposed law is sent to a congressional committee where there are hearings and debate. These proceedings provide a written record that details both alterations in the initial wording of the proposed law, and rationales that underlay aspects of the measure. Some judges pay close attention to these congressional records for help in interpreting a law. Other judges follow an approach called *textualism* and insist that the sole responsibility of a judge is to read as precisely as possible the words of the law and their plain meaning, and not to seek to determine how it came about, as members of Congress who voted for the law may not have been acutely aware of debated points or been in agreement with them.

Once voted on and passed by the Congress, the bill goes to the president, who can sign it into law or exercise the power of veto. The administrative agencies of government are entrusted the duty of developing detailed regulations and sometimes of enforcement, and the courts are given the task of interpreting and applying the law.

In this system, a law, such as the ADA, may be vague. From the viewpoint of those members of the House of Representatives and the U.S. Senate sponsoring the bill, it often is advantageous to keep the language of a law imprecise, lest they lose voting allies who may object to more specific mandates and injunctions. Additionally, legislators often lack the detailed knowledge, the level of expertise necessary, or the foresight or premonition of potential issues that might arise to address them definitively.

Not all parties who participated in the drafting and the passage of the ADA welcomed this state of affairs. They foresaw that the formidable press for equality for persons with disabilities could be eviscerated by skilled and well-financed legal battles waged by those who had financial interest in seeing the law's reach truncated. One congressman pinpointed the problem:

> The legislative history reveals that some members of Congress were uncomfortable with this [subsequent rulemaking and adjudication] feature of the ADA. Rep. Douglas protested that "Congress is abrogating it[s] constitutional duty by writing vague laws which must be clarified by the Federal courts. Our *responsibility* is to write laws which can be clearly understood when reading them—not have another branch of government do our job" (H. R. Rep. No. 101-485(III) at 94 (1990), *reprinted at* 1990 USCCAN 511 {emphasis in original}). His objections did not carry the day (*ILR v. OAC,* 1997f: 737–738, Opinion, 12 November 1997: 63–64).

Intentional or not, this imprecision and the consequent conflicting interpretations in part has resulted in more than 500 volumes of U.S. Supreme Court decisions, close to 1500 volumes of federal courts of appeal rulings, and nearly 800 volumes of federal district court opinions (Acker, 1995).

There is, however, a more subtle implication in the remark of the member of Congress, one that played out in terms of building requirements set out in Title III of the ADA. If the agency charged with promulgating guidelines also fails to do its job with precision, then the level of uncertainty inevitably will prompt lawsuits and judicial vexation. For example, when the Gallatin Airport Terminal in Butte, Montana, underwent an expansion, the restaurant was moved to a mezzanine level located approximately 3 feet above the second floor. Wheelchair access to the restaurant was by an elevator to the second floor and then a lift for the remaining 3 feet. The judge, having to decide whether this arrangement was legally acceptable (he decided it was not), expressed his irritation at the Department of Justice (DOJ), which had promulgated the guidelines: "I am asked to enforce demanding, confusing and controversial design requirements that the Department of Justice itself has never championed in any court or in any rule making procedure, even when invited to do so" (*Coalition of Montanans Concerned with Disabilities, Inc. v. Gallatin Airport Authority,* 1997: 1167).

Businesspersons and professionals, such as architects, typically are not comfortable with uncertainty, unless they can safely employ it to their own advantage. Architects, though they may prefer otherwise, can live with a design mandate if it does not overly interfere with aesthetic concerns. Any additional costs involved can be passed on to those paying the bill. The bill payer might be dissuaded by the additional cost, but, since the requirement binds all architects, competitors will not be advantaged. If, however, architects see that an interpretation of a legal dictate, such as the ADA guidelines, can prove notably expensive for them, they are likely, after some inexact cost-benefit calculus, to decide to litigate the matter. In the ADA cases, the architects' position was two-pronged: that the law did not apply to them, and that, even if it did, it had not been laid down in accord with rules dictated by the federal Administrative Procedures Act (APA).

18.4 THE ROLE OF THE COURTS

The primary role of the courts is to interpret and apply the law. They may also enforce the law in the cases brought before them.

It is an axiom—and a very important one—that American courts can rule only on matters that are properly brought before them. This means that they cannot on their own locate an issue that seemingly needs judicial clarification. It also means that those seeking remedies must have *standing,* that is, that they must have a personal stake in the outcome. In the ADA cases examined, appellants were not only organizations whose members included persons using wheelchairs, but sometimes included individuals using wheelchairs who could claim that they personally were unable to access or enjoy the facilities perhaps to make sure they met the requirements of the law, which stated, "No individual shall be discriminated . . ." (42 U.S.C. §12182(a)). In addition, there are matters of jurisdiction, justiciability, and ripeness that might disqualify cases.

Judges rule on matters of fact and on matters of law, though the latter seems to take up much of their attention. Prior rulings and interpretations become precedent and de facto law as, in their rulings, judges must give deference to rulings by higher courts in the jurisdiction, though sometimes they will try to guess and preempt the effects of judgments by higher courts (see *ILR v. OAC,* 1997f). Although not required to, judges will often pay attention, and sometimes defer, to rulings by equivalent-level courts in other states. In this sense, in matters of law, judges' decisions are not independent, but are linked to create a body of law. For this reason, lawyers work hard to differentiate their cases from similar cases elsewhere with unfavorable rulings.

In their judgments and interpretations of law, judges may seek to apply the spirit of the law or the letter of the law. They also bring in their own predilections, moral, political, ideological, or other, in arriving at decisions.

Judges and courts have limited resources, and may be unable to research questions or obtain expert testimony or help in technical areas in which they lack expertise.

The courts, in the past, have not held architects liable for buildings designed by them when the architects claimed privity of contract, transferring liability for the building to those who could modify or reject the designs. In recent years, however, this common-law doctrine concerning architectural liability has been breached and battered (Nischwitz, 1984; Flatt, 1990; Colgate, 1999; Fritts, 1998). The doctrine that the customer first had to take the contractor to court and then the contractor would have to try to recover from the architect, if the architect were believed to be at fault, is being abandoned. The ADA sports arena cases seemingly dealt a near-fatal blow to this doctrine, though hardly categorically, and perhaps not in regard to other building arrangements outside the scope of the ADA statute.

18.5   *THE AMERICANS WITH DISABILITIES ACT*

The Americans with Disabilities Act was signed into law by the president on 26 July 1990. Its intention was to end discrimination against persons with disabilities and to seek to ensure that the estimated 42 million persons with stipulated disabilities would be moved more effectively into the mainstream of social and business life.

Since the ADA was viewed primarily as a civil rights law, Congress entrusted enforcement to the DOJ. Being considered civil rights law also meant that money considerations generally were not to be regarded as determinative. That is, the justice of the situation was to take precedence over the cost of conformity, even if very high.

Some components of the ADA provide clear guidance. The date for effectiveness of the law, 26 January 1993, is not ambiguous. The number of accessible locations for assembly areas is mandated by ADA regulations:

(a) in places of assembly with fixed seating accessible wheelchair locations shall comply with . . . the following table:

Capacity of seating in assembly areas	Number of required wheelchair locations
4–25	1
26–50	2
51–300	4
301–500	6
Over 500	6, plus 1 additional space for each total seating capacity increase of 100.

(US DOJ 1994a, *28CFR Part 36,* App. A, Ch. 1, §4.1.3(19) 01 July 1994).

This is commonly referred to as the *one percent plus one* formula for fixed seating capacity over 500.

But, many sections of the ADA are vague, unclear, and at times incompatible (see Mazumdar and Geis, 2000). For example, in regard to accessibility, §302(a) of the ADA states:

> No individual shall be discriminated against on the basis of disability in the full and equal enjoyment of the goods, services, facilities, privileges, advantages, or accommodations of any place of public accommodation by any person who owns, leases (or leases to), or operates a place of public accommodation (42 U.S.C. §12182(a)).

This bold declaration is followed by an attempt to define the term *discrimination,* for public accommodations and commercial facilities, as:

> (1) a failure to design and construct facilities for the first occupancy later than 30 months after July 26, 1990 [i.e., 26 Jan 1993], that are readily accessible to and usable by individuals with disabilities . . . (42 U.S.C. §12183(a), §303(a)).

This clarification not withstanding, practitioners and architects attempting to follow the requirements set out in the legal language of this law still run into difficulties even if they consult learned counsel. For example, in the requirement "readily accessible and usable," how readily need access be to meet the stipulation? If a person using a wheelchair must enter Building A and progress over a second-story crosswalk to get into Building B, is this to be deemed to meet the requirement of the law?

But, perhaps to the surprise of those unaccustomed to some arcane aspects of the working life of lawyers, it was not "readily accessible" that would arouse intense dispute, with millions of dollars at stake on the outcome, it was the phrase "design and construct." Were the two verbs to be read conjunctively to refer only to those entities that both designed and constructed the facility, or did they embrace designers and constructors? And were owners and financiers, too, to be held responsible for failure to meet standards enunciated in the ADA? These three words—*design and construct*—provided the basis for protracted litigation and a considerable challenge for attorneys and judges forced into the role of grammarians.

Other unclear items included whether the number of required wheelchair locations could include locations not providing a line of sight unobstructable by spectators in front standing up, requirements for dispersal, integration, and companion seats.

Regulations and guidelines required by the ADA to be developed by the DOJ turned out to be problematic. One example was the DOJ's guideline standard 4.33.3. It provides in part (with awkward diction):

> Wheelchair areas shall be an integral part of any fixed seating plan and shall be *provided so as to provide people with physical disabilities a choice of admission prices and lines of sight comparable*

to those for members of the general public." [US DOJ 1994a, *28CFR Part 36,* App. A, 01 July 1994, p. 547 (italics in original)].

The guideline also stipulates that these seats:

shall adjoin an accessible route that also serves as a means of egress in case of emergency. (US DOJ 1994a, *28CFR Part 36,* App. A, 01 July 1994, p. 547).

and that:

at least one companion fixed seat shall be provided next to each wheelchair seating area. [DOJ 1994a, *28CFR Part 36,* App. A, 01 July 1994, p. 547 (italics in original)].

Before these guidelines were adopted, they were published, with an invitation to interested parties to offer comments. The agency noted that an "overwhelming majority" of responses believed that unobstructed lines of sight (ULOS) for persons using wheelchairs ought to be made a specific item in the rule. But the issue was never formally addressed, beyond a comment that it would be considered subsequently in regard to recreational facilities, which never happened. It was generally agreed that, though somewhat ambiguous, Rule 4.33.3 did *not* call for ULOSs.

The only public comment from the DOJ on the rule came in 1992 during a talk to major-league baseball stadium operators in which the DOJ's chief of the public access section declared that the guideline did not contemplate enhanced sightlines (i.e., ULOS) for wheelchair customers. Judges later made short shrift of this declaration: One dismissed the comment as an offhand observation by a person in a relatively minor position in the department (*PVA v. DCA,* 1997d; 587), and another disposed of it as "post-election remarks made by a mid-level official in a lame duck administration at a convention of baseball stadium operators do not constitute a binding interpretation of agency regulations" (*ILR v. OAC,* 1997f: 737). Had they been so inclined, the judges could well have considered the remark as an enunciation of DOJ policy, providing definitive guidance to architects and builders regarding what was expected of them. But the judges were not so inclined. The rule only meant, one judge observed by way of illustration, that if there were seats behind home plate in a baseball stadium, there also would have to be accessible wheelchair locations in the same general area.

Then, in December 1994, the Department of Justice issued a supplement inserting the following provision in its *Technical Assistance Manual* of November 1993:

. . . in assembly areas where spectators can be expected to stand during the event or show being viewed, the wheelchair locations must provide lines of sight over the spectators who stand. This can be accomplished in many ways, including placing wheelchair locations at the front of a seating section, or by providing sufficient additional elevation for wheelchair locations placed at the rear of seating sections to allow those spectators to see over the spectators who stand in front of them (U.S. DOJ 1994b: *The Americans with Disabilities Act Title III Technical Assistance Manual,* 1994 Supplement III-7.5180, p. 13 insert in *TAM,* November 1993, §III-7.5180, p. 64).

But was this guideline legitimate? The most vexing issue was whether the added guideline was merely an acceptable extension and clarification of a prevailing rule, and therefore not bound by APA procedures. Or was it a new rule that met the promulgation demands of the Administrative Procedures Act, which called for publication of intended rulemaking and the opportunity to submit comments. Had it, as one commentator would note, been inserted in an unacceptable and "backhanded" manner (Conrad, 1998a).

Even if granted legitimacy, this guideline left matters unclear. Could ULOS be achieved by posting signs warning spectators not to stand if they were located in front of a wheelchair spectator? Might the guidelines be met by not selling seats located in front of spectators using wheelchairs? Could wheelchair locations be such that if not sold they could readily be replaced by ambulatory in-fill seating? Approval of any of these options, which were rejected in litigation, would have meant several million dollars' additional income to the arena owners over the course of a season.

18.6 ADA-RELATED ARENA COURT CASES

It is against this background that a number of ADA cases were adjudicated by federal courts in diverse judicial districts. To understand the nature and functioning of the law now, it will be useful to look at a few of the significant court cases. This review of the legal life history of the ADA wheelchair-arena cases will set out some of the substantive issues.

Parties Involved. The parties in these court cases varied, but it is useful to describe some of the primary parties and their intentions.

Perhaps most important were individual persons with disabilities, who felt discriminated against, or their organizations, such as the Paralyzed Veterans of America (PVA), who fought for the rights of persons with disabilities. Their intention was to have the ADA applied and ensure that environments provided equal access and enjoyment to persons with disabilities.

The defendants were the designers, owners, and operators of the facilities. They sought clarification of the law and to protect their profits by avoiding having to expend thousands of dollars to change the designs or make alterations to built structures. The architects wanted to be exonerated as they felt the ADA did not hold them liable, a position they had enjoyed in the past.

The courts and judges were expected to interpret what the law required and rule on the disputes brought before them. Although they were expected to be just in their decisions, they seemed to have the moral approach of providing for persons with disabilities.

The Department of Justice (DOJ), the entity charged with the responsibility of developing detailed regulations and guidelines lacking in the law, was in court mostly as friends and supporters of the plaintiffs. In these arena cases they refused to join the lawsuits even when invited by the judge to do so, although they did try to enforce the law by filing cases against the architects, owners, and operators in a few instances (e.g., *U.S. v. Physorthorad Associates,* 1996).

The American Institute of Architects (AIA) declared that architects were interested in providing appropriate environments for persons with disabilities and claimed to seek clarification of the law. But they filed amicus curiae briefs in support of the architects. Looking after the best interests of their members, they wanted to maintain the lack of liability architects had enjoyed in the past.

Issues. At the heart of the major contested issues in the arena cases was the DOJ's guideline standard 4.33.3 and whether it was promulgated properly. Other issues brought before the courts included the appropriate method of counting for the total number of required wheelchair locations, ULOSs of the main event, dispersal, and integration of accessible locations.

The MCI Center Case (1996)

District Court. The MCI Center case, involving a sports arena in downtown Washington, D.C., was the first major ADA adjudication dealing with assembly facilities. The complainants were the Paralyzed Veterans of America (PVA) and four persons with disabilities, and the defendants included Ellerbe Becket Architects and Engineers (EBAE), the firm that had designed the MCI Center, and seven others: owners, engineers, constructors, and operators

[*Paralyzed Veterans of America* (*PVA*) *v. Ellerbe Becket Architects and Engineers* (*EBAE*), 1996a,b].

The PVA argued that there were insufficient wheelchair locations, that most of them were isolated in the upper "nosebleed" portions of the Center, that they were not acceptably integrated, or dispersed, and that only a small percentage offered unobstructed sightlines.

The lawyers for Ellerbe Becket, the firms responsible for the largest number of new stadia, claimed that the ADA did not hold architects liable for errors of commission or omission and that their client should be dropped from the suit:

> In general, an architect or engineer does not have ultimate control over design and construction unless it explicitly is provided for by his contract. Just as an architect of a private home must verify decisions with the homeowner and comply with the homeowner's direction, the architect or engineer on a public project rarely has absolute control over the design. All decisions are subject to an owner's taste, vision, and pocketbook (*PVA v. EBAE*, 1996b, Defendant's Motion to Dismiss: 8t).

EBAE's position was endorsed in an amicus curiae brief filed by the AIA. The AIA attorneys declared that architects support the ADA and accessible design; an AIA member serves on the Architectural and Transportation Barriers Compliance Board (ATBCB, also known as the Access Board), which developed the ADA Accessibility Guidelines (ADAAG) adopted by the DOJ, that three AIA members serve on the ATBCB-appointed advisory board, and that AIA members apply accessibility guidelines in their designs. It claimed that "the drafters . . . intentionally omitted language . . . which might have encompassed architects and design professionals within the list of responsible parties" (*PVA v. EBAE*, 1996f, AIA Amicus Curiae: 4t).

The plaintiffs in return insisted that the phrase "design and construct" clearly embraced the architects. To say that it included only those who both designed and constructed would be a forced and incorrect interpretation. And since constructors rarely design the building, a rule forbidding them to do so would be "essentially meaningless" and an unacceptably narrow reading of the law.

District Court Judge Thomas F. Hogan agreed with EBAE's argument that the firm was not liable for any ADA-related shortcomings in the MCI Center design, granted two motions, and dismissed them from legal liability in the case (*PVA v. EBAE*, 1996g; *PVA v. EBAE*, 1996h: 2t).

Of key importance, the defendants insisted that the guideline cited by the plaintiffs was not operative because it had not been properly promulgated. EBAE also argued that to provide wheelchair sightlines over standing spectators gave those using wheelchairs a superior or enhanced view. The judge in the MCI Center case was unimpressed with this argument and was in favor of comparable views, even if that meant a design that was slightly superior (*PVA v. EBAE*, 1996j: 400 fn. 16).

Subsequently, in the Portland case, Judge Ashmanskas pounced on this point that ULOS constituted not equal but preferential treatment, noting sarcastically that people were not hurrying to have amputations in order to qualify for such preferential treatment (*ILR v. OAC*, 1997f: 734, Ashmanskas Opinion: 56–57t).

Judge Hogan ruled that the ULOS requirement was an interpretive rule, a clarification of a guideline issued earlier, that it therefore was proper for the DOJ to unilaterally add it to its manual without calling for public comment, and that it had been incumbent on the remaining defendants to adhere to that guideline. Therefore, the defendants were required to redesign the Center to make it accord with ADA standards for wheelchair seating.

Even so, the judge had kind words for the defendants. The case was all the more difficult, he observed, because the defendants had acted in good faith. But good faith alone was not adequate: They should have been wiser and used better judgment about what was required of them (see *PVA v. EBAE*, 1996a–k; *PVA v. DCA*, 1997a–d, 1998).

United States Court of Appeals.   Both the plaintiffs and the defendants were not satisfied with the decision and appealed to the U.S. Court of Appeals.

The specific issues raised were whether the regulations called for ULOSs, whether DOJ had promulgated a substantive rule following the notice-and-comment mandate of the APA, and whether the substantial compliance ruling was valid.

Judge Silberman writing for the D.C. Court of Appeals, ruled that "lines of sight comparable" referred to ULOS "view no more obstructed than would be available to nonwheelchair users" (*PVA v. DCA*, 1997d: 583). The justices also ruled that 4.33.3 is open to the interpretation of ULOS and so the DOJ regulation is valid and receives deference. They upheld Judge Hogan's ruling of substantial compliance stating that the regulation did not call for every seat to have ULOS.

United States Supreme Court.   The case was taken to the Supreme Court, but in March 1998 the Court declined to hear the matter, thus letting stand the decision of the Court of Appeals and the District Court.

The MCI Center ruling in the first ADA sports arena case stands as a landmark decision, but its major findings soon conflicted with those of several other federal courts in the country.

The Rose Garden, Portland, Oregon, Case (1997)

Among the numerous issues brought up in the Rose Garden case were whether placing 33 wheelchair spaces on Level 7, the highest level of the upper section, where there were no other fixed seats, violated the dispersal requirement; whether clustering wheelchair seating in the corners of the end zones met the horizontal and vertical dispersal requirement; whether selling permissible in-fill seats on a permanent basis violated the number of wheelchair locations required; whether required companion seats needed to be fixed, and whether they could be in front and behind rather than side-by-side; whether the DOJ requirement of lines of sight over standing spectators was a valid regulation; whether executive suites were required to have accessible seats; whether visual alarms were necessary in suites; whether camera operator areas were required to have accessible seats; and many questions dealing with passages, concessionaire stands, toilets, and others (*ILR v. OAC*, 1997a). Attention focused primarily on dispersal of wheelchair seating and companion seating (see *ILR v. OAC*, 1995, 1997a–f, 1998a,b).

Judge Ashmanskas, the judge in the Oregon case, reiterated that each ticket category was required to have a proportionate number of wheelchair spaces, and that placing a disproportionately large number of wheelchair spaces on Level 7 of the Rose Garden "makes mockery of the ADA's dispersal requirement" and that this does not meet "the requirement that wheelchair spaces must be an integral part of the overall seating plan" (*ILR v. OAC*, 1997f: 712). He concluded that on realizing that their design did not come close to meeting ADA requirements, OAC and EBAE had decided to take this action "out of desperation" in order to meet the required number of accessible seats (*ILR v. OAC*, 1997f: 710). Regarding dispersal, he ruled that absolute proportionality of wheelchair spaces in each level or section was not required, that some flexibility was necessary, but because neither the DOJ nor the Access Board defined the leeway acceptable, he decreed that a variation of 10 percent would be permissible.

Ellerbe Becket had also placed seats for ambulatory companions of wheelchair patrons in the row in front or behind them, having determined that a next-to-each-other arrangement would result in a loss of 790 ambulatory seats. The judge decided that fixed companion seats "next to each other," required according to DOJ's (1994a) 28 CFR Pt. 36, App. A §4.33.3, was a reasonable regulation. But he declared that these companion seats need not be bolted to the floor, and that they could be Clarin padded folding chairs.

Complaining about the actions and inactions of the DOJ, he ruled, in contrast to Judge Hogan, that the wheelchair sightlines regulation had not been promulgated properly by the DOJ and that the defendant was not required to provide lines of sight over standing spectators at the Rose Garden (*ILR v. OAC*, 1997f: 746–747).

Even though the defendants argued that the "one percent plus one" should not be applied because few persons using wheelchairs attend events at the Rose Garden, the judge's edict was that the rule applies and that the Rose Garden's design violated it.

Judge Ashmanskas ruled that suites are subject to the requirements of Title III of the ADA and needed to provide the formulaic number of accessible seats. He clarified that the law mandated visual alarms in suites. He decreed further that camera operator areas also must comply with the ADA. Later, the judge decided a large number of issues regarding passages, toilets, and others.

Anticipating the probable judgment if the case was appealed, the judge clarified that if on appeal the ninth circuit judges disagreed with his judgment, the defendants would, as a matter of law, be liable to supply unobstructed sightlines for wheelchair spectators. The case never got to the ninth circuit, however. After negotiations, the owners agreed to provide satisfactory wheelchair sightlines, while the plaintiffs abandoned their quest to have the suites in the arena retrofitted to provide satisfactory wheelchair accommodations.

Judge Ashmanskas, in concluding that EBAE had acted calculatedly, watchful of expenses and loss of profits, differed significantly from Judge Hogan, who believed that the architects had acted in good faith.

Broward County Arena, Sunrise, Florida (1997)

This case involving the Broward Stadium, home of the Florida Panthers ice hockey team, was brought before construction had begun. The plaintiffs maintained that they had adequate reason to believe, based on its earlier performances, that Ellerbe Becket would not adhere to ADA mandates (*Johanson v. Huizenga Holdings, Inc.*, 1997). Ellerbe Becket filed a motion to dismiss claiming that architects are not liable under the ADA.

Judge Jose A. Gonzalez found the plaintiffs' argument—that if architects were not liable because they do not "design and construct," then no one would be liable for new commercial facilities—more credible and dismissed Ellerbe's motion (see *Johanson v. Huizenga Holdings, Inc.*, 1996, 1997).

Blockbuster–Sony Music Entertainment Centre, Camden, New Jersey

District Court.   Enhanced lines of sight in the Blockbuster–Sony Entertainment Centre (BSEC) lay at the heart of this lawsuit (*Caruso v. BSEC*, 1997). Judge Joseph E. Irenas ruled, similar to his colleague in Florida, that the enhanced view doctrine for wheelchair spectators had not been properly promulgated, that it was adopted after the construction for BSEC had commenced, and therefore was not binding on the builders. He further declared that "comparable" lines of sight could not, given the history of the ADA, be interpreted to mean lines of sight over the heads of standing spectators (see *Caruso v. BSEC*, 1997, 1999a,b).

Court of Appeals.   Writing for the Court of Appeals, Judge Alito agreed with the earlier ruling that the ADA was vague and could not require for wheelchair users clear lines of sight over standing spectators. But, disagreeing with the lower court, the judge wrote that accessible routes to assembly areas outside was mandated (*Caruso v. BSEC*, 1999a,b).

The Minnesota Case (1997)

Though it had escaped being scathed in the MCI case, Ellerbe Becket's problems with arena wheelchair issues were far from over. On 10 October 1996, the federal DOJ filed suit against Ellerbe Becket in Minneapolis, alleging that it had engaged in a pattern of violations of ADA regulations in the designs of a half-dozen sports arenas. The suit alleged that none of them provided wheelchair patrons with guaranteed unobstructed sightlines (*U.S. v. Ellerbe Becket,* 1996a, Plaintiff's Complaint). Again, the AIA filed a supporting brief, this time in conjunction with the Associated General Contractors of America. The arguments made were essentially those that succeeded in the MCI Center litigation. The AIA brief included a sentence that may have epitomized why this time it failed to carry its case. It read: "Throughout its 138-year history, the AIA has represented the interests and concerns of the architectural profession in every state and legislative body in the country." (*U.S. v. EBI,* 1997a: 3 t). Therein lay the crux of the matter: The AIA was interested in protecting its members; Congress's interest, as read by the courts, lay in enabling persons with disabilities. When the two interests conflicted, there was no impetus for the courts to reach for an interpretation that favored the architects (see *U.S. v. EBI,* 1996, 1997a,b, 1998).

Judge Tunheim rejected the argument that ADA did not apply to architects (*U.S. v EBI,* 1997b). Faced with being placed in an insupportable and expensive position, Ellerbe Becket entered into a negotiated settlement with the government.

In its settlement with the government, Ellerbe Becket did not concede legal responsibility, only that it would meet the dictates of ADA in the future, including the stipulation that it would provide unobstructed seating in all of its future designs (*U.S. v. EBI,* 1998). For its part, the government agreed not to take any action against the firm for projects designed before 1998. In the consent decree, the DOJ specified the dimensions of wheelchair seating and for obtaining ULOS that the architect had to follow (see Fig. 18.1). The Building Owners and Managers Association, pleased with the decision, responded by noting that architects are in the best position to make certain that laws pertaining to construction are obeyed (Winston, 1997c).

FIGURE 18.1 Lines of sight for wheelchair seating. Dimensions specified in Exhibit B:

Average eye height of person using a wheelchair: 47.45 in

Horizontal distance from eye of an average person in wheelchair to edge of tier on which wheelchair rests: 30 in

Average head height of standing spectator: 67.65 in

Average eye height of standing spectator: 63.45 in

Average shoulder height of standing spectator: 55.65 in

3Com Park, Candlestick Point, San Francisco, California (1993–1996)

In a settlement agreement to resolve a case filed by the DOJ, the City of San Francisco agreed to install for persons with disabilities who use wheelchairs an elevator, 61 new wheelchair and companion seats (not folding ones), renovate the restrooms, install 25 assistive listening devices, upgrade signage and accessible parking spaces, and train staff in nondiscriminatory service of persons with disabilities, and take several other actions (*U.S. & DREDF v. San Francisco,* 1996).

18.7   NONARENA CASES

Gallatin Airport, Butte, Montana (1997)

During an expansion of Gallatin airport, a restaurant and bar had been moved from the ground floor. The plaintiffs, the Coalition of Montanans Concerned with Disabilities, Inc., claimed that provision of a separate platform lift for gaining access to the restaurant located on a mezzanine approximately 3 feet above the second floor, failed to meet ADA requirements. Judge Molloy ruled that they were correct in seeking an elevator that would provide ready access to persons with disabilities (*Coalition of Montanans Concerned with Disabilities, Inc., v. Gallatin Airport Authority,* 1997) and required changes.

Physorthorad Case, Hershey, Pennsylvania (1996)

The DOJ filed a lawsuit against the owners, Physorthorad Associates, and the architects of the Parkside Professional Center, an office building, for failure to follow the ADA. In the consent order, the owners agreed to provide an elevator, accessible parking, handrails, and make several changes to internal fixtures (*U.S. v. Physorthorad Associates,* 1996).

Days Inns of America (1996–1998)

The DOJ launched court battles in several states charging the Days Inns of America, a motel chain, for failure to follow the ADA (for background and history of this litigation, see *Days Inn of America v. Reno,* 1996). These have gone both ways in regard to whether architects were liable under the "design and construct" language of ADA. They were deemed not liable in Kentucky (*United States v. Days Inns of America,* 1998b), although an Illinois district court judge thought otherwise (*United States v. Days Inn of America,* 1998a).

18.8   CONCLUSION

This review of court cases dealing with arenas leads to a number of important conclusions about the designs of sports arenas and other buildings so that they do not discriminate against persons with disabilities.

The tripartite division of responsibility in the United States for law- and rulemaking through policy formulation by Congress, detailed regulations and enforcement by the DOJ, and interpretation and application by the courts has produced awkward results. None of these governmental entities declared unequivocally what the law required.

The preamble of the ADA was ringing: "To provide a clear and comprehensive mandate for the elimination of discrimination against persons with disabilities." But, political necessities, it appears, led the Congress to draft the law imprecisely and incompletely even though this was not the first attempt by Congress to address this issue (Null and Cherry, 1996).

Congress left the development of detailed regulations and guidelines to the DOJ. Lack of clarity in the regulations by the DOJ and procedural inconsistencies still left regulatory requirements uncertain. These were, in part, the reasons for the parties to seek the help of the courts through lawsuits. Ultimately, the task of clarifying the law fell to judges, who, with considerable grumpiness, chastised law- and rulemakers for their failure to issue clear and comprehensive guidelines. But the courts did not completely clarify all requirements, either, with their acceptance of the unspecified "substantial compliance" and contradictory rulings (*PVA v. EBAE,* 1996j).

The tripartite division is having another important effect: The law keeps changing. The regulations and standards change when DOJ makes or interprets regulations. After the court rulings, the law in Washington, D.C., now sees ULOS as a legitimate requirement. But in other jurisdictions, such as Portland, this is not so. These changes and differences will continue until the U.S. Supreme Court, which tends to be sensitive to patently divergent edicts from courts immediately below it on the judicial hierarchy, accepts a case (*certiorari* is the legal terminology) and through its ruling reestablishes one version. But the DOJ could mend its ways and follow accepted procedure in declaring what is necessary in regard to wheelchair sightlines. This means that architects have to be aware of the law, regulatory changes, and latest rulings by courts in the jurisdiction and the Supreme Court. Although the changes so far have been delineated here, future rulings are likely to change the way the law stands now. This leaves matters related to accessible buildings still uncertain, in flux, and varying.

In this system, the courts have to rule on the basis of unclear laws. What makes these cases interesting and instructive is that different judges, reviewing much the same set of facts, came to opposite conclusions. Most concluded that the law required such wheelchair viewing lines over seated spectators, but others disagreed.

Judges have to clarify the laws and pass decrees sometimes quite arbitrarily. Some relied on textualism, others tried to understand Congress's intent.

Judges have to deal with procedural regularity issues, a concern that was notably prominent and controversial in the ADA cases that bore upon architectural liability. There are rules specifying how guidelines are to be promulgated. Judges must decide whether these rules were adequately followed or whether unacceptable detours were taken. Opinions differed sharply on this issue in the ADA cases regarding the guidelines concerning wheelchair sightlines.

Judges sometimes have to deal with matters unforeseen by the policy makers. They have to determine whether multiple requirements of the law could be concurrently and practically feasible. For example, Judge Hogan concluded that it was not feasible to meet requirements for ULOS, dispersal, and integration at the same time and that some flexibility in the law was necessary. He rather arbitrarily established what seemed to him to be a reasonable compromise of "substantial compliance" if approximately 75 percent of the accessible locations had ULOS.

The judges must, at times, make decisions on matters about which they possessed only lay insight and partisan briefing by the litigating parties. They may be technical architectural matters, such as the compatibility between wheelchair seat placements and satisfactory sightlines. For some time now, debate in legal circles has been concerned with how well judges are served by partisan briefs on scientific and aesthetic issues that are well beyond their training. Many years ago, John Wigmore, a giant figure in the field of evidentiary law, suggested that impartial expert witnesses ought to be hired by the court and responsible to it, and not to one side in a dispute (Geis and Bienen, 1998). Such an approach might have infused professional concerns more significantly than were reflected in the ADA sports arena judicial rulings.

The federal court decisions focused almost exclusively on matters of law. The cases turned on what is called *statutory interpretation*. Interpretations rely not only on the way in which words are to be defined but also on precedent, a key ingredient in Anglo-American law, and one that is not found in continental jurisprudence. Precedent demands that, in similar-fact situations, judges follow binding rulings rendered by the U.S. Supreme Court and earlier decisions by higher courts in their own district. Though they are not required to do so, federal judges sometimes give "deference" to the wisdom of their colleagues on equivalent-level courts, especially if the courts are in typically well-respected sites, such as Manhattan and the District of Columbia, or, if a particular jurist is highly regarded. But fact situations are rarely identical. This is why lawyers devote much energy and imagination to differentiating convincingly (to the judges) the details of their case compared with others with seemingly similar situations in which rulings have gone against what the lawyers are aiming for. This is one reason different decisions on essentially the same question were forthcoming on apparently similar cases. Facts, perhaps, are less interesting and receive relatively less attention from jurists than commonly assumed. The emphasis on matters of law, which leads to the construction of a body of law, takes law out of the reach of common people and makes it necessary to hire, and be dependent on, lawyers.

Judges are required to be impartial and blindly just. But the judges in these cases showed a concern for aiding persons with disabilities, the goal that had prompted the ADA legislation. Readers of the diverse court decisions will be impressed with the judges' predilection and moral sympathy in favor of persons with disabilities, for finding support, legal or other, for their ideological predisposition, and for locating legally sound reasons for their decisions.

Jurists' egos become evident in some decisions. In one case, the judge declared that he did not believe ULOS seating was mandatory because he thought that the rule had not been properly adopted. But he wanted it to clarify his belief that if the court at the next level overruled him, the defendants would have no other worthy defense against liability. This judicial pronouncement may have been prompted by the jurists' awareness that the ninth circuit court of appeals was notably proplaintiff and might well favor those using wheelchairs (*ILR v. OAC,* 1997f).

For persons with disabilities, these split governmental responsibilities, murkiness of the law and regulations, and uneven efforts meant loss of time and expenditure of resources to achieve what they believed they had been promised by the lawmakers. Persons with disabilities had to file many lawsuits: the PVA brought suits regarding the MCI Center in Washington, D.C., the Fleet Center in Boston, the HSBC arena in Buffalo; the Eastern PVA brought suit against the CoreStates arena in Philadelphia, while other organizations represented them in the Rose Garden in Portland, and the Broward County arena in Sunrise, Florida. Instead of being accepted by larger society, as was the law's aim, the lawsuits cast them as adversaries and detail-oriented taskmasters.

For the architectural firms and their attorneys, the uncertainties of the law and the guidelines at times forced them into court with the claim that DOJ regulations were invalid, and so the architects were not required to follow them. The claim by Ellerbe Becket, supported by the AIA, that it bore no legal responsibility for adherence to the ADA regulations was not an attractive position, both from moral and public relations viewpoints. The firm's position that it serves primarily in a servant role seems demeaning for a profession that justifiably takes credit for some of the world's most remarkable buildings.

How did the architects fare in these lawsuits? Their good-faith efforts at providing more accessible seats than required by the law at the MCI Center were ignored and their designs declared inadequate. On the other hand, the architects were found to be taking self-serving and unflattering positions that disrespected the decent demands of persons with disabilities and their need to participate more fully and effectively in mainstream life. Ellerbe Becket's consent order capitulation was likely based on a desire to do the right thing, as well as a shrewd calculation that, if it did not do so, it very likely was going to find itself involved in highly expensive juridical hot water.

18.9   *GENERAL LESSONS LEARNED FOR UNIVERSAL DESIGN*

In the United States, the preference for the use of law to bring about changes in behavior has several problems. The lawmaking process cannot be relied on for precision and clear directives (Mazumdar and Geis, 2000). The courts can only address matters brought before them. Judges have to decide matters outside their expertise, and often unfamiliar to them, and frequently do not have much help in the law. And, judges are more interested in matters of law than matters of fact. Uneven efforts by those entrusted the job of developing rules, regulations, standards, guidelines, and for enforcement were not helpful. Implementing universal design principles through legislation may lead to problems with the operation of the law as described earlier. Educating and socializing stakeholders should be seriously considered.

Even with the increasing specification of final design outcomes, including detailed anthropometric dimensions (as in the Minnesota consent order, see Fig. 18.1), the needs of people with disabilities may not be addressed adequately. People with disabilities may still be unable to view the main event in buildings designed to those exact DOJ standards if their anthropometric dimensions vary from those selected by the DOJ or if the people standing in front are very tall. The substantial compliance rule also can lead to the same result. This means that people with disabilities will need to be alert in reviewing designs and perhaps will need to continue to influence lawmakers, architects, and owners regarding their needs.

For people with disabilities, even though laws helped bring about some changes to the designs of arenas, these problems with the legal approach were distracting. Their struggle for equality may have to be seen as a long one, requiring patience and vigilance.

For architects, perhaps the most significant outcome of the ADA lawsuits is that their centuries-old immunity from liability for their designs is eroding. Architects will be able to protect themselves in instances where their designs were law-abiding but alterations were made by others. It will become sound practice to keep a meticulous evidentiary trail that demonstrates that their work followed the law, and proves that others modified their designs.

To follow the law, designers will need to know not only the law in its original form, but also rulings by courts in the jurisdiction that become law by precedent. If the law is unclear, architects can refer to research on disability to understand the needs and rights of persons with disabilities (see, for example, Fine and Asch, 1988; Crewe et al., 1983; Gutman and Gutman, 1968; Murphy, 1987; Zola, 1982a,b, 1993). Many environmental design research and architectural writings describe anthropometric and other requirements of persons with disabilities (see, for example, Null and Cherry, 1996; Goldsmith, 1967, 1984; Leibrock and Behar, 1993; Leibrock and Terry, 1999; Pheasant, 1986, 1996; Raschko, 1982; Wilkoff and Abed, 1994). When the law fails to provide clear guidance, these writings may.

Architects need not have waited for laws to be enacted and litigation to be mounted in order to make buildings universally accessible. Rather than fight about the letter of the law, or let financial considerations control, they could have adopted the humanistic spirit of designing environments that serve the needs of all, including persons with disabilities, as a moral imperative and a professional duty. Accessible design questions in the ADA cases were not framed as aesthetic or moral matters, but could be.

Hopefully, the arena cases carry a lesson, one that will impel architects to make life more satisfactory for those who could benefit from their assistance and ingenuity.

18.10   NOTES

1. We are grateful for the help of Daniela Pappada, Julia Gelfand, Paralyzed Veterans of America, Lawrence B. Hagel, American Institute of Architects, Ellerbe Becket Architects and Engineers, and Mark Conrad.

18.11   BIBLIOGRAPHY

AIA, "Justice Department Agreement with Ellerbe Becket Clarifies ADA Compliance," News Release, AIA, 1998; www.e-architect.com/media/releases/04_28_98.asp (checked 27 January 2000).

AIA, "Architects and the Americans with Disabilities Act Accessibility Guidelines (ADAAG)", AIA, 1999; www.e-architect.com/gov/ada/home2.asp (checked 27 January 2000).

AIA, "AIA Files Amicus Brief in Texas ADA Suit," AIA, 2000; www.e-architect.com/gov/govnews/amicusbrief.asp (checked 27 January 2000).

AIArchitect, "DOJ, Ellerbe Becket Agreement Clarifies ADA Compliance," *AIArchitect* (June 1998): 3, 1998.

Acker, James R., "The Law of the Future," in John Klofas and Stan Stojkovic (eds.), *Crime and Justice in the Year 2010*, Wadsworth, Belmont, CA, 1995, pp. 62–83.

Colgate, James P., "If You Build It, Can They Sue? Architects' Liability Under Title III of the ADA," Note, *Fordham Law Review,* LXVIII(1)(Oct): 137–164, 1999.

Conrad, Mark A., "Stadiums Grappling with Disabilities Act," *New York Law Journal* (May 23): 5, 7, 38, 1997a.

Conrad, Mark A., "Sports Arenas Grapple with Disabilities Act," *New York Law Journal* (May 30): 5 (from LEXIS-NEXIS Web), 1997b.

Conrad, Mark A., "Wheeling Through Rough Terrain—The Legal Roadblocks of Disabled Access in Sports Arenas," *Marquette Sports Law Journal* 8(2)(Spr): 263–288, 1998a.

Conrad, Mark A., "Disabled-Seat Pact May Unify Standards for New Stadiums," *New York Law Journal* (May 8): 8–10, 1998b.

Conrad, Mark, personal communication with authors, 1998c.

Crewe, Nancy M., et al. (eds.), *Independent Living for Physically Disabled,* 1st ed., Jossey-Bass, San Francisco, 1983.

Fain, Constance F., "Architect and Engineer Liability," *Washburn Law Journal,* 35: 32–49, 1995.

Fine, Michelle, and Adrienne Asch, *Women with Disabilities: Essays in Psychology, Culture, and Politics,* Temple University Press, Philadelphia, 1988.

Flatt, William David, "The Expanding Liability of Design Professionals," Note, *Memphis State University Law Review,* 20: 610–627, 1990.

Fritts, Jonathan C., " 'Down in Front!': Judicial Deference, Regulatory Interpretation, and the ADA's Line of Sight Standard," Note, *Georgetown Law Review,* 86: 2653–2675, 1998.

Geis, Gilbert, and Leigh B. Bienen, *Crimes of the Century: From Leopold and Loeb to O.J. Simpson,* Northeastern University Press, Boston, 1998.

Goldsmith, Selwyn, *Designing for the Disabled* (2nd ed.), McGraw-Hill, New York, 1967.

Goldsmith, Selwyn, *Designing for the Disabled* (3rd ed., fully revised), RIBA Publications, London, UK, 1984.

Gutman, Ernest M., and Carolyn R. Gutman, *Wheelchair to Independence,* Charles C. Thomas, Springfield, IL, 1968.

Hagel, Lawrence B., "Arenas Are Failing the Disabled," *ESPN SportsZone,* no date.

Leibrock, Cynthia, and Susan Behar, *Beautiful and Barrier-Free: A Visual Guide to Accessibility,* Van Nostrand Reinhold, New York, 1993.

Leibrock, Cynthia, and James Evan Terry, *Beautiful Universal Design: A Visual Guide,* John Wiley & Sons, New York, 1999.

Mazumdar, Sanjoy, and Gilbert Geis, "Stadium Sightlines and Wheelchair Patrons: Case Studies in Implementation of the ADA," in Barbara M. Altman and Sharon N. Barnartt (eds.), *Expanding the Scope of Social Science Research on Disability, Research in Social Science and Disability,* vol. 1, JAI Press, Stamford, CT, 2000, pp. 205–234.

Murphy, Robert F., *The Body Silent,* Holt, New York, 1987.

Nischwitz, Jeffrey L., "The Crumbling Tower of Architectural Immunity: Evolution and Expansion of the Liability to Third Parties," Note, *Ohio State Law Journal,* 45: 217–263, 1984.

Null, Roberta L., and Kenneth F. Cherry, *Universal Design: Creative Solutions for ADA Compliance,* Professional Publications, Belmont, CA, 1996.

Pheasant, Stephen, *Bodyspace: Anthropometry, Ergonomics, and Design,* Taylor & Francis, Philadelphia, 1986.

Pheasant, Stephen, *Bodyspace: Anthropometry, Ergonomics, and the Design of Work,* 2nd ed., Taylor & Francis, Bristol, PA, 1996.

Raschko, Bettyann Boetticher, *Housing Interiors for the Disabled and Elderly,* Van Nostrand Reinhold, New York, 1982.

U.S. Department of Justice, "28 CFR Part 35 Nondiscrimination of the Basis of Disability in State and Local Government Services: Final Rule," U.S. DOJ, Civil Rights Division, Public Access Section, Washington, DC, in *Federal Register* (26 July 1991), 1991a.

U.S. Department of Justice, "Nondiscrimination on the Basis of Disability by Public Accommodations and in Commercial Facilities: Final Rule," *Federal Register,* 56(144): 35544–35961, 1991b.

U.S. Department of Justice, *The Americans with Disabilities Act Title III Technical Assistance Manual, Covering Public Accommodations and Commercial Facilities,* Washington, DC, U.S. DOJ, Civil Rights Division, Public Access Section, 1993a.

U.S. Department of Justice, "Fargodome Will End Discriminatory Ticket Pricing for Persons with Disabilities Under Department of Justice Settlement," (24 November 1993), 1993b; www.usdoj.gov/crt/foia/nd1.txt (checked 12 July 1999).

U.S. Department of Justice, *28 CFR Part 36: Nondiscrimination on the Basis of Disability by Public Accommodations and in Commercial Facilities,* Code of Federal Regulations (Rev. 01 July 1994), incorporating ADAAG for buildings and facilities, 1994a.

U.S. Department of Justice, *The Americans with Disabilities Act Title III Technical Assistance Manual, 1994 Supplement.* Washington, DC, U.S. DOJ, Civil Rights Division, Public Access Section, 1994b.

Wilkoff, William L., and Laura W. Abed, *Practicing Universal Design: An Interpretation of the ADA,* Van Nostrand Reinhold, New York, 1994.

Winston, Sherie, "Disabilities Lawsuit Dismissed Against Arena Designer," *ENR* (*Engineering News Record*) (29 July 1996): 12, 1996a.

Winston, Sherie, "Justice Sues Arena Architect," *ENR* (*Engineering News Record*) (21 October 1996): 11, 1996b.

Winston, Sherie, "Judge Rules Architect Is Liable," *ENR* (*Engineering News Record*) (17 February 1997): 16, 1997a.

Winston, Sherie, "New Guidelines Could Clarify ADA," *ENR* (*Engineering News Record*) (23 June 1997): 8–9, 1997b.

Winston, Sherie, "Accessibility: Disability Law Tests Architects," *ENR* (*Engineering News Record*) 239(13 October 1997): 10, 1997c.

Zola, Irving Kenneth, *Missing Pieces: A Chronicle of Living with a Disability,* Temple University Press, Philadelphia, 1982a.

Zola, Irving Kenneth (ed.), *Ordinary Lives: Voices of Disability and Disease,* Apple-wood Books, Cambridge, MA, 1982b.

Zola, Irving Kenneth, "In the Active Voice: A Reflective Review Essay on Three Books," *Policy Studies Journal,* 21(4): 802–805, 1993.

Court Cases

Baltimore Neighborhoods, Inc. v. Rommel Builders, Inc., 1998, 3 F. Supp. 2d 661 (D.MD).

Caruso, William v. Blockbuster-Sony Music Entertainment Centre (*BSEC*), 1997, 968 F. Supp. 210 (D.NJ).

Caruso, William v. Blockbuster-Sony Music Entertainment Centre (*BSEC*), 1999a, *U.S. Lawweek,* pp. 9–10; lw.bna.com/plweb-cgi/fastweb?getdoc+view6+lawweek+1725+0++.

Caruso, William v. Blockbuster-Sony Music Entertainment Centre (*BSEC*), 1999b, COA Opinion—Alito 174 Federal Reporter 3d 166 (3rd Cir. D.NJ 1999) cf. 166–180, 97-5693, 97-5764, USCOA Opinion Alito, 06 April 1999.

Coalition of Montanans Concerned with Disabilities, Inc. (*CMCDI*) *v. Gallatin Airport Authority,* 1997, Opinion and Order District Judge Molloy 957 F.Supp. 1166 (D. MT 1997) cf. 1167–1171, CV 94-84-BU-DWM, 27 March 1997.

Days Inns of America, Inc. (*Texas*) *v. Reno,* 1996, 935 F. Supp. 874 (W.D. TX 1996).

Independent Living Resources (*ILR*) *v. Oregon Arena Corporation* (*OAC*), 1995, Deposition of Gordon Wood, 982 F. Supp. 698, Summary 753–754, USDC DOR CV95-84-AS Dep, 04 October 1995.

Independent Living Resources (*ILR*) *v. Oregon Arena Corporation* (*OAC*), 1997a, Plaintiff's Complaint, 982 F. Supp. 698 (D.OR) cf. 706–770, USDC DOR CV95-84-AS PC, 12 November 1997.

Independent Living Resources (*ILR*) *v. Oregon Arena Corporation* (*OAC*), 1997b, Defendant's Motion to Dismiss, 982 F.Supp. 698 (D.OR) cf. 770–784, USDC DOR CV95-84-AS DMD, 12 November 1997.

Independent Living Resources (*ILR*) *v. Oregon Arena Corporation* (*OAC*), 1997c, Amicus Curiae brief by U.S. DOJ mentioned at: 982 F. Supp. 698 (D. OR.), USDC DOR CV95-84-AS AC, 12 November 1997.

Independent Living Resources (*ILR*) *v. Oregon Arena Corporation* (*OAC*), 1997d, Plaintiffs' Rebuttal to Defendants' Motion to Dismiss, USDC DOR CV95-84-AS PRDMD, 12 November 1997.

Independent Living Resources (*ILR*) *v. Oregon Arena Corporation* (*OAC*), 1997e, Judge Ashmanskas' Order, 982 F. Supp. 698 (D. OR.) cf. 698–786, USDC DOR CV95-84-AS Order—Ashmanskas, 12 November 1997.

Independent Living Resources (ILR) v. Oregon Arena Corporation (OAC), 1997f, Judge Ashmanskas' Opinion 982 F. Supp. 698 (D. OR.) cf. 698–786, USDC, DOR CV95-84-AS Opinion Ashmanskas, 12 November 1997, pp. 1–158 typescript.

Independent Living Resources (ILR) v. Oregon Arena Corporation (OAC), 1998a, Findings of Fact and Conclusions of Law, 1 F. Supp. 2d 1124 (D. OR) cf. 1124–1159, USDC DOR CV95-84-AS FFCL, 26 March 1998.

Independent Living Resources (ILR) v. Oregon Arena Corporation (OAC), 1998b, Supplemental Findings of Fact and Conclusions of Law, 1 F. Supp. 2d 1129 (D. OR) cf. 1159–1173, USDC DOR CV95-84-AS SFFCL, 08 April 1998.

Inman v. Binghamton Housing Authority, 1957, 3 N.Y.2d 137, 143 N.E.2d 895, 164 N.Y.S.2d 699.

Johanson v. Huizenga Holdings, Inc., 1996, Memorandum of Law of Amicus Curiae United States in Opposition to Defendants Ellerbe Becket Architects and Engineers, Inc.'s Motion to Dismiss, USDC SDFL 96-7026-CIV-Gonzalez, 31 October 1996; www.usdoj.gov/crt/foia/fl14.txt.

Johanson v. Huizenga Holdings, Inc., 1997, Order, 963 F. Supp. 1175 (S.D. Fl.), USDC SDFL 96-7026-CIV-Gonzalez Order, 27 January 1997; www.usdoj.gov/crt/foia/fl14.txt.

McPherson v. Buick Motor Co., 1916, 217 N.Y. 382, III N.E. 1050.

Paralyzed Veterans of America (PVA) v. Ellerbe Becket Architects & Engineers P.C. (EBAE), 1996a, Plaintiff's Complaint for Declaratory Judgment and Injunctive Relief, 945 F. Supp. 1 (D.DC) summary cf. 1–3, USDC DC 1:96CV01354 Plaintiffs Complaint, 14 June 1996.

Paralyzed Veterans of America (PVA) v. Ellerbe Becket Architects & Engineers P.C. (EBAE), 1996b, Defendants' Motion to Dismiss Counts I, II & III, 945 F. Supp. 1 (D.DC 1996) summary cf. 1–3, USDC DC 96CV01354 (TFH), DMD, 28 June 1996.

Paralyzed Veterans of America (PVA) v. Ellerbe Becket Architects & Engineers P.C. (EBAE), 1996c, Plaintiff's Opposition to Defendants' Motion to Dismiss Counts I, II & III, 945 F. Supp. 1 (D.DC) summary cf. 1–3, USDC DC 96CV01354 (TFH), 8 July 1996.

Paralyzed Veterans of America (PVA) v. Ellerbe Becket Architects & Engineers P.C. (EBAE), 1996d, The American Institute of Architects' Brief Amicus Curiae in support of Ellerbe Becket Architects & Engineers, USDC DC 96CV01354 (TFH) AIA AC, 10 July 1996.

Paralyzed Veterans of America (PVA) v. Ellerbe Becket Architects & Engineers P.C. (EBAE), 1996e, Judge Thomas F. Hogan's Preliminary Temporary Injunction Order, 19 July 1996, 945 F. Supp. 1 (D.DC 1996) cf. 2–3, USDC DC 96CV01354 O, 19 July 1996.

Paralyzed Veterans of America (PVA) v. Ellerbe Becket Architects & Engineers P.C. (EBAE), 1996f, Amicus Curiae Brief by DOJ 945 F. Supp. 1 (D.D.C. 1996) summary cf. 2, USDC DC 96CV01354 DOJ AC.

Paralyzed Veterans of America (PVA) v. Ellerbe Becket Architects & Engineers P.C. (EBAE), 1996g, Judge Thomas F. Hogan's Order/Opinion, 21 October 1996, 950 F. Supp. 389 (DDC 1996) summary cf. 389, USDC DC 96CV01354 O, 21 October 1996.

Paralyzed Veterans of America (PVA) v. Ellerbe Becket Architects & Engineers P.C. (EBAE), 1996h, Judge Thomas F. Hogan's Memorandum Opinion explanation of Order, dated 21 October 1996, 950 F. Supp. 389 (D.DC 1996) cf. 389–393, USDC DC 96CV01354 TFH MO, 20 December 1996.

Paralyzed Veterans of America (PVA) v. Ellerbe Becket Architects & Engineers P.C. (EBAE), 1996i, Judge Thomas F. Hogan's Order dated 20 December 1996, 950 F. Supp. 393 (D.DC 1996) cf. 405–406, USDC DC TFH O, 20 December 1996.

Paralyzed Veterans of America (PVA) v. Ellerbe Becket Architects & Engineers P.C. (EBAE), 1996j, Judge Thomas F. Hogan's Memorandum Opinion explanation of Order dated 20 December 1996, 950 F. Supp. 393 (D.DC 1996) cf. 393–405, USDC DC 96CV01354 TFH, 20 December 1996.

Paralyzed Veterans of America (PVA) v. Ellerbe Becket Architects & Engineers P.C. (EBAE), 1996k, Judge Thomas F. Hogan's Order dated 19 February 1997, USDC DC 96CV01354 TFH O, 19 February 1997 (typescript).

Paralyzed Veterans of America (PVA) v. D.C. Arena L.P., 1997a, Brief for Appellants, USCOA, DC. Cir. 117 Fed. Rep. 3d. (DC cir. 1997) 579, USCOA (D.DC. Cir). Case # 97-7005, 03 March 1997.

Paralyzed Veterans of America (PVA) v. D.C. Arena L.P., 1997b, Appellees filing United States Court of Appeals, 117 Fed. Rep. 3d. (DC Cir 1997) 579, USCOA D.DC. Cir. 97-7005, March 1997.

Paralyzed Veterans of America (PVA) v. D.C. Arena L.P., 1997c, Reply brief for appellants and brief for cross-appellees, 117 Fed. Rep. 3d. (DC Cir 1997) 579, USCOA D.DC. Cir. 97-7005, 31 March 1997.

Paralyzed Veterans of America (PVA) v. D.C. Arena (DCA) L.P., 1997d, Opinion—Judge Lawrence H. Silberman, United States Court of Appeals, *Affirmed* 117 F3d (DC Cir 1997) cf. 579–589, USCOA D.DC. Cir. 97-7005, and 97-7017, O, 01 July 1997.

Paralyzed Veterans of America (PVA) v. DC Arena L.P., 1998, U.S. Supreme Court, 117 F3d 579 (USSC *Certiorari Denied*), 118 S. Ct. 1184 (1998)/523 U.S. 1003.

United States v. Days Inns of America, Inc., 1996, 935 Fed. Supp. 874 (W.D. TX).

United States v. (Illinois) Days Inns of America, Inc., 1998a, 997 F. Supp. 1080 (C.D. IL 1998).

United States v. (Kentucky) Days Inns of America, Inc., 1998b, 22 Fed. Supp. 2d 612 (E.D. KY).

United States v. (California) Days Inns of America, Inc., 1998c, 8 Am. Disabilities Cas. (BNA) 491 (E.D. CA).

United States v. Days Inns of America, Inc., 1998d, 151 Fed. Rep. 3d 822 (8th Cir. DIA). cert. denied, 119 S. Ct. 1249 (1999).

United States v. Ellerbe Becket, Inc., 1996a, Plaintiff's Complaint (USDC DMN 4-96-995, 10 October 1996).

United States v. Ellerbe Becket, Inc., 1996b, Ellerbe Becket Inc. Motion to Dismiss (USDC DMN 4-96-995, November 1996).

United States v. Ellerbe Becket, Inc., 1997a, Amici Curiae the American Institute of Architects and the Associated General Contractors of America's Brief in support of Ellerbe Becket, Inc.'s Motion to Dismiss, mentioned at: 976 F. Supp. 1262 (USDC DMN), USDC DMN 4-96-995, 20 January 1997.

United States v. Ellerbe Becket, Inc., 1997b, Order/Ruling on Motion to Dismiss, Judge Tunheim 976 Fed. Supp. 1262–1269 (USDC DMN), USDC DMN 4-96-995, 30 September 1997.

United States v. Ellerbe Becket, Inc., 1998, Consent Order, Judge Tunheim, USDC DMN 4th Civil Action No. 4-96-995, 27 April 1998; www.usdoj.gov/crt/ada/ellerbe.htm.

United States v. Physorthorad Associates, J. Wylie Bradley, and Bradley Chambers and Frey, 1996, Consent Order, USDC DPA 4: CV-96-1077 CO, 25 June 1996.

United States and the Disability Rights Education and Defense Fund (DREDF) v. City and County of San Francisco, etc., 1996, Settlement agreement among the United States of America, the DREDF, and the City and County of San Francisco, the San Francisco Forty Niners Limited, and San Francisco Baseball Associates L.P. (01 September 1996); www.usdoj.gov/crt/foia/ca7.txt (last checked 01 July 1999).

Washington Sports and Entertainment, Inc. v. United Coastal Insurance Co., 1998, Memorandum Opinion Thomas F. Hogan 7 F.Supp.2d 1 (D.DC 1998) cf. 1–14, CIV 97-400 TFH, 26 February 1998.

18.12   RESOURCES

Accessology, Inc.   www.accessology.com

ADA Information Center for the Mid-Atlantic Region.   www.adainfo.org

Adaptive Environments.   www.adaptenv.org

American Association of People with Disabilities.   www.aapd.com

American Disability Association.   www.adanet.org

American Institute of Architects.   http://aiaonline.com/

Americans with Disabilities Act Technical Assistance.   www.adata.org

Building Owners and Managers Association.   www.boma.org/adaup.html

The Center for Universal Design.   www2.ncsu.edu/ncsu/design/cud

Department of Justice—four-page document on Stadium and Arena Design.   www.usdoj.gov/crt/ada/stadium.txt

Disability Statistics Center.   http://dsc.ucsf.edu/UCSF/spl.taf?_from=default

DOJ New or Proposed ADA Regulations.   www.usdoj.gov/crt/ada/settlemt.htm

DOJ's Enforcement of the ADA.   www.usdoj.gov/crt/statrpt.htm

Ellerbe Becket Architects Engineers.   http://ellerbebecket.com/

IDEA—Center for Inclusive Design and Environmental Access.   www.arch.buffalo.edu/~idea

National Center on Dissemination of Disability Research.   www.neddr.org/doorways/univd

National Council on Disability.   www.ncd.gov

National Rehabilitation Information Center (NARIC).   www.naric.com/naric

Pacific Disability and Business Technical Assistance Center.   www.pacdbtac.org

Paralyzed Veterans of America.   www.pva.org

The United States Law Week.   http://lw.bna.com/

The U.S. Architectural & Transportation Barriers Compliance Board.   www.access-board.gov

US DOJ: ADA Home Page.   www.usdoj.gov/crt/ada/adahom1.htm

CHAPTER 19
UNIVERSAL DESIGN IN OUTDOOR PLAY AREAS

Susan Goltsman, F.A.S.L.A.
Moore Iacofano Goltsman (MIG), Berkeley, California

19.1 INTRODUCTION

A quality play and learning environment is more than just a collection of play equipment. The entire site with all its elements—from vegetation to storage—can become a play and learning resource for children with and without disabilities. This chapter discusses how to create a universally designed play area, one that integrates the needs and abilities of all children into the design. An integrated play area allows children with disabilities to participate in the play experiences that other children take for granted. Included are general elements and guidelines as well as specific performance criteria for a variety of play settings. The chapter also provides a case study of Ibach Park in Tualatin, Oregon, a universally designed play area whose theme reflects the unique history of the area.

19.2 BACKGROUND

Play is more than just about having fun. It is a process through which children develop their physical, mental, and social skills. It is value laden and culturally based. In the past, most play experiences occurred in unstructured child-chosen places. Children with disabilities, depending on the type and severity of the disability and the attitudes of their parents, generally have less access to these free-range play settings found around the neighborhood. They also have very limited choices within most structured play settings. It is possible, however, to create well-designed play areas that successfully integrate the needs of children with and without disabilities. The key is to provide diverse physical and social environments so that children with disabilities are a part of the overall play experience.

Designing an Integrated Play Area

A good play area that is designed to integrate children with and without disabilities consists of a range of settings carefully layered onto a site. They contain one or more of the following elements: entrances, pathways, fences and enclosures, signage, play equipment, game areas, land forms and topography, trees and vegetation, gardens, animal habitats, water play, sand play, loose parts, gathering places, stage areas, storage, and ground covering and safety surfacing. In any play area design, each play setting varies in importance, depending on community values, site constraints, and location. The way these elements are used will also determine the degree of accessibility and integration possible in that environment. However, in designing a play space of any size, the full range or settings should be considered.

Diversity of the play setting and opportunities within the setting are key to integration and access in a play area. Since play, to be developmental, must present a challenge as part of its value, not every part of the environment should be physically accessible to every user. Therefore, a play area must support a range of challenges, both mental and physical. Physical challenge within the play area must be part of a progression of challenges that promote an individual's skill.

On the other hand, the social experience available must be accessible to all. Unlike a physical challenge, which to be developmental must be "earned" through a child's efforts, the opportunity for a social interaction should be easily accessible. Social integration is the basic reason why a play area must be accessible to children of all abilities. If the play area truly serves the range of children who use it, then it is considered to meet the definition of universal design.

Creating a universally designed play area requires integrating the needs and abilities of all children into the design of play areas. As mentioned earlier, the diversity of both physical and social environments is the key to accommodating the variety of users in a play area (see Fig. 19.1). Obtaining physical diversity means placing a broad range of challenges within the play setting. Such an environment will allow more children to participate, make choices, take on challenges, develop skills, and, most important, play together.

FIGURE 19.1 Universally designed play areas provide diverse activities.

Social diversity is very much linked to physical diversity. Contact between children of different abilities will naturally increase in play areas that are open to a wider spectrum of users. This interaction is particularly critical for children with functional limitations, who so often are denied these social experiences. Again, while it is often not feasible to make every part of a play area physically accessible to everyone, the social experience must be accessible to all.

Elements of an Integrated Play Area

- Consider the many ways in which children with disabilities can interact. When arranging the play area, integrate accessible play equipment with the rest of the play setting. Placing less challenging activities directly next to those requiring greater physical ability will encourage interaction across all ability levels.

- Provide an accessible route that connects every activity area and every accessible play component in the play setting. (A *play component* is defined as an item that provides an opportunity for play; it can be a single piece of equipment or part of a larger composite structure.) Even though not every play component will be physically accessible to everyone, simply enabling all children to be near the action provides the opportunity and choice, and it promotes the possibility of communication with others. This is a major step toward integration.

- Ensure that at least one of each kind of play component on the ground is accessible and usable by children with mobility impairments. Likewise, at least half of the play components elevated above ground should be accessible. Access onto and off equipment can be provided with ramps, transfer platforms, or other appropriate methods of access. Remember that ramps and transfer systems can also serve as physical challenges and should be designed so that they add to the diversity of the environment.

- Make sand play accessible by providing raised sand areas or installing a transfer system into the sand. Note that raised sand areas, which provide space underneath for children using wheelchairs, provide severely limited play experiences because the necessary clearance keeps the sand depth to a minimum. Providing a transfer system into the sand area will allow users to enjoy full-body sand play.

- Make portions of gathering places accessible to promote social interaction. These areas are important areas of interaction and allow groups of people to play, eat, watch, socialize, and congregate. Include accessible seating, such as benches without backrests and arm supports, so people of varying abilities can sit together.

- Don't forget safety guidelines, which outline important parameters such as head entrapments, safety surfacing, and use zones. At times, however, provisions for safety and accessibility can conflict. For example, a raised sand shelf could be considered hazardous because the shelf is more than 20 in off the ground. If you strictly followed the safety requirements, you would need to construct a nonclimbable enclosure along the edge of the shelf, which would defeat the whole purpose of the design. In such cases, seek solutions that provide other means of access or that mitigate the safety hazard, such as installing rubber safety surfacing on the ground below the shelf.

19.3 *PERFORMANCE CRITERIA FOR INCLUSIVE PLAY SETTINGS*

A universally designed play setting is not high tech; it is *design tech*. In order to accommodate the needs of children with varied abilities, the overall design or individual components need to be reconsidered so that there is an inclusive system. The focus needs to be on good anthropometric data and user-based design guidelines and performance criteria. The accessible components that are created, such as transfer systems for manufactured equipment, are not stigmatizing by their appearance.

The options available for creating a universally designed play area are all based on low technology, innovative thinking, and problem solving. In some instances, certain design solutions, especially new ones, may require fabrication and manufacturing in a state-of-the-art factory. Ideas should be shared in the manufacture of play components. Users, designers, and manufacturers working together will advance this relatively new area of environmental design. The following subsections present performance criteria for creating accessibility and integration in play areas.

Accessible Route of Travel Within a Play Setting

Because play is primarily a social experience, accessible routes through a play setting must connect all types of activities. A path that connects the accessible play elements within a play setting is essential. Without this connection, children with disabilities can too easily find themselves isolated from their friends without disabilities. The accessible route within a play setting avoids problems caused by circulation design flaws and satisfactorily promotes social interaction.

A good play setting has many routes through the space. A route itself may be the play experience. Pathways can be a play element in themselves, supporting wheeled toys, running games, and exploration. To be accessible, pathways require firm and stable surfaces and correct grades (1 up to 20) and cross slopes (2 percent or less). The quality of the pathway system sets the tone for the environment. Pathways can be wide with small branches, long and straight, or circuitous and meandering. Each creates different play behaviors and experiences. Minimum routes or auxiliary pathways through a play experience are exempt from the strict requirements of the primary accessible route of travel to promote the range of challenges necessary for a variety of developmentally appropriate play experiences. The following criteria for accessible route design apply:

- An accessible route to and for the intended use of the different activities within the play area setting must be provided.
- The accessible route should be a minimum of 60-in wide but can be adjusted down to 36-in if it is in conjunction with a bench or play activity.
- The cross slope of the accessible route of travel shall not exceed 1:50.
- The slope of the accessible route of travel should not exceed 1:20.
- If a slope exceeds 1:20, it is a ramp. A ramp on the accessible route of travel on the ground plane should not exceed a slope of 1:16.
- If the accessible route of travel is adjacent to loose-fill material or there is a drop-off, then the edge of the pathway should be treated to protect a person using a wheelchair from falling off the route and into the loose-fill material by beveling the edge with a slope that does not exceed 30 percent. A raised edge will create a trip hazard for walking children. If this route is within the use zone of the play equipment, the path and the edge treatment shall be made of safety surfacing.
- Changes in level along the path should not exceed ½ in.
- Where egress from an accessible play activity occurs in loose-fill surface which is not firm, stable, and slip-resistant, a means of returning to the point of access for that play activity shall be provided, and the surfacing material should not splinter, scrape, puncture, or abrade the skin when being crawled upon.

Play Equipment

Most equipment settings stimulate large-muscle activity and kinesthetic experience, but they can also support nonphysical aspects of child development. Equipment can provide opportunities to experience height and can serve as landmarks to assist orientation and wayfinding. They may also become rendezvous spots, stimulate social interaction, and provide hideaways in hiding and chasing games (see Fig. 19.2). Small, semienclosed spaces support dramatic play, and seating, shelves, and tables encourage social play.

In choosing manufactured play equipment for a recreation area, the following considerations must be made:

- Properly selected equipment can support the development of creativity and cooperation, especially structures that incorporate sand and water play. Play structures can be converted

FIGURE 19.2   A play village that provides access through and around makes the dramatic play social experience available to all.

to other temporary uses, such as stage settings; loose parts can be strung from and attached to the equipment, such as backdrops or banners for special events and dramatic play activities. Equipment settings must be designed as part of a comprehensive multipurpose play environment. Isolated pieces of equipment are ineffective on their own.

- Equipment should be properly sited, selected, and installed over appropriate shock-absorbing surfaces. Procedures and standards for equipment purchase, installation, and maintenance must be developed. A systematic safety inspection program must be implemented.

- There are a number of well-documented safety issues related to manufactured play equipment: falls, entrapments, protrusions, collisions, and splinters. All of these issues must be addressed in the design, maintenance, and supervision of play equipment.

- Equipment should be accessible, but must be designed primarily for children, not wheelchairs. Transfer points should always be marked both visually and tactually. The most significant aspect of making a piece of equipment accessible is to understand that children with disabilities need many of the same challenges as children without disabilities. Using synthetic surfacing can provide access to, under, and through the equipment for children who use wheelchairs. Getting in the center of action may be as important as climbing to the highest point for some children.

- Play equipment provides opportunities for integration, especially when programmed with other activities. Play settings should be exciting and attractive for parents as well as children—adults accompany children to the park or playground more often today than in the past. It is equally important to design for parents using wheelchairs who accompany their able-bodied children.

Play equipment is designed to provide a physical challenge as well as social interaction. For equipment to be appropriate for different skill levels, it must be graduated in its challenge opportunities. Access up to, onto, through, and off equipment should also provide a variety of challenge levels appropriate to the age of the intended users.

Water-Play Areas

Water in all its forms is a universal play material because it can be manipulated in so many ways. You can splash it, pour it, use it to float objects, and mix it with dirt to make "magic potions." Permanent or temporary, the multisensory quality of water-play areas make a substantial contribution to child development (see Figs. 19.3 and 19.4). Water settings include a hose in the sandpit, puddles, ponds, drinking fountains, bubblers, sprinklers, sprays, cascades, pools, and dew-covered leaves.

FIGURE 19.3   Water sprays installed below a rubber surface provide water play for everyone.

If water play is provided, a part of the play area must be wheelchair accessible. If the water source is manipulated by children, then it must be usable by all children. If loose parts such as buckets are provided and children have access to the equipment storage, then the storage must be usable by all children. When water is provided for play, the following dimensions apply:

- Forward reach: 36 to 20 in.
- Side reach: 36 to 20 in.
- Clear space: 36 × 55 in. The clear space should be located at the part of the water-play area where the most water play will occur. If the water source is part of the active play area and children turn the water on and off, then it must be accessible. If the water source is part of a spray pool, the area under the spray should be accessible. Accessibility should involve the dimensions for both clear space and reach.
- Clearance ranges: top height to access water, 30 in maximum; underclearance, 27 in minimum.

Sand-Play Areas

Children will play in dirt wherever they find it. Using props such as a few twigs, a small plastic toy, or a few stones, children can create an imaginary world in the dirt, around the roots of

FIGURE 19.4   Giant nautilus water-play element.

a tree, or in a raised planter. The sandbox is a refined and sanitized version of dirt play. It works best if it retains dirt's play qualities. You should provide small, intimate group spaces, adequate play surfaces, and access to water and other small play props.

If a sand-play area is provided, part of it must be accessible. Important elements are clear floor space, maneuvering room, reach and clearance ranges, and operating mechanisms for control of sand flow. When products such as buckets and shovels will likely be used in the sand-play area, storage places should be at accessible reach range.

Raised sand play is a very limiting play experience because of the way a raised area must be constructed. To provide a place for the wheelchair user under the sand shelf, there is very little depth of sand available for play. Therefore, a raised sand area by itself is not a substitute for full-body sand play. All sand-play opportunities on the site should be made usable by all children to the greatest extent possible.

If the sand area is designed to allow children to play inside the area, a place within the sand-play area should be provided where a participant can rest or lean against a firm, stationary back support in close proximity to the main activity area. Back support can be provided by any vertical surface that is a minimum height of 12 in and a minimum width of 6 in, depending on the size of the child. Back support can be a boulder, a log, or a post that is holding up a shade structure. A transfer system into a sand area may also be necessary if the area is large and contains a variety of sand activities. A transfer system would be appropriate if there are no areas of raised sand play in the primary activity area, or if the sand area is over 100 ft^2 and the raised sand area would tend to isolate accessible sand-play activities.

When raised sand is provided, the following clearance ranges apply:

- Top height to sand: 30 to 34 in maximum
- Underclearance: 27 in minimum
- Side reach: 36 to 20 in
- Forward reach: 36 to 20 in
- Clear space for wheelchair: 36 × 55 in

Depending on the site conditions and the amount of sand play, shade may be required. If shade is required by site conditions, it may be provided through a variety of means, such as trees, tents, umbrellas, structures, and so forth. This advisory requirement for shade is based on the site context, program, and users. Some shade in or around sand is usually desirable.

Gathering Places

To support social development and cooperation, children need comfortable gathering places. Parents and play leaders need comfortable places for washing up, sitting, socializing, and supervising. If gathering places are provided, a portion of them should be accessible (see Fig. 19.5). A gathering place contains fixed elements to support play, eating, watching, talking, or assembling for a programmed activity. Gathering places should serve people of all ages.

- *Seating.*   A variety of seat choices should be made available. At least 50 percent of fixed benches should have no backs and arms.
- *Tables.*   Where tables are provided, a variety of sizes and seating arrangements should be provided.
- *Game tables.*   Game tables provide a place for two to four people to play board games. Where fewer than five game tables are provided, a minimum of one four-sided game table should include an accessible space on one side.
- *Storage.*   If storage is supplied and a part of the gathering area and the storage is used by children, accessible shelves and hooks should be a maximum of 36 in above the ground. The amount of storage is dependent upon program requirements.

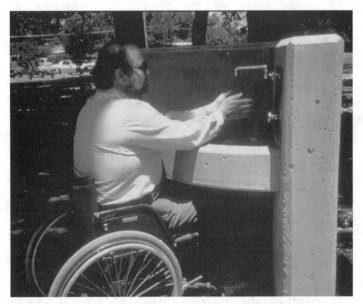

FIGURE 19.5   Accessible wash-up sink serves people of all ages and abilities.

- *Shade.*   Shade may be desirable for gathering areas where people will be participating in activities over a long period of time. Shade can be provided by a variety of means such as trees, canopies, or trellises, depending on site context.

Garden Settings

A powerful play-and-learn activity, gardening allows children to interact with nature and each other. Garden beds and tools can be adapted for use by children with disabilities. Scent gardens are attractive to all children and are of special appeal to children with low vision.

Gardens in play areas are primarily used to provide a program with the activities of planting, tending, studying, and harvesting vegetation. Depending on the type and height of plantings, planter boxes may require a raised area for access or a transfer point. A garden must provide a minimum of one accessible garden plot.

- *Raised gardens.*   If a raised area is provided, it should have the following features:

  The raised area should be located as part of the main garden area. The amount of raised area is determined by the program, but a minimum of 10 percent of the garden should be raised.

  The edge should be raised above the ground surface to a minimum of 20 in and a maximum of 30 in.

  The garden growing area should require access either by side or by forward reach 12 to 36 in above the ground.

- *Transfer systems.*   If children are required to sit in the dirt to garden, a transfer point should be provided which enables a participant to transfer into the garden.

- *Potting and maintenance areas.*   Potting and preparation areas should require access either by forward or side reach. The amount of area to be made accessible depends on the program. At least one workstation for potting should be made accessible.

- *Storage.*   Storage areas for the garden should provide access for children who use wheelchairs. Hooks and shelves should be a maximum of 36 in off the ground.
- *Circulation.*   Aisles around the garden (36 to 44 in) should be provided on a main aisle so a child using a wheelchair or walker can get to the garden. This larger aisle (48 to 60 in) should also provide access to the accessible gardening spaces.

Vegetation, Trees, and Landforms

Vegetation, trees, and topography are important features in a play setting. These features should be integrated into the flow of play activities and spaces, or they can be play features in themselves.

Landforms help children explore movement through space and provide for varied circulation. Topographic variety stimulates fantasy play, orientation skills, hide-and-seek games, viewing, rolling, climbing, sliding, and jumping. "Summit" points must accommodate wheelchairs and provide support for children with other disabling conditions.

Trees and vegetation comprise one of the most ignored topics in the design of play environments. They are two of the most important elements for integration because everyone can enjoy and share them. Vegetation stimulates exploratory behavior, fantasy, and imagination. It is a major source of play props, including leaves, flowers, fruits, nuts, seeds, and sticks. It allows children to learn about the environment through direct experience.

Designers and program providers should emphasize integrating plants into play settings rather than creating separate "nature areas." For children with physical disabilities, the experience of being in trees can be replicated by providing trees that a wheelchair user can roll into or under. You can create an accessible miniforest by planting small trees or large branching bushes.

If vegetation, trees, or landforms are used as a feature, a means should be provided for access up to and around the feature. Tree grates and other site furniture that support or protect the feature must also be selected so as not to entrap wheels, canes, crutch tips, and so forth.

Animal Habitats

Contact with wildlife and domestic animals stimulates a caring and responsible attitude toward other living things, provides a therapeutic effect, and offers many learning opportunities. Play areas can provide opportunities to care for or observe domestic animals. Existing or created habitats of insects, aquatic life, birds, and small animals should be protected. Planting appropriate vegetation will attract insects and birds.

Entrances and Signage

Entrances are transition zones that help orient and inform users and introduce them to the site. They are places for congregating and for displaying information. Not all play areas, though, have defined entrances. Sometimes entry to a play area can be provided from all directions.

Signs can be permanent or temporary, informative or playful. Expressive and informative displays use walls, floors, ground surfaces, structures, ceilings, sky wires, and roof lines on or near a play area to hang, suspend, and fly materials for art and education. Signage is a visual, tactile, or auditory means of conveying information, and it must communicate a message of "*All* Users Welcome." You can use appropriate heights, depths, colors, pictures, and tactile qualities to make signs accessible. Signs should primarily communicate graphically. Talking signs are also effective.

Fences, Enclosures, and Barriers

An enclosure is a primary means of differentiating and articulating the child's environment. For example, fences can double back on themselves to provide small social settings. Fences, enclosures, and barriers protect fragile environments, define pathways, enclose activity areas, and designate social settings. Low fences could be play elements and must be considered as such. The entrance to an enclosure should be clearly visible and wide enough for wheelchair passage (36 in minimum, 48 in preferable).

To fully support child development and integration of children with and without disabilities, a well-designed play environment should be augmented by a risk management program and professional play leadership. Part of the play leader's job is to set up, manipulate, and modify the physical environment to facilitate creative activity. Since the environment can either support or hamper play, designers must learn how they can best empower children through design. Compensating for an inadequate environment is a drain on the leader's time and energy.

19.4   CASE STUDY: IBACH PARK

Ibach Park is a 19.8-acre neighborhood park which is owned and operated by the City of Tualatin Parks and Recreation Department. The design and development of the park reflects the benefits desired by community residents who were active participants in the design process. Park features include a ¾-acre play area that reflects the rich heritage of Tualatin from prehistory to early settlement days, a preteen play area, a soccer field, a softball field, tennis courts, a basketball court, picnic areas, an open turf area, Hedges Creek greenway trail development and restoration, interpretive signage, restrooms, and parking. Although unique in its specific elements, the play area is a typical Moore Iacofano Goltsman (MIG) project in three major respects:

- The design was developed with input from the local community.

- The play area integrates children of varying abilities.

- The goal was not simply to provide a space for physical stimulation, but to create an environment that encourages social and intellectual development as well.

Design Process

To ensure that as many community members as possible would have an opportunity to contribute their ideas, visions, and goals to the site design, a series of five workshops was held. Participants included residents, members of the Tualatin Parks and Recreation Advisory Board, and members of the Tualatin Disabilities Advisory Board. As part of the public participation process, a design charrette was held to involve residents in identifying the future benefits that will be provided at the park (see Fig. 19.6). From these public meetings, the design concept for the site was identified based on the archaeological context of the site. This concept was then overlaid with the functional requirements for a neighborhood park, including active and passive recreational opportunities. Other overall goals for the park included safety, security, accessibility, environmental preservation and education, and ease of maintenance.

The design team worked closely with the maintenance staff during the design process to reduce operation costs. Some features designed to reduce costs include special court surfacing on tennis and basketball courts, timed release of water for water-play features, the use of synthetic safety surfacing in the play area, plant selection, and the design of mow strips to facilitate equipment use. Restroom doors have timers that can be set automatically to lock and

FIGURE 19.6   An interactive design process allows the community to help make design tradeoffs.

unlock, again reducing personnel costs. Sand-based softball and soccer fields are designed for year-round use, with complete underground drainage and irrigation to increase playing time and user safety.

Children's Play Area

Based on the desired benefits identified by community members and the results of the community design charrette, the design team developed an interpretive design concept for the Ibach Park play area. The accessible play area reflects the rich heritage of Tualatin, from prehistory to early European settlement days. Children learn about the history of Tualatin through interactive play, reenacting historic events in play settings. The play area was designed to meet the developmental needs of preschool and school-age children, and to allow children with and without disabilities to have equal access to integrated play opportunities. The final design provides play area users with an interactive tour through the history of Tualatin, allowing people to experience a bit of the city's past. The play area design includes three distinct areas reflecting significant historic periods: the Prehistory area; the Native American area; and the Early European Settler area. A water course of real and simulated water runs through the entire play area, symbolizing the historic impact the Tualatin River has had on the life of the city and its people. In areas where real waterflow is not feasible, the river flow is continued, represented by blue synthetic safety surfacing. A segment of the river is a water-play element that allows children to turn on the river flow by pressing against a bollard. The water flows into the riverbed, which is at the ground level on one side and at a raised height on the other side to allow water play while using a wheelchair or while standing (see Fig. 19.7).

Prehistory Area.   The Prehistory area begins at a high, rocky area, simulating a mountain, such as Mt. Hood, which can be viewed from this site. It contains a child-sized version of the

FIGURE 19.7   Water-play channel at Ibach Park provides wheelchair user access on one side.

meteor that landed on earth millions of years ago and was carried west to Tualatin by the surging waters of the Bretz Floods over 15,000 years ago. In addition, an archaeological dig allows children to unearth fossil rocks and fern prints and explore a giant mastodon rib cage, providing them with a wonderful sense of discovery. Not only do children get to see what happens when you scrape away a layer of earth to discover fossils, but they can develop a better sense of the natural world and the history of life. The mastodon rib cage also provides an excellent climbing structure. One missing rib permits wheelchair access (see Fig. 19.8).

FIGURE 19.8   Water element, mastodon rib cage, and sand play at Ibach Park. The transfer platform (left) and missing mastodon rib (right) facilitate access.

Native American Area.   Replicas of Native American petroglyphs signal the change from prehistory to Tualatin's Native American history. The Native American area contains a circle of drums that allow children to create music. Each drum is at a different height so that children can play either standing or sitting down. A dugout canoe (cut out to accommodate wheelchair users) allows children to experience how the Native Americans might have traveled down the river to catch fish or look for a better hunting ground.

Early European Settler Area.   This area provides an opportunity to learn what it was like for the early settlers of the region. Young children can play in log cabin–style play houses, prepare breakfast at the sand tables, harness a team of horses to pull a covered wagon, ride a ferry across the Tualatin River, or care for the family milk cow.

Children can view the play environment from the swing area, which is entirely covered with accessible safety surfacing and includes a bucket swing for young children or children with disabilities. Located at a distance from the play area for younger children is a separate area that allows both active play and hanging-out space for youth ages 10 to 14. Both play areas meet state-of-the-art safety standards.

The final design for the Ibach Park play area symbolizes the collective memory of generations of Tualatin inhabitants. What makes this area unique is its translation of history into an interactive educational resource, providing play and learning opportunities through direct interaction with the natural and built environments. This results in a diversity of play opportunities and environments that allow children of all abilities to discover, learn, and have fun.

Since its grand opening on May 18, 1996, Ibach Park has attracted users from throughout the region. The park has been featured in the *Oregonian* as an example of creative play area design, and in *InterpEdge* magazine for the play area's unique interpretation of area history. Ibach Park won the Oregon Parks and Recreation Design Award.

19.5   RECENT LEGISLATION

Laws recently enacted in the United States to ensure and promote the rights of persons with disabilities have now created market demand for and social acceptance of all-inclusive, universally designed play environments. In 1998, the Architectural and Transportation Barriers Compliance Board (Access Board: www.access-board.gov) created a committee to amend the Americans with Disabilities Act Accessibility Guidelines (ADAAG) (Access Board, 1991) by adding a section for play areas. The guidelines were developed by a regulatory negotiation committee composed of representatives from a variety of interest groups. The final guidelines proposed for the new rules (ADAAG 36 CFR Part 1191) have gone through public review and were published in the *Federal Register* in October 2000 (Access Board, 2000). The new guidelines specify the minimum level of accessibility required in the construction and alteration of play areas covered by the ADA. The rule provides scoping requirements, which indicate what is to be accessible, and technical requirements, which explain how to achieve access. The guidelines cover play areas provided at schools, parks, some childcare facilities and other facilities subject to the ADA. The requirements of these rules will become part of the enforceable ADA guidelines once they are adopted by the Department of Justice.

19.6   CONCLUSION

Play is learning. Play helps children to express, apply, and assimilate knowledge and experience. A rich play environment encourages all children to grow and develop into healthy adults.

Most children today have very sanitized, packaged play experiences. There are 12-year-olds who live in urban areas who have never built a fort or "claimed" a piece of an outdoor environment as their own private adventure area.

One day the author was walking through a housing development that was under construction and saw the following scene: There were piles of dirt and boards around—great fort-building materials! On one of the dirt piles was a 5-year-old boy, and at the bottom of the pile was his grandfather watching him play.

The construction foreman soon came over to the grandfather and told him that the boy must get off the dirt because his liability insurance would not cover allowing the child to play on the site. The grandfather called to the boy to tell him what the foreman said. The child looked around and could not understand why he could not play there. He knew he was doing nothing wrong. The grandfather tried to explain to the child. Finally the boy got up off the pile of dirt, looked around, and said to the grandfather, "A kid's got to do what a kid's got to do." It is the responsibility of every adult and every institution that serves children to create places so a kid—*any* kid—can do what a kid's got to do.

19.7   BIBLIOGRAPHY

U.S. Architectural and Transportation Barriers Compliance Board, *Americans with Disabilities Act Accessibility Guidelines (ADAAG): Accessibility Requirements for New Construction and Alternation of Buildings and Facilities Covered by the ADA,* Access Board. Washington, DC, 1991.

U.S. Architectural and Transportation Barriers Compliance Board, *Americans with Disabilities Act Accessibility Guidelines (ADAAG) for Buildings and Facilities: Play Areas,* Access Board, Washington, DC, 2000.

19.8   *RESOURCES*

Access Board.   www.access-board.gov.

International Association for the Child's Right to Play.   www.ncsu.edu/ipa.

American Society for Testing and Materials, *Standard Consumer Safety Performance Specification for Playground Equipment in Public Use,* ASTM, West Conshohocken, PA, 1995.

Center for Accessible Housing, *The Recommendations for Accessibility Standards for Children's Environments,* North Carolina State University, Raleigh, NC, 1992.

Moore, R. C., *Plants for Play: A Plant Selection Guide for Children's Outdoor Environments,* MIG Communications, Berkeley, CA, 1993.

———, and H. H. Wong, *Natural Learning: Creating Environments for Rediscovering Nature's Way of Teaching,* MIG Communications, Berkeley, CA, 1997.

———, D. S. Iacofano, and S. M. Goltsman, *Play for All Guidelines: Planning, Design, and Management of Outdoor Play Settings for All Children,* MIG Communications, Berkeley, CA, 1992.

PLAE, Inc., *Universal Access to Outdoor Recreation Areas: A Design Guide,* MIG Communications, Berkeley, CA, 1993.

U.S. Architectural and Transportation Barriers Compliance Board, *Recommendations for Accessibility Guidelines: Recreational Facilities and Outdoor Developed Areas,* Access Board, Washington, DC, 1994.

CHAPTER 20

GUIDELINES FOR OUTDOOR AREAS[1]

Gary M. Robb, M.S.
National Center on Accessibility, Indiana University,
Bloomington, Indiana

20.1 INTRODUCTION

Outdoor or natural areas offer special challenges as well as special opportunities for planners, engineers, and designers. The natural environment does not always lend itself to manipulation to the same extent as built structures. The character and enjoyment of natural settings are often inextricably linked to land formations, topography, or landscape that nature has chosen. These environments are not always conducive to use by a large cross section of people with diverse physical characteristics. For example, caves may be more conducive to exploration by those of short stature, cliffs only by the most adept and skilled climber, and sandy beaches only by those who are highly ambulatory. Given these challenges, many unique natural environments can be opened up to many more potential users with designs and modifications that do not detract from the natural characteristics of the site. While some natural environments may never be manipulated by humans, a vast number are routinely altered to accommodate user recreational interests and movement of a high volume of people. When natural areas or features are manipulated or altered, opportunities are presented to include universal design principles that will accommodate a larger spectrum of potential users.

20.2 BACKGROUND

People love the outdoors! The recent *National Survey on Recreation and the Environment* (Teasley et al., 1998) showed that over 78 percent of the people in the United States are active in outdoor-related pursuits. While participating in outdoor recreation activities is a matter of personal preference, some of the more popular activities appear to be day hiking, sightseeing, camping, swimming, boating (canoeing, kayaking, sailing, and motorboating), bird-watching, horseback riding, and biking (Teasley et al., 1998). Throughout the world, the number of people participating in ecotourism and adventure-tourism programs has increased at a rate of 60 percent in the last 10 years and is expected to increase at a rate of an additional 20 percent per year over the next 10 years (Furze, Lacy, and Birckhead, 1996).

Regardless of country or location, the very nature of our outdoor environment provides a multitude of opportunities. With the growth of ecotourism and outdoor adventure programs that use all areas of the outdoors, opportunities range from a stroll through an urban park, to physical challenges such as mountain climbing and sea kayaking. Most people choose their outdoor recreation experiences based on their skills and ability in the activity and their own comfort level with the outdoors. The key issue is that our global society has to create opportunities for its collective citizenry to experience natural wonders such as Yellowstone National Park in the United States, the Arctic International Wilderness Area, or the jungles of South Africa's Folozi Game Reserve, where the only companions are likely to be of another species. In the United States alone, there are more than 260 million federal, 42 million state, and 10 million local acres of land available for recreational use. These areas include mountains, deserts, prairies, forests, lakes, beaches, and rivers (Cottrell and Cottrell, 1998). It is also important that we not overlook the many and varied opportunities for "local outdoor recreation experiences" as well, such as a picnic in a local park in America or the country gardens in England.

Outdoor recreation provides benefits to the self, the local community, and society (Driver, 1999). Without the ability to participate in outdoor recreation, we stifle our ability to grow. There have been many studies that have looked at the benefits of outdoor recreation in recent years (Driver, Brown, and Peterson, 1991; Sefton and Mummery, 1995; Witt and Crompton, 1996). These studies all point to the fact that outdoor recreation provides society with many benefits that historically were not commonly associated with recreation in the outdoors. These benefits can be classified as personal, including better mental health, personal development and growth, personal appreciation and satisfaction, and psychophysiological. There are other benefits associated with outdoor recreation. These may be classified into the following categories: social, cultural, economic, and environmental (Driver, 1999).

Although it is possible to participate in outdoor recreation in a city park or even in a small backyard, this chapter addresses outdoor recreation that generally takes place on large tracts of land that are usually in undeveloped areas, and where the character of the environment is not lost. These benefits need to be available to all people, but this requires proper planning and programming of resources. Providing access to challenges that can only be found in undeveloped areas is not the same as developing the outdoors and taming the challenge to an acceptable level for visitors, irrespective of their abilities. The benefits of outdoor recreation are maximized by the individual who has the ability to select the level of challenge, the activity, the place, and who they wish to participate with, and not in having these factors dictated to them. Using universal design concepts allows both the user and the manager to maximize their abilities and resources.

While the concept of *challenge* is often central to the outdoor experience, *challenge* is a relative term. One individual may be challenged only until she or he reaches the peak of the mountain, whereas for another person, the ultimate challenge may be to get off the road and into places never traversed by motorized vehicles before. Still for others, significant challenge may be represented by a 15-minute stroll through a relatively flat contoured trail or getting from the parking lot to the picnic table. While the relative solitude of a backcountry hike may provide a variety of sensory and physical challenges, the leisurely stroll through the urban park or the overnight camping experience immediately adjacent to the super highway can provide others with just as stimulating of an outdoor recreation experience. Challenge is only one reason why people participate in outdoor recreation experiences. The outdoor environment provides restorative powers and spiritual experiences that are often more difficult to achieve in more developed hustle-bustle environments.

There are significant populations around the world that would relish the opportunity to have even the previously described roadside experiences. In the United States, almost 20 percent of the population has some type of permanent disability that makes even the least challenging of outdoor recreation activities sometimes very difficult to access (U.S. Bureau of the Census, 1994). In other countries, such as England and Wales, outdoor recreation (or countryside recreation, as it is referred to) is a secondary land use and access to it is typically by license (Newman, 1996). Unlike in many other countries, the national parks in the United Kingdom are owned privately, therefore even further restricting access to the citizenry.

The basic fabric of the UK countryside is not particularly access-friendly. Footpaths, historic landscapes and buildings established centuries ago made no allowance for people with disabilities. (Newman, 1996)

A recent study (Brown, 1999) revealed that people with mobility impairments rated going on "nature outings" as important. The findings also suggested that preferences of people with mobility limitations are no different than anyone else in the kinds of settings they would like to experience. The study also revealed that there were much stronger preferences for "forest settings versus open areas."

As technology and medical advances continue to assist people in living longer, more and more people can no longer experience the challenges of their youth. Families with young children, mothers to be, and people with temporary impairments also experience these situations, although often briefly.

If universal design is of significance in the built environment, then it surely has a role to play in the natural environment as well. The challenges of applying universal design to the natural environment can be far greater than those of the built environment, as the author will explore in this chapter.

20.3 HISTORICAL EVOLUTION OF THE ACCESSIBILITY MOVEMENT

In the United States, it was not until the passage of the American's with Disabilities Act (ADA) in 1990 that much thought was given to ensuring that people with disabilities had the same types of opportunities to participate and take full advantage of the vast participatory opportunities available in society. Other countries have passed similar laws in recent years, such as the Disability Discrimination Act in Australia in 1992 and the similarly titled legislation in the United Kingdom in 1995. In 1993, Sweden passed the Act concerning Support and Service for Persons with Certain Functional Impairments and The Assistance Benefit Act. In Sweden, decision making is largely decentralized. While the state is responsible for legislation, local authorities enjoy great freedom to decide the specific nature of their practices. In 1994, they created The Office of the Disability Ombudsman, which is responsible for monitoring issues relating to the rights and interests of persons with disabilities (1998). Canada has long emphasized the rights of people with disabilities and, while they have not passed specific disability legislation, they do have a Charter of Rights and Freedoms that states that one cannot be discriminated against due to disability. Canada has had building codes that include regulations regarding accessibility since 1985 (Ringaert, 2000). Similar to Sweden, Canada has elected to work through its provinces, local authorities, and private partnerships to address disability issues.

In 1996, First Ministers agreed to make addressing the needs of people with disabilities a priority. Federal, provincial, and territorial governments have been working together to better coordinate their efforts in addressing disability-related issues. In the fall of 1998, the Government of Canada and the provinces and territories jointly released *In Unison: A Canadian Approach to Disability Issues.* This framework document provides a shared vision, principles and objectives to guide future action on disability. The Social Union Framework Agreement, signed in February 1999, now provides us with shared principles and approaches to advance social policies for all Canadians, including Canadians with disabilities.   [Mallory-Hill and Everton (Sec. II, Chap. 10, "Canadian Design Codes") (http://socialunion.gc.ca/pwd/unison/values_e.html)

A number of countries are beginning to specifically address recreation and disability. The ADA specifically references recreational activities as being covered under the law (PL 101-

336, Title III, Section 301, 7.I). Many organizations in the United States and other countries have been providing outdoor recreation experiences for people with disabilities long before the 1990 legislation. Wilderness Inquiry has been providing integrated wilderness expeditions for participants with and without disabilities since 1978. Easter Seals and other private organizations have been providing outdoor recreation and camping experiences for people with disabilities since the 1930s, many of which have focused on inclusion. Similar organizations in the United Kingdom, Canada (Canada's Allied Living Alliance and the Canadian Recreational Canoeing Association), and the Scandinavian countries, to name a few, have also focused on the provision of accessible outdoor experiences for people with disabilities for many years.

Environmental Settings Approach

In the United Kingdom, The Fairfield Trust, an independent, consumer-lead charity in conjunction with other organizations, including British Telecommunications, formed the BT Countryside for All Advisory Committee in 1993. Unlike other countries, such as the United States, where the central government is the primary catalyst in the development of outdoor recreation guidelines, the BT Countryside for All Advisory Committee is made up of many organizations that are guiding the context in which countryside access is developing in the United Kingdom. Their approach to universally designing countryside environments is to look at the characteristics of different environments in considering the level of universality that will be sought. These environments include: formal landscapes, the managed countryside, seminatural environments, and wild country (Newman, 1996).

Parks Canada has developed a *spectrum* approach to considering people with disabilities. For example, the Parks Canada program entitled "Best Practices for Parks Canada Trails" has implemented a spectrum of *environment types*. Parks Canada (1994) established the Access Program in 1990, which produced 118 access plans in 1990 and 1991.

In the United States, various governmental land managing agencies have developed individual approaches to providing access to outdoor environments for people with disabilities. The USDA Forest Service has probably been the most vigorous in pursuing the acceptance of an approach to universal design to outdoor recreation environments in the United States. The USDA Forest Service has chosen a specific approach and published, *Universal Access to Outdoor Recreation: A Design Guide* (USDA Forest Service, 1993). While the Forest Service's guidelines have no legal basis, they have provided guidance for their constituency in addressing accessibility in their vast national land holdings. The USDA Forest Service has used their broader recreation management approach, called the *recreation opportunity spectrum* (ROS), to develop their approach to accessibility. The ROS framework is based on a continuum of possible combinations of recreation settings, activities, and experiential opportunities, as well as the resulting benefits that can accrue to the individual (by improving physical and mental well-being) and society (1993). As with the Canada Parks and U.K. schemes, the USDA Forest Service uses *environments* as the cornerstone of decision making. These environments broadly include: urban/rural, roaded natural, semiprimitive, and primitive.

The major problem that has faced these and other countries and organizations worldwide has been the lack of acceptance on the part of societies to view the need for full inclusion of all people into programs and facilities that are universally designed to facilitate a transparent participation. Instead, more often than not, programs that provide outdoor recreation experiences for people with disabilities have encountered hostility, indifference, and facilities and areas that are not universally designed or usable.

Using the environment as the primary criterion in determining how much accessibility should exist, the concepts of universal design become somewhat secondary in importance, and often it is considered important only in highly developed areas. This approach may lessen the potential of universally designing areas and facilities in vast areas that could possibly be designed for more potential users. The assumption that considering the concepts of universal design in areas other than those in urban or developed settings will create a threat to the environment or destroy the integrity of the experience eliminates up front potentials for providing greater access.

The Universal Design Approach

The Principles of Universal Design should be applied across the board when considering access to outdoor environments. The Center for Universal Design (1997) at North Carolina State University defines *universal design* as "The design of products and environments to be usable by all people, to the greatest extent possible, without adaptation or specialized design." The Center identifies the seven Principles of Universal Design as:

1. Equitable Use
2. Flexibility in Use
3. Simple and Intuitive Use
4. Perceptible Information
5. Tolerance for Error
6. Low Physical Effort
7. Size and Space for Approach and Use

An example of the integration of these principles can be seen in the amphitheater at Bradford Woods Outdoor Learning Center, a program of Indiana University. When these principles are considered in diverse outdoor environments, including wilderness areas, the likelihood of achieving greater access for a greater number of people will result. It does not mean that significant natural features will or must be disregarded or that "we are going to pave the wilderness." That is a psychological barrier that must be overcome prior to looking at the application of a comprehensive approach to establishing guidelines.

20.4   GUIDELINES OVERVIEW

The development of standards and guidelines for outdoor environments, regardless of country or specific geographic location, is challenging at best. In the built environment, earth can be moved and construction methods may be altered to accommodate certain unfavorable site characteristics. In the outdoor environment, however, such methods are more likely, in some areas, to destroy the very nature and characteristic that is essential to the desired user experience.

The development of guidelines or standards for universal usability of outdoor environments is still in the early stages in most countries. As was mentioned earlier, Canada and the United Kingdom are perhaps ahead of other countries in these regards, since various agencies in those countries have been free to develop their own guidelines without specific legal mandate. While the United Kingdom does have The Disability Discrimination Act of 1995, there is not a regulatory body responsible for developing specific standards or guidelines related to outdoor recreation. Since Canada has no specific disability law, their efforts regarding access to outdoor environments has been voluntary. In the United States, an independent federal agency, The U.S. Architectural and Transportation Barriers Compliance Board (Access Board), is responsible for developing guidelines and standards for all environments, including the outdoor environment. The Access Board has been developing standards for recreation environments since 1993. In 1994, a Recreation Access Advisory Committee (RAAC) issued a report in which it addressed the various types of recreation facilities and identified the features of each facility type that are not adequately addressed by current accessibility standards in the United States (U.S. Architectural and Transportation Barriers Compliance Board, 1994). A lack of consensus on some major issues regarding outdoor environments was revealed through public comment, and in 1997 the Access Board created a Regulatory Negotiation Committee on Accessibility Guidelines for Outdoor Developed Areas. This Committee to the U.S. Access Board submitted a final report in July 1999, and the Board subsequently accepted the report in September 1999 (U.S. Architectural and Transportation Barriers Compliance Board, 1999).

In the United States, any newly proposed regulations go through a series of steps, sometimes taking years to actually become final standards. Proposed standards must go through a Notice of Proposed Rulemaking, followed by a period (usually from 60 to 120 days) of public input. The U.S. Office of Management and Budget must make a regulatory assessment of the financial impact of the regulations, and finally, the final rule is developed and signed off by the U.S. Department of Justice. It is expected that finalization of these guidelines will not be completed until late in 2001.

In 1997, standards and guidelines were published in the United Kingdom. The publication, *BT Countryside for All—Standards and Guidelines—A Good Practice Guide to Disabled People's Access in the Countryside,* was printed (British Telecommunications PLC and The Fieldfare Trust, Ltd., 1997). This publication provides extensive information and guidance on making outdoor environments more usable by all people. The guidelines and standards are developed under the following two premises:

1. The standards for access in towns and cities

2. The concept that people expect different levels of accessibility away from the towns and less easy access is acceptable in the countryside (p. 2)

The publication provides extensive suggestions and detailed information on how managing agencies can decide on the amount of accessibility that they provide. The standards follow the *settings approach,* including: (1) urban and formal landscapes, (2) urban fringe and managed landscapes, (3) rural and working landscapes, and (4) open country, semiwild, and wild land, in suggesting the degree of access to be considered. It also provides extensive support and guidance in working with various local and national organizations. It also goes well beyond physical access in providing guidance for developing programmatic access. This includes guidelines for information to be included on signs and maps and how to make them more accessible. Transportation, interpretation, and event planning guidelines are provided with extensive graphics and information sheets.

In 1994, Parks Canada published a document entitled *Design Guidelines for Accessible Outdoor Recreation Facilities* (1994). Similar to the approach in the United Kingdom, the Canadian approach has been to follow the ROS as has been used in the United States by the USDA Forest Service in determining the appropriateness of accessibility levels. The Parks Canada guidelines cover a wide range of outdoor recreation facilities, including bridges and boardwalks, campgrounds, picnicking sites, outdoor amphitheaters, playgrounds, equestrian activities, beaches, swimming pools, fishing, and docks.

The Canadian trails system probably most exemplifies the ROS type of approach in establishing accessibility guidelines. Canadian trails are classified as class 1: primitive; class 2: minor; class 3: major; class 4: walks; and class 5: special-purpose. This latter classification includes provisions for people with disabilities. The Canadian classification system follows.

Parks Canada Trail Classification System Summary

- *Class 1.* Primitive trails are marked, but generally unimproved except for clearing and some work on dangerous areas. Tread widths should be as narrow as possible, up to 70 centimeters (cm). The overall grade should be less than 25 percent, except for short distances, where it should not exceed 40 percent.

- *Class 2.* Minor trails are signed and improved to accommodate foot, horse traffic, or both. Tread widths should be 50 to 150 cm. The overall grade should generally be less than 25 percent, except for short distances, where it should not exceed 40 percent.

- *Class 3.*   Major trails are signed and improved for foot, horse traffic, or both. A major trail usually reaches many of the main visitor attractions and serves as a terminus for minor and primitive trails. Tread width is usually 60 to 200 cm. The overall grade is less than 20 percent, except for short distances, where it should not exceed 30 percent.
- *Class 4.*   Walks include boardwalks and bituminous trails that interconnect developed areas, or serve as short scenic walks. Tread widths should be 100 to 300 cm. Major boardwalks in urban settings may exceed these widths. The overall grade is less than 15 percent, except for short distances, where it should not exceed 25 percent. Stairways should be used on steep sections.
- *Class 5.*   Special-purpose trails include bicycle, cross-country ski, and accessible trails for disabled persons (excerpted from the Parks Canada Internet site entitled "Best Practices for Parks Canada Trails" at http://parkscanada.pch.gc.ca/library/trails/english/trailc_e.htm).

The following provides some detail as to how accessibility fits into the Parks Canada system:

- *Special preservation.*   The special preservation zones generally do not permit trails or their components in order to isolate and protect threatened or endangered resources.
- *Wilderness.*   Trails for experienced hikers and horseback riders only are permitted in this zone, depending on management plan direction. Trails should be natural in character and installed to the most difficult level of use. These trails have the lowest maintenance priority.

Trails should be very unobtrusive and rest gently on the landscape. Their width should be for persons walking single file. Their surfaces should be composed of native or native-appearing materials. Disabled access is allowed, although there is a high degree of difficulty and risk involved. Trails and components are totally unmodified to accommodate persons with disabilities; therefore, some users must have assistance.

- *Natural environment.*   Trails for hikers, day users, cyclists, and horseback riders are permitted. Trails should be somewhat rustic in character and installed to more difficult levels of use. These trails have a lower maintenance priority. Generally, class 2 and class 3 trails are suggested in this zone.

Trails should be unobtrusive and rest gently on the landscape. Their width may be for one or two persons. Surfaces should be built primarily of native or natural-appearing materials. Disabled access may be provided. However, trails and components are difficult for persons with disabilities. Some special modifications to trails, or sections thereof, are made as per management direction. Some persons with disabilities will require assistance.

- *Outdoor recreation.*   Trails in this zone permit hiker, nonmotorized, and limited-motorized access, depending on management direction. Trails should be rural in character and installed to somewhat difficult levels of use. These trails have a medium maintenance priority. Generally, trails of classes 3, 4, and 5 are suggested in this zone.

Trails should be obvious and fit comfortably into the landscape. Their width may be for two or three persons. Their surfaces should be built primarily of natural or natural-appearing materials. Disabled access is provided. Trails and components are moderately difficult for persons with disabilities. Some special improvements or modifications to trails or components are made as determined by the public access plan. Some users may need assistance.

- *Park services.*   Trails in this zone permit nonmotorized and limited motorized access. All trails and components should be sympathetic to the broader surroundings and yet complement the characteristics of the immediate site. They may be refined and somewhat urban in character. Trails should be installed to least difficult levels of use and maintained at a high priority. Generally, class 4 and class 5 trails are suggested in this zone.

Trails should be very obvious and complement their immediate surroundings. Their width should accommodate two to four persons and their surfaces may be built using native, natural, and synthetic materials. Disabled access is provided and trails and components should be easy to use. They should be fully accessible and practical for persons with disabilities. Special improvements or modifications to trails or components are made, as determined by the public access plan. Most users should not require assistance (Parks Canada Internet site, "Best Practices for Parks Canada Trails").

In Finland, the Finnish Association of Sports for the Disabled published guidelines and recommendations for authorities and organizers involved in the exercise, health care, tourism, and environments sectors (Verhe, 1995). This extensive resource includes sections devoted to physical activities outdoors and in ancillary spaces, daily outdoor recreation and exercise, outdoor recreation and nature activities, physical exercise, outdoor activities and their locations and special issues. Specific recommendations and specifications are made for dozens of environments and activities.

20.5 U.S. PROPOSED GUIDELINES FOR OUTDOOR DEVELOPED AREAS

In the United States, the proposed guidelines for outdoor developed areas are in the rulemaking process. The 2-year work by the Regulatory Negotiation Committee provides a preview of what accessibility standards will look like in the United States. The U.S. approach is quite different than that taken by other countries, including the United Kingdom and Canada. The principles, which the committee established in 1997, guided their deliberations over the course of 2 years and nine full committee meetings. This group, which represented a diverse cross section of interests, believed that accessibility guidelines should:

1. Protect resource and environment
2. Preserve experience
3. Provide for equality of opportunity
4. Maximize accessibility
5. Be reasonable
6. Address safety
7. Be clear, simple, and understandable
8. Provide guidance
9. Be enforceable and measurable
10. Be consistent with ADA Accessibility Guidelines (ADAAG) [U.S. built environment standards (as much as possible)]
11. Be based on independent use by persons with disabilities (Final Report, Regulatory Negotiation Committee on Outdoor Developed Areas, 1999)

A final report to the Access Board (including recommendations) was submitted in July 1999. The proposed guidelines are currently in an extensive regulatory review process. It is not expected that they will become final rules until late 2001 at the earliest. The provisions of the report include recommendations for campgrounds, picnic areas, beaches, and trails. By far, the most difficult task of this committee was to agree on standards for newly constructed or altered trails. Trails come in all shapes, sizes, and locations. They have purposes covering a diverse range of interests, which include: accessing primitive environments, providing challenge, enjoying the aesthetics of the outdoors, and simply for taking leisurely walks. Developing a single standard to cover all trails was impossible.

The committee considered many different approaches to developing standards for newly constructed and altered trails and other outdoor recreation sites. Each approach balanced accessibility with the uniqueness of the outdoor environment. Examples of the approaches developed for trails throughout the committee's deliberations included the following:

1. *Requiring a percentage of the miles of trails provided to be accessible.*   Using this approach, it was agreed that some trails, such as paved urban and suburban transportation routes, should usually be accessible. But the committee could not agree on the types of trails, other than the aforementioned type, that should be accessible and to what percentage. The committee determined that this approach would be too arbitrary and too difficult to follow.

2. *Dividing trails into different categories and requiring certain accessibility guidelines to be followed.*   The committee could not agree on the categories, nor could it agree that a trail in one category would always be different than a trail in another category. A fear in this regard was that only easy trails would be accessible, thereby eliminating the option for people with disabilities who can use more difficult trails of the opportunity to do so.

3. *Requiring a certain level of access dependent on the location of the trail in terms of the type of setting.*   Definitions must be agreed upon and understood by the trails community, people with disabilities, and land management agencies that are a part of the federal government, states, and local entities. The committee could not find acceptable definitions for a *settings* approach.

Since one of the goals established by the committee was to provide maximum accessibility while maintaining the integrity of the environment, none of the approaches above worked in total. Primitive trails, for example, are usually found in remote locations or in a natural state with limited development. Providing access in some cases may change the experience or result in a significant environmental impact. Even providing universally designed trails in a highly developed setting could result in all trails beginning to look alike. Committee members did not want the proposed guidelines to impede the creativity of planners or designers (Final Report, Regulatory Negotiation Committee on Outdoor Developed Areas, 1999).

Exceptions

Because of the complexity and diversity of outdoor environments, the one-size-fits-all scheme does not seem to work when considering the design of these environments to include all people. As a result, but with the intent to consider universal design to the greatest extent possible, the U.S. proposed guidelines start from the premise that all new and altered campgrounds, beaches, picnic areas, and trails should consider access for all possible participants. Realizing that it is not possible, and in some cases not even desirable, to require changes in some environments, some general exceptions need to be made when certain environmental conditions are present.

The best example of this can be found in the proposed U.S. standards for trails. In the proposed guidelines, there are four specific conditions where trail construction projects can depart from technical provisions of the standards. These departures are allowed for the duration of the existence of the condition, or unless that condition is such that it makes it impractical to make the remainder of the trail accessible. Some of these conditions may also exist in areas where camping, picnicking, and beach facilities are being constructed. In most cases, these are limited to technical provisions for clear space, surface slope, and accessible surfacing. Conditions that may exist that would not require accessibility would include:

• *Where compliance would cause substantial harm to cultural, historic, religious, or significant natural features or characteristics.*   For example, a significant natural feature may include a large boulder, outcrop, or tree, or some water feature that would block or interfere with trail construction to the extent that the trail could not, at that point, be made accessible. This includes areas protected under federal or state laws, such as areas with designated, threatened, or endangered species or wetlands that could be threatened or destroyed by full compliance with the technical provisions. Or, where trail conditions would directly or indirectly harm natural habitat or vegetation. Significant cultural features include areas such as archaeological sites, burial grounds and cemeteries, Indian tribal protected sites, and so on. Significant historical features include properties on or eligible for the National Register of Historic Places or other places of recognized historic value. Significant religious features include Indian sacred sites and other properties designated or held sacred by an organized religious belief or church.

- *Where compliance would substantially alter the nature of the setting, the purpose of the facility, or a portion of the facility.*   Examples include a trail intended to provide a rugged experience such as a cross-country training trail with a steep grade or a challenge course with abrupt and severe changes in level. If these types of trails were flattened out or otherwise made to comply with the technical provisions for accessible trails, they would not provide the intended and desired level of challenge and difficulty to users. Other examples include trails that scramble over boulders and rocky outcrops. The purpose of the trail is to provide people the opportunity to climb the rocks. To remove the obstacles along the way would destroy the purpose of the trail.

  Trails are designed to provide a particular opportunity for the user. Throughout the discussions regarding trails and accessibility, many committee members were concerned that complying with the technical provisions could change the nature of some recreation opportunities. Further, compliance could negatively impact the unique characteristics of the natural setting, the reasons why people choose to recreate in the outdoor rather than the indoor environment. People using primitive trails, for example, often experience the outdoor environment in a more natural state with limited or no development. In these settings, people are generally looking for a higher degree of challenge and risk where they can use their outdoor/survival skills. This condition is included to address these concerns.

- *Where compliance would require construction methods or materials that are prohibited by federal, state, or local regulations or statutes.*   For example, federally designated and some state designated wilderness areas prohibit use of mechanized equipment, limiting construction methods to hand tools. Imported materials may be prohibited in order to maintain the integrity of the natural ecosystem. Construction methods and materials employed in designated wetlands are strictly limited. For traditional, historic, or other reasons, many trails are built using only the native soil for surfacing, which may not be firm and stable. Compliance with provision for firm and stable might conflict with the prevailing construction practices by requiring the importation of a new surfacing material that would not otherwise have been used. Federal, state, and local governments are often restricted by statute and/or regulations imposed to protect or address environmental concerns, through environmental quality reviews.

- *Where compliance would not be feasible due to terrain or the prevailing construction practices.*   For example, complying with the technical provisions, particularly running slope, in areas of steep terrain may require extensive cuts or fills that would be difficult to construct and maintain, or cause drainage and erosion problems. Also, in order to construct a trail on some steep slopes, the trail may become significantly longer causing a much greater impact on the environment. Certain soils are highly susceptible to erosion. Other soils expand and contract along with water content. If compliance requires techniques that conflict with the natural drainage or existing soil, the trail would be difficult, if not impossible to maintain. This condition may also apply where construction methods for particularly difficult terrain or an obstacle would require the use of equipment other than that typically used throughout the length of the trail. One example is requiring the use of a bulldozer to remove a rock outcropping when hand tools are commonly used.

There are also two general exceptions to designing for universality where severe conditions exist. They include:

1. *Where one of the aforementioned four conditions are present.*   The trail does not have to revert back to the standards after the first point of departure, if the trail is less than 500 feet in length; or, if there is no prominent feature less than 500 feet from the trail head (Final Report, Regulatory Negotiation Committee on Outdoor Developed Areas, 1999, p. 11).

2. *Where one of the four conditions is present for over 15 percent of the length of the trail.*   In this case, the standards do not apply after the first point of departure.

In designing outdoor recreation areas, consideration needs to be given to both the technical aspects of the design as well as the scoping needs (i.e., the *when, where,* and *how many* of the design). The technical provisions provide specific measurements, and the scoping information provides important information as to the circumstances that need to be considered and applied to the design.

Technical Provisions in the United States

Outdoor Recreation Access Routes (ORAR)

> *An outdoor recreation access route is a path or walk for pedestrians that connects developed spaces and elements that support the primary activities offered within the recreational area.*   (Final Report, Regulatory Negotiation Committee on Outdoor Developed Areas, 1999, p. 11)

The provision of an outdoor recreation access route in the proposed guidelines is included since many outdoor recreation areas provide for transportation access or other improvements at a designated site (e.g., parking, visitor centers, concessions, restrooms, etc.). While the actual recreation experience desired may be more remote, the committee felt it important that areas close to the trailhead, parking lot, and such should be accessible. The ORAR differs from the U.S. Access Board's ADAAG standards in that it assumes that grades and cross slopes will not always be able to meet the rigid built environment standards. However, an ORAR should be considered more like an access route than a trail. Therefore, the ORAR proposal calls for a minimum of 36 in wide but allows for a 3 percent maximum cross slope (versus 2 percent in ADAAG). The ORAR also allows for longer distances between designated passing spaces where the width of the route is less than 60 in wide.

Picnic Areas.   Typical picnic areas may include tables, fire rings, grills, benches, and trash containers. Other areas might include woodstoves and fireplaces, overlooks and viewing areas, telescopes/periscopes, utility-cleanup sinks, storage facilities, and/or pit toilets. In areas where multiple elements of any type are provided, the proposed guidelines require that 50 percent, but no less than 2 percent, shall be accessible. Of the 50 percent required to be accessible, a minimum of 40 percent, but no less than 2 percent, shall be located along an ORAR as defined (Final Report, Regulatory Negotiation Committee on Outdoor Developed Areas, 1999, p. 27). For example, if 10 fixed picnic tables are provided in an area, 5 (50 percent) would be required to be accessible, and 2 (40 percent of the 50 percent) would be required to be located on an accessible ORAR.

Some of the major features of accessible picnic elements include the following:

- All elements must be dispersed among available sites.
- Surfaces around accessible elements must be firm and stable.
- Picnic tables must provide a minimum of one accessible seating space [30 in wide, 19 in deep, and 27 in high (leg clearance)]. If the tabletop is over 24 linear ft, additional spaces are required.

- Fire rings must provide a clear space of 48 × 48 in around all usable portions of the fire ring. Fire surface height must be a minimum of 9 in above the ground.
- All cooking surfaces (grills) must be between 15 and 34 in above the ground/floor. All controls and operating mechanisms must comply with current ADAAG standards. The clear floor space requirements identified for fire rings are also applicable to grills as well as to wood stoves and fireplaces.
- Benches, where provided and accessible, must provide armrests that meet current ADAAG standards. Bench seats shall be a minimum of 17 to 19 in above the ground to facilitate transfers.

TABLE 20.1   Proposed Scoping for Accessible Picnic Tables: United States

Tabletop linear feet	No. of wheelchair spaces
25–44 lf	2 spaces
45–64 lf	3 spaces
65–84 lf	4 spaces
85–104 lf	5 spaces

Picnic areas provide excellent opportunities for designers to practice universal design. Since most picnic areas are easily accessible by transportation systems, are often located in urban parks, and are generally designed for family outings, universally designed picnic areas greatly enhance the opportunities for the majority of people to take full advantage of the outdoor environment.

Camping Facilities.   Typical camping facilities may include tent spaces, camping shelters, RV and trailer spaces, kiosks, tent pads, and fee depositories. Some of the major features of accessible camping facilities include the following:

- All elements must be dispersed among available sites.
- Surfaces around accessible elements must be firm and stable.
- Where there are camping spaces (RVs/trailers or tent spaces), 50 percent must comply with specifications for width (RVs/trailers: 20 ft, tent spaces: 16 ft in width).
- Elements (spaces) must have a clear space of 48 × 48 in.
- Tent pad surfaces must be firm and stable and must not exceed 2 percent slope (3 percent under some conditions).
- Curbs, walls, railings, or projecting surfaces that provide people from slipping off a tent platform shall be 4 in high, minimum.
- Where one or more rinsing showers are provided, at least one must have a showerhead usable between 48 and 54 in and one a minimum of 72 in above the floor/ground.

Beaches.   A beach has been defined in the proposed guidelines "as a designated area along a shore of a body of water providing pedestrian entry for the purposes of water play, swimming, or other water shoreline related activities" (Final Report, Regulatory Negotiation Committee on Outdoor Developed Areas, 1999, p. 17). The proposed guidelines define mean high-tide level, mean riverbed level, and normal recreation pool level in differentiating among types of water sources. Both new beaches and existing beaches are addressed and are treated differently. When creating a new beach, there is a much better opportunity to

TABLE 20.2 Proposed Scoping for Accessible Camping Spaces: United States

No. of camping spaces	No. of accessible camping spaces (tent, RV, shelters)
1	1
2–25	2
26–50	3
51–75	4
76–100	5
101–150	7
151–200	8
201–300	10
301–400	12
401–500	13
501–1000	2 percent of total
1001 and over	20 plus 1 for each 100 over 1000

create a universally designed environment. In new beaches the proposed guidelines recommend a minimum of one accessible route each ¼ mile. The definition of an accessible beach route extends to the mean high-tide level (coastal beaches), mean riverbed level (river beaches), and normal recreation pool level (lakes and reservoirs). Tidal differences may vary 10s or even 100s of feet at different latitudes and, therefore, the mean designation is significant. In existing beach areas, where an entity provides a pedestrian route, the route must extend to the mean water level. The proposed guidelines do provide the option to use a temporary surface as, in some cases, constraints of the environment may make the construction of permanent structures impracticable and/or undesirable. In either case, the route surface must be firm and stable, and must be a minimum of 36 in wide. As with the ORAR, if the width of the access route is less than 60 in, passing spaces will be required. For beach access routes, these spaces would be required at intervals of 200 ft maximum. Additionally, the route shall not include any obstructions, and the average maximum slope can't exceed 5 percent for any distance and a maximum cross slope of 3 percent. Because of the surrounding sand surface, edge protection (curb) must be provided where there is a drop-off of 6 in or higher. The edge projection must be a minimum of 2 in. In cases where the drop-off is more than 1 in but less that 6 in, the edge must be beveled.

Trails.   For a trail to be considered accessible, it needs to conform to several specifications. The U.S. proposed guidelines require that trails contain the following:

- Surfaces must be firm and stable.
- Clear width of 36 in must be provided.
- Surface opening cannot be greater than ½ in in diameter (¾ in in certain circumstances).
- Must have no protruding objects (same requirement as in built environment).
- Tread (wheelchair) obstacles no greater than 1 in.
- Passing space at every 1000 ft if trail is less than 60 in.
- Average running slope not to exceed 7 percent.
- Maximum cross slope not to exceed 5 percent (12 percent allowed for a distance not to exceed 5 ft—running slope must be 5 percent or less in such cases).
- Resting intervals are required.
- Where provided, edge protection of 4 in is required.

- Signs are required that will identify accessible trail segments and total distance of the accessible segment.

The design process for all new trails and those being considered for alteration should begin with the premise that they will be accessible.

There are also situations that are so severe that it would not be practicable to consider accessibility further. These situations include the following:

- The combination of running slope and cross slope exceeds 40 percent for over 20 ft.
- A trail obstacle measures 30 in or more in height.
- There is an unstable surface for 45 ft or more.
- A trail width measures less than 12 in for 25 ft.

20.6 EXAMPLES AND IDEAS—UNIVERSAL DESIGN IN OUTDOOR ENVIRONMENTS

The Access Challenge program of the British Columbia Mobility Opportunities Society in Canada is another example of an organization that is using initiative and creativity in getting people with disabilities into the wilderness. They have designed and constructed the Trail Rider. Built with lightweight aluminum, the Trail Rider allows people with severe disabilities to access rocky, rough terrain; tree roots; and streams (Paralyzed Veterans of America, 1998). The limitation in the construct of universal design in the wilderness is the individual's ability to work with others—if people work together and utilize each other's strengths, then all can enjoy wilderness areas. This type of universally designing for access to the wilderness is being used throughout the world—from the Boundary Waters of Minnesota to the great wilderness of Siberia. While it may not be universal design in the architectural sense, programmatic design using universal design concepts can accomplish many of the same results. By applying the concept of universal design in the manner previously described, coupled with education about the needs of people with disabilities, practitioners have the ability to provide the opportunity for people who have wanted to experience the challenges, benefits, and rewards of being active in a wilderness adventure, but felt they could not due to the constraints of the environment. By thinking about universal design in this manner, users both with and without disabilities are allowed to seek the adventure that they desire and enjoy all the benefits of these experiences, while at the same time protecting wilderness for others to share the same rewards.

Many times amphitheaters are constructed on naturally sloping terrains. Often, if accessibility is considered, the last or first rows are likely to be targets for accommodating people with disabilities. However, such an accessible design is not a universal design. A universally designed amphitheater will maximize the options for seating by people who use mobility devices, children in strollers, and people who are elderly or who have unsteady gaits. Additionally, a universally designed amphitheater will make it easier for production companies, equipment movers, and other production or service personnel to provide appropriate services. A universally designed amphitheater will have an assistive listening system that allows for anyone, sitting anywhere, to auditorily participate. It will have retractable or removable seats that will allow anyone to sit anywhere, and it will have access routes to every level that will allow any wheeled device to independently access. One such amphitheater has been constructed at Indiana University's Bradford Woods Outdoor Center in the United States.

A fishing pier that provides for lowered railings for people who are seated, also provides better fishing visibility for children or others of short stature. A fishing pier that provides independent access to people who use wheelchairs also provides access to parents with young children.

FIGURE 20.1   A universally designed amphitheater at Bradford Woods, Indiana University, Bloomington, Indiana. The universal design of this amphitheater allows anyone to sit in any seat at any level.

FIGURE 20.2   Fishing can be enjoyed by all if the pier is designed for universal use. Universally designed outdoor areas can provide solitude and relaxation for all.

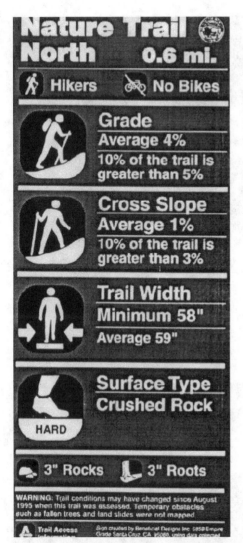

FIGURE 20.3 Trail signs that include pertinent trail characteristics assist all users in making decisions as to the level of difficulty of the trail for them.

In working toward universally designed trails, an assessment process should be conducted on both existing and proposed trails. This process should provide information about specific features on the trails such as trail length; maximum grades and cross slopes; location and size of obstacles; minimum trail width and locations; and surface characteristics (e.g., hardness, stability, etc.) (Axelson, 1998). This information then will provide the data and criteria required to construct or retrofit the trail so that it is usable by more people. Information in the form of specific signage and auditory and/or printed literature detailing the trail characteristics allows people with all types of characteristics to make decisions as to if or how much of the trail they wish to hike. Pictorial and/or graphic information makes the trail even more universally designed for people of all nations and languages.

Picnic tables can easily be universally designed by adding an extension to one or both ends of the table and/or designing a seating arrangement that allows space(s) around the table for strollers, movable chairs, scooters, larger persons, and wheelchairs. Such designed tables also provide more flexibility for users to place picnic baskets, portable grills, ice chests, or other equipment. Costs for the construction of these types of tables should be no greater than the traditionally designed table when produced in quantity.

Many surfacing materials for paths and trails can serve multiple purposes. In addition to providing a firm and stable surface necessary for easy access of wheeled devices, such as wheelchairs, strollers, trolleys, hand trucks, and such, stabilized surfaces can reduce erosion, divert water runoff, and halt unwanted encroaching vegetation. Built wooden trails can, for example, not only provide easy access, but also eliminate human destruction of wetland areas or other sensitive natural features by routing human traffic over and/or around them.

20.7 GENERAL LESSONS LEARNED FOR UNIVERSAL DESIGN

- All types of people derive benefits from outdoor recreation! There is no typical face or body of an outdoor user, and because of this it is necessary to design our outdoor facilities and programs in a universal manner.

- Historically, few countries have focused on and given consideration to the concept of universal design in the outdoor environment. This is a relatively new concept, and because of that, there is not one set of guidelines or standards that can be used.

- The seven Principles of Universal Design can be used when approaching any type of development in the outdoors:

  1. Equitable Use
  2. Flexibility in Use
  3. Simple and Intuitive Use

FIGURE 20.4 Picnic tables can easily be designed and constructed to accommodate a multitude of users.

4. Perceptible Information
5. Tolerance for Error
6. Low Physical Effort
7. Size and Space for Approach and Use

- Universal design includes changing programs and developing programs that find ways to get around barriers and constraints without creating new ones. The goal of universal design is to include the opportunity for a maximum number of people, regardless of ability, to participate in outdoor recreation activities.

FIGURE 20.5 Testing soil stabilizers will assist managers in making decisions on trail surfaces that may be used to provide access to a wide range of potential users.

FIGURE 20.5   *(Continued)*

20.8 *CONCLUSIONS*

While countries around the world have dealt with and are approaching the development of guidelines and standards for outdoor recreation access differently, most seem to be moving toward a common goal—that is to include people with disabilities and others who may be excluded more fully into all aspects of the society in which they live. It must be understood that it is not always necessary to make major physical changes to the environment in order to make it more accessible to people with disabilities, even in remote or wilderness environments. For the most part, people with disabilities do not want the natural environment altered any more than people without disabilities. By employing the principles of universal design, managing agents can look at each environment as an occasion to create opportunities for participation for more people. Once universal design concepts are incorporated into the incipient planning stages, many more people will benefit from the final product. To develop these concepts into working models that are employed throughout the world there needs to be a concerted effort to educate all people that have any role in planning, design, and development of outdoor recreation areas. This process of education has already begun in developed countries such as the United States; the United Kingdom, Sweden, Australia, and others, and it includes such things as classes in universal design at the college level, symposia for the park designer, and international conferences for the practitioner.

20.9 *NOTE*

1. *Special acknowledgment.*   The author wishes to recognize the contributions of Dr. Todd Paxton to the contents of this chapter. Dr. Paxton contributed to the discussion on wilderness accessibility issues and assisted in other parts of the chapter. Dr. Paxton was Director of Research at the National Center on Accessibility and is currently on the faculty of Ferris State University.

20.10   BIBLIOGRAPHY

Axelson, P., *Universal Trail Assessment Process Training Guide,* PAX Press, Santa Cruz, CA, 1998.

British Telecommunications PLC and The Fieldfare Trust, Ltd., *BT Countryside for All—Standards and Guidelines—A Good Practice Guide to Disabled People's Access in the Countryside,* British Telecommunications PLC and The Fieldfare Trust, Ltd., 1997.

Brown, T. J., R. Kaplan, and G. Quaderer, "Beyond Accessibility: Preference for Natural Areas," *Therapeutic Recreation Journal,* XXXIII (3), 1999.

Canadian Standards Association, *Barrier Free Design: CSA-B651 Barrier Free Design,* Canadian Standards Association, Etobicoke, ON, Canada, 1994.

Cordell, K., *Outdoors Recreation in American Life: A National Assessment of Demand and Trends,* Sagamore, Champaign, IL, 1999.

Cottrell, S. P., and R. L. Cottrell, "What's Gone Amok in Outdoor Recreation?" *Parks and Recreation Magazine,* National Recreation and Park Association, Alexandria, VA, 1998.

Driver, B., "Outdoor Recreation and Wilderness in America: Benefits and History," in K. Cordell (ed.), *Outdoors Recreation in American Life: A National Assessment of Demand and Trends,* Sagamore, Champaign, IL, 1999, pp. 1–26.

Driver, B., P. Brown, and G. Peterson, (eds.), *The Benefits of Leisure,* Venture, State College, PA, 1991.

Furze, B., T. Lacy, and J. Birckhead, *Culture, Conservation, and Bio-diversity: The Social Dimension of Linking Local Level Development and Conservation Through Protected Areas,* John Wiley & Sons, Chichester, NY, 1996.

Hitzhusen, G., and P. Thomas, (eds.), *Global Therapeutic Recreation IV,* Columbia, MO: University of Missouri, Columbia, MO, 1996.

MIG Communications, *A Design Guide: Universal Access to Outdoor Recreation,* Plae, Inc., Berkeley, CA, 1994.

Minister of Canadian Heritage, *Design Guidelines for Accessible Outdoor Recreation Facilities,* 1994.

Ministry of Health and Social Affairs International Secretariat: Sweden, Act Concerning Support and Service for Persons with Certain Functional Impairments and The Assistance Benefit Act, Ministry of Health and Social Affairs International Secretariat, Stockholm, Sweden, 1993.

Newman, I., "BT Countryside for All Accessibility: Standards for Countryside Recreation," in G. Hitzhusen and P. Thomas (eds.), *Global Therapeutic Recreation IV,* University of Missouri, Columbia, MO, 1996, pp. 119–127.

Paralyzed Veterans of America, "Blazing Trails," *Sports 'n Spokes,* 25(8), 1998.

Ringaert, L., Personal communication, Canadian Institute for Barrier Free Design, Winnipeg, MB, Canada, 2000.

Sefton, J., and W. Mummery, *Benefits of Recreation Research Update,* Venture, State College, PA, 1995.

The Swedish Ombudsman, Swedish Institute, *Fact Sheets on Sweden,* 1998.

Teasley, R. J., J. C. Bergstorm, H. K. Cordell, C. J. Betz, and M. G. Edwards, *National Survey on Recreation and the Environment,* Venture, State College, PA, 1998.

U.S. Architectural and Transportation Barriers Compliance Board, *Final Report,* Recreation Access Advisory Committee, 1994.

———, *Final Report,* Regulatory Negotiation Committee on Accessibility Guidelines for Outdoor Developed Areas 1999.

U.S. Bureau of the Census, Statistical Brief, U.S. Bureau of the Census, Washington, DC, 1994; www.census.gov/prod/3/97pubs/p7c-61.pdf.

Verhe, Irma, *Outdoor Recreation for Everyone: The Adaptation of Outdoor Activity Areas for the Use of the Disabled,* The Finnish Association of Sports for the Disabled, Helsinki, Finland, 1995.

Witt, P., and J. Crompton, (eds.), *Recreation Programs That Work for At-Risk Youth: The Challenge of Shaping the Future,* Venture, State College, PA, 1996.

20.11   RESOURCES

Peter Axelson, President, Beneficial Designs, Inc.
5858 Empire Grade
Santa Cruz, CA 95060-9603
831-429-8447 (phone)
831-423-8450 (fax)
E-mail to mail@beneficialdesigns.com
Internet: www.beneficialdesigns.com/

Beneficial Designs is an engineering and design firm specializing in technologies that enhance access for people of all abilities.

Peggy Greenwell, Designated Federal Officer for Recreation,
  United States Architectural and Transportation Barriers
  Compliance Board
1331 F Street, NW
Washington, DC 20004
800-872-2253 or 202-272-5434 (phone)
E-mail to: info@access-board.gov
Internet: www.access-board.gov/

The Access Board, created in 1973, has served the nation as the only independent federal agency whose primary mission is accessibility for people with disabilities.

Ian Newman, Director, Field Fare Trust
67a The Wicker
Sheffield, South Yorkshire S38HT
United Kingdom
0114-270-1668 (phone)
E-mail to: Fieldfare.Scotland@osprey.force9.co.uk
Internet: www.fieldfare.org.uk/

Greg Lais, Director, Wilderness Inquiry
1313 Fifth St. SE, Box 84
Minneapolis, MN 55414-1546
612-379-3858 (phone)
1-800-728-0719 (toll free)
E-mail to: info@wildernessinquiry.org
Internet: www.wildernessinquiry.org

Wilderness Inquiry (WI) creates outdoor adventure for people of all ages, abilities, and backgrounds. WI's travels to over 30 destinations by canoe, sea kayak, dogsled, horseback, and backpack.

David C. Park, Chief, Accessibility Management Program,
  National Park Service
Office of Accessibility
P.O. Box 37127
Washington, DC 20013-7127
202-565-1255 (phone)
E-mail to david_park@nps.gov

The NPS Office of Accessibility provides technical assistance and training for NPS personnel in all aspects of program and physical accessibility.

Parks Canada
25 Eddy Street
Hull, Quebec
Canada
K1A 0M5
E-mail: parks webmaster@pch.gc.ca
Internet: http://parkscanada.pch.gc.ca/parks/main e.htm

Laurie Ringaert, Director, Canadian Universal Design Institute,
   Assistant Professor, Department of Landscape Architecture, University of Manitoba;
   Chair of the Canadian Centre on Disabilities Studies
The Canadian Institute for Barrier-Free Design
201 Russell Bldg.
University of Manitoba
Winnipeg, Manitoba R3T 2N2
204-474-8588 (phone)
E-mail: universal_design@umanitoba.ca

The Canadian Institute for Barrier-Free Design's mandate is research, education, and promotion of universal design.

Gary M. Robb, Executive Director, National Center on
   Accessibility, Indiana University
5020 State Road 67 North
Martinsville, IN 46151
765-349-9240 (phone)
E-mail: nca@indiana.edu
Internet: www.indiana.edu/~nca

The National Center on Accessibility is an organization committed to the full participation in parks, recreation, and tourism by people with disabilities. The NCA conducts research, provides technical assistance, and offers training and education programs.

20.12   WEB CITATIONS

Canadian Charter of Rights and Freedoms.   http://socialunion.gc.ca/pwd/unison/values_e.html

North Carolina State University, The Center for Universal Design, 1997.   www.design.ncsu.edu:8120/cud/

U.S. Bureau of the Census, Statistical Brief, 1994.   www.census.gov/prod/3/97pubs/p70-61.pdf

CHAPTER 21

JAPANESE GUIDELINES FOR UNIVERSAL DESIGN IN PARKS: HARMONY BETWEEN NATURE AND PEOPLE

Mamoru Matsumoto, J.C.E.A.
*City and Regional Development, Ministry of Land,
Infrastructure, and Transport, Tokyo, Japan*

21.1   INTRODUCTION

In recent years, universal design has been employed increasingly in various fields and is gradually being introduced in the urban parks of Japan. In 1999, the Ministry of Construction published *Japanese Guidelines for Universal Design in Parks* to serve as a design guide for urban parks that will meet the needs of all users. These guidelines are the first design guide in Japan that illustrates the universal design concept in outdoor settings. This chapter presents an outline of these guidelines.

21.2   BACKGROUND

It is estimated that the aging population in Japan will account for over 25 percent of the total population in 2015. The declining total fertility rate (i.e., the number of children to be born per female in a lifetime), which is the major factor of an aging society as well as the increasing longevity, was reported to be 1.39 in 1997 and continues to decline every year. Fifteen years ago, *Design Guidelines for Urban Parks/Accessible Facilities for People with Disabilities* was published by the Ministry of Construction for the purpose of removing barriers that hinder park use by people with disabilities. However, when focusing on barrier removal in order to make a place functional for people with disabilities, these modifications would often involve destruction of the natural environment, resulting in a number of uniformly designed, barrier-free parks all over the country.

Thus, based on the universal design concept, *Japanese Guidelines for Universal Design in Parks* was developed to improve the previous guidelines, responding to the demographic changes affecting lifestyles and recreation opportunities and considering the broadest spectrum of potential park users. The new guidelines intend to realize an urban park where people of all ages and physical abilities are able to gather in nature and interact in a spontaneous manner.

Considerable time and the efforts of many people were required to complete the guidelines. The preparatory period to collect relevant information from the United States, the United Kingdom, Canada, and Finland, as Robb states in Chap. 20, "Guidelines for Outdoor Areas," was followed by a 3-year period of drafting. The Urban Parks Access Committee that was organized for the project held several meetings to review the previous guidelines and to obtain input and new ideas during the drafting period. In particular, comments and opinions of various user groups, including those with disabilities, were reflected in the guidelines. Much of the work reported in this chapter is based on the publications on universal design in Japan and the United States (Miyake, Kameyama, and Miyake, 1996; PLAE, Inc., 1993; Recreation Access Advisory Committee, 1994).

Urban parks are nurtured by nature and by human beings—that is, they mature by the blessing of nature, which causes the growth of vegetation, and by increased efforts to incorporate user needs. The guidelines have been prepared based on the concept that urban parks should continue to grow—that is, they should be improved constantly to meet all user needs—which is in agreement with the concept of creating universally designed urban parks.

21.3 ORGANIZATION OF THE GUIDELINES

This section briefly introduces the organization of the guidelines, along with the table of contents (see Fig. 21.1). Basic concepts contained in Chaps. 1, 2, and 4 of the guidelines are incorporated in the design specifications, which are presented in Chap. 3. Each conceptual description focuses on the creation of urban parks with the objective of meeting diverse user needs, the principles of circulation planning and access in harmony with nature, and the introduction of recreation programs in urban parks, respectively.

Japanese Guidelines for Universal Design in Parks is composed of four chapters, summarized as follows.

Chapter 1: Creation of Urban Parks Meeting Diverse Needs of Users

In order to meet diverse user needs in the changing society, park planners and designers should create urban parks based on the following concepts:

- A variety of purposes for the use of urban parks and a variety of user needs should be taken into consideration.
- Park facilities should be designed to accommodate a wide range of individual ability, providing choice in adaptation.
- Urban parks should be improved constantly, reflecting voices of diverse users in the community.
- Urban parks should provide recreation programs that assure participation by all people.
- Park planners should improve their understanding of user priorities through a public campaign.
- Urban parks should be managed by signaling hospitality in order to make users feel welcome.

Chapter 2: Planning Urban Parks for All

This chapter focuses on park planning concepts such as circulation and access in harmony with nature. Planning of accessible paths for all, nature preservation, and alternative methods

Contents

Chapter 1   Creation of Urban Parks Meeting Diverse Needs of Users

1-1 Concepts of urban parks meeting diverse needs of users

1-2 Functional limitations of people who are elderly and disabled, and necessary design considerations

1-2-1 Range of users

1-2-2 Functional limitations of people who are elderly and disabled, and necessary design considerations

Chapter 2   Planning Urban Parks for All

2-1 Planning urban parks for all

2-1-1 Key points of planning

2-1-2 Procedures from planning to project completion

Chapter 3   Design Guidelines

3-1 Classification of park facilities

3-2 Access system
Parking areas / Entrances / Paths / Stairs and ramps / Rest areas

3-3 Communication system
Signage / Emergency system / Public telephones and fax machines

3-4 Landscape elements
Planting, flower beds, and trees / Fountains and ponds / Monuments and sculptures

3-5 Recreation facilities
Play lots / Beaches, docks, fishing areas, and swimming areas / Campgrounds / Amphitheaters / Sports facilities

3-6 Other facilities
Toilets / Information desks / Restaurants / Shops, vending machines, and ticket machines / Pergolas, benches and picnic tables / Drinking fountains / Fences / Trash receptacles / Lighting

Chapter 4   Recreation Programs

4-1 Recreation programs for all

4-1-1 Needs of recreation programs for all

4-1-2 Examples of recreation programs in urban parks
Tennis / Horseback riding / Golf / Archery / Bicycling / Fishing / Athletics / Camping / Beep baseball / Canoeing / Boating and yachting / Water skiing / Scuba diving / Alpine skiing / Cross-country skiing / Ice sledge hockey / Gardening

4-1-3 Considerations for program implementation

FIGURE 21.1 Table of contents from *Japanese Guidelines for Universal Design in Parks* (Ministry of Construction, 1999).

of access are discussed, and key considerations in each stage of the planning, design, construction, and post-occupancy phases are explained.

Chapter 3: Design Guidelines

Design specifications of park facilities are presented in this chapter. Compared to the previous guidelines, features of recreation facilities such as play lots and sports facilities are added, reflecting the expanded use of urban parks in recent years. Basic design concepts and design guidelines or recommended design considerations for each facility feature are explained with examples.

Chapter 4: Recreation Programs

Recreation programs that assure participation by all people should be provided in urban parks. This chapter includes examples of recreation programs implemented mainly in the United States and key considerations following the introduction of recreation programs in urban parks in Japan.

21.4   *CREATING URBAN PARKS TO MEET DIVERSE USER NEEDS*

Following are the basic concepts to be borne in mind by planners, designers, and managers when planning urban parks which are to be usable by all people. They serve as the core concepts for planners and designers to determine both physical and nonphysical aspects of park design to meet all user needs.

Meeting a Variety of Purposes and User Needs

There is a substantial difference in the use of urban parks compared to that of other public facilities in cities. People use train stations or shops for the sole purpose of travel or shopping, while they use urban parks for a variety of purposes. As values and lifestyles change, the spectrum of park users is widened, and people who are aging or have disabilities use urban parks more than before, and for different purposes. Accordingly, the role of urban parks should be changed to serve as a multipurpose recreation facility with the intent of satisfying diverse needs of people of all ages and abilities. The following includes possible purposes regarding urban park use by people who are aging or have disabilities:

- Interacting with nature
- Engaging in a variety of exercises to stay healthy
- Fostering interaction among people of different generations living in the community
- Engaging in group activities (e.g., recreation or hobby)
- Working as a volunteer (e.g., in park maintenance or as an interpreter)
- Relaxing and resting

However, it has been pointed out by some park users that park facilities, such as entrances, paths, playgrounds, campgrounds, rest rooms, and signage are not usable by them. As usability is the primary criterion in enhancing the enjoyment of parks, it is essential that park planners and designers research the shortcomings of existing designs that hinder access from the users' point of view in the initial stage of each project, and reflect on the research findings for improvements based on carefully considered planning and performance criteria.

Diversity and Choice in Park Facilities

Until several years ago, people with disabilities rarely visited urban parks alone; they were generally accompanied by attendants. With the growing number of elevators and accessible rest rooms provided in public facilities in cities these days, and with improved public transportation, more people with disabilities are going out and visiting urban parks independently. It is not unusual to see people using wheelchairs or a group of people with visual impairments visiting urban parks without any attendants. Therefore, it is necessary to create urban parks that are independently usable by all people.

To maximize usability by people with different abilities, it is necessary to provide choice in adaptation. Provision of uniform adaptation cannot accommodate a wide range of individual abilities. For example, one type of bench that is usable by the majority of people does not fully satisfy all users because each user may have slight difficulty in use. Instead, provision of different types of benches allows each individual to choose the one best suited to satisfy his or her needs. Diversity in design brings choice in adaptation and will accommodate individual users' needs.

Examples of diversity in design are shown in Figs. 21.2 to 21.4.

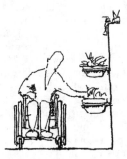 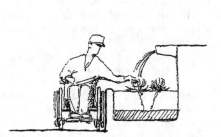

FIGURE   21.2 Planters placed at different heights.   **FIGURE 21.3** Aquatic plants touched by a wheelchair user.

FIGURE 21.4   Planting taking advantage of different ground levels.

Continuous Improvement of Urban Parks

Urban parks should be nurtured as they mature naturally. Park managers should continue to examine whether user needs are met after a project is completed. In case any shortcomings of physical or nonphysical aspects of design are observed (i.e., in any system or program intended to facilitate smooth park operation), proper countermeasures should be taken, or new approaches should be introduced.

When planning a park with input from community members, the public's ideas and intents can be identified through surveys using questionnaires or interviews, workshops, and meetings. A community park cannot be created and nurtured by park management and design professionals alone. Involvement of people living in the community is essential to start the project and to ensure that continuous improvements are realized toward the goal of a park for all, including those who are aging or have disabilities. User participation, starting in the planning stage, will help heighten public awareness regarding the park, and as a consequence, it will bring about interactions between people focusing on the community park.

Recreation Programs for All

Reflecting the advance of aging in society, there is a growing demand for urban parks as a space to recover or improve health. Moreover, aging people are utilizing urban parks to engage in their hobbies or recreational activities independently or in groups, as well for purposes of resting or walking. According to a government survey in 1995, 42.3 percent of those who were over 60 had engaged in some activities in parks during the past year, either alone or with friends, and the average number of activity types engaged in was 1.6 per person. The most popular activity was sports and health exercise (18.9 percent) followed by hobbies (17.9 percent).

This report suggests that elderly people are interested most in improving their health and in making friends. Urban parks are suited to realize this need. Furthermore, parks can also serve as a place for interaction among young people, little children, and aging people. This mitigates the potential isolation that may result from the increasing number of nuclear families in urban areas, where opportunities to communicate with people of different generations are decreased these days. Thus, the role of urban parks becomes even more important.

From the viewpoint of vitalization of urban parks, it is necessary to devise positive means to fulfill the user needs mentioned here. The physical and nonphysical aspects of design should be planned for from the initial stage of the project, including recreation programs for all people and recruitment of aging people as park volunteers. Both will contribute not only to vitalize urban parks, but they will also make park users' life more enjoyable and worthwhile.

Campaigns for User Priorities

In a park design which maximizes usability, and which at the same time is appealing in aesthetic quality, some design considerations are essential only to those who need them. Therefore, park users should understand the notion that priority in the use of certain facilities will be given to people who are aging or have disabilities. Rest rooms with enough space to accommodate a person using a wheelchair are also usable by other people, who often find the extra space desirable, but this may make it difficult to keep the space available for the people who really need it.

Furthermore, there are some prioritizations that need to be heeded by people with disabilities. For example, a parking area should be reserved with a width of 3.5 m or more, at least 1 m wider than the usual 2.5 m, for loading and unloading wheelchairs—that is, for the exclusive use of drivers using wheelchairs. Drivers who wish to use the area should present a permit for identification. It should be noted that in Japan, a permit to allow parking wherever needed is issued specifically for drivers using wheelchairs. There are also priority parking areas for people with other disabilities. Therefore, people with disabilities who do not have the permit for wheelchair users should be encouraged to use these other areas. Needless to say, able-bodied drivers should not park their cars in these areas.

Urban parks provide amenities and pleasure. To retain the intrinsic nature that attracts various user groups, park management should inform all users of the park's available amenities by means of posters, brochures, and signage.

Hospitality in Urban Parks

Once an urban park that meets all users' needs is completed and opened, park operation and maintenance should be conducted properly with *hospitality* in mind. Hospitality is intended to welcome users. A wide range of services can be rendered in a park, and each service should be offered hospitably to provide comfort, safety, and reliability for users. For example, tactile displays and Braille panels are cleaned so that the fingers of people using them do not become dirty; rest rooms are always clean enough to prevent trouble when used by people with visual impairments; and trees planted along a bench are trimmed to cast shade. Continual efforts to implement appropriate park operation and maintenance will expand the park's use by more people, and will heighten its value as a park for all.

To deepen the relationship between a park and the community, it is essential to invite community participation in park operation. Active participation in events and recreation programs beginning at the planning stage, which goes beyond conventional modes of participation such as cleaning and gardening, is encouraged to expand the park's potential. A system to support participation will enable the park to become the core of community activities, and it will further enhance the park's significance as a worthwhile park for all.

21.5   PLANNING URBAN PARKS FOR ALL

Circulation is a primary issue in planning urban parks to meet all user needs, in order to provide access to all facilities constructed on the site. However, the question is raised as to whether equal access should be provided to all users if the existing natural environment is destroyed in the process. This section discusses principles of circulation planning and solutions for incorporating access in harmony with nature.

Circulation Planning

In many urban parks, major facilities are inaccessible by people who are aging or have disabilities, due to various topographical and geographical restrictions. *Major facilities* are buildings and outdoor facilities that attract many users. In most cases they are the highlights of the park, and frequency of use is therefore very high. To solve this problem, top priority in circulation planning should be given to providing a fully accessible path linking those facilities. In the case of new construction, the creation of a fully accessible path and the placement of major facilities along it should be considered in the initial planning stage. This should address various conditions, including geographical features, the natural environment, and the purpose of the facility. With existing parks, modification of a path connecting major facilities may be required to provide full access. However, if large-scale modification involving destruction of the natural environment is unavoidable and modifications seem to be unfeasible due to various conditions, an alternative method of access should be introduced, based on the equal access concept. This is discussed in the following section.

A *path for all* is defined as a path with a gentle slope and safe surface materials, which allows smooth travel by people who are aging or have disabilities. The following principles should be applied in planning a path for all in urban parks:

- The main route should be identical with the path for all.
- Major facilities should be located along the path for all.
- Parking areas and entrances usable by all people should be connected with the path for all.
- Rest rooms and rest areas should be located along the path for all.
- A uniform type of signage should be used to guide people with visual and hearing impairments to ensure smooth travel on the path for all.

An example of circulation planning is shown in Fig. 21.5.

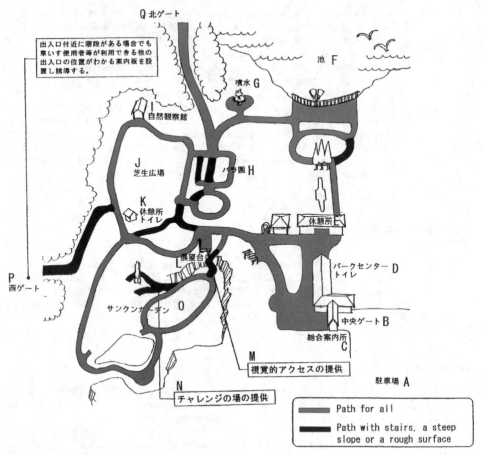

FIGURE 21.5   Example of circulation planning: (A) parking area, (B) central gate, (C) information, (D) park center, (E) rest pavilion, (F) pond, (G) fountain, (H) rose garden, (I) hide, (J) lawn, (K) rest place, (L) observatory, (M) provision of visual access, (N) provision of a challenge opportunity, (O) sunken garden, (P) west gate, and (Q) north gate.

Access in Harmony with Nature

The natural environment is a major target for people to visit and enjoy. Therefore, large-scale modification involving destruction of the natural environment should be avoided. To conduct park planning in harmony with nature, planners should investigate various site conditions and carry out access planning based on the following concepts:

- Respect the goal that people visit parks in order to enjoy the natural environment.
- Avoid large-scale modifications which are contradictory to the goal.
- Preservation of the natural environment helps fulfill user expectations.

Park planning based on these concepts may result in a situation in which a path for all cannot be provided throughout the park because of existing geographical features. In such a case, the following countermeasures should be taken, depending on site conditions.

Provide Access Information.   People who use wheelchairs often complain that inaccessible places are unpredictable—for example, there may be stairs or unpaved surfaces in the middle of a path that is seemingly accessible, but that has insufficient access information posted. Therefore, sufficient access information should be provided at park entrances and at the starting point of each path to prevent problems and difficulty in using paths.

Provide a Challenge Opportunity.   The ability to use a path differs from person to person, depending on the type and degree of disability. Physical strength and the kind of assistive equipment used vary among individuals. Some people may choose to proceed along a path which is not fully accessible, while others may not be able to do so independently. It is important to provide the options for such challenge opportunities to people with disabilities. For this purpose, circulation should be planned to provide choices in routes by degree of challenge. Signage becomes important to provide information about route accessibility, including the path's width, slope, and surface materials.

Provide an Alternative Method of Access.   If there is a facility or area that is not accessible, some other method should be sought to provide alternative methods of access. These alternatives include the following:

- Provision of visual access, that is, viewing from an accessible vantage point (see Fig. 21.6)

FIGURE 21.6   Visual access.

- Concentration of park highlights in an accessible area
- Communication of park highlights information by means of photos, videotapes, or push-button audio systems
- Provision of parking areas for exclusive use by users with wheelchairs near park highlights, allowing vehicle use all the way to the destination

21.6   RECREATION PROGRAMS FOR ALL

As mentioned in Sec. 21.4, a government survey shows that many people who are aging are interested in improving their health and in making friends. Urban parks are suited to realizing these needs. It is necessary to introduce recreation programs for participation by all people in urban parks in order to provide opportunities to enhance enjoyment and life experiences.

Recreation Programs in Urban Parks

Recreation has effects on health improvement and disease prevention. Recreation programs to be introduced in urban parks go beyond this, aiming to help all park users enjoy a fulfilling life and a restoration of confidence. The following are expected outcomes of recreation programs in urban parks:

- Urban parks become a place to enjoy and a reason to go out.
- Program participation helps park users gain confidence, leading to the opportunity to participate in society and to have a fulfilling life.
- Park users develop progressive involvement in recreation programs, as participants in the beginning and as instructors in the next stage.
- Solidarity, a sense of achievement, and satisfaction are attained by all participants, regardless of age or disability.
- Interaction between participants helps form a good community.

A spectrum of recreation programs implemented in urban parks is shown in Fig. 21.7.

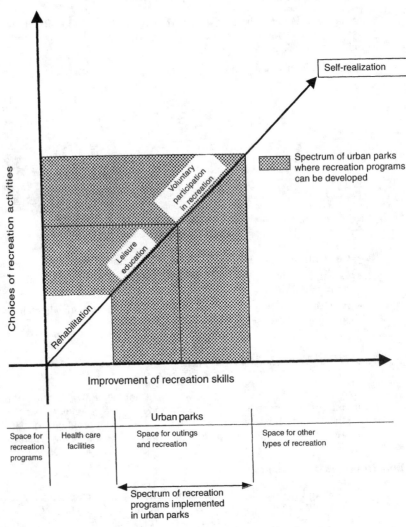

FIGURE 21.7   Spectrum of recreation programs implemented in urban parks.

Examples of Recreation Programs

The guidelines include examples of recreation programs, implemented mainly in the United States, that are considered feasible for introduction in urban parks in Japan. Each program is explained, with information on adaptive tools, design considerations of outdoor settings, and sports rules adapted for people with disabilities.

Although space does not permit description of each program here, the types of recreation programs covered in the guidelines include tennis, horseback riding, golf, archery, bicycling, fishing, athletics, camping, beep baseball, canoeing, boating and yachting, water skiing, scuba diving, alpine skiing, cross-country skiing, ice sledge hockey, and gardening.

Considerations for Implementing Recreation Programs

Unlike the situation in the United States, where there is an established system to cultivate and certify recreation specialists, there are no professional recreation specialists to serve as program instructors in Japan. Although adoption of such a system may be needed in the future, it is planned to use volunteers who have special skills and experience, and proper volunteer training is to be conducted.

In the implementation of recreation programs, it is necessary to understand each participant's type of disability, physical abilities, and any special considerations that may be needed. In the case of sports programs, health conditions should be checked in advance. Medical equipment for emergencies, medical volunteers, and means of communication with emergency hospitals may also be required. Adaptive tools for sports and gardening should be prepared depending on the physical abilities of the participants.

21.7   GENERAL LESSONS LEARNED FOR UNIVERSAL DESIGN

In 1995, when the concept of universal design was not well known in Japan, the Urban Parks Access Committee was established by the Ministry of Construction to develop the guidelines formalized in *Japanese Guidelines for Universal Design in Parks*. Park users with disabilities were included among the committee members, and several meetings were held for input of new ideas and suggestions. A number of opportunities were also provided to hear opinions from various user groups. As a result, it was recognized that the key consideration should be creating a park where all people with diverse abilities can enjoy activities and rest in a natural setting. As there is no specific goal for universal design, creating a universally designed park means providing diversity in design in order to allow people to make choices in the use of facilities depending on their individual abilities. Therefore, it is essential to make special efforts to provide a wider range of choices in order to maximize usability for all park users. This should be the spirit of creating people-friendly parks in the new century.

21.8   CONCLUSIONS

Japanese Guidelines for Universal Design in Parks (Ministry of Construction, 1999) is the first design guide for outdoor settings that incorporates the universal design concept in Japan. The goal of the guidelines is to provide all people with access to the attractions of outdoor settings, while balancing this with preservation in the creation of the park environment. In view of the expectation that universal design will be introduced in numerous settings in the twenty-first century, the guidelines include not only physical design specifications but also nonphysical aspects of design, such as the introduction of recreational programs targeting an integrated society, the ultimate objective of universal design. People have different abilities. Natural environments in the urban parks of Japan are diversified. When all people with and without disabilities are able to fully enjoy nature in urban parks, one can imagine the realization of the concept of a fully integrated society. It is hoped that the guidelines will help in the realization of such a universal society in Japan.

21.9   BIBLIOGRAPHY

Ministry of Construction, *Japanese Guidelines for Universal Design in Parks,* Ministry of Construction, Tokyo, Japan, 1999.

Miyake, Yasuyo A., Hajime Kameyama, and Yoshisuke Miyake, *Creating People-Friendly Parks: From Barrier-Free to Universal Design,* Kajima Institute Publishing, Tokyo, Japan, 1996.

PLAE, Inc., *Universal Access to Outdoor Recreation: A Design Guide,* PLAE, Inc., Berkeley, CA, 1993.

Recreation Access Advisory Committee, *Recommendations for Accessibility Guidelines: Recreational Facilities and Outdoor Developed Areas,* Access Board, Washington, DC, 1994.

21.10   RESOURCE

Parks and Greens Division, City Bureau
Ministry of Construction
2-1-3 Kasumigaseki
Chiyoda-ku, Tokyo 100-8944, Japan
Fax: 03-5251-1916

CHAPTER 22

DESIGN FOR DEMENTIA: CHALLENGES AND LESSONS FOR UNIVERSAL DESIGN

Margaret Calkins, Ph.D.
IDEAS, Kirtland, Ohio

Jon A. Sanford, M.Arch.
Atlanta VAMC, Decatur, Georgia

Mark A. Proffitt, M.Arch.
Dorsky, Hodgson & Partners, Cleveland, Ohio

22.1   INTRODUCTION

Universal design has most commonly been applied in connection with physical or sensory impairments, and thus, at least in practice, does not specifically address the needs of individuals with significant cognitive impairments. Yet there are increasing numbers of people who suffer from cognitive impairment who could also benefit from having environments be more usable. The past decade has seen a dramatic increase in the understanding of several basic principles that can help to create more supportive environments for people with dementia. This chapter explores these principles and examines the extent to which they are congruent with or contradict universal design principles. It provides guidelines that expand the utility of the Principles of Universal Design (Center for Universal Design, 1997) for people with deteriorating cognitive abilities.

22.2   BACKGROUND

Universal design is "an approach to creating environments and products that are usable by all people to the greatest extent possible" (Mace, 1991, in Welch, 1995). While the concept of universal design is laudable, unfortunately the application of universal design principles has not fulfilled the potential of this definition. The vast majority of universal design applications have focused on people with physical impairments—that is, on issues that affect mobility, reach, and motor skills. Wide doorways, level thresholds, and lever handles, along with the popular Good Grips kitchen utensils, rocker light switches, and curb cuts are all hailed as best-practice exemplars of universal design. While these features certainly support the Principles of Universal Design, they are only pieces of the puzzle, or elements in a larger environment. They are not the totality of universal design. As Welch (1995) explains, "An accessible building implies that a person using a wheelchair can get into the building, but the notion that the building is convenient to public transportation, has an easily located front door, and provides good directions for wayfinding is not usually part of the image of accessibility that comes to mind for designers. These features, however, are the essence of a universal design approach."

If universal design is to achieve its goal of making settings that are usable by all people, it must go beyond simply or primarily eliminating physical barriers for people with sensory and physical impairments. It must also address the cognitive barriers that affect the decision-making processes of individuals with memory impairments. This chapter explores the extent to which universal design principles have been—and should be—applied to people with Alzheimer's disease or related dementias. It is worth noting that much of the work on creating supportive settings for people with dementia has focused on group residential settings: nursing homes and assisted living centers. While the majority of people with dementia live at home in the community, they also account for between 40 and 70 percent of all residents in long-term care. The term *long-term care* in this chapter refers to any group residential setting in which individuals are expected to reside for significant periods of time, and includes nursing homes, assisted living, group homes, and the like. Hence, much of the information in this chapter refers to long-term care settings, in which an organization must make decisions that address both the organization's needs and responsibilities as a care provider and the needs and desires of the people it cares for.

Within this framework, the basic universal design principle of making products and spaces more usable by people, specifically people with cognitive impairments, is only part of the problem. In fact, the basic underlying assumption of universal design, that design should be independently usable to the greatest extent possible, must be carefully examined, and even questioned, when being applied to settings which include people with cognitive impairments. This leads to two questions. First, one must question to universal design or not to universal design? In settings used by people with cognitive impairments, there may be situations where the goal is not to promote and facilitate access and use by the individual, but to actually minimize or restrict access and use. For example, access to the outside might be restricted to prevent a person from wandering away. In such situations, the design must be usable by people with dementia to the least extent possible. Second, universal design for whom? Universal design must recognize and accept that in some situations, the goal needs to be to create settings that support the provision of assistance by a caregiver. For example, extra space around the toilet and movable grab bars to facilitate an assisted transfer may be more desirable for some people with dementia than a toilet designed to help someone transfer independently. As a result, designs must be usable by caregivers to the greatest extent possible.

This chapter examines the basic universal design principles as they do or do not apply to design for people with cognitive impairments—specifically people with Alzheimer's disease and related dementias. The goal in exploring this issue is to broaden the range of populations universal design is typically applied to. If one thinks of the total population as being represented by a bell curve, traditional design tends to focus on the middle eightieth percentile— what is thought of as "normal" individuals (Fig. 22.1). Universal design and universal design principles can be seen as a way to increase the portion of the bell curve that is well served by design, up to approximately the ninetieth percentile. This chapter suggests that by incorporating some lessons learned from dementia design into universal design, it may be possible to increase the percentile of people who are well served by the designed environment even further.

This chapter begins with a brief overview of the cognitive changes and behavioral implications that are associated with dementia as well as how and why design affects these behaviors. This is followed by a discussion of basic principles of design for dementia. Finally, examples of typical applications of universal design principles are examined for their applicability to people with dementia. Explanations for why certain principles may need to be considered in a dif-

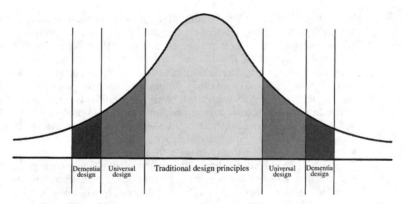

FIGURE 22.1 Expanding the population served by universal design principles.

ferent light, or why a different principle may be more applicable, are provided. When possible, specific research studies are cited. Unfortunately, environmental research is sparse for both universal design and design for dementia, and thus recommendations are often based on anecdotal evidence or clinical knowledge of the populations being considered.

22.3 UNDERSTANDING PEOPLE WITH DEMENTIA

Alzheimer's disease is a progressive, irreversible disease, which gradually impairs the cognitive intelligence and functional abilities of oftentimes otherwise healthy individuals to the point that they require total and complete care by others. Although Alzheimer's disease is only one of over 100 forms of dementia, it accounts for over 50 percent of all cases of dementia (U.S. Department of Health and Human Services, 1987). The Alzheimer's Association estimates that 4 million Americans have Alzheimer's disease, and that this may rise to 14 million by 2050 (Alzheimer's Association, 1998).

The decline which results from dementia is global, affecting all areas of one's functioning. In the early stages, losses are most noticeable in complex tasks, such as balancing a checkbook; planning an event, such as a dinner party or a meeting; or working at a demanding job. As the dementia progresses, and the neuropathological damage becomes more pervasive, it becomes increasingly difficult for afflicted individuals to make sense of the world around them. They may not recognize a spouse of 40 years, understand how to use a knife and fork, or be able to discern the potential hazards present in the typical kitchen.

It is important to note that, given that the majority of people with dementia are older, these cognitive changes are most often experienced in addition to the normal physical and sensory changes that occur with age—decreased mobility, limited reach, reduced range of motion, difficulty sitting down and standing up, diminished visual acuity, altered perception of colors, increased accommodation time, and differentially diminished hearing. These physical and sensory changes can greatly exacerbate the problems someone with dementia experiences in trying to function in a world of increasing environmental demands.

Environmental Models

There have been several attempts to develop models to increase our understanding of many of the behaviors that are typically seen as difficult in this population. The best-known model is

Lawton and Nahemow's environmental docility hypothesis (Lawton, 1975; Lawton and Nahemow, 1973), which suggests that each individual has certain levels of competence, and that negative consequences occur when the environment (broadly defined) exerts more press than an individual can manage. In addition, Lawton and Nahemow suggest that as individual competence declines, the environment assumes increasing importance in determining well-being.

Hall and Buckwalter's (1987) progressive lowered stress threshold (PLST) model is a reformulation of this, focusing exclusively on cognitively impaired individuals; it suggests that as the dementia progresses, the individual's threshold or ability to receive and interpret stimulation from the environment decreases—that is, what Lawton and Nahemow refer to as *environmental press*. The point at which press exceeds competence is called the *stress threshold*. When a person crosses this threshold, dysfunctional behavior or catastrophic reactions occur.

Hall and Buckwalter submit that a person with dementia experiences stress when coping with the demands of the environment, and that this stress accumulates, pushing the individual over the threshold, which results in the manifestation of a variety of negative and catastrophic behaviors unless specific interventions are made to reduce the stress level (see Fig. 22.2). Thus, according to Hall and Buckwalter, the dynamics of the care setting, particularly those aspects of the setting related to stimulation, play a critical role in influencing the behavior of individuals with dementia.

Although both environmental models focus on the amount of stimulation in the environment, the major difference between the two is that Lawton and Nahemow's model argues that understimulation can also have negative consequences, while the PLST model does not suggest any negative consequences of too little stimulation. As a result, the latter appears to argue for environments that have less stimulation, and consequently less press, as people continue to deteriorate.

Information-Processing Theories

While the theories of Lawton and Nahemow (1973) and Hall and Buckwalter (1987) provide interesting rationales for explaining how the environment affects the behavior of people with dementia, they provide only limited insight into why the environment causes these behaviors to occur. Such an explanation is critical in developing design principles for people with dementia and in understanding the impact of universal design on this population.

There is a substantial body of research that repeatedly demonstrates that elderly people perform less well on a variety of cognitive tasks than their younger counterparts (Baltes and Dixon, 1986; Baltes and Kliegl, 1992; Bandura, 1977; Hess and Tate, 1991; Lachman, 1983;

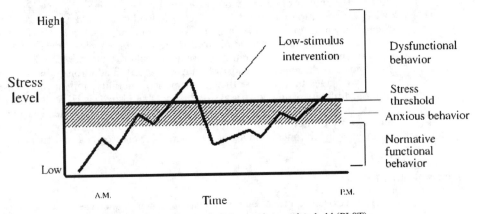

FIGURE 22.2   Hall and Buckwalter's progressively lowered stress threshold (PLST).

Salthouse and Babcock, 1991; Salthouse, Babcock, and Shaw, 1991; Salthouse, 1990). Much of the early research in memory was based on a linear stimulus-response information-processing model in which information is thought to be stored first in an *iconic store,* which holds this information for less than a second. Information is then transferred to *short-term memory,* where it is available to be used. Some information may then be transferred to *long-term memory,* while the rest is lost (Schlesinger and Groves, 1976).

An expansion of the information-processing theory suggests that there is a separate memory function, called *working memory,* which is different from the other forms of memory because of the assumption that it reflects both processing and storage (Salthouse and Babcock, 1991; Salthouse, Babcock, and Shaw, 1991). It is hypothesized that the working memory is primarily responsible for age-related changes. In this extension of the theory, working memory is composed of structural components and operational components. *Structural capacity* is the number of distinct items of information that can be remembered at one time, while *operational capacity* is the number of processing operations that can be performed while still preserving the product of earlier operations. Research based on this theory suggests that older people have a decreased ability to simultaneously store and process information (Kausler, 1992). Applying this theory to people with dementia, it appears that the greatest problems are likely to be encountered when the environment requires the individual to both rely on memory and process simultaneous and competing demands—that is, sources of environmental press.

An alternative theory of memory is the *neural network theory.* This theory is based on a framework developed by Cerella (1990) and Greene (1983), which represents the brain as a series of linked neural networks. Grossly simplified, this theory suggests that the cognitive process is the propagation of a signal (i.e., information) from input to output. This signal travels along a series of short neurons, which transmit to each other to pass the signal (information) along. Each step takes a certain amount of time. If there is a breakdown in the order of the steps (i.e., one of the neurons breaks), then the signal must find a new path, which involves additional steps, taking additional time. This accounts for the research results that suggest that aging people require increased processing time. The theory also suggests that the number of neuronal paths a bit of information travels depends on how many other bits of information it is related to: information with many associations will travel along more paths than information with fewer associations.

Beyond explaining differences in processing time for normal elderly people, the neural network theory provides a useful model for exploring information processing in individuals with dementia. The neural process described here applies equally well to both encoding and decoding information. In the encoding process, as related bits of information are repeatedly presented (i.e., someone becomes more familiar with something), new paths are created from stimulus input to where it is encoded in the brain. Thus, when trying to recall this information, there are many different paths that can be called upon. Hence, the breaking of some portion of these neural networks may not cause a significant delay or impairment, because alternative paths are readily available. This may explain, in part, why long-term memory is less impaired in people with dementia than is short-term memory (Mace and Rabins, 1981). Long-term memories (information) have been processed (i.e., recalled) so many times over the course of the lifetime that there are virtually an unlimited number of paths that lead to this memory (information). However, recent memories, such as what was eaten for breakfast, may be associated with significantly fewer neural network paths, and thus are harder to recall. (In Chap. 8, "Universal Design to Support the Brain and Its Development," John Zeisel relates brain functions with environmental characteristics to support people with Alzheimer's disease.)

22.4   *DESIGNING FOR DEMENTIA*

While the models of memory and cognitive processing described in the preceding section provide some insight into the cognitive changes associated with dementia, their relationship to specific strategies for creating supportive environments is less clear. The basic premise of environment-based strategies is that achieving a proper environmental fit can optimize the abilities, health, and morale of the person with dementia. Thus, the environment is used as a therapeutic tool to achieve person-based goals. Since the manifestation of dementia in each person is unique and changes over time, design principles often are applied using multiple strategies that overlap and reinforce other principles.

The following sections identify a set of seven goals for creating successful settings for people with dementia and five basic principles for using the environment to meet these goals that are based in models of memory and cognition. Each principle has design strategies that are presented as a series of design guidelines for structuring the environment.

Person-Centered Therapeutic Goals

Early design paradigms for people with dementia mostly focused on specific deficits—for example, wandering and unauthorized or unattended exiting from the care setting, agitation, combativeness, and disorientation. More recent work, however, has taken a more positive or therapeutic approach. Over the course of the past two decades, a number of researchers and other experts have recommended their own sets of "therapeutic" person-based goals (Calkins, 1988; Cohen and Weisman, 1991; Lawton, 1983; Regnier, 1997; Zeisel, Hyde, and Levkoff, 1994) that serve as directives for creating a successful setting for people with dementia. Despite differences in how and when these various sets of goals and directives were developed, there is a significant amount of congruence between them (see Weisman and Calkins, 1994, for more details). For the purposes of this chapter, the following therapeutic, person-based goals, which are common among the existing recommendations, are used:

Awareness and Orientation

Safety and Security

Privacy

Functional Abilities

Personal Control

Continuity of Self

Social Contact

Therapeutic Goals and the Environment

The environment can play a major role in achieving these person-based goals as well as in supporting caregivers and family members while they provide care. Due to the progressive nature of most forms of irreversible dementia, frequent assessments should be conducted to ensure that the environment continues to support the individual and his or her caregiver. Accordingly, some goals may be more salient at certain times than other goals. In the early stages of dementia a caregiver may be more concerned with promoting functional abilities, while in latter stages maximizing safety and security may take a higher precedence. The progressive nature of this set of diseases also means that regardless of what environmental supports are provided, at some point caregiver support will be required, and thus the needs of the patient/caregiver dyad should also be considered. Each of the person-based goals is manifest in different environmental interventions.

Awareness and Orientation.   This goal relates to the ability to know where one is in relation to where one wants to go. It is affected by the structure of the environment (e.g., overall building layout and direct views to destinations), as well as by the placement, design and use of signage, landmarks and other cues.

Safety and Security.   These are fundamental concerns in any dementia care setting since cognitive and physical impairments can make any environment hazardous. Features such as rounded edges, level surfaces, appropriate mobility, and balance supports impact basic physical functioning, while automatic shut-off or power control devices (e.g., on kitchen appliances), security systems, and wander-monitoring systems address changes in cognitive functioning. Moreover, as dementia progresses, concern about safety often results in greater levels of caregiver support. For example, the environment may be used to create additional barriers or restraints that limit access, such as the use of an 8-ft-high fence to create a secure courtyard.

Privacy.   This relates to being able to control interactions with others. One environmental application of this goal is the provision of private rooms in a sheltered care setting. Organizational policies can also reinforce these goals by requiring staff members to knock before entering the room.

Functional Abilities.   People with dementia typically lose the ability to complete complex functional tasks, including most self-care or activities of daily living. Supporting functional abilities in people with dementia requires a competence-inducing environment and social supports that promote the continued use of everyday skills. For example, the labeling of drawers may provide enough support for some residents to independently find clothing. Yet, as cognitive skills diminish, providing environmental support for caregiver assistance may be more critical (e.g., having adequate space in the toilet room for a two-person assisted transfer from a wheelchair to a toilet).

Personal Control.   This goal is highly prized and desired by most individuals, yet is something that is often taken away from people with dementia without serious consideration of the implications or of alternative options. Opportunities for personal control are related to how an environment offers options and the degree to which policies allow a person to make meaningful choices. Having a variety of social spaces is one environmental example of promoting this goal. However, promoting independent autonomy may eventually be reduced when the person with dementia is unable to make wise choices.

Continuity of Self.   This goal compensates for the progressive erosion of personal identity, one's knowledge of self, and self-memory associated with the disease. Environmental applications of this goal promote the concept that familiar places and environments that contain elements from a person's past provide appropriate contexts for behavior. Providing shelves, bulletin boards, and other display features in each resident's room reinforces the goal of continuity of self.

Social Contact.   With the loss of executive function and some basic language skills, many people with dementia experience a decreased ability to initiate and sustain interactions with others. This goal focuses on the extent to which the environment supports social interaction. One example of applying this goal is the provision of small seating alcoves in the care setting that promote one-on-one conversations.

22.5   DESIGN PRINCIPLES FOR DEMENTIA

These therapeutic, person-based goals are manifest in five basic design principles that relate specifically to the symptoms of dementia: Minimal Negative Stimulation, Maximal Positive Stimulation, Familiarity, Continuity, and Regulated Access. These principles have made the most significant impact on the design of therapeutic dementia care settings and have been applied using a variety of design guidelines. The application of these guidelines frequently correlates to a number of the person-based goals.

Minimal Negative Stimulation

The extent to which the environment is free of noxious or competing stimuli is *Minimal Negative Stimulation*. Reducing negative stimulation is based on the ecological model that establishes a relationship between competence and the demands or press of the environment (Lawton, 1975). Stimulation is negative if it causes stress or competes with other more relevant stimuli. There are four design guidelines that address negative stimulation: reducing scale, controlling ambient conditions, limiting unnecessary choices, and providing a place for retreat.

Reducing Scale. Reducing the scale of the environment through controlling size has been found to have a positive impact on people with dementia (Day, Carreon, and Stump, in press). This concept has been applied through segregating small care units and creating small group dining and activity areas. Reducing the scale of the environment relates to the goals of Awareness and Orientation, Functional Abilities, Social Contact, and Safety and Security.

Controlling Ambient Conditions. This guideline addresses the removal of unnecessary auditory and visual background stimuli that create distractions, including overhead paging, loud televisions, glare, and busy wallpaper patterns. Although little research has been done in this area, Namazi and Johnson (1992b) found that people with dementia had an increased ability to focus on tasks when visual distractions were reduced in activity rooms. Controlling ambient conditions relates to the goals of Awareness and Orientation, Functional Abilities, and Social Contact.

Limiting Unnecessary Choices. This guideline entails reducing the complexity of decision making by limiting, but not eliminating, choices. Too many choices may create confusion and cause people with dementia to be more dependent than they actually should be. For example, people with dementia have been found to be able to select clothing and dress with increased independence when closets are modified to expose only one or two garments in the sequential order a person typically dresses (Namazi and Johnson, 1992a). However, providing no choice may also cause negative affect. For example, Sloane et al. (1995) found that the provision of a handheld shower in addition to a standard showerhead reduced combativeness during assisted bathing because it enabled an individual to choose when his or her hair got wet. Functional Abilities and Personal Control are the key goals that relate to this guideline.

Providing a Place for Retreat. If a person with dementia is feeling overwhelmed, the best option may be to provide a place for retreat, such as a quiet room or secure outdoor area. For example, Zeisel (1998) found significant decreases in occurrences of agitation when residents of a dementia special care unit were allowed access to secure outdoor areas. Retreat spaces promote the goals of Privacy and Personal Control.

Maximal Positive Stimulation

The extent to which the environment provides positive, engaging characteristics is *Maximal Positive Stimulation*. These engaging characteristics are intended to promote the morale, dignity, and health of the person with dementia. There are three design guidelines that address positive stimulation: augmentative cues, focusing, and nonstandardization.

Augmentative Cues. Promoting full sensory experiences through the use of augmentative cues encourages the use of situational environmental cues that are intended to orient people to time and place. For example, setting the dining room table is a visual cue that can be used to orient people with dementia to when and where a meal will occur. Similarly, striking a dinner gong is an auditory cue, and introducing the smell of fresh-baked bread is an olfactory cue. The use of augmentative cues can be especially important in settings where the same room is used for multiple functions. Functional Abilities is the key goal application for augmentative cues.

Focusing.   Focusing attention on the most salient environmental features is another important design guideline to maximize positive stimulation. For example, Namazi and Johnson (1991a) found that increasing the prominence of the toilet in residence rooms by removing the bathroom wall and using a dark background to contrast with the white toilet fixtures significantly increased continence levels among people with dementia in a special care unit. Attention can also be focused creating important central spaces in a dementia care facility. For example, arranging resident rooms around a large activity and social space focuses attention on the central space, encouraging participation in activities and socialization. This type of building arrangement was associated with increased time spent in social spaces and increased focus on the activity at the Weiss Institute in Philadelphia (Lawton, Fulcomer, and Kleban, 1984). Finally, focusing attention can be used to highlight important design characteristics. For example, contrasting colors are often used on residence room doors to make the doors more visually prominent. Functional Abilities and Social Contact are the primary goals to which this goal relates.

Nonstandardization.   The use of nonstandardized environmental elements promotes differentiation of like design features and avoidance of repetitive architectural forms and spaces. For example, Cohen and Weisman (1991) encourage the use of bright colors, landmarks, and prominent signs to differentiate between areas of a dementia care facility and promote wayfinding. Nonstandardization applies to the goals of Awareness and Orientation, Personal Control, Continuity of Self, and Functional Abilities.

Familiarity

The extent to which the environment facilitates and replicates typical life patterns is *Familiarity*. Familiarity is based upon the theory that information in long-term memory may be easier to recall, due to the greater number of network paths associated with these memories. Therefore, the more familiar an environment is, the more supportive it will be. This principle has typically been applied through three design guidelines: use of noninstitutional imagery, provision of a hierarchical spatial continuum from public to private, and reduced learning of new concepts.

Use of Noninstitutional Imagery.   According to Goffman (1961), institutional environments are settings in which people live separated from society, under the control of supervisory staff, and in which the realms of sleep, play, and work are combined in one place. Clearly, these types of settings are not where most people spend their lives. Therefore, institutional imagery is unlikely to be either familiar or therapeutic for most people with dementia. The provision of familiar domestic imagery is the most common application of this design guideline (Cohen and Weisman, 1991). Day, Carreon, and Stump (2000) found a variety of studies that showed that noninstitutional characteristics improved residents' agitation levels and pleasure, as well as their functional abilities. Noninstitutional imagery relates to the goals of Continuity of Self and Awareness and Orientation.

Provision of a Hierarchical Spatial Continuum from Public to Private.   This guideline is closely linked to domestic imagery. Most homes are set up with a progression from public zones to private zones; that is, the entry is near the public living room, and private areas, such as bedrooms, are deeper in the home. In contrast, many care settings are designed with entries onto private bedroom wings, with public rooms being located at the end of the hall. This arrangement is unfamiliar and can cause confusion. Accordingly, building designs for care settings that replicate the arrangement of a house are hypothesized to be more supportive of person-based goals. Another application of this guideline is the use of private rooms or the design of shared rooms that promote territoriality. Most people do not live in a communal setting. Therefore, private rooms are considered more familiar and supportive. Privacy, Personal Control, and Continuity of Self are the most salient goals for this guideline.

Reduced Learning of New Concepts.   Elements in the environment should not be foreign to a person's frame of reference. Examples include the use of signs and familiar symbols that are simple, clear, and large enough to see, such as using a picture of a toilet to assist people with dementia in finding the bathroom. Functional Abilities, Continuity of Self, and Awareness and Orientation are the key goals that relate to this guideline.

Continuity

The extent to which the environment supports past patterns in a person's life and provides places to permit expression of an individual's past is *Continuity*. This design principle strongly relates to the Continuity of Self goal and goes beyond just familiarity to design that relates directly to the individuals in the setting. This principle has been promoted through three design guidelines: personalization, recreating past patterns and places, and individualized cueing.

Personalization.   This guideline promotes using meaningful objects from an individual's past to encourage well-being and orientation. For example, personal objects placed in a display case have been found to promote orientation to the appropriate bedroom among people with moderate levels of dementia (Namazi, Rossner, and Rechlin, 1991). Continuity of Self, Personal Control, and Awareness and Orientation are the primary goals that correspond to this guideline.

Recreating Past Patterns and Places.   This guideline supports continuity by promoting the recreation of domestic, work, and leisure settings that occur in the everyday lives of individuals. For example, a therapeutic kitchen provides a setting for domestic activities such as baking or folding clothes for someone who spent a lot of time engaged in these activities. Alternatively, an office space might be meaningful to an individual whose memories of a past work role are still prominent. Similarly, a living room might be recreated to provide individuals with familiar places to socialize. Such spaces are postulated to allow people with dementia to fulfill their own personal behavior agendas, which may decrease agitation or wandering behaviors. The creation of these realms relates to the goals of Continuity of Self, Functional Abilities, and Social Contact.

Individualized Cueing.   Multisensory cues should also be based on past patterns and traditions which are familiar to individuals in the setting. One example is using the familiar pattern of smelling food before a meal to cue a person with dementia that it is time to eat. A person with dementia who never has an opportunity to smell food before it is served because it is covered on trays may have a difficult time understanding that it is time to eat. Therefore, preparing familiar foods and serving individual plates from steam tables located on the care unit can be effective in promoting appetites. Individual cues may also relate to using familiar traditions to orient people to special days, such as serving fish on Fridays. Awareness and Orientation, Continuity of Self, and Functional Abilities are the critical goals for this guideline.

Regulated Access

The degree to which the environment controls, limits, or facilitates the monitoring of access to potential hazards is *Regulated Access*. This principle relates to design features that promote the goal of Safety and Security. This principle has been applied using three design guidelines: selective access, shielding hazardous areas, and supervision.

Selective Access.   Access to hazardous elements and potential hazards are typically selectively provided based upon individuals' cognitive abilities. Examples include locking doors, using complicated latches, or using fences to create secure courtyards. As the disease progresses, a person with dementia may not be able to function independently without a caregiver's support. Accordingly, environmental features that facilitate supportive caregiving practices and promote the safety and health of people with dementia are significant aspects of this guideline. For example, Hutchinson, Leger-Krall, and Wilson (1996) found that continence levels increased with toilet rooms that accommodated the staff in providing assistance with a transfer. This guideline relates to the goals of Safety and Security and Functional Abilities.

Shielding Hazardous Areas.   This guideline typically capitalizes on an individual's cognitive deficits to create barriers to hazardous areas and items. For example, doors that lead to unsafe places are often camouflaged or disguised to discourage use, while those that lead to safe areas are highly differentiated to encourage use. Namazi, Rosner, and Calkins (1998) found that doorknobs that were hidden from view were effective in keeping residents from eloping from a care setting. Interestingly, disguising exit doors also has been found to reduce agitation in people with dementia (Zeisel, 1998). Shielding hazardous elements can also be accomplished by controlling use, such as providing a hidden gas control switch for a stove, or a timer to automatically shut off an appliance after a certain amount of time. Other examples of shielding hazardous elements include the use of environmental barriers to reduce mobility, such as the use of lowered beds to prevent people from rising independently and to reduce the impact of a fall. The critical goal for the application of this guideline is Safety and Security.

Supervision.   The arrangement of spaces and plans has been found to strongly inhibit or facilitate the ability of caregivers to monitor the caregiving setting (Morgan and Stewart, 1999). Supervision may also be applied through technology, such as the use of wander-monitoring systems. Safety and Security is the key goal that relates to this guideline.

22.6   COMPARISONS BETWEEN DEMENTIA DESIGN AND UNIVERSAL DESIGN PRINCIPLES

Having examined the design principles for people with dementia, it is clear that there are a number of basic conceptual similarities as well as differences between universal design and dementia design approaches. What are the implications of these differences for universal design? By focusing primarily on loss of cognitive functioning as opposed to physical and sensory challenges, dementia design can suggest new ways of interpreting and even expanding the universal design principles to make them more useful to designers. The last part of this chapter compares and contrasts the two design approaches in order to explore how universal design can learn from dementia design. In addition, specific suggestions will be made for ways in which universal design can be expanded to make it more dementia-friendly through the adoption of dementia design principles, guidelines, and design strategies.

The Principles of Universal Design were developed by a multidisciplinary team (including one of the authors of this chapter) at the Center for Universal Design, North Carolina State University (see Chap. 10, "Principles of Universal Design," by Story). These principles are intended to promote independence and ease of use for all people, regardless of ability or disability. The comparison between dementia design and universal design maintains a strict interpretation of the universal design principles to preserve their original intent. This highlights sometimes subtle yet important differences between the two and permits the development of a broader, more holistic understanding of universal design.

A summary of comparisons between the two sets of principles is presented in Table 22.1. Circles are used to indicate where there is an overlap between dementia design and universal design principles. A black circle indicates that a dementia design guideline is explicitly stated in the universal design principle; a gray circle indicates that the dementia design guideline is implicit within the universal design guideline, but the universal design guideline has not generally been interpreted in that manner; and a white circle indicates that one or more of the universal design guidelines appears to specifically contradict the dementia design guideline. Whereas design guidance for dementia that is explicitly stated in the universal design principles is useful, clearly, the important insights for reinterpreting, redefining, and expanding the universal design principles will result from the differences between the two sets of principles.

Minimal Negative Stimulation

Of the four dementia design guidelines to minimize negative stimulation—reduce scale, provide a place for retreat, control ambient conditions, and limit choices—universal design principles are at least partially supportive of the first three.

Explicitly Stated in Universal Design Principles.   Principle 3, Simple and Intuitive Use, advocates design that is easy to understand and minimizes complexity. This understanding of the universal design principle supports the guideline to control ambient conditions by promoting the removal of unnecessary stimuli that could add complexity and confusion. Specifically, minimizing visual complexity might entail eliminating extraneous signage (Fig. 22.3) or eliminating design elements that obscure important information, as Namazi and Johnson (1991a) demonstrated by the reduction in continence when a curtain blocked a toilet from the view of residents with dementia (Figs. 22.4 and 22.5).

Implicit in Universal Design Principles.   Principle 7, Adequate Size and Space for Use, is necessary to reduce scale and provide places for retreat. The intent of this universal design principle is to ensure that enough space is provided for use (e.g., Fig. 22.6 illustrates a clear floor space with a 5-ft radius, which is needed to turn a manual wheelchair around in a room). However, the dementia design principle suggests that other attributes of the space, not merely how large it is, are also important in promoting use. The universal design principle, as it is written and typically applied, does not address these other attributes. Thus, when smaller spaces are appropriate to reduce scale, or to provide retreat, the universal design principle really does not apply. For example, research shows that when Creekview, at Evergreen Retirement Community in Oshkosh, Wisconsin, added a new unit designed as four households for nine residents each (Fig. 22.7), residents with dementia exhibited fewer agitated behaviors than did those who lived in a larger, traditional 36-bed unit (Fig. 22.8) (Calkins, Meehan, and Lipstreuer, 1999).

Contradiction Between Universal Design and Dementia Design Principles.   Both universal design Principle 1 (Equitable Use) and Principle 4 (Perceptible Information) potentially increase rather than limit choice. As a result, these two principles conflict with this dementia design guideline. Although choice and redundancy are major themes of universal design, and are central to accommodating use of products and environments by people with a wide range of abilities, the provision of choices can potentially be confusing to individuals who are cognitively impaired and who have difficulty discriminating between alternatives. It should be noted, however, that Equitable Use would be appropriate for dementia design when only one, rather than equivalent choices, is provided for all users. For example, a gently sloped pathway that can be used by everyone entering a building (Fig. 22.9) may not be as confusing to a person with dementia as the equivalent alternatives of both stairs and a ramp.

TABLE 22.1 Relationship Between Universal Design and Dementia Design Principles

Universal design principles and guidelines

Dementia design, principles, and guidelines	Equitable Use: Provide same means of use for all users—identical when possible, equivalent when not.	Flexibility in Use: Provide choice in methods of use; accommodate right- or left-handed use; facilitate user's accuracy and precision.	Simple and Intuitive Use: Eliminate complexity; be consistent with expectations and intuition; provide effective prompting of sequential actions and timely feedback.	Perceptible Information: Use redundant presentation; contrast between information and surroundings; maximize legibility.	Tolerance for Error: Make most used elements most accessible; isolate, eliminate, or shield hazards; provide warnings of hazards; discourage unconscious action.	Low Physical Effort: Allow user to maintain natural body position; use reasonable operating forces; minimize repetitive actions and sustained physical effort.	Size and Space for Approach and Use: Provide adequate space for approach, reach, manipulation, use, and assistance; provide clear line of sight; accommodate variations in hand and grip size.
Minimal Negative Stimulation							
Reduce scale.							●
Control ambient conditions.							
Limit choices.	○		●	○			
Provide place for retreat.							●
Maximal Positive Stimulation							
Augmentative cues.				●			
Focusing.				●			
Non-standardization.				●			
Familiarity							
Noninstitutional imagery.			●				
Public to private continuum.			●				
Reduced learning.	○	○	○			○	
Continuity							
Personalization.	●						
Familiar patterns.	●						
Individualized cueing.	●	◐					
Regulated Access							
Selective access.	○		○	○	●	○	○
Shielded hazardous elements.	○	○	○	○	●	◐	○
Supervision.	○	○	○	○	●	◐	◐

Note: ● = Explicitly stated in universal design principle; ◐ = implicit in or compatible with universal design principle; ○ = contradiction between universal design and dementia design principles.

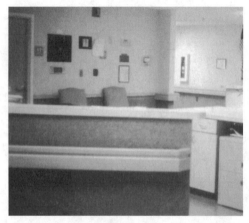

FIGURE 22.3 Too much signage can be confusing.

FIGURE 22.4 Toilet at Corinne Dolan Center, curtain closed.

FIGURE 22.5 Toilet at Corinne Dolan Center, curtain open.

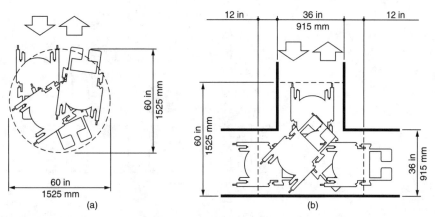

FIGURE 22.6 Example of a 5-ft turning radius: (*a*) 60-in (1525-mm)-diameter space, and (*b*) T-shaped space for 180° turns.

Similarly, the provision of redundant information, such as the use of a text sign with the word MEN as well as icons of a man and an international symbol of accessibility on a men's room door, may be too much information for an individual with dementia to actually perceive.

Maximal Positive Stimulation

Principle 4 (Perceptible Information) is the only universal design principle that directly impacts positive stimulation.

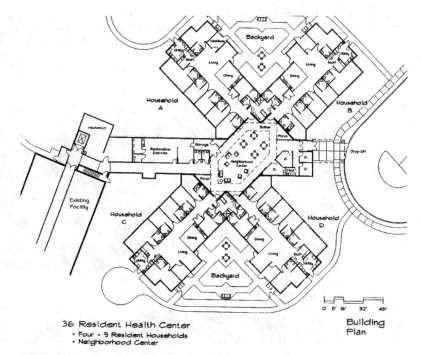

FIGURE 22.7 Household plan of smaller-scale spaces.

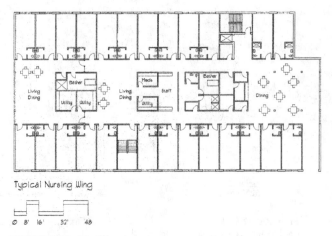

Typical Nursing Wing

Ø 8' 16'   32'   48

FIGURE 22.8   Traditional nursing home unit.

Explicitly Stated in Universal Design Principles.   Principle 4 (Perceptible Information) directly supports focusing attention. Adhering to the guidelines of heightened contrast between information and surrounding and maximizing legibility, Principle 4 can be used to focus attention on essential information while extraneous information is hidden from view. For example, Namazi and Johnson (1991b) found that colored, highly contrasting arrows on the floor of a dementia care unit (Fig. 22.10) were effective in focusing the attention of residents so that they could locate the restrooms.

Implicit in Universal Design Principles.   Principle 4 (Perceptible Information) does not explicitly refer to the dementia design guidelines of providing augmentative cues and non-standardization. Rather, these dementia design guidelines are implicit in an expanded definition of this universal design principle. First, augmentative cueing is similar to the universal design concept of redundant cueing, but augmentative cueing goes a step further by provid-

FIGURE 22.9   Sloped walkway provides the same means of access for everyone.

FIGURE 22.10   Graphics on the floor at Corinne Dolan Center.

ing additional, often multisensory and noncompeting information that enhances information provided by other means. Thus, rather than potentially increasing negative stimulation, augmentative cues can en-hance positive stimulation by providing supplemental and complementary information to communicate to an individual with dementia. For example, by placing personal memorabilia in a trophy case beside an individual's room to augment an individual's name on the bedroom door, Namazi, Rosner, and Rechlin (1991) found that more people with dementia were able to locate their own rooms (Fig. 22.11).

Second, although there is no specific mention of nonstandardization in the universal design guidelines for Perceptible Information, this principle is relevant because the more unique a design element is, the more perceptible it will be. This is particularly important for people with dementia who have difficulty differentiating between the same element used repeatedly, such as doors to rooms. Although not empirically studied, many dementia care facilities paint doors or door frames different colors, with the hope that at least some residents will be able to use this differentiation to help them find their rooms independently. Thus, even though Perceptible Information does not explicitly cover this dementia design guideline, a broader interpretation illustrates the connection between the two.

Contradiction Between Universal Design and Dementia Design Principles.   None of the universal design principles contradict Maximal Positive Stimulation.

Familiarity

Familiarity is perhaps the most important dementia design principle, as it underlies many of the other guidelines and directly addresses the use of environments and products. However, none of the universal design principles specifically attributes ease of use with familiarity.

Explicit in Universal Design Principles.   Although four of the universal design principles, including Principle 1 (Equitable Use), Principle 2 (Flexibility in Use), Principle 3 (Simple and Intuitive Use), and Principle 6 (Low Physical Effort), directly address use of the environment and products, none of these principles explicitly attribute use to familiarity.

Implicit in Universal Design Principles.   Principle 3 (Simple and Intuitive Use) promotes design that is consistent with users' expectations. As noninstitutional imagery and a public-to-private continuum are design practices that meet expectations of most users, Simple and Intuitive Use can be interpreted as at least implicitly supporting the dementia design principle of Familiarity. For example, noninstitutional elements and a progression of spaces from public to private in an institutional setting may be consistent with experiences in an individual's own home as well as in the homes of others.

Contradiction Between Universal Design and Dementia Design Principles.   Four of the universal design principles, including Principle 1 (Equitable Use), Principle 2 (Flexibility in Use), Principle 3 (Simple and Intuitive Use), and Principle 6 (Low Physical Effort), potentially contradict the principle of Familiarity. As previously described, design that is consistent with users' expectations is compatible with familiarity. However, being consistent with users' expectations may only be applicable to sensory input, not to intuition and rationalization. In other words, if

FIGURE 22.11 Display case at bedroom entrances at Corinne Dolan Center.

it looks like a telephone (e.g., black with a rotary dial), sounds like a telephone (e.g., has a loud ring and clicks when the dial rotates), and feels like a telephone (e.g., finger moves the dial in a clockwise direction), it is more likely to be recognized and understood as a telephone and used appropriately. Therefore, when Simple and Intuitive Use relies on rational expectation (i.e., when familiar design elements are used in new ways), the result may be novel to a person with dementia even when the elements are consistent with how they would be expected to work. For example, it may be difficult for a person with dementia to associate a numbered photograph on a telephone with the corresponding numbered button that is used to automatically dial the person in the photograph.

Three other universal design principles—Principle 1 (Equitable Use), Principle 2 (Flexibility in Use), and Principle 6 (Low Physical Effort)—may also promote novelty even though they are intended to facilitate use. In fact, providing flexibility and minimizing the amount of strength and precision required to physically use a design are qualities that are highly desirable for frail older individuals. However, when learning is necessary for a design to be used by an individual with dementia, it may be more difficult to use than one that requires greater effort, but is familiar. For example, caregivers have reported that some people with dementia have a difficult time regulating the temperature of water when using a single-lever faucet (Olsen et al., 1993). Although the single-lever faucet requires less strength and precision to operate than separate hot and cold handles, the lack of familiarity with this type of faucet makes it more difficult to understand than ones with which people with dementia are familiar.

Continuity

Whereas many of the dementia design principles are relevant to individuals with other cognitive or learning disabilities, Continuity of Self is a strategy unique to dementia design that seeks to enhance functioning through environments that have meaning to a particular individual.

Explicitly Stated in Universal Design Principles.   None of the universal design principles specifically address the issue of continuity. This is not surprising, since universal design is design for everyone, and continuity is person-specific design.

Implicit in Universal Design Principles.   At first glance, Principle 1 (Equitable Use), which advocates the same or equivalent means of use for all users, seems to be inconsistent with personalization and individualization to facilitate use. However, a broader interpretation of this universal design principle suggests that the same or equivalent design elements can be used to support continuity. The same display cases adjacent to bedroom doors in a dementia care facility can be personalized with memorabilia to assist individuals in finding their rooms (Namazi, Rosner, and Rechlin, 1991). In addition, equivalent design elements such as the recreation of an office for an individual who had been an accountant or a kitchen for someone who enjoyed being a homemaker can be used to promote continuity through the creation of familiar patterns and places.

Contradiction Between Universal Design and Dementia Design Principles.   None of the universal design principles contradict Continuity.

Regulated Access

In universal design, independence within the limits of an individual's physical abilities is the underlying presumption upon which all of the principles are based. In contrast, in dementia design, caregiver intervention, supervision, and assistance are increasingly necessary as the disease progresses even if an individual is physically able to complete tasks without assistance. For example, providing universally designed controls on a bathtub that require little effort to operate may enable or even encourage individuals to try to bathe themselves. However, at later stages of dementia, independent bathing is often discouraged due to the risk of falls and injury.

Explicitly Stated in Universal Design Principles.   None of the universal design principles explicitly address regulated access.

Implicit in Universal Design Principles.   Two of the universal design principles, Principle 5 (Tolerance for Error) and Principle 7 (Size and Space for Approach and Use), implicitly address regulated access. On paper, Tolerance for Error—which addresses cognitive processing and safety concerns through eliminating or shielding hazardous elements, providing warnings of hazards and errors, providing fail-safe features, and discouraging unconscious actions in tasks that require vigilance—would appear to be entirely supportive of promoting safe environments for people with dementia. However, the compatibility between safety concerns in the universal design and dementia design principles looks better on paper than it does in practice.

In practice, Principle 5 (Tolerance for Error) is inconsistent with dementia design for several reasons. First, it frequently relies on cognitive processes (e.g., such as clearly marking those features that can cause error), rather than on physical or sensory ones, to deter error. This strategy works only if an individual with dementia recognizes the demarcation (perhaps a skull and crossbones). Second, it presumes that error is inadvertent. For people with dementia, actions that cause error may be intended even if the error is unintended. For example, someone may leave the house or care setting to take a walk (intended) but not recognize the risk of getting lost (unintended consequence).

In addition, Tolerance for Error advocates that the most-used design features should be the most accessible. Unfortunately, many of the safety risks in dementia design arise from the design elements that are most accessible, and (based on the six other universal design principles) easiest to use. For example, the front door to a home is usually the most convenient, accessible, and easiest to use. It is also the one that leads directly to the street and provides the easiest opportunity for an individual to wander off and get lost.

To provide a safe environment for a person with dementia, it is likely that access to or through the front door will have to be regulated. Ironically, a switch-operated powered door opener, which requires low physical effort for an individual in a wheelchair, can be an effective means of regulating access for people with dementia, particularly if the switch is not highly visible. Alternatively, disguising the door so it is not obvious to a person with dementia (Fig. 22.12) can regulate access. Of course, the door could simply be locked and selective access permitted only when supervised by a caregiver.

Regulating access through supervision is also implicit in Principle 7 (Size and Space for Approach and Use). Although this universal design principle explicitly provides space for caregiver assistance, the need for and types of assistance necessary in dementia design differs considerably from that anticipated in universal design. Whereas universal design presumes that the need for assistance by adults with disabilities is relatively static over time, dementia design is based on the assumption that people with dementia have constantly changing needs for assistance. For example, someone with dementia who requires the assistance of only one caregiver to get out of a wheelchair and into a tub on one day might require the assistance of several people and a mechanical lift on another day. Clearly, this has great implications for the size and layout of the space.

FIGURE 22.12 Example of a disguised door.

Contradiction Between Universal Design and Dementia Design Principles. The changing concept of independence for people with dementia clearly affects the level of compatibility between universal design and dementia design. At the early stages of the disease, when facilitating independent use to accommodate an individual's physical frailty is important, the two sets of design criteria are quite compatible. However, at the later stages of dementia, all of the universal design principles except Principle 5 (Tolerance for Error) and Principle 7 (Size and Space for Approach and Use) become largely contradictory.

Lessons Learned for Universal Design

What are the practical applications of these lessons? There are several overarching principles from dementia design that impact all of the universal design principles. Most important, while safety is a major concern for all individuals, it is perhaps the major concern for people with dementia, who often are unable to make appropriate decisions. As a result, regulated access must be embedded in each of the universal design principles. To accomplish this, universal design must be broadened to recognize that for certain users, determining what is "usable to the greatest extent possible" is often not an individual decision. Rather, the decision is generally made by organizational policy makers, caregivers, and family members who determine when independent use is no longer safe. As a result, universal design must accommodate independent use when possible, but also reduce access and usability when necessary. Under these circumstances, design may, in fact, be more for caregivers' use than for care recipients'.

In addition, familiarity is extremely important. Everyone, not just people with dementia, become accustomed to doing things in certain ways. This is referred to as *procedural mem-*

TABLE 22.2 Recommendations for Expanding the Universal Design Principles

Universal design principle and guidelines	Additional guidelines	Strategies for implementation
Equitable Use Provide same means of use for all users—identical when possible, equivalent when not.	Avoid confusion. Promote continuity.	Limit options. Use design elements, patterns and cues that are expected to be familiar to users.
Flexibility in Use Provide choice in methods of use; accommodate right- or left-handed use; facilitate users accuracy and precision.	Avoid learning new concepts or means of use.	Limit choices; use design elements that are familiar to users.
Simple and Intuitive Use Eliminate complexity; be consistent with expectations and intuition; provide effective prompting of sequential actions and timely feedback.	Innovation and intuition are often incompatible.	Use expected imagery and standardized elements when possible. Use familiar and expected spatial continuum.
Perceptible Information Use redundant presentation; contrast between information and surroundings; maximize legibility.	Avoid confusion. Enhance distinctiveness of important cues. Reinforce important information.	Limit choices. Use nonstandardized design elements. Use noncompeting augmentative cues.
Tolerance for Error Make most used elements most accessible; isolate, eliminate, or shield hazards, provide warnings of hazards; discourage unconscious action.	Error is unpredictable and can occur at any time.	Accommodate selective access. Recognize that the most useable elements are potentially the most hazardous. Recognize that error may result from intended as well as unintended actions.
Low Physical Effort Allow user to maintain natural body position; use reasonable operating forces; minimize repetitive actions and sustained physical effort.	Minimize learning.	Use familiar behavior patterns when possible.
Size and Space for Approach and Use Provide adequate space for approach, reach, manipulation and use; provide clear line of sight; accommodate variations in hand and grip size.	Size should be appropriate for the type of use. Quality of space is as important as size.	Reduce scale when necessary; provide private places for retreat.

ory. While changing these routines may enhance ease of use, there may be benefits to working to maintain as much of the familiar patterns as possible—to rely on procedural memory. Therefore, it may be easier to partially modify how one does something rather than learning a whole new way of doing it. This particularly applies to products or aspects of design that might be used in an emergency, when procedural memory is much more likely to be relied upon. Procedural memory is also better preserved in dementia than are other aspects of memory. For example, to facilitate entering a building, universal design principles might suggest that a push-button automatic door opener be installed so individuals using wheelchairs or other mobility assistive devices do not have to hold the door open while pushing themselves through. People with dementia, however, might not understand that the way to open the door is to push a button, as this is not how they are used to opening doors. Indeed, this very principle is sometimes applied to keep people with dementia from leaving a place, typically a care center. Many facilities have installed systems that require pushing one or two buttons in order to open the doors, and have found that this keeps residents from leaving the unit unattended.

Finally, the comparison between dementia design and universal design suggests that there are specific recommendations for new guidelines and strategies that will expand the utility of the universal design principles to environments for people with dementia. These are listed in Table 22.2.

22.7 CONCLUSIONS

On the surface, it would seem that the basic dementia design principles are not that dissimilar from universal design principles. Minimizing Negative Stimulation, Enhancing Familiarity, and Continuity all seem to fit nicely with Tolerance for Error, Simple and Intuitive Use, Equitable Use, Perceptible Information, and so on. However, the overlap is much less than might be expected. There are only two relationships where the overlap is explicit: (1) Controlling Ambient Conditions/Simple and Intuitive Use and (2) Focusing Attention/Perceptible Information. In other cases, the relationship between the two sets of principles is only implicit, such as between the dementia design principle of Regulated Access and universal design Principle 5 (Tolerance for Error). Finally, dementia design and universal design appear to be contradictory in the majority of comparisons, such as the former recommending that choices be limited, whereas the latter proposes providing options.

The importance of these differences is not to illustrate that universal design is not really universal, but rather to broaden the universal design principles in an effort to make them more universal. Thus, lessons can be learned from dementia design. More important, those lessons have been applied to the universal design principles in the form of expanded interpretations, new guidelines, and proposed strategies to broaden their applicability and their relevance to new populations and environments.

This begs the question of whether it is appropriate to suggest changes to universal design for what is a relatively small group of people in a limited number of environments. The answer is that these revisions may be applicable not just in relation to people with dementia—they may be equally applicable to any person or population that is cognitively compromised. This includes young children whose cognitive abilities are not fully developed, people with developmental disabilities or mental retardation, and people with brain injuries. While it is not possible to explore specific principles for each of these populations here, it should be recognized that there is a much larger population that could benefit from these revisions.

Finally, there are times and situations in which anyone might find him- or herself cognitively challenged. In such cases, lessons learned from dementia design might benefit everyone. For example, in a busy airport the provision of Perceptible Information generally results in a plethora of signs overhead and on the walls. Sometimes it seems as if there are too many signs and too much information to sort through easily. In such situations, everyone might benefit from a little less sensory stimulation and a little better understanding of how environmental complexity impacts all users. For people with dementia, who find even simple environments confusing or unintelligible, dementia design is intended to reduce complexity and promote safety. The application of this concept to all environments for all users is the real lesson to be learned from design for dementia.

22.8 BIBLIOGRAPHY

Alzheimer's Association, *Breaking Through: Alzheimer's Association 1998 Research Report,* Alzheimer's Association, Chicago, 1998.

Baltes, P., and R. Dixon, "Multidisciplinary Propositions on the Development of Intelligence During Adulthood and Old Age," in A. B. Sorenson (ed.), *Human Development and the Life Course,* Lawrence Erlbaum Associates, Hillsdale, NJ, 1992, pp. 467–508.

———, and R. Kliegl. "Further Testing of Limits of Cognitive Plasticity in Old Age: Negative Age Differences in a Mnemonic Skill are Robust," *Developmental Psychology,* 28: 121–125, 1992.

Bandura, A., "Self-Efficacy: Toward a Unifying Theory of Behavioral Change," *Psychological Review,* 84(2): 191–215, 1977.

Calkins, M. P., *Design for Dementia: Planning Environments for the Elderly and the Confused,* National Health Publishing, Owings Mills, MD, 1988.

———, R. Meehan, and E. Lipstreuer, *Creekview: Its History and Evaluation: Final Report* [unpublished report], I.D.E.A.S., Inc., Kirtland, OH, 1999.

Center for Universal Design, *The Principles of Universal Design,* Center for Universal Design, North Carolina State University, Raleigh, NC, 1997.

Cerella, J., "Aging and Information Processing Rate," in J. Birren and W. K. Schaie (eds.), *Handbook of the Psychology of Aging,* vol. 3, Academic Press, San Diego, CA, 1990, pp. 200–221.

Cohen, U., and J. Weisman, *Holding on to Home,* Johns Hopkins University Press, Baltimore, MD, 1991.

Day, K., D. Carreon, and K. Stump, "The Therapeutic Design of Environments for People with Dementia: A Review of Empirical Literature," *The Gerontologist,* 4(4): 397–415, 2000.

Goffman, E., *Asylums: Essays on the Social Situation of Mental Patients and Other Inmates,* Anchor Books, Garden City, NY, 1961.

Greene, V. L., "Age Dynamic Models of Information-Processing Task Latency: A Theoretical Note," *Journal of Gerontology,* 38(1): 46–50, 1983.

Hall, G., and, K. Buckwalter, "Progressively Lowered Stress Threshold: A Conceptual Model for Care of Adults with Alzheimer's Disease," *Archives of Psychiatric Nursing,* 1(6): 399–406, 1987.

Hess, T., and C. Tate, "Adult Age Differences in Explanations and Memory for Behavioral Information," *Psychology and Aging* 6(1): 86–92, 1991.

Hutchinson, S., S. Leger-Krall, and H. Wilson, "Toileting: A Biobehavioral Challenge in Alzheimer's Dementia Care," *Journal of Gerontological Nursing,* (October): 18–27, 1996.

Kausler, D. H., "Intelligence," in *Experimental Psychology and Human Aging,* vol. 2, John Wiley & Sons, New York, 1992.

Lachman, M., "Perceptions of Intellectual Aging: Antecedent or Consequence of Intellectual Functioning?" *Developmental Psychology* 19(4): 482–498, 1983.

Lawton, M. P., "Competence, Environmental Press, and the Adaptation of Older People," in P. G. Windley and G. Ernst (eds.), *Theory Development in Environment and Aging,* Gerontological Society of America, Washington, DC, 1975, pp. 33–59.

———, "Environment and Other Determinants of Well-Being in Older People," *Gerontologist,* 18: 556–561, 1983.

———, and L. Nahemow, "Ecology and the Aging Process," in C. Eisdorfer and M. P. Lawton (eds.), *Psychology of Adult Development and Aging,* American Psychological Association, Washington, DC, 1973.

———, M. Fulcomer, and M. Kleban, "Architecture for the Mentally Impaired Elderly," *Environment and Behavior,* 16(6): 730–757, 1984.

Mace, N., and P. V. Rabins, *The 36-Hour Day,* Johns Hopkins University Press, Baltimore, MD, 1981.

Morgan, D. G., and N. J. Stewart, "The Physical Environment of Special Care Units: Needs of Residents with Dementia from Perspective of Staff and Caregivers," *Qualitative Health Research,* 9(1), 1999.

Namazi, K. H., and B. D. Johnson, "Environmental Effects on Incontinence Problems in Alzheimer's Disease Patients," *The American Journal of Alzheimer's Care and Related Disorders & Research,* 6(6): 16–21, 1991a.

——— and ———, "Physical Environmental Cues to Reduce the Problems of Incontinence in Alzheimer Disease Units," *The American Journal of Alzheimer's Care and Related Disorders & Research,* 6(6): 22–28, 1991b.

——— and ———, "Dressing Independently: A Closet Modification Model for Alzheimer's Disease Patients," *The American Journal of Alzheimer's Care and Related Disorders & Research,* 7(1): 16–28, 1992a.

——— and ———, "The Effects of Environmental Barriers on the Attention Span of Alzheimer's Disease Patients," *The American Journal of Alzheimer's Care and Related Disorders & Research,* 7(1): 9–15, 1992b.

———, T. T. Rosner, and L. Rechlin, "Long-Term Memory Cueing to Reduce Visuo-Spatial Disorientation in Alzheimer's Disease Patients in a Special Care Unit," *The American Journal of Alzheimer's Care and Related Disorders & Research,* 6(6): 10–15, 1991.

———, ———, and M. Calkins, "Visual Barriers to Prevent Ambulatory Alzheimer's Patients from Exiting Through an Emergency Door," *The Gerontologist,* 29(5): 699–702, 1989.

Regnier, V., "Design for Assisted Living," *Contemporary Long Term Care,* (February): 50–56, 1997.

Salthouse, T. A., "Cognitive Competence and Expertise in Aging," in J. E. Birren and K. W. Schaie (eds.), *Handbook of the Psychology of Aging,* 3d ed., Academic Press, San Diego, CA, 1990, pp. 310–319.

——— and R. Babcock, "Decomposing Adult Age Differences in Working Memory," *Developmental Psychology,* 27(5): 763–776, 1991.

———, ———, and P. Shaw, "Effects of Adult Age on Structural and Operational Capacities in Working Memory," *Psychology and Aging,* 6(1): 118–127, 1991.

Schlesinger, K., and P. Groves, *Psychology: A Dynamic Science,* Wm. C. Brown, Dubuque, IA, 1976.

Sloane, P., J. Rader, A. Barrick, B. Hoeffer, S. Dwyer, D. McKenzie, M. Lavelle, K. Buckwalter, L. Arrington, and T. Pruitt, "Bathing Persons with Dementia," *The Gerontologist,* 35(5): 672–678, 1995.

U.S. Department of Health and Human Services, *Losing a Million Minds: Confronting the Tragedy of Alzheimer's Disease and Other Dementias,* U.S. Government Printing Office, Washington, DC, 1987.

Weisman, G., M. Calkins, et al., "The Environmental Context of Special Care," *Alzheimer's Disease and Associated Disorders,* 8(suppl. 1): S308–S320, 1994.

Welch, P. (ed.), *Strategies for Teaching Universal Design,* Adaptive Environments Center, Boston, and MIG Communications, Berkeley, CA, 1995.

Zeisel, J., J. Hyde, and S. Levkoff, "Best Practices: An Environment-Behavior Model for Alzheimer Special Care Units," *The American Journal of Alzheimer's Care and Related Disorders & Research,* 9(2): 4–21, 1994.

———, "A Review of Special Care Setting Design," presentation at the Seventh National Alzheimer's Disease Education Conference, Indianapolis, IN, July 29, 1998.

CHAPTER 23

LIFE SAFETY STANDARDS AND GUIDELINES FOCUSED ON STAIRWAYS

Jake Pauls, C.P.E.
Consulting Services in Building Use and Safety, Silver Spring, Maryland

23.1   INTRODUCTION

Safety is an integral part of universal design. Safety, even in the limited context of buildings, includes many considerations, indeed too many to cover in a single chapter. Thus, to help illustrate some key considerations in safety, this chapter focuses on a relatively dangerous product—stairways. Stairways pose major usability problems for persons with mobility limitations. As the presented epidemiological background makes clear, people who are elderly are relatively vulnerable and overrepresented in terms of fatal injures and seriously disabling nonfatal injuries. Moreover, persons who suffer an injury resulting in a disability (even a relatively minor one) will be more vulnerable to future injury and further disability in addition to heightened risk of other morbidity. Altered gait is a prime example. Societal costs as well as individual costs of injury are also important. Thus, prevention of injuries (which should not be termed "accident prevention") is an important, generally applicable, universal design objective. Recommendations presented here cover design, modification, and behavioral considerations.

23.2   BACKGROUND

An environment or product that is the result of a universal design process should be reasonably safe for foreseeable users. "Safety" in this context means the prevention and control of injuries, both fatal and nonfatal. The emphasis here is on nonfatal injuries and their prevention through environmental and product design, construction, plus retrofit. One reason for this emphasis is that many nonfatal injuries result in permanent disabilities and reduced quality of life. Thus inadequate design, construction, and retrofit of an environment or product might result in a greater need for subsequent special features in an environment setting or a product to mitigate a disability.

The current increase in average life span is exacerbating this problem. A central dilemma faced by people who are aging is the need to maintain activity so that both physical and mental ability are preserved to the greatest degree possible for the individual. Unfortunately, for the safety problems highlighted in this chapter, increased mobility tends to increase injuries—if all other factors remain fixed. To alter the association of exposure and injuries, the other factors must be altered. Chief among these other factors (at least from the perspective of this book) is modification of the environment and other products that affect the prevention of injuries.

The underlying principle behind many recommendations, especially in the first part of this chapter (and, indeed, this entire book) is that "To err is human. To forgive, design." A related statement of this principle is "To lose capability is human. To compensate, design." Here "design" is used in a broad sense encompassing the devising of environmental or product solutions and constructing or retrofitting those solutions. The emphasis on building design is not intended to ignore the contributions that people bring to preservation of their own safety. People must be a central part of any systematic approach to safety.

23.3 SEMANTICS

Safety is not "accident prevention." Within the community of injury prevention professionals, especially in the United States, use of the term *accident* is strongly discouraged. "Accident" has the connotation of an unpredictable, unpreventable event; a related term is *act of God.* The significant problems raised by the use of these terms in fields such as public health and law are well discussed by Loimer and Guarnieri (1996). Thus "accident prevention" is an oxymoron.

One of the annoying defects with the common use of the term *accident* is that it covers such a large variety of events, ranging from the dog soiling the carpet to a nuclear power plant meltdown and including, in between, falls on stairs and collisions involving motor vehicles. Furthermore, in addition to the term's vagueness in nature and extent of consequence of an event, there is the typical ambiguity about the cause of the event. Are one or more of the injured parties responsible? Or are they hapless victims of an unforeseeable, unmanageable event?

It is far more useful to state clearly what the event is, along with its consequence, without prejudging or oversimplifying fault. For example, compare how much more helpful it is to say that a car occupant was injured in a multivehicle collision on the highway than to say there was an accident on the highway. Similarly, it is immediately clear if one describes, as a stairway-related injurious fall, the fairly common kind of incident focused on in this chapter.

Moreover, the need for semantic precision in this treatment of safety goes beyond use of the term *accident.* For example, rather than accept as physically accurate the common expression, "I slipped on the stairs," readers will come to appreciate that the term *slip* is often misused. In actuality, there was a misstep—not necessarily a slip—that led to a complete loss of footing and balance. A slip is one kind of misstep, but there are others that are both common and amenable to prevention through design. For example, other types of missteps include an *overstep,* an *air-step,* and a *trip.*

Moreover, "missteps" should not be confused with "mistakes." Even "human error" might not be the fault of the victim in a fall, especially when there are system flaws. Environmentally triggered human error is not the fault of the victim. An alternate term unfortunately too applicable to many stairways is "architecturally triggered human error." Within the field of ergonomics or human factors a realistic understanding of how error occurs—and how it can be prevented, reduced, or mitigated—is of paramount importance. Indeed, with this field's focus on designing for human use, we see that universal design is subsumed by ergonomics. As previously noted, To err is human. To forgive, design.

23.4 HOW ARE PEOPLE INJURED?

Surprisingly, when compared with other leading causes of death in the United States—cancer and cardiovascular diseases—injuries exact a comparatively large toll but receive comparatively little attention. Quoting from Rice et al. (1989):

> Cost of Injury in the United States: Injury deaths represent 36 life years lost per death and a productivity loss of $334,851 per death. Life years lost per death for the three other leading causes are 12 years for cardiovascular diseases (heart disease and stroke combined) and 16 years for cancer. Cost per death is $51,000 for cardiovascular diseases and $88,000 for cancer.

The cost estimates provided in this quotation are based on what is termed a "human capital" method for characterizing injury cost; it focuses on medical care costs and costs of lost working time but does not include pain and suffering. Also using the conservative human capital figures from this cost-of-injury report, for injuries in the year 1985, Table 23.1 lists the leading causes of injury with data on the number of fatalities and nonfatal injuries plus their lifetime costs.

In the injury prevention field a distinction is often made between intentional and unintentional injuries. For this one-year set of U.S. injuries, one-third of the deaths were intentional, with 21 percent suicide, 14 percent homicide. For firearm fatalities, 95 percent were intentional, with 56 percent suicide, 39 percent homicide. Nearly half of all poisonings were ruled as suicide. Whether intentional or not, injuries are increasingly treated as public health problems amenable to solution in the same way that infectious diseases have been successfully addressed by a public health approach that is marked by its emphasis on prevention.

Admittedly, the data presented in Table 23.1 are somewhat dated, though still informative. For the best collection of a wealth of information on injury control, the Internet Web site to visit is that of the Injury Control Resource Information Network (ICRIN). (Its URL, along with URLs for several other relevant Web sites, is provided in a Resources section at the end of this chapter.) For those preferring conventional print sources, one of several helpful books on injury control is that of well-known injury-prevention professionals Christoffel and Gallagher (1999).

Falls

Falls are identified as a major injury problem by epidemiological data, such as those authoritatively presented by Baker et al. (1992). In the United States, as in some other countries, falls

TABLE 23.1 Injuries Sustained in 1985 in the United States and Their Total Lifetime Costs

Causes	Fatal injuries	Hospitalized injuries	Injuries not hospitalized	Total lifetime costs, billion $
Motor vehicles	45,923	523,028	4,803,000	48.7
Falls	12,866	783,357	11,493,000	37.3
Firearms	31,556	65,129	171,000	14.4
Poisonings	11,894	218,554	1,472,000	8.5
Fires/Burns	5,671	54,397	1,403,000	3.8
Drownings	6,171	5,564	26,000	2.5
All Others	28,487	696,707	35,001,000	42.4
Totals	142,568*	2,346,736	54,369,000	157.6
Total lifetime cost, billion $	49.4	80.0	28.2	157
% Lifetime cost	31	51	18	100

* An additional 13,097 deaths occurred in later years due to injuries sustained in 1985.

are the leading cause of nonfatal injuries, exceeding those related to motor vehicles. Based on Quinlan et al. (1999), Table 23.2 presents estimates of hospital emergency department treatments for the five leading causes of nonfatal injury, a tabulation in which falls are most prominent. Falls are the second-leading cause of spinal cord and brain injuries. After motor vehicle incidents, falls are the second-leading cause of unintentional fatal injuries.

Elderly persons are at much higher risk of death due to falls: about three-fourths of deaths due to falls occur to the population aged 65 or more. About two-thirds of "accidental" deaths among older adults are due to falls. For people 75 years and older, falls are the leading cause of fatal injuries and the sixth-leading cause of death; falls account for twice as many deaths as motor vehicle incidents. A recent estimate of fall-related injuries treated in hospital emergency departments has about 1.8 million annual treatments for falls in the 65-and-over age group; although this age group only accounts for about 13 percent of the population, this is 26 percent of the total of about 7 million fall-related injury treatments in emergency departments across all age groups. Each year about one third of community-dwelling, generally healthy elderly persons experience a fall; about 5 percent of these sustain a fracture or require hospitalization, a proportion much higher than among younger fallers. Falls are the cause of 87 percent of all fractures in the elderly. Fractures of the hip, 85 percent of which are suffered by persons over 65, led to over 250,000 hospitalizations in 1988; these resulted in an average hospital stay of over 13 days. According to Baker et al. (1992): "These figures exclude the many days of nursing home care and do not convey the tragic changes in lifestyle and loss of independence that commonly ensue." Additional statistical background on the age-related impact of falls comes from the Institute of Medicine (1999) and Rubenstein (1999).

As is well understood, especially in litigation, there are fates worse than death; therefore the toll of nonfatal, often disabling falls suffered disproportionately by the elderly is a large and growing concern for individuals, families, and society generally. Moreover, disabilities from falls or the fear of falls are a leading reason for seeking nursing home care, an economically crippling situation for many elderly people and their families. Thus the prevention of falls is a major concern in universal design, especially of home environments.

Falls Involving Stairways.   Stairways are generally a leading site for serious, injurious falls, especially in homes. U.S. statistics are used in the following discussion. When exposure, or extent of usage, is taken into account, stairs are clearly a relatively hazardous environmental feature. For example, although we tend to spend much more time—more by one or two orders of magnitude, up to one-hundred times greater—standing and walking on floors than using stairs, in the United States there are generally more injuries treated in hospital emergency rooms involving stairs than there are involving floors. See Table 23.3, which covers architectural elements and furnishings. The absolute number of serious fall injuries by the elderly is about one and a half times greater for floors than for stairs, but when the reduced exposure of the elderly to stairs is factored in, stairs are much more dangerous.

Intuitively, one can understand at least two reasons why stairs are such a hazard. It is relatively easy to misstep and fall on stairs; second, a stair is a very bad place to land on in a fall because of the increased fall heights, and hence impact speeds, and the angular surfaces that focus injurious forces. Prominent stair researcher John Templer has referred to this phenomenon by using the term "the unforgiving stair."

U.S. Consumer Product Safety Commission (CPSC) National Electronic Injury Surveillance System (NEISS) data on injuries involving (but not necessarily caused by) stairs reveal that stairs were involved, annually, in an estimated 1 million hospital emergency department-treated injuries in the United States beginning in the mid-1990s. In the United States there is approximately a one-in-three chance that an individual has a hospital-treated, stair-related injury during his/her lifetime. This makes the likelihood of a stair-related injury approximately 35 times greater than a fire-related injury to a civilian nonfirefighter in the United States. Note that NEISS data are sometimes reported as "stairs, ramps, landings and floors"— accounting for an estimated 1,975,000 hospital emergency-department-treated injuries in 1998 (CPSC, 1999). Many stair-related falls terminate on a floor, thus a stair-related fall could

TABLE 23.2 Five Leading Causes of Nonfatal Injuries Treated in U.S. Hospital Emergency Departments by Age Group

Rank	0–4	5–9	10–14	15–19	20–24	25–34	35–44	45–54	55–65	65+	Total
	Estimated Annual Number of Injuries in Thousands										
1	Fall 854	Fall 600	Struck by/against 592	Struck by/against 556	Struck by/against 457	Over-exertion 820	Fall 760	Fall 609	Fall 413	Fall 1,797	Fall 7,007
2	Struck by/against 393	Struck by/against 486	Fall 454	MV, traffic 514	MV, traffic 450	Struck by/against 814	Over-exertion 639	Struck by/against 327	MV, traffic 196	Struck by/against 242	Struck by/against 4,575
3	Cut/Pierce 205	Cut/Pierce 288	Cut/Pierce 276	Fall 379	Over-exertion 420	Fall 761	Struck by/against 622	MV, traffic 314	Struck by/against 186	MV, traffic 189	Cut/Pierce 3,401
4	Natural/ environment 166	Pedal cyclist 216	Over-exertion 230	Over-exertion 372	Cut/Pierce 396	MV, traffic 697	Cut/Pierce 555	Cut/Pierce 299	Cut/Pierce 184	Cut/Pierce 185	MV, traffic 3,256
5	Poisoning 109	MV, traffic 118	Pedal cyclist 220	Cut/Pierce 328	Fall 371	Cut/Pierce 684	MV, traffic 527	Over-exertion	Over-exertion 284	Over-exertion 123	Over-exertion 3,156 139

TABLE 23.3 Hospital-Treated Injuries Related to Products in the United States, 1994 (CPSC/NEISS)

Stairs or steps	1,030,000
Floors and flooring materials	940,000
Bicycles	605,000
Beds	392,000
Tables	327,000
Doors	314,000
Chairs	293,000
Glass doors, windows, and panels	210,000
Bathtubs and showers	174,000
Ladders	144,000
Porches, balconies, open-sided floors	134,000
Rugs and carpets	129,000

also be coded as involving a floor in the NEISS data which allow for multiple-product coding. Notably, NEISS statistics are based on injuries treated in U.S. hospital emergency departments; however, 55 percent of ambulatory injury treatments were at physicians' offices, not hospitals, and, increasingly, there are additional options for treatment including outpatient departments and walk-in clinics, according to the National Center for Health Statistics (NCHS, 1995). Thus CPSC's NEISS statistics are indicative rather than complete.

Stair-Related Fall Time Trends. In contrast to most other types of injury, those from stair-related falls are not decreasing. Although detailed trends are difficult to estimate from NEISS data, these data suggest that since the 1970s the increase in estimated number of stair-related injuries treated in U.S. hospital emergency departments has averaged about 2 percent per year. During this time the U.S. population growth averaged 1 percent per year and stair use likely did not increase even by this amount. See Fig. 23.1, which shows for a 25-year period the 95 percent confidence intervals and central trend for estimated number of stair-related injuries treated in U.S. hospital emergency departments. This is based on CPSC/NEISS data adjusted for sampling changes in 1997. By contrast, as also shown in Fig. 23.1 using National Fire Protection Association (NFPA) data, since the mid-1970s the decrease in fire-related civilian injuries in the United States has averaged about 3 percent per year. According to statistics from the National Safety Council (NSC), during approximately the same period the rate of motor-vehicle-related fatalities dropped by about one half.

Astute observers of both the fire safety and traffic safety scenes can explain the laudable reductions in injury rates, which are corrected for population and usage. For example, roadways and automobiles have incorporated significant design improvements—achieved at a time of reduced car sizes and greatly increased usability and comfort, plus fuel efficiency.

For another comparison, correcting for exposure, for stair use in the United States there is approximately 1 hospital emergency room-treated injury for every 5000 hours of stair use, based on 3 million flight uses for each such injury and one death in 1 million hours of use. For motor vehicle use in the United States, there is approximately 1 injury in 20,000 hours of use and 1 death in about 3 million hours of use. The death rate for scheduled airliners in the United States is on the order of 1 death in 100 million hours of use (NSC, 2000). At a time of improving safety for motor vehicles manufactured since the 1960s, the quality and safety of stairs were reduced. Ironically, this occurred despite major growth in the understanding by researchers of where, how, and why stair-related falls and injuries occur. For example, counterintuitively, most stair-related falls occur on stairs with which the fall victim is assumed to be familiar.

Stairway Injuries Ranked by Comprehensive Injury Costs. Recent estimates available for product-related injury costs come from the work of Lawrence et al. (1999), who used a CPSC injury cost model that takes into account a wide range of costs from 18 large data sets including, prominently, pain and suffering—the largest component of what is termed "comprehen-

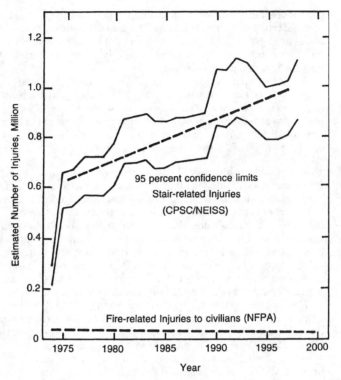

FIGURE 23.1 Approximate time trends of two sources of injuries: stairs and fires.

sive injury cost." Pain and suffering account for 76 percent of total comprehensive injury cost and, notably, are not included in the early cost-of-injury data for 1985 reported in Table 23.1 that are based on the relatively conservative human capital estimating method, which ignores pain and suffering. Using the comprehensive cost method for CPSC/NEISS-coded consumer products (not including guns, food, and motorized vehicles, which are not addressed by CPSC), consumer product injuries were estimated to cost approximately $517 billion, expressed in 1997 dollars, in 1996 in the United States.

Ranked by comprehensive injury cost, stairs are the leading product category for injuries covered by the NEISS data. According to Lawrence et al. (1999), stairs or steps account for approximately $47 billion of comprehensive injury cost for 1996 in the United States; this is 9 percent of the comprehensive cost of injuries related to NEISS-coded consumer products. In the same ranking, floors account for 8 percent of comprehensive injury costs.

It is helpful to put the stair-related injury costs into perspective. Lawrence et al. estimate nonfatal injuries due to motor vehicle crashes, excluding crashes with motorcycles, pedal cycles, and pedestrians, at $85.5 billion of comprehensive cost for 1996 in the United States. Comparing the injury costs to some major causes of property damage, one notes that the toll of natural disasters is often reported extensively by mass media. The tornados that devastated Oklahoma City in 1998 caused about $1.5 billion in property losses. Hurricane Andrew caused about $15 billion in insured property losses; however its toll of life loss was about the same as the toll of fatalities in U.S. stair-related fatalities on the same days that Andrew was devastating Florida and the other Gulf states. In 1997 in the United States, fires caused $8.5 billion in direct property damage according to John Hall of the National Fire Protection Association (NFPA), and when all costs are taken into account, including the kinds of indirect costs assumed in comprehensive injury cost estimates previously noted, fire is responsible for losses as high as $200 billion in 1997 in the United States (North American Coalition for Fire and Life Safety Education, 1999). Similarly, there are significant nonproperty losses for the previously noted natural disasters.

The most remarkable statistic is that annual, 1996-based comprehensive injury costs related to stairs are estimated to exceed annual construction costs of new stairs by a factor of ten! Very few products (e.g., guns and cigarettes) have this extraordinary ratio of injury costs relative to construction or manufacturing costs. Even gun-related injuries in the United States have medical treatment costs of $2.3 billion for 1994 according to Cook et al. (1999); these costs are only half those of stair-related injuries.

Stair-Related Problems in Homes. For those stairs where the fall location is known, 85 percent occur in residential settings, according to NEISS data. Clearly, home stairways provide a very fertile ground for achieving significant reductions of injuries. Potential injury cost reductions likely exceed greatly the costs of improvements to stairway features such as step geometry, handrails, and lighting. Moreover, with our rapidly aging population and the propensity of elderly persons to be somewhat overrepresented in nonfatal stair-related injuries and greatly overrepresented in fatal injuries related to stairs, any benefit-cost analysis shows increasing value in prevention of injuries, especially through environmental intervention. There are also serious usability problems for elderly persons who need to use stairways. Overwhelmingly, persons over 50 years of age want to stay in their homes and never move (AARP, 1999).

Unfortunately, as noted by Ulria and Sven (1995), "Unintentional injuries occurring in the home are regarded as natural consequences of the victims' inappropriate behavior. . . . With this perspective, systematic preventative efforts are not perceived as relevant." A person's familiarity with an environment or with an activity does not mean that its hazards are understood or eliminated. Hence it is crucial that environmental factors in successful (safe) and unsuccessful (injurious) use of stairs be better understood by not only professionals but consumers as well. This means attention to much more than simply urging people to "watch your step" or "be careful when using stairs." In fact, as well established in the public health field and as stressed by the Committee on Trauma Research (1985):

> Of three approaches—persuading behavior change, requiring behavior change, and providing automatic protection—the most effective is the automatic protection through product and environmental design. Especially effective is doing this with laws and regulations.

From John Archea's insights we also learn of the key priorities for improved safety and usability of stairways (Asher, 1977):

> The key to stair safety lies not so much in the hazard itself as it does in the users' awareness of their vulnerability to it. . . . So long as the users know they are coming to a stair, where to place their feet on the treads, and where to grab if they should lose their balance momentarily, then they are most likely to make it to the top or bottom of the flight without an accident.

Rephrasing these three criteria, as stressed by Pauls (1982, 1984, 1998a, 1998b) we need:

1. Visibility of the stair flight and its individual steps, especially when viewed in descent
2. Adequacy and uniformity of step dimensions in relation to human gait
3. Availability of reachable, graspable handrails which also provide accurate visual cues about the presence and location of steps

The recommendations that follow are discussed in greater detail in references such as Archea et al. (1979), Pauls (1982, 1998a, 1998b), and Templer (1992).

Lighting and Stair Visibility. The International Building Code™ 2000 edition includes an illumination requirement for a minimum of 10 fc (107 1x) (measured at the treads) for dwelling unit stairs. Illuminating Engineering Society (IESNA, 1996) recommendations are for twice as much light for the kinds of stair environments typical in homes where there is a strong likelihood that the stairs will be used by persons who are elderly. However, there are several other considerations beyond absolute light levels—including environmental surfaces around the stair, relative levels, gradients, glare, controls, and source redundancy. (Refer to Fig. 23.5 in the handrail section, which also illustrates glare and gradient problems on a stairway.) These are important factors relative to the age of users, with older users needing far more light and being more susceptible to glare. With the lack of any apparent correlation between residential fixture light output and its cost, and with the growing availability of efficient light sources, lighting holds much promise as a very cost-effective means of improving stairways. However, home builders in the United States have vigorously opposed inclusion of detailed requirements in the new International Residential Code™, which has very little by way of illumination requirements for stairs. Aside from lighting, there are other important factors in visibility of steps, including careful choice of stair covering materials to avoid patterns that tend to camouflage the step nosings, the critical leading edges of treads (Archea et al., 1979; Archea, 1985; Asher, 1977; Templer, 1992).

Dimensional Uniformity of Steps. Although there has been much debate on the issue of riser/tread dimensions, this has focused on the absolute dimensions, not the differences between and among rise and tread dimensions within a stair flight or within multiple-flight systems. Laboratory studies, field observations, plus case studies from litigation identify dimensional nonuniformities as a common, extremely potent factor in stairway-related falls. It appears that a great deal of improvement on this matter could be achieved by better educating builders as to the critical importance of close tolerances—preferably 3/16 in (5 mm)—between adjacent treads and between adjacent risers, with no more than twice this tolerance within a stair flight. Moreover, code inspectors need to watch this far more carefully and need to enforce the codes diligently. Mitigation efforts should also be developed for existing stairs with serious nonuniformities. Hazard marking and warnings might be adequate for small nonuniformities. Larger nonuniformities [e.g., greater than ¾ in (20 mm)]—a relatively common defect on new home stairs in the United States, especially at the first steps at the top of a flight—warrant rebuilding. For example, the very common defect of a nontypical or absent nosing projection at the top nosing of a stair flight can be fixed for a dollar or so if the defect is spotted by an inspector before carpet installation.

Carpeting. Carpeting typically increases the effective rise height by a dimension equal to the compression of the carpet and pad; carpeting also decreases the effective depth of the tread by the uncompressed thickness of the carpet and pad (see Fig. 23.2). For carpeted stairs that are already deficient in tread depth with excessive riser heights (and this includes most home stairs), the cheapest, most effective fix is simply to remove any pad and possibly also the carpet. Securing the carpet tightly to the steps helps improve matters. There is a minor countervailing risk with carpet removal; cushioning might help mitigate some falls by slightly absorbing impact, especially for hip and pelvic impacts (Maki and Fernie, 1990; Templer, 1992). However, this minor benefit might be more than offset by a increased number of missteps and falls due to the soft, short, unstable footing provided by the thickly carpeted steps. One thing that should be initiated as soon as possible by building regulatory authorities is to measure the stair dimensions as the users' feet actually experience them! That is, inspection measurements should be made with the carpet in place. Alternatively, code-required minimum tread dimensions should be increased by 1 in (25 mm) and code-required maximum rise reduced by 0.5 in (13 mm).

Minimum Tread Depths. Based on the U.S. national standard for usable buildings and facilities (ICC/ANSI A117.1) as well as national safety standards and model codes, both fall-prevention and usability considerations clearly identify 11 in (279 mm) as the minimum tread depth, exclusive of projecting nosing, and permitting only a small nosing rounding or bevelling amounting to less than 0.5 in (13 mm). Requiring the same for dwellings makes practical

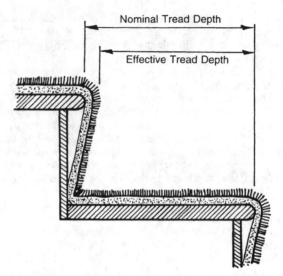

FIGURE 23.2 Thick carpeting on stair reduces effective tread depth.

and philosophical sense; more vulnerable home stair users require at least as good a design standard as is commonly required on other stairs.

Maximum Riser Height. As with minimum tread depths, the recommendation here is well founded philosophically and otherwise—especially as many home stair users have limited joint movement capability and balance due to age. Safety and preference studies point to a maximum of 7 in (178 mm). With step dimensions not easily retrofitted, we need to get this right during initial construction.

Slip Resistance. There is much evidence that many stair-related falls are initially blamed on "slipping," simply because of the inadequacies of commonly used language (i.e., people who have fallen often say they "slipped"). Typically, stair-related falls are due to a *misstep* (i.e., a departure from normal gait) such as overstepping, where, in descent the ball of one's foot ends up ahead of the step nosing). Generally, a simple rule of thumb applies to necessary slip resistance of treads: it does not have to be any greater than what works for level floors; moreover, there should be intertread and intratread consistency.

Handrails. Both in terms of their usability and their contribution to safety, functional handrails provide excellent value for cost; they are the only "safety net" when using stairs. However, builders have perfected a number of ways of making this potentially useful component of a residential stairway worse than useless, by specifying expensive, pseudotraditional railing systems rather than functional handrails that cost only a fraction as much. Important safety enhancing aspects of handrails include their provision on all stairs—including stairs with only one or two risers, the site of a disproportionate number of injurious falls. See Fig. 23.3, showing the prudent, prominent provision of handrails to help warn of the presence of a two-riser stair in a shopping complex. Also important are configuration, continuity, height, terminations, graspability, clearance, and fixing (Feeney and Webber, 1994; Maki, 1988; Maki, et al., 1984, 1998; Pauls, 1989, 1998b; Templer, 1992). Despite the biomechanical evidence, the relatively simple issue of handrail graspability has been especially badly treated by residential stairway manufacturers, home builders, and building code authorities who fail to adopt and enforce meaningful requirements. In the United States, a commonly used residential railing section allows only an ineffective pinch grip. In Fig. 23.4 this railing section is contrasted with a functional handrail section facilitating a power grip by a wide range of hand sizes. Notably the more functional railing section is considerably less costly than the larger one.

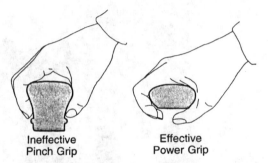

Ineffective
Pinch Grip

Effective
Power Grip

FIGURE 23.3   Graspability difference with common railing and functional handrail.

Handrails are universal design elements that are critical on all stairways as well as being helpful on ramps and some corridors. Although not everyone can use stairs, for many persons facing extra difficulty with stairs, handrails are crucial for both usability and safety. This is well depicted in Fig. 23.5, where a boy with multiple sclerosis, and usually a crutch user, can negotiate a stairway provided with a functional handrail placed high enough to provide adequate stability. This photograph also shows an unfortunate lighting problem due to the strong sunlight at the upper part of the stair flight and the deeply shadowed area lower down. This would pose extra problems for aging eyes, which are unable to adjust rapidly to greatly varying light conditions or gradients.

FIGURE 23.4   Prudent provision of handrails on two-riser stair.

FIGURE 23.5 Boy with multiple sclerosis using functional handrail

Winding, Curved, and Spiral Stairways. Such stair configurations optimize space usage for a stairway and, if designed properly, with attention to the previously stated criteria, they are relatively safe, usable, and cost-effective. For example, winders are best used only at the base of straight stair flights to minimize tread depth nonuniformity effects; handrails should be continuous at turns. Spiral stairs should provide, at the "walking lines," the same tread depths needed on stairs generally; handrails should be provided on the inner, post side as well as the outside. Essentially, one set of criteria can be applied generally to curved, helical, and winding stairs and that set of general criteria utilizes the concept of an inner and outer "walking line," each of which is typically about 10.5 in (270 mm) from a handrail (Templer, 1992).

Summary of Recommendations for Safer Stairways. Table 23.4 distills many recommendations into a single-page checklist applicable to both new stairways and existing stairways. Note that it is largely organized according to the three basic criteria: *step geometry, visibility,* and *handrails.*

The Traditional Double Standard Behind Residential Stairway Design. Virtually all building codes and standards maintain a traditional double, lower design standard for dwellings

TABLE 23.4 Checklist for Stairways, Especially for Homes

General

[ ] Locate steps only where they are necessary and logical.
[ ] Avoid small changes of levels, especially single steps.
[ ] Use safety glazing for glass that could be impacted in a fall on a stairway.
[ ] Arrange for any doors adjacent to steps not to swing over any steps.
[ ] Install gates to prevent infant and toddler use.
[ ] Avoid sudden changes of views and visual distractions from stairways.
[ ] Make steps visually prominent so that their presence is obvious.

Steps

[ ] Build step riser dimensions no higher than 180 mm (7 in).
[ ] Build step tread dimensions at least 280 mm (11 in) front to back.
[ ] Keep risers and treads consistent in size (to within 5 mm or 3⁄16 in).
[ ] Conspicuously mark flights having inconsistently sized steps; e.g., paint a contrasting stripe (25 to 50 mm, 1 to 2 in wide) on all step nosings (right at the leading edges of the treads) so that the nonuniformity stands out.
[ ] Remove thick (10 mm, ⅜ in, or more) carpets and underpads on treads.
[ ] Fix tread coverings securely; coverings must be tight against the nosings.
[ ] Remove/repair tripping surfaces and projecting screws or nails on treads.
[ ] Provide slip-resistant (rough) finish on stair treads subject to wetting.
[ ] If stairs are subject to wetting, slope treads approximately 1 percent (⅛ in per foot) to drain water away from nosings.
[ ] If nosings project (and they do not need to if treads are 280 mm, 11 in deep), keep nosing projections uniform in the flight and configure them so that there is no tripping hazard to persons with poor foot control.

Visibility

[ ] Provide slightly rounded nosings (maximum radius 13 mm, 0.5 in) for visibility and injury reduction.
[ ] Avoid tread materials and coverings with visually distracting patterns.
[ ] Mark nosings permanently (not with tape) if they are not distinctly visible. Note that a painted stripe even works very well, and looks good, on carpet.
[ ] Provide lighting that makes tread nosings distinctly visible.
[ ] Provide lighting that does not cause glare or strong shadows.
[ ] Illuminate stairs with no less than two bulbs (especially if incandescent).
[ ] Have light levels on stairs at least as high as on adjacent areas.
[ ] Unless continuously lit or automatically switched on, provide light switches at each point of stairway access.
[ ] Install permanently illuminated, small light sources (e.g., night lights) on stairs.

Handrails

[ ] Provide at least one handrail on each stair regardless of the number of steps.
[ ] On wide, monumental stairways provide at least one handrail at the normal path of travel. If used by crowds, space handrails 1525 mm (60 in) apart.
[ ] Continue handrails between stair flights at the shortest path of travel.
[ ] Extend the handrail, without a break, the full length of the stair between floors.
[ ] Augment any decorative stair railing system with a functional handrail.
[ ] Install a handrail around which fingers and thumb can encircle. If one wraps a measuring tape around the railing, the loop should be shorter than 160 mm (6¼ in) for adult hands. A smaller railing is desirable for children.
[ ] Maintain adequate hand clearance between the handrail and nearby surfaces.
[ ] Position handrails at about adult elbow height, 900–1000 mm (36–39 in).
[ ] Provide handrails that are visually prominent.
[ ] Repair or replace broken handrails.
[ ] Fix handrails securely to walls and posts. One should be able to bear the entire weight on the handrail without damaging the handrail.

relative to the design standard applicable to other buildings. Indeed, in recent decades, with the improvement of step geometry and handrail requirements for stairways other than in homes, the gulf between design of stairways for homes and for other buildings has grown. The maintenance and exacerbation of the double standard, especially for stairs and railings, flies in the face of universal design and public health generally. The double standard is partly due to the disproportionate power of the homebuilding industry in the development, adoption, and enforcement of codes for dwellings. In the United States the National Association of Home Builders (NAHB) has been especially effective in not only keeping national model code requirements for stairways from matching those for other buildings, the NAHB and its local subsidiaries have been very effective in preventing even the newer but compromised requirements from being enforced at state and local level. As a result, in the United States, new home stair treads, which are typically covered with thick carpet and pad, are effectively about 3 in (76 mm) shorter than treads of other building stairs, and risers are effectively about 1.5 in (38 mm) higher than risers of other building stairs. Also, home stair railings are commonly too large, too short, and too low to be functional—for usability and safety—as handrails or guardrails. For more details, see the checklist in Table 23.4. Most of these shortcomings, relative to universal design, will continue to hinder, handicap, and harm users for many decades to come.

This issue of the double standard has been very prominent in model code-change deliberations in the United States; however, there has been little participation in the debate by those prominent in universal design and home modification fields. At the heart of the debate is the basic question: If the so-called "7-11" standard—with 180 mm maximum step riser height with 280 mm minimum tread depth—is the widely accepted, required minimum standard for those stairs having the minority of injuries (i.e., stairs not serving dwellings), then why should it not be the minimum requirement for stairs serving dwellings, where most injuries occur and where the needs by the young and elderly are greatest?

Bases for the so-called "7-11" standard (addressed, for example, by Templer, 1992) include:

Expert judgement based on experience

Studies of actual fall/injury circumstances:

- Research investigations
- Investigations related to litigation

Field and laboratory studies of gait on stairs

Field and laboratory studies of missteps

Foot/shoe size in relation to tread size

Laboratory studies of energy expenditure

Preference studies

Sometimes countering, but usually ignoring such technical bases in their campaign against improvements to residential stairs, the homebuilders in the United States have:

- Denied that the safety problems exist
- Denied any responsibility for the problems
- Argued that solutions are too costly to implement
- Coaxed or intimidated officials to prevent effective intervention through building code development, adoption, and enforcement

The importance of the roles of builders and regulators is not unique to the United States, although defects in code development, adoption, and enforcement in the United States are more openly perpetrated (Pauls, 2000a, 2000b). Some of the problems of flawed model code development relative to stairways in Canada have been addressed by Pauls (1998b, 1998c).

Affordability of Safe Stairs.   In focusing so much on their limited conception of "affordability," the builders fail to recognize that housing affordability is a function of lifetime costs. Comprehensive injury costs exceeding original construction cost by an order of magnitude and the ongoing costs of reduced usability, during the many noninjurious uses, contribute to huge lifetime costs of stairways that should not be ignored. Moreover, currently in the United States there is little or no relationship between the selling price or size of a house and the safety or functional quality of its stairs. For an average new home built in the United States in 1998, the cost of stairs accounted for only 0.4 percent of the total construction cost. Since 1950, average home size in the United States has nearly doubled, increasing by about 1200 ft^2 (111 m^2) or about 100 times more than the additional area needed to provide, in a dwelling, the "7-11" step geometry as opposed to the inferior stairs builders continue to provide.

With few exceptions, home builders appear blind to the growing market—including many people who are elderly—who need, want, and can afford dwelling units designed with greater attention to function and safety, especially for stairs. Perhaps the most notable exception was a home builder in Ohio who made a selling point of the fact that the improved stair geometry often requires at least one additional riser; the builder of Dominion Homes™ called attention to this in sales literature recommending, "Take these 14 steps to a safer home." For further information on these marketing and feasibility issues see Pauls (1998a, 1998b).

What Can Be Done to Improve Existing Stairways?   The checklist provided as Table 23.4 has a comprehensive list of considerations for both new stairways and existing stairways. In prioritizing recommendations, the addition of a functional handrail—which an average hand can completely encircle and which is positioned at the elbow height of the shortest adult user—is the first-priority, prudent, cost-effective retrofit to be considered for all residential stairs (including those with one or two risers!). Figures 23.3, 23.4, and 23.5 illustrate these handrail features.

Lighting should be improved, including doubling the wattage capability of fixtures or converting to more efficient lamp types for stairways used by persons over 50 years of age. Install permanent night lights so that stairs are never in total darkness; inexpensive, automatic switching, light-sensing units are available for this. Strive to keep illumination levels on stairs and landings at least as high as in immediately adjoining areas; this means that bright stairways in daytime situations, and dim ones at night, are appropriate.

Many stairways in older homes come closer to the basic step geometry criteria previously noted than do stairways in homes constructed in recent decades in the United States; treads are deeper; risers are lower; and better workmanship—including seemingly forgotten skills in joinery—is seen in the critically important consistent step dimensions. Moreover, treads are finished so that they do not need to be covered with wall-to-wall, thickly padded carpet. Thinner carpets should be securely fixed so that they are tight at the nosings. Thickly padded carpets should be removed to optimize effective step dimensions (see Fig. 23.2.) Uniformity should be improved as much as possible, for example by modifying the nosing at the top if it does not project and others in the flight do; consistent nosing-to-nosing dimensions are a goal. Nonuniformities exceeding ¾ in (20 mm) should be eliminated by rebuilding if possible; remaining nonuniformities of between ³⁄₁₆ and ¾ in (5 to 20 mm) should be marked. Painted nosings 1 to 2 in (25 to 50 mm) wide at the front edge of each tread work well even on carpeted stairs! Remove metal strips at nosings; remove flimsy rubber tread mats; remove friction "nonslip" tape—all of these can easily become tripping hazards.

For additional general information on retrofitting guidelines, refer to the comprehensive recommendations of Archea et al. (1979) or the reiteration of key retrofitting guidelines and general stair safety recommendations provided by Pauls (1982, 1998b). As in medicine, the first rule for home modification (and generally) is "do no harm" (i.e., do not make a bad situation worse).

Falls: General Observations

Common kinds of missteps occurring because of inadequate or unexpected inconsistent walking surface conditions can be categorized as follows:

- Air step—Stepping onto unexpected depression or step down.
- Overstep—Ball of foot placed too far forward on short stair tread.
- Heel scuff—Back of shoe scuffs riser of stair with short tread.
- Unstable footing—Ankle injured, in roll, as one side of foot is not supported by an uneven walking surface.
- Trip—Stubbing toe or catching heel on an unexpected projection; bottom of shoe locks on surface with excessive friction.
- Slip—Losing traction due to unexpected friction reduction.

Factors in slipping incidents include:

- Walking surface selection
- Walking surface maintenance
- Shoe geometry and surfaces
- Human gait
- Expectancy

Given the large role of users' common expectations of consistent walking surfaces, one should ask the following key questions when attempting to reduce fall-related injuries:

- What is the expectation of the user?
- What perceptual testing is performed by the user?
- Is the user's expectation realistic?
- What environmental deceptions affect the user's expectation?
- Are the user's capabilities adequate to deal with the situation?
- How can the danger be eliminated?
- If not eliminated, how can the danger be mitigated?
- How can perceptual errors be reduced?

Guidance Programs and Materials to Help Everyone Prevent Falls

As part of the Pauls (1998b) video program, *Stairways and Falls,* a few relatively good videos are excerpted: *Home Safe Home* from the Johns Hopkins University Injury Prevention Center (1990); *Why Move? Improve,* from the American Association of Retired Persons (1995); and *Stepping Out,* from the University of Victoria School of Nursing (1996). A relatively new program, developed by the National Fire Protection Association (NFPA) and the U.S. Centers for Disease Control and Prevention (CDC) is titled *Remembering When;* it focuses on safety for persons who are elderly. Information on the program is available from NFPA. Such programs focus on both environmental factors and behavioral factors, such as are highlighted in the following recommendations.

- Have eyes checked regularly.
- Review medications with physician to eliminate combinations or reduce dosages affecting alertness and balance.

- Exercise to improve balance, upper body strength, and grip.
- Wear flat, thin-soled shoes with wide bases.
- Get out of bed and chairs slowly.
- Install and use handrails and grab bars.
- Permanently mark—with contrasting paint—edges of steps, thresholds, and other hurdles.
- Move unstable furniture from traveled areas.
- Use bath mats or slip-resistant stickers in bathtubs.
- Remove thick throw rugs and other clutter from traveled areas.

Curiously, given the major role of stair-related falls in homes, there is generally no mention in guidance materials regarding the most common adaptation that people make when using typical home stairs with undersized treads and high risers. This is the unconscious tendency to twist one's feet, even one's entire body, to one side to maximize footing on the treads, in what is called a "crablike" gait. Moreover, there appears to be a consistent direction of such twisting by individuals, and, in relation to handrail provision, it appears biomechanically to be more effective for the twist to be toward the handrail. This common, partial adaptation to typical stairs, encountered especially in homes, should be addressed in guidance materials intended to reduce falls. Also, research is needed to explore this prevalent adaptation to the lack of universal design in relation to home stairs, including the negative aspects of the related biomechanics as people have to continually step over their feet when descending a stair with this twisted posture and crablike gait.

23.5 PROGRAM AND ORGANIZATIONAL TRANSITIONS

The NFPA/CDC program "Remembering When" deals only with falls and fires as significant dangers for people who are elderly. In focusing on this one population group, the program does not take a universal design approach. Whether such tailored programs are appropriate or whether a program should not isolate or single out older adults was one of the areas of professional disagreement in a symposium, "Solutions 2000," organized by the North American Coalition for Fire and Life Safety Education and held in April 1999 to examine fire safety challenges of people who cannot take lifesaving action in a timely manner. However, there was consensus that when dealing with persons who are elderly, programs should promote life safety, not just fire safety. NFPA took this to heart in 1999, not only in cosponsoring the *Remembering When* program and at least one other, *Risk Watch,* focused on children's safety, but in significantly modifying its mission statement to deal with fire *and other* hazards, not only fire *and related* hazards. Falls were specifically noted when NFPA announced this shift in its mission statement.

NFPA's important mission transition came at the same time that NFPA announced its intention to produce a model building code to compete with the International Code Council's (ICC) International Building Code™ (IBC) and other new model code documents produced for adoption, as a complete family of codes, in the United States beginning in 2000. Thus far, the ICC has rejected both procedures and technical requirements for the IBC and its companion International Residential Code™ that would promote universal design. Whether NFPA exploits its expanded mission statement to deal with universal design remains to be seen as this chapter is prepared in late 2000. During 2000 the NFPA dealt with the controversial matter of mainstreamed stair step geometry requirements, using the "7-11" rule, for dwellings addressed within its prominent, internationally used standard, NFPA 101, the Life Safety Code, and a relatively new standard, NFPA 501, on manufactured housing.

For both of these NFPA standards, the National Association of Home Builders (NAHB) continues to be a reactionary force opposing the author's championing of both the "7-11" and a more universal design approach generally as part of an enlargement of focus to deal with safety and usability of buildings. Providing a potential new counterbalance to the NAHB in the model codes and standards arena is the American Public Health Association (APHA), which, in late 1999, adopted a public policy dealing with the public health role of codes regulating design, construction, and use of buildings (APHA, 2000). This APHA public policy took a critical view of the ICC's policies and procedures, which have given too much power to the NAHB in the development of the International Residential Code™ (IRC) while keeping off the committees anyone representing the public health community. While one of NAHB's main reasons for having a significant number of committee positions for the new IRC was clearly the defeat of attempts to bring in improved stairway requirements for dwellings, there also might be a continuation of NAHB's traditional opposition to some improved fire safety features in dwellings, beginning with opposition to home smoke detectors and, more recently, with home fire safety sprinklers. Home fire sprinklers and, significantly, a universal design approach to homes (including stairway standards similar to those applied to public buildings) were recommended by a follow-up policy resolution adopted by APHA in November 2000. This policy was especially directed to NFPA and its broadened scope of addressing all safety problems, not only fire.

23.6   GENERAL LESSONS FOR UNIVERSAL DESIGN

The same design features that make stairways safer will also make them more usable. Deeper tread depths reduce the chance of overstepping and suffering an injurious fall; they also reduce the need for the awkward crablike gait that entails repeatedly stepping over sideways-pointed feet. A proper handrail section will allow greater forces and moments to be exerted to help arrest a fall and to help pull oneself up a stair when leg muscles lose some of their capability. A higher handrail height, installed at elbow height, will be more effective in arresting a fall, while making all of the other uses of the stair easier by allowing the arm to rest on the handrail to share some of the load while descending the stair; as is illustrated in Fig. 23.5. Similarly, improvements to stairway lighting and marking will be appreciated with every use as well as reducing the chance of a misstep. Generally, the eventual elimination of the double standard for stairway design in homes relative to other buildings will reduce users' need to continually reprogram themselves when traversing from a public building to a home. They will not have to repeatedly alter their gait to account for significantly higher risers and much shorter treads in the latter settings.

Universal design would be better served than currently is the case if there were greater use of ergonomics or human factors knowledge and approaches to problem solving in the design, construction, management, and regulation of buildings. Rather than leave tasks as important as these largely to manufacturers and builders, who lack the training in and sensitivity to human factors, there needs to be a broadened involvement of professionals with education, training, and experience dealing with the interaction of people and their environments. It is critical to focus more on making the environments fit (for) people than expecting people to adjust to the shortcomings of the built environment. Again, to err is human, to forgive design.

23.7   CONCLUSIONS

The stairway focus in this chapter should help to spur renewed consideration of how to best exploit all building features in a universal design approach to reasonably safe, usable buildings. The landmark report, *Injury in America* (1985) put it well: "The most successful injury prevention approaches have involved improved product designs and changes in the man-made environment that will protect everyone." It is hoped that with the inclusion of this chapter in the first universal design handbook, designers will embrace safety as an integral part of universal design, just as the injury prevention community recognized universal design as the most successful approach to injury prevention.

23.8 BIBLIOGRAPHY

American Association of Retired Persons (AARP), *Why Move? Improve,* 1995 (13-minute video available from the Housing, Consumer Issues Section, American Association of Retired Persons, 601 E Street, NW, Washington, DC 20049).

———, *Fixing to Stay: A National Survey of Housing and Home Modification Issues.* AARP, Washington, DC, 1999.

APHA, "Policy Statement 9916: Public Health Role of Codes Regulating Design, Construction, and Use of Buildings," *American Journal of Public Health,* 90(3): 467–469, 2000.

Archea, J. C., "Environmental Factors Associated with Falls by the Elderly," *Clinics in Geriatric Medicine* (Symposium on Falls in the Elderly: Biologic and Behavioral Aspects), 1(3): 555–569, 1985.

———, B. L. Collins, and F. I. Stahl, *Guidelines for Stair Safety,* NBS-BSS 120, National Bureau of Standards, Gaithersburg, MD, 1979.

Asher, J. K., "Toward a Safer Design for Stairs," *Job Safety and Health,* 5(9): 29, September 1977.

Baker, S. P., et al., *The Injury Fact Book (second edition),* Oxford University Press, New York, 1992.

Christoffel, T., and S. S. Gallagher, *Injury Prevention and Public Health: Practical Knowledge, Skills and Strategies,* Aspen Publishers, Inc., Gaithersburg, MD, 1999.

Committee on Trauma Research, *Injury in America: A Continuing Public Health Problem,* National Academy Press, Washington, DC, 1985.

Consumer Product Safety Commission (CPSC), *Consumer Product Safety Review,* U.S. Consumer Product Safety Commission, Washington, DC, 1999.

Cook, P. J., et al., "The Medical Costs of Gunshot Injuries in the United States," *Journal of the American Medical Association,* 282: 447–454, 1999.

Illuminating Engineering Society of North America (IESNA), *IESNA Lighting Ready Reference,* Illuminating Engineering Society of North America, New York, 1996.

Institute of Medicine, *Reducing the Burden of Injury: Advancing Prevention and Treatment,* National Academy Press, Washington, DC, 1999.

Johns Hopkins University Injury Prevention Center, *Home Safe Home,* 1990 (17-minute video distributed by Injury Prevention Program, Bureau of Public Health Nursing, Baltimore County Department of Health, One Investment Place, 11th Floor, Towson, MD 21204-4111).

Loimer, H., and M. Guarnieri, "Accidents and Acts of God: A History of the Terms," *American Journal of Public Health,* 86(1): 101–107, 1996.

Maki, B. E., *Influence of Handrail Shape, Size and Surface Texture on the Ability of Young and Elderly Users to Generate Stabilizing Forces and Moments,* NRC 29401, National Research Council of Canada, Ottawa, 1988.

———, and G. R. Fernie, "Impact Attenuation of Floor Coverings in Simulated Falling Accidents," *Applied Ergonomics,* 21(2): 107–114, 1990.

———, et al. "Influence of Stairway Handrail Height on the Ability to Generate Stabilizing Forces and Moments," *Human Factors,* 26(6): 705–714, 1984.

———, et al., "Effect of Stairway Pitch on Optimal Handrail Height," *Human Factors,* 27(3): 355–359, 1985.

———, et al., "Efficacy of Handrails in Preventing Stairway Falls: A New Experimental Approach," *Safety Science,* 28(3): 189–206, 1998.

National Safety Council (NSC), *Injury Facts,* National Safety Council, Itasca, Illinois, 2000.

Pauls, J., *Recommendations for Improving the Safety of Stairs,* Building Practice Note No. 35, National Research Council of Canada, Ottawa, 1982.

———, "What Can We Do to Improve Stair Safety?" *Building Standards,* May-June 1984, pp. 9–12, 42–43; July-August 1984, pp. 13–16, 42; *Southern Building,* April-May 1984, pp. 14–20; June-July 1984, pp. 22–28; *The Building Official and Code Administrator,* May-June 1984, pp. 30–36; July-August 1984, pp. 10–15.

———, "Are Functional Handrails Within Our Reach and Our Grasp?" *Southern Building,* September/October 1989, pp. 20–30.

————, "Benefit-Cost Analysis and Housing Affordability: The Case of Stairway Usability, Safety, Design and Related Requirements and Guidelines for New and Existing Homes," *Proceedings of Pacific Rim Conference of Building Officials,* Maui, HI, 1998a, pp. 21–38.

————, *Stairways and Falls,* 1998b. (Extensive notes, along with a two-hour video, based on a workshop for the British Columbia Injury Prevention Conference. For sales information contact the author.)

————, "Stair Safety and Accessibility: Standards Development in the Building and Housing Industry," in Gutman, G.M. (ed.), *Technology Innovation for an Aging Society: Blending Research, Public and Private Sectors,* Simon Fraser University, Vancouver, BC, Canada, 1998c, pp. 111–129.

————, "How Well Are U.S. Building Codes Serving Universal Design?" Presented at Designing for the 21st Century II: An International Conference on Universal Design, Providence, RI, June 2000a. (Paper is posted at conference Web site, www.adaptenv.org.)

————, "Representation of the Elderly in Premises Liability Cases with a Focus on Falls," *Reference Materials Volume II, Convention of the Association of Trial Lawyers of America,* August 2000b, pp. 2613–2626. (Paper is also on Westlaw, and a one-hour video on the presentation is distributed by Jake Pauls.)

Quinlan, K. P., et al., "Expanding the National Electronic Injury Surveillance System to Monitor All Nonfatal Injuries Treated in U.S. Hospital Emergency Departments," *Annals of Emergency Medicine,* November 1999, Table 4, p. 641.

Rice, D. P., E. J. McKenzie, and Associates, *Cost of Injury in the United States: A Report to Congress,* Institute for Health & Aging, University of California, San Francisco, CA, and Injury Prevention Center, The Johns Hopkins University, Baltimore, MD, 1989.

Rubenstein, L. Z., "Preventing Falls and Other Injuries in Older Adults: Research and Policy Priorities," International Conference: Promoting Independence and Quality of Life for Older Persons, December 1999, Arlington, VA.

Templer, J. A., *The Staircase: Studies of Hazards, Falls and Safer Design,* MIT Press, Cambridge, MA, 1992.

Ulria, T., and B. Sven, "Unintentional Injuries: Attribution, Perceived Preventability, and Social Norms," *Journal of Safety Research,* 26(2): 63–73, 1995.

University of Victoria, *Stepping Out,* 1996 (26-minute video available from Elaine Gallagher, School of Nursing, University of Victoria, P.O. Box 1700, Victoria, BC, V8W 2Y2, Canada).

23.9 RESOURCES

Adaptive Environments Center.   www.adaptenv.org

American Public Health Association.   www.apha.org

Human Factors and Ergonomic Society.   www.hfes.org

Injury Control Resource Information Network.   www.injurycontrol.com/icrin/

International Code Council.   www.intlcode.org

National Association of Home Builders.   www.nahb.org

National Fire Protection Association.   www.nfpa.org

National Resource Center on Supportive Housing and Home Modification.   www.homemods.org

National Safety Council.   www.nsc.org

U.S. Consumer Product Safety Commission.   www.cpsc.gov

P · A · R · T · 4

PUBLIC POLICIES, SYSTEMS, AND ISSUES

CHAPTER 24
UNIVERSAL DESIGN
IN MASS TRANSPORTATION

Edward Steinfeld, Arch.D.
RERC on Universal Design, University at Buffalo, Buffalo, New York

24.1   INTRODUCTION

Universal design in transportation is not just more accessibility in the traditional sense; it is actually a redefinition of goals for planning and design. Everyone who uses a mass transportation system stands to benefit from universal design. Thus, everyone shares in the investment, not just people with disabilities. This chapter describes how mass transportation can facilitate convenience and safety for all, including people with disabilities, elderly people, and many others who are often "handicapped" by conventional planning and design. Many ideas and examples described here demonstrate how universal design can be achieved in simple ways and without the need for extraordinary capital investments, if it is considered from the start of a capital project and as part of long-range planning. In today's global economy, a convenient, safe, and effective mass transportation system is an essential feature of any city or community that expects to be healthy over the long term. Not only does universal design benefit the citizens of the community but it also benefits visitors from near and far, encouraging trade, tourism, and future visits. Universal design in mass transportation makes economic as well as social sense.

24.2   BACKGROUND

Transportation systems are a key element of our urbanized and industrialized world. Without them, our life experience would be radically different. These systems define the contemporary world by knitting disparate and far-flung cultures together and expanding the scale of urban areas. At the local level, advanced transportation has transformed our cities from a geography characterized by central cities with outlying suburbs and towns dependent upon them into the deconcentrated urban geography of the multicentered regional network (Gottdiener, 1994). While the central city still has great economic and cultural significance, it now has healthy competition from suburban development. In some parts of the world, the major cities cannot even be viewed as distinct entities anymore. In fact, high-speed transportation systems often make it possible for journeys between destinations in different cities hundreds of miles apart to be more convenient and faster than journeys within each city itself.

The rapid evolution of high-speed mass transportation systems, including airways, rail systems, ships, automobiles and highways, provides both challenges and opportunities for universal design. On one hand, older transportation systems can be very difficult and expensive to change. For example, subway systems like those in New York, London, and Paris include deep tunnels primarily served by stairways and escalators. Surface rail systems, like Japan Railways, one of the most sophisticated in the world, have systemwide features like differences between train floor and loading platform heights that create major obstacles (Kanbayashi, 1999).

On the other hand, throughout the world, governments are recognizing the importance of good transportation in the economy of the new millennium. Moreover, in many countries, the emphasis of urban transportation planning has shifted from highway infrastructure to mass transport systems. New mass transit projects are conceived regularly as part of showcase urban developments such as were undertaken in Barcelona prior to the Olympics. Cities like São Paulo, Brazil, are rapidly expanding their public transportation systems to overcome the massive congestion caused by unplanned growth. In the United States, the prototypical automobile-oriented country, the environmental problems caused by overreliance on automobiles and fossil fuel consumption are driving an expansion of public mass transportation systems like those in Los Angeles, Atlanta, and the Bay Area of Northern California. New airports are being built all over the globe and are usually connected to rail or highway systems at multimodal transportation hubs.

These new developments offer opportunities to make strategic improvements in the accessibility of infrastructure and ensure that the entire population will benefit from more efficient and easier-to-use transportation. But, most important, it is critical that new systems do not discriminate against underrepresented groups such as people with disabilities. Also, ways must be found to upgrade the older infrastructure and vehicle stock to ensure that the full benefits of regional and global transportation are available to all.

This chapter includes attention to the design of mass transport terminals and vehicles and the related information and communication systems. In the space available, there is not enough room to do justice to all the details of technology, architecture, and planning that need to be addressed. Thus, the focus is on some key issues and several examples of how the Principles of Universal Design can be applied to this complex topic. A basic concept uniting all the issues discussed in this article is the concept of a universally accessible "travel chain." The travel chain includes the information used to understand and gain access to the transportation, the stations, and the vehicles. A systems approach is needed to ensure full continuity of access in the travel chain (Naniopoulos, 1999, p. 11).

24.3   *SPATIAL ORGANIZATION AND WAYFINDING IN TERMINALS*

In a complex transportation terminal, finding one's way to ticketing areas, vehicles, and services is perhaps the most important concern of the traveler. Most trips are tied to arrival and departure times so that the pressure to avoid mistakes is high. The impact of a mistake is greater when a passenger is toting heavy baggage. In airports, especially, there is a time-consuming process for obtaining boarding passes, checking baggage, and clearing security, tasks that are usually completed in places with high levels of congestion. In railway and bus terminals, the check-in process is less involved, but the need to sort out the location and schedule of departures makes negotiating these terminals complex as well. Large terminals are often complicated by fast-moving flows of commuters. In some cities, multimodal transportation terminals present a daunting challenge, particularly for the first-time or infrequent user or one who does not speak or read the local language. Consider the Shinjuku station in Tokyo, one of the busiest in the world. It serves commuter rail, subway, and bus lines, and about 3 million travelers pass through it every day.

Most travelers can understand the basic organization of terminals, with their common spatial syntax of entry area, ticketing area, main circulation halls, and gate areas. Generally, terminal designs channel people effectively through the building—for example, with pathways from arrival and ticketing areas through security checkpoints or ticketing gates to the departure gate areas. But the larger the terminal, the more difficult wayfinding becomes. A terminal that reduces the number of decision points for the traveler between the entry area and the gate areas reduces the potential for decision-making error.

For example, consider an airport terminal with a *treelike* structure (Fig. 24.1). As travelers proceed deeper into the system, they must make a series of decisions about which direction, left or right, to select at each decision point. Only the links encountered are visible to them. If a mistake is made, it becomes hard to orient oneself to the whole and recover from the wrong choice. Moreover, recovering from an error that leads a traveler deep into the wrong branch can be very time consuming and increase physical effort substantially, because one has to retrace the path back down the trunk and then out the new branch. Moving from the end of one branch to another is also time consuming and strenuous, particularly if the traveler has a short time period to change planes and is carrying baggage. Although the overall system is easy to comprehend, actually experiencing the plan is not the same as viewing a diagram of the whole.

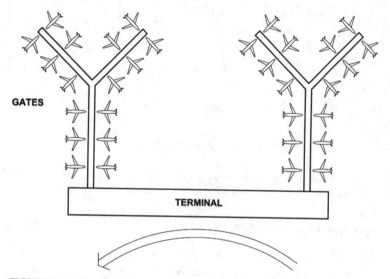

FIGURE 24.1   Treelike terminal plan.

A terminal design with one or more *islands* has a very different spatial structure (Fig. 24.2). The only choice is to decide from which building one's flight leaves. All the gates in each section are relatively close to one another and visible to the traveler from a central space. Going the wrong way is not as serious a mistake in this layout.

Most wayfinding aids found in terminals are visual. This excludes people with visual impairments and makes them dependent on others, unless they are extremely familiar with these buildings. A promising technology for helping people with visual impairments find their way in transportation systems is the *talking sign* system. These systems use small infrared (IR) transmitters that are fastened to ceilings, signs, and other locations. They beam a continuous message. People with visual impairments carry receivers that pick up the signals broadcast from the transmitters and convert them into a spoken message. Testing and evaluation of this technology indicates that it has great promise in a transportation environment (Golledge, Marston, and Costanzo, 1998; Crandall et al., 1999). A set has been installed permanently in a commuter rail station in downtown San Francisco and is working quite well (Fig. 24.3).

There are constraints to the IR talking-sign technology. For example, there must be a clear and uninterrupted line of sight between receiver and transmitter, and the receiver must be

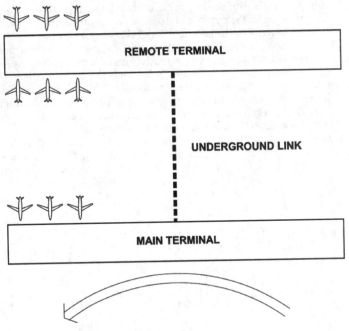

FIGURE 24.2 Island terminal plan.

pointed roughly in the direction of the transmitter to pick up the signal. However, experience with the technology has identified solutions to these problems.

Guidelines for Spatial Organization

- Reduce the number of decision points to each destination.
- Make all services and destinations as visible as possible.
- Provide a guidance system for people with visual impairments.

FIGURE 24.3 Talking-sign installation in a Caltrans terminal, downtown San Francisco.

24.4 INFORMATION SYSTEMS

All transportation terminals must include information resources to make travelers aware of schedules, where and when specific vehicles are arriving or departing, the location of amenities such as restrooms and food services, late-breaking news on schedule changes, and paging and other announcements. In public transit systems, terminals, stations, vehicles, and stops have information on routes served, schedules, and wayfinding. The ability to use these resources effectively is critical for the traveler.

Sign Systems

All terminals today have sign systems that identify the key locations for passengers, such as the ticket counters, security checkpoints, information services, and gates. These systems should have fonts that are large enough to be seen at a distance and are easy to read, have-contrasting text and background, and be located overhead, the most visible position. In general, contemporary sign systems fulfill all these criteria. A trend toward dynamic signage, however, suggests that additional criteria be identified that would ensure that the rate of information presented is within the abilities of people with sensory and cognitive limitations to perceive. Other topics related to dynamic information displays that need research include glare reflected off displays, resolution of fonts on displays, and color and flicker issues.

FIGURE 24.4 Tactile map that benefits all users.

The use of international symbols for common destinations is a major aid to the traveler who has sight but cannot read the languages available; it also reduces the amount of information that needs to be displayed by condensing words into pictures. Of course, visual information alone does not help people who cannot see it. Talking signs, tactile maps, and tactile tiles used as guide strips are useful in helping people with visual impairments find all these important destinations. These devices could be designed to benefit all travelers. For example, the tactile map provides more information than a two-dimensional map and can include color coding and other visual features to make it useful for everyone (Fig. 24.4).

In most international airports, signs are provided in more than one language. Extending this concept to rail, bus, and rapid transit systems would benefit tourists and business travelers from other countries. Two strategies are useful in overcoming the inability to read signs due to lack of ability in the language, illiteracy, or impairments of sight. In large transportation terminals, information clerks and ticket agents can provide personalized information on gates and schedules. Unfortunately, access to such services is often limited and available only by waiting in line, something that is often not possible when the information is most desperately needed. In smaller terminals or where congestion limits access to people, either telephone hot lines with direct connections to a central information service or automated systems can be used to augment or replace information assistants in the terminal itself. Buffalo, New York, has free telephones that provide schedule information at each transit station. As another example, the

Japan Tourist Office offers a toll-free hotline in foreign languages for tourists, through which one can obtain information on train and bus schedules from anyplace in the country 24 hours a day. Automated information kiosks can provide very useful assistance for people with visual and auditory impairments as well as for sighted and hearing travelers.

Additional Information Strategies

Additional sources of information include public address announcements, message boards and video/computer monitors that display arrival and departure information as it becomes available. It is these systems that often create barriers for people with sensory impairments. For example, message boards do not always have the same late-breaking information that is provided by public address systems. The message boards and monitors may not scroll and update fast enough to keep up with recent announcements. People with hearing impairments may miss information that is not available on monitors or message boards, like reasons for delays and track or gate changes. This is particularly a problem when using restrooms and waiting at gates in airports where flight information is not made available except through public address systems.

Technology for in-vehicle information has been enhanced through digital technologies. Newer vehicles provide information on upcoming stops prior to arrival in both visual mode, using an electronic display board mounted at the front of the passenger compartment, and as verbal announcements on a public address system. These systems can be automated and tied to a geographic positioning system (GPS) so that the vehicle will broadcast the right signal even though routes are changed. Moreover, GPS technology allows the automatic transmission of information about delays from the vehicle to stations and bus stops.

The advent of Web-enabled cellular phones and wireless Internet connections could provide a technology for notifying individuals with hearing impairments about changes in schedules in large terminals. The potential of personalized *software agents* that collect specific information and notify the user automatically could be harnessed to keep notifications up to the minute. Such a system would need to include wireless Internet connections in the terminal area. A less advanced technology would be to install terminals or message boards displaying real-time captioning of announcements at each gate. Such a system could be implemented now, but it would not reach individuals who were out of visual range (e.g., in the restroom).

Guidelines for Information Strategies

- Text on signs should be in a large easy-to-read font and should contrast with its background.
- Signs should be overhead and should be directly related to the location described or the path to it.
- Use international symbols wherever possible.
- Use multiple languages.
- Provide information kiosks or human information clerks.
- Provide text-based information for public address announcements.

24.5 NEGOTIATING LEVEL CHANGES IN TERMINALS

A key aspect of mobility in terminals is negotiating level changes. Level changes are often necessary to efficiently process arriving and departing passengers and their baggage in large terminals. Even in small buildings and individual transit stops, the boarding platforms are often above or below ground, out of the way of surface traffic. Terminals that serve intersecting rail lines must have more than one level in order to allow one line to pass over the other. Level changes within transportation terminals are accommodated the same way they would be in any multilevel structure—with elevators, stairs, and ramps. In terminals where access to some zones is restricted, however, providing an accessible path of travel is complicated by the need to provide accommodations in the areas on either side of the security perimeter.

The security perimeter can take two forms: (1) the boundary of the area secured by the checkpoints and the screening system for weapons and contraband or (2) the boundary of the area within which all riders must have valid tickets. In airports, changes in level may be needed both outside and inside the security perimeter. In local transit systems and some commuter rail systems, this perimeter is defined by the fare gates. Once individuals pass through a fare gate into the secure area, they should not have to go back outside that boundary to use vertical circulation systems unless they can do so freely without paying twice, without assistance, and without inconvenience.

One of the most serious problems faced by people with mobility impairments is being forced to rely on assistance to negotiate changes in level or to circumvent security perimeters to use accessible routes. Forced dependency results in inconvenience, embarrassment, anger, and exposure to injury by poorly trained attendants (Kawauchi, 1999). This is especially the case when people who use wheelchairs must use an entirely different circulation path from that used by other travelers. Careful design can reduce the need for duplicate elevators or ramps—for example, a system that keeps all passengers at the same level once they are within the secure perimeter. From the perspective of management, universal design reduces the need for special training for all new staff and improves productivity because workers do not have to leave their posts as often to assist an individual traveler.

The geometry of rail terminals is governed by the tracks. Access to them requires underpasses or overpasses to reach any platforms not adjacent to the entry hallway. The level changes need to be at least one complete story. Since large terminals have many tracks, providing wheelchair access to platforms can be very involved. Each platform has to have an elevator or ramp access. To minimize elevators, a transverse link can be provided that is served by both an elevator and escalators or stairs. From this link, ramps can feed each platform for all travelers. But such a system has to be planned into the design from the early stages. The two critical factors to be addressed in planning are the horizontal length required for the ramps and the difference in elevation between the transverse link and the loading platforms. Without considering these factors in the basic planning of the terminal, there may not be enough room available for ramps, and an elevator for each platform will be required. This circulation problem is less severe if the rail terminal is at the end of the line or on a spur. In such a case, access to each platform can be provided from the end of the tracks at the same level, and an overpass or underpass is not necessary. However, most railway terminals have to traverse the path of the train lines.

Guidelines for Level Changes

- Plan terminals to reduce level changes to minimum.
- Provide wheelchair accessibility to all vehicle loading areas.
- Maintain accessible circulation within and without the security perimeter.

24.6 NEGOTIATING LONG DISTANCES IN TERMINALS

Regardless of how well a terminal is planned, the sheer size of large terminals results in the need to traverse long distances. Some prototypical terminal designs are more compact than others—

for example, compare a terminal that clusters gates together in a node to one that spreads the gates out in long fingers. In Reagan International Airport in Washington, D.C., the plan has a long pedestrian hallway with the gates all distributed in fingers off the base; the passenger who has to rush from one gate to another in different fingers has a long route to travel in a short time. In large terminals like this, it is often impossible to avoid excessively long routes.

Most new terminals provide assistance through mechanical means to reduce the burden of these distances and increase the speed of traffic flow through the terminal. At Atlanta's Hartsfield Airport, the distance between gates has been reduced by planning several smaller terminals linked by an automated rapid transit system. Each remote terminal is an island that can have gates on all sides. This reduces overall corridor length. At Newark International Airport and at Chicago O'Hare International, new transit links have been added to the original multiterminal complexes. Another helpful strategy is the addition of moving walkways. A third approach is the use of small electric shuttles to transport individuals with limitations in mobility. Each of these mechanical systems needs to be designed for universal access.

Rapid transit links should provide some seating for people who have limitations in balance or low stamina. Their use has to be easily understood by people with sensory limitations and those who do not speak the local language. And they need markings to ensure priority access for people with disabilities and elderly people. In the United States, only English is used for the verbal instructions on most automated transit links. In fact, there are often not even any text instructions describing the operation of the systems, which restricts understanding to only those people who can perceive and understand spoken English.

Moving walkways should be wide enough to accommodate a wheelchair user, with additional space so that a pedestrian can pass by. Wheelchair users should not be prohibited from using these walkways since they are most likely to benefit from their use.

Although electric courtesy carts and trams are very beneficial for many people with disabilities, they can also cause problems in the terminal environment. In crowded conditions, they can be dangerous to pedestrians sharing the same path. In addition, the crowds of pedestrians can thwart the efficiency of the carts themselves. To make pedestrians aware of their approach, the carts are equipped with loud warning signals, but deaf people in the terminal are still not able to hear them. Moreover, there are many wheelchair users who cannot stand and climb into the carts. Thus, the people who could benefit most from their use do not have access to them. Clearly, the use of shuttles is a band-aid solution used to overcome the poor design of terminals. As the population ages, we can expect that demand for such assistance will increase. Thus, creative solutions to terminal design should be developed to integrate such vehicles into the overall circulation scheme or, better yet, provide a more universal transit system designed for all passengers.

Guidelines for Long Distances

- Plan terminals to minimize the distance of trips within the terminal.
- Integrate transit systems into terminal design.
- Provide automated walkways wherever there are long pedestrian paths in terminals.
- Plan terminals to separate courtesy carts from pedestrian traffic.

24.7   VEHICLE LOADING

One of the most difficult issues in universal design of transportation systems is accommodating level changes to board and exit vehicles. There are two basic accommodation strategies that can be used separately or in tandem. The first focuses on design of the vehicle and the second on design of the transit station or terminal. In addition to accommodating the level change, there is also a need to protect people from falls and other safety hazards in the loading area, especially when there is a loading platform. A uniform design approach throughout a system eliminates surprises for travelers, especially travelers who have visual impairments.

Rail Systems

In new transportation systems for interurban and commuter rail, loading platforms should be at the same level as the vehicle floor. There must also be a narrow gap between the platform and the vehicle to keep the front wheels of wheelchairs, walking aids, and people from falling into the gap. Consistency across the system and over time is important, so standards have to be established for both rolling stock and terminal construction. In existing systems where there are inconsistencies between levels, vehicles will have to be fitted with lifts or small ramps. The latter is preferable since all passengers can use them. But where there are great differences in levels, lifts may be needed because ramps would be either too long to fit on the platforms or too steep to negotiate. A small ramp (drawbridge) can be used to overcome small level differences between vehicle and platform. Such ramps may also be necessary to span large gaps between platform and vehicle that trap the front casters of wheelchairs or are hazardous for users of walking aids and people with visual impairments (Kanbayashi, 1999, p. 23). Although it is desirable to reduce such gaps to a safe size, larger gaps may be necessary to accommodate the lateral tolerances of rolling stock, especially if the vehicles used on one track have different widths.

One major obstacle to ensuring continuity in the travel chain in rail transport is the need to coordinate different companies. On a continent like Europe where crossing international borders is common, the need to ensure compatibility of systems across international borders requires a high level of coordination.

Street-Level Transit

Many systems, especially transit and shuttle buses but also trolleys, must load and unload on street surfaces. There are several technologies available to reduce or compensate for the level change between vehicle and street or walkway surface. The best approaches reduce the difference in height to a minimum. A good example is the *low-floor* bus that has a lower floor level than conventional buses and a "kneeling" hydraulic feature that brings the floor level within one step of the ground when activated. Such vehicles reduce the difficulty of entering and exiting significantly. This helps children, parents with children, older people, and people with limitations of mobility. It also helps people with visual impairments, because it reduces the complexity of entering and exiting a strange vehicle. A wide doorway and handholds add another level of convenience.

The low-floor kneeling bus does not eliminate the step up entirely, so it is not sufficient to provide a truly accessible vehicle. The most common solution to this problem is the wheelchair lift. Several types of wheelchair lifts are available for transit buses. They include lifts at the front or rear doors, lifts that slide out from under the bus floor, and lifts that are integrated into the stairs of the bus and fold out when the lift is needed. Another solution that works particularly well is the *stairless bus* (Fig. 24.5). This bus combines the low floor and kneeling features with a ramp that serves all passengers. Compared to the lifts, this solution is a far better example of universal design. Lifts benefit only wheelchair users. They are time consuming to operate and expensive to purchase and maintain. The person using the lift becomes a spectacle. Some riders and drivers may even become annoyed with the extra time required to load or unload someone with the lift, reinforcing the stigma associated with disability. With the stairless bus, however, all passengers enter the same way when the ramp is used. No one becomes a spectacle, and there is no delay caused by the need to accommodate

FIGURE 24.5   Stairless bus.

a person with a disability. Where the ramp can be extended onto a curb, the incline is eliminated as well, adding great convenience for all passengers.

Changing Platform Levels

Raising the entire secure area of transportation terminals off the ground to the same level
as vehicle floors is a good universal design strategy. It requires passengers to negotiate only
two level changes—at the end and at the beginning of the trip. Moreover, this strategy benefits all riders by increasing safety and convenience and improving service response by
reducing the time of loading and unloading a vehicle. Finally, it makes it easier to maintain
security in the system since no one can get on vehicles unless they are on a platform. All
entries to platforms can be controlled easily. Typically, platform loading is used only in rail
systems, but bus systems can also benefit from the use of this strategy. In Curitiba, Brazil,
all express bus terminals and stops are raised off the ground and are accessible by ramp or
lift as well as stairs. Once riders enter the secure area, they stay at the higher level and never
have to use stairs or any other vertical circulation. All vehicles are loaded with great ease
from the platform. In Buffalo, New York, a variation of this strategy was used for a light-rail
rapid transit system that has both underground and aboveground stations (Fig. 24.6). In the
belowground stations, trains are loaded directly from a platform as in all subways. All the
aboveground stations have a small raised platform level with the vehicle floor, and they all
have ramps. The platforms at the aboveground stations serve only the first car in the train,
however, to keep the platform length to a minimum. Thus, if someone with a wheelchair or
with a carriage boards a rear car, they cannot get off the train independently in the aboveground area.

FIGURE 24.6 Aboveground transit stop in Buffalo, New York.

Access onto Airplanes

Most airports in major cities have direct ramp access to airplanes, although the slope can be an obstacle. At commuter gates of these airports, where smaller planes are boarded from the pavement, elevators are needed to bring people who cannot manage escalators or stairs to the boarding level. This can result in the need for many elevators, each serving one or two gates. Although not common in the United States, in many airports in other countries, waiting areas are not directly connected to the gates. There is a separate circulation system for access to gates, using perimeter corridors. This system can be used to reduce the number of elevators serving gates that need access from the ground, because one elevator can serve a corridor connecting many gates.

Access to Water Transportation

Ferries designed to carry automobiles and trucks generally are easy to access because vehicles have to be driven on from the dock, although the boat needs an onboard elevator to provide full access if the passenger seating compartment is on a different level from the automobile garage area. Even when land vehicles are not accommodated, boarding from docks can usually be negotiated with ramp systems, although the ramp may have to be very long when the deck of the boat is considerably higher than the dock. The design of ramps serving boats is complicated by two factors. First, many docks serve several different kinds of boats with different entry conditions and deck levels. Second, tidal action results in the need to have an adjustable system that accommodates differences in the water height. There are many examples of adjustable ramps. One approach is the *telescoping ramp* that can adjust in length to accommodate different height decks. More than one can be available at a dock to accommodate a wide range of boats (Fig. 24.7). Such ramps can even serve more than one deck of a single boat. Another is the use of a *floating dock*. The ramp landings float up and down as the tides change and the ramp segments are hinged so that they move with the dock sections (Fig. 24.8).

Guidelines for Vehicle Loading

- Reduce the change in level between vehicle floor and loading surface to a minimum, or eliminate it entirely.

FIGURE 24.7 Telescoping ramp system.

FIGURE 24.8 Floating dock system.

- Provide mechanical loading systems when level changes cannot be eliminated.
- Eliminate the gap between platform and vehicle either by initial design or through mechanical means when a vehicle stops.

24.8   SAFETY ISSUES AT PLATFORMS

Avoiding falls off loading platforms is a major safety concern for individuals with visual impairments and children. There are several methods that can be used to protect the traveler from falling. One is the use of a gate-and-barrier system. This is by far the safest strategy, although it constrains the location of loading and unloading. Other safety measures include the use of warning signals and tactile warnings on floor surfaces. Warning signals provide advance notification that vehicles are about to arrive at the platform. Visible, audible, and tactile signals should all be used at once to provide redundant modes of information. In Washington, D.C., Metro stations, the platform edge is marked by a row of lights embedded in a granite strip about 18 in wide along the platform edge. When a train is about to arrive, the entire row of lights starts flashing. There is also a companion voice announcement of the arriving train. Many rapid transit systems have one mode of warning but not the other.

Detectable Warnings

Platform edges can be marked visually with a contrasting color and also tactually with a change in texture. Many different textures have been used by transit authorities. They include rough concrete or stone, applied resilient plastic materials, and contrasting paving materials like concrete against brick. There is great concern as to whether any of these methods are effective in warning pedestrians, particularly people with visual impairments (Richmond and Steinfeld, 1999). Several materials have been developed specifically to provide an effective warning for this latter group of people (Fig. 24.9). It is manufactured as a modular plastic tile with a pattern of small raised domes or ridges on it, usually with a bright yellow color. Such tiles were recently installed in new and existing Washington, D.C., Metro stations, augmenting the existing signaling system. This material has been used extensively, particularly in Japan.

There are many professionals and consumer advocates who question the relative reliability and safety of tactile tiles compared to other approaches. The textures may have to be very obtrusive to work with a high degree of reliability. However, exaggerated textures could cause tripping hazards for pedestrians. An important consideration in utilizing tactile tiles is to ensure that people with visual impairments can distinguish between an edge warning and a guide path. Guide paths on platforms should keep travelers away from the dangerous part of the platform (the edge). Kanbayashi (1999, p.23) points out that tactile tiles along a platform are not very useful as a navigation aid because people and luggage create obstacles along them. He also reports that people with visual impairments still fall onto the track despite the presence of these tiles at the edge of the platform.

Platform Barriers

The most secure system for protecting waiting passengers at the platform edge is a physical barrier along the entire platform. In Curitiba, Brazil, the express bus stations are constructed of a prefabricated plastic tube system in a circular section that protects all passengers from falling off the platform. Bus trains have automated systems that open the doors on both buses and stations when sensors on the platform detect that the bus is in the proper position. At that time, travelers are free to load and unload. In Curitiba, bus stations provide such a barrier (Fig. 24.10). They have sliding door openings where the doors to the vehicles will be when the

FIGURE 24.9   Tactile tiles used to mark platform edge.

FIGURE 24.10   Protected access at station platform in Curitiba, Brazil.

FIGURE 24.11   Guard rails with sliding gate at transit station platform.

buses stop at the stations. In new subway stations, guard rails and even complete glazed enclo-sures with automated sliding doors are being introduced (Figs. 24.9 and 24.11). Not only do such barriers protect people with visual impairments; they also protect the general population from being pushed off the platform, and they prevent people from committing suicide by jumping in front of trains. Current technology supports the precise docking of vehicles that enables this approach to platform safety.

Guidelines for Platform Safety

- Protect people from accidents at loading platforms by means of barriers or warning devices.
- Warn passengers on the platform of arriving vehicles, using both visual and audible means.

24.9   VEHICLE DESIGN

Three of the links in the transport chain that need a lot of attention with respect to universal design are taxis, shuttle buses, and intercity buses. There are some examples of accessible taxis, but the idea has not caught on in a major way. Small modifications in taxi design, includ-ing wider and higher doors, better grips, more leg room, and swivel seats, would be a start toward universal design (Short, 1999, p. 7). But, ultimately, a fully universally accessible taxi fleet would include only taxis that could load a wheelchair and its occupant without the need for transfer. The London taxis have high doors, wide openings, and a ramp that pulls out to allow wheelchair boarding. A new taxi in a Japanese city uses a minivan with rear wheelchair loading. So, hopefully, universal taxis will become more common. The provision of universal access to shuttle buses like those that connect hotels with air terminals is an even more neglected area. The same methods used to make transit buses accessible can be used on shut-tles, but where accessible shuttles exist, they usually are special vehicles that need to be sched-uled in advance, and service may require a longer than normal wait. In the United States, there has been a long battle to bring intercity buses into compliance with the Americans with Disabilities Act (ADA). In the fall of 1999, the Department of Justice negotiated an agree-ment with Greyhound, the leading intercity bus line in the United States, to provide accessi-bility features on all its buses.

Seating and Circulation in Vehicles

Universal design of mass transit vehicle interiors starts with ensuring that wheelchair users can negotiate from the entry to the wheelchair seating area. Wheelchairs are often wider than the aisles in buses, trains, and airplanes. Wider aisles can be provided in two ways. The first is to use perimeter benches, rather than seating in rows, and the second is to remove some seats. Both these strategies also provide more space for standing passengers during rush hours, thus increasing overall system capacity. Perimeter seating does not have to be continued throughout the vehicle. It may be limited to the area close to the accessible entry.

If seating in rows is provided, space for wheelchair users can be left at the entry areas. These spaces can double as locations for storing luggage, carriages, or bicycles or for standing passengers when not accommodating a passenger using a wheelchair. Another approach is to provide convertible seats that can be folded up against the wall when a passenger using a wheelchair needs to be accommodated. In Curitiba, the express buses have very few seats in order to maximize their capacity during rush hours. This has created a very convenient interior for wheelchair access.

Tie-Down Mechanisms

Spaces for passengers using wheelchairs should be equipped with tie-down devices to prevent injury to both the wheelchair user and other passengers if an emergency stop is necessary. Some types of tie-down devices are easier to use than others. One type can be used without assistance. The wheelchair is simply backed up into clamps that automatically grab the wheels and hold them still. The device can be disengaged easily with a release lever within reach of the passenger. These devices do not prevent an individual from being launched out of the chair in a sudden stop. Moreover, to be universally effective, the clamp system needs to be flexible enough to accommodate a wide range of wheelchair sizes and tire widths. Another type of device uses a web belt and clamp to hold the wheel from moving. However, it requires that an attendant fasten it around each wheel. Other tie-down devices utilize a combination wheelchair tie-down and passenger restraint system (e.g., shoulder or seat belt). These systems provide more comprehensive protection. However, in bus systems, their complexity can require a lengthy delay while the driver fastens them in place. This delay can effectively negate the time savings of the stairless design, and it puts wheelchair users in a position of dependency and makes them objects of unwanted public attention.

Currently there are no wheelchair tie-down systems that accommodate all wheelchair designs, are easy to use without assistance, and are safe for all passengers. It should be noted that there are many rail systems that do not provide tie-down devices, particularly older systems. One could argue that there is no difference in leaving luggage, a carriage, a bicycle, or a wheelchair user loose in the vehicle. But the absence of tie-downs and restraint systems could leave a transit authority vulnerable to legal action if passengers were to be injured during an emergency stop since accepting the responsibility of transporting a person using a wheelchair brings with it the obligation to protect the passengers, both the wheelchair user and others. This is particularly the case for some wheelchair users with low strength and limited ability to maintain their balance. These passengers may encounter injury when others would not, and they may not be able to keep a wheelchair from moving using their own strength. Clearly this design issue requires much more research and product innovation.

Airplane Seating

The most difficult problem of wheelchair seating is on airplanes, where removing seats has a significant impact on costs. The current approaches for loading and unloading passengers using wheelchairs are demeaning, dangerous, and fraught with inconvenience and difficulty for the passenger. The best current strategy is to provide independent wheelchair access at least to the entry of the plane, where a passenger is then assisted to his or her seat (often with a transport chair), and the wheelchair is removed to the baggage compartment. This strategy, however, separates the passenger from his or her wheelchair and creates dependency on airline staff, who are often under great time pressure. An individual who needs the custom designed seating system in his or her wheelchair will not be comfortable during the flight. Moreover, if the individual cannot walk at all, he or she may be effectively imprisoned in the seat until arrival at the destination. If the chair does not reach the same city at the same time, serious personal problems can result. During plane changes, the passenger has to be assisted by airline personnel in the terminal. In addition, putting a wheelchair into baggage means that it will not be treated the same way as if it were within sight of the owner. Many people who use wheelchairs complain that their chairs are often damaged during airline trips. Innovative design solutions are needed to develop a system that can accommodate wheelchair users in their own chairs without taking up valuable floor space when not in use.

Wheelchair users are not the only passengers that need accommodations while in the passenger compartment. Frail older people, people with disabilities who don't use wheelchairs, pregnant women, and others who need to sit should have access to priority seating. This seating should be near a door of the vehicle and well marked.

Onboard Toilet Facilities

On long-distance vehicles like trains, intercity buses, and aircraft, access to toilet facilities is needed. This, of course, is one of the most challenging design problems in universal design of transportation. Access to toilet facilities on trains has been successfully accommodated in many locations. The approach is similar to that used in single-user toilet rooms in buildings. To conserve space, such facilities can be designed so that turning a wheelchair around inside the toilet compartment is not possible as long as the individual can back in or enter forward, reach all the fixtures and equipment, and open and close the door. Automated door opening and closing with a sliding door facilitates access significantly. Wheelchair access in airplanes is now available on some newer equipment. Larger toilet compartments are provided in the front of the plane near the boarding area, and wheelchair users are seated close to the compartment.

Guidelines for Passenger Seating

- Provide space for wheelchair users in the vehicle with tie-down devices.
- Develop tie-down devices that can be used independently.
- Provide priority seating for people who have difficulty standing during a trip.
- Include accessible toilet compartments on long-distance vehicles.

24.10 *BARRIERS IN TICKETING AND SECURITY*

For many riders, with disabilities or not, the most annoying parts of a transit system to negotiate are the ticketing and security areas. This is especially the case for first-time users of the system who do not understand how it works. The problems are compounded if the rider has a visual disability or does not read or speak the local language. In many cases, there are unusual rules that have to be followed, with significant penalties for transgressors. In others, ticket prices vary based on the length of the trip and are difficult to determine without prior experience with the system. In most cases, the rider must purchase a ticket or token and then use it in a fare gate. This can often require the use of two different machines. Security systems at airports and other terminals require complex maneuvering under great social pressure, especially at rush hours. At the destination, users often find that the ticket they purchased is not sufficient to leave the system, and they need to add fare or upgrade tickets to exit. And transfers to other lines and vehicle types often have to be obtained. Universal design can help to reduce the stress and increase convenience for all riders.

Design of Ticket Machines

The operation of fare and ticket machines is often incomprehensible, even if one reads the local language. But, through design that provides strong "affordances" and utilizes intuitive methods of operation (Norman, 1988), the steps for using these machines can be conveyed easily without complicated instructions and with illustrations to overcome language and literacy limitations (Fig. 24.12). To be usable by people with visual impairments, fare and ticket machines should be standardized in design and include tactile or audible cues to their operation.

The most direct way to increase usability of ticketing systems is to simplify the task of purchasing tickets. For example, many systems separate the change machine from the ticket machine. While it may be useful to have separate change machines for convenience in purchasing food or beverages, there is no reason to separate those functions for purchasing tickets. Ticket machines can also provide change. This eliminates a source of congestion and reduces the number of tasks necessary to use the system. A second strategy is to combine the ticket machine with the access gate. For example, money can be used to get access instead of a token or fare card. Fare cards could be issued as money is inserted in a combined ticketing/access gate machine.

FIGURE 24.12 Fare machine with simple instructions for use.

Additional simplification can be achieved by the use of fare cards that one can purchase for varying denominations and that can be used throughout a system, even on different modes of transport. Passengers only have to pay once to buy the card and can use it for many trips until it runs out. Daily, weekly, or monthly passes eliminate the need to obtain transfers, so such options greatly facilitate the use of transit systems. Finally, the actual task of buying a fare card or token can be simplified through the use of card-swipe systems for purchasing tickets with credit cards, or sensor systems with prepaid or automated credit card billing, like those now used on highways.

Options for Negotiating Barriers

Ideally, aside from local buses or light-rail vehicles that stop on the street, the only time one should have to pay is when entering the security perimeter of a transit system. Once in the system, there should be a continuity of path within the perimeter without having to pass through and exit and reenter. This may not be possible, however, when changing modes (e.g., from train to bus) or lines. In these cases, intermodal passes or transfers should be available to avoid the need to pay again. Transfer dissemination should be obvious and evident to the user. In many cases, one has to ask the driver of a vehicle for a transfer. This can be impossible or difficult for people who have communication limitations.

Every bank of gates should have at least one gate that is sufficiently wide for wheelchair passage. Turnstiles are impossible to negotiate with a wheelchair and also are difficult to negotiate with a baby carriage or baggage. Retracting barriers are an accessible alternative to turnstiles. Another alternative is to have one larger gate supervised by a fare collector. In such situations, this option should be available to any passenger and be within close proximity to the fare-collection station. If the alternative gate becomes difficult to supervise, it increases the potential that the person in charge will resent the inconvenience of accommodating passengers using wheelchairs, and breakdowns in accessibility will likely result.

The most universal solution is to have no fare gates at all. Buffalo, New York, utilizes such an honor system. Roving transit police make random checks for valid tickets. The penalty for being caught without one is severe, like a traffic ticket. The result is that few riders take the chance. However, to make this honor system work, there have to be sufficient fare machines to purchase tickets even when there is a mechanical breakdown. If there is only one machine at a stop and it is broken, the rider is forced to choose between the inconvenience of walking to the next station with a working fare machine or taking a chance on not being caught. Buffalo has instituted a policy of free access for people who use wheelchairs that effectively reduces the burden of finding a working fare machine, but it does not address the significant problem this can create for people with other disabilities. Luckily, malfunctions are not common and passes are available for frequent users.

Guidelines for Ticketing

- Provide intuitive fail-safe ticket machines.
- Simplify the ticketing system by reducing steps and offering optional payment plans and methods.
- Reduce the number of payments to a minimum, preferably one.
- Provide access for passengers using wheelchairs, pushing baby carriages, and carrying luggage at all security and fare perimeters.
- Fare- and ticket-collection machines should be designed for use without having to read and without vision.

24.11 CASE STUDY: CURITIBA'S INTEGRATED TRANSPORTATION SYSTEM

Curitiba, Brazil, has a worldwide reputation for its accessible urban transit system. Starting in 1970, the city planned and implemented a modern system that was designed from the ground up to replace a system of many poorly coordinated private bus lines. By 1992, the system was serving 600,000 people daily while the city had grown to a size of about 1.5 million people. A main planning goal was to provide public transportation that would be so effective that citizens would find little need for private transportation, thereby creating more concentrated growth, protecting the environment, and reducing congestion and urban sprawl in the future. Many elements of the system are described as examples in the preceding sections. This section focuses on the overall planning concepts and how the system works as a whole.

The system was designed to provide full accessibility for people with disabilities, but, in addition, the planners perceived similar benefits for elderly people, children, and others with limitations in mobility and improved efficiency for the general population. The design adopted combined low cost with a high level of performance and accessibility for all. The components of the system include express *busways* with dedicated rights-of-way on radial routes into the city core, conventional local bus routes that join each other or the busways at key terminals, interline connector buses that circumnavigate the city, and paratransit vans for door-to-terminal service for those who require it.

All transportation terminals are designed to be fully accessible. There are three basic types. The first is an on-grade local terminal, where transferring passengers cross from one loading area to another at grade. Accessibility is provided by buses with lifts at these terminals. The second type is the multimodal terminal, where local buses deliver passengers to the stops on the express busway system. Busway vehicles are large *bus-trains* consisting of 2- or 3-unit articulated buses carrying 250 to 350 people each (Fig. 24.13). The bus-trains load and unload directly to raised platforms. This allows safe and rapid movement of passengers as well as full accessibility like in a subway. All express bus terminals have ramps and/or lifts. The *tube stations* are smaller intermediate stops along the busways and are also sometimes linked together to construct larger stations at heavily used destinations. The tube stations are constructed of large prefabricated plastic and steel tubes with lifts at one end (Fig. 24.14). By clever planning of circulation, transfers take place entirely within the secure areas of terminals, avoiding the necessity of multiple fare collections and transfer tickets at mode changes.

The planning, design, and construction of the system and rolling stock was coordinated by a new agency, URBS, set up for this purpose. This agency centralized purchasing and coordinated standards so that all the pieces of the system would work together. However, management of each district of the system was contracted to the existing private bus lines. Private individuals operate the parataxi vans. Originally, they were designed specifically to provide people with disabilities home-to-station access, but experience demonstrated that there were not enough customers among the population of people with disabilities to maintain such a system on fares alone. These vehicles are now available for all riders.

Terminals and tube stations were constructed of low-cost prefabricated systems. The small size of the prefabricated structural components reduced construction costs significantly by

FIGURE 24.13   Bus-train.

FIGURE 24.14 Tube stations.

allowing fabrication without expensive equipment, using low-cost hand labor that is available in Brazil. The rolling stock was designed and bid out to specifications that included accessibility features and environmentally conscious fuel systems. Volvo, the winning bidder, established a production facility in Curitiba that is now a major employer. A surface bus system was chosen because it reduced the cost of developing a modern mass transit system significantly and because the bus-train concept provided the same capacity as a fixed-rail system.

The system is a good example of universal design. It is remarkable in that it was conceived before the concept was well established. It provides a high level of access, especially for a bus system. The tube stations and platform loading create great convenience for all riders. A surface system is generally more pleasant to ride than a subway and facilitates wayfinding. Moreover, the integrated system of local routes, interline routes, and express routes with a single fare purchase provides a seamless experience with great convenience.

The vehicles for each type of line are color-coded, which makes it easy for even a rider who does not read to understand the type of service that is being provided at each stop. Moreover, the protection of platform edges on the express lines by the tube-station design removes the safety hazard for visually impaired travelers. As of this author's visit in 1994, all the route information provided was in visual mode, so users with visual impairments did not have full access to information about the routes. Not all the express stations used the tube-station construction, so the design of the platform edge was not addressed everywhere. Moreover, all stations had a human fare collector, so automation of fare collection was not an issue. Nevertheless, Curitiba is an excellent model for developing economies and also for urbanizing areas anywhere that are seeking to improve their mass transportation systems without investing in a fixed-rail system.

24.12 CONCLUSIONS

Universal design in transportation systems is in its infancy. While many of the lessons learned in the general design of accessible buildings are applicable to transportation terminals, stations, and stops, there are many unique concerns that cannot be addressed by existing guidelines and standards. For example, current guidelines and standards do not provide any guidance on access to dynamic scheduling information, and many of the universal design concerns are beyond issues of minimal accommodation for disabilities such as inability to read the local language. Moreover, high-technology applications are emerging that could increase the options available for universal design.

There is very little ergonomic research on usability of mass transportation systems by diverse passengers. These systems are designed using a knowledge base derived for the broader, nondisabled population. It is time to reexamine this knowledge base with the new demographics in mind. In the field of transportation ergonomics, there is a developing body of research and information that has identified problems with both the physical design of vehicles and the information systems used in conjunction with them. However, there is clearly a need for research on the effectiveness of new technologies in addressing the needs of older people and people with disabilities. Some specific design issues for accommodating people with disabilities still have not been resolved adequately, such as access to airplanes, tie-downs in buses, and solving the information needs of people with sensory impairments.

As the demand for transportation at all scales continues to increase, the issues identified here are likely to become more and more important. Research is needed both to identify the severity of problems and the priorities of the traveling public and also to address the gap in information needed for design decisions. It is important that this research give attention to the differences and similarities between travelers who are older or have disabilities and those who are not. The value of including universal design features in transportation systems will be increased if research can demonstrate their widespread value to all travelers. But the specific needs of people with disabilities need to be identified and prioritized as well, so that they do not become neglected in the pursuit of greater convenience and usability for the broader population. It is likely that those specific needs can be addressed in ways that will benefit everyone. Some examples include better information on scheduling changes in terminals and greater accessibility of vehicles.

The challenge of universal design in transportation is not only to eliminate discrimination in access but to ensure social integration in use. While the existing mass transport infrastructure is difficult to change, particularly old underground systems in deep tunnels and interurban rail systems, there are many opportunities for practicing universal design in developing countries and the rapidly growing suburbs of major cities. Reducing environmental pollution and depletion of dwindling fossil fuel resources are key goals of mass transport planning. The city of Curitiba is an excellent example of how the goal of increasing convenience for all can be added to the planning strategy to attract commuters away from automobiles.

24.13   BIBLIOGRAPHY

Crandall, W., J. Brabyn, B. L. Bentzen, and L. Myers, "Remote Infrared Signage Evaluation for Transit Stations and Intersections," *Journal of Rehabilitation Research and Development,* 36(4), 1999.

Golledge, R. G., J. R. Marston, and C. M. Costanzo, *Assistive Devices and Services for the Disabled: Auditory Signage and the Accessible City for Blind or Vision Impaired Travelers,* Working Paper UCB-ITS-PWP-98-18, California PATH Program, Institute for Transportation Studies, University of California, Berkeley, CA, 1998.

Gottdiener, M., *The Social Production of Urban Space,* University of Texas Press, Austin, TX, 1994.

Kanbayashi, A., "Accessibility for the Disabled," *Japan Railway and Transport Review,* 20, 1999.

Kawauchi, Y., "Railway Stations and the Right to Equality," *Japan Railway and Transport Review* 20, 1999.

Naniopoulos, A., "European Approaches to Accessible Transport Systems," *Japan Railway and Transport Review* 20, 1999.

Norman, D., *The Design of Everyday Things,* Doubleday, New York, 1988.

Richmond, G., and E. Steinfeld, "The Usability of Tactile Warning Surfaces for People with Visual Impairments," in Edward Steinfeld and G. Scott Danford (eds.), *Measuring Enabling Environments,* Kluwer Science/Plenum Press, New York, 1999.

Short, J., "Transport Accessibility," *Japan Railway and Transport Review* 20, 1999.

24.14 RESOURCES

The following resources can be helpful in learning more about universal design in transportation systems:

TELSCAN.   European Commission research project that develops resource information on all aspects of travel for people with disabilities. Web site: http://hermes.civil.auth.gr/telscan/telsc.htm.

European Conference of Ministers of Transport (ECMT).   Publishes many reports on these issues.

Access Exchange International. Publishes a newsletter, *Accessible Transportation Around the World.* Address: 112 San Pablo Ave., San Francisco, CA 94127-1536.

International Centre for Accessible Transportation.   Supports the development of new technologies and other methods to make transportation more accessible to all travelers. Address: World Trade Centre, 380 Saint-Antoine St. West, Suite 3200, Montreal, Quebec, PQ, Canada H2Y 3X7. Web site: www.icat-ciat.org.

CHAPTER 25
THE BOTTOM-UP METHODOLOGY OF UNIVERSAL DESIGN

Selwyn Goldsmith, M.A., R.I.B.A.
London, United Kingdom

25.1 INTRODUCTION

This chapter reviews the concept of universal design as it relates to public buildings from a British perspective. In this context, universal design, with its aim of making buildings that are convenient for everyone and its principle that architects ought constantly to be looking to expand the accommodation parameters of normal provision, is incompatible with prescriptive design standards. It has a bottom-up methodology, as compared with the top-down process of legislative controls based on mandatory design standards. For regulating the accessibility of public buildings, the design standards in Britain, as in America, are based principally on the capabilities of independent wheelchair users. These are discriminatory. They come with cutoff points, meaning that those above the line, such as severely handicapped people with disabilities, are excluded. Nor is it the case that all people with disabilities below the line are suitably catered for by arrangements that are appropriate for independent wheelchair users. Relatedly, others than people with disabilities can be vulnerable to architectural discrimination when they use public buildings and are not protected by mandates for those with disabilities, noting, for instance, women when they use public toilet facilities. The views that this chapter expresses are personal and should not be understood to reflect prevailing opinion in Britain.

25.2 BACKGROUND: PRODUCTS VERSUS BUILDINGS

Broadly, universal design means that the products that designers design are universally accommodating, that they cater conveniently to all their users. The process by which this is achievable is usually evolutionary. Initially, the product may be designed for the mass market of normal able-bodied people. Subsequently, it is refined and modified, the effect, with accommodation parameters being extended, being that it suits all its other potential users as well, including people with disabilities.

Four examples of this process are cited, none of the products concerned being ones that in previous forms had been geared to suit people with disabilities: first, the television remote control; second, the personal computer—as word processor, e-mail communicator, and, through the Internet, information provider; third, the mobile telephone; and fourth, the microwave oven.

These are products of a kind that are amenable to the accomplishment of universal design. Buildings, the products that architects design, are for four reasons not so amenable. First, technology: The products just cited are all ones whose design has owed much to modern technology—most important, electronic technology. And while electronic technology has a role to play in buildings—manifested, for example, by automatic-opening doors in public buildings—it has little effect on how buildings are planned and designed so that people can use them. Second, market forces: Manufacturers and retailers compete for customers for the products, and customers have a choice. The way that public buildings present themselves to their potential users is not subject to their necessarily being attractive to their consumers, who in effect are not given a choice. Third, design standards and costs: Buildings, when they are designed and built, are subjected to minimum standards (including those for access for people with disabilities) with which architects are statutorily obliged to comply, and if the minima are exceeded, costs—always a vital consideration—will rise. Fourth, relatedly, the architect has no material incentive to make buildings universally accommodating.

Tackling Architectural Discrimination

None of these factors ought, however, to prevent architects from looking to apply universal design principles to the buildings they design. When designing new buildings or alterations to existing ones, the aim of architects who work to the precepts of universal design is to cater to all the users of their buildings and ensure, so far as they can, that none will be threatened by architectural disability—by being unable to use the building or a feature of it on account of the way the architects designed it, or, meaning in effect the same thing, being subjected to architectural discrimination that makes the building difficult or impossible to use. Broadly, those who are discriminated against are those who are affected by the depth, breadth, or height of spaces in and around a building or the steepness of its surfaces. With respect to people with disabilities, a consequence is that architects can do a great deal to assist those with locomotor disabilities—people with disabilities who are ambulatory or use wheelchairs—but relatively little for people with sensory or cognitive disabilities (Goldsmith, 2000).

Access Standards

The vulnerability of people with disabilities to architectural discrimination is now widely acknowledged, and in Britain, America, and other developed countries around the world, architects who design public buildings are required by law to make them accessible to people with disabilities. To do this they have to apply minimum design standards—in America, those prescribed in the Americans with Disabilities Act Accessibility Guidelines (ADAAG) (U.S. Access Board, 1991), and in Britain, those prescribed for meeting the requirements of Part M, the national building regulation, which covers access and facilities for people with disabilities (Department of the Environment and Welsh Office, 1999).

Whether for newly designed public buildings or alterations to existing ones, the realization of universal accommodation can for three reasons be hindered when architects apply statutorily prescribed accessibility standards for people with disabilities. First, attention to the needs of people with disabilities does not encompass many other building users who are prone to architectural discrimination. Second, access standards come with cutoff points, with the result that those people with disabilities who need particular kinds of provisions that the access standard does not embrace are excluded. Third, the methodologies of designing for people with disabilities versus designing for everyone are conflicting.

Design for people with disabilities, as represented, for example, by the Americans with Disabilities Act access standards or the British Part M building regulations, has a top-down methodology. The presumption is that people with disabilities are abnormal, are peculiar and different, and that, in order to make buildings accessible to them, they should be packaged together and then, with a set of special accessibility standards, have their requirements presented, in top-down mode, as add-ons to the unspecified provisions that are made for normal people.

25.3   THE UNIVERSAL DESIGN PYRAMID

The bottom-up methodology of universal design is demonstrated by the universal design pyramid (Fig. 25.1). It corresponds with that of the European Concept for Accessibility, which for the accommodation of people who are disabled advocates "integral" (i.e., normal) provision rather than "categorical" (i.e., special) provision (CCPT, 1996). The principle is the same whatever kind of building the architects may be working on; it is that they should be looking to expand accommodation parameters so that, as best as can be, their building will cater conveniently to the needs of all its potential users. Those needs are variable according to the character of the building and its users, and to judge how architects in Britain have tended to perform over the last 50 years or so, the buildings under review are assumed to be public buildings such as theaters, department stores, pubs, hotels, and restaurants, ones which have public toilets for the benefit of their customers among their amenities.

Rows 1 and 2 are at the foot of the 8-level pyramid. In row 1 are fit and agile people, those who can run and jump, leap up stairs, climb perpendicular ladders, dance exuberantly, and carry loads of heavy baggage. In row 2 are the generality of normal adult able-bodied people,

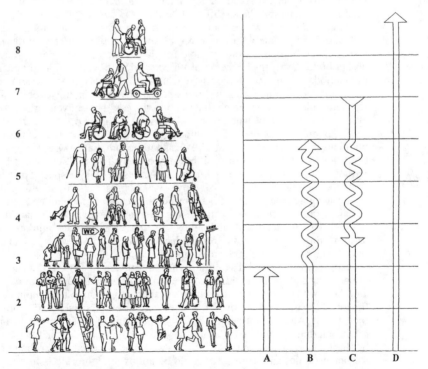

FIGURE 25.1   The universal design pyramid demonstrates the bottom-up methodology of universal design.

those who, while not being athletic, can walk wherever needs or wishes may take them, with flights of stairs not troubling them. Architects do as a rule cater well enough for these people. It needs, however, to be noted that there are no small children in rows 1 and 2.

In row 3 are people whose needs when using public buildings are not covered by the application of access standards for people with disabilities, even though some among them may have disabilities. Frequently, the architects fail them. These are women, the users of public buildings who are regularly subjected to architectural discrimination because, unlike men, they can be obliged to join a long queue or abandon the quest when they attempt to use public toilets, and along with that find that the toilet compartments—if they are able to reach them—are extremely inconvenient to use on account of their small size.

In row 4 are older people who, although perhaps going around with a walking stick, are not perceived as being "disabled." By comparison with younger adult people, however, these are people who are commonly prone to discrimination caused by steep steps or stairs without handrails.

Also in row 4 are people with infants in strollers, who—men as well as women—can be architecturally disabled when looking to use public toilets. Three common impediments feature here. First is the approach to public toilet facilities—in public buildings in Britain, it is common for this to be by way of stairs to an upper or lower floor. Second is the lack of adequate turning space in enclosed restroom lobbies. Third is that the very confined space of normal men's or women's toilet stalls, along with in-opening doors, precludes stroller accessibility.

Not only with respect to public toilet usage, building users who have infants in side-by-side double strollers can be particularly vulnerable to architectural disability. The width of the strollers necessitates wide doors, ones that are wider than those which are prescribed in the British Part M access standard for meeting the needs of wheelchair users (Department of the Environment and Welsh Office, 1999).

Ambulant people with disabilities are in row 5. Broadly, the building users who are in rows 3, 4, and 5 are people who would not be subjected to architectural discrimination if normal provisions in buildings were suitable for them, if it were standard practice for architects to design buildings to the precepts of universal design, with public toilet facilities being more accommodating and conveniently reachable, and steps and stairs being comfortably graded and equipped with handrails to both sides. Across Britain, however, this is not by any means a general rule, the effect being that people who could be conveniently accommodated by suitable normal provisions when new buildings are designed often are not.

The people in row 6 are independent wheelchair users. In the years since 1985, new public buildings in Britain have had to be designed in compliance with the Part M building regulations, meaning that access provisions for people with disabilities have to be made in them. The Part M process operates top-down, and it focuses on making special provisions in buildings. It is independent wheelchair users who govern its prescriptions, an effect of which is that when the design guidance in Part M is followed, the needs of independent wheelchair users may be satisfied, but not necessarily those of ambulant people with disabilities or people using wheelchairs who need to be helped by someone else in public buildings. The outcome of this selective top-down procedure is that the needs of people in rows 5, 4, and 3 may be only partially taken care of when they use public buildings.

People with physical disabilities whose particular needs are not fully covered by Part M are at the top of the pyramid. In row 7 are people with disabilities who drive electric scooters and those who use wheelchairs who need another person to help them in public buildings. At the top of the pyramid, in row 8, are wheelchair users who need two people to help them when they go out. A need that people in row 8 and many of those in row 7 could have when using public buildings would be for a suitably planned unisex toilet facility where a wife could help her husband, or a husband his wife. This would be a special rather than normal provision, but for universal design purposes it would be admissible; the rule is that supplementary special provisions may be made where normal provisions cannot cater to everyone.

Height of Fixtures and Fittings

The ideal outcome would be buildings that are entirely convenient for all their users. As has already been noted, however, the universal design pyramid does not show children, and for them the important factor in the context of universal design is the height of fixtures and fittings.

The issue is exemplified by lavatory basins. In restrooms in public buildings where there is a single basin, and also where two or more basins are at the same level, it is customary for the bowl rim to be at about 820 mm (31.5 in) above floor level. This is not convenient for young children. Nor is it convenient for standing adult people, for whom 950 mm (37.5 in) is more suitable. There is no single level at which a lavatory basin can be fixed so that it suits all users.

Relatedly, urinal bowls cannot be at a common height to suit all users. For elevator controls, the advice for meeting the Part M requirements is that they be placed between 900 and 1200 mm (35.5 and 47.25 in) above floor level. This is convenient for independent wheelchair users, but not for standing people, particularly those with sight impairments.

By comparison with height dimensions, horizontal spaces need not be problematical for architects. When they are working on their screens or drawing boards, their plan drawings of a new public building may illustrate convenience for all users: horizontal dimensions show spaces that are wide enough for wheelchair users and others; at building entrances there are no steps; elevators are there as well as stairways; unisex toilets are there to supplement normal men's and women's facilities, and so on.

The Principles of Universal Design are not compromised by the fact that a wash basin cannot be fixed at a height which will be convenient for all its users. By expanding the accommodation parameters of normal provision, with supplementary special provisions being added on where appropriate, the architect's objective is to make buildings as convenient as can be for all their potential users. The operative condition is *as convenient as can be*. There are times when architectural discrimination is unavoidable.

Alterations to Existing Buildings

The expansion of accommodation parameters of normal provision is a principle that applies equally when alterations are made to existing buildings. Automatic-opening doors can replace heavy side-hung doors. Steps can have handrails to help ambulant people who have disabilities. In-opening doors to toilet stalls can be rehung to open out, facilitating access for people using wheelchairs or pushing strollers. The altering of existing public buildings may not, however, afford the same scope as newly constructed buildings. Entrance steps which cannot be substituted or bypassed will preclude accessibility for independent wheelchair users. And, more frequently for alterations than for new buildings, it may be appropriate to incorporate supplementary special provisions—for example, the installation of a platform lift to carry wheelchair users where there are steps on circulation routes.

Set against the current rules for accessibility minimum design standards, universal design represents a paradigm shift. Had its virtues been recognized and acknowledged at the start of the accessibility era, from around the mid-1950s, it could have become the construct to which architects around the world would now have been working. Had this occurred, access for everyone would have informed legislative instruments, not access for people with disabilities. The social model of disability would have been operational, not the medical model. Bottom-up methodology would have been in order, not top-down. And design standards for "accessible" building features would not have been circumscribed.

For regulating the accessibility and usability of the built environment on behalf of all building users, the question that is prompted is whether, either in America or Britain, it would now be practicable to make the leap, to shift from the access-for-people-with-disabilities to the universal design paradigm. Notionally, it could be, but in its way there is an obstacle, the solid entrenchment in both countries of the former brand of accessibility.

25.4   *THE 1961 AMERICAN STANDARD*

Tim Nugent was the visionary idealist who began it all. As director of rehabilitation services on the Champaign-Urbana campus of the University of Illinois, he recognized in the early 1950s that architectural barriers stood in the way of his students with disabilities being able to realize their potential for achievement and compete successfully with others for the material rewards which America offered. Having demonstrated how the university and public buildings in Champaign and Urbana could be altered so that students with paraplegia could use them independently in their wheelchairs, he was asked to prepare the draft of what was to be the seminal document in accessibility history, the initial American standard, Standard A117.1-1961: *American Standard Specifications for Making Buildings and Facilities Accessible to, and Usable by, the Physically Handicapped* (American Standards Association, 1961).

Self-help, not welfare, was Nugent's guiding axiom, and that was reflected in the terms of the A117.1 design specifications, most notably those for toilet facilities. Integral provision for people with disabilities was to be the rule, and within normal toilet rooms at least one toilet stall was to be designed to cater to wheelchair users. Its internal dimensions were cramped like others, but Nugent rejected complaints that it was too small for wheelchair users to manage; those who were well motivated and properly trained could, he maintained, use it independently.

By the tenets set out in this chapter, Nugent, we may now observe, was the first exponent of the Principles of Universal Design as they apply to buildings. He was an advocate of access for everyone; he did not want his students to be singled out for special treatment. But it was perhaps unavoidable that the 1961 American standard was presented as being for people with physical disabilities; to be successfully promoted across America, it was essential for it to have the emotionally appealing cachet of "helping the handicapped."

Standard A117.1-1961 was acclaimed across America. With it was launched the national movement to eliminate the architectural barriers which confronted people with disabilities. In 1963, state governments began enacting legislation requiring that new public buildings be designed to be accessible in conformity with Standard A117.1. "Access rights" were demanded by consumer groups advocating for people with disabilities, and in the mid-1970s, when the campaign for civil rights legislation for people with disabilities took hold, it was access rights that were at the head of the agenda. In 1990, the U.S. federal government passed into law the civil rights Americans with Disabilities Act (ADA), augmented in the following year by the Americans with Disabilities Act Accessibility Guidelines (ADAAG) (U.S. Access Board, 1991). These guidelines, displaying inheritance from the 1961 American standard, set rules for accessible buildings and building features, ones which for new buildings (and for alterations where "readily achievable") were held to constitute rights for people with disabilities. That it was only people with disabilities, and not women or other building users who could also be subjected to discrimination, and who were protected under civil rights legislation, served to reinforce the impression that people with disabilities were a distinctive subgroup of the population, one that deserved exclusive special-status privileges.

Britain—Welfare Culture and Public Toilets

In Britain in 1961, the idea that there ought to be rules for making public buildings accessible to people with disabilities was unheard of. Britain had a welfare culture, the established principle being that the only way to help people with disabilities was to treat them as individuals, and respond to the special needs that each had for health and welfare services. The idea that they could be bunched together and treated congregately was virtually unheard of. The message that Tim Nugent brought from America was, however, greeted with enthusiasm, and Britain, it was decided, must have an access standard like America's, one that the British Standards Institution would produce; in its first version it was issued in 1967 as Standard CP96: *Access for the Disabled to Buildings* (British Standards Institution, 1967).

For toilet facilities, the specifications in CP96 differed markedly from those of A117.1. In Britain, research had shown that for people with severe disabilities there was a vital unmet need—nowhere were there any public toilets which were designed and equipped to cater to wheelchair users who could not use a toilet unaided, and who needed, as a rule, to be helped by their spouses. Here was a design issue that CP96 had to deal with, the result being specifications for a unisex toilet for people with disabilities—one which, because it was unisex, would have to be set apart from normal men's and women's toilets.

In 1979, a revised access standard, BS5810 (British Standards Institution, 1979), replaced CP96, and it came with specifications for a standard unisex toilet that was more spacious and more suitably equipped. When the Part M national building regulations were introduced in 1987, one requirement was that buildings should have at least one BS5810-type unisex toilet. The current 1999 Part M regulations retain that requirement. But as research findings have shown (Department of the Environment, 1992), the BS5810-type toilet does not suit all wheelchair users; an independent wheelchair user who enters forward may find it difficult to close the door, and there is insufficient space inside for a wheelchair to be easily turned around. Research has confirmed that most wheelchair users (whether managing independently or assisted by a helper) prefer to transfer laterally and have the toilet in a corner position with securely fixed grab rails, the proviso being their need for the facility to be larger than the BS5810-type toilet. There are, however, some wheelchair users who, with their helpers, need a toilet stall that has a peninsular layout with open space on three sides of the toilet.

In America, the rule is that a toilet compartment suitable for independent wheelchair users should be a feature of all toilet rooms in public buildings; it is a normal provision. In Britain, the BS5810-type unisex toilet is special for people with disabilities—the one that has to be provided in each public building is separated from the normal provisions for men and women, alongside or elsewhere. And for normal public toilet facilities, there are no statutory rules—no minimum standards for the size of toilet compartments and no conditions aimed at preventing discrimination against women.

25.5 DISCRIMINATION AGAINST WOMEN

When architects plan toilet facilities in public buildings, a common practice is to map out two approximately equal areas of space on plan drawings, allocating one to men and the other to women. Urinals occupy much less space than toilet compartments, the effect when the areas are filled being more amenities for men than there are for women.

Table 25.1 lists a selection of the findings of a survey made in 1992 of toilet amenities for men and women in public buildings in London and elsewhere in Britain (Goldsmith, 1992). It indicates that the number of urinals and toilets men are given in public toilet facilities is typically twice as many as the number of toilets that women get. The effect in terms of discrimination—or being "disabled" on account of not being able to use the facilities—is that women are 4 times more vulnerable than men. This is because, along with being given only half as many toilet facilities, there is the time factor—women take longer than men when using them, relevant research indicating that on average they take twice as long (Goldsmith, 1997). To achieve parity where the incidence of usage is roughly the same, women ought therefore to have twice as many amenities as men. Given the disproportionate amount of space that a toilet compartment occupies as compared to a urinal, the rule when restrooms are planned is that the zone for women should be 3 times the area of that for men (Goldsmith, 2000).

TABLE 25.1   Public Toilet Facilities for Men and Women in Public Buildings in London

| | Men | | | |
Building	Urinals and toilets	Urinals/ toilets	Women, toilets	Total ratio, men:women
Theaters, concert halls				
National Theatre				
Serving Olivier and Lyttleton Theatres	64	53/11	28	
Serving refreshment facilities	12	10/2	6	
Serving Cottesloe Theatre	7	6/1	2	
All facilities	83	69/14	36	2.3:1
Barbican Centre, in basement, serving concert hall and theater	43	37/6	22	2.0:1
Royal Festival Hall, all facilities	64	45/19	28	2.3:1
Museums, art galleries				
British Museum, all facilities excluding those in special exhibition areas	41	25/16	19	2.2:1
National Gallery, all facilities	33	24/9	24	1.4:1
Department stores				
Harrods, all facilities	55	33/23	60	0.9:1
Selfridges, all facilities	51	28/23	42	1.2:1
Hotels				
Langham Hilton, Portland Place, basement facilities	23	17/6	12	1.9:1
Copthorne Tara, Kensington, facilities serving first-floor conference suite	31	24/7	7	4.4:1
Railway stations				
Euston, concourse facilities	42	27/15	20	2.1:1
Liverpool Street, below-concourse facilities	49	31/18	20	2.5:1

Note: The buildings listed were surveyed in 1992. The figures for toilets exclude unisex facilities for people with disabilities.
Source: Goldsmith, 1997, p. 179.

A further aspect of discrimination against women comes from the architects' customary practice of planning public toilets so that toilet compartments for women are identical in size to those for men. But for three reasons women are more commonly disadvantaged than men by the constricted size and awkward configuration of a typical toilet compartment in a public lavatory in Britain. The first is that women's clothes are more prone to sweeping the toilet seat and hence to contamination. The second is that women always have to sit down or squat, which involves the adjustment of clothing (and sometimes taking off an overcoat) in a confined space. The third is that a sanitary waste disposal bin placed to one side of the toilet restricts the maneuvering space available.

Liverpool Street Station

The effects of currently prevailing design standards for public toilets in Britain are manifested at Liverpool Street Station, the railway terminus which serves destinations in East Anglia, to the northeast of London (Figs. 25.2 to 25.5). When the station was comprehensively reconstructed in the late 1980s, all parts of it were designed to be accessible to people with disabilities in conformity with the Part M building regulations, meaning that with respect to public toilet facilities, the only requirement was for a BS5810-type unisex toilet.

Tucked away to one side of the ticket office at concourse level at Liverpool Street Station are two BS5810-type toilets. They are directly accessible only to the few people with disabili-

FIGURE 25.2 Liverpool Street Station, London, main concourse. Train platforms are off to the left. The approach to public toilets is beneath the information board; the ticket office is to the right.

FIGURE 25.3 Liverpool Street Station, London. The approach to the public toilets is below the concourse.

ties who carry with them the national key which, at railway stations and many public toilet venues around Britain, unlocks the doors of the special unisex toilets. The public toilets for everyone else are below the concourse, reachable only by way of two flights of steeply graded steps, and it is these which serve families with small children, people encumbered with heavy baggage, mothers with infants in strollers, pregnant women, and all the people with disabilities who do not have the key to the special unisex toilets.

By the precepts of universal design, the public toilet facilities at Liverpool Street Station score poorly. As well as not catering to the needs of the great majority of people with disabilities, they do not cater to the needs of many of those who do not have disabilities. None of the toilet stalls in either the women's or men's toilet rooms below the concourse are accessible to people with strollers, and, as Table 25.1 shows, the number of amenities provided for women as against men manifests overt discrimination against women.

FIGURE 25.4 Liverpool Street Station, London. Another view of the approach to the public toilets below the concourse.

25.6 ACCESS STANDARDS

In America, and then in Britain and elsewhere around the world, A117.1-1961 set the mold for access standards. It drew on four premises, ones which we may now perceive to be flawed, but which in the context of official accessibility regulatory controls have effectively remained unchallenged. They were: (1) that architectural barriers in and around buildings are a threat to people with disabilities, but not to able-bodied people; (2) that all people with disabilities—anyone with a physical, sensory, or cognitive impairment—can be disadvantaged by architectural barriers and can be emancipated where they are removed; (3) that what for accessibility purposes suits wheelchair users will generally serve for all other people with mobility impairments, and that, with design criteria giving priority to independent wheelchair users, rules for accessibility can come in a single package of prescriptions with a common set of design specifications; and (4) that the design specifications can be precise and definitive—there are "right" solutions.

An effect of regulations set in this form is the "what can we do for?" syndrome—what, for example, can be done for ambulant people with disabilities; what for people with mental impairments; what for people who are hard of hearing; what for people with sight impairments; what for people who are blind? In Britain, in anticipation of what the consequences of the Disability Discrimination Act might be, it is a syndrome which has become increasingly charged in recent years.

The Disability Discrimination Act

The Disability Discrimination Act, Britain's version of the Americans with Disabilities Act, was passed into law in November 1995. Most of its provisions have been brought into force, but not yet those that are to give people with disabilities access rights. These are due to take effect in 2004, and when they do, a person with a disability who comes across a building that he or she feels discriminates against him or her unreasonably will be entitled to complain—and, if necessary, to pursue the matter through the courts.

Following consultations on how the act's access provisions might be applied in practice

FIGURE 25.5   Liverpool Street Station, London. The door to the side of the ticket office which leads to special toilets for people with disabilities. The notice advises that a key to it can be obtained from station staff.

(Disability Rights Commission, 2000), an official guidance document advising service providers and architects on what may be reasonable under the terms of the act is to be issued in 2001. Its focus will predictably be on situations where alterations should be made to existing buildings to provide wheelchair accessibility, of particular concern being small high street shops and office buildings where there is a step to the entrance door or two or three steps. Two examples are illustrated of building entrances where alteration work has already been undertaken.

Figure 25.6 shows a shop where an existing step has been replaced by a ramp, a method feasible only where the footway outside is sufficiently wide for the ramp not to impede the passage of pedestrians along the footway. In this case the ramp is graded at about 1:7 (as against the maximum 1:12 which is prescribed for new buildings), and there is no platform in

FIGURE 25.6   Altered entrance to a shop in London.

front of the door for wheelchair users to secure themselves while opening the door. The outcome is convenient for neither wheelchair users nor ambulant people with disabilities.

Figure 25.7 shows the entrance to an existing office building. With the entrance door being recessed, the three steps up to the platform in front of the door are associated with a ramp that has been placed within the limited space available; it has a length of about 2.5 m (8.4 ft) and a gradient of about 1:6. For independent wheelchair users, a ramp as steep as this would be difficult or impossible to manage, but in the context of universal design the arrangement ought not to be rejected. A wheelchair user with a strong helper could be pushed up, and helped down backwards. For ambulant people with disabilities, the steps are manageable. In addition, the ramp will benefit people with strollers.

25.7   PEOPLE WITH SIGHT IMPAIRMENTS

People who are blind could be among the complainants when access rights are enforced, on occasion their dissatisfaction perhaps being caused by a lack of tactile pavings to help them get into and around the building concerned. In Britain, the presumption has been that tactile floor surfaces are valuable for people who are blind, and across the country tactile pavings have been laid at street crossings (Figs. 25.8 and 25.9). There is no substantive evidence which supports the proposition, whereas it is apparent that such pavings can be troublesome, uncomfortable, and sometimes hazardous for other street users. But the generality of British people are, it seems, convinced that tactile pavings are essential for people who are blind; that anything that can be done to help people who are blind ought to be done; that it is immaterial how tiny the number of people who are blind in the total population may be (Goldsmith, 1997); and that, while it is apparent that they are a nuisance and a hazard to others, tactile pavings must be commended because they help people who are blind, and it is right for large sums of public money to be spent on them.

A universal design principle is that special provisions may be made for people with disabilities where normal provisions do not suit; to test their appropriateness in any particular circumstance, two conditions would be in order. One would be that there must be substantive data or secure prima facie evidence confirming that the provisions will genuinely be valuable to the people with disabilities concerned. The other, noting the convenience-for-all precept,

FIGURE 25.7   Altered entrance to an office building in London.

would be that the provisions not inconvenience other users of the buildings; the proviso here would be that the disadvantages to others would have to outweigh the benefits there would be for the people with disabilities concerned. On both counts, it is clear that tactile pavings for people with sight impairments fail. On the footways of streets around Britain they are now commonplace; they are normal provisions. Pedestrians can be inconvenienced by them, and interestingly, the effect has been to curtail rather than extend the accommodation parameters of normal provisions.

Elsewhere in and around buildings, people with special needs may be accommodated by provisions which supplement the normal provisions that are convenient for others. Unisex toilets are an example. So also is the American practice whereby in the men's and women's zones of toilet rooms the normal toilet cubicles are supplemented by one which is wheelchair-accessible. Relatedly, "special" can become normal. Where in a small public building the single toilet room for staff and visitors is designed and equipped to cater to everyone, including

FIGURE 25.8 Tactile pavings on a London street.

FIGURE 25.9 Close-up of tactile pavings on a London street.

wheelchair users who need to be helped, that is a normal provision. The line between *normal* and *special* becomes blurred.

25.8 CONCLUSION

Universal design can be obtained only by working from the bottom up, by looking to make normal provisions suitable for everyone. It aims to be socially inclusive and is compromised if "accessibility" is defined in terms of provision for people with disabilities only (*see* Chap. 28). Three examples serve. First is hotel guest rooms—universal design is not achieved by ignoring normal guest rooms and decreeing that only a selected few should be "accessible," meaning accessible to people with disabilities. Second is public toilet facilities—universal design is not achieved by ignoring normal men's and women's toilet provision and decreeing that only special provisions for disabled people should be "accessible." Third is seating provision in sports stadiums and buildings such as cinemas, concert halls, and theaters—universal design is not achieved by disregarding normal seating areas and decreeing that only spaces suitable for wheelchair users should be "accessible."

In Britain, America, and elsewhere around the world, statutory controls for making public buildings accessible to people with disabilities have been beneficial; whatever their shortcomings, they have been instrumental in massively extending the accommodation parameters of the generality of public buildings. That is the plus factor. The downside is that the universal design ideal remains elusive. An impediment in the way of developing regulatory controls for its operation is that with its bottom-up methodology there cannot be prescriptive design standards for its realization. But for architects who are keen to take it on board, informative design guidance would help. A contribution is *Universal Design: A Manual of Practical Guidance for Architects* (Goldsmith, 2000). To explain the implementation of the principles and practice of universal design it has some 370 diagrams.

25.9 BIBLIOGRAPHY

American Standards Association, A117.1-1961, *American Standard Specifications for Making Buildings and Facilities Accessible to, and Usable by, the Physically Handicapped,* American Standards Association, New York, 1961.

British Standards Institution, CP96:1967, *Access for the Disabled to Buildings,* British Standards Institution, London, UK, 1967.

———, BS5810:1979, *Access for the Disabled to Buildings,* British Standards Institution, London, UK, 1979.

Central Coordinating Committee for the Promotion of Accessibility (CCPT), *European Concept for Accessibility,* CCPT, Rijswijk, Netherlands, 1996.

Department of the Environment, *Sanitary Provision for People with Special Needs,* vol. 1, part 1: "The Practicalities of Toilet Usage"; part 2: "Population Needs Estimates"; vol. 2, part 3: "Tabulated Project Data," Department of the Environment, London, UK, 1992.

Department of the Environment and Welsh Office, *Access and Facilities for Disabled People,* Building Regulations 1999, Part M Approved Document, Her Majesty's Stationery Office, London, UK, 1999.

Disability Rights Commission, *The Disability Discrimination Act 1995: New Requirements to Make Goods, Facilities and Services More Accessible to Disabled People from 2004; Consultations on a New Code of Practice, Regulations and Practical Guide,* Department for Education and Employment, London, UK, 2000.

Goldsmith, S., "The Queue Starts Here: A Raw Deal for Women," *Access by Design* 57 (Journal of the Centre for Accessible Environments, London, UK), 1992.

———, *Designing for the Disabled—The New Paradigm,* Architectural Press/Butterworth-Heinemann, Oxford, UK, 1997.

Universal Design: A Manual of Practical Guidance for Architects, Architectural Press/Butterworth-Heinemann, Oxford, UK, 2000.

U.S. Access Board, *Americans with Disabilities Act Accessibility Guidelines for Buildings and Facilities,* U.S. Government Printing Office, Washington, DC, 1991.

CHAPTER 26
ACCESSIBILITY AS UNIVERSAL DESIGN: LEGISLATION AND EXPERIENCES IN ITALY

Fabrizio Vescovo, Dr.Arch.
Architect, Rome, Italy

26.1 INTRODUCTION

This chapter explores the evolution from barrier-free design to universal design in Italy and presents its implementation through case study examples. Accessibility and, beyond that, universal design is a quality issue of the constructed environment that affects a broad range of users. As such, it includes safety and comfort. Article 2 of the 1989 Law DM 236 on "technical rules to guarantee accessibility, use and adjustability of areas and buildings" defines *architectural barriers* as physical barriers impeding mobility for anybody, particularly people with temporary or permanent mobility impairment; barriers that prevent comfortable and safe use of parts, mechanisms, or components of buildings; and lack of orientation and wayfinding aids to identify potential sources of danger for anybody, especially people who are blind, have visual impairments, or are deaf.

26.2 BACKGROUND: ITALIAN LEGISLATION

The target of the Italian law is clearly extended to all people, with a specific and appropriate emphasis on those with reduced independence and mobility, namely about 20 percent of the total population (Vescovo, 1990).

Universal design is a cross-discipline whose far-reaching human, social, and economic goals must be pursued gradually and constantly, at all levels and with all means. This is possible only if the general public is made aware of the need to achieve certain standards which help improve the overall quality of construction and of the environment, bringing them more in line with the real needs of citizens, especially those with reduced mobility or sensory ability. The aim is thus the maximum enhancement of everybody's independence, regardless of any temporary or permanent psychophysical conditions, leading to a general improvement of the environment in which their everyday lives take place (Vescovo, 1992).

Improving accessibility goes hand in hand with a more general urban quality, with an across-the-board cut of resources and situations of danger, difficulty, and fatigue. Therefore, such a qualitative issue, so important in the design of the built environment, must be taken into account together with other relevant issues. This should be initiated right from the beginning of the various and normal organizational and mental processes necessary for any given design (Vescovo and Antoninetti, 1998).

This is indeed a new way of looking at the design and management of urban areas, industrial buildings, and products. It calls into question several features and performance standards in order to make their use comfortable, safe, and accessible to the largest possible number of people. All this must be brought about easily and naturally, without the need for special adjustments or any customized design. The goal is making everybody's lives easier through the construction of spaces and devices that can be used by people with a wide range of different needs, people of any age, of diverse conditions and skills, at a cost that is identical or slightly higher than that of traditional construction (Norwegian State Council on Disability, 1997).

The Italian legislation on public as well as private spaces makes it possible to further investigate such important principles, in line with the pursuit of equal opportunities and with the enhancement of individual freedom. More specifically, laws DM n. 236/89 and DPR n. 503/96 indeed contain design criteria for the accessibility, visitability, and adjustability of different environments. Also illustrated are the technical standards to be complied with in order to meet the specified requirements.

Yet, in the design phase it is possible to put forward alternative solutions to what said technical standards provide for, as long as they meet the requirements specified in the design criteria. In such cases, the designer has to express his or her motivations and clearly illustrate with graphs and technical reports the proposed alternative, as well as the equivalent or better quality of results that can be attained. In addition, a technical expert must certify the conformity and compliance of what has been designed with the intended performance, not the standards, specified by the decree itself.

Building licenses can only be issued after a check of project conformity by the municipality's technical office, which is specifically responsible for this kind of activity. Local authorities, universities, and individual professionals may propose alternative technical solutions to a standing committee at the Ministry of Public Works, which, in case of certified conformity, can utilize them to update the previously stated rules by means of a subsequent decree.

The existing legislation is thus flexible and enables designers and producers to pursue their interests in the final result and to compare several technical solutions with the purpose of raising the satisfaction level of end users. Therefore, said technical rules must not be regarded statically but as an important starting point for a continuous and proficient approach of research, experimentation, and verification of the solutions adopted by technicians and users.

In other words, the idea was that of doing more than just establishing absolute constraints and measures, together with excessively rigid dimensional standards, defined once and for all and specifically directed at those who use wheelchairs (Vescovo, 1997). There can be no doubt that an excessive number of constraints and specific technical rules, imposed on top of other rules related to different issues, is frustrating for the designer, makes research impossible, and often eliminates imagination.

One of the reference points in the Italian legislation on accessibility is that of performance features of spaces and objects, which in all cases must guarantee anybody comfortable use of environments and related facilities. These legislative provisions are indeed a positive element in that they allow considerable margins of flexibility, sometimes necessary in the implementation stage. They have been enforced for over 10 years now, and thanks to them, designers can develop good spatial solutions that are also safe, comfortable, and expressive for anybody. Furthermore, they can stimulate the search for innovative solutions that keep pace with the unceasing progress made by technology, as well as encourage the use of new materials.

Unfortunately, the important option of creating accessible spaces is hardly ever considered by most designers and producers, even with external contributions of expertise and imagination. Just the opposite, they seem to be more comfortable following the rigid rules contained in the enormous number of regulations, which continue to proliferate in all directions.

Here is an example. In designing new environments—even more so when it comes to the alteration of existing structures, especially if open to the public—the secret lies in modifying bathrooms, so that they can also be used by people using wheelchairs. In such cases it is advisable to resort to alternative solutions, because the space available may be rather limited. This is particularly true for historical buildings, where structural constraints do not leave room for creating bathrooms in keeping with the usual standards, nor for complying with the sizes normally specified by regulations and handbooks.

In addition, conformity with the legislation does not necessarily require a bathroom specifically dedicated to people with disabilities. What is needed is a bathroom that may also be used by people who are impaired in their mobility or are using wheelchairs.

The alteration of an existing bathroom to achieve the desired accessibility objective sometimes requires only very simple solutions, such as changing the way a door opens, moving a sink elsewhere, or taking out the lower part of a wall. All this can also be attained by identifying alternative solutions, but only based on in-depth knowledge of the users' real needs.

In spaces open to the public, compromise solutions must be found to fulfill the users' different needs, while customized rooms can be provided in individual housing units to meet unique requirements in specific situations. Every possible opportunity must be used to communicate knowledge about the real needs of citizens, and to do away with negative stereotypes that hinder the creation of an urban environment accessible to all. It is thus necessary that experts and administrators be informed and made aware. Synergies have to be found between the various efforts, even if they are the result of rules related to different areas of intervention.

The rules on accessibility are not exclusively directed at people with disabilities; they are also aimed at improving the quality of construction, means of transport, spaces, and equipment used by everybody. The goal is thus to offer to all citizens the benefits resulting from a comfortable use of all spaces and buildings.

The reason for this is that often environments and equipment made for people with disabilities only engender negative attitudes in the general public, which sometimes rejects them. Special environments and devices made for people with disabilities are thus a source of discrimination against all those who have special needs, including people with reduced mobility or sensory abilities (e.g., older people, little children, people with heart conditions, people who have had accidents, etc.).

A number of experiences and verifications of widespread attitudes observed in different parts of the world have led to the abandonment of the concept of spaces and objects that are designed exclusively for people with disabilities. These people must always be regarded as part of the world and not as a world apart. Hence, decision makers or administrators for urban and architectural projects must also take into account the psychological issues involved. People with mobility limitations must be able to easily utilize cities and buildings in their entirety, with a minimum price to pay in terms of anxiety, humiliation, and frustration. For these reasons, it is necessary to create environments and equipment that can be used normally by a broad range of people. This is one of the key elements of the concept of universal design.

This perspective of planning projects is based on the seven Principles of Universal Design, identified by a group of experts at the Center for Universal Design (1997) in the United States. They are as follows:

- *Equitable Use.* The design is useful and marketable to any group of users.
- *Flexible Use.* The design accommodates a wide range of individual preferences and abilities.
- *Simple and Intuitive Use.* Use of the design is easy to understand, regardless of the user's needs.
- *Perceptible Information.* The design effectively communicates necessary information to the user, regardless of environmental conditions and the user's sensory abilities.
- *Tolerance for Error.* The design minimizes hazards, adverse consequences of accidental or unintended actions.
- *Low Physical Effort.* The design can be used efficiently and comfortably with a minimum of fatigue.

- *Size and Space for Approach and Use.* Appropriate size and space specifications are provided for approach, reach, manipulation, and safe use, regardless of the user's height, posture, and mobility.

Unfortunately, such simple but essential concepts still meet with a lot of difficulty and resistance in the consciousness of the general public.

When designers and builders are required to bear in mind the needs of all people when designing urban areas, housing units, furniture, and so on, it helps not only people with disabilities, but also the public in general, including manufacturers.

26.3 UNIVERSAL DESIGN

Universal design takes into account multigenerational needs, namely the needs of children, the needs of elderly people, and the needs of those who, for whatever reason, have sensory or mobility impairments (Mannucci, 1998). In particular, residential environments should be designed in order to respond to the real needs of people, from childhood to old age, with all the adjustments required by the changes over time.

This way of thinking implies a higher level of social awareness, as well as a business obligation to reach as many market segments as possible. It is a very significant issue, particularly in Europe, where an aging population plus a significant number of people with disabilities account for a large proportion of consumers, with enormous purchasing power. However, and unfortunately, new technologies and new products are rarely devised and developed based on this perspective.

Rather than demanding products that are logically more in line with this philosophy, people have now become accustomed to coping with spaces and objects whose specifications do not meet their real needs. This can be explained by looking at designers and producers, who are in general people in good health, young and able, gifted with perfect sight, and unable to imagine that life can be different. In order to bring about a real change and to work together to achieve such an objective, it is therefore necessary to make architects and designers aware and prepared, much in the way it is done at the Royal College of Art in London (see Chap. 4).

As potential users, it is important that older people, as well as people with disabilities, be consulted and involved in the design and subsequent testing of manufactured products (Wilkins, 1996). Furthermore, the idea of cooperation is backed by the European Institute for Design and Disability. It has members in 11 countries of the European Union and has been very active in promoting universal design.

Technological solutions can be designed so as to take into consideration, right from the start, the requirements of people with disabilities, in their capacity of ordinary customers. It is far more expensive to make product adjustments at a later stage. With the growing awareness of the need in Europe for nondiscriminatory legislation for people with disabilities, manufacturers will have to ensure that they do not infuse any discrimination into the products and services they offer. Thus, one has to acknowledge the overall significance of accessibility in buildings and urban areas, while applying the principle of equal opportunities and the right to participate at all levels of human society. There is a need for operational programs that can ensure full access to the environment, information, and communication for everybody. The principle must be acknowledged, and the appropriate actions taken, whereby people at a disadvantage are put in a position to exert their rights, to strengthen their personal autonomy, and to reach their financial independence.

Furthermore, as far as residential units are concerned, the Italian legislation is particularly comprehensive and advanced, since it contains clearly articulated and modern provisions. Unfortunately, to date it has not been fully understood and adopted by most designers and entrepreneurs. Among other things, the Italian law provides for all housing units to be visitable, whether they are new construction or renovations. The requirement of visitability is considered complied with if the living or dining room, one bathroom, and related passage-ways inside the housing unit are accessible. Furthermore, each housing unit for which accessibility and/or visitability are not required must be adjustable in all its parts.

These rules are of a performance nature, and, based on the identification of certain requirements, they establish the criteria and features that designers must comply with. In other words, they do not express a fixed list of absolute dimensions in terms of size or volume. The rationale of the law is that by fostering studies, research, and experiments on the part of professional experts, ever more innovative design proposals will result, including alternative solutions that favorably meet the real needs of people with reduced mobility. Consequently, it is appropriate to put forward new ideas and to stimulate experts and users alike. A recent experience in this regard is described in the following case study.

26.4   THE FRIENDLY HOUSE

Under the auspices of the MOA Cooperative, the Twenty-Fourth Furniture and Interior Decoration Expo, MOA Casa, was held in Rome in October 1998 in the area called Fiera di Roma. Over and above the usual exhibits of furniture and various household items, a new project was presented in this exposition: an experimental housing unit called Friendly House. It was designed not only for commercial purposes, but with the aim of attracting ordinary people, as well as experts and users. Furthermore, an important objective was to draw the audience's attention to a number of issues and solutions for the design of housing units that were little known to developers, producers, sales agents, and also to observers and consumers (Vescovo, 1999).

The project idea and its implementation can be attributed to the author. In this case, business was combined with pleasure through the presentation of a new and compact type of housing unit, with a total area of little more than 60 m^2 (645 ft^2), divided into several rooms: hallway, kitchen, living room, two bedrooms, and two bathrooms. The rooms were designed and equipped in a simple but functional manner. Yet, thanks to the elimination of architectural barriers, they were also in compliance with the law.

The rooms were designed and organized in order to facilitate all everyday domestic operations, even for people with limited mobility, such as older people, those affected by arthritis, people with disabilities, and so on. The fundamental objectives pursued in the experiment were accessibility (i.e., the extent to which places and equipment are reachable), user safety, environmental and psychological comfort, and good aesthetics.

A number of significant and smart solutions in terms of size were critical in the design of the Friendly House, coupled with equally smart choice and appropriate location of furnishings. It was intended as a pleasant environment where any person could live, and certainly not as a house specially designed for people with disabilities.

The use of specific components was coupled with the rooms' characteristics in terms of size and spatial distribution, such as sliding doors, folding doors, bathroom fixtures, technical devices, and furnishings that are not custom-made, but mass-produced. This resulted in a better housing unit in which the furnishings, chosen for their functional as well as aesthetic value, were provided by several firms that cooperated in this initiative.

In view of its social and cultural significance, this initiative was promoted by the Municipality of Rome, which organized a special information desk for the occasion pertaining to current activities and proposals. This service was very useful in that it provided a large number of visitors with reliable information on the many technical aspects of this initiative.

The Friendly House was finalized in the framework of, and in full compliance with, all applicable laws. However, being an experimental unit, it was also intended as a sort of grab bag of ideas that might suggest to experts and users new potential design solutions. It might also provide some orientation when it comes to choosing furnishings on the market that are capable of meeting anybody's needs, including all necessary and tailor-made adaptations.

This was a way of demonstrating how careful design can make it unnecessary to opt for housing units that are larger than average and also more expensive, especially with appropriate choices of standard furnishings, fixtures, and accessories. Neither is it necessary to have those complex and marginalizing custom-made assistive devices that guarantee full utilization of facilities to any person, including those with reduced mobility. The main idea was to raise the quality of the housing unit's spaces, resulting in a very compact spatial configuration that allowed for a variety of innovative solutions. Based on the author's extensive previous experience, the unit's design subsequently underwent real-world tests. The tests were favorably reviewed by the press and by the great number of visitors who came to see them. Among the visitors were project designers and many people with disabilities using wheelchairs, some of them electric. Furthermore, doors, passageways, and small hallways were tested, and they proved to be user-friendly for everybody.

For bathroom specifications, and in compliance with applicable regulations, a number of smart solutions were adopted as far as space was concerned. This was to guarantee free movement for people using wheelchairs, an indispensable element for user-friendly sanitary facilities. One of the two bathrooms proposed for the housing unit was about 2×2 m (6.5×6.5 ft), and it was completely usable and accessible via a small hallway [2×1.6 m (6.5×5.25 ft)] that connected the entrance hall with the main bedroom. The bathroom had a sliding door that could be powered. Part of the free space at the end of the small hallway could be used for a chest of drawers, or for other storage.

The second bathroom was deliberately designed as a very small space [about 1.5 m^2 (16 ft^2)] and was directly connected to one of the bedrooms. It was meant as an alternative solution, as envisaged by the previously mentioned regulations, and could be used in cases where the space available was very limited.

The small hallway also contained a large suspended wardrobe with sliding doors to keep the area underneath [70 cm (27.5 in) from the floor] clear. In spite of the limited space available, this solution made it easier for everybody to move through the three doors, and to use the wardrobe itself (see Fig. 26.1).

In practice, once the large folding door was fully opened, even people using wheelchairs could make full use of the bathroom. They would leave the wheelchair outside the small room, and would shift from the wheelchair to a seat with grab bars installed nearby, from which it was easy to use the four available fixtures: toilet, bidet, shower, and sink.

The main bedroom included a wardrobe fitted with sliding doors, which made it possible for any person to get inside, even with a wheelchair, thus ensuring full and comfortable use of the various spaces for all (see Fig. 26.2).

Like the two bathrooms, the kitchen, too, was designed so as to occupy a rather small area. This was to demonstrate that full use can also be achieved by resorting to a number of smart choices in terms of space, and by using furnishings normally available on the market. Mass-produced furnishings, sometimes without their lower parts, are a good solution, even for people who have to cook while leaning on a stool. For those who cannot bend over, the refrigerator was simply placed on a 20-cm (8-in)-high base for improved accessibility.

It is noteworthy to stress how the Friendly House, presented at one of the regular furniture fairs, has given a small but important impetus towards universal design in Italy. It has created a greater awareness of accessibility as a specific issue, while at the same time being in full compliance with the objectives and regulations contained in the current law (Scott, 1996).

26.5   ALTERNATIVE SOLUTIONS

All figures in this section refer to alternative solutions and project schemes of spaces and equipment that are accessible to all. They were designed by the author. Each of the proposed ideas has been successfully implemented in prototypes or models. Universal design was the inspiring, underlying philosophy and approach.

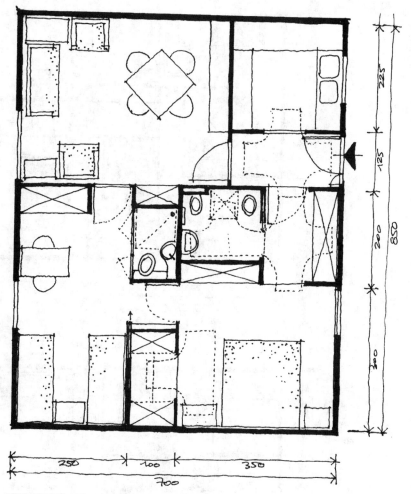

FIGURE 26.1 Floor plan of accessible apartment.

Bathrooms

In order to move about the built environment and to use related services and facilities, people with reduced mobility need a number of simple but critical elements. These elements can be realized using design techniques and methods that comply with appropriate standards and are compatible with the movements necessary for those who use a wheelchair. This is the reason why all turning points with a diameter of 1.5 m (5 ft) must be clearly identified in the design stage and specified in technical manuals and legislative provisions aiming at removing architectural barriers.

Often elevators, bathrooms, corridors, and kitchens are oversized compared to their actual use. Adjoining spaces like bedrooms, small connecting hallways, and living rooms are therefore downsized and compromised, while the total area of the dwelling remains unchanged. If greater care is used when considering the average volume occupied by a person using a wheelchair, and the number of possible movements the person can perform, it becomes appar-

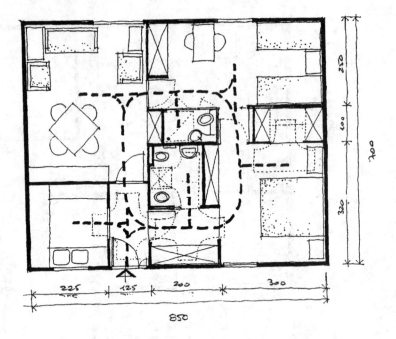

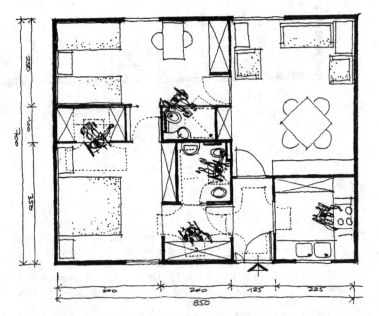

FIGURE 26.2 Wheelchair routing in accessible apartment.

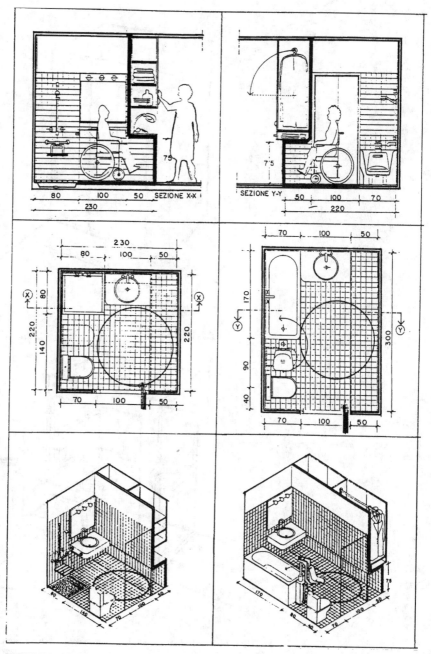

FIGURE 26.3   Basic concepts in bathroom accessibility.

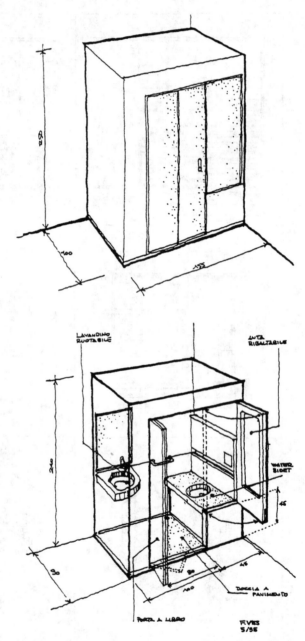

FIGURE 26.4   Alternative solution for an accessible bathroom.

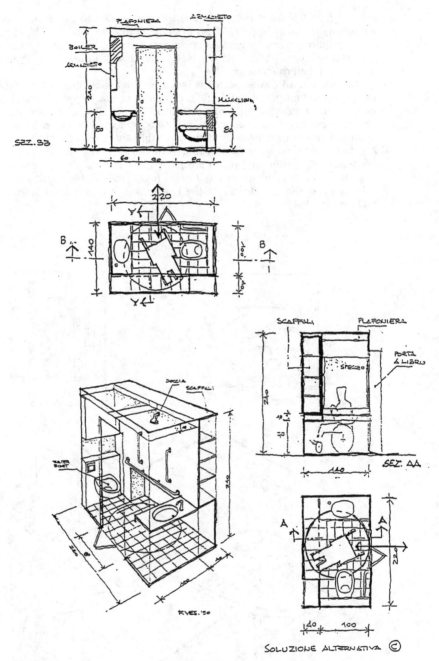

FIGURE 26.5   Minimum bathroom alternative solution.

ent that only the lower portion of any space is needed to make these movements possible, approximately up to 70 cm (27.5 in) from the floor. The head and the upper part of the body do not require a similar amount of open space, even in the turnaround maneuver, because the head is closer to the axis of rotation. Hence, the idea illustrated in this chapter for a more functional use of bathroom space.

This method of design, which is based on the logical use of constructed spaces and on the principle that people come first, can be effectively applied to any environment, both indoors and outdoors: kitchens, corridors, small hallways, bedrooms, living rooms, balconies, elevators, and even relationships of spaces that make up a housing unit. This approach to the organization of the available space ensures tangible results in terms of a better utilization of the built unit, with obvious cost savings. The technical solutions adopted in the proposed project are fairly easy to implement. It can be used in new construction, with the constructed unit being highly modifiable over time. Furthermore, these solutions can also be adopted in the renovation of existing buildings in order to remove architectural barriers.

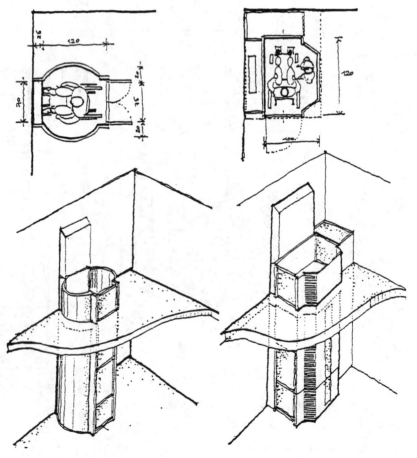

FIGURE 26.6   Elevator.

FIGURE 26.7 Support handrails.

One can imagine a housing unit with a cross section of the wall that divides bathroom and bedroom. The lower portion of space, up to 70 cm (27.5 in) from the floor, can be dedicated to the bathroom, allowing people with disabilities to perform rotating movements, while the upper part, going up to the ceiling, can be part of the bedroom, and can efficiently be used for a cupboard, a bookshelf, or the like (Figs. 26.3 to 26.5).

Emergency Egress

In case of emergency, people with reduced physical abilities, or people with disabilities, are usually unable to quickly leave premises. So their only chance is to promptly reach a static or dynamic safe place. In the case of a dynamic safe place (internal or external stairs or flights of stairs), a safe haven must be provided on each floor. This will allow people who are unable to move quickly to stop and wait, without creating impediments to the orderly movement of those who want to use the stairs going up or down, and to exit from places that can be potentially dangerous. Such a safe haven is generally found in a place such as a staircase landing. Alternative solutions are also acceptable, provided that the safe haven still provides conditions of survival for an appropriate period of time. Such technical solutions are extremely simple, and they are not very expensive. It is thus advisable that they also be installed in residential buildings that are not subject to fire prevention inspections.

Elevator Platforms

Some prototypes developed by the author on behalf of Ceteco Srl, Pisa, contain alternative solutions for elevator platforms. The goal is to propose alternative solutions for such types of elevator systems, in line with existing regulations (Ministerial Decree 14.6.1989 n.236, articles 4.1.13 and 8.1.13). When building or renovating in open or closed spaces, it can be very useful

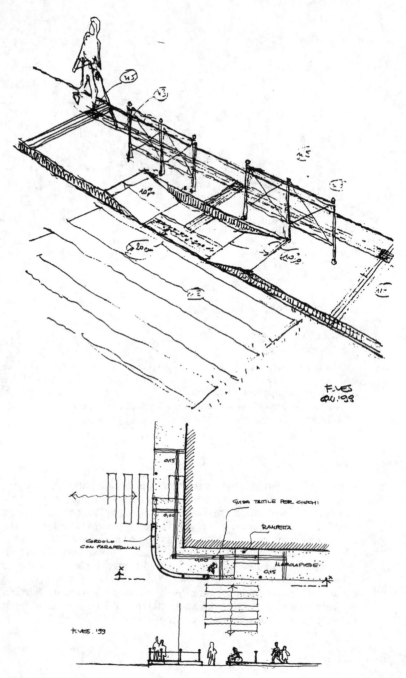

FIGURE 26.8 Street crossings.

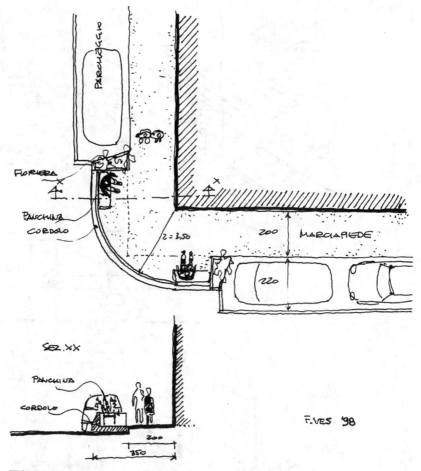

FIGURE 26.9 Street corner with benches and planters.

to rely on elevator systems that have not only functional but also aesthetic attributes. Several solutions were developed as part of a modular set, and with a number of different optional elements (Fig. 26.6).

Supportive Handrails

The upright position causes fatigue in many people. The solution shown in Fig. 26.7 is the installation of a handrail fitted with a supportive element for people to lean on, thus providing relief while standing up. This simple and inexpensive solution can also prove to be very useful for older people, or for people who have had accidents. It can be applied inside buildings or outdoors at bus stops, taxi stands, and the like.

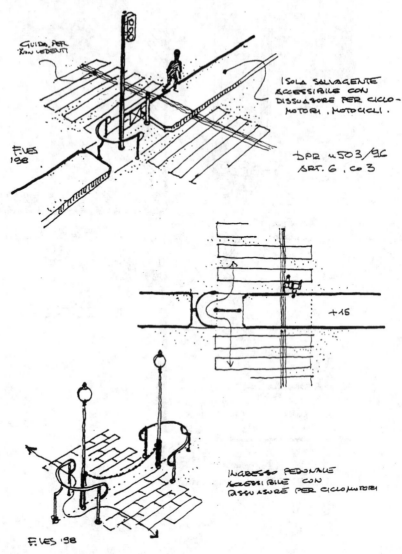

FIGURE 26.10 Pedestrian access only.

26.6 URBAN DESIGN ELEMENTS

From the very beginning, the design, construction, and management of urban areas, buildings, and industrial products is planned with those features and services in mind that make them safe and comfortable to use by the largest possible number of people, including people of different ages and different levels of individual abilities. Each project complies with the various laws and governing standards, such as removal of architectural barriers, safety, and traffic regulations.

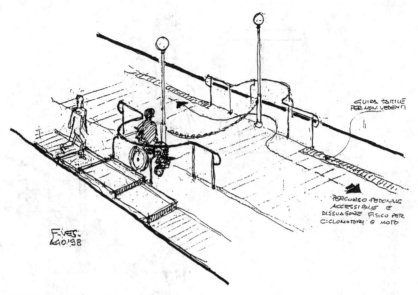

FIGURE 26.11   Keeping out vehicles.

Street Crossings and Park Access

Street crossings that are safe for people who are older or have disabilities are shown in Figs. 26.8 and 26.9. Figures 26.10 and 26.11 show an access system to parks and areas that are closed to vehicles. It provides easy passage for everybody, including people with disabilities or using wheelchairs. At the same time, the system prohibits access for mopeds and motorcycles. In some urban areas, there is an increasing need to ban road traffic from those areas which are reserved for pedestrians. The areas between roadways and sidewalks are very dangerous, because signs and barriers are insufficient to prevent intrusions by motorcycles and mopeds to such areas or paths, which are intended for pedestrian use only. The proposed solution meets several requirements. Thanks to the shape, size, and positioning of the various tubelike metal elements, there are no barriers for pedestrians. At the same time, they prevent dangerous intrusions by mopeds.

Bus Shelters

Figure 26.12 is related to project proposals of multipurpose elements for urban design or decoration. Pedestrian-related elements are generally used to make footpaths safer and to separate them from car-only areas. After adding to the previously mentioned pedestrian-related devices a simple horizontal structure, they can also be effectively used for people standing in line and waiting for a long time, especially at bus stops.

Other elements include transit systems, from driveways to footpaths, which are comfortable to use by anyone, as they do not have architectural barriers. Nevertheless, the particular shapes and sizes of these elements can turn them into real physical barriers which can prevent possible intrusions of motorcycles and mopeds into pedestrian areas. A few urban areas can thus be reserved for pedestrians, who can use them more comfortably and safely.

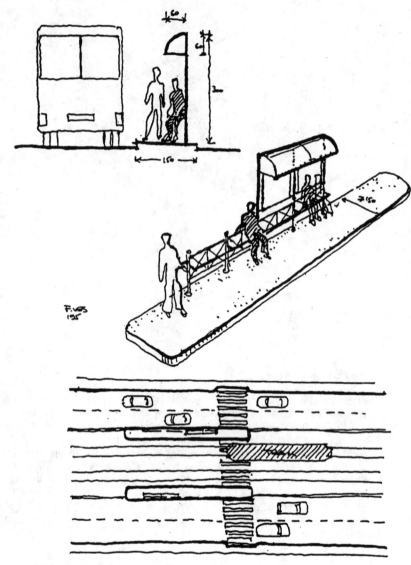

FIGURE 26.12 Bus shelter.

26.7 CONCLUSIONS

This chapter indicates that legislative measures, changing public attitudes, and a facilitating role of planners and designers can move universal design forward. Examples of residential and urban design for all show that with minimal modifications, spaces and places can often be made accessible and usable for all. Critical in advancing universal design, even in industrialized nations, is better public awareness of the benefits and solutions which can enhance the quality of life without necessarily incurring extra costs.

26.8 BIBLIOGRAPHY

Center for Universal Design, "The Principles of Universal Design," Center for Universal Design, Raleigh, NC, 1997.

Mannucci, Nancy, "Accessible Design Can Be Beautiful," *Inside MS—The Magazine for Members of the National MS Society,* 16(3), 1998.

Norwegian State Council on Disability, *Universal Design Planning and Design for All,* Oslo, Norway, 1997.

Scott, Andrew, "Guida Europea di Buona Prassi—Verso la parità di opportunità delle persone disabili, C.E.," Helios II, 1996.

Vescovo, F., *Accessibilità e barriere architettoniche,* Maggioli Press, Rimini, Italy, 1990.

———, "L'accessibilità urbana: Considerazioni di base e concetti introduttivi," *Paesaggio Urbano* 1, 1992.

———, *Progettare per tutti senza barriere architettoniche,* Maggioli Press, Rimini, Italy, 1997.

———, "Accessibilità come progettazione Universale. La difficile strada per raggiungere l'obiettivo," *Paesaggio Urbano* 1, 1999.

——— and M. Antoninetti, "La nuova linea del trolley in San Diego, California," *Paesaggio Urbano* 5, 1998.

Wilkins, Rosalie, "Il Design Universale e la nuova tecnologia," *Helioscope* 7, 1996.

CHAPTER 27
THE EVOLUTION OF DESIGN FOR ALL IN PUBLIC BUILDINGS AND TRANSPORTATION IN FRANCE

Louis-Pierre Grosbois, Architect, D.P.L.G.
Ecole d'Architecture, La Villette, Paris, France

27.1 INTRODUCTION

The recent concept of universal design, also known as *design for all,* is in fact an extension of *commoditas,* the ancient historical concept of use value. Vitruvius, the architect of Roman antiquity, specified that *commoditas* was the third element of architectural creation, along with beauty and firmness (or *voluptas* and *firmitas*). Nowadays, if one wants the *commoditas* concept to adapt to all ages and all situations of life, one must take physical, sensorial, and mental abilities and/or disabilities into account in order to exercise optimal use value. It used to be that a large part of population was excluded by dubbing people with disabilities "handicapped." These people were reintegrated into social life when the diversity of uses became one of the purposes of architectural creativity. This chapter presents examples of public buildings and transport systems in France (Paris, Lille, and Grenoble) from 1975 to 1995 to analyze how this evolution has revolutionized society. All of these were designed to comply with the concept of universal design. This evolution has helped change attitudes by creating a culture of comfort of use that takes the diversity of individuals into account in its aesthetic, technical, and economic choices.

27.2 DESIGN FOR ALL AGES AND LIFE SITUATIONS IN PUBLIC BUILDINGS

What is a public building? The word *public* comes from the Latin word *populus,* which means that it has something to do with the people, with the state, and with the public realm. The expression *publicus usus* can be translated as *public use,* or use for all. A public building is therefore, by definition, a building that must accommodate everybody, at whatever age in life. But there are many differences between the ways one approaches mobility and perception in infancy, teenage years, maturity, and old age. How can architectural spaces adapt to these differences? How should these spaces be conceived, and on which criteria should they be based?

In addition, society is now living through an unprecedented demographic revolution, which is actually the second of two revolutions that took place during the twentieth century; the first was an industrial revolution at the beginning of the century. In 1900, life expectancy in societies with the highest standards of living was around 40. By 2000, life expectancy had shot up to 75 for men and 85 for women. In 200 years (from 1750 to 1950), life expectancy has increased three-fold for men and six-fold for women. In the past 100 years (from 1900 to 2000), men have gained an extra 35 years and women an extra 45 years in life expectancy. This will continue, as each child born today has a 50-50 chance of living to be 100 years old.

The effect of this demographic revolution is particularly evident in public buildings, as the way buildings are used is changed by people with age-related disabilities.

Disabilities, Limitations, and Handicaps

In its Resolution 48/96, on December 20, 1993 (Kallehauge, 1995), the General Assembly of the United Nations adopted the main rules of equality of opportunity for people with disabilities. These rules include, among others, easy access to public buildings, to means of transport, and to the exterior spaces of the built environment. People with intellectual, sensory, or physical disabilities, must not encounter any obstacles when trying to access information or facilities for education, transportation, and so on. Access is a right for people with disabilities, and this point of view complies with the analysis of handicaps proposed by the World Health Organization in 1980.

Disabilities are defined in terms of the activities in which a person wants to engage; disability appears when the built environment cannot provide any compensation for a person's disabilities. In other words, the architecture of buildings creates or suppresses the disability.

The first theorem of accessibility therefore states: *A disabled person in an accessible building is an able-bodied person.* The corollary of this theorem is that an able-bodied person in a nonaccessible building is a disabled person.

27.3   ARCHITECTURE, HUMAN REQUIREMENTS, AND ACCESSIBILITY

In the first century B.C., the Roman architect Vitruvius (translated, 1965) said that architecture is based on three qualities: *"firmitas, voluptas, commoditas"*:

- *Firmitas*—solidity of construction
- *Voluptas*—the experience of pleasure, aesthetics
- *Commoditas*—adaptation to use, suitability

Specifically, Vitruvius recommended "laying the building out so ingeniously that nothing could hinder its use."

Today, one has to acknowledge two facts. First, these three criteria are still the pillars of architectural creation—good architecture depends on achieving the right compromise and balance between them. The second is that adapting to people and their uses is an inescapable consequence of achieving this compromise—the architect can discover directions in this process of adaptation and can therefore qualify the building and its aesthetics. As Otto Wagner (1984) concluded in *Modern Architecture,* first published in Vienna in 1895: "Architecture is both an art and a science based on human requirements. If architecture is not inspired by life and men's requirements, it will lose part of its spontaneity, of its vitality and of its freshness. It will sink to the level of a sterile and simple reasoning and will even cease to exist as an art. The artist must never forget that art exists for the sake of man, rather than man for the sake of art."

Historical Background and Attitudes

Since antiquity, the European culture that developed around the Mediterranean Sea has always used a universal graphic representation of the human body that obeys a set of proportions, in which all the parts are geometrically related. By reducing the body, as an ideal figure, to a mathematical model, Greek and Roman architecture identified the forms of the human body with geometrical figures.

Renaissance artists, such as Francesco di Giorgio, Leonardo da Vinci, and Albrecht Dürer, tried to apply the proportions of the human body to the layout of buildings, and even cities. When Le Corbusier created Modulor in 1946, he also took his due place in an historical continuity of thought. Here is how he defined it: "Modulor, a range of dimensions on the human scale, universally applicable to architecture" (Fig. 27.1).

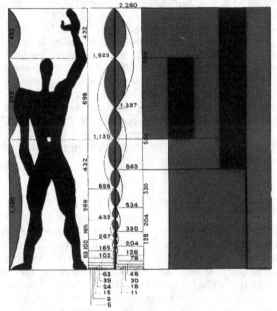

FIGURE 27.1   Le Corbusier's "Modulor Man" with universal dimensions (1946).

Models of social behavior have been changing since the 1960s, particularly in Northern Europe and Great Britain. The representation of the human as "well formed, well proportioned, an ideal image", an anthropometric model, has been replaced by the model of the pyramid that describes a person's evolution through the phases of the life cycle.

Henceforth, the architectural notion of a unique use related to an individual at the peak of his or her physical and sensorial capacities disappeared, to be replaced by the diversity of individuals, or of the same individual throughout the phases of the life cycle, or after an accident or an illness. At the same time, population studies carried out in the framework of such new disciplines as biochemistry have shown that diversity is found more often within one population than between different populations, leading to the abandonment of the concept of races, among others. This progress was crucial to countering architectural standards of construction and module notions based on human models (i.e., the average individual) that branded people who were different as physically disabled.

Comparative Anthropometrics

One of the first books written by an architect to take age, sex, capacity, and posture into account through anthropometric measurements was Selwyn Goldsmith's *Designing for the Disabled,* published in London in 1963. The book shows how the gestures made by a wheelchair user, for example, can be taken into consideration in architectural planning. The *Human Scale* anthropometric booklets by Dreyfuss (1981) added children's dimensions to adult ones and included standing, seated, and wheelchair positions, as well as furnishing each body dimension with upper and lower variations. The notion of the diversity of individuals is obvious and the recommendation to designers is clear: "Try to accommodate everyone."

When the Cité des Sciences et de l'Industrie was built, it was designed to be a model of a building accessible to everyone. The *Critères d'accessibilité aux présentations du musée* study (Grosbois and Araneda, 1982) defined the public as a whole, both anthropometrically and ergonomically, in a graph showing the distribution of the French population.

The study compared human dimensions, from children in their strollers to adults (men and women), either upright or using a wheelchair, to older people. These anthropometric data enabled people with various abilities to use the building by means of a series of plaques, including the rules of visual and tactile accessibility for people who are blind, with more favorable areas of reading and seeing for all; and the rules of anthropometric accessibility, with access areas compared with the planned museum accommodation types.

As a guide to architectural conception and to the control of the project, this methodological tool made it possible to have the different spaces of the building play a compensatory role for people with a mobility or perception deficiency, as well as for all other people. The objective Vitruvius defined 2000 years ago, "to lay out the building so ingeniously that nothing could hinder its use," was therefore achieved as a consequence of a new concept, dating from this end of the millennium: "Design for all."

Beyond Formalism: Design for All

Between the three qualitative criteria of architecture—*firmitas* (construction and solidity), *voluptas* (aesthetics), and *commoditas* (usability)—a compromise has to be found in terms of design. One needs to remind the reader that a compromise is an agreement in which each participating party makes concessions. This means that the range of constraints and freedoms of each criterion must be known exactly in order to link all the phases of designing a building. Unfortunately, there are examples of buildings in which the accent was on the construction and aesthetics, while the building's use was forgotten. There is no really comprehensive design when one of the three fields of construction, aesthetics, and usability is either ignored or predominant. Nowadays, the formalist doctrine, based on aesthetics and construction, carries too much weight, leading to more constraints of use for people. The architect Alvar Aalto (1970) already said so in 1935: "Today, when standardization is a principle of production, we realize that formalism has an utterly inhuman nature." He then followed up on this acknowledgment by giving himself this principle of creation: "An object must be produced so that man and all the individual laws controlling him can complete its form." The idea that people complete form by their gestures is obvious in Aalto's modular door handle. Several superimposed handles are laid out for people with different statures. The curve of each handle respects the open angle between the hand and the arm of each person, so that it can be easy and harmonious to push the door open (Fig. 27.2).

This idea is also obvious in the form of the sofa made by the architect Gaudi for the Casa Calvet in Barcelona. Attention is drawn to the beautiful curl of the armrest on which the bent elbow rests (Fig. 27.3).

Without any dialogue or encounter between solidity, usefulness, and beauty, only inhuman constructions can be produced. In order to establish a good compromise between these three criteria, and thus achieve the concept of design for all, two objectives have to be met:

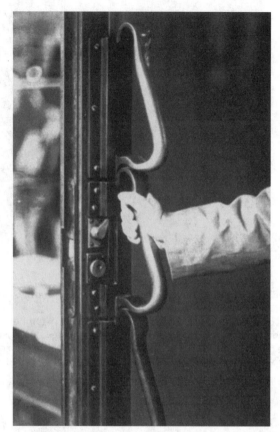

FIGURE 27.2   Aalto's modular door handle: several handles for persons with different heights.

- Reinforce the morphological and anthropological knowledge of humankind.
- Always start from the most architecturally demanding limitations, or those which require the most extensive alterations. As Aalto said: "Each solution is, one way or the other, the result of a compromise, which was found more easily when the human limitations were studied."

This approach enables people with a permanent or temporary disability (be it related to age, disease, etc.) to be included in society. It also brings more comfort to everyone.

Fiskars: The First Product Application of Design for All

The intention in applying the concept of design for all is from the point of view of people's human limitations; i.e., from our disabilities rather than our abilities. During the 1960s, a period marked historically by the international leadership of Scandinavian design, this was found in the production of a pair of scissors by a Finnish industrialist, Fiskars. This tremendous development is not extraneous to the essential role played by those architects who were also designers, particularly Alvar Aalto. Characteristic of this pair of scissors is the ergonomic study of the rings that the fingers have to grasp (Fig. 27.4), which are designed to guarantee a

FIGURE 27.3 Gaudi's sofa: dialogue between the curl of the armrest and the bent elbow.

good grip by either the right or the left hand, regardless of whether the fingers grasp them with a lot of strength or not.

The solution consisted mainly of proposing two models of rings for right-handed or left-handed people, and making two asymmetric rings, one for the thumb and the other for 2 or 3 fingers (the forefinger, the middle finger, and the ring finger). As a consequence, left-handed people are not put in situations where they are disabled, nor are people who do not have

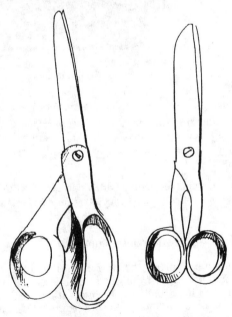

FIGURE 27.4 Fiskars scissors: ergonomic design of the ring grasped by the fingers (*left*). Traditional scissor design (*right*).

enough strength to press the ring with only one finger due to a disease, an accident, or because of old age. More than 300 million pairs of this type of scissors have been produced since. They are used all over the world and have been defined by their creator as "an element of the process of democratic sharing."

27.4 DESIGN FOR ALL IN FRANCE: FOUR EXAMPLES

An architect, like any designer, is a person who needs to compromise. A compromise has to be made between all technical, economic, social, and aesthetic data and human requirements. Moreover, the strongest human requirements have to be envisaged, namely those linked to disabilities. This designer attitude was well defined by H. Dreyfuss when he proposed trying to accommodate everyone. Cities will become accessible only when substantial human requirements are included in the compromises of architectural design. Furthermore, when this design methodology materializes in reality, it becomes obvious that either mobility or the uses of spaces are more comfortable for all people. In fact, this implies a passage from a point of view considered as marginal to an overall vision—deriving accessibility from technical and statutory regulations in order to include it at the core of a society's culture and in all the spaces being built by that society. The argument is that it is a key to the development of other decisions and designs that generate behavior, ensuring that no actor and no use is forgotten. Currently, a process is under way of evaluating how this change of behavior can be felt in French society, considering that other countries in Northern Europe and North America have already integrated them. Why is there a necessity for a new culture called the *culture of accessibility?*

Criteria for Analysis

In 1998, a research project entitled *Living in an Accessible City—From Uses to Design* was carried out on behalf of the Ministère de l'Equipement et du Logement by a multidisciplinary team comprised of the author, the architect Paul Sautet, and the sociologist Isaac Joseph. The study tried to define the nature of problems encountered in design for all and came to the conclusion that seven points have to be dealt with, as differently emphasized ways of seeking out compromises:

- *Language.*   What are the attitudes in play, and how did they develop? What mental blocks can occur, and how can one remove them when carrying out a project? All the vocabularies in use must be studied, as they reveal attitudes.

- *Legal requirements.*   To which legislative framework does each project belong, considering that projects are developed at different periods and that legislation evolves in time? Has the law been respected strictly or not?

- *Advocacy.*   How can community associations defending people with disabilities influence decisions towards a nonsegregation policy during the course of urban planning? Different situations can be compared in order to see how groups can influence planning, and how they evolve toward accessibility for all.

- *Planning.*   This is the decisive phase when the criteria of accessibility for all are integrated into designers' specifications. How far can intentions and facts be extended, considering that specifications enable results to satisfy needs and assess the project manager's will?

- *Technical traditions.*   Once the obstacles of attitudes have been overcome, obstacles raised by technical choices have to be dealt with. It is always more convenient to use well-known techniques rather than to search for others.

- *Accessibility follow-through.*   When strong intentions are expressed at the beginning of a project, it is necessary to make sure that a good accessibility level is actually reached and subsequently maintained.

- *Communications.*   In light of what is at stake for society as a whole, a breakthrough in accessibility for all, whatever its importance, amounts to nothing if all decision makers, designers, and the general public do not know about it. Attitudes evolve mainly through examples. Signaling accessibility takes part in this communication.

The study used these seven points to analyze several types of urban spaces—those that pay particular attention and those that are particularly innovative with regard to design for all:

- Two transportation systems—the Lille metro system and the Grenoble tram system
- Two public buildings with cultural features in Paris—the Cité des Sciences et de l'Industrie and the Grande Galérie of the Muséum d'Histoire Naturelle

The Lille Metro System (1973–1983)

When Lille began discussing the town's mass transit at the beginning of the 1970s and became involved in urban research and high technology, Paris was the only city in France with a metro system. A new mass transit concept was subsequently created: a metro was thought better suited to what the town had become. The detailed examination of all data, human and technical, was undoubtedly what made the operation so successful.

A new mass transit concept, including new human and technical data, represents the most favorable opportunity for changing social behavior, thus enabling sweeping accessibility to be introduced into a developing town.

Language.   The negative terms in use in 1981, such as *architectural barriers,* have now been replaced by the positive introduction of the people involved: "disabled people, people using sticks, crutches, wheelchairs . . . old or tired people, people pushing buggies, carrying loads, pregnant women . . . blind or deaf people." Terms such as *metro for all* reveal the nature of the objectives. A strong image was also suggested when the term *horizontal elevator* was coined to describe a means of horizontal movement offering both security and easy accessibility.

Legal Requirements.   When the Communauté Urbaine de Lille (CUDL) wanted to create its own mass transit network, France did not yet have any legislative framework of reference. The law governing issues about people with disabilities was only voted in 1975, while the metro prototypes were being perfected as early as 1973 and 1974. Furthermore, the law did not favor specific transportation for people with disabilities over transport accessible to all. Indeed, one can say that as far as transportation accessible to all is concerned, the Lille metro, Véhicule Automatique Leger (VAL), was a pioneer in implementing the legislation of reference.

Advocacy.   In 1982, the inhabitants of Lille were invited to get to know their new metro for 4 months, and to take part in evaluation surveys. The accessible-to-all solution was appreciated by everybody. In 1989, a law required the creation of a users' committee. When a good level of accessibility had been obtained for people with reduced mobility and visual disabilities, those with hearing deficiencies also expressed their needs. This case brings to mind the legislation that introduced the obligation of providing accessibility for people with motor deficiencies, in which no provision was made for persons with sensorial deficiencies.

Planning.   Toward the end of the 1960s, while drawing up its main urban planning concept, the CUDL decided to create a mass transit network with several lines and using equipment that had yet to be defined. The elements of specification submitted to the rolling-stock builder, together with the VAL light-rail system's technical response, were of course essential in the city's choice of an accessible means of transport. For the accessibility and architecture of the underground stations, specifications were drawn up stating that better installations for persons with reduced mobility would benefit all travelers, and specifying that the reverse was not true. Station planning work in 1975 involved providing access according to a process of adaptation—when immediate financing was not available, specifically for elevators, cavities were reserved for installations to be made later.

Technical Traditions.   As far as French mass transit was concerned, the Paris metro and its latest, more comfortable versions were well known. But the CUDL wanted something lighter and better suited to urban settings. The quantity of travelers to be transported on each train was less important than the frequency of those trains. Mass transit was just then evolving from the railway tradition into a more intimate, comfortable mode of transport. The metro itself had to be thought out in a different way, and urban issues reconsidered to their very foundations. Designing both the rolling stock and the stations at the same time was an opportunity in which the accessibility challenge could be finally taken up.

  Rolling Stock.   Accessibility was mainly achieved through the rolling stock. The objectives were small size, automation, and frequent trains. They led the manufacturer, Matra, to develop technical solutions that required a very fast transfer between platforms and trains. A sliding-door system was thus a determining factor, shutting off access to the platforms after a train has departed, and synchronizing the opening and closing of the train doors upon arrival or departure (Fig. 27.5). Perfect synchronization and reliable leveling were thus essential, de facto bringing about excellent accessibility. It can be said that a mode of public transport described as a "horizontal elevator" enabled an important phase in the accessibility to be achieved.

FIGURE 27.5   The Lille metro: perfect synchronization of train and station sliding-door opening.

Stations.   Accessibility to the stations is essential to the completion of the transport system. There are two types of access, underground and aboveground. Each underground station is equipped with elevators linking the sidewalk to the platforms. The emerging elevator structures on the sidewalk are covered with transparent or colored panels so that they are easy to spot. Elevators and escalators are built according to an equity principle, so that all people can truly choose their way to or from the trains according to their abilities. The stations above ground are located on two levels. The ticket office is at the sidewalk level, and elevators provide access to both platforms. This metro is not yet fully equipped for people with sensory deficiencies, nor for people who are blind or deaf. Some specific work is still necessary to improve these people's comfort, such as voice messages in elevators, visual signals for people who are deaf when doors close, and so on. Technical traditions have to evolve in order to tackle the issues raised by sensory deficiencies.

Elevators.   The policy, already in use when work on the transport system started, of postponing elevators when financing had not yet been secured is the right solution to protect future accessibility. Such a preventative policy would be also be necessary for apartment blocks where elevators are not mandatory. The added cost would be minor, and planning for the aging population and the future would be secured. This postponing policy, which disrupts technical traditions, is also applied in Sweden and the Netherlands. It enables buildings to be made accessible by adapting to the lower cost.

Accessibility Follow-Through.   The project manager coordinates visits and meetings in order to improve access to network transportation. Continuity is therefore ensured. The old tramway network was not accessible but started to change as a result of the influence exerted by the VAL. Buses are also in the process of changing their accessibility, and it can be hoped that the "accessibility chain" will work seamlessly in a few years in the whole mass transit network in Lille.

Communications.   The CUDL insists that the Lille metro is safe and accessible to all: "All movements from the area around the stations to the station exits, including access into trains, were devised with the same goal in mind: 'a metro for all.' " The new design-for-all metro led to a general improvement and appreciation of mass transportation.

The Grenoble Tram System (1978–1988)

Grenoble is a city with 400,000 inhabitants, including its suburbs, in a rather small area surrounded by mountains. The city spread considerably at the end of the 1970s. Car traffic then became saturated, and bus lines did not extend beyond the town center, so that a new mass transit network became necessary. Considering the nature of the subsoil, only tramways could be used. They were to be accessible to the whole population. Planned in 1983, after a popular referendum, the first Transport de l'Agglomération de Grenoble (TAG) line, which stretches nearly 9 km (5.4 mi) and called for the entire urban area to be thoroughly replanned, opened in 1987. A second 6-km (3.6-mi) line was completed in 1990, and a third one is currently under consideration.

Language.   In addition to the town's intentions to "question its development and the alterations to its infrastructure . . . symbols of massive changes . . . ," the language used in documents about the TAG consists of mass transit terminology—that is, passengers, clients, and users. *Accessibility* is the generic term found most frequently with regard to people with disabilities. "Accessibility for travelers in wheelchairs, persons with reduced mobility, buggies, strollers, exceptional accessibility for all."

Legal Requirements.   When preliminary studies started in 1979, legal provisions governing the accessibility of buildings and mass transit systems were practically complete, with the 1975 law and two 1978 enforcement decrees. The law covered new buildings open to the public, while the decrees dealt with existing buildings that were not yet accessible and with the adaptation of mass transit services. The distinction between new and existing buildings expressed the basic idea of implementing accessibility through adaptation, a process that was to be used for transportation systems. By implementing trams accessible to all and minibuses suitable for people with disabilities, the city of Grenoble applied the new legislation to the letter.

Buses and Regulations. When the tram project was planned, accessible buses had not yet been envisaged. Moreover, by limiting access to one person using a wheelchair, which has to be secured by means of a bar that can be pulled down, the French regulations for buses were an obstacle to accessibility. These regulations now have to be modified. In France, the carrier is responsible for the people carried, unlike in Germany, for example.

Advocacy. The first demonstration by people with disabilities to be held in France took place in Grenoble in 1971. The same year also saw a person with a serious disability elected to the city council. These two events defined the context of the time, where the trend in opinion ran against all forms of social exclusion. "A handicap is not due to the chances of nature but to deficiencies in social legislation, to industrial injuries, to the welfare state. . . . Cities have to be changed" (Comité de lutte des handicapés, 1978). This trend of opinion, opposed to traditional associations, represented strong pressure for a better development of a city for all.

Among the pressure groups were several rehabilitation centers in this urban area. In one of them, the Centre Médico Universitaire D. Douady, classes and medical care are given along with primary and secondary education. It is also the only place where disabled people can receive a university education in France. Because they needed to break down their isolation, the centers acted to open up to the nearby town, and the new tram helped them. In 1983, after the positive referendum result in favor of the accessible tram, elected representatives reacted to pressure from these groups.

Planning: Accessibility as an Urban Criterion. In 1976, an architect and town planner, R. Herbin, was given a consultancy assignment to promote accessibility operations in the town's highway maintenance, and to assist the highway maintenance operatives in legal issues.

Tramway. The project first required a long process of technical research and further thought about the urban settings. The industrialists involved had to respect a strict plan covering comfort, accessibility, security, and aesthetics. These constraints led research departments to completely think out technical solutions again.

Technical Traditions. The French standard tramway had then been in use for a short time but, although this equipment could meet technical criteria, it could not meet accessibility criteria. Bogies and electric motorization equipment in the low part of the car were already at such a height that the floor had to be heightened too, and could not be compatible with the level of the platforms at the tram stops (Fig. 27.6a).

For a time, the possibility of overcoming this height difference by adding some structural modifications, such as elevator platforms or huge ramps, was envisaged. But conclusions led toward solutions within the rolling stock itself. Was this a technical feat or a cultural revolution?

The associations were under pressure from the technicians and the managers, who argued that a great deal more money than usual would have to be spent. But a technical solution was finally found by reversing the usual logic: the electrical drive elements, such as shift-batteries, were placed on the tram's roof (Fig. 27.6b), enabling a lower floor to be put in the tram.

A retractable door step completed the leveling with the sidewalk (Fig. 27.7). This example is quite telling: from a very precise criterion of use (i.e., accessibility for the whole population), technical constraints described as obligatory were revised. According to Donald A. Norman (1996), "People have to be the priority issue, before technology, in order to ensure that the final result will meet the needs of the persons who are expecting it."

Buses. Would it be possible to improve buses like tramways? Trams had introduced such an access quality to the town that equipment had to be faultless and highway maintenance

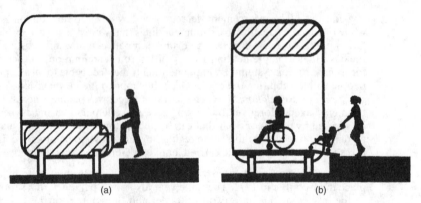

FIGURE 27.6   Principle of tram accessibility: moving batteries up. (*a*) Classical tram with low technical elements—3 steps up. (*b*) Low-floor tram with raised technical elements—near-level entry.

FIGURE 27.7   Horizontal accessibility from the platform.

had to be reorganized. Low floors had been chosen, together with heightened platforms (Fig. 27.8). The first accessible bus was bought in 1995. In the end, 20 bus lines and three trolley bus lines were accessible.

Accessibility Follow-Through. The first important action was undoubtedly the City of Grenoble's December 1976 decision to commission an accessibility consultant. This assignment contributed to the extension of the accessibility concept to the whole population.

Associations and Administrations. Associations have been playing a watching game. They were consulted and have been suggesting improvements. Whenever the tram's accessibility is not followed up, it is fair to say that associations and the entire public end up controlling the tram's use. However, good coordination is always difficult to establish between scattered administrative services and associations.

Communications. Communications worked quite well in this project, which was pioneering a new type of accessible mass transit. The TAG won the European Community Helios award for best transport achievements in 1989. The cities of Strasbourg and Bordeaux followed the example of Grenoble, and the Grenoble accessible tram model was then chosen abroad by the cities of Turin and Brussels.

(a)

(b)

FIGURE 27.8 Accessible bus station: (*a*) the station, and (*b*) the bus entry ramp.

Education.   Why not make the most of the strong educational action of the presence within cities of numerous people in wheelchairs, sometimes with very severe disabilities, but all moving, studying and working, as they are fully entitled to?

Signage and Signals.   Recognizing accessibility signs does not seem to be a problem for users. Easily identifiable directions are indicated at both ends of each tram. The button to be pushed for the retractable doorstep opening is highly visible and accessible. Voice announcements are made at each station on the tramway and bus lines, and tactile signs are installed for users who are blind.

The Cité des Sciences et de l'Industrie de la Villette, Paris (1980–1987)

The idea of a Museum for Sciences, Technology, and Industry was born in 1975, as a use for the remodeled La Villette slaughterhouse. This very important facility had been completed in 1960, and because of an enormous planning error had never been used. This was so scandalous that it was decided to give a new function to the huge sales hall with a floor surface of 3 ha (7.4 acres).

Building Refurbishment Plan.   In 1980, Adrien Fainsilber won the architectural competition and created an important structural design to adapt the building to his ambitious program. During the design process, Paul Delouvrier, president of the Établissement Public de la Villette, decreed as an inescapable constraint that it had to be accessible to the entire public, including people with disabilities. The consultant's work started at the very beginning of the project, and the architect's design thus integrated the accessibility proviso. All public vertical circulation was grouped in a central core (Fig. 27.9).

Horizontally, all levels, elevators, and escalators were connected with a continuous, accessible path from the square to the Park gardens. On the entire 55-ha (135.8-acre) site, accessibility became a requirement too. The consultancy commission was then extended to the park, designed by the architect Bernard Tschumi.

Museum Layout.   After the general budget had been decided and when the 7-ha (17.3-acre) building was completed with its networks and circulation spaces, new design teams started working on the exhibition space, including layouts, choosing furniture, graphic design, and directional signs. This was to be a place where discoveries and interactive manipulation could be experienced by numerous visitors of any age, coming from different cultures and with different abilities. In 1987, the Cité des Sciences et de l'Industrie (CSI) was finally opened to the public.

Language.   As the integration of persons with disabilities was considered from the beginning of the project, it was necessary to thoroughly examine the concept of integration and to evaluate the various disabilities of the persons who make up a museum's visiting public—for example, older people, children in strollers, wheelchair users, people who are blind or deaf, and so on. In the preliminary study for specifications, *Accessibility Criteria to the Musée National dès Sciences et de l'Industrie Présentations* (Grosbois and Araneda, 1982), it was suggested using the term *handicap* rather than the expression *disabled persons:* "the handicap concept is misinterpreted because it is often applied to persons who could be treated like competent persons if adapted changes were made. The study therefore aims to propose adjustments that could balance each visitor's possibilities."

Legal Requirements.   The legislative framework for the accessibility of public buildings was achieved in 1980. However, as the project manager was aware of the complexity of the issue, he asked the architect who had won the competition to reinforce his team by commissioning the author as its consultant. In 1983, the Exhibitions Graph incorporated the criteria of accessibility in the museum's accommodation principles (Grosbois and Araneda, 1982). Full-size mockups incorporating accessibility took place in 1985, thus largely exceeding the requirements of the normative framework.

Advocacy.   The most efficient pressure for achieving exemplary accessibility came from the mandate of Paul Delouvrier, president of the Etablissement Public de la Villette: "No one

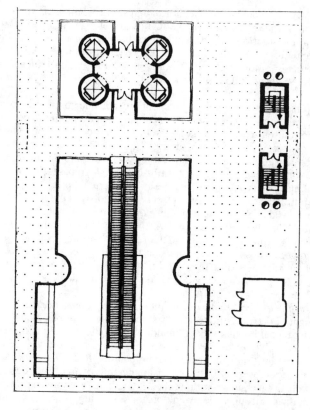

FIGURE 27.9   Cité des Sciences et de l'Industrie (CSI) core: vertical building elevators and escalators.

had constructed a large technical and scientific museum for a very long time. Therefore, there was no expert and no vast know-how to refer to. Strained relationships became unavoidable, but the solution did not finally mar the building's beauty and practicability, nor did the drastic requirements that had to be imposed for disabled people's use!" A committee for people with disabilities met subsequently in order to establish an accessibility chart (Musée National des Sciences et de l'Industrie, 1984).

Planning.   The building's accessibility became possible, in excess of the standards provided in legislation, as a consequence of cooperation between the project's architect and its consultant. For museum accommodation, another method was used. Rules had to be defined in order to organize, lay out, furnish, identify, and devise signage for exhibition spaces, which will accommodate as many active visitors as possible. Ergonometric principles were suggested during the layout of spaces and their furniture. For example, seating able-bodied people in front of handling devices puts wheelchair users in an equal position. As another example, more extensive ability to move within the museum and improved independence for visitors who are blind was achieved by setting up a guideway within the building, where visitors can find their way with a stick, tactile maquettes, and Braille inscriptions within the exhibition rooms. In the Louis Braille Room in the multimedia reference library, visitors can read through a scanner with a voice synthesizer. This room's design, by Grosbois and Sautet, is based on nonvisual perceptions, such as sound, tactile, odor, and kinaesthetic perceptions (Fig. 27.10). Access using sign language was also planned for visitors who are deaf.

Technical Traditions.   It is interesting to consider traditions when designing a museum. In light of their experience, the first thing that the public planners asked themselves was whether there could actually be a "handicapped public." One recurrent question was "How many of them will there be?" People with disabilities were usually seen as marginal and limited, and no one had yet thought about accommodating the largest possible number of people with disabilities. There were also the professional habits of architects, designers, and graphic designers, who were not inclined to rethink their immediate reactions to problems. For example, if cables could not pass through, they would install a floating floor, or a platform—both nonaccessible solutions. As another example, exhibitions are created by people working with their sight (video, computer screen, etc.), and vision is the dominant sense they will refer to. Designs for all perceptions are simply not common. In fact, the technical obstacles raised at the CSI often came from behavior obstacles with regard to disability.

Accessibility Follow-Through.   At the CSI, this was quite innovative. There were analyses, proposals, visits to the construction site, and a final assessment of the accessibility to exhibitions. After the museum had opened, the consultant's commission was extended to work with the accessibility team, an internal CSI service. This was quite necessary, as accessibility can often gradually disappear if there is no strict follow-up, even when it was thought to be permanent.

Communications.   The information circuit for associations and the public works very well. One only has to walk around the CSI spaces and in the Parc de la Villette to see that the challenge for social integration has been taken up. There are many people—mostly young people—using wheelchairs. They are very happy to be in a large public space with other people. The public adapted quite well to facilities that place side by side people who are usually separated. Is this accessibility easy to perceive? It is not obvious at all. In fact, such an achievement implies that comfort becomes commonplace, which is precisely the goal of design for all.

The Grande Galérie in the Muséum d'Histoire Naturelle, Paris (1989–1994)

This building, which belongs to the Muséum d'Histoire Naturelle, was built in 1889. It is located in the Jardin dès Plantes, between the Austerlitz Station and the Latin Quarter. Closed to the

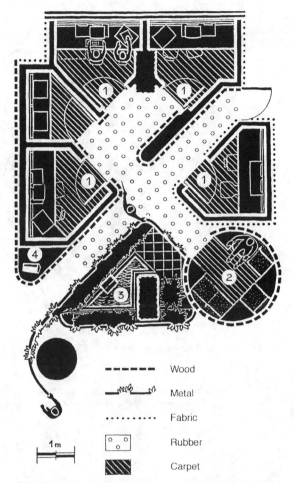

Wood

Metal

Fabric

Rubber

Carpet

1 m

FIGURE 27.10 Louis Braille Room at the CSI (Grosbois and Sautet, architects): (1) reading booths, (2) relaxation room, (3) basin with fountain, and (4) tactile sculpture.

public in 1965, the Galérie has always been rather secret and could have been defined as a place where knowledge and research had been gradually accumulated through the years.

Refurbishment Project.   An architectural competition organized in 1989 was won by Paul Chemetov and Borja Huidobro. Of all the building's functions (i.e., research and collection storage), only the exhibition function survived. But the architects had a double goal—to create a contemporary museum, to be visited by the general public, and to restore a nineteenth-century building. The public entrance was originally located on the façade that overlooks the garden, up an inaccessible monumental staircase (Figs. 27.11 and 27.12).

The architects decided to move the entrance to the side of the building, at street level, thus allowing easy access to the surrounding area. Panoramic glass elevators made the three gallery levels surrounding the central space accessible. As the space was not large enough to comply with the plan as envisaged, the basement was converted into exhibition space. This project actually involved redesigning a building, which was done by identifying the old and the new parts, and by integrating accessibility into the new parts from the start.

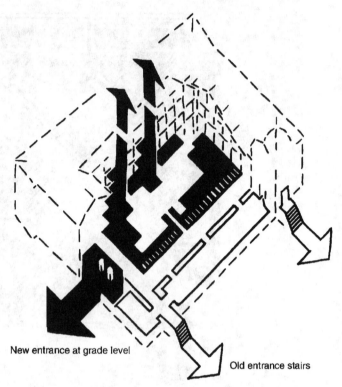

New entrance at grade level

Old entrance stairs

FIGURE 27.11 Grande Galérie du Muséum (Chemetov and Huidobro, architects): new accessible entrance after renovation.

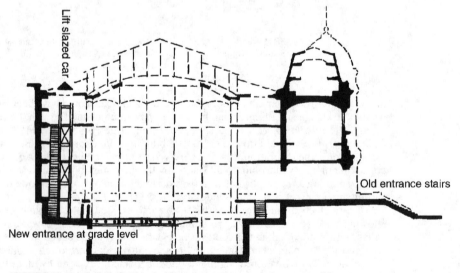

Lift slazed car

Old entrance stairs

New entrance at grade level

FIGURE 27.12 Grande Galérie du Muséum: another view of new accessible entrance.

Language.   *Amplitude* was developed by the consultancy department in the quest for accessibility to the public, whose aim was to "include every handicap, physical, sensory and intellectual accessibility to spaces and objects for all visitors." This was to "include reduced mobility and sensitivity in the definition of the criteria used to establish the conditions in which visitors might experience disability for themselves," and the public was "to be able to move through the gallery's entire exhibition space without having to be segregated through any specific route" because "comfort and social interaction have to be sought." Designed in the 1990s, this project illustrates how the vocabulary has evolved significantly as a consequence of the CSI experience.

Legal Requirements.   The Grande Galérie was opened in 1994, the very year when the regulations were strengthened. The project proved that regulations were only a minimal aspect that could be broadened by the architect, Paul Chemetov, who argued that "cities today attribute a dominant role to information and reception" and "locating the entrance on the side, i.e., the accessible façade, was thus a key option." Architecture critics perceive the concept of the free visitor flow as part of the concept of accessibility: "Nothing has changed but everything is different. A direct entrance at street level leads to the museum, since now the street and the side aisle are connected and are on the same reference level" (F. Lamarre, 1994). This case shows that the concept of the free visitor flow went beyond the literal application of legal requirements.

Advocacy.   As part of the refurbishment project of the Grande Galérie, consultancy for accessibility for disabled people was commissioned from a design office. In this case, there was no advisory committee of associations.

Planning.   The Laboratoire de Sociologie de l'Éducation had been developing a research program with the Québec University CREST (Évolution, 1993) since 1989, whose purpose was to study the Jardin dès Plantes and the general public attending exhibitions in the museum. This program consisted of three parts: a sociodemographic survey, a survey of knowledge and competence, and a survey of visitors' ways and habits. Accessibility studies were subsequently integrated into the specifications to take the variety of the public's motor and sensory abilities into account. Several teams of contributors shared the work based on their competence: the team of architects worked on refurbishing the building, the film-maker René Allio designed visual and sound spaces, and the A.D. Sign Agency worked on designing the exhibition spaces.

Technical Traditions.   The A.D. Sign Agency was created by the architects involved in the museum project as an independent organization in charge of the museum spaces and the development of display designs. The beautiful nineteenth-century industrial display cabinets were kept, but adapted to contemporary museographic requirements, such as optical fibers. The 3-m (10-ft)-high display cabinets provided all visitors, including children and wheelchair users, with excellent visual access.

   This case shows that it is possible to combine traditions with technical evolution to preserve a building's identity. René Allio's stage-set approach combines sound and light, and he says that animals "will prompt visitors to complete the images they see with their feelings and sensitivity, so strong is the animals' evocative power." As the animals are set on the floor, persons with visual deficiencies can make tactile explorations. According to Sandrine d'Eggis, "the association of visual, sound and tactile possibilities provided by the scenography explains a great deal of the Grande Galérie's success."

Accessibility Follow-Through.   A reception unit for people with disabilities is currently following up on accessibility and has made the whole museum staff aware of this issue. Guided visits for the deaf are accompanied by deaf lecturers and actresses, who adapt their tours to the needs of their audiences. Workshops have been created to complete training for the blind, so that they can sense the exhibits better by touching them—for example, forms, textures, animal furs, skeletons, and dimensions (Fig. 27.13). This accessibility follow-up policy is essential to "attract an audience that does not always fit in cultural places" (S. d'Eggis).

Communications.   The magazine *Musées et Collections publiques de France* published an issue entitled "How to Receive People with Disabilities," showing that communications about all these new accessible places are developing ("Recevoir les handicapés," 1997). The community of associations is also very strongly involved in this communications network.

27.5   GENERAL LESSONS LEARNED FOR "DESIGN FOR ALL" IN FRANCE

Having analyzed two public transport systems and two public buildings, it is interesting to note the evolution of various elements, especially because these achievements have been made during the past 20 years.

Language.   In 1975, the expression *architectural barriers,* a negative vision, was gradually being replaced by *accessible measures,* a positive goal. The adjective *handicapped* was also being replaced by the expression *person with reduced mobility.* Since 1985, the concept of specific use "for handicapped people" has been dropped and replaced with the concept of *use for all* or *accessible to all.*

Legal Requirements.   A few years before 1975, in the case of the Lille metro, politicians began to support the implementation of accessibility laws. By 1980, in the case of the Grenoble tram, the laws had been fully implemented and comfort was afforded all users. Since 1985, in the cases of the Cité des Sciences and the museum, the law has needed amendment because it provides

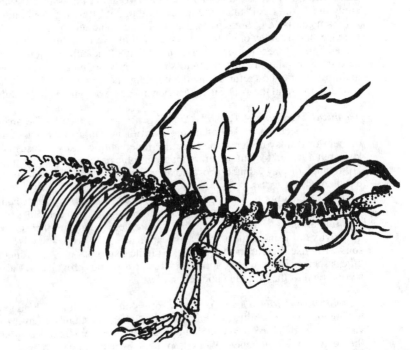

FIGURE 27.13   People who are blind can sense objects better by touching them.

only a minimum set of regulations. In the case of the Grenoble metro and tram, a popular referendum on accessibility approved the system chosen by the city. It now appears clear that law is only a tool, but political determination and democratic referenda set the various objectives and pursue their implementation.

Advocacy. The evolution of the role of associations for people with disabilities is structured in two phases:

- An incentive role to develop awareness of the issue in 1975 (Lille and Grenoble).
- A consulting role, associated with a commission for technical consultancy awarded to an expert in 1982 (Cité des Sciences).

Planning. This reflects the evolution of the users' demands and the evolution of technical solutions. In Lille and Grenoble, this entailed the demand for accessibility and the definition of new transportation equipment, meaning an automatic metro and new, comfortable, safe, and elegant tramways. In public buildings, it entailed accessibility for all users, after having defined their various abilities.

Technical Traditions. Humankind has to be considered as the key figure, with technology playing the supporting role. Overturning technical traditions by insisting on accessibility for all in the Lille metro and the Grenoble tramway was a significant development from 1975 to 1985. Professional habits prove to be symbolic rather technical obstacles.

Accessibility Follow-Through. The Grenoble experience in 1975 and the Cité des Sciences in 1985 created a permanent team called an *accessibility cell.*

Communications. As a consequence of the aforementioned developments, the following can be said:

- The best communications are guaranteed by everyone's presence (Lille, Grenoble, Cité des Sciences, and the Muséum d'Histoire Naturelle).
- There is added value to the image of achievement for transport and public buildings, as shown in the preceding examples.
- The design-for-all concept is becoming increasingly significant, and the public wants it to be developed everywhere to ensure comfort.

A Culture of Accessibility

The objective of design for all could be achieved by encouraging a new culture to emerge: a culture of accessibility that considers human beings in their diversity as key figures in economic, technical, and aesthetic choices. French examples have provided an awareness of the progress made in the area of transportation and the design of public buildings—a real breakthrough in a 20-year period. Accessibility is now an unavoidable principle; it is feasible, economically reliable, and even quite aesthetic.

This overall progress resulted from partial evolutions that have come about in various ways. These were designated as key processes, leading to the following proposals for action:

- The issue of disability must become a commonplace issue, so that people with disabilities are entitled to control their environment.
- The legal requirements governing accessibility could be developed into regulations governing the convenience of diversified uses.

- Associations should be reminded that, although they have specific goals, they must also take the general public's interest into account: comfort for the whole population will ensure social cohesion.

- Managing complex situations in the evolution of techniques and uses is part of the art of building. This implies that design-for-all design concepts have to be taught.

- Professional traditions must be able to evolve with the evolution of uses.

- The practice of maintaining and perfecting diversified uses in managing transportation and buildings should be reinforced.

- Communication should be developed by increasing the number of accessible places and by setting up multisensory information supports.

27.6 CONCLUSIONS

The French examples of design for all reveal issues of compromise based on cultural, technical, and economic data. Such a concept is a route to accessibility to be developed by each country, each making different social and technical compromises. A universal goal that deserves everyone's respect, and one that respects everyone's right to be part of social life through the use of buildings and transportation, must not be confused with a policy of compromise to be defined according to each culture, even though the experience of other cultures may increase one's own knowledge.

This is actually the way this concept has been introduced in recent years. The Design for All—Toward the Mainstream conference held in Barcelona in January 1998 established the objective of "Design for All as a new concept introduced in the disciplines responsible for an environment's configuration, including social planning, town planning, architecture, design and industrial production. The aim is to achieve an environment which is adequate for human diversity, and in such a way that all can enjoy and participate in the building of society." This had the following consequences: "The new Design for All concept will generate a friendlier and less aggressive environment and will provide benefits to all social sectors; i.e., users and consumers, enterprises, professionals responsible for product configuration and the environment, administration, as well as the organizations representative of civil society."

Similarly, in the December 1997 survey *Universal Design, Planning, and Design for All,* conducted for the Norwegian State Council on Disability (Alslaksen, et al., 1997), the seven Principles of Universal Design, developed by the Center for Universal Design (1997), were recommended based on a use for all:

- Equitable Use
- Flexibility in Use
- Simple and Intuitive Use
- Perceptible Information
- Tolerance for Error
- Low Physical Effort
- Size and Space for Approach and Use

Focusing on diversified uses, these recommendations do not go into the social and technical compromises that have to be made. Starting in the 1960s, the Scandinavian culture of democratic sharing facilitated these compromises, as can be seen with the creation of objects, transportation, public buildings, and dwellings that are accessible to all. Imposing a universal design model is therefore out of the question within the person/culture/techniques interfaces. The suggestion is to share a humanist vision described by D. A. Norman (1996): "In the 1933 Chicago

World Fair, there was such a craze for the wonders of modern technology that one of the slogans was 'Scientists discover, Industry applies, and Man gets adapted.'

"On entering the 21st Century, this slogan should be changed into 'Man proposes, Scientists or Architects create, and Technology gets adapted.' "

27.7 BIBLIOGRAPHY

Aalto, A., *Synopsis,* Birkhäuser Verlag, Basel, Switzerland, and Stuttgart, Germany, 1970.

Alslaksen, F., S. Berger, O. R. Bringa, and E. K. Heggem, *Universal Design, Planning, and Design for All,* Norwegian State Council on Disability, Oslo, Norway, 1997.

Beckman, M., *Building for Everyone,* Ministry of Housing and Physical Planning, Stockholm, Sweden, 1976.

Center for Universal Design, *The Principles of Universal Design,* Center for Universal Design, Raleigh, NC, 1997.

Comité de lutte des handicapés, "Abrogation complète et immédiate," *Habitat et vie sociale,* 22: 36–37, 1978.

d'Eggis, S. and Y. Girault, "Politique d'accueil des personnes handicapées visuelles et audititives," *Musées et Collections publiques de France,* 214: 41–42, 1997.

"Design for All—Toward the Mainstream," conference, Barcelona, Spain, January 1998.

Dreyfuss, H., *The Human Scale,* MIT Press, Cambridge, MA, 1981.

Évolution 93, *Lettre d'information de la cellule de préfiguration de la Grande Galérie,* 11, Museum National d'Histoire Naturelle du Jardin dès Plantes, Paris, France, 1993.

Goldsmith, S., *Designing for the Disabled,* 1st ed., Royal Institute of British Architects, London, UK, 1963.

Grosbois, L. P., *Handicap physique et construction,* 1st ed., Le Moniteur, Paris, France, 1983.

———, and Araneda, A., *Critères d'accessibilité aux présentations,* Musée National des Sciences et de l'Industrie, Paris, France, 1982.

———, I. Joseph, and A. Araneda, *Habiter une ville accessible,* Ministère de l'Equipement et du Logement, Paris, France, 1982.

Kallehauge, H., *Summary of Report from the 50th UN General Assembly,* Copenhagen, Denmark, 1995.

Lamarre, F., *Revue d'architecture* 47, 1994.

Le Corbusier, *Oeuvres complètes 1946–52,* Editions Boesiger, Zurich, Switzerland, 1995.

Musée National des Sciences et de l'Industrie, *Charte des personnes handicapés,* Paris, France, 1984.

Norman, D. A., "Grandeur et misère de la technologie," *La recherche* 285: 22–25, 1996.

"Recevoir les handicapés," *Musées et Collections publiques de France 214* Paris, France, 1997.

Vitruvius, *Les dix livres d'architecture,* Les Libraires Associés, Paris, France, translated in 1965.

Wagner, O., *Architecture moderne et autres écrits,* Edition Pierre Mardaga, Bruxelles, Belgium, 1984.

CHAPTER 28
THE DUTCH STRUGGLE FOR ACCESSIBILITY AWARENESS

Maarten Wijk, M.Sc.
Delft University of Technology, Delft, the Netherlands

28.1   INTRODUCTION

This chapter describes the efforts the Dutch made in raising accessibility awareness. It traces the evolution of the accessibility movement over the years, starting in the mid-1960s. The chapter identifies the origin of the low priority given to accessibility in daily building practices. It questions the words, symbols, and arguments that have been used to convince the building trade, but failed. Insight into these Dutch historic precedents help one understand what should be done to finally bridge the gap between the environment and its use: There is a need for a new approach, which is based on ergonomic differences within the common human needs. In its marketing, this approach should not be given a name like *inclusive design, design for all,* or *universal design.* Instead, good design helps provide integral accessibility for all and does not require a label.

28.2   THE GAP BETWEEN THE ENVIRONMENT AND ITS USE

Inconveniences

Through the ages, man has put energy into arranging his environment to facilitate his everyday activities. Man built houses to live in, buildings to work in, streets to move around in, and parks for leisure. And, as if the world is not enough, man even set foot on the moon in a first attempt to explore the universe. For man, it seems, there is no mountain high enough. So why do people still trip over thresholds? Why can't people find the dressing room, when they need it most? Why do people catch their sleeves on the doorknob and spill coffee on the floor in the process? Why are people getting annoyed by a little girl rattling the flap of the mailbox because she cannot reach the doorbell? Why does the luggage get stuck in a turnstile? Why do people lose their high heel in a floor grating or bump their heads against an awning because their glasses are steamed up? Is it because they think that these, as Fig. 28.1 shows, are just problems caused by the designed and built environment, or that these are just everyday inconveniences caused by clumsiness or by not paying attention? Do they believe it is just not their day?

FIGURE 28.1 Hole in ramp: gap between the environment and its use.

It is other people, after all, who have to cope with children, who are careless on the road, who sprain their ankles, or who do not realize the dangers of a ski slope. That is when pushing a stroller, wearing a sling, ending up in a cast, or spending some time in a wheelchair makes the surrounding built environment take on a less friendly character. Later on, they can laugh about it and tell their friends dramatic stories, that it was not so funny: into the bus with the stroller and back out; having to camp out in the living room for a month because they could not use the stairs; not being able to enter the street because of busy traffic; and not going to work because of the lack of an elevator. Everything should have been made easier, but it simply was not.

Still, others have been stricken with a disease affecting the heart, eyes, ears, legs, hips, hands, nervous system, or lungs. This is when the environment becomes an enemy, day in and day out. These people are literally excluded, becoming dependent on a friendly passerby, faithful friend, or a surly professional. They are supposed to be grateful for this, no matter how ineptly assistance may sometimes be rendered. Nothing is so frustrating as having to depend on others for things that everyone should be able to do as a matter of daily routine. All people wish, and have the right, to move about freely, but reality does not allow this.

People

How is it possible that, after arranging and rearranging the designed and built environment generation after generation, people still make streets, buildings, and houses that do not completely facilitate a comfortable and safe use for everyone? Is it because there is too little knowledge about what people actually need while trying to function in their environment? Is it because those who are responsible just do not care? Or is it because people only seriously wish to invest in pushing forward frontiers, instead of facilitating the ordinary?

FIGURE 28.2 Three of the 16.5 million different Dutch individuals.

The ordinary—and universal—truth is that people differ in every single aspect of their abilities. People differ in the way they perceive their environment, the way they understand it, the way they act in it, and the way they just are in it. They are thin or fat, short or tall, strong or weak, or anything in between. (See Fig. 28.2.) Some people cannot see or hear well; people's minds can change gradually or suddenly from being clear-minded to confused. Unacquainted with a city, people can get lost. Whether this happens occasionally or all of the time, it will happen to everybody at one time or another.

It would seem only logical to take human diversity as a self-evident starting point in the arrangement and rearrangement of the environment—not only because human diversity is a universal fact, but also because doing it any other way is morally unacceptable, that is, if one considers that everyone should have equal opportunities to function in society. It sounds so logical, and yet, it did not and still does not occur to those who are responsible for arranging the environment: the building trade. Instead, there is a mismatch, a gap between human needs and man-made reality, a gap between the environment and its use. This is shown in Fig. 28.3, in which the circle represents the total sum of human needs, and the square represents a design that is implicitly based on a uniform presumption of human abilities. The gap is the space between the circle and the square.

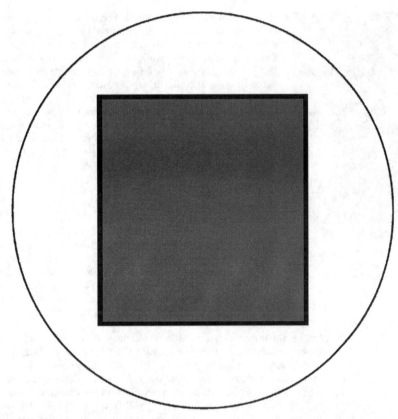

FIGURE 28.3   Traditional implicit design.

28.3   TRAPPED IN THE GAP

Accessibility

It is not that human needs are ignored in general. Through the ages, life has become more comfortable for all—more complex, but still more comfortable. Smoke signals changed to a World Wide Web and mobile phones, horses to magnetic trains, feathers to laser printers. And now, only a few decades after the first Model-T Ford was produced, people have cruise control, air conditioning, and air bags in their cars, equipped any way they want.

And yet houses are built the way the Romans did, brick by brick, with doors sometimes too hard to handle, stairs sometimes too steep to climb, forcing their occupants to move when they get older, keeping them dependent when they are struck by fate. Somehow, the drive to innovate for comfort does not fully apply to the making of the built environment. Innovation in the building industry is basically form driven. This leaves the users of the built environment with a gap. The question is how to bridge this gap. (See Fig. 28.4.)

In the Netherlands, the term *accessibility* is used when referring to bridging the gap. Although the term has been used in the Netherlands since 1962 (see the following section, "What Became of the Term?"), it is only since 1991 that the term was actually defined explicitly as "the feature of built facilities that enables people to reach and use those facilities"

FIGURE 28.4 Unique housing project in the Netherlands: a form-driven innovation.

(Rijksgebouwendienst, 1991). According to this definition, accessibility can be interpreted as the feature that enables people to live in a house, to work in a building, or to make use of the services offered inside. It is also the feature that enables people to walk on the street and to make use of the park. A built facility is accessible when people can get in, and once inside, can reach the place they want to be in, which might be left, right, up, or down. It is when people can perceive what has to be noticed, understand what has to be understood. It is comfort—not too cold, not too hot, and not too windy. It is fresh air. It is safety, or the mere perception of it. It is limiting the chance of slipping, tripping, or being hit by a car. Just following the definition, accessibility is everything that has to do with quality of use. Unfortunately, the term was already in the vocabulary of another agenda.

What Became of the Term?

As mentioned previously, the term *accessibility* was used for the first time in the Netherlands in 1962. This happened during a marathon TV spectacle in which Mies Bouwman, a famous Dutch TV moderator, appealed to the Dutch public to reveal its generosity to build a barrier-free residential complex for people with disabilities: Het Dorp in Arnhem. After that legendary broadcast, everyone became aware of the existence of people with disabilities, the fact that they have problems with the environment, and that this concept could be labeled *accessibility*.

Ever since, the term has been in the vocabulary of everybody who wants to improve life of all those people who find themselves, more than anybody else, trapped in the gap: the disabled. And when the relation between disability and accessibility became a rehabilitation issue, the term *accessibility* got stuck in the gap. Even in 2000, one could ask anyone and hear, "Accessibility has something to do with people who are disabled." Ask any architect, any real estate owner, and any legislator, and one will find the term is "trapped in the gap." This is shown in Fig. 28.5, in which one can see that focusing on the specific needs of categories (squares) beyond the implicitly presumed uniform needs, hardly covers the total sum of human needs (circle), and in fact, is stigmatizing and isolating the issue.

The exclusive relationship between the needs of people with disabilities and accessibility would not be a problem if it would actually lead to an environment that could be used by all. But this is not the case. The building trade is not really interested in solving the problems of a relatively small group of users. The general attitude of Dutch designers and builders is to let welfare take care of the unfortunate ones, if they think about it at all.

28.4 THE FIRST EFFORT: DESIGNING FOR PEOPLE WITH DISABILITIES

The Book

It is historically completely understandable that accessibility got trapped in the gap. After all, it is people with disabilities who are the first to suffer from inaccessibility. Accessibility is their issue. However, this does not necessarily mean that organizations for people with disabilities should have the overall control over the term and its contents. Due to the lack of interest of others, however, these organizations learned to think they had the right to have the control.

It is easy to conclude afterward that organizations for people with disabilities could have made their job so much easier by making use of other marketing strategies. But no, instead of promoting accessibility to be a universal issue, these organizations made it their personal

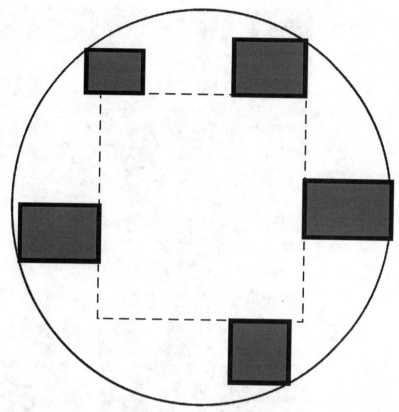

FIGURE 28.5 Designing for the disabled.

agenda, publishing handbooks with titles like *Designing for the Disabled* and labeling this accessibility. The Dutch had an equivalent handbook with a very ambiguous and, in either way, stigmatizing title: *Geboden Toegang* (Gehandicaptenraad, 1973/1993), meaning both *access provided* and *access demanded*.

The first interpretation of the title *Geboden Toegang* (Access Provided) is marginalizing accessibility to the entrance of buildings, and this is still what the members in the building trade associate the issue with. The second interpretation (Access Demanded) refers to the needs of revolting outsiders, who obviously want to get in. And, naturally, people who are already inside the building can not associate with this request. The subtitle of the book emphasizes what everybody already had learned from the title *Handbook for Accessible Building and Designing for Handicapped People,* and so did the cover design: a white wheelchair on a blue background (Fig. 28.6).

The Symbol

It is not just a word game, it is the impact that words can have on the attitude and perception of the intended readers, who are not only architects, but also real estate owners, policy makers, occupational therapists, and—last but not least—locally operating organizations of people with disabilities. And just to show how big the impact was, some figures are presented

FIGURE 28.6   Cover of *Geboden Toegang.*

here. The first edition of the handbook was published in 1973. The last and 11th edition was published in 1993. In 20 years' time, 25,000 copies were sold to national and local governments and the building trade. It became the national standard reference work on accessibility.

The title of the book fully covered the contents and the many publications, brochures, and leaflets that were derived from it. The guidelines were primarily based on the anthropometrics of an average wheelchair user, some for a blind male with an average stature. So it seemed that not only was accessibility marginalized, but also the group of people it was meant to serve.

To promote the issue, actions were taken to convince real estate owners to comply with the dimensions in *Geboden Toegang*. If they did, they were allowed to put the international symbol of accessibility next to the front door, which is the well-known white wheelchair on a blue background, a legally protected logo in the Netherlands and Belgium (filed under registration number 158.302, classes 37 and 42). The symbol indicated that the building was accessible to people with disabilities. But the only thing it guaranteed, in fact, was that the presumed essential parts of public spaces were accessible for wheelchair users. What else could the building trade do than to think accessibility is not their problem?

So the building trade kept on building like it always had done: designing implicitly for a majority, sometimes adjusting public spaces for people who are disabled (e.g., a platform lift at the backdoor entrance; a wheelchair toilet in the basement; and the blue sign at the front door, with an arrow to show the way to the back). However, in 1992, the categorical approach promoted by *Geboden Toegang* celebrated a major success. In that year, a couple of accessibility requirements became part of the Dutch National Building Code (Bouwbesluit). Though these requirements only apply to a certain part of the public areas of newly built buildings, the Dutch Council of The Disabled, publisher of *Geboden Toegang*, was very proud. Also, given the fact that quite a number of public buildings actually have the international accessibility symbol next to the entrance door, one could say that *Geboden Toegang* and its categorical approach had success. But what is success, if accessibility is limited to wheelchair access to the public areas of buildings and, therefore, people still bump their heads at an awning when their glasses are steamed up?

28.5 SECOND EFFORT: THE INTEGRAL APPROACH

Beyond People Who Are Disabled

While at the beginning of the twenty-first century the building trade still considers accessibility to be the adjustment of public spaces to the needs of people who are disabled (i.e., a rehabilitation issue), in the mid-1980s a new concept was launched: the *integral approach*. To really integrate the needs of people with disabilities in the building process, the Dutch promoters of accessibility awareness had to break with their categorical way of thinking. No longer should the needs of people who are disabled be emphasized. No, the message became that also others would benefit from accessibility: women with strollers, people with luggage, people who are elderly, and children. With this new insight, the inner circle of accessibility promoters started to change their marketing strategy. Seemingly out of the blue, the term *accessibility for people who are disabled* was replaced by *integral accessibility for all*.

Adaptable Housing

The first impulse for the integral approach was initiated by the Dutch National Federation of Housing Associations (Nationale Woningraad). In 1987, this organization published a package of requirements for the experiment known as *adaptable housing* (Nationale Woningraad, 1987). This concept was a new approach for dealing with disability and housing. The idea was to build "ordinary" houses in such a clever way that, with minor adaptations, these houses could be adapted to individual disabilities, whenever the need to do so occurred. The concept was a promising alternative for the traditional Dutch way of providing a special housing stock for people with disabilities. *Adaptable housing* quickly became a popular term, thanks to the argument that the concept was an effective way to reduce the budget for special care facilities, while adaptable housing itself was said to be cost-neutral.

In 1997, five requirements with regard to adaptable housing became a part of the Dutch National Building Code. In theory, these requirements created favorable conditions for

FIGURE 28.7 A typical Dutch stairway: steep, narrow, and twisted.

the implementation of adaptable housing. The traditional dangerous steep, narrow, and twisted Dutch stairways, like in Fig. 28.7, are, for example, no longer prohibited. However, in Dutch building practices the emphasis is put on building as cheaply as possible. In this culture, any quality aspect gets overshadowed by doubts about its cost neutrality. Owing to cost pressures, municipalities and housing corporations often discard ad hoc requirements, and real estate developers skim off requirements even further. Nevertheless, a positive development has been set in motion. Building special housing facilities for people with disabilities has at least a competitive approach.

European Concept

A second development of importance started also in 1987. A European conference, "Access to Public Buildings for the Handicapped," organized by the Dutch Council of the Disabled, was held in Utrecht in October 1987. The aim was to generate new initiatives to improve access to the built environment in the European Community. One of the recommendations of the conference was to harmonize the different national accessibility dimensions within Europe. The Utrecht conference developed this recommendation further by advising the European Commission to compile a European Manual.

The Dutch Governmental Coordinating Commission for the Promotion of Accessibility (CCPT) subsequently took the initiative for the development of this manual. In November 1990, the first edition of the *European Manual for an Accessible Built Environment* (Wijk, 1990) was published. (See Fig. 28.8.) Not everybody in Europe was happy with the book. In fact, people in the field were a bit shocked. Although they embraced the approach, it seemed the manual contained too many details for which there was no common European ground. And, indeed, the Dutch developers had been a little too rushed and a little bit too naive to presume the Dutch could provide European guidelines in only 2 years' time.

In the years that followed, more European experts in the field got involved, and the CCPT published a new draft with fewer details. The draft, entitled *European Concept for Accessibility* (Wijk, 1996), was presented in March 1996, supported by a steering committee of 35 people representing governmental and nongovernmental organizations from all over Europe. The *European Concept* was subsequently distributed all over Europe.

In itself, the development of the *European Manual* and the *European Concept* might not have had this great impact, if it was not for the European and Dutch governmental money involved. The budget to develop the European products made it possible to invest in a new concept, instead of merely interpolating the categorical approach into a European context. Especially for the Dutch, the making of the *European Manual* was an opportunity to explore the merits of the integral approach, its consequences, and its promotion. But not only the Dutch benefited from it. Other countries took elements to revise their own points of view. The most significant success story is, without any doubt, the Hungarian one: In Hungary, the *European Concept* became an integral part of the Hungarian Building Code.

Rgd Accessibility Guide

A third major event in the development of the integral accessibility approach was initiated by the Dutch Governmental Building Agency (Rijksgebouwendienst, in short Rgd), which is the facilitator of state building projects. Impressed by the approach presented in the *European Manual* of 1990, the Building Agency started research on the consequences of integral accessibility for new governmental buildings in 1991 (Rijksgebouwendienst, 1991), and existing buildings in 1992 (Rijksgebouwendienst, 1992).

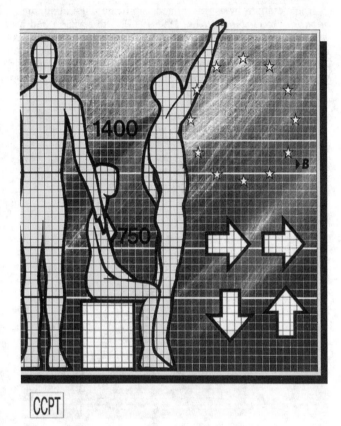

FIGURE 28.8   Cover of the *European Manual for an Accessible Built Environment.*

The budget for this research was considerable and the output quite significant. The accessibility requirements were no longer a more or less ad hoc set of wishes from the outside (the criteria for the international symbol of accessibility), but a serious set of performance specifications to be promoted and enforced by governmental organizations themselves. These specifications—based on the definition of accessibility mentioned earlier—did not only apply to public space in buildings, but also to private workplaces. Thanks to the work of the Building Agency, accessibility became a measurable performance for the public and employees, laid down in the *Rgd—Toegankelijkheidsgids* (Rgd Accessibility Guide) (Rijksgebouwendienst, 1996).

Handbook for Accessibility

Throughout the 1990s, the Dutch Council of The Disabled, publisher of the handbook *Geboden Toegang,* was fully aware of the new developments. In fact, the Council participated very constructively in every committee involved. To the Council it became obvious that it was time for changing its own marketing of accessibility. At the beginning of 1994, the Council started a quite remarkable revision of their "bible," *Geboden Toegang.* The new book was published at the end of 1995.

The metamorphosis was indeed significant. First of all, the revision had a new title: *Handbook for Accessibility* (Wijk, 1995/1998). The revision had a new cover: no longer a white wheelchair on a blue background, but a colored picture of a group of "ordinary" people on a green background. (See Fig. 28.9.) The book also had a new publisher: no longer the Council of The Disabled, but a commercial publisher, very well known in the building trade. The book was immediately a commercial success. At the end of 1999, it sold over 8,000 copies.

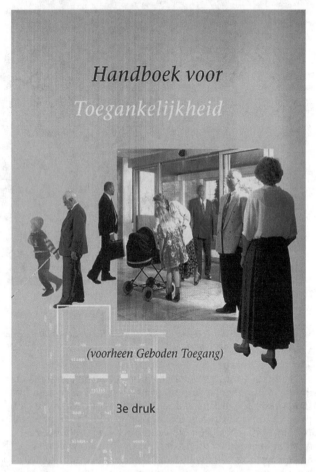

FIGURE 28.9   Cover of the Dutch *Handbook for Accessibility.*

Furthermore, the contents were also completely changed. The requirements now matched the performance specifications as laid down in the *Rgd Accessibility Guide* of the government. Besides, the requirements for adaptable housing were fully incorporated. The anthropometrical criteria were no longer based on the average wheelchair user, but on the more universal concept of people derived from the *European Concept for Accessibility.*

28.6   TOWARD THE THIRD GENERATION

The Revised Dutch Standard

In 2000, the completely revised Dutch standard on accessibility, entitled *Toegankelijkheid van buitenruimten, gebouwen en woningen* (Accessibility of Outdoor Environments, Buildings and Houses) (Nederlands Normalisatie Instituut, 2000) was published. This *Standard* was fully in line with both the *Rgd Accessibility Guide* and the *Handbook for Accessibility*. And so it seems the mission to change from the categorical approach to an integral approach was completed successfully. This was quite an achievement in only 10 years' time. But then again, the Netherlands is a small country with only 16.5 million people living together on 41,000 square kilometers, which is 225 times smaller than the United States. The Netherlands is small but crowded with a positive side effect—namely, that communication lines are relatively short. Everyone who wants to be involved and who has the opportunity to find out what is going on is, in principle, able to participate in the discussion.

The Dutch inner circle of accessibility promoters, consisting of national consumer organizations and some converted architects and academic researchers, has always been one big family. In fact, they are calling themselves "The Family." Like any family, The Family argues once in a while, but it has a common goal: harmonization of requirements and the promotion of the integral approach. At the end of the twentieth century, the aim of harmonizing accessibility requirements seemed to work out just fine.

Continuing Lack of Acceptance

However, at the same time that The Family was tormenting the building trade with arguments to comply with the requirements of integral accessibility, there were other groups in the arena. First of all, there were the local organizations of people with disabilities, with their own local ways of promoting accessibility. The national proclamation and implication of the integral approach was not fully getting through at the local level. It seemed the local organizations, which are mostly small groups of volunteers, were still pursuing the categorical approach.

Second, there were other consumer groups, which had their own strategies for promoting their needs, basically in the field of housing and the domestic environment: elderly people with their *Senior Housing Label* (ANBO, 1995), and the Women's Advisory Committees with their *Housing Quality Guide* (Hilhorst, 1997). The needs of these groups included accessibility requirements. These requirements were not fully in line with the performance requirements of the integral approach. The building trade became confused, not only because of The Family, but also local groups of people who are disabled, elderly people, and women were promoting their specific needs. And, as a rule, confusion is a perfect alibi to stick to traditional routines.

Though the request for a better, accessible environment was getting louder and louder, at the beginning of 2000 it was still not loud enough to really make a difference. As mentioned, the building trade was not taking the issue too seriously, and the consumers were not yet united to enforce their wishes. Was there still something wrong with the message? Was the integral approach as a marketing strategy really effective, or was it leading the issue even deeper into the gap? There is a reason to think it was.

When one takes a good look at the requirements of the integral approach, and all instruments that were based on it, one will notice a blind spot. Something was just not right. Was the integral approach nothing but a cosmetic operation—in essence, not different from the categorical approach? Was the integral approach building for all? For mothers with strollers? For older people? For children? Were these people actually ever asked for their specific needs? And even if they were asked, is it not a fact that these people were also categories by themselves, just like people with disabilities are, and thus still creating a gap between the environment and its use?

The integral approach of combining the special needs of categories of people into one package of requirements, and by emphasizing that not only people with disabilities, but also other categories (e.g., children, elderly people, and people with strollers and luggage) benefit from the measurements, still leads to a mismatch between the environment and its use.

To put it boldly, one could say that people who are disabled were sucked into the average, creating a new kind of average. To make this possible, people with disabilities had to compromise, because if special needs become normal needs, the needs of those who have severe handicaps cannot be part of the package. On the other hand, the so-called normal users will not really recognize themselves in the package, because there is not an absolute urge to do that. This is shown in Fig. 28.10. As one can see, the square has grown, covering parts of the special needs, but not covering the whole circle. There is still a gap, which is logical: A square will never cover a circle.

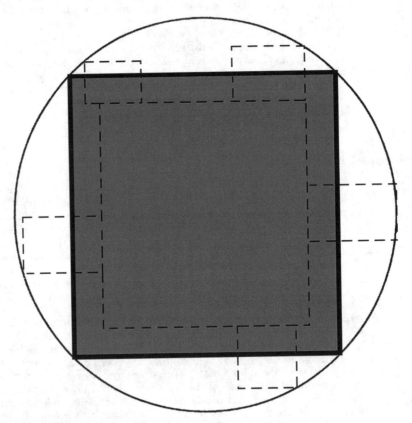

FIGURE 28.10 The integral approach.

This conclusion was addressed during the installation of the academic Chair of Accessibility at the Department of Architecture of the Delft University of Technology. In his inaugural speech, entitled "Differences We Share" (Wijk, 1998), the author pointed out that the promotion of accessibility awareness would not have a future if the need for an accessible environment was not truly based on the common needs of all people. To move the promotion of accessibility awareness into the mainstream of building practices and architectural training, it is the common needs of people in general, not the specific needs of specific groups, that should be addressed. Without the restricting ties to tradition and its representatives, the existence of the academic chair made it possible to set this new approach into motion.

Back to Basics

To understand the difference between the specific needs of groups—even if they are integrally incorporated into one set of coherent requirements—one just has to enter the mind of an 18-year-old who starts his education to become an architect. As a rule, any starting architecture student has no idea of the dimensions of things, despite the fact that she has been exploring her environment from the moment she could walk. To him, as to anybody else, the environment is just the way it is: Streets are streets; pavements are pavements; buildings have roofs, windows, doors, and stairways. In this respect it is only logical that a student takes herself as the starting point in the first designs she makes. With the existing environment as natural reference, he refers to his own physical dimensions to determine the required width of a door, the height of a window, and the size of a bathroom.

In one of the first lectures that a student receives, he will be introduced to the principles of human dimensions illustrated by, for example, Le Corbusier's *Le Modulor,* which is a romantic way of looking at human dimensions fitting the golden section but, in essence, meaningless. (See Fig. 28.11.) Pictures like *Le Modulor* fail to tell what the student actually should know: the adequate widths, lengths, and heights of things.

The student will, however, be introduced at some stage to *Bauentwurfslehre* (Neufert and Neufert, 1936/1992). The starting points for proper dimensions can be found in the first chapter of the book. In this chapter, a standard man and a standard woman are introduced. These standards, with all their physical features and dimensions, are thoroughly incorporated into

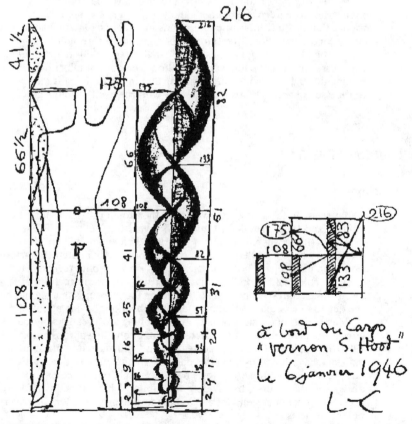

FIGURE 28.11   Le Corbusier's *Le Modulor.* Does it resemble a human being?

various types of facilities presented in the rest of the book. For a student, there is no reason to doubt the Neuferts, as long as the stature of the standard man or standard woman resembles his or her own stature—that is, until the student is asked to design, for example, a primary school or a child care facility. The student will find out that the anthropometrics of children are quite different from the standard. And, if being confronted with the aging of his or her grandparents, the student will also discover the importance of creating separate facilities for people who are elderly. In due course, the student will learn that besides the average person, children and senior citizens also exist, which is good.

It is good that students—and the building trade—are aware of different user categories, but it is not good enough. First, a home for the elderly is not just for old people, just as a primary school is not just for children. Children, adults, and elderly people go everywhere as residents, as staff members, as the public, or as guests. Second, within all those categories, one will find tall adolescents, elderly tall people, small men, and sturdily built women.

Just as no "standardized" man, woman, child, or elderly person exists, neither does, for example, any "standardized" person with disabilities. The essence of the new approach is to let go of the idea of standard categories. People cannot be pigeonholed, and no norms for separate categories exist. Individually, people are undefinable and diverse. In architectural training, and in the building trade, attention should be paid to the ways in which individuals can be diverse. It is of no use to identify the common needs within categories and then to combine them into collective needs. The opposite is necessary: to look for the needs that, in principle, every individual has, and then to identify the differences in every aspect of functioning.

People differ individually, but people also have the same needs. All users of the environment want to perceive, to understand, to act, and to stay healthy. Everybody needs pavements, sign posting, lifts, air conditioning, and doors. One individual may want a wide door, the second a high door, the third a door without a threshold. But people all need doors to enter a building. It is not a question of what category this door should be for, nor who benefits most from specific dimensions: It should be one for all and all for one. If this principle is getting through in architectural training and in the building trade, everyone will be involved in the issue of accessibility. Accessibility will free itself from the gap.

28.7 BRIDGING THE GAP

Human Functioning

It may seem quite complex to investigate the proper provisions for all 16.5 million Dutch individuals, and the half million tourists who visit the Netherlands every year. Nonetheless, in principle, this is what should be done. But, then again, it is not the individuals about whom one should have knowledge, but the environmentally related characteristics of human functioning in its diversity. For example, seeing is a facet of perceiving. To see, people need light, contrast, color, and so on. What is needed are proper criteria for these environmental parameters, while addressing poor sight, including the measures that should be taken if one cannot see due to illness, glasses that are steamed up, or not paying attention.

Many facets of human diversity in functioning are already explored. However, most knowledge is fragmented and is formatted for neither building purposes nor architectural training. In 1997, the academic Chair of Accessibility in Delft started to make an inventory of ergonomic data, and has been translating this data for architectural training. This is an ambitious project, but promising. Figure 28.12 shows how the gap between the environment and its use can be bridged in due course, by dividing the circle of human needs into separate sectors for all aspects of functioning, and then searching for the proper criteria to cover the extremes in each of these aspects.

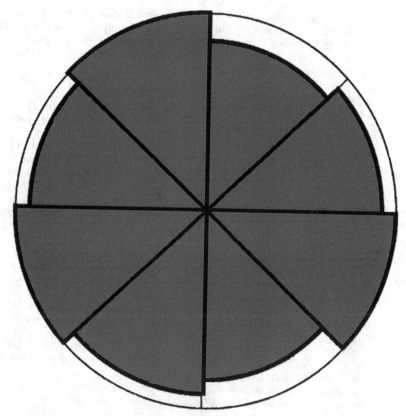

FIGURE 28.12   Design based on ergonomic diversity.

Changing Attitudes

Still, if all ergonomic data are available for architectural training and building purposes, one still has to cope with existing attitudes. To make a change in schools of architecture and the building trade toward a better accessible environment for all, one could devise or revise various instruments. For example, one could devise theories, handbooks, standards, and legislation. However, these instruments are only effective if the attitudes of all participants involved (e.g., building professionals, legislators, governments, and consumer organizations) change. Historic precedents may help one understand how these attitudes can be changed effectively.

First of all, one should realize that things only change when those who benefit most will enforce the change. The first to benefit from an accessible environment is, of course, the consumer. It is only if the consumer is actively involved in the agenda that those who develop the environment, the real estate developers, and clients, have to follow. Once they set their priorities straight, architects will have no choice but to do what they are expected to do: facilitating human activities.

The consumer is primarily responsible for changing building traditions, but until now, his voice was too weak to be heard, and most people were not very interested. Those who were interested were marginalizing themselves and were just not professional enough to interfere in building processes. It is the responsibility of nationally operating consumer organizations to unite and professionalize consumer groups at local levels. Consumer organizations should concentrate on what they ought to do best: monitor the state of the art and address their needs in functional requirements. The translation of these requirements into performance specifications and solutions should be left to the professionals on the provider side, like governments and relevant organizations.

Once consumers become a professional and united force, real estate developers and clients will begin to understand that they, too, will benefit from an accessible environment. Usually, accessibility is the last item of a client's brief, if it is mentioned at all. This low priority makes the issue lose on the battlefield of cost reduction or of other requirements that are deemed more important. Research has already demonstrated that integral accessibility is cost-neutral if the requirements are taken into account from the start (Rijksgebouwendienst, 1991). However, the marketing of accessibility should not focus on costs, as this would suggest that the right to accessibility to exist depends on cost neutrality. In the long run, proving the cost neutrality of accessibility is a battle that cannot be won. Those who fight it lose their awareness of quality. Instead, one should turn the question around and seriously investigate what an inaccessible environment costs society. The marketing of accessibility awareness should focus on the advantages of good accessibility, and it is these advantages that should be promoted.

Another option to enforce accessibility requirements is legislation. This can be done in the constitution and in building codes. At the beginning of 2000, the Dutch still did not have a clause in the constitution that prohibits discrimination against people with disabilities. There should be such a clause, and thanks to European law, there will be one some day. However, building legislation may have a negative side effect, which may spoil its advantages. First, any legislation with concrete measures is an act of discrimination in itself, since it excludes everyone outside the law. Second, legislation can be an alibi for the building trade not to develop quality awareness on its own and to look for solutions beyond the minimum. Therefore, building legislation should only prescribe some firm functional requirements. The industry can, and should, take it from there.

There are three reasons for architects and other designers not to comply with accessibility requirements: ignorance, unfortunate conditions, and obstinacy. Ignorance can be tackled by providing good handbooks. Unfortunate conditions (e.g., a complex building location or conflicting requirements) can be tackled by the client, who has to put his or her priorities straight. The promoters of good accessibility should also provide architects inspirational, ingenious, and imaginative examples. Obstinacy can also be tackled by the client, who can be selective in the architect he or she contracts with. A major concern is that to some clients, not the least of which is local governments, building is just an opportunity to show off their artistic awareness or economic potency, instead of adequately facilitating the activities of building users. This is to show that the consumer must gain power!

However, and quite ironically, artistic, conceptually form-centered architects play a very constructive role in the promotion of building for human diversity. In their competition to display the most eye-catching form concepts, they produce objects that can be used by hardly anyone. It seems that, especially during the 1990s, Dutch architecture deliberately disregarded very basic functional principles, and so did the foreign signature architects who were invited to compete for design commissions in the Netherlands. This process widens the gap between man and his environment to such an extreme, that even the general public is getting annoyed, thus making accessibility an issue for all and of all. (See Fig. 28.13.) It is a logical reaction to decadence, and it will lead at some stage—without any doubt—to a new age of design for all people.

FIGURE 28.13 Perverted form fetishism in the Netherlands.

28.8   CONCLUSIONS

Since its introduction in 1962, the term *accessibility* has suffered from a stigma, because it was exclusively attached to the needs of relatively small and nonunited consumer groups. In fact, as long as the promoters of accessibility awareness need to use buzz words to spread their message, they are talking from inside the gap. It does not matter whether the issue is called *integral accessibility, inclusive design, universal design,* or *design for all;* these terms are only euphemisms for the struggle to get out of the gap.

If and when human diversity becomes a natural starting point for architectural design, all special terms will vanish, and so they should. In the end, one could say that there are only two types of design: bad design and good design. Bad design can be anything between form fetishism and the result of ignorance. Good design is free of failures that could have been avoided if the designer had just used the knowledge available about human needs and differences.

Even if man knows everything there is to know about human functioning, the existing environment and attitudes will only slowly change. Energy and patience are the only tools to finally emerge from the gap into the mainstream of daily architectural routines. And to demonstrate that help can come from any direction, the Dutch struggle can be concluded with an optimistic hypothesis: The worse architecture gets, the more one learns about what one wants to know. It is just like Nietzche said: "Only from chaos a star is born."

28.9   BIBLIOGRAPHY

ANBO, *Seniorenlabel* (Housing Label for the Elderly), Stuurgroep Experimenten Volkshuisvesting (SEV), Rotterdam, Netherlands, 1994.

Gehandicaptenraad, *Geboden Toegang,* Federatie Nederlandse Gehandicaptenraad (Dutch Council of The Disabled), Utrecht, Netherlands, 1973/1993.

Hilhorst, H. C. C., *VAC—Kwaliteitswijzer* (Housing Quality Guide), Landelijk Contact van de VAC's (National Coordination Point for the Women Advisory Committees), Utrecht, Netherlands, 1997.

Nationale Woningraad, *Eisen voor Aanpasbaar Bouwen* (Requirements for Adaptable Housing), Nationale Woningraad (Dutch National Federation of Housing Associations), Almere, Netherlands, 1987.

Nederlands Normalisatie Instituut, *Toegankelijkheid van buitenruimten, gebouwen en woningen* (Accessibility of Outdoor Environment, Buildings and Houses), NNI (Dutch Standardization Institute), Delft, Netherlands, 2000.

Neufert, Ernst, and Peter Neufert, *Bauentwurfslehre,* Vieweg, Braunschweig/Wiesbaden, Germany, 1936/1992.

Rijksgebouwendienst, *Integrale Toegankelijkheid van Bestaande Rijkshuisvesting* (Integral Accessibility of Existing Governmental Buildings), Ministerie VROM (Ministry of Housing, Spatial Planning, and Environment), The Hague, Netherlands, 1992.

———, *Integrale Toegankelijkheid van Nieuwe Rijkshuisvesting* (Integral Accessibility of New Governmental Buildings), Ministerie VROM (Ministry of Housing, Spatial Planning, and Environment), The Hague, Netherlands, 1991.

———, *Rgd—Toegankelijkheidsgids* (Rgd Accessibility Guide), Ministerie VROM (Ministry of Housing, Spatial Planning, and Environment), The Hague, Netherlands, 1996.

Wijk, Maarten, "Differences We Share" (original Dutch title: Niets menselijks is ons vreemd, 1997), Publicatiebureau Bouwkunde (Publishing Office of the Department of Architecture, Delft University of Technology), Delft, Netherlands, 1998.

———, *European Concept for Accessibility,* CCPT, Rijswijk, Netherlands, 1996.

———, *European Manual for an Accessible Built Environment,* CCPT, Rijswijk, Netherlands, 1990.

———, *Handboek voor Toegankelijkheid* (Handbook for Accessibility), Elsevier, Doetinchem, Netherlands, 1995/1998.

CHAPTER 29
NORWAY'S PLANNING APPROACH TO IMPLEMENT UNIVERSAL DESIGN

Olav Rand Bringa, M.Sc.
Ministry of the Environment, Oslo, Norway

29.1 INTRODUCTION

Although high-level objectives and international policies on accessibility exist, the common practice in many countries is to relegate the implementation to a technical, ineffective, and poorly coordinated effort with little prestige. In 1997, the Norwegian government began an effort to find a high-level strategy to address this problem, and to develop more integrated and comprehensive planning. This chapter describes the policies of the Norwegian government, detailing the objectives and actions that support accessibility in general planning, the core of this program, at all administrative levels. Universal design plays a vital part in this policy. The basic and normative high-level objectives are the foundation of international and national policies, providing directions of how fundamental values should be translated into practice.

The norms in the UN Charter, national constitutions, and elsewhere create a moral, ethical, and legal basis for making and implementing national and international policy, and there is hardly any dispute over these basic values. In implementing the policy, governments use a variety of means: laws, regulations, economic programs, as well as information supported and controlled by public administrations at the various administrative levels.

29.2 BACKGROUND FOR NORWAY

Every country chooses a way of conducting policy based on political priorities, political tradition, and the economic situation. When it comes to implementing the policy of accessibility for people with disabilities, there appear to be some general characteristics shared worldwide:

- Focus on high-level national issues
- Lack of visible and effective objectives at other levels
- Concentration of responsibility in the social sector
- Lack of responsibility in other sectors

There will, of course, be variations in this typical political and administrative picture around the world. The Norwegian political and administrative picture may be typical. It includes the following actions:

- The UN Charters and UN Rules, which fit in well with the basic ideas of the constitution, are accepted.
- The parliament formulates national policies and approves white papers with high-level objectives and places the main responsibility on the Ministry of Social Affairs.
- Other ministries like transport, housing, and planning follow up by including considerations for people with disabilities in specific areas. Most common are specific regulations for the design of trains and buses, building regulations for buildings, and guidelines for sidewalks.

This picture looks much the same at the regional and municipal levels. The social sector has the overall responsibility, and the technical sector with its departments handles accessibility as a fragmented end-product issue in some predefined areas. In the technical sectors, accessibility is generally a low-priority and low-prestige issue.

The consequences of this are evident and have been pointed out repeatedly: insufficient quality, poor coordination of efforts, and a lack of strategic thinking. In short, solutions prescribed include the following: more regulations, more user participation, better professional qualifications, and taking accessibility considerations into account at an early stage in planning.

The Norwegian government started to look into this situation in 1997, trying to find a strategy that would accelerate the implementation and quality of accessibility in the built environment on a national basis. Having used most common means (e.g., national guidelines, regulations, and special programs for development), interest turned to the political and professional planning processes in counties and municipalities. The master plans for these administrative levels refer to the directions for how major political issues shall be implemented and financed in the coming years. These plans must include national objectives and local priorities, and comply with all legal and regulatory requirements. Surveys showed that objectives concerning accessibility were practically nonexistent, leaving both politicians and professional planners without any guidance and support to address this aspect. A survey conducted in 190 municipalities, out of a total of 420, showed that 17.1 percent had accessibility objectives for people with disabilities in their plans for the physical development of the municipality. Just over 3 percent had made these objectives binding by law (Hanssen and Stokke, 1999). This percentage is much lower than that for national objectives like industrial development, general housing policy, and sustainable development. But it is also lower than another critical aspect, namely, considerations for the living conditions of children.

29.3 PLANNING PROCESS

User participation in the planning process, which is statutory in Norway, was almost nonexistent regarding people with disabilities. Only 42 percent of the municipalities had organized user participation when it came to people with disabilities. Most of this participation took place in the last hearing phase of the planning process. Only 3.4 percent of the communities invited the advocacy organizations for disabled people to take part in the important early phases, where the objectives of the plan were discussed (Hanssen and Stokke, 1999). This is lower than for almost any other important population group. These results were disappointing, but they pinpointed clearly one of the reasons for the problems in the built environment. At the same time, an area with a great potential for improvement was identified.

Suggested Improvements

In searching for possible obstacles and ways to improve the results of the planning processes, community planners and advocacy organizations were both asked to give their opinions and suggest solutions. It became evident that planners were positive about including accessibility, but they considered accessibility to be a question of details, and not an issue for master plans, as well as local plans and development plans. When asked to define what they needed to incorporate universal design into master plans, the planners rather surprisingly rejected the idea of producing more guidelines (Høyland and Nistov, 1999). They argued that they already had too many manuals on accessibility and universal design from too many sources. They pointed out that their primary requirements for doing a better job with universal design were the upgrading of skills and better planning tools. They suggested that both planners and representatives of user organizations should be given the opportunity to improve their skills and qualifications locally through seminars, and also at colleges and universities. Their preferred tools were concise information on planning and universal design, as well as references to the most important information. They emphasized that the authorities should be more precise in describing policy, guidelines, and solutions, and that examples of good practice were preferred and useful information.

The user organizations with their local members played a key role in improving user participation in the planning process. A report developed by a research institute served as the basis for the strategic discussions with the organizations. It pointed out some important issues to be considered (Saglie, 1999). The responsibility of the local communities to take initiatives and marshal the user participation processes to suit all users was underlined, and the need for training and information was established. Based on the evaluation of previous pilot projects, the organizations were urged to give priority to subjects that they felt were important, and to choose to be represented in processes where their influence would be as effective as possible. Earlier experience showed that advocacy organizations had had problems in meeting demands for user involvement in public decision-making processes. This caused frustration among public administrators, who did not get the response they needed, and among advocacy organizations, which were unable to meet the demand for user input. There were also strong reactions from users to not having the influence they expected to have on the end results.

It was evident that the planning arena, where strategic priorities are established and work is coordinated, did not recognize accessibility as an issue. At the same time improvements could obviously be achieved, and many parts of the strategies for obtaining better results were clear.

These aspects were carefully considered to make accessibility a key issue in planning in the 4-year program launched by the Norwegian government in 1998.

29.4 PILOT PROGRAM: ACCESSIBILITY, A KEY ISSUE IN PLANNING

The Ministry of the Environment, which is the planning authority, was responsible for the program. The objective was to introduce accessibility as a high-level goal in plans, and to integrate the issue in all the relevant parts of the plans. Using any means available, laws should be evaluated, directives issued to clarify national policy, research launched, development of methods carried out, pilot projects conducted, and a series of educational programs supported.

A survey of the existing laws, regulations, and guidelines from the national authorities showed that most of these could serve as basic tools for improving the environment and providing better conditions for all. The general obstacle for effectiveness was the inability to identify and involve people with disabilities as part of the population, given a broadened perspective of functionality that is necessary to support the development of better practical solutions. The national planning code requires in its present version that considerations be given to the needs of all citizens, and it has been evaluated to find better ways to clarify with precision what is meant by "all citizens," as seen in Fig. 29.1. New laws and support for interpretation of the text

FIGURE 29.1   People of all ages will be considered in the new master planning program.

will be issued in 2001 at the earliest. Therefore, efforts were made to meet the immediate need for better planning tools and political instructions for counties and communities.

Accessibility-for-All Directive

Using the existing laws, relevant political resolutions, and available results from research and pilot projects, a directive was issued in late 1999. The directive, "Accessibility for All," was based on universal design thinking and clarifies national policy in the field of planning for all. The main issues include: the objective of accessibility for all should be included in master plans and supportive plans; user participation should begin with dialogue in the early stages of the planning processes; and high-level authorities may stop plans that do not fulfill these requirements. In Norway, all municipalities are obligated to have an approved master plan covering priority activities for the coming 4-year period.

In addition to these formal and political instructions for the planning process, the directive defines accessibility in a broad context covering national policies and regulations for land use, road building, city design, pollution, and noise control. The goal is to provide basic, easy-to-handle areas of concentration within the existing mainstream planning concerns. In all these aspects, universal design is introduced with a focus on the technical requirements concerning people with disabilities.

Previous national initiatives, which introduced new priorities and aspects in planning, provided both methodical and strategic experience regarding accessibility and universal design. In recent years, consideration for children and concern for women's priorities have been emphasized by the planning authorities.

29.5 PILOT PROJECTS

Important information and knowledge on how the Ministry of the Environment should conduct the work and move forward will be supplied by 20 pilot projects in counties and communities. The pilot projects all have the same objective: introduction of accessibility and universal design thinking in the actual planning. Various strategies are used to test the best practical way to achieve this in the pilot projects.

The projects use, in general, one of two main approaches:

1. Universal design applied directly to the ordinary planning process.
2. Universal design introduced in the general plan. This is done through a parallel plan developed with a special focus on accessibility, universal design, and people with disabilities.

It is important to establish knowledge on which of these approaches is more effective and reliable in view of the objectives that are to be achieved. So far, the latter approach is the one that has been used more, and many argue that a plan with a special focus is necessary to bring the issues of universal design on the agenda. Universal design applied directly to the general planning process is considered the next step. Others believe that the general planning processes can handle universal design directly and effectively, given that the right information is available and that user participation is present. It is also noted that the principles of equality point toward an inclusive process.

The pilot projects deal with different types of plans and a variety of focused themes. Four projects involve county plans. Most of the projects deal with community master plans, development plans, and thematic plans for recreation and historic areas. One project is testing the best ways of combining universal design with the national Local Agenda 21 program (LA 21)—the environmental and development action plan adopted in 1992 at the United Nations Conference on Environment and Development in Rio de Janeiro. Under the slogan "Think globally, act locally," the purpose of LA 21 is to encourage broad public participation and specific steps to improve the environment. LA 21 has spawned activities in communities all over the world. One pilot project is creating a strategic plan for the city of Bergen on the Norwegian west coast. Bergen, with its hilly terrain and historic sites, is a challenge for universal design, as seen in Fig. 29.2.

FIGURE 29.2   A bird's-eye view of Bergen in the winter.

Evaluation

Based on 2 years' experience, the pilot projects tend to be more successful if the responsibility for including universal design is assigned to the community's planning department, not the social affairs department. It has also been observed that special plans developed with a focus on people with disabilities lose momentum and do not easily find their way into the mainstream political process and general plans. This tends to be the case even if political and administrative agreements have been made prior to the planning process. Follow-up is a crucial point, whether the plan developed is a special action plan for people with disabilities or a general plan for the municipality. A plan will not produce results before the intentions and actions that have been adopted are implemented. This requires follow-up of plan elements and the inclusion of specific measures and economic means in the municipality's action plan, economic plan, and annual budgets. Much of the success of the pilot projects will depend on achieving these objectives.

29.6 CURRICULA FOR PLANNERS

The Ministry of the Environment uses various methods to achieve the aims set for the priority area. As a long-term strategy, initiatives are made to upgrade the curricula for planners. In cooperation with the Norwegian State Housing Bank, the Ministry of the Environment manages a project to introduce "Universal Design/Design for All" as a part of the basic education of planners. In this project, planning is an integrated part of a program that covers 11 universities and colleges. There are more than 20 universal design activities. Five universities and colleges with planning programs are involved along with two of Norway's three architectural colleges and colleges offering degrees in product design, engineering, and occupational therapy. In addition, the comprehensive upper secondary schools are working to strengthen their building-related vocational programs.

Traditionally, accessibility has been an occasional part of the education of designers, architects, and planners. Representatives from organizations for people with disabilities or research institutes were invited to give a single lecture on the subject. This was not satisfactory, as these were random lectures covering only parts of the field, and they were usually not adapted to the audience. Students understandably lost interest and fell asleep when comprehensive building regulations were presented in detail or research reports were introduced.

In planning the Universal Design Curriculum project, the Norwegian State Housing Bank sought to involve university personnel with pedagogical backgrounds and experience in the project, and to offer them the best resources available. This meant financial resources, professional experts, and expert users. It also meant forging a cross-professional network of universities, different educational programs, organizations, and authorities. The Ministry of the Environment assumed a special responsibility for both the funding and the professional contact with universities with planning programs. A total of four ministries, five of the largest user-based advocacy organizations, and two research institutes were involved in the project, in addition to the Norwegian State Housing Bank. Valuable information from foreign initiatives and projects was used (Welch, 1995). (See Chap. 51, "Teaching Universal Design in the United States," by Polly Welch and Stanton Jones, and Chap. 52, "Infusing Universal Design in Interior Design Education," by Louise Jones.) All universities and colleges with educational programs covering product design, architecture, planning, or occupational therapy were invited to launch pedagogical pilot projects with the aim of including universal design in their curricula. They were at the same time offered funding and the resources in the network. The independence and autonomy of universities required this approach, whereas comprehensive schools educating carpenters, plumbers, electricians, and masons were treated a little differently. The curricula of these schools were stipulated by the Ministry of Education, Research, and Church Affairs, and they already had a usable basis for teaching universal design. For example, the curriculum for electricians contains a general introduction to universal design from a social perspective. In addition, the students have to be able to choose the correct switches and other equipment and place them correctly so that they can be used by everyone, including people with disabilities. For these schools, better teaching aids were all that was needed. In particular, they expressed a need for examples of successful usage of universal design and upper secondary-level textbooks, compendiums, CD-ROMs, and videos.

In general, more than 50 percent of the universities and colleges with relevant educational programs joined the project, choosing an impressive variety of strategies to reach the objective. Conferences, seminars, competitions, and exhibitions have been held. The broad network has proven useful in connecting activities in different environments and sectors, giving the students more interesting challenges. To some extent it has been possible to look across professional barriers. This has been encouraged by the project. Cooperation between occupational therapists and engineers on the one hand and designers and architects on the other has been fruitful, and has emphasized the connection between function and design.

Examples of Planning Curricula

The two universities and three colleges that are involved in the integration of universal design into their planning programs may serve as examples of the activities and results of the Universal Design Curricula Project. The two universities that have participated in the project are the leaders in educating land use planners in Norway. Both universities wanted to address basic curricula as well as postgraduate programs. One of these schools started the process with a 2-day conference debating the issues of including universal design with both the students and staff in the department (Ridderstrøm, 1998). These subjects were covered on a regular basis within a year, and the university continued the process of developing readers to fit their basic courses and to establish postgraduate courses. The second university, having previously used guest lecturers from research institutes and user organizations for a number of years, chose to try other methods. It turned to learning by experience and to compiling compendia on universal design to fit into the obligatory curriculum.

As part of the learn-by-experience approach, the students were sent in groups into the city, accompanied by representatives of various user organizations. Their task was to isolate and record difficulties with regard to pollution, noise, visual and tactual information in the environment, and accessibility. In addition, the students were to pay special attention to children and elderly people. Their job was to isolate the problems of the user groups and propose specific steps in a written report. The students quickly identified the specific problems together with the advisers from the user organizations and pointed out solutions. The university set few guidelines regarding how accessibility for people with disabilities could be solved, but the strategy of universal design was reviewed. In this way, the students were not only given a relatively clear focus for the assignment, but also ideas about how to incorporate a broad perspective in their analyses. Linkages to other disciplines such as safety, social contact opportunities, traffic engineering, and pollution were obvious in the answers. The students' choice of analysis methods also showed that they used knowledge from other disciplines. Map analyses were frequently used to show contexts, and interviews were conducted with residents and users of the areas to identify characteristics and problems.

Discussions and evaluation indicated clearly that this way of learning was both more engaging and more effective than lectures. Student design competitions and exhibits were another way for students to address design problems. Figure 29.3 is the model of the building that was the winner in a design competition sponsored by the Universal Design Curriculum Project. The university plans to pursue this line in the future and will attempt to expand the potential of learning by experience.

FIGURE 29.3   Tonje Loevdal's winning model for dwellings in downtown Hamar.

The university made an attempt to improve the traditional method of learning with seminars where guest lecturers from user organizations and research institutes spoke on universal design and accessibility for people with disabilities. Better planning and coordination of the various presentations elicited a more favorable response from students, but the university still found it necessary to do a better job of adapting and concentrating the academic material and to include and highlight it in the obligatory part of the curriculum. The academic compendium by Nistov (2000) emphasizes strategies and ways and means of promoting universal design. Students are presented with the main problems faced by people with disabilities. Parallels are drawn between other population groups, and the social implications of a lack of accessibility are addressed. Considerable emphasis is placed on giving the students an overview of national political goals and planning guidelines. The presentation of this subject also provides a clear illustration of how interwoven accessibility is with other sectors and disciplines. Documents from the UN and national public institutions are used. Basic land use policy and policies concerning children, adolescents, the elderly, outdoor recreation, and so on are represented.

In this way, the strategy of universal design is illustrated through existing public documents on planning. It provides a thought-provoking presentation in support of the idea that universal design is not a conflict-creating approach, and that it does not need to be distant from today's methods. The compendium concludes by specifying the factors that should be taken into consideration at a master planning level. These include topography, building density, siting, transportation and communications systems, and air quality.

Of the three colleges, one educates engineers and the other two planners. The engineering school quickly identified development plans as the most important issue, and prepared a reader defining accessibility for all user groups in this context.

The other two found it necessary to develop basic ideological thinking to achieve what they considered to be the best approach in their colleges: to develop a module in a master degree program and to define social distribution policy in sustainable development thinking. The master degree module is based on two axes: lifespan and cross-professions. The idea is that the planner should be able to see the factors that influence the population's lives in all stages, and to see the relationships between the different professional sectors.

The relationship between social distribution policy and the policy of sustainable development is explored in environmental health care and environmental and energy engineering courses at one college. Sustainable development is one of the factors considered in universal design (Center for Universal Design, 1997). The issue is not only interesting for that reason, but also because it is often claimed that there are conflicts between taking people with disabilities into consideration and the means necessary for achieving sustainable development. (See Chap. 33, "Sustainable Human and Social Development: An Examination of Contextual Factors," by C. J. Walsh.) Among other things, it has been pointed out that the use of hiking and walking paths by people in wheelchairs ruins outdoor recreation areas, that the dispensation given to people with disabilities to operate their vehicles in car-free areas causes pollution, among other things. Some factions of the environmental protection movement embrace ideologies that give more consideration to animals, larvae, and microorganisms than people. Consequently, finding common ground in the development of a universal design standard cannot be taken for granted. By juxtaposing social distribution policy where those who need the most social and health benefits receive the most allocations with sustainable development, the challenges are made clear to the students. For example, would it be right to bar people with disabilities from wilderness areas because they have the fewest opportunities to enjoy the great outdoors? We know that people with allergies get sick at pollution levels lower than those tolerated by the average population. Is this acceptable or should pollution levels be lowered to protect those who need it most?

Initial Findings from Curricula Projects

An evaluation of the Universal Design Curriculum Project was presented in May 2000 (Halvorsen, 2000). The goal of the project is to achieve a permanent place for universal design and accessibility in the curricula of universities and colleges. It remains to be seen whether the project will succeed. The evaluation will provide answers as to which steps should be taken to ensure a solid platform over a long period. From experience we know that this is necessary because universities and colleges must constantly respond to changing needs and trends. The universities themselves seem interested in anchoring the concept through the development of competencies (i.e., that educational institutions be given resources to immerse themselves in the subjects through master's theses, doctorate degree work, and research).

Of the preliminary conclusions that can be drawn from the project, the following are significant:

- Academically sound initiatives by public bodies, such as the Norwegian State Housing Bank and the Ministry of the Environment, have been a factor in the positive response of colleges and universities.

- The far-reaching across-the-board commitment of public bodies, user organizations, and research institutions has been an important factor in the involvement and efforts of the universities and colleges.

- New access to resources and funding possibilities have been an important, but not crucial, factor in the initiative taken by the colleges and universities.

It is also worth noting that none of the universities or colleges believe the project has improperly challenged their autonomous role. In fact, they would like to see more initiatives executed in this manner.

It is, nevertheless, paradoxical that many colleges with relevant programs rejected the idea of a cooperative educational project at the start. This means that the educational sector as a whole has far to go in acknowledging the need for an academic approach to accessibility for people with disabilities.

It appears that the projects conducted by the educational institutions have generated greater interest among students. One indication of this is that there has been a significant increase in the number of graduate students writing theses about universal design. There is a clear trend among universities and colleges to use more educational methods than before in studying accessibility and universal design. Projects and seminars are being utilized better in addition to the traditional practice of guest lecturers. In addition, the concept has been clearly embraced in curricula, often through the use of compendia compiled by the universities and colleges. As mentioned earlier, the challenge will be to ensure that these changes are permanent and that they also are discussed by more colleges and universities.

Besides the results from universities, colleges, and comprehensive schools, the curricula project has also provided input for developing accessibility in master planning through discussions with students and professors. The following issues have been pointed out as important:

- The link between housing areas, city centers, and transportation intersections
- Strategic and preferential accessibility development plans
- The need to understand the relationships between accessibility, aesthetics, and historic preservation in the development of centers
- The connection between the consideration for people with disabilities and sustainable development, preventive health measures, safety, and local security plans

29.7   GENERAL LESSONS LEARNED FOR UNIVERSAL DESIGN

In these programs universal design serves as a fresh and good strategy for developing integrated thinking and to make students and professors explore new viewpoints. The universal design concept offers a framework for handling more accessibility features simultaneously, taking into account reduced mobility, hearing, vision, and learning abilities, as well as allergic reactions. These factors represent a challenging diversity of planning conditions. At the same time, these conditions fall in line with the classical planning premises like high-quality pedestrian areas, acoustics and noise, perception and design of the urban environment, and pollution control.

Universal design is a very demanding way of working, and the distinct way in which the concept addresses a general approach to meeting challenges seems to simplify logical planning considerations. Developed for the design of products and buildings, the remarkable Principles of Universal Design as defined by the Center of Universal Design can be applied just about anywhere, but they also have some shortcomings when used in large-scale planning. This is probably one application that could contribute further to filling the gap between national high-level objectives for accessibility, and the fragmented accomplishments in detailed implementation.

Based on the experiences so far, it is obvious that the fundamental ideas behind a strategy of universal design provide a better point of departure for discussing and explaining how taking people with disabilities into account can be developed within a broader and higher-quality framework. At the same time, the Principles of Universal Design are a little too concerned with design detail to provide the right academic associations in master planning. Experience has also shown that it is extremely important to underline that both the special and general aspects must be included in universal design. Taking people with disabilities into account has to be a point of departure for developing general solutions. One can see where it is easy to fall back on an unconscious for-all formulation if this is not done consciously.

29.8 CONCLUSIONS

The results achieved in this priority area by the Ministry of the Environment will be evaluated on a national basis in 2001, and this will hopefully give more detailed information on how universal design can be realized in planning in Norway.

The venture aims to develop better accessibility through master plans in order to fill the gap between master ideology and goals and detail execution. In theory, there is fundamental agreement on master international and national goals. This means that a municipality should have no problem in drafting strategies to implement these goals. Municipalities do not operate that way, however. Consequently, a national process of change has been mapped out. The process uses a number of ways and means that have shown themselves to be effective, and there are plans to make simultaneous use of these means for greater effect.

The main inertia in the system is not the unwillingness of the players: politicians, planners, users, and educational institutions, but rather their inability to understand that accessibility for people with disabilities is relevant to their work and discipline. The strategy of universal design has shown itself to be useful in simplifying and identifying priorities and contexts, but it has also become obvious that each player and each subject requires precise follow-up in addition to general initiatives. The education sector cannot make immediate use of reports and materials prepared by research institutions: They have a need to develop their own forms. Public planners give the impression that they need other tools and information about universal design than those developed by user organizations and research institutions. Planners ask for clear public guidelines, precise and succinct professional methods, and examples of legal obligations. These are clearly not excuses but real needs for executing the work efficiently and correctly in a busy work situation.

The pilot projects in municipalities and counties so far show that it is possible to place accessibility for people with disabilities higher up on the agenda through systematic planning. It is also clear that there is considerable potential for improvement on a national basis in Norway. The planning processes give users the right and opportunity to get their viewpoints across to public authorities and politicians. To succeed in these processes, the user organizations have pointed out the lack of information and training adapted to their needs. It remains to be seen how the interests of advocacy organizations for people with disabilities come across relative to the strong commercial interests at stake in the planning process. Not least, it will be a challenge for the public authorities to ensure practical implementation of the good intentions and programs described and approved in plans and an improvement in the quality of life for people with disabilities. The latter will be the decisive measure of the success of the Ministry of the Environment's program. The approach used by the Ministry of the Environment is often referred to as the classic European model for bringing about change. It will be interesting to see if these processes result in permanent change.

29.9 BIBLIOGRAPHY

Aslaksen, Finn, Steinar Bergh, Olav Rand Bringa, and Edel Kristen Heggem, *Universal Design: Planning and Design for All,* The Norwegian State Council on Disability, Oslo, Norway, 1997.

Aslaksen, Finn, *Handlingsplan for økt medvirkning fra funksjonshemmede i plansaker,* Vista Utredning A/S, Oslo, Norway, 2000.

Center for Universal Design, *The Principles of Universal Design,* Center for Universal Design, North Carolina State University, Raleigh, NC, 1997.

Halvorsen, Halvor Kr., *Universell utforming av bolig, bygning og utemiljø. Evaluering av Husbankens utdanningsprosjekt,* Asplan Analyse AS, Sandvika, Norway, 2000.

Hanssen, Martin A., and Knut Bjørn Stokke, *Funksjonshemmedes interesser i planlegging etter plan- og bygningsloven,* Norsk institutt for by- og regionforskning, Oslo, Norway, 1999.

Høyland, Karin, and Sverre Nistov, *Hensynet til funksjonshemmede i planleggingsprosessen,* SINTEF Bygg og miljøteknikk, Trondheim, Norway, 1999.

Joranger, Pål, and Kari Bjerke Karlsen, *Forslag til innarbeiding av temaet "universell utforming" i tre studier ved Høgskolestiftelsen på Kjeller,* Høgskolestiftelsen på Kjeller, Lillestrøm, Norway, 1999.

Ministry of the Environment, *Planning for All: Introduction to Priority Area,* Ministry of the Environment, Oslo, Norway, 1999.

Ministry of the Environment, *Directive T-5/99E: Accessibility for All,* Ministry of the Environment, Oslo, Norway, 1999.

Nistov, Sverre, *Planlegging for alle: Strategier og virkemidler for å fremme universell utforming,* Norges teknisk- naturvitenskapelige universitet, Trondheim, Norway, 2000.

Ridderstrøm, Gunnar, *Universell utforming: Rapport fra arbeidskonferanse,* Norges landbrukshøgskole, Institutt for landskapsplanlegging, Aas, Norway, 1998.

Saglie, Inger-Lise, *Medvirkning for funksjonshemmede: Deltakelse i plansaker etter plan- og bygningsloven,* Notat 1999:111, Norsk institutt for by- og regionforskning, Oslo, Norway, 1999.

Welch, Polly, *Strategies for Teaching Universal Design,* Adaptive Environments, Boston, MA, and MIG Communications, Berkeley, CA, 1995.

29.10 RESOURCES

Planlegging for alle (Web site): www.miljo.no/pfa. All publications in English that have been used as references may be found on this site. Most of the documents published by the Ministry of the Environment's priority area project may also be found here.

CHAPTER 30

FROM ACCESS FOR DISABLED PEOPLE TO UNIVERSAL DESIGN: CHALLENGES IN JAPAN

Gihei Takahashi, Dr.Eng.
Toyo University, Kawagoe City, Japan

30.1 INTRODUCTION

This chapter focuses on three principal objectives: First, the intent is to clarify the development and characteristics of the move toward barrier-free and universal design that has taken place in Japan through the advocacy of people with disabilities. The origin and characteristics of this movement will be described. A second objective is to clarify the characteristics of Japanese laws concerning accessible and usable structures as well as the problems entailed in implementing universal design. Another objective is to introduce to the reader two examples of universal design: (1) Saitama New City Center, which was completed in May 2000, and (2) Takadanobaba District, Tokyo. Since universal design attaches great importance to the development and design stages of the planning of cities and buildings, these two cases, with their different historical backgrounds, indicate how universal design has been and is being accepted in Japan. Finally, future challenges related to universal design will be identified.

30.2 BACKGROUND: ORIGINS OF POLICIES CONCERNING PEOPLE WITH DISABILITIES

The origins of policies for people with disabilities started in Japan 30 years ago and have developed under the influence of the civil rights and independent living movements in the West.

Advocacy for and by People with Disabilities

Advocacy for and by People with Disabilities, the actual civil rights movement by people with disabilities in Japan, was born in 1969 in the Nishitaga Work Campus, a facility for people with disabilities located in the suburbs of Sendai City. Under the welfare policy of the time, such facilities were often built in the mountains, far from the urban areas. A severely handicapped individual living in the Nishitaga facility went to town with a volunteer, where they encountered numerous physical barriers. As a result, they voiced the need for improved accessibility, but the government considered the demands for change to be extremely aggressive. The resulting movement was based on the idea that people with disabilities should not be segregated in a remote facility but should be allowed to become useful and effective members of society.

By that time in northern Europe a movement toward "normalization"—the development of community-oriented services and the abolishment of isolated, institutional facilities exclusively dedicated to people with disabilities—had already started. Although the Japanese movement did not use the label *normalization,* the same trend was beginning in Japan. This epoch-making movement was taken over by the Civic Meeting for Barrier-Free Environments for all Citizens in Sendai, in 1971. This was a group aiming to popularize the international symbol of access, which had been established in Dublin in 1969. Thereafter, the movement spread quickly to major cities all over Japan.

In 1970, the Organic Law for Mentally and Physically Disabled People was established. Clause 2, Art. 22 of the law required improvements in transportation and public facilities. The law was established rather early compared to those in other countries, but it did not immediately gain societal consensus. As for city environments, however, by the mid-1970s the government had announced guidelines for improving access to buildings and for smoothing uneven sidewalks to facilitate access for people using wheelchairs. The report of the Welfare Council for the Disabled, published by the Ministry of Health and Welfare in 1970, clearly emphasized that these improvements would be helpful not only for people with disabilities but also for children, pregnant women, and elderly people, even though the anticipated, very large population of elderly persons did not yet exist. This report suggested a step toward universal design.

In 1973, The First All-Japan Civic Meeting for Wheelchair Users was held in Sendai, the birthplace of the civil rights movement by people with disabilities. The meeting, which was chaired by wheelchair users themselves, provided an opportunity for them to recognize that they are members of society and to discover their right to be independent. Since then, similar bimonthly meetings have been held all over Japan, with the participation of many Japanese experts on causes for people with disabilities. Leaders who have emerged from these meetings have played an important role in realizing improvements in buildings and streets, as well as in passing new laws.

In Japan, the first welfare model city project started in 1973. In 1974, Japan's first guidelines for barrier-free designs, which advocated a wheelchair-accessible city, were issued in Machida City. In the same year, the United Nations Expert-Group Meeting on Barrier-Free Designs was held. Through this meeting, the idea of barrier-free designs was spread to the whole world. The delegates reported their epochal goal of making barrier-free buildings available to every person, disabled or not, in the world. This report had a great impact on subsequent development of barrier-free designs in Japan (Rehabilitation International, 1975).

An examination of the development of the civil rights movement by people with disabilities in Japan indicates that, from the start, their activities focused on improving city conditions not only for people with disabilities but for all citizens. The following are examples of goals held by activists: "Citizen-centered town planning" (Sendai City); "A wheelchair-accessible city" (Machida City); "A subway accessible to all" (Kyoto City, mid-1970s); and "Citizen-centered town planning for citizen unity" (the All-Japan Civic Meeting for Wheelchair Users). In reality, however, all activities directed toward the improvement of environments for people with disabilities were considered by city administrators and building designers to be for the benefit of people with disabilities alone, and barrier-free designs became synonymous with welfare planning for people with disabilities.

In 1976, 60 disabled demonstrators in Kawasaki City occupied a bus, protesting the bus company's discrimination against them. This became a noteworthy event because it was the first time that a transportation company faced a problem involving people with disabilities. In fact, in the same year, a similar event happened on a bus in Sweden, a country that is well known for its advanced welfare policies. It seems inappropriate to call it a coincidence that this kind of demonstration for transportation rights by people with disabilities took place almost simultaneously in two countries.

The issues central to the civil rights movement for people with disabilities during the 1980s included improving the transportation system, city environment, housing, and social security. The International Year of People with Disabilities, observed in 1981, played a very significant role in Japan. People with disabilities took action and demanded further progress in the civil rights movement for people with disabilities, emphasizing a barrier-free subway system, and elevators in train stations. In 1985, the Ministry of Construction established a law to provide Braille pavers on sidewalks to guide visually impaired pedestrians. In the latter half of the 1980s, as the growing number of senior citizens in Japan came to be even more widely anticipated and discussed, previous efforts to improve the quality of life and environments for people with disabilities started bearing fruit.

In the 1990s, laws related to accessible and usable designs were finally enacted in response to Japan's aging society and the enactment of the ADA in the United States. The local government of Osaka Prefecture was the first community in Japan to enact a disability-related bylaw, which included the concepts set forth in the ADA. In 1994, the Ministry of Construction, in preparation for the predicted aging of the population, established the Accessible and Usable Building Law, which encouraged the construction industry in Japan to focus more on elderly people, people with disabilities, children, and women. (Kose discusses this in more detail in Chap. 17, "The Impact of Aging on Japanese Accessibility Design Standards," by Kose.) However, this law was not fully satisfactory for people with disabilities for the following reasons:

- Though the law intended to make the environment more accessible and usable to people with disabilities, penalties for noncompliance were unclear.
- The law itself did not ensure the rights of people with disabilities.
- The number of buildings to which the law was applied was limited and specific.
- The most serious issue of all was that this law did not apply to schools, transportation facilities, or other institutions essential to everyday life.

In March 2000, nearly 25 years after the bus occupation incident in Kawasaki City, a law was passed regarding barrier-free designs for transportation facilities, such as train stations and airports, watercraft, and road vehicles, as well as for urban areas with certain levels of traffic. The law also addresses the design of new buildings in light of universal design. Once again, however, sufficient improvement is not stipulated for existing train stations and other transportation facilities. Also, people affected by design, such as people with disabilities and other citizens, are not expected to participate with construction companies in the planning and design processes. Nor is there any organization to defend the rights of people with disabilities when the law is violated. However, there is an increasing demand by disabled people for equal rights and opportunities.

The Civil Rights Movements by People with Disabilities in Japan

The civil rights movement for people with disabilities in Japan was established by disabled persons themselves but had no legal authority. However, today most local governments in Japan consider the civil rights movement for people with disabilities as well as normalization to be significant policy objectives. As history indicates (Takahashi, 1996a), Japan's civil rights movement for people with disabilities aims, using barrier-free design, to achieve a society or community in which all citizens have the right to participate in various kinds of social activities, and to improve living conditions so that the environment will be more accessible to people with disabilities, elderly people, and children. The civil rights movement for people with disabilities seeks to remove four different types of barriers: (1) discrimination and prejudice, and any barriers to (2) communication, (3) the social system, and (4) the physical environment.

One distinctive characteristic of Japanese society is that nondependence on the legal system is considered a virtue, and it is no exaggeration to say that many of the various civil rights movements for people with disabilities have received support primarily through the generosity of the people. That is, welfare movements are mostly volunteer activities. On the other hand, the civil rights policies for people with disabilities implemented by the national government, by definition, aim at not only improving living environments, but also at providing a guaranteed pension, home care, employment, transportation, and housing. However, the scope of each goal is not well defined.

As stated thus far, while Japan's welfare movements have characteristics in common with the worldwide goal for barrier-free design, the movements themselves have not yet become a part of the fabric of Japanese society. In this respect, Japan is different from the United States, where implementation of universal design has been based on the idea that people with disabilities have civil rights. With citizen participation, universal design is being expanded in Japan, but the real civil rights movement for people with disabilities has just begun.

30.3   CHANGING ACCESSIBILITY LAWS

The following section illustrates the changing characteristics of Japan's accessibility laws over three decades, beginning in the 1970s.

Characteristics of the Early Laws: 1970s and 1980s

The system that was established in 1974 in Machida City, Tokyo, was mainly concerned with wheelchair users, and it called for a consultation between the design engineers and the city when a new building was being planned. The system, called "barrier-free design guidelines," was a breakthrough idea at that time. By the end of the 1980s, the guidelines had been adopted by nearly 90 local governments throughout the nation. This was slow progress in view of disabled persons' demands for legal measures, but, considering the limited consensus of the Japanese people and society at that time, it was epochal progress. Meanwhile, the Ministry of Construction initiated national guidelines to provide accessibility for people with disabilities at public facilities. Unfortunately, the actual implementation of the guidelines was ambiguous and unsuccessful (see Kose, Chap. 17). Problems arose in regard to the laws established by local governments from the 1970s through the 1980s. The four primary problems were:

1. The guidelines for accessibility were ambiguous, and each local government set up its own criteria; some simply imported guidelines from overseas.
2. No legal regulations were implemented for the construction of buildings.
3. Each local government made its own laws, but no consistent administrative guidance was given and no discussion was held among neighboring cities and communities.
4. The guidelines were not enforced to ensure the accessibility rights of people with disabilities and elderly people.

As a result, two important issues remained unsolved. First, it remained unclear as to what degree architects and business owners should ensure accessibility. Second, if the accessibility guidelines were violated, there was no one against whom disabled and elderly people could file suit and no place where they could plead their case. These variations in guidelines and guidance showed the necessity of federal laws.

Later Laws: The Inclusion of Elderly People

As Japan became a nation of increasingly older people in the 1980s, there was recognition of the necessity of laws for accessibility and usability, as Kose discusses in Chap. 17. At that time, a financial aid system was established to promote accessibility and usability in urban public spaces, buildings, and so forth, but this program did not apply to private residences. In 1986 the national government outlined a civil infrastructure to serve the aging population of the future. However, even then the government did not agree with normalization—the idea that everyone, including people with disabilities and elderly people, should be recognized as members of the local community. In addition, the issue of accessibility for people with disabilities was not solved at all.

The turning point finally came in the 1990s. At that time, the idea of Japan as simply an aged society changed when it was revealed that not only was the population getting older but the birth rate was declining. In 1990, Kanagawa Prefecture revised its construction regulations based on Art. 40 of the Building Standards Law and introduced guidelines to provide accessibility to elderly people, people with disabilities, and others. The actual contents of the guidelines were simple (i.e., the same as the standards of the symbol of access). However, for the first time there were some requirements. For one to obtain permission to construct a building, these guidelines had to be satisfied. In 1992, Kanagawa Prefecture also started a fund to pay for the installation of elevators at train stations. The prefecture thus worked toward assuring accessibility in transportation.

Influenced by the actions taken by Kanagawa Prefecture, in 1992 the cities of Yokohama and Osaka, and then Tokyo in 1993, took a major step in revising construction regulations. However, the regulations only focused on some public buildings. To create an infra-structure that would more significantly improve citizens' lives, a broader law was necessary. Thus the Local Law for Accessibility and Usability was created.

Osaka Prefecture, the initiator of the Local Law for Accessibility and Usability, used the spirit and ideas of the ADA, enacted in the United States in 1990, as its reference. The process used by Osaka Prefecture to establish accessibility and usability laws became known all over Japan as a model legislative process in which all citizens, including people with disabilities, could participate. The law targeted all facilities necessary to everyday life, such as public and private buildings, roads, parks, and train stations. Highly influenced by these movements and by the national laws of other countries, such as the ADA, the Standard Law for People with Disabilities was established in 1993. Six months later, the first national access law for buildings in Japan, called the Accessible and Usable Building Law, was finally enacted. As of this writing, more than 95 percent of Japan's 47 prefectures have enacted the local accessibility laws. In 1999, six years after the passing of the Accessible and Usable Building Law, the Mobility and Barrier-Free Transportation Law was established.

30.4 NATIONAL AND LOCAL LAWS

As has been observed so far, Japan currently has three major laws related to accessibility and usability: the Local Law for Accessibility and Usability, the Accessible and Usable Building Law, and the Mobility and Barrier-Free Transportation Law. If these laws are used in association with one another, Japan can ensure a high level of accessibility for all people. It will also be possible to achieve universal design for cities. Although the difference between accessibility and universal design is not yet clearly defined in Japan, the design guidelines given in the laws just mentioned are characterized by aspects of both accessibility and universal design. Table 30.1 compares the Local Law for Accessibility and Usability with the Accessible and Usable Building Law.

National Laws and Issues

The Accessible and Usable Building Law was established separately from the Building Standard Law for the following reasons: First, the Building Standard Law is a minimum set of requirements that have not been significantly changed since the law was written in 1950, at a time of much different awareness of the needs of older people and those with disabilities. The

TABLE 30.1   Comparison of the Accessible and Usable Building Law and Local Laws

		Local laws for accessibility and usability	
		Accessible and usable building law	
Local laws for accessibility and usability	Buildings to which the regulations of the Building Standard Law apply	Museums, art galleries, libraries Public halls, assembly halls Hospitals, clinics Senior citizens' centers, children's centers Welfare centers for people with disabilities Shopping facilities, eating and drinking establishments Hotels, lodges Theaters, movie cinemas, exhibition halls, auditoriums Entertainment halls, video arcades, dance halls Public bath houses, Gymnasiums, ice skating rinks Swimming pools, bowling alleys Parking lots	Schools Crematoriums Social welfare facilities Apartment houses, dormitories Industrial plants
	Buildings	Public utilities, telephone, and revenue offices Post offices, banks, financial institutions Hair dressers Dry cleaning establishments, pawn brokers, clothing rental shops Public restrooms Platforms or waiting rooms at train stations, airports	Public service offices Business offices Residences and residential areas Train stations (concourses, platforms, etc.)
	Nonbuildings	City parks, zoos, amusement parks Roads, sidewalks, driveways, and other paved areas Ball parks Trains, buses, taxis, etc.	

Source: Takahashi, G., *Design Handbook for Accessibility and Usability for All,* Shokokusya, Tokyo, Japan, 1996a.

1994 Accessible and Usable Building Law changed the way in which construction policies were administered, in order to handle the significant aging of the population. Second, the Accessible and Usable Building Law laid down, for the first time in Japan, upgraded design standards that took accessibility and usability into consideration. Third, the improvement guidelines were not mandatory but were left up to the will of the people. The new law could be used in the place of the Building Standard Law, as an alternative procedure to obtain a building permit. Now, one unified set of national accessibility and usability guidelines was legally implemented in regard to architects and designers of construction companies. The guidelines were divided into two sections; the architects were allowed to choose which section to follow. Table 30.2 displays the basic guideline for Minimum Standard of accessibility as well as the Recommended Standard, with its higher and more extensive improvements. When the

TABLE 30.2 Accessibility Guidelines Set by the Accessible and Usable Building Law

Areas in and around buildings	Minimum standards	Recommended standards
Ground-floor doorways, exits leading to parking lots	• Internal width of more than one doorway shall be 80 cm or wider. • Automatic doors or easily accessible doors. • No steps shall be allowed.	• Internal width of doorways shall be 90 cm or wider; more than one of the doorways shall be 120 cm or wider. • Automatic doors. • No steps shall be allowed. • These rules shall be applied to indoor doorways.
Hallways, etc.	• Width of hall shall be 120 cm or wider; it must be nonslippery. • End of hallway shall accommodate turning of wheelchairs. Space shall be clear for wheelchairs to turn within a 50 m distance. • If there are different levels, a ramp or elevator shall be installed. [Ramp structure]: This applies to other parts of buildings. Internal width shall be 120 cm or wider (with stairs, 90 cm). Incline shall be ½ or less (⅛ if the height is 1 cm or less). • More than one hallway shall have floor guidance material or a voice-guidance system for visually impaired persons (if an aide is available, this need not apply).	• Internal width of hallway shall be 180 cm or wider; if a space for two wheelchairs to pass each other is available within a 50 m distance, the internal width of hallway may be 140 cm or more. • If there are different levels, ramps or elevators shall be installed. • No projecting objects shall be allowed on the walls. • Resting benches shall be available. [Ramp structure]: This applies to other parts of buildings. Internal width shall be 150 cm or wider (with stairs, 120 cm). Incline shall be ½ or less. • Each end shall have a handrail. • More than one hallway shall have floor guidance material or a voice-guidance system for visually impaired persons.
Stairways	• Handrail shall be available (one side only). • Circular stairs shall be avoided. • Difference in color intensity between the flat surface and the vertical surface of steps shall be assured. • Attention-attracting floor material shall be used at top and bottom landing areas of stairways.	• Internal width shall be 150 cm or more. • The height of each step shall be 16 cm or less; the width shall be 30 cm or more. • Both sides shall have a handrail. • Other items shall be the same as listed on the left.
Elevators	• Total area shall be 2000 m² or larger. • Floor area of car shall be 1.83 m² or larger (11 persons or more). • No obstacles preventing a wheelchair from being turned shall be allowed. • Internal width of doorway shall be 80 cm or wider. • Voice-guidance or stop-floor indicator, etc. shall be available.	• Floor area of car shall be 2.09 m² or larger (13 persons or more). • No obstacles preventing a wheelchair from being turned shall be allowed. • Internal width of doorway shall be 90 cm or wider. • Area for getting on/off the elevator shall be 180 × 180 cm or more. • Voice guidance or stop-floor indicator, etc. shall be available.
Restrooms	• More than one restroom for wheelchair users shall be available (if separated for female and male use, accommodate each sex). • Sufficient floor area for wheelchair users shall be assured. • Internal width of doorway to each stall and restroom shall be 80 cm or wider. Door structure shall be accessible. • If urinals are installed, more than one floor-set-type urinal shall be used.	• Number of restrooms for wheelchair users shall be 2% of the restrooms on the floor if the total is 200 or fewer; if more than 200, 1% + 2. • Internal width of doorway to each stall and the restroom shall be 80 cm. Any restroom without wheelchair accommodations shall be located near a wheelchair-accessible restroom. • If urinals are installed, more than one floor-set-type urinal shall be used.

(Continued)

TABLE 30.2   Accessibility Guidelines Set by the Accessible and Usable Building Law (*Continued*)

Areas in and around buildings	Minimum standards	Recommended standards
Parking lots	• More than one parking space for wheelchair users shall be available. • Width of the parking space shall be 350 cm or wider. • Distance to exit/entrance shall be as short as possible.	• 2% or more of the total if 200 or fewer; if more than 200, 1% + 2. • Same as the list on the left. • Same as the list on the left.
On-site walkways	• Width of the path shall be 120 cm or wider. • If the path is not on one level, ramp or elevator shall be installed. • More than one path shall be equipped with floor guidance material or a voice-guidance system to assist visually impaired persons. • Attention-attracting floor material shall be used at crosswalks, ramps, and top landing areas of stairways.	• Width of the path shall be 180 cm or wider. • If there are different levels, ramp or elevator shall be installed. • Floor guidance material or voice-guidance system, etc. shall be used on the path leading to each entrance/exit to assist visually impaired persons. • Attention-attracting floor material shall be used at crosswalks, ramps, and top landing areas of stairways.

Table compiled by G. Takahashi, 1996.

latter guideline was followed, the government provided preferential tax treatment, larger spaces for construction, and subsidies for construction costs.

However, the Accessible and Usable Building Law has many problems. The first is that the law applies only to new buildings, and it does not require that existing buildings be made barrier-free. The second problem is that the law targets only certain buildings used by the public. Third, the rationale behind some guidelines is not based on any solid researched evidence. Given that the Accessible and Usable Building Law will gradually evolve into universal design in Japan, the following points must be taken into account.

First, the scope of application of the law must be extended. For instance, the law must be applied to schools, public housing, business offices, and government office buildings. Second, laws for accessibility in existing buildings need to be considered and implemented. Third, currently the law applies only to buildings that are larger than 2000 m²; this should be lowered to 300 m². Fourth, an agency is needed through which citizens can communicate their complaints and suggestions to the government as quickly as possible. As of March 1999, 1062 high-quality buildings nationwide have been surveyed and rated as meeting the higher standard of the Accessible and Usable Building Law (Ministry of Construction, 1999). This is but a small fraction of the many facilities built in the period since the new law was enacted, and does not include such important buildings as schools, public housing, and business and government offices, as they are not yet covered under this law. If—and only if—the issues just mentioned are solved can the Accessible and Usable Building Law fully promote the concept of universal design.

Local Laws and Related Issues

What distinctively characterizes the local laws is the scope of facilities that need to be reported to the examining organization of administration of local governments. This issue has to do with the size of the facilities to which each local law applies; the criteria vary from one prefecture to another. Table 30.3 identifies the types of facilities and the minimum sizes covered by the laws in four cities.

TABLE 30.3   Minimum Size of Facilities to Which Local Regulations Shall Apply

Facility to which local law applies	Osaka	Hyogo	Kyoto	Saitama
Schools, institutions		All		
Museums, art galleries, libraries				
Hospitals, clinics (medical facilities)		300 m² or more		
Public halls, assembly halls		All	All	All
Senior citizens' centers, children's centers	All			
Public service offices				
Public utilities and telephone offices				
Public restrooms	—		—	
Banks, cooperative banks, credit associations, agricultural cooperatives, etc.	All		All	
Crematoriums		—		500 m² or more
Eating and drinking establishments	Exceed 500 m²	300 m² or more	Exceed 500 m²	200 m² or more
Shopping facilities				
Gymnasiums, bowling alleys, ski areas, swimming pools, etc.			Exceed 1000 m²	500 m² or more
Theaters, movie cinemas, auditoriums	Exceed 1000 m²	All		
Exhibition halls		1000 m² or more		All
Video arcades, dance halls		500 m² or more		500 m² or more
Movie studios, TV studios	—	—	—	
Public bath houses	Exceed 1000 m²	300 m² or more	Exceed 1000 m²	200 m² or more
Hotels, lodges		Exceed 1000 m²		
Hair dressers, dry cleaning establishments, pawn brokers, etc.	—	300 m² or more	Exceed 500 m²	
Marketplaces	—	—	—	500 m² or more
Facilities for ceremonial functions	Exceed 1000 m²			—
Business offices	5000 m²	3000 m² or more	Exceed 3000 m²	500 m² or more
Industrial plants			Exceed 5000 m²	
Apartment houses, dormitories train stations, bus terminals	Exceed 50 rooms	51 rooms or more	Exceed 3000 m²	1000 m² or more
Passenger terminals at airports, ports, and harbors	All	All	All	All
Underground shopping arcades			—	—
Temples, shrines, churches		—	Exceed 500 m²	—
Parking lots (off-road)		500 m² or more	50 vehicles or more	Automobile garage
(Parking areas regulated by road laws)	All	30 vehicles or more		500 m² or more
Roads				All
City parks		All	All	
City parks, zoos, botanical gardens		—		

(*Continued*)

TABLE 30.3   Minimum Size of Facilities to Which Local Regulations Shall Apply (*Continued*)

Facility to which local law applies	Osaka	Hyogo	Kyoto	Saitama
Natural green spaces at ports and harbors		All		—
Airports		—	—	All
Implementation date of each local law	April 1, 1993	October 1, 1993	October 1, 1995	April 1, 1996

Note: The number in the list is the minimum amount that shall be reported to local government. "Exceed" means the area of the facility that exceeds the specified number. "—" indicates that nothing is specified.
Table compiled by G. Takahashi, 1996.

Most local laws apply to buildings that are smaller than 2000 m^2, the standard size to which the Accessible and Usable Building Law applies. In addition, the local laws have more control over citizens' lives and the availability of facilities. Yet almost all of the local laws conform to the basic guidelines of the Accessible and Usable Building Law, so the parts of buildings, elevators, restrooms, and so forth to which improvement standards apply and the level of accessibility required are regulated consistently across the nation.

However, the local laws differ from the Accessible and Usable Building Law in that the local laws are concerned with all people, including children through elderly people and people with disabilities and they focus on improving roads, houses, and other buildings. The local laws also pursue improvement not only on the physical level but also on the level of social services, possibly involving health care and the welfare system. Local laws also attempt to make rules allowing elderly people and people with disabilities to participate in community improvement activities.

The major issue regarding these laws is the cost of improving existing buildings. To reduce the economic burden for private businesses, projects for facility improvement were supported by grant and fund allocation, and have been implemented all over the nation. Examples of projects in two regions that received this assistance are shown in Table 30.4.

Such financial assistance for city planning and the improvement of facilities is important. However, with the current stagnant economy in Japan, it is very difficult to secure sources of revenue for the future. Of course, this kind of assistance will not directly lead to universal design of cities, but it will help to build public interest in universal design.

TABLE 30.4   Example of Public Aid to Make Buildings, etc. Accessible

Tokyo Metropolis	Assistance in planning a welfare community and model-district development projects by wards, cities, towns, and villages.
	Model-district development projects carried out by wards, cities, towns, and villages are funded in part by the metropolis.
	Expenses supported by the metropolis: Planning, developing public structures, public relations.
	Maximum amount of assistance: 20 million yen.
Tottori Prefecture	Assistance in promoting a welfare-community project.
	Support for private builders who develop public facilities.
	Objective: Accessible restrooms and elevators in existing public facilities.
	Maximum amount of assistance:
	Restrooms, 1.5 million yen,
	Elevators, 10 million yen (existing), 1.5 million yen (new).

30.5 SAITAMA NEW CITY CENTER PROJECT

The Saitama New City Center project started in the early 1980s in central Saitama, near Tokyo, in order to relocate some business activities from Tokyo to Saitama. The project was completed on May 5th, 2000. The development covers 47.4 ha, and the total number of employees is expected to be 50,000. Facilities include governmental and private business offices, a post office, a sports arena, hotels and other lodgings, a train station, a park, and parking lots.

In 1997, the governor of Saitama Prefecture issued the Declaration of a Barrier-Free City, the first project implemented in an area as big as the New City Center in Japan. During the next year, practical plans were developed with the involvement of people with various disabilities (Takahashi, 1998). The number of participants with disabilities in this project was 40, and they included a paralyzed person, visually and hearing impaired persons, persons with muscular dystrophy, and parents of people with mental and severe physical disabilities. Many of the participants were those who previously had insisted on the right of people with disabilities to participate in the accessibility project for a new city environment. The project was imperfect in its process but was implemented on the basis of rules of universal design. For instance, meetings with people with disabilities were held on a regular basis. Opportunities were provided for people with disabilities to experiment with potential New City Center technology using small models, including a tactile model for visually impaired people. People with disabilities were encouraged to visit the construction site, and experiments were performed on the actual Braille pavers. This project is the latest experiment regarding an accessible, universally designed city; the scale of the project can be seen in the photograph in Fig. 30.1. However, difficulties in the universal design process were discovered during this project. These will be explored later in this chapter.

FIGURE 30.1 Panorama of the Saitama New City Center project.

Improvement Concepts

The first goal of this project was to implement universal design in creating a city where anyone, from children to elderly people, regardless of gender, homeland, or disability, could visit freely, meet various types of people, and move around comfortably.

Second, the project aimed at providing a place and opportunities for the business owners involved in the project, people with disabilities, other citizens, and the local government to work together. At the planning meeting, models and 3-D maps for visually impaired persons were provided. Almost all groups of people with disabilities within Saitama Prefecture participated in this project, including those who are mentally disabled, visually and hearing impaired, and wheelchair users.

Third, a full support center was established at the site of the new train station in the New City Center. This support center was designed to assist elderly and disabled people to enter, exit, and transfer between trains. It was also designed to facilitate communication. To achieve an environment in which any citizen can participate, volunteers are currently being trained to assist and guide outside visitors.

Fourth, public sidewalks were designed with safety, continuity, and comfort in mind. The main sidewalk is covered to provide protection from rain and other inclement weather conditions. Elevators were installed to transfer people from the ground or road level to higher levels of the facilities. In addition, Braille pavers and appropriate voice-guidance systems were installed to help visually impaired persons get around easily and safely.

The fifth concept concerns measures to help people with both visual and hearing impairments. Pictograms, voice-guidance systems, and electric illumination indicators were integrated into signs used at intersections and on sidewalks. This is standard in universal design. The signs include directions to facilities, shopping information, and event information. These help the city become more accessible to all.

The sixth concept is concerned with making evacuation routes barrier-free in case of a natural disaster. New City was designed to play an important role as an evacuation center for citizens from neighboring cities when natural disasters occur (Takahashi, 1996b). Evacuees may include elderly people, people with disabilities, and children. Thus rescue methods, distribution of evacuation information, living assistance, and nursing care are important. The importance of preparedness has been given higher priority since the Hanshin earthquake, when the lack of information provided to people with communication disabilities, including visually and hearing impaired persons, became a major issue.

Universal Design Features

A plan for an accessible second-floor pedestrian deck (marked "Pedestrian Deck" in left center of accessibility planning map) is shown in Fig. 30.2. This is the most significant part of the New City project. In this plan, the deck level was designed to be exclusively a pedestrian area giving all people maximum accessibility to all facilities. The following other plans have been implemented:

- *Continuous mobility.*   The route from the New City train station to each facility will enable wheelchair users and visually impaired persons to reach their destination without discomfort. Also, many types of information can be accessed at the train station's main information center.

- *Braille pavers.*   To achieve accessibility for visually impaired persons, Braille pavers run continuously through the main sidewalk leading from the train station to each facility. Yellow, the most recognizable color for visually impaired persons, was selected for the Braille pavers since luminance contrast between the pavers and the flooring material is important in order for people who are amblyopic to walk safely. During experiments with visually impaired persons, the luminance contrast was shown to be more than 1.7. In addition, Braille pavers with built-in LED displays are now being considered for use at steps, junctions of the main sidewalk, and other dangerous areas so as to guide visually impaired people at night. Various patterns of the pavers with the LED displays are shown in Figs. 30.3 and 30.4.

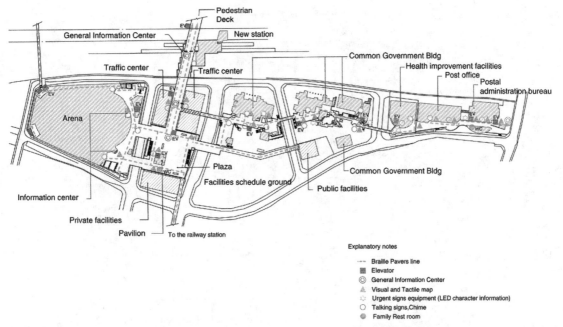

FIGURE 30.2 Accessibility planning map of Saitama New City project, showing accessible second-floor pedestrian deck.

- *Sign system.* For visually impaired people, a magnetic sensor able to transmit and receive communication with the use of a portable device was mounted onto the normal voice-guidance system. It can be activated within a 15 to 20 m radius of the voice-guidance sign and indicates the individual's current location. The base of the main sign information system (shown in Fig. 30.5), has an LED variable display which provides everyday information and emergency information to users, including children, elderly persons, and hard-of-hearing people. On the main guidance signboard, a three-dimensional map and Braille directions are included, as shown in Fig. 30.6.

- *Sidewalks and driveways.* These are level at intersections, and Braille pavers with built-in LED displays line the sidewalks at intersections. Visually impaired persons can rely on the Braille pavers in all areas and are enabled by this system to walk from one place to another.

- *Restrooms.* Designated areas inside restrooms are, in principle, available for use by males, females, wheelchair users, and mothers with nursing infants. Also, a family restroom, as seen in Fig. 30.7, is available for disabled persons who are with a helper of the opposite sex. For major events at the super arena, the number of restrooms for each sex can be changed depending on the expected male-to-female ratio.

Issues Facing the Saitama City Project and Universal Design

There are still several issues facing the project at this time:

- Ideally, people with disabilities should be able to participate fully in this project, from the design phase all the way through to construction. However, such a large-scale project involves many individuals, so it is not easy for people with disabilities to participate in all of its aspects. For example, it is nearly impossible for wheelchair users to visit the construction site.

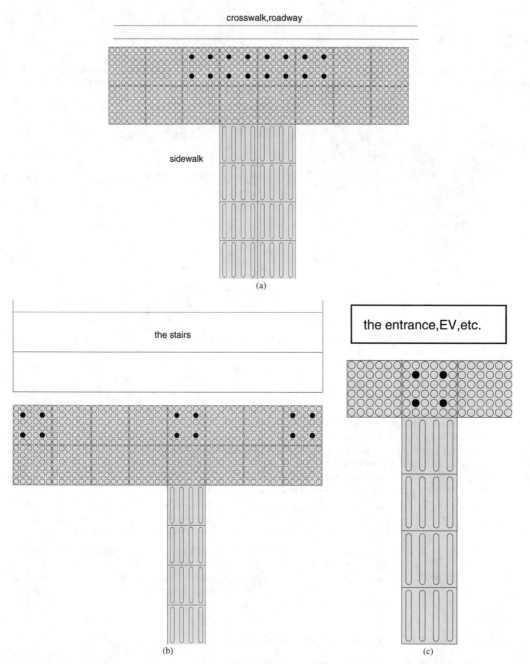

FIGURE 30.3   Base patterns of Braille pavers with built-in LEDs: (*a*) in front of roadway crosswalk; (*b*) in front of steps; (*c*) in front of entrance.

FIGURE 30.4   Sidewalk with Braille Pavers with built-in LED.

FIGURE 30.5   Sign system with LED display.

- A coordinator is needed who can facilitate communication among the architect, the owner of the construction company, and people with disabilities. This position could be filled by an architect. Japan's government administrators tend to lack technical knowledge; therefore, a specialist such as a town planner or architectural designer would be ideal for this position.

- As yet, many architects are satisfied with minimal accessibility design. They do not have a mental picture of a city that incorporates universal design to facilitate common use. Some people hold the mistaken belief that universal design is meant to provide minimal improvements useful to all people. There is disagreement as to what the ultimate goal of universal design should be. Because no satisfactory goal in universal design has been reached, improvements must continue with maximum effort and continuous trial and error.

- Lack of green space was pointed out at the beginning of this project. In Japan, nature in harmony with technology is another basic principle of universal design. In this regard, both Braille pavers for visually impaired persons and green plants lining the sidewalks are essential to people's physical and psychological comfort.

- Universal design is said to be invisible design, but when accessibility is first implemented, the improved portions of facilities should be clearly indicated to citizens. For the sake of the development of universal design, it is necessary to let society know of the existence and the rights of such diverse citizens as people with disabilities.

FIGURE 30.6 Sign system with three-dimensional map.

The case of New City Center described here has helped to solve various problems that Japan faces on its way from mere accessibility to universal design. To achieve universal design on a larger scale, such as on a city level, it is important to train citizens engaged in city construction and social activities. Even if facilities are accessible, real universal design cannot be accomplished without mutual communication among citizens.

30.6 TAKADANOBABA DISTRICT: IMPROVEMENT OF THE SIDEWALK ENVIRONMENT

Since the late 1940s, many public facilities that provide information to visually impaired persons (including the Helen Keller School of the Tokyo Helen Keller Association, Inc.), have been built and established in the Takadanobaba District of Tokyo. At present, the Japan Braille Library and other organizations for blind people are located in this district. In other words, this area has a history with visually impaired persons. Thus much can be learned by studying how people with disabilities in this district have been able to become an integral part of the community. The basic Principle of Universal Design is the exercise of citizens' right to be involved in society, especially citizens who have been victims of discrimination from the community. The case that follows outlines an actual situation.

Universal Design Community with Braille Pavers

In the mid-1970s, Braille pavers, which had been invented in 1965 and were being increasingly used across the nation, were installed for visually impaired people in the Takadanobaba District. In 1975, an accident occurred in which a blind person fell off the platform of the Takadanobaba train station and died. Braille pavers were not in use at this train station, and

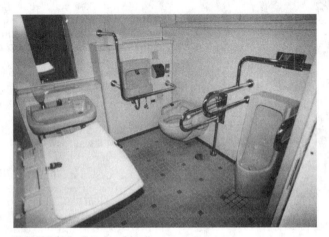

FIGURE 30.7 Interior of family restroom.

FIGURE 30.8 Visually impaired person walking on a Braille-paved sidewalk.

this was said to be the cause of the accident. The case was later disputed in court, and the Takadanobaba District decided to provide Braille pavers for installation on all roadways, even those as narrow as 4 m, passing through residential areas of the district. This kind of extreme example had not been previously seen in other parts of Japan, and at first, citizens were confused and puzzled by it. However, today the Braille pavers give meaningful information to all of the citizens as well as to first-time visitors in the area. Braille pavers are now functioning very effectively to help people with disabilities find their way to main public facilities and to get directions to their destinations. Figure 30.8 shows a visually impaired person using the sidewalks with pavers. Today, nobody feels uncomfortable with narrow sidewalks that are fully covered by Braille pavers.

On the other hand, one will often encounter a dangerous situation on Braille pavers at the train station, a scene that is seen everywhere in Japan. Figure 30.9 shows illegally parked bicycles that cover Braille pavers and may be dangerous to pedestrians. This is a serious and seemingly unsolvable issue that city and transportation planners have been grappling with for the past 30 years. This problem may be peculiar to Asian countries, especially Japan. Universal design in cities must pursue the construction of a safe walking environment. Manley, in Chap. 58, "Creating an Accessible Public Realm," also discusses the challenge of making safe urban environments for pedestrians.

FIGURE 30.9 Illegally parked bicycles covering Braille pavers.

FIGURE 30.10 Speaker and voice chime for visually impaired people.

Figure 30.10 shows a voice chime set up at an intersection. This chime sounds off regularly to help people with visual disabilities confirm their current location. When the chime was first installed, neighborhood residents were bothered by it, but now it is a familiar characteristic of the city. All citizens in this town wake up every morning and go to sleep every night to the sound of the chime.

30.7   FUTURE CHALLENGES FOR UNIVERSAL DESIGN IN JAPAN

Following are the future challenges surrounding universal design in Japan:

- *Measures for natural disasters.*   These include countermeasures for issues regarding elderly and disabled people in case of a natural disaster, issues revealed during the Hanshin earthquake of 1995. In a disaster situation, everyone becomes vulnerable. Thus it is important to set forth a plan (Takahashi, 1996b) for disaster that involves many citizens. To achieve universal design in a city, preparation for natural disasters is essential.

- *Alignment of and collaboration among diverse fields.*   To accomplish universal design, people from all different fields and specialties must cooperate with elderly and disabled people. These experts should include public administrators, social workers, economists, medical personnel, rehabilitation engineers, architects, civil engineers, and precision engineers. Though many experts in Japan are aware of the need for such cooperation, its actual implementation has not yet been achieved.

- *Introduction of information/communication technology.*   The rapid changes in communication devices and information technology (IT) facilities have technologically raised the level of accessibility in many places. With improved city environments, significantly disabled persons have become able to participate in social activities. IT/communication devices can change participants' attitudes. People with significant disabilities are now expected to do more than just be present. Obviously, the use of such technology has a great impact on city planning and building design. If universal design cannot be implemented in a city or house, an IT/communication device can partly compensate for the problem.

- *Use of educational institutions.*   Schools and institutions for compulsory education must be most accessible, and universal design concepts must be used in these places. Japan's low birth rate will enhance the quality of education by lowering the student-to-teacher ratio. It will also change the role that schools play in the community. Perhaps, universal design in Japan will spread and become known among children and young people rather than becoming the property of the business world, as it appears in the United States. The starting point for real universal design is providing an environment where children without disabilities can make normal friendships with elderly people and children with disabilities.

Finally, based on the cases observed here, it can be said that universal design cultivates history— and history builds up universal design. In other words, a design that is effective in everyday life is sustainable. The two examples discussed here had different starting points, the past and the present, but both have shown that if accessibility is based upon citizens' needs, it can easily evolve into universal design. An independent movement by a person with disabilities led to the establishment of new laws, and the movement toward universal design is now under way. Considering this trend, accessibility in Japan will continue to progress and eventually become universal design.

30.8   CONCLUSION

As pointed out in this chapter, the Japanese movement for the rights of people with disabilities, with the active participation of citizens, has made continuous progress. Also, despite

Japan's small land area and narrow sidewalks and roadways, barrier-free designs have been implemented quickly as the aging of the population has become a growing concern. On the other hand, some underlying discrimination remains because of many people's lack of awareness about human rights. Even though businesses say that they cannot construct accessible buildings for financial reasons, the real reason is that they are unaware of human rights. Although these are real issues, the possibility of Japan's wholeheartedly adopting the universal design concept for its cities and buildings is still strong.

With elderly people currently making up more than 17 percent of Japan's population, universal design is starting to be taken more seriously throughout the country. Universal design emphasizes the involvement of all citizens, including elderly people and people with disabilities. With all citizens participating, universal design will eventually be implemented, and the required design evaluations and inspections will be carried out.

30.9   BIBLIOGRAPHY

Ministry of Construction, *Investigation of Buildings Constructed Under Accessible and Usable Building Law,* 1999; www.jaeic.or.jp/hyk/ninteiitiran.htm.

Ministry of Health and Welfare, *Report of the Welfare Council for the Disabled, Tokyo,* Tokyo, Japan, 1970.

Rehabilitation International, *Barrier Free Design,* Report on United Nations Expert Group Meeting, United Nations, New York, 1975.

Takahashi, G., *Design Handbook for Accessibility and Usability for All,* Shokokusya, Tokyo, Japan, 1996a.

———, "Town Planning with an Emphasis on Protecting Elderly and Disabled People in Time of Disaster," in Bachir Mekibes (ed.), *Domotic Environment and Users, Proceedings International Symposium 2, IAPS 14 Conference,* 1996b.

———, *Basic Study on Accessible and Universal Design in Saitama New City Center, Saitama Prefecture,* 1998.

CHAPTER 31

ISRAEL: A COUNTRY ON THE WAY TO UNIVERSAL DESIGN

Avi Ramot, Ph.D.
Israel Center for Accessibility, Jerusalem, Israel

31.1 INTRODUCTION

Universal design, or the design of environments and products that can be used by most people, and accessible design, or design that allows those with disabilities to use facilities and products, are not new concepts in Israel. As early as 1965, the Planning and Construction Law required public buildings to be accessible to those with disabilities. In addition, there has always been great sensitivity to the special needs of disabled soldiers, whose injuries were received in the service of the country. Despite laws and sensitivity, the reality of Israel's public places is far from reflecting the ideals of universal design. This chapter will describe the history of legislation and regulations concerning accessibility in Israel, current efforts to expand the application of universal design in this country, and the challenges that are faced by architects and designers, planners and advocates, in creating a country that is accessible to all.

31.2 ACCESSIBILITY LAWS AND REGULATIONS

The State of Israel was established in 1948, and legislation as early as 1949 took into account the needs of people with disabilities (Veteran's Law—Returning to Work, 1949; Law for the Disabled—Payments and Rehabilitation, 1949). The Planning and Construction Law of 1965 set out requirements for public buildings to make arrangements for people with disabilities, and specified that local authorities may not grant building permits for public buildings that are not accessible. Existing public buildings that cannot be renovated because of engineering considerations may be entitled to dispensations, as long as people with disabilities are not prevented from using the building. According to the 1988 Law of Local Authorities, sidewalks need to be accessible with ramps and without obstacles, and crosswalks need to be accessible to those in special vehicles (e.g., wheelchairs).

In 1993, the Law of Parking for the Disabled was instituted, which established specially marked parking places for use by people with disabilities, and in 1994 and 1995 regulations were drafted that provide for the safe transportation of disabled children to and from school. In addition, the schools themselves must be accessible, and legal provision is made for people to accompany children with disabilities, if necessary.

Because the Israeli legal system is based on precedent or case law, following the British system of justice, it is possible to trace the changes in attitude by reviewing the legal decisions made in court cases regarding people with disabilities. For example, in 1996, the Supreme Court ruled that the local authority of Maccabim-Reut was required to make the changes that would allow a 13-year-old resident who used a wheelchair to have access to public buildings. The ruling specified that the school must provide accessible toilets for people with disabilities, and that the toilets must be located in reasonable proximity to the parts of the building used by such people. The ruling also required the local authority to provide wheelchair access to the schoolyard, to the local health care provider's building, to the town library, to the sports auditorium, and to the synagogue. Other changes were to the sports auditorium and to the synagogue. Other changes were required, including creating an accessible water cooler and crosswalks. The court noted that the main purpose of the 1965 law is to allow a person with disabilities to take part in normal, public activities. To achieve this goal, the local authority must provide for accessibility to those using wheelchairs, including installing accessible toilets and ramps.

In 1998 the Equal Rights for People with Disabilities Law was passed, specifying that people with disabilities have the right to equal and active participation in society, and prohibiting discrimination in employment and requiring public transport services to be fully accessible. The new law is very similar to the 1990 Americans with Disabilities Act. A second law, called "Equal Rights for People with Disabilities Law (part 2)," relates to the accessibility of public buildings. It would replace the 1965 Planning and Construction Law, is currently before the parliament, and is expected to be passed in 2001.

31.3 HISTORY OF ACCESSIBILITY ADVOCACY IN ISRAEL

In 1977 the Umbrella Organization of Associations for the Disabled in Israel was formed. This organization was a coalition of all the organizations in Israel for people with disabilities and those with special needs, and it had two major goals: (1) creating committees on accessibility that would talk about how to deal with the concept of accessibility in Israel, and (2) lobbying at the national level for legislation concerning accessibility issues. The Umbrella Organization of Associations for the Disabled in Israel was instrumental in setting up the position of advisors to local municipalities on issues of accessibility. These were volunteer positions, and the goal was to increase awareness of the problems faced by people with disabilities, and to create accessible environments at the local level. In two cities, Tel Aviv and Jerusalem, the volunteer positions eventually became paid positions, and issues of accessibility of public places were addressed by the municipality.

Lobbying in the Parliament in Israel was also successful and led to provisions in the 1965 Planning and Construction Law for accessibility of public buildings for people with disabilities. Other legislation aimed at creating accessibility for people with disabilities was also passed, including the 1985 addition to the Regulations on National Parks and Nature Sites, which created special half-price discounts for people with disabilities, and the 1988 Law of Local Authorities (Arrangements for the Disabled), which provided for accessible sidewalks and crosswalks. In 1993, the Law for Parking for the Disabled was passed, and in 1994 and 1995, regulations were written for the transportation of disabled children.

In the last decade, the Umbrella Organization of Associations for the Disabled in Israel has reduced its activities, and today is primarily involved in publishing a quarterly newsletter, *Tadmit* (Images of the Disabled). Other organizations have taken over many of its activities. One of these organizations, *Bizchut* (the Israel Human Rights Center for People with Disabilities), is an independent, nonprofit organization working to advance the rights of people with disabilities, physical or mental, and to make possible their full integration into society. Bizchut was established in 1992, and since then has become known for legal work, community work, and education. The staff of Bizchut drafted the Equal Rights for People with Disabilities Bill, and played a key role at every stage of its progress through the Knesset, Israel's parliament.

Over time, projects were undertaken by different organizations to improve accessibility. For example, in the 1970s, the Israel Museum, a large complex with many buildings and outdoor exhibits, including the Shrine of the Book, worked actively to improve its accessibility, especially to those in wheelchairs, by adding ramps, elevators, and wheelchair-accessible toilets. The Biblical Landscape Reserve in Neot Kedumim was developed in the 1980s and 1990s. It re-creates the physical setting of the Bible, including plants and trees, vineyards, wine and olive oil presses, cisterns, and threshing floors. It is a 625-acre site, with miles of wheelchair-accessible paths and a "Tour Train" that gives groups access to the reserve. The park made special efforts to create paths and exhibits accessible to wheelchairs.

Until recently, accessibility has been defined as wheelchair accessibility, and little attention was paid to other kinds of disabilities, such as vision and hearing impairments. In the past decades, awareness of issues of accessibility has grown, and from 1994 to 1997, JDC-Israel, a branch of the American Jewish Joint Distribution Committee, allocated resources to developing projects for people with disabilities, and created the Division for Populations with Special Needs. The Division was involved in many projects, three of which involved accessibility. These included increasing accessibility in national parks and nature reserves, working with local authorities to encourage them to create policies and budgets to improve accessibility at the local level, and making hotels more accessible to the hearing impaired through the provision of special kits. In each of these areas, the Division worked with government agencies and other partners to create policies on accessibility and to develop tools and projects for people with disabilities.

As an outgrowth of these activities at JDC-Israel, SHEKEL, the Israeli Center for Accessibility, was formed in 1998 as a partnership between JDC-Israel and the Shekel Association, a nonprofit organization that provides a network of community-based services for individuals with disabilities. SHEKEL's mandate is to promote the process of creating more accessible environments for people with disabilities in Israel. The organization works to remove physical and communication barriers, helping people with disabilities to move easily and safely from place to place, to have access to buildings and other facilities, to take advantage of available services, and to have access to all kinds of information, through means adapted to their abilities. While SHEKEL aims to enable more independent functioning for people with special needs, the organization's brochure notes that the ultimate beneficiaries of universally designed environments will be society as a whole. SHEKEL has four primary goals:

1. The creation and operation of a knowledge and information resource center. This center will encompass existing and developing information in Israel and throughout the world.

2. The initiation of unique projects designed to enhance the day-to-day quality of life of people with special needs.

3. The promotion of awareness of accessibility issues on the part of policy makers, planners, professionals, and the general public.

4. The initiation and development of ongoing mechanisms to deal with the issue of accessibility.

SHEKEL's main areas of activity include initiating projects to increase accessibility; collecting information and knowledge to promote accessibility in Israel; and professional activities, which include publishing policy papers and manuals and offering workshops, seminars, and conferences. SHEKEL works in partnership with government ministries, local authorities, client organizations, and academic institutions.

Two other organizations that serve people with disabilities deserve to be mentioned here: (1) MILBAT, which was established in 1981 as a demonstration center for assistive devices for people with disabilities, and (2) GeronTech, the Israeli Center for Assistive Technology and Aging, which was established in 1998 to facilitate the development of assistive products. Other organizations in Israel, such as Yad Sarah, also devote resources to assisting those people with disabilities.

31.4 RECENT INITIATIVES

National Parks and Nature Reserves

During the 1990s, JDC-Israel's Division for Populations with Special Needs encouraged the Ministry of Tourism to target two sites for renovations: (1) the Tel Dan National Park in the north and (2) the Hula Nature Reserve north of the Sea of Galilee.

Tel Dan is a nature reserve with plants, trees, and streams near Kibbutz Dan. JDC-Israel's Division for Populations with Special Needs, together with the Ministry of Tourism, decided to improve Tel Dan, making a significant part of the area accessible to people who use wheelchairs. An external consultant on accessibility was hired, who recommended building a 1200-meter wooden pathway right into the heart of the reserve, and making the visitor center wheelchair-accessible. The changes cost about $150,000 and have succeeded in making about one-third of the park available for those who use wheelchairs. Today everyone who visits the park takes advantage of the walkway, which has made it easier for elderly people and families with small children in strollers to take advantage of the views. What was designed specifically for people who use wheelchairs has turned out to be an attraction to everyone else as well.

The Hula Nature Reserve was completely redone, with the renovations also originating with JDC-Israel's Division for Populations with Special Needs and the Ministry of Tourism, again using an external consultant. This time the goal was to base the renovations on concepts of universal design, and they were planned to facilitate the use of the whole reserve by all visitors, including those with disabilities. All of the park except one observation tower for birdwatching was upgraded, taking the approach that every aspect of the site should be accessible to everyone, including those people using wheelchairs. The paths, walkways, and observation points were constructed in such a way that a visitor to the site will not necessarily be aware that it is accessible unless it is pointed out. As a result, the park is comfortable for people using wheelchairs, for people who are elderly, and for families with strollers, as well as for people without any problems of mobility. These changes cost about $180,000, and are considered well worth the time and expense. Both in the case of the Hula Nature Reserve and in Tel Dan, people with disabilities are made aware of these changes through organizations and publications aimed at this population. (See Figs. 31.1 through 31.6.)

The most successful aspect of these two projects was improving accessibility for people with difficulties in mobility. In both cases, the solutions were less satisfying for people with visual impairments, and future projects should pay more attention to the particular needs of this population.

In addition to specific projects, at the end of 1995 the Ministry of Tourism enacted regulations restricting support by the Ministry for renovations and expansion in National Parks and Nature Reserves to those projects that improve accessibility. Not only is accessibility indicated for people using wheelchairs, but for visually and hearing-impaired people as well. One outgrowth of this work with the Ministry of Tourism was the development of a tool to improve accessibility for people with hearing impairments—a kit that uses transmitters and earphones and allows a guide to address a group of hearing-impaired visitors at a tourist site.

Municipalities and Local Authorities

A second major thrust of the Division for Populations with Special Needs was encouraging local authorities to develop policies and allocate funds to improve accessibility. This effort has been continued by SHEKEL. In two cities, Haifa and Eilat, real progress was made, and in several others, support from the mayor and from the local authority has given an impetus to changes that increase accessibility.

FIGURE 31.1 An observation point on the covered, nonslip walkway in the Hula Nature Reserve.

- In Haifa, a steering committee on accessibility was established, reporting directly to the mayor. Its role was to determine which projects in Haifa would receive priority. Three projects have already been implemented:
  1. Installing stoplights with audio signals at key intersections for people who are blind
  2. Renovating a large, central community center to make it wheelchair-accessible
  3. Adapting the "German Colony" (an older, central neighborhood of the city) to make the crosswalks and public walkways wheelchair-accessible

- In Eilat, the mayor declared that the city would be an "accessible tourist city." Planning is underway to ensure that two beaches are accessible, including opportunities for diving, with the goal of keeping the changes as unobtrusive as possible. Hotels are in the process of improving their accessibility, and 17 public buildings are being renovated.

- In Tel Aviv, there is a paid position to deal with issues of accessibility. A survey of public buildings is being carried out, and a budget has been prepared to carry out necessary repairs. A long-term planning process is underway to improve accessibility in public buildings.

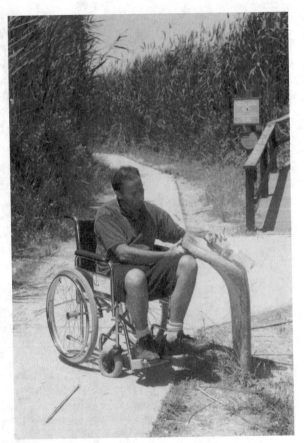

FIGURE 31.2   A water fountain in the Hula Nature Reserve, designed to be easily used by everyone.

FIGURE 31.3   A walkway in the Hula Nature Reserve over the swamplands.

FIGURE 31.4   A view of the walkway over water and swampland in the Hula Nature Reserve, with a path and the mountains of the Galilee in the background.

- In Petach Tikva, the municipality decided to repair all crosswalks to make them wheelchair-accessible, without curbs, and with special texture for the benefit of the visually impaired.

Accessibility for Hearing-Impaired People in Hotels

The Division for Populations with Special Needs also developed a kit to improve accessibility for hearing-impaired people in hotels. The kit, which fits into a briefcase, includes small amplifiers and a remote control unit that allow a person with hearing problems to use the telephone, hear the television, and hear if someone is knocking at the door. The remote control unit has both tactile (vibrating) and visual signals, and can even be used in the bathroom. Special earphones allow a person with impaired hearing to listen to the television. The kits are self-contained and are easy to use. A person with hearing problems simply requests the kit on registering, and a member of the hotel staff will bring the kit to the room and demonstrate its use. These kits are currently being used in hotels in Eilat and Jerusalem, with the goal of encouraging distribution to all hotels in Israel. Promotion and distribution of these kits has continued through SHEKEL.

FIGURE 31.5   A gently sloping path in the Hula Nature Reserve is accessible to everyone.

31.5   *ACCESSIBILITY AND UNIVERSAL DESIGN IN ISRAEL TODAY*

Accessibility and universal design are becoming more widespread in Israel, with increasing awareness in the past few years of the problems of people with disabilities, and greater pressure for change coming from many directions.

Government Ministries and Private Corporations

Government ministries and private corporations are beginning to incorporate ideas of accessibility into their policies. Examples include the following:

- The Ministry of Transportation has introduced the first "kneeling buses" in Tel Aviv as a pilot project.

FIGURE 31.6   The entrance to the Hula Nature Reserve has eucalyptus trees and paved paths.

- The Israel Railroad Authority is planning to make their train stations accessible, especially to people who are visually impaired, through the use of large, high-contrast signs, audio messages, and other technology. Most train stations, especially the newer ones, are already wheelchair-accessible. This is not an instance of universal design, but a response to the needs of specific populations.

- El Al, Israel's national airline, is working with SHEKEL to create new regulations and a training program for flight attendants and ground personnel to improve the quality of service to people with disabilities. Again, this represents improvements in accessibility rather than the application of concepts of universal design.

- Israel's largest health insurance fund, Kupat Cholim Clalit, has requested information from SHEKEL on how to improve accessibility in their hospitals and clinics.

National Financing

Resources have been allocated at the national level to improve accessibility. The National Insurance Institute decided in 1998 to allocate $1.5 million to improve accessibility. In 1999, the second year of the program, the director decided to allocate $5.5 million for this purpose. These funds can be accessed by local municipalities and other organizations, which may apply to the National Insurance Institute for up to 80 percent of the cost of renovations and improvements that increase accessibility. This program does not have a specific universal design orientation, but rather, is an example of the commitment the government is making to improve access to people with disabilities.

Professional Education for Designers

There is also increasing awareness among professionals of issues of accessibility and universal design. At the School of Architecture at the Technion in Haifa, a special course is offered on accessibility. There is an annual competition, with prizes to students of architecture for the best accessibility project, or new ideas for universal design in construction. At the University of Tel Aviv, the School of Architecture has integrated material on accessibility into the regular course work.

Seminars and conferences on accessibility are being held. A large conference for engineers and architects who work in local authorities was held in the spring of 1999. A seminar for architects working in the north of the country was held at the Technion in the summer of 1999. Also, a seminar for museums on improving accessibility, as well as applying concepts of universal design, was held in the summer of 2000. A course for advisors on accessibility to local municipalities is being offered by SHEKEL together with the Technion.

Increased Public Awareness

Today, the issue of accessibility and universal design is becoming more prominent in the awareness of professionals and the public as a whole. In December 1999, people with disabilities staged a strike for increased financial benefits. One outcome of the strike was increased visibility of people with disabilities in the media and increased awareness of issues affecting them. A second outcome was empowerment of people with disabilities themselves, who were successful in fighting for their needs in a public forum.

In addition to an increasing awareness of accessibility and universal design in architecture and other built environments, there is also an increasing emphasis in Israel on new technology. GeronTech, established in 1998, is working to develop products that embody principles of universal design, and SHEKEL is working with start-up companies, some based in Europe and the United States, to develop new technologies for accessibility.

The increasing visibility of people with disabilities and the acceptance of those using wheelchairs are raising public awareness of issues of universal design. There are two judges in wheelchairs, and some Knesset members have assistants in wheelchairs. This increasing visibility raises awareness, which, in turn, will lead to more pressures for solutions that are acceptable to all.

31.6   CHALLENGES AND OPPORTUNITIES

Israel is a country on the way to implementing principles of universal design. Legislation to extend the 1998 Equal Rights for People with Disabilities Law to include accessibility to public buildings is before the Knesset. This new law requires all new public buildings to be fully accessible, and it supersedes all previous legislation.

The most important challenge facing Israel today in becoming truly accessible is to ensure that this legislation is passed, and to provide the legal framework to begin the implementation of the Principles of Universal Design. Following this, the next steps will be to ensure that these rights become part of policy statements, regulations, and general principles of local municipalities, public institutions, and private corporations.

The second challenge is to ensure that Principles of Universal Design are made part of the educational process of those who will be involved in implementing them. This means that students in architecture, landscape architecture, engineering, city planning, and urban design must be aware of the importance of universal design and how to plan and build for everyone, including those with disabilities. Students in hotel management and tourism need to be made aware that not all visitors can climb stairs, hear the phone ring, or see and understand signs.

The third challenge is to ensure that the Principles of Universal Design and accessibility are implemented throughout Israel—in local authorities, in public institutions, and in private companies. This requires political and organizational mechanisms that will ensure that existing problems of accessibility be rectified, and that future construction and development incorporate the principles of universal design.

The fourth challenge is to encourage high-tech companies in Israel and abroad to devote resources to developing assistive devices and solutions to problems of accessibility for people with disabilities. Israel is known as a country that actively supports start-up and high-tech companies. SHEKEL, MILBAT, and GeronTech are working with these new companies to encourage the development of new products, which will make universal design a reality and not just a dream.

Once Israel is able to take these four challenges and turn them into opportunities, the country will be well on the way to implementing concepts of universal design. The reality of Israeli public places is still far from reflecting the ideals of universal design, but awareness is increasing, both in the public and private sectors. Many improvements have been made during the past 10 years, and as the visibility of people with disabilities increases, these changes will take place even more rapidly.

31.7   GENERAL LESSONS LEARNED FOR UNIVERSAL DESIGN

SHEKEL has found that two principles of operation have been particularly successful in inducing local authorities, public institutions, and private companies to consider issues of accessibility and universal design and to follow through on implementing changes.

The first principle is to begin with those projects that are of particular interest to the organization, rather than trying to impose a specific project on the organization. In the case of municipalities, where differing political interests can impede the implementation of projects, this principle is especially effective. For example, Tel Aviv began with a survey of public buildings, whereas Petach Tikva preferred to begin by upgrading all their sidewalks, and Kiryat Shemona decided that they should begin by developing a 5-year plan for total accessibility. When the impetus for accessibility comes from the municipality or the organization rather than from an external body, the prospects for success are much greater.

The second principle is not to begin projects without the public support of the mayor and the city council, or the director or president in the case of public organizations and private companies. Once the mayor and city council are prepared to muster political support behind a project, or the director or president is prepared to assign staff to a project, it is much more likely to be implemented.

31.8 CONCLUSION

Great strides have been made in Israel, marked by the passage in 1998 of the Equal Rights for People with Disabilities Law. Together with the law currently in the parliament concerning the accessibility of public buildings, the legal framework for accessibility in Israel is assured. The law establishes a commissioner who is responsible for enforcement, which will help bridge the often substantial gap between the law and its implementation. In fact, Israel has had legal backing for many aspects of accessibility of its public areas for some time now, but the reality seldom matches the ideal. One of the problems in this country has to do with cultural differences that allow for considerable leeway in law enforcement. This is true in many areas of life in Israel, including, for example, traffic regulations and building codes, where stringent laws and regulations are often not enforced with uniformity.

In tandem with the legal changes being implemented, awareness of issues of accessibility and universal design is increasing, and these concepts are being introduced into the educational process of professionals who will implement these concepts in their work. This means that planners, designers, architects, and politicians are all becoming more aware of the needs of their disabled clients and constituents.

In some cases, principles of universal design are used without specific reference to people with disabilities. For example, some cities in Israel, Bnei Brak among them, have a high percentage of large families with many small children. In these cities, design considerations often include mothers with strollers, which, incidentally, also include people in wheelchairs and elderly people.

Taken together, these trends paint a bright future for real changes in Israel's public facilities and recreation areas over the next decade.

31.9 BIBLIOGRAPHY

City of Eilat, Declaration of the Mayor of Eilat: "Eilat, An Accessible City for Tourists," Conference on Accessibility, Eilat, Israel, 1995.

Government of Israel, Veteran's Law—Returning to Work, 1949.

Government of Israel, Law for the Disabled—Payments and Rehabilitation, 1949.

Government of Israel, Planning and Construction Law, 1965.

Government of Israel, Law of Local Authorities, 1988.

Government of Israel, Law of Parking for the Disabled, 1993.

Government of Israel, Equal Rights for People with Disabilities Law, 1998.

JDC-Israel, *The Concept of Working with Local Authorities* (working paper), 1996.

Leibrock, C., and S. Bahar, *Beautiful Barrier Free,* Van Nostrand Reinhold, New York, 1993.

Ministry of Tourism, *Nature Sites: Regulations of the General Director Concerning the Improvement of Nature Sites,* 1995.

SHEKEL, *Survey on the Accessibility of Public Buildings in Five Local Authorities,* 1998.

SHEKEL, *Accessing Train Stations: A Survey of Train Stations in Israel,* (working paper), 2000.

Steinfeld, Edward, *The Concept of Universal Design,* State University of New York at Buffalo, Buffalo, NY, 1994.

CHAPTER 32
DEVELOPING ECONOMIES: A REALITY CHECK

Kenneth J. Parker, M.Sc., FCIBSE
National University of Singapore, Singapore

32.1 INTRODUCTION

There are vast differences between the East and West; between developed and developing economies; between the financial, social, governmental, and legal circumstances of different countries. Yet, the fundamental needs of elderly persons and persons with a disability remain constant; it is the different ways to serve these needs and aspirations to improve the quality of life of all people through the application of universal design that requires attention. Universal design solutions require the use of appropriate technology, moderated by many, including social, constraints. There is a large, and expanding, market in Asia for products and systems that meet local wants and needs and which preserve, or enhance, the dignity of the user. The purpose of this reality check is to demonstrate the contextual nature of universal design and the diversity of needs and wants, and other factors, that influence appropriate universal design solutions in developing economies with a particular focus on Asia.

32.2 BACKGROUND

It is difficult to give a short review of attitudes, achievements, and developments in Asia—a diverse continent. Asia is a glorious melting pot of cultures, religions, and social practices, and a source of deeply enriching and moving experiences; modern and populous cities are found, along with remote and undeveloped villages, wherein the gap between the haves and have-nots is often wide. It is also dangerous to generalize, and contradictory examples to those given can be found readily. It is not the intention to be critical but to illustrate differences between the West and the East to broaden understanding.

Globally, strategic interests and history have influenced infrastructural differences, in terms of state and private involvement, between societies—especially in the case of the United States and China following the Cold War. This is reflected in the private and public ownership and development of assets, from housing and personal transport through to streets and urban planning. With such differences, the question then becomes: How can technology and information transfer be established to benefit all societies and individual users, regardless of their personal abilities and impairments?

There are many differences between the peoples of the world, and although television, communications, and the Internet have all served to improve understanding—the differences are still significant. These differences can be in many forms: cultural, social, financial, technical, technological, attitudinal, legal, and so on. In the West, groups representing the elderly population have worked with disability groups and found that they have many common agenda items; this form of cooperation is rare in the East. Americans, possibly more than many other people, appreciate these differences as they live in a multiracial, multicultural society. Yet, even this set of life experiences does not prepare Americans for the stark differences encountered in other parts of the world, for example, in India and China, which have different languages, climates, balances of religions, and political and economic systems. Thus, a particular set of design/living options for a particular situation in the West is often inappropriate for adoption in the East. Yet, with an understanding of these differences, appropriate universal design solutions can be developed.

Awareness of the need for equitable environments and full participation by all citizens in Asia has been heightened by the efforts of the United Nations: the International Year of the Disabled (1982), the Decade of People with Disabilities (1983–1992), and the Asian and Pacific Decade of Disabled Persons (1993–2002).

32.3   DEMOGRAPHY IN ASIA

Asia is the largest of the world's continents, and China is the world's most populous nation; since the 1980s, China's one-child policy, with fines and other measures for those who break the law, has been enacted to limit the population to 1.6 billion in the middle of the century (1.4 billion in 2010). India has, presently, 1 billion citizens, and various studies suggest that it will surpass China as the world's most populous nation by 2040; the population is growing annually by 2 percent.

A recent newspaper article states:

> Poverty remains a staggering social challenge for Asia despite the phenomenal growth in the 1980s, but the Asian Development Bank (ADB) believes the region could be free of its yoke by 2025 if economic growth continues and there is no substantial increase in inequality. About 900 million Asians—representing two-thirds of the developing world's poor—live on less than US$1 a day. The overwhelming majority of the poor live in South Asia. Efforts to reduce poverty over the past four decades have met with little success. While abject poverty has been virtually eradicated in some Asian countries, particularly Hong Kong, Singapore, South Korea and Taiwan, South Asia lags behind.   (*Straits Times*, 2000)

More than 300 million of the 900 million Asians who live on less than US$1 a day live in India, and, of these, about 130 million are jobless.

Asia has the fastest-growing elderly populations in the world. There is a tremendous, and expanding, market for products and services to support normal aging. For example, there is a need to provide adaptable housing to promote "aging in place"; the "lifetime homes" concept and elderly-friendly villages are examples. Yet, local authorities and businesses are often hesitant to provide a new accommodation type until others have "tested the water."

China has the fastest-aging population, in terms of average age increase, in the world. Recent headlines read: "Grim future for 60 million disabled Chinese" and "The Government estimates that more than 7 million impoverished handicapped people have no guarantee of a basic standard of living" (*Straits Times*, 1999).

In recent years, key changes have come into being, that is, the promotion of smaller families (with one child per couple in China), the trend toward independent aging, and increased longevity. Recently, the eldest living person in the *Guinness Book of Records* died at 119 years; yet, a common, attainable age of 120 years has been suggested for future generations of seniors (Fischetti, 2000). With people living longer and less children being born, attention is being given to the elderly members of societies, who have voting power and more disposable income than previous elderly generations. It is hoped that decisions to make and adapt environments to become elderly-friendly will widen to become people-friendly and, hence, inclusive of all users in the true spirit of the application of universal design.

32.4 GENERAL CONDITIONS IN ASIA

In Asia the issue of independence is different for a member of a large family living in a high-rise, high-density apartment than for someone living alone in a rural setting. Although this is an extreme contrast of lifestyles and conditions, it serves to illustrate the vast variances within Asia. More people are living in megacities, which attract migrant workers and temporary shelters, and even slum communities; these are dangerous, unhealthy, with poor water quality and sanitation, and inaccessible. This disadvantaged, poor population often has a higher-than-average percentage of persons with disabilities gathered together in a setting with scant regard to accessibility and mobility. The colloquial expression "no money, no talk" is often heard; yet, shouldn't all individuals be entitled to have certain basic needs and rights regardless of wealth? For example, access to water, food, sanitation, security, freedom of thought and movement, and freedom from discrimination and persecution should be human rights. In 2000, the lake in Rajsamand village in India's Rajasthan State dried up for the first time in 300 years—thousands of villages in the drought-stricken state face an acute water shortage from the worst drought in a century. In the face of severe water shortages, the provision of non-handicapping environments is relegated to a lower priority.

Some forms of disability can be prevented; polio and cataracts, though preventable, are still present, due to a lack of awareness, will, and resources. However, great strides have been taken to tackle polio and the problem is nearly solved. In April 2000, the U.S. Centers for Disease Control and Prevention (CDC) reported that occurrences of polio "remain high" in Northern India and Bangladesh. Timely and adequate intervention can reduce the problem, and teachers and the medical profession need to increase sensitivity and awareness. Physiotherapy, speech therapy, and occupational therapy can go a long way to reducing disability. The level of funding for assistive technology is low, so there is a need for affordable intermediate technology; disability is related to poverty.

An excellent example from Singapore of how a code was made necessary is that of the *Code on Barrier-Free Accessibility in Buildings* (PWD, 1995), launched in 1990, which was given greater importance when the projected numbers of seniors was added to the statistical numbers of those with physical disabilities. Without this addition, the voluntary *Code* might not have been made mandatory so rapidly. However, the *Code* and its revision in 1995 didn't address the needs of older persons; instead, it addressed the needs of wheelchair users and the ambulant disabled. It is hoped that the forthcoming third edition of this *Code* will provide the necessary improvements to the sensory environment and address the needs of elderly persons using the built environment.

Malaysian guidelines (Jabatan Kerajaan Tempatan, 1999) reflect the way that wheelchair and mobility issues are given more initial attention than sensory issues. The guidelines contain matrix tables with features listed against building types and are categorized as "mandatory," "encouraged," or "optional," and many of the physical features of the environment, such as ramps, accessible toilets, and lifts, are classified as mandatory. However, visual signs are mostly mandatory and encouraged, whereas audio signals are mostly encouraged and optional. Hiroshima, a port in southwestern Japan and the site of an atomic explosion in 1945, has become very accommodating of the needs of vision-impaired people. This is one of the very few examples where, in recent times, sensory issues have been given more attention than physical issues.

Dignity and independence are not tenets of the usually accepted Principles of Universal Design (Center for Universal Design, 1997). Nowadays, independent living centers are found next to hospitals, to aid those who are recovering and to help prepare them to return to the community. In Singapore, the Disabled People's Association (DPA) has both an independent living center and an occupational therapist on their premises, and it has developed a virtual independent living center on the Web. The DPA takes a more proactive approach to home adaptations and changing lifestyles, whereas in Malaysia, an independent living center is a relatively new concept, following the medical model.

32.5 *LEGAL AND POLITICAL CLIMATE IN ASIA*

Discrimination is still seen in Asia, where most countries do not have antidiscrimination legislation, and even if it is in place, it often is not enforced rigorously. Australia, a country bordering Asia, is one of the leading countries in the Asian Pacific region where the recent changes in legislation have impacted on more accessible public and private buildings, streets, transportation systems, and employment opportunities. But the whole society benefits: For example, safer streets have less accidental falls; buses and trains are easier to use; and the economy benefits with less strain on medical services, safer tourism; and so forth. The Australian experience is an example of how properly enforced and suitable codes and legislation can lead to an improvement in the quality of life of the whole population.

In Asia, the government and the people are generally quite distant from each other in evolving policies, and it is quite rare to witness action groups, open discussions, or forums. Most Asian governments believe in individual and family responsibility to cope, but realizing that this is not enough, there have been recent initiatives such as long-term-care insurance and the encouragement of voluntary groups and nongovernmental organizations (NGOs).

Demonstrations and aggressive campaigning are rare, and social progress is often evolutionary. Recently, user groups have started to campaign for disability matters to be part of developmental issues, alongside poverty alleviation, housing development, and human resource development. The national budget in India on disability is about 20 rupees per person per annum, assuming that there are 70 million people in India with disabilities. The vast majority of persons with disabilities in India do not have the means and opportunities to optimize their faculties and realize their potential. Economic deprivation should not affect social responsibility, social acceptance, or the presence of a helping hand.

It is difficult to know the number of persons with disabilities in a country, or in a region of a country. Registering as disabled has a certain stigma attached, and there are rarely benefits to be gained from such registration. In Thailand, registration as a disabled person with provincial public welfare offices has recently been made compulsory, yet is being handled in a positive manner with advice and consultation. This takes the form of medical advice, health guidance, advice on the government's nonformal educational system and Internet services, job placement and career guidance, and so on.

Government intervention in India is slow: The first disability legislation took 25 years to come on the statute books after its first announcement. The Persons with Disabilities Act of 1995 was passed by the Indian parliament and implemented in February 1996. This provides for a 3 percent reservation in vacancies of identified posts in every establishment (a 1 percent reservation each for visually impaired, hearing impaired, locomotor disability, or cerebral palsy). The Act states:

> Government agencies within the limits of their economic capacity and development shall provide incentives to employers both public and private sectors to ensure that at least 5% of their work force is composed of persons with disabilities.

If disability discrimination is experienced in India, a person has recourse to the law.

Little, if any, social security, safety net, or pension provision is to be found in developing economies, and exceptions to this are more commonly enjoyed by civil servants and employees of public corporations rather than within the private sector. Thus, the support and coherence of the family unit is of high importance in developing economies. Large families were, to some extent, insurance and a support network for parents as they age through grandparents to great-grandparents.

Multigenerational families are still very prevalent, and family bonds remain strong in Asia. A recent law in Singapore permits elderly parents to bring legal action against their siblings for support funds; this is an interesting reversal of the usual filial family support arrangements commonly found in Asia, and a legal confirmation of the government's recognition of the importance of, and rights to, family support.

Although the working hours are long, and efficient and frugal societies have driven rapid development in Asia and provided much economic success, there are no tight safety nets for people in their 60s, 70s, and 80s. This is a problem that will become greater before the generations with pensions and insurance policies come to the fore in the next 20 to 30 years.

32.6 SOCIAL CONDITIONS IN ASIA

Street life is very important in Asia, where many do literally live on the street and use street stalls and markets in a very pedestrian, rather than automobile-oriented, lifestyle. In Asia, there is a dilemma in that persons with disabilities cannot use inaccessible streets (Fig. 32.1), and others perceive their lack of presence on the streets as either "persons with disabilities don't want to come out" or as "the problem of a lack of accessibility is small in scale and of low significance."

FIGURE 32.1 Persons with disabilities rarely visit inaccessible shopping centers.

In Asia, the standing-squat type of toilet is commonly provided, because it is relatively fast and easy to clean, provides the option of a low squat (which is anthropometrically better for evacuating the bowels), and is hygienic because there is no contact with the body. Yet, the squat toilet is dangerous for elderly persons when a foot slips into the center and they fall, damaging their ankle. These toilet types are also not wheelchair-user friendly.

Although developing Asian economies have been experiencing rapid social, cultural, and economic changes, women's traditional roles as nurturer and housekeeper remain the norm. Throughout life they are identified as daughters, wives, or mothers, regarded as subordinate to men (United Nations, 1995). Nowadays, many women enter higher education, are graduated from overseas universities, and pursue a job career. If the financial resources of the family are limited, priority for schooling is gender-based rather than based upon intellectual potential (William, 1991). In Indonesia, in 2000, the number of women aged 45 and above was estimated to reach 20 million, which was about 19 percent of the country's 104 million women population (*Jakarta Post*, 1999). Habitual body positions of squatting and kneeling and the dynamic outdoor life of developing Asia are both physical and cultural barriers in social life for elderly people and for people with disabilities. Women are often more involved and active in socializing and celebrating traditional ceremonies, which require squatting, kneeling, and sitting on the floor (Fig. 32.2). Women also use the environment beyond the house more frequently than men. As they get older, habitual body positions and outdoor life become distressing if they still want to participate fully in social life.

FIGURE 32.2   Sitting on the floor is common in Asia.

Many day-to-day activities and traditional ceremonies are done in positions of squatting, kneeling, and sitting on the floor. In conventional trade vegetables, fruits, knickknacks, food and beverage, in bamboo baskets and tins are carried on shoulder poles. The seller displays the goods on the ground where bargaining and transaction proceed in squatting positions. Conventional manners of cooking, eating, washing up, laundry, body cleaning, wedding and funeral ceremonies, religious activities such as praying, listening, and reading verses, kowtowing to older persons, gatherings to celebrate events such as the seventh month of pregnancy, are traditionally done on the floor. With respect to these habits elderly women with mobility problems find it oppressive to participate, and to sit on chairs or benches will make them feel embarrassed. In Asian culture, people's conduct is ruled by external norms rather than by their conscience. This induces shame when their behaviour does not meet society's standard of what is perceived as proper. While cultural barriers cannot be removed overnight, taking into account the needs of elderly women in the planning and design of built environments is essential to make life more convenient. The western style of performing household chores, celebration, personal care and gatherings are done standing or sitting on chairs. These body positions are more appropriate for elderly women, because this style protects them from unnecessary worries and embarrassment.   (Komardjaja, 1999)

In Thailand, often called "the land of smiles," the people are seen to respect and help elderly persons by giving up seats on transportation vehicles and helping them cross roads, for example. However, some Buddhists believe that one should not help a person with a disability, because that person has been reincarnated as disabled to atone for their sins or failings in a previous life. Others believe that they can catch a disease or affliction from a person with a disability, or they treat them as lesser beings. It is an unenviable situation—people with disabilities are able to receive less education, can find lesser employment, and have to endure appalling stereotyping. Examples of this limited thinking are that a blind person can be a telephone operator, a deaf person can operate a keyboard, and little else!

In developing Asia, the street is the most significant site of the day-to-day life (Harrison, 1998). (See Fig. 32.3.) Dynamic street life, however, does impede the mobility of less able-bodied persons. Along the street one can find a variety of eating stalls, vehicle repair facilities, street vendors, parked vehicles (e.g., tricycle pedicabs, motorbikes, taxis), and people casually waiting for public transport at any point of the street. People are more oriented to outdoors rather than indoors, which leads to crowded streets, but this does not mean they are enthusiastic walkers.

FIGURE 32.3   Asian streets and markets are busy and lively places.

In some developing economies, there is a wide rich-poor divide. For the rich, servants provide care and assistance (e.g., chauffeur, cook, nurse, maid). (See Fig. 32.4.) Those with sufficient funds can overcome the hazards of the street, being unaffected by busy public transport because of personal private transport and occupying better designed, and located, housing.

Over the last decade, there have been tremendous changes in affluence and lifestyles in Asia. One of the greatest changes that the author salutes is the move away from ignoring, or playing scant regard to, people with disabilities to providing high-quality, centralized facilities. But there has been a further, more mature, move to bring persons with disabilities back into the community, encouraging full participation in all aspects of living, such as education, recre-

FIGURE 32.4   Relatives and servants can provide assistance.

ation, and employment. This can be surmised as a move from a concentration on growth, per se, to focusing upon quality-of-life issues, which reflects a more caring and equitably aware society.

Disability affects social attitude and domestic harmony. There are no schemes in India, operated by the state, which take care of parentless children. India is a poor country, yet policies and awareness can be increased. The government handles disability as a charity issue rather than as a developmental one, and all problems associated with disabilities are directed to the welfare departments. "India requires a shift of focus from 'disability' to 'ability in disability,' " according to Major General K. M. Dhody, Hon. Secretary General, Cheshire Homes. A major challenge is to change society's perception of people with disabilities from liabilities to assets. This should then lead to changes in the mind-sets of governments and civil servants.

Hirotada Ototake, born without arms or legs, describes his life story, and how people with disabilities are treated in Japan, in his book which has sold over 4 million copies in Japan.

Sad to say, in Japan today it is hard for people with disabilities to move about freely, and it's not easy for us to live on our own. So there's no denying the fact that we need a great deal of help. But it's the environment that forces us into that position. I always think that with the right environment, a person with physical handicaps like mine would not be disabled.   (Ototake, 2000)

32.7   RECENT APPLICATIONS OF UNIVERSAL DESIGN

The Principles of Universal Design, when established and recognized as prerequisites, will get away from the notion that the problems faced by those with disabilities can be tackled, primarily by funding as opposed to appropriate design and implementation. The United Nations Economic and Social Commission for Asia and the Pacific (UN ESCAP) has worked tirelessly in the region to promote accessibility and nonhandicapping environments and has reported on pilot projects in Bangkok, Beijing, and New Delhi (United Nations, 1999).

An example is the use of tactile pavements by the Bangkok Metropolitan Authority (BMA) in Thailand, where the larger blocks used are more expensive yet faster to lay, so the overall installation cost is similar to that of standard paving systems. In New Delhi, India, universal design modifications have been carried out as part of routine maintenance (e.g., the addition of ramps, handrails, and grab bars). This is often funded from within maintenance budgets, which is important, because items to improve accessibility cannot be readily identified, separated, or pruned from budgets by the unenlightened.

The upgrading of the Hong Kong China Mass Transit Rail (MTR) subway system is a positive example of the retrospective application of universal design, where all stations are being upgraded for independent personal access by those using wheelchairs and with vision and auditory impairments. The cost has been considerable, yet the expenditure has been accepted as necessary to meet the needs of the population in terms of social equality. Singapore, to some extent, has begun to follow this lead with the recent announcement to make all of its Mass Rapid Transport (MRT) system elderly-friendly by installing lifts, ramps, and such: "*Wherever possible we will install lifts in existing stations*" (*Straits Times*, 1998). This is close to accommodating wheelchairs, but there remains an unjustifiable fear that wheelchair access delays stops at stations and that evacuation of wheelchair users from tunnels is problematic. It is hoped that further schemes will be implemented in Asia and other developing economies because they are deemed essential and just.

32.8   GENERAL LESSONS LEARNED FOR UNIVERSAL DESIGN

In promoting and applying universal design, it may be valuable to emphasize that universal design is not a global standard, it is a collection of principles that are adjustable for local needs, customs, economic factors, and other reasons. As such, universal design is not out of reach to developing economies or certain sectors of society. Many of the examples of universal design (e.g., of assistive devices) in developed economies are seen as rather sophisticated, expensive, or too comprehensive a solution to meet the present and projected needs of those in less developed economies.

The challenge then is for professionals and users to become aware of, and demonstrate, the appropriate solutions for problems encountered. These solutions can be low-cost, or even of no cost; can use available and appropriate technology; and can apply methodologies to increase access, mobility, and usability for everyone. An example is that access may be required to a higher floor, and this can be achieved by means other than the installation of an elevator (e.g., through the use of ground topography, building a bridge to an adjacent building, etc.). The Internet is a wonderful information source: For example, the Independent Living Centre of South Australia has an extensive database at www.ilc.asn.au; however, this is of no use if Web addresses are unknown or an Internet-connected computer is unavailable.

An analogy of the variety of appropriate solutions that are commonly available is how buildings may appear similar in different countries, yet they still have to comply with local regulations and be designed to cope with local conditions, such as climate and available fuels, funds, and materials. The contextual nature of universal design, and the solutions that it generates, is an important lesson for all involved in shaping and using built environments.

32.9   CONCLUSION

The rate of change of almost everything in Asia is astonishing. With the most rapidly aging populations in the world, it is possible that environments for people with disabilities will rapidly improve, providing that elderly-friendly design is viewed as a subset of universal design. There is a pressing need to nurture a disabled-friendly society through accessible pavements, buildings, and transport, and a holistic approach to rehabilitation needs. It is important to combat stigmas and prejudices toward greater social assimilation. Educational, vocational, and professional needs should be serviced, and social integration to bring disability issues into the mainstream is another high priority.

Also, one should be mindful of the effect of market forces such as economies of scale driving products and services, and a future trend toward people-friendly tourism. With this in mind, it is anticipated that the great differences affecting persons with disabilities between East and West will be eroded with time, as Asian societies catch up and take the lead in some areas. Technology will have great impact on people with disabilities, but there is also the fear that knowledge-based economies do not have to be as physically and sensorily accessible as the nonvirtual environments that we presently, and always will, inhabit.

32.10   BIBLIOGRAPHY

Baguioro, Luz, "900m Asians Live on Less Than US$1 a Day," *Straits Times,* 26 April 2000.

Center for Universal Design, *The Principles of Universal Design,* Center for Universal Design, North Carolina State University, Raleigh, NC, 1997.

Fischetti, M., and G. Stix, "When Life Knows No Bounds," *Scientific American,* 11(2): 2000.

"Grim Future for 60m Disabled Chinese," *Straits Times,* 16 October 1999.

Harrison, J. D., and K. J. Parker, "Pavements for Pedestrians: Asian City Streets for Everyone," in *Proceedings: Designing for the 21st Century,* Hempstead, NY, 17–21 June, 1998.

"How to Pay Special Attention to Family Harmony," *Jakarta Post,* 3 January 1999, Jakarta, Indonesia. (Copyright, *Jakarta Post.*)

Jabatan Kerajaan Tempatan, *Guidelines on Building Requirements for Disabled Persons,* Ministry of Housing and Local Government, Kuala Lumpur, Malaysia, 1999.

Komardjaja, I., and K. J. Parker, "Mobility and Accessibility for Elderly Women in Developing Asian Countries," in *Proceedings Women's Health—The Nation's Gain: An International Conference with a Special Focus on Older Women in Asia,* Singapore, 5–7 July, 1999.

Ototake, Hirotada, *No One's Perfect,* Kodansha International, Tokyo, Japan, 2000.

Public Works Department, *Code on Barrier-Free Accessibility in Buildings,* Building Control Division, Public Works Department, Singapore, 1995.

Toh, Su Fen, "More 'disabled friendly' features added to MRT," Forum letters page, *Straits Times,* 2 May, 1998.

United Nations, *Hidden Sisters: Women and Girls with Disabilities in the Asian and Pacific Region,* ST/ESCAP/1548, United Nations Economic and Social Commission for Asia and the Pacific (UN ESCAP), New York, 1995.

————, *Promotion of Non-handicapping Environments for Disabled Persons: Pilot Projects in Three Cities,* ST/ESCAP/2005, United Nations Economic and Social Commission for Asia and the Pacific (UN ESCAP), New York, 1999.

William, Walter L., *Javanese Lives—Women and Men in Modern Indonesian Society,* Rutgers University Press, London, UK, 1991.

CHAPTER 33

SUSTAINABLE HUMAN AND SOCIAL DEVELOPMENT: AN EXAMINATION OF CONTEXTUAL FACTORS

C. J. Walsh, B.Arch., M.R.I.A.I., M.I.B.C.I., M.I.F.S.
Sustainable Design, Dublin, Ireland

33.1 INTRODUCTION

The new World Health Organization's (2000) International Classification of Functioning and Disability heralds not only a dramatic change in the language and philosophy of what constitutes disability, which is both positive and liberating, but the possibility of creating a more tangible, interactive relationship with mainstream society, because of its reference to both *environmental* and *personal contextual factors*, as defined later.

This chapter examines disability and contextual factors from the broad perspective of sustainable human and social development. It remains, then, for the reader to decide whether the concepts of *universal design* or *design for all* are sufficiently elastic to remain on the European disability agenda for the medium term (i.e., until the year 2040) or even the short term (i.e., until 2010).

33.2 BACKGROUND: A CHANGING GLOBAL CONTEXT IN THE TWENTY-FIRST CENTURY

In the United Nations *Human Development Report* (1998), it was estimated that 20 percent of the world's population in the highest-income countries consumed 58 percent of total energy, while the poorest one-fifth consumed less than 4 percent, and that the burning of fossil fuels had almost quadrupled since 1950. Even as the process of globalization has continued to gather pace, with nation states being crisscrossed and undermined by transnational corporations (TNCs), the mass media, nongovernmental organizations (NGOs), the Internet, and enormous flows of investment and private capital, one sees an increasingly fragmented built environment, with a social fabric that is threadbare and torn in many areas. It is also clear, at the time of writing, that the developed regions of the globe will not successfully meet the first major environmental performance target of the new millennium [i.e., the legally binding Kyoto Protocol to the United Nations Framework Convention on Climate Change (UNFCCC) (1998), which was agreed upon at the third meeting of the Conference of the Parties (COP 3) in December 1997]. The challenges for an integrated society are seen in Fig. 33.1.

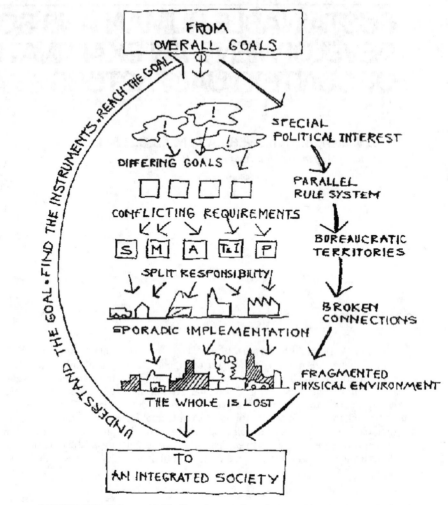

FIGURE 33.1   A fragmented built environment.

Within this existence that human beings themselves are shaping, individuals and even large groups of people no longer feel that they have the freedom to remain apart from the rest of society. Furthermore, the health and happiness of everybody else now depends very much on which model of capitalism is being advocated by economic gurus in a small number of geographic locations. Is the emphasis being placed on capital and shareholders' value, or on human resources and added value? These and other existing economic models, however, are extremely wasteful of resources.

Should a more person-centered approach to how society is ordered and organized be adopted? A welcome addition to the European Plan of Integration, a Charter of Fundamental Rights, is currently being drafted in the European Union (EU). Social justice and the protection of human rights are founding principles of the EU, and these are the single most essential prerequisite for its growing legitimacy. In fact, an extensive body of EU legislation already exists relating not only to social concerns, but also to environmental, energy, economic, institutional, and political matters, all of which fall precisely within the zone of pertinence for the concept of *sustainable development*. Where EU legislation exists, it is superior to (i.e., takes precedence over) the national legislation of its member states.

The construction of an interim, rational understanding of sustainable development, which makes sense for Europe and nowhere else, is rooted firmly in the elaborate legal base of the EU, international agreements and treaties to which it is a signatory, and the aims and objectives of international organizations in which it agrees to participate. It has been possible to make rapid progress and to propose within that unique context how, for example, a more coherent and comprehensive program of action concerning people with activity limitations might be implemented. This exercise also allows the robustness of the interim understanding to be tested as it is developed and evolves.

33.3 SUSTAINABLE HUMAN AND SOCIAL DEVELOPMENT

Significant progress was made in the last treaty addition to the primary legislation, or constitution, of the EU—the *principle of sustainable development* was explicitly incorporated in the EU Amsterdam Treaty (1997), although it was not defined or expressed in the clearest terms. This concept was first presented internationally at the beginning of the 1970s. It was elaborated in a readily understandable form only at the end of the 1980s in the report, *Our Common Future,* produced by the World Commission on Environment and Development (1987).

It is convenient, therefore, to go back to that report and to start with the following definition of *sustainable development,* also known as the *Brundtland definition,* after the Commission's Norwegian chairperson, Gro Harlem Brundtland: "Development which meets the needs of the present without compromising the ability of future generations to meet their own needs."

However, with the benefit of many years' hindsight since then, a more thorough explanation of sustainable development must also embody a resolution of some further issues:

- The place of human beings in the environment, and the relationship between both
- The nature of human, social, cultural, economic, institutional, and political development, their current imbalances and inequities, and their future course
- The healing of existing harm and injury to the natural environment

It is important, here, to distinguish between the *natural environment* and the *built environment* (i.e., anywhere that there is, or has been, an intrusion or intervention by a human being in the natural environment). This may be urban, suburban, rural, or marine, and it includes not only buildings, but also civil engineering projects, infrastructural networks, service and support systems, transport, and so on.

Principle 1 of the 27 principles in the Rio *Declaration on Environment and Development* (United Nations, 1992) states: "Human beings are at the centre of concerns for sustainable development. They are entitled to a healthy and productive life in harmony with nature."

And the World Health Organization, in the preamble to its constitution, defines *health* as: "A state of complete physical, mental and social well-being, and not merely the absence of disease or infirmity."

The European Energy Charter Treaty (European Union, 1994) also provides some necessary definitions:

Environmental impact. Any effect caused by a given activity on the environment, including human health and safety, flora, fauna, soil, air, water, climate, landscape and historical monuments or other physical structures or the interactions among these factors; it also includes effects on cultural heritage or socio-economic conditions resulting from alterations to those factors.

Energy cycle. The entire energy chain, including activities related to prospecting for, exploration, production, conversion, storage, transport, distribution and consumption of the various forms of energy, and the treatment and disposal of wastes, as well as the decommissioning, cessation or closure of these activities, minimising harmful environmental impacts.

A definition of *energy cycle* is required because, more and more, the potential life cycle cost of future actions that modify or influence the built environment has to be considered and carefully examined. EN ISO 14040 [European Committee for Standardization (CEN), 1997], an international standard also adopted as the European standard, defines *life cycle assessment* as follows: "Compilation and evaluation of the inputs, outputs and the potential environmental impacts of a product and/or service system throughout its life cycle."

At the start of the twenty-first century, a growing consensus in Europe is finally acknowledging that in order to accommodate further human and social progress, with an assured minimum quality of life and health for all peoples, harmony between global regions, and world economic stability, it will be necessary to convert from current irresponsible patterns of human development, with their attendant wasteful environmental destruction and societal stresses.

And so, the long-term goal over the next century and a half in Europe will be to convert to sustainable human and social development (i.e., the creation of a sustainable built environment within a flourishing natural environment, each coexisting with the other in harmony and dynamic balance, and each in their own way capable of providing for responsible and equitable human, social, cultural, and economic development). Previous injury to the natural environment must be healed in order to arrive at this outcome, and initial damage repair by human intervention, sufficient only to promote a process of natural self-healing and self-management, is suggested.

33.4   *EUROPEAN CHARTER ON SUSTAINABLE DESIGN AND CONSTRUCTION*

Intended primarily for senior policy and decision makers within the Commission (i.e., the sole EU institution with the power of initiative), the *European Charter on Sustainable Design and Construction* (Sustainable Design International Ltd., 1998) was a logical next step following a major study project coordinated by Working Commission 82 of the International Council for Research and Innovation in Building and Construction (CIB). The project culminated during May 1998, with the production of CIB Publication 225—*Sustainable Development and the Future of Construction*—comprising national reports from 14 different countries, and accompanied by an international synthesis.

The structure of the *European Charter on Sustainable Design and Construction* reflects the fundamental view that sustainable construction is the response, in built form, to the concept of sustainable development (as diagrammed in Fig. 33.2), and its initial formal elaboration at the global level. The 27 principles of the Rio *Declaration on Environment and Development,* with Agenda 21 as detailed supporting guidance, was agreed upon at the United Nations Conference on Environment and Development, held in Rio de Janeiro, Brazil, on June 3 through 14, 1992.

However, it is sustainable design, a quantum leap in design philosophy, which will direct the future course of this innovative approach to construction.

For the first time in the *European Charter,* a comprehensive scope of concern, relating to ethics and social values, is outlined for the subject: A rational decision-making framework is presented; human development, social justice and inclusion, environment, and energy issues are discussed in a coherent format; and finally, technical terms are defined for better communication.

Preserving intact the legal intent of the Rio *Declaration on Environment and Development,* the *European Charter* also comprises 27 principles:

- Each principle of the original Rio *Declaration* was closely examined and redrafted to suit more closely a European context. On the basis of existing EU primary and secondary legis-

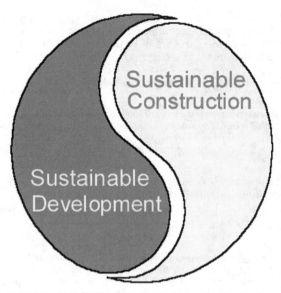

FIGURE 33.2 Relationship between sustainable development and sustainable construction.

lation, and international agreements and treaties to which it is a party, it was strengthened considerably in expression.

- Where appropriate, a clause relevant to sustainable planning, design, construction/deconstruction, and maintenance was added.

The *European Charter* also places special emphasis on implementation through the informed application of construction-related sustainability performance indicators (i.e., short statements of targeted, real performance in the built environment).

The *European Charter* and People with Activity Limitations

The long-term importance of the 1997 Amsterdam Treaty for Europe is the emphasis it places on the individual citizen in society. It brings the institutions of the EU and their services closer to people, and it requires that the baseline for public health and consumer protection be set at a high level. Prior to the national referendum on Ireland's ratification of the treaty in 1998, therefore, a public seminar, entitled "Citizens in the E.U.—Equality of Opportunity for Minority Groups in Society," was organized on behalf of the Dublin Branch of the European Movement to promote awareness of these treaty changes.

For the first time, many representatives of the unemployed, refugees, travelers (indigenous nomadic people), gays and lesbians, the homeless, people with activity limitations, elderly people, and single parents gathered in one room for a day of interactive discussion and an exchange of views. The day's proceedings were a revelation to everyone who attended. It was more fully appreciated that there are many minority groups in society and that they must cooperate with each other and not compete. Restrictions on participation in society are numerous, rigid, and certainly do not always take the form of physical barriers. It was also important to accept that not all people wanted to be included in mainstream society. Details of the seminar are contained in Appendix 4 of the Irish National Report (Sustainable Design International, Ltd., 1998), CIB Publication 225, *Sustainable Development and the Future of Construction.*

In addressing the wider context of sustainable human and social development, therefore, it was a purposeful intention to deal positively and directly in the *European Charter* with issues that concern, and are of concern to, people with activity limitations, as one minority group, among many, in society:

- Principle 20 of the *European Charter* states:

  Women have a vital role in environmental management and development. Their full participation is essential to achieve sustainable development. The experience and wisdom of the elderly should be valued; the abilities of every person should be cherished; and the creativity, ideals and courage of youth should be mobilized to forge a European partnership in order to ensure a better future for all.

  The original Rio *Declaration* did not contain any reference to either people with activity limitations, or elderly people.

- Among a small number of relevant extracts from the 1997 Amsterdam Treaty, Appendix I of the *European Charter* quotes the new antidiscrimination clause:

  Without prejudice to the other provisions of this Treaty and within the limits of the powers conferred by it upon the Community, the Council, acting unanimously on a proposal from the Commission, and after consulting the European Parliament, may take appropriate action to combat discrimination based on sex, racial or ethnic origin, religion or belief, disability, age or sexual orientation.

  This isolated clause will be added to in successive treaties, and the currently extensive texts devoted to gender equality in EU primary legislation will be matched.

- Appendix III of the *European Charter* provides a useful list of terms and definitions. A principle must be that everyone has the right to choose exactly what they wish to be called. There should be, accordingly, much greater tolerance and acceptance of the use of the different terms (e.g., "people with activity limitations," "people with disabilities," and "disabled people" and "disABLED") in different regions of the world.

- The Preamble to the *European Charter* makes reference to a coherent European "Guideline Framework on Social Justice and Inclusion," which is later outlined in Appendix II.

Guideline Framework: Achievement of Equality of Opportunity and Social Inclusion for Every Person in the European Union (EU)

Direct and meaningful consultation with people, partnership between all sectors of society, consensus, transparency, and openness are essential elements in social well-being. Following are a number of areas that should be actively considered by the institutions of the EU and implemented and effectively monitored by relevant authorities in each member state, through the informed application of sustainability performance indicators.

Empowering People for Participation in Society

- Respecting autonomy and independence
- Readjusting education and training programs to facilitate participation
- Readjusting welfare and other supports to facilitate participation
- Moving toward a person-centered approach in the design/implementation of support services
- Mainstreaming
- Ensuring seamless provision of services
- Ensuring the principle of participation

Removing Physical Barriers to Participation

- Viewing access/egress/evacuation and health/safety/welfare issues in the light of equality of opportunity and the right to participate
- Developing effective legislation, standards (nationally transposed ENs), and technical guidance to eliminate all forms of barrier
- Monitoring and controlling compliance with legislation
- Moving toward a person-centered approach in the planning/design/construction/maintenance of a sustainable built environment

Opening Up Various Spheres of Society

- Upholding the equal civic status of every person
- Promoting employment for people as a key to social inclusion

Nurturing Opinion of the Public, Government Administrators, and Design Professions to Be Receptive to "Person Centeredness" of the Built Environment

- Concerted programs of awareness raising and education at all levels

Origins of Person-Centered Design

The concept of person-centered design is introduced, and referred to, throughout the document. Principle 1 of the *European Charter on Sustainable Design and Construction* (Sustainable Design International Ltd., 1998) states:

> Human beings are at the centre of concerns for sustainable development. They are entitled to a healthy and productive life in harmony with nature. Movement towards a person-centred and socially inclusive approach in the planning/design/construction of a built environment; i.e., placing real people, their needs and responsible desires at the centre of creative endeavours, should be encouraged and fostered by every key sector in society.
>
> The method of work in the various processes of planning/design/construction should be widely multi-disciplinary. An active dialogue between practitioners, researchers and end-users, based on meaningful consultation, partnership, and consensus should become the standard.

33.5 DEVELOPMENT OF PERSON CENTEREDNESS IN DESIGN

The concept of *person-centered design* is also defined in Appendix III of the *European Charter*. It states:

> . . . that design process which places real people at the centre of creative endeavours and gives due consideration to their health, safety and welfare in the built environment. It includes such specific performance criteria as a sensory rich and accessible (mobility, usability, communications and information) environment; protection from fire; air, light and visual quality; protection from ionizing and electromagnetic radiation; thermal comfort (EN ISO 7730) (1995); unwanted or nuisance noise abatement; etc.

An important person-centered design aid is the questionnaire survey, carried out by an independent, competent, nonthreatening individual on a person-to-person basis, and which comprises both open- and closed-format questions. These surveys are not only valuable sources of information, but they formalize the process of meaningful consultation between practitioners and end users. Two key architectural projects, in particular, furthered the development of this design concept.

The Sessions House: A Major Civic Building in Dublin

A detailed energy survey was carried out on The Sessions House, a historic building built in 1797. The survey was based on the knowledge gained in the extensive literature review carried out from May to September 1995. This review led to the development of sustainable energy-efficient environment-friendly development (SEED) (Environment 2000 Ltd. and Sustainable Design International Ltd., 1996), an acronym that brought together, for the first time, three separate areas of concern and that crystallized the single idea that building performance cannot be evaluated in isolation from interrelated human factors. In practice, SEED gave shape and resolution to an array of new methods of work in construction.

Part of the process of working with EN ISO 7730 [European Committee for Standardization (CEN), 1996], using a questionnaire survey, proved to be a very useful design aid because of the reliable information that it uncovered. It had also facilitated, in a formal way, meaningful consultation with the building's occupants and other users. In what might have been described as a "problematic workplace," morale improved dramatically following the survey. From then on, the employees were working with the project team, instead of resisting its every action.

In the model proposed by the European and international standards, thermal comfort is dependent on air speed (drafts), relative humidity (damp walls), radiant temperature (the ability of thick walls to heat up quickly), and air temperature. During the survey, it was found that if comfort was lacking in one aspect, it was overcompensated for in another. For example, electrical misuse in one office could finally be explained by the need to compensate for uncomfortable drafts with higher room temperatures ($27°$–$28°C$). The uneconomical use of portable electrical heaters as heat sources throughout the building then became understandable. In this case, the standard allowed the team to comprehend what was happening, but it was the questionnaire survey that permitted the building user to be seen as a real person. Once the problem was examined from the point of view of that person, practical and effective design solutions were quickly formulated. Recommendations in the completed *The Sessions House—Energy Survey Report* (Environment 2000 Ltd. and Sustainable Design International Ltd., 1996) for this historical building reliably identified potential savings in energy costs of approximately 51 percent, while working properly within the constraints of the International Council on Monuments and Sites (ICOMOS) Venice Charter (ICOMOS, 1964).

Local Housing Authority in Dublin: Public Housing

Toward the end of a difficult summer in 1996, the most extensive cross-sectional energy study of existing public dwellings ever carried out in Ireland was completed. Its objective was to develop a more precise understanding of the issue of fuel poverty among low-income groups in the urban built environment of Dublin. As project coordinator and technical controller, the author reported the study's principal findings (SEED) (Environment 2000 Ltd. and Sustainable Design International Ltd., 1996) to a Paris conference coorganized by CIB/CSTB, "Buildings and the Environment," from June 9 to 12, 1997.

In demanding to observe the real energy performance of real buildings, the use of long-wave (8–12-μm waveband) infrared thermography opened up a world that hitherto had been closed off to full exploration, because of an unquestioning reliance on reference documentation and computer software, which were either limited in scope and/or not properly validated, as shown in Fig. 33.3.

Apart from the obvious benefit of reliable energy flow information, thermal imagery also showed areas of dampness, porosity, and poor maintenance of a building's fabric, factors that not only influenced energy consumption, but also had a direct impact on human thermal comfort and health within the building.

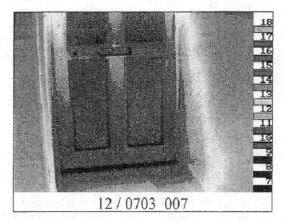

FIGURE 33.3   Door to exterior. Thermal image of interior showing cold surfaces.

The Sustainable Building

Slowly, a profile is emerging of what sustainable building of the future may entail. It will relate, in harmony and balance, with its natural environment, as shown in Fig. 33.4, and will be:

- Person-centered and socially inclusive
- Flexible and adaptable
- Spatially complex, with some ambiguity, and yet understandable (e.g., easy for building users to find their orientation and to connect with the exterior)
- Electronically mature, yet energy-efficient throughout
- A place of stimulation and encouragement to human creativity

FIGURE 33.4   Design concept of the future sustainable building.

- Always at the limits of economic viability and technical feasibility
- Beneficial with regard to overall environmental impact

At the international level, further research and development work is already being undertaken that focuses on sustainable planning of the built environment, and sustainable civil engineering, service, and infrastructural construction.

33.6   REAL IMPLEMENTATION OF A SUSTAINABLE BUILT ENVIRONMENT

As previously stated, the *European Charter on Sustainable Design and Construction* places emphasis on real implementation, which is reproducible, by means of the informed use of construction-related sustainability performance indicators (i.e., some understanding and competence is required).

In particular, Principle 26 of the *European Charter* states:

> Harmonized short, medium and long-term strategies in the policy areas of energy efficiency, environmental protection and sustainable development should be planned for implementation in the European Union over the following time frames: up to 2010; between 2011 and 2040; and between 2041 and 2100.
>
> Such is the threat to quality of life and human progress caused by current environmental degradation, and such is the great time lag between implementation of corrective actions and resulting beneficial environmental impacts, that sustainability performance should be benchmarked at year 1990 in the Member States of the E.U. Detailed performance indicators for all stages of design, construction/deconstruction, maintenance and disposal should be used to target improvements in sustainability performance, verify target attainment, and continually readjust targets at appropriate intervals thereafter.

The primary purpose of construction-related sustainability performance indicators is to commence, in earnest, the practical task of implementing a sustainable approach to the future development and modification of the built environment in Europe, while also ensuring a flourishing future for the natural environment by carrying out sufficient repair to past, present, and potential future damage directly or indirectly caused by construction.

- Principle 26 of the *European Charter on Sustainable Design and Construction* signaled that a futures scenario should be developed that would cover the short, medium, and long terms, at least until the end of this century.

- Using this futures scenario, incremental improvements in construction performance required to achieve a sustainable built environment within a flourishing natural environment may then be plotted. The focus of attention must always be on real, rather than theoretical, performance.

- Construction-related sustainability performance indicators, harmonized for application in the EU, will allow targeting, reliably quantifying, and monitoring construction performance in the built environment, which, by general international agreement, has been benchmarked at 1990 levels. Rigorous procedures will be required to process the data and statistics generated in order to ensure that they, too, are reliable.

A secondary, short-term purpose in Ireland will be to develop a *sustainability rating system*, initially for buildings, which will be a major departure from existing methods of energy and/or environmental rating systems. Based on an understanding of sustainable development that is current and generally held at any particular time, an objective sustainability performance statement may be made about a specific building, the performance of different buildings may be

FIGURE 33.5 Harmonized European sustainability label for buildings.

compared, or more favorable working methods in the building design process itself may be identified. Figure 33.5 shows the rating label developed by Sustainable Design International Ltd.

At the global level, the implementation plan for the United Nations Commission on Sustainable Development (UNCSD) Work Program on *Indicators of Sustainable Development* was in its third phase: January 1998 until January 2001. An initial Working List of Indicators has already been produced, which is intended for global application. It is, therefore, necessarily general in nature, and not at all construction related. These indicators cover four aspects of sustainable development (i.e., social, economic, environmental, and institutional) and are presented in a "Driving Force—State—Response" framework. Trial application of initial indicators is taking place in four global regions: (1) Africa, (2) Asia and the Middle East, (3) Europe, and (4) Americas and the Caribbean. In Ireland, work is being undertaken on a fifth aspect of sustainable development (i.e., political).

Supported by a value system, there are five essential requirements. Therefore, in order to achieve sustainable human and social development, these aspects need to be considered:

1. Social justice and inclusion
2. Economic equity
3. Beneficial environmental impact
4. Institutional openness
5. Political accountability

To manage all of this information on performance indicators, it was necessary to develop a fundamental matrix, as shown in Table 33.1.

Following are three examples of relevant indicators:

- *Equality of opportunity for every person in society.* To advance the principle of sustainable development, and in order to combat discrimination and remove restrictions on participation in society by the year 2010, every new building shall be fully and independently accessible with regard to mobility, usability, communications, and information.

- *Fire protection in buildings.* To advance the principle of sustainable development, and in order to provide a high level of protection and improvement of the quality of the environment by the year 2010, every new building shall be designed, constructed, and managed so as to ensure the least adverse environmental impact in the event of fire.

- *Indoor climate/air quality in buildings.* To advance the principle of sustainable development, and in order to provide a high level of human health protection by the year 2010, radon activity (including Rn-222, Rn-220, RnD) in every new building shall, on average, fall within the range of 10 to 40 Bq/m^3, but shall at no time exceed 60 Bq/m^3.

The underlined text in each indicator is a direct quotation from the 1997 EU Amsterdam Treaty.

33.7 IMPORTANCE OF TARGETED RESEARCH AND DEVELOPMENT

Without a concerted program of innovation, research, development, demonstration, and precise observation of results, the desired end condition in Europe (i.e., a sustainable built envi-

TABLE 33.1 Fundamental Matrix: Construction-Related Sustainability Performance Indicators*

Design, construction, and logistics			Setting targets and monitoring real				
			Social			Economic	
			Driving force	State	Response	Driving force	State
1. Design	a. Planning	i. Urban					
		ii. Rural					
		iii. Marine					
	b. Architectural						
	c. Engineering						
	d. Industrial						
2. Construction							
3. Use							
4. Maintenance							
5. Adaptation							
6. Deconstruction							
7. Disposal	i. Reuse						
	ii. Recycle						
	iii. Waste						
8. Products (Dir. 89/106/EEC)							
9. Services							
10. Incentives							

* Matrix prepared for the International Council for Research and Innovation in Building and Construction (CIB).
Source: © Sustainable Design International Ltd., 1999–2000.

ronment) cannot be attained. Sustainable design is tailor-made as the most cogent concept and the most relentless driving force for this task in construction of the future. Following are proposed stages of research.

- *Stage I.* The realistic end condition, or reality, is a sustainable built environment (i.e., the response, in built form, to the concept of sustainable human and social development). It may take another 7 to 10 years before this concept is fully understood.

  Literature dealing with reality is reviewed. Relevant hypotheses are extracted, and as many variables as possible are identified. Inputs also include the futures scenario(s), and initial construction-related sustainability performance indicators. Use of statistics is limited to those that can be shown to be impartial, reliable, objective, scientifically independent, cost-effective, and statistically confidential [see the Amsterdam Treaty, New Article 213a in the treaty established by the European Community (TEC)].

- *Stage II.* An artificial reality is designed so that it is complex enough to permit testing of the hypotheses formulated in Stage I. Observations must be capable of description in quantitative terms.

- *Stage III.* Artificial reality is broken down into simple experimental situations at small and medium scale [e.g., advanced energy surveying of buildings or groups of buildings using

| performance in the built environment | | | | | | | | | | |
| Response | Environmental | | | Institutional | | | Political | | | |
	Driving force	State	Response	Driving force	State	Response	Driving force	State	Response

infrared thermography, detailed analysis of air quality in buildings and at external locations, real-time monitoring of thermal comfort (EN ISO 7730) conditions in buildings, etc.], which generate test results under controlled conditions, like a laboratory in the real environment.

Special attention is paid to measurement/calculation uncertainty and test method precision. Computer models must be transparent to practitioners and validated by an independent, competent individual and/or organization.

Questionnaire surveys are carried out with real users of buildings, civil engineering works, and infrastructural networks (e.g., transport). To be effective, it is essential that a survey is carried out on a person-to-person basis by an independent, competent, nonthreatening individual using open- and closed-format questions. These surveys formalize the process of meaningful consultation between practitioners and end users.

- *Stage IV.* A simple theory, or microtheory, is developed to explain the test results (e.g., person centeredness of the built environment). When this microtheory is tested and found valid, it is expanded to contain test results in more complex situations. This process is repeated until a macrotheory is formulated, which explains the artificial reality.

- *Stage V.* Artificial reality is modified in the direction of reality, and Stage IV is repeated yielding a fresh macrotheory. The process is repeated again and again.

- *Stage VI.*   When a macrotheory is sufficiently developed, it can be used to extrapolate an explanation of reality.

It is essential that such a theory of reality be accessible to all concerned with the implementation of sustainable design, construction/deconstruction, and maintenance in the built environment and, therefore, a boundary to the use of terminology is delineated. Terminology must focus on, and be always directly related to, the realistic end condition.

33.8   CONCLUSION: A REGIONAL DISABILITY AGENDA FOR EUROPE

On May 12, 2000, the Commission of the European Communities (the European Commission) issued *Towards a Barrier Free Europe for People with Disabilities.* This document proposed to declare the year 2003 as the European "Year of Disabled Citizens," with the objective of strengthening the concept of citizenship for people with disabilities. It included text devoted to a renewed approach to disability, detailing major steps forward and a new impetus.

The reality of present-day Europe, however, is far from being ideal. Notwithstanding the fact that different types of EU legislation that require that buildings and places of work be accessible have existed for many years, a lack of political will on the part of European politicians and controlling authorities at national and local levels has ensured that, even today, countless barriers to that accessibility are still being erected in the built environment. In particular, the nonexistence of comprehensive technical guidance on protection from fire in buildings has resulted in the creation of a far more pervasive form of barrier to the full inclusion of people with disabilities into the economic, cultural, and social life of the general community. As just one example, access to accessible buildings is regularly being refused in Ireland, for reasons of fire safety.

Sustainable design, on the other hand, represents a quantum leap in the evolution of design philosophy. Accessibility of buildings and the built environment is approached, and effectively solved, on a person-centered basis. There is no competition between different demographic groups, because every person is an individual. This approach is directly supported by Principle 1 of the Rio *Declaration,* further reinforced by the more elaborate Principle 1 of the *European Charter on Sustainable Design and Construction,* and placed within a coherent and comprehensive agenda on social justice and inclusion, a core value of sustainable human and social development. In turn, this value is rooted, at a fundamental level, in elaborate EU legislation.

The language of sustainable design is positive, progressive, and facilitating with regard to the future provision of a sustainable, sensory-rich built environment, which will stimulate and encourage human creativity in an environment where the abilities of every person will be cherished.

Aided by an examination of contextual factors, a European Disability Agenda would contain the following elements:

- A coherent action program, initially covering the short term, up to 2010
- A properly harmonized and multilingual EU vocabulary of disability
- A reliable and properly harmonized EU database of disability-related statistics
- Targeted research and development that answer the real needs of people with activity limitations, and those who plan and design an accessible built environment
- A comprehensive array of disability-related performance indicators
- The positioning of social justice and inclusion within the area of responsibility for action coordination by the Secretary-General of a reorganized European Commission
- An effective EU regime of performance monitoring and technical control

Furthermore, this agenda simply cannot be limited any longer to considering technical issues of accessibility in splendid isolation (e.g., what should be the clear width of a door opening). It must now be extended, under legal imperative in the European Union, to deal comprehensively with the social well-being of people with activity limitations in the built environment. A reasonable concern about the *universal design/design for all* concepts is that their dated and inadequate scope hinders this new approach.

33.9   BIBLIOGRAPHY

Commission of the European Communities, *Towards a Barrier Free Europe for People with Disabilities,* Commission of the European Communities, Brussels, Belgium, 2000.

Environment 2000 Ltd., and Sustainable Design International Ltd., *"SEED" Housing Agenda 1996—The 12th Report,* Environment 2000 Ltd., and Sustainable Design International Ltd., Dublin, Ireland, 1996.

———, *The Sessions House—Energy Survey Report,* Environment 2000 Ltd., and Sustainable Design International Ltd., Dublin, Ireland, 1996.

European Committee for Standardization (CEN), *Moderate Thermal Environments,* EN ISO 7730, CEN, Brussels, Belgium, 1995.

———, *Environmental Management—Life Cycle Assessment—Principles and Framework,* EN ISO 14040, CEN, Brussels, Belgium, 1997.

European Union, The Amsterdam Treaty, "Treaty of Amsterdam amending the Treaty on European Union, the Treaties establishing the European Communities and certain related acts," 97/C 340/01, Office for Official Publications of the European Communities, Luxembourg, Luxembourg, 1997.

———, "The European Energy Charter Treaty," *Official Journal of the European Communities,* Ser. L, No. 380, 1994-12-31, Office for Official Publications of the European Communities, Luxembourg, Luxembourg, 1994.

International Council on Monuments and Sites (ICOMOS), *International Charter for the Conservation and Restoration of Monuments and Sites,* ICOMOS, Paris, France, 1964.

International Council for Research and Innovation in Building and Construction (CIB), *Sustainable Development and the Future of Construction,* CIB Publication no. 225, CIB, Rotterdam, Netherlands, 1998.

Sustainable Design International Ltd., *European Charter on Sustainable Design and Construction,* produced in cooperation with the Commission of the European Communities and the International Council for Research and Innovation in Building and Construction (CIB), Dublin, Ireland, 1998.

United Nations, *Declaration of the United Nations Conference on Environment and Development,* 1992; Internet: www.un.org/esa/sustdev/agenda21.htm.

———, *The Kyoto Protocol* [agreed upon at the third meeting of the Conference of the Parties (COP 3) to the United Nations Framework Convention on Climate Change], Kyoto, Japan, 1997.

United Nations Commission on Sustainable Development (UNCSD), *Indicators of Sustainable Development;* Internet: www.un.org/esa/sustdev/isd.htm.

United Nations Development Report, *Human Development Report,* Oxford University Press, New York, 1998.

World Commission on Environment and Development Report, *Our Common Future,* Oxford University Press, New York, 1987.

World Health Organization, *International Classification of Functioning, Disability and Health,* Prefinal Draft, Full Version, WHO, Geneva, Switzerland, 2000.

33.10   RESOURCES

Commission of the European Communities. *Towards a European Research Area,* Brussels, Belgium, 2000-01-18, COM (2000) 6

European Concept for Accessibility (ECA) Network, Luxembourg, Luxembourg. www.eca.lu/

Legislation of the European Union—Eur-Lex. www.europa.eu.int/eur-lex/en/index.html

Statistics of the European Union—Eurostat. www.europa.eu.int/eurostat.html

Sustainable Design International, Ireland. www.sustainable-design.com/

United Nations Commission on Sustainable Development. www.un.org/esa/sustdev/

United Nations Development Program—Sustainable Human Development. www.undp.org/indexalt.html

World Health Organization. www.who.int/

P · A · R · T · 5

RESIDENTIAL ENVIRONMENTS

CHAPTER 34
THE NEXT-GENERATION UNIVERSAL HOME

Leslie C. Young, M.S., and Rex J. Pace, B.E.D.A.
*Center for Universal Design, North Carolina State University,
Raleigh, North Carolina*

34.1 INTRODUCTION

The next-generation universal home is an example of how housing may evolve over the next 15 to 20 years in response to present and future demographic and marketing trends. The concepts and solutions it presents are based on existing technologies and coincide with popular design approaches. As such, the next-generation universal home's innovation is not based on manufacturing or technological advances, but rather on a change in thinking about how people actually live in their homes. This change is called *universal design,* and this chapter lays out the need, range of solutions, and possible future of this concept when applied to housing.

34.2 DEMOGRAPHIC TRENDS AND LEGISLATIVE BACKGROUND

Single-family houses in the United States today are built much as they were 50 years ago, with the exception of better insulating, tighter windows, and electronic devices. The housing industry, slow to change, did not respond to the predictions of the 1960s that suggested homes in the year 2000 would "[be made of] wonderful new materials far stronger than steel but lighter than aluminum [and] be able to fly" (Arthur C. Clarke, *Vogue Magazine,* 1966). And "in the year 2000 we will . . . cook in our television sets and relax in chairs that emit a private, sound-light-color spectacular" (*New York Times,* 7 January 1968). Fantastic though these projections may sound, the characteristic common to these two futuristic statements proving true is that housing in the future will exhibit a much greater degree of flexibility, with spaces and products offering multiple functions or doing double duty.

The housing industry must adapt to the growing diversity in family types in the United States. Tom W. Smith, director of the General Social Survey and researcher at the National Opinion Research Center, states, "It means that employers, schools, the government—all the institutions in society—can't assume there's . . . one-model family." Unmarried and no children has become the most common living arrangement in the United States, according to a study by the National Opinion Research Center at the University of Chicago. In 1972, the most common type of household (i.e., 45 percent) consisted of married couples with children.

But by 1998, the number had dropped to 26 percent, a dramatic shift from a generation ago. And the population is aging. By 2030, there will be close to 70 million older adults, more than twice the number in 1996 (cited in "A Profile of Older Americans," American Association of Retired Persons, 1997).

Concurrent with these trends, initially brought into the national social consciousness by the civil rights movement for people with disabilities, is the recognition and acceptance that being human means that there is no one-model individual whose characteristics remain static through their lifetime. The self-identified population of people with disabilities is around 49 million. But, in truth, the figure is probably higher. If older American adults, who may or may not have a disabling condition, are aggregated with people with disabilities, as well as the rest of the population in the throes of the life process where common limitations are normally experienced due to stature, pregnancy, temporary injury, or even a circumstantial limitation such as hands full or low light, the magnitude of the range of human diversity is immense.

A growing community of designers, researchers, and educators worldwide is recognizing that the built environment cannot be designed for one specific population, but for a dynamic range of people and abilities. Housing of the future must possess, in addition to being environmentally sensitive and sustainable, an ability to adapt to the differing needs and requirements of the users, no matter their age or strength or agility. This design approach, known as *universal design,* strives to make the practical day-to-day tasks involved in living possible and safer for everyone.

34.3   *UNIVERSAL DESIGN*

Universal design was a philosophy and goal coined by Ronald L. Mace, F.A.I.A., in the early 1980s. It was shaped by his recognition that the community of people with disabilities had been relegated to a parallel universe, and through a new design approach to the built environment, he and other members of the disability community could take their rightful place as integral members of a common universe (Mace, Hardie, and Place, 1991).

As understood today, universal design respects, values, and strives to accommodate the broadest possible spectrum of human ability in the design of all products and environments. It encompasses and goes beyond the accessible, adaptable, and barrier-free design concepts of the past. It helps eliminate the need for special features and spaces for a specific group of people, eliminating the stigma and additional expense that specialness and special products often generate. Universal design works to proactively address human needs within the mainstream.

Universal Design and Housing

Few areas in the United States have any design requirements to make single-family housing or other forms of private housing barrier-free or accessible. Change within the private housing industry is historically exceedingly slow. In fact, most accessible housing is built by and for people with disabilities on an individual basis. Very little accessible housing is available on the open market and housing opportunities for people with disabilities continue to be extremely limited. Realtors, citing stigma and customization that are too geared to a specific individual, generally view accessible homes as not marketable.

Universal design in housing grew out of the recognition that most of the features needed by people with disabilities were useful to all members of a family. The inclusion of universal features could become common practice. This was recognized in the passage of the federal Fair Housing Act Amendments of 1988, which offer an increased, albeit still low, level of accessibility to people with disabilities in multifamily housing and condominiums. But these same requirements also provide more flexibility and enhanced usability to the general population (see Chap. 35, "Fair Housing: Toward Universal Design in Multifamily Housing," by Edward Steinfeld and Scott M. Shea).

Universal Design Features

Several cities in the United States have in the last few years passed legislation requiring new single-family houses, and some two- and three-unit buildings that receive public money, to be "visitable" (see Chap. 14, "Lifetime Homes: Achieving Accessibility for All," by Richard Best). A few changes in design and construction significantly expand housing options for much of the population when the following are included:

- One no-step entrance
- All doors with at least a 32-in-clear opening
- One half- or full bathroom on the entry level that a person with a mobility aid may enter and use
- Reinforcing installed in bathroom walls around toilet and bathing fixtures for future installation of grab bars

These and other universal features make day-to-day activities possible for some and easier for many. For example, moving day is less arduous in houses with a stepless entrance and wider doors and hallways, the same entrance a person with a mobility aid may use to visit a friend. When well designed, bathrooms with a little extra floor space are perceived as luxurious. The extra space allows the bathroom to be personalized with the addition of a chair, bookcase, towel rack, or étagère, and provides a marketable spaciousness and elegance. Such items can be removed if the space is ever needed to accommodate a family member or friend who uses a mobility aid.

Universal design in housing far exceeds the minimum specifications of legislated barrier-free and accessible mandates. Universal design can be applied to all spaces and elements to create housing that is usable by and marketable to a broad cross section of potential buyers. It addresses buyers' future and often unrecognized needs. It minimizes and may even eliminate tragic experiences such as older adults who are forced to leave the security and comfort of familiar surroundings and a community of friends because they are no longer able to climb stairs in their own home. Being able to remain in one's home through life-stage changes is more psychologically supportive and cost-effective than being forced to move or, due to insufficient resources, to forgo necessary large-scale modifications and "make do" in a potentially dangerous environment. Because features in a universal home have more than one function or can be easily adapted later, costly renovation often may be avoided, fostering a sense of security and helping to maintain a sense of place and community.

34.4 THE NEXT-GENERATION UNIVERSAL HOME

This project, one of the last in which Ron Mace actively participated, represents the collective experience of the staff at the Center for Universal Design, affiliated with the School of Design at North Carolina State University, and identifies specific universal features and design elements that can be incorporated into most houses. The *Wall Street Journal,* while planning a feature article describing future housing trends for the retiree market, sought the Center's assistance to define the house of the future. The Center helped to shift the newspaper's thinking away from the retiree-only niche into a broader and more universal approach. Thus was created the concept illustration showing how housing could evolve to satisfy the needs of a changing market, while retaining features that many home buyers request (see Fig. 34.1).

The term *next generation* recognizes that new home design is not accomplished in radical shifts, but as part of an evolution in thinking. The original illustration appeared in the 14 September 1998 issue of the *Wall Street Journal* (Winokur, 1998) with the express goal of raising awareness of possibilities using current building methods and technology. The house was conceived as a holistic environment (i.e., living spaces designed to benefit multiple users in a flexible format).

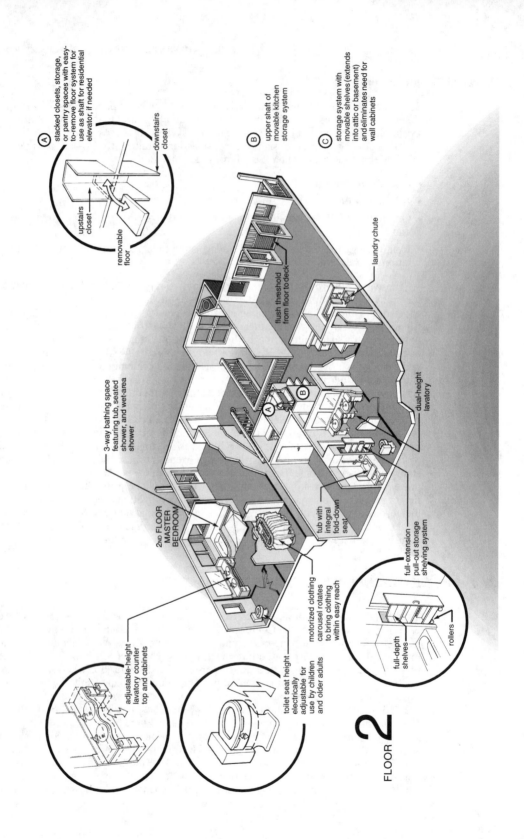

A stacked closets, storage, or pantry spaces with easy-to-remove floor system for use as shaft for residential elevator, if needed

upstairs closet

removable floor

downstairs closet

B upper shaft of movable kitchen storage system

C storage system with movable shelves (extends into attic or basement) and eliminates need for wall cabinets

laundry chute

flush threshold from floor to deck

dual-height lavatory

tub with integral fold-down seat

3-way bathing space featuring tub, seated shower, and wet-area shower

2ND FLOOR MASTER BEDROOM

adjustable-height lavatory counter top and cabinets

toilet seat height electrically adjustable for use by children and older adults

motorized clothing carousel rotates to bring clothing within easy reach

full-extension pull-out storage shelving system

full-depth shelves

rollers

FLOOR 2

34.6

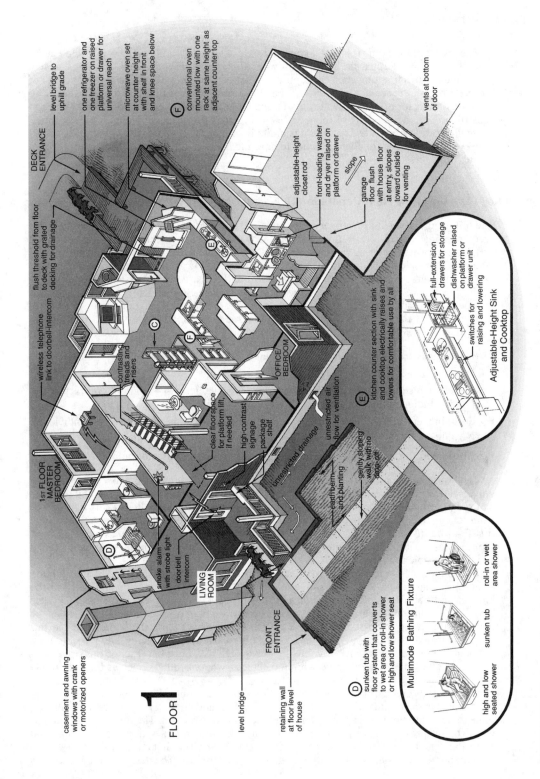

FIGURE 34.1 Next-generation universal home.

casement and awning windows with crank or motorized openers

wireless telephone link to doorbell-intercom

flush threshold from floor to deck with grated decking for drainage

level bridge to uphill grade

one refrigerator and one freezer on raised platform or drawer for universal reach

microwave oven set at counter height with shelf in front and knee space below

conventional oven mounted low with one rack at same height as adjacent counter top

DECK ENTRANCE

1ST FLOOR MASTER BEDROOM

contrasting treads and risers

clear floor space for platform lift, if needed

high-contrast signage

package shelf

OFFICE/ BEDROOM

adjustable-height closet rod

front-loading washer and dryer raised on platform or drawer

slope

garage floor flush with house floor at entry, slopes toward outside for venting

vents at bottom of door

smoke alarm with strobe light

doorbell intercom

LIVING ROOM

unrestricted drainage

earth berm and planting

gently sloping walk with no drop-off

unrestricted air flow for ventilation

kitchen counter section with sink and cooktop electrically raises and lowers for comfortable use by all

full-extension drawers for storage

dishwasher raised on platform or drawer unit

switches for raising and lowering

Adjustable-Height Sink and Cooktop

FLOOR

level bridge

retaining wall at floor level of house

FRONT ENTRANCE

sunken tub with floor system that converts to wet area or roll-in shower or high and low shower seat

Multimode Bathing Fixture

high and low seated shower

sunken tub

roll-in or wet area shower

The layout of the home is based on typical home designs currently being constructed in the United States. Middle-market house plans were analyzed and synthesized to create a design that demonstrates that universal concepts can be effective in popular home styles and do not require futuristic space-age approaches. For example, features like no-step entrances and reinforced walls for grab bars, while easy to implement, run counter to traditional design and construction methods. Many of the features included in this discussion and in the accompanying illustration presently exist, but it is unlikely to find them all included in any one home. However, it is possible for all new homes to have incorporated these and other features within the next 25 years.

Overview of Key Features in the Next-Generation Universal Home

The following discussion is an overview of the key universal design features presented in the accompanying concept illustration. Careful space planning and fixture placement, combined with necessary maneuvering clearances, are reflected in the floor plan. Deliberate and careful attention was paid to details to ensure that this hypothetical design incorporates the full intent of universal design. However, all universal features are not discussed in the narrative, as the focus here is on the critical conceptual approaches. Other features and details not included are possible and also may work. For a more exhaustive listing of universal features as developed by the Center for Universal Design, see the article "Universal Design in Housing" (Mace, 1998). Key features in the next-generation universal home include the following:

- All entrances are stepless with low thresholds.
- Construction incorporates future access to second floor.
- Some elements and features are located at more usable heights—locations contrary to traditional building industry norms.
- Adjustable features are provided throughout, so the environment adapts to the individual.
- Adjustable bathing fixture in each bathroom allows for multiple bathing options.
- Features and elements are included to improve warning and aid in orientation.

The house layout reflects some current trends in house design that support the concept of universal design, such as open and flexible spaces. A first-floor master bedroom and bath can be used as a suite for care of an elderly parent, a relative, an older child making a transition from college to employment, or a family member who is unable to climb stairs. An additional first-floor bedroom can be designed to accommodate a guest or to double as a home office, especially as the largest growing workforce are employees who work from home, estimated at 40 million people in the year 2000.

In comparison to most of the fixed features of the past, this house relies on adjustability to accommodate the widest range of users. Height-adjustable countertops are used in the kitchen and bathrooms. Rotating and adjustable-height shelves maximize storage. Toilets have height-adjustable seats. Bathing spaces allow more than one method of use, such as the multimode bathing fixture, shown in the first-floor bathroom. (See the section later in this chapter entitled "Bathrooms.") All of the illustrated features are technically possible now, although some are not widely available. Features such as adjustable countertops or vertical rotating shelves are available in very limited choices, whereas others, such as the multimode bathing fixture, are not yet being manufactured.

One obvious element of the next-generation universal home is its upper floor. This may run counter to the popular concept of a universally designed home. Two-story homes will continue to be built, due to cost savings in both land and construction. The next-generation universal home provides several invisible and carefully integrated options for efficient and cost-saving modifications to gain access to the second story. The first is a stairway designed to accommodate the installation of either a chair or platform lift. The design of the stairway is critical as lifts often cannot be installed if the stairway turns or if there is insufficient space at the top and bottom landing.

Second, the house contains stacking storage closets, with removable floors, in the center of the house. (See Fig. 34.2.) The flooring, when removed, exposes an elevator shaft. This allows the installation of a residential elevator without disrupting either the layout or the aesthetics of the house and saves significantly on retrofit costs of adding such a feature. (See Fig. 34.3.)

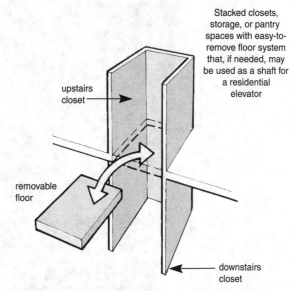

Stacked closets, storage, or pantry spaces with easy-to-remove floor system that, if needed, may be used as a shaft for a residential elevator

upstairs closet

removable floor

downstairs closet

FIGURE 34.2 Stacking storage spaces can be adapted for an elevator shaft.

Additional noteworthy features are described in the following sections. Their inclusion in a careful and thoughtful way adds significantly to the usability of any living configuration. Spaces and rooms described include entrances, kitchen and laundry areas, bathrooms, controls, alarms, home automation, stairs, and storage.

Entrances

Each entrance shown in the next-generation universal home is universally usable. Not only is this philosophically significant, but having more than one means of accessible egress is an important safety consideration. Reflecting conditions common to most houses, the house incorporates three significant methods for creating universal entrances. For each to be successful, some unconventional, although not necessarily difficult, construction details may be required. (See Fig. 34.4.)

Creative landscaping can be an effective way to resolve differences in elevation between the exterior grade and interior floor level. Using soil from excavation for the foundation or having additional soil brought onto the site allows the creation of an earth berm and bridge—a combination landscape and hardscape feature. (See Fig. 34.5.) This strategy, shown at the front entrance, uses soil pushed up against a new retaining wall leaving a "dry moat" just in front of the house foundation for drainage and air circulation. A bridge spans the moat from the gently sloping walk to the house entrance. A planter replaces the handrail on one side of the bridge and acts as a landscape transition from the hard-surface walk to the house porch.

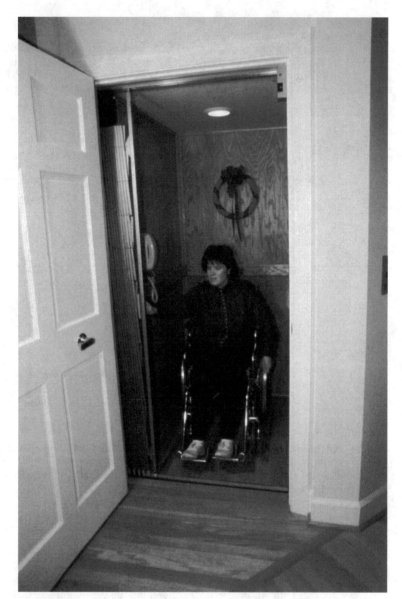

FIGURE 34.3   Residential elevator.

The front porch is covered to provide weather protection and has sufficient maneuvering space for a person using a mobility aid. Other universal features include a package shelf, high-contrast signage, carefully placed night lighting to illuminate the house numerals and keyless lock, lever hardware, doorbell intercom or video entry system, as well as sidelights for viewing approaching visitors.

At the rear entrance is shown another important method. When site conditions allow, the existing grade may be used to advantage. At the back of the house a level entrance is created

FIGURE 34.4   A front entrance similar to that on next-generation universal home.

by extending a bridge from the deck to an uphill point. All decks for the house are set at the same level as the interior floor to eliminate steps and provide smooth transitions. Slotted decking to provide adequate drainage is incorporated at all entry doors. (See Fig. 34.6.)

The relationship of parking to the house is critical and must be adequately addressed for an entrance to be truly universal. For the homeowners' own use, the most important consideration may be level changes between a garage and interior living spaces. In this design, the garage floor at the house entry matches the interior floor level of the house. However, to prevent gas and toxic fumes from seeping into the house, the entire garage floor is sloped gently away from the entry door. Venting is provided at the bottom of the garage doors for additional protection. (See Fig. 34.7.) Still meeting some resistance because many U.S. building codes require a step down from the house to the floor of an attached garage, waivers are being issued in some areas of the country.

Kitchen and Laundry Area

A home of this size may have several people participating in food preparation either together as part of a family activity or solo. Their needs can be accommodated in a kitchen that is still conventional in overall appearance. Features that increase usability for all, such as work surfaces at multiple heights, raised appliances, easy-to-reach storage, and an efficient layout can be seamlessly integrated into the overall design. Key to usability are work surfaces at a variety of heights, ranging from 28 to 42 in. Varying counter heights ensure that the differing needs of users—short, tall, adults, children, seated, or standing—are accommodated. A variety of methods are shown in this kitchen: an adjustable-height countertop, an island with fixed dual-height work surfaces, and a peninsula with multiple functions.

An electrically operated countertop, shown here containing a sink and cooktop, allows the user to set the counter at the exact preferred height for any given task. Knee space is provided under the sink and cooktop to facilitate use by a seated user. Rolling carts can be placed in this open space when knee space is not desired.

(a)

(b)

FIGURE 34.5 Earth-berm-and-bridge concept. (*a*) Conventional entrance with steps. (*b*) Earth berm and bridge used to create a stepless entrance.

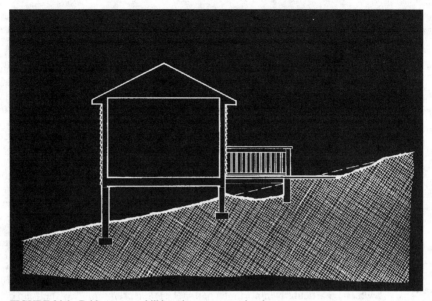

FIGURE 34.6   Bridge to an uphill location to create a level entrance.

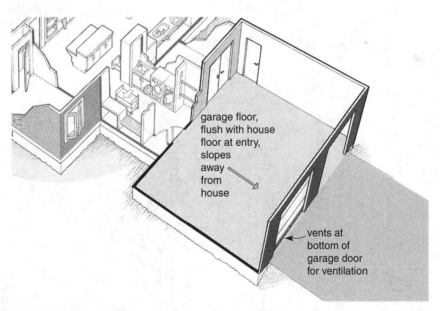

garage floor,
flush with house
floor at entry,
slopes
away
from
house

vents at
bottom of
garage door
for ventilation

FIGURE 34.7   Gently sloping garage floor to eliminate step at entry door. At the entry door, the garage floor is flush with the house floor and slopes away from house. There are vents at the bottom of the garage door for ventilation.

The peninsula between the kitchen and family room doubles as both a lowered work surface and an informal eating area. The secondary, or bar, sink has retractable doors that cover storage space that doubles as knee space. The faucet at this sink is mounted at the side of the basin to minimize reach distance.

Located adjacent to the primary sink, the raised dishwasher is positioned at a height generally easier for all to load and unload. Several appliances in the kitchen and laundry have been elevated to minimize the need for bending and stooping for standing people and to place items inside the appliance within easier reach of wheelchair and scooter users.

Counter edges contrast with the surrounding countertops and help define the counter edge, an important cue for people who have reduced vision. Contrasting edges or strips also can be used to demarcate the sink basin, or the color of the basin could be selected to sharply contrast with the countertop itself.

Ovens are located to maximize adjacent counter space and minimize the need to lift heavy and/or hot dishes and pans. The conventional wall-mounted oven is set with one of the oven racks at the same height as the adjacent work surface, allowing the user to easily slide hot dishes from the rack onto the counter surface. The microwave oven is also set at counter height with a shelf in front, so hot or heavy dishes may be slid easily to a safe resting area.

Cold-food storage is provided in two independent units (i.e., a refrigerator and a freezer placed side by side on an elevated platform so more of the storage is within a comfortable reach). Dry goods are stored in a mechanized vertically rotating shelving system running between floor levels. (See Fig. 34.8.) This system provides a large volume of storage using very little floor area, which is particularly important when space is at a premium. Any shelf can be brought to the optimum convenient height. The programmable system, by entering a number on a keypad, automatically brings the desired item within reach. A more conventional alternative, although not providing as much storage, is a floor-to-ceiling pantry with storage at all reach ranges.

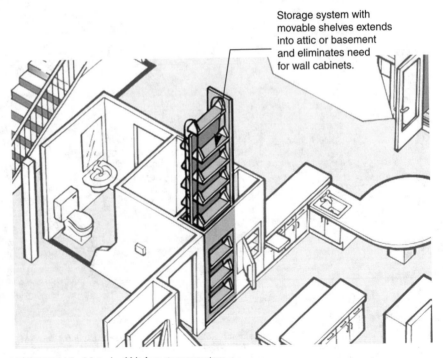

Storage system with movable shelves extends into attic or basement and eliminates need for wall cabinets.

FIGURE 34.8 Motorized kitchen storage system.

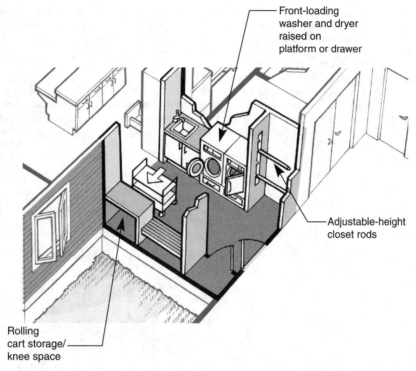

Front-loading
washer and dryer
raised on
platform or drawer

Adjustable-height
closet rods

Rolling
cart storage/
knee space

FIGURE 34.9   Laundry/"mud room."

The laundry area creates a transition space between the garage and kitchen, sometimes called a "mud room." (See Fig. 34.9.) The space is equipped with a bench on which to sit and change shoes, and a coat/recreational equipment storage closet. The washer and dryer are front-loading with front-mounted controls and, like the refrigerator/freezer and the dishwasher, are raised on a platform to reduce the need for bending and stooping. Adjustable-height hanging rods are provided for both dry and wet clothes. Rolling cart storage can be moved easily to a different work location to free up space beneath a counter, if the counter is needed for someone who prefers to sit while performing tasks.

Bathrooms

Having universal features as a standard part of at least one and preferably all bathrooms of any new home expands the likelihood that family members may confidently remain in their homes as long as they wish and their friends may visit regardless of any changes in physical condition. Each bathroom in the next-generation universal home is carefully arranged to allow approach and use by someone who may, at some point, need to rely on a mobility aid.

All walls around the toilet and bathing fixture include reinforcing for the addition of grab bars, not only at the minimal positions as suggested in many design standards, but whole wall reinforcing. A person who may desire a vertical grab bar can screw it in place and be confident that the bar is secure.

The bathrooms contain many of the commonly recommended universal features, such as lever-handle faucets, mirrors over the lavatory that are long enough for both a seated and a standing person to view themselves, and offset tub/shower controls. This concept illustration promotes a high level of usability by incorporating greater flexibility in the design of the plumbing fixtures. All toilets and lavatories are adjustable in height. Even more uniquely, each bathing fixture allows multiple bathing options, offering a person choices to meet their individual need or preference without disadvantaging others.

The first-floor bathroom suite contains a bathing fixture still in the conceptual stage. In some respects, it is like the more traditional bathtub in that the options are provided within the same fixture, and it occupies a similar shape and area. However, the movable segmented floor of the fixture can be positioned or removed to create four bathing options. The user may be submerged in the water, seated on the bench, stand, or use a shower wheelchair. (See Figs. 34.10 and 34.11.)

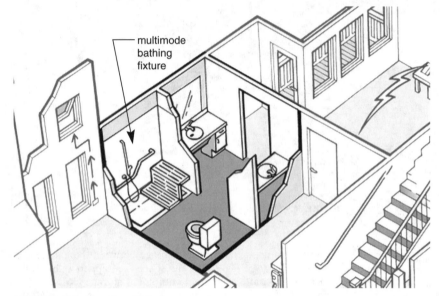

FIGURE 34.10   First-floor full bath.

The second bathroom upstairs off the master bedroom also offers at least four bathing options. Unlike the fixture downstairs with the segmented floor, fixed features are used here to create choices. The area just outside the tub is the shower and a seat spans the back of the tub and extends across the shower area. The shower has a sloped floor to contain water and no curb or lip, thus allowing someone in a shower wheelchair to enter the shower area.

The second upstairs bathroom is shown with what appears to be a standard shower/bathtub. With the addition of a fold-down seat and hand-held shower, the traditional arrangement offers a third bathing option for users who prefer not to stand, but may not wish to get down into the tub. The seat offers an alternative for a person who wishes to sit at a height close to standard dining room chair seat height. This is a particularly helpful feature for some older adults who may have difficulty with balance. (See Fig. 34.12.)

Controls, Alarms, and Home Automation

Careful consideration of the design, selection, and installation of environmental controls, alarms, and home automation systems is critical. If done successfully, it will not only benefit

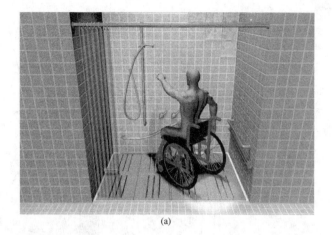

(a)

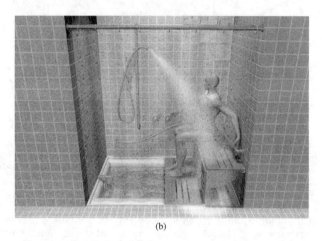

(b)

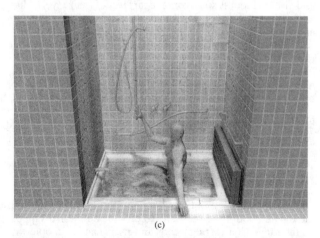

(c)

FIGURE 34.11   Multimodal bathing fixture concept design. (*a*) Roll-in or wet area shower. (*b*) High- and low-seated shower. (*c*) Sunken tub.

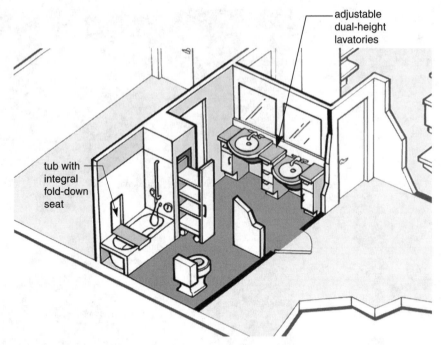

adjustable
dual-height
lavatories

tub with
integral
fold-down
seat

FIGURE 34.12 Second-floor shared bath.

everyone in the household, but it will allow people with perception and communication disabilities to more fully participate in home life. This next-generation universal home illustration integrates several of these concepts into the overall design.

All environmental controls are located in convenient locations and mounted at easy-to-reach heights. Light switches and electrical outlets are within reach of a wide range of people (i.e., standing adults, children, people in wheelchairs, and those who may have trouble bending or stooping). The thermostat is mounted at a convenient height and provides both audible and visual readouts so anyone, regardless of their level of hearing or vision, may control their environment. (See Fig. 34.13.) Smoke alarms and fire detectors emit both visual (strobe light) and audible signals.

Home automation is an important concept that is incorporated into the illustration. Features include a doorbell intercom that can be linked to a wireless telephone for use in remote locations, remote controls for lighting and window operators, and a computer network running throughout the house. A video version of the intercom system can be installed with optional preprogrammed responses to visitors at entry doors. Motion-sensor lighting is incorporated into the exterior landscape to provide safe entry to the house at night and alert the homeowner of a visitor's arrival. A computer can be centrally located and programmed to access doorlocks, regulate mechanical systems, and set appliances.

Stairs

Appropriate inclusion of certain stair features ensures the user's safety. Handrails extend at the top and bottom of the stairs to provide additional support for users who may be unsteady on their feet. Contrasting treads and risers can aid people with low vision or when the lighting is dim. An electrical junction box is fitted with safety lighting to illuminate the stair treads, even

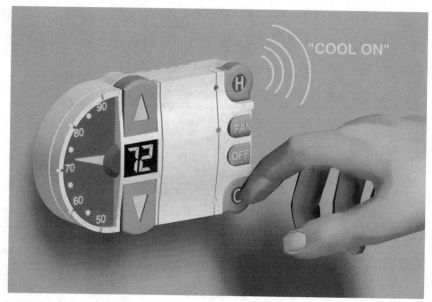

FIGURE 34.13 Thermostat (concept design) with visual, audible, and tactile cues.

when the other house lights are off. And a new and innovative strategy to incorporate at the time of initial construction is an electrical connection that can later be used as the power source for a platform lift. (See the section entitled "Overview of Key Features in the Next-Generation Universal Home" earlier in this chapter for a discussion of access to the upper floor.)

Storage

Storage throughout the house is provided in as flexible a way as possible. Since retrieving items from drawers is generally easier than pulling them out of cabinets, deep drawers on full-extension glides are used extensively so contents are easily seen and reached by all. In the upstairs front bathroom, this concept is interpreted so the entire cabinet pulls into the room. A clothing carousel that rotates to bring the clothing into reach is placed in the master bedroom on the second floor. Where floor space is at a premium, adjustable-height closet rods can be installed in a child's bedroom or in the laundry area. A more elaborate design, mentioned earlier, is the vertical rotating storage as shown in the kitchen.

34.5   CONCLUSION

The concept of universal design is increasingly finding its way into the single-family housing market. Several high-profile demonstration houses, which feature universal design, have been built across the country in recent years. These include Lifestages 99, an innovative modular home design, which was the center floor exhibit at the 1999 International Home Builders' Conference; *Better Homes and Gardens'* Blueprint 2000 House; and the Universal Design House sponsored by the American Association of Retired Persons, which appeared on national media. These projects have ranged from just the inclusion of prominent universal features to true holistic universal designs. However, all of these designs reflect a belief in the marketing advantage of universal design. As a sign of even further progress, Home Planners, LLC, a Division of Hanley-Wood, Inc., recently published the first book of universal home plans, called *Products and Plans for Universal Homes,* which is widely distributed across the country.

As one progressive builder, Ron Wietzel of Cincinnati, Ohio (Uhlenbrock, 1999), states: "The future for universal design looks bright. I see the senior-oriented market adopting the idea first, but as people see how well it works for everyone, universal design will expand to all markets." And with the advent of the Internet and the need to stay competitive, builders may find they must adapt more quickly as potential homebuyers have access to more information. Buyers even now are requesting changes (i.e., changes that include features of universal design).

However, many challenges remain for industry-wide adoption of universal design. While many necessary construction details are not difficult, they are unconventional and, as such, are perceived as expensive, and passed over in favor of more traditional building methods. Product availability is still an ongoing problem. Often, choices for universal products are limited and must be specially ordered. Many manufacturers are unaware of the need for and advantages of universal features. Builders are often unable or unwilling to invest the time into creating new designs, changing over to new construction methods, or making the best product selections. Ultimately, many of these problems stem from a lack of understanding about or even an awareness of universal design.

Furthering the understanding of what universal design is and could be makes projects like the next-generation universal home critical. As the concept becomes more widely understood and embraced, construction techniques will reflect these universal design elements and features, making them commonplace and, soon, the new paradigm. Reportedly, there are only slightly increased costs of incorporating universally designed features during initial home construction, and these, too, may decrease. Products will become mass produced and available at a lower cost for all.

This century promises new innovations in housing resulting from advances in computer technology, manufacturing processes, materials applications, and continuing ergonomic and anthropometic studies. With a thoughtful and knowledgeable application of the concept of universal design and its inherent commitment to the end user, even more effective solutions than those presented here are anticipated.

34.6 BIBLIOGRAPHY

Mace, R. L., "Universal Design in Housing," *Assistive Technology,* 10: 21–28, 1998.
———, G. J. Hardie, and J. P. Place, "Accessible Environments: Toward Universal Design," in W. F. E. Preiser, J. C. Vischer, and E. T. White (eds.), *Design Intervention: Toward a More Humane Architecture,* Van Nostrand Reinhold, New York, 1991.
Uhlenbrock, D., "Homes 2000: Accessibility and Convenience Highlight the Future of Homes in the 21st Century," *Cincinnati Magazine,* 32(4), January 1999.
Winokur, L. A., "Down with Doorknobs! Homes Based on Universal Design Are Attracting More Builders and Buyers," *Wall Street Journal,* p. 10 (Encore section), 14 September 1998.

34.7 RESOURCES

Articles

Mace, R. L., "Universal Design in Housing," *Assistive Technology,* 10.1: 21–28, 1998.
Pace, R. J., "Future Housing Now: The Next Generation Home," The Center for Universal Design *UD Newsline,* 2, 1999.

Trachtman, L. H., R. L. Mace, L. C. Young, and R. J. Pace, "The Universal Design Home: Are We Ready for It?" *Physical and Occupational Therapy in Geriatrics,* 16: 1–18, 1999.

Books

Barrier Free Environments, Inc., *Adaptable Housing: Marketable Accessible Housing for Everyone,* U.S. Department of Housing and Urban Development, Washington, DC, 1987.

———, *The Accessible Housing Design File,* Van Nostrand Reinhold, New York, 1991.

———, and NAHB Research Center, Inc., *Residential Remodeling and Universal Design: Making Homes More Comfortable and Accessible,* U.S. Department of Housing and Urban Development, Washington, DC, 1996.

Dobkin, I., and M. J. Peterson, *Gracious Spaces: Universal Interiors by Design,* McGraw-Hill, New York, 1999.

The Editors of Home Planners, The Center for Universal Design, and The Philip Stephen Companies, Inc., *Products and Plans for Universal Homes,* Home Planners, LLC, Raleigh, NC, 2000.

Peterson, M. J. *Universal Kitchen Planning: Design That Adapts to People,* The National Kitchen and Bath Association, Hackettstown, NJ, 1995.

———, *Universal Bathroom Planning: Design That Adapts to People,* The National Kitchen and Bath Association, Hackettstown, NJ, 1996.

Steven Winters Associates, *Accessible Housing by Design: Universal Design Principles in Practice,* McGraw-Hill, New York, 1997.

Wylde, M., A. Baron-Robbins, and S. Clark, *Building for a Lifetime: The Design and Construction of Fully Accessible Homes,* The Taunton Press, 1994.

Web Sites

The Center for Universal Design
www.design.ncsu.edu/cud

IDEA Center
www.arch.buffalo.edu/~idea

National Resource Center on Supportive Housing and Home Modification
www.homemods.org

Universal Design and Home Accessibility
www.exnet.iastate.edu/Pages/housing/uni-design.html

CHAPTER 35

FAIR HOUSING: TOWARD UNIVERSAL DESIGN IN MULTIFAMILY HOUSING

Edward Steinfeld, Arch.D.
University at Buffalo, Buffalo, New York

Scott M. Shea, M.Arch.
Scheuber & Darden Architects, Aurora, Colorado

35.1 INTRODUCTION

The disability rights movement in the United States has been successful in achieving legislation that guarantees people with disabilities the right of free and equal access to the physical environment. As universal design becomes standard practice, products and environments will be accessible from the beginning; reducing the need for legislating access. However, the legislative process can encourage universal design by providing a mandate for innovation. Although the 1988 Fair Housing Amendments to the Civil Rights Act of 1968 are not as well known as the Americans with Disabilities Act (ADA), they are equally as important, guaranteeing the rights of people with disabilities in the residential setting. By mandating adaptability and minimum accessibility in all new multifamily projects, this law is a major step toward achieving universal design in housing. This chapter describes the Act and its impact, and, using bathroom design as an example, illustrates how the practice of universal design relates to such legislation.

35.2 THE FAIR HOUSING AMENDMENTS

In 1988 Congress passed the Fair Housing Amendments to the Civil Rights Act of 1968. These amendments extended coverage of the fair housing law to people with disabilities and established design and construction standards for all new multifamily housing with four or more units built for first occupancy on or after March 13, 1991. To ensure compliance with the provisions of the Act, the U.S. Department of Housing and Urban Development (HUD) issued the Fair Housing Accessibility Guidelines. The Guidelines contain seven requirements:

- An accessible building entrance on an accessible route
- Accessible and usable public and common use areas

- Usable doors
- An accessible route into and through the dwelling unit
- Environmental controls in accessible locations
- Reinforced bathroom walls for grab bars
- Usable kitchens and bathrooms

The law is intended to ensure that multifamily housing will be accessible to people with disabilities. Each of the seven requirements has specific technical design criteria that provide a basic level of accessibility. It is important to note, however, that as minimum design criteria the Fair Housing Accessibility Guidelines do not serve all of the needs of people with disabilities, particularly inside the dwelling unit. Unless designers go beyond these guidelines, their building specifications will not result in accessible housing for people with severe impairments or those with sensory disabilities. The Guidelines are not as comprehensive as the ADA Accessibility Guidelines, but the latter do not have requirements specifically for dwelling units.

35.3   FAIR HOUSING AND UNIVERSAL DESIGN

Universal design is more than just a new buzzword for accessibility. Terms like *barrier free* or *accessible design* refer to special design features that specifically accommodate people with disabilities. While this solves the accessibility problem, it separates people with disabilities. This "special" approach can create stigma. Universal design does not separate people, but seeks to accommodate everyone at all times with the same design or provide choices for different needs.

A good example of a universal design approach is the Oxo Good Grips line of cooking utensils (see Chap. 49, "Universal Design of Products," by Story and Mueller). The original idea for the handles on the Good Grips utensils came from the practice of rehabilitation therapy. Therapists built up handles to allow people with disabilities to grip them easily. Because of this, the utensils are more comfortable and easier to use by anyone. The handles are larger, softer, and more slip-resistant than those on standard utensils. Utensils with this handle are sold all over the country at competitive prices. Universal design, then, is design for usability and consumer acceptance by everyone.

The idea that homes should be more convenient and safer has been around a long time. It is something everybody wants. In seeking to provide a minimum level of usability in the home, the Fair Housing Accessibility Guidelines move beyond being a set of codes to provide access for people with disabilities. They begin to benefit all households, and become an example of universal design. The fact that the Guidelines apply to all buildings and all units except those that are above or below the ground floor of walk-up buildings reinforces this approach.

The Guidelines do not produce buildings designed specially for people with disabilities. They produce housing that is accessible and convenient to all at a basic level. They incorporate features that allow the units to be easily adapted to an individual's specific needs. For example, the Guidelines specify that reinforcing must be installed in the walls of bathrooms to allow grab bars to be installed easily in the future, should they be needed by the current resident of a dwelling or some future resident. Furthermore, although the Guidelines do not specifically require it, there is an incentive for vanities to be removable or adaptable in order to provide knee space under a sink in the kitchen or bath. Both of these measures save money by reducing modification costs, and provide added value to the design of the home.

Because of its adaptability, housing designed to the Guidelines is attractive to a wide variety of people, including people with chronic pain or temporary disabilities, visitors with a disability, and older people. Incorporating adaptability into the design also reduces the cost of making modifications to accommodate an individual with a disability. This type of change is not limited to a new tenant moving in. As people age in their homes, their ability to carry out many activities of daily living changes. Disability can also occur to previously healthy people as a result of accident or disease. In either case, modifications to the home may be required in order for the occupant to maintain independence.

35.4 IMPLEMENTING THE GUIDELINES IN PRACTICE

It is important to note that compliance with the Fair Housing Act, like the ADA, is a complaint-driven process. Many of the states in the United States have adopted accessibility codes that are generally equivalent to the ADA. This means that compliance with ADA requirements is integrated into the local building regulatory process. The Fair Housing Accessibility Guidelines, however, do not have an equivalent local method of implementation in regulations. This situation may change rapidly in the future as the United States adopts a uniform model building code. A move is underway to include the Guidelines in the model code documents.

While the Guidelines are generally easy to implement in comparison with the far more extensive ADA Accessibility Guidelines, information about the Fair Housing Act and its impact on new construction has not been nearly as easy to obtain as information about the ADA requirements. In fact, there is evidence that architects and builders are not very knowledgeable about the Fair Housing requirements. Some builders who prefer to construct the same design over and over have made few changes in their designs to address compliance. The result has been the beginning of a wave of litigation by the U.S. Department of Justice to raise the profile of the law and ensure that the building industry understands that it must comply. Several cases either have been settled or are in the process of litigation. Settlements can include requirements to renovate common facilities, renovate units already constructed, provide opportunities for owners of condominiums to renovate their apartments if desired, and/or provide funding for other home modifications in the community if renovating the non-complying units is infeasible.

From experience in litigation on compliance with the Act by the authors and others, these are the most common noncompliant features found in the field in newly constructed housing:

- Sidewalks and curb ramps that are too steep
- Inaccessible common facilities like picnic areas, swimming pools, and community centers
- Nonaccessible mailbox areas and garbage/recycling facilities
- Lack of adequate accessible parking
- Inaccessible laundry facilities
- Lack of adequate clearances at appliances in the kitchen
- Inadequate clearance spaces at bathroom fixtures
- Thresholds at doors, especially patio doors that are too high

In states where the accessibility code covers housing, the ICC/ANSI A117.1-1998 Standard is usually used for the technical criteria. Until recently, the requirements exceeded the Fair Housing Accessibility Guidelines. However, starting with the 1999 version of the Standard, both the old requirements and requirements equivalent to the Guidelines are now incorporated as two independent options (Type A and Type B, respectively) that can be adopted by state code officials. It remains to be seen if the states that have already adopted the equivalent of the Type A designs will reduce their current requirements for accessibility in housing to the less stringent Fair Housing level (Type B). Advocates are promoting the idea that the Type B units are the bottom line for all multifamily housing and that the more stringent Type A requirements should be used in addition for a designated percentage of units. However, the building industry is lobbying for only the Type B requirements. Thus, there is currently much conflict and confusion about this issue and the future is uncertain.

35.5   THE IMPACT OF THE GUIDELINES IN BATHROOM DESIGN

The best way to understand the value and limitations of the Fair Housing Accessibility Guidelines is through a demonstration of how they improve usability. To do this, a simulation system was constructed (Fig. 35.1) that allowed the Center for Inclusive Design and Environmental Access (IDEA Center) to evaluate four different bathroom designs (Fig. 35.2). A series of demonstrations was implemented to compare Fair Housing Designs to typical bathroom configurations. The focus was on the bathroom because bathrooms must comply with five of the seven requirements of the Act—more than in any other room in the apartment unit.

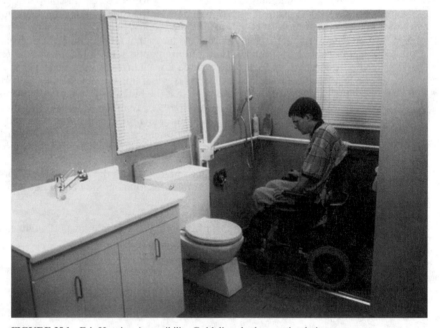

FIGURE 35.1   Fair Housing Accessibility Guidelines bathroom simulation system.

Volunteers simulated use of each bathroom in a standard way. Each person was asked to:

- Enter the bathroom
- Close the door
- Use the grooming center
- Get on and off the toilet
- Use the bathtub to bathe or shower

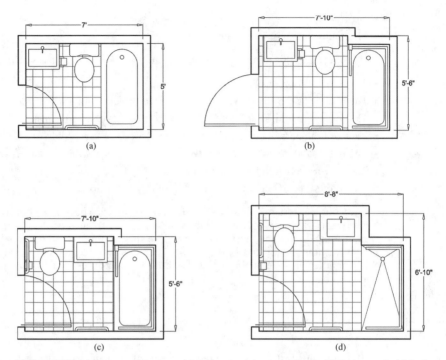

FIGURE 35.2   Demonstration plans: (*a*) Bathroom 1 (noncompliant); (*b*) Bathroom 2; (*c*) Bathroom 3; and (*d*) Bathroom 4.

- Retrieve and store a towel
- Select items in the medicine cabinet or lavatory

Many individuals with different disabilities (and some with no disabilities) took part in these demonstrations. There were old, middle-aged, and young people; some of the volunteers used walking aids, some used wheelchairs, and some simply had difficulty walking. Some had only minor or moderate limitations, while others were more severely disabled. Several volunteers brought family members who usually helped them out in the bathroom. In short, the needs and preferences of a wide variety of people were evaluated. After they were finished using each room, participants explained what they liked or disliked about each of the designs.

Bathroom 1

The first bathroom was a typical bathroom plan similar to what a homebuilder would use to keep the floor area of the bathroom as small as possible. Its features include:

- Dimensions approximately 5 by 7 ft
- Fixtures all located against one wall
- A door that was 28 in wide, opened inward, and had a standard threshold
- A sink that was enclosed underneath by a lavatory cabinet

- No grab bars
- Minimum clearances between fixtures

As a result of this minimum configuration, there were many accessibility problems:

- The narrow door was difficult to pass through.
- The room was too small to close the door for privacy with a wheelchair user inside.
- The room had limited space for caregivers to assist others in transfers.
- Lack of grab bars made the toilet and tub difficult or impossible to access.
- The faucet and medicine cabinet were difficult or impossible to reach.
- The mirror was often too high or too low.
- The limited space between fixtures created a tight fit for all.

Any one of these problems could mean the difference between a bathroom that is usable by a person with a disability and one that is not. When such problems appear in combination, it typically means that a bathroom like this is completely inaccessible. Furthermore, people without disabilities find bathrooms of this type to be cramped and difficult to use. Such facilities provide the basic fixtures, but they do not respond to what people really need and prefer.

Bathroom 2

Bathroom 2 meets the minimum requirements of the Fair Housing Guidelines. Changes from the first bathroom include:

- The room was 9 in longer and 6 in wider (8 ft^2 larger).
- The door opened outward and was 6 in wider.
- Grab bars were added on the walls in the tub area and next to the toilet.
- The vanity cabinet was removed to provide knee clearance (this was not a requirement).

As a result of these changes, our participants found that the bathroom was much easier to use. The wider door was easier to pass through, and with the vanity removed, knee space under the sink allowed easier access to the faucet and medicine cabinet. The grab bars also provided much greater access to the tub and toilet, and participants felt safer using them.

Although Bathroom 2 was an improvement over Bathroom 1, there were still some problems. Despite the larger size, the bathroom was still too small to turn a wheelchair around when the vanity was in place. The removal of the base cabinet in a bathroom vanity is one of the most vital features of any accessible bathroom. With the vanity cabinet removed, the area under the sink can be used as maneuvering space for wheelchairs. Not only can people who use wheelchairs get closer to the sink and medicine cabinet, but it is also easier to close a door that swings in and to turn around inside the bathroom. This could be critical for someone who can only transfer to a toilet from one direction. When the cabinet is removed, it is important to cover the plumbing either with a panel or insulation in order to prevent burns and other injuries.

Another problem with this bathroom was that it was too small to maneuver a wheelchair alongside the tub, forcing people to do a 180-degree pivot transfer. This is a difficult and sometimes dangerous maneuver to perform. Many people with disabilities cannot do this without the help of a caregiver, and the size of this bathroom made it difficult for someone to provide safe assistance. While generally more accessible than Bathroom 1, Bathroom 2 is not a good option for people with more severe disabilities.

Bathroom 3

There are many ways to meet the Fair Housing Guidelines. Bathroom 3 demonstrated a different approach than Bathroom 2. While the bathroom was exactly the same size, the arrangement of fixtures was slightly different. The sink was located next to the tub, which allowed enough clearance for the door to open inward. This is preferable if a bathroom opens off a perpendicular corridor. The new positions of the sink and toilet also provided some additional benefits:

- It was easier to close the door and turn wheelchairs.
- There was more room in the toilet area.
- The wall next to the toilet was long enough for a standard grab bar, which everyone found useful.
- As in Bathroom 2, the wider door made it easier for everyone to use.

While Bathroom 3 was clearly more accessible than the previous examples, there were still problems. For example, one participant found the grab bar next to the toilet too short. He had to use the door handle for support so that he could reach the bar. If the bathroom had been wider, there would have been room for a longer grab bar.

Bathroom 4

By increasing the length of the bathroom another 10 in and the width 15 in, a very accessible space was provided. The improvement in usability for all of our participants was considerable.

- There was plenty of space to maneuver and close the door, even for electric scooters and wheelchairs with long wheelbases.
- There was enough room for a parallel approach to the tub. This is safer for both the individual and caregivers.
- There was enough space for a longer grab bar at the toilet, and more elbow room.
- The side-mounted medicine cabinet was easier to reach.
- The roll-in shower was much more convenient for bathing.
- Lever-style handles made it easier for all users to open the door.

Bathroom 4 provided one feature that significantly increased the usability of the bathroom—a sink recessed in an alcove. Not only did this provide greater clear floor space for general wheelchair maneuvers, but it also provided the room for a person using a wheelchair to pull up parallel to the tub. This enabled easier side transfers because people can slide instead of standing and pivoting. It also makes it easier to reach the controls and adjust the water before getting into the tub. There is more room for caregivers, and medicine cabinets can be side-mounted for easier reach.

Summary of Findings

It is clear that the Fair Housing Guidelines result in more usable bathrooms than are provided for in typical designs.

- The wider door provides a more convenient entry into the bathroom. In our focus groups, even people without disabilities preferred wider doors.
- Knee space under the sink allows easy access to the medicine cabinet and faucet.

- The usable floor area under the sink and the wider room improves maneuvering inside the bathroom, increasing independence.
- With grab bars, the toilet and tub are easier and safer to use by people with and without disabilities.

Adding slightly more space, and incorporating some innovative products, can provide even greater accessibility for little additional cost.

- The added floor space makes the bathroom fully accessible for people with severe disabilities.
- More space is definitely a great help for caregivers as well.
- A recessed sink improves access to the tub.
- The recessed sink also allows side access to the medicine cabinet. Consumers find this convenient and desirable. They also like the additional mirror.
- A roll-in shower instead of a bathtub makes bathing and transfers easier because it is not necessary to climb over a tub rim.

These demonstrations show that extra attention to bathroom design can improve access and convenience for everyone. More space is not always the answer. Simple changes like providing knee space and alternative fixture layouts can significantly improve the safety and usability of a design. This is what universal design is all about.

35.6   *UNIVERSAL DESIGN IS COST-EFFECTIVE*

The National Association of Home Builders (NAHB) argued that the cost of meeting the Guidelines is significant. When the Guidelines were being formulated, the Association mounted a substantial lobbying campaign to reduce the "burden" of the Act by watering down the regulations. At the time, NAHB was aided in its efforts by several other organizations, including national disability organizations like the Paralyzed Veterans of America,

design professional organizations like the American Institute of Architects, and real estate and financial interests like the Mortgage Bankers Association (NCCSCI and NAHB, 1989; Paralyzed Veterans of America, 1990). Grass roots disability advocates, mobilized by the Disability Rights Education Defense Fund, developed a countercampaign, and the result was a negotiated compromise.

It doesn't necessarily cost more to meet the Guidelines. In response to the opposition to the Guidelines, the U.S. Department of Housing and Urban Development, the agency responsible for implementing the Fair Housing Act, commissioned a careful study that demonstrated that the additional cost for new construction is minimal (Steven Winter Associates et al., 1993). The HUD-sponsored study reported estimated cost increases of 0.07 to 0.85 percent, with an average of 0.34 percent. This nominal increase in costs can be offset easily by cost reductions in other features.

There are many ways to keep the cost of meeting the Guidelines low. For example, by swinging the door out, less floor space is required in an accessible bathroom. The out-swinging door improves accessibility without any additional cost. Although the Guidelines do not require knee space under a vanity, a bath-

FIGURE 35.3   Pivoting grab bars and adjustable toilet.

room with a closed vanity has to be 18 in longer. This is because the Fair Housing Accessibility Guidelines specify that a sink without knee space underneath must be centered in a clear space 30 by 48 in wide to allow a parallel approach. Sinks that do have knee space underneath however, require only a 30-in-wide clear space. This allows the bathroom to be 18 in shorter and about 8 ft^2 smaller—a significant cost savings. An adaptable vanity allows the use of the smaller clearance, even when the knee space is not open from the start.

Cost-effectiveness is not always measured in first costs. The advantages of accessibility, such as the elimination of stairs to first-floor units, have economic value in themselves. Moreover, features that add value to the design can improve marketability. For example, the lack of stairs to unit entrances is preferred by most tenants; and lever handles on faucet controls and doors make them easy for everyone to use (particularly people with limited gripping abilities). Adding universal design features beyond the letter of the law can increase value even further. An example is grab bars at the toilet that lift out of the way (see

FIGURE 35.4 Adaptable vanity: (*a*) Closed position, before adaptation; (*b*) Open front, base extends for child step; (*c*) Base removed for wheelchair access; (*d*) Sides removed to increase turning radius.

Fig. 35.3). When such bars are used, the toilet can be located away from a side wall. The bars do not get in the way of others using the toilet or the bathroom. For many people (especially older individuals), these bars provide a greater level of usability than bars on side walls.

Universal design makes good marketing sense. In a highly competitive world, producers have to reach out to the broadest market. Safety and convenience are things that everyone—and especially the older consumer—wants. As the population ages, the market for accessibility will increase substantially. Older people have the largest disposable income of any segment of the population, and housing is the largest expenditure they make. In a rapidly aging society, accessible housing will have lasting value.

Developers should anticipate changes in housing preferences now; the housing being built today will be more marketable in the future if it is built to be accessible from the start.

FIGURE 35.5   Adjustable shower.

35.7   *UNIVERSAL PRODUCTS*

One of the most immediate ways that housing can be made more universal is through the inclusion of innovative products. There are many products that can help create housing that everyone can use with greater ease and safety now and in the years to come.

- The adaptable vanity (Fig. 35.4) is an innovative approach to accessibility. A prototype design created by the IDEA Center in conjunction with KraftMaid has several built-in accessibility features. With the base removed and doors open, knee clearance for wheelchair users is provided. The base could also be used as a step for children. With sides and door removed, the vanity is even more accessible. A wheelchair can be turned under it. If it is wall-hung, the counter height can be adjusted to an individual's needs.

- An adjustable shower (Fig. 35.5) is a feature that everyone can appreciate. A tall or a short individual can set it for their best height. Those who are unable to stand can hold the handle and use it independently. Caregivers can use it to bathe someone without getting wet themselves.

- Tubs made from a new soft material (Fig. 35.6) help to prevent injury. This is a feature that will please families with children as well as older adults.

FIGURE 35.6   Soft bathtub.

- Adjustable lavatory sinks (Fig. 35.7) provide more flexibility and accessibility. They make it possible for people of widely different statures to use the same sink without difficulty. As the baby boomer generation advances through middle age, products like this that reduce back pain will be very popular.

- Adjustable toilets provide convenience for residents of all ages, from children to elderly family members. Although not yet available to the mass market in the United States, such toilets are now available in Europe and Japan (see Fig. 35.3).

In addition to products, interior design concepts can also provide universal design. For example, a shower stall—particularly one without a curb—is one of the most accessible bathing environments. When an apartment has two bathrooms, equipping one with a curbless shower stall improves marketability.

FIGURE 35.7 Adjustable lavatory.

35.8 CONCLUSION

The Fair Housing Act can create more enabling environments for people with disabilities, and at the same time, increase safety and convenience for all people. The provisions of the Act are not costly to implement. Although initial experience with the Act demonstrates that architects and homebuilders still have a lot to learn about how to meet its provisions, it can be expected that over time they will incorporate accessibility as a normal part of doing business. HUD is, as of this writing, planning a technical assistance program to improve knowledge about the Fair Housing Accessibility Guidelines in the building industry.

As our population ages and more people with disabilities enter the housing market, building accessible housing now makes financial and marketing sense. Building universal design into housing today will ensure its marketability in the future and reduce the need for expensive modifications as people's abilities and needs change. It is clear that by following the Fair Housing Accessibility Guidelines, greater access and ease of use by almost everyone can be achieved. Perhaps more important, however, is the Fair Housing Act's role as a catalyst for a new thinking in the future. Through bringing issues to the attention of designers and adopting an approach that provides access through flexibility, the Guidelines are educating designers about the Principles of Universal Design. By continuing down the road opened by the Fair Housing Accessibility Guidelines, designers and architects will establish universal design as the standard in the future.

35.9 NOTES

The work that provided the basis for this chapter was supported by funding under a grant with the U.S. Department of Housing and Urban Development. The substance and findings of the work are dedicated to the public. The authors are solely responsible for the accuracy of the statements and interpretations contained in this chapter. Such interpretations do not necessarily reflect the views of the government.

This chapter is based on a videotape entitled *Fair Housing Means Universal Design,* produced by the Center for Inclusive Design and Environmental Access (IDEA Center). The Center produces many other additional resources on fair housing with an emphasis on universal design. Center for Inclusive Design & Environmental Access, 378 Hayes Hall, SUNY at Buffalo, Buffalo, NY 14214, 716-829-3485, (fax) 716-829-3861, idea@ap.buffalo.edu, www.ap.buffalo.edu/~idea.

The *Cost of Accessible Housing,* available from the U.S. Department of Housing and Urban Development, describes in detail how to meet the provisions of the Fair Housing Act Accessibility Guidelines. A series of before-and-after case studies using real-world examples shows how to design sites, buildings, and units to meet the provisions of the Act. Detailed cost analyses accompany the case studies.

35.10   BIBLIOGRAPHY

Anonymous, *Recommendation for Implementation of the Handicapped Accessibility Provisions of the 1988 Fair Housing Amendments Act,* National Coordinating Council on Spinal Cord Injury and National Association of Home Builders, Washington, DC, 1989.

————, *The Impact of Proposed Fair Housing Guidelines on Multi-Family Housing Design and Costs,* Paralyzed Veterans of America, Washington, DC, 1990.

Steven Winter Associates, Inc., Tourbier and Walmsley, Inc., E. Steinfeld, and Building Technology, Inc., *The Cost of Accessible Housing,* U.S. Dept. of Housing and Urban Development, Washington, DC, 1993.

CHAPTER 36

PROGRESSIVE HOUSING DESIGN AND HOME TECHNOLOGIES IN CANADA

Mary Ann Clarke Scott, M.Arch., M.A.I.B.C., Sylvia Nowlan, B.Sc., and Gloria Gutman, Ph.D.
Gerontology Research Centre, Simon Fraser University, Vancouver, British Columbia, Canada

36.1 INTRODUCTION

Universal design is an inclusive design philosophy that spans age, gender, and ability. Universal design is the design of products and environments to be used by all people, to the greatest extent possible, without the need for adaptation or specialization (The Center for Universal Design, 1997). The overall goal of universal design is to minimize barriers to individual independence by reducing or eliminating the need for special adaptation to the physical environment. The seven fundamental Principles of Universal Design (see Chap. 10, "The Principles of Universal Design," by Story) primarily address the universality of the usability of a particular design. It is important to note that the practice of design involves much more than usability issues; these seven design criteria for seniors draw attention to the need to address livability issues as well.

This chapter highlights projects and design solutions that have been used recently in Western Canada to address the special needs of different user groups. While they do not, arguably, satisfy all of the criteria and goals of universal design, they do illustrate the state of the art in housing design in Canada and draw attention to emerging trends and future directions. The chapter concludes by pointing out that changing attitudes and acceptance of new technologies will facilitate the trend toward integration.

36.2 BACKGROUND AND OVERVIEW

It is valuable to review and discuss trends in seniors' housing, integrated community housing, as well as housing built for special populations because the universal design philosophy is relevant to all of these approaches and applications. Its fundamental principles are inclusive, thereby promoting the independence, safety, and security of all residents, regardless of their characteristics now or in the future.

Many other design philosophies complement or coexist within the greater universal design framework, such as transgenerational, life-cycle housing, accessible, barrier-free, Flexhousing

(CMHC, 1997), and user-friendly (Simpson, 1998) design. All of these have a common design objective, which is to promote independence by removing and minimizing barriers in the built environment, and contributing to safety and ease of use. As James Pirkl points out, *transgenerational design* is also referred to as design for the lifespan, because it is sensitive to the changes that occur as people age (Pirkl, 1994). However, transgenerational design does not speak to the needs of those with disabilities that result from injury or illness. The Center for Accessible Housing defines *accessible design* as "design that meets prescribed code requirements for use by people with disabilities" (Center for Accessible Housing, 1991). *Barrier-free design* is similar in that it seeks to remove physical barriers to those with various disabilities. *User-Friendly Homes' design* concepts embody many of the principles already discussed and outline many practical guidelines and building techniques that also take into consideration standard building practice and cost. These user-friendly characteristics allow planning for future needs as well as current ones, thereby allowing housing projects to be designed to meet the needs of the widest range of users. The language used, however, is inclusive and nonstigmatizing, broadening its market appeal.

According to Vanderheiden, there are three ways to intervene to enhance an individual's capabilities: (1) change the individual, (2) provide the tools that he or she can use, or (3) change the environment (Vanderheiden, 1997). In many cases, it is extraordinarily difficult and/or unethical to entertain changing the individual. Providing the tools can include assistive technologies to support an individual in his or her built environment. The final option, changing the environment, involves designing the built environment to better meet the needs of the individual. Universal design and user-friendly design seek to provide an open-ended and flexible environment that adapts to meet different and changing user needs. A universally designed environment minimizes the need for either assistive devices or adaptation of the individual user's behavior.

36.3   CANADIAN CONTEXT

In Canada, accessibility in building design is legislated only to the extent that the National Building Code of Canada and related provincial and municipal building codes require (see Chap. 16, "Accessibility Standards and Universal Design Developments in Canada," by Mallory-Hill and Everton). Section 3.7 of these codes, relating to accessibility, applies to the design of new buildings excluding the following: some buildings of less than 600 m^2 (6459 sq ft); all dwelling units and boarding homes; apartment buildings except for access to the street, to parking, and to an elevator; shops, stores, and supermarkets with an area of less than 50 m^2 (538 sq ft); and high-hazard industrial occupancies (City of Vancouver, 1998). As can been seen, residential design and construction is omitted from this legislated requirement.

There are many Canadian organizations, however, that advocate universal design and provide information and education about it. Examples include such university-based centers as the Gerontology Research Centre at Simon Fraser University and the Universal Design Institute, formerly the Canadian Institute for Barrier-free Design, at the University of Manitoba, and departments within provincial Ministries of Housing, such as the British Columbia Ministry of Municipal Affairs and Housing's Accessibility Program.

The Ministry of Municipal Affairs and Housing's Accessibility Program provides information to planners, municipalities, and the public through regular newsletters, an *Annotated Bibliography on Universal Design* (BC Ministry of Municipal Affairs and Housing, 1998), coordination with public libraries, and organization of events such as Access Awareness Day.

As part of the Faculty of Architecture at the University of Manitoba, the Universal Design Institute is actively involved in providing educational support to five departments of the Faculty and to the broader community. It engages in research, training, and advocacy on behalf of accessibility issues in Canada. It also has produced *Access: A Guide to Accessible Design for Designers, Builders, Facility Owners and Managers* (Finkel, 2000) to assist with interpretation of the National Building Code of Canada.

The Gerontology Research Centre at Simon Fraser University in Vancouver conducts research in several thematic areas, one of which is aging and the built environment. Applied research on aspects of universal design as well as other environmental design and environment-behavior research questions is conducted in the Dr. Tong Louie Living Laboratory, a full-scale simulated residence whose flexible wall system can be reconfigured in various ways. The Living Lab is also a venue in which assistive devices and other products and equipment can be tested with volunteers from the community who are elderly or disabled.

36.4 SENIORS' RESIDENCES

There is a broad range of housing options available to older Canadians, offering varied tenures, physical designs and social features, services, supports, and amenities. When appropriately designed for changing needs, preferences, and technologies, most housing types can provide supportive environments that promote independent living and facilitate aging in place.

The key terms or aspects of universal design philosophy that best characterize effective housing for the elderly are *flexibility* and *adaptability*. This applies equally to the functional programming, detailed design of the physical environment, as well as to the support services that are included, such as social activities, health care services, and assistance with activities of daily living.

To provide an environment that is suitable for aging in place and for individuals with varied physical capabilities, the physical and social environment must be adaptable to the changing and variable needs and preferences of end users. Specifically, there is a concern that many accessible design solutions aimed at people who use wheelchairs can pose problems for frail elderly people. A current research project being conducted at Simon Fraser University's Dr. Tong Louie Living Laboratory addresses this question by testing a range of accessible design solutions for the residential kitchen and bath with people who vary in mobility and strength.

As the Canadian population rapidly ages and the seniors' housing market increases, it is important to understand market demands in order to meet the needs and preferences of older consumers. Consultation with end users can provide insight into the design features that older people value as the physical environment begins to pose greater challenges to independent living.

Current housing options for older Canadians illustrate several important features that facilitate aging in place and meet a wide range of user needs. While not traditionally viewed as aspects of universal design, such features include a range of tenure options and support services, as well as flexible, adaptable, and supportive physical design.

Seniors 75 years and over, who are the major occupants of specially targeted seniors' housing, benefit not only from an accessible, flexible, and supportive physical setting, but also from a supportive social network in a safe, secure, and stable neighborhood. These work together to help them maintain continuity of role and self-identity as changes to health and mobility over time challenge their independence. Also important are services such as meal programs, housekeeping and building maintenance services, and social activities.

In addition to resident involvement in management issues and autonomy in affairs of daily living, factors that contribute to a sense of safety and security for older people include: the quality of the neighborhood, 24-hour security systems, on-site management, and closed-circuit TV channels with resident-controlled building access.

Other physical design features that are important for older residents include:

- Nonslip flooring in the bathroom, showers with seats, low thresholds, safety bars in appropriate locations, wall reinforcing for flexible grab bar installation, generous and accessible medicine cabinets and hand-held flexible shower nozzles

- Kitchen design features such as pull-out drawers, wall ovens at the appropriate height with heat-resistant surfaces beneath, full-height pantry storage, lowered light switches, raised electrical outlets, and appliance and environmental controls with large and illuminated lettering
- Lever door, window handles, faucet handles, and easy-to-use light switches
- Maintenance-free finishes and details inside and out
- Sheltered balconies or solaria with glassed-in and operable enclosures
- Wide hallways and doors, flush thresholds, shelving at the entry to units, storage rooms with adequate space and electric outlets for a variety of uses including appliances and scooter battery charging
- Large windows with lower sills; ample overhead lighting in every room with supplementary task lighting; user-friendly door, window, cabinet, plumbing fixture, and appliance hardware
- Sheltered and well-lit building entries and accessible landscaped courtyards and gardens with ample seating, gentle gradients, adequate lighting, and shelter from excesses of sun, rain, and wind

36.5   ACCESSIBLE GROUP LIVING PROJECTS

William Rudd House, although, strictly speaking, is a licensed care facility, demonstrates a trend toward specially designed residential settings for people who might otherwise have to live in larger, institutional environments (Fig. 36.1). This unique residence for younger adults with disabilities has separate bedrooms for 11 permanent residents, plus one room for respite use. Each room has its own accessible bathroom and shares a shower with another. Bedrooms are tastefully decorated, have adjustable storage for personal belongings, and a tracking and sling system that allows residents to transfer independently in and out of bed. A cheerful and spacious dining room, living room with fireplace, guest lounge, and staff office/storage area complete the

FIGURE 36.1   William Rudd House—exterior.

8,600-square-foot building. A Medinorm® snack area enables the raising and lowering of counters and provides access to cupboards for residents who use wheelchairs (Fig. 36.2).

FIGURE 36.2   William Rudd House—Medinorm® snack kitchen.

This single-story freestanding residence, located adjacent to Queen's Park Hospital, a 600-bed extended-care facility, includes landscaped pathways and outdoor patio areas. Residents have access to many services available at the hospital. These include hairdressing, dentistry, podiatry, pastoral care, social work, occupational therapy, diet counseling, physiotherapy, gym and hydrotherapy pool, plus hospital gift shop.

Consistent with the philosophy of William Rudd House, residents have formed a partnership with the staff, the community, and Queen's Park Hospital to shape the care environment within the residence.

Benefits from this partnership include:

- Participation in determining and directing their own care
- Social interaction in a nongeriatric, nonhospital setting
- An environment in which to practice acceptable social behavior
- Participation in an enriched program where ability, physical, mental, and/or social functioning is maximized
- An environment where family and friends feel welcome, but where the resident has a choice in the degree of family involvement
- An increase in the amount of time a resident with deteriorating physical abilities can remain in a noninstitutional setting

William Rudd House residents, whose average age is 45 years, exhibit disabilities associated with spinal cord fractures, multiple sclerosis, muscular dystrophy, cerebral palsy, and Parkinson's disease. They are encouraged and supported to actively further their education or to work outside the home. Participation in community, recreational, and social activities is also encouraged and facilitated by a wheelchair van donated by the community.

In response to a request from the Ministry of Health and in consultation with its Continuing Care Division and the Multiple Sclerosis Society, the proposal for William Rudd House was prepared in 1991 by the Pacific Health Care Society. At that time, the society operated Queen's Park Hospital in New Westminster and Felburn Hospital in Burnaby. In April 1996, it became one of the partners of the Simon Fraser Health Board.

The project was consistent with Queen's Park Hospital's mission "to support the independence and well-being of seniors and the disabled, living in its facilities or in the community." Under the leadership of Board Chair Marilyn Cassady and President/CEO Keith Anderson, the planning group wanted to create a noninstitutional, homelike atmosphere, and yet keep within the cost parameters specified by continuing care. At the same time, they needed to ensure that the house had access to a broad range of professional support services. A key concept was that this housing option had to be affordable.

William Rudd House was officially opened in October 1995. It demonstrates that an affordable model of residential-based care can be provided that improves quality of life for younger adults with deteriorating physical abilities who might otherwise have to live in a larger institutional environment. While younger people with severe disabilities have been housed in unlicensed group homes in residential neighborhoods, as well as in apartments in integrated buildings in previous years, funding for these options, available through other ministries (such as Social Services) and other programs within the Ministry of Health, have since disappeared. The determined vision, available land, and resources of this sponsor group are responsible for the reintroduction of a smaller residential solution for these clients, despite its location on a hospital site. In addition, other housing (for seniors) has been and continues to be developed on the site, creating a more mixed-use community than previously existed. This dilemma between the perceived negative or inappropriate image of institutional settings continues to be a challenge in a context where funding for public health care remains available and, increasingly, funding for public housing is not.

Capital construction costs of approximately C$1.39 million were funded on a 60-40 basis by the Ministry of Health and the Greater Vancouver Regional Hospital District. The Ministry of Health covers operating costs.

In its naming, this residence commemorates the significant contribution made by William Rudd to the development of community-based health services in the Simon Fraser Health Region. Mr. Rudd was a former Chairman of the Board of the Pacific Health Care Society and was a prominent corporate citizen in New Westminster. He had a special understanding of the unique needs of younger adults with disabilities.

36.6 MULTIFAMILY HOUSING PROJECTS

Some of the most innovative and unique housing options incorporate universal and user-friendly design philosophies to promote community integration and independent living, and all that independence implies (safety, security, and environmental adaptability and flexibility.) Each of the following examples incorporates design innovations and state-of-the-art technologies.

Lions Millennium Place is the result of a partnership between the British Columbia Paraplegic Association, the Lions Paraplegic Lodge Society, and British Columbia Housing, with collaboration from many other stakeholders, including the G. F. Strong Rehabilitation Centre. The project, which opened in March 1999, provides 39 rental units targeted to low- to middle-income families and individuals, as well as people with spinal cord injuries (Fig. 36.3).

Of the 39 units, 12 are fixed-term rentals for those making the transition from hospital back to their community, and as many as 6 are for residents with high-lesion spinal cord injuries.

Many user-friendly and universal design features are incorporated into Lions Millennium Place. The project is prewired to allow home automation technologies (i.e., autodrapes; home security; remote control of doors, TV, stereo, lights, fireplace), which promote independence,

FIGURE 36.3   Lions Millennium Place—exterior during construction.

safety and security (Fig. 36.4). Other design features include safety bars, offset shower/bath faucets (Fig. 36.5), flush thresholds, low windowsills, frontset counter A/C outlets, lowered light switches, and raised A/C outlets.

In addition to the features previously mentioned, Lions Millennium Place is designed with mobility in mind, incorporating flush thresholds, wider doors and hallways, and adequate turning and transfer space in all rooms.

Although there are no preprogrammed assisted living services, the physical design and layout of the project accommodates any of the required assistance that individual tenants may arrange. A few units include spare bedrooms located between two adjacent units, designed to accommodate a live-in care aid.

St. George's Place is another integrated housing project that has implemented several universal and user-friendly design strategies (Fig. 36.6). St. George's Place evolved from a partnership between the Anglican Church and the BC Rehabilitation Foundation and their common desire for a socially progressive housing project that could facilitate supportive community living.

According to Bert Forman, project facilitator, St. George's Place attempts to house people with disabilities "within a socioeconomically and diagnostically integrated family housing project." Residents are, therefore, intended to be both low income (subsidized) and middle income (able to pay market rents), singles, couples, seniors, and families with children, as well as include both able-bodied tenants and those with a wide range of physical disabilities. The project consists of 19 units. Sixteen were designed to accommodate residents who use a wheelchair. Although St. George's Place does not offer support services, the objective of the project is to provide a physical environment that facilitates individual supports. However, it is the responsibility of the individual resident to secure the necessary support services.

St. George's Place offers a range of floor plans and unit sizes. Of the 19 units, there are 9 one-bedroom units (64.4–66.1 m^2, or 693–712 ft^2), 6 two-bedroom units (72.3–76.5 m^2, or 779–824 ft^2), 3 three-bedroom units (96.9 m^2, or 1043 ft^2), and 1 four-bedroom unit (125.7 m^2, or 1353 ft^2).

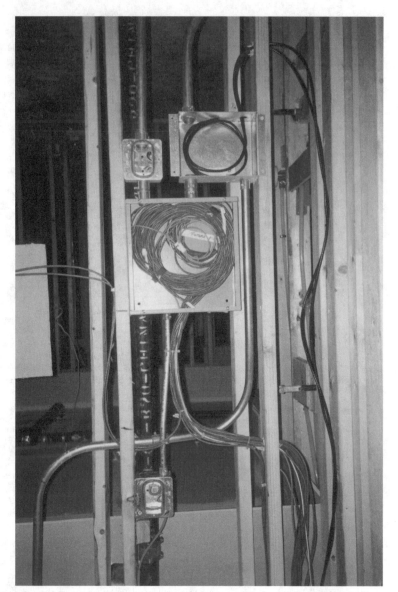

FIGURE 36.4   Lions Millennium Place—Node Zero panel and smart wiring.

Because St. George's Place aims to provide integrated community housing, 8 units are rented at near-to-market value, and 11 units are available to low-income individuals via the BC Benefits Shelter Allowance program (rent is 30 percent of their gross household income).

Design features vary from unit to unit and include: wall-mounted ovens, front-mounted A/C outlets on counters, flush thresholds and showers, knee space under sinks, raised A/C outlets and lowered switches, offset shower/tub controls, lowered medicine cabinets and low-pile carpet to assist people who use wheelchairs (Fig. 36.7).

FIGURE 36.5 Lions Millennium Place—bath and shower with offset controls, hand-held shower nozzle, and safety bars.

Zajac Norgate House is a third example of housing that integrates a variety of users, including seniors and adults with multiple sclerosis in an independent living setting (Fig. 36.8). Like the previous two cases, it incorporates a number of universal and user-friendly design features. The initial impetus for the design of Norgate House was the designer's personal experience dealing with her husband's rapidly progressing multiple sclerosis. Her frustration about being unable to care for her husband at home, and the inappropriate institutional (geri-

FIGURE 36.6   St. George's Place—exterior.

atric long-term care) environment where he eventually lived out his life, fueled her determination to create a more appropriate, supportive living environment.

The project, although designated seniors' housing, permits rental by residents under 55 years of age if they have a disability. BC Housing Management Commission provides a rental subsidy for 10 units, while the remaining units are rented at market rates, starting at about

FIGURE 36.7   St. George's Place—typical unit kitchen.

FIGURE 36.8   Zajac Norgate House—exterior.

C$900 per month. All 36 units are one bedroom, and each unit is designed with adaptable features to meet the diverse and changing needs of tenants. The building is designed to be completely barrier-free, based on universal and CMHC Flexhousing principles.

All units have wheelchair turning radii at the entry, kitchen, bathroom, and bedroom closet areas. Adaptable features include flush thresholds and 3-ft-wide suite doors with two peephole heights. Interior doors are 3-ft-wide heavy-duty pocket doors with D-handles. Rocker-type light switches have been placed at 3-ft-6-in and electrical outlets at 1-ft-8-in heights. Prewiring for assistive devices throughout and automatic opening devices at all suite entry doors have been incorporated for future use.

Kitchens have adjustable-height countertops and upper cabinets, with removable base cabinets under the sink and range. Pullout work surfaces have been provided, and appliances include a range top with side controls, a wall oven with side-opening door, a dishwasher, and side-by-side refrigerator. Bathrooms can accommodate a wheelchair with removal of the vanity base cabinet. Half the units have side-transfer space and six units have wheel-in showers. Other features in the bathroom include reinforced walls for installation of safety bars, telephone-type adjustable showerheads with shutoff controls and nonslip flooring.

In May 1997, the design won the BC and Yukon division of Canadian Mortgage and Housing Corporation (CHMC's) Flexhousing national design competition. This competition, entitled "Flexhousing: Homes That Adapt to Life's Changes," set out to raise awareness, tap into talent, and set national goals toward a more universal design standard for all new housing. Initial development funds of $75,000 were contributed by CMHC's Public-Private Partnership Program.

The challenging development process began with the formation of the Norgate House Society, a partnership between Bev Nielsen, the designer, and Innovative Housing, a housing development consulting firm, in response to a municipal proposal call. Significant development challenges included establishing project financing, negotiating with a management group, obtaining charitable status, and obtaining municipal approval. Key aspects of the project's success include a land lease from the city, funding from the Zajac Foundation and the BC Real Estate Foundation, and the BC Housing rent subsidy.

Providing supportive housing with a care component was a significant part of the vision. The goal has been to achieve this without creating an institution. The local home support agency is working with the board to provide more efficient care to residents in their own homes, and more appropriate and flexible service delivery. Common areas in the building, such as a kitchen and amenity room, allow for future potential linkages with community programs such as recreational, therapeutic, or other services that may meet tenants' needs.

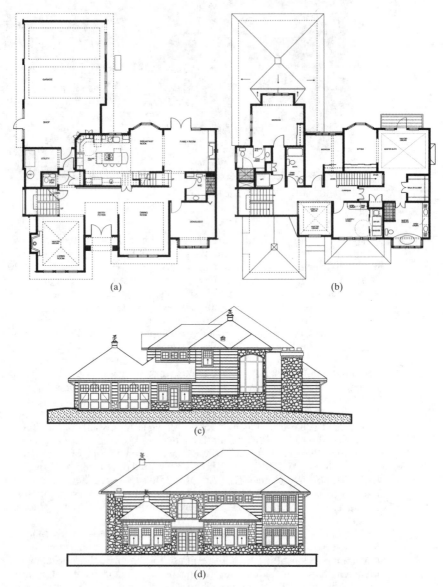

FIGURE 36.9   User-Friendly Demonstration Home—plans and elevations. (*a*) User-Friendly Home second-floor plan. (*b*) User-Friendly Home main-floor plan. (*c*) User-Friendly Home side street elevation. (*d*) User-Friendly Home front elevation.

32.7 THE USER-FRIENDLY DEMONSTRATION HOUSE

This project is intended to raise industry and consumer awareness, and to provide information and training and, most important, first-hand experience with a universally designed single-family home.

The project is a by-product of the City of Vancouver Planning Department's Accessible and Usable Dwellings Project (AUDP), which ran from 1995 to 1997. The City commissioned its planners, together with designers, experts, advocates, and stakeholders from the community "to work together to re-examine design standards for universal access as well as strategies and tools for implementation" (City of Vancouver, 1997).

Participating User-Friendly Homes' President Patrick Simpson envisioned a vehicle for implementing the new standards of housing design identified in the AUDP final report. Not only a life-size, hands-on demonstration of the design ideas, principles, and technology as advocated by User-Friendly Homes through its consulting and publication, it is also a base from which further dissemination of ideas can flow. The demonstration home was open to the public from January to November 2000 and housed professional and industry workshops aimed at designers, builders, manufacturers, realtors, and the trades.

The 344-m^2 (3700 ft^2) house, designed by Generations: Architecture, Planning, Research of Vancouver, includes a living room, dining room, large kitchen, breakfast nook, and family entertainment area (Fig. 36.9). As well, there is a multifunction office/guest room on the main floor (Fig. 36.9a), together with a utility area where "Node Zero" and other important system controls are located. This area links to an at-grade two-car accessible garage, and opens onto a large landscaped terrace and garden area designed jointly through a design charette by University of British Columbia Landscape Architecture students. On the second floor (Fig. 36.9b), there are three bedrooms and three bathrooms, including a generously proportioned master suite, and an accessible laundry and hobby room. The two floors are linked by two carefully designed staircases and a two-story electric lift installed into stacked storage closets that demonstrate the principal of design for flexibility and change to accommodate different users' needs at different times.

Other features incorporated into the User-Friendly Demonstration Home include: additional electrical receptacles at exterior doors, in the bedrooms, washrooms, on the front face of cabinets, in closets, and at Node Zero; additional telephone and cable outlets in every room to accommodate home automation and information technologies; 6-in-wider doors throughout the house (42 in); reinforcing in bathroom walls; lowered rough-in for waste pipes; knee and transfer space in all bathrooms; offset shower controls; zero shower thresholds; flush exterior door thresholds; and wider staircases with tread lighting and flush nosings. It also incorporates a universally designed kitchen with full-height pantry, rollout storage, and counters at different heights to maximize flexibility of use, along with many other features that contribute to safety and ease of use.

36.8 CONCLUSION

Universal design is catching on in Canada. From September 5 to 9, 1999, one stream of the Fourth Global Conference of the International Federation on Ageing focused on universal design. Interestingly, in the conference brochure, Professor Joseph A. Koncelik was quoted as saying, "The consumer products industry has no future if it neglects the aging market . . . designers have no future if they fail to understand the aging process" (International Federation on Ageing, 1999).

From September 30 to October 2, 1999, the Universal Design Institute and the Faculty of Architecture at the University of Manitoba presented a conference entitled "Universal Design in the City: Beyond 2000," featuring important speakers from across North America.

Universal design concepts are beginning to infiltrate the consciousness and everyday language of Canadians. Average consumers, including older adults and their adult children, are becoming aware of design features that promote independence via adaptability to future needs, preferences, and technologies. Aging baby boomers are among those predicted to "dig in their heels" and insist on changes to housing, health care, and support services to allow them to live out their years in the community, rather than in an institutional setting. They are also a group that is well educated, informed, and influential enough to insist that these significant changes occur.

Home automation technologies are entering a new phase of awareness and acceptance as Canada, the most wired and willing country in the world to embrace information technology, integrates the multitude of functions and potential inherent in the home wiring and computer systems that seemed only yesterday to merely facilitate use of the Internet, movie videos, and home shopping. Research in the area of health telematics is also state of the art in Canada, and reforms in the health care system may find consumers ready to embrace this new, high-tech form of health care delivery.

As discussed in the overview, universal design and its related theoretical frameworks strive to promote individual independence through the sensitive and flexible design of the built environment. Many of the design concepts employed in the cases discussed here illustrate the fundamental Principles of Universal Design. By preplanning for a variety of residents' needs and preferences, universal design facilitates independence. There is a vital relationship between universal design and home automation technologies. As an integral part of a universal housing design strategy, home automation wiring enhances the built environment by accommodating current and future technologies.

Home automation includes telephones, computer equipment, cable, fiber optics, audio-visual equipment, home security systems, and automated control devices.

Increasingly, in Canada, there is a move away from the stigmatizing language historically associated with design for special needs populations. Housing designed for and marketed to older people is beginning to change to better serve the needs of an increasingly more educated and technologically sophisticated population.

36.9 BIBLIOGRAPHY

BC Ministry of Municipal Affairs and Housing, *Annotated Bibliography on Universal Design: A Selection of Books on Creating Accessible Environments,* Queen's Printer, Victoria, BC, Canada, 1998.

Canadian Mortgage and Housing Corporation (CMHC), *Flexhousing: Homes that Adapt to Life's Changes: Design Options for Barrier Free and Adaptable Housing,* CMHC, Ottawa, ON, Canada, 1997.

Center for Accessible Housing, *Definitions: Accessible, Adaptable and Universal Design,* (fact sheet), Center for Accessible Housing, North Carolina State University, Raleigh, NC, 1991.

———, *Definitions: Accessible, Adaptable and Universal Design (Fact Sheet),* North Carolina State University, Raleigh, NC, 1991.

Center for Universal Design, *The Principles of Universal Design,* Center for Universal Design, North Carolina State University, Raleigh, NC, 1997.

City of Vancouver, *City of Vancouver Building Bylaw,* Vancouver, BC, Canada, 1998.

———, unpublished staff report, 1997.

Finkel, G., *Access: A Guide to Accessible Design for Designers, Builders, Facility Owners and Managers,* Universal Design Institute, Winnipeg, MB, Canada, 2000.

Gutman, G., "Technology Innovation for an Aging Society: Application to Environmental Design," *GRC News,* 18(1): 3–5, 15, 1998.

International Federation on Ageing, Conference brochure, 1999.

Pirkl, J. J., *Transgenerational Design: Products for an Aging Population,* Van Nostrand Reinhold, New York, 1994.

Simpson, P., *User Friendly Homes: Building for Your Future.* Self-published, Vancouver, BC, Canada, 1998.

Vanderheiden, G. C., "Universal Design vs. Assistive Technology," in G. Gutman (ed.), "Technology Innovation for an Aging Society: Application to Environmental Design," *GRC News,* 8(18): 3–5, 15, 1997.

CHAPTER 37
THE CARE APARTMENT CONCEPT IN SWITZERLAND

Matthias Hürlimann, Architekt E.T.H.
archi-netz, Zürich, Switzerland

37.1 INTRODUCTION

This chapter reports on recent developments in universal design in the area of care apartments for persons who are elderly. In Switzerland, for the last 10 years, care apartments, or decentralized care stations, have offered an alternative to living in a care home or hospital. They were opened with the intention of widening the range of options for elderly care patients. Today, there are approximately 75 care apartments in Switzerland. What is remarkable is the fact that the care apartments are integrated into regular housing without standing out or causing potential stigma.

37.2 BACKGROUND

The care apartment is a local or community care station for senior citizens with medium to high levels of care requirements. The apartment is normally in a building with other apartments and offers facilities for between five and eight people to be cared for, with the care staff living elsewhere. The inhabitants take part in daily work and join each other for excursions. They either have a single room, or may share a twin room at their own request. Confused persons can live in mixed or special care apartments.

Today, care apartments offer an alternative to care for patients who are elderly and who are seeking something more than hotel-like care, or isolation in a conventional care home. There are other alternatives, such as apartments shared by senior citizens, accommodation of those in care in unrelated families, or independent groups within a care home.

Within this framework, the care apartment offers its own advantages: The local model permits patients to maintain contact with relatives and existing friends. Frequently, elderly persons are visited by these people and taken to their homes. This way, the previous lifestyle is not interrupted and permits continuity. All residents can remain for the rest of their lives.

Care apartments allow the residents to keep up previously learned skills. Everyday tasks and work can be continued, according to individual circumstances and interest, and can replace ergotherapy as well as physiotherapy, respectively.

For care personnel, care apartments are attractive workplaces, offering a wide range of tasks and the opportunity for personal development. Medical factors take a backseat, and living together locally becomes the center of activity. In economic terms, both the low level of investment required and the flexibility in planning make the care apartment the ideal accommodation and form of care of the future.

Decentralized care stations should provide just one of the future living options for elderly persons. Important experience is being gathered currently in this field with regard to living together with those in care, with the aim of connecting better to the individual life experiences of elderly persons. In addition, the goal is to move away from the medical-technical criteria and to facilitate a higher quality of life together in the space provided.

37.3 THE SWISS CONTEXT

In Switzerland, the attempt has been made to provide various choices for housing elderly persons, one of which are the care apartments, which will be described in detail later. One current model for housing elderly persons could be called "family care," especially under the assumption that families are capable of accommodating additional persons in their apartments. For example, when children move away, there might be underused space in a single-family home. The Swiss Red Cross places persons who are interested in this type of accommodation. They train a family, who will then take care of the residents, who are given their own room, and who can pursue leisure activities on their own.

Another model is the accommodation of young people in apartments of elderly persons. The assumption here is that elderly persons often live in apartments that are too large for their needs. Therefore, for them not to have to move to another apartment, younger persons who still do not have a need for their own apartment, can be placed into those of elderly persons. In return, the young people are asked to provide certain services, care, and assistance. In return for their services, the elderly person would provide a low-cost rent or even a stipend. This type of placement of young and older persons has been quite successful in the city of Zürich.

In general, no special housing arrangements for elderly persons have existed in Switzerland in the past. Elderly persons with only a few limitations move together with the expectation of mutual assistance and care. In the long run, certain services can also be provided from the outside. This is a model, which has been successful all over Switzerland already, and it can be expected that with the aging trend of the younger generation, this model will be used even more.

Also, in the classical "retirement homes," changes can be observed. Restructuring of these homes can change them from a hospital or hotel-like character to even smaller living clusters with about 5 to 10 persons each. Each group can live in the area independently, and they can prepare their meals and pursue their leisure-time activities individually, while also being integrated into the larger structure and organization of the home. This is a method where one can use existing resources and existing buildings, which can be improved by using new models, thus creating group living environments.

The third model is the so-called decentralized care station, which is the official term for the care apartment. In Switzerland, there are approximately 75 of these, both in cities and rural areas. Key features of the care apartment include being located in a regular multifamily, multiunit apartment house, and, if possible, at ground level. That way, one can have access to the garden, so that the residents can enjoy both the inside and the out-of-doors, including the changing seasons, weather, and such.

Typically, five to eight persons live in such an apartment, which has about six to nine rooms. A professional care team will take care of a group of elderly persons. Often, members of the care team are still in training. Between 7:00 P.M. and 7:00 A.M., a night staff member stays in the apartment (i.e., after the day staff has left). The night person is allowed to sleep after all the residents are in bed, but will only get up in serious or urgent cases. For example, this can involve changing the bed linens or assisting an ill person.

37.4   A CASE STUDY IN ZÜRICH

The key features of a care apartment will be described later using the example of a six-room unit, located in a relatively new apartment building in Zürich (see Fig. 37.1) dating to the late 1980s. The entire building has 72 apartments. The care apartment is on the second floor, and the possibility exists that it will be moved into the ground floor in the near future.

What is remarkable is that the apartment looks quite normal (i.e., it does not stand out, and neither does it provide any stigma or clues that it houses people who require medium to intensive care) (see Fig. 37.2). It is interesting how the care apartment came about (i.e., by combining a two- and a four-room apartment into the large six-room unit). The apartment has a contemporary character; it is a unit without any special features, and without special funding support. However, it is fully accessible and barrier-free, meeting the current standards. Furthermore, this building in which elderly persons requiring care are living was built without any extra construction costs. Meanwhile, yet another apartment, which was next door, has been incorporated, adding two more residents requiring care who now also belong to the group.

It is important to point out that, at a very moderate cost, this could become a model for new construction (see Fig. 37.3). In this case, an additional 10,000 Swiss francs were invested by the sponsor of the original units. The owner did not need to invest any more. A sum of 10,000 Swiss francs was spent on the shower bathroom, with the shower costing approximately 5,000 francs. The remainder was spent on smart features in the kitchen. This example shows how contemporary housing can be created in a barrier-free manner without any significant cost increases.

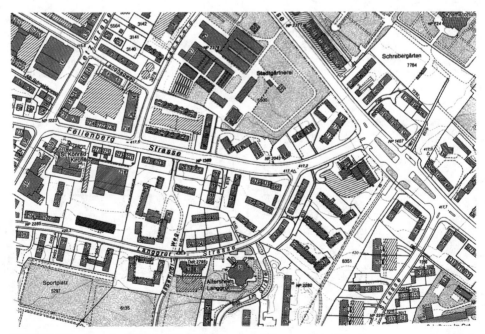

FIGURE 37.1   Area plan with location of the care apartment.

FIGURE 37.2   Front façade of the care station Langgrütstrasse.

37.5 *SPECIAL FEATURES OF A CARE APARTMENT*

What is the difference between a care apartment and general housing? Critical preconditions for housing that is subsidized by the federal government apply in principle to the creation of a care apartment. Architects are helped by the respective requirements, which are stated in the brochure entitled *Barrier Free and Adaptable Housing of the Swiss Consortium on Barrier Free Building.* Several particular measures should be considered, especially in new construction, if possible:

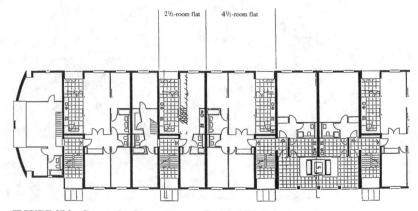

FIGURE 37.3   Care station Langgrütstrasse Zürich-Albisrieden: overview.

General Space Criteria

- Bedrooms should be no less than 12 to 14 m² and, if possible, a room should be big enough for two persons.
- It is desirable to have a second lockable living room in addition to a staff accommodation.
- The kitchen should have places to sit and/or have a direct connection to the living room.
- It is desirable to have storage space inside the apartment (in publicly subsidized housing).

A typical floor plan for the care apartment described here can be found in Fig. 37.4.

Sanitary Facilities

- It is desirable to have a wet room with shower, toilet, and sink, and another one with a bathtub, toilet, and sink.
- Often, a shower/toilet room is requested.

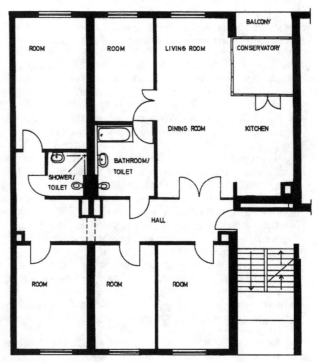

FIGURE 37.4 Second-floor plan.

- A staff toilet room is not necessary; however, it is often requested.
- It is desirable to have adequate space in the wet rooms for storage, small furniture, and so on.

Kitchen

- Stove with glass top and five burners, perhaps also a separate main switch.
- Double sink.
- Adequate cool storage with separate freezer in the kitchen and the basement.
- Adequate daylight is desirable.

Electrical

- Bedrooms should have three electrical outlets and telephone and television connections.
- Lighting in the hallways should have a sliding switch, which can be switched to a regular light switch.
- Electrical outlet should be adjacent to the toilet in the shower/toilet room.

Doors

- To be opened inside with a code number key, and from the outside with a general key.

37.6   A TYPICAL DAY IN A CARE APARTMENT

To illustrate life in a typical care apartment for elderly persons, activities during the course of a day are highlighted. For example: dispensation of the daily pills, using a shopping cart for going shopping, active participation and involvement of elderly persons in their daily lifestyle. Life in a care apartment can be illustrated as follows:

At 7:00 A.M., the night staff will be relieved by the day staff. They will be together with the residents until about 9:30 A.M. The residents can get up when they wish, and breakfast will be ready. They come in smaller groups, two or three, to have breakfast. At 9:30 A.M., the second day-shift staff member starts the day service, and the last residents get up and eat at the breakfast table. When one of the service staff workers goes shopping for lunch, he/she is typically accompanied by one or two of the residents of the care apartment. Then the preparation of lunch starts, and the residents (in this case, five women) can contribute to the preparation of the meal based on their life experience. The percentage of women in the care apartments is very high and is expected to go up. It is a fact that it is a real advantage that women, based on their life experience, can contribute to the household, especially cooking. They can use their know-how to participate in preventative therapy for residents (e.g., by cleaning up after their meal or by putting their dishes into the dishwasher), thereby increasing their physical activity. Also, they may help carry laundry to the laundromat in the basement and help with other chores, which are necessary in daily life.

After lunch, the residents rest until about 3:30 P.M. Typically, they go to their rooms, which may be shared, and they rest. After 3:00 P.M., or sometimes as late as 4:30 P.M., the early staff shift ends, leaving one staff member in the apartment who will take care of preparing dinner. Individuals determine when they are going to go to bed, but typically, by around 10:00 P.M., everybody is asleep. Some residents are dependent on some assistance when going to bed. At 7:00 P.M., the night staff arrives and takes over from the day team.

In summary, the ambiance in the care apartment clearly serves the well-being of the residents. It is a place that is much more interesting, lively, and humane when compared with the hotel or hospital model. There is comfort in knowing that individual needs, activities, and people's own experiences can be accommodated.

37.7 LESSONS LEARNED

The key points in this case study relate not only to urban design and architectural issues, but also to the assistance and care of elderly persons, as well as the role of planners in the design itself. These points are also relevant to the "design for aging" network, which has dealt with this topic for the past 3 years. Critical to future design and construction and an improved approach to care apartments will be the better utilization of existing resources. In the existing building stock, there are lots of highly specialized buildings, which are not barrier-free and, therefore, do not meet expectations.

It is important to recognize existing assets regarding the ambiance and image, the infrastructure, and the technical systems providing comfort; it is also important to facilitate barrier-free design and opportunities for choices. Existing resources are not only buildings and infrastructure, but also people (i.e., the residents) who can collaborate and contribute. In any case, one should plan for and facilitate the collaboration of persons with disabilities, something that has been attempted in recent projects in Switzerland. Critical for the quality of new approaches and building solutions will be the inclusion of the experience and potential of the end users of care apartments. It is important to create opportunities for choices and to utilize models that can adapt to the changing needs of the residents. These choices are very important, and they can only be implemented using relatively small, decentralized housing units. Building networks and fostering better solutions in this area of concern should be supported. However, planners will also have to rethink their role, which can be compared with that of a moderator. As moderator, one cannot predict the course of things, one can only encourage participants to utilize their own resources and to seek new solutions through a combination of knowledge and experience.

37.8 CONCLUSIONS

Switzerland is characterized by a social order that functions from the bottom up. Communities and many regional private bodies carry out very important tasks, both in the health and care domains.

Therefore, decentralized structures support the care of elderly people. Citizens, including elderly people, participate, and the existing potential of ideas, competence, and financial resources are well utilized. In the last few years, this has lead to development away from centralized, standardized solutions to very diverse and innovative housing models.

On one hand, the bottom-up movement has meant that new standards in the building industry are only slowly being implemented; on the other hand, there are increasingly better buildings and solutions, which act as examples for other regions. Decentralized solutions make possible greater participation of the end users, their relatives, and neighbors. That way, housing opportunities can be realized that are locally supported and socially integrative.

Interesting in this context is that planning personnel participate with great motivation, making possible a great deal of creativity, despite the high degree of responsibility that has to be shouldered.

In the past, buildings were focused on specialized designs: homes for elderly persons, for people who are disabled, for those with behavior problems, or otherwise specialized forms of housing and workplaces. In this context, a critical change has occurred. Both in the organizational structures and the building solutions, one can find more and more integrated concepts.

A significant part of this is the so-called *adaptable housing*. It creates the building foundation based on which individual adaptations for specific requirements can be carried out without significant investment. This is the premise upon which a number of care apartments were made possible.

In Switzerland, universal design is not well developed, yet. There are very progressive regions and those that lag behind. The network of counseling services, meanwhile, is very dense, and progress, which has been made both in the public consciousness and also in the realization of designs and buildings, is remarkable, indeed.

CHAPTER 38

AGING WITH A DEVELOPMENTAL DISABILITY AT HOME: AN AUSTRALIAN PERSPECTIVE

Susan Balandin, Ph.D. F.P.S.A.
University of Sydney, Sydney, Australia

Robyn Chapman, B.App.Sc.Phty.
Spastic Centre of NSW, Sydney, Australia

38.1 CHAPTER OVERVIEW

In Australia, the majority of people with a developmental disability are now living in the community. In common with their nondisabled peers, they are living longer and experiencing many of the problems commonly associated with the aging process, such as, for example, loss of visual acuity and increased incidence of falls. In this chapter, issues related to aging with a disability that may impact on housing requirements are explored. It can be argued that if the concepts of universal design were incorporated into Australian housing, those who are aging with a disability would be able to remain in their homes and maintain their preferred lifestyle with a minimum of disruption. Indeed, universal design features help ensure that the housing would be suitable for all members of the community. Thus, housing that incorporates universal design features helps to reduce discrimination within the community, as it is suitable for people both with and without a disability. In Australia, adaptable housing includes many features that are compatible with the concept of universal design. In this chapter, the features of adaptable housing in Australia relevant to people who are aging with a disability are examined and the current adaptable housing standards are discussed. The potential benefits of adaptable housing are illustrated by two case studies.

38.2 BACKGROUND

Individuals with a developmental disability now have the chance to enjoy a long period of old age similar to that of the rest of the community. Within Australia, there is recognition that good design has the potential to enhance quality of life not only for those with a developmental disability but also for all members of the community. However, to date no Australian legislation enforces the concepts of universal design. On the other hand, there is growing agreement across many sectors of the community that good design features such as those incorporated into universal design will support equitable use of the built environment by all members of the community. Government agencies, councils, and communities are being called on by community groups and those involved in planning and design to consider the concepts of universal design in conjunction with legislation that is already in place, namely the Commonwealth Disability Discrimination Act, 1992. Nevertheless, despite agreement that a coordinated national approach to design of housing, public buildings, and accessible transport is desirable, to date the Australian environment is still not inclusive (National Access Working Group, 1999). Indeed, there is still limited understanding of what features might constitute universal design. Although federal, state, and local governments as well as nongovernment agencies are contributing to improving the accessibility of the built environment across Australia, there is still no integrated approach for building on universal design initiatives to create a more equitable society (National Access Working Group, 1999).

"Adaptable housing" (known outside Australia as *mobility housing, universal housing, and/or lifespan dwellings*) is being promoted as the housing of the future, particularly by groups concerned with modifying or designing housing for people with a developmental disability or those who are aging (Pratt, 1998). Adaptable housing is, very simply, housing that can be adapted to the changing needs of the occupants. It is different from modified, accessible housing that is usually developed to meet a specific need, such as to accommodate an individual with a spinal cord injury in a wheelchair. Such modified, accessible housing is often expensive due to its "special" character, and is usually not revertible without considerable expense. Adaptable housing facilitates those with a disability who are aging to (a) age in place, (b) remain in the community, and (c) maintain their natural support networks and regular social contacts. It can be argued that the Adaptable Housing Standards AS 4299 (1995) are a forerunner to universal design in Australia. Sadly, within Australia Adaptable Housing Standards AS 4299 (1995) are not mandatory for all public housing.

When considering appropriate planning for those who are aging with a developmental disability, planning for adaptable housing is as important as financial and legal planning. If the housing needs of individuals who are aging with a disability are to be met, Australia must have a commitment to the provision of adaptable housing and the resultant changes to building practice. Both state and local governments need to legislate for the provision of adaptable housing and include adaptable housing in policy and planning, as has occurred in other countries including the United States, Sweden, and the United Kingdom.

38.3   BACKGROUND AND GENERAL HEALTH TRENDS

"Increased longevity represents one of the great social achievements of the twentieth century" (Walker and Walker, 1998, p. 125). In Australia, as in many other western countries, people are living longer (Ministerial Council of Review on Aging, 1994). Included in the aging population are people with a disability, as the life expectancy of adults with a developmental disability continues to improve (Hawkins, 1993). Indeed, Hogg (1997) noted that people with a developmental disability are now a well-defined population who are living in the community and are likely to remain in it. The life expectancy of those who do not have a profound level of disability mirrors that of individuals without a disability (Krauss and Seltzer, 1994), although individuals with Down's syndrome have an approximately 10 to 15 years lower life expectancy than their nondisabled peers (Hogg, 1997).

In 1993, 18 percent of the total Australian population and 51 percent of the population over 60 years of age were identified as having a disability. As the population continues to age and people with a disability live longer, it is expected that by 2001 the population aged over 65 years will increase by 21 percent and the population aged over 85 years will increase by 61 percent (ABS 1993). Given the disability rate in the population over the age of 60 years, Australia can expect a large increase in its aged population with a disability, all of whom should be planning for their retirement and old age now. The Australian Bureau of Statistics (ABS) makes no distinction between different types of disability and includes both acquired disabilities (for example, spinal cord injury) and sensory disabilities (such as visual impairment or hearing loss). However, this chapter will focus on individuals who are aging with a developmental disability. "Developmental disability" can be defined as an intellectual disability and/or a physical disability that is congenital or acquired in the first 18 years of life—for example, cerebral palsy.

There is some evidence that health conditions associated with aging seem to manifest earlier in people with a developmental disability, compared with nondisabled members of the general population (Devenny et al., 1996; Jenkins, Brooksbank, and Miller, 1994). Individuals with cerebral palsy may experience the effects of aging at an earlier age than their nondisabled peers (Balandin and Morgan, 1997; Schwartz, 1994; Turk, Overeynder, and Janicki, 1995). Although some authors (Turk, 1994; Turk, Geremski, Rosenbaum, and Weber, 1997) have suggested that other issues, such as level of fitness, may impact on the individual's ability and be mistaken for early aging, there are reports that individuals with cerebral palsy experience problems commonly associated with aging (for example, increased fatigue and/or loss of mobility) from as early as 25 years (Balandin and Morgan, 1997; Murphy, Molnar, and Lankasky, 1995; Willner and Dunning, 1993). Despite differences in reports on the onset of problems with health and well-being associated with the aging process, there is agreement that individuals with cerebral palsy are living longer mainly due to advances in medicine and improved health care (Crichton, Mackinnon, and White, 1995). Kailes (1993), who herself has cerebral palsy, noted that the issue of concern is no longer how long individuals with a disability will live but rather how well they will live. The ability to age in place—in other words, age in their own homes—is also an issue of concern for many individuals who fear that a nursing home is the only alternative if their health and mobility decline with age.

Australia has a high rate of home ownership, with approximately 80 percent of people over the age of 65 years living in homes that they own or are in the process of buying (Kendig and Gardner, 1997). The bulk of residents in nursing homes are women over the age of 70, though, despite the move to close institutions and provide accommodation for people with a disability in the community, approximately 7000 individuals with an intellectual disability or cerebral palsy are still living in institutions in Australia (AIHW, 1999). Most older people want to age in their own homes; consequently the Australian housing industry must develop housing that will meet the changing needs of occupants across the lifespan.

In the early 1970s, Wolfensberger recommended that people with a disability should live in typical community housing, that is, regular houses (Wolfensberger, 1972). Since the early 1980s, there has been a move in Australia to close institutions for individuals with a developmental disability and accommodate them in small group homes, hostels, boarding houses, and independently supported accommodation within the community (Intellectual Disability Services Council, 1994). A recent Australian study of 279 adults with cerebral palsy (Balandin and Morgan, 1997) indicated that 32 percent of the respondents lived in their own or their parental home; a total of 44 percent lived in group homes in the community, 21 percent still lived in institutions, and only 3 percent lived in nursing homes.

The preferred residential option for older individuals with a disability who live in the community (whether independently, in a group home, or cared for by their families) is that they should age in place, that is, in their home within the community, and not be moved to a nursing home or back to an institution as they become older (Ashman, Suttie, and Bramley, 1993; Intellectual Disability Services Council, 1994). If they are to remain in the community as they grow older, housing must be adaptable in order to meet residents' changing needs. Some of the changes people with a developmental disability may experience are pertinent to older people in general and some are specific to individuals with a disability.

38.4 POLICIES FOR ADULTS WHO ARE AGING WITH A DEVELOPMENTAL DISABILITY

Walker and Walker (1998) suggested that the notion that aging is a "natural process" has hidden the needs of older people with a disability within the ideology of the principles of normalization (Walker and Walker, 1998). This view, when applied to the aging of people with a developmental disability (for example, those with intellectual disability or cerebral palsy), requires the individual to adapt to the norms of society, which for the aged in our community often means a limiting and dependent lifestyle. It also effectively absolves policy makers from recognizing the specific needs of older people with a disability and from developing appropriate policies and action plans (Walker and Walker, 1998). Walker and Walker (1998) suggested that the support and care provided to people with a developmental disability "is more driven by expectations of dependence rather than the goals of independence or inter-dependence." Older people in Australia tend to be concentrated in the suburbs (Kendig and Gardner, 1997) where support facilities, including health services, transport, and leisure options may not be accessible for people without cars. Many group homes have also been purchased in suburban areas, which present problems in terms of visitability and socialization for the residents. To date, planning adaptable housing that will foster independence and facilitate individuals with a disability to remain in the community has not been a priority on either policy makers' or the disability movement's agenda. In addition, there has been limited, if any, planning that considers the specific health conditions experienced by many individuals who are aging with a disability that may impact on their accommodation as they age.

38.5 HEALTH CONDITIONS RELEVANT TO ADULTS WITH A DEVELOPMENTAL DISABILITY

In Australia, the main causes of death in individuals aged 65 years and over are ischemic heart disease, cancer, and cerebral vascular disease. The main disabling conditions are arthritis, circulatory diseases, and hearing loss (McFee and Bray, 1995). However, adults with a developmental disability frequently have untreated simple medical conditions, and untreated or undiagnosed health problems related to their disability (Beange, 1996). Such problems can result in injury in the home or in a belief that individuals with a disability and additional health problems cannot be cared for adequately in the community home. There are some reports that health problems are dismissed by the medical profession as being a facet of the condition rather than a disorder that requires further investigation and treatment (Turk et al., 1995). Consequently, poor health is a major barrier for individuals with a developmental disability aging in place. Those who are frail or are in poor health require more care and more equipment, and are more likely to sustain falls or injuries that impact on their ability to live at home. Housing should be easily adaptable to accommodate an individual's changing needs throughout the aging process, thus facilitating the opportunity to remain at home within the community.

There are a number of design features that should be incorporated into housing for people with a developmental disability. For example, corridors should be wide enough to accommodate wheelchairs, and bathrooms should have enough circulation space not only to facilitate the use of shower chairs and handrails but also to enable a caregiver to assist with bathing if necessary. Kitchens should have adjustable counters and water faucets that are easy to operate. Shiny, slippery surfaces should be avoided. In addition, housing should be close to transport, community support services, and social and leisure activities (Gething, 1990; Kendig and Gardner, 1997).

Dementia

The incidence of Alzheimer's disease, particularly among women, increases with age, with approximately 10 percent of individuals over 65 years affected (Gething, 1990). Individuals with Down's syndrome have an increased likelihood of developing Alzheimer's disease in midlife, with approximately 40 percent of people with Down's syndrome developing Alzheimer's disease by the time they are 60 years old (Moss and Patel, 1997). The incidence of Alzheimer's disease in individuals without Down's syndrome but with an intellectual disability is thought to be similar to that of the nondisabled population (Janicki, Heller, Seltzer, and Hogg, 1996).

Housing should accommodate the changing needs of residents who, as a result of dementia, require increased support and additional equipment to conduct routine activities, such as personal care, cooking, and recreation. The design features for wayfinding and orientation as well as security features become increasingly important and should be incorporated into the plans for the home (Lewit, 1999). Some of these features, including double locks on all doors leading to the outside and locks on windows, are now considered desirable for all metropolitan housing in major Australian cities.

Vision and Hearing

Janicki and Dalton (1998) evaluated the extent of vision and hearing impairment for adults with an intellectual disability in New York State. They found that these sensory deficits occurred more frequently in individuals with an intellectual disability than in the general population. In common with the general population, individuals with a developmental disability also experience an increased likelihood of hearing and visual impairments as they age. Regular vision and hearing screening and treatment will assist older individuals with a developmental disability to maintain optimum function.

Housing should be constructed to ensure that there is adequate lighting, clear signage, and appropriate interior decoration. This would include avoidance of reflective paints and surfaces, and floor covering that is safe and that contrasts with walls. Such design features will also benefit visitors who have sensory impairments and will facilitate the likelihood of people remaining in their own home as they age rather than having to relocate to a nursing home or specially adapted environment.

Cancer

Although comparatively low levels of cancer-related deaths have been reported in individuals with an intellectual disability or cerebral palsy, Turner and Moss (1996) and Uehara, Silverstein, Davis, and Geron (1991) noted that there is evidence that death from cancer is on the increase in this group. A study by Bohmer, Klinkenberg-Knol, and Niezen-de Boer (1997) of the incidence of cancer of the esophagus indicated that the incidence of this form of cancer increased with age for those with a developmental disability, as compared to the general population.

Individuals are more likely to be able to remain at home during illness related to cancer if the home can be easily adapted to accommodate changing needs and to facilitate home nursing. Such adaptations include bathrooms that can accommodate wheelchairs and shower chairs, corridors that are wide enough for a wheelchair, and preferably no steps (or at least steps that can be ramped easily).

Osteoporosis

In a recent study conducted in the northern region of Sydney, Beange and McElduff (1995) examined the prevalence and risk factors for osteoporosis in a community population of young adults with a developmental disability. The results showed that this group had lower bone mineral density compared to an age-matched reference population. Individuals who are unable to participate in weight-bearing activities (for example, people with a severe physical disability) are also at risk for osteoporotic changes from an early age (Dorval, 1994). It is important that the houses and surrounding areas of individuals at risk of osteoporosis are safe and built to minimize the incidence of falls.

Polypharmacy

Polypharmacy is an issue for older individuals with a developmental disability. Individuals with a disability are likely to take more medication than their nondisabled peers. Changes in metabolic rates that occur in normal aging may result in health problems that are a result of polypharmacy and overmedication—for example, confusion or depression. Individuals and their caregivers should be encouraged to access one pharmacist and to consult with the pharmacist on both prescription and nonprescription medication (Balandin and Kerse, 1999). All older individuals who take medication are at risk of suffering effects of polypharmacy and consequently are more likely to fall or injure themselves at home. Therefore, it is important that housing be constructed to minimize injury from falls and accidents in and around the home.

Fall Prevention

Preventing falls in the aging population is of importance not only to individuals but also to communities, due to the high economic and social costs associated with these injuries. Significant economic costs are incurred by health and community sectors to provide acute care, rehabilitation, resultant mobility equipment, and personal care to individuals injured in falls. Moreover, victims of falls incur the loss of independence and mobility that inevitably lead to social isolation and reduced quality of life. These in turn decrease the individual's ability to participate fully and equally in society (Chapman, 1997).

As already noted, people with a disability may appear to age earlier as a result of their already reduced functional levels. In addition, they may be more likely to fall because of concomitant problems, for example epilepsy, unsteady gait, visual problems, and too much medication. Therefore, preventing falls becomes an important issue for them at an earlier age. Successfully preventing falls requires that all causes of falls be addressed. These causes can be divided into two major components—environmental issues and medical issues. The two major environmental issues are safe access within the home and safe access in the community. The medical issues include inappropriate medication, poor vision, poor footwear, and poor health status including inadequate fitness.

To prevent falls within the home there should be:

- A continuous accessible path of travel from sidewalk and car garaging areas to the entries and around the internal spaces of the home, including the provision of handrails at stairs and ramps
- Practical flooring, including slip-resistant surfaces and removal of unfixed rugs
- Adequate lighting, especially for those with a vision impairment
- Well-designed work spaces, especially in the kitchen to reduce the amount of bending, stretching, carrying, and lifting
- Well-designed accessible bathrooms with doors that open outward
- Adequately sized, well-designed laundries
- Appropriate aids to assist in activities of daily living
- Space to move, with a mobility aid if necessary, throughout the home

- Appropriately contrasted color schemes that facilitate good visual perception
- Adequate community-based support services, including gardening, laundry, housework, and home maintenance

Recent Australian studies have shown that when changes such as those just listed are made to the built environment, the incidence of injury is effectively reduced (Clemson, Cumming, and Roland, 1996; McLean and Lord, 1996).

38.6   BENEFITS OF ADAPTABLE HOUSING IN AUSTRALIA

Adaptable housing provides a choice in housing. Older people (particularly those who are frail) and people with disabilities currently have little choice in housing. Adaptable housing will increase choice, especially if applied to all sections of the housing market, private and public, rented and owned. Increased choice in public housing stock will increase the viability of that stock and facilitate early placement of older people and those with disabilities into appropriate dwellings. Adaptability allows tenants to remain in place instead of being rehoused at times of crisis.

As previously noted, adaptable housing is housing that can be altered to suit the needs of the occupants. According to the Adaptable Housing Standard guidelines in AS 4299 (1995), it has many benefits as it aims at good design for everyone. It provides for all ages, levels of ability, and degrees of mobility impairment. However, to date there is very little adaptable accommodation in Australia. Most accessible or adaptable housing is provided by public housing authorities that currently manage less than 40 percent of the Australian housing stock. Consequently, it can be argued that there is a large untapped market for adaptable housing within the Australian community (National Access Working Group, 1999).

Under the Building Code of Australia (BCA), new public buildings are required to be accessible for people with a disability, but there is still no legislation to enforce accessibility of existing buildings or of access issues beyond the building, including streets or parks—nor does the BCA include interiors or fittings. Neither is there to date any legislation to ensure that public housing does not discriminate against people with a disability. However, the Adaptable Housing Standard, A.S. 4299–1995 provides guidelines for the design for Australian residential accommodation that can be utilized by architects, planners, and homeowner/builders to ensure that housing is both accessible and adaptable and will consequently meet the owner's/resident's needs across his/her lifespan.

AS 4299 (1995)—The Australian Adaptable Housing Standard

This Standard was developed by Standards Australia to provide guidelines to architects, builders, developers, and owners on adaptable housing. It draws on AS 1428 (1993) parts 1 and 2—Design for Access and Mobility. AS 1428.1 (1993) provides the minimum requirements for access based on a wheelchair footprint of 1250 mm × 740 mm that applies to 80 percent of adults, that is, people aged 18 to 60 years.

AS 1428.2 (1993) provides enhanced guidelines based on a wheelchair footprint of 1300 mm × 800 mm that apply to 90 percent of adults. There is currently no research available on the wheelchair footprint of chairs used by those in the over-60 age group. In some states in Australia (for example, South Australia), providers of public housing are now building public housing to Adaptable Housing Standard AS 4299 (1995). Adaptable Housing Standard AS 4299 (1995) contains four sections that lead the reader logically through the Standard. The four sections provide information and guidance on the potential for adaptation, issues relating to siting and access, and finally specific design features. Table 38.1 provides a summary of the main sections in AS 4299 (1995).

TABLE 38.1   A Summary of the Main Sections in Adaptable Housing Standard AS 4299 1995

Section one	Section two	Section three	Section four
Scope	Objectives	Scope	Design of the housing unit
Application	Performance requirements	Siting	Scope
Reference documents	Potential for adaptation	Access within site	Floor level
Definitions		Building location	Entrances, doorways, circulation spaces
		Landscaping	Sanitary facilities
		Security	Kitchen areas
		Car parking	Bedrooms
		Mailboxes	Living areas
		Signage	Floors
			Lighting
			Ancillary items

Classification of Adaptable Housing.   One important aspect is the classification of Adaptable Housing suggested in AS 4299 (1995). Adaptable housing can be certified *Class A, Class B,* or *Class C.* The features discussed in the four sections of AS 4299 (1995) have been further designated *Essential, First Priority Desirable,* and *Desirable* status. The class of certification of the building depends on the status of the features included. There are 55 Essential features, 42 First Priority Desirable features, and 22 Desirable features. Class A adaptable housing incorporates all Essential, First Priority Desirable, and Desirable features. Class B incorporates all Essential and 50 percent of the Desirable features, including all First Priority Desirable features. Class C incorporates all Essential features. Examples of the features that are included in each class are given in Table 38.2.

Aging in Place—Planning for Old Age.   All people, including those with a disability, need to plan for retirement and old age. Planning usually involves financial and legal issues and may also include planning for personal care needs. However, currently in Australia, very few individuals with a disability plan for their future housing needs. In the recent past, if a high level of care is required, options for the elderly were limited to leaving the family or community home and entering retirement living options or a nursing home. Those with a disability who could no longer manage at home, or who no longer had family to care for them, had one option—to enter institutional care. The result of this lack of choice was social isolation, including isolation from family, friends, and the wider community.

There is now evidence to suggest that individuals who are able to age in place enjoy a better quality of life and live longer. Pratt (1998) reported that people living in adaptable housing live two to three years longer, are happier, have fewer accidents, and have better general health than those who have to move into institutions, including long-stay hospitals or nursing homes, because of conditions associated with their aging process. Both Todd (1990) and Ashman and Suttie (1996) stressed the importance and value for individuals with a disability to live in the community and be part of it. It seems unfair if such hard-won community status is lost as a result of the aging process and the reward of living longer is a return to an institutional lifestyle. The emphasis on aging in place for individuals with a disability means that planning for changing need in the home environment is now as vital as financial planning. If dwellings are built that are adaptable to the changing needs of the occupants, successful, healthy inclusive aging in place is more likely to result. Many people with a developmental disability have a limited capacity to forward plan and therefore require support to plan for the future. Families, case managers, and social workers can assist them if they are aware of the need to consider adaptable housing in the planning process. In Australia many disability ser-

TABLE 38.2 Examples of the Features That Are Included in Each Class of Adaptable Housing

Essential features	First priority desirable features	Desirable features
Continuous accessible path of travel from street frontage and car parking space to accessible entry.	Covered access to car parking space to and from dwelling, including a minimum level of lighting of 50 lux.	All pathways to provide a continuous, slip-resistant, accessible path of travel.
Mailbox on a hardstand connected to an accessible path.	Common use facilities within a residential complex to be accessible.	Storage for wheelchair and external battery charging.
Accessible car parking space.	Switches located 900 to 1100 mm above the floor.	Folding shower seat.
Level, accessible entry with an 850 mm clearance at doorways.	Level or gently sloping site not steeper than 1:14.	Guide dog accommodation.
Corridors with minimum width of 1000 mm.	Accessible entry with combined security door and a lighting level of 300 lux.	
Circulation spaces of 2250 mm diameter in living areas.	Electrical outlets located at a minimum height of 600 mm.	
Slip-resistant floor surfaces.	Slip-resistant surfaces externally.	
At least one kitchen work surface adjustable in height or easily replaceable.	Kitchen includes a thermostatic mixing water faucet and task lighting over kitchen sink.	
Internal door spaces to have 820 mm minimum clearance.		
Visitable toilet complying with AS 1428.1 with provision to become accessible to AS 1428.2.		

vice providers employ workers specifically to assist consumers in identifying and planning for their life goals. These professionals need to be aware of the advantages of adaptable housing so that they can advise and advocate for the people they are supporting.

Accessibility in Australia.   The Australian Urban and Regional Development Review identified principles of barrier-free design that would help ensure that housing is built to meet the needs of people across their lifespan. However, these principles are not yet incorporated into legislation. A few councils across Australia include adaptability and visitability in their local development control plans, but currently the lack of legislation prevents councils from enforcing these building practices in either private or public housing development.

Development of age-specific housing has led to older people moving into smaller homes with smaller rooms and space. A better option for older people would be homes of the same size or smaller, with fewer rooms, larger room space, and smaller gardens. Adaptable housing would provide this. In addition, the incorporation of adaptable features at the time of building will add little to the cost of the housing but may save expensive modification costs at a later date.

Inclusion and Safety.   There are a number of features included with adaptable housing plans that facilitate social inclusion and safety rather than isolation due to social or health problems. A continuous accessible path of travel into and around the home facilitates a safer environment. Such a path ensures that it is easier and safer to maneuver furniture, children in strollers, shopping and laundry trolleys, and of course wheelchairs and other mobility aids. In combination with nonslip surfaces, falls are prevented. Accessible electrical outlets, switches, water faucets, door and window latches, and well-designed bathroom, kitchen, and laundry space prevent further injury and promote independence. A safer work environment is provided for employed personal care attendants.

Economic and Marketing Benefits. Adaptable housing is economically sensible. Pratt (1998) stated that the extra cost of adaptable housing is not more than 5 percent higher than traditional housing, whereas the added cost of modifying a standard dwelling in order to make it fully accessible is 12.5 percent (Campbell, 1997). Increased marketability also makes good economic sense, and the Australian Standard for Adaptable Housing (AS 4299—1995) recommended the certification of dwellings according to adaptability. A vendor could promote a dwelling according to the level of adaptability and purchasers could select the level of adaptability that meets their requirements. Currently, a model for a house that is adaptable, energy efficient, and suitable for people with a disability, including those with allergy problems (including asthma sufferers), has been developed by a group in South Australia comprised of the Department of Human Services, the South Australian Housing Trust, the Office of Energy Policy, the City of Onkaparinga, and the building firm A. V. Jennings Ltd. The house is designed by A. V. Jennings Ltd. in South Australia. The house incorporates design features that would allow older people with a disability to remain at home. These features are summarized in Table 38.3. A version of the house plan designed by A. V. Jennings Ltd., is reproduced in Fig. 38.1.

Houses such as this are an exciting innovation in Australia. It incorporates many features that fit well with the concepts of universal design. It could be used by a wide sector of the community, including those who are aging, those who have a disability, or those who are allergy sufferers. It would also be suitable for people with no disability who are looking for a family home that will accommodate them even if their independence changes due to their aging process or unforeseen trauma (for example a motor vehicle accident). Housing such as this is a flexible option for people that breaks down barriers of discrimination and access and provides scope for a wide range of individual preferences.

38.7   ADAPTABLE HOUSING—A QUALITY OF LIFE ISSUE

The following case studies provide examples of the importance of adaptable housing to people who have a disability and live in the community. Participating in the community, maintaining friendships, and working are all quality of life issues that are often taken for granted. Yet it is traumatic for individuals and families if family members must relocate to residential care, such as a nursing home, because they can no longer manage in their own home, or because the caregivers, usually parents, can no longer provide enough support. If people live in adaptable housing they are more likely to be able to remain within their community, even if their level of independent functioning and support requirements change.

Case Study 1

John and Sharon, an independent married couple, both have cerebral palsy. They have lived all their married life in the community in their own three-bedroom home. The house was built in 1935 and they bought it when they married 25 years ago. John is 55 years old and recently took early retirement from his job as a receptionist as he was finding it increasingly difficult to physically manage at work. Sharon is 54 years old and did not work as they had

TABLE 38.3 Features of an Adaptable, Energy Efficient House Suitable for People with a Disability, Including Those with Allergy Problems

Features	Features	External environment
Sloped architraves and skirting minimize dust catchment.	Minimal eastern windows with sun control glazing.	Allergy-minimizing garden design.
Hard floor surfaces adjacent to northern windows.	Central ducted vacuum system.	Drip watering system.
Pergola to northern windows allows winter sun entry, screens summer sun.	Smooth hard flooring minimizes dust collection.	Low-fragrance lawns and plants.
Adjustable-height kitchen counters.	Wide corridors and doors for wheelchair access.	Paving areas adapted to avoid steps.
Kitchen design allows wheelchair access.	Electric cooking appliances.	
Evaporative cooling of whole house.	Open shelving to cupboards.	
Extractor fan in cooking area.	No knobs in showers, controls at edge of shower.	
Cupboards built to ceiling.	Reinforced walls to allow installation of handrails.	
Doors allow zoning of living spaces.	In-floor heating to bathroom	
Radiant ceiling heating.	Wet areas vented to outside— one-way vents.	
Solar-convertible hot water service.	Lever-type door handles.	
	Deadlock and peep hole to front door.	
Low-irritant paint throughout.	Wall and ceiling insulation to Australian standards.	
Visual-vibration smoke alarms.	Solar-controlled glazing to western window.	

a child who is now living in another state. Although Sharon has always been independent apart from needing handrails in the bathroom, she is now experiencing back pain and arthritis in her knees and hands. She is finding it increasingly difficult to access public transport and relies on John to take her shopping in the car. They receive help with domestic duties, for example cleaning, for two hours a week. Apart from this, they do not receive support from any other services. However, John relies on Sharon to assist him with some skills such as putting on his socks.

   John was hospitalized with a hip fracture due to osteoporosis and a bad fall. He is unlikely to walk independently again and thus will need to use a wheelchair for mobility. He will need a power chair, and increased personal care support. While John was in hospital their home was assessed for modifications needed to meet John's new needs. Both the toilet and bathroom are too small. The wall between these two rooms is a structural wall and cannot be removed without considerable expense. Furthermore, the combined area still would not have enough circulation space to accommodate John's new chair. The doorways and corridors within the house need to be widened and the landings at the front and rear entrances need enlarging. The house has external steps, and the garage is too narrow to fit a power chair beside the car. If John and Sharon are to remain at home, major home modifications are

FIGURE 38.1 Floor plan of an adaptable, energy-efficient house suitable for people with a disability, designed by A. V. Jennings.

required. These are financially prohibitive. John and Sharon are unable to fund the building and must now consider alternative accommodation. They are unlikely to qualify for public housing as they own their own home. In the meantime, John will have to rely on home support services, as Sharon cannot give him enough physical support because of her own level of disability. Ultimately, they will have to sell the house and either move to cheaper accommodation, possibly in another state, or consider moving to a retirement village where they will be able to have an accessible unit but will be living with older people and away from their current social networks.

If their house had been adaptable, the garage, doorways, entrances, and the bathroom would be accessible. Reduced numbers of structural walls would enable rooms such as the toilet and bathroom to be enlarged without difficulty. The accessible car space would have allowed John to access the car from his chair. The expense of modifications would be reduced, and consequently John and Sharon could remain in their own home and maintain much of the lifestyle they had prior to John's accident.

Case Study 2

Anne has cerebral palsy and a moderate intellectual disability. She walks using two sticks and has limited strength in her hands. She works in supported employment and has access to some support from the nongovernment organization that provides service to people with cerebral palsy. She lived with her family until she was 35 years old, but was keen to have her own home. Her family considered that it would be ideal if Anne could settle into her new home while both her parents were still young and well enough to give her support rather than her having to relocate at a time of crisis, for example, if her parents became too frail to support her or died. In planning for Anne's later years, her mother approached an architect who understood the objectives and principles of adaptable housing. The architect consulted an occupational therapist, and together with Anne and her mother, they designed Anne a home that would meet her future needs even if her physical condition changed and she became less independent. The design incorporated wide doorways and corridors, a spacious bathroom with enough circulation space to accommodate a wheelchair or personal-care lifting hoist. There were few structural walls internally, so that should the need arise, Anne would be able to rearrange the internal spaces of the home to suit her requirements. The kitchen contained adjustable-height benches and all the switches were placed at accessible heights. Lever water faucets were installed. The plumbing was placed within the wall cavity, so this too could be easily relocated as needed. The architect provided a certificate of compliance with the Adaptable Housing Standard (AS 4299—1995) and a full set of plans showing the house in several states of adaptability.

One immediate advantage to Anne was that her visitable house enabled friends in wheelchairs to visit and to stay if they wished. Anne's next of kin were pleased with this arrangement, not only because they were assured that Anne would have an optimal chance of remaining in the community as she ages, but also were she to sell the home, the home's adaptability would mean that any new owner could revert the internal spaces of the home to suit their own needs.

Summary

Both John and Anne have cerebral palsy and live in the community, yet the future outcomes for them in terms of quality of life are very different. John, despite having worked and been independent, is having to come to terms with major changes that will affect both his and his wife's quality of life. When John and Sharon bought their house, there were few couples with cerebral palsy living independently in the community, and there was no planning process to help people with a disability or their families consider the future and the possible implications of their aging process. Anne, on the other hand, has the advantage of new information on housing and the funds to have a house built that will adapt to her needs should they change. In addition, there is now a greater understanding of aging with a disability and the challenges that must be faced. These challenges include parental death, and increased physical dependence associated with the aging process. Thus, it is likely that she can remain in the community with little, if any, disruption to her life. This is reassuring for Anne and also for her family who have planned for her future.

Nevertheless, adaptable public housing was not available for either Anne or John. These case studies highlight the need for adaptable public housing within Australia. Until such housing is available, sectors of the community, including people with a developmental disability, will continue to battle discrimination and segregation, not only in community attitudes but also in housing.

38.8   *GENERAL LESSONS LEARNED FROM UNIVERSAL DESIGN*

Services for people with a disability throughout the world are now starting to deal with issues that arise from the aging process. Many community homes are not able to accommodate the needs of people with changing physical support needs or other disabilities that are associated with the aging process. Such disabilities include loss of visual acuity, hearing impairment, and dementia. In Australia since the mid-1980s, many people with a developmental disability have moved from large residential services into community housing and are now experiencing improved quality of life and greater community participation (Young, Sigafoos, Suttie, Ashman, and Grevell, 1998). Individuals with a disability, families, caregivers, service providers, and government departments are now having to consider how best to accommodate people who are no longer able to manage easily in their community housing. Universal design concepts, if put into practice in Australia, would facilitate individuals remaining in their home and maintaining their newfound improved quality of life. In addition, universal design concepts help foster community integration of people with a disability and thus help break down barriers of discrimination and exclusion.

38.9   CONCLUSION

Within Australia, adaptable housing is still a fairly new concept that is not covered by any housing or planning legislation. Nevertheless, it is clear that as many Australians (including those with a disability) age, the benefits of adaptable housing will become increasingly apparent. The additional 5 percent building costs incurred in building adaptable housing are more than made up for by the cost benefits to the community. Such cost benefits relate to reduced care costs for the individuals and the state (for example, a reduction in need for nursing home placement) and also relate to quality of life, visitability, and the opportunity to age in place. To plan for the future, policy makers, service providers, and all of those involved in supporting individuals with a developmental disability and their families need education and information not only on the advantages of adaptable housing but also on where they can obtain help in planning such housing. Architects and planners should be aware of the need for individuals with a disability living in the community to have adaptable housing that will meet any changing needs throughout their lifespan.

38.10   BIBLIOGRAPHY

Ashman, A. F., J. Suttie, and J. Bramley, *Older Australians with an Intellectual Disability,* Fred and Eleanor Schonell Special Education Research Centre, Brisbane, Australia, 1993.

Ashman, A. F., and J. N. Suttie, "The Social and Community Involvement of Older Australians with Intellectual Disabilities," *Journal of Intellectual Disability Research,* 4: 120–129, 1996.

Australian Institute of Health and Welfare (AIHW), "Disability Support Services Provided Under the Commonwealth/State Disability Agreement," *National Data 1998 AIHW Cat. No. DIS 16,* AIHW, Canberra, Australia, 1999.

Balandin, S., and N. Kerse, "Aged Care," in N. Lennox and J. Diggens (eds.), *Management Guidelines. People with Developmental and Intellectual Disabilities,* Therapeutic Guidelines Ltd., Melbourne, Australia, 1999, pp. 47–60.

Balandin, S., and J. Morgan, "Adults with Cerebral Palsy: What's Happening?" *Journal of Intellectual and Developmental Disability,* 222: 109–124, 1997.

Beange, H., "Caring for a Vulnerable Population," *Medical Journal of Australia,* 1645: 159–160, February 1996.

Beange, H., A. McElduff, and W. Baker, "Medical Disorders of Adults with Mental Retardation: A Population Study," *American Journal on Mental Retardation,* 996: 595–604, 1995.

Bohmer, C. J., E. C. Klinkenberg-Knol, R. C. Niezen-de Boer, and S. G. Meuwissen, "Cancer of the Esophagus," *Journal of Gastroenterology Hepatology,* 9: 589–592, 1997.

Campbell, C., "Benefits of Adaptable Housing," *Proceedings of the Local Government and Shires Association of New South Wales Workshop,* Sydney, Australia, September 1997.

Chapman, R., "A Collaborative Approach to Fall Prevention," paper given at *"They Said We'd Never Make It" Asian and Pacific Cerebral Palsy Conference,* Sydney, Australia, May 1997.

Crichton, J. U., M. Mackinnon, and C. P. White, "The Life Expectancy of Persons with Cerebral Palsy," *Developmental Medicine and Child Neurology,* 37: 567–576, 1995.

Devenny, D. A., W. P. Silverman, A. L. Hill, E. Jenkins, E. A. Sersen, and K. E. Wisniewski, "Normal Aging in Adults with Down's Syndrome: A Longitudinal Study," *Journal of Intellectual Disability Research,* 403: 208–221, 1996.

Dorval, J., "Achieving and Maintaining Body Systems Integrity and Function: Clinical Issues," paper presented at *Preventing Secondary Conditions Associated with Spina Bifida or Cerebral Palsy,* Crystal City, VA, 1994.

Gething, L., *Working with Older People,* W. B. Saunders, Sydney, Australia, 1990.

Hawkins, B. A., "Health and Medical Issues," in E. Sutton, A. R. Factor, B. A. Hawkins, T. Heller, and G. B. Seltzer (eds.), *Older Adults with Developmental Disabilities,* Paul H. Brookes Publishing Co., Baltimore, MD, 1993, p. 1.

Hogg, J., "Intellectual Disability and Aging: Ecological Perspectives from Recent Research, *Journal of Intellectual Disability Research,* 412: 136–143, 1997.

Intellectual Disability Services Council, *Having a Place in the Community.* Disability Services Office, Adelaide, Australia, 1994.

Janicki, M. P., and A. J. Dalton, "Sensory Impairments Among Older Adults with Intellectual Disability," *Journal of Intellectual and Developmental Disability,* 231: 3–12, 1998.

Janicki, M. P., T. Heller, G. B. Seltzer, and J. Hogg, "Practice Guidelines for the Clinical Assessment and Care Management of Alzheimer's Disease and Other Dementias Among Adults with Intellectual Disability," *Journal of Intellectual Disability Research,* 404: 329–339, 1996.

Jenkins, R., D. Brooksbank, and E. Miller, "Aging in Learning Difficulties: The Development of Health Care Outcome Indicators," *Journal of Intellectual Disability Research,* 38: 257–264, 1994.

Kailes, J. I., "Aging with a Disability: Educating Myself," *The Networker,* 71: 6–9, 1993.

Kendig, H., and I. L. Gardner, "Unravelling Housing Policy for Older People," in A. Borowski, S. Encel, and E. Ozanne (eds.), *Aging and Social Policy in Australia,* Cambridge University Press, Cambridge, UK, 1997.

Krauss, M. W., and M. M. Seltzer, "Taking Stock: Expected Gains from a Life-Span Perspective on Mental Retardation," in M. M. Seltzer, M. W. Krauss, and M. P. Janicki (eds.), *Life Courses Perspectives on Adulthood and Old Age,* American Association on Mental Retardation, Washington, D.C., 1994, 213–219.

Lewit, J., "Choosing the Right Architect," *AGENDAS—Aged and Community Services in Australia,* 12: 12–13, Summer 1999.

McFee, G., and G. Bray, "Older People in New South Wales: A Profile," *Government Report 4108.1,* Australian Bureau of Statistics, 1995.

Ministerial Council of Review on Aging, *Positive Aging: On Our Agenda,* NSW Government Printers, Sydney, Australia, 1994.

Moss, S., and P. Patel, "Dementia in Older People with Intellectual Disability: Symptoms of Physical and Mental Illness, and Levels of Adaptive Behaviour," *Journal of Intellectual Disability Research,* 411: 60–69, 1997.

Murphy, K. P., G. E. Molnar, and K. Lankasky, "Medical and Functional Status of Adults with Cerebral Palsy," *Developmental Medicine and Child Neurology,* 37: 1075–1084, 1995.

National Access Working Group, *Accessible Design in Australia: A discussion Paper,* The University of Sydney, Sydney, Australia, 1999.

Pratt, R., "Adaptable Housing—Better Homes for Everyone," paper given at the *Conference of Elegant Access Solutions,* Adelaide, South Australia, May 1998.

Schwartz, B., "Aging for Persons with Cerebral Palsy," *Communicating Together,* 123: 14–15, 1994.

Standards Australia, *AS 1428 Design for Access and Mobility,* Standards Australia, Homebush, NSW, 1993.

Standards Australia, *AS 4299 Adaptable Housing,* Standards Australia, Homebush, NSW, 1995.

Todd, S., G. Evans, and S. Beyer, "More Recognised than Known: The Social Visibility and Attachment of People with Developmental Disabilities," *Australia and New Zealand Journal of Developmental Disabilities,* 163: 207–218, 1990.

Turk, M. A., "Attaining and Retaining Mobility: Clinical Issues," paper presented at the *Preventing Secondary Conditions Associated with Spina Bifida or Cerebral Palsy,* Crystal City, VA, 1994.

Turk, M. A., C. A. Geremski, P. F. Rosenbaum, and R. J. Weber, "The Health Status of Women with Cerebral Palsy," *Archives of Physical Medicine and Rehabilitation,* 78: S-10–S-17, December 1997.

Turk, M. A., J. C. Overeynder, and M. P. Janicki, (eds.), *Uncertain Future—Aging and Cerebral Palsy: Clinical Concerns,* New York State Developmental Disabilities Planning Council, Albany, NY, 1995.

Turner, S., and S. Moss, "The Health Needs of Adults with Learning Disabilities and the Health of the Nation Strategy," *Journal of Intellectual Disability Research,* 40(5): 438–450, 1996.

Uehara, S. E., B. J. Silverstein, R. Davis, and S. Geron, "Assessment of Needs of Adults with Developmental Disabilities in Skilled Nursing and Intermediate Care Facilities in Illinois," *Mental Retardation,* 294: 223–231, 1991.

Walker, A., and C. Walker, "Normalisation and 'Normal' Aging: The Social Construction of Dependency Among Older People with Learning Difficulties," *Disability & Society,* 131: 125–142, 1998.

Willner, L., and D. Dunning, *Aging with Cerebral Palsy,* SCOPE, London, UK, 1993.

Wolfensberger, W., "Additional Implications of the Normalization Principle to Residential Services," in W. Wolfensberger (ed.), *The Principle of Normalization In Human Services,* National Institute on Mental Retardation, Toronto, ON, Canada, 1972.

Young, L., J. Sigafoos, J. Suttie, A. Ashman, and P. Grevell, "Deinstitutionalisation of Persons with Intellectual Disabilities: A Review of Australian Studies," *Australian and New Zealand Journal of Public Health,* 22: 155–170, 1998.

38.11   RESOURCES

Australian Bureau of Statistics.   www.statistics.gov.au/websitedbs/d3310114.nsf/Homepage

Australian Institute of Health and Welfare.   www.aihw.gov.au/

Australian Urban and Regional Development Review.   www.ahuri.csiro.au/othersites/plandevt.htm

Building Code of Australia.   www.dlg.nsw.gov.au/bran.htm

CHAPTER 39

HOUSING POLICY AND FUNDING MECHANISMS FOR ELDERLY AND DISABLED PEOPLE IN GERMANY

Karin Piltner, M.A.
KOOB Agentur für Public Relations, Mülheim a.d. Ruhr, Germany

with Brigitte Halbich
HEWI Heinrich Wilke GmbH, Bad Arolsen, Germany

39.1   INTRODUCTION

At this time, the topic of universal design is synonymous with barrier-free building in Germany. In 1992 and 1996, revised versions of planning guidelines and standards for public buildings and workplaces [Deutsche Industrie Norm (DIN18024)] and housing (DIN18025, Part 1, barrier-free housing for people in wheelchairs, and Part 2, barrier-free housing in general) were put into effect.

These DIN standards constitute the basis for planning, especially in the area of building for elderly people and people who are disabled. Based on demographic changes and the growing age pyramid in Germany, there is increased demand for barrier-free housing, which permits people with advanced age and mobility limitations to lead an independent life. Therefore, different ways were developed to subsidize the remodeling and the construction of housing. This chapter describes policies and funding agencies in Germany for barrier-free construction, home modification, and technical assistance, and how they relate to new product development regarding universal design.

39.2   BACKGROUND

Over the past 5 years, barrier-free design has become a much-debated topic in Germany. Many valuable and easy-to-understand brochures were issued by organizations for seniors, associations serving disabled people, and industry. The discourse about barrier-free living in this context is focused at the perspective of building for elderly people and people who are disabled. Because of this, the term *barrier-free* is generally understood by the public and experts as being for people who are elderly and disabled. However, guidelines and regulations for barrier-free living clearly differentiate between the standards for people in wheelchairs (DIN18025, Part 1) and the standards for barrier-free living (DIN18025, Part 2). The implications of Part 2, as published in the DIN standards of December 1992, read as follows:

This standard is applicable, depending on individual need, to planning, construction and furnishing of barrier-free new, improved or renovated buildings, as well as modernization of apartments, housing complexes, as well as homes. The apartments have to be usable by people, the residents have to be able to be independent of outside assistance. This is particularly true for blind persons, persons with other disabilities, elderly persons, children, short and very tall people.

The fact that the standard is concerned with elderly people and people who are disabled has evolved into a misguided conception that often reads: "Barrier-free living is for people who are elderly and disabled." This is also reflected in the regulations governing funding for housing in the various states. However, the situation of people who are elderly and who are disabled requires particular sensitivity and tact. Unfortunately, this is lacking all too often, whether in the context of writing laws and regulations and consulting, or in advertisements. In the real world, this stigma tends to make these people who need such help the most reluctant to apply for it, or even admit they need it. In today's society, people rarely wish to be considered "old." If, however, the removal of barriers in apartments leads to the perception that persons are limited or inferior, then it appears necessary to change the focus of discussion or discourse, both in public and in expert circles. In this context, it is possible that the term *barrier-free* in itself constitutes a barrier. While it may be well intended to treat people who are elderly and those who are disabled as equal to socially weak persons or families with many children, as in the context of funding programs, this could be interpreted as stigma. In spite of this, the idea to subsidize home modifications constitutes a particularly valuable development in Germany.

39.3   THE NEED FOR HOME MODIFICATIONS

Only few elderly people are willing to give up their comfortable environment and move into an apartment that serves seniors' needs, and even less so into a public retirement institution for elderly people. This trend can be observed, not only in Germany, but also in other western industrialized countries.

The Europeans, however—in contrast to U.S. citizens—are, as a whole, less flexible concerning changing living locations, or living space, and this even applies to younger people. Thus, in Germany, the desire to remain independent is, especially for elderly people, strongly linked to the desire to stay in the present living environment.

Compared with the United States, more people in Germany rent than own a home. But more and more people who are elderly are deciding to modify their homes—to make them barrier-free. This allows them to preserve their independence at a much more advanced age without giving up their familiar environment. In many cases, people who are elderly and people with disabilities qualify as recipients for special funding programs from the state or from long-term-care insurance for these home modifications.

Generally, subsidies for the remodeling and construction of housing are linked to certain conditions such as living in a rented house or apartment (i.e., living in a condominium). In addition, each federal state in Germany provides grants and funds for the remodeling or construction of senior homes and assisted living facilities, according to German industrial standard DIN18025.

Interestingly enough, the system of funding agencies in Germany differs from that in the United States. In Germany, the subsidies are often provided to individuals as part of the government insurance programs, whereas in the United States, funds and grants are mainly provided through special programs by the Department of Housing and Urban Development (HUD) or state housing and finance agencies. These programs are primarily for remodeling or new construction of multifamily dwellings, for people who are elderly, or people with disabilities. In the United States, the programs to provide or fund home modifications for individuals are very limited, but there is growing advocacy for systems to adapt homes in order to help people remain in their homes.

39.4 *HOUSING FOR PEOPLE WHO ARE ELDERLY AND PEOPLE WHO ARE DISABLED IN GERMANY*

Demographic Trends

The demographic trends in Germany lead to rethinking in many areas as the percentage of the total population of people who are elderly continues to grow. As of 1998, out of a population of over 82 million, about 13 million were older than 65.

According to information from the German Census Bureau (Statistisches Bundesamt in Wiesbaden), by the year 2050 people between the ages of 58 to 63 will represent the largest group. Today, the largest group is between 35 and 40. At the same time, Germany's population will decline in the next 50 years by at least 12 million.

These changes in the age pyramid have led architects, designers, and gerontologists to rethink building construction and focus on barrier-free design. For instance, public buildings as well as many facilities designed and built for elderly people in the past 15 to 20 years fulfill the guidelines for barrier-free construction, according to German industrial standards (DIN18024 and 18025, Part 1). Beyond this, the changes in the age pyramid have also caused changes in the insurance system and in product development.

Changes in the Social System

From the perspective of society as a whole, however, the barrier-free construction and remodeling of public and private buildings is just beginning. To accelerate this process, the government has developed a variety of programs to subsidize barrier-free houses and apartments, especially for people who are elderly and people who are disabled. Helping this target population to live independently is also seen as a chance for significantly lowering the costs for institutional long-term care.

In this context, long-term-care insurance was introduced in Germany. On May 26, 1994, a law was passed in order to ensure against the risk associated with the need for long-term care. The purpose of long-term-care insurance is to provide help in many different ways. One form of supporting people who need care is to facilitate caregiving through home modifications.

Besides this, the government health insurance program offers assistance for independent living. For example, not only are costs for medicine paid for through health insurance, but technical aids such as assistance for climbing stairs, grab bars, or shower seats may also be prescribed by the doctor, and thus be paid for.

Furthermore, industry is making efforts to adjust to the changing demographic situation by developing innovative ergonomic products. A benchmark in the development of such products was recently set by the internationally active technical control board, the TÜV Rheinland Product Safety GmbH, in Cologne. In cooperation with the German Society for Geronto-Technology (GGT) in Iserlohn, Germany, the TÜV developed a special test certificate for products to ensure safety and comfort for all users.

Living to Advanced Age

The question of how people who are elderly live is often associated with the subject of "living in a public or private home for elderly people." Contrary to widespread opinion, only a minority of people who are elderly live in senior housing facilities or in nursing homes in Germany.

As can be seen in Table 39.1, in Germany 11.6 million people who are elderly live at home. In other words, 93 percent of those over 65 live either in an apartment or in their own home. Only a small number of these people live in a nursing home.

According to Klaus Grossjohann of the German Foundation for Elderly Care (Kuratorium Deutsche Altershilfe), quantitatively, models such as assisted living are not yet playing a

TABLE 39.1   Housing Choices for Elderly People in Germany, 1998

Housing choices	Total no. of elderly people	Percentage of elderly population*[†]
Standard apartments and houses	11.6 million	Approx. 93
Senior living facilities	661,000	5.3
Nursing homes	375,000	3.0
Senior homes	204,000	1.6
Independent living facilities (including those provided by charitable institutions)[‡]	82,000	0.7
Hospice	100	
Housing designed for elderly people		
Housing for elderly people	200,000 (to 250,000)	1.6 (to 2.0)
Assisted living facilities	Approx. 30,000	0.25
Total no. of housing for elderly people	861,000 (to 911,000)	6.9 (to 7.3)
Innovative housing choices	No. of units	
Assisted living group apartments	Approx. 100	
Group homes for elderly people	Approx. 1000	
Integrated/multigeneration living	Approx. 1000	

* 12,451,773 elderly people age 65-plus.
[†] The values listed in this column are percentages.
[‡] The selection is not clearly defined. Placements from housing provided by charitable institutions might be found under independent living facilities and also under assisted living facilities.
Source: German Census Bureau (Statistisches Bundesamt, Wiesbaden/Germany), 1998.

large role in Germany. However, there is no reliable statistical information on this issue at this time. Models with optional services such as assisted living residences are therefore not included in the statistics of Table 39.1. These barrier-free models of the most current type target primarily wealthy seniors, and have shot up everywhere in the past few years.

The Need for Care

Elderly people between the ages of 60 and 75 generally enjoy relatively good health. The life expectancy of a 60-year old man in Germany is an additional 18.3 years and for a 60-year old woman it is 22.6 years.

In fact, the probability of the need for long-term care before the age of 80 is quite low. Only 0.5 percent of people under 60 years of age require caregiving. Between ages 60 and 80, this percentage grows to 3.5 percent. For those over 80 years, 28.0 percent require caregiving.

In absolute numbers, 1.81 million people presently require caregiving in Germany. Due to demographic changes, this number will increase by about 333,000 by the year 2010.

Living in a Nursing Home

The desire to remain at home can also be seen among those depending on caregiving at an advanced age. Fewer between the ages of 80 and 84 live in a nursing home: less than 10 percent.

Due to the lack of single rooms, the thought of moving to a nursing home is a horrifying thought for many. During the International Year of Seniors in 1999, the German Commission on the Aged recommended that a room for one person should have at least 14 square meters. Furthermore, they pointed out that almost two-thirds of all occupants of nursing homes must share a room. In the reunited states of the former East Germany, this number is somewhat higher.

A lack of single rooms can be observed in senior homes. In the western states of Germany, 27 percent of elderly people have to share a room in a senior home, whereas in the reunited states this figure is over 40 percent. Thus, existing building stock does not meet the need for privacy, nor does it fulfill the requirements of barrier-free design.

Home Modification—The More Economical Choice

For those who prefer living at home, barrier-free design and home modification will help to maintain a maximum level of independence and enhance their ability to stay in a comfortable environment. Almost a third of the homes with people over 60 are presently not barrier-free and have deficiencies in interior design. In particular, bathrooms and shower tubs have serious deficiencies; in some cases, there is no bathroom at all. Again, the situation is worse in former East Germany: About 26 percent of the housing for elderly people is classified as being deficient.

Making home modifications earlier has been recommended for over 12 years by the German Foundation for Elderly Care. This seniors' organization has persistently pointed out in their brochures that not only is the quality of life of people increased, but home modifications are significantly more economical than maintaining a nursing home. For the cost of care of 5000 DM (German marks) per month, the government will pay about 3500 DM. Reducing the cost for institutional care through savings from 10 modified homes will allow home modification for about another hundred elderly people. This is the argument presented by the German Foundation for Elderly Care.

This approach may be adequate. However, the cited example, which is used to appeal to local politicians and politicians that are concerned with care for the aged, may be too optimistic (i.e., 25 to 50 modifications of homes appear to be more realistic). Supporting home modification and technical assistance should pay off for the government.

39.5 LIVING CONDITIONS FOR ELDERLY PEOPLE IN THE EUROPEAN UNION

Developing subsidy programs for home modification and barrier-free construction of private homes requires that all parties of the ownership structure be examined. In Germany, this structure is different from the United States, and occasionally even so within the European Community, according to the report of the Census Bureau of the European Community, Eurostat.

In 1995, about 60 percent of the EU households lived in their own home. (See Table 39.2.) Compared with the total population, elderly couples represented the largest group of home owners. In this context, all EU countries present a similar picture when differentiating between elderly couples and elderly persons living on their own: Elderly people living by themselves are much less frequently home owners than couples.

The living conditions for people who are elderly may vary significantly within the individual member states of the European Union. While the number of elderly couples living in their own homes is 93 percent in Ireland, in the Netherlands it is only 40 percent, and in Germany 52 percent. Among those elderly people living by themselves, about 20 percent own their homes in the Netherlands, and in Germany this number is 31 percent.

With respect to quality of life, problems can be primarily found with those living by themselves, as Table 39.3 indicates. For 12 percent of this group within the European Union, at least one basic comfort is lacking (i.e., a bathroom). The percentage is 14 in Germany, 21 in Ireland, and 54 in Portugal.

TABLE 39.2   Households of Elderly People Owning Homes and Total Percentage of Households Living in Own Home, 1995*

	Households of elderly people			Total no. of households
	Average	Couples	Singles	
EU15[†]	59	68	50	60
Belgium	68	78	56	69
Denmark	51	74	33	57
Germany	41	52	31	42
Greece	88	91	84	81
Spain	80	83	76	81
France	65	77	52	56
Ireland	87	93	81	83
Italy	75	80	70	74
Luxembourg	74	82	64	69
Netherlands	30	40	21	49
Austria	44	57	35	49
Portugal	63	67	55	62
United Kingdom	59	71	50	68

* All values are percentages.
† No data are available for Finland and Sweden.
Source: EUROSTAT.

39.6   FINANCIAL SUBSIDY POLICY IN GERMANY

During the International Seniors' Year Expert Conference in 1998, the German Commission on the Aged strongly recommended general implementation of barrier-free construction and design. According to the Commission, the use of public funds should be coupled with the general meeting of the German industrial standard (DIN18025, Part 2). Many of the companies who are members of the German Federal Association of Housing (Bundesverband Deutscher Wohnungsunternehmen) are already applying barrier-free guidelines as a standard for remodeling and modernization. With a total of 7 million apartments, these companies may plan improvements better and implement them more clearly than private-sector owners of rental apartments.

In addition to general programs for subsidizing housing, each federal state in Germany has developed special programs providing funds and grants for both the barrier-free rehabilitation, as well as for the construction of new buildings. Although many programs do not consider barrier-free construction explicitly, more and more home owners choose barrier-free design and architecture for additions and modernization.

39.7   FUNDING AGENCIES

Federal Subsidies for Barrier-Free Housing in Germany

Each of the 16 federal states of Germany, including the city states of Hamburg and Bremen, has developed numerous subsidy programs for housing, which are tailor-made to the respective situations. Typically, the support of barrier-free construction and modernization of housing in respective states has the following in common:

• The support of apartments is a focus of support within the context of social housing construction.

TABLE 39.3   Percentage of Households in Europe Lacking at Least One of the Three Basic Comforts, 1995*[†]

	Households of elderly people			Total no. of households
	Average	Couples	Singles	
EU15[‡]	9	6	12	5
Belgium	15	10	21	7
Denmark	3	1	4	3
Germany	11	8	14	7
Greece[§]	—	—	—	—
Spain	11	10	12	4
France	12	7	16	5
Ireland	14	8	21	6
Italy	8	4	13	3
Luxembourg	8	5	10	4
Netherlands	2	2	2	2
Austria	14	10	16	8
Portugal	41	34	54	23
United Kingdom	1	0	2	0

* Lack of comfort is defined as the following: (1) lack of bathtub/shower, (2) lack of hot water, (3) lack of toilet within the apartment/house.
[†] All values are percentages.
[‡] No data are available for Finland and Sweden.
[§] Data on comfort are not available for Greece.
Source: EUROSTAT.

- The support of private houses and condominiums is tied to certain income limits.
- The support may include new construction, as well as home modification.
- More than one program for funding support may be available for the construction/ modification of barrier-free housing.
- Special funding support for barrier-free housing is tied to certain use conditions; for example, utilization for people who are elderly or disabled.
- Support for interest-free loans extends over a certain period—typically, 15 years.

The following examples from the old federal states of Germany, in addition to the new reunited states of the former East Germany, may give a good idea of the most important subsidy programs for barrier-free housing.

EXAMPLE 39.1: LOWER SAXONY

Item 1: Funds and grants for private homes for new construction, modification, remodeling, and acquisition
In the state of Lower Saxony, building projects for private ownership are supported for personal use. This includes private homes and apartments. Selection of projects that are to be supported is principally based on social priorities.

For example, people who are disabled, or households with dependents who are disabled, are target groups. This includes people with exceptional mobility problems, people in wheelchairs, and people who are severely disabled or ability-impaired, as well as people suffering from multiple sclerosis. Other criteria include:

- *Total income, including the income of all persons belonging to a household, which may not exceed a certain limit. This limit is set in paragraph 25 of the second housing law.*

- *Living conditions of the applicants need to be inadequate, according to the guidelines for barrier-free building.*
- *Additional building space is required, because of a disability.*

Item 2: Rental apartments

The support of apartments is focused on future-oriented projects, as well as on construction or remodeling of rental apartments for people of limited income. For example:

- *Projects with apartments for people who are elderly or people who are disabled, including the necessary service support*
- *Multigenerational housing*
- *The adaptation of existing housing to meet the needs of people in wheelchairs*
- *The modernization of existing housing*

The preconditions for funding support, as far as building is concerned, is the need to meet the DIN standards for barrier-free living, in addition to creating certain apartment sizes that lawmakers determined, and that are considered, adequate: For a single person, up to 50 m²; for two members of a household, up to 60 m²; for three members of a household, up to 75 m²; for any additional members of the household, up to 10 m² more. Funding support of housing for people who are severely disabled includes, in addition to the two types of basic mortgages (see Table 39.4), another loan of up to 25,000 DM per apartment.

The size of the mortgage is established in the German housing law (WoBauG). The following mortgages are offered for the construction of apartments for elderly people:

- *60,000 DM per apartment for eligible persons, according to paragraph 25*
- *45,000 DM per apartment for eligible persons, according to paragraph 88a*

Regulations are applicable to the adaptation of apartments, which can be determined on an individual basis.

In the state of Lower Saxony, a total of 96,000 apartments was supported/funded between the years 1990 and 1999. The situation varies in the respective individual states based on their size, and because states have their own laws and regulations and can have variations not just in programs but also in definitions. For example, the state of Bavaria has funds for home modification meeting

TABLE 39.4 Funding of Private New Construction in the Federal State of Lower Saxony, 2000

	First mortgage up to*:	Second mortgage up to*:
For families including		
3 and more children (with 2 children age 14 and younger)	80,000	—
4 children (with 3 children age 14 and younger)	80,000	40,000
5 and more children (with 4 children age 14 and younger)	80,000	60,000
For people who are disabled/families including a person who is disabled and		
Up to 2 children	60,000	—
3 and more children (with 2 children age 14 and younger)	80,000	—
4 children (with 3 children age 14 and younger)	80,000	40,000
5 and more children (with 4 children age 14 and younger)	80,000	60,000

Note: For the construction of housing for people who are disabled, generally an additional loan of DM 25,000 is being granted as a second mortgage.

* The values listed in this column are German marks (DM).

Source: Landestreuhandstelle of Lower Saxony, 2000.

the requirements of the DIN standard 18025, under the category of modernization, *whereas Lower Saxony uses the term* modification. *Beyond that, the state of Bavaria, for example, has special programs to support nursing homes. In the five new Federal states of the former East Germany, however, there are more programs for the modification of apartments, which are described as* modernization, remodeling, improvements, *or* expansion of existing buildings.

EXAMPLE 39.2: THURINGIA

Thuringia, which belongs to the so-called five new Federal states, has a total of 14 subsidy programs for social housing. The requirements of the DIN standard 18025 (barrier-free housing) are specifically mentioned in the programs under the term WO1, WO7, *the descriptions of which follow.*

Item 1: Private homes and apartments (WO1 Modernization and Remodeling)
The target group, among others, are people who are ill or disabled. The goal of these measures is, among others, the improvement of:

- *The layout of the apartment and functional relationships*
- *The sanitary fixtures and installation*

   Specially mentioned in the program WO1 are these items, in connection with DIN18025:

- *Barrier-free entrances*
- *Wheelchair-accessible toilets, bathrooms, and kitchens*
- *Ramps*
- *Stairway lift and floor covering*
- *Grab bars in bathrooms and toilet rooms, shower seats, and benches or seats for bathtubs*

Item 2: Modification of housing for persons who are ill or disabled (WO7)
This program is similar to the preceding one. At the same time, however, it is intended to serve the removal of "exceptional housing needs of low-income households," who live in their own private home or condominium.

Item 3: Private home or apartment, based on the acquisition of existing housing (WO8)
This program is targeted at people who are severely disabled, among others, and enables this group to obtain support for the purchase of a home or apartment, or to obtain support for modifications, according to DIN18025.

All states maintain, though, that in principle nobody has a right to funding support. Funds are only made available according to need and fiscal-year budget limit.

Other programs of the state of Thuringia also support modernization and remodeling. However, in this case, barrier-free design/building is not considered a necessity. Furthermore, without specifically targeting DIN18025, subsidies are provided for:

- *Ground-level parking lots*
- *Improvement of entrances and access to the building*

The requirement to link public funds to meeting DIN18025 standards appears reasonable, considering the aforementioned examples, and was strongly recommended during the International Seniors' Year Expert Conference in 1998. It makes particular sense to underscore the topic of barrier-free building in the context of modernization, in order to create housing that is acceptable to all people on a long-term basis, a central goal of universal design in the United States.

Taxation Subsidy in Germany

According to current legislation, tax deductions/amortization and special funds to support the cost of remodeling can be obtained. Although these are not linked to the requirements of the

DIN standards for barrier-free building, they can still be used for that purpose. The income limit in this case is 165,000 DM for single and 320,000 DM for married persons. That way, qualifying homeowners can obtain the so-called home improvement funds for expansion or remodeling projects. This amounts to 5000 DM per year, and can be distributed over a period of 8 years. Maximum support (i.e., 5 percent of 100,000 DM in building costs) can amount to up to 40,000 DM for a total project cost of 100,000 DM.

Special Community Programs

Many communities have established special programs for the improvement or modification for the purpose of barrier-free buildings. Conditions vary significantly from one community to another and extend from mortgages to full coverage of costs. Typically, the maximum support is 10,000 DM, and it is tied to the income limit of the applicant. Applicants can also be renters.

Long-Term-Care Insurance and Home Modification

Even long-term-care insurance can be used for subsidizing home modifications. This form of insurance is relatively new and was introduced in the context of demographic change in the year 1995.

About 90 percent of the population in the Federal Republic of Germany has health insurance by law and, therefore, is automatically insured in case long-term care is needed. Health insurance and long-term-care insurance are offered by the same insurance company, although these are two different types of insurance with separate fee structures. Even those who are not insured by law, or are privately insured because they are independently self-employed or have a relatively high income, are insured by law. They pay their premiums to private insurance companies.

With the introduction of this additional type of insurance, government has created a system that covers everybody. This is in order to keep pace with the growing age pyramid and respective need for care that is socially fair, not only for home, outpatient, and institutional care, but this also provides for expenditures of up to 5000 DM for home modifications furthering barrier-free design.

By law, the insurance consultants are obliged to correctly and completely inform about home modifications and independent living kits. The topic of barrier-free design has become a central theme for health maintenance organizations. Beyond home modifications, which could include such things as installation of a shower at floor level as shown in Fig. 39.1, there are certain provisions that cover the cost of so-called technical care systems and devices. The insurer covers these systems and devices under the following conditions:

- To facilitate caregiving
- To reduce pain
- To facilitate independent living

Typical technical care systems and devices include:

- Special beds that facilitate caregiving
- Emergency call systems that facilitate independent living
- Special devices that enhance personal hygiene
- Special bedding/support systems that reduce pain

Basically, the same provision of services applies to private insurers, just like the public insurance. Based on the long-term-care insurance law, both will scrutinize the specific situa-

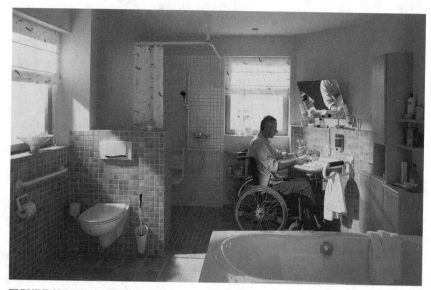

FIGURE 39.1   Shower installation, an example of home modifications funded by long-term-care insurance.

tion of an applicant who is elderly or disabled. According to the law, there are no upper limits for the cost of technical care systems and devices. However, insurers have agreed with the providers of various technical care systems and devices as far as fixed prices are concerned. If someone selects a more expensive item, then that person will have to bear the extra cost. In principle, it is possible to make almost all living areas barrier-free using technical care systems and devices. In that context, based on safety concerns, the bathroom is the most frequently modified area.

Some of the devices listed in Table 39.5 can also be covered by health insurance (i.e., if a person does not depend on caregiving, but needs some sort of technical assistance after surgery or because of an accident). Thus, the definition of technical assistance and technical devices may vary.

Besides providing the costs for technical systems and devices, long-term-care insurance covers home modification to a certain extent, as mentioned before. The maximum amount that long-term-care insurance will pay for each home modification is 5000 DM. The areas in which home modification can be accomplished are shown in Table 39.6.

Generally, renters have to first ask the owner for permission to have their home modified. Furthermore, there is no right to obtain provisions in every situation; every single case has to be examined by a medical expert who decides on the basis of clearly defined criteria whether a person needs caregiving.

Overall, the long-term-care insurance differentiates between three levels of caregiving. Home modifications and technical devices shall help to either avoid caregiving at all, or to make caregiving at home easier for nurses. The general goal is to keep costs as low as possible. Home care is far less expensive than institutional care, and home modifications that help maintain independent living as long as possible are less expensive than home care. Also, each individual person has to share the costs for his/her home modification to a certain amount, as shown in Table 39.7.

In some cases the provisions of long-term-care insurance might be neither sufficient nor fair (i.e., when people who receive lower retirement have to share the costs according to the funding system described in Table 39.7). If the costs for a home modification exceed the

TABLE 39.5   Technical Care Systems and Devices That May Be Funded by Long-Term-Care Insurance

Living area	Technical care systems and devices	Commentary
House entrance	Entrance ramps Walking supports/handgrips Reflective colored stair edges Collapsible wheelchair	
Stairway	Handrails Stair lift Seating and handgrips in elevators	
Bedroom	Special ergonomic beds Support springs Standing help devices	
Bathroom and toilet	Grab bars Handrails Faucets with pullout spray head Tub seats Shower seats	Regarded as a technical assistance being paid by health insurance if it is not mounted to the wall.
	Lifting cranes Toilet seat adapters	Regarded as a technical assistance being paid by health insurance if it is flexible and not fixed to the toilet.
	Adjustable-height toilet	
Kitchen	Faucets with pullout spray head Rolling cabinets	
Emergency	Emergency call facilities (panic buttons)	

Source: Ministerium für Arbeit, Soziales und Stadtentwicklung, Kultur und Sport des Landes Nordrhein-Westfalen, 1999.

amount of 5000 DM, self-payment may be equal to 50 percent of a person's gross income. However, the long-term-care insurance will cover the costs of additional home modifications, if needed in the future, each up to 5000 DM.

Financing of Technical Assistance Through Health Insurers

Generally speaking, technical assistance is provided for those who do not depend on caregiving. While long-term-care insurance provides technical care systems and devices, health insurance covers the costs for so-called technical assistance. For example, if a disability is certified by a doctor, then the health insurance will cover the costs, and only a small amount has to be paid by the patient.

If someone is not contributing to private health insurance or long-term-care insurance (e.g., he/she is unemployed or lives on welfare), then they can apply to their welfare or assistance office for technical assistance and technical care systems and devices, in addition to home modifications.

Often, people need help and assistance for a period of time (i.e., after surgery, after treatment in a rehab due to an accident, etc.). As long as they are able to continue living independently at

TABLE 39.6 Home Modifications/Improvements That May Be Funded by Long-Term-Care Insurance

Improvements outside an apartment/house	
Elevator	Adaptation according to the DIN standards for people in wheelchairs: entrances at floor level, widening of doors, light switches and lever handle within hand's reach
Orientation systems	Installation of technical devices for blind people (i.e., on each floor)
Entrances	Installation of pneumatic doors; installation of even surfaces without barriers
General modifications/improvements within the living area	
Moving space	Improvements to create barrier-free space for movement
Floor covering/surfaces	Removal of barriers (i.e., elimination of multiple floor levels or slippery surfaces)
Reorganization of the living area	Adaptation of the architectural plan to meet the changing needs of elderly people (i.e., switching room functions within a two-level house, such as remodeling and redesigning rooms on ground-floor level into bedrooms and bathrooms)
Modifications/improvements of bathroom	
Creation of a new bathroom/toilet	Adaptation/redesigning of the living area to create space for a bathroom as well as construction of a bathroom/toilet
Adaptations within an existing bathroom	Installation/mounting of ergonomic faucets (i.e., faucets with pullout spray head)
Bathtubs	Installation/mounting of special technical device to assist in entering bathtub
Floor covering/surfaces	Removal of barriers (i.e., elimination of multiple floor levels or slippery surfaces)
Shower stall	Installation/creation of a shower on floor level if entering bathtub is not possible with technical devices and without caregiving
Furniture and fixtures	Adaptations of heights (i.e., cabinets, sinks, toilets according to the standards for people in wheelchairs)
Modifications/improvements of bedroom	
Entrances	Removal of barriers to ensure an easy and safe entrance to the bed
Floor covering/surfaces	Removal of barriers (i.e., slippery surfaces)
Light switches	Installation of light switch within arm's reach of bed
Modifications/improvements of kitchen	
Plumbing/faucets	Installation of hot-water systems and ergonomic faucets with pullout spray head
Floor covering/surfaces	Removal of barriers (i.e., slippery surfaces)
Kitchen cabinets	Modification of furniture height (i.e., height of cabinets, refrigerator, sink, etc., and creation of rolling cabinets according to the standards for people in wheelchairs)

Source: Verbraucher-Zentrale Nordrhein-Westfalen, 2000.

TABLE 39.7   Maximum Costs That Long-Term-Care Insurance Covers for Each Home Modification, and Minimum Deductible Based on Income*

Costs for home modification	6200	5200	6200
Monthly gross income of person depending on care giving	2000	800	6000
Self-payment	1200	400	1200
Amount covered by long-term-care insurance	5000	4800	5000

* All values are German marks (DM).
Source: Verbraucher-Zentrale Nordrhein-Westfalen, 2000.

home, with the help of technical assistance, they are not defined as dependent on caregiving. In all these cases, health insurers cover 90 percent of the costs for technical assistance.

As Table 39.8 indicates, the definition of technical assistance may depend on whether a device is mounted to a wall or is flexible. However, experiences show that insurers do not apply the definition according to the book, but try to keep procedures flexible in order to maintain a person's self-independence.

The basis for prescribing this type of technical assistance is the general technical assistance directory (*Hilfsmittelverzeichnis*) that is used by all insurers. Companies that have developed ergonomic and barrier-free products have the opportunity to have their products evaluated by the medical services of the insurers. If the product meets the established criteria, then application for inclusion in the technical assistance directory can be made and can be prescribed by the doctor through a prescription or certification.

For example, technical assistance devices in the kitchen can consist of a walker or a slip-resistant mat. Typical assistance devices in the bathroom are shower seats and grab bars in the sink or toilet area.

TABLE 39.8   Technical Assistance That May Be Funded by Health Insurance

Living area	Technical assistance	Commentary
House entrance	Walking supports/handgrips	
	Collapsible wheelchairs	
Stairway	Handrails	
	Chair lift/walking supports	
Bedroom	Special ergonomic beds	
	Standing help devices	
Bathroom and toilet	Grab bars	
	Handrails	
	Tub seats	Defined as technical assistance
	Shower seats	if it is not mounted to the wall.
	Lifting cranes	Defined as technical assistance if a full bath is required.
	Toilet seat adapters	Defined as technical assistance if it is flexible and not fixed to the toilet.
Kitchen	Standing help devices	
	Walkers	
	Technical assistance for the housework (i.e., trays for one-handed people, antislip mats, special ergonomic gripping tongs)	Devices for daily use might be defined as technical assistance, depending on a person's situation.
Living room	Special ergonomic arm chairs/seats	

Source: Ministerium für Arbeit, Soziales und Stadtentwicklung, Kultur und Sport des Landes Nordrhein-Westfalen, 1999.

As the bathroom is the area at home that is most frequently modified to make independent living possible and ensure safety, even bathroom designers can directly bill the health insurance under certain conditions. They can do so if their services and fixtures meet the criteria of the general technical assistance directory, which is provided by the board of health insurers.

39.8 PRODUCT DEVELOPMENT AND CERTIFICATE: "COMFORT & QUALITY"

People who are elderly or disabled and who are dependent on barrier-free products have a need to select products that are appropriate for the given situation. Often, people in this situation have difficulties because even relatives or advisors who assist cannot truly comprehend the range of products that are on the market. Many of these are low-cost imitations that do not meet the necessary safety regulations and, therefore, present yet another source of accidents in the household.

In March of 1999, the seal of "Comfort & Quality" was introduced in Germany, which constitutes an important aid in implementing improvements in barrier-free homes and apartments, but which also helps with the purchase of quality products. This certificate of quality was developed in cooperation with the German Society for Geronto-Technology (GGT) and the nationally active TÜV Rheinland Product Safety GmbH (TÜV). It is intended to assist

FIGURE 39.2   Example of a grab bar that may be prescribed by the doctor.

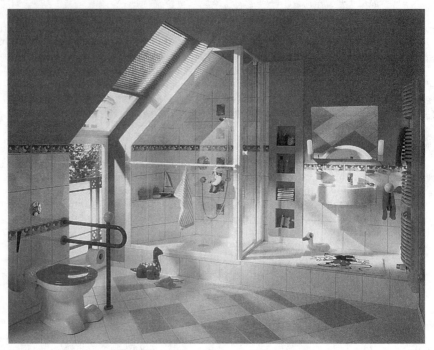

FIGURE 39.3   Children's version of grab bars representing the idea of universal design.

planners and buyers in assessing the quality of products, especially when modifying homes. The seal "Comfort & Quality" signifies a high degree of user comfort and technical safety.

The test persons doing a practical test are primarily elderly people. The specifications for the products go way beyond just designing for elderly people, although much emphasis is placed on people who are elderly. The experience of the TÜV Institute has proven to be advantageous for other age groups as well. For the first time in Germany, not only has an official institution paved the way for barrier-free products for people who are elderly or disabled, but for universal design as well.

The first company in Germany that had their barrier-free product line tested by the TÜV and the GGT was Hewi, based in Bad Arolsen. Hewi, a company that is well known for its sophisticated barrier-free design in Europe, applied for the procedure with different versions of grab bars (see Figs. 39.2 and 39.3) and a splash protection curtain unit (see Fig. 39.4).

Designed for multifunctional purposes, both products can be used in different situations. The grab bars ensure safety in the area of toilets and sinks, and also serve as towel holders or as towel rails. A smaller version, developed especially for children, was fitted with fun elements, which may also play a therapeutic role. Thus, these grab bars more closely resemble a toy rather than a handle meant to be used by a child with disabilities. Furthermore, the little grab bars may help children to become more self-reliant when using the bathroom.

The second Hewi product that passed the test to qualify for the "Comfort & Quality" seal was first used in hospitals to protect caregivers from splashing water and to provide privacy for patients. Meanwhile, many people use the splash protection curtain unit when caring for their elderly parents at home, or even may use it while sitting in a shower seat.

While the curtain unit folds flexibly upward and its outer end can fold around, it is also designed for the care of people in wheelchairs or for people with disabilities. To ensure the frame rods can also be seen by those with poor eyesight, the colors were chosen on a psychological basis.

FIGURE 39.4   Shower with a splash protection curtain unit that can fold upward when not in use.

In Germany, Hewi is regarded as the first company that embraces the idea of universal design in the field of sanitary fixtures. Until several years ago, that company called their design "social design" to point out that products have to be more than just functional, and that they should not be contributing to stigmatization when meeting the needs of people who are elderly or disabled.

At various symposia, Hewi invited designers, architects, gerontotechnologists, and experts from the field of community development, as well as people with disabilities, in order to exchange ideas about barrier-free housing and living in general. For example, the designer Roger Coleman was one of the speakers at the first Hewi Forum in 1996 in Bad Arolsen, Germany. In 1998, Beverly Jones, a member of the board of directors of the National Easter Seal Society in Chicago, informed an international audience at the second Hewi Forum about her first-hand experiences as an American. Thus, the company promoted both—the issue of barrier-free design, as well as of universal design in Germany.

Besides the universal design–oriented "Comfort & Quality" seal, Hewi has received many international design awards for its barrier-free product line. Meanwhile, it also offers products that qualify as technical assistance (see Fig. 39.2), based on the general technical assistance directory, referred to previously. Many of the products are for multipurpose use. The seat, as shown in Fig. 39.5, can also be used as a shower seat and may be funded by long-term-care insurance.

FIGURE 39.5   Seat for multipurpose use in the bathroom.

FIGURE 39.6   Bathroom with tilting mirror and rolling cabinets beneath the sink.

FIGURE 39.7   Award-winning ergonomic lever door handle, designed by Hewi in 1970.

As the company's overall goal is a "design for all," the interior design of a barrier-free bathroom may even resemble a living room, as shown in Fig. 39.6. With its rolling cabinets beneath the sink, a tilting mirror, and lots of space, this room complies with German DIN standard 18025. Its barrier-free character is noticeable at best at a second glance.

Another example of nonstigmatizing, ergonomic design is the lever handle No. 111, which Hewi designed in 1970 (see Fig. 39.7). Meanwhile, this award-winning door handle can be seen all over the world in private homes, as well as in public buildings, such as hospitals and airports. According to the Center for Universal Design at Raleigh, North Carolina, this lever handle is regarded as an excellent example for universal design.

39.9   CONCLUSIONS

From a long-term perspective, barrier-free design needs to be thought of as the ultimate goal and benchmark for building activity, which has been recognized in the United States through the now-established term *universal design* (i.e., to design for all).

Regarding building laws, this has so far only been implemented in a few states. For example, the state of Schleswig Holstein changed its state building code on March 1, 2000, and the law was changed to be effective on June 1, 2000, in the state of Nordrhein-Westfalen. The new building laws of these states provide that, in principle, public places have to be barrier-free. Even in the private housing sector, these standards will be considered the basis for future planning. For example, in a building with more than two levels, two of the apartments will have to be accessible to people in wheelchairs. Beyond that, all apartments have to be designed barrier-free.

The demand for a "design for all" was presented in the second HEWI Forum in October 1998, entitled "Barrier-Free Living: Suggestions and Visions for Living." The resolution that was passed requires that product development and building modifications in the spirit of design for all, or universal design, requires inter-disciplinary collaboration and cooperation among the different relevant industries. Closer collaboration between seniors' organizations, home modification consultant agencies, and trade and industry on the other hand appears to make sense. The way from barrier-free design to universal design will create new professional pathways in the future, which require a new competence beyond the knowledge of the DIN standards. For example, those that specialize in the design of kitchens and bathrooms need to get more involved with ergonomics and gerontology, design, and marketing. To be accepted by elderly people, technical assistance will need to have reduced technical emphasis and features. However, if such products look like they have been designed specifically for people who are elderly and people who are disabled, they risk not to be accepted, and their marketing will fail from the start.

39.10   BIBLIOGRAPHY

Bayerisches Staatsministerium des Innern (ed.), *Wohnen ohne Barrieren* (Arbeitsblätter Bauen und Wohnen für Behinderte Nr. 5), Oberste Baubehörde im Bayerischen Staatsministerium des Innern, Munich, Germany, 1995.

Behindertenbeauftragter des Landes Niedersachsen (ed.), *Selbstbestimmung bis ins hohe Alter—Wie behinderte Menschen im hohen Alter leben wollen,* Behindertenbeauftragter des Landes Niedersachsen, Hannover, Germany, 1998.

BGBL, *Sozialgesetzbuch SGB* (Becktexte im dtv 5024), 25th ed., Verlag C. H. Beck, Munich, Germany, 1999.

Blonski, Harald (ed.), *Wohnformen im Alter. Ein Praxisberater für die Altenhilfe* (edition sozial), Weinheim: Beltz, Weinheim, Germany, 1997.

Bundesarbeitsgemeinschaft Wohnungsanpassung e.V. (ed.), *Wohnberatungsstellen für ältere Menschen in Deutschland,* Düsseldorf: Verbraucher-Zentrale, St. Wendel, Germany, 2000.

Bundesgeschäftsstelle Landesbausparkassen (ed.), *Altersgerecht Wohnen. Umgestalten oder umziehen* (LBS-Ratgeber Bd. 5), Deutscher Sparkassen Verlag GmbH, Stuttgart, Germany, 1999.

Bundesministerium für Raumordnung, Bauwesen und Städtebau (ed.), *Wohnungen für ältere Menschen. Planung Ausstattung Hilfsmittel,* Bonn-Bad Bundesministerium für Raumordnung, Bauwesen und Städtebau, Bonn-Bad Godesberg, Germany, 1995.

Coleman, Roger (ed.), *Design für die Zukunft: Wohnen und Leben ohne Barrieren,* DuMont Buchverlag, Cologne, Germany, 1997.

Deutsches Institut für Normung e.V. (ed.), *Barrierefreie Wohnungen—Wohnungen für Rollstuhlbenutzer, Planungsgrundlagen, DIN 18 025, Teil1,* Deutsches Institut für Normung, Berlin, Germany, 1992.

———, *Barrierefreie Wohnungen—Planungsgrundlagen, DIN 18 025, Teil 2,* Deutsches Institut für Normung, Berlin, Germany, 1992.

Geschäftsstelle zum Internationalen Jahr der Senioren (ed.), *Vorbereitung Internationales Jahr der Senioren 1999: Fachtagung der Nationalen Kommission 12.–14. Mai 1998,* 2nd ed., BAGSO/IJS, Bonn, Germany, 1999.

Grossjohann, Klaus, "Wohnen—Bedingungen für ein aktives Alter." *Vorbereitung Internationales Jahr der Senioren 1999: Fachtagung der Nationalen Kommission 12.–14. Mai 1998,* BAGSO, 202–229, Bonn, Germany, 1999.

HEWI Heinrich Wilke GmbH (ed.), *Barrierefreies Wohnen. Planungsempfehlungen für Grundriss, Ausbau und Einrichtung,* HEWI Heinrich Wilke GmbH, Bad Arolsen, Germany, 1990.

———, *Dokumentation HEWI-Forum Barrierefreies Wohnen—Leben ohne Ausgrenzung,* HEWI Heinrich Wilke GmbH, Bad Arolsen, Germany, 1997.

———, *Wohnungsanpassung. Empfehlungen für Ausstattungen, Nachrüstung und Umbau,* HEWI Heinrich Wilke GmbH, Bad Arolsen, Germany, 1998.

———, *Dokumentation HEWI-Forum Barrierefreies Leben. Ansätze und Visionen für ein Miteinander,* HEWI Heinrich Wilke GmbH, Bad Arolsen, Germany, 1999.

HUD Department of Housing and Urban Development (ed.), *Housing Our Elders.* HUD, San Diego, CA, 1999.

IKK-Bundesverband, *Zuschüsse für Massnahmen zur Verbesserung des individuellen Wohnumfeldes des Pflegebedürftigen nach § 40 Abs. 4 SGB XI,* IKK Bundesverband, Bergisch Gladbach, Germany, 1995.

Kliemke, Christa, Heide Knebel, and Erhard Böttcher, *Wohnungsanpassung—Anpassung an die Wohnung,* TU Berlin, Institut für Krankenhausbau, Berlin, Germany, 1988.

KDA Kuratorium Deutsche Altershilfe (ed.), *Rund ums Alter. Alles Wissenswerte von A bis Z,* Verlag C. H. Beck, Munich, Germany, 1996.

———, *Hilfe und Pflege im Alter zu Hause,* Kuratorium Deutsche Altershilfe, Cologne, Germany, 1997.

Landesarbeitsgemeinschaft Wohnberatung NRW (ed.), *Wohnberatungsstellen in Nordrhein-Westfalen, Adressen, Materialien, Informationen,* Verein für Gemeinwesen—und Sozialarbeit Kreuzviertel e.V., Dorkmund, Germany, 1999.

Loeschcke, Gerhard, and Daniela Pourat, *Barrierefrei und integrativ,* Verlag Das Beispiel, Darmstadt, Germany, 1994.

———, and ———, *Wohnungsbau für alte und behinderte Menschen,* Kohlhammer Verlag, Stuttgart, Germany, 1995.

Ministerium für Arbeit, Soziales und Stadtentwicklung, Kultur und Sport des Landes Nordrhein-Westfalen (ed.), *Altersfreundlicher Wohnungsumbau,* Verbraucher-Zentrale, Düsseldorf, Germany, 1999.

———, *Ohne Stufen und Schwellen. Wohnungsanpassung. Ein Beispiel aus dem ländlichen Raum,* Verbraucher-Zentrale, Düsseldorf, Germany, 1999.

———, *Zu Hause älter werden. Leitfaden: Wohnungsanpassung und Finanzierung,* Gemeinnützige Werkstätten Neuss GmbH, Neuss, Germany, 1999.

Ministerium für Bauen und Wohnen des Landes Nordrhein-Westfalen (ed.), *Neue Wohnformen für ältere Menschen,* 2nd ed., Ministerium für Bauen und Wohnen, Düsseldorf, Germany, 1997.

Neufeld, Hildegard, *Der ältere Mensch als Wirtschaftsfaktor,* 2nd ed., Universität des 3. Lebensalters an der Johann Wolfgang Goethe-Universität e.V., Frankfurt am Main, Germany, 1999.

Pack, Jochen, *Zukununftsreport demographischer Wandel. Innovationsfähigkeit in einer alternden Gesellschaft,* Bundesministerium für Bildung und Forschung, Bonn, Germany, 1999.

Philippen, D. P., *Wohnen ohne Barrieren: Leitfaden zum Planen, Bauen, Einrichten barrierefreier Wohnungen,* Sozialverband Reichsbund, Bonn, Germany, 1992.

———, *Der barrierefreie Lebensraum für alle Menschen, Leitfaden nach DIN 18024 Teil 1 und Teil 2* (Schriftenreihe des Sozialverband Reichsbund—Folge 59), Sozialverband Reichsbund, Bonn, Germany, 1998.

———, *Spaziergang durch einen barrierefreien Lebensraum,* 4th ed., Der Beauftragte der Bundesregierung für die Belange der Behinderten, Bonn, Germany, 1999.

Stolarz, Holger, *Wohnungsanpassung. Kleine Massnahmen mit grosser Wirkung* (vorgestellt 57), 2nd ed., Kuratorium Deutsche Altershilfe, Cologne, Germany, 1998.

Thieler, Volker, and Axel Eichhorst, *Tatort Altenbetreutes Wohnen,* Carl-Hendrik Bäumler Verlags GmbH, Ingolstadt, Germany, 1999.

Verbraucher-Zentrale Nordrhein-Westfalen e.V. (ed.), *Die Pflegeversicherung. Informationen und Tips für Betroffene und Pflegepersonen,* 3rd ed., Verbraucher-Zentrale, Düsseldorf, Germany, 1999.

———, *Pflegefall—Was tun?* 4th ed., Verbraucher-Zentrale, Düsseldorf, Germany, 2000.

39.11   RESOURCES

Web Sites

e-mail: **kontakt@altenarbeit.de**.   www.altenarbeit.de

BAGSO Bundesarbeitsgemeinschaft der Senioren-Organisationen.   www.bagso.de

BMFSFJ Bundesministerium für Familie, Senioren, Frauen und Jugend.   www.bmfsfi.de

BMG Bundesministerium für Gesundheit.   www.bmgesundheit.de

The Center for Universal Design, Raleigh, North Carolina.   www.design.ncsu.edu

DIN Deutsches Institut für Normung e.V.   www.din.de

Feuer TRUTZ Bauvorschriften und Brandschutzinformationen.   www.feuertrutz.de

HEWI Heinrich Wilke GmbH.   www.hewi.de *and* www.hewi.com

HUD Department of Housing and Urban Development.   www.hud.gov

IKK-Bundesverband.   www.ikk.de

KDA Kuratorium Deutsche Altershilfe Wilhelmine-Lübke-Stiftung e.V.   www.kda.de

Spezialsuchmaschine für Soziales, Gesundheit und Pflege; e-mail: info@onlinenetzwerk.de
www.onlinenetzwerk.de

Statistisches Bundesamt.   www.statistik-bund.de

TÜV Rheinland/Berlin-Brandenburg e.V.   www.tuev-rheinland.de

BAG Bundesarbeitsgemeinschaft Wohnungsanpassung e.V.   www.wohnungsanpassung.de

Das wohnungswirtschaftliche Forum im Netz.   www.wowi.de

dito; Förderinstitute und Wohnungsbauförderung der Bundesländer.   www.wowi.de/info/
foerderung/bundeslaender/menu.htm

Addresses

Arbeitsgemeinschaft der Verbraucherverbände e.V. (AgV)
Heilsbachstrasse 20
D- 53123 Bonn/Germany
e-mail: mail@agv.de

Arbeitsgemeinschaft Wohnberatung e.V. (AGW)
Bundesverband Bau- und Wohnberatungen
Buschstrasse 85
D- 53113 Bonn/Germany

Bundesarbeitsgemeinschaft Hausnotruf e.V.
c/o Frankfurter Verband für Behindertenhilfe e.V.
Vorsitzender
Alfred Viola
Mainkai 43
D- 60311 Frankfurt a. Main/Germany

**Bundesarbeitsgemeinschaft der Senioren-Organisationen
e.V. (BAGSO)**
Pressestelle
Ursula Lenz
Schedestrasse 13

D- 53113 Bonn/Germany
e-mail: kontakt@bagso.de

Bundesarbeitsgemeinschaft Wohnungsanpassung (BAG)
c/o Wohnberatungsstelle Stiftung Hospital
Ulrich Weissenauer
Hospitalstrasse 35–37
D- 66606 St. Wendel/Germany

Bundesministerium für Familie, Senioren, Frauen und Jugend (BMFSFJ)
Glinkastrasse 18-24
D- 10117 Berlin/Germany
e-mail: info@bmfsfj.bund.de

Bundesministerium für Gesundheit
Am Propsthof 78a
D- 53121 Bonn/Germany

Bundesministerium für Raumordnung, Bauwesen und Städtebau
Postfach 20 50 01
D- 53170 Bonn/Germany

Deutsches Institut für Normung e.V (DIN)
Postfach
D- 10772 Berlin/Germany

Gesellschaft für Geronto Technik mbH (GGT)
Pressestelle
Bettina Wilmes
Max-Planck-Strasse 5
D- 58638 Iserlohn/Germany
e-mail: mail@gerontotechnik.de

HEWI Heinrich Wilke GmbH
Postfach 12 60
D- 34442 Bad Arolsen/Germany
e-mail: info@hewi.de

HEWI, Inc.
2851 Old Tree Drive
Lancaster, PA 17603/USA
e-mail: hewi@hewi-inc.com

IKK-Bundesverband
Technologie Park
Friedrich-Ebert-Strasse
D- 51429 Bergisch Gladbach/Germany
e-mail: Christiane.Mahnke@ikk-nordrhein.de

Kuratorium Deutsche Altershilfe (KDA)
Wilhelmine-Lübke-Stiftung e.V.

An der Pauluskirche 3
D- 50677 Köln/Germany
e-mail: info@kda.de

Sozialverband Reichsbund e.V.
Bundesvorstand
Beethovenallee 56-58
D- 53173 Bonn/Germany

Statistisches Bundesamt
Postfach
D- 65180 Wiesbaden/Germany
e-mail: info@statistik-bund.de

TÜV Rheinland/Berlin-Brandenburg e.V.
Am Grauen Stein
D- 51105 Köln/Germany
e-mail: presse@de.tuv.com

Verbraucher-Zentrale Nordrhein-Westfalen e.V.
Mintropstrasse 27
D- 40215 Düsseldorf/Germany
e-mail: vz.nrw@vz-nrw.de

CHAPTER 40
HOUSING FOR OLDER PERSONS IN SOUTHEAST ASIA: EVOLVING POLICY AND DESIGN

James D. Harrison, B.Arch.
National University of Singapore, Singapore

40.1 INTRODUCTION

Most people hope to live long and fulfilling lives, but few would admit to wanting to grow old when doing so involves deterioration in their physical or mental abilities. Where this is aggravated by inconsiderate design in the built environment, personal mobility can easily be reduced, and fear of accidents—particularly of falling over—will justifiably inhibit many peoples' lifestyles as they age. The habitat with which individuals have become familiar may present increasing hazards as they age, made even more dangerous because they are unaware or unwilling to admit their limitations. Universal design, by definition, anticipates the problems encountered in aging and can provide for safer and more amenable surroundings, especially for the more vulnerable sectors of the population.

Although universal design as a concept is still largely unacknowledged in Asia, some of its basic principles are beginning to emerge in policy making and design of built environments and housing in the faster-developing countries in the region. This is a relatively intuitive approach, resulting from concerns for the future needs of growing numbers and proportions of older people. In East and Southeast Asia, population trends will include a dramatic increase in those of 80 years and over, who will tend to be more frail and at greater risk of personal accidents. Design of suitable habitats for such older people must be considered as a natural subset of universal design, supporting the maintenance of personal independence and dignity as people age. In Singapore, whose high-rise housing ranks among the most envied in the world, a pattern can be traced in recent developments, showing how universal design principles are beginning to change the accepted policy and standards for public housing.

40.2 BACKGROUND

Recent policy studies in Singapore have led to significant proposals that could be the cue for universal design as a key design factor in housing and habitat. This could also become true in many facets of public provision—provided that these opportunities are backed up by informed design skills and well-considered legislation to ensure their application. As in other parts of

Asia, with increasing average age and longevity of populations, the integration of nonhandi-capping environments can make the difference between whether many people live rich and fulfilling lives or whether they are handicapped and trapped by their surroundings. Historical and geographical context, humid tropical climates, enduring cultural and family values and patterns of support, all distinguish the Asian context from the West, despite the apparent signs of rapid urbanization and growing economic prosperity. The idea of a greater degree of inclusion for everyone is gaining understanding, however, as the benefits for other citizens, of all ages and abilities, become manifest. Moving from the disability-led approach to a more universal one will take time, especially where attitudes and availability of resources inevitably lag behind more developed economies.

Changes are taking place in the region to provide for aging populations, with certain unique qualities. Previously, building codes and standards in the Asian region, whether legally enforced or not, have generally regarded barrier-free access as the ideal, based on the needs of wheelchair users. Increasing awareness, particularly among policy makers and designers, about the imminence of a growing population of older people, is having an impact in a number of ways. Various initiatives to disseminate information, both on awareness and on technical realities, have already been seen in different parts of Asia, each with their own priorities and imperatives. This may be due not only to increasing ease of communication, but also to interventions from agencies, and by personalities of strong conviction. Slowly but surely, those involved in providing for the needs of people with disabilities are appreciating the value of broadening their approach, by including a wider range of people than before, rather than making a case only for those with special needs (see Fig. 40.1).

40.3   RECENT INITIATIVES IN ACCESSIBLE DESIGN IN THE UN ESCAP REGION

The United Nations "Asian and Pacific Decade of Disabled Persons" commenced in 1993, and since that time a number of significant moves have been brought about in the United Nations Economic and Social Commission for Asia and the Pacific (UN ESCAP) region, as discussed in Chap. 59, "Promoting Nonhandicapping Environments in the Asia-Pacific Region," by Sato. Many of the initiatives inaugurated during the Decade are still gaining momentum, but the seeds have been sown. In many instances these involve moves toward the adoption of building codes on nonhandicapping environments. But in the less prosperous countries in the region, the issue of rights for people with disabilities is likely to be low on national agendas. Even where a country can be motivated to adopt a code on access, the application and enforcement of that code is a completely different matter. Many of these countries are experiencing increases in the proportion of older people, as a result of the post–World War II baby boom, improved health care standards, and subsequent longevity. Inevitably the new generations of older people will need greater consideration, if they are not to become a burden on society at large, and the design and provision of supportive human habitats will play a part in giving these coming generations more independence in their daily lives.

A great diversity of levels of development may be seen in Southeast Asia. As a model, Singapore is not typical of many aspects of urban development in Asia, but, along with Japan and Hong Kong, it has developed sophisticated housing policies and technical solutions to deal with rapid urbanization and growing populations. Moreover, because of Singapore's limited resources, especially its land area, social cohesion is a strong political aim, with housing forming one of the cornerstones of this strength. Other areas of Southeast Asia, which have less concerted programs of rehousing and less ambitious levels of social care by their governments, still aspire to levels of development seen in these leading countries.

As a major part of the UN ESCAP "Asian and Pacific Decade of Disabled Persons 1993–2002," a survey of the situation of people with disabilities in the region was undertaken by that agency (UN Case Studies 1995). Probably for the first time, a full picture of the acces-

FIGURE 40.1 As people age they deserve better environments for their continued well-being.

sibility situation in the region was gathered, including reviews of existing access policy legislation and standards in ESCAP member countries, as well as the role of educational systems and professional bodies in bringing about barrier-free environments. From the findings of this study, Expert Group Meetings of interested parties from member countries in the Asian Pacific region were convened in Bangkok in 1994. Among other initiatives the *Promotion of Nonhandicapping Physical Environments for Disabled Persons: Guidelines* was published (UN Guidelines 1995). These guidelines, and related workshops and training sessions, have been instrumental in bringing about a variety of significant changes in awareness, legislation, and standards of barrier-free design, even in the less-well-developed economies in the region.

At the time the guidelines were published, however, less priority was given to the status of elderly people than to persons with disabilities. Perhaps significantly, the term "universal design" was not then in common use in the area, although the principles were anticipated in many of the recommendations, which advocate holism in accessibility. Even though representatives of a range of disability groups articulated their needs to the Expert Group Meetings, these did not specifically include advocacy on the needs of older people. In hindsight, the opportunity to justify environmental accessibility and safety for the benefit of everyone, not just people with disabilities, was not given particular stress. Since then, growing awareness of the potential magnitude of the aging problem in the area shows that this aspect could have been given more emphasis, giving added weight to arguments for the need for inclusive environments.

More recently, the "International Year of Older Persons, 1999" focused on a range of topical "elderly issues." Particular concerns have been expressed about the changing role of older people in society and the possible problems of supporting an increasing number of them in the future, if they are not able to live with some degree of independence. For housing, questions have been raised, such as how to integrate aging citizens back into the community after retirement, or of how to avoid ghettos of older residents. Clearly, there is a need to design housing, public buildings, streets, and transportation facilities and services for increased opportunities for personal independence, and thereby respect the individual's dignity in all walks of life.

While many Western models are available, their appropriateness in an Asian context is not clear-cut, especially when the perception of aging is in a state of flux. Education on aging issues, and exemplars of good design provision and appropriate solutions are needed for professional designers and for policy makers, as well as for developers in the private sector. In Singapore, for example, there continues to be reluctance on the part of private developers and organizations to build appropriate apartments for retirees, whether integrated into existing housing areas or in special "retirement villages." In the public sector, however, paradigm shifts in intention have been made recently, following major reappraisals of policy.

40.4   THE BACKGROUND TO HOUSING POLICY IN SINGAPORE

The Republic of Singapore has a population of over 2.7 million people living on an island of 646 km^2. Since land is a very expensive commodity, high-rise living is accepted as the norm. Most "public" housing is in the form of slab blocks of up to 16 stories, in developments which are based on British new-town models, with distinct town centers, social amenities, and transport interchanges (Wong 1985). Smaller neighborhood center precincts have essential amenities, with grocery stores, primary schools, and community health centers (known locally as *polyclinics*) clustered around this hub, within easy walking distance of each other.

During the almost 40 years since Singapore gained its independence, the Housing and Development Board (HDB) has been the major provider of housing. Currently about 87 percent of the population live in high-rise blocks, designed and built to good but economical standards by this "statutory board." HDB not only constructs and maintains these apartment blocks, but also controls the sale and financing of them, so that although they are generally referred to as "public housing," they may not be directly comparable with public rental housing projects in other countries. Some 95 percent of these properties are sold to their occupants, usually on 99-year leases. Various financial incentives are offered to buyers to encourage positive social patterns through family cohesion. For instance, financial rebates are available to attract two or more generations of a family to buy units nearby or in the same block. Buyers may use a proportion of their Central Provident Fund (CPF) savings as a mortgage, and this can be taken out jointly by several wage-earning members of a family. As only a small proportion of public housing stock is for rental, generally to very low-income groups or more transient tenants, the idea of a "stakeholder society" generally pertains (Chua 1997). In the private sector a very small proportion of residences are "landed properties" (that is, built on a single plot), whether freehold or leasehold. The remainder of the housing stock comprises privately-developed condominium-style apartments, which sometimes include communal social and recreational facilities, swimming pools, or squash courts.

HDB housing in Singapore, while being adequate and well-constructed, has never been generous in space or quality of finish, being aimed at providing affordable homes for the masses. Since apartments are expensive to buy, even by most Asian urban standards, it is common for many younger people to continue to live with their parents, even after marriage. It is also relatively uncommon for older people to move away from the family home—although this trend is now changing. The extended family and multigenerational living are traditional and continue as the norm among the main ethnic groups that are found in Singapore, whether they be Chinese, Malay, or Tamil Indian. It is generally assumed that there will be family members on hand to care for the elderly or the disabled members. Equally, it is very common for grandparents to look after preschool infants during the day, when their parents are out earning a living. Although this will continue among the majority in the foreseeable future, there will also be a growing incidence of people wanting to live independently as they age, and housing policy will have to allow for this trend.

40.5 AGING AND ACCESSIBILITY ISSUES IN SINGAPORE

Since 1997 a number of housing initiatives have been announced by the HDB, while increased media coverage on the growing aging population has helped to raise awareness about issues such as independent living, accessibility, and home safety. Statistics indicate that more people will live independently in old age, either by necessity or choice. The implications of this on the design and development of appropriate dwelling forms are thus of pressing concern, as both the demography and socioeconomic trends are changing. Currently, Singapore has some 235,000 people aged 65 and above, representing 7 percent of the total population, and this is predicted to rise to 796,000 (or 19 percent) by the year 2030. Falling birth rates, as well as increased longevity due to improved health care and better levels of nutrition contribute to these statistics.

> The nuclearisation of families will probably continue to rise. Between 1990 and 1997, the fastest growing type of living arrangement for older persons was that of the older person living alone or with their spouse only—away from their children. In 1997, such households formed 15 percent of all households with older persons aged 60 and above, compared with 9 percent in 1990. However, it is important to note that a high percentage of older Singaporeans still live with their children, in some 78 percent of the households with older persons. (IMC Report 1999, 31)

In revisions to the Concept Plan, the Population and Housing Subcommittee chaired by the Urban Redevelopment Authority (URA) adopted the assumption made by the Ministry of Health (MOH) that 25 percent of people aged 60 years or more would be living on their own in the long term (IMC 1999). Here the application of universal design principles to increase accessibility and safety as a fundamental of any planning and design, will be crucial and require consistency of approach. At present, provisions for older residents in HDB housing are not particularly conducive to independent living, since this has been the exception rather than the rule; although for the last 10 years or more public spaces and hard landscape in HDB estates have been designed to barrier-free standards. Some older slab blocks, however, have elevator configurations with a lobby only at alternate staircase half-landings, or even every third floor. The original concept was that residents of intermediate floors should be able to walk up, or down, one story, thus cutting construction costs. In the HDB's current extensive upgrading of older housing stock, such blocks prove difficult to convert to provide access to each floor, depending on the configuration of the elevator shaft relative to access decks (see Figs. 40.2 and 40.3).

For over 12 years HDB has had its own in-house codes on accessibility in the public areas of its housing. The Singapore "Code on Barrier-Free Accessibility in Buildings, 1995," first

FIGURE 40.2 Some older apartment blocks can have elevator lobbies only at intermediate floors.

introduced in 1990 and revised in 1995, makes no provision for the interior of dwellings (PWD 1995). This mandatory code has had a significant impact on buildings to which the public has access, even when privately owned. This is manifest in most nonresidential developments designed and built since 1990, and also applies to many buildings that have been retrofitted since then. Today, new commercial developments, stores, and entertainment buildings are generally accessible, while many older buildings are not. Although HDB has a policy of adapting existing apartments for people with disabilities, as well as providing barrier-free public spaces, there are consistent problems with this, in that almost all units have stepped thresholds or even full flights of steps at their front doors (see Fig. 40.4).

FIGURE 40.3 Some older blocks can be upgraded to have elevator lobbies at each floor.

40.6 *TRENDS TOWARD INDEPENDENT LIVING*

Some of the practical problems of making older housing blocks more livable for elderly inhabitants on low incomes have been addressed as part of recent upgrading schemes for older HDB housing estates. The joint venture scheme to modify old, single-room flats as "congregate housing" was set up by the Ministry of Community Development and the Housing and Development Board in 1996, acknowledging the fact that not all older people are able to live with their families (Harrison 1997). In this program, existing blocks of one-room rental apartments have been improved to allow for their use by elderly people who are generally of limited financial means and sometimes supported by subsidies from welfare organizations. These rather basic units have a double-loaded interior corridor, protected from the weather, and thus with-

FIGURE 40.4   Many apartment blocks have flights of steps at front doors.

out the need for steps at the threshold. Though small in area, they are self-contained with a separate shower and toilet—although neither is accessible for a person using a wheelchair. All apartments are equipped with alarm cord-pulls for use in emergency, connected to a superintendent's room at "void-deck" or ground-floor level, where provision is made for communal day-center facilities, provided by voluntary welfare organizations (VWOs). At detail level, lever-handles for doors and cupboards and similar elements are now standard.

So far, many of the measures to make housing units accessible have tended to take a "check-list approach," rather than an understanding of the potential for integrating the piece-meal adaptations into a consistent whole, when it could be regarded as universal design. Consequently many of the provisions, while functional, are obtrusive and may stigmatize the occupants as being old and therefore possibly incapable. These units cater to an existing need, but one that may be superseded as following generations of independent older people have more savings and higher expectations of their habitat. As such, they are not very useful models for the apartments for senior citizens in the future.

40.7 PILOT PROGRAM: STUDIO APARTMENTS

In 1998 HDB proudly announced a proposal to construct 580 "studio apartments" in a pilot phase located in existing "new town" areas, close to social and commercial facilities. This move was in answer to growing comment that no future provision had yet been made for the needs of many elderly people who may wish to live socially and financially in a more independent way. Newspaper articles and letters had been asking for more scope in housing for older people, now living with their families, to be able move away and live independently if they so wish. With the complex controls that HDB imposes on the financing, buying, and selling of apartments, it has always been much easier to upgrade to a larger and more expensive unit than to downsize to a smaller one. But many people approaching retirement would benefit from a more compact unit, once their children have grown up and left. Smaller units are generally easier to fund and to maintain, and might be designed to be safer and more convenient, if purpose-built for aging occupants. Anticipating such demand, HDB planned some 580 'studio apartments' to be completed in 2000 (Harrison 1998; see Fig. 40.5).

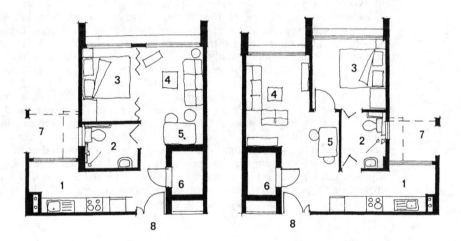

35 m. sq. Apartment   45 m.sq. Apartment

Legend:
1 Kitchen
2 Bathroom (shower & W.C)
3 Bedroom / Bedspace
4 Living Area
5 Dining Area
6 Household Shelter
7 Laundry Drying Racks
8 Access Corridor / Entrance

FIGURE 40.5   Plans of studio apartments, with floor areas of 35 m^2 and 45 m^2 respectively.

The units are clustered together in 17-story blocks, sharing communal facilities for services such as medical care, located at ground level. Apartments are available in two sizes, with floor areas of either 35 or 45 m^2. The smaller of these apartments has a bed alcove, separated from the living room by a folding door, while the larger unit has a conventional double bedroom. Each unit incorporates a Civil Defense "household shelter," which is now mandatory for all new housing in Singapore. These take up valuable space, and, although they can be used for storage, would be inaccessible to a person using a wheelchair because of the raised threshold.

The design approach is generally well-considered in regard to accessibility and incorporates a number of "elderly-friendly" and barrier-free features, without these being overtly conspicuous. Nonslip floor surfaces are provided throughout, with support handrails and lever faucets in the bathroom, which is spacious enough to take a wheelchair. Doorways are generous in width, and bathrooms have a folding door arrangement to save space. In the larger unit, the bathroom has two doors, allowing it to be approached from either the kitchen or the bedroom. This allows for privacy, so that a person in the bedroom would not have to go through the living space if they needed to go to the bathroom. Switches and sockets are located at waist-height, while lever door handles and rocker-type light-switches are specified, to allow for ease of use by stiff fingers. There are alarm cord pulls in the bedroom and bathroom, which are linked to a superintendent's room at the ground level. Front entrances are without steps, and the elevators stop at every floor for ease of access. Only in the bathroom—where the grab rails and alarm pull are somewhat conspicuous—are these obvious clues that this is specifically designed for older people (see Fig. 40.6).

To be eligible to buy one of these studio apartments, which are purchased on a 30-year lease, applicants must already be HDB owners aged 55 years or over, and have children living close to the intended location. If the buyer's offspring are willing to take over the parents' existing flat, or intend to buy another flat in the same electoral division, they will also qualify for priority allocation. These moves are to counter fears that families will be irrevocably disrupted if the elder parents move away, or even that families may try to abandon elderly relatives, making them become a burden on the state or the charitable voluntary welfare organizations that now provide for destitute elderly people. Some criticism has already been leveled at the fact that complete 17-story blocks will be devoted to housing older people, which may cause these to be seen as ghettoes. HDB has implied that future housing developments may incorporate such apartments mixed in with family-sized units. Such apartments could also appeal to younger people or couples without children, hence allowing for a more multigenerational mix, rather than just for retirees. This is, however, a pilot project, and the current conditions of purchase are overtly tight in order to avoid speculation by nondeserving buyers, as well as to sound out the demand by older people.

40.8 PLANNING FOR AGING: THE INTERMINISTERIAL COMMITTEE REPORT

Aware of the looming social problems that the increasingly aging population may bring if relevant issues are not fully addressed, the "Interministerial Committee on the Ageing Population" (IMC) was established in September 1998, to discuss the future needs of the graying population and to make positive proposals to provide for these. Six workgroups were tasked with investigating the major areas of concern: Social Integration of the Elderly, Health Care, Financial Security, Employment and Employability, Housing and Land Use Policies, and Cohesion and Conflict in an Aging Society.

The IMC workgroup on Housing and Land Use Policies met during June and July 1999, with interim draft proposals being presented at feedback sessions, attended by interested parties—users, as well as professionals, sociologists, architects, planners, and others. From these sessions a number of issues emerged, including housing and the nature of Asian family patterns in Singapore, which, although still relatively traditional, are likely to change to

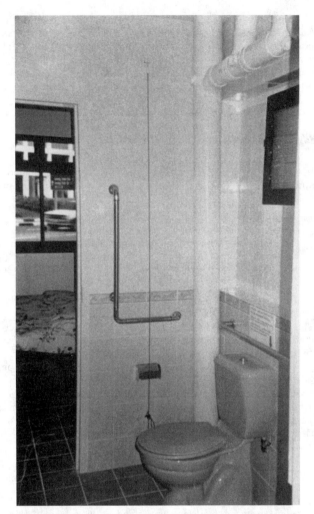

FIGURE 40.6 Studio apartment bathrooms have handrails, alarm cord-pulls, and lever faucets.

some degree in the future. Multigenerational living is still regarded as highly desirable, but so far the implications of this on planning and dwelling layout are not widely discussed outside of HDB. Particular discussion was generated by the aspect of "independent living," a topic that is not completely accepted or even fully understood by many older people. There is still a feeling that for a person to live alone, either by choice or from necessity, is an unnatural thing. Many members of feedback sessions associated living away from the family with being "put in an institution." Indeed, this sometimes tends to be the case at the present, when the older people become unable to live in the family home, or are without skills or resources to support themselves—perhaps after the death of their spouse. Older people from lower-income groups may be housed in adapted, rented apartments, of adequate but basic standards, shared between two or more people. But the idea that older people of the emerging generation will tend to have higher expectations than their parents, who were often poorly educated immigrants, seems to have been difficult for many to grasp. The fact is that many of tomorrow's older people will be better educated and will ask for more independence. Access to modern communications and more familiarity with these mean that actual physical separation may not cause the complete breakup of the multigenerational family, particularly in a small island state such as Singapore, as its already efficient transportation systems are made more usable and accessible for older people.

40.9   RECOMMENDATIONS: AGING IN PLACE

In the "Key Strategic Thrusts" of the IMC Report published in November 1999, one important recommendation is facilitating aging in place as the key principle in housing and land use policies:

> Aging in place refers to enabling people to continue living in the community they are familiar and comfortable with well into old age, amidst family and community support. Aging in place entails provisions which make it possible for senior citizens to stay put in the homes which they possess at the point of retirement, or to move if they wish, within their own community (IMC 1999).

The IMC report also proposes "that Government review and extend its policies and programmes that cater to the two-generation family to the three-generation family" (IMC 1999). This last issue is discussed in the IMC Report under the sections on "Social Cohesion," but there is an obvious relationship to housing accessibility and the "lifetime-home" approaches—"where older people can live a lifetime in their own homes." The report also points out the "need to raise the awareness of architects, developers and homeowners of the need to create safe home environments for older persons. Building codes and standards should also be reviewed to ensure new homes are elder-friendly. The curriculum of courses in our tertiary institutions should also include planning and building for an aging generation" (IMC 1999). These salient recommendations are made within the report's section on "Housing and Land Use Policies," which states:

> An elder-friendly built environment is a key determinant of the extent to which older people have access to services, and are integrated into the wider community. With the increasing trend towards nuclearisation, family care and support for senior citizens as well as their social integration could be affected. Community-based services should be made easily available to enable older people to stay in their communities, and not in enclaves of older people. . . . For public housing, the Government should retrofit Housing & Development Board (HDB) flats and multi-storey car parks progressively with lift landings on every floor, where feasible. Elder-friendly features should be provided for households in older HDB blocks. HDB should also consult para-medical professionals in designing and modifying flats to understand and meet the needs of senior citizens better (IMC 1999, 20).

Although the IMC report calls for the further development of building codes or standards on elderly-friendly homes, Singapore's Building and Construction Authority (BCA) has requested that this should be not be compulsory for housing. Currently, the *Code on Barrier-Free Accessibility in Buildings, 1995* (PWD 1995) applies only to buildings used by the public, even where these are privately owned, while HDB incorporates a range of access features in its own design standards. As well as asking for improved access standards in new-built housing, the IMC report recommends a "standard package" of detailed improvements of existing HDB apartment blocks (IMC 1999, 165). Although comprehensive, these do not break any new ground, but propose the removal of barriers such as thresholds at front entrances and bathroom doors by providing ramps wherever necessary. Inside the apartments, nonslip floor finishes are specified throughout and handrails are provided in the bathroom. Lever-type handles for doors and faucets at sinks and showers are standard, with rocker-type switches providing controls that are easy to use by arthritic hands. Sockets and other controls are mounted at reachable height, but this has been standard practice in HDB apartments for some time (for reasons other than being elder-friendly).

In public circulation areas, elevators will have lobbies on every floor, wherever possible, and elevator cars will have support bars, lower control panels with clear floor-numbering systems, safety light sensors, and alert alarm systems to the superintendent's office. Better lighting levels are recommended in corridors, support handrails and ramps provided wherever needed, and door-closers should have slower actions to accommodate older people's needs.

As far as new housing is concerned, the studio apartments set a new standard for safety and barrier-free design features, which approach universal design in their scope. The IMC Report urges that in order to raise awareness of developers and consumers as to the value of these standards and thereby achieve environments that are more than just "barrier-free," there is a need to focus on these issues in the education of planners, architects, and others in the building industry. This recommendation is most significant, since it recognizes (quite rightly) that "friendly" built environments are more than the mere application of codes, no matter how well these are drafted. One may begin to recognize that here is the cue for universal design principles to be followed, as the only way to achieve something that is "more than just barrier-free." The limited level of awareness currently prevailing among professionals, both as building providers and designers, is a key issue. Many practitioners may be well motivated but simply unsure of how to proceed, or find the code difficult to use. In the past, accessibility features have more often than not been seen as "add-on" items which could easily be removed if budgets were too tight. Having to integrate these into the design in a seamless way has few precedents in the region, and building owners may actually doubt the credibility of doing so. It is also true that any good example of a universally-designed building or environment would probably not look anything out of the ordinary, if integration has been carried out effectively.

Methods to foster and promote good design are urgently needed within the professions and design schools. The creation of a database of exemplars, for instance, or initiatives to acknowledge well-designed buildings and environments, can all play a part. In Singapore, such moves are already under way. In January 1998, the biennial "Award for Handicap-Friendly Buildings" was instituted jointly by the Singapore Institute of Architects (SIA) and the Handicaps Welfare Association (HWA) for buildings that are friendly toward disabled and elderly users. These awards recognize and publicize good examples from which clients and professionals can learn, as "standards of excellence in accessible built environments." Using the five qualities of Convenience, Usefulness, Innovation, Transferability, and Integration, works in the four categories of Housing, Commercial Development, Community/Public Building, and Landscape were evaluated to decide which were worthy of award, merit, or mention.

In the field of design education, the "Tsao Ng Yu Shun Awards for Excellence in Aging Study," the precursor of future regular annual awards on this theme, were set up in 1999, concurring with the United Nations "International Year of Older Persons." The donor, the Tsao Foundation, is a not-for-profit organization aimed at promoting healthy aging and enhancing quality of life for elders. Each year the award is given to students of the School of Architecture at the National University of Singapore, for design work that "most ably demonstrates an understanding and sensitivity towards making buildings friendly to elderly people." Entries for the award are evaluated on the sensitivity of the design, which should extend beyond standards already required by regulations and codes, in providing solutions that are practical, safe, and user-friendly for all, and in particular, for elderly users. The design should integrate provision for the needs of the older user into the overall design, without overtly stigmatizing any part of the building as being solely for use by the elderly, or designing only for "special needs." Hence the building type should, preferably, be for a wide spectrum of users and not only for older citizens.

40.10   PUBLIC AND PRIVATE INITIATIVES

Public Sector

So far, the HDB studio apartments are the only purpose-designed units to be under construction, and these are still of limited availability. One recommendation of the IMC Report is that:

> . . . senior citizens of the future may aspire to have a wider range of housing options. To address this, Government should allow the private sector to provide housing for senior citizens on HDB terms, through "Design-and-Build" schemes and the development of Studio Apartments. Housing for senior citizens should be integrated into existing housing estates to allow senior citizens to downgrade to smaller flats in familiar surroundings (IMC 1999).

The most recent innovation in apartment design in Singapore is the HDB's "White Block" pilot project. The first move is to create 36 units, on the upper floors of apartment blocks in Punggol New Town, in the north of Singapore Island. These will have basic facilities and structure, but no partitions to rooms, other than toilet and CD shelter, presumably. So far, few details have been announced, but the intention is to allow their occupants to configure the plan to their own needs as the family changes in size or circumstances, or as they age-in-place (*Straits Times* 1999). What is not yet clear is whether the occupants will be given some guidelines as to how best to create their home. It may be assumed that the designers could not have evolved this scheme without studying at least some of the feasibility aspects of possible layouts and later adaptations.

Private Sector

Certainly there is a need for meaningful research into multigenerational living arrangements, and for improved design and facilities for independent living, in both the public and private housing sectors. When it comes to property and financial affairs, Singaporeans are essentially pragmatic people. Property buyers relate the value of the apartment directly to the floor area, just as much as to the location or amenities, so that aspects of "added value" in housing, such as Lifetime Homes, have not been given much consideration (Brewerton 1997). In the public sector, HDB can deliver housing for sale on 99-year leases, or 30 years for studio apartments, at very competitive prices. So far, the private developers have been wary about building smaller units to appeal to the older buyers, perhaps believing that the public would not be attracted to these, especially in competition with HDB units—even though many potential buyers would not be eligible for these, on various grounds. A research project begun in 1999 at the Faculty of Architecture, Building and Real Estate at the National University of Singapore is investigating possible housing forms that would appeal to older buyers in the more expensive end of the housing market. This survey assesses potential buyers' demands in order to propose prototype design models for retirement housing, including apartment configurations, and external factors, such as siting and communal amenities. The study is intended to give private developers confidence about the kinds of housing that will be in demand in the near future.

40.11   RESEARCH AND FUTURE DEVELOPMENT ISSUES

HDB is a very large organization, employing many professionals, architects, planners, sociologists, and specialized researchers of many kinds. It has been the provider of the majority of Singapore's housing stock, to high standards, in the years since Independence. Housing is a political element in keeping Singapore as a stable and buoyant economy, but HDB tends to be reluctant about releasing information on its future plans until they are already signed and sealed. The impetus started by the IMC report may have prompted their designers to study ways to provide more flexible solutions in the future, as well as to look for a wider variety of housing solutions for tomorrow's population. Such initiatives might integrate units for older persons into existing estates and apartment blocks alongside family-sized units.

The IMC recommendations also call for a revision of codes to include a wider range of people who will benefit. These include those elderly people, who although not medically classifiable as having disabilities would certainly benefit from safer and more convenient environments. Significantly, the request to include someone as an advocate for older people came from members of the Handicaps Welfare Association (HWA), previously the only local user group represented on the reviewing committee. This may have been stimulated, in an indirect way, from members of HWA having attended training sessions promoted by UN ESCAP, as follow-up to the pilot project workshops on accessible environments. These sessions were set up by UN ESCAP to give the user groups more confidence in expressing the needs, not only of their own groups, but of the those whose needs are not yet represented in the legislation, including older people and people who are sight-impaired.

40.12 CHALLENGES OF UNIVERSAL DESIGN

Recent policy moves in Singapore have highlighted the need to provide appropriate settings that will enable older people to live in social harmony, security, and safety. That a significant number of these people will have varying degrees and forms of disability is implicit, despite increasing longevity and health care standards. The Interministerial Committee Report has focused attention on many related issues, and will have far-reaching effects in housing and many other areas of the built environment. For instance, the Land Transport Authority (LTA) is already appraising many of its policies on transportation, in the light of the demands for increased accessibility, and these generally suggest moves in the right direction.

> The key recommendations from the IMC on housing and planning clearly have a number of positive implications. "Aging-in-place," whether in the family home or a studio apartment or other form of independent unit, will require standards of design to be constantly appraised. Recommendations to integrate older people into the community, to create more choice in housing arrangements, and to make homes and environments "Elder-Friendly" should be clarion calls to professionals, but professional awareness and design skill will need to be increased if this is to become reality. Along with other initiatives in many other walks of life, the picture on accessibility is getting better all the time, and this is to be greatly applauded. (Harrison 1999).

But in order to sustain this momentum, better education on design issues—especially on the relative value and ease of application of universal design principles—is required, so that these become an integral way of designing, rather than applied as an afterthought. This was underpinned in the IMC Report, which urges that curricula in tertiary institutions should also include design, planning, and building for an aging population (IMC 1999). Such information and awareness-raising is also needed by practitioners, through such channels as Continuing Professional Development courses and locally applicable publications.

40.13 AGING AND DESIGN CHALLENGES IN AN ASIAN CONTEXT

For over 10 years, the Asian Training Centre on Ageing (ATCOA), an initiative of HelpAge International based in Chiang Mai, Thailand, has run workshops on "Buildings and Environments for Older and Disabled Adults" in various centers in the ASEAN Region. The value of this initiative is beginning to bear fruit in various ways. Architects, designers, managers, and health care personnel have benefited from these courses, and a number of the most active personalities in bringing about accessible environments received their initial motivation and skills from the expert workshop trainers.

By prolonged immersion in workshop training, professionals are better able to project to what aging will mean for themselves, in order to empathize with the needs of people in older age. It is useful for designers to consider that most people will hope to continue to do the things that they like to do for as long as possible, without having to rely on others, and so keeping their self-respect, gaining fulfillment and feeling part of a community, not a burden on it. But increased longevity is worthless unless individuals can live safely and independently into ripe old age, even with the likelihood of a reduction in faculties, sensory, physical and mental, as well as increasing frailty. Designers may not appreciate just how much older people can become fatigued and frustrated by all kinds of barriers, including distance and exposure to tropical sun and rain. Even though a combination of physical and sensory impairments may not constitute full disability in the medical sense, these will inevitably impose limits and increase vulnerability to accidents—particularly to tripping and falling over. This has serious implications on the design of the built environment, since accidents that would be mildly painful or embarrassing for a young person can be disabling or possibly even fatal for an older person. In this respect, universal design can help professionals play a responsible role in providing environments that do not disadvantage or endanger "our future selves" through thoughtless design, while also providing barrier-free and accessible places for people using wheelchairs, strollers, or trolleys.

Removal of some of the main obstacles is vital in providing inclusive environments, but it may take some effort to convince engineers or architects of the viability and advantages of possible alternatives. The world is full of barriers and hazards. For instance, in many parts of tropical Asia, where high rainfall is commonplace, traditional buildings usually stand well above the ground surface, with steps leading to the entrance. Entrances to colonial-style buildings have always had steps at the threshold, in order to keep out storm water from violent tropical downpours. In everyday building maintenance, the washing of floors of public hallways, as well as kitchens and bathrooms, inevitably involves copious quantities of water. In order that adjacent floor finishes are not damaged by excess cleaning water, the floor surfaces of such wet areas are often set at a slightly lower level, thus creating a step, or are provided with a raised bar about 2 in high. Many public buildings, such as cooked food centers or canteens and "wet" markets selling fresh vegetables, fish, and meat, also use these devices to contain water on their floor surfaces. But for buildings and public places to become safer and more user-friendly, these barriers must be removed. Consequently there is a need to raise awareness about the dangers and the barriers that even one single step can present to any older person, as well as developing technical solutions on alternative drainage methods. Best-practice solutions for improving safety and access standards are not yet well developed in Southeast Asia, although exemplars in the region are being appraised by several bodies.

40.14   REMOVING BARRIERS AND PROMOTING UNIVERSAL DESIGN PRINCIPLES

Surprisingly few people think objectively about retirement and old age, but early planning can make life more convenient as well as preempting accidents, which could precipitate the need to move from the family home. Preparation for old age in a more systematic way could be at an individual level, or, for professionals, to ensure safe, accessible and appropriately designed environments for everybody—in housing, in public and commercial buildings, in services and transportation, and in affordable products. Sadly, lack of foresight about aging appears to be endemic: For instance, at a recent exhibition in Singapore to promote services and products for people entering their "Golden Age," exhibitors included real estate agents and country clubs. But, when asked how the places or services they offered could accommodate a person in his/her old age, such as "life membership" in a club, and apartments considered suitable for retirees, the sales staff had no answers. For example, no consideration had been given that anyone might need a device to help them get in and out of the swimming pool if they developed some disability—although swimming can be a most beneficial exercise for people who may not be able to walk so well. It is significant to note that the promotional brochures showed no gray-haired people sitting at the poolside. The same held true for the apartments on sale, with no suggestions as to their suitability as a place to grow old in, or whether a wheelchair could be used inside the flat. Some of this apparent lack of awareness may be attributable to the fact that potential buyers do not ask for or expect such standards, as a result of which the situation does not improve (see Fig. 40.7).

An enduring misconception, observed in Southeast Asia but probably widespread, is that of considering older people as "the elderly" and people with disabilities as "the disabled"[1]—

FIGURE 40.7 The lack of accessible features in a residential home for older people may limit aging in place for users.

as though they are separate groups, each with one set of needs, for whom separate amenities should be provided. This approach inevitably makes for environments that are exclusionary, as well as often duplicating facilities unnecessarily. If an additional ramp for those who cannot climb stairs has to be provided alongside the steps at a building entrance, additional cost is incurred. Moreover, unless building codes are enforced, when costs are to be cut that ramp may well be omitted on the grounds that not many people with wheelchairs will use it. Similarly, merely providing special housing units for the so-called "elderly" is not the sole solution to providing for the housing needs of the older generation, and is exclusionary in practice.

Since Asian values question the idea of putting old people in special homes unless this is absolutely necessary, there is a very good case for universal design, to ensure that the family home is always truly accessible, safe, and supportive of the needs of older residents. Assistance by family members is widely taken for granted and often expected by many older people in more traditional Asian families. Where this is so, the needs of the caregivers should also be taken into account, in terms of space and facilities provided, particularly by providing step-free and nonslip floor surfaces. If all of the built environment were to be made accessible and safe, residents could go on using their familiar habitat for much longer, without having to move if they can no longer get about in safety. In an ideal world, all housing, whether multigenerational or for independent living, should be habitable and visitable by people even if they become disabled. If this happens, well-designed "adaptable" apartments can be easily and economically modified to suit the occupants' changing needs.

Regrettably, many components that are provided for "independent living" lack sensitivity in their design and look more as though they belong in the hospital than in the home, thereby discouraging many people from using or installing them in order to prevent accidents before they happen. This is true both of assistive devices like wheelchairs and fixtures such as handrails, which are often only available from "independent living centers" attached to rehabilitation units in hospitals, and principally available only to older people after an accident or hospitalization.

40.15 CONCLUSION

Universal design, by definition, may usefully be applied in any situation, but there are some which cry out for an integrated, holistic approach. Housing must be high on the list of priorities, since it is fundamental to everyone's life and should support people throughout their life-course. Multigenerational living, currently the preferred housing solution in Singapore, presents just as valid a case for universal design as do the smaller, purpose-designed units, such as the studio apartments. To allow greater personal independence in the family home will actually encourage aging in place, since the older resident will not be obliged to move to a more "suitable" place if and when he or she becomes less mobile. Increased independence, for mobility and for carrying out personal daily activities, such as using the bathroom, will also relieve the caregivers of some of the attention that they now have to expend to ensure, for example, that the elderly person does not fall. Patently, here is a very strong case for universal design, where every family apartment can provide the standards of safety and support for any member of the family, regardless of age or ability. But elderly friendly apartments and environments still have to be recognized for their worth, among the policy makers, developers, and by professional bodies. Awards for well-designed exemplars can demonstrate how universal design is appreciated by design professionals as well as users.

Meeting the needs of those sectors of the population that have so far been marginalized by barriers in the built environment is possible by considerate design, without great expense or loss of aesthetic quality. All buildings—and in particular the home environment—should allow their inhabitants to do what they came to, and not overawe or inhibit the occupants by the architecture. But if one component of the access system is inadequate it can make the whole place inaccessible for many users. Design details of many everyday things like street furniture, hard landscaping, legible routes, and usable road crossings should form part of the overall pattern of accessibility, safety, and convenience. The dwelling itself should be the core of this system, allowing independence and safety well into old age, whatever the cultural patterns or economic status of the occupants. Particularly where demographic and social patterns are changing, as in developing Asian countries, universally designed environments can allow for future adaptation with minimal disruption or expense. These should be the trends for a better future, making the designed world friendly for as many people as possible through consistent approaches that accept that users may be young or old, or have different capabilities.

A built environment that fully integrates the older person into the community will, in the words of the IMC Report, "be critical in determining the extent to which older people can be integrated into the wider community and can lead active lives. Aging in place, as a strategy for housing senior citizens, can help promote the desired integration. For this to be successful, the built environment and transportation system must be elder friendly, and have barrier-free features which facilitate safe access to services" (IMC 1999).

40.16 NOTE

1. The insistence on politically correct terminology is not common in Asia, and even causes confusion in some circles. But these niceties will surely come along with increasingly positive attitudes and awareness of issues of designing for people with disabilities.

40.17 BIBLIOGRAPHY

Brewerton, J., and D. Darton, *Designing Lifetime Homes,* Joseph Rowntree Foundation, York, UK, 1997.

Chua, Beng Huat, *Political Legitimacy and Housing: Stakeholding in Singapore,* Routledge, London, UK, 1997.

Harrison, James D., "Housing Singapore's Frail Elderly in the Next Millennium," *Singapore Medical Journal,* vol. 38, 1997.

———, *Housing Provision for an Aging Population in Singapore.* CIB/TG 19 University of Reading, UK: Proceedings, Designing for the Ageing Society.

———, and K. J. Parker, "Getting it Right: Housing Design for an Aging Society in a Changing World," *Proceedings of the XXV IAHS Housing Congress,* Lisbon, Portugal, 1998.

IMC, *Report on the Interministerial Committee on the Ageing Population,* Ministry of Community Development, Singapore, 1999.

PWD, *Code on Barrier-free Accessibility in Buildings, 1995,* Building Control Division, Public Works Department, Singapore, 1995.

Straits Times, "HDB to try out Flexi-Flats in Punggol," *Straits Times,* Singapore, 18 September 1999.

UN, *Promotion of Non-Handicapping Physical Environments for Disabled Persons: Guidelines,* UN (ESCAP), New York, 1995.

UN, *Promotion of Non-Handicapping Physical Environments for Disabled Persons: Case Studies,* UN (ESCAP), New York, 1995.

Wong, Aline K. and Stephen H. K. Yeh, *Housing a Nation: 25 Years of Public Housing in Singapore,* Maruzen Asia for Housing and Development Board, Singapore, 1985.

CHAPTER 41
UNIVERSAL KITCHENS AND APPLIANCES

Abir Mullick, M.A., M.C.R.P., and Danise Levine, M.Arch.
RERC on Universal Design, University at Buffalo, Buffalo, New York

41.1 INTRODUCTION

The kitchen is one of the most-used places in the home. It is both private and public. The use of the kitchen is essential to preparing food, carrying out household activities, maintaining family contact, and fostering social interaction in the home. Modern society has linked independent use of the kitchen to independent living, and those unable to use the kitchen on their own are dependent on others for nutritional assistance. Many older people and individuals with disabilities require assistance because they have difficulty using the kitchen. These individuals have problems accessing storage cabinets, using appliances, and maneuvering inside the kitchen.

Most current kitchen designs are noninclusive and they do not allow equal participation by all members of the family. They are poor designs and do not work well for anyone, including able-bodied persons as well as people with disabilities. The universal kitchen is an inclusive approach to kitchen design. This chapter will discuss the fundamental basis for universal kitchens and appliances, present design guidelines and examples, offer a design discussion on the future of universal kitchens and appliances, and conclude with strategies to market them.

41.2 BACKGROUND

Every member of the family shares the residential kitchen, which consists of appliances and cabinets. These kitchens come in standardized designs such as galley style and L-shaped and U-shaped plans. All people, regardless of their physical condition, are limited by the design of kitchens—some more than others. While able-bodied individuals usually experience only minimal difficulties, children, older people, and those with disabilities are often greatly challenged by kitchen designs. People with disabilities have great difficulty moving around in the kitchen, gaining access to storage, and operating appliances. Mobility-impaired individuals such as wheelchair and walker users are often handicapped by kitchen layouts and have difficulty functioning in the kitchen. Seated users and people of short stature have difficulty reaching inside refrigerators, washers, dryers, and ovens. Many people are unable to operate cooktops and rear-mounted controls on ranges, washers, and dryers. Those with arthritic hands or hand deformities find it difficult to grasp round knobs and controls; they may be unable to use handles on refrigerators, ovens, and microwave ovens. Visually impaired people have difficulty reading and understanding written information on control panels. Safety, convenience, and independence in the kitchen are not only concerns for the elderly and people with disabilities. All users, regardless of their physical condition, value them greatly. They are all interested in kitchens that meet their functional, social, and cultural expectations.

New Considerations in Kitchen Design

Today's kitchens are significantly different from 1940s kitchens, when their layout was first established to promote work efficiency in the home and the mother was the exclusive user of the kitchen (Hayden, 1982). The demand for universal kitchens has increased as society has been transformed by demographic change. Advances in medical technology and health care not only enable people who might have died from illness or accidents to survive; they have also lengthened human life spans and improved the quality of life in old age. People are living longer, and they are living longer with disabilities. Historically speaking, there are more people now with disabilities than ever before. Moreover, society is finding ways to accommodate and integrate people with disabilities with the rest of the population. The independent living movement affirms the right of people with disabilities to live an independent life in residential settings of their own choice. Health care reforms have emphasized community integration over institutionalized care. The demand for accessible homes will increase as more and more people grow older, acquiring some disabilities but choosing to age in place or wherever they wish—but not in a nursing home (AARP, 2000). It is important to note that people with disabilities often live in households in which not everyone has a disability. Thus, the same kitchens and appliances must be usable by people with varying physical capabilities.

Today's kitchen users are no longer only women; all members of the family use the kitchen. While most kitchens were designed primarily for preparing food, it is very clear that the use of the kitchen today is diverse, and that people perform many different activities in the kitchen beyond food preparation. For example, the kitchen has become a place for social and family interaction, where many people read, write, watch TV, and look after children. Moreover, the usability of the kitchen varies widely with age, income, physical condition, and family structure. As a result, today most kitchens do not meet individual or collective needs, and they will require expensive modifications to accommodate their users' present needs and their changing needs over time.

Even though today's kitchen users and activities have changed significantly, the appliances and cabinets that make up the kitchen have not changed much since the 1940s. They have only been transformed in appearance and improved technologically to become more efficient. A few major appliances, such as dishwashers and microwave ovens, are relatively new additions to the kitchen. The majority of kitchens today do not meet the social and collective goals of universal design, as they continue to focus exclusively on efficient food preparation, with little attention being paid to user variations and other kitchen activities. They ignore the fact that today the kitchen is the nucleus of family activity, where young children play, teenagers watch TV, adults meet and discuss family matters, and the elderly interact with children and grandchildren. Today's kitchen must be a place for all family members, and it needs to accommodate a variety of family activities.

41.3   UNIVERSAL KITCHENS

The universal kitchen is an equal opportunity kitchen; it should benefit all users and meet their present and future needs. Universal kitchens are not the same as accessible or "barrier-free" kitchens, as these kitchens incorporate design features specifically for accommodating people with disabilities. While accessible kitchens have the potential to solve the accessibility problem, they also separate people with disabilities, and create social stigma and personal alienation. The universal kitchen is a kitchen for all people; it considers the range of users, provides choices for different needs, and accommodates everyone at all times. The universal kitchen is a design for usability and consumer acceptance by everyone. A universal kitchen can not only offer a high degree of safety, security, and independence to all users; it can also provide individual satisfaction, family interaction, and long-term economic benefits to home owners.

The universal kitchen is not one design for all people. A single universal kitchen would be the same as "one-size-fits all," and this approach contradicts the pluralistic aspirations of universal design. Supporters of universal design argue that the one-kitchen-design for all is exactly the problem with kitchen designs in existence today. Due to their limited flexibility in appliance design and placement layout, these kitchens neither provide choices nor do they allow individuals to adapt the kitchen environment so they can participate on their own or with other members of the family. The universal kitchen supports the idea of individualization and personalization through design flexibility and diversity; that is, different designs for different users within the same system, or adaptability and adjustability that can accommodate all users. A universal kitchen is a place where all members of the family can participate.

Some people believe that due to a tremendous variation in user populations and their needs, a true universal kitchen may be an unachievable goal. This is because universal design does not prescribe a final state, as nothing is truly universal, nor is the process of designing a kitchen ever finished. Many advocates of universal design realize that the lofty social and inclusive goals of universal design are almost unattainable, as it is nearly impossible to design kitchens for "all" people. The goal, therefore, is to approach the objectives of social inclusion through an ongoing process, best termed as "universal designing" (Mullick and Steinfeld, 1997). The process of designing a universal kitchen must be viewed as incremental and ongoing, the primary goal being individual and social inclusion through design.

41.4   *KITCHEN DESIGN GUIDELINES*

The Center for Inclusive Design and Environmental Access (IDEA) at Buffalo, New York, developed kitchen design guidelines as part of a demonstration study called, "Fair Housing Means Universal Design." This study compared Fair Housing Designs with typical kitchen designs and it involved a variety of volunteers who demonstrated the use of four kitchen designs: two typical kitchen designs (one galley-style and one U-shaped) and two kitchens that met and exceeded the Fair Housing guidelines (one modified galley-style and one modified U-shaped). All four kitchens were full-scale, simulated environments, equipped with real but nonworking appliances. Volunteers were asked to perform usual kitchen activities such as opening the refrigerator, using the counter workspace, accessing the cabinets, operating the range, oven, and dishwasher, and using selected products that make activities in the kitchen easier. Many individuals with different abilities, and some with no disabilities, took part in these demonstrations. There were old, middle-aged, and young people; some of the volunteers used walking aids, some used wheelchairs, and some simply had difficulty walking. Some volunteers had minor to moderate limitations while others were more severely disabled. Several volunteers brought family members who usually helped them in the kitchen.

Kitchen A was a small galley-style kitchen of the type found in most affordable apartments. Some of its features included: an 8 by 9 ft floor area, 36 in of space between countertops, and a refrigerator with the freezer on top. Kitchen B, the modified galley-style kitchen, met minimum Fair Housing requirements. Differences from Kitchen A included: 40 in between counters; clear floor space in front of appliances allowing for parallel wheelchair parking; knee clearance under the sink; and a lever faucet instead of a conventional two-knob faucet. The U-shaped Kitchen C demonstrated a different approach from Kitchen B and had several new features: 60 in between counters; a side-by-side-type refrigerator/freezer; and rotating-base corner cabinets. In comparison to Kitchen C, the modified U-shaped Kitchen D was 10 in longer and 15 in wider; it had a motorized device for adjusting the sink height; larger toe kick space; and larger wall cabinets, with the lower shelf at a height of 48 in. (See Figs. 41.1 through 41.4.)

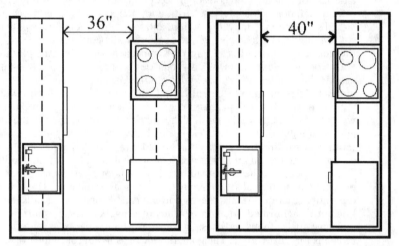

FIGURE 41.1   Galley-type Kitchen A.   **FIGURE 41.2**   Modified galley-type Kitchen B.

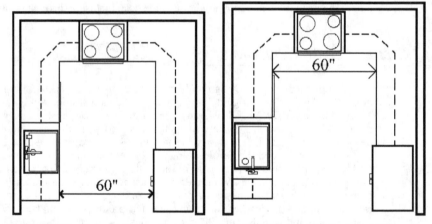

FIGURE 41.3   U-shaped Kitchen C.   **FIGURE 41.4**   Modified U-shaped Kitchen D.

IDEA Center Guidelines

The demonstration study outlined many important kitchen design guidelines, such as (Steinfeld and Levine, 1997):

- Knee clearance under work surfaces, adjacent to storage cabinets and appliances
- Enough clear floor area to accommodate someone using a wheelchair or walker, or to allow more than one person to work efficiently
- Electrical controls, switches, and receptacles in easy-to-reach places

- Enough counter space near appliances and storage to enable users to transfer items easily
- A spill-proof countertop edge that contrasts in color
- Task lighting in key work areas
- Storage at optional heights, and adaptable storage systems
- Storage options within comfortable reach ranges, between 24 and 48 in above floor level, or mechanical devices that bring storage within that range
- Storage system that offers high visibility and exposure
- Side-by-side refrigerator for easy access to both cooling and freezer areas
- Accessories such as sliding shelves, racks, and drawers that reduce bending and reaching
- Raised dishwasher that minimizes bending
- Shallow sink that reduces bending
- High-contrast, low-glare fixtures
- Controls that are easy to see, grip, and operate

National Kitchen and Bath Association Guidelines

The National Kitchen and Bath Association (NKBA) is a trade and industry organization consisting of kitchen and bathroom designers, manufacturers, and trade professionals. It is dedicated to the kitchen and bath industry, and is a leading source of industry information and professional education. NKBA certifies kitchen and bathroom designers, publishes educational materials, and promotes kitchen and bathroom design standards. It is committed to universal design and promotes universal design values to its members. NKBA promotes 20 kitchen planning guidelines, many of which are universal-design-focused. With consultation by Mary Jo Peterson, NKBA has developed universal strategies for selection and placement of countertops, cabinets, and appliances to better accomplish universal access in the kitchen (NKBA Web site, 2000). These guidelines include the following recommendations:

Countertops

- 28 to 32 in high for standing and seated users; at this height they are comfortable as chopping or baking centers.
- 36 in high for general use by standing users; at this height the countertop works well as a snack bar.
- 42 to 45 in high for taller users.

Cabinets

- Use motorized or mechanical systems to adjust wall cabinets.
- Install dividers in drawers, and roll-out shelves to provide access to interior storage systems such as drawers.
- Provide doorless open shelves for quick detection of items; this eliminates the hazard posed by open doors.
- Include drawers instead of doors; this eliminates the need to get around the open door.
- Install lever handles, as opposed to knobs; this allows for easy opening of doors.

Appliances

- Install side-by-side refrigerators or bottom-freezer models; these provide ideal access for all users, including the young and old.
- Separate cooktop and oven; this accommodates height differences.

- Select appliances with front-end-placed controls; this offers safe and easy access.
- Select appliances that have easy-to-read numbers and touch-pad controls.
- Place microwave within reach and sight of the individuals who will use it.

41.5   *KITCHEN DESIGNS*

Kitchens have traditionally been a constructed environment that requires assembling a variety of components such as cabinets, fixtures and appliances. While these components are carefully designed products, they have a strong relationship to one another, and their interrelated use makes up the kitchen. Kitchen design fundamentals, therefore, deal with the design, selection, and placement of these components and how their integrated use can offer safety, utility, and work efficiency. Most kitchen designers are architects or certified kitchen designers; they are familiar with the various cabinets, fixtures, and appliances available in the market and how to organize them to make a kitchen environment. These professionals do not design cabinets, fixtures, and appliances; such components are designed by industrial designers who have expertise in human factors, work efficiency, product design, and production technology. In recent years, industrial designers have become involved in kitchen design; their designs show that the kitchen is a product environment, and that integrating cabinets, fixtures, and appliances with one another can improve work efficiency. The two examples of kitchens that follow suggest two very different design approaches: one is a one-of-a-kind environment created by a kitchen designer, and the other is a generic, mass-produced environmental product assembled by an industrial design team.

General Electric's Real Life Design

Sponsored by General Electric, the Real Life Design project is a kitchen designed by Mary Jo Peterson, a certified kitchen designer, for use by a variety of people living at home (see Fig. 41.5). It uses standard appliances and stock cabinets in a nontraditional way to produce an innovative kitchen design (General Electric, 1995). This kitchen considers the range of users and supports its use by all members of a family, including, for example, able-bodied adults, older people, wheelchair or walker users, children, and tall people.

There are numerous design details that make this kitchen an inclusive family environment. For example, it incorporates three counter heights: 30 in for seated users, the traditional 36 in, and 45 in for taller individuals. Varying heights of countertops address stature variations among family members, accommodate both seated and standing individuals, and provide a useful variety of work surfaces. The clear floor space in this kitchen is generous and offers sufficient room for maneuvering a wheelchair, assisting another person, or simply working with someone else. Cabinet heights have also been varied to make their contents more accessible to a wide range of users. All storage in the base cabinets rolls out for clear view and easy access. All base cabinets and many floor-mounted appliances have a 9-in toe kick for improved wheelchair or walker access; this also facilitates detecting fallen objects and cleaning the kitchen floor. Many wall cabinets have open shelving, and base cabinets have shatterproof glass doors; this offers a clear view and easy access to stored items. A selected base cabinet has a built-in step stool to facilitate reaching hard-to-access spaces; the stool folds and locks into place. The sink is height-adjustable, has clear knee space, and features a shallow basin. Important work surfaces, such as a microwave table, also have clear knee space.

Appliances such as the dishwasher are raised 9 in off the floor. Height adjustability, clear knee space, and raised appliances and shallow sink basins reduce bending and overextension by all users. There are many pullout shelves on which to set hot food items and to provide

FIGURE 41.5 General Electric's "Real Life" kitchen with variable-height countertops and height-adjustable sink.

additional and variable-height work surfaces. They also benefit seated users or those of shorter stature. Some pullout shelves have cutouts to hold bowls for one-handed mixing operation. There is a raised oven at one end of a cabinetry run; this offers clear and unobstructed access to seated users, and reduces bending by all standing users. There is a rolling table underneath the microwave oven; this offers additional work surface to seated or standing users. A rolling cart fits neatly beneath the rolling table, which can be easily moved around for food preparation and serving. The recycling bins are located in the corner base cabinet; they slide out, rotate, and are easily removed, thereby providing easy access and eliminating fatigue-related heavy lifting. The countertop adjacent to the flat cooktop is lined with heat-resistant ceramic tile; this facilitates sliding heavy hot pots and thus also eliminates the need for heavy lifting. Other countertops have raised, contrasting color inserts to contain spills and offer visual and tactile cues to people with visual impairments. Cabinet doors, such as the one under the cooktop, fold open to provide clear knee space for seated users.

Rhode Island School of Design Universal Kitchen

In 1993 a team of 100 students and faculty under the direction of Marc Harrison, Professor of Industrial Design at the Rhode Island School of Design, embarked on a project to design a universal kitchen. This project, which was in response to the Cooper-Hewitt' National Design Museum's plans for a design exhibition, undertook an interdisciplinary research on universal kitchen and a program to rethink the kitchen standards. The project drew many industry sponsors such as Frigidaire, Kohler, Dow Plastics, Masco Corporation, International Paper, Schott Glass, Lightolier, Hafele, and Boran (a Nortek company), and many national advisors such as Julia Child, George Covington, Niels Diffrient, Cynthia Leibrock, Ron Mace, Patricia Moore, Elaine Ostroff, and James Pirkl. The primary goal of this project was to develop an integrated kitchen system that would adjust to meet the needs of a wide range of kitchen users. The project produced three designs, of which two were made into prototypes. The two prototypes were displayed at the Cooper-Hewitt's "Unlimited By Design" exhibition in New York City in November 1998.

The faculty/student team conducted design research dealing with time and motion studies on a wide range of users in food preparation and analyzed existing kitchens in the local community. The research showed that there are more that 400 discrete steps in preparing a simple dinner, and that routine kitchen tasks require continuous bending, stooping, reaching, and lifting. Much unnecessary travel and many gross inefficiencies happen due to bad kitchen layout and inappropriate placement of appliances and cabinets. The faculty/student team used the research to identify the limits of comfortable reach, called the *comfort zone,* and developed kitchen designs in which all critical activities are located within this zone (RISD, 1998).

The universal kitchen is an integrated kitchen system that aims to involve all members of the family. It minimizes the effort involved in and the time taken to move between the traditional work triangle in the kitchen: the sink, refrigerator, and cooktop. Conceptually, this kitchen design is an embodiment of a minimum-size work triangle, and it does so by integrating appliances and cabinets and accommodating them within the easy reach or comfort zone of all users. Unlike most of today's kitchens, where appliances and cabinet placement make up a kitchen environment, the universal kitchen is a synthetic unit designed to fit in a wide variety of architectural settings. It is compact, nonobtrusive, and it can be easily transported to any location.

The universal kitchen is made up of a "kit of parts," with interchangeable modular components for refrigeration, cooking, water delivery, and storage. Users can select and arrange these components, and adjust their heights and depths, manually and automatically, to be within their particular comfort zone. All three kitchen models, MIN, MID, and MAX, are equipped with a baking center, multiple washing and cooking stations, and innovative cleaning systems. At the present time, two kitchen prototypes have been constructed: MIN, a small-size kitchen for studio apartments, dorm rooms, hotel suites, and independent living centers; and MAX, a gourmet kitchen and dining area that accommodates a large family (see Figs. 41.6 and 41.7). They each have the potential of being stand-alone, mass-manufactured units for wide distribution and sale.

The universal kitchen is not only user and technology focused; it suggests a paradigm shift in kitchen design and production, from kitchens as constructed environments to kitchens as manufactured systems. Such a paradigm shift can have a huge impact on design, construction, manufacture, sale, and maintenance of future kitchens. There are tremendous commercial opportunities surrounding this kitchen design. In fact, Maytag Corporation has signed a technology transfer agreement with the Rhode Island School of Design (RISD) with exclusive worldwide rights to the Universal Kitchen.™

41.6   *UNIVERSAL APPLIANCES*

Like universal kitchens, universal appliances are designed to benefit a wide range of users. They are an important aspect of universal kitchens, and together with cabinets they make up the kitchen environment. Universal appliances conform to many universal design principles, such as providing the same means of use for all users (identical when possible, equivalent when not); providing privacy, security, and safety to all users; not segregating and/or stigmatizing users by offering unequal social and personal benefits; appealing to all users; free of unnecessary complexities; consistent with user expectations and intuition; accommodating a wide range of users; prompting users for sequential actions; and providing timely feedback (Story, 1997). A universal appliance offers a high degree of safety, security, and independence to all users, and has the potential to provide individual satisfaction, family interaction, and long-term economic benefits.

Universal appliances are different from mass-marketed appliances retrofitted for people with disabilities. For example, washer/dryers with retrofitted controls and information panels

FIGURE 41.6   Rhode Island School of Design's "MAX" kitchen for the family.

are not universal appliances. Retrofitted appliances are similar to assistive technology products, made adaptable for people with disabilities. These products increase, maintain, or improve functional capabilities of individuals with disabilities. Although indispensable for helping people with disabilities to achieve independence, retrofitted appliances suggest "special-purpose-design" and they draw attention to people's disabling conditions. They are clinical-looking and stigmatize those who use them (Mullick and Steinfeld, 1997).

There are no universal appliances, only appliances with universal features. Due to tremendous variation in user population and the complex nature of appliance design, it is nearly impossible to design a universal appliance for "all" people. In a way, true universal appliances may be an unattainable goal, and what are realizable are universal features. The approach to appliance design, therefore, must focus on social and personal inclusion. The process of developing a universal appliance is an incremental and ongoing process, more and more universal features, so individual and social inclusion can happen through design. Several corporations have developed universal design strategies for appliance design, since they view these strategies as a means to improve competitiveness by both making a better product and also enhancing marketability of existing features. Some new appliances are even being developed directly in response to the universal design philosophy.

In this new millennium, there seems to be a shift toward more pluralistic values and actual performance rather than just superficial styling and "featurism." While it is difficult to accurately identify the forces behind this shift, it could be attributable to: (1) the demographic change that is creating a more diverse population; (2) the aging of the baby boomers and their parents, which is making everybody aware of the fact that the young of today are the aged of tomorrow; and (3) long experience and frustration with products that are difficult to use and even understand.

Universal design features in appliances are becoming commonplace as corporations realize the power of universal design to develop better appliances, improve competitiveness, and enhance marketability. Side-by-side refrigerators, water/ice dispensers in refrigerators, flat-top

FIGURE 41.7   Rhode Island School of Design's "MIN" kitchen for apartment, dorm, and hotel suites.

radiant cooktops, front-end controls in cooktops, combination washer/dryers, and subzero-drawer refrigerators are all examples of universal design features in appliances. Appliance design can greatly benefit from adopting a universal design approach to product development, since this will produce innovative, user-centered appliances that a wide variety of people can personalize. The universal design approach can also help appliances reach a broad market, and make products and services economical and accessible to all people. In spite of all the benefits associated with following a universal design focus, there are many barriers restricting innovation in universal appliance design. The removal of these barriers is essential for more rapid appliance innovation.

41.7 APPLIANCE DESIGN GUIDELINES

The appliance industry is continually developing new features in response to consumer demand and to obtain a competitive edge. The following strategies for the development of appliances offer a high degree of usability for everyone.

Refrigerators

- Refrigerators that are nearly flush with the base cabinets provide a wider and more accessible clear floor area.
- Folding shelving systems in refrigerators allow rapid adjustment of space to accommodate storage containers of various sizes.
- Continuous door handles on refrigerator doors provide infinite grasping positions.
- Water dispensers improve visibility for those with low vision or in low levels of lighting.
- Front-mounted control systems offer easy access, increased visibility, and easy operation.
- A built-in water filtering system saves loading bottled water.
- A bottom-mounted freezer with shelves that pull out reduces reaching.
- Electronically operated systems that will adjust shelves at the touch of a button greatly facilitate storage and retrieval.
- Toe clearances can reduce the strain associated with reaching.

Cooktop

- Front- or side-mounted control panels are safer than rear-mounted panels because they eliminate the need to reach over a flame.
- Feedback regarding which burners are hot to touch even after they have been turned off can prevent burns.
- Spatial "mapping" of controls to burner location makes it easier to understand which controls operate which burners.
- Backlighted control panels are more legible for people who have difficulty reading print and graphics.
- Glass cooktops are the easiest to clean.
- Controls shaped to facilitate gripping can improve access and convenient use.

Conventional Ovens

- Wall-mounted ovens are safe and easy to use for standing users because they require no bending.
- Side-mounted control panels or a control panel between two stacked ovens improve convenience and use.
- Simplified systems that are easy to operate and understand greatly improve the usability of ovens.
- Backlighted control panels increase readability and comprehension.
- Rollout shelves or electrically operated shelves facilitate sliding of hot and heavy items.
- Fold-down doors offer an intermediate resting surface for hot and heavy items.
- Side-mounted doors could facilitate oven use by seated users—but none are currently produced.

Microwave Ovens

- Microwave ovens located on a counter are most usable for the broadest population.
- If microwave ovens have counter space adjoining them, it facilitates the transfer of hot items.
- Most microwave ovens are equipped with side-opening doors, but fold-down doors could provide a work surface useful for transferring hot items.
- A simplified system that is easy to understand greatly improves the operation of microwave ovens.
- Backlighted control panels and symbols increase comprehension of information.

Dishwashers

- A higher mounting location prevents bending and overextension and reduces back pain.
- Newer models have sophisticated controls to ease decision making.
- Interior illumination helps the user ascertain the condition of dishes and also helps those who have low vision.
- Backlighted control panels and iconic signs increase comprehension.
- An easy-to-load system would greatly increase convenience and facilitate use.
- A compact, portable model for smaller households can be installed in many different locations, making it highly accessible and usable.

Washers/Dryers

- Higher bases raise the opening height and minimize the need for stooping over.
- Front-loading washers and dryers are useful because they can be used from a seated position.
- Some washers/dryers can be installed side-by-side or up-and-down, based on user preference and available space.
- Side-by-side installation greatly facilitates transferring laundry.
- Combination washer/dryers eliminate the need to transfer wet laundry.

41.8 EXAMPLES OF KITCHEN APPLIANCES

Students in the Art Center College of Design in Pasadena, California, worked on a kitchen of the future project for Whirlpool Corporation, called Kitchen 2000. They were asked to be architecturally innovative, consider nontraditional kitchen layout, and conceive how the kitchen of the future might look and work. Students were required to produce appliance designs that were real "machinery," and to search for and apply future technology that might make life a little easier and more fun in the kitchen. In a concept paralleling a contemporary workstation for information processing, the students visualized the future kitchen as domestic workstation for food preparation. They saw the kitchen as a place for processing food and an environment that supports interaction between people and appliances. Instead of creating a room with built-in cabinets, plumbing, and appliances, the students created freestanding pieces that perform many complex tasks, and these pieces could be arranged in numerous ways for maximum mobility and to reflect the preferences of the individual user. Even though not directed to design for universal access, these student projects have many inclusive design features. They are convenient to use for all people, do not segregate users based on age or ability, and address the seven Principles of Universal Design (Center for Universal Design, 1997).

The appliances by students show a strong link in design thinking to the universal kitchen by the Rhode Island School of Design. They are the work of industrial designers who focus greatly on work efficiency and convenience. Like the RISD kitchen, the appliances are future

FIGURE 41.8 The doorless "Cool-Column" refrigerator provides an uninterrupted view of the contents plus easy access.

technologies, stand-alone environmental products, designed to be mass-produced for mass consumption, have an independent spatial presence, and resonate strong industrial aesthetics. The RISD kitchen and appliance designs are based on a systems approach, which focuses on user-environment interaction to provide a high degree of comfort, work efficiency, and time-saving. The student projects are innovative examples of universal design; their design incorporates many of the appliance design guidelines mentioned earlier and many new features not part of the guidelines. Their strong focus on user convenience makes them universal.

"Cool Column" Refrigerator. This doorless refrigerator by Darrek Rosen opens on two sides (see Fig. 41.8). A cool air tornado swirls around the circumference of the interior, keeping the inside cool, while a vertical curtain of air, much like those at store entrances, moves down over each large opening through jets in the top and bottom of the frame, keeping warm air out. The upper portion has chilled, rotating clear shelves for conductive cooling and visual access to food and drink. The freezer compartment is located at the bottom, and retracts on drawer slides. A modular heat exchange is located on top for energy efficiency and would be completely removed and replaced for service. The cool column can be easily moved around in the kitchen and can serve as a room divider that happens to hold food. People from the adjoining room can reach inside for ingredients to make a sandwich or get a drink.

The lack of a door is the primary reason why this refrigerator is universal. It provides an uninterrupted view of, and hassle-free access to, the contents inside. Two-sided access into the refrigerator offers choice in approach and interaction. Because the refrigerator can be easily rolled to places of food consumption, it eliminates the need for carrying several items of food, thus reducing movement and inconvenience associated with food transportation.

"Jet Stream" Burner. This freestanding burner by Ed Hawkins has many high-octane sources called candles (see Fig. 41.9). The burner rests on a freestanding portable base that makes it possible to cart around; it can stay in a kitchen or be rolled outdoors for barbecues. The cooking panhandle can be captured by non-slip coated "fingers" of various lengths, allowing custom pan design. Special fittings can be added to adapt the burner capability for many types of cooking vessels, such as a wok or other shaped pans.

This design recognizes problems many people have with cooktops that are situated in constricted spaces or in awkward locations. It can be easily moved around in the kitchen, and taken outdoors if necessary, allowing users the freedom to make locational decisions based on personal convenience and adjacency to related kitchen items. Burner design in this project is very accommodating of pots and pans; it allows situating many types of cooking vessels, which is often a huge problem in existing burners. Improved burner design offers better fit between the burner and cooking vessels, providing everyone with a high degree of safety, security, and cooking convenience.

FIGURE 41.9 The portable "Jet Stream" burner offers moving convenience and secure pan grip.

"Food Closet" Computerized Pantry. This freestanding smart unit by Marde Burke stores and dispenses dry goods that are stored in climate-controlled rotating cylinders (see Fig. 41.10). Double doors can be opened to take goods out directly from the shelves, or the desired product can be automatically dispensed

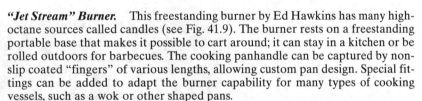

FIGURE 41.10 The temperature-controlled "Food Closet" pantry uses an interactive screen for ordering and retrieving groceries.

when called up on a removable interactive screen. To replace goods, an inventory microprocessor uploads to the supermarket's computer, and indicates necessary items for purchase. The storage unit can be placed against the exterior wall of the house so that goods can be loaded as they are brought from or delivered by the supermarket. Using the bar codes on canned or dry goods, the unit reads where food should go, keeps track of the inventory, and maximizes shelf life by maintaining optimum temperature and humidity. It also comes with a planner that presents menus based on what items are stored in the unit.

This futuristic food storage offers many universal design benefits associated with inventory maintenance, loading and unloading of groceries, and storage and retrieval of food items. Especially helpful for busy people and those who have difficulty remembering information, this appliance communicates directly with the supermarket's computer and orders delivery of items that need to be replenished—a huge convenience to everyone. Because it is located against the exterior wall, it facilitates delivery of food items, freeing everyone from the drudgery of grocery loading and unloading. The ability to call for desired products eliminates problems associated with reaching into hard-to-access spaces in storage closets.

Microwave "Dome" Oven.   This portable, rechargeable warming device by Takanori Kawai uses microwave technology to warm food right on the plate just before eating (see Fig. 41.11). It offers three important benefits: first, being portable, it can be easily transported to the place of eating such as dining tables or outdoors for backyard picnics. Second, on-plate warming eliminates the need to transfer food between containers for eating and warming. Third, it reduces the inconvenience associated with transporting and handling food in and out of microwave ovens and with cleaning dishes.

FIGURE 41.11 The portable "Dome" microwave oven conveniently transfers and warms food.

"Cylinder" Dishwasher.   This dishwasher by David Zimberoff features over/under compartments with on-board water heater, gray water reclamation, and a foot pedal to operate the door when the operator's arms are full (see Fig. 41.12). The washing compartment has a two-tiered rotating carousel that eliminates the difficulties associated with reaching inside a dishwasher to load and unload dishes. A separate open tub on the top of the main washing

FIGURE 41.12 The computerized "Cylinder" dishwasher provides easy access and recycles gray water.

cylinder cleans smaller items ultrasonically after they are loaded on a submersible rack. This method assures hygienic cleaning of utensils, which can often trap food particles that escape ordinary washing.

41.9 THE FUTURE OF UNIVERSAL KITCHENS AND APPLIANCES

The current approach to kitchen development has focused on designing "one appliance at a time." This approach has produced a diverse set of products that do not work well with each other, except in appearance. Unlike most machines for contemporary industrial production and office work, which are designed for optimum interface and work flow, home appliances are independent pieces of equipment that, when used together, do not make up a universal kitchen environment. As a result, the workflow in the kitchen is intermittent and needs supervision. This may be fine for most people, but it can be very demanding for many others. Better integration of appliances could produce improved kitchen environments and significant benefits for everyone. Combination washer/dryers, for example, eliminate the need to transfer clothes and can eliminate human intervention during the process. They would benefit not only people with disabilities, but also many other user groups such as pregnant women, people with arm and back problems, those who hate housework, and the time-impaired. Everybody would benefit from such appliances because they minimize human attention and facilitate workflow.

The future of universal kitchens and appliances lies in recognizing that smart technology can ease housework and add convenience to daily living for everyone. The kitchen, which is a microenvironment consisting of appliances and cabinets, has to relate to a larger macroenvironment, the home. Successful universal kitchens will employ smart digital technology to establish intracommunication between kitchen appliances, as well as intercommunication with outside appliances. This will produce nonintermittent workflow within and outside the kitchen, which otherwise needs constant monitoring. Better integration of appliances with smart technology can eliminate unnecessary human intervention, facilitate workflow, produce improved kitchen environments, and offer significant benefits for everyone.

Digital technology enables machine-to-machine communication, such as the interaction of television and videotape recorder. The same thing happens in many burglar alarm systems where clock, radio, and lights are programmed to send and receive signals and they go on and off as planned. They recognize human presence, automatically connect when programmed to do so, and record programs or turn on lights and radio. Digital synchronization eliminates human presence and offers power and convenience to do complicated tasks easily. Nicholas Negroponte, director of MIT's media lab, suggests that linking appliances to better serve people will be a major emphasis of the digital revolution (Negroponte, 1995). He calls this *unification* or *intercommunication*. Linking coffee makers with grinders already brews freshly ground coffee from beans. If a clock is added, the three devices can intercommunicate and brew freshly ground coffee without human intervention to anticipate the user's needs. Negroponte suggests developing networks of information between appliances and other components of the household, such as between refrigerators and cabinets. Refrigerators and cabinets can automatically maintain an inventory and inform the user about what and when to replenish. Better yet, the two can be linked with a car. The car can remind the user to pick up food items on the way home, or even as the user passes the supermarket, if a global positioning system is included.

Today's homes each have an average of more than one hundred microprocessors, but they are not yet communicating with each other. As a result, they require a great deal of human attention. They cannot talk with each other about what food is available and when and how to prepare it. Intercommunication between appliances could produce revolutionary universal kitchens. Just like the coffee grinder/maker, the freezer could transfer food in the microwave oven, set cooking conditions, prepare the food, and deliver it when desired—all without human intervention. Digital technology will reduce the need for users to constantly communicate with machines and assist everyone in performing tasks involved in everyday life. Everyone will benefit from digitally based universal kitchens.

41.10   CONCLUSIONS

The kitchen is one of the most important places in the home, and all members of the family share it. The use of the kitchen is essential to preparing food. Modern society has linked independent use of the kitchen to independent living, and those unable to use it independently are dependent on others for nutritional assistance. All people, regardless of their physical condition, are limited by the design of kitchens, some more than others. While most able-bodied individuals experience minimal difficulties with kitchen designs, children, older people, and those with disabilities are greatly challenged by them. The kitchen is a place for social and family interaction, and many people read, write, watch TV, and look after children in the kitchen. The usability of the kitchen is varied, and it changes rapidly with age, income, physical condition, and family structure.

Universal appliances are an integral part of a good kitchen (i.e., universal appliances make universal kitchens). For a kitchen to be universal, it must be user-oriented, address people's individual and collective needs, and allow people to make changes to it as their needs change over time. Universal kitchens are not the same as accessible or "barrier-free" kitchens, as the latter embody special design features that have the potential to solve the accessibility problem but also separate people with disabilities, create social stigma, and promote personal alienation. The universal kitchen is an equal opportunity kitchen, a kitchen for all people. It considers the range of users, provides choices for different needs, and accommodates everyone at all times.

Universal appliances benefit a wide range of users, and they conform with the Principles of Universal Design. Universal appliances are different from mass-marketed appliances retrofitted for people with disabilities. Retrofitted appliances suggest "special-purpose design," and they draw attention to people's disabling conditions. A universal appliance offers a high degree of safety, security, and independence to all users, and has the potential to provide individual satisfaction, family interaction, and long-term economic benefits.

Design guidelines are available for developing universal kitchens and appliances; for example, the Center for Inclusive Design and Environmental Access, and the National Kitchen and Bathroom Association offer universal kitchen guidelines. These guidelines are user-focused and can help in producing environments for all people. Two universal kitchen prototypes, General Electric's Real Life Kitchen and Rhode Island School of Design's Universal Kitchen, show that kitchens can be safe and secure environments where all members of the family can participate. What makes them universal is that they are designed to be convenient and have many multi-purpose features that benefit everyone.

The future of universal kitchens and appliances lies in recognizing that smart technology can ease housework and add convenience to daily living for everyone. Successful universal kitchens will employ smart digital technology to establish intracommunication between kitchen appliances and intercommunication with outside appliances. This will produce nonintermittent workflow within and outside the kitchen that otherwise needs constant monitoring and supervision. Better integration of appliances with smart technology can eliminate unnecessary human intervention, facilitate workflow, produce improved kitchen environments, and offer significant benefits for everyone.

Finally, universal design in kitchens and appliances can serve as a marketing theme, and producers can use it as an opportunity to increase demand. To be effective, universal kitchens and appliances must have a broad appeal and not focus only on accessibility. Universal kitchens and appliances should emphasize use by a diverse population and the universal value of increased safety, convenience, and usability.

41.11   BIBLIOGRAPHY

American Association of Retired Persons' Web site, 2000; www.aarp.org/.

Hayden, Dolores, *The Grand Domestic Revolution,* MIT Press, 1982.

Center for Universal Design, *Principles of Universal Design,* Center for Universal Design, Raleigh, NC, 1997.

General Electric, *Real Life Design,* General Electric Appliances, 1995.

Mullick, Abir, and Edward Steinfeld, *Innovation,* Industrial Designers Society of America, 1997.

Rhode Island School of Design (RISD), *The Universal Kitchen: Research, Analysis and Design,* Rhode Island School of Design, 1998.

Steinfeld, Edward, and Danise Levine, *Fair Housing Means Universal Design 2: With Emphasis on Kitchens,* Center for Inclusive Design and Environmental Access, University at Buffalo, Buffalo, NY, 1997.

Story, Molly, "Is It Universal?" *Innovation,* 1977.

Negroponte, Nicholas, *Being Digital,* Vintage Books, New York, 1995.

National Kitchen and Bath Association Web Site, 2000; www.nkba.org/.

41.12   RESOURCES

Accessible Appliances, a technical report by Abir Mullick and Danise Levine (published by the Center for Inclusive Design and Environmental Design, University at Buffalo).

Accessible Cabinetry, a technical report by Edward Steinfeld (available from the Center for Inclusive Design and Environmental Design, University at Buffalo).

Designs for Independent Living: Kitchens and Laundry, 1986 (available from the Whirlpool Corporation, Benton Harbor, MI).

Elder Design: A Home for Later Years, by Rosemary Bakker (published by Penguin Books, USA).

Fair Housing Means Universal Design 2, a video presentation to create awareness of the Fair Housing Guidelines and how they make housing usable by everyone (available from the Center for Inclusive Design and Environmental Design, University at Buffalo).

Kitchen Industry Technical Manuals by Ellen Cheever, Annette Depaepe, Nick Geragi, and Marylee McDonald, 1966 (available from National Kitchen and Bath Association).

Lighting Kitchens and Baths by Jane Grosslight (available from Durwood Publishers, Tallahassee, FL).

Strategies for Kitchen Design (posted on the Center for Inclusive Design and Environmental Access Web site).

The Essential Kitchen Design Guide (available from the National Kitchen and Bath Association).

Tools for Independent Living: Appliances Information Service, 1986 (available from the Whirlpool Corporation, Benton Harbor, MI).

Universal Design Newsletter (available from Universal Designers and Consultants, Rockville, MD).

CHAPTER 42
UNIVERSAL BATHROOMS

Abir Mullick, M.A., M.C.R.P.
RERC on Universal Design, University at Buffalo, Buffalo, New York

42.1 INTRODUCTION

Universal design is a commitment to designing for all people. It implies a belief that all persons, regardless of their physical condition or limitations, can benefit from the same environment or product if it is designed appropriately. Universal design of residential bathrooms, therefore, should benefit all people living in a home. Most residential bathrooms fall short of this goal. They do not support all users equally, and while their designs may improve the functional independence of some people, others experience functional deprivation.

This chapter will focus on universal bathrooms and how they provide access to all people. It will discuss functional performance and environmental fit, critique accessibility standards, outline bathroom development, present bathroom design guidelines, and offer examples of universal bathrooms. It will conclude with a discussion on the future of universal bathrooms.

42.2 BACKGROUND

A person's ability to function independently in the bathroom environment changes over time. Bathroom independence is minimal at a young age, and children must rely on parental assistance as they learn proper hygiene and bathroom practices. With growth and maturity, functional dependence changes to reliance on environmental factors for support. For example, a child may need to use a stool to use the toilet or sink. Most young and middle-aged adults operate independently in the bathroom. They can adapt to a wide range of conditions. With advancing age, independence diminishes, first requiring environmental support, such as bath seats and grab bars, and then requiring assistance by care providers.

While independence in the bathroom is largely based on functional ability, the environment plays a crucial mediating role in determining the level of independence possible. In rehabilitation practice, the environment has been conceptualized as a prosthetic support for functional independence. Standards and codes are used to establish how environmental interventions should be designed. The prevailing view is that an environment is either accessible (i.e., meets standards) or inaccessible (i.e., does not meet standards). This conceptual model takes an extreme position about accessible environments and has limited usefulness because it does not consider the complexity of person-behavior-environment relationships. Some people need far more support than others, and sometimes only one small detail may make the difference between independence and dependence. Moreover, this conceptualization ignores the array of in-between conditions that may offer a range of functional independence.

Transactional Perspectives

According to the literature on person-behavior-environment relationships, how a person behaves in a particular situation does not reflect either the person alone or that person's environment, but rather the interaction between the two (Mead, 1934; Cronberg, 1975). This perspective on the nature of person-behavior-environment relationships is the defining characteristic of a transactional perspective (Altman and Rogoff, 1987; Moore, 1976; Wandersman, Murday, and Wadsworth, 1979; Stokols, 1981). According to Steinfeld and Danford (1997), the transactional model is an important concept for understanding functional independence in the bathroom. It allows us to understand the wide range of conditions that contribute to functional independence, rather than being wedded to simplistic normative models. The National Institute on Disability and Rehabilitation Research expresses a similar view as it describes the changing definition of disability as a dynamic interaction between a person and the environment rather than a condition that resides in the person (NIDRR, 1999).

A term used to characterize the appropriateness of a particular person-behavior-environment transaction is *congruence* or *fit*. Fit is a state of equilibrium where an individual's capabilities are in balance with the demands of the environment. Equilibrium may not be a specific pivot point, but rather "zones of adaptation" within which individuals are sufficiently challenged, yet not so challenged or deprived that they are under pathological stress. Perception of users plays a role in fit. Enabling environments, designed to achieve the best fit, should be congruent with the functional requirements of users. An accessible bathroom, for example, can be overaccommodating and not provide the level of challenge necessary to maintain an individual's ability to adapt. Consumers do not appreciate the reflection of self in an accessible bathroom environment, for such an environment will imply that they have a lower level of functional ability than actually exists. On the other hand, an inaccessible bathroom can be overly demanding and may put that person under serious stress due to unsafe conditions. It is essential that the level of fit in enabling environments be neither too little nor too much. By fitting the individual, the environment will be both supportive and attractive.

Accessibility Standards May Limit Independence

Accessibility standards, such as the Americans with Disabilities Act Accessibility Guidelines (ADAAG), are perceived as important means of achieving a person-behavior-environment fit for persons with a disability. Steinfeld and Danford (1997) argued that the transactional model offers at least three reasons why such a strategy might not yield a desirable outcome. They wrote about standards in general, not the transactional model. First, accessibility standards, viewed as a socially formulated definition of fit, do not always represent the latest knowledge and social values. For example, the inertia of the regulatory process tends to work against timely application of emerging technologies such as foldable grab bars or height-adjustable fixtures in the bathroom. As a result, the standards could actually be restricting potential for independence. Also, standards are typically written for achieving a minimum level of accessibility. This could lead to less-than-optimum responses for a particular individual's needs. Second, accessibility standards are largely insensitive to individual differences. They prescribe unvarying responses in the face of varying individual needs. For example, standards prescribe fixed grab bar or toilet heights in the bathroom even though the stature of users varies greatly. Third, standards are often not based on empirically derived performance models. If empirical data are used as a basis for evolving standards, it is usually obtained from studies that seek to define "boundaries" of accessibility rather than seeking to understand the relationship between the individual and the environment. For example, it is unreasonable to assume that there is one fixed "accessible" toilet height for people with disabilities as currently suggested in all standards. The accessibility of a toilet is a function of an individual's stature, ability, and perception of convenience or risk. It is clear that standards are often not based on empirical research at all. Accessibility standards provide a benchmark to evaluate compliance with a law. Unfortunately, they do not correctly reflect the reality of everyday life.

42.3 *HISTORIC DEVELOPMENT OF THE BATHROOM*

The fundamental basis of the bathroom, as a place of relaxation and regeneration, has its roots in the social and cultural history of the bathtub. The present-day bathtub originated as a washbowl for external cleansing, and its design dates back to the Minoan dynasty in Crete, 1800–1450 B.C. The Minoans not only had bathtubs; their bathrooms were equipped with hot and cold water, sewer systems, and water closets. According to archaeologists, the queen's apartment in the palace of Knossos in Crete had a painted terracotta tub, shaped much like the bathtubs that are in use today (Evans, 1921–1935). In later years, the Minoan tradition of bath was adopted by the Greeks of the Mycenaean period, and Homeric heroes bathed in Cretan-tub-like bowls.

The ancient world, like Islam and to some extent the Middle Ages, considered human regeneration as a basic social responsibility. This concept withered away during the Renaissance, and bodily care in the seventeenth and eighteenth centuries was condemned to the point of total neglect. The nineteenth century reintroduced the idea of regeneration, and bathing reappeared around 1830, as a natural act, laying much stress upon cold-water treatments. Around 1850 the Islamic concept of regeneration regained support, and the home vapor bath became popular. Its popularity continued until the end of the century. Home vapor baths, together with the shower and the sunbath, prospered successively, and side by side. Finally, the bathtub emerged as a popular option, and it has remained as the fundamental device for body cleansing.

The development of the bathroom has closely followed the history of social attitudes and cultural values. This is not to recognize that technology had no role to play in bathroom development. The Minoan and Roman eras produced many advanced bathroom technologies, which were sources of pleasure for the rich Minoans and enriched the lives of everyday Romans. Community values that followed the Minoan and Roman periods set aside these technologies and kept them out of reach for many subsequent centuries. As a result, these advanced bathroom technologies were inaccessible to everyone (including the rich and powerful) and people were deprived of the benefits of using the bathroom. It was not until the late 1800s that a somewhat improved version of Minoan bathroom technology reappeared in the United States and was made minimally available to the general public. Bathrooms finally became part of American homes as recently as the early 1900s, when indoor plumbing and running water became available in common buildings.

The basic design of the bathroom as a room for three containers—the tub and the sink for body washing, and the toilet for disposing of body wastes—has remained practically unchanged since its inception many centuries ago. Depending on the social milieu and community support, these containers were either integrated into the building architecture and supplied with running water and sewage disposal systems, or they were placed away from the home and provided with pumped water and a simple gravity-assisted waste disposal system. The Minoans were the first to have private bathrooms; these were followed by the early monasteries, which operated as protectors of culture and social values. Bathrooms for nomadic and rural cultures were greatly different from the Minoan bathrooms: They were open to the sky and relied on natural systems such as locally available water supplies for washing the body, the earth for disposing of waste, and the air for ventilation. The establishment of permanent communities and the growth of cities triggered the development of centralized water supplies and efficient methods of waste disposal. This led to the development of modern-day plumbing and sewage disposal technology, which allowed connecting the bathtub, sink, and toilet to the overall building system, so that water can be supplied uninterruptedly and waste disposed of in a sanitary fashion.

Bathrooms in middle-class American homes first appeared in the urban areas. Fixture placement in these bathrooms was driven by plumbing and drain technologies and cost conservation, not by usability or human considerations. The present 5- by 7-foot bathrooms first appeared in 1920s, when plumbing in private bathrooms became available in American homes. Even though a great number of technological and design improvements (including prefabricated designs and integrated systems) have been suggested during recent years, this bathroom design has remained basically unchanged up to this day. None of these new technologies has received acceptance within the plumbing industry. This is because the plumbing industry maintains a peculiar relationship with the home-building industry, and continues to support on-site bathroom construction, which requires assembling thousands of unrelated parts. Not only has the plumbing industry denounced new bathroom designs; it has employed very restricted codes to limit new bathroom and technological innovations. People's attitudes toward privacy, security, and independence in the bathroom, and their adaptability to deal with difficult environmental conditions, have allowed the plumbing industry to not address many pressing bathroom problems. Consequently, the modern bathroom has failed to serve as a place of comfort for most people, and users have compromised convenience for efficient water supply and waste disposal.

42.4 THE NEED FOR A UNIVERSAL BATHROOM

Life expectancy in 1900, when bathroom technology first found its way into the home, was 44 years (i.e., a shorter life expectancy meant that most bathroom users were young people and they did not experience physical limitations associated with old age; Dychtwald, 1990). Most people, then, made use of bathroom facilities independently, children being the only recipients of bathroom assistance. Assistance in the bathroom did not become a factor of the early design since most users did not live long enough to become old and acquire physical limitations. The concept of early bathroom design, therefore, centered on independent use and use in complete privacy, and those who were unable to use the current design under these conditions had to seek assistance, thus compromising privacy, dignity, and self-reliance.

Advances in medical technology and health care have prolonged life, and this has contributed to demographic changes in society. Within a longer lifespan, therefore, people are encountering a greater number of disabling conditions, for which they must develop new coping strategies. Furthermore, while survival rates for previously fatal injuries and diseases are now much higher, chronic disabling conditions often result. Bathroom dependence among older people and those with disabilities escalates with age and severity of physical limitations, and this often compels them to move from residential settings to institutions. However, social trends are also changing the demographics of people with disabilities. The Independent Living movement encourages all people, including people with disabilities, to live as independently as possible. Everyone wants to be independent in the bathroom. Deinstitutionalization is being encouraged by government agencies seeking to save money, as well as to improve the quality of life of former institutional residents. Older people repeatedly assert their preference to remain in their homes as they age (AARP, 2000). Demographic changes, together with the government's desire to keep people in their homes, means that the number of people with functional limitations is ever-increasing, and that there is more of a need for bathroom independence than ever before.

Bathroom users today are vastly different from the users when the bathroom first originated. They are a mixed population consisting of independent users, dependent users, and care providers of dependent individuals—they are tall and short people, young and old people, large and small people, able-bodied people and people with disabilities. People with a range of disabilities, then, are far from comprising the only demographic group that is hampered by the lack of a change in bathroom standards. For example, standard fixture design and location is outside the reach and strength of most children, and they need assistance. Cabinets are too removed for many individuals who are short in stature, and they often overextend to maintain bathroom independence. Tall people are greatly inconvenienced by fixture height and placement, and they have to stoop down to operate them. Wheelchair and walker users are restricted by the bathroom layout, and they have difficulty moving around and operating fixtures. People with arthritic hands are unable to grasp controls and operate faucets, and they are limited by design. Visually impaired people have difficulty comprehending fixture design and placement, and they often misjudge their location and bump into them.

Clearly, most bathrooms are not enabling environments, and they present serious imbalance between an individual's capabilities and the demands of the environment. Current bathrooms continue to primarily support independent users because they are based on the premise that all users will operate independently, and they do not support dependent users and their care providers, or those using assistive technology products. Even though providing care is a normal aspect of bathroom life, it is nearly impossible to care for children and dependent adults in the bathroom. Most care providers are seriously inconvenienced by bathroom design and they operate in very unsafe conditions. The average bathroom is an unsafe and inconvenient place for all users, and it is inconsistent with the needs and requirements of most users.

At no time in the history of the bathroom has the need for a better bathroom been more urgent than now. A universal bathroom, an equal opportunity environment, stands for a better bathroom for everyone, and it should benefit all users and meet their individual and collective needs. Universal bathrooms are not the same as accessible or "barrier-free" bathrooms, as these terms primarily refer to bathrooms with special design features that accommodate people who use wheelchairs, separately from those who do not use wheelchairs. Therefore, even as such bathrooms "solve" a problem, they create social stigma and personal alienation, which some see as an even greater problem. Universal bathrooms must be for all people, must achieve consumer acceptance by everyone, must consider the range of users, must provide appropriate choices for different needs, and must accommodate everyone at all times. They must embrace new technology to provide a dynamic bathroom environment that adapts to people's changing conditions; allow users to customize their environment; offer a high degree of safety, security, usability, and independence to all users; provide individual satisfaction; and support the offering and receiving of assistance. A universal bathroom is not a finite product. Like universal design, it is a concept that symbolizes a movement for a better bathroom, which requires the support of users, manufacturers, design professionals, and government officials to foster a climate of innovative bathroom development.

The universal bathroom is not one design for all people. A single universal bathroom would be the same as "one size fits all," and this approach contradicts the pluralistic aspirations of universal design. Supporters of universal design argue that the one-bathroom-design for all is exactly the problem with bathroom designs in existence today. Due to their limited flexibility in fixture design and placement layout, these bathrooms neither provide choices nor do they allow individuals to adapt the bathroom environment so they can participate on their own or with their care providers. The universal bathroom supports the idea of individualization and personalization through design flexibility and diversity; that is, different designs for different users within the same system, or adaptability and adjustability that can accommodate all users. A universal bathroom is a place for all members of the family, and it will offer many different designs.

Some people believe that due to a tremendous variation in user populations and their needs, a true universal bathroom may be an unattainable goal. This is because the universal design does not prescribe a final state, as nothing is truly universal, nor is the process of designing a bathroom ever finished. Many advocates of universal design realize that the lofty social and inclusive goals of universal design are almost unattainable, as it is nearly impossible to design bathrooms for "all" people. The goal, therefore, is to approach the objectives of social inclusion through an ongoing process, best termed as *universal designing* (Mullick and

Steinfeld, 1997). The process of designing a universal bathroom must be viewed as incremental and ongoing, the primary goal being individual and social inclusion through design.

42.5   BATHROOM RESEARCH AND DESIGN GUIDELINES

The Center for Inclusive Design and Environmental Access (IDEA) at Buffalo, New York, with the support of the National Institute on Disability and Rehabilitation Research, U.S. Department of Education, conducted two bathroom studies, "Listening to People" and "Measuring Universal Design." These studies produced important research findings and design recommendations. The National Kitchen and Bath Association (NKBA), a trade and industry organization consisting of bathroom designers, manufacturers, and trade professionals, developed 41 bathroom design guidelines based on the Uniform Federal Accessibility Standards (UFAS) and the American National Standard Institute's ANSI A117.1. These guidelines are widely circulated by the NKBA for use by practicing kitchen and bath designers.

Listening to People

"Listening to People" interviewed 30 consumers about bathroom use and bathroom products; they were independent users, dependent users, and care providers of different ages, sexes, races, and physical conditions—a wide variety of users (Mullick, 1997). The study examined the need, importance, and benefits of universal design in the bathroom, generated qualitative information about the bathroom, and evolved common design features to benefit all users. Selected findings include:

Doors
- Doors must be wide enough for easy passage; narrow doors often force wheelchair users to leave the chair outside the bathroom, compromising safety or making assistance necessary.
- Door weight is a safety issue; heavy doors require more force to open/close and can be difficult to use without assistance.
- A wide door that opens outward increases its appeal, offers space benefits, and is easier to operate than an inward-opening door.
- Pocket doors are a design solution to problems associated with door swing and aesthetic appeal.
- Lever-handled door latches make opening the latch easier to operate.

Grooming Area
- Flexibility in the layout of the bathroom fixtures supports customization.
- "Flexible" or "adjustable" fixtures, such as movable cabinets/mirrors and adjustable-height sinks, alter the space, allow for the presence of an attendant, promote safety, and enable easy access and use.
- Adjustability in bathroom fixture mount height and location helps everyone.
- Storage of toiletries adjacent to the sink, in a wall-cabinet or vanity near the sink, or on shelves offers convenience.
- Wheelchair users object to a vanity under the sink, stating that this eliminates legroom, thereby causing extended reach.
- A cabinet mounted on a sidewall is easier to access because it does not require a forward reach.

Toilet Area

- A wall-mounted toilet frees floor space, allows easier access, offers ease of transfer, and makes the bathroom floor easier to clean.
- No consensus could be reached concerning the most desireable location and height of a toilet paper holder or the number and location of grab bars.
- Towel holders near the lavatory and bathtub should also function as grab bars.
- A fold-down grab bar on the sides of a toilet or near the bathtub offers choice of use.
- There is a stigma associated with fixed grab bars due to their "institutional" aesthetic.

Bath and Shower Area

- There is a strong personal preference regarding the use of a bathtub or shower.
- The use of a shower bench eases transfer into and out of bathtubs or showers.
- Grab bars mounted on each wall of the tub surround increase safety; recessed grab bars are aesthetically pleasing.
- Showers are easier for people to adapt to their changing needs.
- An adjustable and removable shower spray with flexible hose is easy to operate.
- Shower and bath controls closer to the outside edge of the tub or shower reduce over-extension.
- A nonslip bathroom floor surface eliminates fear of falling.
- There is a need for storage near the bath/shower or a desire for a better system for dispensing soap, hair care products, or other accessories.

Selected design recommendations include:

Safety

- Provide supporting devices such as wall and floor supports throughout the bathroom. Include unobtrusive devices such as wall systems and fixtures that people can hold on to when they need to use them.
- Provide seating possibilities throughout the bathroom; incorporate foldable seats, pull-out seats, or fixtures that allow seating possibilities.
- Eliminate protruding objects such as faucets, soap dishes, or towel bars; recess these fixtures or build them into the wall.
- Round all edges and soften all corners.
- Provide a nonslip floor surface such as textured floor tiles to prevent chances of fall.
- Furnish the bathroom with soft surfaces such as soft floor tiles or a soft tub.
- Provide more storage space; incorporate shelves into the wall to eliminate projecting surfaces.
- Include antiscalding devices in the faucet controls.
- Provide better illumination in the bathroom; build task lighting into the fixtures.

Usability

- Incorporate fixtures that are usable by all users; this creates an inclusive bathroom.
- Install fixtures that respond to the expectations of users; such fixtures provide a good match between user capabilities and usability requirements.
- Incorporate simple-to-operate fixtures; eliminate the need to learn new technology.
- Include easy-to-understand designs; fixture designs should have built-in cues for use.

- Install fixtures that have multiple uses, such as a tub wall that allows seating, or a nonslip floor that drains water.
- Locate storage within convenient reach.
- Include fixtures that are easy to maintain and service.

Appearance

- Install fixtures that have a safe appearance; recessed or rounded fixtures appear safe.
- Include fixtures that have general appeal; neutral color and nonspecific designs have a mass appeal.
- Provide stigma-free fixtures; user-specific fixtures have stigma attached to them.

Personalization

- Provide a flexible bathroom layout for performing multiple activities and use throughout a person's lifetime.
- Enable making easy adjustments of fixture location for independent use, dependent use, and care-providing needs.
- Include adjustable fixtures such as adjustable-height sinks, toilets, grab bars, cabinets, mirrors.
- Adopt a systems approach to design; include systems such as modular bathtubs, showers, or shelves so that their designs can be altered as people's needs change.
- Provide adaptable supporting devices that allow adjusting size and location.
- Providing adjusting storage; make it possible for people to add, reduce, or alter storage location.
- Provide adjustable control for illumination level.
- Make it possible to easily alter the appearance of the bathroom, such as by changing color or adding decorations.

Measuring Universal Design

"Measuring Universal Design" utilized a full-scale modeling system to study how accessible and conventional bathrooms, both code-complying and not, benefit many groups of users (Mullick, 1997). It utilized two full-scale simulated bathrooms, as shown in Figs. 42.1 and 42.2, to identify differences in functional ability that are related to design differences and benefits of accessible bathrooms for people with and without disabilities. The study also evaluated consumer perceptions of bathroom features and determined features for a universal bathroom. "Measuring Universal Design" asked participants to simulate the performance of several activities of daily living (ADL) like grooming, toileting, and bathing in two full-scale bathroom designs, designated "challenging" and "supportive."

The challenging bathroom, which resembled a small apartment bathroom, was not in compliance with any current accessibility standards. Its key characteristics were as follows:

- 15 ft^2 of clear floor space.
- 32-in-wide (30-in clear opening) entry door equipped with conventional doorknob handles, with a swing that swept across that open floor area inside the room.
- Enclosed vanity with a small (20- by 20-in) countertop and traditional dual-knob faucets and high-mounted (47-in) mirror cabinet.
- 16-in-high toilet with no attached grab bars or wall-mounted grab bars.
- 15-in-high bathtub with a fixed showerhead, no wall-mounted grab bars, and a single-knob-handle faucet mounted on-center.

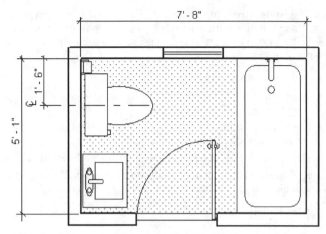

FIGURE 42.1   Challenging bathroom.

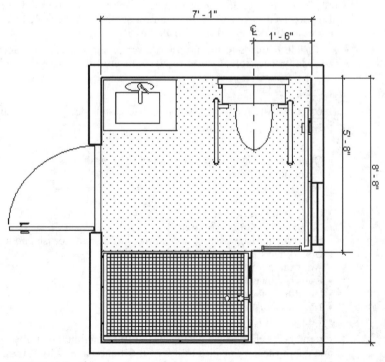

FIGURE 42.2   Supportive bathroom.

The supportive bathroom had many accessible features, was in compliance with the current accessibility standard (ANSI A117.1), and included a few universal fixtures such as a foldable grab bar near the toilet and a large-size toggle light switch. The characteristics of this bathroom were as follows:

- 25-ft^2 open floor area.
- 34-in-wide (32-in clear opening) entry door equipped with lever handles and a swing away from the inside of the room.
- Open vanity counter with a larger (20- by 30-in) countertop, single-lever handle faucet, and a low-mounted (40-in) mirror cabinet.
- Higher (18-in) toilet equipped with both a wall-mounted grab bar and fold-down grab bars.
- Roll-in shower with a height-adjustable/hand-held shower head, wraparound grab bars, and a single-lever faucet mounted on-center.

Key findings include:

- An accessible bathroom based on current codes greatly improved functional independence for people with disabilities; marginally improved the functional independence of all users; and did not improve the functional independence of able-bodied users.
- Accessible fixtures, primarily designed to provide access for people with disabilities, result in a higher level of performance or no difference. They did not diminish the performance of any user group.
- Functional performance is not always directly related to perception of difficulty/ease. For the most part, accessible features that improved performance were perceived as "easy" to use. On many occasions, however, accessible features that did not improve function were also perceived as "easy" to use. On rare occasions, accessible features that improved function were still perceived as "difficult" to use.
- Accessible features result in either lower or similar level of effort required to perform a task. Many accessible features reduced the effort level to near minimum.

Selected design recommendations include:

1. Incorporate the functional benefits of accessible design as a baseline.
2. Provide at least the following features:
   - 25-ft^2 open floor area
   - Generous amount of wall supports such as grab bars
   - Lever-handle faucets
   - Wide toggle switch
   - 34-in entry door with outside swing
   - Lever-handle door latch
   - Large (20- by 30-in) countertop
   - Convenient place to store accessories such as toothbrush and hair dryer
   - Clear knee space under lavatory
   - Adjustable mirror
   - Roll-in shower stall
   - Hand-held shower
   - 18-in-high or height-adjustable toilet
   - Convenient location for toilet paper
3. Extend accessible design principles to offer functional benefits to people with disabilities and adapt these principles to offer functional benefits to other people. The Principles of Universal Design (Center for Universal Design, 1997) could accomplish this goal.

4. Include features that create a high level of convenience for everyone.

5. Develop a modular approach to design so that each individual can purchase only the components they need to maximize functional performance and minimize expenditure.

National Kitchen and Bath Association

The National Kitchen and Bath Association (NKBA) certifies kitchen and bathroom designers, publishes educational materials, and promotes design standards. It is committed to universal design, and uses bathroom design guidelines that are based on the Uniform Federal Accessibility Standards (UFAS) and the American National Standard Institute's ANSI A117.1. Selected design guidelines (Peterson, 1998) emphasize:

- The clear space at doorways must be at least 32 in wide and not more than 24 in deep in the direction of travel.
- Provide a minimum clear floor space of 48 by 48 in in front of the toilet.
- Provide a minimum clear floor space of 60 in wide by 30 in deep in front of the bathtub for parallel approach and 48 in deep for perpendicular approach.
- Plan clear floor space of 60-in diameter or 36 by 36 by 60 in for turning mobility aids.
- Incorporate a kneespace at the lavatory, minimum of 27 in above floor.
- Install lavatory mirrors, flat mirrors 40 in above ground and tilted mirrors 48 in above ground.
- Showers should include a bench or seat, 17 to 19 in above ground, and a minimum of 15 in deep.
- The width of shower door openings must consider the interior space and maneuvering. The entry needs to be 32 in wide for a 60-in-deep shower and 36 in wide for a 42-in-deep shower.
- The shower door must open into the bathroom, and not into the shower.
- Install safety rails near the bathtub and shower area to facilitate transfer.
- Equip all showerheads with a temperature-limiting device.
- Install shower and tub controls so that they are accessible from inside and outside; locate tub controls 33 in and shower controls between 38 and 48 in above the floor.

42.6 UNIQUE BATHROOM DESIGNS

Most bathrooms are constructed environments, which require an integrated effort by several trades such as plumbers, electricians, general contractors, and cabinetmakers to offer safety, privacy, and independence for the users. Furthermore, bathroom construction involves selection and installation of bathroom products, so work efficiency can be maintained. While bathroom designers supervise construction of most "designed" bathrooms, the fixture design and installation details are designed by architects and industrial designers, who have expertise in human factors, work efficiency, product design, and production technology. Presented in chronological order, the following bathrooms by architects and industrial designers have been designed to provide safety, independence, and work efficiency. They suggest a unique design approach based on modular parts and mass production, which shifts the idea of the bathroom from a constructed environment to a manufactured environment. This allows for the introduction of new plumbing and drainage technology, necessary to incorporate important universal design features related to adaptability, adjustability, and personalization.

Bathroom for Elderly People

Robert Graeff, an architecture professor at Virginia Polytechnic Institute and State University, developed a bathroom that allows elderly people to maintain personal independence and prolongs their ability to stay at home (Singer, 1988). This bathroom, which can be located adjacent to (or as part of) the bedroom to provide easy access, safety, and privacy, reduces the walking distance, especially at night, to the customary bathroom. It incorporates many functional features such as wraparound hand-rails, skid-proof flooring, enclosed storage and open counter space, shallow washbasin, spacious shower booth, equipment for perineal cleaning, and effective lighting. There are unobtrusive supports that guide users through the enclosure. Being next to the bedroom, removed clothing can be stored dry in the bedroom. The storage is designed to keep medications and related supplies organized in the cabinets, and the shelves have magnifying shields to make fine print on drug labels readable.

Two standardized volumetric modules, a 78- by 36-in sink/toilet module and 52- by 36-inch shower module, separated by a 7-in-thick integral plumbing wall, make up the bathroom. These modules, shown in Fig. 42.3, fit within most existing closet spaces and can be assembled on site, offering flexibility in design, layout, and placement. Graeff acknowledges that some users may feel uncomfortable about sleeping in the same room with the toilet and drains. A ventilation system has been planned to supply continuous negative airflow to keep the bedroom freshened. Care has been taken to keep the toilet out of sight of viewers, even with the entry slide door opened. To avoid costly plumbing installations, Graeff suggests incorporating a grinder pump to carry wastewater through 1-in pipework to the centralized plumbing core. The design facilitates self-help and potentially will improve the quality of life and psychological well-being of users, especially those who would otherwise have to be institutionalized. According to Pirkl, this bathroom, which has the potential to preserve privacy, independence,

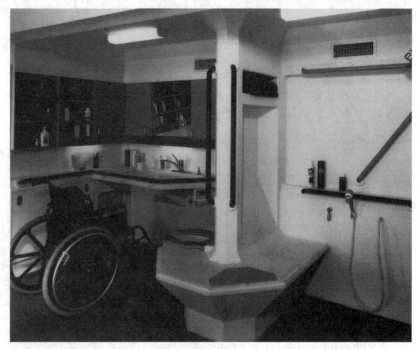

FIGURE 42.3   The Graeff toilet and shower modules.

and dignity, and offer safety and comfort to many elderly people, could well postpone institutionalization on an average of 10 to 15 years" (Pirkl, 1994). The Department of Veterans Affairs is currently supporting this bathroom project, and Aquarius of Praxis International, Inc., in Tennessee, is considering it for production.

Metaform Bathroom

The Metaform bathroom system was designed to provide greater flexibility, safety, convenience, and independence for people of all ages and abilities, including children, able-bodied adults, older people, and people with disabilities (The Metaform Personal Hygiene System, 2000; Zacci, 1993; Raver, 1998). Gianfranco Zaccai of Design Continuum, Inc., designed it for the Herman Miller Research Corporation (HMRC) and the Herman Miller Corporation (HMI). The parent company of HMRC was expected to commercialize it. Universal aesthetics are used in this bathroom to eliminate social stigma, appeal to a larger population, integrate user groups, expand the potential market, and offer economic feasibility associated with volume production. Made up of modular components and moving parts, the bathroom system can be easily transformed as people's needs change over time. It has been designed to blend with existing architecture and details, and to require little labor and cost to install, maintain, and repair.

The Metaform system consists of associated yet independent components that form three bathroom activity nodes—the lavatory, toilet, and bathing nodes. These components can be assembled to form activity centers and construct variable-size accessible bathrooms with the minimum size being 5 by 5 ft. The lavatory node, shown in Fig. 42.4, consists of a self-contained, height-adjustable assembly, and it has a sink, work surface, storage, lighting, and mirrors. This node can be installed within the thickness of a standard 4-in studded wall, or retrofitted to the outside of any existing wall that provides for hot and cold water and waste lines and electricity. The entire assembly is driven by an electrical motor actuated by an "up-down" button switch located at the front edge of the sink. At the touch of a button, the entire assembly can be repositioned to accommodate the needs of standing or seated users of all sizes and ages within the same household. Wheelchair riders have unimpeded knee access, and can use the clear space under the sink for the necessary turn radius.

The toilet node, as shown in Fig. 42.5, consists of a toilet that adjusts automatically in height to facilitate transfer and use by small children, tall adults, and people with disabilities. It incorporates optional features such as foldaway arms to facilitate transfer, bidet wand with dryer, and automatic self-cleaning/sanitizing. When not in use the toilet bowl can be rotated into a cavity and out of the way. The toilet unit can be installed into a 9-in-deep wall cavity, and its height can be motor adjustable. The toilet bowl rotates into the wall cavity for storage, opening up floor space in the bathroom. In this position it can be automatically washed and sanitized.

The bathing node includes four basic components: (1) a "water column" assembly, (2) the shower floor pan/drain system, (3) resilient bathtub with optional hydraulically powered transfer chair, and (4) a support bar/accessory rail system. The "water column" is a self-contained, preassembled unit, which contains the shower/hand wand controls, a range of showerheads, ambient and foot lighting, an integral support bar to facilitate safe transfer, and integral forced ventilation for the shower or shower/tub. The pan/grill system channels shower water into a trough underneath and drains into the bathroom drain. Because it has no threshold, it eliminates all obstacles that pose mobility barriers to wheelchair or walker users. The system can be installed or retrofitted over any existing floor joist pattern, as it occupies only the depth required for a subfloor and finished floor. It can be integrated into one-piece shower pans of modular sizes, or be installed in conjunction with custom floors such as tile or stone. The tub, which is made up of a resilient outer surface, is designed to facilitate transfer, while providing safety and comfort. The rim of the tub is bowed outward to facilitate seated transfer from a wheelchair. It incorporates optional whirlpool jets for massage and hydrotherapy and a

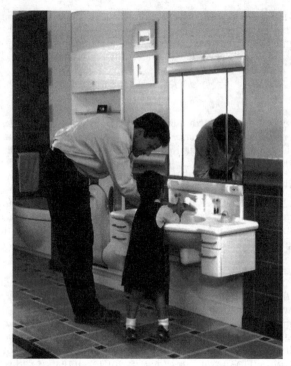

FIGURE 42.4   The Metaform sink node.

hydraulically operated *transfer chair* that hangs on the handrail. The support bar/accessory rail, as shown in Fig. 42.6, is a modular component. It can be installed over any existing stud wall and can support 1000 pounds of load at any point. Made of a tubular steel core with a resilient urethane skin, it offers a secure and natural gripping surface for the hand, and allows attaching and repositioning accessories, such as a folding shower chair, soap and shampoo dispenser, soap dish, shaving caddy, baby's bath bassinet, and such, thus enabling users to customize the bathroom environment. Optional features include an electronic temperature control valve with large, easy to see and operate controls and a radio/hands-free telephone unit that can serve as an automatic call-for-help system if needed.

The Metaform bathroom system was exhibited at the Unlimited by Design show in 1997, sponsored by Cooper-Hewitt Museum of Design in New York. It was well received by all viewers—able-bodied people, people with disabilities, and care providers. The national and international press took note of this bathroom, and it appeared in many prestigious publications. Since its design development, the Herman Miller Corporation, which was expected to commercialize it, has had a change of mind and has decided not to enter into the bathroom fixture business. This bathroom, which has several innovative technologies, has received many patents. The Industrial Designers Society of America recognized it with a gold award and published it as a good example of universal design (Zacci, 1993).

IDEA Center Bathrooms

The Center for Inclusive Design and Environmental Access, with the support of the National Institute on Disability and Rehabilitation Research, U.S. Department of Education, has

FIGURE 42.5   The Metaform toilet node.

developed two adjustable bathrooms, called the "Movable Fixtures" bathroom and "Movable Panels" bathroom (Mullick, 2000). These designs, which are the outcome of two research studies mentioned earlier, are based on the premise that bathroom use must prolong independence, allow the offering of care, and assist care providers. Unlike most current bathrooms, which are designed primarily for independent users, these bathrooms consider the needs of the human life cycle, and address dependent use and care providing alongside independent use in the bathroom. In these bathrooms (as seen in Figs. 42.7 and 42.8) two fixtures, the sink and shower, move around the bathroom wall to adjust for use conditions. Through fixture movement, the bathrooms reorganize and open up spaces for both independent and dependent use, to suit various body sizes and preferences, and for specific care-providing situations. The fixtures also adjust in height to accommodate variations in stature and for standing users, sitting users, and children, in relation to user needs and capabilities, and the demands of the environment. Fixture movement is a result of technological innovation in existing plumbing and drain technology used unconventionally. Both the Movable Fixtures and Movable Panels bathrooms let bathrooms adapt to people, and not the other way around, so there is a best "fit" between users and their environment. They are excellent examples of how the universal design philosophy can create a flexible environment and produce innovative designs that are high in usability, convenience, aesthetics, surprise, and fun. They allow fixtures and spaces to adapt, provide the highest degree of design flexibility, allow user personalization, and offer multipurpose use of features to minimize stigma associated with user-specific products. They are also designed to be attractive.

The Movable Fixtures and Movable Panels bathrooms are each made of four units: the lavatory unit, toilet unit, and shower and support unit. These units can be used independently or in combination with each other, and they can be used in place of existing fixtures or in new

FIGURE 42.6 The Metaform shower node.

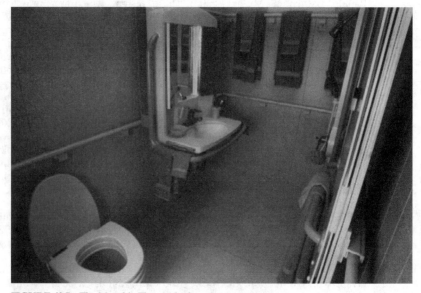

FIGURE 42.7 The Movable Fixtures bathroom.

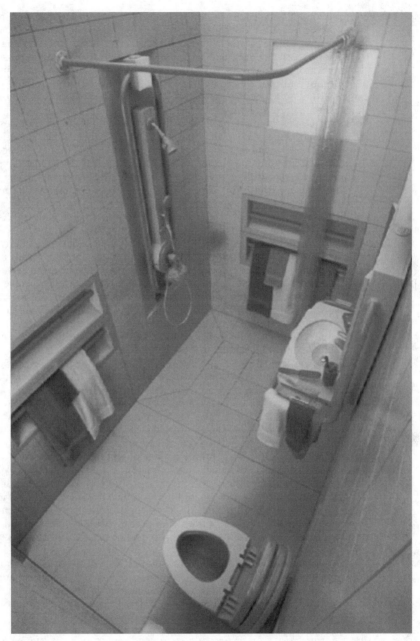

FIGURE 42.8   The Movable Panels bathroom.

construction. In both cases, they offer superior consumer benefits while reducing total installation costs and space requirements. Two working bathroom prototypes have been fabricated, and they have running water, a drain, and electricity. These prototypes have been constructed to fit the smallest-size everyday bathroom of 5 by 8 ft to assess their benefits.

The lavatory unit has the sink, work surface, storage, lighting, and mirrors. The sink is fitted with hot and cold water, drain, and electricity. The countertop has pull-out shelves and can be extended to increase the work surface when needed. It has built-in lights for additional illumination and wraparound grab bars for support and storage. Specially designed hooks can be mounted on the support bars to store daily-use accessories. There are two tilting mirrors for improved viewing conditions. The lavatory unit is height-adjustable; its height adjustment is achieved manually or by using a motorized unit. This accommodates a range of users including standing and seated users as well as children and tall persons. During use it can be locked securely in position to limit unwanted movement. The sink, as seen in Fig. 42.9, has knee clearance for wheelchair users and a clear space underneath for wheelchair turning.

The toilet unit, which uses conventional water and drain technology, is wall-mounted, and consists of a multilayer seat and a flusher. The toilet height adjusts to facilitate use by a range of users, including small children, able-bodied adults, tall persons, elderly people, and people with disabilities. The multilayer seats can be stacked to achieve a variety of seat heights. In the raised position, the toilet offers easy transfer by wheelchair users and older people, and in lower position it is convenient for children and short-statured people. The wall-mounted installation allows easy floor cleaning and maintenance, and the clear space underneath helps to make wheelchair turns. The flusher, which is large-size, wall-mounted, and located close to the user, is designed for easy detection, access, and operation. The toilet can be fitted for low water use technology and includes optional features such as bidet wand with dryer, and automatic self-cleaning and sanitizing.

The shower unit shown consists of three basic components: the shower column, drainage floor, and the shower door—together they provide an optimal bathing environment. It has optional features such as resilient floor tiles, transfer chair, shower seat, and emergency communication system. The shower column, which serves as the water source, is fitted with hot and cold water and water control. It has built-in lights for additional lighting, and wraparound grab bars for support and storage. Specially designed hooks can be mounted on these bars to hold daily-use accessories. The shower is height-adjustable, achieved manually or by using a motorized unit. The shower door is a lightweight-telescoping screen that is stored inside, and that can be extended out to make the shower enclosure. When not in use, the screen stores flat against the wall in a locked position. The drainage floor consists of specially designed tiles that are gently tapered to drain water along their edges into a trough underneath, and they drain water quickly and keep the floor dry to prevent the slips and falls that happen on wet, slippery floors. The entire bathroom floor could be made up of the drainage floor. This will allow the bathroom to be hosed down for easy cleaning and maintenance.

The support unit is an architectural solution, which consists of a multipurpose grab bar system that can be used as structure, support, and storage. The grab bar system supports the lavatory and shower units, and it has been designed to be attractive to eliminate the stigma associated with grab bars. It allows the storage of everyday accessories and the mounting of specially designed accessories such as a fold-down grab bar.

The concept for the Movable Fixtures bathroom is based on the premise that the bathroom needs to be regularly adjusted to meet the changing demands of a variety of users living in the home. Furthermore, the technology must allow instantaneous and easy movement of fixtures to create larger spaces for bathroom activities based on user preference, as well as independent and dependent use. In this bathroom, the sink and shower roll along the bathroom wall to create larger toilet, grooming, and showering areas. Fixture movement, in principle, instantly creates three "large" bathrooms in one small space and opens up the bathroom for easy operation during dependent and independent use. As shown in Fig. 42.10,

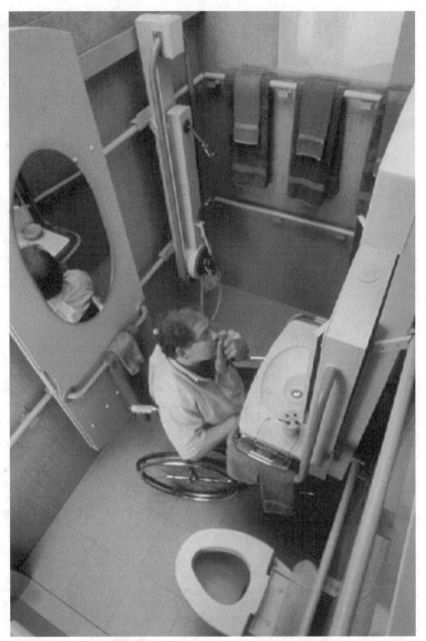

FIGURE 42.9   Sink in Movable Fixtures bathroom has knee clearance and moves around the bathroom to open up floor space for wheelchair movement.

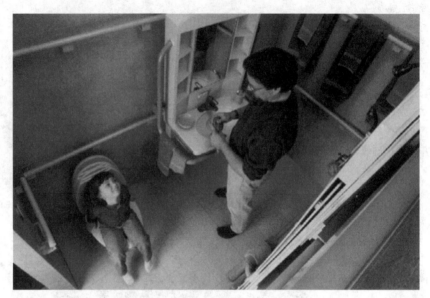

FIGURE 42.10   Fixtures in Movable Fixtures bathroom reconfigure for independent use, dependent use, and for providing care.

fixture movement instantly reconfigures the bathroom for simultaneous use by couples, for easy use by large persons, or to provide space for caregivers to assist children and dependent adults. The sink and shower units, which rest and roll on the wall-mounted bars, also adjust in height for stature variations and usage. When not in use, the fixtures lock securely in place to prevent unplanned movement. Plumbing, drainage, and electricity are supplied through a flexible assembly, and these services are designed to support fixture movement. The shower area consists of a fast-draining shower floor and a telescoping shower screen that moves with the shower to provide various-size enclosures. The screen stores flat against the bathroom wall when not in use.

The concept for the Movable Panels bathroom is based on the premise that users will only make major bathroom changes periodically, not every day. They will make environmental changes when there is a major change in the family composition, such as the birth of a child, temporary disability, or when an elderly parent moves in to live with his or her children. In this bathroom, shown in Fig. 42.11, the sink, shower, and storage units are on panels, and they can be moved anywhere in the bathroom. Panel movement will require assistance of trained professionals or experienced do-it-yourselfers, just as when installing an air-conditioning unit, or a radio in an automobile. The fixture designs are the same as those in the Movable Fixtures bathroom and allow the same height variation and movement. The shower area consists of fast-draining shower floor and a shower curtain. There are other supportive and multipurpose features in this bathroom, such as the panels for accessory storage. These panels also serve as recessed grab bars for support and assistance. They are available in different lengths and can be installed in horizontal or vertical position. When in production, panels will be available in a range of colors so they can be mixed and matched to produce a variety of attractive bathrooms, thereby making it easy and less expensive to redecorate over time.

The bathroom designs provide an opportunity for manufacturers to address the needs of a wide variety of users, including children, adults, older people, people with disabilities, and even the baby boomer generation. They will provide a competitive advantage and an increase in market share. The modular nature of the designs, which allows continued service and production of replacement parts, will eliminate the need to manufacture many types of bathroom

FIGURE 42.11 Fixtures and storage units in the Movable Panels bathroom are on panels that can be moved anywhere in the bathroom.

technologies, preventing overproduction and reducing waste associated with manufacture, use, and disposal. The bathrooms have been designed for ease of operation, maintenance, repair, and recycling. In case of fixture breakdown, they can be easily removed and replaced, while the old unit is sent in for repair. This will prolong their use cycle and reduce waste. The Movable Panels design is amenable to an innovative marketing approach such as leasing panels. This will greatly increase use cycle of fixture panels and fascia wall, and minimize disposal associated with planned obsolescence.

These bathrooms reflect the social and inclusive philosophy of universal design. They have the potential to unify diverse population groups, so no one user group is excluded by their design. Existing bathrooms, with permanently installed fixtures, symbolize one bathroom-design-for-all users. They tend to homogenize through uniformity. These bathrooms, on the other hand, encourage individualization through flexibility and choice, so users can adapt their bathrooms to suite their personal taste and preferences.

42.7   THE FUTURE OF THE UNIVERSAL BATHROOM

The current approach to bathroom development has focused on designing task-oriented fixtures—sink, toilet, and tub-shower—with little consideration for their use, placement, and how they make up the bathroom environment. These fixtures require people to use them individually and move around to perform bathroom activities. Emphasis on task-oriented fixture design, rather than on the bathroom as a place to carry out daily hygiene, undermines the relationships between fixtures, and forces moving around to perform bathroom activities. For example, shower designs do not support shaving and brushing when showering, even though it is quite common among bathroom users to shave and shower at the same time. Users are forced to groom at the sink and travel to the shower for showering. Similarly, many people, including those who catheterize themselves, seek access to the sink when using the toilet. This access is denied because of the fixtures' distant placement. The task-oriented fixture design approach has produced distinctly different products that do not relate well to each other. Unlike the machines for contemporary industrial production, these fixtures do not interface well with each other, and bathroom activity requires time, movement, and effort. This may be fine for most people, but it can be very demanding for many others. Fixtures that allow performing multiple and related tasks can produce improved bathroom environments and provide significant benefits for everyone. Better fixture design will diminish time in the bathroom, reduce unnecessary movement, and benefit not only people with disabilities, but also many other user groups. Everybody will benefit from well-designed fixtures.

Design and placement of early bathroom fixtures was driven by plumbing and sewer technologies and cost conservation, not by usability and human considerations. This is true of today's bathrooms as well. As a result, the bathroom is unsafe, inconvenient, and not user-friendly for most users. Fixture design and installations are regulated by national and local codes, which seem to favor plumbers rather than designers, manufacturers, and users. Consequently, many innovative bathroom designs and improved bathroom technologies have not met with much success of being accepted. The future of universal bathrooms, as an environment for a diverse population, is dependent on the cooperation and support of code-making and enforcement agencies. These agencies need to understand the pressing needs of users and how current bathrooms are crippling them. The future of the universal bathroom is dependent on revising age-old plumbing codes, without which it is impossible to foster design innovation and market new technologies.

The future of the universal bathroom lies in recognizing that smart technology can ease bathroom use and add convenience to daily living for everyone. The successful universal bathroom will employ smart digital technology to establish communication within and between fixtures, so the bathroom environment would adapt to users and not the other way around. For example, smart technology in the IDEA Center's Movable Fixtures bathroom has the potential to make instantaneous fixture adjustment for stature variations and move them for better bathroom layout and fixture relationships. By activating coded information, it is possible for digital technology to customize the Movable Fixtures bathroom to suit people's individual and collective needs. Better integration of smart technology in the bathroom will increase convenience, eliminate unnecessary human intervention, facilitate hygiene maintenance, produce improved bathroom environments, and offer significant benefits for everyone.

42.8 CONCLUSIONS

The bathroom is one of the most important places in the home and all members of the family need to be able to use it. Modern society has linked independent use of the bathroom to independent living, and those unable to use the bathroom on their own are dependent on others for assistance, thus compromising their privacy and self-reliance. All people, regardless of their physical condition, are limited by the design of the bathroom, some more than others. While able-bodied individuals experience minimal difficulties, children, older people, and those with disabilities are greatly challenged by bathroom designs.

Contemporary residential bathroom design generally follows a two-stream approach—bathrooms designed for people with disabilities, and bathrooms designed for able-bodied people. In fact, even accessible bathrooms are based primarily on the needs of wheelchair users and do not address the needs of people with other disabilities. This approach assumes that accessible bathrooms improve functional independence for people with disabilities and that able-bodied people do not have the need for improved functional performance.

A universal bathroom is an equal opportunity environment. It should benefit all users and meet their individual and collective needs. Universal bathrooms must be for all people, offer consumer acceptance by everyone, consider the range of users, provide choices for different needs, and accommodate everyone at all times. They must embrace new technology to provide a dynamic bathroom environment that adapts to people's changing condition, allows users to customize their environment, offers a high degree of safety, security, usability, and independence to all users, provides individual satisfaction, and facilitates offering and receiving assistance.

Three universal bathroom prototypes—Graeff's bathroom for the elderly, the Metaform bathroom system, and IDEA Center's bathrooms—demonstrate that bathrooms can be safe and secure environments that all members of the family can use. They suggest a unique design approach that allows introducing new plumbing and drainage technology necessary to incorporate important universal design features related to adaptability, adjustability, and personalization.

The future of universal bathrooms lies in recognizing that fixture design must address placement and use; bathroom codes must foster design innovation and new technology; smart technology needs to be incorporated to facilitate bathroom use; and users must be able to customize their environment. Improved bathroom design will increase everyday convenience, eliminate unnecessary human intervention, facilitate hygiene maintenance, and offer significant benefits for everyone.

Finally, universal bathrooms have the potential to open up new marketing opportunities, and generate increased market share. For universal bathrooms to be effective in the market, they must have a broad appeal and not focus only on accessibility. They should highlight use by a diverse population and the value of increased safety, convenience and, usability.

42.9 BIBLIOGRAPHY

Altman, I., and B. Rogoff, "World Views in Psychology: Trait, Interactional, Organismic and Transactional Perspectives," in D. Stokols and I. Altman (eds.), *Handbook of Environmental Psychology,* vol. 1. John Wiley & Sons, New York, 1987.

Center for Universal Design, *Principles of Universal Design,* Raleigh, NC, 1997.

Design Continuum, Inc., "The Metaform Personal Hygiene System: Universal Accessibility," Information supplied by Design Continuum, Inc., 2000.

Evans, Arthur, *The Palace of the Minos at Knossos,* London, UK, 1921.

Mead, G., *Mind, Self and Society,* University of Chicago Press, Chicago, 1934.

Moore, G., "Theory and Research on the Development of Environmental Knowing," in G. Moore and R. Golledge (eds.), *Environmental Knowing: Theories, Research, and Methods.* Dowden, Hutchinson and Ross, Inc., Stroudsburg, PA, 1976.

Mullick, Abir, "Listening to People," unpublished report to the National Institute on Disability and Rehabilitation Research, 1997.

———, "Measuring Universal Design," unpublished report to the National Institute on Disability and Rehabilitation Research, 1997.

———, "Universal Bathroom Design," unpublished report to the National Institute on Disability and Rehabilitation Research, 2000.

National Institute on Disability and Rehabilitation Research; Long-Range Plan, 1999.

Peterson, Mary Jo, *Universal Kitchens and Bathroom Planning,* McGraw-Hill, New York, 1998.

Pirkl, James, *Transgenerational Design,* Van Nostrand Reinhold, New York, 1994.

Raver, Anne, "A Bathroom That Makes You Want to Stay Home for the Day," *New York Times,* 1998.

Singer, Len, (ed.), *A Bathroom for the Elderly,* Virginia Polytechnic Institute, Blacksburg, VA, 1988.

Steinfeld, E., and G. S. Danford, "Environment as a Mediating Factor in Functional Assessment," in S. Dittmar and G. Gresham (eds.), *Functional Assessment and Outcome Measures for the Rehabilitation Health Professional.* Aspen Publishers, Inc., Gaithersburg, MD, 1997.

Stokols, D., "Group x Place Transactions: Some Neglected Issues in Psychological Research on Setting," in D. Magnusson (ed.), *Toward a Psychology of Situations: An Interactional Perspective.* Erlbaum, Hillsdale, NJ, 1981.

Wandersman, A., D. Murday, and J. Wadsworth, "The Environment-Behavior-Person Relationship: Implications for Research," in A. Seidel and S. Danford (eds.), *Environmental Design: Research, Theory and Application.* Environmental Design Research Association, Inc., Washington, DC, 1979.

42.10 RESOURCES

Accessible Plumbing, a technical report by Edward Steinfeld and Scott Shae (available from the Center for Inclusive Design and Environmental Design, University at Buffalo, Buffalo, NY).

Accessible Appliances, a technical report by Abir Mullick and Danise Levine (published by the Center for Inclusive Design and Environmental Design, University at Buffalo, Buffalo, NY).

Bathroom Lifts by Abir Mullick (available from the Center for Inclusive Design and Environmental Design, University at Buffalo, Buffalo, NY).

Bathroom Seats and Benches by Abir Mullick (available from the Center for Inclusive Design and Environmental Design, University at Buffalo, Buffalo, NY).

Elder Design: A Home for Later Years by Rosemary Bakker (published by Penguin Books, New York).

Fair Housing Means Universal Design (1), a video presentation to create awareness of the Fair Housing Guidelines and how they make housing usable by everyone (available the Center for Inclusive Design and Environmental Design, University at Buffalo, Buffalo, NY).

Lighting Kitchens and Baths by Jane Grosslight (available from Durwood Publishers, Tallahassee, FL).

Strategies for Bathroom Design (posted on the Center for Inclusive Design and Environmental Access Web site).

The Accessible Bathroom by Lori Jablonski and Karen Nickels (available from Design Coalition, Inc., Madison, WI).

Transgenerational Design by James Pirkl (available from Van Nostrand Reinhold, New York).

Universal Design Newsletter (available from Universal Designers and Consultants, Rockville, MD).

Universal Kitchens and Bathroom Planning (available from National Kitchen and Bath Association, Hackettstown, NJ).

P · A · R · T · 6

UNIVERSAL DESIGN PRACTICES

CHAPTER 43
UNIVERSAL DESIGN PRACTICE IN THE UNITED STATES

Elaine Ostroff, Ed.M.
Adaptive Environments Center, Boston, Massachusetts

43.1 INTRODUCTION

This chapter presents the origins of universal design in the United States from a civil rights perspective. The precedent of "Separate is not equal," which was established by a crucial Supreme Court decision, is introduced as an underpinning for universal design. With examples from recent architecture and landscape architecture practice, this chapter illustrates some of the differences between universal designing and meeting minimum accessibility requirements of federal standards. As universal design is less documented in the larger-scale environment, the emphasis in this chapter is on the regulated environment. The chapter also describes initiatives that promote universal design by government, private industry, professional design, and advocacy organizations. A timeline charting milestones that lead to universal design identifies key individuals and events.

43.2 U.S. BACKGROUND IN CIVIL RIGHTS

The civil rights history in the United States that confirms the right of every individual to participate fully in society provides the background to the values-based philosophy of universal design.

Civil Rights Origins—Separate Is Not Equal

The U.S. Constitution provides the legal basis for the civil rights movement. The Constitution was ratified by the first states in 1791, and it formed the basis of law in the United States in its guarantee of individual rights. The 10 amendments added to the Constitution at the time of ratification are referred to as the Bill of Rights. These included: religious freedom; freedom of speech; freedom of the press; and the right of the people to peaceably assemble. The 14th Amendment to the Constitution also assured that the government will not enforce any law that deprives any person of their privileges or liberty as citizens.

In the early 1950s, racial segregation in public schools was the norm across the United States. Although all schools were supposed to be equal in the quality of education that they provided, most schools for African Americans were far inferior to their counterparts for white children. In Topeka, Kansas, an African-American third-grader named Linda Brown had to walk 1 mile through a railroad yard to get to her segregated elementary school, although the white school was only seven blocks away. The National Association for the Advancement of Colored People (NAACP) aided her father and 13 other parents in a class action suit against the Board of Education of Topeka Schools.

This suit finally reached the Supreme Court. On May 17, 1954, the Court ruled in a unanimous decision that the "Separate but equal" clause was unconstitutional because it violated the children's 14th Amendment rights by separating them solely by the color of their skin. Chief Justice Warren delivered the Court's opinion, stating, "segregated schools are not equal, and cannot be made equal, and hence they are deprived of the equal protection of the laws." This *Brown v. Department of Education* ruling was a major step toward civil rights (Cozzins, 1998).

Civil Rights Based on Race

In March 1963, decades of organizing and struggles by African Americans and many other dedicated civil rights workers culminated in the historic march to the U.S. national capital in Washington, D.C. Martin Luther King spoke at the Lincoln Memorial and defined the civil rights vision of a just and inclusive society. Some fragments of his remarkable speech included: "I have a dream that one day this nation will rise up and live out the true meaning of its creed. We hold these truths to be self evident, that all men are created equal . . ." (King, 1963).

U.S. President Lyndon Johnson was able to move the civil rights agenda forward with new legislation. The Civil Rights Act of 1964 was the first of the major civil rights statutes passed by the U.S. Congress. It established the foundation for Sec. 504 of the Rehabilitation Act of 1973 and, later, the Americans with Disabilities Act (ADA). It prohibited discrimination by employers, public accommodations, and recipients of federal funds on the basis of race, religion, national origin, and sex. Although the 1964 Civil Rights Act was broad in its definition of protected classes, it did not cover people with disabilities.

In 1965, the Voting Rights Act was passed. This was the second major antidiscrimination statute. In 1968, following the assassination of Dr. King, Congress quickly passed the Fair Housing Act. This Act prohibited discrimination in the sale and rental of housing based on race, religion, national origin, and sex. Palames (1995) provided a strong link between race-based civil rights and disability-based civil rights.

Civil Rights Based on Disability

From 1919 to 1964, federal rehabilitation law provided benefits for people with disabilities. Initially focused on rehabilitation services for veterans, federal rehabilitation laws were amended to include workers who were disabled on the job. In 1954, the Vocational Rehabilitation Amendments Act widely extended services and benefits to people with disabilities who were not veterans. Lusher in 1989 and Peterson in 1998 provide detailed background on the federal legislation and policies that benefited people with disabilities. Lusher cites early reports that although veterans and other people who are disabled might be ready to take part in society, society was not ready for them. This led to efforts in the late 1950s to create federal accessibility standards and legislation.

The Rehabilitation Act of 1973 included Sec. 504. This was the first federal law that incorporated civil rights prohibiting discrimination toward people with disabilities by recipients of federal funds. The law states (1973):

> No otherwise qualified handicapped individual in the United States . . . shall solely by reason of his handicap, be excluded from the participation in, be denied the benefits of, or be subjected to discrimination under any program or activity receiving federal financial assistance.

For the first time, access to employment, education, buildings, and society were taken out of the realm of a government benefit and put into the context of a civil right (Bonny, 1997). Although the law was signed in 1973, the regulations that would give guidance to its implementation were not signed for 4 years. During the 4-year period, there was increasing pressure from people with disabilities all over the United States. The American Coalition of Citizens with Disabilities delivered an ultimatum to the responsible federal official, Joseph Califano, Secretary of Health, Education, and Welfare. They demanded that he sign the regulations by April 4, 1977, threatening that protests would be held all over the United States.

When the regulations were not signed, people with disabilities occupied federal buildings around the country. The "sit-in" in the San Francisco office of the U.S. Department of Health and Welfare lasted 25 days and, ultimately, led to the signing of the Sec. 504 Regulations. This was a stunning, well-organized effort, representing the combined participation of many people, both those inside the building and those on the outside. Over 100 people with many significant disabilities lived in the office building, under difficult circumstances. The governor of the state, the congressional delegation, organized labor, the churches, the Black Panthers, gay rights groups—all offered assistance.

The media covered the remarkable event very closely. When a group of the disability leaders from the San Francisco sit-in went to Washington to try to meet with Secretary Califano, a reporter and a photographer accompanied them. Although Califano would never meet with the protesters, the regulations were finally signed on April 28, 1977. Historic documentation, including videotapes, audiotapes, and extensive photographs of that milestone time, was prepared for the 504 20th-anniversary celebration in 1997 (Bonny, 1997).

The signing of the 504 regulations of the Rehabilitation Act of 1973 was a tremendous victory for people with disabilities. It showed the power of the growing disability movement, the power of independent living, and the power of a well-organized cross-disability coalition. The regulations recognized that people with disabilities had been treated as second-class citizens (Breslin, 1997).

Although the regulations were limited to organizations that received federal funding, they established the basic principles that would become the basis for the ADA. The 504 regulations identified the need to remove architectural and communication barriers, and provide accommodations in order to remove barriers to participation. It established a new legal definition of disability, not just a medical one. It also balanced the right of the individual with the cost of the accommodation to society, and it established the right of an individual to seek a remedy or to go to court.

43.3   FEDERAL ACCESSIBILITY REGULATIONS

Federal regulations on accessibility began with the Architectural Barriers Act of 1968 (Public law 90-480); Salmen traces the development of the regulations and standards in Chap. 12, "U.S. Accessibility Codes and Standards: Challenges for Universal Design." This section addresses only Sec. 504 of the 1973 Rehabilitation Act, the Fair Housing Amendments Act (FHAA), and the ADA—the three civil rights laws with major architectural implications.

Section 504 of the Rehabilitation Act

Section 504, as noted earlier, only applied to entities that received federal funds. Each federal agency that provided funding to external activities or that operated programs that served the public was required to issue regulations specific to those programs. There was a potential to lose federal funding if they were not in compliance with regulations. For example, the lead agency, the Department of Education (DOE), developed regulations that affected both public and private educational institutions for which they provided any funding. The DOE regulations prompted major training and technical assistance efforts for colleges and universities. The Department of Housing and Urban Development (HUD) developed regulations relating to the housing programs they funded. The language of the 504 regulations was the basis for much of the ADA, as seen in the excerpt that follows, a guide written by Coppelman (1977). The concept of "program accessibility" was introduced; emphasizing that the purpose of the architectural access was to ensure that people could use the service within a facility—a concept seen later in the ADA. The requirement to develop a transition plan, with input from people with disabilities, was also repeated in the ADA, for state and local government agencies.

Subpart (C)—Program Accessibility

Different rules apply regarding accessibility, depending upon whether we are dealing with new construction or alteration of existing facilities. The requirements are straight-forward and direct: any facility or part of a facility constructed after the effective date of the regulations must be readily accessible to handicapped persons. In addition, any new construction which is performed on an existing facility or part of a facility which could affect the usability of the facility by the handicapped must, to the maximum extent feasible, be altered in such a way as to make the facility accessible to handicapped persons.

As to existing facilities, recipients have up to three years to make facilities accessible. However, within six months of the effective date of the regulations, these recipients have to develop a transition plan that will analyze their current facilities and set forth the necessary steps to make the facilities accessible within three years. In developing the transition plan, once again, the recipients are mandated to enlist the assistance of interested persons "including handicapped persons or organizations representing handicapped persons." There is an exception for recipients who have fewer than 15 employees. If this small provider of services cannot make his facility accessible by any means other than significant alterations, then the recipient may, as an alternative, refer a handicapped person to another provider of services whose facility is accessible.

Section 504 was limited to recipients of federal funding, and in 1984, a Supreme Court decision, *Grove City College v. Bell,* further limited the application of the regulations to the department or unit of the institution that received the federal funds—not the entire institution. However, in 1988, the U.S. Congress passed the Civil Rights Restoration Act, overriding President Reagan's veto and overturning the 1984 Supreme Court ruling. The 1988 act asserted that when a program receives federal financial assistance, civil rights laws apply to the entire institution (Palames, 1995).

Fair Housing Amendments Act

The Fair Housing Amendments Act (FHAA) of 1988 built on the 1968 Fair Housing Act. The 1988 amendments added people with disabilities and families with children as two new protected classes. It was revolutionary in that it covered all multifamily housing, including private developments, even if there was no federal financing involved.

The U.S. Congress understood that people with disabilities needed choice in their living arrangements and that access to housing was a fundamental right. Many people with disabilities had been denied the right to buy or rent housing in neighborhoods of their choice. In addition to providing regulations that addressed housing policies and administrative practices, the FHAA created far-reaching accessibility guidelines. The new levels of accessibility were required only in new multifamily housing, with four or more units. Although the access guidelines of the FHAA were not as stringent as the full accessibility required in limited numbers of units under other federal or state laws, the impact was in the scope of these guidelines. All units in a building with an elevator had to include the Fair Housing Design Guidelines in every unit as well as in all common areas. In a nonelevator building, only those dwelling units on the ground floor were covered. The intent was "to build for now and for the future." Units that can meet the need of people with changing abilities need not be stigmatizing. Also, the FHAA guidelines supported more social integration for people with disabilities. Steinfeld and Shea, in Chap. 35, "Fair Housing: Toward Universal Design in Multifamily Housing," discuss the universal characteristics and efficacy of the FHAA Guidelines.

The FHAA guidelines became effective for buildings first occupied on or after 13 March 1991. The timing of the effective date was after the passage and signing of the ADA. There was great confusion—the publicity about the ADA was so extensive that many people thought that everything was covered by ADA requirements—and the FHAA was often overlooked. There was also strong resistance to the FHAA guidelines from the organized homebuilding community, especially the builders of multifamily developments. Their publications emphasized the high cost of compliance, pitting the affordable housing advocates against the disability advocates. Although HUD provided researched studies on the low costs associated with the guidelines, there was limited compliance. FHAA did not have an ongoing, extensive educational effort nor any significant technical assistance as was later organized to assist the voluntary compliance with the ADA.

Multifamily housing is the most complicated area for architects and builders. There are state, local, and federal regulations. A public housing authority can be subject to Sec. 504, as a recipient of federal funds, to Title II of the ADA as a state or local entity, to FHAA in relation to building four or more units—and there are different state regulations as well. Private developers have slightly fewer conflicting regulations, but it is still complex.

More recently, the large builders have discovered the market potential of universal design. Mcleister, writing in the *Professional Builder* (1999), reviews the positive shift in some of the country's largest home builders. He notes the combination of recent strong FHAA enforcement efforts that culminated in fines and settlement in 1998 and the growing awareness of the market needs as instrumental in this shift. In the agreement between Pulte Home Corporation, the nation's largest home builder, and the U.S. Department of Justice, Pulte will provide seminars about accessibility requirements of the FHAA. The corporate counsel for Pulte notes that they will go beyond the requirements of the legislation and apply the broader concept of universal design (Mcleister, 1999).

However, in the 106th Congress, FHAA continued to be a controversial issue. The Consortium for Citizens with Disabilities Task Force (2000) reported,

> During the 106th Congress, opponents of the Fair Housing Act's accessibility guidelines attempted to have the effective date of those rules pushed back from March 1991 to 1999 or later. The result of such a move would have been to immunize against prosecution many properties that have been built since March 1991 in violation of the Act. Fortunately, efforts to undermine the FHAA were unsuccessful. In the meantime, many disability advocates, housing industry representatives, and building code officials have worked together to promote the incorporation of the FHAA accessibility guidelines into building code language. Including the FHAA accessibility guidelines in building code language is viewed as a positive way of assuring compliance with the law at the state and local level.

Americans with Disabilities Act

The ADA is the most significant and far reaching of all civil rights acts for people with disabilities. Building on the momentum of the FHAA, and with tremendous national coalition work, the ADA was signed in 1990. The coalitions included people of every disability, along with their families and advocates. Their solidarity withstood every attempt to omit one or another group from coverage by the ADA. People with psychiatric disabilities, people with AIDS, alcoholics, and recovering drug addicts were often threatened during the passage of the ADA.

The ADA is a very complex and comprehensive law; it is often referred to as a "patchwork quilt" of regulations as it identified and attempted to remedy all the areas where people with disabilities still faced discrimination. Other laws had addressed these discriminatory areas for other citizens, but they were still unresolved for people with disabilities. The ADA includes employment—all aspects of the employment relationship including hiring, retention, benefits, and firing. It includes businesses that provide services to the public, such as retail stores, restaurants, hotels, service industry, recreation facilities, cultural settings—there are 12 categories of this type of private entity that serves the public. It includes all aspects of state and local government, related government authorities, public and private transportation, and some telecommunications services.

Throughout all of these areas, there are architectural implications; all of the aforementioned areas exist in some physical setting. The ADA reinforced the concept that the lack of architectural access was a form of discrimination. The regulations for accessibility under the ADA were defined in the ADA Standards for Accessible Design. Salmen, in chap. 12, "U.S. Accessibility Codes and Standards: Challenges for Universal Design," discusses the issues that these standards present for universal design.

The ADA Standards for Accessible Design are a comprehensive standard that includes technical requirements for all building components and the related site of the facility or the place of any program that is used by the public. The intent is that people with disabilities can enjoy the goods and services of any program that is offered. Throughout the ADA Standard, the phrase "in the most integrated manner" is used frequently. For example, in seating in an assembly hall, the accessible seats should be distributed throughout and not just contained in a single area. Beasley and Davies detail this concept in Chap. 47, "Access to Sports and Entertainment."

In new construction, the Standards are the most comprehensive and easy and are not expensive to incorporate. However, if this accessibility is not considered until late in the design stage, it can be difficult and expensive. The intent of Congress was to assure access for the future and to avoid costly renovations. When building alterations are planned, there are extensive requirements, but fewer than in new construction. These requirements are carefully balanced with the idea of maximum feasibility and also a cost limitation. The accessible route is a critical component of the alterations requirements.

In existing buildings, even when there is no plan to make alterations, the building owner or the government still has an obligation to assure the ability of people to use the services. This was one of the most controversial aspects of the ADA; never before did private business have an obligation to address accessibility in such settings. It was an area of compromise in the drafting of the ADA. Congress had to decide between grandfathering existing buildings or requiring that existing buildings be renovated to a new construction standard. They established this new expectation, that private businesses would remove barriers if readily achievable. The process of determining if barrier removal was possible for a business was detailed in the regulations. Cheap and easy was the idea, and the Department of Justice provided technical assistance with recommendations for prioritizing barrier removal.

The 10th anniversary of the ADA was celebrated throughout the summer of 2000. The Department of Justice issued a report, *Enforcing the ADA: Looking Back on a Decade of Progress* (DOJ, 2000), highlighting the range of agreements, court settlements and consent decrees, and mediation results affecting the daily lives of people with disabilities. The architectural changes can be seen everywhere—in historic town halls and new airports, in courthouses and motels, and in shopping centers and theaters. Former Attorney General Janet Reno noted in the report, "As this report illustrates, the ADA has made a difference in the lives of so many. But there are many others who still face barriers—barriers that man-made structures create . . . these barriers took generations to create. It will take continued vigilance and dedication to remove them."

The publicity surrounding the ADA more frequently focuses on negative stories that highlight some media-worthy aspect, such as Clint Eastwood's complaints about harassment of inaccessible businesses in the town where he is mayor. The National Council on Disability, the small federal agency that provided the initial background on the discriminatory processes that led to the ADA, reported in its 10th anniversary report to the president, "It is journalists who have run amok—not the ADA" (June 2000).

Mazumdar and Geis, in Chap. 18, "Interpreting Accessibility Standards: Experiences in the U.S. Courts," review the heavily publicized sports and entertainment law suits, which included the roles of architects and owners and focused on who was liable in ensuring that what was designed met ADA requirements.

Unfortunately, the published articles on buildings by high-profile architects in the professional magazines rarely identify the thoughtful ADA-inspired solutions that make prestige buildings work well for everyone. The silence on these aspects of well-designed facilities leaves uncontested the criticisms of the ADA and how it stifles good design. Of equal interest to this author would be more insight into why some buildings by name architects ignore some basic accessibility concerns as integral to overall considerations.

Several authors in this book address the problem by asserting the need for more and better education in degree granting programs. This is an essential objective but architectural education is a hotly contested issue. The education of architects has been debated for some time, not only concerning gaps relating to the understanding of diverse users but its overall relevance to architectural practice. The Boyer and Mitgang Report, *Building Community,* was commissioned to examine the state of the art in architectural education and make recommendations. (Boyer and Mitgang, 1996). The persistent and ongoing pedagogical arguments about what should happen in the academy is beyond the scope of this chapter, but it is noteworthy to consider the changes in the requirements for architectural programs by the National Architectural Accrediting Board (NAAB) as mentioned in Chap. 51, "Advances in Universal Design Education in the United States," by Welch and Jones. Furthermore, based on her experience in Ireland and England, Morrow, in Chap. 54, "Inclusion as Tool in Design Education," provides important insight into the professional ideology of architects that conflicts with issues of diversity, inclusion, and the mundane context of daily life.

Unlike earlier legislation, the ADA includes extensive technical assistance that helps practicing architects. The technical assistance that is available to architects through toll-free hotlines is a contributing factor to their ability to incorporate the ADA Standards into their work. The two major sources are (1) the 10 regional Disability and Business Technical Assistance Centers (DBTACs), funded by the National Institute on Disability and Rehabilitation Research (NIDRR), and (2) the Department of Justice (DOJ). Over 500 publications were created through these federally funded technical assistance systems. Some were produced by the DOJ both through grants awarded to professional associations and through DOJ staff, providing easy-to-use and industry-specific information that clarifies some of the more confusing aspects of the ADA (DOJ, 2000).

However, getting the needed document in the hands of the right person when needed is not a simple task. The DBTACs report that architects are among their regular callers with whom they build relationships. Andy Washburn, the information specialist in the New England ADA Technical Assistance center says, "About one in four callers are architects. There are 2 tiers of architects who call us. There are the very experienced callers: people whom I have gotten to know quite well, who want to discuss a complex issue, and there are the newer people who just discovered us, and who have straightforward questions about how to use the ADA Standards. Soon they become part of the experienced tier" (Washburn, 2001). However, many first-time callers express surprise and appreciation for the service, "You should publicize your services to the AIA, both locally and nationally" (Gattis, 2001).

That comment and other responses from architects are tabulated and analyzed as part of a federally mandated system that NIDRR has instituted to begin to evaluate the effectiveness of the DBTACs. Robert Gattis, ADA Impact Measurement (AIM) project manager, reports that more than 14 percent of callers to all the centers are architects, and that 96 percent report that the technical assistance that they received has helped their ADA-related work.

Another important source of technical assistance is the U.S. Architectural and Transportation Compliance Board (Access Board); they have had ongoing cooperation with the AIA to create educational opportunities for architects. The most recent is an online course that can be used at no charge on the AIA website (Access Board, 2000).

Challenges to the ADA

Although the ADA was enacted by a bipartisan congressional effort in 1990, there are continuing challenges to the ADA, both in Congress and the Supreme Court. The most recent was the Supreme Court case, *Alabama v. Garrett,* involving two state employees with disabilities—Patricia Barrett, a registered nurse with breast cancer, and Milton Ash, a security officer with asthma and sleep apnea, both of whom claimed discrimination in their jobs beacause of their disabilities. The case was brought by the state of Alabama and challenged the constitutionality of the ADA. The Alabama Attorney General argued that parts of the federal law violated states' rights. In their briefs to the Supreme Court, lawyers for individuals with disabilities and friends of the court contended that states' history of discrimination based on disability was so egregious that Congress had the power to override state sovereignty.

The 21 February 2001 decision was 5–4 in favor of Alabama and highlighted the deeply divided interpretations within the Court. However, the decision was narrowly defined and only covers Title I, the section of the ADA that prohibits discrimination on the basis of disability in employment; Title II prohibits discrimination on the basis of disability by public entities. Although the parties to Garrett had originally asked the Supreme Court to decide both Title I and Title II, the Court only decided Title I, and explicitly stated that they were not disposed to decide the constitutional issue regarding Title II. This means that the result of the Garrett decision ony impacts state employment practices. All other facets of state (and local) government activity are unaffected by this opinion (Robertson, 2001).

A week later, the Supreme Court, without comment, declined to accept a case that could have made states immune from lawsuits alleging discrimination against people with disabilities in access to public services, programs, and buildings.

43.4   RELATIONSHIP OF UNIVERSAL DESIGN TO ACCESSIBILITY REGULATIONS

The ADA Standards and the FHAA provide minimum standards that are a basis for universal design. Many architects persist in defining "universal design" as synonymous with the ADA Standards for Accessible Design. They consider the requirements a needless complication, to meet the needs of a limited number of people. They do not treat the regulations as opportunities to increase the good design and usability of a building. They are not aware of or perhaps choose not to see the design opportunities in going beyond the minimum standards to a more universal design. The negative attitudes sometimes create new buildings that have had accessible features added late in the design stage. For example, the requirement to have at least one accessible entrance can be met without the accessible entrance being the one that most people use. If the initial planning has sited the building so that the main entrance needs stairs, the ramped solution may be exhausting to use, unattractive, and stigmatizing.

43.5   EXAMPLES OF UNIVERSAL DESIGN

Universal design is inclusive design. It considers the needs of as many users as possible, at the outset of the design. Ordinary people with diverse abilities will be able to participate and use what the designer has created. The designer must be mindful of the range of users and incorporate these needs throughout all the design stages. Usability and good design are the emerging themes for universal design. They build on a hard-won legacy of civil rights victories that assert that "Separate is not equal," and that all people have the right for equal opportunity in society. This section includes three projects that illustrate a universal design approach. They include a historic building renovation on the campus of a private university, a newly constructed state courthouse, and the planning and site design of a new bank in a historic district.

Patton Hall, Princeton University, Princeton, New Jersey

The Context.   Patton Hall is one of Princeton University's older collegiate Gothic residence halls, built in 1906. It is located amid a group of dormitories that constitute a residential precinct, as seen in Fig. 43.1. The University decided to renovate Patton as the first in a planned 33-year campus-wide dormitory upgrade program. The building has 48,000 ft^2 on four floors. It steps down the sloping site, with four separate sections, and multiple levels in each section. Each wing's floors connect via stairs to each other, making continuous access very difficult. The existing layout included a stair-entry system of circulation, with nearly all bathrooms in the basement.

Client Priorities.   Princeton's goals for the historic building included providing a reasonable degree of accessibility, including creating corridors on some floors while minimizing the loss of beds. The interior had a considerable amount of original detail and woodwork that was to be repaired and retained. Since there are other dorms of the same vintage as Patton, and which would be renovated in subsequent years, decisions made on Patton were likely to establish standards/precedents for the future. Overall, the solutions had to respect the original character of the building.

Solutions.   The renovation by Goody Clancy & Associates (GC&A) included the following:

- Adding new student rooms in attics and ground floor areas
- Providing accessibility both horizontally and vertically
- Restoring or replacing interior finishes
- Creating new shared bathrooms on each floor
- Replacing mechanical, plumbing, and electrical systems completely (including provision of full Internet access)

The project also included adding a new pedestrian passageway cut through the building, as part of the larger campus master plan. This included related site work to the east and west of the building. Exterior restoration included new roofing and windows.

FIGURE 43.1   Patton Hall multilevel dormitory, on sloping site.

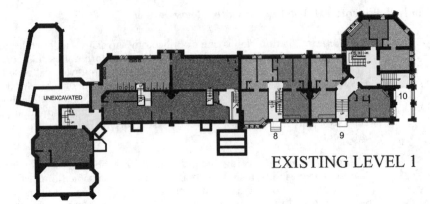

EXISTING LEVEL 1

FIGURE 43.2 First level of dormitory, before renovation—none of the levels have any interior connection.

Approach and Entrance. The building was previously entered at multiple points, with separate stairs serving suites on each floor, as seen in Fig. 43.2 (the plan view of existing level 1). The new configuration is illustrated in Fig. 43.3 (the renovated first level). It incorporates corridors serving most parts of the building, and is accessed via an elevator at the southern end of the building. The elevator lobby is served by a new accessible building entrance, created by subtle regrading of the landscape, as shown in Fig. 43.4. The new portal, which cuts across the building at grade as seen in Fig. 43.5, is on an accessible pathway and aligns with a new campus pedestrian route. It was designed in the architectural vocabulary of the original building, using the same or similar materials.

Getting Around the Building, Wayfinding. Wayfinding is much simplified now that corridors connect most parts of the building, as shown in Figs. 43.2 through 43.3. Several options were developed that would have resulted in varying degrees of accessibility, but Princeton opted for a hybrid approach in which some suites were maintained as is, without being accessible to the corridors in the other parts of the floors. Floor level changes from wing to wing made this necessary.

Key Features of the Major Functional Areas. Many student rooms on each level and all the new student rooms in the attic are accessible. Student dorm rooms are arranged in three-room suites that include two bedrooms and a study room. Given the irregular floor plan, some suites are larger than others. Some suites have the original fireplaces and other unique

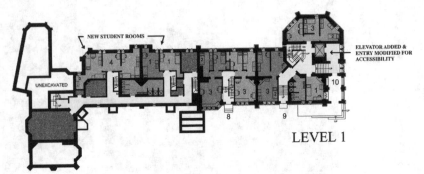

LEVEL 1

FIGURE 43.3 First level of dormitory after renovations; every entrance is modified for accessibility.

FIGURE 43.4 Landscaped and accessible new entrance to dormitory.

FIGURE 43.5 Courtyard, with new portal on accessible route.

features. All of the common areas provided in Patton, other than the shared study rooms within suites, are fully accessible. These include two lounges, a laundry room, and a small kitchen for general use by the students. New shared bathrooms are provided on every floor and are accessible.

The client contact at Princeton was John Hlafter, Director of Physical Planning. GC&A's project manager was Deborah Robinson, A.I.A. There was a client team that included students, physical plant staff, administrators, and others. The University played a very hands-on role in reviewing all aspects of the design and documents.

User Comments. The students who live in Patton are very pleased that the significant renovation was accomplished without losing the historic detail that defines Patton's unique character. They appreciate the look of the new portal as it blends in with the original building and facilitates pedestrian connections across the campus. Princeton has engaged GC&A to renovate two additional historic dormitories (Robinson, 2000). (See Figs. 43.4 and 43.5.)

Architect's Comments. Deborah Robinson, Project Manager, explained their approach: "Accessibility must be considered from the outset as an integral part of the design process. In renovation projects, existing conditions often prevent full access from being achieved without damaging the historic fabric of the building. This demands more creativity from us as architects, and sometimes, requires that compromises be made on all sides. At Patton Hall, one of the original drivers for the project was accessibility; thus every decision about corridors, bathrooms, quality of student life and cost was balanced against access. We stay closely in touch with the maintenance staff as well as the facilities planners, to learn what works and what doesn't."

Firm Description. Goody, Clancy & Associates, a Boston-based firm, provides complete architectural, planning, and urban design services for a variety of project types. Services include feasibility studies, master planning, construction administration, and interior design and space planning. They have applied these capabilities to a broad spectrum of projects across the country.

Edward W. Brooke Courthouse, Boston, Massachusetts

FIGURE 43.6 Site plan highlights the irregular parcel of land used for the new courthouse.

The Context. The Edward W. Brooke Courthouse, designed by Kallmann McKinnell & Wood Architects (KMW), is part of a statewide courthouse improvement program. Built on a partially abandoned and irregular parcel of land from Boston's Government Center urban renewal plan of the 1960s, the courthouse had to solve several major urban design problems. The site had been used for surface parking and was a desolate area avoided by pedestrians. It had to mesh with the adjacent Paul Rudolph Government Services complex, as well as link to other nearby government structures and the Bulfinch Triangle district, as seen in Fig. 43.6, the site plan.

The building is home to four specialized court groups and all of their related functions. It also accommodates facilities for alternative dispute resolution, central detention, a child care center, and a cafeteria. The building also houses the Suffolk County Registry of Deeds and the statewide court administrative offices. An urban park connects the Courthouse with the neighboring government service building.

Client Priorities. Providing equity in the courthouse design so that everyone could use the building with dignity was considered early on by the client, the state agency that manages all aspects of capitaldevelopment in the state. Lark Palermo, former Commissioner of the Department of Capital Asset Management (DCAM), is a lawyer who knows a great deal about the multiple users of courtrooms. "Architects often find it hard to know the needs of all the clients, and some are more demanding than others" (Palermo, 2000). Having practiced law and clerked in courtrooms all over the state, she was able to bring a fresh perspective to the varied needs of lawyers, judges, defendants, plaintiffs, family members, and others. As the client of record, with financial control, she was able to exert leadership in establishing some design criteria that had budgetary implications.

First, she insisted that there be two major entrances so that no one would have excessive distances to walk. Security is a major component of design decisions in courthouses, and in partnership with the Supreme Judicial Court, she was able to assure the additional staffing for security, making it possible to have two, secure entrances. Palermo also required that every courtroom have natural light. In addition, she established the precedent that all 18 courtrooms would be fully and invisibly accessible for people with mobility or hearing limitations. "The scheduling problems of trying to work with only a few accessible courtrooms would be an administrative nightmare."

Solutions.   Both entrances are very visible and clearly accessible. The main entrance is very dramatic, there is no ambiguity that this is how one enters the building. Once inside, wayfinding is extremely easy, with major functional areas arranged around a sky-lit four-story atrium. Anticipating many first-time users, the building is completely comprehensible and easy to orient oneself in. There are nearby elevators; also, there are clear, high-contrast, legible signage and a low central reception desk from which the security guard can direct visitors by pointing. All these facilitate getting around the building, and easily viewed open staircases provide a form of security for the occasionally emotional participants in family, probate, and housing courts.

Deep clerestory windows from each courtroom to the exterior wall bridge over the private back-of-the-house secure corridor, filling the courtrooms with natural light, as shown in Fig. 43.7. The acoustics within courtrooms and the spacious atrium work well, according to the building superintendent. Built into the wall surface are preformed acoustic panels that absorb sound while providing surface detail. The air quality issues have been addressed by Cosentini Associates, which included ducted return air, sealed ducts during construction, and state-of-the-art building management systems.

Each of the 18 courtrooms has the following amenities, including the lift at 1, many of which can be seen in the floor plan (Fig. 43.8):

- All judges have access to their benches from either three steps or a built-in lift that is an integral part of the millwork enclosure to the bench and witness box. All witnesses use the same lift that has three stopping points: the court floor (0 in), the witness stand (+6 in), and the judges' bench (+18 in). The elegant lift has no resemblance to the familiar mechanical devices often added to address level changes. As shown in Fig. 43.7, it has a hardwood sliding gate that is operated for all witnesses and the judge by the courtroom security officer.

- The jury box is on two or three levels, the first at floor level, has flexible seating as well as space for a person using a wheelchair, as shown in Fig. 43.7.

- The lawyer's lectern is height adjustable, and can be raised or lowered by a rocker switch on the lectern.

- All tables and benches are fully accessible with 19 in of knee space clearance under all tables.

- Assistive listening systems are wireless and consist of a mixer, transmitter, and antenna. Headsets are available upon request, and allow for separate frequencies between courtrooms and six-channel frequencies within each courtroom.

- The spectator benches have been shortened. The flexible space accommodates people using scooters or wheelchairs. When not needed, a courtroom chair can be placed there.

FIGURE 43.7   Clerestory windows provide natural light in every courtroom.

User Comments.   Commissioner Palermo was very appreciative of KMW and their numerous design solutions that made the Edward W. Brooke Courthouse an award-winning and elegant building. "They had the imagination and problem-solving skills to make it all work." The neighborhood association had been a long-time advocate for the site to be well used. "Anything would have been an improvement, but they have gone beyond all our expectations," said Robert O'Brien, Executive Director of Downtown North, the neighborhood association that promotes economic growth for the area. "The courthouse has been a shot in the arm, and is revitalizing the area" (O'Brien, 2000). Robert Campbell (1998) architecture critic of the *Boston Globe,* wrote about the courthouse:

> A single building, if it is the right one, can bring order out of chaos and transform an urban mess into a civic symphony.

Lawyers and other users give high marks to the building and tout it as a model. "It's an unusually well designed and comfortable building, it is very easy to navigate and use" (Dietrich, 2000).

Architect's Comments.   Bruce Wood, principal-in-charge, said, "It was primarily common sense. We never felt that the access requirements drove the design. We dealt with the overall urban design and complex program needs of the courts, and in the development of the design incorporated the dimensional provisions for access with the goal of making all spaces accessible without visible distinctions."

Firm Description.   Kallmann McKinnell & Wood Architects, Inc. (KMW) is a Boston-based, 75-person firm offering comprehensive design services including feasibility studies, programming, master planning, architectural design, interior design, and landscape architecture.

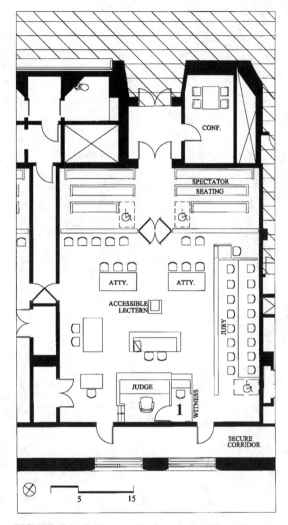

FIGURE 43.8 Courtroom plan shows built-in lift at 1 for judge and witnesses, height-adjustable lectern, accessible jury box.

The Federal Reserve Bank of Minneapolis, Minneapolis, Minnesota

The Context.   The Federal Reserve Bank of Minneapolis (FRBM) overlooks the Mississippi River from a historic site—the location of Minneapolis' original settlement and the first public square. As the city grew, the central part of downtown Minneapolis had been largely cut off from the river for more than a century. The 9.4-acre site is at the junction of the city's two major street grids: Nicollett Mall and Hennepin Avenue. The design returned the site to its traditional role as the gateway to the city. It represented one of the last opportunities to provide public pedestrian access from downtown to the river. The site, as shown in Fig. 43.9, with its 22-ft elevation change accommodated the two bank-related structures connected by a courtyard: an eight-story office tower and a four-story operations center. The office tower is set back from Hennepin Avenue in order to frame a public square, providing direct access to

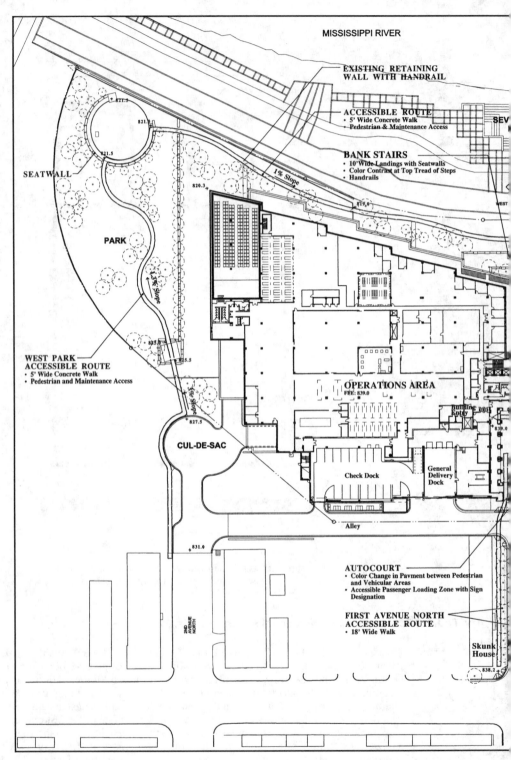

MISSISSIPPI RIVER

EXISTING RETAINING WALL WITH HANDRAIL

ACCESSIBLE ROUTE
• 5' Wide Concrete Walk
• Pedestrian & Maintenance Access

SEV

BANK STAIRS
• 10' Wide Landings with Seatwalls
• Color Contrast at Top Tread of Steps
• Handrails

SEATWALL

1% Slope

WEST

PARK

1.5% Slope

WEST PARK ACCESSIBLE ROUTE
• 5' Wide Concrete Walk
• Pedestrian and Maintenance Access

1.5% Slope

OPERATIONS AREA
FRE: 839.0

Building Entry

839.0

CUL-DE-SAC

Check Dock

General Delivery Dock

Alley

2ND AVENUE NORTH

AUTOCOURT
• Color Change in Pavment between Pedestrian and Vehicular Areas
• Accessible Passenger Loading Zone with Sign Designation

FIRST AVENUE NORTH ACCESSIBLE ROUTE
• 18' Wide Walk

Skunk House

838.2

FIGURE 43.9 The site plan documents the accessible routes around the bank to the river and parkway.

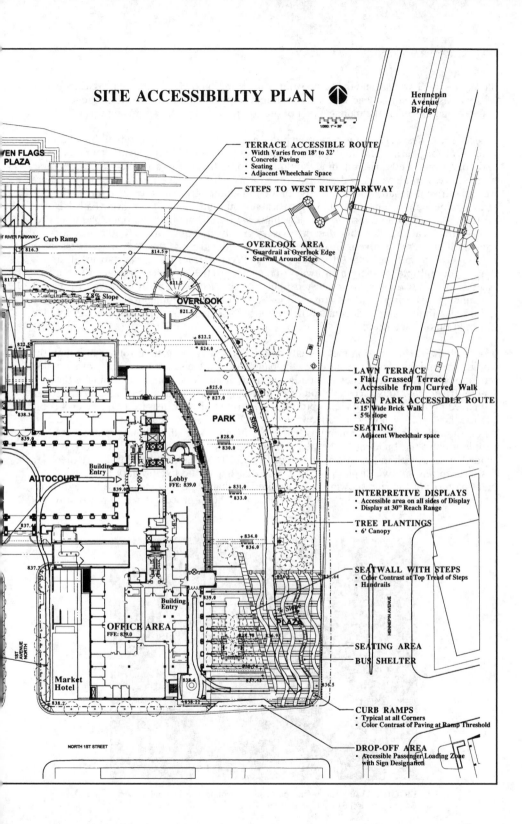

SITE ACCESSIBILITY PLAN

Hennepin
Avenue
Bridge

1/360; 1" = 30'

SEVEN FLAGS
PLAZA

TERRACE ACCESSIBLE ROUTE
• Width Varies from 18' to 32'
• Concrete Paving
• Seating
• Adjacent Wheelchair Space

STEPS TO WEST RIVER PARKWAY

WEST RIVER PARKWAY   Curb Ramp

816.3

814.5

817.9

811.5

2.8% Slope

OVERLOOK AREA
• Guardrail at Overlook Edge
• Seatwall Around Edge

OVERLOOK

821.5

822.5

822.2
824.0

LAWN TERRACE
• Flat, Grassed Terrace
• Accessible from Curved Walk

825.0
827.0

EAST PARK ACCESSIBLE ROUTE
• 15' Wide Brick Walk
• 5% slope

838.34

5% Slope

SEATING
• Adjacent Wheelchair space

839.0

PARK

828.0
830.0

Building
Entry

AUTOCOURT

Lobby
FFE: 839.0

839.0

831.0
833.0

INTERPRETIVE DISPLAYS
• Accessible area on all sides of Display
• Display at 30" Reach Range

837.0

834.0
836.0

TREE PLANTINGS
• 6' Canopy

837.7

SEATWALL WITH STEPS
• Color Contrast at Top Tread of Steps
• Handrails

839.0

Building
Entry

OFFICE AREA
FFE: 839.0

Stop
PLAZA

SEATING AREA

BUS SHELTER

Market
Hotel

838.4

838.2

838.22

837.45

836.5

CURB RAMPS
• Typical at all Corners
• Color Contrast of Paving at Ramp Threshold

NORTH 1ST STREET

DROP-OFF AREA
• Accessible Passenger Loading Zone
with Sign Designation

1ST AVENUE NORTH

HENNEPIN AVENUE

the West River Parkway and the Mississippi Waterfront Park. Hellmuth, Obata & Kassabaum (HOK) was the architect and landscape architect responsible for site selection, site and landscape design as well as the management of the complex design and approval process.

Project Priorities.   The bank wanted its new Ninth District headquarters to be a building with great long-term value and a setting that would support their growing administrative and operations needs. The building would need to be flexible to accommodate bank functions and essential, very rigid security operations without becoming a "fortress." There were significant urban design and environmental challenges. The reconnection to the city was extremely important. As a public institutional building in the St. Anthony Falls Historic District, a high-profile riverfront site, the bank wanted to maintain an open planning process. At the local level, a public design advisory group and approximately 30 public and private entities were involved, including neighborhood groups, the Mayor's Accessibility Review Group, and the State Office for Historic Preservation. There was a strong desire by the city for multiple pedestrian pathways to the river. In the early design stages, the Mayor's Accessibility Review Group requested plans for the accessible path of travel, to discuss maximum access for people with disabilities. As part of the planning process, HOK created a site accessibility plan to document the options. As seen in Fig. 43.9, the plan has detailed information on each route.

The Solution.   The solution met the city's desire for three pedestrian routes; although the sloping site and the security required by the bank complicated access for people with mobility limitations, HOK created pedestrian pathways from three access points, all leading to the Mississippi River. One fully accessible route, at the far right of the site plan (Fig. 43.9) is through the paved public plaza along Hennepin Avenue, through the greensward to the river. People can use the curved walkway or walk on the grass. The plaza, designed for ease of use and clarity of wayfinding, has many amenities. There are highly visible directional graphics that lead people around the building to the plaza; there is stonewall seating along the way, both in shaded areas as well in the sun, as shown in Fig. 43.10. Along the curved walkway there are cast-bronze interpretive displays—sculptures—that depict the five primary historical settlement patterns of the site, showing how the site has changed over time. These large displays were created in part to comply with the federal mitigation process that allowed removal of

FIGURE 43.10 The public plaza leads around the Federal Reserve Bank through the greensward to the river.

buildings within the historic district. They also address accessibility needs of users who are blind. All of the detailed displays were highly tactile and one sculpture displays a tactile map of the city as it was 100 years ago, as shown in Fig. 43.11.

Another access point is at the center of the site, at the extension of First Avenue. It goes through the auto court and downstairs to the main path leading to the overlook. On the far left of the site plan is the third access point from the western edge. This meandering walkway has native trees, wildflowers, and grasses all the way to the river. The three routes all connect along the river, leading to the overlook at the upper right section of the site plan.

Architects' Comments.   Bill Palmer, Group Vice President, HOK Planning, reflected, "The timing of the project was ideal—it coincided with the start of the firm-wide training and technical assistance in response to the ADA. There was high awareness about access needs. HOK designated two people, Mary Ann Lazarus, and myself, for quality control on accessibility. We created many accessibility planning tools and 'best practices.' We are a lot bigger now and it's harder, with growth and turnover, to be able to reach out to everyone."

Jim Fetterman, Vice President, HOK St. Louis Planning Group, said, "It was such a difficult site and there was so much to be communicated that we documented the planning of accessible routes on the site plan. The accessible routes drove the design of the site." He recommends that all planners and landscape architects create such planning tools.

User Comments.   Sue Lasoff, occupational therapist, access consultant, and one of the members of the Mayor's Accessibility Review Group, noted, "The architect was very cooperative, and eager to go beyond minimum requirements. I love going there, they have done a great job making the pedestrian paths along the river accessible."

Sheldon Azine, Senior Vice President, FRBM, was the senior bank officer in charge of the new building project. "We were anxious to be good neighbors as there was some opposition to the acquisition of property in a historic district. We focused on the landscape, to make it as open and inviting as possible."

The Firm.   Hellmuth, Obata + Kassabaum (HOK), Inc. is now the largest full-service architectural, engineering, interior design, and planning firm in the world as ranked by *World Architecture* magazine, in the January 2000 issue. They have offices worldwide and have worked for the Federal Reserve Bank system since 1976. The St. Louis office led the design at the FRBM.

FIGURE 43.11   This bronze sculpture with its tactile map is one of five sculptures depicting the historic area.

43.6 *SUPPORTS FOR UNIVERSAL DESIGN*

There are numerous interlocking relationships, between individuals, the federal government, university-based centers, nonprofit organizations, professional and trade associations, businesses, and corporate foundations. This will describe the efforts of key federal agencies and selected private businesses and organizations.

Federal Government

The federal government continues to be the most influential player in promoting universal design. Numerous agencies are involved in making laws and regulations for the underlying accessibility requirements, as well as providing funding for research, education, and promotion of universal design. Recently, the federal government introduced market incentives related to the purchase of information technology and equipment. Vanderheiden, in Chap. 65, "Fundamentals and Priorities for Design of Information and Telecommunication Technologies," describes the incentives as "pull legislation" and the requirements as "push legislation." Once Congress passes a law, there are different agencies named in the law that have responsibility to develop regulations for implementation and for enforcement. Some agencies have broadly interpreted their mandates; these are the agencies that have contributed to the development of universal design. As discussed earlier, the ADA, and its enforcement by the United States Department of Justice, has had the most powerful impact of all the federal laws. This was confirmed in the research conducted by the Trace Center that identified factors motivating private business to adopt universal design (Vanderheiden and Tobias, 2000). Many cited the ADA, even though their specific industries were not covered by any regulations. Following are descriptions of some of the other agencies that have had a significant impact in universal design.

National Endowment for the Arts.   The National Endowment for the Arts (NEA) is a small federal agency whose guiding principle is that the arts should be made available to all Americans. Although it has limited funding, this was the first federal agency to show leadership in promoting universal design. In 1990, they convened a national group of experts to make recommendations for a Leadership Initiative on Universal Design and, also in 1990, provided seed money to the Adaptive Environments Center for the Universal Design Education Project. In 1992, they funded the first national conference on universal design at Pratt Institute. The NEA has also convened meetings with other federal agencies to stimulate interest in universal design. The NEA continued to follow the recommendations of the 1990 expert group, including the funding of two programs to identify examples of excellence in universal design. The first project funded Universal Designers and Consultants (1996) in a national search of excellent examples, leading to the *Images of Universal Design* slide collection. The NEA funded the Center for Universal Design (2000) to collect more examples, and to create a CD-ROM, *Exemplars of Universal Design*. In 1999, the NEA convened another national group to advise on future directions to ensure universal design. The report (NEA, 1999) emphasizes mainstream promotion, expanded efforts in design education, and expanded collaboration with policy-setting organizations.

National Institute on Disability and Rehabilitation Research.   The National Institute on Disability and Rehabilitation Research (NIDRR) is part of the U.S. Department of Education. Its mission is to improve the rehabilitation outcomes for people with disabilities and to support the goal that every adult American possesses the skills to compete in a global economy. Through their funding for research, NIDRR has played an extremely important role in the development of universal design in the United States. They are the major funders of the Rehabilitation Engineering Research Center (RERC) at the three largest university-based centers on universal design: (1) the Center for Universal Design at North Carolina State University, Raleigh, North Carolina; (2) The RERC on Universal Design at the University at Buffalo, Buffalo, New York; as well as (3) the RERC on Telecommunications at the Trace Center at the University of Wisconsin, Madison, Wisconsin. They have funded a range of projects beginning in 1992, including the collaborative research at the Center for Universal Design that led to the Principles of Universal Design and the universal bathroom project at the State University of New York at Buffalo. Recently, the NIDRR long-range plan identified universal design as an important approach for the inclusion of people with disabilities in community life.

U.S. Access Board.   The U.S. Access Board, also known as the Architectural and Transportation Barriers Compliance Board, is the only independent federal agency whose primary mission is accessibility for people with disabilities. The Access Board has developed most of the federal accessibility guidelines that provide the minimum requirements leading to universal design. The Board develops its accessibility guidelines and standards through a process that provides an opportunity for public comment. It has become standard practice for the Board to establish advisory or regulatory negotiation committees to help develop or update its accessibility guidelines and standards. These committees allow interested groups, including those representing designers, industry, and people with disabilities, to play a substantive role in the Board's development of guidelines that are then proposed for public comment.

These committees produce consensus reports that become the basis for the Board's guidelines, that are then proposed for public comment. Once the Board's final rule is published, it must be adopted by an enforcing agency such as the DOJ or Department of Transportation before it becomes an enforceable standard. However, the published results of the committees are valuable tools for architects and other designers. These constitute informed opinions that can be used when architects want to innovate and create solutions that differ from the standards. This opportunity to provide alternatives is part of the "Equivalent Facilitation" section of the ADA Standards.

In 1998, the Board issued final guidelines on state and local government facilities, building elements designed for children's use, telecommunications equipment, and over-the-road buses. In 2000, *Play Areas, Passenger Vessels,* and *Electronic and Information Technology* were published. Various guidelines and standards are in development and scheduled for publication by the Access Board in 2001. These include: *Recreation Facilities; Outdoor Developed Areas;* and *Public Rights-of-Way.* Research is planned on at least two other areas of emerging concern, including acoustics and environmental air quality (Access Board, 2000).

U.S. Park Service and the U.S. Forest Service.   These two federal agencies have been influential in the universal design of outdoor environments. The U.S. Forest Service has addressed the outdoor, natural environment in a very thoughtful way. They have collaborated with a design firm, Moore Iacofano and Goltsman, to publish their *Universal Access for Outdoor Recreation.* These well-illustrated guidelines provide a way of describing natural settings that is not limited to detailed specifications. The U.S. Park Service supports a major education program at the University of Indiana, the National Center for Accessibility, as noted in Chap. 20, "Guidelines for Outdoor and Wilderness Areas," by Robb. The U.S. Park Service Web site provides excellent communication about national parks and their access.

Federal Communications Commission.   The Federal Communications Commission (FCC) is especially important as the world moves further into universal design of telecommunications. The mission of this independent government agency is to encourage competition in all communications markets and to protect the public interest. In response to direction from Congress, the FCC develops and implements policy concerning interstate and international communications by radio, television, wire, satellite, and cable. Vanderheiden discusses the sweeping legislation initiated and enforced by the FCC in Chap. 65. "Universally Designed Telecommunications." Former FCC Chairman William E. Kennard's statement on the ADA anniversary included the following (Kennard, 2000):

> . . . But we also see that the world today is very different than the one that existed ten years ago. We see the jobs that once required physical access now requiring virtual access. We see the educational tools that were once contained in textbooks now contained on software and over the Internet. We see children once limited to board games now enthusiastically greeting the challenges of surfing the world's resources through the Web. The ADA focused on the world made of bricks and mortar. But now we are presented with a different world—a world of networks, of fiber optics, a world of billions of digital bits that are becoming indispensable to our daily lives. And this world has presented the Commission with a unique challenge—a challenge to ensure that all Americans, including Americans with disabilities, have full and equal opportunities to access and enjoy this virtual world. Technology has the power to unleash access to jobs, education, and information in ways undreamed of ten years ago. And our challenge is to make sure that all Americans, regardless of ability, have equal opportunities to enjoy this access . . .

Marketing, Advocacy, and Education in Private Organizations

American Association of Retired Persons.   The American Association of Retired Persons (AARP) reports the stunning response of their huge membership to universal design as a way for people to "age in place." More than 100,000 people requested information on universal design when first introduced in *Modern Maturity,* the organization's monthly periodical. Harper, at the NEA June 1999 meeting, emphasized AARP's intent to educate the consumer and to promote a universal design approach through mainstream media. He illustrated his point with a segment about universal design that was seen on broadcast news, *Good Morning America* in June 1999, sending a popular message about graceful and practical features that many people would want in their homes.

The AARP in-house communication tools have tremendous reach. Their Web site has seen an increase in use during the last 2 years' activity from about 400,000 hits per month to the latest figure of 750,000 hits per month and continues to grow. New features on universal design are planned for both the periodicals and the Web site (Harper, 2000).

Concrete Change.   Concrete Change is a small nonprofit organization based in Atlanta, Georgia. Founded by Eleanor Smith in 1988, this grassroots organization is the driving force behind the visitability concept in single-family housing in the United States. (This term is seen elsewhere in the book; the use in Europe and South America has a different meaning as it applies to public buildings. Both D'Innocenzo and Morini, in Chap. 15, "Accessible Design in Italy," and Guimaraes, in Chap. 57, "Universal Design Evaluation in Brazil: Development of Rating Scales," describe the application in their countries.) Smith's underlying objective is simple: People with disabilities want to be part of community life and should be able to visit with their friends and family. She has advocated for two modest items in single-family housing construction that make this possible: (1) the zero-step entrance and (2) wider doorways throughout the house.

Smith describes her early efforts with Habitat for Humanity in Atlanta; in trying to encourage the local Habitat board to include these two details in their low-cast volunteer built housing. She said, "Finally, a few board members were open and their knowledgeable construction engineer checked it out. He assured them that it wasn't hard and could be done for under $200 per house. Smith noted that portions of Atlanta are hilly and the houses are not built on slab construction, "but they found several creative ways to accomplish this." Soon, 30 houses were built and the evidence of their acceptance and affordability helped when Concrete Change introduced a local ordinance to require visitability in new single-family construction that used any government funding. Variations of this ordinance are now in place or under consideration in a number of towns and states throughout the country.

The visitability concept was included in the Department of Housing and Urban Development (HUD) Hope VI program, after conversations with former Secretary of Housing Henry Cisneros, Smith, and other members of the advocacy community. Hope VI is a well-funded program to revitalize distressed urban neighborhoods. Applicants for funding get extra points if they include visitability. Visitability is defined in the application kit as applying standards to "allow a person with mobility impairments access into the home" (this is in addition to the required accessibility features).

An article in the Hope VI online newsletter (McGovern, May/June 2000) illustrates the ease and affordability of visitability. The article is full of quotes from housing authority directors who note how ordinary and simple it is. John Hiscox, executive director of the Macon Housing Authority (MHA) explains, "In 1994 the MHA built a 91-unit, single-family public housing development in which the concept of visitability emerged almost naturally as a byproduct of the design. This is not rocket science. In terms of a no brainer, it ranks right up there with not locking your keys in the car . . . just take ten seconds and think about what needs to be done." McGovern quotes another developer, Jack Morse, developer of Atlanta's East Lake Commons, who puts it succinctly when he says, "There was just no reason not to do it." All 67 units in this private townhouse development are visitable. The Hope VI material emphasizes that the key is to plan for visitability from the outset to avoid incompatible retrofits. The Congress for New Urbanists (CNU) is involved with these urban projects and is also incorporating the visitability idea in its principles for inner-city neighborhoods (CNU, 2000).

The National Association of Home Builders. The National Association of Homebuilders (NAHB), an outspoken opponent of accessibility by regulation, is now promoting *amenities* (all of which meet the needs of people with limitations as well as everyone else's for convenience and style) in new housing as a way to attract the baby boomer market. With funding from the Administration on Aging, the NAHB National Center for Seniors' Housing Research sponsored a student design competition for "Aging in Place: A Smart-Aging Residential Design Competition for Students" (NAHB, 2000). AARP is working closely with NAHB's Senior Housing Council, the Center for Seniors' Research, and the Remodelers' Council to develop an educational program for builders and contractors to prepare them to meet the remodeling needs of aging customers. At the community level in Atlanta, Georgia, the local NAHB affiliate is working with the local AARP chapter and Concrete Change to create a model home for "Easy Living" (February 2001).

The Industrial Design Society of America. The Industrial Design Society of America (IDSA) has been extremely active among the professional design organizations in their attention to universal design. Their professional interest group in universal design chaired by one of the handbook authors, Jim Mueller, has grown from 9 members in 1997 to over 500 members in 2000 (Mueller, 2000). Their annual Industrial Design Excellence Awards (IDEA) program explicitly includes usability and attention to diversity of users in the criteria for selection. In their "Designs of the Decade," the Good Grips products described by Moore in Chap. 2, "Experiencing Universal Design" were honored. In January 2001, IDSA and Motorola sponsored a Wireless Universal Access Competition.

The Foundation for Interior Design Education and Research (FIDER). This group, which evaluates and accredits interior design programs in colleges and universities in the United States and Canada, also sets the standards for their educational content (see Jones, Chap. 52, "Integrating Universal Design into the Interior Design Curriculum"). Building on their user-centered design orientation, the FIDER 2000 Professional Standards explicitly state that

> Student work must demonstrate understanding of universal design concepts and principles as the basis of their design work . . .

No other accreditiation program or discipline has made this commitment for universal design in professional design education. There are 137 interior design education programs, as of this writing, that will respond to these revised standards.

GE Real Life Kitchen.   The GE Appliances approach to universal design was a business decision. They believed that they would be more competitive if they promoted their individual kitchen products within an overall kitchen design that had broad appeal. Their goal was truly universal in that they promoted existing products in widely usable applications, rather than products and concepts designed for specialized, disability-related access. In this way, they broadened their market base by focusing on inclusive concepts that accommodated diverse users. They identified contractors who build and remodel kitchens as a key target audience for their marketing and unveiled the model kitchen designed by Mary Jo Peterson and described by Mullick and Lavine in Chap. 41, "Universal Kitchens and Appliances," at the huge annual show of the NAHB. More than 30,000 builders and contractors were able to walk through and examine the model kitchen. In the planning for the kitchen, they responded to a key problem often noted by contractors, that they can't find universally designed products. In addition to GE appliances, this kitchen uses many off-the-shelf products and cabinets that are readily available. GE also created an illustrated booklet that contractors could show to customers. The publication, *Real Life Design* (GE Appliances, 1995), also provides product sources to assist contractors. In collaboration with another promoter of universal design, the National Kitchen and Bath Association (NKBA), they created a seminar that Peterson teaches for contractors and builders who want to learn how to use universal design in their business.

Milestones in Universal Design: A Timeline

There are many sectors as well as individuals in the United States that have contributed to and continue to sustain universal design. The following list, which reflects the author's perspective and is not comprehensive, cites events from 1954 to 1998 that have been catalysts for universal design.[1]

1954: Brown Versus Board of Education.   U.S. Supreme Court decision establishes that "Separate is not equal." Separate facilities prevent equal opportunity in education.

1959: Gunnar Dybwad introduces concept of humane, functional environments.   He emphasizes needs of individual residents in planning residential care facilities at the annual convention of the National Association for Retarded Children.

1961: The American National Standards Institute.   A117.1 provides the first voluntary standards for accessible design.

1964: The Civil Rights Act of 1964.   The first of several major pieces of civil rights legislation in the U.S., this was the foundation for future civil rights laws such as Section 504 and the ADA.

1968: Architectural Barriers Act.   U.S. Congress passes the first law requiring accessibility for people with disabilities in federal buildings.

1973: Section 504 becomes law.   First civil rights legislation prohibits discrimination against people with disabilities in programs that receive federal funding.

1974: Ron Mace founds Barrier-Free Environments, Inc.   Mace and colleagues found research, design, and educational firm in Raleigh, North Carolina. Their initial work in writing and illustrating access codes in North Carolina puts a human face on accessibility.

1974–79: Ed Steinfeld leads research for revised ANSI A.117 Standard.   State-of-the-art report includes survey of state and federal regulations, human factors and social science research. Laboratory research on the functional abilities of 200 people leads to recommendations for accessibility standards.

1975: National Center for a Barrier-Free Environment founded.   First U.S. center established, to create and distribute educational materials on accessible design. In 1982, they host a conference, "Designed Environments for All People," at the UN in New York.

1977: Marc Harrison redesigns consumer product.   Industrial design faculty member at Rhode Island School of Design redesigns Cuisinart food processor based upon limited hand function and vision deficits, improving the ease of use for everyone.

1977: People with disabilities demand 504 regulations.   More than 100 people stage sit-in at federal offices in San Francisco for 28 days until 504 regulations are signed. This was the birth of the nationwide disability rights movement.

1978: Adaptive Environments Center founded.   Elaine Ostroff and Cora Beth Abel cofound the Adaptive Environments Center at Massachusetts College of Art in Boston. Emphasis is on teaching a participatory design process to teachers, to redesign their classrooms to include disabled children.

1979: Ray Lifchez teaching students about design for someone unlike themselves.   Ray Lifchez, Professor of Architecture at the University of California, Berkeley, brings people with disabilities into the design studio so that students would learn to design for people unlike themselves.

1979: PLAE is founded, integrating children with disabilities in outdoor play.   Robin Moore's Environmental Yard, Berkeley, is the setting where parents of disabled children worked with MIG to initiate integrated program.

1979–82: Patricia A. Moore, disguised.   Industrial designer tours the U.S. disguised as an older woman, to experience the discrimination faced by elderly people.

1982: Go beyond the codes, design for all people.   Adaptive Environments funded by the National Endowment for the Arts to convene symposium of designers and educators. Recommendations: teaching that goes beyond barriers, infused in curriculum, involves the users in real world projects.

1982: Gregg Vanderheiden discusses "electronic curb cuts."   Engineer, cofounder of the Trace Center at the University of Madison, Wisconsin in 1971, he introduces the term to identify communication access solutions that are useful to all users.

1984: Adaptive Environments Laboratory at Buffalo State.   Research center established with prototype settings to test usability. In 1992 it becomes the IDEA Center.

1985: Ron Mace introduces term *universal design*.   In an article in *Designers West,* Ron Mace explains universal design; the first documented use of the term.

1986: Apple computer builds easy access into operating system.   Alan Brightman works with Gregg Vanderheiden to integrate accessibility into Apple products.

1988: The Fair Housing Amendments Act.   People with disabilities and children are added to the 1968 civil rights law that prohibits racial discrimination in housing. It establishes guidelines for universal design in new multifamily housing.

1989: The Center for Accessible Housing.   National Institute on Disability and Rehabilitation Research funds center at North Carolina State to support the accessible and universal design of housing for people with disabilities, families, and older people. In 1994, it becomes the Center for Universal Design.

1990: The Americans with Disabilities Act.   The most comprehensive civil rights legislation for people with disabilities establishes that the lack of access to programs, employment and facilities is discrimination, in public and private settings. It establishes a baseline for universal design.

1990: National Endowment of the Arts and universal design.   NEA convenes a panel on universal design and recommends actions to encourage universal design in education and practice.

1990: James Mueller produces *Workplace Workbook*. Graphic workbook uses functional limitations as basis for design. The 1992 edition includes universal design.

1990: Oxo International debuts Good Grips products. Sam and Betsey Farber work with Smart Design, Inc.; Pat Moore to create the model for universally usable kitchen products.

1990: Classic Tupperware redesigned for usability and style. Norman Cousins redesigns products into museum-quality objects, inspired by his mother's need for easier-to-use tabs.

1991: Universal Design Education project. Adaptive Environments begins national project for design faculty to infuse universal design in the teaching of design. *Strategies for Teaching Universal Design* documents pilot program and is published in 1995.

1992: Access to Daily Living. Pratt Institute's Centre for Advanced Design Research produces the first U.S. conference on universal design.

1993: *Beautiful Barrier Free: A Visual Guide to Accessibility*. Cynthia Leibrock, with Susan Behar, show that aesthetics can be part of accessible design.

1993: *The Universal Design Newsletter*. The first periodical dedicated to universal design, begins quarterly publication.

1994: Raynes rail provides tactile wayfinding and communication. Coco Raynes creates railing system initially for Braille users. Audio additions provide universal usage in museums.

1995: Principles of Universal Design. The Center for Universal Design develops first edition of performance criteria with group of U.S. experts.

1996: Telecommunications Act. New U.S. law extends universal access to communications for people with hearing, speech, and vision disabilities.

1998: Designing for the 21st Century international conference. First international meeting on universal design involves 450 people from 21 countries. Sponsored by Adaptive Environments with Center for Universal Design, Eastern Michigan University, Hofstra University, and the *Universal Design Newsletter*.

43.7 CONCLUSIONS

This chapter highlighted the development of the civil rights legislation that provided a minimum set of accessibility standards for the practice of architecture and landscape architecture and a baseline for the practice of universal design.

The development of the legislation and key architectural requirements in the United States was traced back to the civil rights history of the 1950s. "Separate is not equal" provided the conceptual basis for the practice of universal design in all of the design fields. The struggles of African Americans for equal opportunity in education, employment, voting, and housing inspired the growing disability rights movement. In 1977, that nascent movement showed its power in the demonstrations and sit-ins that led to the signing of a strong set of regulations to implement Sec. 504 of the 1973 Rehabilitation Act. Section 504 was the first civil rights law for people with disabilities and provided the basis for future laws, especially the ADA. The challenges and continuing implementation issues for Sec. 504, the FHAA, and the ADA are detailed.

Three case studies illustrated the process and results of a universal design approach, and highlights of selected government agencies and private organizations are noted for their significant efforts in the development of universal design. A timeline of milestones identifies people, organizations, legislation, and events that have been instrumental in developing and sustaining universal design.

Over the past 15 years, there has been a tremendous change in both the consciousness of universal design as well as in the practice and experience of universal design. What began as a conviction by people with disabilities and their advocates now has form and reality in the built environment. In reflecting on what the future holds for universal design in the United States, key issues emerge from an examination of the history.

- There is responsibility and opportunity for every individual to use their democratic privilege for collective action to keep universal design viable. The history of the civil rights legislation highlights the constant political action that is required to educate and to advocate for universal design. Although there is a growing marketplace demand for universal design, without the underlying legislation that is shaping the larger-scale environment, the integration into professional practice may be subverted.

  Chairman of the National Endowment for the Arts, William Ivy, stated (Garfield, 1999):

  When we look at the accomplishments in universal design since 1990, it's obvious that the concept is finding growing support, but as we all know, there is still much to be done. We need to talk about ways we can infuse the concept of universal design into the thinking and practices of those who plan and build communities, own businesses, and teach in the important field of design.

- The leadership role of clients in demanding universal design needs to be appreciated. In the examples that were presented in the chapter, the clients wanted universal design features incorporated from the very beginning. Architects have been blamed for not delivering more universally designed settings, but it is sometimes very difficult to sell the concept to clients when they are negative about accessibility and only want some minimum response that will avoid any liability. Ivy's statement speaks to the education of decision makers. Harper's statements at the NEA meeting in 1999 emphasized that mainstream promotion about universal design is the way to reach the broadest client population. It is noteworthy that many of the industry leaders who have incorporated universal design into their work have had some personal experience that motivated their innovations.

- Architects and other designers do have responsibility, to educate themselves and their clients, bringing their imagination, sensibilities, and sensitivities to each project. The challenge in large public buildings is that the client of record is rarely the user of the building. It is essential that the design programming stage find ways to bring users into the process, and use them at key junctures. The design and construction of large buildings involves many firms, many decision makers. Without some organized way of getting and documenting user input, the details that ensure usability will not be addressed.

  The role of users in the teaching, learning, and practice of universal design was explored and documented in the Universal Design Education Project (UDEP) and acclaimed as the essential component that changed students' attitudes about how and for whom they designed (Welch, 1995; Ostroff, 1997).

  Education is essential, but the process to create a significant impact in design curricula needs more attention and involvement beyond the "usual suspects" who have been involved to date. The 1982 meeting of design practitioners and educators illuminated some issues that have still not been addressed (Ostroff and Iacofano, 1982). Robert Shibley, speaking to the group of faculty involved in the UDEP in 1994 reflected on the problems of discourse, and the need to avoid righteousness in communication with colleagues. He optimistically called universal design a "border pedagogy" that has the potential to address both modernist and postmodernist concerns.

- Donald Norman, in the *New York Times* (2001), reminds everyone about the centrality of usability in the designed environment. His partner, Dr. Nielsen, said, "Humans are an incredibly error-prone species. It's very hard to change human nature. It's really easy to change design, if you bother doing so."

43.8 NOTE

1. Adapted from E. Ostroff, *Global Universal Design Timeline,* 1998.

43.9 BIBLIOGRAPHY

Access Board, Reports, 2000; www.access-board.gov/index.htm.

AIA and Access Board, AIA / Access Board Accessibility Guidelines Education Course, www.access-board.gov/news/AB-AIAcourse.htm.

Azine, S., Phone interview with author, December 2000.

Bonny, S., "Update" in 504 Sit-In 20th Anniversary Celebration and Commemoration, 504 Sit-In 20th Anniversary Committee, DREDF, Berkeley, CA, 1977.

Breslin, M. L., "From Section 504 to the Americans with Disabilities Act," in 504 Sit-In 20th Anniversary Celebration and Commemoration, 504 Sit-In 20th Anniversary Committee, DREDF, Berkeley, CA, 1977.

Campbell, R., *Boston Globe,* 29 November 1998.

Center for Universal Design, *Exemplars of Universal Design,* Center for Universal Design, North Carolina State University, Raleigh, NC, 2000.

Chang, K., "From Ballots to Cockpits, Questions of Design," www.nytimes.com/2001/01/23/science/23USEA.html.

Congress for the New Urbanism, www.cnu.org/otherresources.html.

Consortium for Citizens with Disabilities, "Opening Doors," June 2000; www.c-c-d.org/od-junr00.htm.

Coppelman, P., "A Layperson's Guide to Section 504" (reprinted from *The Independent,* Summer 1977; www.dredf.org/504guide.html.

Cozzens, L., *"Brown v. Board of Education," African American History,* 25 May 1998; http://fledge.watson.org/~lisa/blackhistory/early-civilrights/brown.html.

Department of Justice, "Enforcing the ADA: Looking Back on a Decade of Progress," Department of Justice, Washington, DC, 2000.

Dietrich, E., Phone interview with author, September 2000.

Fetterman, J., Phone interview with author, December 2000.

Garfield, D., *Report of Universal Design Meeting,* National Endowment for the Arts, June 1999; www.arts.gov/explore/ud/contents.html.

Gattis, R., *Architects' Use of the DBTACs in 1999, 2000, ADA Impact Measurement Project,* (unpublished preliminary report), Rocky Mountain Disability and Business Technical Assistance Center, Colorado Springs, CO, 2001.

GE Appliances, *Real Life Design,* GE Appliances, Louisville, KY, 1995.

Harper, L., Phone interview with author, December 2000.

King, M. L. Jr., *Address at March on Washington for Jobs and Freedom,* Martin Luther King, Jr. Papers Project at Stanford University, Stanford, CA, 1963; www.stanford.edu/group/King/.

Lasoff, S., E-mail interview with author, September 2000.

Lusher, R. H., "Handicapped Access Laws and Codes," *Encyclopedia of Architecture,* John Wiley & Sons, New York, 1989.

McGovern, J., "Visitability Improves Accessibility for All," *Hope VI Newsletter,* no. 42 (May/June 2000).

Mcleister, D., "An Open Door for Universal Design" in *Professional Builder,* Cahners Business Information, Des Plaines, IL, March 1999.

Mueller, J., "Universal Design: Growing Up Without Growing Old," *Innovation, Quarterly Publication of the Universal Designers Society of America,* Winter 2000.

National Association of Home Builders Research Center, www.nahbrc.org/.

National Council on Disability, "Promises to Keep: A Decade of Federal Enforcement of the Americans with Disabilities Act," 27 June 2000; www.ncd.gov/newsroom/publications/promises_1.html#3.

O'Brien, R., Phone interview with author, September 2000.

Ostroff, E., "Mining Our Natural Resources: The User as Expert," *Innovation, the Quarterly Journal of the Industrial Designers Society of America,* 1(1), 1997.

Ostroff, E., and D. Iacofano, *Teaching Design for All People: The State of the Art,* Adaptive Environments Center, Boston, 1982.

Palames, C., "Historical Context" in *ADA Core Curriculum,* Adaptive Environments Center, Boston, 1995.

Palermo, L., Phone interview with author, September 2000.

Palmer, W., Phone interview with author, September 2000.

Peterson, W., "Public Policy Affecting Universal Design," *Assistive Technology,* 10(1), 1998.

PLAE, Inc., and USDA Forest Service, *Universal Access to Outdoor Recreation,* MIG Communications, Berkeley, CA, 1993.

Robertson, A. (for the Colorado Cross Disability Coalition), in F. Fay (ed.), *Justice for All,* 27 February 2001; www.jfanow.org.

Robinson, D., Written interview with author, September 2000.

Smith, E., Phone interview with author, January 2001.

Steven Winter Associates, Inc., Tourbier & Walmsley, Inc., Edward Steinfeld, and Building Technology, Inc., *Cost of Accessible Housing: An Analysis of the Estimated Cost of Compliance with the Fair Housing Accessibility Guidelines and ANSI A117.1,* U.S. Department of HUD, Washington, DC, 1993.

Universal Designers and Consultants, *Images of Universal Design Excellence,* Universal Designers and Consultants, Takoma Park, MD, 1996.

Vanderheiden, G., and J. Tobias, "Universal Design of Consumer Products: Current Industry Practice and Perceptions," in *Proceedings of the XIVth Triennial Congress of the International Ergonomics Association and 44th Annual Meeting of the Human Factors and Ergonomics Society,* 6, 19–22, Human Factors and Ergonomics Society, Santa Monica, CA, 2000.

Welch, P. (ed.), *Strategies for Teaching Universal Design,* Adaptive Environments Center, Boston, and MIG Communications, Berkeley, CA, 1995.

CHAPTER 44

PROJECT BRIEFING FOR AN INCLUSIVE UNIVERSAL DESIGN PROCESS

Rachael Luck, B.A.(Hons.), Dip.Arch., M.Sc., A.R.B.
Hans Haenlein, M.B.E., Dip.Arch., RIBA, F.R.S.A.
and Keith T. Bright, M.Sc., F.R.I.C.S., F.B.Eng., M.A.S.I.
University of Reading, Reading, United Kingdom

44.1 INTRODUCTION

This chapter describes a process being used at the University of Reading to design an inclusive building. The case for the introduction of a universal approach to project briefing is presented, first describing different methods for project briefing, then a discussion of why more user involvement is necessary. Based on these findings, the universal approach is presented. The building users are an integral part of the briefing process; voicing their needs and being part of the project organization's decision-making process, they filter ideas as the design is refined.

44.2 BACKGROUND

The change in language associated with universal design has been noticeable, and it not only reflects semantic preferences but also different meaning and responsibility. The shift in terminology from *barrier-free, design for all, design for disabilities,* and *accessible design* to *universal design* parallels the change from *the handicapped* to *people with disabilities.* This has been well documented and summarized by the Center for Universal Design, and others.

This semantic issue, discussed elsewhere within this book, is raised within this chapter because of the specific use of the term *inclusive environments.* The work of the Research Group for Inclusive Environments at the University of Reading, United Kingdom, centers on the belief that if a building cannot be satisfactorily used by all, the fault lies with the building, not the user. An inclusive environment is one that meets the needs and requirements of all users, as well as addressing the problems experienced by people who have reduced physical, sensory, or cognitive ability. This parallel's Jim Sandhu's definition that universal design extends beyond the issues of accessibility for people with disabilities to respond to the broad diversity of users who have to react with the built environment. The politics of inclusion are an integral part of universal and inclusive design, and the terms are used interchangeably within this chapter.

In common with Susan Goltsman's work in Chap. 64, "The Ed Roberts Campus: Building a Dream," this chapter describes the design of an inclusive building. The Ed Roberts Campus project is much further developed than the building at Reading, and is able to report on the site and physical aspects of the building. This chapter focuses on the process used to capture project ideas and to structure these so they can be incorporated within the project brief, without constraining or losing the meaning intended by the user representative.

44.3   DIFFERENT APPROACHES TO PROJECT BRIEFING

It is difficult to generalize about the briefing process; the methods used will vary from project to project, and people have preferences and preferred modes of working. Although briefing is not a homogenous, unified activity throughout the construction industry, recent developments in briefing research have highlighted some deficiencies that have led to the development of a more inclusive approach.

The *brief* is the term used in the United Kingdom to describe one of the legal documents that encapsulates the client's requirements for a building. The RIBA *Plan of Work* (RIBA, 1967) linked the client's brief to be developed between stages A (inception) and D (sketch design). Later editions of the *Appointment of an Architect* (RIBA, 1992, 1999) and *The Process Protocol* (Kagioglou et al., 1999) research at the University of Salford proved what many in the industry were already conscious of: The sequential staged model used to plan projects did not describe what happened in reality. The process model shows stages of work that are in progress concurrently, and it shows that briefing is not a distinct project phase that starts at one time and is completed before design commences. If the needs of the commissioning and user client could be established at a distinct phase early in the project's progress, this would increase the certainty of the project and limit the number of design changes in the procurement process. Reality, unfortunately, is different from this. Acknowledging that briefing is a process introduces some flexibility in the timing of design decisions—some can be made later without delaying the progress of the project. In this way, the rich information that users suggest, as their understanding of the building becomes more refined, can be included with a positive effect on their satisfaction with the building.

The information needed to develop a project brief was another area identified for attention. There is a need to include both overt and tacit information (Cooper and Press, 1995). This is information that not only describes the functional requirements for the building, but also background information to better meet the needs of those who will eventually occupy the building. The importance of "hard-" and "soft-"people-focused information was mentioned in Barrett's work (Barrett, 1991). His inclusion of the Johri window as a method for revealing unknown information through discussion follows the same line of thought as the approach proposed at Reading. These methods support Mintzberg and Water's (1985) management studies, which show that discussion can reveal the unknown and also enhance better understanding among those involved (Mintzberg and Water, 1985). The message coming through was that understanding the human dimension was an integral part of the briefing process, equally important to knowing the functional requirements.

To bring the universal approach to briefing in line with current thinking, the outmoded views of the brief as a document and of briefing as a fixed stage of the project were addressed. The issue of the type and quality of information gathered during the briefing process introduces the universal component to the briefing process. This provoked thought about appropriate user involvement.

44.4   WHY IS MORE USER INVOLVEMENT NECESSARY?

Greater user involvement in the design and briefing processes of buildings was considered necessary because of critical attention since the 1980s, which showed that the lack of inclusion of the user perspective in the design process was considered an oversight. The lack of user participation and designer accountability in the design process were considered to be the two main reasons for inadequacies in design decisions. It was thought that the inclusion of the end user's views in the design and briefing processes would result in greater user satisfaction with the building. This inclusive approach also acknowledges that giving users a voice reflects an organization's consideration for its employees' working conditions. This has been explained in terms of power within organizations, where greater participation in design is linked to the empowerment of those whose views are sought.

There was also a movement toward more user-centered methods for product design and development in the 1980s. The term *user-centered* came to mean a number of different things: *design for users,* where analysts study user behavior to design a system to match user characteristics, and *designed by users,* where users become involved in the design exercise. The movement toward actual user involvement whenever possible was popular as many were critical of empirical anthropometric models, finding the data unreliable. This is especially true when designing for people with disabilities, because generalized anthropometric data is rarely useful.

A participatory building design and management approach was developed to create good relationships among a project team. This approach was concerned with enabling people to complete a project together, rather than the power of one party over another. The approach was of particular interest to the design of this building as it is concerned with the representation of views from different parties and gaining experiential knowledge through dialogue. It is considered that knowledge is socially constructed through dialogue. Knowledge is generated and confirmed in a dialectic between concrete and abstract knowing. This is a key observation, which coincides with the approach of this project, that knowledge and new ideas are generated through the process of discussion.

The movement toward greater user involvement has also been driven by changes in societal values and the politics of inclusion. Linked with a need for user participation in the design process is the issue of hegemony and representation of disabled building users. Hegemony concerns the location of power and decision making by a majority on behalf of others. There is a feeling among people with disabilities that many decisions that have a strong influence on their lives are made by others on their behalf. This includes the design of environments, even those with specific access concerns, where design decisions are made with little or no consultation with people with disabilities. Much has been written about design for disabilities, which is uncomplimentary about the process. A study of architects' understanding of building needs of people who are disabled, based on their relationships with building users, found that many designers were assuming a knowledge and understanding of environmental issues for people who are disabled, which they do not have. This presumptive designing has led to dissatisfaction. Research by Mason has shown that when people with disabilities are able to choose for themselves, their decisions differ from those made on their behalf (Mason, 1992). This reflects that their interests have not accurately been represented in the past.

The issue of user participation and representation of views is particularly sensitive when designing for people with disabilities, because activity in the field of disability theory and the politics of disability has increased since the 1970s. There has been a rejection of the hegemonic labeling of a person as "handicapped" or "disabled" among disability theorists, and an introduction of disability as a social construct. Disability studies look at impairment from a sociological perspective, where a social model of disability considers that societal values determine what is considered abnormal, and this defines normality.

The politics of disability and the social model have brought about a movement among people who are disabled for greater involvement and empowerment rather than representation of their needs and values. Although there are mechanisms for the inclusion of various disability perspectives to influence the design of environments (e.g., access committees to influence local authority access policies), there is criticism of the effectiveness of these methods. It has been observed that the advice given to a local authority by an access committee can be ignored, but the fact that an access committee exists meets the authority's consultative duties. This illustrates the main thrust of this chapter that, although design exercises may involve a user consultation process, ultimately there is an autocratic decision made on behalf of people who are disabled.

Taking a social model of disability perspective, to address this imbalance, people with disabilities should be emancipated and their views should be considered from an empowered, rather than a consultative, position. This would be possible if people who are disabled were designing environments for themselves, but few architects and planners are disabled. Even this position would be flawed if a person with one disability had to presume the needs of a person with a different disability, and the approach assumes that the views of one individual will represent those of a larger population. The nature of design makes an emancipatory approach difficult. Design is a specialist activity, with liabilities, knowledge, and expertise that cannot arbitrarily be passed on to someone else.

44.5   A UNIVERSAL APPROACH TO PROJECT BRIEFING

The approach to project briefing presently being used to design a universal building attempts to address some of the drawbacks of previous approaches to project briefing and to impartially include end users' views in the briefing process. This approach uses three mechanisms to collect data and knowledge from the client-user about the future building: (1) semistructured interviews, (2) an information base for the feasibility study, and (3) steering group meetings.

These mechanisms formalize the social processes within the project to develop the brief. The mechanisms are discreet events to gather information about building users' preferences and to elicit the commissioning client's requirements for the building. The approach is systematic without being a checklist. There is a process for the collection of this information, but each person presents their information and their requirements in their own terms.

Semistructured Interviews

The method used to gather information for the feasibility study was to individually interview people to better understand their experience of working and communicating in buildings. An interview approach was preferred because this allowed people to express their perceptions of the built environment in their own terms. This will generate a theoretical model of the user based on experience and scenarios of use.

The interviews were concerned with the ease with which people use the built environment and encouraged people to illustrate their responses with examples of buildings or environments and to describe these settings. The semistructured approach used a series of headings on an *aide-mémoire* (Table 44.1) to steer the discussion and to better understand the experience of people working and communicating in different environments.

The aide-mémoire prompts were developed through experience in architectural practice, developing an understanding of the information needed to design different types of projects. The prompts relate to aspects of a potential building and were used to encourage the respondent to generate ideas for the building being designed. The headings also prompted comments on other buildings and the experience of other environments, as well as qualitative judgments on the relative merit of these environments. It was this rich information that provided invaluable data for the next stages of the briefing process.

The aide-mémoire prompts were trigger statements used by the person conducting the interview. The response to a prompt was the raw data for collection and analysis. The responses to

TABLE 44.1 Aide-Mémoire, in Part

Aide-mémoire (part)
Needs of users
Relationship of functions
Activities and needs of occupants/users (compare with current conditions)
Duration and frequencies of activities
Special provisions (e.g., for elderly people, people who are disabled, patients with strollers)
Special items to be accommodated [e.g., machines, animals, vehicles (give key dimensions, weights, etc.)]
Safety and health risks (e.g., noise, heat, chemicals, plant, etc.)
Flexibility
Future trends/changes, need for flexibility
Phasing, any occupancy retained during contract works
Area usage
Zoning, permanent and temporary
Conflicting activities
Communication routes, systems
Access and escape
Fire and security
Flexibility of use, partitioning, etc.
Rooms, workstations, work spaces
Staff priorities, confidentiality
Special occupational needs
Health and safety
Amenities
Internal environment
Energy policy
H&V, air conditioning
Special services
Noise and lighting criteria
Fenestration, building envelope, and structure

each heading were included in the notes taken by the interviewer, usually verbatim. The notes were very detailed, and the interview text was unmodified within the "information base" document produced in the next stage of the briefing process.

The interviews were conducted by the same project researcher, who used the same prompt sheet for each person interviewed. This allowed consistency across the sample of people interviewed, but did not impose a structure as to how an individual should respond. The responses were the unprompted, unbounded ideas of the person interviewed. This approach generated rich data of personal perceptions of their experience of buildings and suggestions for improvements to the built environment. Interviewing people individually had the advantage that their ideas were personal and not affected by group pressures and influences.

A key concern was selecting who should be interviewed. The research team felt that it was important to consult a range of people with different disabilities, not only to satisfy sociopolitical pressures of representation but also because of the universal ethos of the project, that the built environment should be accessible for all. The individuals selected were people who would occupy the building, when complete. These people had a range of different disabilities and experience of the built environment and were interviewed so they could have direct influence on the design of the project. This group included people with visual impairments, wheelchair users, and people who were profoundly deaf and hard of hearing. This method of inclusion of views directly empowered those involved, as their unsolicited opinions were included in the exercise. The decision of whether to include an idea within the scheme was subject to the filtering processes of the steering group at a later stage rather than the decision of a single person.

One of the interviews was conducted with an interpreter using British sign language to enable the discussion between the researcher and a deaf person. This changed the dynamic of the interview and introduced another level of interpretation. It became more of a three-way discussion of ideas rather than a prompt-and-response conversation. Although the words and terms referred to in the checklist were easy to understand, they were all relevant to construction and their use needed to be clarified to the interpreter so they were not misrepresented to the person being interviewed. The response to the prompt was also influenced by the person's knowledge of construction and a keenness to respond in an informed manner. A key factor to get across to the respondent was that there was no correct response. Each response was an expression of his or her personal wish list for the building based on his or her own experience. In this way, the response was individual and was not assumed to represent a disability. The wealth of information gathered when communicating with a person with a different experience of the built environment is captured in the written notes from the interview.

The criticism of interviews that a person may attach more than a single meaning to his or her experience was addressed within this process by grouping a person's response, which may be applicable to more than one heading of the aide-mémoire, within several headings. This meant that the textual transcript of the interview included some repetition, but that the response was included with the prompt that triggered that response, as well as the heading where comments on a related subject from other people interviewed would be included. The interview process encouraged the interviewee to discuss around the prompt, so this type of response was acceptable and easily accommodated within the data-gathering process.

Developing an Information Base

Developing an information base for the feasibility study was the term given to the second stage for gathering user and commissioning client expectations from the built environment. The *information base* is the preliminary briefing document for the project developed by combining each of the notes from the semistructured interviews into a single document. Each interview was written as a separate document, then a combined document was generated, structured around the aide-mémoire headings. Each person's response to an item was presented in a section so it was possible to read the different responses together. This document was an unedited version of the views represented by those interviewed and contained some conflicting items. This approach was the cumulative representation of the responses without interpretation or judgment on the content of the response. Developing the information base is a consolidation process allowing all views to be equally represented within a single document.

The interview text was unmodified during the information base stage of the process. The response to a prompt was treated as an individual's view or experience that would unconditionally be included in the information base document. It was accepted that the information base would include personal judgments that may not be representative of a wider population, even for people with a similar disability. This stage of the process, allowing conflicting and unreasonable views to be expressed, was a method that allowed "unknown knowledge," at the project outset, to be found out during the interview process. Filtering opinions and addressing any conflict of need or preference would be considered at the steering group stage.

The Steering Group

The steering group is a management device within the project organization introduced to formalize decision making and to ensure the representation of different user groups. The steering group is a useful mechanism for tradeoff and negotiation of problems. It is a forum for problems to be raised and resolved. The intention was for the steering group to be at regular

2-month intervals; in practice, however, the meetings were less frequent and were called when there was sufficient new user requirements to discuss, approximately every 4 months.

A key to a successful tradeoff is the composition of the steering group and the understanding that each member of the group has of his or her role within the group. A consideration was that the representatives on the group were of sufficient status, part of the decision-making unit, so views could be expressed without reference back to others. Each person on the steering group totally represents an organization, but acts in a personal capacity. In this way, each person is encouraged to have his or her own head and generate ideas. If there are different views among members of the steering group, in order to resolve disputes and move forward and make a decision, the members are reminded of their role as a representative of their own organization and the wider objectives of the project. This shift in perspective and reminder of the broader responsibilities is a useful mechanism of the steering group structure for group problem solving. The steering group should include representation of the building user group as well as the commissioning client. For this project, this includes the involvement and representation of various disability groups. The minutes of the steering group meeting represent a paper trail recording the decisions made on the project.

The ability to generalize the findings from one person to a larger population was a representation issue to consider. The concern was that the needs and views expressed by one person with one type of disability may not represent the needs and views of others with the same disability. It also applies that one person who is disabled may not be able to represent the views of a person with a different disability. Although the time constraints of the project meant that a small range of commonly occurring disabilities were represented during the structured interviews, these were considered adequate because of the structure of the interview and the way the data were interpreted.

The steering group resolved any differences of opinion within the information base and feasibility study through discussion. Careful examination of the language used and the emergence of ideas were the main thrust of this way of working. It was recognized that metaphors illustrate views and subtly encapsulate and communicate complex ideas. The synergy of the group was a good platform for generating and developing ideas. An advantage of this method is that ideas are discussed and more fully understood by all concerned, which may lead to a more valid and thoroughly negotiated result. The process is transactional and relies on the shared understanding of language to quickly arrive at a valid result.

44.6   GENERAL LESSONS FOR UNIVERSAL DESIGN

Applying the inclusive briefing approach to this project was a useful exercise that demonstrated how complex the project organization can become. It becomes complex in terms of the number of people that are involved in the briefing process and, as a consequence of this, there is an increased number of differing views and perspectives expressed.

Transcribing the user discussions into text documents using the aide-mémoire headings meant that much paperwork was generated. The information for the feasibility study produced a lengthy document combining the views of everyone interviewed. This took over a month to produce, because of people's availability to be interviewed and editing the interview notes into a single document—the information base. It took the steering group 1 to 2 weeks to read and digest the lengthy document before they could comment on any conflicting items suggested by the users. Developing a method to quicken this process would improve the ease of using this approach.

This exercise demonstrated that although many user representatives were interviewed, and this increased the number of suggestions for the project, in this instance only one item was in potential conflict and needed further clarification. This may not always be the case, and using the steering group, drawing upon the members' role as a representative of an organization, may be necessary.

An interesting issue to consider is defining the boundaries of the project team. The issue of commissioning client and user-client discrepancy is something that has been raised within briefing literature and is also an issue for this approach to project briefing. The issues of hierarchy that may already exist within an organization are transferred to the project and become a project issue. Should there be any difference in the way the views, wants, and needs of the commissioning client, who may also be an end user of the building, are handled from the other users of the building? For the process to be truly universal, there should be no hierarchy reflected within the process, the ethos is equal representation and equitable use. In practice, the commissioning client's requirements control the budget for the project. How can this ultimate command control position be avoided in project briefing?

A concern for designers may be that although ideas have been "universally" suggested, designers will assume design responsibility for these suggestions. If a designer refines an idea to make the design work in practice, has the designer breached the ethos of inclusive design? At what point does design modification become presumptive designing?

Using the inclusive briefing approach raised many issues for further investigation; many are common to the universal design of the built environment:

- Who should be interviewed, whose views should be gathered?
- Who should conduct the user consultation exercise?
- Who should act as the intermediary, the interface between the construction team and the building users?
- What is the legal position of user representative information?
- How should this information be incorporated within the design of the scheme?
- Should there be a cutoff point for the inclusion of ideas?

44.7 CONCLUSION

The approach described here was developed to address some of the user representation imbalances discussed. Current briefing methods do not per se cause an imbalance, but the manner in which user information is gathered and decisions are made within the briefing process can influence whether these views are represented. This was a prototype, with the potential for further development as knowledge and experience are gained. The method is a series of decision-making mechanisms and social processes that regulate the capture of ideas and filter these among the project team. Taking a consultative approach was the only reasonable game plan to arrive at common understanding among the project team.

44.8 BIBLIOGRAPHY

Barrett, P., "The Client's Brief: A Holistic View," *Management, Quality and Economics in Building,* E&FN Spon., 1991.

——, and C. Stanley, *Better Construction Briefing,* Blackwell Science, Oxford, UK, 1999.

Bejder, E., "From Client's Brief to End Use: The Pursuit of Quality," in P. S. Barrett and A. R. Males (eds.), *Practice Management: New Perspectives for the Construction Professional,* E&FN Spon, London, UK, 1991.

Cooper, R., and M. Press, *The Design Agenda—A Guide to the Successful Management of Design,* John Wiley & Sons, Chichester, UK, 1995.

Kagioglou, M., R. Cooper, G. Aouad, and J. Hinks, "The Process Protocol: Improving the Front End of the Design and Construction Process for the UK Industry," *Journal of Construction Procurement,* 5(2), 1999.

Mace, R., *Housing Definitions: Accessible, Adaptable and Universal Design,* Fact Sheet No. 6 OPFS.4.91, The Center for Universal Design, North Carolina State University, Raleigh, NC, 1990.

Mason, P., "The Representation of Disabled People: A Hampshire Centre for Independent Living Discussion Paper," *Disabled, Handicap & Society,* 7(1), 1992.

Mintzberg, H., and J. A. Water, "Of Strategies Deliberate and Emergent," *Strategic Management Journal,* Vol. 6, 1985.

Royal Institute of British Architects (RIBA), *Plan of Work,* RIBA, London, UK, 1967.

———, *Appointment of an Architect SFA./92,* RIBA, London, UK, 1992.

———, *Appointment of an Architect SFA./99,* RIBA, London, UK, 1999.

44.9 RESOURCES

The Web site of the Research Group for Inclusive Environments is part of the University of Reading's Department of Construction Management & Engineering site at www.construct.rdg.ac.uk/.

CHAPTER 45

OFFICE AND WORKPLACE DESIGN

James L. Mueller, M.A.
J. L. Mueller, Inc., Chantilly, Virginia

45.1   INTRODUCTION

The unemployment rate among people with disabilities has remained at an appalling 67 percent for many years, despite passage of the Americans with Disabilities Act (ADA) of 1990, which assured the rights of individuals with disabilities in the workplace and in the community. The Ticket to Work and Self-Sufficiency Act was passed in 1999 to remove financial disincentives to work for people with disabilities. As the workforce ages and the cost of work disability rises, demographic and economic trends have combined with legislation regarding employment of people with disabilities to make universal design in the workplace a powerful issue. Workplace design that considers age-related changes in vision, hearing, posture, and mobility will be critical to an aging workforce expected to work even further into their senior years than previous generations (Anders et al., 1999).

This chapter will discuss how these trends impact designers and manufacturers of furniture, equipment, materials, and other workplace products. This chapter will also present specific examples of how designers, employers, and manufacturers have responded by implementing the concept of universal design.

45.2   ECONOMIC BACKGROUND

"What do you do for a living?" This is one of the first questions asked when people meet. Especially in the United States, one's personal identity depends heavily on what he or she does for a living. But approximately two-thirds of Americans with disabilities are unemployed. This is an enormous burden on them and on their families. Hundreds of thousands of employees become disabled each year and leave the workplace permanently. Their former employers must bear the burden of replacing them as well as paying disability benefits, and taxpayers must help fund public benefit programs for them such as Social Security Disability Income (SSDI).

The SSDI program is the primary source of income for millions of Americans considered too disabled to work. Between 1985 and 1994, SSDI payments doubled from $19 billion to $38 billion (U.S. General Accounting Office, 1995). Realizing that continuing increases like this could destroy the Administration's budget, Congress passed the Ticket to Work and Self-Sufficiency Act of 1999 to provide greater vocational rehabilitation services and financial incentives to enable more Americans with disabilities to work (Social Security Administration, 2000). This law, combined with the ADA, makes job and workplace design that considers the needs of workers with disabilities more important than ever.

The workplace is the site of millions of injuries per year. The average permanently disabled employee costs his or her employer $154,400 in benefits, insurance costs, and lost productivity through age 65 (Farrell et al., 1989). But not all disabilities are caused at work.

Seventy percent of all people with disabilities are not born with them, but develop them during the course of their lives (Louis Harris and Associates, 1994). As more people live longer lives, the likelihood of experiencing a disability during one's lifetime increases.

Medical progress has had a profound effect on treatment of illness and accidents that a short time ago were fatal. More than 3 million Americans each year survive severe auto accidents, sports injuries, strokes, and heart attacks (Lowery, 1994). From 1970 to 1997, the survival rate from strokes more than doubled and the survival rate from traumatic brain injury improved from 10 percent to 90 percent (Jones and Sanford, 1996).

Historically, both government and business have been much more willing to pay cash benefits than to provide assistance to help disabled workers return to productive employment. For every $100 in cash benefits paid to disabled persons, the Social Security Administration spends only about a dime for rehabilitation services (U.S. General Accounting Office, 1995). Among private businesses, the total of insurance costs, replacement expenses, and workers' compensation and other disability benefit payments due to work disability reached $160 billion in 1992 and was expected to reach $200 billion per year by the turn of the century (Farrell et al., 1989).

Both the ADA and its predecessor, the Rehabilitation Act of 1973, prohibit employers from discriminating against individuals with disabilities who are qualified and able to perform the essential duties of an available job, with or without reasonable accommodation. Though these laws have boosted the employment rights of people with disabilities, they have had little effect on the level of unemployment among people with disabilities. But increasingly common occupational injuries such as repetitive upper-extremity stress and back pain and the steadily aging workforce assure that disability will continue to be a common concern among American workers and their employers.

Compared with the enormous cost of paying disabled employees not to work, making accommodations to bring them back to the job are cheap. According to the Job Accommodation Network, 71 percent of accommodations cost $500 or less. For every $1 spent on job accommodation, the employer gets back $26 in savings (Job Accommodation Network, 1999). Yet, about one-quarter of the ADA Title I complaints filed with the Equal Employment Opportunity Commission (EEOC) allege failure to provide reasonable accommodation.

45.3 JOB ACCOMMODATIONS FOR EVERYONE EQUAL UNIVERSAL DESIGN IN THE WORKPLACE

Job accommodations usually benefit coworkers without disabilities as well as the worker requesting accommodation. It is rare that on-site job accommodation needs analysis does not reveal risks of reinjury to the returning disabled worker that are also hazards to other employees. Accommodations developed with this in mind bring employers the double benefit of accommodating as well as preventing disability.

At the very least, job accommodations for workers with disabilities should be transparent, or have no effect at all on coworkers or customers. This is not as difficult as it may sound. For employers with little experience with disabilities, it can be very difficult to imagine how an employee with very different abilities from his or her coworkers might share similar needs. But the same barriers to productive and safe work faced by an employee with a significant disability are usually barriers to nondisabled coworkers as well, though to a lesser degree.

For example, an individual with limited manual strength and coordination was hired by a window manufacturer to insert weatherstripping into 12-foot sections of window frame channel. Previous workers had used a pair of pliers to tightly grasp the end of the weatherstripping in order to pull it the length of the channel. Even for workers with a very strong grip, this was a tiring job that often caused considerable hand pain by the end of a workday.

The individual with manual limitations was unable to exert adequate grip on the pliers to pull the weatherstripping through the channel without slipping. Instead, the author designed a simple tool, shown in Fig. 45.1, shaped to fit the channel, with a large hand loop and a toothed gripping surface for the weatherstripping. The worker was able to hold the tool without gripping tightly and was able to use his body weight on the tool to supply adequate pressure on the toothed gripping surface. The rest of the task simply required him to walk the length of the channel.

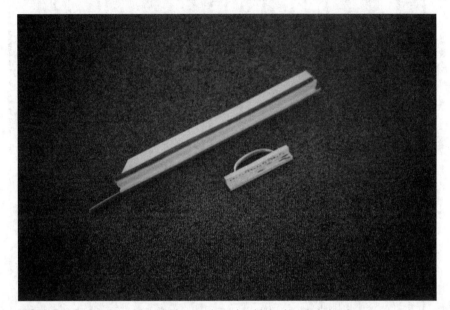

FIGURE 45.1   Simple tool for gripping weatherstrip with section of window frame.

Both his coworkers and his supervisor were surprised at the ease with which he was now able to perform a task that had been difficult for even the strongest employees. Not surprising, the supervisor suggested that all workers use this tool.

There are many examples like this of successful job accommodation benefiting all workers. Employers in these situations commonly ask, "Why didn't we do this in the first place?" With growing emphasis on reducing risk of cumulative and repetitive stress injuries, the supervisor in the aforementioned example might well have asked this very question. He realized the importance of the simple tool in preventing injuries to other employees, as well as in accommodating the worker with the disability. In situations like these, employees formerly seen as "different" due to their disabilities help to identify job and workplace design problems affecting all workers. Their ergonomic needs become effective templates for improvements in job and workplace design for all.

45.4   *WORKING TOWARD A UNIVERSAL WORKPLACE*

At the request of the U.S. Department of Defense, the author applied this principle to the development of a workbook for workers and supervisors to assess the fit between the employee and his or her workplace. This resource helped to identify workplace design factors that might be barriers to workers with disabilities, as well as risks to workers not yet experiencing a disability. The result was *The Workplace Ergonomics Workbook,* illustrated in Figs. 45.2 and 45.3.

Ergonomic Needs Assessment   Visual and Auditory Information

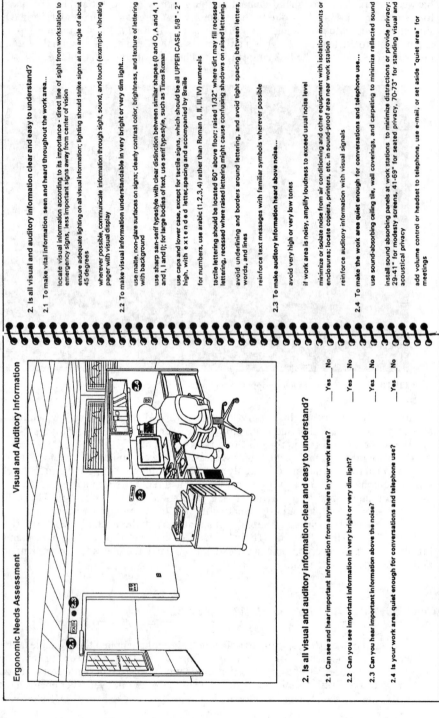

2. Is all visual and auditory information clear and easy to understand?

2.1 Can you see and hear important information from anywhere in your work area? _Yes _No

2.2 Can you see important information in very bright or very dim light? _Yes _No

2.3 Can you hear important information above the noise? _Yes _No

2.4 Is your work area quiet enough for conversations and telephone use? _Yes _No

2. Is all visual and auditory information clear and easy to understand?

2.1 To make vital information seen and heard throughout the work area...

locate visual information according to its importance - direct line of sight from workstation to emergency signs, less important signs away from center of vision

ensure adequate lighting on all visual information; lighting should strike signs at an angle of about 45 degrees

wherever possible, communicate information through sight, sound, and touch (example: vibrating pager with visual display

2.2 To make visual information understandable in very bright or very dim light...

use matte, non-glare surfaces on signs; clearly contrast color, brightness, and texture of lettering with background

use sharp san-serif typestyle with clear distinction between similar shapes (0 and O, A and 4, 1 and I, I and l); for large bodies of text, use serif typestyle, such as Times Roman

use caps and lower case, except for tactile signs, which should be all UPPER CASE, 5/8" - 2" high, with e x t e n d e d letter, spacing and accompanied by Braille

for numbers, use arabic (1,2,3,4) rather than Roman (I, II, III, IV) numerals

tactile lettering should be located 60" above floor; raised 1/32" where dirt may fill recessed lettering, recessed where raised lettering might cause confusing shadows on raised lettering.

avoid underlining and borders around lettering, and avoid tight spacing between letters, words, and lines

reinforce text messages with familiar symbols wherever possible

2.3 To make auditory information heard above noise...

avoid very high or very low tones

if work area is noisy, amplify loudness to exceed usual noise level

minimize or isolate noise from air conditioning and other equipment with isolation mounts or enclosures; locate copiers, printers, etc. in sound-proof area near work station

reinforce auditory information with visual signals

2.4 To make the work area quiet enough for conversations and telephone use...

use sound-absorbing ceiling tile, wall coverings, and carpeting to minimize reflected sound

install sound absorbing panels at work stations to minimize distractions or provide privacy: 29-41" for modesty screens, 41-69" for seated privacy, 70-73" for standing visual and acoustical privacy

add volume control or headset to telephone, use e-mail, or set aside "quiet area" for meetings

FIGURE 45.2 Illustration from *The Workplace Ergonomics Workbook*.

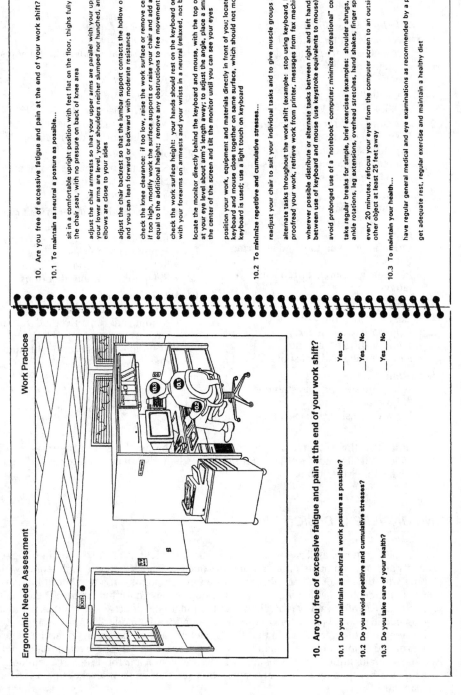

FIGURE 45.3 Illustration from *The Workplace Ergonomics Workbook.*

As shown, this workbook asks simple questions about common workplace features and provides direct answers, based on a broad compilation of ergonomic resources addressing the following:

1. *Workplace Accessibility*
   - Accessible routes of travel
   - Accessible doorways and door hardware
   - Changes in level
   - Flooring
   - Navigation
   - Emergency evacuation

2. *Visual and Auditory Information*
   - Positioning and lighting
   - Typeface selection
   - Ambient noise and auditory signals
   - Sound attenuation

3. *Lighting*
   - Illumination without glare
   - Materials and positioning to reduce glare
   - Orientation to natural and artificial light
   - Personal measures for reducing eye fatigue

4. *Storage*
   - Easy access at the workstation
   - Appropriate containers
   - Identification of materials
   - Safe handling and transport

5. *Seating*
   - Adequate support and stability
   - Adjustability features
   - How to adjust seat, back, and arm support

6. *Work Space Layout*
   - Adequate space
   - Work surface materials and adjustments
   - Air quality

7. *Computer Displays*
   - Monitor placement
   - Minimizing glare
   - Locating source documents

8. *Computer Inputs*
   - Keyboard and mouse positioning
   - Hand and wrist support
   - Software options

9. *Telephones and Other Office Equipment*
   - Electrical supply
   - Equipment controls
   - Telephone location
   - Telephone peripheral options

10. *Work Practices*
   - Maintaining a neutral posture
   - Minimizing repetitive and cumulative stresses
   - Maintaining general health and productivity

Significantly, this resource covers not just ways the employer can improve workplace ergonomics, but also how workers can minimize their risk of workplace-related injury by adopting healthy work habits.

45.5 ENABLING UNIVERSAL DESIGN

The Enabler system was developed in the 1970s (Steinfeld et al., 1979) to aid designers of products and environments in integrating the needs of elderly people and people with disabilities with the rest of the population. This approach points out a number of human functional characteristics that are important to consider in design for human use. The Enabler, as seen in Fig. 45.4, offers a way of dealing with the functional effects of disabilities without getting tangled in medical jargon or compromising confidential medical information.

It is important to keep in mind that the impact of each of these functional characteristics is determined as much by the demands of the environment as by the level of functional ability. For example, limitation of balance is far more significant for a high-rise building construction worker than for a data entry operator, even though their level of limitation may be very similar.

THE ENABLER

DIFFICULTY INTERPRETING INFORMATION   A

SEVERE LOSS OF SIGHT   B1

COMPLETE LOSS OF SIGHT   B2

SEVERE LOSS OF HEARING   C

PREVALENCE OF POOR BALANCE   D

INCOORDINATION   E

LIMITATIONS OF STAMINA   F

DIFFICULTY MOVING HEAD   G

DIFFICULTY REACHING WITH ARMS   H

DIFFICULTY IN HANDLING AND FINGERING   I

LOSS OF UPPER EXTREMITY SKILLS   J

DIFFICULTY BENDING, KNEELING, ETC.   K

RELIANCE ON WALKING AIDS   L

INABILITY TO USE LOWER EXTREMITIES   M

EXTREMES OF SIZE AND WEIGHT   N

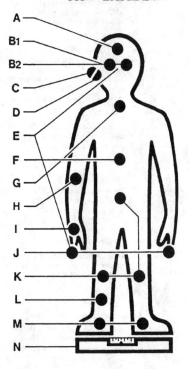

FIGURE 45.4   The "Enabler."

Since its development in the 1970s, the Enabler system has been widely adapted. This concept was adapted for use in *The Workplace Workbook 2.0* (Mueller, 1992), an illustrated guide to reasonable accommodation and assistive technology for employers. As seen in Fig. 45.5, the explanatory text clarifies each functional limitation.

This approach has also been used by office furniture system manufacturer Herman Miller, Inc., in helping their customers comply with the ADA and gain control of disability management. Herman Miller, Inc., based in Zeeland, Michigan, and second in size only to Steelcase among office furniture manufacturers, had a strong reputation among its customers and wanted to maintain it (Mueller, 1997).

In response to their customers' concerns, Herman Miller, Inc., produced a videotape and an illustrated guide to office planning and design entitled *Designing for Accessibility* (Herman Miller, Inc., 1995). These materials include recommendations for creating a workplace that was as usable for workers with disabilities as for those without disabilities. This workplace also includes flexibility for making specific accommodations for employees who need reasonable accommodation.

In the early 1990s, Herman Miller's customers felt the impact of the ADA in several ways. Public businesses and state and local government facilities were required to ensure that their facilities were accessible to people with disabilities. This meant that interiors, office systems, and furniture designed and supplied by Herman Miller to these customers had to comply with the accessibility guidelines of the new law.

Furthermore, the ADA required that employers make "reasonable accommodation" for employees with disabilities. This meant that a Herman Miller workstation might have to

Difficulty in Processing Information

This characteristic is defined as an impaired ability to receive, interpret, remember, or act on information. Among Americans experiencing this characteristic are approximately 614,000 workers ages 18-69 with learning disabilities, mental retardation, or senility.

Limitation of Sight

This characteristic is defined as a difficulty in reading newsprint-size copy, with or without corrective lenses, and extends to "legal blindness" (but not TOTAL blindness). Among Americans experiencing this characteristic are approximately 829,000 workers ages 18-69 with glaucoma, cataracts, or other eye disorders.

Total Blindness

Total blindness is the complete inability to receive visual signals. It is experienced by approximately 164,000 American workers ages 18-69.

Limitation of Hearing

Limitation of hearing is defined as a difficulty in understanding normal speech (but not TOTAL deafness). It is experienced by approximately 320,000 American workers ages 18-69.

Total Deafness

Total deafness is the complete inability to receive auditory signals. It is experienced by approximately 78,000 American workers ages 18-69.

Limitation of Stamina

Limitation of stamina is defined as fatigue, shortness of breath and/or abnormal elevation of blood pressure due to mild exercise or sensitivity to chemicals. Among Americans experiencing this characteristic are approximately 6,935,000 workers ages 18-69 with heart disease, emphysema, or other respiratory or circulatory conditions.

Difficulty in Lifting, Reaching, Carrying

This characteristic is defined as impaired mobility, range of motion, and/or strength of one's upper extremities. Among Americans experiencing this characteristic are approximately 9,522,000 workers ages 18-69 with arthritis, bursitis, tendonitis, loss/paralysis/deformity of extremities, back impairment, hernia, or quadriplegia, paraplegia, or hemiplegia.

Difficulty in Manipulating

Difficulty in manipulating means impaired hand or finger mobility, range of motion, and/or strength. Among Americans experiencing this characteristic are approximately 2,833,000 workers ages 18-69 with arthritis, carpal tunnel syndrome, cerebral palsy, or multiple sclerosis.

Inability to Use Upper Extremities

This characteristic is defined as complete paralysis, severe incoordination, or bilateral absence of upper extremities. Though not specifically itemized in the National Health Interview Survey data, this characteristic is experienced by Americans ages 18-69 with severe cases of conditions such as multiple sclerosis, spinal cord injury, or cerebral palsy, as well as by those without arms as a result of amputation or congenital loss.

Limitation of Speech

This characteristic is defined as a capability of only slow or indistinct speech, or non-verbal communication. Among Americans experiencing this characteristic are approximately 280,000 workers ages 18-69 with cerebral palsy, a distinct speech impairment, or total deafness.

Susceptibility to Fainting, Dizziness, Seizures

This characteristic may be spontaneous or inducible by environmental factors such as sudden sounds or flashing lights, resulting in loss of consciousness, balance, or voluntary muscle control. Among Americans experiencing this characteristic are approximately 2,094,000 workers ages 18-69 with epilepsy, diabetes, or cerebrovascular disease.

Incoordination

Incoordination is defined as limited control in placing or directing extremities, including spasticity. Among Americans experiencing this characteristic are approximately 442,000 workers ages 18-69 with multiple sclerosis, cerebral palsy, Parkinson's Disease, quadriplegia, paraplegia, or hemiplegia.

Limitation of Head Movement

This characteristic is defined as a difficulty in looking up, down, and/or to the side. Among Americans experiencing this characteristic are approximately 1,732,000 workers ages 18-69 with curvature of the spine or intervertebral disc disorders.

Limitation of Sensation

Limitation of sensation means an impaired ability to detect heat, pain, and/or pressure. Among Americans experiencing this characteristic are approximately 1,789,000 workers ages 18-69 with diabetes, multiple sclerosis, or full or partial paralysis.

Difficulty in Sitting

Difficulty in sitting is defined as excessive pain, limited strength, range of motion, and/or control in turning, bending, or balance while seated. Among Americans experiencing this characteristic are approximately 4,367,000 workers ages 18-69 with curvature of the spine, deformity or impairment of the back, intervertebral disc disorders, complete or partial paralysis, or quadriplegia, paraplegia, or hemiplegia.

Difficulty in Using Lower Extremities

This characteristic is defined as slowness of gait, difficulty in kneeling, sitting down, rising, standing, walking, and/or climbing stairs or ladders. Among Americans experiencing this characteristic are approximately 1,915,000 workers ages 18-69 with cerebral palsy, multiple sclerosis, deformity/absence/impairment of one or both lower extremities, or quadriplegia, paraplegia, or hemiplegia.

Limitation of Balance

Limitation of balance means a difficulty in maintaining balance while standing or moving. Among Americans experiencing this characteristic are approximately 939,000 workers ages 18-69 with cerebral palsy, cerebrovascular disease, complete or partial paralysis, or Parkinson's Disease.

* LaPlante, M.P. (1988). Data on Disability from the National Health Interview Survey, 1983-85. Washington, D.C.: National Institute on Disability and Rehabilitation Research. Data for 1990 are unpublished tabulations provided by the Disability Statistics Program, Institute for Health & Aging, School of Nursing, University of California, San Francisco.

FIGURE 45.5   Illustration from *The Workplace Workbook 2.0.*

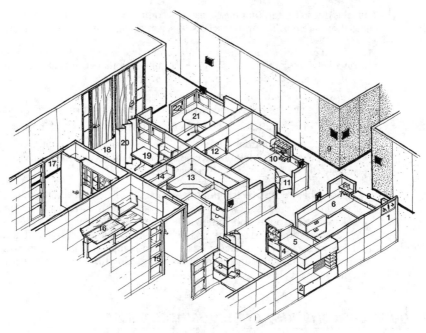

FIGURE 45.6   Illustration from *Designing for Accessibility.*

accommodate an employee who could be blind, deaf, a wheelchair user, or limited in a variety of other ways.

At the same time, a recession was causing American businesses to postpone investments in new facilities and equipment. Seeking to make the most of their investments, Herman Miller's customers began to preface their contacts with sales reps with "How do your products comply with the ADA?" Since the ADA does not include standards for products like office furniture, Herman Miller products by themselves could not comply with this law. Instead, the law was written for compliance by organizations. Herman Miller set out to address their customers' question by developing a program to communicate its philosophy of complying with the ADA through workplace design that eliminates barriers that all workers face.

For example, a worker using a wheelchair might have difficulty retrieving a thick file folder from the top drawer of a four-drawer file cabinet. But a coworker with short stature might have similar difficulties, as would a worker with wrist fatigue after long hours of keyboard work. The philosophy also included the capability to make "reasonable accommodation" for disabled employees without undue effort or expense.

The company also developed an *Applications Guide* to help customers understand the requirements of the ADA, the incentives for returning employees with disabilities to the job, and the features of office furniture products that made reasonable accommodation possible. The drawing shown in Fig. 45.6 is keyed to a list of features. Both the video and the *Applications Guide* were made available on request through Herman Miller's sales network.

45.6 *EFFECTIVE JOB ACCOMMODATION*

Through Herman Miller's efforts, customers learned the best approach to effective job accommodation. When approaches such as that suggested by Herman Miller are applied to the overall design of workplaces, most job accommodations are very simple and involve minimal cost.

But this doesn't mean that inexpensive accommodations are reasonable and expensive accommodations are not. The most successful accommodations are those that are:

1. *Effective.*   The solution enables the individual with the disability to do his or her job productively and safely. An effective accommodation does not substitute for the individual but enables the individual to use his or her own abilities.

2. *Transparent.*   The solution either has no effect on coworkers, customers, and other aspects of the business, or it has a positive effect in improving productivity and/or safety.

3. *Timely.*   The solution can be implemented within a reasonable time frame.

4. *Durable.*   The solution is useful and flexible enough to remain effective throughout the employee's service. Maintenance, as well as modifications necessary due to business or technology changes, can be readily accomplished.

Reasonable workplace accommodations are likely to be a compromise among these criteria. For example, it may be less expensive for a business to relocate an employee who uses a wheelchair to a ground-floor office than to invest in an elevator to the usual workplace. Or it may be less disruptive to coworkers to invest in a document scanner than to restructure jobs so that a coworker can read documents to a blind employee. Each employer must select from a number of accommodation alternatives that solution that best suits the needs of the individual and the business.

45.7   *GOOD DESIGN IS JUST GOOD BUSINESS SENSE*

Thorough analysis of Herman Miller products and their usefulness to workers with disabilities was needed for production of their *Application Guide.* This effort revealed that Herman Miller's traditional strong attention to established ergonomic principles was a good beginning. The flexibility of their products in meeting a variety of needs also helped customers to meet unique ergonomic needs of workers with disabilities.

Significantly, Herman Miller's competitors also realized the importance of providing guidance to their customers regarding workers with disabilities. Haworth, Inc., developed an *ADA Handbook,* which described the requirements of the law and, like Herman Miller, illustrated ways in which Haworth products could be used effectively to comply with these requirements. The Knoll Group produced a guidebook entitled *Workplace Issues: Universal Design and the ADA.* Steelcase, in its award-winning design for a self-contained workspace called "Personal Harbor," incorporated accessibility guidelines into the parameters of the project.

45.8   *LESSONS LEARNED FOR UNIVERSAL DESIGN*

Given the ever-increasing rate of technological advances, it seems that just about whatever can be imagined may very well be technologically possible. Through the efforts of citizens with disabilities, Congress, and rehabilitation technologists, major manufacturers of telecommunications and computer hardware and developers of software seem to have gotten the message that people with disabilities are a major portion of the workforce and are likely to remain so. This awareness must be constantly reinforced and expanded to other fields of workplace access and technology, including landscape and building architects, facility managers, interior designers, product designers, and graphic designers.

It has become clear that "disabled" no longer means "unable to work." This attitude has been rendered obsolete by law, as well as by population demographics and the economic realities of disability in business. Accommodation of workers with disabilities through job and workplace design is here to stay.

By instilling a universal design approach among those responsible for the development of work environments and products, the incidence of work disabilities can be reduced. And those accommodations that are required for workers with disabilities will be much more likely to be reasonable accommodations.

45.9 BIBLIOGRAPHY

Andres, R., M. Redfern, and S. Wilker, "Ergonomics and the Aging Worker," in *Compendium of the National Ergonomics Conference and Exposition,* pp. 86–96, Anaheim, CA, 7–9 December 1999.

Computer/Electronic Accommodations Program, *The Workplace Ergonomics Workbook,* U.S. Department of Defense, Washington, DC, 1998.

Farrell, G. P., S. K. Knowlton, and M. C. Taylor, *Second Chance: Rehabilitating the American Worker,* Brandeis University, Waltham, MA, 1989.

Herman Miller, Inc., *Designing for Accessibility: Beyond the ADA Applications Guide and Video,* Herman Miller, Inc., Zeeland, MI, 1995.

Job Accommodation Network, *Second Quarterly Report,* Job Accommodation Network, Morgantown, WV, July 1999.

Jones, M., and J. Sanford, "People with Mobility Impairments in the United States Today and in 2010," *Assistive Technology,* 8(1): 43–53, 1996.

Louis Harris and Associates, *The New Competitive Advantage,* National Organization on Disability, Washington, DC, 1994.

Lowery, C., "The 2 Year Track in Rehabilitation," *The Washington Post,* 23 January 1994, p. M5.

Mueller, James, *The Workplace Workbook 2.0,* HRD Press, Amherst, MA, 1992.

———, *Case Studies on Universal Design,* The Center for Universal Design, Raleigh, NC, 1997.

Social Security Administration, 16 February 2000, online; www.ssa.gov/legislation/legis_bulletin_121799.html.

Steinfeld, Edward, S. Schroeder, et al., *Access to the Built Environment: A Review of Literature,* U.S. Department of Housing and Urban Development, Washington, DC, 1979.

U.S. General Accounting Office, *Social Security Disability,* U.S. General Accounting Office, Washington, DC, 1995.

45.10 RESOURCES

Haworth, Inc.
Phoenix, AZ 85004
www.haworth.com

Herman Miller, Inc.
8500 Byron Road
Zeeland, MI 49464
www.hermanmiller.com

The Knoll Group
105 Wooster Street
New York, NY 10012

Steelcase, Inc.
901 44th Street, SE
Grand Rapids, MI 49508
www.steelcase.com

CHAPTER 46

EDUCATIONAL ENVIRONMENTS: FROM COMPLIANCE TO INCLUSION

Fred Tepfer, B.Arch.
University of Oregon, Eugene, Oregon

46.1 INTRODUCTION

This chapter reports on universal design in educational environments and the importance of universal design to educational institutions. It also discusses the influence of educational institutions on the growth and development of universal design. It tracks the evolution of accessibility and universal design in education from barrier removal, through concern with physical features, to other areas such as technology and curriculum. Universal design is especially important in education, in particular because of the role of educational environments as examples that students can draw on later in life, and because of the importance of educational institutions as environments in which inclusion is taught. This chapter also provides recommendations for best practices and the universal design rationale behind them.

46.2 BACKGROUND

Schools, colleges, and universities are ideal environments for fostering universal design. Compared with other types of uses, education has the most extensive experience with the broadest range of diverse needs. By comparison, commercial environments are used by large numbers of people from a broad spectrum of the population, but typically in a brief, transitory way. At the other end of the spectrum, employment settings must be adapted to the permanent needs of each individual's disabilities, but most employers do not experience accommodating many different individuals. In educational settings, people with disabilities require individualized semipermanent accommodation, yet this population is much more numerous and more transient than that in employment settings. This diversity of experience creates a valuable knowledge base and constituency for universal design.

Educators are also beginning to realize that their responsibility to foster diversity extends beyond racial and cultural issues to physical needs. In the same way that a multicultural curriculum is needed to create racial and cultural tolerance and diversity, universal design is needed to encourage inclusion and acceptance of all abilities. Young people are educated as much by example as by teaching. Environments that segregate teach acceptance of segregation, and inclusive environments teach inclusion. If all students are taught the benefits of inclusive environments through experiencing inclusive education, an inclusive society will eventually be created.

Four Stages of Accessibility in Education

Historically, accessibility and inclusion in education divide broadly into four stages, or eras. These reflect changing attitudes toward disabilities and inclusion, so it is natural that different regions and different organizations have moved through these periods at different rates. Programs and buildings of the first stage had no access provisions, as shown in Fig. 46.1. Students with disabilities were often prevented by law from being educated in contact with the general population. In the exceptional instances that students or teachers with disabilities were included at all in public schools and colleges, there was no support from the built environment. Segregated facilities were the norm.

In the second stage, students were given at least a theoretical right to be included and a right to accommodation for their disability. Federal law provided the springboard for disability rights, and states supported the sweeping change by incorporating at least limited accessibility provisions into building codes. Parents and students fought for the right to education and the right to be included with others. Many barriers were removed, but many others became apparent (U.S. GAO, 1995). Although students with disabilities were being educated in the same building as other students, segregation continued, in some cases due to physical barriers and in other cases due to commonly accepted practices of special education as well as discrimination (Ansley, 2000). It became clear that more work was needed on the theoretical framework for inclusion, as well as on the standards for accessibility, but that the relatively limited amounts of major construction in the 1980s limited how much could be accomplished.

The third stage (which is the present stage), depending on location and institution, moves thinking about accessibility in education from a focus on barrier removal and barrier prevention to the challenge of creating broadly inclusive physical environments. The emphasis has shifted from finding some accessible facilities for students with disabilities to identifying and finding the most appropriate way of dealing with the remaining inaccessible facilities. With an increased volume of school and university construction, administrators and designers are beginning to find opportunities for integrated design solutions through the conceptual guidance of universal design, as shown in Fig. 46.2. In this stage, integration of students with physical disabilities has been successful, but integration of students with developmental disabilities has lagged behind.

FIGURE 46.1   Whiteaker School, Eugene, Oregon: entrance from the 1920s.

FIGURE 46.2   Grants Pass High School, Grants Pass, Oregon: main entrance.

Signs of the fourth stage of educational accessibility are just beginning to emerge, in which concepts of inclusion and universal design that were learned in the built environment are now being applied to other areas, largely through technology. "Electronic curbcuts," pioneered by Vanderheiden and the Trace Center, are informing the design of telecommunication and other electronic devices to remove barriers (Trace Center, 1999). The Center for Applied Special Technology (CAST) is designing digital-based curricula, which are as broadly inclusive as possible (CAST, 2000). Previous thinking about physical and sensory barriers is being applied to the broadest range of human abilities, moving beyond concepts of physical disabilities and hidden disabilities to include the full range of ages, sizes, and other factors, and beyond physical disabilities to inclusion in all areas.

K–12 Education and State and Federal Legislation from 1968 to 1990

The concept of inclusive educational environments is new, relative to the age of most school and university buildings. Thirty years ago, it was unheard of and often illegal to integrate children with disabilities into the school systems. The passage of the Rehabilitation Act in 1973, with its significant Section 504, and the Education of all Handicapped Children Act in 1975, which later became the Individuals with Disabilities Education Act (IDEA), brought children with all types of disabilities into school systems throughout the country. These laws required programs receiving federal funding to make their programs accessible, and forbade discrimination on the basis of disability. School districts had to figure out how to adapt their programs, their school buildings, and their design and construction practices within a relatively short time. However, the reality of the implementation often only achieved limited access into parts of schools and into programs limited to students with disabilities.

This was followed up in many states by the addition of accessibility provisions to building codes and adoption of disability rights legislation at the state and local level. Although there was wide variation during the 1970s and 1980s in code requirements and enforcement, depending on which state or locality was involved, the inclusion of these provisions across the nation helped set the stage for more comprehensive approaches to the creation of inclusive environments.

K–12 Education and Universal Design

K–12 curriculum trends in individualized instruction, in combination with the 1970s mandate to accommodate individuals with disabilities, exposed many educators to the broadest range of needs among their students. Parents and students fought for their rights, and slowly, sometimes grudgingly, teachers and administrators who worked with students with disabilities also became advocates for disability rights in the schools. This was especially true in schools serving larger, more diverse populations, where individual educators were exposed to a wide range of needs. The frustrations of accommodating these individuals taught them that the prescriptive federal and state mandates for barrier removal and accessible construction were not always delivering effective accessible environments for the full range of individual needs (Ansley, 2000).

For example, even after remodeling or constructing buildings to comply with Section 504 of the 1973 Rehabilitation Act construction standards used at that time, students and employees with disabilities often needed a greater degree of accessibility beyond that found in the newly built or altered environments. Furthermore, the continuing evolution of accessibility standards led educators, parents, and students to realize that they had to think beyond the minimum standards and codes, especially as the 1980 revision of the American National Standards Institute (ANSI) accessibility standard (ANSI A117.1-1980) and the Uniform Federal Accessibility Standards (UFAS) that are based on it, came into general use. This new ANSI standard was significantly different from earlier standards such as the 1961 and 1974 versions of ANSI A117.1. Much of the barrier removal done in the late 1970s and early 1980s before the widespread use of ANSI A117.1-1980 did not comply with the dimensional requirements of the 1980 standard, which led many people in education to start thinking beyond the minimum standards in the design of educational environments. In later years, when the concept of universal design was introduced, these earlier experiences helped educators and facilities planners understand the basic concepts of and the need for universal design.

Higher Education and Federal Legislation from 1968 to 1990

The passage of the Architectural Barriers Act (ABA) in 1968 had a small but noticeable effect on facilities in education. It required federal construction projects to meet certain accessibility standards. Many larger colleges and universities had at least one federally funded building built between the passage of the ABA and the implementation date of Section 504 of the Rehabilitation Act in 1977, and had to begin to learn how to plan environments for people with disabilities. However, in higher education, the major triggering event for disability awareness was the return of veterans from the Vietnam War, many with war injuries, who were being educated with federal assistance. Universities quickly discovered that this vocal and growing population demanded education but could not be served. The passage of the Rehabilitation Act in 1973 and the issuance of Section 504 implementation regulations in 1977 were largely in response to this frustration (DREDF, 1997). Figure 46.3 is an example of an ABA-influenced design from 1970. Note the new ramp under construction in the summer of 2000 to the right of the front entrance.

Section 504 of the Rehabilitation Act of 1973 (Section 504) prohibited federally assisted programs from discriminating on the basis of disability. This included making their programs accessible to people with disabilities. The Rehabilitation Act led to an outpouring of manuals, training sessions, self-evaluations, transitions plans, and, eventually, at least limited barrier removal in nearly all colleges and universities. This extensive examination of the physical barriers to education in the United States was, in many ways, a rehearsal for the passage and

FIGURE 46.3   Grayson Hall, University of Oregon: front and side entrances.

implementation of the Americans with Disabilities Act nearly 2 decades later. Large research universities, being dependent on federal grant support for their research functions, were especially sensitive to Section 504 compliance.

However, misfortune placed the implementation of Section 504 in a period of financial retrenchment and low construction volume in higher education. This further reinforced the emphasis on barrier removal instead of creation of inclusive new environments, and on accommodation of individuals instead of creation of inclusive organizations and institutions housed in inclusive physical settings. Many of the barrier removal projects were done in the hastiest and most expedient manner possible, resulting in a token level of service to people with disabilities and, in some cases, resentment on the part of other facility users due to the sometimes shoddy or ugly changes for accessibility. Figure 46.4 shows one example of well-meant but ineffective barrier removal constructed in response to Section 504 requirements.

The mixed effectiveness of Section 504 barrier removal efforts became evident very quickly as the first generation of mainstreamed students made their way through schools, then colleges and universities, and medical technology allowed more survivors of traumatic brain injuries to be reintegrated into society. Unlike the returning veterans, many of whom had full use of their upper bodies, this newer population brought a much wider variety of needs and more challenges to the built environment. Many of these individuals' needs, such as for power-wheelchair users and people with low vision, deafness, or multiple disabilities, were not met in facilities that appeared to comply with federal mandates, leading those institutions that experienced this disconnect to begin to think much more broadly about the creation of inclusive environments. Perhaps the earliest and most visible change was growth in the numbers of users of power wheelchairs, scooters, and other mobility aids. People using these devices often found that facilities designed for manual wheelchairs were difficult or impossible to use. They often experienced buildings designed with very accessible restrooms, yet with inaccessible classrooms and laboratories. Nothing in the standards prepared colleges and universities for the needs for safety considerations for the deaf, amplification systems for people who were hard of hearing, or wayfinding and safety issues for blind people.

FIGURE 46.4   Ramps at Stella Magladry School, Eugene, Oregon.

The Americans with Disabilities Act

Despite its limitations, Section 504 and the effect it had on school facilities placed many educational institutions at the forefront of creating accessible environments in the 1980s. Passage of the Americans with Disabilities Act (ADA) effectively raised the bar for the rest of the nation to the level that education had been experiencing. Educational institutions could no longer get by with relatively better but inadequate access. Educators and administrators again reexamined their programs to weed out barriers to access, or they were forced to do so by disability advocates who now expected meaningful action on barrier removal. In this reexamination, the limited success of many earlier barrier removal efforts gave great impetus to the concept of universal design.

One large effect of the implementation of the ADA was to further standardize many accessibility provisions, by requiring states and local governments to use the more stringent federal standards when confronted with state and local building codes. This led in many states to outright incorporation of ADA accessibility guidelines for construction into state and local codes. The minimum standard of compliance provided by these codes often represented a major improvement in access for students and teachers with disabilities, but they also became ingrained in some designers' minds as maximums as well as minimums. See Chap. 12, "ADA Standards and Challenges for Universal Design," by John Salmen, regarding how minimums become maximums. This is shown by the barrier removal plan for the three buildings and four building levels of Colonel Wright School in Fig. 46.5, which provided an accessible route to all ground-floor spaces, but not a route that would be effective for either a student or a teacher at a school in that cold, wet climate. Nor does the accessible route at Colonel Wright School effectively integrate people with mobility impairments with the rest of the school population.

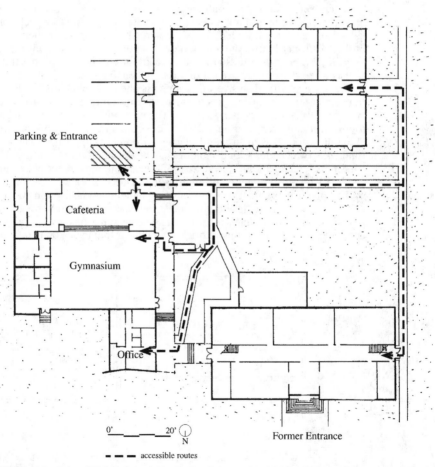

FIGURE 46.5   Colonel Wright School, The Dalles, Oregon: site plan showing accessible routes built in the mid-1990s.

46.3 INCLUSIVE EDUCATIONAL ENVIRONMENTS

Reasons for Universal Design

Three main factors brought many educators and administrators to begin to think about the concepts that are now called *universal design*. First, it became clear that the variety of needs in the population they served was not met fully by the federal accessibility standards such as ANSI A117.1, UFAS, and the ADA Standards for Accessible Design (ADA Standards). In some cases, the architects and administrators interpreting the standards were unfamiliar with the reasons behind them, and failed to follow them adequately. In other cases, the standards themselves failed to serve a broad enough population base.

Second, those codes and standards were changing fairly rapidly, so an attitude favoring minimal compliance quickly became a liability. An elevator that was built to the minimum state and federal requirements in 1979 was no longer large enough in 1985, as shown in Fig. 46.6.

Third, they found that where barriers had been removed in a more thoughtful and inclusive way designed beyond the minimum requirements, there were unanticipated benefits to other segments of the population. For example, elevators, which can carry bulky and heavy loads, are very useful for maintenance staff, but mechanical platform lifts and stair-climbing lifts are not. Full-length mirrors are appreciated by people in wheelchairs, but they are also valued by anyone who needs to make sure they are looking their best. Anyone who has pushed a stroller knows the location of all of the curb cuts and entrance ramps, but if these only allow circuitous routes and serve backdoor entrances, they frustrate stroller pushers as well as wheelchair users. The list of multiple benefits goes on and on.

Experience with inclusive environments showed educators further advantages. Educational institutions are examples to their students. During these formative years, children and young adults learn from many sources. The environment within which learning occurs can be a powerful educator, and an inclusive environment can provide a basis of understanding about inclusion. This inclusive physical environment supports an inclusive educational environment in which kids with disabilities can be educated side by side with other children. That concept is fundamental to the success of the current movement of genuine mainstreaming of children with developmental disabilities to the greatest extent possible.

This thinking about inclusion extends beyond the physical realm. Most schools make a large effort to educate students about the need to be accepting of other cultures and other races. A physical environment that is inclusive of all needs supports the broader aims of a multicultural curriculum by demonstrating in bricks and mortar that inclusion helps everyone.

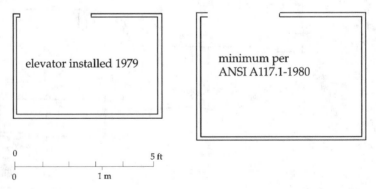

elevator installed 1979

minimum per
ANSI A117.1-1980

FIGURE 46.6 Minimum elevator sizes, pre-1980 and post-1980 ANSI A117.1 requirements.

By the same token, universal design now influences the design of curricula. The beginnings of curriculum approaches that are suited to multiple ability levels are beginning to be seen. Although a very new field, universal design of curricular offerings is getting a great deal of attention and has the potential to become very influential in the future. The CAST has been a leader in this field (CAST, 2000).

Advocacy for Universal Design

The constant, day-to-day realities of accommodating students and employees, as required by Section 504 and by IDEA within facilities that demonstrated the partial successes of Section 504 barrier removal, have led many education professionals, teachers, administrators, and facilities planners to become proponents of universal design. It is more likely in design of educational facilities that the clients and users will be the proponents for more accessibility, not the architects who are more likely to view accessibility as a code compliance issue. In this way, participatory design of education facilities is important in creating inclusive environments. Ideally, a broad spectrum of individuals should be represented in the design process. But even if they are not, many educators have experience with the broad range of needs and will advocate for it in the design process.

To a certain extent, the recent trend toward hiring prominent architects to design signature buildings for educational institutions works against this trend. As physical distance between users and architects increases, communication links can grow thin and opportunities for interaction can decrease. As these communication problems stretch or break the user collaboration, which assures consideration of the full breadth of population needs, inclusive design suffers.

46.4   RECOMMENDATIONS FOR BEST PRACTICES

Universal design continues to be very important in educational environments. As discussed previously, schools experience intensive use by a wide range of the population representing the full spectrum of needs, so schools need a built environment that supports and integrates all of these individuals without special intervention. In addition, by creating an inclusive environment that encourages the active, equal participation by people with a variety of needs, the entire population is educated about the advantages of inclusion and inclusive environments, which should embrace the following considerations.

General Design Standards and Processes

Anthropometric and ergonomic considerations of children and young adults.   These are of paramount importance in designing educational facilities. Standard dimensions for accessibility may not be suitable for the larger range of needs encountered in schools. Furthermore, since buildings typically far outlast the programs that they contain, the building site and envelope should be designed for the full range from a small child to an adult. These issues influence design of elements ranging from window sill heights to workstations to handrails.

Participatory design.   This provides the forum for the broad range of needs to be expressed, either directly through people with specific environmental needs or through the teachers and administrators who work with them on a daily basis. Where used, participatory design has proven its value by providing a way for users to insist on the creation of broadly accessible environments through the design process. It is important for facilities professionals to represent the school or college to ensure that the design reflects a broad-based view and is not tailored to the needs of the last student who needed to be accommodated.

Designing beyond the standards.   This is essential for the creation of inclusive educational environments. When the minimum requirements become the de facto maximums, it is a certainty that the user needs will not be met, either due to errors in construction, or because the standards do not accommodate a broad enough range of needs. The best practice is to take on the needs of the whole population as a design challenge rather than relying on minimum standards.

Accessible to all.   All building elements and program activities should be accessible to all of the anticipated building users without assistance or intervention, or else a clear justification as to why not should be written and agreed to with the school users. If certain areas such as libraries or shops will require staff assistance, the school should be made aware of that during the design phase. In addition to applying to people with disabilities, this logic also applies to environments for young children.

Flexibility and choice.   The most successful design solutions integrate the activities of all elements of the population while also providing choice and flexibility. The redundancy that may result is also an opportunity for a more richly varied design palette. For example, some individuals cannot climb stairs. Others have trouble with ramps, but can manage stairs if proper handrails are provided. The best solution integrates both stairs and ramps into the design so that neither appears to be an afterthought. In Fig. 46.7, a former loading dock and service area was reconfigured to provide accessible slopes (nearly all at 5 percent or less) and stairs connecting four buildings. The fountain (by Alice Wingwall, Berkeley, California) was part of Oregon's Percent-for-Art program, and was designed specifically to enhance the experience of blind and low-vision individuals.

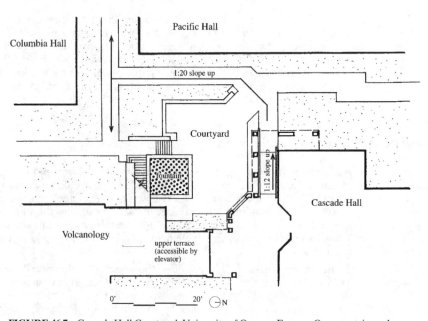

FIGURE 46.7   Cascade Hall Courtyard, University of Oregon, Eugene, Oregon: stairs and ramps.

Site and Building Planning

Careful site selection and site planning.   These are the foundation of creating inclusive places. If the approach to development of a site adds access considerations to the design instead of including them from the beginning, the places that are built are almost certain to be unsatisfactory. Accessible routes limited to slopes of 5 percent or less guarantee inclusion with the rest of the population. If the site is sloping, major level changes along accessible routes should be made within buildings via elevator, as the travel time and effort demanded by long ramps create de facto barriers.

Connections.   Connections within and between buildings need to allow for quick, convenient travel. Schools and colleges work on schedules with fixed travel times. If these times are not adequate because of excessive route length, someone is probably being denied access to education.

Ergonomics.   Design appropriate to size and physical function of users is essential to universal design. In educational environments, many designers forget that the requirements for children are different from those for adults. The Access Board has published guidelines that although not yet enforceable, provide an initial level of guidance in this regard (ATBCB, 1998).

Power doors.   In keeping with providing an environment that fosters everyone's independence, maneuverability should ensure that everyone can enter the building without assistance. The best-practices design provides power door operators on at least some exterior building entrances and at key internal doors. These should be designed realizing that many users have limited or no use of one side of their body. If button door actuators are used, clearances should be provided to allow use from either the right or the left hand or by mouth stick without blocking the door operation. This may require enough space for the wheelchair user to operate the button and then turn 180° to move through the door, as shown in Fig. 46.8. Alternatively, motion sensors, multiple buttons, or buttons that can be bumped by a footplate can be provided.

Orientation and wayfinding.   Clear, "imageable" building organization and floor plans are essential. Legible environments are often easier to move around in, and much easier for blind people and those with cognitive or psychiatric disabilities to navigate. Even better are those that provide tactile cues at major intersections of circulation systems. A clear approach to building layout has major advantages for other purposes such as supervision and security. Figure 46.9 illustrates one approach to providing a layout that ties together existing buildings in a way that works well for blind and sighted building users.

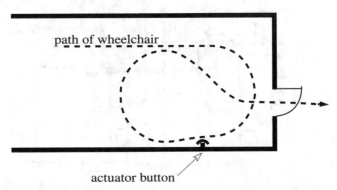

path of wheelchair

actuator button

FIGURE 46.8 Path of a left-handed wheelchair user at a right-handed power door actuator.

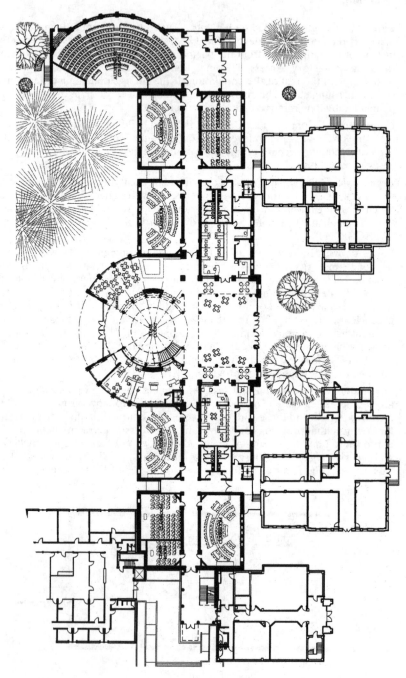

FIGURE 46.9   Gilbert Hall Addition/Lillis Building Complex, University of Oregon: main floor plan, schematic design.

Signs are also an important part of solving wayfinding problems. Directional signage, directories, and signs identifying major spaces should be large and clear so that they are visible to people with low vision and also visible at a distance. At the entrance to each room, signs identifying the space, including raised letters and Braille, must be placed in the precise location that blind people will reach out to in order to read them: 5 ft above the floor on the latch side of the door. Even small deviations in location make these room identification signs useless to those who cannot see. Going beyond the requirements, room identification signs can contain more information than just room numbers, especially in large buildings with many rooms and complicated floor plans. They should also identify the permanent function of the room, such as library, laboratory, or auditorium, in raised letters and in Braille. Room numbers alone in raised letters and Braille are not enough information for blind people who have no visual cues to guide them. However, most other signs are pointless to provide as tactile signs, because blind people have no way to find the sign in order to read it. Some concepts, such as handrails that provide information in Braille, have been used successfully in certain applications such as museums, but these are only effective if used in consistent ways as an addition to standard locations for room identification signage (Raynes, 1996).

Travel distances.   Travel distances within buildings and between buildings should be tested during design with the hourly and daily routine schedules of the students and staff in mind. Many people with disabilities travel more slowly than average people and, therefore, need more time to get from place to place. Their routes should be shorter than inaccessible alternatives, or slightly longer if necessary, and in total should provide an entire educational environment that works well for them. This thinking should be applied at a detailed planning level, as well as in site and building planning. For example, accessible stations in labs should be close to exits and safety equipment, such as showers and fire extinguishers. Similarly, accessible toilet stalls should be the nearest to the entrance to a toilet room, not the farthest away, which is commonly the case.

Building Systems

Acoustics.   Most designers are not aware of the importance of room acoustics in educational environments. Most small children experience at least temporary hearing loss due to allergies and infections at the very age that auditory comprehension is most important (i.e., while they are learning to read). Even in the higher grades and in higher education, research shows that many students are auditory learners, and if they cannot hear well or if other sounds are too distracting, learning suffers. Many students with learning disabilities suffer from noise distractions. Good practice includes careful acoustical design in all learning environments. Material choices that have and will maintain the appropriate acoustical qualities are the main vehicle to ensure that this important characteristic is maintained. Beyond basic comprehension of teachers by students, good acoustical design also provides for clear interaction among presenter and audience in a lecture situation, and allows for breakout into smaller groups of students even within a large group setting, such as a lecture hall. In addition to the acoustics of the room itself, the technology that is used for teaching must also be integrated.

In the lower grades, there are also spaces specifically designed for speech therapy. Proper acoustical design of these spaces is particularly important. Not only should they be acoustically isolated from distracting noises, but they should also be relatively acoustically absorbent.

Amplification systems that can amplify all auditory media for individual users are fundamental to providing a learning environment for those with hearing loss. User involvement in the selection and installation of these systems is critical to ensure the following:

- That they are interchangeable with other systems used by the same school or college
- That they can accommodate future technologies
- That they are robust enough to survive the hard use that teachers and students impose on equipment
- That they are simple enough for everyone to understand

Some schools are installing amplification systems with cordless microphones for teachers connected to a room amplifier and built-in speakers. Although this technology is effective in ensuring that students can hear presentations from the teacher, it does nothing for communication between students. Curricula are moving more and more toward concepts of cooperative learning in which students work in small groups and learn from each other as well as from teachers, books, and other sources. Good room acoustics that help everyone hear each other is an essential element of a learning environment for this type of curriculum. [For more information, please refer to Grondzik (1999).]

Indoor air quality.   Indoor air quality and multiple chemical sensitivity (MCS) are increasing concerns for educators. A hidden disability, MCS is sometimes overlooked during the design process, but it is primarily in the design of buildings that it can be accommodated. Best-practice design makes available large amounts of outdoor air under the control of the users, most often through operable windows. Careful site planning and building layout are needed to ensure outdoor air entering the building is not contaminated with vehicle exhaust, such as from loading areas, kitchen or laboratory exhaust, or other pollutants. Attention should be paid to interior materials to avoid those that release volatile organic compounds (VOCs) over a long period of time. Release of VOCs may be controlled by "cooking" the building at a relatively high temperature before the building is occupied, or by "breathing" it with plentiful ventilation to allow materials to off-gas before occupants move in.

It is equally important to plan a building so that maintenance and remodeling activities can happen in the future without needing to shut down the entire facility because of air contamination issues. A modular floor plan that allows part of the school to be separated and that uses windows for local ventilation of individual spaces during construction goes a long way toward preventing future management problems.

Designing fire and life safety in schools and colleges for the entire population.   This is a new and changing field. Current standards require *areas of rescue assistance* (ARA) to provide a refuge from fire from which people can signal their need for help. While this is probably a major improvement over previous efforts to provide fire and life safety for the entire population, best-practice design would first install fire sprinklers, which prevent fires from growing to a dangerous size, and which have an excellent track record for providing a high degree of fire safety for all elements of the population from infants to nursing homes. Unlike fire alarms, notification devices, and other electrical systems, fire sprinklers have proven their reliability in almost every imaginable situation. Where fire alarms are used, strobes for visual signaling devices are now required in certain areas by the ADA, and should be designed to be easily extended into rooms used extensively by deaf individuals.

In educational institutions, it is also wise to provide training for any part of the population who may have special needs for fire and life safety, such as people with mobility impairments who cannot use stairs, deaf people who cannot perceive pre-ADA fire alarm systems, blind people and small children who may not be able to find exits, and so forth. By providing training for the target population in addition to the building staff, skills can be provided that have life-long use, with special emphasis on the need for advance planning, use of all possible means for calling for help such as telephones, and identification of ARAs, as well as people to assist in an emergency. For more information, please refer to Chap. 23, "Life Safety Standards and Guidelines" by Jake Pauls.

Functional Areas

Classrooms.   At all levels of education, classrooms are remade daily and weekly. Furniture is rearranged, equipment is set up and then removed. A universal design by its nature must accommodate uses that can change quickly, so perhaps the most critical elements are the furniture and the equipment. Tables, chairs, projectors, chalk, or marker boards—all must be designed to accommodate a wide range of body types, adaptive equipment, and sensory disability. The basic tools are to provide adjustability, especially vertical adjustability of work and equipment surfaces; variety of choice (e.g., furniture of different sizes, left and right tablet arms, high and low sinks, etc.); and multisensory equipment, including captioning, materials that are also available on accessible Web sites, and so on.

Lecture halls.   In a conventional college lecture hall with tablet arm chairs on a steeply raked floor, the conventional solution is to put a little extra floor space at the front and, if the designers are thoughtful, also at the rear. This leaves the users of these spaces without a writing surface and limited to one or two prominent locations. Even if a writing surface is provided, it is probably provided with the ADA minimum 27-in knee clearance [which, in this case, became a maximum (see Chap. 12, "U.S. Accessibility Codes and Standards: Challenges for Universal Design," by Salmen)], ignoring the needs of users of power wheelchairs, scooters, and technologies other than manual wheelchairs.

A better solution would provide continuous desks for all at a height appropriate for manual wheelchairs, as illustrated in Fig. 46.10, and with horizontal clearances that allow for wheelchair use. With the appropriate design effort, an accessible route could be created to the front, near the rear, and at least some points in between. This provides locations for students with mobility impairments who need front access, possibly due to multiple disabilities, and also provides them the same range of choice that other students experience. At a certain number of locations, students would find desks that allow for vertical adjustment by the users, allowing for use by students with power wheelchairs and others whose needs are not met by the standard accessibility compliance dimensions. The front of the room would have an accessible teaching station or lectern, with pullouts for adaptive technology or laptop computers, and all controls would be within easy reach for all.

Laboratory and other special environments.   Easily adaptable laboratory and other special environments are a key component of universally designed schools and colleges. Minimum standards that architects rely on, such as the ADA Accessibility Guidelines, have not yet addressed these situations. Teachers and administrators know that they will be responsible for teaching all students who are enrolled, so if these places are designed to be accessible or at least adaptable, then the school will face fewer problems in the future.

FIGURE 46.10   Knight Law Center, University of Oregon: lecture hall.

The best-practices solution is to design for a seating level that is easy to incorporate accessible features into. Standard high-bench science labs, for example, do not allow adequate reach across the bench tops. Low lab benches can be designed so that a drawer unit can be removed or a cabinet door unscrewed to reveal a knee space, which is adequate for wheelchairs and which allows the users to reach to the full depth of the work surface. The lower height gives the teacher or lab manager the advantage of being able to see farther across the lab while standing, improving supervision and safety for all.

Theaters and stages.   These can be a design challenge, and because they are often used for graduation ceremonies, public presentations, and other major events, they have the potential to be troublesome embarrassments. Figure 46.11 shows one solution. The best-practice approach integrates them to the fullest extent into the fabric and, perhaps, the topography of the building or campus so that movement through, within, and across them is planned to be accessible for all. In particular, the movement of people through this environment for public ceremonies, such as graduation, should be carefully planned to ensure that all students will use the same path throughout the ceremony. Beasley, in Chap. 47, "Access to Sports and Entertainment," discusses the issues in assembly area seating and offers good examples.

Eating areas.   Eating areas with proper knee clearance and movable chairs permit a wide variety of choice for all. This less formal, less institutional setting can also encourage more social interaction, and it allows for a wide variety of uses, from daily dining to presentations and performances.

Outdoor play areas.   These are a fundamental part of the education of children. For more information on this subject, refer to Chap. 19, "Play for All Guidelines," by Goltsman.

Toilet rooms.   These are covered extensively in federal standards and local codes. There are a few aspects that are especially relevant to schools. One of these is making provisions for attendants of the opposite sex. Children in the United States are typically very sensitive to this issue, so providing a reasonable number of unisex single-user toilet rooms is recommended.

FIGURE 46.11   Grants Pass High School, Grants Pass, Oregon: auditorium.

Recreational facilities.   Recreational facilities are often overlooked because of mistaken assumptions about the physical needs of people with disabilities. Fitness and regular exercise are very important to people with disabilities, and the opportunities for exercise are typically fewer. Properly designed accessible recreation facilities are very popular (see Fig. 46.12), and often require only moderate extra effort to provide for everyone's activities (e.g., swimming pools that provide a variety of ways to enter the water, fitness centers designed for transfer from wheelchair to weight-lifting equipment, running/rolling tracks with accessible routes from locker areas, and many others). Locker areas should have facilities that provide for attendants of the opposite sex, which are also greatly appreciated by families with small children.

Equipment.   All too often, schools and universities build buildings that are accessible and inclusive, and then install equipment that is not in any way accessible. Inclusive environments extend to the furniture, to the electronic equipment, even to the software that runs on the computers. A few examples include the following:

- Software, and in particular Web sites, should be designed to be accessible via screen readers for people who are blind. This technology reads text, making computers very valuable tools for the blind, but it does not interpret graphics. The top of Web pages should have a link to a text-only alternative. For more information, see Chap. 66, "Access to the World Wide Web: Technical, Policy, and Implementation Issues," by Brewer.
- Equipment such as photocopiers should have control panels that are readable by all who may wish to use the equipment.

FIGURE 46.12   Student Recreation and Fitness Center, University of Oregon: wheelchair user on elevated indoor track accessible by elevator.

- Adaptive pointing devices, text input devices, screen magnifiers, and many other aids are available to bring computer technology to as many people as possible.
- Laboratory equipment should be selected that allows the fullest range of users access to the technology.
- Computer technology should be available at a reasonable number of stations equipped with extremely adjustable work surfaces, which the users can adjust to suit their individual needs.

As with any best-practices list, the aforementioned items represent the current level of understanding and design. Future research and innovation will undoubtedly improve on our current practices as well as infuse universal design into additional aspects of schools and colleges. However, the basic lessons and fundamental principles previously discussed will continue to be useful as guidance in the evolving environment of universal design.

46.5 GENERAL LESSONS LEARNED FOR UNIVERSAL DESIGN

The most important lesson learned from educational environments is the importance of learning from the diversity of experience that is available in educational environments. Large educational institutions often have a wealth of experience and expertise in this area simply because they have experienced a broad range of issues and needs.

This concept relates to the importance of user involvement in design of places and products. For educational settings, users are an essential voice in creating good places as well as inclusive places, a notion that translates to many other areas of design.

In addition, education is discovering the beneficial spin-offs of universal design in other areas. Fields beyond education will also find interesting and useful analogues to universal design far beyond the built environment.

46.6 CONCLUSIONS

Education, with its unique position of dealing with large numbers of people as individuals, has been the breeding and testing ground for many advances in accessible design, and universal design is no exception. Educators are becoming strong advocates for inclusion of people with disabilities, and have learned the advantages of applying the Principles of Universal Design not only to their physical environment, but also to the equipment that they use, to the software on their computers, and even to their curricula. Where participatory design provides them with an appropriate voice in the design process, they can help architects and designers anticipate the widest range of possible needs and design for them.

Where universal design has created successful learning environments, many schools and universities have found that the resulting inclusion helps support other areas such as multiculturalism and acceptance of others.

As society moves into the fourth stage of accessibility education, universal design concepts are essential to educators and designers. These concepts ensure that educational environments continue to inspire a vision of an inclusive society as an example for the rest of the world.

46.7 BIBLIOGRAPHY

American National Standards Institute (ANSI), ANSI A117.1-1980, *American National Standard: Specifications for Making Buildings and Facilities Accessible to and Usable by Physically Handicapped People,* American National Standards Institute, Inc., New York, 1980.

Ansley, J., *Creating Accessible Schools,* National Center for Educational Facilities (NCEF), Washington, DC, 2000; www.edfacilities.org/ir/irpubs.html.

Architectural and Transportation Barriers Compliance Board (ATBCB), *Americans with Disabilities Act: Accessibility Guidelines for Buildings and Facilities; Building Elements Designed for Children's Use,* ATBCB, Washington, DC, 1998; www.access-board.gov/adaag/kids/child.htm.

Center for Applied Special Technology (CAST), *Concepts and Issues in Universal Design for Learning,* CAST, 2000; www.cast.org/concepts/.

Disability Rights Education and Defense Fund, Inc. (DREDF), *504 Sit-in Commemoration and Anniversary,* Berkeley, CA, 1997; www.dredf.org/504home.html.

Grondzik, A., "Rethinking Classroom Acoustics," in *Proceedings of ASHRAE Winter Meeting,* Florida Design Institute, Tallahassee, FL, 1999.

Lubman, D., "America's Need for Standards and Guidelines to Ensure Satisfactory Classroom," in *1334d Meeting Lay Language Papers,* Acoustical Society of America, Melville, NY, 1997; www.acoustics.org/1334d/2paaa1.html.

Raynes, C., et al., *The Raynesrail, a Braille and Audio Handrail System,* 1996; www.raynesrail.com/applications.htm.

Trace Center, *General Concepts, Universal Design Principles and Guidelines,* Trace Research and Development Center, Madison, WI, 1999; http://trace.wisc.edu/world/gen_ud.html.

U.S. General Accounting Office (GAO), *School Facilities: Accessibility for the Disabled Still an Issue,* U.S. General Accounting Office, U.S. Government Printing Office, Washington, DC, 1995; http://frwebgate.access.gpo.gov/.

46.8 RESOURCES

Center for Applied Special Technology (CAST).   *www.cast.org*

ERIC Clearinghouse on Disabilities and Gifted Education (ERIC EC).   http://ericec.org/

National Center for Educational Facilities (NCEF), "Accessibility Topics."   www.edfacilities.org/ir/accessibility.cfm.

Trace Research and Development Center.   http://trace.wisc.edu/world/gen_ud.html.

Bar, Laurel, and Judith Galluzzo, *The Accessible School: Universal Design for Educational Settings,* MIG Communications, Berkeley, CA, 1999.

McGuinness, K., "Beyond the Basics," *American School & University,* 69(11), 1997.

Moore, D., "ADA Means All Children Can Have a High-Quality Education," *School Planning and Management,* 36(10), 1997.

Rydeen, James E., "Universal Design," *American School and University,* 71(9), 1999. www.asumag.com/magazine/Archives/0599ada.html.

Sydoriak, D., "Designing Schools for All Kids," *Educational Facility Planner,* 31(5), 1993.

Tepfer, F., "Fred Tepfer's Home Page," University of Oregon. http://darkwing.uoregon.edu/~ftepfer/.

CHAPTER 47
ACCESS TO SPORTS AND ENTERTAINMENT

Kim A. Beasley, A.I.A.
Beasley Architectural Group, Alexandria, Virginia

Thomas D. Davies, A.I.A.
Paralyzed Veterans of America, Washington, D.C.

47.1 INTRODUCTION

Access to sports and entertainment facilities for persons with disabilities has been the objective of many new design practices and technical innovations. The most significant advance in the design and construction of these facilities is the extent to which accessibility has been provided in virtually every element and space. This chapter provides insight into some of the most recent design innovations that enable persons with disabilities to participate in all aspects of sports and entertainment activities—performing, watching, working, and virtually every other aspect of this fast-growing industry. In most cases, design innovations and facility features intended for persons with disabilities have produced more accessible environments for everyone. This trend toward universal design appears to be increasingly embraced by the sports and entertainment industry.

47.2 BACKGROUND

The last decade witnessed a proliferation of new sports stadiums, arenas, and entertainment facilities in the United States. This construction is primarily attributed to professional and amateur sports' tremendous growth stimulated by television and other entertainment media coverage. Intercity bidding wars to attract and retain professional sports teams added further impetus. For team owners, new marketing opportunities for corporate suites and other luxury amenities offered an enhanced revenue stream compared with traditional sports facilities. Because of their vast magnitude and high construction costs, public attention has focused on the design of these facilities. For local disability advocates, their accessibility has also been subjected to intense scrutiny, particularly since the passage of the 1990 Americans with Disabilities Act (ADA). As a result, many recent accessible design innovations emerged in new sports facilities.

The 1990 ADA was the most far-reaching civil rights law for people with disabilities. Since its enactment, design advances have been incremental, but each small victory builds upon the next, slowly achieving a built environment that provides more access, integration, and employment opportunities for this new protected class. Many important legal and political victories have emerged in the design and operation of new sports and entertainment facilities.

Historically, sports and entertainment access for persons with disabilities has evolved slowly. Wheelchair users' first challenge was simply to gain access to these facilities. This success was followed by better and more integrated wheelchair seating choices. More recently, wheelchair locations have been designed to provide comparable sight lines at all times, even when spectators in front stand [*Paralyzed Veterans of America (PVA) vs. Ellerbe Becket Architects & Engineers, P.C. (EBAE),* 1996a]. (See Chap. 18, "Interpreting Accessibility Standards: Experiences in the U.S. Courts," by Mazumdar and Geis.) Finally, full access has been provided to other facility users such as employees, entertainers, and athletes with disabilities.

The 1996 Atlanta Olympic Games was the first major sports facility construction program to address the ADA's design and operational requirements. For the first time in Olympic history, new competition and noncompetition venues were designed and constructed to serve persons with disabilities including spectators, athletes, employees, and volunteers. In every 1996 Olympic sports facility, all wheelchair seating was fully integrated and provided sight lines over standing spectators, as seen in Fig. 47.1. Purchasers of wheelchair seating had a broad choice of locations throughout every seating bowl. In addition, these facilities provided

FIGURE 47.1   The 1996 Olympic Games became the first Olympic site to comprehensively address the needs of persons with disabilities.

full access to athlete areas such as lockers, showers, toilet rooms, and the field of play. Furthermore, the new Olympic structures left Atlanta with a lasting accessibility "legacy," including facilities like the former Olympic Stadium (now Ted Turner Field) and the former Olympic Village (now Georgia Tech dormitories). For accessible design, Olympic facilities became the accessibility benchmark for stadiums, arenas, and other sports facilities constructed both in the United States and worldwide.

Innovative facility designs of any type often result in the development of new standards. These same innovations also test the application of accessible design standards. Accessible movie theater design, for example, must respond to the new stadium-style seating concept that significantly improves movie patrons' sight lines and thereby enhances their enjoyment of movie presentations. Unfortunately, the theater industry often fails to meet the ADA's design intent because wheelchair users do not always have comparable seating. Disability organizations' advocacy efforts, however, have raised public awareness and thereby improved recent stadium-style movie theater designs. In addition, new case law on sight line comparability has increased industry sensitivity to the ADA's requirements for full and equal enjoyment of the goods, services, and privileges that are afforded other patrons in these types of movie theaters.

Access to sports and recreation is important for persons with disabilities and their families. With the removal of physical barriers to these facilities and the provision of more enabling environments and new technologies, opportunities are opened to people with disabilities to lead more active and enriched lives. In all sports and entertainment facilities, quality accessible spectator seating is the primary key to design success. For individuals who use assistive devices such as a wheelchair, the architectural design implications are extensive but critical to equal access. For individuals who have hearing or sight impairments, technologies and services such as assistive listening systems or interpretive programs are equally important, although they are not typically achieved through architecture. Accessible architectural design, when coupled with new innovative technology, enables users with a wide range of capabilities and limitations to participate in sports and entertainment activities.

47.3 KEY ACCESSIBILITY ISSUES IN SPORTS AND ENTERTAINMENT FACILITIES

For all sports and entertainment facilities, access to and throughout the facility is the primary objective for full and equal enjoyment of the associated activities. Current design trends and innovations reflect a growing industry awareness and need for full access to spectators, performers, athletes, employees, and volunteers with disabilities who use these facilities. In many cases, sports and entertainment industry representatives have begun to see accessible design features as a means to enhance enjoyment, usefulness, and effectiveness of the facility for everyone. The most significant innovations in recent years have focused on spectator seating. Because of the large numbers of fixed seats typically designed for sports and entertainment facilities, improvements to spectator seating have had a significant effect on the overall facility design. In particular, the quality and location of wheelchair seating, the adjacent companion seating, and spectator sight lines for mobility-impaired individuals are key elements for successful design.

The Americans with Disabilities Act Accessibility Guidelines (ADAAG) contain specific requirements for fixed seating in stadiums, movie theaters, and other assembly areas (U.S. Access Board, 28 CFR Pt. 36, Appendix A, dt. 01, July 1994). These requirements can be met in several different ways, although some designs are better than others. In the past, failure to meet these requirements has resulted in owners paying for costly redesign, demolition and reconstruction, and litigation (*PVA v. EBAE*, 1996b). The quality of seating for persons with disabilities is shaped by several factors, including provisions for companions and the lines of sight to the focal point or field of play.

Companion Seating

Appropriate companion seating is important to wheelchair users' enjoyment of sports and entertainment facilities. The numerical ratio of companion seats to wheelchair spaces and their spatial interrelationship are two factors that must be integral to the overall facility's design. ADAAG specifies that "at least one companion seat shall be provided next to each wheelchair seating area." Department of Justice interpretations and also federal court decisions have indicated that a companion seat is required to be directly adjacent to each wheelchair space (U.S. DOJ, 1993/1994). One common arrangement is the "two-two-two distribution" (i.e., two companion seats, next to two wheelchair spaces, next to two companion seats, etc.).

While the two-two-two plan satisfies the law, it is not ideal because it limits the available seating combinations and arrangements. For example, only two wheelchair users can sit together, unless management removes a companion seat. For families or groups with three or more members attending, the two-two-two configuration reduces the possible seating arrangements. For example, a father in a wheelchair may not be able to sit *between* his wife and child, but rather *beside* either one or the other. The seating alternative that can address these two issues is *adaptable* seats for both wheelchair spaces and companion seats. Both plans are seen in Fig. 47.2.

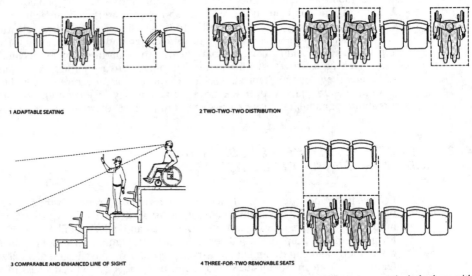

1 ADAPTABLE SEATING　　　　2 TWO-TWO-TWO DISTRIBUTION

3 COMPARABLE AND ENHANCED LINE OF SIGHT　　　　4 THREE-FOR-TWO REMOVABLE SEATS

FIGURE 47.2 Flexible seating arrangements provide more options for all spectators, particularly those with mobility impairments who use assistive devices such as a wheelchair or scooter.

Adaptable and Removable Seats

Many major seating manufacturers now make stadium seats that can either be folded up to accommodate wheelchair users or used as conventional seats. Adaptable seating's flexibility allows spectators to choose any combination of seating arrangements within a family or group. It also allows last-minute adjustments to previously purchased tickets. Adaptable seats' drawbacks are the higher initial unit costs and horizontal space limitations. These seats are several times more expensive than conventional seats, and they are typically slightly wider so more row space is required. In some stadium designs, these differences can be critical. In other instances, however, extra space is available and within the overall project cost, the difference is negligible. To save on either space or construction cost, a mix of different accessible seating types, some infill (or removable) seats and some adaptable seats, for example, can be used within a single facility (Beasley and Davies, 1995).

Current ADAAG criteria permit the installation of removable seats in unsold wheelchair spaces. Different facilities have addressed this option in different ways, and the choices made can impact the integration of the wheelchair users into the viewing audience, the facility's ticket sales policy, and the potential revenues to the owners. The installation of removable seats in unoccupied spaces makes the wheelchair user and their companions more integral members of the audience instead of being visually isolated from conventional seating. Removable seats also better define the wheelchair spaces in terms of their precise location, and they prevent disruptive ticket holders from informally congregating in unclaimed locations. Some infill seating arrangements are designed to maximize the owner's revenues but have operational and qualitative disadvantages for wheelchair users. Large, inflexible infill arrangements like multitiered platforms fail to provide wheelchair users with the desired visual integration and encourage unfair ticketing sales policy. In some facilities, for example, the sale of a single wheelchair seat could displace as many as 27 conventional seats. A better and more flexible system is a "three-for-two" configuration, where two unoccupied wheelchair spaces are infilled with three conventional seats.

Comparable Lines of Sight

ADAAG requires accessible seating to provide wheelchair users with comparable lines of sight to the performance area. Though strongly implied in the ADA's statutory language, ADA design standards do not specifically state that wheelchair users must be able to see over a spectator who stands up in the row in front. The Department of Justice has published interpretations, however, that indicate that this is necessary at facilities where designers can expect spectators to stand and thereby block wheelchair users' view at critical periods during the event (U.S. DOJ, 1998). Recent U.S. court rulings (*Paralyzed Veterans of America v. Ellerbe Becket Architects, P.C.,* 1996a,b) have supported this interpretation. Where wheelchair users are not seated in the front row, a vertically enhanced seating tier will elevate the seats to provide the necessary sight line.

47.4 ASSEMBLY AREA FACILITY TYPES

All types of assembly seating are different, and, therefore, each individual facility needs to be evaluated in order to determine the appropriate features for spectators with disabilities. In different sports facilities, for example, different seat locations are considered ideal. Midfield or 50-yard-line seats are considered the best football seats. Corner quadrant seats are the best for ice hockey. Many baseball fans like the first- or third-base-line seating. Wheelchair seating is required to be dispersed in every type of facility and offer spectators with disabilities a full range of seat locations, including the prime locations.

Optimum sight lines also vary with different types of assembly facilities. Sports facilities, for example, require good side-to-side visibility so those spectators can follow the action up and down the field of play. A theater stage has a much more limited performing area with the most critical events set in center stage. A movie theater's focal point is the projection screen that can be designed in a manner that partially compensates for sight line deficiencies. ADAAG's comparability requirement, therefore, must be addressed in a different way in each facility type.

Audience behavior also varies in different facility types. In a football stadium, for example, fans can be expected to stand up at the contest's exciting moments. If you are unable to stand, standing spectators may completely block your view unless you have front-row seats or other provisions have been made. On the other hand, tennis spectators never stand up during a match because of the sport's rigid social etiquette. Each specific facility needs to be examined on a case-by-case basis in order to determine appropriate accessibility provisions for their specific seating configuration and anticipated use.

47.5 *ACCESSIBLE SPORTS FACILITIES*

Two projects stand out in the development of accessibility in American sports facilities: Oriole Park at Camden Yards and the 1996 Centennial Olympic Games in Atlanta, Georgia. In both projects, disability advocates were actively involved in the design process and they were pivotal in the outcome. Through this involvement, they ensured that spectators, athletes, and employees with disabilities could enjoy these facilities from spectator viewing areas, on the field of competition (as seen in Fig. 47.3), and in all employment areas.

FIGURE 47.3 Accessible sports facilities such as the Atlanta Olympic Tennis Center provided state-of-the-art access for world-class disabled athletes during the 1996 Paralympic Games.

Camden Yards

Oriole Park at Camden Yards, in Baltimore, Maryland, was the first new stadium to capture "old-time" ballpark magic. Beneath this traditional exterior, however, this park incorporates state-of-the-art technology for a modern major league baseball facility. The nostalgic exterior was a response to its unique site, Baltimore's Inner Harbor with its historic transportation and industrial buildings. The modern design features were the result of hard work by the owners, the state of Maryland, HOK Sport, and the other professional and volunteer members of the design team.

In the late 1980s, when the Maryland Stadium Authority was formulating its first plans, Governor William Donald Schaeffer invited disability advocates to participate on the governor's advisory committee for the stadium design. As members of the committee, architects from the Paralyzed Veterans of America provided stadium designers with a written design program that outlined the specific requirements and design objectives established to enable persons with disabilities to enjoy the completed facility. Comprehensive design reviews were conducted periodically by disability advisors throughout the project's design.

One Camden Yards innovation was the *adaptable seat,* which allowed unlimited seating arrangements to accommodate wheelchair users' families and friends (Fig. 47.4). Developed and patented by the PVA, this invention sparked other accessible innovations throughout the seating industry. A version of PVA's convertible seat, for example, was installed in the 1996 Centennial Olympic Stadium, and later became a permanent seating option for the Atlanta Braves' Turner Field.

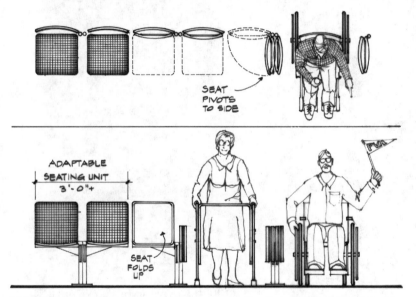

FIGURE 47.4 The PVA adaptable seat was first proposed during the design of Camden Yards and later was developed by American Seating Company as the equal-access seat.

Although it was designed and constructed prior to the ADA's enactment, Camden Yards' accessibility established a benchmark for the new ballparks that followed throughout the United States. The 1991 adoption and publication of new ADA design guidelines showed that, in spite of its innovations, Camden Yards' design fell slightly short in some areas. The most notable deficiency was line of sight to the field for wheelchair users on the lower level when spectators stand. This singular design issue proved to be pivotal in future stadium and arena design, particularly the 1996 Centennial Olympic Games.

1996 Olympic Games

The United States hosted the 1996 Centennial Olympic Games in metropolitan Atlanta and outlying venues in Georgia and Tennessee. This international event was the first sports facility construction program initiated after the ADA. The U.S. Department of Justice helped guide the planning efforts that created these model facilities of fully accessible new construction. The

PVA again played a key role in helping make the Atlanta Games the most accessible games in Olympic history. The Atlanta Committee for the Olympic Games (ACOG) retained PVA architects to consult on 12 new sports venue designs and on adaptations of numerous existing Atlanta sports facilities. In addition, the PVA consulting team helped prepare operational plans and spectator information booklets that assisted visitors with disabilities from all over the world to utilize the access that was designed and planned for the Games (Fig. 47.5).

FIGURE 47.5   A variation of the PVA adaptable seat was created by Camatic Seating Company from Melbourne, Australia, and supplied for the 1996 Olympic Stadium.

The Atlanta Paralympic Games were held immediately following the Olympics, an arrangement that allowed these athletes with disabilities to take full advantage of the new and accessible Olympic sports venues. The Olympic facilities included not only conventional sports stadiums, but also more unique competition venues such as rowing and canoeing, bicycle racing, and rifle shooting. These unusual specialized facilities posed new accessibility challenges such as the temporary seating platforms that were floated in the middle of Lake Lanier in order to view rowing and canoeing competition from the best vantage point possible. All together, the accessibility that was planned and provided for the 32 separate Olympic sports competitions established new benchmarks for future sports venues.

The Atlanta Olympic Stadium, the Games' flagship facility, was where the 1996 Olympics and the 1996 Paralympics held their opening and closing ceremonies and also the site of some of the Games' most exciting athletic competition. For spectators using wheelchairs, this facility was the first of its kind, allowing almost all wheelchair seating (840 spaces) to have an unobstructed view of the field of play even when the audience stood. In addition, for the first time, athlete areas such as locker rooms, showers, toilet facilities, and other related functions were fully accessible. World-class Paralympic athletes used these amenities as did Olympians like Jackie Joyner-Kersee and Michael Johnson, who suffered injuries during competition.

As part of the Olympics' legacy, the stadium was reconfigured after the Paralympic Games to be the new home of the Atlanta Braves baseball team. Accessible spectator seating, athlete lockers, showers, and even the new dugouts are notable reminders of the Olympics' broad commitment to accessible design excellence.

A unique service offered to Olympics spectators was a continuous color commentary on each competition event that was broadcast in real time in several different languages. Recognizing this service as a potential benefit to spectators with sight impairments, ACOG elected to offer receiver headsets free of charge to all spectators with impaired vision so that they could better enjoy the Games.

47.6 STADIUM-STYLE MOVIE THEATERS

New stadium-style movie theaters are currently being constructed at locations throughout the United States. Disability advocates have often been dissatisfied with the wheelchair seating many of these new theaters provide. This issue has been raised at public forums for local building codes and also by federal legal action. Theater seating has much different sight lines than the seating in live assembly facilities such as sports arenas, music and lecture halls, playhouses, and legitimate theater. The appropriate wheelchair seating design specifically for movie theaters, therefore, needs to be carefully considered in order to provide full and equal enjoyment of the facility.

A movie theater's theoretical focal point, for example, is always horizontally centered and is vertically located above most viewers' heads because film images are projected onto an elevated (and sometimes tilted and curved) screen. In stadium-style theaters, the middle and rear seats are installed on steep tiers, while the front seating is installed on a gently ramped floor. This front area is typically where the wheelchair seating is located. For these front seats, the upward viewing angle is steep. The rear and middle theater seating is arranged on flat tiers (like sports arena seating), so those viewers have an unobstructed view over the preceding row's occupants. The tiered seats have ever-decreasing vertical sight line angles as the viewer moves up and farther away from the screen.

Comparable Sight Lines

For wheelchair seating, several design considerations must be addressed for full and equal enjoyment of this sophisticated new technology. The most critical issue is the provision of comparable sight lines. Seats located in the theater's front seating area provide viewers with less comfortable seating because of the steep viewing angle to the screen. Industry guidelines indicate that many viewers find uncorrected sight line angles in excess of 35° to be uncomfortable. Conventional theater seats partially mitigate steep sight angles by installing chairs with a built-in seat recline of approximately 12°. Most rigid-frame wheelchairs, however, have much less recline, so the wheelchair user's view angle is typically steeper than that from the immediately adjacent companion seat. In some instances, wheelchair spaces' sight lines are further compromised because the seating area floor has a forward slant of approximately 1° (2 percent slope). (See Figs. 47.6 and 47.7.)

For theater owners, the nontiered front seating is almost a free bonus because it makes use of the otherwise unproductive space. When the performance or show is sold out, the front seats will often be occupied. They are typically less popular choices, however, in partial-capacity audiences. As marginal seating, this is not a satisfactory location for wheelchair seating, particularly when this is the facility's only wheelchair seating.

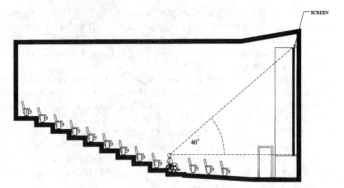

FIGURE 47.6   Many of the earlier-designed stadium-style movie theaters placed wheelchair seating areas in less-than-desirable locations with limited choices for seating.

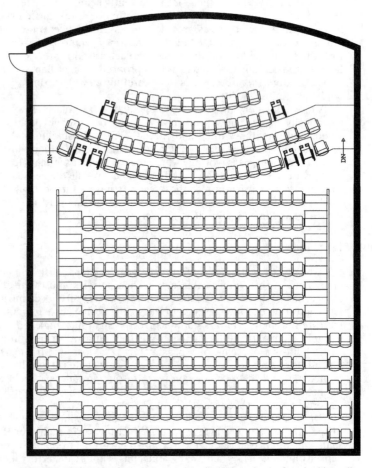

FIGURE 47.7   Wheelchair users are given limited seating options in this stadium-style theater with viewing angles that cause discomfort.

Several stadium-style theaters are typically clustered around a common lobby with food concessions and public restrooms. The individual theaters are entered through rear vomitories from the common lobby. Larger complexes may include theater access from both upper and lower lobbies.

Seating Choices

Another accessibility issue is the range of seating choices that are offered to wheelchair users. In larger-capacity theaters and in those complexes with two lobby levels, it is possible to provide two or three good wheelchair seating options in each facility. In many smaller theater facilities, practical wheelchair seating options may be more limited. (See Figs. 47.8 and 47.9.)

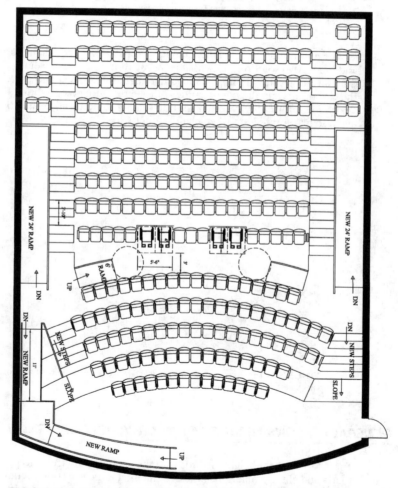

FIGURE 47.8 PVA architects proposed this modification to a stadium-style movie theater in order to more fully integrate wheelchair seating into the seating plan and to provide comfortable, and comparable, lines of sight.

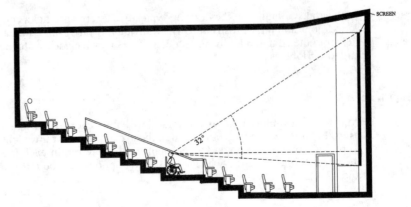

FIGURE 47.9   Redesigned for better access, integration, and sight lines, the movie theater now more fully accommodates patrons who use wheelchairs.

The appropriate accessibility standards for stadium-style theaters require careful analysis and innovative solutions. Discussions between industry representatives, public officials, and disability advocates are important so that all parties have an accurate understanding of both the constraints and the opportunities offered by these new facilities. For theater owners, new design ideas for easily convertible infill seating can partially offset their economic concerns and also better integrate wheelchair seating. (See Fig. 47.10.)

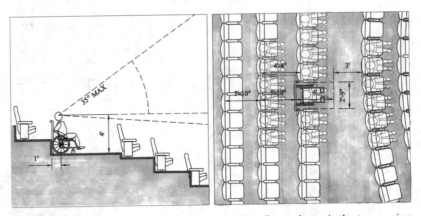

FIGURE 47.10   Accommodating wheelchair access in stadium-style movie theaters requires careful attention to detail.

47.7   *GENERAL LESSONS LEARNED FROM UNIVERSAL DESIGN*

The ADA brought heightened focus to the need for full access to persons with disabilities to sports and entertainment facilities. As the first large event that required the design and construction industry to respond to the new ADA requirements, the 1996 Centennial Olympics was an unarguable success. For the first time in sports history, facilities, spaces, and elements were uniquely configured to address the needs of the broadest range of users, including athletes, entertainers, spectators, employees, and volunteers, with or without disabilities. The universal application of many accessible features was apparent during the Olympics when world-class athletes experienced sports-related injuries and thereby became the first beneficiaries of accessible lockers, showers, and toilet facilities. Speakers, entertainers, and former athletes such as Muhammad Ali, Christopher Reeve, and others were able to participate and contribute to this worldwide event because of the full access provided to virtually all areas of Olympic facilities. Thousands of fans with disabilities, along with their families and friends, were full participants in the Olympic experience. The PVA adaptable seat, first unveiled in Baltimore's Camden Yards, evolved into yet another variation in the Atlanta Olympics, and more recently in the Sydney 2000 Olympics.

Another initiative that captured universal interest was the *tactile stadium model,* created by PVA's National Architecture Program in cooperation with Pratt Institute (Fig. 47.11). This scaled model of the Centennial Olympic Stadium provided visually impaired individuals with a "vision" of the facility's design through their touch. It also enabled children to use their tactile sense, along with their visual sense, to more fully experience architectural design.

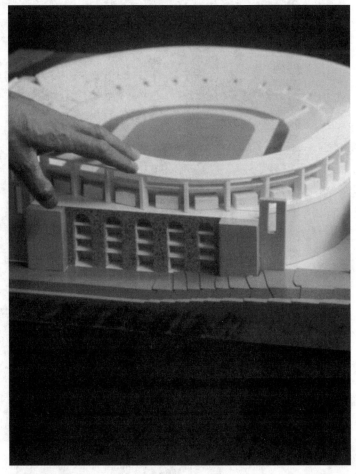

FIGURE 47.11   The tactile stadium model enabled visually impaired individuals, as well as children, to experience the Olympic Stadium through touch.

A special Olympic spectator guide (the *Centennial Olympic Games Guide for Guests with Disabilities*) was created and published by the PVA and the Eastern Paralyzed Veterans Association to familiarize Olympic visitors with the accessibility that was provided at each venue and also the fully accessible Olympic transportation system (Fig. 47.12). In the hundreds of hot-line calls that the Atlanta Committee for the Olympic Games (ACOG) received daily during the events, this guide proved to be indispensable. A typical phone call would, for

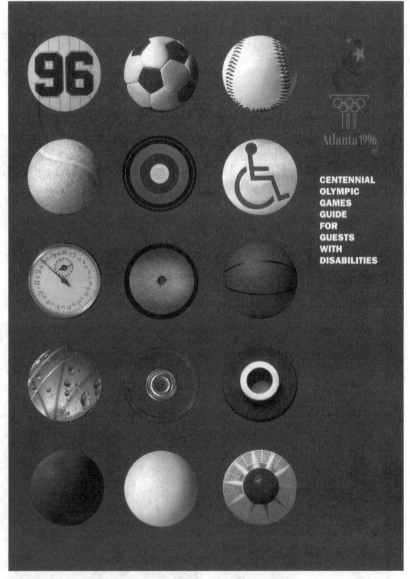

FIGURE 47.12   The *Centennial Olympic Games Guide for Guests with Disabilities* was not only a valuable resource for persons with disabilities, but proved universally popular among many Olympic visitors.

example, cite a family member that had just broken a leg and needed to exchange the tickets for conventional seating that they had purchased many months before. With the accessible spectator guide, the visitor could find their accessible seats, locate an accessible venue entrance, and plan their transportation to the event. All the visitor needed to do was stop at any information kiosk and ask for the free guide. Advance purchasers of wheelchair seating were mailed a copy of the guide with their tickets.

47.8 CONCLUSION

Accessibility is achieved through careful coordination between the facility owners, the designers, the operators, and the end users. For unique events like the 1996 Olympics, the accessibility challenges were significant. Design considerations included the need for strict security provisions, the use of many older and less accessible venues, the hybrids of temporary and permanent structures used in almost every venue, the huge spectator crowds from hundreds of countries, and most important, the requirement to *get it right* the first time. In spite of these challenges, the event was successful because of ACOG's strong commitment to accessibility, their use of professional ADA technical assistance, and their proactive efforts to engage the disability community throughout the planning, design, and construction process.

In recent years, compliance with the minimum standards for accessible design in sports and entertainment facilities has helped this growing industry so that design for persons with disabilities has many universal benefits to a broad range of facility users. Seating options, alternative communication techniques, and other disability-related accommodations enhance the effectiveness of the facility for everyone. In many ways, design diversity that directly responds to the need for "required" access is representative of the richly varied population of end users who benefit from universal design.

47.9 BIBLIOGRAPHY

Beasley, Kim, and Thomas Davies, "Seating for Assembly Areas," in *Progressive Architecture,* Penton Publishing, Cleveland, OH, 1995.

Paralyzed Veterans of America (PVA) v. Ellerbe Becket Architects & Engineers P.C. (EBAE), Plaintiff's Complaint for Declaratory Judgement and Injunctive Relief, 945 F. Supp. 1 (D.DC) summary: 1–3, USDC DC 1:96CV01354 PC 14, 1996a.

Paralyzed Veterans of America v. Ellerbe Becket Architects & Engineers, P.C., Judge Thomas F. Hogan's Order 19 July 1996, 945 F. Supp. 1 (D.DC 1996) 2–3, USDC DC 96CV01354 O 19 July 1996, 1996b.

Peredo-Lopez, Carol, *ADA Compliance Study for Stadium-Style Movie Theaters,* Paralyzed Veterans of America, Washington, DC, 1999a.

——, *ADA Compliance Study for Stadium-Style Movie Theaters,* Paralyzed Veterans of America, Washington, DC, 1999b.

——, *ADA Compliance Study for Stadium-Style Movie Theaters,* Paralyzed Veterans of America, Washington, DC, 1999c.

——, *ADA Compliance Study for Stadium-Style Movie Theaters,* Paralyzed Veterans of America, Washington, DC, 1999d.

——, *ADA Compliance Study for Stadium-Style Movie Theaters,* Paralyzed Veterans of America, Washington, DC, 1999e.

U.S. Access Board, *ADA Accessibility Guidelines,* 01 July 1994 28 CFR Pt. 36, U.S. Access Board, Washington, DC, 1994.

U.S. Department of Justice, *The Americans with Disabilities Act Title III Technical Assistance Manual* (covering public accommodations and commercial facilities), U.S. Department of Justice, Washington, DC, 1993.

———, *The Americans with Disabilities Act Technical Assistance Manual, 1994 Supplement, Architecture Magazine,* U.S. Department of Justice, Washington, DC, 1994.

———, "Justice Department Reaches Settlement With Architect of New Sports Arenas" (press release), U.S. Department of Justice, Washington, DC, 1998.

47.10   RESOURCES

www.pva.org
www.pvapub@aol.com
www.paraplegianews.com
www.psa.org

CHAPTER 48
LANDSCAPE DESIGN

Yoshisuke Miyake, J.C.E.A., J.I.A.
SEN, Inc., Osaka, Japan

48.1 INTRODUCTION

This chapter focuses on how universal design is incorporated in landscape design of public parks, which is the author's primary field of work. It describes the social background for universally designed parks in Japan, presents findings from a user opinion survey on what is needed by people who are disabled and elderly, and it highlights design considerations for universal use of parks. Based on the author's recent work, two case studies are presented: Sensory Garden as well as Rinku Park/Symbolic Green Park/South Zone, both in Osaka.

48.2 SOCIAL BACKGROUND FOR UNIVERSAL DESIGN IN PUBLIC PARKS IN JAPAN

Because of the limited land area for housing, the per capita living space in Japan is rather small when compared with the United States and Europe. It is only natural that Japanese people should seek a space for relief outside of the house. In particular, many people visit public parks to enjoy a spacious setting with green and flowers. Serving as an oasis in the dense urban areas, public parks are playing an important role in Japanese society. The changing demographics and growing orientation toward community-based care, as well as changing attitudes, all add to the importance of universal design in public parks.

Until recently, Japanese people received the benefits of a national health insurance system through which all Japanese nationals could obtain medical treatment at low cost. However, with the emergence of an aging society, this system has begun to deteriorate in recent years, resulting in an imbalance between expanding medical expenses for the aged population, and a sharp drop in revenue from a decreasing number of people in the younger generation. Although a national health care insurance system is being introduced to respond to the aged population with disability, it is assumed that in the future medical institutions will be incapable of providing the types of quality treatment provided now, and that more people will receive home care and spend plenty of time in the community. This means that the use of public facilities by people who are elderly and disabled will increase further; therefore, universally designed public facilities including parks will be in great demand, as Matsumoto states in Chap. 21, "Japanese Guidelines for Universal Design in Parks: Harmony Between Nature and People," of this handbook.

On the other hand, the cultural attitude toward people with disabilities has changed, and more people with disabilities are going out to enjoy community life. Also, the Japanese media tend to showcase more advanced social environments surrounding people with disabilities, as well as their outstanding achievements in foreign countries. With similar cases being reported in Japan, and a growing number of people with disabilities participating in society nowadays, universally designed public parks are urgently needed to serve diverse user groups.

48.3 CREATING PUBLIC PARKS WITH THE UNIVERSAL DESIGN CONCEPT

The author's work of creating landscaping with the universal design concept usually follows four steps:

1. Survey and needs analysis for universal design
2. Design and field inspection
3. Verification by users
4. Feedback

The first step is to heighten people's awareness of the need for universal design and to incorporate the universal design concept in the design process. Then, a user opinion survey is conducted through interviews to identify needs and to obtain comments from people who are disabled and elderly.

The second step is to add design considerations for requested items found in the survey to the accumulated design data, and to propose the most appropriate and best-suited design for the site. It is also important to provide aesthetic value. In the construction stage, in order to be implemented based on the design, the design intent needs to be reflected with the utmost care.

The third step requires participation of prospective users, including people who are disabled and elderly, for a trial use to check that the completed work is usable by people with different levels of ability.

Based on the results of verification, the final step involves addition to and modification of the completed work and research data. When incorporating universal design principles in landscape design, it is very important to repeat these steps in order to attain the greatest extent of usability for diverse user groups. In the section that follows, findings from an opinion survey on public park usage are reported.

Findings from a User Opinion Survey

The key point of universal design is *diversity*. As each individual is different, there is a difference from person to person in type of disability, daily activity patterns, the way individuals advance in age, how one enjoys, or how one feels uncomfortable. However, diversity is the most difficult subject to give form to in design, and it is clarified solely by a survey of a number of people and their individual needs. The author and his staff conducted a survey through interviews of park users (i.e., approximately 600 people who are disabled and elderly) to find out what is needed for the universal use of public parks. Requests and comments expressed in the survey became useful material and a valuable resource for the author's design work. Some of the survey findings follow.

People in a Group Home.   The people in a group home varied in age and types of disability. Their age ranged from the 30s to the 50s, and the types of disability included mobility and dexterity limitations, and hearing impairments. Some of them used wheelchairs. First, they complained about poor maintenance of park facilities. For example, in some cases, wheelchairs provided in parks were dirty or had flat tires, and rest rooms were often locked to prevent vandalism. They were disappointed with their inability to use these facilities when needed. Second, they commented that accessibility in parks is a matter of absolute necessity, and further, that rain shelters should be provided in large park areas because people with disabilities generally have difficulty in exiting promptly in case of a sudden change of weather. Other requests to improve the use of parks included wider pathways, so that two people using wheelchairs are able to proceed side by side, and provision of human assistance for recreational activities such as fishing. Last, regarding the fundamental issue of access to parks, they said they would appreciate it if volunteers were provided to take them to parks.

Wheelchair Users.   A comment that was common for many wheelchair users was their desire to enjoy parks either in a group or alone on their own. They were interested in path width, paving materials for smooth travel, as well as the height of handrails that permitted a vista from wheelchairs. Different types of benches were also desired, such as one where a wheelchair user and an able-bodied person could sit side by side, or a movable bench or table, depending on the circumstances. Wheelchair users were mostly troubled with insufficient information on accessibility; therefore, maps and other information devices were strongly requested.

People with Visual Impairments.   This survey revealed that people with visual impairments had the most difficulty in parks when compared with people with different types of disabilities. Therefore, their requests and comments covered quite a wide range of design details. Basically, where they felt most uneasy were very large spaces, such as parks lacking orientation so that it was almost impossible to move about alone. They placed high emphasis on feelings of anxiety when visiting a very spacious playground, fear of stairs lacking uniformity, or a tree branch protruding into their path at the height of their face. In short, difficulties can arise when visually impaired park users have to respond to unpredictable situations.

On the other hand, they desired flower beds with distinctive colors and play equipment providing physical challenges. Also, they suggested that help in orientation be given by slightly sloping ground surfaces and that chairs with backrests be provided to make them feel more at ease. Regarding signage, they requested tactile signage and maps on the assumption that many blind people are unable to read Braille.

Examples of undesirable features they noticed were flower beds with strong fragrances, as well as the loud sound of cascading water, both of which result in adverse effects of unpleasantness and uneasiness. The majority of those who were interviewed for the survey mentioned that they had their own way to enjoy the natural environment, which was different from that of able-bodied people (e.g., listening to rough-and-tumble children's play or to little birds singing). It was their hope that able-bodied people would understand that no special provisions were necessary for visually impaired people to enjoy nature.

In addition to the aforementioned, valuable survey findings were obtained from children with developmental disabilities and elderly people in nursing homes.

Design Considerations for Universal Use of Public Parks

Basic design considerations for universal use of public parks are listed in Table 48.1.

48.4   CASE STUDIES

Two case studies from the author's recent work with universal design concepts are presented in the following sections.

TABLE 48.1   Basic Design Considerations*

Item	Basic design considerations	W	E	V	H	D
	Path					
Configuration	• The walkway width is enough to allow a wheelchair user to turn a right-angled corner with ease.	●				
	• The walkway width is enough to allow two wheelchair users to pass each other.	●				
	• The straight walkway is easier to walk.			●		
	• The level difference at a sloped curb is less than 1 cm.	●	○			
	• Walkway edges are provided, where necessary, to protect wheelchairs from dropping off.	●				
	• Other methods of directional guidance than detectable warnings are provided depending on the condition of the walkway.			●		
	• A boardwalk is provided with raised edges with a height of 10 cm to protect wheelchairs from dropping off.	●		○		
Surface material	• Crushed stones, gravel, etc., are not used so that wheels are not stuck or get flat tires.	●				
	• Concrete causes glare and uncomfortable light levels. Surface material with less glare is used.	●	●	▲		
	• The joint of surface materials is finished smooth with minimum recess and minimum width.	●	○			
	• When dust is used, proper drainage is necessary to prevent a muddy condition.	●	●			
	• Slippery materials are not used. A level surface is recommended.		●	○		
	• It is preferred to use a surface material that makes desirable sounds when walking on it.			●		
	• Surface materials of the main walkway and those of other walkways are different.			●		
	• Use of too many different types of surface materials may cause confusion.			●		
	• The surface material of both sides of the walkway is distinctly different from that of other portions to protect wheelchairs from dropping off.			●		
	• The color of walkway edges is in contrast with that of other portions of the walkway.			▲		
	• Surface material is selected so that people can walk with less fatigue.		●			
Handrail	• Braille or other tactile signage is provided at the top end and the bottom end of handrails.			●		
	• Wooden handrails feel comfortable when used.	○	●	○		
Slope	• The sloped board is grooved to prevent slippage.	○	●			
	• Handrails are provided for a steep slope without exception.	○	●			
	• A walkway with a slope, rather than with steps, is desirable for main access.	●				
Stairs	• The surface material immediately before the first step and immediately after the last step is different from that of other portions to warn users.			●		
	• In the case of a wide step, the surface material of the edge is different from that of other portions to warn users.			●		

TABLE 48.1 Basic Design Considerations (*Continued*)

Item	Basic design considerations	W	E	V	H	D
	Path					
Stairs	• Handrails are provided for stairs without exception. A landing is provided for long consecutive stairs. • Effective lighting is provided so that each step can be recognized with ease.		●	▲		
	Entrance/exit					
	• The door at the entrance/exit is easy to open and close.	●	○			
	• All entrances/exits on a level ground allow a wheelchair user to pass through.	●				
	• An entrance must not allow motorcycles and cars to pass through.		○	●	●	○
	Rest room					
	• Accessible rest rooms are provided separately for men and women.	●		○		
	• The door of an accessible rest room is kept open, when not used, in order to prevent vandalism.	●		○		
	• Heavy doors are not used. A sliding door has a doorstop to protect user's fingers from injury when it is opened with a grip.	●				
	• Braille and tactile panels are provided to give useful information, such as the location of toilet paper, a push-type flush button, etc.			●		
	• A push-type flush button is provided on the wall and on the floor for people with limb disabilities.	●	○			
	• The sink and doorknob are at an appropriate height for wheelchair users.	●				
	• Toilet seats are always clean to prevent problems in use.	●		●		
	• Colorful design is desirable for accessible rest rooms to provide a pleasant atmosphere.	●				
	Parking area and others					
	• Parking areas for exclusive use by drivers in wheelchairs are provided near the park entrance/exit and major facilities of the park.	●	○	○		
	• Smooth and safe travel, without a change in level, is secured from a parking area to the park entrance/exit.	●	○	○		
	• The park entrance/exit is located in the vicinity of the nearest train station or bus stop.	●	○	○		
	Rest area and others					
Trellis	• A clear space is provided to allow a wheelchair user to turn around in a trellis, and easy access is provided.	●				
	• A wind-resistant shelter is recommended, instead of a trellis, for people who are sensitive to temperature changes.	○	●			
	• Plants with fragrant vines or fruits are recommended.			●		
	• A trellis with foliage is recommended to provide effective shade.	○	○	●	○	○
Bench and others	• The bench has an adjacent clear space to allow a wheelchair user to sit right next to other, seated persons.	●				
	• A space for wheelchair users is provided for a set of picnic tables and benches.	●				

(*Continued*)

TABLE 48.1 Basic Design Considerations (*Continued*)

Item	Basic design considerations	W	E	V	H	D
	Rest area and others					
Bench and others	• A bench is placed in the sun in a windless setting. (A shade area is desirable in summer.)	○	●	○	○	○
	• A backrest and armrests are provided with a bench.		●	●		
	• A rest area with a bench is provided at an interval of 100 m.		●	○		
	• The bench is provided with a device to hold a cane.			●		
Drinking fountain	• Clearance of more than 69 cm is provided under the drinking fountain for a power wheelchair user.	●				
	• The ground surrounding a drinking fountain has a level area for access of wheelchair users.	●				
	• The spout is set at an appropriate position for a wheelchair user.	●				
	• Two types of controls, a faucet and a pedal, are desirable for each drinking fountain.	○	●	○		●
	• Waterflow in a trajectory is desirable.	●				
Others	• The telephone booth is accessible and has sufficient floor area for wheelchair users.	●				
	• A vendor is stationed at least at one shop in a park.	○	○	●	○	○
	• Vending machines are placed in a park, usable by people in wheelchairs and with visual impairments.	●		●		
	• Placing a telephone along a walkway is desirable for emergency use.		●			
	Signage and guidance					
	• Major park entrances have signs with a park guide and accessible routes.	●	○	●		
	• The main gate has signs showing a map and functions of major facilities of the park.	●	○	●▲	○	○
	• Accessible routes information includes estimated time required for travel.	●	●			
	• The main gate has a tactile panel showing the park layout with Braille.			●		
	• At the main gate, an appropriate audio system is desirable for information on park facilities.			●		
	• Water sound, taped singing bird, or other natural sounds are used for directional guidance.			●		
	• A large label is used for identification of a tree or a flower.		○	▲		
	• Use of Braille and relief tiles is desirable for identification of plants.			●		
	• A tactile information panel and Braille panel are placed where temperature does not rise too high in summer.			●		
	• A sculpture, which allows people to touch, helps orientation.			●		
	• For orientation, a bell or other sound device is provided along a path.			●		
	• Fragrant plants are planted in places as a landmark.			●		
	• If a walkway has stairs, a sign is provided in advance to inform of its existence.	●				
	• Pictograms are used for signs.		●			●

TABLE 48.1   Basic Design Considerations (*Continued*)

Item	Basic design considerations	W	E	V	H	D
	Campground					
	• The shower room and a rest room are usable by people in wheelchairs and with visual impairments.	●	○	●		
	• A kitchen is accessible by people using wheelchairs and canes.	●	●	●		
	• A log or a bench is provided around a fire to sit in a circle, instead of sitting on the ground.	○	●	○		
	• The cabin is clean and usable by a wheelchair user. A rest room is provided inside the cabin.	●	○	○		
	• At a campsite, a push-button audio system, Braille, and a tactile panel are provided for information on camp facilities.			●		
	• The barbecue grill is usable by a wheelchair user.	●				
	Playground and others					
	• The play area is designed so that children can feel safe.	○		○	○	●
	• There is a rubber mat in the fall zone of the swing.			●		○
	• A swing is provided with a backrest.	●		●		
	• A trellis designed in children's scale is desirable, and fragrant flowers are used for the trellis.	○		●	○	○
	• A small zoo, which allows children to touch animals, is desirable.	○		●	○	●
	• The sandbox is raised for access of children in wheelchairs.	●				
	• Knee clearance is provided under a raised sandbox.	●				
	• A composite play structure is provided with a slope and handrails.	●		●		
	• A passage within a composite play structure is wide and accessible by children using wheelchairs.	●				
	• Access to a water feature is provided.	●	○			
	• Scattered play equipment in a large area is not recommended.			●		
	• Play equipment provoking physical challenge or composite play structures are preferred by children.			●		●
	Planting					
	• Tree branches are trimmed correctly so that they do not hurt people on a walkway.			●		
	• An accessible rest space with a bench is provided under a large tree.	●		●		
	• A raised lawn area is provided to allow a wheelchair user to transfer.	●				
	• As people with visual impairment have a desire to touch trees, those with a distinctive characteristic are planted to allow people to touch a trunk, branches, and leaves.			●		
	• Trees and flowers are planted in consideration of the wind direction so as to avoid a very strong fragrance.			●		
	• Trees and flowers having aesthetic quality are planted near a bench or a trellis.		●			
	• A flower bed with distinctive colors, such as orange, white, and yellow, is provided for people with poor eyesight.			▲		

(*Continued*)

TABLE 48.1   Basic Design Considerations (Continued)

Item	Basic design considerations	W	E	V	H	D
	Planting					
	• Trees and flowers with unique texture and fragrance are planted for directional guidance.			●		
	• A raised flower bed is provided to allow people to touch.	●		●		
	• Plants with thorns or sharp leaf edges are used for planting with care.			●		
	• Ends of tree stakes are trimmed properly so that they do not hurt people.			●		
	• Use of a flower bed fence is minimal so that people can touch flowers.			●		
	• Plants that cause allergy are not used.	○	○	○	○	○
	Building in park					
	• A shower room and other facilities in a sports clubhouse are accessible by people using wheelchairs and with visual impairments.	●		●		
	• A space where a wheelchair user removes dirt on wheels is provided in front of the gymnasium entrance.	●				
	• All facilities in a lodge are accessible by wheelchair users.	●	○	○		
	• The inside of the building is designed so that people with disabilities are able to proceed freely and safely.	●	○	●		
	• For people with visual impairments, right-angled corners are sometimes easier to follow than curvilinear or odd-angled corners.			●		
	• A warning sign is provided where transparent glass is used.			▲		
	• Dining tables in the restaurant are accessible by wheelchair users.	●				
	• In the restaurant, a menu using large print is provided. Dining tables are placed so as not to reflect sunlight.			▲		
	• The cafeteria is accessible by wheelchair users.	●				
	Landscape					
	• For a vista from a wheelchair, a handrail height of less than 90 cm is desirable.	●				
	• Visual access is provided for inaccessible areas for aesthetic reasons, such as stepping-stones in a Japanese garden.	●				
	• For a railing in a preserved area, natural materials, such as arranged stones, are used instead of a safety fence.	○	○	○	○	○
	Water area					
	• At a shoreline, access is provided as close to the water as possible.	●				
	• A railing and a fishing rod holder are provided in a fishing area.	●				
	• An adaptable wheelchair is provided for access into water.	●				
	• Swimming pool facilities are accessible by wheelchair users.	●				
	Amphitheater					
	• In a park landscape, an amphitheater is designed to be inclusive.	○	○	○	○	○

TABLE 48.1   Basic Design Considerations (*Continued*)

Item	Basic design considerations	W	E	V	H	D
	Amphitheater					
	• An amphitheater is accessible by both the audience and performers.	●				
	Others					
	• An open space is provided for new sports such as disc golf and pétanque.			●		
	• For horseback riding and canoeing, a transfer deck is provided for wheelchair users.	●				

* D: people with developmental disability; E: elderly people; H: people with hearing impairments; V: people with visual impairments; W: wheelchair users; ○: consideration is needed; ●: consideration is particularly needed; ▲: consideration for people with poor eyesight.

Source: Excerpts from Miyake, Yasuyo A., Hajime Kameyama, and Yoshisuke Miyake, *Creating People-Friendly Parks: From Barrier-Free to Universal Design,* Kajima Institute Publishing, Tokyo, Japan, 1996, pp. 110–113.

Sensory Garden

Project Background.   Initially, this project was planned to renovate the Garden for the Blind in Osaka, situated in the southwest corner of the approximately 90-hectare (ha) Oizumi Ryokuchi Park. The Garden for the Blind was opened in 1974, designed to focus on the special needs of people with visual impairments. However, after a quarter-century, it had become a secluded space where very few people visited. Considering the fact that such a segregated design concept for a particular group of people is no longer accepted in public parks, the initial renovation plan was eventually changed to the creation of a 2000-m² new garden at another location in the park, to be opened in 1997. The following were the key considerations of this project in the planning stage:

• Select the best location in terms of usability and view.

• Serve all users, so that people with and without disabilities, people who are elderly, as well as small children, can equally enjoy the place.

• Provide for aesthetic value, a pleasant experience, and joy of life.

Illustrations of Key Features of Universal Design.   The tactile gate door in Fig. 48.1 mirrors the inner atmosphere when the Sensory Garden is closed. In Fig. 48.2, each of the relief tiles installed in the entrance wall is shaped in the form of a plant that can be found in the Garden. The handrail underneath has each plant name in Braille on its back face. These two designs and the information display in Fig. 48.3 are universal ways of communicating information on the Garden atmosphere, image, and layout, respectively.

The hedge picture window in Fig. 48.4 peers into the Garden, revealing an image of the inside. Distinctive colors of flowers in Fig. 48.5 are for people with poor eyesight. In Fig. 48.6, a pair of pillars at each corner is crowned with bright-colored ornaments and, thus, indicates in which direction to proceed. By appealing to the senses, these elements serve as a guide to visitors. Visitors are gently introduced to the world of the Sensory Garden by the picture window with light and shade, then guided naturally to the distinctive color garden and to a pair of pillars provided at each corner for spatial orientation. Fig. 48.6 also shows two stainless-steel railings, which are installed parallel with the path for directional guidance.

The bench in Fig. 48.7 is set back so as not to hinder the circulation of visitors. Figure 48.8 shows the bench with flowers behind it, to which wheelchair users can transfer.

FIGURE 48.1 Tactile gate door.

FIGURE 48.2 Relief tiles in the entrance wall and handrail with Braille underneath.

FIGURE 48.3 Information display with letters, Braille, a tactile map, and a push-button audio system.

FIGURE 48.4   Hedge picture window.

Figures 48.9 through 48.12 show examples of not only access to water through placement of seating and positioning of running water, but also designs that maximize the aesthetic value of water features.

Figure 48.13 shows a universal way of appreciating art. A senior high school student, Kohki Sakata, who uses a wheelchair and also has visual and hearing impairments, is touching his own creation, the sculpture *Hands*.

FIGURE 48.5   Flowers in distinctive colors for people with poor eyesight.

FIGURE 48.6   A pair of orientation pillars at each corner.

FIGURE 48.7   Bench adjacent to the path.

Involvement of Specific User Groups in the Design.   The conventional type of plant labels in Braille often required people with visual impairments to bend their wrists when reading Braille. This problem was solved by improvement of the plant label configuration through consultation with a school teacher for blind students. The final design of the plant labels enabled Braille reading with low physical effort.

Other elements were designed for universal use, based on the findings of our survey of about 600 people with disabilities.

FIGURE 48.8   Bench with raised flower bed.

FIGURE 48.9   Raised pond.

Design Challenge.   The major challenge for the designer was to what extent accessibility could be introduced, in light of the beauty of the site, which faced a serene pond. Although aesthetic considerations and accessibility may often conflict with each other, this project succeeded in providing as much accessibility as possible. As shown in the examples of water access, this was accomplished by dealing with this problem as an integral design issue.

Benefits to a Variety of Users in the Final Design.   The final design enabled wheelchair users to tour the entire Garden on a smooth path surface while enjoying a variety of textures

FIGURE 48.10 Wheelchair-accessible cascade.

FIGURE 48.11 Bench in a nook created in the raised pond.

such as bricks, natural stones, and boardwalks. Raised flower beds, a raised pond, and a sculpture were accessible by wheelchairs. The guide railings installed along the path enabled people with visual impairments to proceed unhindered and to access all the highlights of the Garden. The user experience of the Garden was enriched by a variety of tactile and audio information, opportunities to touch and smell flowers, as well as to feel water or a sculpture. Small children were able to enjoy the feeling of cascading water on their own.

People who are disabled and elderly, having been kept from access to public gardens in the past, are now able to visit the Sensory Garden and enjoy time with others. The Sensory Garden not only created a space for outdoor recreation for people with disabilities, but it also provided a significant setting for the integration of all users.

FIGURE 48.12   Access to aquatic plants.

FIGURE 48.13   Sculpture touched by the artist, Kohki Sakata.

Lessons Learned from the Project.   Because the top priority in the planning stage was to select the best location in Oizumi Ryokuchi Park, the Sensory Garden was created in the central part of the park facing the serene and quiet pond, which is some distance from the park entrance. Although the walkway from the park entrance to the Sensory Garden is totally accessible and assures smooth travel by all visitors, it is probable that this distance may cause a problem to some people. However, it is believed that the site selection was made properly due to the following two reasons:

1. It is obvious that there has always been a variety of visitors since the Sensory Garden was opened, and all of them enjoy the unique waterscape at the site. This is exactly as expected in the planning stage (i.e., to create a garden for all people rather than creating a special place for people with disabilities).

2. The site selection necessitated a modification of the path connecting the park entrance with the Garden to improve accessibility, which resulted in a positive impact on the overall park design.

Comments Received from Others.   Since its opening in 1997, the Sensory Garden has attracted much attention in the media as an excellent example of universal design in an outdoor recreational setting. Design details of the Garden were introduced in NHK (Japan Broadcasting Corporation) TV programs, major newspapers, and magazines for landscape professionals. Also, people with disabilities (in particular, those with visual impairments) were highly appreciative of the diversity of sensory experiences in the Garden.

Rinku Park/Symbolic Green Park/South Zone

Project Background.   With the opening of Kansai International Airport in Osaka in 1994, new construction of a commercial and residential area named Rinku Town was planned on reclaimed land on the side of the bay that is opposite to the Airport. As a leading image symbolic of Rinku Town adjacent to the sea, this project was to create 8.5-ha Rinku Park/Symbolic Green Park/South Zone, which was to be opened in 1996. Following are the key considerations of the project in the planning stage:

• Create a gateway park to Japan that conveys a message of the new style of Japanese culture.

• Provide full use of the park location adjacent to the sea.

• Provide access for all people, including those with different nationalities, people with and without disabilities, elderly people, and small children. Provide various types of enjoyment regardless of length of stay in the park.

Illustrations of Key Features of Universal Design.   All people appreciate the impressive scene of the Fountain of Four Seasons (*Shiki-no-Izumi*) and Inner Sea (*Uchiumi*), shown in Fig. 48.14, as well as the beauty of nature reinforced by the sunset (Fig. 48.15). The information display, descriptive panels, minidisk audioguides, and the guide pipe embedded in the path (depicted in Figs. 48.16 to 48.18) show several ways of informing visitors of the park layout, major sights, and wayfinding. All of these helped minimize the number of people who did not obtain essential information about the park.

The raised flower beds in Fig. 48.19 enabled visitors using wheelchairs or small children to touch flowers. Also, the handrails in Fig. 48.20 did not restrict but helped provide a view for those people. Both facilitated access to landscape beauty by enhancing the design.

Adaptive wheelchairs for water use, shown in Fig. 48.21, increased the number of people who could share the enjoyment of the Inner Sea. The observatory on the top of the arched bridge, shown in Fig. 48.22, is within a short distance from the park entrance, so that people with limited physical strength can also enjoy the park as viewers in a less demanding way. The observatory is accessible by an elevator or a ramp.

FIGURE 48.14   Fountain of Four Seasons and Inner Sea viewed from the arched bridge.

FIGURE 48.15   Sunset scene with the Fountain of Four Seasons.

FIGURE 48.16 Descriptive panel with letters and Braille at each major sight of the recommended route.

FIGURE 48.17 Descriptive panel showing the route direction, a minidisk audio-guide channel, and the maximum slope.

FIGURE 48.18   Aluminum guide pipe embedded in the recommended route.

FIGURE 48.19   Raised flower beds conveying the image of the Japanese garden.

FIGURE 48.20 Handrail that provides views for people using wheelchairs.

FIGURE 48.21 Adaptive wheelchair for access to the Inner Sea.

The handrails for the arched bridge (Fig. 48.23) were provided with Braille on the back face so that they can easily be read by people with visual impairments.

Figure 48.24 shows a wheelchair space adjacent to each bench so that all people can enjoy sitting together side by side.

Involvement of Specific User Groups in the Design. As the guide pipe embedded in the recommended route was the first example of its use in Japan, a trial use was conducted and use-

FIGURE 48.22   Observatory on top of the arched bridge.

FIGURE 48.23   Handrail for the arched bridge with Braille on the back face.

FIGURE 48.24   Benches with side-by-side wheelchair space.

ful advice was given by the staff of Osaka Institute for the Blind. Also, information to be included in minidisk audioguides, as well as information and descriptive panels, along with valuable comments and advice, were received from a school teacher for blind students. Other elements designed for universal use were based on the findings of our earlier-mentioned survey of about 600 people with disabilities.

Design Challenge.   The major challenge for the designer was how to introduce the universal design approach to such a large area (8.5 ha). Many problems arising from this issue were solved by providing visitors with choices of destinations they could visit depending on their physical strength. Examples of choices included the observatory, the rest pavilion, and the recommended route, each of which could be reached with relatively little physical effort.

Benefits to Various User Groups in the Final Design.   Wheelchair users are able to visit the entire park, wheeling on the smoothly finished path surfaces. Flower beds are raised so they can be touched with ease, and views for people using wheelchairs are provided at any location in the park.

The guide pipe embedded in the recommended route enables the visually impaired people to experience all the major sights. Necessary information is provided by a tactile map, Braille, the push-button audio system incorporated in the panels, and minidisk audioguides.

Not only small children, but also people in adaptable wheelchairs can safely enjoy the beach and the Inner Sea for which water is let in through the stone embankment from the open sea.

People who are disabled and elderly, having had limited access to public parks, are now able to visit this park and enjoy the experience with others. Rinku Park/Symbolic Green Park/South Zone not only creates a space for outdoor recreation for people with disabilities, but provides a significant setting for the integration of all park users.

Lessons Learned from the Project.   Since Rinku Park/Symbolic Green Park/South Zone is built on reclaimed land, the ground is not stable and has been sinking. Significant subsidence in part of the grounds was remedied temporarily by pavers, to be followed by final finishing in several years. Considering that amenity is an essential factor in attracting people to the park, it is necessary to develop quality pavers in order to meet the needs of flexible use and aesthetic quality.

Comments Received from Others.   Rinku Park/Symbolic Green Park/South Zone was much appreciated by people with disabilities for having expanded their opportunities for outdoor recreation. The 1997 Design Honor Award was awarded to this project by the Japanese Institute of Landscape Architecture for excellence in universal design and aesthetic quality.

48.5 GENERAL LESSONS LEARNED FOR UNIVERSAL DESIGN

Universal design is a concept that integrates all aspects of life and behaviors of users. For example, a universally designed park alone does not meet the needs of the users, unless access to the park, as well as information of its existence, are provided to the users. As shown in the case of Sensory Garden in the preceding section, the distance from the park entrance to the Garden may pose a problem in access to some users. It resulted from the initial planning, which covered only a part of the entire park within a large area. It is essential in universal design that access from where users live to the destination should be provided continuously. There are many issues that need to be carefully examined, such as transportation and communication, to realize total access to the parks, and they should be solved on a governmental basis.

48.6 CONCLUSIONS

As landscape design cannot be attempted without consideration of the natural environment, incorporating universal design in landscaping means providing access to nature. For people who are disabled and elderly, gaining access to nature goes beyond mere physical access. It involves exploration of the realm of healing of their soul and spirit, plus the therapeutic effects of nature. It has been said that our society is in transition from a century of materialistic values to a new century of spiritual values. Therefore, the importance of the role of landscape design professionals in society will become even greater in the future. Regarding the design approach for a physical design, an approach that creates tension or unfamiliarity is not suited for a public park. Many people visit parks to seek relief in peaceful and familiar settings, rather than expecting the tension in unfamiliar places. This concept is associated with the classic arguments as to what *paradise* means in landscape design. In a park landscape, consideration of the people who use it is the fundamental principle for the first and most important step of design. When such careful considerations are given and aesthetic quality is incorporated in the design, their existence may only be perceived by those who need them, which is the essence of universal design.

Presently, the number of universally designed public parks is very small in Japan. Although there is a growing movement for universal design in the field of landscape design, it involves difficulty in heightening clients' awareness toward its necessity. Although in Japan it may take longer to disseminate the universal design concept compared with the United States and Europe, it is hoped that universally designed parks will increase further in the future in this country.

48.7 BIBLIOGRAPHY

Miyake, Yasuyo A., Hajime Kameyama, and Yoshisuke Miyake, *Creating People-Friendly Parks: From Barrier-Free to Universal Design,* Kajima Institute Publishing, Tokyo, Japan, 1996.

48.8   RESOURCE

SEN, Inc.
4-11 Tsuruno-cho
Suite 1106
Kita-ku
Osaka
530-0014 Japan
Phone: 06-6373-4117
Fax: 06-6373-4617
E-mail: post@sen-inc.co.jp

CHAPTER 49
UNIVERSAL DESIGN OF PRODUCTS

Molly Follette Story, M.S., I.D.S.A.
*Center for Universal Design, North Carolina State University,
Raleigh, North Carolina*

James L. Mueller, M.A., I.D.S.A.
J. L. Mueller, Inc., Chantilly, Virginia

49.1 INTRODUCTION

The term *products* describes a wide range of objects, from the tiny (e.g., miniature toys) to the very large (e.g., trade show exhibits). In this chapter, *products* can be defined as encompassing objects that can be found:

- In the home (e.g., appliances, furniture, cabinetry)
- At work (e.g., office furniture, equipment, tools, dispensers)
- In public spaces (e.g., elevators, kiosks, rest room fixtures)

While each product presents its own unique challenges, many of the issues of universal usability are common.

This chapter clarifies the relationship between assistive technology and universal design, and it highlights the potential of universal design to increase social integration for people with disabilities into their communities.

49.2 DEFINING UNIVERSAL DESIGN

> Universal design is the design of all products and environments to be usable by people of all ages and abilities, to the greatest extent possible.   (Mace, 1991)

While this definition makes universal design an unattainable goal, it remains an ideal well worth striving to achieve. Universal design accommodates the range of variation inherent in diverse populations and varied situations.

The variations in functional characteristics that universal design addresses can happen to anyone at any time in their lifetime—even young, nondisabled people. Limitations can be caused by situation and environment as much as by personal abilities. For example, noisy environments impair anyone's hearing; dimly lit rooms impair anyone's vision; and having the flu reduces anyone's stamina.

The U.S. Department of Education's National Institute on Disability and Rehabilitation Research (NIDRR) recognized this in the new paradigm of disability they published in their long-range plan of 1998 (National Institute on Disability and Rehabilitation Research, 1999). In the plan, they postulate that disability is not a personal trait but rather, a characteristic of the complex and dynamic relationship between an individual and his or her environment. NIDRR's new paradigm builds on the growing appreciation of the impact that the environment has in enabling or disabling individuals and the shift from the rehabilitation/medical model of care that focused on treating only the person to the independent living model that emphasizes societal factors. (Laurie Ringaert, in Chap. 6, "User/Expert Involvement in Universal Design," discusses the independent living model.) This approach is, by nature, variable with personal conditions and environmental and programmatic circumstances. It has significant implications for designers, businesses, and governments, who should anticipate and accommodate a wider range of individuals than they typically do.

Only someone who has personally experienced disability can fully understand the challenges of living with a disability in a world designed for the most part as if such persons did not exist. However, individuals without disabilities can get a modicum of first-hand knowledge by participating in exercises that simulate the effects of having a functional limitation. For example:

- Zip up your jacket—with one hand.
- Carry on a conversation—while a train goes by next to you.
- Count the money in your wallet—in the dark.
- Write your name using your nondominant hand—and then ask someone else to read your signature.
- Read the instructions for your new VCR—in another language.

These situations point out that unusual circumstances can cause functional limitations that make even common, simple tasks difficult for persons of average abilities. Everyone who lives long enough will experience situations like these—at least sometimes, at least temporarily, and often, eventually permanently.

Traditionally, the needs of individuals with disabilities have been addressed by providing assistive technologies. These devices are designed for limited populations with specific special needs and generally are neither useful nor appealing to other users. In contrast, universal design integrates accessibility concerns within devices that are useful and attractive to as many users as practical. This accommodation of everyone's needs to the greatest extent possible results in better usability for all users, regardless of their personal abilities or temporary circumstances.

The design details of products can profoundly affect people's experiences. Even small changes in features can determine how a product is used or whether it is usable at all.

49.3 EVALUATION CRITERIA FOR ASSISTIVE TECHNOLOGIES

A number of groups have published guidelines for the development of various specific assistive technologies, most notably those contained in the Telecommunications Act of 1996 (Telecommunications Access Advisory Committee, 1997). Several references contain criteria for telephones (Francik, 1996; Pacific Bell, 1995), consumer electronic products (Electronics Industries Association and Electronic Industries Foundation, 1996), accessible Internet Web sites (Vanderheiden and Lee, 1988), or computer software (Vanderheiden, 1994). Some published guidelines address the needs of specific groups of users, such as those published by The Lighthouse, Inc., for legible text for people with low vision (Arditi, 1997a,b). These guidelines are useful for improving the accessibility of products for individuals with specific disabilities.

Staff at Honeywell, Inc. (1990), wrote some preliminary *Human Factors Design Guidelines for the Elderly and People with Disabilities,* which address the following:

- Controls
- Visual displays
- Auditory displays
- Functional allocation and panel layout
- Operating protocol

Although incomplete, the guidelines represent one company's noteworthy attempt to serve a broader market. Vanderheiden and Vanderheiden (1992) developed a set of excellent guidelines for the design of consumer products, addressing:

- Output/displays
- Input/controls
- Manipulations
- Documentation
- Safety

They offer specific recommendations to improve the accessibility of a wide range of products.

In addition, researchers have developed sets of general design evaluation criteria, most notably those developed by Batavia and Hammer (1990) for the evaluation of assistive devices. Their original 17 criteria were subsequently reviewed by the Rehabilitation Engineering Research Center on Technology Evaluation and Transfer (RERC-TET) at the University at Buffalo (New York). Their researchers asked consumers to evaluate the criteria and help amend them for use by the RERC-TET in their formal product evaluation activities (Lane et al., 1996, 1997). Their resulting 11 criteria, condensed from the Batavia and Hammer criteria, are as follows:

- Effectiveness
- Affordability
- Reliability
- Portability
- Durability
- Securability
- Physical security/safety
- Learnability
- Physical comfort/acceptance
- Ease of maintenance/reparability
- Operability

These criteria are helpful in evaluating and developing designs that address issues identified by consumers as being the most important to them when purchasing and living with assistive devices. They cover all aspects of ownership and use.

Similar to these evaluation criteria for assistive technologies, criteria have also been developed for universal design. In 1997, a set of seven Principles of Universal Design was developed by The Center for Universal Design (see Chap. 10, "The Principles of Universal Design," by Story). The purpose of these Principles was to articulate the concept of universal design in a comprehensive way. The Principles of Universal Design and their associated guidelines were intended to guide the design process, allow the systematic evaluation of existing designs, and assist in educating both designers and consumers about the characteristics of more usable products and environments.

49.4   PRODUCTS WITH UNIVERSAL DESIGN FEATURES

In its most successful applications, universal design is inconspicuous: The usability of the design is remarkable, not the accommodation it incorporates. Users simply comment that the product or environment is easy to use.

The quality that makes the critical difference between successful universal design and design for limited populations is integration of the accommodating features. Accessible features must be functionally and visually integrated into the overall design. When accessible features are integrated into products and environments from the beginning of the process, they are not noticeable, and they cost less and benefit more people in a wider range of situations.

The following designs satisfy the Principles of Universal Design and help communicate what is meant by universally usable product design. These images were reprinted from *The Universal Design File: Designing for People of All Ages and Abilities,* published by the Center for Universal Design (Story, Mueller, and Mace, 1998). These designs are not necessarily universal in every respect, but each is a good example of a specific guideline and helps convey its intent.

Principle 1.   Equitable Use
Guideline 1A.   Provide the same means of use for all users: identical whenever possible; equivalent when not.

- Side-by-side refrigerator/freezers with door handles that extend the full height of the doors suit users of all statures and postures.

Guideline 1B.   Avoid segregating or stigmatizing any users.

- A powered door (Fig. 49.1) at the entrance to a building such as a retail store is convenient for all shoppers, especially if they are pushing a shopping cart or a baby stroller or their hands are full.

FIGURE 49.1   Power door at entrance to public building is convenient for all.

Principle 2.   Flexibility in Use
Guideline 2A.   Provide choices in methods of use.

- Computer hardware devices and software features (Fig. 49.2) offer users a choice of input and output options. Input can be controlled via a mouse, trackball, touch pad, graphics pad, joystick, keyboard arrow keys, sip-and-puff switch, even a tongue-activated touch pad, or eyegaze software. Output can be presented through a monitor (with custom appearance, such as specific colors, fonts, and graphics), speakers, printer, Braille printer, dynamic Brailler, and other devices.

FIGURE 49.2   "Easy Access" software offers computer users choices.

Guideline 2D.   Provide adaptability to the user's pace.

- Some audiocassette players and dictation machines have a control that will adjust the speed of playback. This suits individuals with different needs, such as transcriptionists who may need to listen slowly to understand an indistinct passage, or talking-book users with and without vision impairments who may listen at accelerated rates.

Principle 3.   Simple and Intuitive Use
Guidelines 3A and 3C.   3A: Eliminate unnecessary complexity. 3C: Accommodate a wide range of literacy and language skills.

- Cross-culturally meaningful icons on devices can reduce the visual complexity of a product and communicate quickly and effectively, regardless of individual abilities or circumstances.

Guideline 3B.   Be consistent with user expectations and intuition.

- The automobile power seat control (Fig. 49.3), which is located on the inside face of the car door, is the same shape as the seat itself, which makes it easy for the driver or passenger to make seat adjustments intuitively. The user simply pushes the control in the same direction as he or she wants the seat to move.

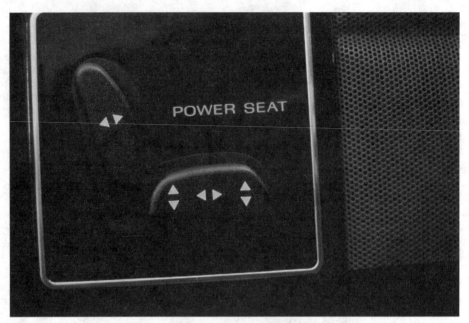

FIGURE 49.3   Automobile seat control mimics the shape of the seat.

Principle 4.   Perceptible Information
Guideline 4A.   Use different modes (pictorial, verbal, tactile) for redundant presentation of essential information.

- Appliance manufacturers should supply use and care manuals in large print, Braille, and audio and video formats (Fig. 49.4) to suit individual user abilities and preferences.

FIGURE 49.4   Manufacturer's care and use manuals should be provided in multiple formats.

Guideline 4B.   Maximize legibility of essential information.

• The issue of legibility applies to information in all modes. One of the primary considerations for legibility is contrast between the information and the surroundings. For visibility, this means contrast between object and background; for audibility, it means contrast between signal and ambient noise; for tactility, it means contrast between raised letters or symbols and surrounding surface texture.

Principle 5.   Tolerance for Error
Guideline 5B.   Provide warnings of hazards and errors.

• The red tip on a bottle of contact lens cleaner warns the user not to confuse it with the bottle of eyedrops, which has the same shape but a white tip (Fig. 49.5).

FIGURE 49.5   Different contact lens solution bottles are distinguished by color.

Guideline 5C.   Provide fail-safe features.

• Ground-fault interrupter (GFI) electrical outlets reduce the risk of electrical shock when installed in bathrooms and kitchens. When a GFI senses a current surge, it trips a circuit breaker that stops current flow.

Principle 6.   Low Physical Effort
Guideline 6A.   Allow the user to maintain a neutral body position.

• A computer keyboard that is split in the middle and has its two sides angled outward to align with the user's forearms allows a computer operator to maintain a neutral body position from elbows to fingertips (Fig. 49.6).

Guideline 6D.   Minimize sustained physical effort.

• Free-wheeling casters on luggage greatly reduce the physical effort involved in traveling.

FIGURE 49.6   Ergonomic keyboards avoid awkward postures while typing.

Principle 7.   Size and Space for Approach and Use
Guideline 7B.   Make reach to all components comfortable for any seated or standing user.

- Installing multiple vending machines or drinking fountains at varying heights offers controls at comfortable locations for seated or standing individuals of any stature.

Guideline 7C.   Accommodate variations in hand and grip size.

- Open-loop door handles (Fig. 49.7) and cabinet and drawer pulls do not require grasping and accommodate hands of all sizes and shapes.

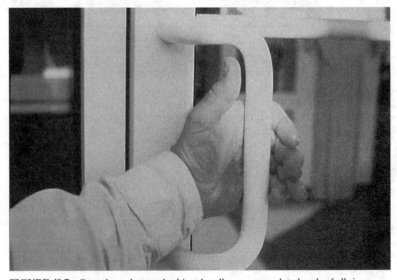

FIGURE 49.7   Open-loop door and cabinet handles accommodate hands of all sizes.

49.5   THE INTERRELATIONSHIP OF UNIVERSAL DESIGN AND ASSISTIVE TECHNOLOGY

The distinction between universal design and assistive technology depends on both the user and the application. Depending on the situation, the same device can be perceived in completely different ways. For example, when a magnifying lens is used for examining tiny objects such as a splinter, it is considered a tool; when it is used by someone to read typical-sized type, it is called assistive technology. When television headphones are used by one member of a household to watch a show without disturbing others, they are dubbed a clever gadget; when they are used by an individual who has a hearing impairment, they are considered assistive technology. When an electric cart is used by an individual while playing golf, it is considered sports equipment; when an electric scooter is used by an individual with limited stamina, it is perceived to be assistive technology. In fact, there is a large area of overlap between universal design and assistive technology, and the name applied to a device is most reflective of the attitude of the person assigning the label.

Ron Mace discussed the distinctions between assistive technology and universal design at the conference "Designing for the 21st Century" (Mace, 1998). Using both his personal and professional experience, he described many examples of personal equipment that some people need, such as eyeglasses and portable oxygen tanks, which will never be attractive to individuals who do not need them.

However, products sometimes move across the dividing line from assistive technology to universal design and become appealing to a larger number of users. Mace mentioned a number of devices that had crossed this line, such as TV headphones and oversized toilet stalls. Other products that started as *accessible* and became *mainstream* include the following:

OXO Good Grips®.   The OXO Good Grips line of kitchen utensils was conceived when founder Sam Farber retired from the kitchenware business he founded (Copco) and moved to France with his wife, Betsey. Betsey and Sam spent much of their free time cooking, and Betsey noticed that the kitchen utensils they used aggravated the pain of the arthritis in her hands. This motivated Sam to develop better handles for these tools, which brought him out of retirement to start a new company, OXO International. Betsey, a designer, became OXO's director of design. The handle designs inspired by Betsey's needs were sold not as assistive technology but introduced without promotion as standard kitchen products sold through numerous retail outlets in the United States and abroad. Today, the popular soft, rounded Good Grips handles can be seen on OXO products ranging from kitchen peelers to garden trowels and have inspired many imitators. (For more information, see Mueller, 1997.)

Voice recognition technology.   Early military research in voice recognition technology was applied to create some of the first systems used by persons with severe disabilities. At the same time, some developers recognized the potential mainstream civilian uses of this technology. Research advanced in both rehabilitation technology centers and in the computer and communications industries. Today, voice recognition systems are available on the open market at a much lower cost than initially and are used by people with and without disabilities.

Telephone volume controls.   Telephone receiver volume controls were an early accommodation for persons with hearing limitations. These and other telephone modifications were made available to customers with disabilities through AT&T's Special Needs Center in the 1980s. How many travelers can say that they have never used the volume control now commonly available on pay phones in airports, train stations, and bus stations? They are a convenience for all.

Vibrating pagers.   Vibrating pagers were a technology introduced as a means of communicating with persons with hearing limitations. As the use of pagers has become widespread, the ability to silently page persons in meetings, theaters, and such has made this feature popular with and highly marketable to mainstream consumers.

Tacky surface materials.   Tacky plastic sheets, marketed as Dycem or Vykem by assistive technology suppliers, were originally used to stabilize telephones, writing pads, dinner plates, and other items that might be difficult to use for persons with limited coordination. Today, this material is used for many applications, such as table mats used in boat galleys that keep plates and glasses from slipping off the table in rough seas.

At the same time, some products start as *mainstream* and become *accessible.* The need for affordable and easily available assistive technology has caused persons with disabilities and rehabilitation technologists alike to creatively adapt mainstream commercial products for unforeseen uses. For example:

Bicycle grips.   Soft foam tubing sold as bicycle grips can be easily purchased in most sports, toy, and hardware stores to replace inferior or worn-out parts. They also provide a readily available and inexpensive way to improve grasp for persons having difficulty holding a variety of items. Smaller-diameter foam tubing sold as pencil grips can be used on smaller utensils to make them easier to grasp.

Rubber sheeting.   For persons with grip limitations due to conditions such as arthritis, opening jars can be an impossibility. Thin rubber sheets sold in kitchenware stores—and even offered as giveaways at trade shows and conferences—make jars easier for anyone to open.

Magnifying glasses.   Magnifying glasses are available in a wide selection of powers and sizes and in any price range. As a result, people with vision limitations can use these in the workplace or at home as an alternative to more complex and expensive assistive technologies.

Cordless headphones.   Cordless headphones are a popular gift "for the person who has everything." They can also be a very useful assistive technology for the person with a hearing impairment who doesn't want to disturb other members of the household by setting the TV or stereo volume at a high level.

49.6   WHY PRACTICE UNIVERSAL DESIGN FOR PRODUCTS?

Product designers and their clients are beginning to understand and take advantage of the business incentives toward universal design created by federal legislation, reduced costs associated with a universal design approach, improved survival rates of trauma and illness, an aging population, and a shrinking globe.

Accessibility Legislation Supports Universal Design

Various pieces of U.S. legislation require access for persons with disabilities. The Trace Research and Development Center conducted a research project to identify the factors that motivated businesses to adopt universal design; for many companies, legislation such as the Telecommunications Act of 1996 provided critical incentives (Trace R&D Center, 1999). The Americans with Disabilities Act (ADA) of 1990 has done much to promote accessibility of the built environment by persons with disabilities. The ADA required the removal of barriers in the built environment, as well as "reasonable accommodation" for qualified employees with disabilities. (For a thorough discussion of the incentives for practicing universal design in the workplace, see Chap. 45, "Office and Workplace Design," by Mueller, in this handbook.)

Universal design can provide cost-effective solutions to environmental access problems. Creative implementation of accessibility guidelines by architects has resulted in environments that are in themselves excellent examples of universal design.

The home of the Baltimore Orioles baseball team is just one example. Orioles Park at Camden Yards was designed to incorporate gradual ramps, handrails, and Braille signs, as well as 426 seats that will accommodate wheelchair users. The Camden Seat folds up and pivots out of the way to one side to create space for a wheelchair user, yet can also be used conventionally by nondisabled fans. These seats offer a choice of locations and seating prices throughout the stadium. (See Chap. 47, "Access to Sports and Entertainment," by Beasley and Davies, in this handbook.)

In some cases, architectural changes can be costly and difficult to make; in some buildings, historic preservation efforts conflict directly with ADA compliance. For those situations where architectural changes are not practical or timely, assistive and universal products may help resolve these problems. One such accessible product is the Evac-Chair, developed by an industrial designer and first made available in 1982. This product enables an able-bodied person to assist a nonambulatory person to evacuate a building via existing stairwells.

However, in some instances, creative designers have found ways to accomplish historic preservation and also respect civil rights. Carol Johnson's landscape architecture work on the Hunnewell Visitor Center at Harvard Arboretum was noted in the Images of Universal Design Excellence collection (Johnson and Jones, 1996). Johnson regraded the site around this historically significant building to support a sloped walkway that led to the landing at the top of the entry stairs. The walkway provided a universally usable alternative to having to climb a long stairway without disturbing an important architectural element.

Several recent laws have drawn attention to the accessibility and usability of electronic data-processing equipment and telecommunications equipment: the Telecommunications for the Disabled Act of 1982; the Hearing Aid Compatibility Act of 1988; the Television Decoder Circuitry Act of 1990; and Sec. 255 of the Telecommunications Act of 1996. In a way similar and complementary to the ADA architectural guidelines, the Telecommunications Act of 1996 contained requirements but also offered designers and manufacturers guidance for producing technology that is more universally usable by a diversity of potential users. (See Chap. 65, "Universally Designed Telecommunications," by Vanderheiden.)

The most recent piece of legislation is Sec. 508 of the 1998 Amendments to the Rehabilitation Act of 1973, which has remarkable incentives for universal design. Section 508 requires that information technology suppliers to the federal government (a not-insignificant customer) make this technology usable for people with disabilities. In her comments to the Federal Open Systems Exhibition (FOSE) conference on 18 April 2000, former Attorney General Janet Reno called on all producers of information technology to consider the implications of this law (Reno, 2000):

> Over the past 15 years, many have realized that making technology accessible does not have to be expensive. It does not have to be difficult, and it is very much the right thing to do. But technology has changed so rapidly that accessibility has oftentimes been an afterthought, if a thought at all. Modifying existing technology to be accessible is much, much more difficult, much more expensive than designing technology right in the first place. Accessible design is good design. . . . Accessibility doesn't have to cost a lot. It doesn't have to be difficult. It is good design. Making this technological revolution work for everyone is what this is all about.

Changing Demographics

Consumer businesses are beginning to understand what rehabilitation professionals have always known: Disability is a common fact of life. Most people with disabilities develop them during the course of their lives (National Organization on Disability/Harris, 1994). People with disabilities make up approximately 20 percent of the U.S. population (McNeil, 1997). And as life expectancy increases, so does the incidence of disabilities. Persons who are 65 years of age and over made up 11.9 percent of the total U.S. population at the end of 1994, but they accounted for 30.0 percent of all persons with any disability and 40.1 percent of all persons with a severe disability (McNeil, 1997). In the United Kingdom, France, Germany, Japan, Sweden, and Norway, the proportion of persons over 65 years of age is greater than in the United States and increases at a greater rate (Ministry of Welfare, Japan, 1998). Importantly, disability also touches the family, friend, and business relationships in which people with disabilities are involved.

Medical progress has had a profound effect on the treatment of illnesses and injuries that a short time ago were fatal. Between 1970 and 1997, the survival rate from strokes more than doubled, and the rate of survival from traumatic brain injuries improved from 10 percent to 90 percent. The survival rate from spinal cord injuries has also increased steadily each decade since the 1950s (Jones and Sanford, 1996). These changes result in more people living with disability now than at any previous time.

That large proportion of the aging population represented by the baby boom generation is a major force driving interest in universal design. With age may come wisdom, but also changes in sensory and physical abilities. As it advances through middle age, this cohort is demanding design that accommodates their needs without stigmatizing them. These customers expect to have car dashboard displays they can read easily, appliance controls they can understand intuitively, and packages they can open comfortably—and they have the money to buy them. They demand design that is desirable as well as usable.

Disability is more a part of mainstream life than it has ever been and, therefore, a more important consideration in design. Although progress in the development of assistive technologies has served to increase the independence of people with disabilities, it is universal design that will increase the integration of people with disabilities in mainstream society. Universal design is technology that reduces the demands of the built environment and enables users to function most efficiently, without emphasizing disabilities.

Universal Design Can Reduce Costs

Beyond legislation and for products unaffected by it, one benefit of practicing universal design is that it can increase the size of markets and reduce the costs associated with manufacturing. Voice recognition technology, as previously mentioned, has come into common use in assistive technologies and in mainstream products. Initially, these units, produced practically one at a time, cost over $10,000. Because of the broad applicability of voice recognition, these systems now cost approximately one-third of this amount, yet are far more effective and easy to use.

Universal design can also increase personal independence in daily living and help control the costs of home care. More Americans than ever before are depending on home health care, which may involve highly skilled medical services as well as basic assistance with homemaking chores. In response to this need, in March 1999, former President Clinton introduced a $6 billion program of tax credits to families providing long-term home care for people with disabilities. Whether provided by professional caregivers or family members, this one-on-one care is costly. Research is just beginning to show the potential for savings in home health care and hospital stays that can be achieved through the application of universal design principles in home and product design, as well as in assistive technologies such as lifts, mobility aids, communication devices, and environmental control systems (Mann et al., 1999).

International Markets: Additional Challenges for Universal Design

Beyond the challenges of language, cultural conventions affect user intuition when faced with a new design. However, when successful, a product can be an international ambassador for universal design, as products such as OXO Good Grips have proven. While examples of universally usable buildings or communities can serve as models only through images and description, one product can directly affect the quality of the daily lives of millions of individuals around the world. It can enhance their capabilities, independence, and self-image.

49.7 CONCLUSION

To make products more universally usable, user-centered design must be embraced by the designers employed by manufacturers, and marketing managers must broaden their definition of *target users*. However, in most cases, the knowledge base of these practitioners is not yet sufficient to practice universal design well. Changing established business patterns will take considerable advocacy, education, and training.

Most important, people's attitudes must change toward people who are different from themselves, particularly toward people with disabilities that everyone has the potential to acquire through disease, injury, or the aging process. The biggest challenge is changing people's attitudes about whom universal design is for and whom it benefits.

Where possible, practicing universal design makes sense for product manufacturers:

- *Legally.* Accessibility is the law in many venues and universal design satisfies but goes beyond the minimums specified within.
- *Politically.* Practicing universal design can attract beneficial publicity and provide good public relations.
- *Economically.* Universal design increases the sizes of target markets and, consequently, through economies of scale, reduces unit costs.
- *Socially.* Everyone in society is enriched when all of its diverse citizens participate.
- *Morally.* Providing maximum access and optimal usability is the right thing to do.

The most significant effect of successful universal design practice is on the social integration of people with disabilities into their communities. The ultimate goal of universal design is the natural inclusion of all members of society in the everyday activities of living, both within their own homes and as they move around their communities. With better designs, individuals with disabilities can manage independently in their own homes, communicate with others over the telephone or through the computer, and seek a wider range of jobs. They can travel along streets, move through buildings, and interact with various environmental features as easily as anyone without a disability. With universal design, individuals with disabilities are able to become active participants in and contributors to the society in which they live, to the enrichment of all.

49.8 BIBLIOGRAPHY

Arditi, A., *Color Contrast and Partial Sight: How to Design with Colors that Contrast Effectively for People with Low Vision and Color Deficiencies,* New York, NY: The Lighthouse, Inc., New York, 1997a; available online at www.lighthouse.org/1lh32a.html.

———, *Print Legibility and Partial Sight: Guidelines for Designing Legible Text,* New York, The Lighthouse, Inc., New York, 1997b; available online at www.lighthouse.org/1lh32b.html.

Batavia, A. B., and G. Hammer, "Toward the Development of General Consumer Criteria for the Evaluation of Assistive Devices," *Journal of Rehabilitation Research and Development,* 27(4), 1990.

Electronics Industries Association and Electronic Industries Foundation, *Resource Guide for Accessible Design of Consumer Electronics,* 1996; available online at www.eia.org/eif/toc.htm.

Francik, E., *Telephone Interfaces: Universal Design Filters,* Pacific Bell Human Factors Engineering, San Ramon, CA, 1996.

Honeywell, Inc., *Human Factors Design Guidelines for the Elderly and People with Disabilities* (Revision 3, Draft), Honeywell, Inc., Minneapolis, MN, 1990.

Johnson, C. R., and A. C. Jones, in *Images of Universal Design Excellence,* Universal Designers and Consultants, Takoma Park, MD, 1996.

Jones, M. L., and J. A. Sanford, "People with Mobility Impairments in the United States Today and in 2010," *Assistive Technology,* vol. 8.1, 1996.

Lane, J. P., D. J. Usiak, and J. A. Moffatt, *Consumer Criteria for Assistive Devices: Operationalizing Generic Criteria for Specific ABLEDATA Categories,* Proceedings of the RESNA '96 Annual Conference, RESNA Press, Arlington, VA, 1996.

———, ———, V. I. Stone, and M. J. Scherer, "The Voice of the Customer: Consumers Define the Ideal Battery Charger," *Assistive Technology,* Vol. 9.2, 1997.

Mace, R. L., "The Evolution of Universal Design," Jan Reagan (ed.), keynote speech presented at "Designing for the 21st Century: An International Conference on Universal Design," 19 June 1998, Hofstra University, Hempstead, NY, 1998; available online at www.adaptenv.org/examples/default.asp?f=4.

———, G. J. Hardie, and J. P. Place, "Accessible Environments: Toward Universal Design," in W. F. E. Preiser, J. C. Vischer, and E. T. White (eds.), *Design Intervention: Toward a More Humane Architecture,* Van Nostrand Reinhold, New York, 1991.

Mann, W. C., K. J. Ottenbacher, L. Fraas, M. Tomita, and C. V. Granger, "Effectiveness of Assistive Technology and Environmental Interventions in Maintaining Independence and Reducing Home Care Costs for the Frail Elderly," *Archives of Family Medicine,* vol. 8 (May/June), 1999.

McNeil, J. M., *Americans with Disabilities: 1994-95: U.S. Bureau of the Census. Current Population Reports, P70-61,* U.S. Government Printing Office, Washington, DC, 1997.

Ministry of Welfare, *The Percentage of Elderly People Who Are 65 Years Old and Over,* Ministry of Welfare, Tokyo, Japan, 1998.

Mueller, J., *Case Studies in Universal Design,* The Center for Universal Design, North Carolina State University, Raleigh, NC, 1997.

National Institute on Disability and Rehabilitation Research (NIDRR), "Long Range Plan," in *Federal Register,* 64(3) (7 December 1999), U.S. Government Printing Office, Washington, DC, 1999.

National Organization on Disability/Harris Survey of Americans with Disabilities, Study No. 942003, Louis Harris and Associates, Inc., New York, 1994.

Pacific Bell, *A Compilation of Interface Design Rules and Guidelines for Interactive Voice Response Systems (IVRs),* Pacific Bell Human Factors Engineering, San Ramon, CA, 1995.

Reno, J., "Remarks of Attorney General Janet Reno: Section 508; FOSE 2000," 2000; available online at www.usdoj.gov/ag/speeches/2000/doc3.htm.

Story, M. F., J. L. Mueller, and R. L. Mace, *The Universal Design File: Designing for People of All Ages and Abilities,* The Center for Universal Design, North Carolina State University, Raleigh, NC, 1998.

Telecommunications Access Advisory Committee, *Access to Telecommunications Equipment and Customer Premises Equipment by Individuals with Disabilities,* Architectural and Transportation Barriers Compliance Board, Washington, DC, 1997.

Trace Research and Development Center, *Universal Design Research Project,* 1999; available online at www.trace.wisc.edu/docs/univ_design_res_proj/udrp.htm.

Vanderheiden, G. C., *Application Software Design Guidelines: Increasing the Accessibility of Application Software to People with Disabilities and Older Users,* The University of Wisconsin, Trace R&D Center, Madison, WI, 1994.

———, and C. C. Lee, *Considerations in the Design of Computers and Operating Systems to Increase their Accessibility to Persons with Disabilities* (Version 4.2), The University of Wisconsin, Trace R&D Center, Madison, WI, 1988.

———, and K. R. Vanderheiden, *Accessible Design of Consumer Products: Guidelines for the Design of Consumer Products to Increase Their Accessibility to Persons with Disabilities or Who are Aging* (Working Draft 1.7), The University of Wisconsin, Trace R&D Center, Madison, WI, 1992.

CHAPTER 50
UNIVERSAL DESIGN IN AUTOMOBILE DESIGN[1]

Aaron Steinfeld, Ph.D.
California Partners for Advanced Transit and Highways,
University of California, Berkeley, California

Edward Steinfeld, Arch. D.
Rehabilitation Engineering Research Center (RERC) on Universal
Design, University at Buffalo, Buffalo, New York

50.1 INTRODUCTION

In most cities in North America, automobile transportation is the means by which most citizens travel in their communities when destinations are too far to walk to. Without access to automobiles, access to community resources and quality of life may be severely restricted. Despite the need for alternative forms of transportation, "automobility" is likely to remain dominant for a long time to come. Currently, major transformations are taking place in the design of automobiles, driven by the need to increase fuel efficiency, to keep the cost of vehicles within the range of low- and middle-income consumers, to address concerns about product safety, and to integrate new safety and information technologies. Thus, this is a good time to consider other improvements that would address demographic shifts toward an older society and the initiative to provide the highest degree of self-determination for people with disabilities. This chapter examines the major issues related to universal design of automobiles. It also identifies and discusses the advent of new technology, both its promise and its potential liabilities.

50.2 BACKGROUND

In recent years there has been increased attention to applying universal design to automobiles. Specifically, there is a significant increased awareness of the aging of populations in developed countries, and the impact that this demographic shift will have on the use of automobiles (Waller, 1991). While the use of the term *universal design* has not been particularly popular in the automotive industry, other related terms like *transparent enablers* have increasingly garnered attention in the press and industry (e.g., Ford, 2000). The use of such indirect language reflects an underlying ambivalence about addressing the needs of older people and people with disabilities directly. In the highly competitive automotive industry, marketing has a very high priority in making design decisions. Too close an association with aging and disability has been perceived as a marketing liability—particularly if such features conflict with styling goals.

Despite their ambivalence and priorities, companies in the industry are now fully aware that the disabled community and aging population are more and more potent as a consumer force in the marketplace. Thus, improvements in designs to accommodate these groups have become more common as the risk of neglecting the concerns of this market increase. Furthermore, new advances in automobile technology may alleviate driving problems and safety risks associated with functional impairments related to disability and aging. This new technology has the promise to make automobiles more universally accessible. At the same time, there are dangers that the population with functional impairments could be denied access to the benefits of the new technology.

Not only is the industry changing its objectives, it is also changing its design process. The Ford Motor Company, in conjunction with the University of Loughborough in the United Kingdom, has developed a "Third-Age Suit" that enables its designers to experience the limitations in performance associated with aging (see Fig. 50.1). The idea is to make them more sensitive to the impact of aging and to experience design concepts as an older driver might (Block, 1999). Other automobile companies have similar efforts underway (Parker, 1999). The power of using "empathic models" (Pastalan et al., 1971) in design is demonstrated by the words of this quote from the manager of the Ford Human Factors and Ergonomics Division: "It's one thing to read customer feedback in a marketing study. . . . It's a whole different thing to feel what they are feeling while driving a car. This has been a real eye opener for our engineers" (Ford, 2000).

Yet, to this point, most developments in universal design of automobiles focus on interiors and have not addressed the more difficult issues where there are conflicts with structural, aerodynamic, performance, and stylistic design goals. These will take longer for the industry to address.

There are many factors that affect the usability of automobiles. Some of these are directly related to driving, while others are more general and affect both drivers and passengers. The key design issues are: entering and exiting the vehicle, seating, perception of the surrounding environment, navigation, safety, and obtaining information external to the vehicle.

50.3   ENTERING AND EXITING

Limitations in motor abilities cause significant problems in entering and exiting from vehicles (James, 1985). It is noteworthy that the human factors research on aging and use of automobiles has emphasized the driver. Among the frail older population, however, a large number only use vehicles as passengers. About 50 percent of one large sample reported difficulties getting in and out of vehicles (Steinfeld et al., 1999). Light trucks such as minivans, vans, sport utility vehicles, and pickups with high floor levels are very popular today. Although older people with disabilities do not necessarily purchase such cars, their baby boomer children and their grandchildren are doing so in increasing numbers. Moreover, senior citizen transportation services use vans and minivans. Many older consumers have difficulty getting in and out of these vehicles. Another popular style is the sport coupe that is usually styled to look sleek and low. These require agility to bend the body under the low ceilings and to rise up out of very low seats.

There are many other barriers to entering and exiting vehicles. Front doors are difficult to use when the distance between the front of the seat and the front of the door opening is too narrow. People who have difficulty bending their knees or lifting their legs find it impossible or painful to get their legs in and out of narrow door openings or over high door sills. Back seats with narrow doors are literally impossible for many older passengers to use because they cannot bend their legs enough to get them through the door. The rear seats of two-door models are particularly difficult for older frail people to use. Drivers who use wheelchairs and can transfer

FIGURE 50.1   Ford Third Age suit.

on their own, on the other hand, find that two-door models are best for entering and exiting and loading and unloading their chairs. The doors of these models are very wide, and having only one door on each side makes it possible to get the chair into the rear of the automobile.

Design Guidelines for Entering and Exiting

- Accommodate passengers with mobility impairments by reduced floor height, high door openings, and wide doors.
- Seats should be high enough to reduce the need to extend legs and reduce the need to push up while exiting and bend down while entering.
- Include adequate handholds and consider improving access by designing new seating systems like swivel seats.

50.4   SEATING

Seating and positioning play major roles in supporting driving tasks. The size of the *useful field of view* (the spatial area within which an individual can be rapidly alerted to visual stimuli) has been linked directly to accident frequency and driving performance in the older population (Owsley and Ball, 1993; Ball and Owsley, 1991). Positioning the body for visibility is clearly constrained by the design of some automotive seating systems and the vehicles themselves. As such, consumers in focus groups reported that many of them use cushions to prop themselves up to get a better view of the road (Steinfeld et al., 1999).

Use of safety restraints has also been identified as a major problem. Physiological changes associated with aging like reduced bone density, muscle mass, and resistance to skin laceration potentially increase the risk of injury during a collision. People over 55 who had been in accidents have a significantly higher proportion of deaths even though a higher proportion of the older group used seat belts (Cushman et al., 1990). One proposed explanation for this finding is that the incidence of inappropriate use of seat belts by older people is much higher than the younger population. Failure to properly fasten a seat belt can put the passenger at great risk of injury in collisions. Unfastened seat belts or improperly adjusted belts can result in "submarining" under air bags and the dashboard.

In focus groups conducted by one of the authors (Steinfeld et al., 1999), older people with disabilities reported that they didn't use seat belts or they pull the shoulder belt over their shoulder because the belts are uncomfortable and cause pain. Individuals who have had recent surgery or chronic health problems in the trunk of the body reported the most discomfort with seat belts. The difficulties of buckling the belt and reaching shoulder belts and pulling them across the body were also common complaints by those with arthritis and limitations in range of motion. People who had paralysis or other limitations on one side of their body reported great difficulty in using seat belt buckles. Although some participants simply viewed seat belts as a nuisance and unnecessary, and stated that they had used cars successfully for a long time without them, others had positive attitudes regarding seat belts and were explicit about their safety value.

In the focus groups already mentioned, participants identified another seating design issue that caused them problems in using automobiles—the location of controls for adjusting the seat position. Many automobiles have the control lever located under the front edge of the seat. Participants reported that this position was very difficult for them to reach. Understanding how the controls operate is also a problem because the control is difficult to see in that location (Steinfeld et al., 1999).

Another issue related to seating is support for interaction between drivers and roadside devices and toll booths. Limitations in range of reach or grasp function can make it very difficult or even impossible to use drive-in banking, drive-in restaurants, or toll roads. One solution for tolls is electronic toll collection (ETC). This usually involves a component mounted in the vehicle's windshield that transfers information with a roadside counterpart to handle payment without driver interaction. Coming to a full stop is not needed (which has the added value of reducing the risk of low-speed collisions while occupying a toll booth queue). While these systems were originally designed to increase efficiency and reduce the potential for non-payment of tolls, they have significant benefit to people with mobility impairments. A driver with limited reach no longer needs to hand money to a toll collector or pull close to roadside infrastructure and risk damage to his/her vehicle.

Design Guidelines for Seating

- Seating controls should be easy to understand and operate. Whenever possible, controls should be located in a visible and easy-to-reach location.
- Seat belt buckles should be located and designed so that they can be easily found and fastened by drivers with limited upper limb mobility, and without favoring one side of the body.
- Roadside devices and toll booths should be designed to accommodate drivers with limited reach and limited upper limb mobility.

50.5   CASE STUDY—LEAR TransG

Lear Corporation, a major supplier of interior components to the global automobile industry, has developed a concept interior that demonstrates how seating, instrument panels, environmental controls, and window and door controls can be made more usable for older people. Lear calls their concept interior the "TransG" for "transgenerational." This concept has features such as swiveling power front seats, enhanced graphic display of instruments, and a storage cart that can be stowed in the trunk (see Fig. 50.2).

A key component of this concept vehicle is the power swivel seat system. These seats rotate out at a 45° angle to facilitate getting in and out and to reduce dependence on good balance or strength. The integrated seat belts have a four-point arrangement with a center-positioned buckle that is easy to latch and see. This design makes the act of fastening the belt much easier for people with limited dexterity or a limited range of head movement. The belts not only secure the occupant more uniformly but are more comfortable because they do not cut across the chest on a diagonal. The restraint systems also include air bags that form a collar upon deployment, thus reducing potential injury to the neck and spine. The seats have many convenience features, some of which are already available on many luxury cars and others that are enhanced in this concept model. These include power adjustable headrests, lumbar support, seat height and recline adjustments, and seat heaters. The TransG instrument pod and pedals move toward and away from the driver rather than having the seat move. A memory control automatically adjusts the seat, instruments, and pedals for each user. The floor of the concept interior has a lower step-up height and a flat-load floor to enhance ease of ingress and egress.

50.6   PERCEPTION OF THE ENVIRONMENT

Aging, sensory disabilities, and in-vehicle factors such as loud car radios all can result in impaired perception of the surrounding environment. For example, when such distracting stimuli are present, older drivers can exhibit reduced performance with respect to visual field size, dynamic visual acuity, and velocity estimation, as well as other perceptual and cognitive characteristics (Hakamies-Blomqvist, 1996).

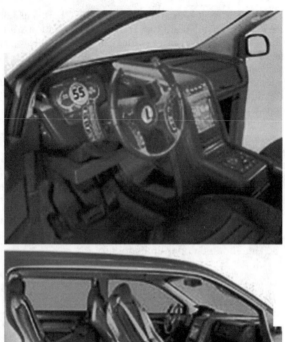

FIGURE 50.2   Lear TransG interior.

The effects of multiple disabilities exaggerate limitations in performance. For example, drivers with restricted neck motion may also have hearing and vision problems. These drivers can have a reduced ability to detect horns, emergency vehicles, and other unusual events. Vehicle adaptations that accommodate these problems include parabolic rear-view mirrors and siren detectors.

Night vision aids are now being introduced via head-up displays (HUDs). These devices, such as Cadillac's Night Vision system, are particularly attractive to drivers who have impaired night vision. As such, it is possible that older drivers, who in the past were unwilling to drive at night, may begin doing so without any increase in risk. However, it is also quite possible that they may make matters worse. A system that utilizes a HUD or in-dash screen can only provide monocular views to the driver, thus reducing depth perception. Furthermore, night vision systems do not provide images that look normal. In addition, drivers are still susceptible to glare from oncoming vehicle headlamps.

HUDs have been increasingly attractive to vehicle designers because they provide a means to superimpose visual information on the road scene (see Fig. 50.3). Theoretically, this will reduce the time that drivers take their eyes off the road to view instruments. HUDs in automobiles have been introduced to the market, yet most have not been successful products.

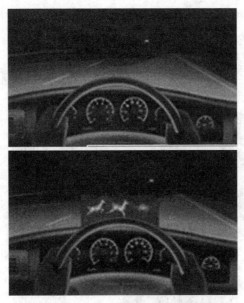

FIGURE 50.3   Head-up display (HUD).

Experiments involving advanced technologies have shown that HUDs have promise (Steinfeld and Green, 1998), but there are serious perceptual and cognitive issues that need to be addressed with additional research (Weintraub and Ensing, 1992; Tufano, 1997). HUDs will probably become more prevalent, but they will be introduced slowly since some key perceptual and cognitive problems have yet to be solved.

One technology that is particularly promising for drivers who have limited neck motion is the parking aid. Parking aids include rear proximity sensors and other parking collision avoidance systems (e.g., Ward and Hirst, 1998). These systems typically utilize audible alerts or iconic displays on the dashboard to indicate that the driver is about to back into an object. Audible alerts should work fine for drivers who have good hearing but may not be effective for drivers with hearing impairments; they should be designed with redundant visual displays on the dashboard and mirrors.

Design Guidelines for Perception of the Environment

- Provide sensory enhancement systems to facilitate night vision and compensate for sensory loss.
- Ensure that the introduction of new technology does not introduce side effects with negative safety impacts that outweigh the benefits.
- Whenever possible, systems should provide redundant information in alternative sensory modalities.

50.7   NAVIGATION

As drivers age, some begin to lose their confidence and/or their ability to navigate in unfamiliar territory. If they acknowledge this difficulty, these drivers only drive on very familiar routes. Thus, in-vehicle navigation systems are very attractive to drivers as a means to compensate for poor or impaired wayfinding abilities. This includes typical drivers in a new or confusing area who are "handicapped" by the unfamiliarity of the environment. These devices provide automated instructions on how to proceed from one address to another (see Fig. 50.4). Some identify and locate points of interest such as libraries, hospitals, and supermarkets, and a means to enter origin and destination selections.

FIGURE 50.4 Navigational aid.

The user interfaces of most navigation systems are small flat-panel displays mounted on the dashboard. Audible directions are included in some systems, and HUDs are being examined as alternative display devices. Navigation directions are provided in the form of turn-by-turn instructions, overhead plan, or bird's-eye views (most high-end systems allow for several choices). Interaction problems can include difficult destination input methods (Steinfeld et al., 1996), visual attention conflicts, and poor understanding of imagery. The most serious problem is misinterpretation of instructions. For example, in a real-world experiment of one system, there were four critical incidents where subjects changed lanes upon instruction from the system without checking for other cars (Katz et al., 1997). Regardless, a variety of automobile manufacturers, rental agencies, and car audio system manufacturers have begun to offer these systems as optional accessories.

Clearly, it is important that these new technologies do not distract drivers from scanning the road ahead or overload their mental capabilities. While systems that use speech output are clearly beneficial in that respect, they exclude use by people with hearing impairments unless visual information is also provided. Information has to be presented in a manner that has little or no significant impact on the safety of the driving task. The task of selecting choices and commands can also distract the driver. Voice recognition systems are often heralded as the safest input method but, to benefit all users, they have to be usable for individuals with speech impediments or accents. Multiple-language capabilities are also desirable.

Design Guidelines for Navigation

- Destination selection should be as easy as possible, without causing distraction from the driving task.
- Destination entry controls should consider voice recognition and shape coding for use by touch. Controls for tactile use should be few and large.
- Systems should be accessible to drivers with hearing impairments both for output and input.

50.8 SAFETY

Crash-protection or "occupant-packaging" features like air bags, whiplash-protection seats, and passive restraint systems will not be addressed here. (This subject is heavily studied by a variety of laboratories and tracked by the U.S. Government via the National Highway Traffic Safety Administration [NHTSA].) Instead, the focus here is on high-tech, precrash safety developments that are just now being introduced to the general public. These include a variety of safety devices such as collision warning and adaptive cruise control.

Collision warning systems have been on the market for several years but have mostly been installed on commercial vehicles like tractor trailers. Companies like Eaton-Vorad market systems that use forward sensors to determine the presence and trajectories of potential obstacles. A user interface indicates to the driver when dangerous scenarios exist, thus prompting corrective action. When collision warning systems are tied to the automobile's cruise control, the vehicle can automatically respond to forward obstacles by releasing the accelerator, shifting to a lower gear, and/or activating the brakes. This system is referred to as adaptive cruise control (ACC). Currently, ACC is an option on some luxury cars and is available in some modern commercial trucks. While collision warning systems and ACC will likely improve safety for the general public, there are certain design concerns regarding ease of use and comprehension. Given the need for research on ACC, the U.S. Department of Transportation has recently finished funding a large-scale field operational test of an ACC system without active braking (Fancher et al., 1998). Additional large-scale studies of ACC and collision warning systems are underway.

Calls for standardized audible tones to reduce confusion—"Did that beep mean the ACC activated or did I get a new e-mail?"—are becoming more common (Kantowitz and Moyer, 1999). However, using only an audible mode to convey safety-related information does not provide sufficient information to drivers who are hearing impaired. Innovative new approaches have been attempted to transmit safety-critical messages to the driver without visual stimuli. Examples are accelerator pedals that push back, brake pulses (Lloyd et al., 1999), and torque or vibrations applied to the steering wheel (Schumann et al., 1996; Hsu et al., 1998). The tactile modality is more direct and, like hearing, does not necessarily depend on selective attention to convey a message. Such systems would definitely be beneficial to people with hearing impairments and to the general population, but would be less useful to drivers with limited tactile sensation or alternative vehicle controls. Thus, a combination of audible and tactile/haptic alerts may be ideal.

Design Guidelines for Safety

- System functionality should be easy to understand.
- Alerts and other interfaces should utilize redundant sensory modalities in a standardized manner.

50.9 IN-VEHICLE INFORMATION

These days consumers are expecting more than a car radio and traveling games like Punch Bug to keep themselves entertained. Unfortunately, the addition of advanced information and entertainment devices can lead to "button overload." Vehicle safety can be compromised when many controls are available for adjustment while driving. One study documented a maximum duration of 3.7 s for a single glance while using a car radio (Ayers et al., 1996), a highly unsafe scenario. Fortunately, auto companies have begun to express interest in multifunction interfaces that offer integration of many in-vehicle features through a limited set of buttons and selections by using a "menu tree" (Sumie, Li, and Green, 1998). Such interfaces will allow designers to significantly reduce the number of buttons on a device and thus permit simpler interfaces, larger buttons, larger text, and more logical grouping of functions—all of which should be beneficial to older drivers and those with mobility impairments. However, the trade-off could be a longer selection period and additional glances to the dashboard. For example, a particular action may require one button of many as opposed to one button only with three selections down a menu tree. The need to switch visual attention several times between the road and the dashboard may be especially problematical for older drivers given the necessity to refocus twice for each glance. Older eyes require more time to refocus as attention shifts from a distant object (road scene) to a close object (control panel).

The main problem with many in-vehicle devices is that they are not accessible to people with hearing impairments. Some traffic safety experts may argue that this is good since there are fewer potential distractions for these drivers, but, there are significant benefits for some of the features these devices offer. For example, as any hearing person who lives in a large city can attest, receiving a warning about the location and reasons for traffic congestion over the car radio is immensely beneficial. Besides the time saved by being able to alter a route before getting mired in gridlock, there is a safety benefit because advanced warning of congestion leads to higher levels of alertness. Furthermore, the lack of entertainment beyond basic driving tasks during long trips can lead to increased boredom due to a lack of mental stimulation.

As companies begin to offer wireless access to the Internet, additional demands for attention will be placed on the driver. One positive characteristic of Internet-equipped cars is that most e-mail, chat, and Web information is in text form. For example, a potential application for drivers with hearing impairments is the ability to access traffic reports available on Web sites. However, the current trend for Internet-equipped vehicles is to use voice interaction, via speech recognition and digitized text to speech, rather than keyboards and screens. An ideal solution would be to allow the use of keyboards and screens when the vehicle gearshift is in park, so that drivers with hearing impairments can utilize the functionality of these systems. Passengers could safely use such systems while a vehicle is in motion, however. Systems could be designed to operate even when not in park as long as they are only oriented toward the passenger side of the vehicle, out of reach and view of the driver. Keyboard and screen interfaces would allow passengers to access information without the driver being distracted. It should be noted that features like text-based traffic alerts are useful to hearing-impaired drivers who have good eyesight, but the small text may lead to difficulty for older drivers (see Fig. 50.5).

FIGURE 50.5   Text-based traffic alert.

Design Guidelines for In-Vehicle Information

- Provide redundant information in a form accessible to drivers with hearing impairments whenever possible.
- Activities that require high cognitive demands should only be permitted when the vehicle is not in motion or if safeguards are provided to only permit passenger use while in motion.

50.10   CASE STUDY—ONSTAR

A relatively new in-vehicle system that has universal design characteristics is integrated live-operator assistance. An example is the OnStar system. Activation is as simple as pressing a blue button, usually placed above the rear-view mirror. Pressing the button opens a cellular phone call to specially trained operators. Using their computers and information relayed from the vehicle through the system, operators can provide a variety of assistance to the driver. OnStar services include convenience services, concierge services, roadside assistance, route support, and emergency services. Other features include remote door unlock, air bag deployment notification, and theft protection. The integration of a live human operator increases the ability of the system to provide assistance during scenarios that system designers may not have envisioned. Furthermore, including human operators by nature also includes characteristics like intuition, curiosity, and compassion that are useful in emergency events. For example, an operator will have the initiative and capability to try to determine if a person in an altered state is experiencing a stroke. Again, the major limitation of this system from a universal design perspective is that it is not accessible to people with significant hearing impairments. It is hoped that future versions will permit a TTY option.

50.11   GENERAL LESSONS LEARNED FOR UNIVERSAL DESIGN

Common themes throughout this chapter are the importance of large, easy-to-operate controls, redundant information for drivers with hearing impairments, and simple designs that do not tax cognitive or perceptual resources. Furthermore, designers should be aware that new systems designed to improve driving may, in fact, lead to undesirable degradations in driver performance.

Additionally, most designers focus their attention on the person who will eventually drive the vehicle and then mimic the features for passengers. This is a faulty approach in that passengers often have impairments that the driver does not have. Simply duplicating the driver's seat and door when designing the passenger areas can quickly lead to problems for friends or family members with mobility impairments. Awareness of this issue is not simply good design; it is also good business. Consumers will base initial and repeat purchases on their experiences.

50.12   CONCLUSIONS

In contrast to mass transport systems (Chap. 24, "Universal Design in Mass Transportation," by Steinfeld), automobile design can change rapidly because of the "loose fit" between vehicles and the highway infrastructure. The marketing focus of automobile design is a vehicle for introducing innovation through universal design. There is clearly a move in the industry to address the purchasing power of the aging baby boomer generation in industrialized countries. Initially, the focus may be on the easy problems of interior features. It is hoped that attention will shift to some of the more difficult issues where usability conflicts with structural, aerodynamic, and styling concerns. For example, the older generation would be well-served by vehicles that have low floors but high roofs and high seats. But this combination neither provides a sleek, sporty appearance or a tough, off-road look, two of the most popular styles of the times. Priorities will have to change before usability improves when it conflicts with other goals, unless creative solutions can identify new approaches that satisfy all goals at once.

Of particular concern for the future is the rapid introduction of new technologies. Many of the early information and telecommunications products introduced do not demonstrate a universal design philosophy. If the designers and manufacturers of these new technologies incorporated more universal design strategies, these new products would likely be safer and easier to use for the whole consumer population. There is still much to learn before all the problems are solved; for example, provision of information in alternative sensory modes. Vanderheiden discusses design decision making that relates to this issue in Chap. 65, "Fundamentals and Priorities for Design of Information and Telecommunications Technologies," in this handbook. Clearly, there are new products that can have significant benefits to the driver with a disability or simply an older person seeking to continue driving safely for as long as possible.

50.13 NOTE

1. Research for this chapter was partially supported by funding from the Rehabilitation Engineering Research Center on Aging, SUNY/Buffalo, which is sponsored by the National Institute of Rehabilitation Research, U.S. Department of Education.

50.14 BIBLIOGRAPHY

Ayres, T., A. Donelson, S. Brown, V. Bjelajac, and W. Van Selow, "On-Board Truck Computers and Accident Risk," *Safety Engineering and Risk Analysis 1996,* SERA-Vol. 6, American Society of Mechanical Engineering, 1–6, 1996.

Ball, K., and C. Owsley, "Identifying Correlates of Accident Involvement for the Older Driver," *Human Factors,* 33(5):583–596, 1991.

Block, D., "A Well-Suited Approach to Auto Design for Older Drivers," *Pittsburgh Post-Gazette,* January 24, 1999.

Cerrelli, E., *Older Drivers: The Age Factor in Traffic Safety,* DOT, NHTSA (DOT HS 807 402) Springfield, VA, 1989.

Cushman, L., R. Good, R. Annechiarico, and J. States, "Effect of Safety Belt Usage on Injury Patterns of Hospitalized and Fatally Injured Drivers 55+," in *34th Annual Proceedings Association for the Advancement of Automotive Medicine,* Scottsdale, AZ, 1990.

Fancher, P., R. Ervin, J. Sayer, M. Hagan, S. Bogard, Z. Bareket, M. Mefford, and J. Haugen, *Intelligent Cruise Control Field Operational Test, Final Report Volume I: Technical Report* (DOT Report No. DOT HS 808 849), 1998.

Ford, R., "Auto Designs for the Ages—Marketers Appeal to Boomers, Young Drivers," *Boston Globe,* A01, March 5, 2000.

Hakamies-Blomqvist, L., "Research on Older Drivers: A Review," *International Association of Traffic and Safety Sciences (IATSS) Research,* 20:91–101, 1996.

Hsu, J.-C., W.-L. Chen, K. H. Shien, and E. C. Yeh, "Cooperative Copilot with Active Steering Assistance for Vehicle Lane Keeping," *International Journal of Vehicle Design,* 19:78–107, 1998.

James, G., *Problems Experienced by Disabled and Elderly People Entering and Leaving Cars,* Transport and Road Research Laboratory Research Report, Institute for Consumer Ergonomics, Loughborough University, Loughborough, UK, 1985.

Kantowitz, B., and M. J. Moyer, "Integration of Driver In-Vehicle ITS Information," in *Proceedings of the Intelligent Transportation Society of America (ITSA) Ninth Annual Meeting and Exposition* (CD-ROM), Intelligent Transportation Society of America, Washington, DC, 1999.

Katz, S., J. Fleming, P. Green, D. Hunter, and D. Damouth, *On-the-Road Human Factors Evaluation of the Ali-Scout Navigation System,* Technical Report UMTRI-96-32, The University of Michigan Transportation Research Institute, Ann Arbor, MI, 1997.

Lloyd, M., G. Wilson, C. Nowak, and A. Bittner, "Brake Pulsing as a Haptic Warning for an Intersection Collision Avoidance (ICA) Countermeasure, in *Transportation Research Board 78th Annual Meeting* (Preprint CD-ROM), Transportation Research Board, Washington, DC, 1999.

Owsley, C., and K. Ball, "Assessing Visual Function in the Older Driver," *Clinics in Geriatric Medicine,* 9:389–401, 1993.

Parker, J., "Innovative Designs Help Aging Baby Boomers and Motorists with Disabilities Drive with Ease," *Detroit Free Press,* December 9, 1999.

Pastalan, L. A., "Empathic Model Project," paper presentation at the Gerontological Society Annual Conference, Houston, TX, 1971.

Schumann, J., J. Lowenau, and K. Naab, "The Active Steering Wheel as a Continuous Support for the Driver's Lateral Control Task," in A. G. Gale, I. D. Brown, C. M. Haslegrave, and S. P. Taylor (eds.), *Vision in Vehicles—V,* Elsevier Science, Amsterdam, Netherlands, pp. 229–236, 1996.

Steinfeld, A., and P. Green, "Driver Responses to Navigation Information on Full-Windshield, Head-Up Displays," *International Journal of Vehicle Design,* 19:135–149, 1998.

Steinfeld, A., D. Manes, P. Green, and D. Hunter, *Destination Entry and Retrieval with the Ali-Scout Navigation System* (Technical Report UMTRI-96-30), The University of Michigan Transportation Research Institute, Ann Arbor, MI, 1996.

Steinfeld, E., M. Tomita, W. Mann, and W. DeGlopper, "Use of Passenger Vehicles by Older People with Disabilities," *Occupational Therapy Journal of Research,* 19(3):155–186, 1999.

Sumie, M., C. Li, and P. Green, *Usability of Menu-Based Interfaces for Motor Vehicle Secondary Functions* (Technical Report UMTRI-97-19), The University of Michigan Transportation Research Institute, Ann Arbor, MI, 1998.

Tufano, D., "Automotive HUDs: The Overlooked Safety Issues," *Human Factors,* 39:303–311, 1997.

Weintraub, D., and M. Ensing, *Human Factors Issues in Head-Up Display Design: The Book of HUD,* (CSERIAC state of the art report), Crew Systems Ergonomics Information Analysis Center, Wright-Patterson Air Force Base, OH, 1992.

Waller, P., "The Older Driver," *Human Factors,* 33:499–505, 1991.

Ward, N. J., and S. Hirst, "An Exploratory Investigation of Display Information Attributes of Reverse/Parking Aids, *International Journal of Vehicle Design,* 19:41–49, 1998.

50.15 RESOURCES

National Highway Traffic Safety Administration (NHTSA).   www.nhtsa.dot.gov

Intelligent Transportation Society of America.   www.itsa.org

P · A · R · T · 7

EDUCATION AND RESEARCH

CHAPTER 51

ADVANCES IN UNIVERSAL DESIGN EDUCATION IN THE UNITED STATES

Polly Welch, Architect
Stanton Jones, M.L.A., M.C.P.
University of Oregon, Eugene, Oregon

51.1 INTRODUCTION

The education of future design professionals is and should be central to any discussion of universal design as a transformative paradigm in environmental design. Clearly, if universal design is going to be made manifest in built form and become second nature in the design responses of designers, communicating this concept to designers in their formative years of development is critical. Yet, the teaching of a value as complex and pervasive as universal design is not accomplished simply by adding a lecture, assigning a reading, or teaching "the code." Design instructors teaching universal design have found that, to effect attitudinal change, the pedagogical strategies must be at once engaging and critical.

Besides developing syllabi and classroom teaching, inculcating design education with universal design values also requires addressing more systemic academic considerations such as program accreditation; continuing education to graduates; knowledge generation and scholarly development of faculty; recruitment and retention of a diverse group of students and faculty; a physical environment conducive to learning; and administrative support of innovation and excellence.

This chapter examines a range of curricular responses that have been utilized at universities in the United States in the past decade, highlights possibilities for structure and content within the design curricula and programs in architecture and landscape architecture, and concludes with some recommendations for future developments.

51.2 BACKGROUND

Universal design is not an entirely new concept in design education. Concern about segregated design and the need for a more integrative concept was identified over 20 years ago (Bednar, 1977; Ostroff, 1982), but the codification of barrier-free design solutions and a market economy that is based on segmentation of users have reinforced separate and unique accommodation (Jones and Welch, 1999).

The development of universal design education is intricately intertwined with the evolution of universal design as a concept for a more equitable world and as a value in designing places that meet the needs of a variety of people. The vision of those who first promoted the concept of universal design was to underscore the mainstream applications and benefits of accessibility standards and assistive devices that had previously been seen as exclusively designed for and used by people with disabilities. The concept was robust because it sought to simultaneously integrate disability issues into the marketplace by showing the benefits to all people of designs previously believed to exclusively serve people with disabilities (Mace et al., 1991) and by expanding the definition of *consumer* to include a greater diversity of people (Anders and Fechtner, 1992).

The opportunity to transform prevailing norms and practices related to accessibility, implicit in the universal design concept, also captured the imaginations of faculty in professional design schools. Some saw an opportunity to expand the population benefiting from accessible design without significantly altering the accessibility content they were already teaching. Others sought to locate design for people with disabilities within a broader context that included a wide array of users, thereby including disability in the critical examination of who is and is not served through placemaking.

The development of universal design has been shaped by the frustration of disability advocates toward design solutions that persisted in segregating people with disabilities from the rest of the world. To Ron Mace, architect and tireless proponent of accessible places and products, universal design meant designing with *all* users in mind—that is, democratic design. As a result of his years of work making the built environment more equitable for people with disabilities, he saw the need for a fundamentally different way of framing the concept of accommodation. In spite of his efforts to distinguish universal design practice from ADA compliance, many teachers and students treat the two as synonymous, casually identifying clearly segregated access as universal design.

The teaching of universal design is confounded by confusion among students and faculty as to what *universal design* actually means. Universal design continues to be misunderstood as a euphemism for *accessible design*. Some design instructors have used the terms *universal design* and *accessibility* interchangeably, describing universal design as a better way to create environments for people with disabilities. The confusion is not surprising because the first proponents of universal design were already leading advocates for accessibility. The plethora of euphemisms in use for identifying people with disabilities reinforces the belief that universal design is just another, more current, term for accessibility. This is further fostered by the presence of a person using a wheelchair in many of the visual examples of universal design. While universal design appeared to grow out of the accessible design arena, its paradigmatic shift from *special design for people with special needs* to *inclusive design for a diverse population* has not been sufficiently clear for people to realize that universal design is fundamentally different from achieving accessibility as outlined in the American with Disabilities Act and building codes.

Proponents of universal design have not been clear until recently about how universal design is connected to and yet distinct from accessibility. Design features that usually make buildings and landscapes accessible, for example, may not be universal design if the solution addresses narrowly the needs of some users over others. The photograph of an apartment doorway in a family housing complex (see Fig. 51.1) illustrates an accessibility feature—the lever door handle—that might be universal design under some circumstances. It meets the needs of a person with poor hand dexterity as well as a resident with an armful of groceries but, in this context, does not sufficiently address the ease with which it can be manipulated by a young child who can run out into the street.

This chapter assumes that the two terms reflect "nested" concepts: Universal design embraces accessibility as one of an array of responses to achieve inclusive design. Proponents of the most inclusive definition of universal design include consideration of broad and overlapping facets of identity, including—in addition to age and ability—size and stature, ethnicity, gender, culture, sexuality, and class (Welch and Jones, 1998).

In addition to the terminology confusion, the incorporation of universal design into school curricula has been stymied by prevailing norms about what constitutes fundamental knowledge in design education. From its inception, universal design has been promoted as an eco-

FIGURE 51.1   Door with lever door handle and baby gate.

nomic idea—design that is usable by the greatest number of people to the greatest degree possible increases market share and makes sense in terms of life cycle cost (Mace, 1991). The argument was made that if an object produced for someone with special needs could accommodate everyone, its cost could be reduced and thus its availability increased. When designing new places, it was argued, it is less expensive to design one solution for all than to add a separate alternative for a subset of people. This argument for universal design, however, was not a particularly powerful motivator for faculty and students to think more inclusively about users, largely because economic considerations have rarely been given much more than lip service in design teaching, especially if they appear to conflict with formal and aesthetic principles. Universal design, as will be discussed later, also needs to be addressed as a social and political idea—the idea that all people have the right to have their needs considered and addressed in the design of the built environment.

51.3   *THE EVOLUTION OF UNIVERSAL DESIGN TEACHING*

Efforts to incorporate universal design into design curricula have been undermined by the historical resistance of designers to privilege social concerns as a critical component in the act of designing. For more than 50 years, social scientists and environmental design researchers have sought to influence the design professions through the publication of positivistic and phenomenological research studies on person-environment relations. Few contemporary designers acknowledge the existence or influence of this form of knowledge; even fewer actually consciously use environment-behavior research as a source of information for their design decisions. For a short period in the 1960s and early 1970s, attention to users' needs was fashionable in design schools and practice. Not surprisingly, this was also the decade when American society was struggling with civil rights and political and cultural imperialism. People with disabilities took to the streets to demand equal access and independence, underscoring the need for more democratic values about the human experience in the built environment. Many of the faculty teaching universal design today participated in these social and political movements as design students or young scholars.

During the last 3 decades, formal instruction about user accommodation has been relegated to a small niche in the curriculum of many schools, often in a single required course about social and behavioral factors in design. These courses have frequently employed a user typology—elderly people, people who are disabled, children, and the homeless—for conveying the needs of key groups of people. This means that the needs of older people, for example, have been well described—but separately from the needs of people with disabilities, and separately from any consideration of designing for the life span.

In response to designers' ambivalence toward user needs, and to a growing body of knowledge about the needs of specific user groups, some efforts were made to bridge the gap between designers and the information that could inform and enhance their design work. One venue was through curriculum development, banking on the idea that design students might be more successfully influenced than practitioners. In 1975 the Gerontological Society of America (GSA), funded by the Administration on Aging, undertook a 2-year project to develop educational materials for design instructors so they could utilize the research on aging and environments in their teaching. The GSA believed that this would prepare future practitioners to design for an aging American population that was growing exponentially and whose needs would impact a range of building types. The project compiled a sourcebook of 30 research articles on environment and aging and pretested this package and strategies for course development at 11 universities. Case studies from these schools showed that: the students read only a few of the research materials, finding them theoretically cumbersome and limited in graphic examples; students responded well to input from visiting experts (usually gerontologists and architects, except in one case where students met with older people); and students had difficulty applying the information they received to design studio problems (*Journal of Architectural Education*, 1977).

Commenting on her teaching with the GSA materials, Sandra Howell of MIT (Howell, 1977) noted that treating age in isolation from other facets of identity "narrows opportunities for illustrating user needs and behaviors relevant to future professionals who will undoubtedly be designing for a range of populations and of settings. . . . Not all students are psychologically or intellectually prepared to explore human aging. Introduction of materials and methods equally appropriate to issues of old and young, workers and retirees and able and disabled, brings the student more slowly into the arena and offers options within a behavioral context."

Formal dissemination of the aging curriculum to a broader constituency never happened. In 1994 the AIA/ACSA Council on Architectural Research was awarded a grant from the

Administration on Aging to develop new curriculum materials, building on the earlier endeavor. A new package of materials was created with a more critical eye toward what students might read and find useful to their design inquiry. While providing a rich resource of materials and syllabi, it primarily served educators who chose to teach the design of institutional settings for old people, and contained little to inform design students about accommodating the needs of people in the latter half of the lifespan across a range of places and building types.

Teaching about disability followed closely behind aging. In 1979, Raymond Lifchez, a professor of architecture at the University of California at Berkeley, was funded by the Exxon Education Foundation for an experimental project, "Architectural Design with the Physically Disabled User in Mind," in which he sought studio teaching methods that would help students "bridge the gap between able-bodied and disabled people." This well-documented project (Bassett, 1984; Lifchez, 1987) has been an inspiration as well as a cogent appraisal of the frustrations and benefits of engaging real users in design studio teaching. Still germane to current issues in universal design education, Lifchez's project addressed disability in the context of user needs accommodation more generally and used the studio venue as the site for strategic engagement between students and real users (see Fig. 51.2). Lifchez's goal was to test teaching methods that would "place clients at the heart of the design process." His perspective was in radical contrast to the prevailing postmodern design philosophy of the 1980s that privileged

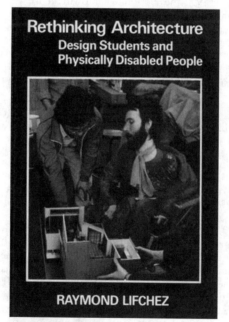

FIGURE 51.2 Student and user/consultant in Raymond Lifchez's studio at the University of California at Berkeley.

form over other considerations: "Client accommodation is not merely the third element in design, alongside aesthetics and technology, but it is in fact the context within which all factors of architectural design must be placed" (Lifchez, 1977).

In 1981, Uriel Cohen at the University of Wisconsin-Milwaukee (UWM) expanded his research on the important distinctions between mainstreaming people with disabilities and barrier-free design by publishing a resource guide for teachers and students of environmental design. He had discovered that the intense focus on removing mobility barriers had failed to address the needs of 90 percent of the affected children—those with non-mobility-related disabilities. This was the first curriculum material to raise the subtle but equally important issue of designing for people with learning disabilities, sensory disabilities, developmental disabilities, and mental illness. The resource guide contains programmed instructional materials and information resources and continues to be available from UWM. The content of the materials is still relevant even if the terminology and bibliography are dated.

A small number of design educators have sustained an interest over the years in addressing the accommodation of people with disabilities and taught classes devoted to barrier-free design. Even since the passage of the ADA and its mandate for equal access to the built environment, these courses have been students' primary, if not only, exposure to substantive information on the needs of people with disabilities. Technical requirements for accessibility are also randomly addressed as discrete bits of information, in technology classes and in studio teaching, at the discretion of the professor. The recasting of accessibility legislation as civil rights law has possibly increased the pressure to teach about the ADA but has had a surprisingly small impact on *how* this material is taught, even though it presents a robust opportunity to discuss the social and political context of design. External forces such as digital technology, seismic risk, and sustainability values, on the other hand, have been a bit more successful at permeating design curriculum.

The role of education in achieving universal design has been a key issue for advocates and activists. In 1990 the National Endowment for the Arts (NEA) organized a gathering of experts in the design fields, education, and disability rights to identify how the NEA might support educational efforts to bring a more "universal approach to the design of buildings, public spaces and products." This group made it clear that fundamental changes in attitudes and practice depended on exposing students to universal design as they were forming their values and developing their skills in school. The NEA meetings resulted in the following agenda to support design teaching by (Welch and Ostroff, 1995):

- Engaging the power of the accrediting organizations to emphasize universal design in curricula
- Funding the development of design problems for universal design instruction
- Helping publicize resources and experts available to educators

The NEA funded several universal design education efforts, including the Universal Design Education Project, proposed in 1989 by Elaine Ostroff of Adaptive Environments (Welch, 1995); the first U.S. conference on universal design, "Universal Design: Access to Daily Living," held in 1992 in New York City (*Metropolis*, 1992); a video, *Towards Universal Design* (Mueller, 1993); and the dissemination of visual examples of good universal design (Universal Designers and Consultants, 1996).

In 1991 the Adaptive Environments Center was awarded the initial funding for its Universal Design Education Project (UDEP) to "challenge existing values in design education and to stimulate innovation in design curriculum that will lead to the development of products and environments which incorporate universal design concepts." In addition to the NEA, the Department of Justice Disability Rights Section, NEC Foundation of America, and several other national foundations later contributed support for UDEP. The pilot project sought to integrate the concept of universal design into the curricula of five design disciplines (architecture, landscape architecture, interior design, industrial design, and urban design). The project, which encouraged innovative engagement of universal design by faculty and students, supported faculty in 22 design programs and resulted in ongoing curricular changes in a number of the schools. It is documented in *Strategies for Teaching Universal Design,* a book of case studies highlighting the teaching process and outcomes (Welch, 1995). UDEP awarded a second round of grants 3 years later to nine design programs, six of which were building on innovations developed in the first grant period.

The UDEP faculty experimented with developing contextually responsive teaching materials, with positioning universal design in a variety of curricular locations, and with bringing user/experts into their classrooms and studios. Although there was a strong effort at dissemination to other design education programs, through the book of case studies, some publications, videos, and numerous presentations at design educators' conferences, universal design has not achieved the visibility or recognition in the design community that one would have expected. Many of the UDEP faculty have had difficulty achieving significant dialogue with their design colleagues about the incorporation of universal design values into design education, in general, and into their programs' design curriculum, more specifically. Students, however, have demonstrated interest and acceptance of the value, as documented in the evaluations that the faculty conducted as part of their grants. How students choose to incorporate universal design values into design practice remains to be seen over the next decade.

51.4 *TYPES OF CURRICULAR RESPONSES*

The UDEP faculty and others teaching universal design have utilized numerous strategies. Determining how best to introduce universal design values into the curriculum requires weighing the unique context of a design program with a variety of possible strategies for introducing a new perspective. Appropriate locations for introducing universal design into the design studio sequence or subject area courses vary, depending on a number of factors that include the pedagogical structure of departmental curricula, faculty interest, knowledge, and commitment, and disciplinary demands. Ostroff envisioned and encouraged the development of diverse strategies that built on faculty strengths and were appropriate to the unique culture of each school (Welch, 1995).

While many educational interventions have been simple injections of new course materials, such as stand-alone assignments worked into preexisting courses, others have entailed more prolonged, infusion-based approaches to teaching design students that have required a greater, if not more sustained, commitment to curricular alteration. In many cases the contextual realities of an educational institution have dictated the nature of the "curricular response" in ways that sometimes enhanced, and sometimes hindered, the delivery of information and ideas to students.

Participants in UDEP taught universal design at different levels. In some cases the level was determined by departmental teaching assignments, and in others the level of engagement was determined by faculty consciously positioning universal design in the curriculum. Iowa State faculty had the foresight and sufficient status, seniority, and departmental involvement to integrate universal design into multiple courses at different levels, with the conviction that students needed repeated exposure at increasingly higher levels of engagement to understand that universal design is an essential consideration in the design process. Most UDEP schools clarified through reflection and evaluation their understanding of when universal design is most effectively introduced and integrated into the curriculum. At SUNY Buffalo, universal design was initially incorporated into the fourth-year studio as a vehicle for teaching good design. The faculty team came to the conclusion, however, that universal design principles needed to be introduced earlier, with frequent subsequent engagements building on the first early exposure. In the second phase of UDEP the SUNY Buffalo faculty was able to further develop this teaching experiment by incorporating universal design into the second-year design studios and made several other critical changes based on their initial experience.

Weighing the benefits of injecting a teaching unit into the curriculum or infusing the curriculum in a more systemic manner is a fundamental decision in developing a strategy for teaching universal design. As illustrated by the programs and schools that participated in the UDEP, many different elements might factor into a decision regarding which direction to take; understanding the programmatic and pedagogical context, as well as the nature of faculty support and expertise, are but two of the many variables that must be weighed in deciding which strategy or strategies to pursue. Figure 51.3 illustrates the spectrum of curricular strategies utilized by faculty at design schools across the country.

A Typology of Curricular Responses

INJECTION

- Inject a unit of teaching into a given course syllabus
- Inject a course devoted to universal design into the curriculum
- Offer a one-time event/workshop

INFUSION

- Infuse universal design into a subject area course
- Infuse universal design into a studio problem
- Infuse universal design into a single year of the curriculum
- Infuse universal design into the entire design curriculum

FIGURE 51.3   Types of curricular responses.

Injection Strategies

Curricular change often starts with small trial injections of new materials. In the design fields, new concepts like ecology and sustainability were first introduced by prescient instructors in the form of small teaching units in traditional courses and over time have become fully integrated in many curricula.

The benefits of injection usually outweigh possible deficits. Faculty can inject new material into a course, offer a seminar on a previously untaught topic, or plan an event with relative autonomy and ease. Injections can be modified and expanded from year to year, creating the opportunity for pedagogical development without major curricular modification. The risks of injection are the potential for superficial engagement by the students, perceived marginality of significance in relation to the rest of the curriculum, and the added responsibility for students to integrate the material into their design work.

Stand-Alone Unit.   In response to practitioners' expectations that students must be familiar with codes, many educators have responded with useful but pedagogically limited lessons geared toward teaching technical information about users, especially in the form of codes and anthropometric data.

The benefit to educators of the stand-alone unit, consisting of a reading, lecture time, and an assignment, is the ease with which it can be worked into a course. Stand-alone units also have portability, enabling an instructor, whose teaching assignments vary from year to year, to incorporate the material into other courses. Universal design teaching that is modeled after teaching accessibility codes is typically taught this way.

The drawbacks to this approach are significant, however, especially if this is the only strategy utilized within the curriculum to teach universal design. One UDEP landscape architecture program developed a basic computer tutorial that introduced universal design as a resource for students. Few students actually used it beyond a few minutes of exploration, and it appeared to have little impact on student design work. With little or no reinforcement from other assignments or courses, the information can appear marginal (even irrelevant) to the main focus of design, to students who are simultaneously wrestling with all of the other ideas and lessons from their other courses. A stand-alone unit or assignment can actually work to reinforce the notion of universal design as an entirely separate design consideration, like accessible design, meant for a special population or a distinct segment of the population. Still, as one strategy used in concert with other strategies, teaching a unit or making an assignment can provide useful lessons, or create interesting opportunities for dialogue that may not happen otherwise in a student's training.

The relative benefits of a stand-alone unit are not measurable by the amount of time allotted to the topic. UDEP schools found that a 1-hour session with a panel of users could alter attitudes more than a reading or writing assignment. An assignment to incorporate universal design into a design problem has enormous potential to engage students in imaging the world differently, but without sufficient time spent on attitudinal change the design may reflect only superficial attention to diversity of use and inadvertently perpetuate formulaic answers.

Over time, the single most frequently used exercise by design faculty to teach accessibility has been the empathic experience, formalized by Lee Pastalan et al. (1973), in which students navigate their environment in a wheelchair or blindfolded. Although the UDEP stressed the participation of users in the teaching and learning process, many of the UDEP faculty employed this empathic technique as one of the methods to communicate the needs of people with disabilities. While the wheelchair exercise is engaging because it so dramatically illustrates for students the physical barriers created by designers and builders, some people believe that it furthers the stereotypes that impede the teaching of universal design values. It also tends to obscure the important concept that many different people are "disabled" by not having their needs met by the design of the physical environment. Empathic techniques create sufficiently robust teachable moments that it is worthwhile to use them, but care and sensitivity are required. Empathic techniques can be implemented responsibly if they are linked to experiences with real users (Welch, 1995). (See Fig. 51.4.)

Discrete Course.   There are far more opportunities in the venue of a full course to expand the dialogue in time and content than are possible in one unit of teaching. Faculty has the luxury of adapting their approach and being opportunistic with classes and events to actually construct a perspective that stimulates attitudinal change. While a single unit seldom allows enough time to address context sufficiently, a course can provide opportunities for students to develop an understanding of applications and rationale as well as history and the broader implications of applying universal design into design practice.

Three UDEP schools taught universal design as a discrete subject course; all of the courses had high enrollment and were determined by faculty to have a significant impact on the students. While the connections with design were not as strong as in a studio setting, the lectures, hands-on exercises, and in-depth discussions provided important opportunities for attitudinal change. Some of the most creative pedagogical developments occurred in these courses. Cal Poly at San Luis Obispo sought to give students "an experiential introduction to the theoretical, social, psychological, cultural, legal and ergonomic issues related to designing for diverse users." After an initial set of introductory assignments to explore the ramifications of diversity, the instructors asked the students to select audiences with whom they might share their newfound knowledge and awareness and to develop outreach projects to teach the benefits of universal design. This included making an educational video, visiting elementary school classrooms, and creating human performance sculptures in public places as shown in Fig. 51.5.

Exercises Used by UDEP Faculty

Learning more about the diversity of people:
- Interviewing someone different from self
- Researching how a particular facet of identity impacts the design of the built environment
- Inviting users into the studio for desk crits
- Inviting users to reviews and juries without telling students in advance
- Selecting studio projects with real client users
- Exploring the different perspectives among students in a class (SUNY Buffalo beautiful building slides)
- Tracing students' bodies and comparing to historical ideals of human body dimensions
- Drawing new diverse entourage figures
- Finding examples of the Seven Principles of Universal Design

Critiquing how well the built environment meets the needs of people
- Empathic exercises
- Building walkthroughs with user/experts
- Field trips
- Mock trial
- Installations

Building personal understanding
- Who Am I? exercise
- Teaching others about universal design
- Reviewing other students' design work from a personal nondesign perspective

---

FIGURE 51.4   Types of exercises used by the UDEP faculty.

One-Time Event.   Single, unique events present opportunities for concentrated time and attention on a particular topic and sustained intensity of experience—rare in the day-to-day classroom environment. In addition to other curricular activities, a number of UDEP schools used events as a strategy for reaching an audience broader than design students and a vehicle to raise awareness about universal design in the larger community. The events attracted students who were not able to devote an entire term to taking a course on universal design; students from the nondesign fields, such as law and health sciences, who were curious about disability and the built environment; and community members (both designers and nondesigners) who were looking for information and resources on universal design.

For the most part, these events used a keynote speaker and panel format combined with a participatory activity, or took the form of a "teach-in." The University of Oregon, for example, created a weekend event, titled "Power and Place," to raise awareness and provide an opportunity for students to engage universal design outside the curriculum. The faculty team from four design disciplines planned a forum for students to examine their attitudes and feelings about people different from themselves, through the lens of power and inclusion. The goal was to provide a participatory experience that would simultaneously engage students' empathy, their critical thinking, and their creativity—what architect Joan Goody

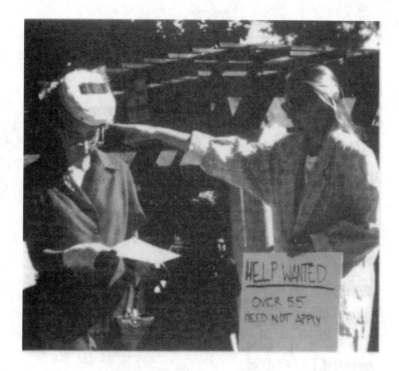

FIGURE 51.5 Photo students at Cal Poly creating human performance sculptures that challenge social stereotypes.

calls *compassionate imagination*. Students were invited to explore their own exclusion experiences, listen to others', and jointly imagine and construct an art installation to reflect their shared understanding of how the university environment supported inclusion (see Fig. 51.6).

FIGURE 51.6   Power and Place installation at the University of Oregon: "A Place at the Table."

The creation of art installations on the campus grounds was chosen over the more typical design intervention workshops that had been used at other UDEP sites, to distinguish this workshop experience from the problem-solving realm in which students are already being well trained. Abstracting the level of engagement was critical to differentiating this event from a diagnostic effort to make the university campus accessible (Welch, 2000). The workshop structure and outcome was conceived as an opportunity for personal reflection and social critique through creative expression that would be transformative for the students and instigate commentary, engagement, and response by passersby and, ideally, the community at large.

The dialogues between individuals and across disciplines raised challenging questions for participants about their personal assumptions, the stereotypes implicit in their training, and the relative absence of ethical discourse in their professions. Participants' written reflections over the subsequent months described a sense of empowerment and engagement that directly impacted their work and their lives. The workshop reinforced the importance of addressing attitudes by combining empathy and creativity in the search for inclusive design (Welch and Jones, 1998).

While these kind of events increase both visibility and participation, and can attract funding to sustain universal design teaching and research, they are also extremely time-consuming and place an enormous burden on individual faculty in terms of planning, coordination, and implementation.

Infusion Strategies

Faculty, teaching material that historically has been absent from mainstream design curricula, routinely face the frustration of the marginal status of their courses. Classes on race, class, and gender are now present in postsecondary education, but the important material and perspective that they cover continues to be separate from the dominant context of design training. More than ever, faculty is examining how these perspectives might be taught within traditional courses, incorporated directly into the curricular core (Dutton, 1991; Ward, 1996; Theodoropoulos and Welch, 1998). Infusion diminishes the potentially marginal status of the course content and introduces new discourse by challenging ableist, gendered, classist, Eurocentric course content.

Infusion, as an approach in general, affords faculty the opportunity to illustrate to students the interconnectedness of factors impacting design, while simultaneously challenging the students to internalize and apply their understanding to projects and tests in much the same manner as they do in a design problem. The difficulty in undertaking infusion, however, is that faculty themselves must be very well versed in the concepts, values, and rationale behind universal design so that they can present universal design as integral to their course material, as opposed to something discrete that can be addressed in isolation from design. It generally requires a critical mass of faculty members with strong commitment to the issues. Although many of the UDEP faculty aimed for infusion and found it challenging to achieve, two of the interior design education projects had more success than most. Both Eastern Michigan University (see Chap. 52, "Infusing Universal Design in Interior Design Education" by L. Jones) and Virginia Polytechnic Institute (Welch, 1995) had significant department-wide support and a history of teaching user-needs-related design.

Infusion of a Subject-Area Course. Different from a discrete course focused on universal design, this strategy utilizes existing courses, such as history or technical courses, as vehicles within which to infuse the information pertaining to universal design in a manner that encourages students to see the links between universal design and design, in general, in a new, more holistic manner. Rather than teaching universal design as a stand-alone subject, this approach embeds the information, and ideally the values and rationale behind the technical aspects, into the curriculum in a way that makes universal design an integral component in design thinking, research, and execution.

The UDEP faculty at the University of Oregon focused their efforts on infusing their respective courses with universal design values. Stan Jones incorporated universal design concepts into his "Land as Media" course 5 years ago. In this technical course designed to teach students the basics of contour manipulation and stormwater management, Jones challenges his students to consistently address, in all of the course assignments, the many social factors inherent within the design of built landscapes. He also expects them to evaluate all of their design work—both in his course and in other concurrent classes—in terms of "who is included" and "who is excluded" by design decisions. Jones includes a lecture on universal design and the ADA and occasionally uses empathic exercises accompanied by in-depth discussions by people with disabilities midway through the term. By linking social factors to the technical factors embedded in every design action they undertake, Jones's course helps students to design inclusive environments that are both technically sound and socially just. Polly Welch, in "Human Context of Design," a required course for architecture students, scrutinizes each lecture for opportunities to critique accepted knowledge from the perspectives of alternative perspectives such as race, class, gender, culture, and ability, demonstrating that the status quo can be interrogated and challenged. She discusses the ADA in the context of the story about how accessibility codes evolved, comparing that evolution to the social and political development of zoning and fire protection. She asks students to interview people different from themselves to learn how well the built environment meets their needs. By the end of 9 weeks, Welch's students demonstrate their understanding of environment-behavior concepts, including universal design, by adding annotations to the drawings of their design studio projects (see Fig. 51.7).

Infusion of a Design Studio Problem. Within design curricula, there are few venues that offer the integrative potential of the design studio (Schon, 1985). Given a specific design challenge, students must be able to draw on internalized knowledge they have been exposed to in previous (or concurrent) courses and be able to pursue a design response that reflects their understanding of the problem at hand—Schon's "reflection in action." While the introduction of new mate-

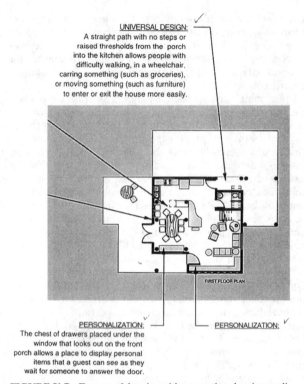

UNIVERSAL DESIGN:
A straight path with no steps or
raised thresholds from the porch
into the kitchen allows people with
difficulty walking, in a wheelchair,
carring something (such as groceries),
or moving something (such as furniture)
to enter or exit the house more easily.

FIRST FLOOR PLAN

PERSONALIZATION:
The chest of drawers placed under the
window that looks out on the front
porch allows a place to display personal
items that a guest can see as they
wait for someone to answer the door.

PERSONALIZATION:

FIGURE 51.7 Excerpt of drawing with annotation showing application of universal design.

rial is common within the structure of a typical studio, it can be a challenge to introduce concepts as complex as universal design. The amount of skill development and design mastery typically expected of students at each level of studio makes them easily overwhelmed when additional expectations are introduced. Several of the UDEP projects found that embedding the concept of universal design in the studio problem made it appear less like an additional expectation but also tended to understate its potential role in producing good design. Others found that students tended to persist with habits learned from earlier studios where they simply added some special features like ramps that they hoped would pass for universal design.

The SUNY Buffalo faculty defined *universal design* as synonymous with *good design* and provided a variety of critical explorations to encourage students to think about what would make the design of a hotel project inclusive. The work produced by some of the most engaged students challenged faculty perspectives on universal design and demonstrated that an exploration of universal design values can produce design innovations. The instructors were particularly pleased that only 3 out of 150 student projects were specially designed for people with disabilities. A few students, however, resented the time occupied by universal design considerations when they would rather be addressing "design concerns" such as structures and aesthetics. This points to the need to introduce universal design in other courses prior to studio if they are being asked to integrate the concept. In the absence of prior exposure, the choice of design problem becomes critical.

Another strategy for embedding universal design into a studio design problem is to assign a problem that requires a critical analysis of exclusion and inclusion. At the University of Oregon, a joint architecture/landscape architecture design studio problem took a critical look at how well existing neighborhoods meet the needs of a diverse and aging population. The students investigated whether permitted increases in density could also enhance livability, by encouraging building innovations that respond to changing household composition and housing needs throughout people's lives; changing uses of and demands on neighborhood open space; increasingly multigenerational and multicultural community makeup; and a more sustainable community. The studio critically examined the homogeneity implicit in neighborhoods developed over time in accordance with traditional zoning and explored the ramifications of designing for greater diversity.

Students worked in multidisciplinary teams that encouraged them to consider multiple perspectives. Design exploration took place at multiple scales from unit design to neighborhood placemaking to urban watershed. Applying universal design at these larger scales was difficult for the students because most of the exemplars illustrate universal design at the object scale. The studio emphasized the use of empirical knowledge and the fruitful involvement of community members. Some students realized that the ultimate engagement with universal design was in the presentation of their work and chose to add Braille and other graphic devices to make their drawings understandable to nonarchitects.

Studio is an excellent venue for students to explore how the design process might be enhanced to bring in opinions and experiences beyond their own, through user participation and student-driven research. The overall challenge of teaching universal design in the studio setting is to help students think creatively and inclusively at the same time, so that their final designs are simultaneously aesthetically "beautiful" and socially equitable.

Infuse Entire Curriculum.   Infusing the entire curriculum requires all faculty members to examine their own courses for how they can reflect the values of universal design at a minimum, and ideally, civil rights and social justice as well. The challenges are immense, given the academic freedom of instructors, and the amount of misunderstanding that exists among faculty, students, and administrators about the value of universal design. The UDEP professors in landscape architecture at Purdue proposed, with the full support of their faculty, syllabi changes to the entire landscape architecture curriculum, showing how universal design concepts could be incorporated into each course. The project was somewhat compromised, however, when the curriculum developer left the project to take another position. The outcome ranged from several instructors creatively incorporating universal design into their courses to others who found it difficult or impossible to modify their established patterns of teaching. The curricular plan, however, remains a template for ongoing change. The Iowa State UDEP team, with senior faculty representing three departments, including landscape architecture, architecture, and interior design, proposed to engage all 1000 students majoring in environmental design disciplines by infusing the curricula with "awareness modules" of increasing intensity. The awareness levels developed by the faculty have helped illustrate for other faculty why no single course can sufficiently cover universal design, and provided a model for discussion on building universal design into the curriculum.

Many faculty interested in universal design teach in an academic context that may not support massive curricular overhaul, even when their own work leads in that direction. In response to the requirement of the second round of UDEP funding "to engage faculty beyond the team" at SUNY Buffalo, Steinfeld, Mullick, and Tauke organized a faculty development workshop that attracted administrators as well as teachers. The agenda was to "search for universal design" and counted on the expertise of faculty to contribute to an interpretation of universal design that would help faculty imagine ways to use universal design in their teaching. The discussion ranged from issues of terminology and technology to ethics and sustainability of cultures and communities. Faculty facilitator Robert Shibley summarized four important points: "design as moral choice, the experience of architecture, the relationship of technology to design, and the politics of design" (Adaptive Environments, 1996). The workshop demonstrated how faculty can engage universal design through discussion of the philosophical and pedagogical aspects of good design.

In spite of the broad array of teaching strategy options, most of the teaching strategies used by design faculty can be categorized as an injection, typically highlighting universal design as one of many issues designers need to address in their work. This approach predominates for many reasons: Foremost among them is certainly the ease with which a discrete unit or problem assignment can be worked into what many perceive as an already overloaded professional curriculum. Discrete units lend themselves to imparting technical information such as the needs of well-defined and well-understood categories of users such as children. The single-unit approach has been less effective at encouraging the conscious integration of a wide array of human needs as originally envisioned by Lusher and Mace. More attention must be paid to the needs of users as real, complex human beings (e.g., a working single mother of three who is also blind) who make multiple demands on the design of places as they go about their daily lives. This complexity only becomes more intense as the spatial scale increases, a fact not lost on design students as they struggle with design problems that ask them to reconfigure parts of the landscape that range in size from a building room to a watershed.

Real infusion will occur when faculty consciously look for opportunities across the curriculum to infuse classes with universal design values. These will include places not normally thought of as venues for universal design, like technology and history/theory courses. There is also a need to examine how students and practitioners might present their ideas in a more inclusive manner: visual accessibility and making visible how building meets needs of diverse users.

51.5   A MODEL FOR INCORPORATING UNIVERSAL DESIGN INTO DESIGN EDUCATION

The strategies described in this chapter raise awareness about inclusive design, but they are not sufficient by themselves. To maximize the impact of these strategies, a comprehensive approach to content must provide students with opportunities for attitudinal change and critical thinking as well as acquisition of process skills and knowledge (Welch and Jones, 2001). Five components, described in the following paragraphs and graphically illustrated in Fig. 51.8, are critical for enabling students to move from general awareness to engagement and integration, and, ultimately, the ability to design inclusively.

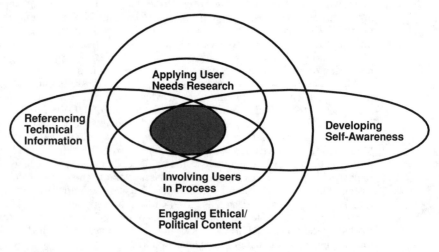

FIGURE 51.8   Model for curriculum development.

Referencing Technical Data That Informs the Fit Between Humans and Their Environments.   This is an essential part of a design education, but instruction in anthropometrics and building codes usually lacks sufficient discussion of the social, political, and ethical agenda underlying the data and standards. Design professionals and students alike often lament the restrictive impacts of the ADA that they believe compel them to design a less-beautiful, "design to the lowest common denominator" type of space (Stilgoe, 1999). The sort of "belligerent compliance" (Jones, 1998) that is practiced today by design professionals has its roots in the well-intended but unimaginative, and sometimes inaccurate, approach that some educators take to addressing the ADA. Codes and dimensions can take on new meaning to students when they are understood within a larger context (i.e., discussed in relation to user needs research and to the social and political realities that surround anthropometric research and the development of most code requirements).

Applying User Needs Research.   This is an essential component in teaching universal design. Students need to be familiar with existing literature on the needs of different users to be able to weigh it against other perspectives, including their own, and to develop a critical understanding of different forms of knowledge. As described at the beginning of this chapter, the user typology that has developed—categorizing and describing users by a single distinguishing characteristic, such as the elderly, single mothers, teenagers, the homeless, or the handicapped—can be misleading without recognition of the other life-shaping facets of users' identities. "The study of special populations has generated important information for designers on how the environment can meet specific needs but special has become another word for separate" (Kailes, 1984).

Involving Users in the Design Process.   This is an important component to consider in crafting educational experience that will support and encourage the creation of innovative, inclusive places. Design educators, however, have not traditionally viewed the user's role to be critical to the teaching of design, even though participatory design has a long and rich history, and has become compulsory in all but a few types of environmental design practice (Hester, 1997). Lifchez and others found that there is novelty in having real people invade the insular setting of the studio. User/consultants provide students with the opportunity to see the world through another person's eyes and to understand how a product or place looks from a different perspective (see Fig. 51.9). For many students it is like seeing for the first time a world they think they know and understand (Welch, 1995). User/consultants also can be very effective at shifting the perception of accommodation beyond the technical focus of codes and at illustrating the variability in how people actually use the environment.

Developing Self-Aawareness in Students.   This is an unresolved and insufficiently debated aspect of design education. To understand the value of universal design, students need to develop a greater awareness of how their backgrounds and world views influence their design as well as how these personal factors impact on how they respond to and interact with people different from themselves. Analyzing and understanding how built form communicates messages about power and control in the environment, and how many people are, in fact, excluded from places or experiences as a direct result of design action, is an essential component of universal design awareness (Hunter, 2000).

Engaging Ethical and Political Content in Design Teaching.   This enables students to see how design may be a tool for perpetuating the status quo and marginalizing minority users (Weisman, 1994; Jones and Welch, 1998). Most advocates of universal design are well versed in techniques for designing more inclusive products and places but are less comfortable with the polemics of this bigger picture. Design departments, however, need to enhance course offerings and broaden degree requirements to ensure that students are exposed to learning opportunities that add an ethical and/or political dimension to their design thinking.

FIGURE 51.9   Mother and baby try a cardboard chair designed by a Kansas State student.

51.6   *INSTITUTIONAL FACTORS*

In addition to the structure and content of the curriculum, the culture of the professions, the nature of academic institutions, and the particular dynamics of departments involved in delivering a design education temper the impact of universal design values throughout the educational experience (Welch and Jones, 2001). Following are some of the key elements that faculty have found influence the adoption of universal design—and that consequently must be discussed and proactively addressed in every design program:

Attitudinal Change.   Universal design teachers generally have found that attitude among students, faculty, and administrators is a far greater barrier to infusion than the actual time and effort required to introduce and elaborate on the materials. Resistance to change is often articulated informally by design faculty with the argument that the design curriculum is already overloaded and that universal design considerations, like client accommodation, can be learned best after graduation in the office setting. Resistance also comes from faculty that equates universal design with ugly ramps and homogenous spaces. Design instructors, in general, are subject to little scrutiny about what and how they teach and may foster stereotypes and discrimination, however inadvertently. Students are far more aware of the values that faculty espouse than are their colleagues, but rarely feel empowered to engage their instructors in discussions that challenge the design field's hegemonic assumptions. For change to occur, attitudes about and understanding of universal design must be shared among the faculty to evaluate how the infusion of universal design values might be incorporated best, given the specific nature of the program or department.

Diversity of Student Body and Faculty.   Design programs in the United States generally suffer from insufficient diversity in both the faculty and the student body. While certainly not a new phenomenon, this is a persistent one in relation to gender, ability, race, and class. The presence of international students presents opportunities for cultural exposure, but does not ensure that culture is thoughtfully addressed in curriculum and instruction. The lack of diversity is becoming more of an issue as the population of the United States becomes more diverse ethnically and racially (Jones, 1996) and as the percentage of old people in the population grows. To address this challenge, design programs must work to attract a more representative body of students so that the professions will come closer to reflecting the makeup of society itself.

Adaptive Environments' "Access to Design Professions" project, developed by Elaine Ostroff, strives to increase the representation of people with disabilities in the design professions. Funded initially by the National Endowment for the Arts, the project is a living memorial to Ron Mace, who died in 1998 (Adaptive Environments, 1999). The project is predicated on the belief that design education will improve when people with disabilities participate, and that the practice of inclusive design will improve with an increase in people with disabilities in the design professions (Hunter, 2000).

Knowledge Generation and Scholarly Development of Faculty.   Teaching universal design in design schools, one might assume, would have resulted in a secondary benefit—exposing the concept of universal design to scholarly scrutiny, critical thought, and knowledge development. While there is some research being conducted on universal design at several nationally funded research centers, these investigations are primarily technology- and data-oriented or dissemination-focused. In the scholarly journals that support architecture and landscape architecture, there has been no scholarly discourse, let alone a special-focus volume. The presence of articles in academic journals and trade magazines is necessary for the dialogue on universal design to grow and mature.

Program Accreditation and Licensing Exams.   The National Architectural Accreditation Board (NAAB), in its most recent (1998) requirements, distinguishes between the levels of accomplishment expected of graduates—*awareness, understanding,* and *ability.* Students are expected to have *ability* to design both site and building to accommodate individuals with varying physical abilities. This means that they "can correctly select the information that is appropriate to a situation and apply it to the solution of specific problems." They are only required to have *understanding* when it comes to their legal responsibilities with respect to accessibility. They need to be able to "correctly paraphrase or summarize without necessarily being able to relate it to other material or see its fullest implications." These two requirements make clear the obligation of architecture schools to ensure that their students can apply the requirements of the ADA even if they cannot "see the fullest implications of the law" and to address the constituency needing accommodation broadly, as "individuals with varying physical abilities."

In the field of landscape architecture, the requirements are considerably less well formulated. The only mention of accessibility is not under an assessment of the curriculum but under educational facilities, where schools are expected to have "safe, convenient, and barrier-free access." While knowledge of the ADA Standards for Accessible Design are incorporated into questions on the landscape architecture licensing exam, there is no specific requirement to teach it in the requirements for accreditation of professional programs.

Physical Environments Conducive to Learning.   Many of the environments in which design is taught are rife with examples of noninclusive settings. While this offers picturesque examples for universal design instructors and powerful experiences of exclusion, it also subtly indicates to students that these characteristics are tolerable aspects of the built environment. Lecture halls may still be inaccessible, especially for teaching staff, and seating may not accommodate left-handed students or students who are larger or smaller than average. Schools should take proactive steps to address these inequities in ways that illustrate beauty and creativity, not functionality and/or failure. Schools of design should also be conscious of the larger landscape around them, paying attention to campus facilities, public spaces, and other elements that compose a typical campus to ensure not only an inclusive school of design, but a universally designed campus as well.

51.7 CONCLUSIONS AND CHALLENGES

Over the last decade the nascent movement called *universal design* has expanded from its origins in accessibility for people with disabilities to become a more expansive and inclusive approach to creating equitable environments for all. While not all those who teach or advocate for universal design agree on its more inclusive definition, most agree that within the curriculum there are definite benefits to examining the built environment through multiple lenses. By challenging students to recognize the power of designers to include and exclude a variety of people through their design decisions and by encouraging them to take a more critical view of how they will practice, *universal design* may become synonymous with *good design,* as Lusher and Mace originally envisioned.

A large body of teaching experiments now exists, some formally documented through programs like UDEP, others shared at conferences and on Web sites. More effort is needed to cross-fertilize design programs with some of these successes, especially for faculty looking for resources at both the course and curriculum scales. Publication of scholarly work in academic journals and documentation of universally designed places in trade magazines will emphasize that this emerging concept of universal design is, in fact, a robust academic and professional topic. At the same time the dissemination of information on universal design as well as visual exemplars will also enhance teaching and research within the academy in the years to come.

There is, however, a constant pressure from grantors to capture the learning curve through prescribed curricula. Previous curricular packages have not fared well, however, and have attracted few design teachers. While dissemination of knowledge, including the sharing of successes and failures, is critical to teaching enhancements, better vehicles are needed to capture the interest of faculty. Interestingly, Lifchez's original project had included producing a primer of fundamentals for others to follow in similar courses. He elected not to, believing that the process could not be "neatly encapsulated." As we have described throughout this chapter, strategies and components of a universal design–based curriculum must vary from one place to the next, due to the inevitable variation in people, place, curricular focus, and in overall acceptance of a new idea such as universal design.

We also believe that a critical discussion of the many ethical and political issues facing designers today is imperative if universal design is to move from the margins into the heart of curricular development. Teaching ethical and political perspectives is not generally effective, however, when transmitted only via syllabi and course readings. Each faculty member and her/his students need to engage attitudes and values with contextual specifics. Participatory design, expanding the range of literature to encompass these issues, and developing a better sense of the designer's own world view are a few of the key components necessary in creating a supportive yet critical discourse about the political and the ethical elements within the design professions.

As American society both ages and diversifies, the values inherent in universal design will only become more critical to understand, address, and apply in the creation of new places and spaces. Design professionals have the opportunity to get out ahead of this change and seek to design environments that are at once inclusive, beautiful, fun, and uplifting. Unfortunately, many current design programs and their curricula do not reflect a sensitivity to this coming wave, limiting consideration of universal design to the realm of technical information or to a subspecialty (such as human factors) within the curriculum. Universal design, and the value of inclusiveness that is at its core, offers a way to reconceptualize design that embraces all members of our increasingly multicultural society who inhabit and use the built environment.

51.8   BIBLIOGRAPHY

Adaptive Environments, *Universal Design Education Project 2.0,* unpublished report, 1996.

———, *Access to the Design Professions: Report to the Task Force,* unpublished report, 1999.

AIA/ACSA Council on Architectural Research, *Design for Aging: Resource Package,* American Institute for Architectural Research, Washington, DC, 1996.

Anders, Robert, and Daniel Fechtner, *Universal Design Primer,* Pratt Institute Department of Industrial Design, Brooklyn, NY, 1992.

Bassett, Bruce, *A House for Someone Unlike Me* (video), National Center for a Barrier Free Environment, Washington, DC, 1984.

Bednar, Michael, *Barrier Free Environments,* Dowden, Hutchinson and Ross, Stroudsburg, PA, 1977.

Blandy, D., F. Tepfer, L. Zimmer, S. Jones, and P. Welch, "Power and Place," in *EDRA 28/1997: Proceedings of the Environmental Design Research Association,* Montreal, Quebec, Canada, 1997.

Cohen, Uriel, *Mainstreaming the Handicapped: A Design Guide,* Center for Architecture and Urban Planning Research, University of Wisconsin—Milwaukee, Milwaukee, WI, 1981.

Dutton, Thomas, *Voices in Architectural Education: Cultural Politics and Pedagogy,* Thomas Dutton (ed.), Bergin and Garvey, New York, 1991.

———, and Lian Hurst Mann, *Reconstructing Architecture: Critical Discourses and Social Practices,* University of Minnesota Press, Minneapolis, MN, 1996.

Cuff, Dana, "Through the Looking Glass: Seven New York Architects and Their People," in *Architects' People,* Russell Ellis and Dana Cuff (eds.), Oxford University Press, New York, 1989.

Goody, Joan, "Moral Architecture," *GSD News,* Harvard University, Cambridge, MA, 1995.

Hester, Randy, "Wanted: Local Participation with a View," in J. Nasar and B. Brown (eds.), *EDRA 27/1996: Public and Private Places: Proceedings of the 27th Annual Meeting of the Environmental Design Research Association,* EDRA, Edmond, OK, 1997.

Howell, Sandra, "The Aged as a User Group," *Journal of Architectural Education,* 31(1):1977.

Hunter, Daniel, "Creeps! Disability in Landscape Architecture," masters thesis in landscape architecture, University of Oregon, 2000.

Jones, Stanton, "Beyond Belligerent Compliance," *Landscape Architecture Magazine,* 1998.

———, "Decolonizing Landscape Architecture: Multiculturalism and the Landscape of Future Possibilities," in Wagner, C. (ed.), *Design For Change: 1996 Annual Meeting Proceedings of the American Society of Landscape Architects,* 162–168, 1996.

Journal of Architectural Education (entire issue devoted to GSA project), 31(1):1977.

Kailes, June Isaacson, "Language is More than a Trivial Concern," *Paralyzed Veterans of America Bulletin,* Virginia Chapter, September 1990.

Lifchez, Raymond, *Rethinking Architecture: Design Students and Physically Disabled People,* University of California Press, Berkeley, CA, 1987.

Lusher, Ruth, and Ron Mace, "Design for Physical and Mental Disabilities," in *Encyclopedia of Architecture,* John Wiley and Sons, New York, 1990.

Mace, Ronald L., Graeme J. Hardie, and Jaine P. Place, "Accessible Environments: Toward Universal Design," in *Design Interventions: Toward a More Humane Architecture,* Preiser, Vischer, and White (eds.), Van Nostrand Reinhold, New York, 1991.

Metropolis magazine ("Access": special universal design report), 1993.

Mueller, James, *Towards Universal Design* (video), Universal Design Initiative, P.O. Box 222514, Chantilly, VA 22022.

Ostroff, E., "Mining Our Natural Resources: The User as Expert," *Innovation, the Quarterly Journal of the Industrial Designers Society of America,* 16(1):1997.

———, and Daniel Iacofano, *Teaching Design for All People: The State of the Art,* Adaptive Environments Center, Boston, 1982.

Pastalan, Leon, R. K. Mautz, and J. Merrill, "The Simulation of Age Related Sensory Losses: A New Approach to the Study of Environmental Barriers," in *Environmental Design Research,* W. F. E. Preiser (ed.), Dowden, Hutchinson and Ross, Stroudsburg, PA, 1973.

Schneekloth, Lynda, and Robert Shibley, "Implacing Architecture into the Practice of Placemaking," *Journal of Architectural Education,* 53(3):2000.

Schon, Donald, *The Design Studio,* RIBA Publications Ltd., London, UK, 1985.

Stilgoe, John, "Lions and Tigers and Stairs, Oh My!" *Boston Society of Architects' Journal,* 1998.

Steinfeld, E. et al., *Accessible Buildings for People with Walking and Reaching Limitations,* U.S. Department of Housing and Urban Development, Washington, DC, 1979.

Sutton, Sharon, *Weaving a Tapestry of Resistance: The Places, Power, and Poetry of a Sustainable Society,* Bergin & Garvey, Westport, CT, 1996.

Theodoropoulos, Christine, and Polly Welch, "Real People, Real Building, Real Sites: A White Paper Concerning Foundation Design Studio," in *Toward a Critical Pedagogy for the Environment: ACSA Western Regional Conference Proceedings,* Berkeley, CA, 1998.

Universal Designers and Consultants, *Images: Universal Design Excellence Project,* Rockville, MD, 1996.

Ward, Anthony, *The Suppression of the Social in Design: Architecture as War in Reconstructing Architecture: Critical Discourses and Social Practices,* Thomas Dutton and Lian Hurst Mann (eds.), University of Minnesota Press, Minneapolis, MN, 1996.

Weisman, Leslie, "Re-designing Architectural Education," in Rothschild, Joan, *Design and Feminism,* Rutgers University Press, New Brunswick, NJ, 1999.

Welch, Polly (ed.), *Strategies for Teaching Universal Design,* Adaptive Environments, Boston, and MIG Communications, Berkeley, CA, 1995.

———, "Taking Space: Locating Disability in Design Education and Practice," in *Confronting Conservative Architecture, ACSA Eastern Regional Conference,* Boston, 2000.

———, and Stanton Jones, "Teaching Universal Design Through Inclusiveness," in *Designing for the Twenty-First Century: Proceedings of an International Conference on Universal Design,* 1998.

———, and ———, "Evolving Visions: Segregation, Integration, and Inclusion in the Design of Built Places," in *EDRA 30/1999: The Power of Imagination: Proceedings of the Environmental Design Research Association,* 1999.

———, and ———, "Designing to Transgress: A Model for Teaching Universal Design Values" (forthcoming article), 2001.

———, and Elaine Ostroff, "The Universal Design Education Project," in P. Welch (ed.), *Strategies for Teaching Universal Design,* Adaptive Environments Center, Boston, and MIG Communications, Berkeley, CA, 1995.

CHAPTER 52

INTEGRATING UNIVERSAL DESIGN INTO THE INTERIOR DESIGN CURRICULUM

Louise Jones, Arch.D., I.D.E.C.
Eastern Michigan University, Ypsilanti, Michigan

52.1 INTRODUCTION

This chapter will define interior design, trace its development as a discipline, and discuss educational preparation for the profession. Discussion will focus on a paradigm shift in interior design practice, namely, the adoption of an all-encompassing, user-centered design philosophy. Adoption of universal design precepts as the foundation for design decisions is facilitating the creation of aesthetically pleasing interior environments that support the functional and psychosocial needs of everyone who uses the environment, regardless of their age, physical stature, or abilities. A case study will be used to illustrate the integration of universal design into the interior design curriculum at one American university. Students in this program came to understand that good design requires much more than compliance with the minimum requirements of laws that mandate a few special features to create barrier-free environments for individuals with disabilities. The chapter concludes with a discussion of the implications of adopting universal design precepts as the basis for design decisions.

52.2 UNIVERSAL DESIGN AND INTERIOR DESIGN EDUCATION

Universal design represents a paradigm shift in interior design practice, namely, the adoption of an all-encompassing, user-centered design philosophy. Historically, interior design focused on the creation of an aesthetically pleasing environment that met the psychosocial and functional needs of the users. However, the characteristics of the users were based on a mythical "average person." With the passage of barrier-free codes and the Americans with Disabilities Act (ADA), features were included in the design to address the needs of special groups (e.g., ramps for individuals using wheelchairs). However, a shift is occurring that negates design for the average person. When universal design is the underpinning for design decisions, the design solution can be aesthetically pleasing, cost-effective, and inclusive of all potential users.

The infusion of universal design precepts into the interior design curricula at colleges and universities is supported by accreditation guidelines, professional certification, and state licensing requirements. The outcome is students' adoption of a personal design philosophy that mandates design solutions that simultaneously address the psychosocial and functional needs of many different user groups. As these designers enter the workplace, they are changing the way interior design is practiced in the United States and around the world.

52.3   EVOLUTION OF INTERIOR DESIGN

Origin of Interior Design

There is no specific time period, person, or place that defines the origin of interior design. A desire for functional and pleasant spaces existed long before the production of buildings. Early cultures practiced interior design when they painted images on the walls of their caves and used animal pelts and indigenous vegetation to make their homes more comfortable. When people began to plan and construct buildings, the design of the interior was an integral part of the total design [i.e., the architecture of that society (Kilmer and Kilmer, 1992)]. Although interior design has always been a part of the built environment, the profession of interior design did not evolve until after World War II.

Science of Home Economics

In the United States, as early as the mid-nineteenth century, women were beginning to analyze the relationship between residential design and efficient completion of household tasks. In 1841, Catherine Beecher published a *Treatise on Domestic Economy,* "addressing every aspect of production in the home and placing household tasks in a broad social context" (Malnar and Vodvarka, 1992). In 1869, she and Harriet Beecher Stowe wrote a sequel, *The American Woman's Home,* a time management study that was later updated to discuss the use of domestic technology. At the beginning of the twentieth century, home economists helped to focus public attention on the emerging discipline of domestic science. Christine Frederick's *Household Engineering: Scientific Management in the Home* (1919) used reliable, recognized, scientific procedures to apply time management studies to the residential environment. "The rational organization of domestic work preceded the new household machines that were to curtail that work . . . [and were] critical to the shaping of interior space" (Malnar and Vodvarka, 1992). During the 1920s, Lillian and Frank Gilbreth conducted time-work studies ". . . to eliminate needless, ill-directed, and ineffective motions" (Griedion, 1948). This early work laid the foundation for theories of interior design that focus on interactions between people and the environment in which they live and work.

Interior Decoration

Concurrent with this interest in domestic science was an emerging interest in the art of interior decoration as a profession distinct from architecture. Candice Wheeler, working with Louis C. Tiffany in the late 1800s, wrote *Interior Decoration as a Profession for Women* (Kilmer and Kilmer, 1992). Elsie deWolfe (1865–1950) is usually credited with being the first professional interior decorator in the United States. Born in New York City and educated in Scotland, deWolfe began her long and successful career in 1904, working primarily with wealthy clientele, borrowing from historic periods to decorate existing, residential spaces. *The House in Good Taste,* her 1913 book, popularized her decorating style, going beyond stylistic imitation to probe more basic questions regarding aesthetic goals (Pile, 1995). Although formal coursework was limited during the first trimester of the twentieth century, numerous books and magazines enabled interior decoration to develop as a profession focusing on the decorative arts (i.e., the selection of ornamentation, finishes, furnishings, and furniture). After World War I, with the increasing wealth of the middle class, interest in professional interior decoration expanded. Department stores such as Macy's and Marshall Fields featured residential vignettes and offered interior decoration services to their clients. Dorothy Draper (1889–1969), whose commissions included hotels, clubs, restaurants, shops, and hospitals, broke the male dominance in the decoration of nonresidential interiors (Kilmer and Kilmer, 1992).

Interior Design Profession

After World War II, during a period of unprecedented economic prosperity, construction of nonresidential spaces flourished. Technological innovations and human factors research caused the focus to shift from decorating spaces to designing functional work environments. Robert Sommer (1969), in his classic text *Personal Space,* contends that "architecture may be beautiful but it should be more than that; it must enclose space in which certain activities can take place comfortably and efficiently." The emerging profession of interior design combined the science and art of design to create aesthetically pleasing interior environments that addressed the needs of the people using the space. Working with the building structure, life support systems, and relevant regulations, interior designers used light, color, materials, furnishings, and human behavior research to create environments that enhanced the quality of life for everyone using the space (Kilmer and Kilmer, 1992). During the 1970s interior designers became members of the team of planners, architects, engineers, and developers responsible for designing the highly complex environments necessary to support a continually changing, technological society.

Although there are multiple definitions of interior design, the one endorsed by the National Council for Interior Design Qualification (NCIDQ), the Foundation for Interior Design Education and Research (FIDER), and the major interior design organizations in the United States is the most widely accepted.

> The Professional interior designer is qualified by education, experience, and examination to enhance the function and quality of interior spaces for the purpose of improving the quality of life, increasing productivity, and protecting the health, safety, and welfare of the public. The professional interior designer . . .
>
> - Analyzes the client's needs, goals, and life safety requirements
> - Integrates findings with knowledge of interior design
> - Formulates preliminary design concepts that are aesthetic, appropriate, and functional, and in accordance with codes and standards
> - Develops and presents final design recommendations through appropriate presentation media
> - Prepares working drawings and specifications for non-load bearing interior construction, reflected ceiling plans, lighting, interior detailing, materials, finishes, space planning, furnishings, fixtures, and equipment *in compliance with universal accessibility guidelines* [emphasis added] and all applicable codes
> - Collaborates with professional services of other licensed practitioners in the technical areas of mechanical, electrical and load-bearing design as required for regulatory approval
> - Prepares and administers bids and contract documents as the client's agent
> - Reviews and evaluates design solutions during implementation and upon completion (FIDER, 1999a)

Interior Design Education and FIDER Accreditation

In 1970 the Foundation for Interior Design Education and Research (FIDER) was established by a coalition of interior design professional organizations, with recognition by the U.S.

government's Council on Higher Education Accreditation (CHEA), to evaluate and accredit interior design programs in colleges and universities in the United States and Canada. FIDER established a set of minimum criteria, based on the technical skills and knowledge necessary for entry into the interior design profession, which served to set the standard for the educational content of interior design programs.

FIDER standards reflect a focus on user-centered design. "The development of space planning and problem solving skills is essential and should relate to a broad range of residential and non-residential projects, including all types of habitation, whether for work or leisure, new or old; large or small; for a variety of populations, young and old, of varying physical abilities, low or high income. Problem solving experiences should follow a theory of design process involving physical, social, and psychological factors and reflect a concern for the aesthetic qualities of the environment" (FIDER, 1996). Specifically, Standard S2.10.3 relates to "Human factors, i.e., anthropometrics, ergonomics" and S2.11.3 relates to "Laws, codes, standards, and regulations, e.g., universal accessibility guidelines, life safety, fire, etc." (FIDER, 1996), thus ensuring that students attending accredited programs understand universal design concepts. The *FIDER Professional Standards 2000* (FIDER, 1999b) revisions require that

> Student work must demonstrate:
>
> - Understanding of theories of design and human behavior including human factors . . .
> - The relationship between the built environment and human behavior . . .
> - The application of: codes, regulations, and standards, barrier-free concepts, ergonomic and human factors data . . .

However, the 2000 standards go beyond previous expectations and explicitly state that "Student work must demonstrate understanding of universal design concepts and principles," thereby mandating that students in accredited interior design programs utilize universal design precepts as the basis for their design work in order to enhance the function and quality of interior environments.

52.4   CASE STUDY: THE INTERIOR DESIGN PROGRAM AT EASTERN MICHIGAN UNIVERSITY

History of the Program

The interior design program at Eastern Michigan University (EMU) can serve as a case study to demonstrate the infusion of universal design precepts into an interior design curriculum. The program at EMU originated as Domestic Science in 1905. Prior to the 1960s, interior design coursework at EMU, as was true at most colleges, focused on design and decoration of the home environment. By 1975, when the interior design program was given full status as a major within the home economics department, the curriculum had been expanded to include nonresidential design. In the mid-1970s, Michigan passed barrier-free regulations mandating that buildings be barrier-free if (1) they were government-owned, (2) employment opportunities existed, or (3) services to the public were available. All new construction was to be barrier-free, and if renovation affected more than 50 percent of the floor area, the entire facility was required to be barrier-free (Michigan Department of Labor, 1987). Michigan quickly gained a reputation as a leader in the development of barrier-free building codes, and subsequently, studio projects in the interior design program at EMU required compliance with these codes in order to address the needs of users with disabilities. As more accessibility research became available (e.g., ANSI W117.1-1980), it was incorporated into human factors lectures that focused on the ergonomic and anthropometric data that was the basis of design for "average" people. In 1987, the interior design major was provisionally accredited by FIDER. With the development of an expanding research base for environment and behavior studies, design for special needs (i.e., elderly people, children, people with disabilities, and people from diverse socioeconomic and ethnic groups) was included in human factors discussions. Thus, even before the passage of the Americans with Disabilities Act in 1990 defined the lack of access in the built environment as discriminatory, the interior design program at EMU was teaching user-centered design theories. EMU documented that they were in compliance with FIDER accreditation standards, and in 1991 the interior design program was awarded full FIDER accreditation. When EMU was selected by the Adaptive Environments Center to be a participant in the first Universal Design Education Project (UDEP) in 1993, the program was ready to move beyond teaching design for special needs and compliance with barrier-free regulations to teaching universal design as the appropriate underpinning for design decisions.

First UDEP Experience

The first UDEP[1] was a collaboration between interior design at EMU and industrial design and architecture at the University of Michigan. Environmental Design for Aging Research Group (EDARG), a multidisciplinary research group, had been working together for several years to improve the design of environments that were used by elderly people who wanted to maintain their autonomy and independence. EDARG members Leon Pastalan (architecture), Ronald Sekulski (industrial design), and the author (interior design) realized that although people who are frail, elderly, or disabled made up an increasingly large segment of the population, little information was available in a format that facilitated students' integration of the information into the design process. For the UDEP project, they decided to modify an assessment game, originally developed by EDARG to collect qualitative research data, for faculty to use as a teaching aid in design studio classes. Therefore, the goal of the UDEP project "was to introduce design students to an experiential, interactive, design research method and to demonstrate that the game/simulation A Day's Journey Through Life© offers students significant insight into the environmental and performance needs of a diverse population, thereby changing their perception of accessibility and design issues" (Jones, Sekulski, and Pastalan, 1995).

The UDEP grant supported activities in two design studios, industrial design at the University of Michigan (under the direction of Ronald Sekulski) and interior design at EMU (under the direction of the author). Senior interior design students at EMU were invited by a nonprofit agency to develop a proposal for the adaptive reuse of a vacant 11-story, 145,000-ft^2 hotel in Ann Arbor, Michigan. The program for adaptive reuse was to incorporate office space; an indoor, year-round park; retail spaces; classrooms and offices for the local community college; senior co-op apartments; management offices; resident activity rooms; an indoor pool and physical fitness center; and a restaurant for both residents and the general public. Industrial design students at the University of Michigan were invited to identify and develop products that could be used in this environment.

To expand the breadth of students' empirical experience, faculty redesigned a game/simulation, A Day's Journey Through Life, for students to use as a design research tool. Students from both universities were assigned to teams to play the game, thereby ensuring multidisciplinary teams. Each student team was joined by a consultant with whom students would play the game. Consultants were selected from the local community to represent multiple user groups: hearing impaired, vision impaired, mobility impaired (both temporary and permanent), frail elderly (i.e., musculoskeletal problems and manipulation, dexterity, and grip problems), and those who fell in the anthropometric extremes (i.e., less than the 5th percentile or greater than the 95th percentile). Some consultants had multiple problems. For example, one consultant was a 48-year-old man with Parkinson's disease whose wife had Lou Gehrig's disease (amyotrophic lateral sclerosis); another, Paul, was a 22-year-old man who lost his eyesight and one leg in a small plane crash; a third consultant, Andrea, was a 40-year-old woman with multiple sclerosis.

A Days' Journey Through Life "provides a multilogue [rather than] a dialogue . . . to sensitize students to disability and lifespan concerns and to the uniqueness of each individual's experiences" (Sekulski, Jones, and Pastalan, 1999). The gameboard and sequence of play (see Figs. 52.1 and 52.2) moved participants through the activities of daily life [e.g., grooming, dressing, cooking, cleaning in a setting familiar to the consultant (i.e., their own home)]. During the game, the consultant was considered the very important person (VIP) because he or she was teaching the students about life as a member of that user group. During the two- to three-hour period of play (including orientation, game play, and debriefing), the student team engaged the VIP and a caregiver (if applicable) in a multilogue to identify the aspects of the micro- and macroenvironment that inhibited the VIP's autonomy and independence.

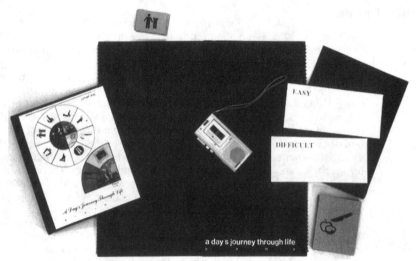

FIGURE 52.1   A Days' Journey Through Life game.

FIGURE 52.2   A Days' Journey Through Life gameboard.

Students initially expressed apprehension about meeting the consultants in their homes, preferring the relative safety of the design studio. But in most instances, the consultants offered insights and coping strategies that would have gone unrecognized if interactions had been limited to the studio setting. One consultant took students to the bedroom to demonstrate an assistive device her husband had constructed to allow her to use the nightstand as a support when transferring from her wheelchair to the bed. Another discovered that the fire extinguisher in the kitchen was out of reach when she was seated in her wheelchair, although accessible if she was using her walker.

After playing the game, students recognized the need for both a broader perspective and for more specific information regarding each impairment. They extended their research to identify the following:

- The underlying characteristics of the impairments (e.g., medical conditions that lead to the use of a wheelchair)
- The prevalence of the condition (e.g., 31 million Americans had mobility impairments)
- The magnitude of the problem (e.g., not all wheelchairs are created equal)
- The relevant codes and legislation (e.g., barrier-free building codes, ADAAG, and the Fair Housing Amendments Act)

This information was essential if students were to understand the full scope of the problems rather than focusing on the narrow perspective narrated by one consultant. Through research and the interactive programming experiences (i.e., participant observations, interviews, and game play), the students came to know, understand, and empathize with the particular user group being investigated. Interior design students created a fact sheet for their classmates regarding the user group (see Fig. 52.3) represented by the consultant with whom they played the game. The students then served as advocates for their user groups for the duration of the semester, working with classmates to resolve design concerns and critiquing design proposals as to their appropriateness for the specific user group.

Working in their respective studios, students moved from design research to conceptual design. To encourage cooperative, collaborative learning, interior design students worked in teams to develop proposals for the public spaces (see Fig. 52.4) and individually to develop proposals for the residential spaces (see Fig. 52.5, a and b). Students were required to recognize the diversity of user needs in the public spaces by employing universal design precepts in their proposals. They used the concept of adaptable design for the housing units to improve the quality of life for all residents. Each student developed a base plan for an apartment using universal design criteria such as wider doorways to facilitate moving furniture or wheelchair access and lever handles to facilitate manipulation by someone with their hands full of groceries or anyone with dexterity/grip problems. Modifications reflecting accessibility guidelines and adaptable design options were then developed for different user groups, such as visual alarms for those with hearing impairments and removable base cabinets for those who used a wheelchair. "When done well, universal design is invisible—that is, it is simply good design" (Jones, Sekulski, and Pastalan, 1995). Therefore, students were asked to document their proposals for implementing universal design and adaptable housing considerations using both traditional design drawings (e.g., sketches, floorplans, and elevations, and annotated floor plans).

Consultants visited the studio for a midsemester project critique and at the end of the semester for project presentations. Many were contacted by phone for additional insights during the course of the project. A few, particularly Paul and Andrea, found that they liked working with the design students and were active guest critics for studio projects for several years following the UDEP project.

Students were overwhelmingly positive regarding their learning experiences during the UDEP project. Responses to open-ended questions on the end-of-term evaluation indicated that students "found the studio experiences to be challenging but rewarding." Many seemed

DESIGN REFERENCE SHEET
HEARING IMPAIRMENT

DEFINITION

There are many degrees of hearing impairment experienced by the 1 in 10 Americans with a hearing loss. The medical and social problems experienced by people with a partial hearing loss are quite different from those experienced by people who have a total hearing loss. The two groups should not be grouped together indiscriminately.

Deafness. A total or severe impairment of hearing. Individuals may use sign language and/or speech reading (i.e., lip reading) to compensate for their hearing loss. *Prelingual deafness* occurs before auditory language skills are developed. Individuals often use sign language as the first language with English (or another spoken language) as a second language. *Postlingual deafness* occurs after auditory language skills are developed. Individuals typically have more advanced speaking skills and a better understanding of spoken language.

Hard of hearing. A partial impairment of hearing, often the result of illness, injury, or aging. Individuals typically use a spoken language to communicate. Individuals may benefit from surgery and/or hearing aids and may read lips to facilitate communication.

INTERIOR DESIGN GUIDELINES

- Specify and/or provide for use of assistive devices such as TDD attachments for the telephone, closed-caption television decoders, vibrating alarm clocks, and blinking-light alarms/timers.
- Keep "visual noise" to a minimum to provide a neutral ground for signing.
- Provide generous, nonglare lighting to facilitate speech reading or sign.
- Specify sound-absorbing materials and finishes to minimize reflected noise and reduce background noise for those with a partial hearing impairment.
- Specify appropriate electrical wiring and controls to permit lights to flicker when phone or doorbell rings.
- Use visual icons for multiple cueing whenever possible. People who use sign as their first language may have difficulty understanding written language.
- Specify supplementary visual alert systems for fire alarms.
- Provide alternate communication systems in locations where emergency phones are used.
- Design furniture arrangements that do not profile people in front of window glazing to assist those who read lips or sign.

BIBLIOGRAPHY

Rezen, Susan, and Carl Hausman, *Coping with a Hearing Loss,* Dembner Books, New York, 1985. The book provides a sensitive discussion of the physical and psychological effects of hearing loss and suggests methods of coping with the related problems.

Ritter, Audrey, *A Deafness Collection: Selected and Annotated,* Rochester Institute of Technology and The National Technical Institute for the Deaf, Rochester, NY, 1985. Bibliography of related readings.

Schein, Jerome, *At Home Among Strangers,* Gallaudet University Press, Washington, DC, 1989. Informative text written by an educator at one of the foremost institutions of higher learning to help others understand the "deaf community."

Suss, Elaine, *When the Hearing Gets Hard,* Plenum Publishing, New York, 1993. A hearing impaired journalist discusses the problems experienced by people with hearing impairments in a "hearing world."

Turkington, C., and A. Sussman, *Encyclopedia of Deafness and Hearing Disorders,* Facts on File, New York, 1992. Text defines words and terms, discusses causes and characteristics of hearing impairments, and identifies assistive devices and support organizations.

Van Itallie, Phillip, *How to Live with a Hearing Handicap,* Paul Ericksson Inc., New York. Author uses his own experiences as a person who is hard of hearing to help others with similar problems understand and adjust to the problems experienced in everyday life.

FIGURE 52.3 Fact sheets were created by students to provide guidelines for their peers.

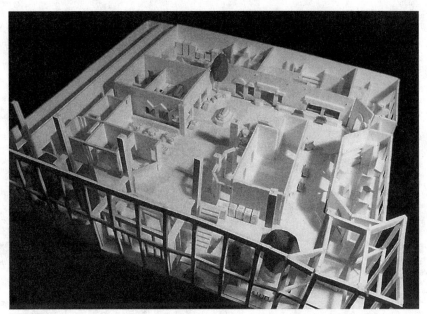

FIGURE 52.4 Design of public spaces in the Ann Arbor Inn project utilized universal design precepts.

to have adopted a universal design perspective. When asked what was the single most important thing learned that semester, one student replied, "to design for everyone, not just the average individual" (Jones, Sekulski, and Pastalan, 1995). A more formal evaluation assessed the changes in knowledge and attitude experienced by students participating in the UDEP studio. Chi-square analysis indicated statistically significant differences in posttest scores between interior design students who participated in the UDEP studio and a control group, suggesting a significant change in knowledge of universal design. The average of the test scores of the participants rose from 35 to 95 percent compared with a consistent 31 percent for the control group. Comparisons of test scores for ADAAG information was not possible since all interior design students scored better than 95 percent on the pretest, probably reflecting the program's mandate that design solutions comply with applicable codes and standards. Faculty concluded that "students were more comfortable interacting with people of different ages and abilities and have internalized pertinent design recommendations, regulations, and codes. Their projects reflect a heightened sensitivity to the design needs of people of different ages and abilities and a proficiency in the development of design criteria reflective of the needs of a diverse population" (Jones, Sekulski, and Pastalan, 1995).

The studio at EMU was team-taught by all four tenure/tenure-track interior design faculty members. This allowed the faculty members not associated with the UDEP proposal to become familiar with universal design precepts and to experience a teaching/learning experience where the precepts were successfully incorporated into interior design problem solving. This positive experience set the stage for full involvement of all faculty members during UDEP 2.0.

Second UDEP Experience

When the request for proposals for the second UDEP project was issued by the Adaptive Environments Center in 1995, interior design faculty at EMU decided the time was right to

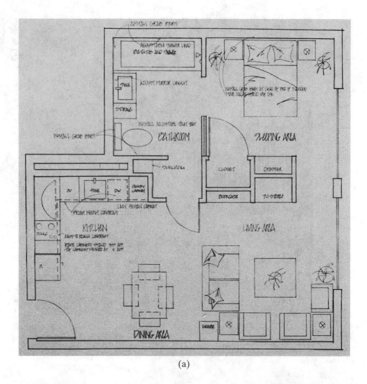

(a)

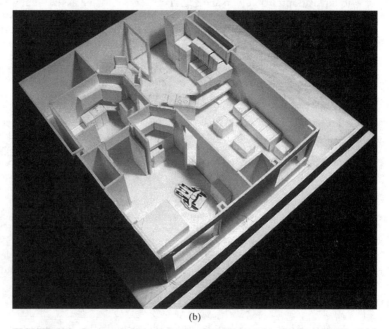

(b)

FIGURE 52.5 Design of residential spaces in the Ann Arbor Inn project utilized (*a*) universal design precepts and (*b*) adaptable design features.

build on the UDEP I experience and introduce universal design precepts as the foundation for interior design decisions (i.e., to be successful, the design of the interior environment must address the needs of all of the users, regardless of their age, physical stature, or abilities). When EMU's proposal was funded, the involvement of all of the tenure/tenure-track interior design program faculty at EMU in the grant activities facilitated infusion of universal design throughout the curriculum.[2]

At EMU, the foundation courses in the interior design program are taught during the first 2 years. Students then utilize the creative problem-solving processes and technical skills that they have learned to address increasingly complex problems during the last 2 years of their studies. Tenure-track faculty are responsible for the design and delivery of the foundation courses. Although adjunct faculty may be invited to participate if multiple sections of a particular studio are required (seldom is a 15:1 student/faculty ratio exceeded), they work closely with, and use the assignments developed by, the tenure-track faculty in order to provide similar learning experiences for all students during the first 2 years. This makes it possible to introduce universal design awareness early in the program of study, expand the exposure to ensure understanding of the precepts, and develop competency through application of the precepts in interior design projects. Most students graduate with a personal design philosophy grounded in a universal design paradigm. For them, user-centered design mandates the inclusion of all users of the space, and universal design precepts ensure that the environment will support users with diverse and multiple needs.

EMU Interior Design Curriculum

The interior design curriculum (see Figs. 52.6 and 52.7) at EMU is delivered over an eight-semester period of time. Each sequential semester has a studio and one or more related lecture classes that are prerequisites for the next semester's work.

During the first semester, universal design precepts are introduced in the Human Factors and Special Needs lecture class. Students study the diverse needs of multiple-user groups, including those who are outside the scope of the 5-95 percentile range, and compare these with the anthropometric data for "average" individuals. As they explore the special needs of diverse groups, they begin to realize that they will inevitably be a member of a special needs group at some point in time as they experience temporary disabilities due to accidents, illnesses, or other life events. A series of empathic experiences enables students to explore the campus from a different perspective than is typical during their day-to-day environmental interactions. This awareness of the design needs of different groups of individuals is enhanced in the Orientation to Interior Design course when universal design is presented as the foundation for design decisions. Discussion follows the video, *Towards Universal Design* (Mueller, 1993), as the philosophy is explored in small group interactions. Upper-level students point out examples of the application of universal design precepts in their portfolio presentations to the orientation class. The students' final assignment in the human factors class asks them to use a check sheet to assess an existing nonresidential environment, identifying both barrier-free and universal design features (see Fig. 52.8 and Goltsman, 1993).

During the second semester, students have an opportunity to problem-solve, using the information they learned during the first semester. When asked to design a granny flat for a frail, elderly female, they must consider the needs of multiple users. Their client is frail but living independently, how can their design facilitate her ability to age in place? With time, she may experience a range of age- and disease-related conditions. How can the interior be designed to be supportive as her needs change with the passage of time? The house will be a modular unit that is moved by truck to her son's home site. He and his wife want to be able to visit with his mother on a daily basis to provide companionship and assistance. The design must accommodate their visits and support their involvement in housekeeping and maintenance tasks while encouraging the client's independence. Grandchildren will be encouraged to visit

EASTERN MICHIGAN UNIVERSITY
INTERIOR DESIGN PROGRAM COURSEWORK

202 ROOSEVELT HALL
YPSILANTI, MI 48197
734-487-2490
www.emich.edu

Number	Course name	Credits	Contact hours
IDE 110	Interior Design Studio I	4	8
IDE 111	Human Factors & Special Needs	3	3
IDE 131	Orientation to Interior Design	1	1
IDE 120	Interior Design Studio II	3	6
IDE 121	Interior Design Materials & Components	3	3
FA 123	Drawing I	3	6
FA 122	Two Dimensional Design	3	6
IDE 210	Interior Design Studio III	3	6
IDE 211	Lighting for Interiors	3	3
IDE 220	Interior Design Studio IV	4	8
IDE 221	Environmental Systems for Interiors	2	2
CNST 201	Construction Systems	3	4
FA 231	Three Dimensional Design	3	6
ATM 235	Textiles for Consumers	3	4
IDE 310	Interior Design Studio V	4	8
IDE 311	History of Interiors: Ancient–1800	2	2
IDE 312	Computers for Interior Design	3	6
IDE 313	Space Planning & Specifications	3	3
IDE 320	Interior Design Studio VI	4	8
IDE 321	History of Interiors: 1800–Present	2	2
MKTG 360	Principles of Marketing	3	3
IDE 410	Interior Design Studio VII	4	8
IDE 420	Interior Design Studio VIII	4	8
IDE 421	Interior Design Internship	2	**1
IDE 422	Professional Practice in Interior Design	3	3
MGMT 386	Organizational Theory & Practice	3	3

** Requires 120 hours of work experience in the interior design field.

See EMU Undergraduate Catalog for course descriptions and prerequisites for each course.

FIGURE 52.6   Eastern Michigan University interior design program curriculum.

frequently, so Grandmother's home must be designed to meet the changing needs of growing children. In the years to come, the client may need a personal care assistant, so the house must be flexible to accommodate this supportive intervention. At some point in the future, the client will no longer need the house, and it will be trucked to another site to meet another family's needs. Therefore, the house design must be flexible enough to reflect the aesthetic preferences and support the physical and psychosocial needs of a sequence of users. Through design, critique, and redesign, students are able to develop comfortable, aesthetically pleasing homes that create a supportive environment for the client and all her visitors and can be easily modified to adapt to changing needs (see Fig. 52.9). During this project, most students come to understand how universal design precepts can be the underpinning for residential design deci-

**INTERIOR DESIGN PROGRAM
TYPICAL COURSE SEQUENCE**

Year 1 fall	Credit	Contact	Year 1 winter	Credit	Contact
IDE 110 ID Studio I	4	8	IDE 120 ID Studio II	3	6
IDE 111 Human Factors	3	3	IDE 121 Materials	3	3
IDE 131 Orient. to ID	1	1	Physical Science*	3	3
FA 123 Drawing I	3	6	FA 122 2-D Design*	3	6
ENG Composition*	3	3	PSY 101/102 Intro Psy*	3	3
COSC 136 Computer Intro*	3	3			
	17	24		15	21

Year 2 fall	Credit	Contact	Year 2 winter	Credit	Contact
IDE 210 Studio III	3	6	IDE 220 ID Studio IV	4	8
IDE 211 Lighting	3	3	IDE 221 Envir. Systems	2	2
CNSC 201 Const. Sys	3	4	ATM 235 Textiles	3	4
FA 231 3-D Design	3	6	General Education Course	3	3
General Education Course	3	3	General Education Course	3	3
	15	22		15	20

Year 3 fall	Credit	Contact	Year 3 winter	Credit	Contact
IDE 310 ID Studio V	4	8	IDE 320 ID Studio VI	4	8
IDE 311 History	2	2	IDE 321 History	2	2
IDE 312 Computers for ID	3	6	MKTG 360	3	3
IDE 313 Space Pl & Spec	3	3	General Education Course	3	3
General Education Course	3	3	General Education Course	3	3
	15	22		15	19

Students typically complete 120 hour field placement in the summer between 3rd and 4th year

Year 4 fall	Credit	Contact	Year 4 winter	Credit	Contact
IDE 410 ID Studio VII	4	8	IDE 420 ID Studio VIII	4	8
MGMT 386	3	3	IDE 421 Field Exper.	2	1
General Education Course	3	3	IDE 421 Prof. Practice	3	3
General Education Course	3	3	General Education Course	3	3
General Education Course	3	3	General Education Course	3	3
	16	20		15	18

2 credits in physical education must be added (PEGN 210)

* Required General Education course

TOTAL CREDITS = 125

FIGURE 52.7   Eastern Michigan University interior design program coursework sequence.

EASTERN MICHIGAN UNIVERSITY   IDE 111 HUMAN FACTORS & SPECIAL NEEDS

INTERIOR DESIGN PROGRAM   RESIDENTIAL ENVIRONMENTAL ASSESSMENT

The following checklist identifies ideal residential environmental characteristics. It is a composite derived from multiple sources to assist you in evaluating residential environments. Note that both universal design and barrier-free characteristics are included to facilitate the design of environments that are supportive throughout the life cycle. Although useful in evaluating generic residential spaces, the design of an environment for a particular individual (or individuals) should reflect that person's current and future needs.

ENTRY

☐ Barrier-free transition space between car and house entrance.

☐ Slope at minimum 1:12 ratio (i.e., 1′0″ rise to 12′0″ run), 1:20 ratio preferred. If rise greater than 6″, ramp must be minimum 36″ width, have grab bars at both sides 34–38″ AFF, extending 12″ beyond top and bottom of ramp, with a flat landing space of 5′0″ × 5′0″.

☐ Nonglare, balanced lighting across entire path of travel.

☐ Sheltered entry with no stairs.

☐ Resting spot for items while unlocking door.

☐ 18–24″ clear space at handle side of door.

☐ Easy-to-use doorlock system, does not require twisting motion.

☐ Door hardware no higher than 48″, 36″ preferred.

☐ Lever handle.

☐ Door requires less than 15# to open.

☐ Doorway provides minimum 32″ clearance.

☐ Level threshold preferred, up to ½″ rise acceptable but must be beveled.

☐ System in place for viewing visitors from inside the home prior to opening door.

☐ Foyer area minimum 4′0″ × 4′0″ for putting on/off coat, 5′0″ × 5′0″ preferred.

☐ Door bell has volume control and can be set to flash.

LAYOUT

☐ Hallways minimum 42″ wide.

☐ Turning spaces for wheelchair (5′0″ × 5′0″) available with only minor modifications.

☐ Option for bedroom and bath on first floor if home has more than one level; provision for stair lift and/or space for residential elevator preferred.

FINISHES

☐ Hard-surface flooring materials are low glare and nonslip.

☐ Carpet is tightly woven; pile height does not exceed ½″.

☐ No more than ¼″ beveled floor transition between rooms, level transition preferred.

☐ Color contrast provided between walls and flooring.

☐ Matt finishes used to avoid glare.

FIGURE 52.8 Checksheet to evaluate residential barrier-free and universal design features.

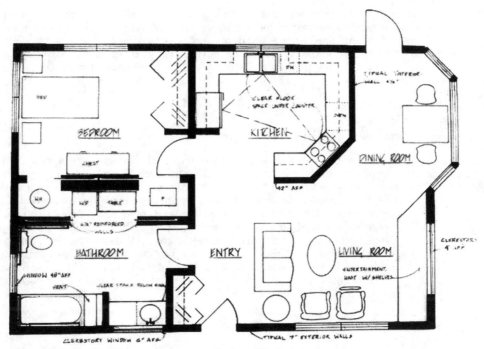

FIGURE 52.9 Design of the granny flat utilized universal design precepts and barrier-free features to address both current and future needs of everyone using the space.

sions. Because universal design is often invisible (being perceived simply as good design), students document their universal design features with an overlay to the floor plan.

During the second year, students explore the design process in nonresidential settings through the development of a small boutique (third semester) and through the renovation of a mainstreet store front into a two-story cafe and night spot. In both projects, students need to consider building structure, life support systems, building codes, and human environment research as they develop interior spaces that respond to their client's goals for the project and address the needs of all of the users of the space (i.e., owners/managers, employees, delivery and maintenance personnel, and patrons of diverse ages, physical stature, and abilities). Through careful programming and extensive design exploration, they create interior environments that are visually exciting and invisibly supportive of everyone who uses the spaces. Students are beginning to understand that universal design is "just good design."

In the third and fourth years, student projects involve increasingly larger and more complex environments. They work with actual clients and/or actual design projects as they explore residential design, historic preservation/adaptive reuse of existing buildings, office design, retail store design, and large-scale nonresidential environments in multistory, mixed-use buildings. Their understanding of the physical and psychological needs of diverse user groups is enhanced during programming by interviewing people of diverse ages, physical stature, and abilities and by inviting users with diverse environmental needs to critique their design solutions during design development. A person of short stature who is less than 4 ft tall and a basketball player who is more than 6 ft tall provide different insights into the appropriateness of the design solution. Through multiple applications of universal design precepts, students develop competency in creative problem solving, resulting in user-centered design that addresses the needs of all users.

By graduation, for most students, universal design is an integral component of their personal design philosophy and has become the underpinning for their design decisions. As one student explained when asked during a critique to point out the universal design components, "It's difficult for me to discuss the universal design features, they are interwoven throughout the design. They are the starting point for the development of my design, that's just the way I work."

General Lessons Learned for Universal Design

The integration of universal design into the interior design curriculum at EMU required the support and involvement of the entire faculty. Each sequential learning experience builds on previous assignments and experiences to enhance students' creative problem-solving abilities. Universal design is understood to be the underpinning for user-centered design. Good design—responsible design—considers the needs of all of the people who use the space and supports those needs in aesthetically pleasing ways. By introducing universal design early in the curriculum and consistently providing reinforcement, user-centered design becomes intuitive. The outcome can be dramatic. During their internships, when students interact with decorators, designers, architects, or contractors who do not share this design philosophy, they frequently indicate a lack of interest in continuing their employment beyond the internship period, "It's not the way I want to do design."

52.5 CONCLUSIONS

In the past 25 years, interior design has evolved from a focus on domestic science and the decoration of residential interiors into a discipline that focuses on user-centered design of both residential and nonresidential interior environments. Interior designers are essential members of the team of professionals responsible for the design of the built environment. Working with architects, engineers, and landscape and product designers, they create aesthetically pleasing interior environments that support the functional and psychosocial needs of all users of the space, regardless of their age, physical stature, or abilities.

Although there are interior decorators who limit their practice to interior decoration, and interior designers who choose to do no more than meet the legal requirements of ADAAG and barrier-free codes, they are becoming a minority in the United States. Graduation from FIDER-accredited schools, NCIDQ certification, and state licensing are becoming the norm. With this movement toward professionalism has come an understanding that good design is both aesthetically pleasing and functional for a wide spectrum of people. Recognition that universal design precepts can serve as the underpinning for user-centered design is changing the way interior design is taught and practiced.

Attendance at Design for the 21st Century: An International Conference on Universal Design, held in 1998 in New York City, by people from 38 disciplines and 26 countries, made it clear that interest in universal design is not limited to the United States. The 2000 conference, held in Providence, Rhode Island, engaged an even more diverse group of participants—700 people from 30 countries. A global movement is under way to ensure that the designed environment supports the psychosocial and functional needs of all people. A paradigm shift is in process. Design solutions that exclude segments of the population are being rejected as unacceptable. Universal design is emerging as the personal design philosophy adopted by those who understand that good design responds to the needs of all users of the designed environment. For these designers, universal design is the basic premise for all design decisions.

52.6 NOTES

1. During UDEP I, the senior interior design studio was team-taught by EMU faculty members. Dr. Deb deLaski-Smith supervised selection and specification of materials, surface finishes, and furnishings; Dr. Abraham Kadushin supervised time management plans and adaptive-reuse considerations, including construction, HVAC, electrical, and plumbing; Dr. Virginia North supervised programming documentation, concept development, lighting design, and design presentation; and the author supervised design exploration and development, including incorporation of universal design precepts as well as compliance with barrier-free building codes, the Fair Housing Act, and ADAAG.

2. EMU interior design faculty involved in delivery of the curriculum during UDEP 2.0 included Deb deLaski-Smith, Ph.D.; Keith Fineberg, J.D.; Abraham Kadushin, Arch.D.; Virginia North, Arch.D.; and the author.

52.7 BIBLIOGRAPHY

Beecher, C., and H. B. Stowe, *The American Woman's Home,* H. A. Brown & Company, Boston, 1869.

deWolfe, E., *The House in Good Taste,* The Century Company, New York, 1913.

Foundation for Interior Design Education and Research (FIDER), *FIDER Accreditation Manual,* FIDER, Grand Rapids, MI, 1996.

———, *Directory of Interior Design Programs,* FIDER, Grand Rapids, MI, 1999a.

———, *FIDER Professional Standards 2000,* FIDER, Grand Rapids, MI, 1999b.

Frederick, C., *Household Engineering: Scientific Management in the Home,* American School of Home Economics, Chicago, 1919.

Goltsman, S. M., T. A. Gilbert, and S. D. Wohlford, *The Accessibility Checklist: An Evaluation System for Buildings and Outdoor Settings,* MIG Publications, Berkeley, CA, 1993.

Griedion, S., *Mechanization Takes Command,* W. W. Norton & Company, New York, 1948.

Jones, L., R. Sekulski, and L. A. Pastalan, "A Day's Journey Through Life©: A Design Education Game," in P. Welch (ed.), *Strategies for Teaching Universal Design©*, Adaptive Environments Center, Boston, and MIG Communications, Berkeley, CA, 1995.

Kilmer, R., and W. O. Kilmer, *Designing Interiors,* Harcourt Brace Jovanovich, New York, 1992.

Malnar, J. M., and F. Vodvarka, *The Interior Dimension,* Van Nostrand Reinhold, New York, 1992.

Michigan Department of Labor, *Barrier Free Design Graphics,* Bureau of Construction Codes, Michigan Department of Labor, Lansing, MI, 1986.

Mueller, J. (producer), *Toward Universal Design* (videotape), Universal Design Initiative, Chantilly, VA, 1993.

Pile, J. F., *Interior Design,* Prentice-Hall, Englewood Cliffs, NJ, 1995.

Sekulski, R., L. Jones, and L. A. Pastalan, "A Day's Journey Through Life: An Assessment Game," in E. Steinfeld and S. Danford (eds.), *Enabling Environments: Measuring the Impact of Environment on Disability and Rehabilitation,* Plenum Press, New York, 1999.

Sommer, R., *Personal Space,* Prentice-Hall, Englewood Cliffs, NJ, 1969.

52.8 RESOURCES

American Society of Interior Design (ASID)
608 Massachusetts Ave NE
Washington, DC 20002
202-546-3480
www.asid.org

**Foundation for Interior Design Education
 and Research (FIDER)**
60 Monroe Circle NW, Suite 300
Grand Rapids, MI 49503
616-458-0400
www.fider.org

International Interior Design Association (IIDA)
341 Merchandise Mart
Chicago, IL 60654
312-467-1950
www.iida.com

Interior Design Educators Council (IDEC)
9202 North Meridian, Suite 200
Indianapolis, IN 46260
317-816-6261
www.idec.com

International Federation of Interior Designers (IFI)
P.O. Box 19126
1000 CG Amsterdam, Netherlands
(020) 276820

National Council for Interior Design Qualification (NCIDQ)
1200 18th Street NW, Suite 1001
Washington, DC 20036
202-721-0220
www.ncidq.org

CHAPTER 53

DESIGNING CULTURAL FUTURES AT THE UNIVERSITY OF WESTERN AUSTRALIA

Annette Pedersen, B.A.(Hons.), Ph.D.
Theory Department, Western Australian School of Visual Arts,
Edith Cowan University, Perth, Western Australia, Australia

53.1 INTRODUCTION

This chapter discusses two major diversity/inclusivity studios, which were part of an on-going experimental project in the School of Architecture and Fine Arts at the University of Western Australia (UWA). Both studios were funded by equity grants from the University's Equity Office. A limited engagement with universal design occurs in the first studio where the brief was to design community housing for a remote indigenous community; the second studio foregrounds universal design within a speculative modernist design paradigm. Thus, in the first instance students worked collaboratively with the Karmulinunga Communities to design houses; in the second they worked with disability consultants to explore interactions between varying human bodies and the built environment. In retrospect, it is possible to argue that the first studio, in designing for a specific cultural community, is limited in terms of design possibilities. It is ultimately, however, a successful model for reconciliation. The second studio, fully incorporating the Principles of Universal Design, produced designs allowing for a play of possibilities of the human body in all its physical, cultural, and historical complexities.

53.2 BACKGROUND

> If you are here to help me, then you are wasting your time. But if you come because your liberation is bound up in mine, then let us begin.
>
> *Lily Walker, an Australian Aboriginal woman*

The teaching/research project described in this chapter involves two experimental design studios taught at the University of Western Australia as part of an ongoing experiment with inclusive pedagogic practices. These studios were built on an ethical premise that by researching and changing the curriculum of universities, and in educating students, it is possible to begin to deal productively with identity and the related range of human rights issues in contemporary society. Both the studios were interdisciplinary (i.e., studio staff included architects, an artist, and an art historian), working with students and community consultants. The first studio also had an interdisciplinary student body. The work is founded on the notion that an understanding of culture and history—especially with a regional aspect; respect for all peoples, indigenous and nonindigenous; and an understanding of the politics of identity—is essential to improving the quality of the built environment.

Knowledge is a powerful tool in the political struggle for equity and social justice. This research project seeks to broaden student understanding and awareness of the human condition in order to respect and attempt to understand the range and diversity of the human experience in Australian culture today. This project is not only original in its brief, but is an important move toward addressing a clearly defined lack in the contemporary field of design pedagogic practices. The ultimate would be, to paraphrase Antonio Gramsci, to improve design for the public such that designed structures cease to be external forces that limit and are limited, but are transformed into a means of freedom, instruments to create new ethical political forms, and a source of new initiatives (Gramsci, 1971).

The School of Architecture and Fine Arts (the School) at UWA provides an ideal site to engage with a pedagogic practice with the potential for creative design problem solving. Design is itself a creative process, but the blend of disciplines at UWA can support the interdisciplinary approach, which is the essential and fundamental strength of inclusive curriculum and universal design.

There are 22 schools of architecture in Australia, New Zealand, and New Guinea. Some institutions, such as Curtin University, combine architecture, construction, and planning; others, such as Deakin University, combine art and architecture; while others combine architecture and building, as does Papua New Guinea University of Technology. However, the School at UWA is unique among Australian universities in housing three studio-based disciplines: architecture, landscape architecture, and fine arts. This union occurred in 1994 through an amalgamation of the departments of architecture and fine arts. The School now offers five undergraduate courses and seven postgraduate courses.

Universal design is a term that has attained currency in the United States of America over the last 15 years to describe a design approach that enfranchises the widest possible population into the use of designed facilities and services. It has been adopted and is being promoted by the social services industry in Australia in the hope that the interests of previously marginalized groups can be represented in the design process. In speaking of *universal design,* the term is used in the sense in which inclusive design explores difference and what it means to be an individual in relation to the particular and the local. The concept of inclusive design goes beyond the provision of special access features for various members of the community, such as wheelchair ramps. Rather, it emphasizes a creative approach that is more inclusive than traditional design methodology, and that "asks at the beginning of the design process how a product, graphic communication, building or public space can be made both aesthetically pleasing and functional for the greatest number of users" (Welch, 1995). Designs resulting from an inclusive approach serve a broad spectrum of the general population, including individuals with temporary or permanent disabilities, parents with small children, elderly people, and individuals from diverse cultural backgrounds, including indigenous peoples.

For example, in Western Australia, Homeswest has struggled for years to house Aboriginal populations. Progress in housing has been impeded by misconceptions, both in the public mind and among design professionals, that Aboriginal communities represent an insignificant portion of the population and that designs for such communities are necessarily restricted by cost, remoteness, inefficient maintenance, and cultural or other problems. Housing tends to be tackled in an ad hoc manner and without due consideration given to the need for what can be rather protracted community consultations. The frequent result is housing that is inappropriate in terms of cultural, physical, geographical, and environmental community needs. This need not be the case. An inclusive design process is actively informed by the diversity of the client population and therefore addresses diverse needs.

In this project, the inclusive approach extends to mean working in interdisciplinary, collaborative design groups, thus involving fine arts practitioners and community consultants in the design process. Although inclusive design can be instrumental in making the arts accessible to people of all ages and abilities, it has the potential for a far more important impact on society and the economy. This potential exists because the practice of inclusive design improves the art of environmental design in all its facets so that all people may contribute to the life of their communities and the ultimate prosperity of their country or community. (See Fig. 53.1.) Thus, the design process is no longer the domain of the architectural "expert."

FIGURE 53.1   Universal Design Studio focus group analysis, October 1999.

The writer Homi Bhabha discusses the "move away from the singularities of 'class' or 'gender' as primary conceptual and organizational categories," which he sees as having resulted in the awareness of a variety of subject positions currently informing ideas of identity. He argues that it is in the negotiation of the range of subject positions in contemporary cultures, from the "periphery," or boundary, to authorized authority, that dominant discourses in Western culture can be challenged (Bhabha, 1994). One can argue, in effect, that architectural and landscape architectural design studios have traditionally been taught within a modernist paradigm in which the architect or studio master occupies the privileged subject position, the position of authority, and the student, or even the client, is deprived of subjectivity. Postcolonial discourse, on the other hand, provides a theoretical position from which each subject may speak. Modernity revisited through colonial discourse provides a space for what Bhabha terms "enunciation," so that design itself can be interrogated and translated in order to value difference.

This project seeks to significantly broaden design educational parameters to explore an inclusive pedagogy, which embraces the notions of coinquiry and collaboration and acknowledges an appreciation of contemporary life experience and diversity (i.e., translating design in order to value difference). This then has the potential to significantly improve the education offered to students, equipping them with an understanding of the diversity of the community they work in, and skills with which to responsibly solve problems in their future workplace. This is done by:

- Proposing and testing a collaborative model of design encompassing an interdisciplinary approach
- Focusing on the cultural and social orientation of the student, consultants, and teaching staff, to further a discussion regarding responsible work practice and ethics

Often, in universities, research and teaching are separate pursuits, with the former eventually informing the latter. This proposal effectively reverses this proposition by suggesting that teaching, carried out in design studios, can form the basis of a research project that has as its ultimate aim a desire to improve education, and thus design, for the benefit of all members of the community. The project has two main components:

1. The first is to investigate and extend the theory and concept of universal design, itself inclusive, for the local and Australian context.
2. The second is to explore the philosophy and theory of architecture and landscape design pedagogy in order to extend its existing perimeters to fully address notions of inclusivity.

One of the dilemmas of design schools in Australian universities is that, traditionally, design has not been regarded by the academic community as formal research in the same way in which clinical research or writing may be. Research at universities tends to be measured by publications. The outcome of studios is often not published, or not able to be published, as it is frequently in the form of a folio of drawings or models. The Committee of Heads of Schools of Architecture of Australasia (CHASA) has for some years been concerned about this problem and in recent years has been rigorously debating this issue. The Committee now formally recognizes the potential for design as a research tool and defines the nature of design in the following:

Design studios are about:

- Entering into a contract not just to digest knowledge, *but also to work to produce knowledge*
- Investigating history, theory, and criticism *through* architectural operations
- *Thinking architecture* rather than applying thoughts to architecture

It is also to be interested in *building* an operative, progressive culture rather than constructing new conceptual systems for theory, or prescriptive methods for design.

This project, whereby teaching becomes part of research methodology, is one way of addressing this academic dilemma. Traditionally, design is a teleological, creative, problem-solving process leading to a constructed or built solution. The teaching of design takes place in a studio, rather than a lecture theater. Design studios are traditionally assessed by a jury, which is a panel of experts who critique student work. The design process has a number of other uses, including the following:

- It provides a logical framework for creating a design solution.
- It helps to ensure that the found solution addresses the specifics of the brief.
- It seeks to find the best use of resources by exploring alternatives.
- The process is the basis for explaining the design outcome to the client.

The design process typically includes a "series of sequential steps from project acceptance, design, construction drawings, implementation, post-occupancy evaluation and maintenance. This basic model is used by landscape architects, architects, industrial designers, engineers and scientists to solve problems" (Booth, 1983). This method of designing tends to lead to multiple separate design solutions. Traditionally, it is taught as a linear process dealing with solving one specific problem. Working with the Principles of Universal Design and taking an inclusive approach acknowledges the students' role in a collaborative design process and, ultimately, the importance of designing for public use.

The pedagogic component of the project began broadly with the general principles of a research project, based in the Department of Education at the University of Liverpool, to introduce Syndicate Group Learning Methods to the School of Architecture. These methods include, for example, collaborative group work. This project was extremely useful as an initial beginning point to review design teaching strategies but, ultimately, was deficient in effecting pedagogic change.

A major element of this British project was monitoring the development and implementation of flexible learning techniques through syndicate group work within undergraduate courses of professional study. The focus of this work was to shift existing power relations within the traditional practice of studio design work, where the studio master is the font of all information, to a perspective of giving students more responsibility for their learning. The Liverpool project was welcomed by some senior staff who believed that teaching needed to reflect recent changes in the architect's role in society—specifically, that less individualistic, more cooperatively based learning would help to equip students for new professional realities (Pearce, Stewart, Garrigan, and Ferguson, 1995). However, the Liverpool project team found their efforts to restructure design studio teaching frustrated by an established and resistant culture. Their research findings then focused on the issues of dealing with transferring a learning system from one discipline to another (i.e., from the Education Faculty to the Architecture Faculty).

53.3 THE WESTERN AUSTRALIAN UNIVERSAL DESIGN STUDIO RESEARCH PROJECTS

The historical production of space has long been a contested space where the exercise of power tends to determine who is advantaged and who is oppressed by the creation of built structures and landscapes. Knowledge then about how space is produced, and for whom, is a vital ingredient of this power struggle. Both of these studios focus on working collaboratively with people who have been largely excluded from the discourses and design practices that shape the physical dimensions of their lives. An inclusive pedagogic practice *and* an inclusive design practice seek to share a valuable conceptual, professional, and practical resource that has the potential to emancipate all.

The Indigenous Design Studio, 1996/1997

This was an equity project conceived in 1996 and eventually carried into the teaching arena in 1997, with a design studio entitled Indigenous Design Studio. As indicated by its title, the studio involved working with Aboriginal communities. (See Fig. 53.2.) Here, universal design was employed as a curriculum strategy in order to design appropriate community housing. This chapter is written from a nonindigenous perspective; although the project was collaborative, it is not ethical to speak for Aboriginal colleagues. Although acknowledging the fundamentally important roles played by the Centre for Aboriginal Programmes at UWA, Homeswest's Aboriginal Housing Board, and the Karmulinunga Aboriginal Communities in Derby in the studio, this chapter is thus focused on an academic teaching practice. Without indigenous collaboration, this work would have simply been reinventing the emperor's new clothes. However, there are important aspects of the project work related to cultural practices that cannot be discussed in this chapter; to do so would be unethical.

In Western culture, when knowledge or theory comprehends the other, then the alterity of the latter is lost as it becomes part of the same. In a refusal to tell the whole story of the project, it is hoped to resist classification within the hegemonic Western colonial system. Those collaborative aspects of the project, key issues for universal design, are aspects that could only be dis-

FIGURE 53.2 Student/community on-site consultation, Derby.

cussed in conjunction with all the people involved. So this writing is effectively about the "edge," the only visible edge of work that can be shared with academic colleagues. It is also "cutting edge," a much abused term these days, but in this case valid, as the teaching model used was both experimental and innovative in relation to the established traditions of teaching design.

It is probably useful to tackle this section of the chapter in a traditional Western fashion by simply beginning at the beginning. A beginning not only creates but is its own method because, as Edward Said understood, "Beginning is making or producing difference which is the result of combining the already familiar with the fertile novelty of human work in language" (Said, 1975). These words of Said's accurately describe the project: working within the established traditions of an academic institution to address issues of equity and diversity in a novel and culturally productive way.

As noted earlier, the School of Architecture and Fine Arts at UWA houses three disciplines: fine arts, landscape architecture, and architecture. Most of the students in the School will, at one stage or another in their undergraduate years, produce work that is exhibited publicly. Occasionally, this work causes controversy and very occasionally is offensive to some. Discussion among teaching staff in the School during 1996, relating to issues of cultural sensitivity and a lack of student awareness of issues of cultural diversity and equity, led to thinking of a way to teach students about responsible work practices—how to teach a course within the School that was based on ethical practice. This issue of ethics, or alterity, was fundamental both to the planned project and to the methodology of carrying out the pedagogic work. It was paramount to find a way in which to "allow the other to remain as other" and not to turn the design process into yet another colonial strategy (Young, 1990).

The School at UWA is characterized by a culturally diverse population of students and an academic staff that is predominantly white, Anglo-Saxon, and male. It is fair to say that while the student population of the School is diverse (approximately 50 percent of the students are women and 10 percent are international students predominantly from Southeast Asia), the culture is not. This is not unusual in Australian educational institutions and has an increasingly negative effect on students from culturally diverse backgrounds. One debilitating effect of organizational failure to effectively recognize diversity and difference can be the emergence of some of the negative legacies of our colonial past such as racism or sexism within the academy. The ongoing problem is that students fail to develop the knowledge, attitudes, and skills necessary to function effectively in a world that is increasingly diverse ethnically, racially, and culturally.

The School is further characterized by the traditional way in which teaching is tackled. By and large, the teacher lectures, and the students take notes. Design studios are still usually run by a studio master. As Donald Schön observes in the introduction to his text *The Design Studio:* "At its best, the architectural studio is an exemplar of education for artistry and problem-solving, architectural studios are prototypes of individual and collective learning-by-doing under the guidance and criticism of *master* practitioners [my emphasis]" (Schön, 1985). The learning process is authoritative, efficient in terms of numbers, not in the least part inter-active, and, of course, Western.

The School perspective is Eurocentric, the teaching process autocratic and directive. This form of educational strategy does not allow for difference and diversity in terms of culture, race, or gender and tends to be ego based (i.e., competitive rather than cooperative). The coordinators of this project believe that this approach to learning disadvantages the students who were encouraged to achieve sometimes at the expense of fellow students. Once qualified as professional artists, architects, or landscape architects, these young people are expected to be able to work collaboratively with other professionals or culturally diverse community groups, but their degree course encourages them to be solo operators. Many graduates struggle to work effectively after leaving the School.

There was a broad agenda to address with the studio, and teaching objectives were documented as follows:

1. To begin to understand race relations in Australia

   To understand the ideological position of Western culture

   To begin to question the idea of race in Western culture

2. To develop an awareness of the history of European contact, the invasion of Australia

   To develop an awareness of ways of thinking other than that of the Westerner

   To gain an appreciation of other concepts of time

   To gain an ability to deal with issues where there may be problems but no solutions exist, or where solutions may exist but can never be enacted

3. To learn to work in a group or community

   To share responsibility

   To shed the need for recognition of individual success

   To understand how to work as a team

   To learn to listen effectively

   To develop problem-solving skills in relation to client needs

   To develop an appreciation of the variety of disciplines in the School

4. To develop a continuation of the critical dialogue on the issues relating to Aboriginal Australia, in particular

   • Native title
   • Land management
   • Environmental and social health
   • Community development models
   • Political power bases

5. To understand desire for the "other" and the need for things to go both ways

   To learn about oneself

   To understand some of the pragmatic issues of working with Aboriginal peoples

6. To undertake interdisciplinary work.

To learn to design with others.

To comprehend the need for respect and etiquette.

To understand inclusion versus exclusion learning models.

Initially, it was planned to enroll students from all disciplines represented in the School. Eventually students from different levels of fine arts, architecture, and landscape were enrolled. The studio teaching program was based on a practical project: Homeswest's Aboriginal Housing Board had offered the studio coordinators a real housing project. This then entailed the students consulting with a remote community and designing houses for that community. To prepare the students, it was decided to subsidize their travel to the community to consult and then to walk the Lurujarri Dreaming Trail. This was a 9-day trek with the Goolarabooloo mob from Broome following a dreaming song cycle. *Mob* is a termed used by Aboriginal peoples in the northwest of Western Australia to describe groups of people. Historically, it derives from the cattle industry. The mob may be diverse and include people outside a particular language group. Thus, the Goolarabooloo mob includes Paddy Roe's family from Broome and a number of Europeans who work with him.

Before leaving Perth, all the students attended 2 days of intensive seminars to prepare them for the design studio. Jane Pearce from the education faculty at Murdoch University ran sessions to introduce everyone to the concept of working as teams across disciplines and to the reality of working in a multicultural situation. Aileen Walsh from the UWA's Centre for Aboriginal Programmes then spoke to the students about their responsibilities as representatives of the University traveling to the Northwest and working with the communities there. The students were introduced to the idea of listening with respect and asking themselves why they needed to know something before they asked a question. If the question was not related to their project, they were told it is inappropriate to ask. They were also given a brief introduction to the violent history of European settlement in the northwest of Australia.

In much academic work related to Australian culture there is an absence or subordination of Aboriginal peoples and their histories. Other writers have described the discourse often produced by the field of Aboriginal studies as *Aboriginalism*. Such discourse in all its manifestations has been complicit with the European invasion of Australia and the consequent dispossession of Aboriginal peoples. Mindful of the ethical considerations with which this project was undertaken, the coordinators wondered if it could be possible to have any worthwhile non-Aboriginal knowledge related to Aboriginal peoples. Is it possible to work together, or is everything Westerners attempt to do inherently flawed either because of the colonial circumstances in which it is conceived or because of epistemological considerations related to issues of representation?

Working jointly on community projects sounds honorable in itself, but any production of knowledge pertaining to the other always involves an act of what can only be termed *translation*. That process risks distorting the lived experience and any understanding of that other. However, in the process of working through these issues and attempting to successfully complete the project brief, the possibility arose that the ethical difficulty of the work lay not in the fact that we speak as Europeans, but in *how* we speak. The nature and content of what is spoken determines whether the work of Europeans is responsible or reprehensible. Trinh T. Minh-Ha writes specifically about this dilemma when she states (Trinh, 1991):

> A responsible work today seems to me above all to be one that shows, on the one hand, a political commitment and an ideological lucidity, that is, on the other hand interrogative by nature, instead of being merely prescriptive. In other words, a work that involves her story in history; a work that acknowledges the difference between lived experience and representation; a work that is careful not to turn a struggle into an object of consumption, and requires that responsibility be assumed by the maker as well as by the audience; without whose participation no solution emerges, for no solution exists as given.

With the notion of a responsible pedagogic practice a foremost concern, a program was written based on the idea that if the students were working on an Aboriginal housing project, it was fundamentally important that their learning experience be grounded in a deeper knowledge and understanding of their own cultural positions. The accepted School practice of a design studio running with weekly 2- to 3-hour blocks of practical teaching was supplemented with a further 3-hour seminar block devoted to historical and theoretical texts.

The seminar series began by examining the notion that all knowledge is socially constructed and that social interactions form the basis of social knowledge. To initiate discussion around these issues, Tracy Moffat's film, *Nice Coloured Girls,* and the 1950s film, *Jedda,* were screened. Within an Australian cultural context, the students explored the notions that because they have different experiences, people have different knowledge, and that knowledge can shift over time. Differences in power result in the commodification of knowledge and a monopoly on knowledge production. They were able to see this amply illustrated within Australian popular culture. Thus, the practical design work was supplemented by an intellectual interrogation of the students' own ideological positions.

The studio coordinators believe that to be involved at any level in working with indigenous peoples requires being fully briefed for that work. Primarily in this case, the students needed to be informed about the colonial history of Australia and the history of racism in Western culture. Racism is not simply a theoretical concept, it continues to be a part of our lives. It is the ongoing legacy of the Enlightenment, woven deeply into the fabric of Western culture. The issue of race is both culturally and socially constructed and structured directly or indirectly by power relations. The legitimation of these power relations is grounded on establishing a hierarchy of differences. It is not what makes racist thought and practice aberrant that is of concern, but what makes it acceptable and legitimate. In Australia, all subjects are equal before the law. Notions of equality or sameness can hide any consideration of how difference may function. Where inequality is actively legislated against, cultural diversity and difference may lead to silent injustice and inequity. In such a climate, racism can flourish because it is hidden behind claims of equality that assume similarity. Educational institutions such as UWA reinforce a Western cultural homogeneity.

Along with coming to terms with Western privilege, it was found that in the teaching itself the Western hunger for racial identity had to be dealt with. Europeans are attracted to gatherings of indigenous peoples because of the obviously strong connection they have to a rich cultural tradition. Students were entranced with the music, stories, and traditions the Aboriginal peoples they met in the Northwest shared with them. These Aboriginal peoples have lives that are imbued with spiritual meaning that seems to be lacking within the School. Students began displaying a deep yearning for cultural connectedness. For example, early in the semester one student painted a huge self-portrait of himself as an Aborigine. When asked why, he replied, "I am reaching deep inside myself, I feel that I have a black heart. An Aboriginal heart. I want to reach deep inside myself to my primitive self." As was quickly realized, the student's quite legitimate search for cultural identity had been misdirected toward Aboriginal culture. It was found that the students not only wanted to be with the Aboriginal people they had met, they actually wanted to *be* Aboriginal. The painful irony of the descendants of the colonialists who had attempted to destroy Aboriginal culture forever trying now to adopt it was not lost on the studio coordinators. Through seminars, students were taught about their own culture and history using an inclusive curriculum. In this way, they were introduced to the idea that there are many voices and, thus, stories in Australian history.

Most courses presented in the School only allow a single Eurocentric cultural perspective. Studio coordinators attempted to unravel this through the following two means. The first was in giving the power over the designs produced by students to the community requiring the housing. This meant that they had the final authority to say what they did and did not want and to direct the students accordingly. It was strongly impressed on the students that they were to listen to these people; in so doing, many had to reconsider their own ideological positions. The students were not experts required to research and document the community in order to produce their own unique designs. They soon realized that they needed to listen, hear, and respond appropriately to what the community told them *they* wanted at studio practical criticism sessions. The students took direction from the community and the community had the final say in which plans they chose for their site. In other words, the students learned not as an anthropologist would learn (i.e., about these people); they learned from them. They learned to understand aspects of housing and environmental design from a non-Western perspective. (See Figs. 53.3 through 53.5.)

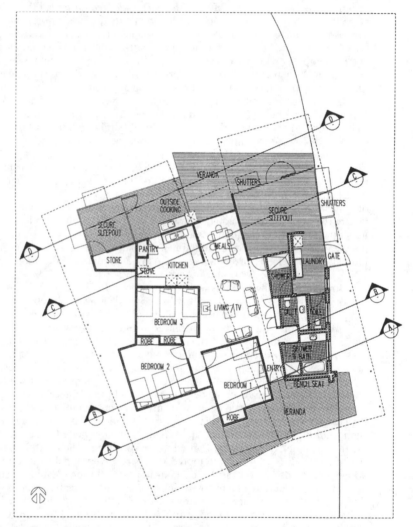

FIGURE 53.3 Student housing design (floor plan), Karmulinunga Community housing.

For many students, they were to learn how the community members actually think rather than how the students thought they should think. Any essentialist or romantic ideas about design for indigenous peoples were swiftly caught up in heated discussion about the pragmatic issues of site, climatic extreme, community needs, and the cost of building. Initially, see-

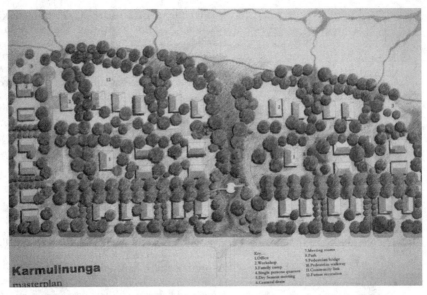

FIGURE 53.4   Student landscape design, Karmulinunga Community housing site, Derby.

ing themselves as urban and the community as somehow connected to nature through the bush, the students swiftly realized the shallowness of these ideas and, more important, where such ideas came from. The binary culture/nature, Western ideas of Aboriginal cultures as timeless and unchanging, issues of authentic and inauthentic were repeatedly challenged and discussed in seminar sessions.

The second way in which a single cultural authoritative perspective was disrupted was by working in interdisciplinary groups following universal design principles and complementing that with the seminars already discussed. This had the effect that fine arts students, for example, attempted housing design, or commented on landscape design. This is a key concept of universal design where cross-disciplinary collaboration can creatively and effectively solve design problems. The students were positively encouraged to work together and to approach design problems creatively. By the end of the semester, architects had designed landscape, landscapers had modeled houses, and so on. Although the students had to meet assessment criteria within their disciplines, they had all ruptured the boundaries of those disciplines to find solutions to design problems.

In the process of encouraging this interdisciplinary collaboration, a profound sense of community developed within the studio. The students spent most of their time together in the studio space working, discussing ideas, and socializing together. By the end of the semester, the students had developed their own distinct culture within the School, and other students were commenting enviously about the positive nature of the group. Odd as this may sound, many people think that *culture* is something someone else has. It is often seen as exotic or foreign. Encouraging a strong sense of group identity within this design studio not only gave the students a challenging learning environment, it also fostered a sense of cultural identity and respect such that, by the end of the semester, the romantic yearnings to be part of Aboriginal culture had shifted to a more productive attitude.

In this process, both consciously and unconsciously, the students, including teaching staff, had learned to embrace a range of interpersonal educational techniques. This follows the work of Kate Grinde, who identifies such valuable group design techniques by emphasizing "the development of trust of oneself and others; the use of barrier free communica-

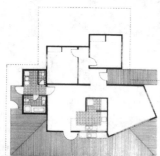

FIGURE 53.5 Student housing design (floor plan for three-bedroom house and two-bedroom house, with photograph of existing housing), Karmulinunga Community housing.

tion techniques; the use of organisational skills; the importance of humour; and separating the individual from the idea" (Grinde, 1998). In this rich pedagogic environment, students learned many skills beyond the simple application of design techniques for a remote and harsh environment. They also learned about their own history and culture to the extent that their design studio became a profound gesture of reconciliation. Many students from the group volunteered to travel to Derby to work for the Aboriginal and Torres Strait Islander Commission during the long vacation. They opted to continue the meaningful work begun in class.

Critical Evaluation of the Project

There was in this instance a clear conflict between the articulated rationale of design tutors in the studio, which involved the promotion of individual creativity and excellence, and the rationale of a project, which included a desire to promote equity, collaborative skills, and teamwork. The introduction of syndicate group learning, one of the fundamental aims of the Homeswest project, in a design culture that can sometimes promote competitiveness and individual achievement, presented a clear challenge. A vital element of this was the fact that while, rhetorically, tutors were able to accept the aims of group work as constructed by the project, there were difficulties for effective action, exacerbated by the need to produce completed designs for a specific Homeswest contract and for formal folio assessment within the School. This had the effect of restricting the field of speculation for most students.

The ultimate success of this studio was as a model of reconciliation. Having evaluated the design project, which, although limited in some respects, was successful in terms of fulfilling a Homeswest contract and achieving excellent design work, it was proposed to refine the teaching model in order to greatly broaden its context to more fully incorporate the Principles of inclusive, or Universal, Design. For the following project it was proposed that design be undertaken in the spirit of speculation, allowing any further work to be taken up as research projects by graduate students with relevant experience. This would allow the studio to remain in the province of what Schön refers to as "reflection-in-action," with the emphasis on design process and speculation related as much as possible to its social context (Schön, 1985).

53.4 *THE UNIVERSAL DESIGN STUDIO, 1999*

This studio's aim was to experiment with further strategies for implementing inclusive/universal design in Australian universities, building on the experience of the previous studio project.

The studio was to practically examine specifically, designing collaboratively, the demographics of diversity, cultural exchange, design for health and ethics, and responsibility in relation to both design and teaching. To achieve this, an equity grant was used to fund the employment of six community consultants with a range of disabilities, who were to work closely with the students for the semester. Again, the studio curriculum was taught by an architect, an artist, and an art historian. The following areas were to be addressed:

- Designing collaboratively
- Design for health
- Community design in Australia
- Ethics and responsibility in relation to both design and teaching

This part of the project was designed to focus on traditional research methodologies throughout its duration and be informed by the design studio both experimenting with theoretical concepts and contributing to them. This, in itself, is a potent innovative nexus that incorporated teaching as part of a research methodology, and it is extremely significant in relation to traditional attitudes to the design studio, which has seen it marginalized as a research source. It is the logical continuation of the work carried out in 1996/1997.

A specific problem of the learning experience of students in architectural schools is the distance they are kept from real interaction with users. Instead, they are required to understand users who are represented through the creative use of imaginary scenarios. Students are required to project onto these scenarios an understanding of the social, political, and economic forces that define and constrain architectural practice outside the university environment. Given that most architectural and design students in the School are Western Australian school graduates, or "leavers" as they are known as in Western Australia, a high percentage of whom are still partially or wholly economically dependent upon parents, government assistance, and other support mechanisms, much of this projection is expectably naive.

Whether consequently or coincidentally, architecture schools have a strong focus and emphasis on the development of design techniques and the exploration of aesthetic outcomes. There is a strong culture within the School, which gives a high value to these. The main aim of the studio was to encourage an active participation with a set of real users with whom genuine relationships can be formed. It is significant that the consultants engaged in this process are all disabled in some way. These relationships, it was believed, would establish in the students a stronger sense of the complex connection between functional, social, and aesthetic concerns, by allowing the definition of the user in the practical scenario of building design to intersect with these concerns. This belief was supported by the evidence of the previous studio.

The term *universal design* is a utopian ideal for the enfranchisement of the widest possible range of users within the brief for building design. This studio did not necessarily attempt to promote the use of the term itself or advance its adoption in design schools. Rather, it attempted to encourage students to interrogate its ideals within the framework of design functionalism and aesthetic outcomes, and also the social responsibility of the design professional. Working with people from outside the design professions and with varying degrees of disability provided a challenge to the students, all young with limited life experience. Such community activity, sharing their studio space with people who were in wheelchairs and who had movement and visual impairments, impressed upon the students the importance of space as well as of the body.

Here was an opportunity for students to understand and map the body as an active or productive occupant of space in ways they had not previously imagined. Not only does the body produce social space through its interaction or material practices, but the social context situation encountered by the body plays a vital role in the production of actual social embodiment. For the student architect designing space, then, the body becomes the site for the production of space, and here there was an exciting diversity of articulations possible from the varying capacities of their consultants. The dynamic dialectical relationship of bodies and social space in the studio inspired participants to begin to envisage a real resistance to spatial oppression and to articulate new ways of creating inclusionary spaces and places. (See Figs. 53.6 and 53.7.)

FIGURE 53.6   View 1—Student design for Universal Design Convention.

FIGURE 53.7   View 2—Student design for Universal Design Convention.

53.5   CONCLUSIONS

These two projects critically hinge on linking theory to practice, which is, in itself, a postcolonial move. The philosophy of praxis is full of contradictions, and in these cases the teaching staff act, again to paraphrase Gramsci, both as individuals and as part of a social group, where teaching, an element of the contradiction, is elevated to a principle of knowledge and therefore of action (Gramsci, 1971). The two design projects thus provide a context for an active, local, site-specific, theoretical discussion. By situating the work within the local, individuals not necessarily in the School of Architecture and Fine Arts, or related to design disciplines, become involved. It is generally accepted that university teaching takes place within a lecture theater or laboratory, but mainly on campus within the university environs. By involving members of the local community off campus, an Aboriginal community in the first instance and consultants with a range of disabilities in the second, in the teaching process, these projects ruptured the boundaries between academia and the world. Therefore, the theoretical issues and philosophy of inclusivity are carried beyond the bounds of theory to the community both on campus and beyond. This emphasis on the local is intellectually related to postcolonial theory.

One of the key issues of postcolonial theory has become that of identity, and the space, or site, of difference. Within this space is a play of private and public, past and present, the physiological, the psychological, and the social. In relation to this, Homi Bhabha argues that he has "attempted to constitute a postcolonial, critical discourse that contests modernity through the establishment of other historical sites, other forms of enunciation" (Bhabha, 1994). In addressing inclusivity both in terms of teaching strategies and design itself, these projects aimed to actively explore ways in which the sense of what it means as an Australian can be transformed: "to live, to be, in other times and different spaces both human and historical" (Bhabha, 1994).

This project entails both theoretical research and then practice in the creation of studio-based projects involving staff and students from the School. The propositional nature of the design project potentially enables the cultural or social orientation of the student and the instructor to be brought into focus. These studios work to develop the unique potential that a combination of the creative arts, architecture, landscape architecture, and fine arts have to offer opportunities for developing alternative, more inclusive pedagogic models and design. This is a significant development from the modernist design process, which follows a mechanistic procedure to solve design problems. Such an initiative then proposes to develop further existing strategies for making explicit culturally based differences in the requirement for space and the perception of form, and to make these an integral part of the design process such that the ethical basis of practice and education will be enhanced.

The two studios specifically sought to address the University's commitment to an inclusive curriculum, which is "about curricula and teaching practices that by their very nature allow for student differences, be they individual, gender, cultural or racial or socio-economic background differences, which all contribute to differences in learning styles" (UWA, 1995). The unique combination of research, teaching and design, and putting theory into practice provides theoretical materials that, in a postcolonial sense, transcends boundaries and provide the material for a possible diversity of inclusive futures.

53.6   NOTE

The author wishes to acknowledge Finn Pedersen, architect; Grant Revell, landscape architect; Jaye Johnson, occupational therapist; Karmulinunga Communities, Derby; Romesh Goonewardene, architect; and students from the School of Architecture and Fine Arts at UWA, UWA Centre for Aboriginal Programmes.

53.7 BIBLIOGRAPHY

Bhabha, H., *The Location of Culture,* Routledge, New York, 1994.

Booth, N., *Basic Elements of Landscape Architectural Design,* Elsevier, New York, 1983.

Gramsci, A., *Selections from the Prison Notebooks of Antonio Gramsci* (edited and translated by Quintin Hoare and Geoffrey Nowell-Smith), London: Lawrence & Wishart, London, UK, 1971.

Grinde, K. (Cited in Pedersen, A. and G. Revell, "Design in Derby, Teaching for Diversity," in Arts on the Edge Conference, Edith Cowan University, Perth, Australia, 1998.

Pearce, J., R. Stewart, P. Garrigan, and S. Ferguson, *The Management of Independent Learning,* Kogan Page, Liverpool, UK, 1995.

Said, E., *Beginnings,* Basic Books, New York, 1975.

Schön, D., *The Design Studio,* RIBA Publications Limited, London, UK, 1985.

The University of Western Australia (UWA), *Inclusive Curriculum at UWA* (discussion paper), The University of Western Australia, Nedlands, Western Australia, Australia, 1995.

Trinh, T. M.-H., *When the Moon Waxes Red,* Routledge, New York, 1991.

Welch, P. (ed.), *Strategies for Teaching Universal Design,* Adaptive Environments Center, Boston, 1995.

Young, R., *White Mythologies,* Routledge, London, UK, 1990.

CHAPTER 54

UNIVERSAL DESIGN AS A CRITICAL TOOL IN DESIGN EDUCATION

Ruth Morrow, B.Arch.(Hons.), RIBA
University of Sheffield, Sheffield, South Yorkshire, United Kingdom

54.1 INTRODUCTION

The DraWare project was a pedagogical research project based in the School of Architecture at University College in Dublin, Ireland. The broad aim of the project was to contribute to the creation of a universally accessible built environment through architectural education. Specifically, this meant experimenting with various teaching methods and awareness-raising techniques both within the existing contexts of the School of Architecture and the profession.

Implementing any type of change in an existing system is a slow process, but it can be significantly aided by an understanding of the causes of resistance to change. This chapter aims to explore and highlight some of the prevalent attitudes and accepted teaching methods in architectural education, which in the course of the DraWare project were identified as hampering the mainstream adoption of universal design. Examples of counterteaching strategies will be suggested, which might form a foundation for the development of a new critical theory and practice of design education.

54.2 BACKGROUND

The DraWare project was part-funded by the European Social Fund of the European Community and was based in the School of Architecture at University College in Dublin, Ireland. The School has approximately 250 students, and the project ran for just over 2 years, between January 1998 and March 2000. The main concern of those teaching universal design is how to ensure that its principles are imbedded and infused right across the architectural curriculum. DraWare, using the model of the U.S. Universal Design Education project (Welch, 1995), quickly adopted a multipronged approach, that is, inputting not only into the design studio, but also environmental lectures, history/theory seminars, technology studios, and staff development. However, those involved in the project were aware that a lasting mainstream acceptance of universal design would be difficult given the following evidence.

1. The forerunner to universal design, designing for disability, despite having a long, pedagogically successful, and well-documented history in architectural education, is still not a standard or accepted part of the architectural curriculum.

2. Many universal design and designing-for-disability courses would rapidly disappear if it were not for the personal commitment and, at times, career sacrifices of the "maverick" teachers who lead and sustain them.

3. Given that one would assume that an understanding of the relationship between people and space would seem to be fundamental to the making of architecture, very few architectural curricula make the study of people's needs an explicit part of the course. An important illustration of this is the document revised in 1997 by the Royal Institute of British Architects (RIBA) and the Architects Registration Board (ARB), which acts as a curriculum outline to all schools that seek professional validation for their courses. This currently includes over 80 schools worldwide, including all schools of architecture in Britain and Ireland. However, this document, *Criteria for Validation,* contains vague and fragmented references to the understanding of people's needs in design, particularly when compared with other areas such as technology, structure, information technology, and even sustainability.

4. The DraWare team, when referring to their own previous experience of teaching universal design in other schools of architecture, realized that it was hard and often frustrating work. This frustration arose from the fact that students electing to join previous universal design courses never seemed to have the skills necessary to do justice to universal design and that, regardless of the hard work, shared conviction, and enthusiasm of both students and staff, their efforts were rarely, if ever, fully valued by the educational system. It seemed that schools of architecture simply did not have the means by which to evaluate the design produced. In the end, it is difficult for those involved in universal design to sustain themselves in a system that does not, even in part, reflect their values and interests.

As a consequence of this evidence, the research project undertook to identify the elements and attitudes in architectural pedagogy that hamper the mainstreaming of universal design.

54.3   *GENERAL BARRIERS TO UNIVERSAL DESIGN*

Implicitness.   The position of the user within design education is so implicit, so central, that it is possible to become complacent about it, much in the same way that it is possible to become complacent about the act of teaching within universities. Those involved in architectural education have to be particularly alert to this, reinforcing in their academic aims that meeting the user's needs is a basic requirement of the course. Course documentation should be examined to draw out and make explicit the reference to developing the appropriate knowledge, skills, and values required when designing for people. Otherwise, this fragmented implicitness will continue to weaken the central position of the user within architecture, negating the fact that architecture only exists to house and support human activity.

Relationship to the End User.   In comparison to other design professions, the relationship that exists between architect and end user is relatively loose. Where the end user and paying client are different is that usually the paying client's needs take preference over the needs of the end user. Often, instead of taking moral and professional responsibility for their actions, architects find it easier to say that it is the clients who will not pay for universal design. The problem is further compounded by the fact that architects, drawn from a relatively select and privileged group in society, have little immediate understanding of those who struggle to access the built environment.

In fact, architects could be seen to represent the *antithesis* of those who are environmentally disadvantaged. For example, it is not uncommon to find that architects in their own homes, offices, and so forth will, in order to maintain architectural purity, tolerate a certain amount of physical discomfort. As a consequence, the underlying premise of universal design—that the built environment excludes and discriminates against certain groups in society—is not universally held within the profession. In other words, many architects just do not see the problem.

There are, however, ways to reduce this "disconnectedness" between the architect and the end user. The most obvious one, though difficult to achieve, is to attract and actively encourage young people from a variety of cultural and social backgrounds to study architecture. Architects can also be reminded that they, too, are people with every chance of experiencing mobility, sensory, or mental impairment at some stage in their life. As Franck and Lepori (2000) say in their book, *Architecture Inside Out,* "Exploring one's subjectivity can be a path to understanding the experiences of others." This is a challenge in particular to education, which tends to deny that students coming into an architecture course have any useful experience or knowledge of the built environment. In an initial attempt to address this problem, the DraWare project organized a workshop for first-year students entitled "Using Your Body." Students were asked to work in groups and compare sizes and proportions of their bodies, as seen in Fig. 54.1. They discussed the differences that arose and their cause (i.e., gender, age, ethnic background, etc.). Recognizing the differences between the individuals of one peer group was seen as the first step to understanding the differences between different groups in society. Discussion focused on the danger of designing for the stereotypical user—fit, healthy, young, able, and male—who, in fact, represented only a minority of the population.

The Culture of Architectural Education. The DraWare project postulated that within architecture and, more specifically, the area of architectural education, elements of the existing culture and many of the resulting pedagogical methodologies have a profound effect on the mainstream acceptance and integration of universal design. This mirrors Manley's (1998) statement in her paper, "Designing a Paradigm Shift in Education and Practice," when she says, "Educationalists themselves need to consider their own attitudes and ensure that the teaching methods they employ contribute to the development of the necessary paradigm shift." Such attitudinal and methodological barriers were more specifically defined during the course of the DraWare project and are laid out in Sec. 54.5, "Barriers in the Culture of Architectural Education." The following section outlines, first, some of the major teaching activities that were carried out during DraWare.

54.4   DRAWARE'S TEACHING ACTIVITIES

As previously mentioned, the DraWare project was structured across the course of architecture, participating in and contributing to all areas of dissemination, discussion, and design. In particular, universal design principles were incorporated into lectures, history/theory seminars, and design studios.

Lectures

As part of the third-year lecture course, "Ecology of Architecture (Conservation and Sustainability)," DraWare presented a series of lectures entitled "Designing for Inclusion." Lecture topics addressed legibility and wayfinding, designing for all senses, the spatial environment and exclusion, the visual environment, the aural environment, the tactile environment, the conceptual environment, the cognitive environment, air quality, molding space and artifacts to bodily movement, sound-sculpturing space, light-sculpturing space, and sensing space.

History/Theory Seminars

DraWare and associated staff ran history/theory seminars for fourth-year architecture students, which led to written dissertations. Three seminar series were developed under the following titles:

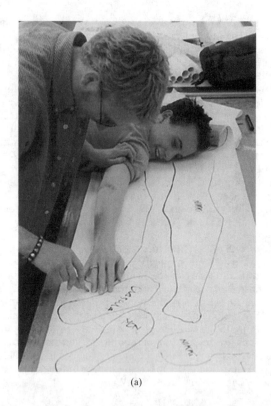

(a)

(b)

FIGURE 54.1   "Using Your Body." Students (*a*) outline and (*b*) compare body measurements.

1. *"Normal People and Their Everyday Lives."* This examined the definition of "normal" and the gap that exists between the general public and the architect's view of architecture. It led to discussions about why architects presently do not design for all people.

2. *"Perception, Representation and Designing for the Senses."* These seminars looked at how the built environment is perceived through the senses, with particular emphasis on the nonvisual processes of the aural, tactile, haptic, and olfactory, and their interaction and interconnection. Representation, meaning, and wayfinding within the built environment, landscape, and gardens were also explored.

3. *"Voices from the Margin."* This explored the secret spaces of a city—the spaces between the everyday. It investigated the position of these spaces in the contemporary and/or historical city, both metaphorically and geographically; their qualities and characteristics; and the nature of the people who inhabit them—outsiders, the homeless, prostitutes, the mentally ill, and those who are excluded on the basis of race, disability, gender, or sexual preference.

Design Studio

Within the Design Studio, DraWare participated in existing projects by simply adding *universal design* to the discussion. On other occasions, it made alliances with other growing interests in the school to create new projects that addressed multiple concerns. At no time did DraWare create a project that only addressed universal design concerns. Examples of these projects follow.

Environmental Barriers—Analysis and Sketch Design Project. Before DraWare's arrival, this project had been designated a landscape project. DraWare chose to site the project in Tullamore, a town in the center of Ireland, surrounded by vast areas of bogland. The students were asked to analyze and understand not only the landscape but also the relationship between the landscape and the people. Links were established with a local umbrella voluntary organization representing people with disabilities, travelers, women's groups, unemployed, and residents of local authority housing estates. With their help, the students involved in the project met a wide range of the local people, including an elderly member of the community, a wheelchair user, a traveler family, a single-parent family, and a local community leader.

Through these meetings, the students uncovered parallel problems of access to town and landscape. In response, the students produced sketch design proposals that sought to address these problems through a mixture of singular interventions and wider planning responses. The completed work was presented to the local umbrella organization, and a video was produced. It is a measure of the students' understanding of the relationship between people and the environment that many of their ideas mirrored similar ideas under discussion among the community of Tullamore.

Residential Centre for People with Severe Physical and Sensory Impairments, and Housing for the Elderly. Both of these projects developed from the work in Tullamore. DraWare was approached by two groups, "Centre for Independent Living" and "Rights for the Elderly," both of which were at the start of the briefing process for accommodations that they wanted to build to meet their group's needs.

DraWare and a group of interested students were asked to submit design proposals for each building. The process was fairly typical of such live projects: an initial set of meetings where the client explains the context and the needs of the user, followed by presentations and roundtable discussions of the students' design work in progress, as seen in Fig. 54.2. The uniqueness of the project arose from the equality of the relationship between client bodies and students.

FIGURE 54.2 The Tullamore project. Students discuss designs in progress with the client groups.

Both groups learned from the process: The students learned about the specific needs of the users, and the client bodies were challenged in rethinking some of their previous ideas. For example, one student proposed that, in order to develop links to the surrounding community, certain of the rooms could be hired out. This drew an overwhelmingly positive response, and the group went on to discuss how renting out a room to someone practicing alternative medicine, such as reflexology, would make a good symbiotic relationship.

Another student altered the clients' views of bedrooms, which, until then, had been seen simply as a space for a bed. It was agreed that bedrooms could vary, that for some people, bedrooms were only for sleeping in, whereas others might prefer to use them as private living/working spaces.

It is certain that the client bodies involved have benefited from working with the students. The discussions that arose from the students' work and proposals not only strengthened many of their previous ideas, but also opened up new possibilities and, overall, raised the level of their architectural expectations.

Sensory Installations. This project initially grew out of fourth-year history/theory seminars entitled "Designing for the Senses." However, it became a collaborative project with the building technology laboratory staff, who saw it as an excellent opportunity for students to experience "designing while making." Seven fourth-year students were involved in designing and building a "Box," seen in Fig. 54.3. The site chosen for its location was the quadrangle around which the School of Architecture was arranged; in this way, the entire process became the center of attention for the school.

Once the external structure was complete, each student used a variety of materials to manipulate the space, exploring one or more aspects of the senses. The sensory installations addressed such issues as how changing light conditions can affect the perception of materials (Fig. 54.4); how sound can suggest architecture; how differing materials, colors, and textures affect the "feel" of a space (Fig. 54.5); how different light conditions can affect the perception of space (Fig. 54.6); how light affects the perception of color (Fig. 54.7); and how moving projection destabilizes solids.

Introduction to People and Space.   DraWare organized an introduction to people and space in the third week of the first-year course. The introduction directly preceded a project where the students visited and analyzed three different architectural spaces—a room, a building, and a part of a city. The aim of the introduction was to encourage students to think more fully about the relationship of the body to architectural space and the factors that affect that relationship.

The introduction was delivered in two workshops. Following the first workshop, "Using Your Body" (described earlier in Sec. 54.3), the students listed the factors that affect the size and proportion of the human body. In this way the conversation, without needing any direction, naturally comes around to issues of ability/disability, race, gender, age, generational and cultural differences, and so on.

FIGURE 54.3 Sensory installation—the finished "Box" in the quadrangle.

FIGURE 54.4 Sensory installation—how changing light conditions can affect the perception of materials.

The second workshop, "Avoiding the Visual," asked students to experience space while wearing a blindfold. This was not a disability simulation exercise, but again in the discussion that followed, students were asked to identify factors that affect sensory perception in architectural space. Their list included temporary and permanent sensory impairments such as fear of reaching out to touch, a head cold, and confusion caused by too many sensory stimuli (noisy room).

At the conclusion of the workshops, the students were given a handout and asked when visiting and analyzing architectural space, not only to look at and think about the spaces but also to note the experience of them, first through themselves, and then by observing others.

Stories of Everyday Living.   This was a second-year studio project that also involved fourth-year students. It is another example of a collaborative project devised by the studio staff of second year, who were interested in alternative methods of representation, and DraWare. Following is an excerpt taken from the project description written for the students.

FIGURE 54.5 Sensory installation—how differing materials, colors, and textures affect the "feel" of a space.

The project focuses on people for whom the ordinary built environment presents problems and inconveniences. Because of their circumstances they have either had to adapt their daily routines to suit their environment, or have adapted their surroundings to suit their needs. In groups of three or four, you are asked to visit these people in their homes to learn how they negotiate the built environment in the course of a normal day. Think of these people as your collaborators and clients. They want you to understand how the built world looks and feels and functions (or not) for them—in the hope that their needs as users will become as important to you as your aspirations as a designer. You will have to ask questions, listen, watch, learn, record, and above all to empathise.

This project should prove useful in the subsequent design project for housing. Having considered the minutiae of domesticity as experienced by a range of individuals you will have to translate what you have learnt into a more general approach to domestic architecture.

Each group was then asked to illustrate one aspect of the knowledge gained through their meeting and present it to a similar-sized group of fourth-year students. The fourth-year students then acted as advocates and described the students' work to a much larger audience made up of second-years, fourth-years, and staff. In total, 98 students were involved. One of the most successful presentations illustrated one small but important frag-

FIGURE 54.6 Sensory installation—how different light conditions can affect the perception of space.

FIGURE 54.7 Sensory installation—how light affects the perception of color.

ment of a person living with multiple sclerosis (MS): getting out of bed in the morning and going to the bathroom. The students presented this in a series of drawings that displayed in detail every aspect of this sequence from gaining balance when getting out of bed to turning on the light switch in the bathroom. As seen in Fig. 54.8, the drawings produced were spaced and hung through a doorway in such a way that slowed the viewer to the pace of a person living with MS.

FIGURE 54.8 Stories of Everyday Living—charting a journey in the day of someone living with MS.

54.5   *BARRIERS IN THE CULTURE OF ARCHITECTURAL EDUCATION*

During the course of the teaching activities, the DraWare team were particularly alert to any attitudinal and methodological barriers that hindered the mainstreaming of universal design principles; those that were identified are discussed in depth in the following sections.

Monologues and Hierarchies

Typically, knowledge is disseminated as finite parcels in monologue form. However, the study of people's needs and aspirations requires other ways of knowing and learning—especially when one considers that people's needs are not only highly individual, but also ever changing, contradictory, and infinite. It is not a knowledge that can be mastered; instead, it requires continuous, open, and respectful dialogue in order that knowledge and understanding are kept up-to-date and relevant.

This then challenges the traditional hierarchical relationship between architect and user and, in particular, teacher and student. In other words, for universal design education to be successful, the teacher's role has to change from disseminator and leader to facilitator and supporter. Students, in turn, have to be encouraged to go out and seek the most current information, in order to develop the skills on which they will rely for the rest of their career. This represents an uncomfortable metamorphosis for both parties and one that neither will easily accept. Fortunately, for universal design, the growing pressure to reduce staff-student contact hours in universities, particularly in Britain and Ireland, means that utilizing strategies to develop students' independent learning skills has become a priority.

The DraWare project set out to develop independent learning skills in students as early as possible. The second-year project, "Stories of Everyday Living," sought to address this by asking the students to investigate and document the lives of a variety of people, each of whom had a disability and was drawn from a wide range of ages and social and educational backgrounds.

The second-year students were asked to illustrate one aspect of these people's everyday lives in a way that would communicate the barriers they faced in the built environment effectively and efficiently to fourth-year students. The tutors, not having participated in the visits, played the role of listener and supporter, while the second-year students played the role of disseminator to the fourth-year students, as seen in Fig. 54.9.

Representation

Representation is both affected by and affects our modes of thinking and evaluation. Traditionally, students concentrate their efforts on representing the formal and organizational qualities of their designs by anticipating such questions as *where is your entrance?* and *what is your spatial concept?* However, this denies the development of a universal design approach in two ways:

1. First, the dominance of design drawings that focus on planning and formal issues leaves little opportunity for the students to develop or illustrate their detailed design skills. This is problematic since accessible space can only truly be guaranteed by accessible detailing. For this reason, the DraWare project specifically ran a series of technology seminars, entitled "Accessible Detailing," for third-year students. The drawings, which students traditionally produce to illustrate their designs, rarely convey how accessible (physically, perceptually, psychologically, etc.) the space is at a detailed level. Design drawings are used to represent almost exclusively the layout, sequence, and dimension of space.

FIGURE 54.9 Stories of Everyday Living. Fourth-year students explain the second-year work to a larger audience.

Detailed construction drawings, on the other hand, illustrate whether the sequence of construction and materials are satisfactory. But somewhere between these two types of drawings is the representation of the physical nature and human scale of space, illustrating more than abstract or technical information.

2. The second problem with traditional forms of representation is that the language used to describe and represent the formal and organizational qualities of space is, by necessity, both abstract and "profession-specific." As a consequence, those outside the profession find it difficult to contribute formative feedback. Universal design draws on the knowledge and input of everyone who uses the built environment not simply for egalitarian reasons, but because long after a building's use has changed and the designer's conceptual ideas have faded, it is the user's *experience* of the space that persists.

Despite the acknowledged difficulty in representing the experience of space, there are certain types of presentations that come close (e.g., large-scale detailed drawings, full-scale mock-ups of parts of space and installations, and multimedia presentations).

The DraWare project, entitled "Designing for the Senses," allowed the students to represent their ideas in an installation rather than through drawing. Although initial drawings were made, the students were surprised how much their ideas changed as they began to physically experience them during construction, as seen in Fig. 54.10.

Creating design projects, which seek both input and feedback from real users, forces students to use alternative, more experiential-based presentation techniques. The UDEP was particularly concerned with this aspect of representation, where "some students started to investigate carrying the values of universal design into the design medium, making their design presentations accessible to a wide range of people" (Welch, 1995). This was demonstrated in several schools, including the work carried out at the University of South Florida where students used dollhouse-scale models to make their design projects accessible to nondesigners.

FIGURE 54.10 Sensory installation. An installation is one of the best methods to test the accessibility of space.

Lopsided Evaluation

The main thrust of architectural endeavor has been and remains *form*. This naturally determines the way in which student design projects are evaluated. Students who produce innovative forms traditionally receive significantly more credit than those who develop new programs or ways of reconceptualizing space/user relationships. In addition, working with form centers the students' efforts on the visual dimension of space. As a result, they create environments that, at best, are lacking the full complement of perceptual stimuli and, at worst, are perceptually excluding. This is often the consequence of professional disciplines in academic institutions, where students are judged against the model of the complete professional and their product. The danger here is that, instead of allowing students the time to develop their design processes, the system implicitly encourages them too quickly to become the complete architect, producing what looks like complete architecture. This is reinforced by judging the students' efforts on the product and not on how they have developed their designs.

The introduction of good learning and teaching practices will be to the benefit of universal design. Many recent pedagogical trends and methods are seeking to place process and product on an equal footing. Two examples are: (1) using an evenly apportioned assessment system that gives students credit for their process of design development as well as the final product or (2) by using learning journals that force students to document and make explicit all of their design and thought processes.

Indulgent Programs

Many of the design programs given to students in the design studio uncritically recycle existing building types, many of which by their inherent nature exclude certain groups of people in society. One such example are the capitol buildings of the 50 state capitals throughout the United States, which are described as buildings that ". . . seem to disempower the public and empower only those already powerful" (Franck, 1994). Students, however, accept such buildings as the status quo and unwittingly become part of a process that perpetuates the production of disabling and excluding environments.

The other aspect of the "indulgent program" is that many of the sites provided for the location for student design projects more often than not reflect the desires of the architect-tutor than the realities of site acquisition. Sites used in design projects could typically be described as "beautiful" and "full of character," yet those sites available to the less privileged in society would best be described using terms such as "suburban" and "bland." Sites of this nature demand good skills and strong motivation to overcome their existing problems and yet students rarely have the opportunity to develop such skills in the course of their studies. In the DraWare project, while carrying out projects for the two client groups in Tullamore it became apparent that working with real clients was much less of a problem for the students than the nature of the sites they had to deal with. The only sites available for the projects, housing for elderly people and a residential center for people with physical and sensory disabilities, were situated in areas of social deprivation. Both sites were also flat, expansive, and left over ground that previously belonged to the health authority. It seemed that reality was just too real, that the students were simply not equipped to deal with the uninspiring nature of the sites; in the end, one of the students refused to participate. Unless students are given more opportunities to face real issues within schools of architecture, they will not be able to combine and sustain a creative optimism while addressing some of the harsher realities of designing for people.

Analysis Under Pressure

Typically, students come to understand analysis in an oppositional relationship to design, equating spending more time on analysis to spending less time on design. Given that they perceive design to be the most valued element of their course, their analytical processes are invariably carried out with little creativity and an overreliance on traditional methods of analysis. It might be argued, however, that excluding environments are as much a product of inadequate and noninclusive methods of analysis as they are of poor design.

The first step to a more inclusive analysis is understanding that objective methods of analysis are influenced by the position and potential biases of both method and method user. The next step is to use a variety of methods to get as close as possible to an understanding of the space and program from the point of view not only of the designer but also of the existing and future users.

Alternative and supportive methods of analysis are available, such as those used by Randolph T. Hester in his analysis of Manteo, North Carolina, and described in the book, *Dwelling, Seeing and Designing: Towards a Phenomenological Ecology* (Hester, 1993). In his analysis of Manteo, a town under threat from tourism, he undertook to discover the "sacred structure" of the town (i.e., the places and landmarks that held the most collective meaning for the town's people) in order to ensure that the town benefited from growth without losing its sense of community.

Missing Reality

The preoccupation with the new, the exotic, and the shiny, which still exists in schools of architecture, means that students rarely have the opportunity to develop the intellectual rigor or motivational skills required to find inspiration in the everyday lives of everyday people. The DraWare project asked students to document the lives of people with disabilities, as seen in Fig. 54.11. It represented one of the few times when the eye of the architecture student falls on such objects as the ordinary lavatory or the humble ramp.

In addition, there is a long-held and, in part, mythical understanding of the source of inspiration, that somehow it is either innate in the designer or appears out of thin air. This, in turn, distances the source of the solution from the reality of the problem. The question has to be

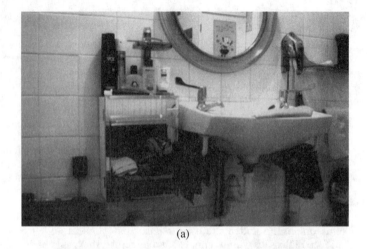

(a)

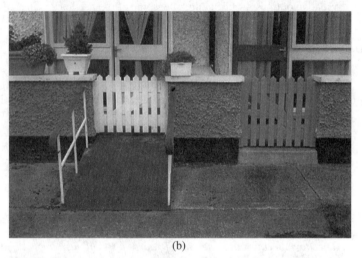

(b)

FIGURE 54.11   Stories of Everyday Living. The students' photographs of (*a*) a lavatory and (*b*) a ramp show an eye for the ordinary.

asked whether innovation stems only from the need to be different and provocative or whether it arises from simply responding critically and laterally to an everyday problem. Universal design demands that it becomes the latter.

A further problem comes to light when one examines the fact that architecture still views universal design as a threat to good design. The definition of *good design* is based on an aesthetic that is centered on beauty; it is an aesthetic that is born of another era and, one could argue, an aesthetic of the privileged.

As schools of architecture start to introduce real programs on real sites for real users, that aesthetic will be forced to expand. Perhaps it will begin to mirror the current trend within the fashion industry where beauty, elegance, and avant-garde exist together with comfort, ease, and wearability to produce truly good design.

54.6 GENERAL LESSONS LEARNED FOR UNIVERSAL DESIGN

Beginning with Raymond Lifchez's work in Berkeley, California (Lifchez, 1987), and more recently, extensively demonstrated in the Universal Design Education project in 1995, putting students in direct contact with users has been essential to the success of both designing-for-disability and universal design courses (Welch, 1995; Ostroff, 1997):

> The faculty found that engaging user consultants in the classroom and studio was the single most valuable strategy for teaching universal design.

Students develop a less stereotypical understanding of people's needs and an increased level of empathy that motivates their sustained involvement.

The DraWare project identified two other consequences of working directly with users. First, the relationship is most successful when it is a two-way learning partnership (i.e., when the users, through working with the students, experience an increase in knowledge of the built environment and the processes that lead to its formation and management but, more important, an increase in their levels of expectation for the built environment). In this way, schools of architecture are not only educating able practitioners of universal design but also increasing the numbers of informed advocates within the general public. Second, involving users in the design studio directly results in the changes in communication, representation, and analysis called for in Sec. 54.5, "Barriers in the Culture of Architectural Education," earlier in this chapter. However, these changes can only be guaranteed when the users' involvement goes beyond consultation to include contributing to the evaluation process.

54.7 CONCLUSIONS

Architecture and architectural education by their nature respond slowly to new influences and movements. Considering how long issues of feminism and sustainability have taken to gain credibility and begin to influence design thinking, it is predictable that universal design will take time to become mainstream.

However, the universal design "movement" benefits from being many faceted. It draws strength and impetus from equality legislation and consumer power, and within architectural design there is a growing awareness that the discipline of universal design can lead to not only fairer but also richer environments.

Identifying and understanding where the difficulties lie in adopting universal design is an important stage in its mainstreaming. As the conclusion in *Strategies for Teaching Universal Design,* Welch (1995) points out, "Critical thinking about universal design is the next important step to considering its place in design education." But in addition, universal design has an important contribution to make to design education. Used as a critical framework, the pedagogical environment of design education can be examined and adjusted to more closely reflect the position and serve the needs of contemporary and future societies.

54.8 BIBLIOGRAPHY

Franck, K. A., "Types Are Us," in K. A. Franck and L. Schneekloth (eds.), *Ordering Space: Types in Architecture and Design,* Van Nostrand Reinhold, New York, 1994.

———, and R. Bianca Lepori, *Architecture Inside Out,* Wiley-Academy, Sussex, UK, 2000.

Hester, R. T., "Sacred Structures and Everyday Life: A Return to Manteo, North Carolina," in D. Seamon (ed.), *Dwelling, Seeing and Designing: Towards a Phenomenological Ecology,* SUNY Press, Albany, NY, 1993.

Lifchez, R., *Rethinking Architecture: Design Students and Physically Disabled People,* University of California Press, Berkeley, CA, 1987.

Manley, S., "Designing a Paradigm Shift in Education and Practice," in *Proceedings of Designing for the 21st Century: An International Conference on Universal Design of Information, Products and Environments,* North Carolina State University, The Center for Universal Design, Raleigh, NC, 1998.

Ostroff, E., "Mining Our Natural Resources: The User as Expert," *Innovation* (quarterly journal of the Industrial Designers Society of America), 16(1): 1997.

RIBA and ARB Joint Validation Panel, *Part 2 Criteria for Validation: Procedures and Criteria for the Validation of Courses, Programmes and Examinations in Architecture,* RIBA Publications, London, UK, 1997.

Welch, P. (ed.), *Strategies for Teaching Universal Design,* Adaptive Environments Center, Boston, and MIG Communications, Berkeley, CA, 1995.

54.9 RESOURCES

DraWare Project.   http://avc.ucd.ie/DraWare/

CHAPTER 55

UNIVERSAL DESIGN RESEARCH COLLABORATION BETWEEN INDUSTRY AND A UNIVERSITY IN JAPAN[1]

Chitose Ikeda, M.B.A.
Noriko Takayanagi
NEC Design, Ltd., Tokyo, Japan

55.1 INTRODUCTION

The NEC Design Group and the Product Design Course of the Department of Product Design at Tama Art University in Japan have long shared this belief: Usable design for people who are elderly and for people with disabilities is ultimately usable for all people. For four years since 1996, NEC carried the educational collaboration and research into the design process to develop universal design products with the third-year students of Tama Art University. The outcome of the research resulted, in the fourth year of the program, as follows: (1) the establishment of a universal design product design process, (2) commercialization of universal design products realized by industry and the university, and (3) completion of a new educational curriculum by applying the collaboration methodology to all of the four-year curricula.

This chapter introduces the actual process of the project, results for both the university and industry, methodologies of universal design education, and future issues.

55.2 BACKGROUND

Social Background

In the near future, Japan is anticipating a predominantly aging society, one never before experienced by any civilization in history. The characteristics and needs of dominant users in all product markets are already going through tremendous changes. Society at large is in urgent need to promptly adjust the environment and services to the new standards of universal design concepts, where designers will have particularly important responsibilities.

NEC and Tama Art University

NEC has been promoting the universal design movement and proposing highly social universal design research under the leadership of its design group, raising awareness within the company, and aiming to adjust actual product development processes to meet universal design standards. Tama Art University had an educational mandate to teach the universal design concept and its methodology of implementation to the students, who will lead the future of industrial design. The reason why both industry and the university chose to collaborate together is not only because both parties wanted to support each other in these objectives but also for two other reasons: (1) industry will be able to accumulate knowledge and methodologies by directly instructing the students, taking one step further than supporting them out of social awareness; and (2) the university can benefit from having access to the latest design processes through long-term, close collaboration with industry, as well as publicizing its high quality of educational programs.

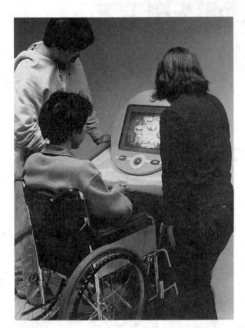

The main significance of this research is that industry and the university worked together to gradually fuse their two separate objectives, "commercialization and profit" and "product design education," in the course of four years and accomplished significant outcomes in order to achieve both aims.

The second notable aspect of the project is its efficient operational process through the cooperation of both industry and the university with regard to: research planning; sharing the definitions of the concept; creating supporting systems; selecting the themes; making a detailed program of curricula; and implementation of design research, presentation, publicity, and commercialization of the designed products (see Figs. 55.1 and 55.2).

FIGURE 55.1 Public terminal for amusement parks (1998).

55.3 COLLABORATION PROCESS

Clarifying Mutual Final Objectives and Sharing Universal Design Concepts

The participants from industry and the university had numerous discussions in order to share common universal design concepts. Through the creation of educational tools for the students, the definitions of terms and concepts were confirmed and reconfirmed. Mutual final objectives were clarified as education and practice of the universal design concept. Universal design was originally defined by Ron Mace at North Carolina State University as: "the designing of products, buildings, and environments to be usable by all people, to the greatest extent possible." Mace's philosophy was made into a graph to visualize this basic definition, and industry and the university came to share the following common belief: that usable design for people who are elderly and for people with disabilities is ultimately usable by all people.

Clarification of Separate Objectives and Benefits for Industry and University

Based on the mutual final objectives just described, separate objectives and benefits for industry and university were clarified. NEC defined its research purposes as (1) raising gen-

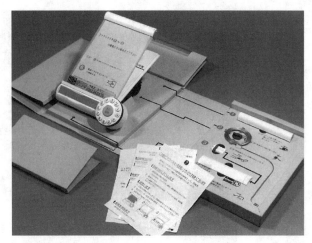

FIGURE 55.2   Senior-user-friendly facsimile (1997).

eral corporate awareness and (2) the development of actual products to achieve corporate profit. Tama Art University defined its objectives as (1) instilling basic understanding of the universal design concept, (2) teaching practical design techniques appropriate for the third-year students, and (3) enhancing the "know-how" of collaboration with industry, based on the results of this project.

Deciding on Themes

It was not easy to decide on the collaboration themes. As for the students, having no restriction in choosing themes can enable them to relate the research to their individual interests and to explore freestyling. For instructors, having certain directions in themes can enable them to plan more effective lessons. Corporationwise, considering desirable profit and approval within the company, it is better to have a theme in accordance with a particular product plan, and to restrict it within possible engineering requirements of actual production, which may conflict with the needs of the university's educational purposes. Considering such conflicting elements, themes were balanced so that, as a whole, both parties could accomplish their objectives. The following are the themes and contents of the projects of each year:

1996: "Supporting People."   Under this unrestricted theme, target users or technological restrictions were not considered. Ten teams of students proposed 10 ideas, for universal design projects, including, for example, musical instruments, public telephones, and scooters. This was an approach to solve the problems of random users (see Fig. 55.3).

1997: "Senior-User-Friendly Information Equipment."   Cellular phones, faxes, and automatic teller machines (ATMs) were selected from NEC's product lines. Two teams made a total of six proposals. Target users were the general public and people who are elderly (see Figs. 55.2, 55.4, and 55.5).

1998: "A New PC for Senior Citizens and First-time Users" and "Easy-to-Use Public Information Terminals."   Three teams made six proposals. These were conceived for a wide range of users, including people with disabilities. The NEC staff joined the teams for each theme. As a result, the proposals became more realistic as corporate activities (see Figs. 55.1 and 55.6).

FIGURE 55.3 Electric scooter unit for a wheelchair (1996).

1999: "Universal Design Public Terminals." Actual product development took place. Based on the previous year's research, a chassis, interface software interactions, and screen were devised with the support of engineering experts. At this point, the corporate objective was achieved.

Establishing an Operative Structure and Assigning Roles

The roles of industry and the university were assigned and repeatedly reviewed in the course of the project period. NEC was responsible for setting themes, reviewing each step of the process, education on the universal design concept through workshops conducting a "virtual senior citizen" experience, and scheduling visits to public facilities. Research guidance, senior user observations, hearings with people with disabilities, midterm evaluation, planning

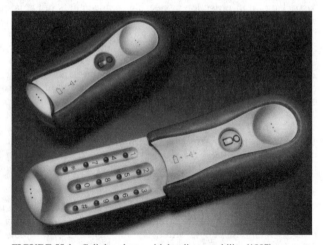

FIGURE 55.4 Cellular phone with hot line capability (1997).

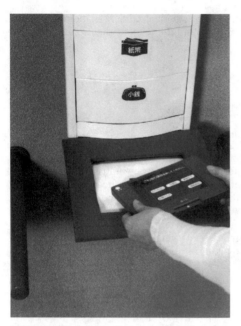

FIGURE 55.5 ATMs that users can use at their own pace (1997).

presentations, publicity, and negotiation with corporate divisions for commercialization of the products were carried out. The university was responsible for putting together teams of students, providing guidance on observation and initial research, setting and managing the overall schedule, design guidance, cost management, presentation guidance, creating publicity tools, scheduling various interviews, and organizing exhibitions.

Evaluation Standards

Industry and the university set separate evaluation standards. The industry's evaluation standard was reviewed each year according to the year's theme. In 1996, originality and viewpoints were assessed. In 1997, six areas, such as quality of observation, interaction ideas, and technical perspectives, were taken into consideration. In 1998, the seven Principles of Universal Design, devised by North Carolina State University, were used as guidelines, as well as assessing how diverse the range of potential users might be. In 1999, accessibility guidelines set forth by the Ministry of Posts and Telecommunications and the Ministry of International Trade and Industry were also taken into consideration. Target users also reviewed the project several times.

The University's Educational Guidelines

The requirements for effective instruction of collaboration between industry and university are: (1) limit the leadership of a project to one person, (2) the leader and subleader should both have experience working in a corporation and be presently working as professional designers, and (3) several advisers should be appointed. The most successful outcomes

FIGURE 55.6 Easy-to-use health maintenance personal computer (1997).

resulted from having advisers guide students with their own ideas while expanding students' perspectives, and eventually organizing the projects around the leaders' decisions.

55.4   *CURRICULAR FEATURES AND RESULTS*

Curriculum Elements

Understanding and Sharing the Universal Design Concept—Experience-Centered Workshops.   In 1996, most students did not even know the term *universal design.* In order to assure common understanding of the universal design concept among industry, the university, and students, various workshops were implemented at the beginning of the project so that the participants could experience and understand the meaning of universal design and become aware of the roles of designers.

Workshop 1: "Who Are the Users with Disabilities?"   The first workshop was dedicated to motivating active engagement in universal design. People who are elderly and people with disabilities are not necessarily "others." Anybody could possibly become a "person with disabilities" under various circumstances.

Workshop 2: "Discovering Barriers in Familiar Environments."   Participants in this workshop practiced brainstorming for universal design. They reviewed familiar environments from various perspectives to see what kind of barriers exist and how they can be removed.

Workshop 3: "Virtual Disability Experience."   The theme of this workshop focused on changing perspectives. Through experiencing disabilities of vision, hearing, and limbs by using eye masks, crutches, and wheelchairs, participants discovered psychological and daily barriers.

Workshop 4: "Virtual Senior Citizen Experience."   This workshop was dedicated to removing prejudice and promoting understanding of the psychological and physical challenges of older citizens. Wearing goggles to simulate cataracts and attaching weights on limbs enabled participants to "experience" old age and review relevant information, equipment, and the indoor and outdoor environments.

FIGURE 55.7   Discovering barriers in currently available products.

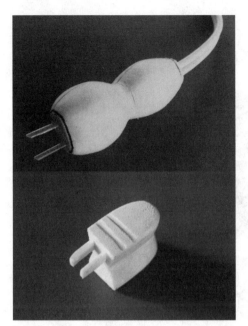

FIGURE 55.8 Redesigning familiar objects with universal design concepts.

Workshop 5: "Discovering Barriers in Theme Products." The focus of this workshop was to enhance the quality of user observations and interview research. Participants used theme products in the present market, analyzed them from the perspectives of users with disabilities, and studied present conditions (see Fig. 55.7).

Workshop 6: "Redesigning Familiar Objects with Universal Design Concepts." This workshop reconfirmed the meaning of universal design and helped prepare for the theme projects. Participants applied universal design perspectives to the redesigning of familiar objects such as doorknobs and electrical plugs (see Fig. 55.8).

Application of Simulation Programs.   Simulating a design business environment, the classroom followed the process of a design office doing a project for a client. Students assumed roles, such as president, engineer, designer, and publicity staff of a design firm, in order to learn about teamwork and leadership.

Observation, Interviews, and On-the-Spot Research of Target Users.   Through hearings with senior citizens and people with disabilities, the participants observed present situations. An all-day task program for people who are elderly, both indoors and outdoors, was designed, in order to observe their physical ranges of motion. Going through this process, the participants learned about the barriers existing presently for senior citizens, and they gained an enhanced understanding of their thinking. Each team planned and implemented street interviews and analyzed the behavior patterns observed on video (see Fig. 55.9, *a* through *d*). This process, which constituted almost 50 percent of the work performed, was a very important part of the project and formed the basis for conceptual planning.

Idea Development, Prototyping, and Reassessment.   In order to develop ideas, various methods were applied, such as the KJ method (a card-based method of organizing and managing information in idea generation sessions); concept mapping through multiple axes; brainstorming and sketching in competition; and, repetitions of creating several 3-D models, by selecting from the models, and by creating multiple versions of models. This ensured flexible development of ideas.

In order to create prototypes, tools such as wheelchairs and canes were consistently utilized to simulate old age. The process was reassessed from various perspectives of many participants, including instructors, NEC staff members, and students. These methods were effective in preventing individual, arbitrary judgments in design, as well as the risk of setting priorities on stylishness rather than user-friendliness, so that universal design quality was ensured before the midterm evaluation by target users.

Technical Guidance, Evaluation, and Promotion

Technical Guidance and Midterm Evaluation.   There were several consultations and evaluations by NEC engineering professionals. This enabled the students to access state-of-the-art technologies, and the engineers to have inspiration through the students' inventive

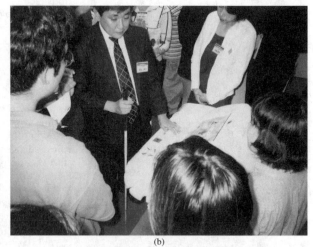

(a)   (b)

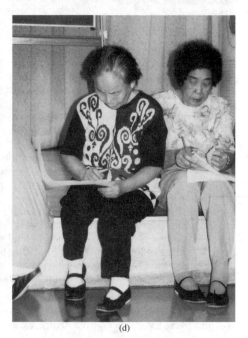

(c)   (d)

FIGURE 55.9   Understanding the target user. (*a*) Observing a target user opening a package. (*b*)A session with people with visual disabilities. (*c*) Observing the daily lives of senior citizens. (*d*) Administering questionnaires/planning.

proposals and ideas. Since 1998, direct and timely guidance using e-mail has become available to students who want to ask the engineers questions.

Project advisors with disabilities reassessed the design process several times, proposed concrete ideas for improvements, and reconfirmed the concept. That helped maintain an accurate direction of the project, and it resulted in substantial design improvement.

Participation of the Second-Year Students.   All second-year students assisted third-year students in all the stages, from model making to presentations every year. This system prepared the second-year students and improved their understanding of universal design, user observation skills, comprehension and market research abilities, and, product proposal skills for when they became the third-year students who would initiate projects.

Presentations and Publicity.   The final presentations were carried out in two steps, which consisted of a presentation in the classroom and then a public presentation at NEC. This enabled both university and industry participants to evaluate and advise in detail, and it served to enhance presentation and communication skills. Furthermore, overall public relation skills were reinforced through making brochures, presenting at international conferences, and receiving journalists for interviews. (See Tables 55.1 and 55.2.)

TABLE 55.1   Schedule of the Collaboration Between Industry and University

---

Preparation Period
  June
    Deciding on themes and clarifying objectives
    Making a budget
    Establishing an operative structure and assigning roles
    Formation of teams

Collaboration Project Period
  July
    Understanding of the universal design concept, socializing
    Virtual senior citizen experience
    Redesigning familiar objects with universal design concepts

  August
    Observation of senior citizens
    Sessions with people with disabilities
    Original research and interviews

  September
    Observation report
    Clarifying the needs and concepts
    Design investigation and making rough models
    Evaluation by target users

  October
    Technical exchanges
    Making mock-ups
    Evaluation within the classroom
    Presentation

Publicity
  November
    Publishing reports
    Publishing brochures

TABLE 55.2   Project Summary

Period	July 1996–July 1999
Participants	NEC Corporate Design Division, other related corporate divisions and NEC Design, Ltd.
	Third and Forth-Year Students in Product Design Course of the Department of Product Design, Tama Art University
Number of Proposals	1999—1 Proposal
	1998—6 Proposals
	1997—6 Proposals
	1996—10 Proposals
Presentations	December 5, 1998: Tokyo Metropolitan Welfare Center for the Disabled (Exhibition)
	December 2, 1998: Yokohama Universal Design International Conference
	November 13, 1998: Toyama Design Exhibition (Presentation at Symposium and Exhibition)
	June 19, 1998: The First NY Universal Design International Conference (Presentation and Exhibition)

55.5   RESULTS IN BOTH INDUSTRY AND THE UNIVERSITY

Results for NEC

Establishment of Universal Design Product Development Process and Design Guidelines.
One of the results that NEC achieved based on this research collaboration was the development of the universal design process, beginning with understanding and sharing of universal design concepts, and finally leading to actual product design. This was directly and indirectly recognized within and outside of the company throughout the four years. There are some plans to develop universal design products incorporating this process in the future.

The workshops developed for universal design education for university and industry collaboration have been modified to apply to educating corporate engineers and designers, and they have been implemented in many sections. Furthermore, they are being introduced as a part of NEC's corporate educational programs this year (2000).

NEC Design Group developed "Five Guidelines" for designers, engineers, marketers, and others to utilize when planning and designing universal design products. Since April 2000, the guidelines have been promoted in NEC groups. Currently, this is only an objective and it is not being enforced. In the future, however, as a part of corporate policy, these are planned to become standard guidelines.

Guidelines Established by NEC Design Group

Intuitive Expressions and Easy-to-Read Panels.   The screens and panels should be designed so that they can be understood and manipulated with anyone's intuition, regardless of culture and degrees of experience, with easy-to-read fonts and colors for people who are elderly and people who have visual disabilities. Braille, touch-to-read symbols, and symbols with sound should also be utilized.

Flexibilities According to Users' Capabilities and Preferences.   Products should offer flexibilities according to the users' capabilities, preferences, and level of comprehensions. Adjustments should be designed in standard models of product; however, in case it is not possible to function with sound and touch guides with one model, a wide selection of models and various options should be available for customers from which to select the best product and service. Operation manuals should have different levels of difficulty and ways of expression according to the needs of various users so that they would be able to select the most suitable for them.

Easy to Use.   Operation methods should be simple and it should be easy to return to the original menu in case of error. Every user should be comfortable and in an easy position to operate the product.

Reasonable Price and Distribution System to Ensure Availability for a Maximum Number of Users.   Prices of products should always be reasonable, making the products available to a maximum number of potential users. Distribution and product information routes should be accessible.

Reliable, Elegant, and Attractive Design.   Elegant design should be offered, with a touch of intelligence, reliability, and without psychological barriers. (See Table 55.3.)

TABLE 55.3   Design Process Chart For Developing a Universal Design Product

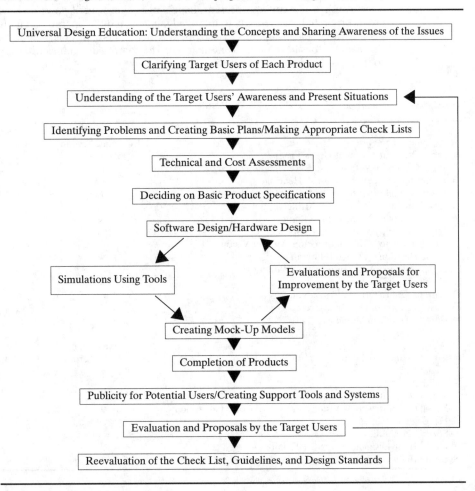

Realization of Universal Design Products

In 1999, a universal design public terminal, developed as a result of industry and university collaboration, was produced and evaluated positively in the market. The hardware and interface software were both designed based on the 1998 research proposals of public terminals. This is an unconventionally universal terminal adaptable to people in wheelchairs and with visual and hearing disabilities, as well as senior users. The hardware has user-friendly details such as adjustable, tilted display; a cane stand; and wheelchair-accessible path. The interface has a pop-up menu with sound and super titles. It has scanning capability for input. It is easy to see, and detailed features like easy button manipulation, dynamic interface development, and easy-to-use icons and sound offer enjoyment and accessibility.

When comparing conventional NEC public terminals and this universal design public terminal with guidelines of the Ministry of Post and Telecommunications, as well as users' range, there are remarkable improvements (see Table 55.4).

More universal design products were developed and initiated by the NEC Design Group. These included a simple remote control for a large plasma display. Since people who are elderly and children in public spaces would use it, it is designed with many user-friendly ideas. The letters on the panels are easy to read, and the material used on the buttons is soft and easy to push. Color-coding, different shapes, high-relief symbols/keys, and easy-to-read layout are incorporated for the users to intuitively understand the operation, and further, to enable people with visual disabilities to also operate them. This control is light and easy to hold, and the main buttons have fluorescent material. The control also comes with a string and magnet so that it will be less likely to be lost.

TABLE 55.4   Comparison of Universal Design Application

Accessibility Check List	Conventional NEC Public Terminal	1999 Universal Design Public Terminal
1. Input Capability Without User's Vision	—	★★
2. Input Capability Without User's Recognition of Color	★★	★★
3. Input Capability Without User's Hearing	★★	★★
4. Input Capability With User's Limited Physical Abilities	★	★★
5. Input Capability With User's Artificial Limbs	—	★★
6. Input Capability Without Time Limit	★	★★
7. Input Capability Without User's Speech	★★	★★
8. Accessibility to Visual Information Without Vision	—	★★
9. Pause Function on Video Information	★	★★
10. Accessibility to Audio Information Without Hearing	—	★★
11. Highly Accessible Buttons	★	★★
12. Easy Return to Original/Any Menu	★★	★★
13. Customization of User Interface	—	★
14. Choice of Control Methods	—	★★

Target Users and Disabilities	Conventional NEC Public Terminal	1999 Universal Design Public Terminal
1. Senior Citizens	★	★★
2. Children	—	★★
3. Visual Disabilities: Blindness	—	★★
4. Visual Disabilities: Low Vision and Lack of Color Perception	★	★★
5. Hearing Disabilities	★★	★★
6. Speech Disabilities	★★	★★
7. Upper Body Disabilities, Artificial Limbs	—	★★
8. Lower Body Disabilities (People in Wheelchairs)	★	★★

★★ Accessible   ★ Partially Accessible   — Not Accessible

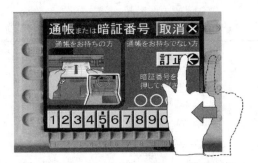

FIGURE 55.11   Barrier-free ATM operator panel.

FIGURE 55.10   Barrier-free automatic teller machine.

Figures 55.10 and 55.11 show an automatic teller machine (ATM) and operator panel, with the latter designed to be highly accessible for people with visual disabilities, as well as for people who are elderly, who would otherwise have difficulties with conventional liquid crystal touch panels. The point of this ATM guide is that it is not a high-cost special model intended only for people with disabilities; rather, all of the standard models can be redesigned to be universally accessible while maintaining the standard price. Without incorporating special numeral buttons or software, the plastic frame attached to the liquid crystal panel provides guidance to users. The added functions to enlarge letters and provide vocal guidance in its input area around the screen make the ATM accessible to people with visual disabilities, to people who are elderly, and to people in wheelchairs. In order to develop this device, interviews with and reviews of prototypes by people who have visual disabilities, people who are elderly, and people in wheelchairs were repeated, in order to improve on the easy-to-read touch-to-read panel, guideline framework, contents of vocal guidance, and clarity of the screen, aiming to perfect the product. Standardization of the accessible design with models of other manufacturers will be the next step.

Results for Tama Art University

Dynamic Curriculum Focusing upon the University and Industry Collaboration.   One result that the university achieved was a major transformation of its educational curriculum through making the most of the collaboration know-how. Through the continuous publicity effects of this project, collaboration projects offered to the school by industries significantly increased. There were more than 10 projects in 1998, compared to 1 in 1996. The university has developed its methodologies to effectively operate collaborations with industries and has established an extraordinarily dynamic educational system incorporating the collaborations in all of the four-year curricula in 1999. Naturally, different collaboration projects are carried out, depending on the levels of the students and educational purposes: Introductory courses including learning about materials are offered to the first-year students, function and form design technique improvements for the second-year students, overall product design process for the third-year students, and the more complete and demanding projects for the fourth-year students (see Table 55.5).

TABLE 55.5   Comparison of Curricula of 1996 and 1999 in the Department of Product Design

1996	First-Year Students	Second-Year Students	Third-Year Students	Fourth-Year Students
April	Theme #1	Theme #1	Theme #1	Theme #1
May				
June	Theme #2	Theme #2	Theme #2	
July	Theme #3	Theme #3		
September	Theme #4	Theme #4	Theme #3	
October	Theme #5	Theme #5		
November	Theme #6	Theme #6		Graduation Project
December	Theme #7	Theme #7	Theme #4	
January				

1999	First-Year Students	Second-Year Students	Third-Year Students	Fourth-Year Students
April	Theme #1	Theme #1	Theme #1	Theme #1
May		Theme #2		
June	Theme #2		Theme #2	
July	Theme #3	Theme #3		
September	Theme #4	Theme #4	Theme #3	
October				
November	Theme #5	Theme #5	Theme #4	Graduation Project
December	Theme #6			
January				

▢ Collaboration classes with industries.

Currently, joint collaborations with other departments of the university have begun. The students' overall abilities—including planning design process, construction of ideas, and communication skills—have improved, and practical education has been established.

55.6 GENERAL LESSONS LEARNED FOR UNIVERSAL DESIGN

One general lesson learned was that a very effective way to promote understanding of the universal design concept is to educate design students—who are not restricted by commercial concerns or conventional business practices. A natural understanding that there are numerous individuals with different capabilities and personal needs is probably best found in those whose environment and experience have allowed them to communicate and collaborate with various types of people from childhood onward.

Early universal design education provides experience with the various needs and limitations of users as part of a designer's basic understanding and encourages depth as well as breadth of knowledge, solutions, and methodologies. Students, given the motivation, learned the concept with great speed and are able to apply it in their designs.

This indicates that it is important to offer the same type of education to young engineers, designers, and marketers. Experience-centered workshops, such as virtual experiences of different types of users as well as observations and feedback sessions with users, are effective. When designers have opportunities to examine barriers experienced by various users in products they have designed, they really learn about the requirements of universal design. It is also important for designers to observe various users using prototypes or products developed with the universal design principle.

It is not easy to change the awareness of experienced designers and engineers who have been designing products for people who are in their prime years and without disabilities. This has been one of NEC's corporate challenges. In order to raise awareness and promote universal design, educating upper management is a key requirement.

It is obvious that the practice of universal design philosophy will lead to long-term social benefits. However, universal design does not provide industries with short-term profits, and it is therefore perceived as a negative factor, leading to higher costs and limitations on design. Much of middle management, seeking short-term profits, is reluctant to make a commitment to universal design. Thus it is crucial to gain top management's understanding and commitment, so that universal design will be seen as a corporate goal and investment in universal design will be implemented.

Examples of competitors' successes in universal design, especially in the marketplace, can be effective sources of information for persuading management. Other helpful information consists of regulations, such as Section 508 of the Rehabilitation Act, and Section 255 of the Telecommunications Act. That way, negative consequences of *not* practicing the universal design philosophy become evident. Useful educational materials include data and knowledge on target users; market size of users who are elderly and have disabilities; regulations in various regions around the world; the movement toward standardization, such as ISO; and, practical methods and realistic figures to realize universal design.

Education is paramount to understanding and promoting universal design. This project has made it clear that the most effective educational methods can be specified depending on the situation, age, and specialties of the people who are given universal design education. The possibilities as to methods and developments are limitless. More study and practice of universal design should be pursued in the future.

55.7 CONCLUSIONS

Universal Design sometimes appears to have many design restrictions and is often rejected not only in the field of education, but also by professional designers. Some even say that "it's hypocrisy that lowers the quality of design," and "it's negative." The fact is that most people involved in design are without disabilities. Teachers and students who focus on unrestricted inspirations for design concepts and styling possibilities criticize universal design as the basic theme because of its many restrictions.

In order to advocate universal design, universal design education should be further researched to allow students to experience users' realities. Inventive ways should be sought to remind professional designers and design students of their social responsibilities. In order to avoid lowering the aesthetic quality of design on behalf of universal design, it is crucial to have a higher level of design quality and stronger direction to inspire new perspectives.

Another challenge is that the trade-offs of developing universal design products, such as additional costs and restrictions in design and interaction, should be balanced with total product quality and resultant social values.

A challenge to implementing collaborations between industry and university lies in communication methods. As few industries have enough personnel and time to closely instruct students, monitoring instruction given to students by professionals is limited to once or twice a month, and in most cases, regarding important points only. On the other hand, students and instructors hope that professionals from industry will get involved more often at every stage. To solve this problem, the Internet is planned to be utilized, by creating an exclusive, information-secure Web site for the project. This will assist in organizing discussion groups online to report on the latest progress, exchanging questions and answers, and discussing ideas. If video conferencing becomes available, collaborations between the university and industry can function as an even more flexible and effective system.

Senior citizens and people with disabilities, who have been conventionally thought of as minorities, are destined to become a majority and active members of Japanese society. It is expected that higher quality of universal design will result in improved environmental design, products, and service. In an aging society, universal design is a crucial concept to promote truly user-oriented product development. The participants of this project will continue to be committed to universal design efforts both in industry and universities, making the most of their research. It is hoped that substantial improvement will result, and that the universal design movement will be extended to other universities and industries in Japan and the world.

55.8   NOTE

1. This chapter is based on the following article:

   Ikeda, Chitose, Noriko Tagayanagi, Tatsuya Wada, and Hideki Tanaka, "Universal Design Research: Industry & University Collaboration Project," *Annual Design Review of Japanese Society for the Science of Design,* vol. 5, Japanese Society for the Science of Design, 1999.

   Setsuko Miura translated the article and this chapter from Japanese into English.

55.9   BIBLIOGRAPHY

Anders, Robert, I.D.S.A., and Daniel Fechtner, M.D., *Pratt Institute Design Primers* (second edition), Pratt Institute Department of Industrial Design, Pratt Center for Advanced Design Research (CADRE), 1998.

Covington, George A., and Bruce Hannah, *Access by Design,* Van Nostrand Reinhold and International Thomsom Publishing Inc., 1997.

Pirkl, James Joseph, *Transgenerational Design Products for an Aging Population,* Van Nostrand Reinhold and International Thomsom Publishing Inc., 1994.

Welch, Polly, *Strategies for Teaching Universal Design,* Adaptive Environments Center, Boston, MA, and MIG Communications, Berkeley, CA, 1995.

55.10   RESOURCES

Important Web Sites

NEC Corporation (English and Japanese)
www.nec.co.jp/

NEC Design Group (English and Japanese)
www.lg.mesh.ne.jp/design/nec/

NEC Design, **Ltd**. (English and Japanese)
www.nec-design.co.jp/

Product Design Course, **Department of Product Design**, **Tama Art University** (English)
www.tamabi.ac.jp/english/pd/

Product Design Course, **Tama Art University** (Japanese)
www.tamabi.ac.jp/product/

Report on the Universal Design Project between NEC and Tama Art University (Japanese)
www.tamabi.ac.jp/product/nec_tamabi/

Tama Art University (Japanese and English)
www.tamabi.ac.jp/

Literature

Ikeda, Chitose, "NEC and Tama Art University: Universal Design Collaboration Project," *Proceedings of Designing for the 21st Century: An International Conference on Universal Design,* June 17–21, 1998, pp. 67–70.

Ikeda, Chitose, and Noriko Takayanagi, "NEC and Tama Art University: Universal Design Collaboration Project," *Proceedings of Universal Design: An International Workshop,* Yokohama, Japan, November 30–December 4, 1998, pp. 89–95.

Ikeda, Chitose, Noriko Takayanagi, Tatsuya Wada, and Hideki Tanaka, "Universal Design Research: Industry & University Collaboration Project." *Annual Design Review of Japanese Society for the Science of Design* (vol. 5), Japanese Society for the Science of Design, 1999.

Moriyama, Ryoko, "Collaboration Research Project between NEC and Tama Art University," *Nikkei Design,* January 1997, p. 25.

Nomura, Masatoshi, Noriko Takayanagi, and Chitose Ikeda, "Activity and Concept for Universal Design," *Proceedings of the 27th Annual Meeting of the Kanto-branch, Japan Ergonomics Research Society,* December 5, 1997, pp. 2–3.

Ogata, Sae, "Thinking Design through Experience," *'99 Barrier Free Guidebook, Japan Economic Journal,* p. 173.

Takayanagi, Noriko, and Chitose Ikeda, "Universal Design Collaboration Research Project between NEC and Tama Art University," *Proceedings of the 29th Annual Meeting of the Kanto-branch, Japan Ergonomics Research Society,* November 20, 1999, pp. 14–15.

Takayanagi, Noriko, and Chitose Ikeda, "Approach from Universal Design," Production in NEC No. 84, March 2000, pp. 24–25.

CHAPTER 56

POST-OCCUPANCY EVALUATION FROM A UNIVERSAL DESIGN PERSPECTIVE

Shauna Corry, M.A.
North Dakota State University, Fargo, North Dakota

56.1 INTRODUCTION

Today, society is poised to begin a transformation. The terms *global diversity, equality, accountability, community, change, social and cultural responsibility,* and *values* are all talked about, taught, and acted upon by people from all walks of life. But society has not yet addressed the needs of all people equally and should continue forward in the move from accommodation to accessibility, and finally to transformation. Diverse needs are still not understood or addressed in the built environment. What does society value? What are the elements of a building that an architect or designer values? What are the elements building users value? These are questions design researchers, architects, interior designers, and facility managers are interested in addressing. Universal design in tandem with post-occupancy evaluation is one answer to these questions that can positively influence the design process and the resulting building. By looking at post-occupancy evaluation (POE) from a universal design perspective, designers, builders, and managers of the built environment can create inclusive workplaces and commercial spaces (i.e., spaces that continue the movement toward transformation of society).

56.2 BACKGROUND

> The idea of a POE is to compare the reality of a building in use to the theory.
>
> *Ziona Strelitz, Principal, ZZA Research*

Post-occupancy evaluation is an important critical analysis tool in enhancing not only function, but overall user satisfaction in the built environment. A well-planned and implemented POE can successfully identify user needs and assess overall building function. Specifically regarding universal design, POE strongly enhances planning and design because it increases the awareness of architects, interior designers, builders, owners, and managers of the needs of all users in all aspects of the built environment, from the air that cools and warms them to the sunlight they crave, the furniture and equipment they depend on, and the sounds they hear.

Universal design is inclusive design (i.e., design that goes beyond meeting life safety and accessibility codes). It is design that positively addresses function, comfort, utilization, satisfaction, and legibility for every building user, ranging from a 60-year-old maintenance worker, to a 32-year-old pregnant manager, to a 6-year-old elementary school student. Building users need built environments to work for them and with them, not against them.

Typical POE (see Chap. 9, "Toward Universal Design Evaluation," by Preiser) methods include facility walk-throughs, user surveys and interviews, behavioral and cognitive mapping exercises, and observation. An additional method (rarely used, but very effective in terms of evaluating the built environment from a universal perspective) is autophotography. Autophotography allows users to capture a visual portrait of the built spaces they encounter throughout their day.

In the past, POE was a one-time service conducted soon after a new facility was occupied, generally conducted by developers, facility managers, or consultants, and the data were gathered from and the analysis was limited to that specific site. POE was not typically budgeted into the project because clients did not want to pay for the extra service. This is still true today; however, the use of POE as a professional service offered by architectural and interior design firms, facility and property management companies, and consultants is increasing.

In *Redefining the Architecture Profession* (1997), a video produced by The American Institute of Architects (AIA), post-occupancy evaluations are noted as an expanded value-added service that architects can offer a client. One of the upsides of this increase is that architectural and design firms are enhancing their involvement in the life of the facility, and as a result they have the ability to then apply the lessons learned to other projects.

There are three levels of service a consultant can offer a client (Preiser, 1999; Haworth, 2000). Indicative POEs are the least expensive evaluation, cover basic positive and negative aspects of the building, and are often conducted using walk-throughs, photographs, and interviews with key building users. Investigative POEs are more expensive to conduct and gather more data using video, physical measurements, and an increased number of interviews and surveys. The diagnostic POE is the most expensive evaluation to conduct, and it gathers detailed data over a longer period of time.

The goal of a POE is to provide hands-on feedback to designers, managers, owners, and users about how a building functions, what works well, and what the problems are. The feedback is then used to refine existing building function and influence future project design. Robert Graves (Tarricone, 1999), a developer of POEs for laboratory facilities, believes:

> The value of POE is the research information that can be applied to the next project, the authorship you gain from individuals who know there will be a POE and that they can raise issues, and the information you get on how to improve the design process from programming through occupancy.

A POE conducted from a universal design perspective effectively addresses the needs of all users, from the commute time experienced by a clerical worker, to the accessibility of a work station for a user who has a mobility impairment, to the thermal comfort of a vice president. Incorporating the Principles of Universal Design as evaluation attributes could expand the definition of and final outcome of a POE. Principles of Universal Design (Center for Universal Design, 1997) include: Equitable Use; Flexibility in Use; Simple and Intuitive Use; Perceptible Information; Tolerance for Error; Low Physical Effort; and Appropriate Size and Space for Approach and Use for all users across their life spans. The principles were created by a team of architects, engineers, environmental design researchers, and product designers to be performance criteria for evaluating existing and designing new products and environments, and each principle has guidelines that help specify performance. Evaluating a built environment from this perspective results in an all-encompassing evaluation, one that can be used to address issues in the existing environment and in future ones.

56.3 PRINCIPLES OF UNIVERSAL DESIGN IN POE

The Principles of Universal Design could be included in a POE not only as guidelines in developing a general evaluation instrument, but also as separate universal design–focused POE. POE instruments such as surveys and mapping exercises should be generated from a universal design perspective. Instruments that are easy to use, intuitive, and available to all will help evaluators collect data and enhance user participation. A separate POE focusing on the Principles of Universal Design would be effective in identifying problems in a facility that would not generally be highlighted in a typical space, thermal comfort, or building systems POE, or even in an ADA (Americans with Disabilities Act) audit.

FIGURE 56.1 Front entrance to the Cooper-Hewitt Museum.

An example of this using Principle 1: Equitable Use (i.e., the design is useful and marketable to people with diverse abilities), can be illustrated with the entrance to the Cooper-Hewitt National Design Museum (see Fig. 56.1). A universal design POE could not only evaluate whether there is an accessible entrance, but whether the entrance works the same for all users (identical whenever possible or equivalent); whether it avoids segregating or stigmatizing any users; whether it provides privacy, security, and safety, equally available to all users; and finally, whether the design is appealing to all users.

The Cooper-Hewitt National Design Museum was built in 1899 for Andrew Carnegie by Babb Cook & Willard. In the mid-1990s the museum was thoughtfully renovated by Polshek and Partners, and one of their principal design goals was to create an accessible front entrance while respecting the historic value of the mansion (Strauss, 1997). The design won the Access New York award in 1997, and is seen to be a successful design by members of the design community, and users who are disabled and disability advocates. The architects designed "a ramp to the Museum's main entrance . . . discretely placed in a planting area, behind the east fence of the original carriage entry on East 91st Street" (Strauss, 1997).

If a universal design POE was conducted of the museum entrance, it would be found that there is an accessible main entrance that is usable by all users. However, the ramp has been seen by some to segregate or hide users who are disabled, because they have to transition from the public sidewalk to the planting area before entering the building. But it is important to note that sensitivity to historical facades is an important consideration in renovations of this type, and although the ramp was "discretely placed," it does provide for privacy, security, and safety, which are important characteristics of equitable use.

The last characteristic that would be evaluated is aesthetics. Is the ramp design appealing to all users? On a tour of the museum during a preconference intensive workshop for Designing for the 21st Century I, attendees were heard saying that they thought the new design was more aesthetically pleasing than the original, and people who were not disabled commented that they preferred entering the building using the ramp because it was located in the planting area and they enjoyed the colors and aroma of the flowers, and the sense of transition from the sidewalk to the entrance. Although the ramp does meet code requirements, the extra care and thought put into the design have elevated it to meeting universal design guidelines— guidelines that are much broader in scope than accessibility codes, and that speak to all users equally.

An example of a product that would not necessarily be evaluated effectively in typical POEs or an ADA checklist is *access flooring*. This product is now being used to address technology needs in office environments. Access flooring is a system that allows for flexible cable management and is composed of floor tiles and spacers that permit floor areas to be built up so that cabling can be hidden. Access flooring is utilized in team spaces and conference areas, and is available through office system manufacturers. However, this type of system requires the addition of small ramps, which in some installations are not accessible. The slope is short, steep, and unsafe. Although longer ramps that meet code are available, some designers and facility managers are installing access flooring without them—the idea being that users who are disabled can be accommodated in other teaming spaces.

If specified and installed correctly, this product can meet the characteristics of universal design Principle 1: Equitable Use. But access flooring is often installed in an unequal and unsafe manner, and as a result it would not be evaluated positively in those cases. A universal design–focused post-occupancy evaluation would help designers, facility managers, and users effectively evaluate contract products and space plans in a relatively easy and successful way.

56.4   POE PROCESS AND METHODS: OVERVIEW OF TYPES

Various post-occupancy evaluation methods have been developed and implemented by researchers, design educators, and practitioners (Preiser, 1988, 1994, 1996; Zeisel, 1981; Bailey, 1987; and Baird et al., 1996). The individual methods can be used alone or combined to develop a set of successful evaluation tools that effectively address the needs of each client. Method types can be categorized as follows: surveys and interviews; behavioral, cognitive, and annotated mapping exercises; filmed visual representations such as video, photographic, and autophotographic exercises; observational methods such as facility walk-throughs using checklists and rating methods; and participant workshops. The method categories do overlap, and it is necessary to develop a set of POE tools based on client needs and facility type.

User Participation

Amy Nadasdi, a consultant and workplaces strategist (Tarricone, 1999), states: "In general, they [post-occupancy evaluations] measure satisfaction with the workplace, and they should anticipate what people might complain about later." In order to do just that, POEs should be developed by consultants in conjunction with the users of a given space. User participation is key to an effective POE. Representatives from various user groups should have the opportunity to help develop the POE instruments, and all users should have some input in the data gathering process. Participation encourages ownership in the process and ultimately allows the POE consultant to gather meaningful data (Baird et al., 1996). The problems that users anticipate may be different from the problems the architect, designer, or facility manager anticipated during the design process.

Surveys

Occupant surveys are common methods that enable the researcher to focus the POE on certain issues in a facility or gather unexpected information from open-ended questions. Surveys gather quantitative results that are easily reported and measured. Successful ones take little time for the respondent to complete, but yield valuable results for the researcher. Surveys can be written or verbal, such as phone interviews and marketing questionnaires; disseminated by mail-out or electronically; or administered to a captive audience. Surveys can be completed by every building user, select users, or a random sample of users. They communicate levels of user satisfaction across a broad spectrum of users at that point in time (Bailey, 1987).

Interviews

Interviews are used to find out what people think about an issue or building, what they know about it, what they are doing about it, and what they expect to happen (Zeisel, 1981). By asking focused or open-ended questions, an interviewer can gather information about any number of issues related to the built environment. There are definite advantages and disadvantages to using the interview method in post-occupancy evaluations. Bailey (1987) states that some of the advantages of interviews are flexibility, response rate, time of interview, and control over the environment the respondent is in. Disadvantages can be cost, time, lack of anonymity, and interview bias. Interviews can be structured or nonstructured. Both styles are effective, depending on the situation and type of data needed. Structured interviews follow a prescribed format and generally gather specific information, whereas unstructured interviews generally have more open-ended questions and build on the respondents' answers.

Structured interviews enable the facility evaluator to explore specific issues identified by the previously outlined POE methods. Furthermore, open-ended or unstructured interviews allow for users to communicate their unbiased perspectives. Interviews are successful in gathering data without assumptions or misunderstandings. The evaluator can ask for immediate clarification and expansion of issues that could only be minimally explained in a written survey.

Annotated Plan

Annotating plans is an effective method of data gathering. Plans of a building are used by the evaluator to document existing conditions by noting areas of concern (Zeisel, 1981). This method is generally used during site visits in conjunction with facility walk-throughs.

Cognitive Mapping

According to Zeisel (1981), cognitive maps provide a mental picture of how users perceive their surroundings. They communicate how users view and interact with a facility. Graphic images provide spatial snapshots that help facility evaluators to understand user responses to the built environment. Lynch (in Zeisel, 1981) defined elements that repeatedly occur in cognitive maps as follows: (1) paths, (2) edges, (3) districts, (4) nodes, and (5) landmarks. In Fig. 56.2, a cognitive map drawn by a first-year student at North Dakota State University, paths are noted as the heavily used streets on campus. The edge is also a street, but it is a main artery in the city that signifies the beginning of the campus. A district is shown between Centennial Boulevard and Administration Avenue, and is the area of the campus that houses buildings commonly used by all students. A major node on campus is the student union building. It is a central gathering place and is noted as the starting point for majority of the students who drew the maps.

In comparison, Fig. 56.3 is a map of the same campus drawn by a fourth-year facility management student and shows a much greater understanding of the site and number of buildings, but it also shows the same edges, nodes, and landmarks shown by the first-year facility management student.

Behavioral Mapping Techniques

Behavioral mapping is a technique that allows the observer to document how people use space and interact with others within a building (Baird et al., 1996; Lang, Burnette, Moleski, and Vachon, 1974). In a health care facility study documenting the influence of design and staff behavior in a psychiatric ward, Proshansky (in Lang, Burnette, Moleski, and Vachon, 1974) used behavioral mapping "to establish by means of systematic and sustained observations what people in the ward are doing, where they are doing it, for how long and with whom." Researchers use observation and annotated plans to track user movement throughout a room, ward, or facility. Numbers of users are counted and locations are mapped on a plan. Researchers note specific and general information about individuals and groups. The information then provides a map or a picture of building use that allows problems to be identified.

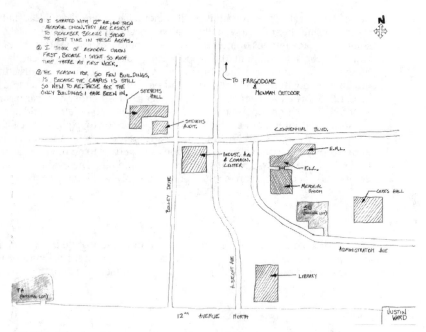

FIGURE 56.2   Cognitive map of campus drawn by first-year student.

Still and Video Photography

Photography has long been used to document existing conditions by designers and architects, because it gives a visual record that can be taken from the site and referred to later to clarify an issue. The advent of video enabled evaluators to make a comprehensive inventory of a building in a short time, and to capture movement patterns and conflicts in public spaces. Video can be used at the micro- or macrolevel, and it can record large gathering spaces or smaller intimate and personal spaces such as work stations. Preiser (in Baird et al., 1996) uses still photographs and/or video during a walk-through evaluation to record building attributes. Corry (2000) used one-time-use cameras and had building users document their daily work and living environments. Participants identified numerous building attributes that were then discussed in personal interviews.

Observations, Walk-Throughs, and Checklists

Observations and building walk-throughs are effective hands-on methods that allow the researcher to see and experience the problems in a given space. Consultants can choose to be hidden or apparent observers in an effort to document behaviors of facility users. A hidden observer documents unrehearsed and natural patterns of behavior. Apparent observers are noticed by the users and can be approached with comments and concerns. Taking notes or diagramming use patterns are important parts of observations.

Walk-throughs can be casual evaluations based on general satisfaction, or they can be more formal in terms of a detailed checklist. Building checklists can be developed for all aspects of the facility and enable the researcher to quantify problems. Checklists can quickly identify perceived satisfaction in terms of issues identified by research such as space usage,

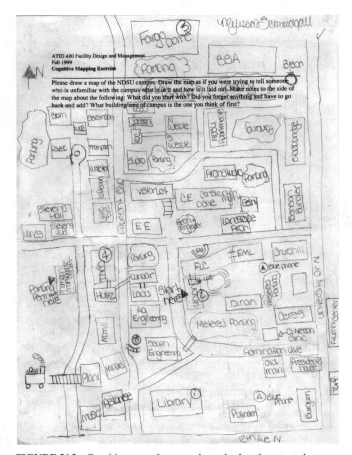

FIGURE 56.3   Cognitive map of campus drawn by fourth-year student.

privacy, user friendliness, facility image, way finding, quality of maintenance, and aesthetics (Baird et al., 1996).

Walk-throughs cover the entire facility, and evaluators use different methods such as photography and direct observation to document existing conditions (Preiser in Baird et al., 1996). Casual walk-throughs can identify unexpected issues. Katherine Klass, project manager for an interior design firm in Boston (Tarricone, 1999), found, "During walk-throughs, we observed things like people with their hand up to their ear when talking on the phone." This highlighted the need for improved acoustics and noise management.

Participant Workshops

Preiser (in Baird et al., 1996) conducts three-day POE workshops that encourage participant ownership of the process, methods, and results. The workshops include facility managers, designers, end users, and decision makers, and create subteams that develop and refine evaluation methods, gather the data, and report their findings in written and verbal reports. Recommendations to address areas of concern are then made according to the immediacy of the situation, future building issues, or issues that require policy changes.

Selecting POE Tools

The methods just discussed can be used alone or combined to effectively evaluate an existing built environment, to address identified problems, and to apply findings and recommendations to future buildings. It is important to specifically tailor the POE tools to the identified users. Methods that work for one group of users may not be as effective for another. Gender, age, cultural and social background, and developmental level of building users combined with the POE budget and completion time frame all influence the selection of evaluation tools. The immediate purpose (e.g., identifying space allocation or thermal comfort problems) and how the outcomes will be implemented are also factors to consider in developing data gathering and analysis tools.

56.5   CASE STUDY: EVALUATION OF ENVIRONMENTAL QUALITIES USING AUTOPHOTOGRAPHY

Autophotography is a qualitative method that results in a richly textured portrait of a user's perception of a space. With this method, users photograph the environments they work, play, and live in for a specified time (e.g., a day, a week, or more, depending on the amount of information the evaluator is looking for). Users can be urged to photograph spaces that they like and that function well for them, and spaces they do not like and that do not function well. Content analysis of the photographs, combined with personal interviews, can reveal the environmental attributes that users prefer and highlight problem areas in a building.

Corry (2000), in a study funded by the International Facility Management Association Foundation, had 21 designers, managers, and users of the built environment evaluate retail spaces in the neighboring communities of Fargo, North Dakota, and Moorhead, Minnesota. Group 1 (i.e., designers, architects, and facility managers) and Group 2 (i.e., users who have mobility impairments) photographed retail stores in the area. Although both groups included spaces that were not functional, users who were disabled photographed functional issues more frequently than did the designers and architects. Figure 56.4 is a photograph of a retail store taken by a user with a mobility impairment, and it highlights the functional issue and universal design principle of Appropriate Size and Space for Approach and Use.

Figure 56.5 shows a retail display that was considered to be aesthetically pleasing by a facility manager. The difference in the perspectives of the user with a mobility impairment and the facility manager is evident in their choice of retail environments to photograph.

One of the goals of the project was to gain an understanding of the differences and similarities in spatial perception by both people who design and manage built environments and users of those environments who have limited mobility. The theoretical base of the study began with Christopher Alexander's "A Pattern Language" (1974). Alexander theorized that:

> A pattern language is, in short, a picture of a culture. And each personal version of the language is a work of art: a personal effort, by each person, to create a single picture of his culture which fits together and makes sense of life. If all of us together, try to create such personal "languages," and share them, then the evolution of our shared language will be a continuous communal effort by all of us, to create an integrated picture of a future way of life, in which all of us can, communally, be whole.

Corry (2000) applied this theory of personal and cultural languages to a group of people who are trained to design and manage the built environment and a group of building users who have limited mobility. The goal was to identify common and discordant themes in each culture's perception of the environments they experience daily. The emerging themes as seen in the photographs highlighted the meaning and symbolism both groups assigned to the built

FIGURE 56.4   Retail environment showing approach and size problem.

environment. Status, discrimination, image, choice, and function were some of the themes that began to develop a pattern language for both groups. Identifying a pattern language for each group and then sharing those languages as Alexander suggests could increase awareness and acceptance of differing needs and ultimately define a language of universal design.

56.6   POE EXAMPLES FROM DESIGN FIRMS

In January of 1999 the San Francisco Public Library commissioned Ripley Architects and Florence Mason with Susan Kent to conduct a POE of the newly completed Main Library (San Francisco Public Library Web site, 1999). Although the library design was impressive and unique in reference to typical libraries of the past, it was not as functional in reality as in theory. The evaluation team identified more than 150 problem items, including concerns with way-finding, circulation, usable area, and equipment malfunctions. Specific problems included inadequate and difficult-to-install-and-read signage; inappropriate selection of flooring materials that made it difficult for staff to roll carts from one area to another; and dark and limited space for popular collection areas. All of these concerns could have been addressed early in the programming process if the Principles of Universal Design (Center for Universal Design, 1997) had been the basis for the design concept.

Routinely incorporating POE into the building delivery cycle and fee structure would decrease the number of avoidable problems and greatly influence the design of future buildings. McLaughlin (1997), director of design at Kaplan McLaughlin Diaz, an architecture and planning firm based in San Francisco, has been conducting POEs for almost 30 years and firmly believes that the impact of POE can be far-reaching: "Over the years, we've discovered that lessons learned from one project can help us even with different kinds of buildings. For example, studies we conducted of housing for the elderly have been applicable to the design

FIGURE 56.5 Retail environment showing aesthetic qualities.

of successful environments for the homeless, the mentally ill, and the incarcerated." There will always be complications and issues to address as society builds bigger, better, and smarter buildings. But owners, designers, and managers would see increased benefits if POE were simply a step in the building delivery cycle, and not just an option suggested by a consultant or requested by the client.

56.7 GENERAL LESSONS LEARNED FOR UNIVERSAL DESIGN AND POE CHALLENGES

Post-occupancy evaluation has not been used as a standard tool in the past. Today, with architecture and design firms expanding the scope of their services, more POEs are being completed. However, they are not often conducted unless there are problems evident to the building users. If POEs were incorporated into the design process as the last phase in the design/build cycle, problems in existing and future buildings could be decreased, with lessons learned communicated to design professionals in periodicals and at conferences.

McLaughlin (1997) stated that "Nearly everyone agrees that post-occupancy evaluations are immensely valuable . . . They benefit the public by contributing to a pool of research on architecture and building types. Simply put, they show us what works and what doesn't." McLaughlin believes that POEs are one tool that can help architects design better buildings, as they enable designers to look at a building through the eyes of the user. A review of POEs completed on existing buildings of similar use and size during the programming phase of the design process would positively enhance the final product.

Another challenge for post-occupancy evaluation is to increase the level of awareness and understanding of POE in design education. If students are routinely taught post-occupancy evaluation as a phase of the design process, when they become professional architects, designers, and facility managers they may be more inclined to plan and budget for a POE. Facility management, architecture, and interior design programs do introduce the concept of POE, but generally in upper division or capstone courses.

At North Dakota State University (NDSU) and Ferris State University, students conduct POEs on campus buildings and regional facilities. At NDSU, interdisciplinary teams composed of facility management, architecture, and interior design students conducted 11 POEs in the fall of 1999. Teams evaluated overall user satisfaction, accessibility, wayfinding, thermal comfort, and indoor air quality in new and renovated buildings such as Ehly Hall, the new architecture and engineering building; the NDSU Alumni Center, a new conference and office center; and Vista, the new research and development building on Great Plains Software's new corporate campus.

The challenge for design educators is to incorporate an understanding of how valuable POE can be early in the curriculum when students learn the design process. The last phase of the design process is evaluation, and although instructors discuss evaluation techniques and often have the students analyze problem environments, observe users in public and private space, and participate in accessible sensitivity training exercises, they rarely detail the process of post-occupancy evaluation. By aiding students in learning that evaluation is an ongoing process in the life span of a building, instructors can positively influence future generations of designers and managers, and by including universal design considerations in POE, designers and managers have the opportunity to strengthen and enhance the effect buildings have on all users.

Preiser (2000, 2001) states that "universal design evaluation (UDE) is the process of systematically comparing the actual performance of universally designed products, buildings, places, and systems with explicitly documented criteria for their expected performance." If designers, design educators, and managers expand the definition of POE to include evaluation of universally designed products and actively begin to address the needs of all users equally, society can ultimately move toward transformation of the built environment.

56.8 CONCLUSION: UNIVERSAL DESIGN AND POE IN THE DESIGN PROCESS

Designers and builders of the built environment commonly create spaces based on their personal perspectives of life, information culled from past projects, and various user groups. Universal design entails an all-encompassing and inclusive perspective that maximizes use and satisfaction with the built environment. Post-occupancy evaluations that incorporate universal design issues can be effective in identifying environmental attributes that improve life quality for all users.

When combined, post-occupancy evaluation and universal design are powerful tools in influencing the quality of the built environment. Individually, both can enhance the function and aesthetics of a building. But together they can begin to provide environments that are supportive, safe, appealing, and equal. By taking the time to conduct a POE from a universal design perspective, architects, interior designers, and facility managers are listening to the diverse needs of users, understanding those needs, and implementing designs that address those needs.

56.9   REFERENCES

Alexander, C., "A Pattern Language," in J. Lang, et al. (eds.), *Designing for Human Behavior,* Van Nostrand Reinhold, New York, 1974.

Bailey, K. D., *Methods of Social Research,* Free Press, New York, 1987.

Baird, G., et al. (eds.), *Building Evaluation Techniques,* McGraw-Hill, New York, 1996.

Center for Universal Design, *The Principles of Universal Design,* NC State University, Version 2.0—1 April 1997.

Corry, S., "Perceptions of Built Environment: Evaluating Environmental Qualities," *Proceedings, Designing For the 21st Century II,* E. Ostroff (ed.), Adaptive Environments Center, Boston; www.adaptenv.org/21century/proceedings.asp.

Haworth Furniture: Facility Resource Center-Facility Planning/Management, "Conducting a Post-Occupancy Evaluation," *Office Journal, 7;* www.haworth.com/resource/fpm/poea.htm.

Lang, J., et al. (eds.), *Designing for Human Behavior,* Dowden, Hutchinson & Ross, Inc., New York, 1974.

McLaughlin, H., "Post-Occupancy Evaluations: They Show Us What Works and What Doesn't," *Architectural Record,* 185(8):4, 1997.

Preiser, W. F. E., "POE Training Workshop and Prototype Testing at the Kaiser Permanente Medical Office Building in Mission Viejo, California, USA," in G. Baird, et al. (eds.), *Building Evaluation Techniques,* McGraw-Hill, London, UK, 1996.

———, "Post-Occupancy Evaluation: Conceptual Basis, Benefits and Uses," reprinted in J. M. Stein and K. F. Spreckelmeyer (eds.), *Classical Readings in Architecture,* McGraw-Hill, New York, 1998.

———, "Post-Occupancy Evaluation Training Workshop, University of Cincinnati," Cincinnati, OH, 1999; www.rware.demon.co.uk/poe.htm.

———, "Universal Design Evaluation." *Proceedings, Designing For the 21st Century II,* Ostroff, E. (ed), Adaptive Environments Center, Boston, 2000; www.adaptenv.org/21century/proceedings.asp.

———, H. Z. Rabinowitz, and E. T. White, *Post Occupancy Evaluation,* Van Nostrand Reinhold, New York, 1988.

San Francisco Public Library Commission, "Main Library Post-Occupancy Evaluation Summary," 1999; www.sfpl.lib.ca.us/www/poe.htm.

Strauss, S., *Renovation and Expansion of the Cooper-Hewitt National Design Museum, Smithsonian Institution,* Architect's Statement, Polshek and Partners, New York, 1997.

Strelitz, Z., "Post-Occupancy Evaluation," 1999; www.rware.demon.co.uk/poe.htm.

Tarricone, P., "The Power of POE," *Facilities Design and Management,* June 1999, 52–54.

Zeisel, J., *Inquiry by Design: Tools for Environment-Behavior Research,* Cambridge University Press, New York, 1981.

56.10   RESOURCES:

American Institute of Architects, *Redefining the Architecture Profession* (video), 1997.

CHAPTER 57

UNIVERSAL DESIGN EVALUATION IN BRAZIL: DEVELOPMENT OF RATING SCALES[1]

Marcelo Pinto Guimarães, M.Arch.
Federal University of Minas Gerais, Belo Horizonte, Brazil

57.1 INTRODUCTION

In this chapter, readers will find the conceptual basis for the Rating Scale of Inclusive Design. Some of these concepts were generated by considering and experiencing Brazilian cultural and environmental circumstances. At the research center of the School of Architecture of the Universidade Federal de Minas Gerais (ADAPTSE, EA-UFMG), Brazil, a case study evaluating accessibility to an existing building demonstrates the application of these rating scales that incorporate universal design in architecture. Universal design is the ultimate outcome of critical design guidelines in the rating scale that describes environmental conditions for use of architectural elements in an inclusive society. The list has an evolutionary stance that links main conceptual definitions to current design practice. The result is an assessment tool for inclusive design that tests design practice and addresses consumers' needs, as far as better cities, buildings, objects, and services are concerned.

57.2 BACKGROUND

Broad use of technology does not assure the existence of adequate responses to social needs. In fact, it is far more important for social and moral values to reflect the cultural context in which technology has to be applied. Otherwise, that technology cannot flourish and result in better quality of life. The main issue is how to develop culturally appropriate design solutions worldwide that can reduce environmental misfit in regarding the needs of people with permanent or temporary disabilities. It is also important to expand awareness about the need to adapt the built environment by use of good design solutions that accommodate everyone's physical abilities and performance (Fig. 57.1).

In Brazil, as in other developing countries, studies of inclusive design began only a few years ago. Therefore, such studies are very different from the experience of other countries, which, since World War II, have been searching for information about barrier-free environments (United Nations, 1986). It is not just a matter of time, but that differential constitutes a

FIGURE 57.1 Technology is just one aspect of different world views about inclusive design.

huge cultural gap about professional respect for user needs as part of designers' challenge of creating better environments.

In Brazil, there is a lot of resistance to the adoption of accessibility design criteria despite strong legislation. The Brazilian Constitution Law of 1988 states clearly that accessibility to buildings, transportation, and communication is one of the basic civil rights. Therefore, there is legislation at federal, state, and city levels that establishes the adoption of Brazilian technical standards (ABNT, 1994) as minimum criteria on ergonomics. Nevertheless, architects, builders, engineers, and clients have not yet explored the development of good inclusive design, since consumers do not regard this worthy of major investments. Consequently, Brazilian standards for accessible buildings and transportation have not led to many alternative solutions to the needs of those who experience permanent or temporary disability problems.

Countries like Brazil have a great but, as yet, unrealized potential to transform existing buildings into examples of good design that really responds to users' needs, despite permanent or temporary disabilities. Many buildings and cities still need to be constructed, while others will soon replace older buildings with better technology. However, administrators, building professionals, and the general population are not prepared to accept accessibility as a concern that promotes quality of life for future generations. Scattered initiatives for accessible design in a few Brazilian cities provide some examples. There is a good design approach in Curitiba for surface public transportation (see Chap. 24, "Universal Design in Transportation," by Steinfeld, and Chap. 69, "Creating the Universally Designed City: Prospects for the New Century," by Weisman). Serious design experiments for tactile sidewalks and parks can be found in Rio de Janeiro. Strong legislation for accessibility has concentrated primarily on adapting buildings for public services in São Paulo. A proposal for informal barrier-free parking next to building entrances has been made in Belo Horizonte. Unfortunately, many outcomes of similar proposals are so inadequate and temporary, that they do not reveal even minimal consideration of the principles and guidelines for universal design.

Thus, the development of systematic approaches to inclusive design is a foreign concept and requirement that has reached the country as an exotic idea. Fortunately, one of the good aspects of Brazilian culture is to accept influences of other cultures, and to develop those influences in such a way as to create innovative Brazilian solutions (Ribeiro, 1995). Today, adaptable design ideas for independent living are rare high-tech solutions seen only in books and magazines. They look so specific and complex that people believe they suit only the needs of wheelchair users and no one else's (Aino et al., 1978; Steinfeld, 1979a,c; EGM et al., 1986). Thus, there is a strong need for designers in Brazil to adopt a conceptual framework that considers a continuous process of adaptation, from poor accessibility to people with mobility problems to full accessibility to everyone (Guimarães, 1999a). Once these concepts are promoted, consumers will demand universal design for environmental settings and for better quality of life. Consequently, there will be great opportunities for Brazilian experiences to be rich in creativity and originality.

Inclusive design represents the entire process of designing for the needs of everyone in an inclusive society. It is a cultural outcome that may be expressed differently around the world. The adoption of inclusive design is the best indication of a general awareness that there is a misfit between a person's abilities and environmental resources for meaningful social interaction (Steinfeld et al., 1977). In a list of environmental conditions that may contain many possible alternatives for full accessibility, universal design ideas are at the top of a new scale of environmental quality assessment, the Rating Scale for Inclusive Design. As designers reach this level, their ideas must take into account different needs simultaneously. All of these needs must be considered, no matter whether a person is part of the aged population or is a small child; no matter whether a woman is pregnant or is oversized; no matter who has an injured limb or lives with permanent disabilities, is blind, deaf, or has minor cognitive disabilities.

Some of these universal design ideas grew out of the Brazilian cultural and environmental circumstances. There are similarities to some aspects of practicing design for accessibility in other developing countries. However, the rule of thumb is for readers never to expect that any specific rationale for alternative solutions used in Brazilian cities can be directly applied to problems of other cities and nations.

57.3   *LINKING EVOLUTIONARY CONCEPTS TO RATING SCALES*

The conceptual basis for the Rating Scale for Inclusive Design is presented here. Its structure contains the evolutionary development of major principles and guidelines for inclusive design arranged in a continuum of concepts. Also, there are slightly modified meanings of some generally accepted ideas, so that the conceptual framework functions as an integrated system (Fig. 57.2).

Inclusive Design Beyond Mainstreaming

In practice, results of inclusive design can provide a degree of accessibility in a wide range of alternatives that include stressful settings on one hand, and, on the other, contain conditions

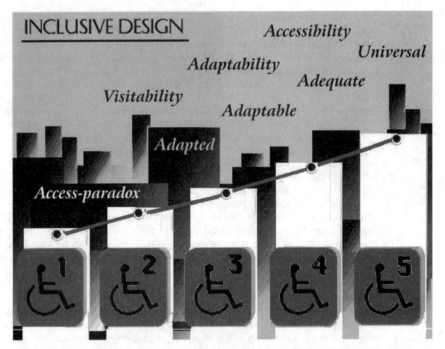

FIGURE 57.2   Conceptual definitions can be arranged in an evolutionary line of development.

supportive of independent performance of social roles (Steinfeld et al., 1977). Inclusive design goes beyond complementary approaches that respond to the needs of people with disabilities, such as aiming at normalization, which "mainstreams" people with disabilities, only to have them end up as a minority group in society. Both normalization and mainstreaming were important concepts in the late 1970s and early 1980s and were responses to institutional treatment and segregated settings for people with disabilities. Their intent was to integrate people with disabilities into the mainstream activities of "normal" community life.

Mainstreaming takes the consideration of special needs into the realm of the needs of the general population. However, it failed to really measure the extent of differences that are inside those groups, the minority and the general population. Accessibility that comes from the mainstreaming approach does not really address a different way of designing things so that the results provide a new, unprecedented architecture. In short, mainstreaming can only be effective if it enables designers to consider special needs in many behavioral conditions that also involve people with disabilities performing different social roles, such as the boss, the employee, the customer, the father, the care provider, and so on.

According to these definitions, accessibility means much more than providing physical access for people with mobility impairments. It is a natural consequence of designers' actions for solving a wide range of disability problems, which means that it constitutes a minimum level of requirements for everyone's full satisfaction and convenience of use of environments.

At the lower end of the Rating Scale for Inclusive Design, there are some environmental situations that illustrate the rehabilitation approach of the medical model (Guimarães, 1991a). Basically, the focus in the medical model is on user efforts for personal adaptation. It does not include the adaptability of design ideas for a variety of human performances. In some cases, adaptability is just an irrelevant aspect to be considered, when compared with others, like aesthetics or economic factors.

The rehabilitation approach makes laypeople believe that there is always a tradeoff decision in the process of creating special accessible settings for people with disabilities. It assumes that the outcome may be of lesser importance than that of designing for people with no apparent disabilities. The general public also believes that only minimum investments should be made in the environment for compliance with codes and legislation (Steinfeld, 1979c). However, technology for personal assistive devices may soon enable users with severe disabilities to reach reasonable performance at standard levels. Without public investment, full accessibility for people with severe disabilities becomes a private investment, a personal struggle to obtain affordable high-tech devices.

Access Paradox

The conceptual definition of the *access paradox,* a term coined by Guimarães in 1991, illustrates the controversy that exists when someone decides to invest only a minimum in environmental resources for accessibility to accomplish the most in terms of financial savings or revenues. The access paradox is the result of minimal design ideas that do not take into consideration the best alternatives for users to perform tasks actively and independently. In such cases, people with disabilities, who have enough self-determination to perform daily activities on their own, are unable to do so and must face circumstances in which the decision to control environmental conditions depends on caregivers, managers, or security or maintenance staff.

Some examples of access paradoxes can be seen in the light-rail train stations at Buffalo's mass transportation system in the state of New York. The author demonstrated how the design of Buffalo's train cars and platform systems represented stigmatizing settings for active people with disabilities (Proshansky et al., 1970). The design decision to adopt elevated miniplatforms in the downtown area, for instance, interfered with the location of foldable seats, which are prioritized for use by people with mobility problems at both ends of light-rail cars. Those miniplatforms could only serve the first door of the first car. Therefore, users with disabilities could not choose any other place to sit, and thus, sometimes suffered from attitudinal and environmental barriers in the metro system. Also, platform waiting areas that are far from train stops prevent users with mobility problems from catching trains in time for departure.

Visitability and Limited Options for Accessibility

Both the European concept of *visitability* as well as the reserved percentage of adapted resources are related ideas that provide accessibility to people with mobility impairments in predictable, but limited, ways. Other definitions and practices of visitability follow. In Europe, visitability is a partial approach to accessibility (Zuylen, 1996). It solves the issue about how to apply small investments in existing environments in order to achieve some accessibility in a short time and at low cost. It is often limited to a few public areas where visitors are encouraged to use facilities, while there are other aspects of buildings, vehicles, or equipment that do not have any means for accessibility for people using wheelchairs.

Promoting visitability as the only way to provide accessibility for people with disabilities is not a good idea. Unfortunately, it is commonly understood as a method to follow in developing countries like Brazil. As a generally accepted definition, it becomes an access paradox. Designers and their clients do not wish to invest in renovation work for public environments. Thus, visitors may have restricted access by means of removable ramps, for instance, as a less expensive alternative to the construction of permanent accessible elements. Also, there may be few existing accessible routes for clients, and in private areas for employees there are no accessibility elements at all. This emphasis on applications of accessibility in public areas only reinforces discriminatory attitudes against hiring people with disabilities. They may gain respect as clients, but they may never have a chance to become ordinary workers anywhere.

In the United Kingdom (see Chap. 14, "Lifetime Homes: Achieving Accessibility for All," by Best) and the United States, there is a growing movement toward the concept of visitability in privately owned housing as promoted by Concrete Change (Smith, 1991). In the United States, this modest approach refers to zero-step entrances and interior doors that are 30 in clear (see Chap. 43 "Universal Design Practice in the United States," by Ostroff). It allows people with mobility limitations, especially those that involve wheelchair use, to enter, move through a home, and access the toilet.

The Differences Between Adaptable, Adapted, and Adequate Design

The application of early requirements for accessibility in multifamily residential units, for instance, often resulted in two problematic situations (Mace et al., 1987). First, the established percentage of required fully accessible units may or may not have matched the need in that particular local area. In some places, wheelchair-accessible units were kept empty because there were not as many wheelchair users as expected. In other places, wheelchair users were not able to find accessible apartments, because other people with no apparent disabilities were greater in number and requested the available units. In the second situation, the numbers of unused reserved accessible units may increase over time. These problems were part of the motivation in the United States that led to the passage of the Fair Housing Amendments Act in 1988 and the requirements for some limited accessibility in all new multifamily units (see Chap. 35, "Fair Housing: Toward Universal Design in Multifamily Housing," by Steinfeld). The intent was to provide both maximum choice as well as marketable units with no stigmatizing features. This also led to the growing movement for visitability in single-family homes. Thus, both the minimally accessible Fair Housing units and the visitable homes had the potential to provide vastly increased residential options. As the population of wheelchair users grows higher, for example, some easy adaptations in such homes could further increase accessibility to meet user needs.

Design ideas that relate to these specific, listed environmental conditions represent another stage of inclusive design. As design solutions present intelligent applications of alternatives to user performance with greater environmental adaptability, accessibility certainly gets better. More people will benefit from it.

Unfortunately, people have tended to mix up the meanings and usage of words, such as *adapted, adaptable,* and *adequate settings,* as they also misunderstood the goals of good inclusive design.

- *Adapted.* In adapted settings, the original design idea is not suitable to respond to everyone's needs. As renovation work takes place, some superficial or even profound changes may occur to achieve compliance with accessibility regulations. The result is completely different from the original environment, the ambience of which cannot be recovered anymore.

- *Adaptable.* In adaptable settings, the original idea already contains arrangements for the development of environmental resources that provide accessibility to people with disabling conditions. The concept of adaptable environments that came about for multifamily housing is also applicable to other types of buildings (Beasley and Davies, Jr., 1988). These arrangements may include enough space for reach and maneuvering tasks, for example. Furthermore, they may offer enough flexibility through assistive devices as accessories for users to explore different functions. Finally, adaptable design of settings may help users to adopt different alternatives for achieving specific goals. In the end, the design idea is reversible to the initial configuration with little effort or personal investment.

- *Adequate.* In adequate settings, the conceptual definition takes into consideration the idea of full adaptability (Barrier Free Environments, Inc., 1984). Adequate design, contrary to simple adaptable design, expresses design ideas that accommodate, simultaneously, different alternatives in the same setting. Nevertheless, every choice in different environmental settings is possible because alternatives of fixtures are there for the users' convenience (Aslaksen, 1997). For instance, one may find stairs parallel to ramps or even stairways close to elevators, all reachable in short distances. Washrooms with adequate design provide more than just a percentile of wheelchair-accessible toilet seats. A roomy toilet stall for side transfer may be available along with long stalls that serve anyone who can make front transfers. Equipment and wall-mounted fixtures at two heights are preferred to those that are at a single, median height.

Benefits of Adequate Design

Adequate design means that environmental settings contain different alternatives for user performance despite temporary or permanent disabilities. It is a natural result of design decisions that develop environmental settings for an inclusive society. Inclusive design refers to the process of considering a number of environmental conditions that respond to the diversity of user needs. Contained in this process are links between concepts that are the basis for universal design. As far as weak links are concerned, so-called access paradoxes exist along with conflicting aspects of the medical model. In addition, while some environmental settings contain a percentage of facilities reserved for accessibility, changes in design configurations can accommodate unexpected needs. That way, they may be the opportunity for a larger number of people with special needs to experience full accessibility, often resulting in greater adaptability of design solutions.

At certain levels of the Rating Scale for Inclusive Design (e.g., a city, a building, or an object), there may be adapted settings, adaptable configurations, or adequate designs for an inclusive society. Adaptable settings are preferred to adapted settings, which do not offer flexible adjustments to various human factors, such as space requirements for moving and reaching tasks, reaction time, and perception. Adequate design offers the most simultaneous alternatives to environmental settings and users' choices. Furthermore, the best design ideas are the ones that suit the diversity of people's abilities, and which require no special arrangements.

Unfortunately, there are very few examples of good, inclusive design at the scale of cities and buildings. Therefore, universal design is a utopian target at the top of the scale, and it represents the goal of improving the entire process. As experiments of inclusive design result in unprecedented responses to user needs, others will surpass existing ideas that designers believe to be universal today. Then, the target quality for universal design will be at even higher levels.

57.4   THE RATING SCALE FOR INCLUSIVE DESIGN

In this section, the rating scales and their use are introduced. First, an overview is presented about the use of the grading scale of inclusive design and how it helps designers organize information for decision making. The grading scale is then applied in a single building (e.g., an elementary school in Belo Horizonte: the Escola Estadual Bueno Brandão, or the BB School). Later in this chapter, several important issues about accessibility in transportation, parking facilities, building entrances, accessible routes, and historical sites are addressed through concrete examples. They illustrate the concepts of inclusive design that serve to create integrated solutions in complex contemporary cities.

Structuring the Flow of Information

The grading scale has a list of building elements that is organized in such a way that early design decisions direct designers to explore different sections of the list. The hierarchical structure for the flow of such information organizes environmental conditions in a treelike configuration that is upside down. Each section contains a number of alternatives. Some alternatives provide conditions that are complementary to one another, while other options preclude later selections. At the main upper end of the scale, there are guidelines about the accessible route, program configuration, human factors, and communication systems. All of these determine general conditions for the quality of environmental settings. Detailed aspects of such conditions may vary according to the designers' decisions. Thus, it is very important that the designer understand basic principles of inclusive design so that he or she can adopt good design decisions, which, in turn, increase the number of important scale items with high scores.

Using the checklist is quite simple, although the design solutions constitute a long list of nearly 420 items. As the designer gets involved in the decision-making process of creating a building element, he or she checks different items in a very personal way. In the end, some items on the list will be deleted, as the designer concludes work on specific programs that may vary from one building to another. As the designer chooses an item on the list with a marking of access grade 1, the score of 1 is placed into a summary. Designers' choices for access grade 2 or 3, respectively, make the scores of 2 or 3 to be part of additional summaries.

As someone reviews the priority order in the scale at higher levels, he or she will recognize better design solutions that may satisfy a variety of needs. The priority order is defined from 0 through 5. Access grade 0 means there are enough architectural barriers that inhibit activities of many people with different mobility, perception, and cognitive disabilities. Then, further information emphasizes better opportunities of accessibility for people who experience severe mobility challenges in existing public environments. Furthermore, the list of environment solutions also includes the needs of people with perception and cognitive problems. Last, the scale of inclusive design reaches access grade 5. This means that solutions fulfill the needs of everyone, and thus, they can be considered universal design. At this level, design solutions will be addressed that may have demonstrated priority attention to lower-grade requirements, as well as those that include the awareness of the simplest problems that may affect people slightly, but still in a very important way.

In the end, the final list of checked items would lead to the construction of a chart with columns that list the percentage distribution of summaries of items. Then, readers may relate the profile of specific column height distributions as the character of building conditions in the overall rating scale at the urban level.

Universal Design Evaluation of an Elementary School

The BB School is an existing building that occupies an entire block in the downtown area of Belo Horizonte, as seen in Fig. 57.3. The site is shaped like a triangle, and the main entrance is located at the vertex of that triangle, which is opposite Getulio Vargas Avenue. There are two other entrances: One is at the avenue side serving the staff parking area; the other entrance is for exclusive use of the administration's staff, teachers, and visitors. Also, the one serving the parking area is the only way for students who are wheelchair users to access the classrooms on the second floor. Other students use the main entrance gate that is at the first-floor level. There, the playground connects open areas to the sheltered plaza and the stairs.

Architecture students of the research center of the ADAPTSE, EA-UFMG used the Rating Scale for Inclusive Design to examine the building and to highlight major problems. In addition to the lack of a ramp connecting the two floor levels of the building, other barely noticed problems were also identified in the checklist. A total of 413 problem items were noted in the list. Some other items were left unchecked since they related only to other aspects of buildings (e.g., bathtub specifications). The percentage distribution showed a concentration of 54.6 percent of environmental conditions that are out of compliance with specifications according to the Brazilian Standards for Accessibility and fire escape safety. These included sharp edges of windows with low awnings that were open to corridors, stairs without handrails, cracked and slippery floor surfaces, bathroom floors that were raised one step

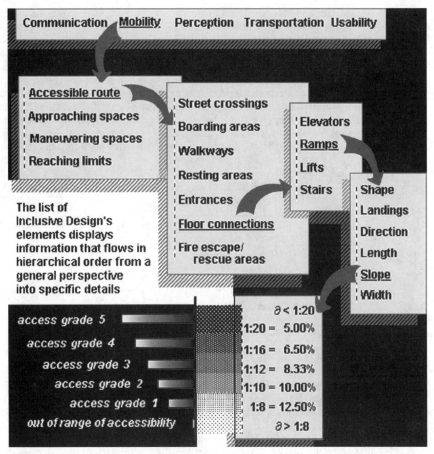

FIGURE 57.3 This chart demonstrates how information is organized in a list of inclusive design elements.

higher than corridors, and so on. Other percentages showed simple solutions such as lever handles for wide doors, blackboards at comfortable reaching heights, and so forth. These met the needs of many people with mobility problems. There were a small number of items on the list that described good overall accessibility conditions. Therefore, the distribution of access grades 1, 2, 3, 4, and 5 were, respectively, 11.7, 2.9, 14.5, 6.7, and 9.6 percent.

After discussions, the architecture students' team decided to adopt design solutions to correct the poor access grades, as well as to eliminate simultaneously the conditions that were out of compliance with official specifications. Fortunately, there was enough space in the building for construction of a ramp that can link the floor levels at comfortable slope, width, and length specifications. That solved the big problem of providing an accessible route for the entire building. With this solution, students and parents who use wheelchairs were able to participate fully in all activities and celebrations of the school.

The students' design did not include any proposals to the school's administration for new spaces and activities. Therefore, a number of design alternatives with high access grade numbers were unable to be used because they required renovation using sophisticated and costly technology.

The students' proposals led to significant changes. As a result, the profile of the percentage distribution of access grades in the checklist was revised in response to the students' proposals. Very few environmental conditions could not fit into the set of proposed requirements (5.3 percent) in the grading scale of accessibility. Those were related to the long distance to bus stops and taxis on the accessible route; lack of resting areas on the sidewalk; lack of sheltered entrances to the building; and finally, steep slopes of surrounding streets, which reduce the ability of pedestrians to travel short distances. These poor conditions prevented students from eliminating all architectural barriers in their proposal. In the outcome, the scored results of the building's assessment were as follows: access grade 1 = 4.8 percent; access grade 2 = 12.2 percent; access grade 3 = 24.8 percent; access grade 4 = 31.2 percent, and access grade 5 = 21.7 percent. The comparison of the results of the rating scale with the initial survey are shown in Table 57.1.

TABLE 57.1   Comparison of Grading Scales Before and After Student Proposals

Environmental conditions	Initial analysis (%)	After proposals (%)
Unsuitable to users' access needs	54.6	5.3
Access grade 1	11.7	4.8
Access grade 2	2.9	12.2
Access grade 3	14.5	24.8
Access grade 4	6.7	31.2
Access grade 5	9.6	21.7

An important observation is that these figures are related to the removal of environmental conditions that did not comply with standards. Other environmental improvements still can alter the distribution of access grade percentages. Thus, some items that describe better quality of solutions can replace existing ones that show low scores. For instance, a water fountain that is operated by hand lever can be replaced by another type with automated infrared operation. Such a qualitative change of a feature can improve scores in the checklist from access grade 3 to access grade 5. Accordingly, the percentage distribution can be different. The importance of those changes is for fine-tuning the profile of the percentage distribution of scores, so that it can be compared with similar ones in a series of patterned sets of access grades. Those patterned sets will give an overall score to the building in terms of accessibility, and the building will be considered as being access grade 1, 2, 3, and so on.

Finally, the use of the grading scale of inclusive design in this study of the elementary school resulted in the overall score for the building in terms of accessibility to be about access grade 3, as seen in Fig. 57.4. Therefore, the existing building has the potential for improved accessibility, especially if investments by the city administration will promote the removal of barriers over time.

57.5   TOWARD MORE EFFICIENT, ACCESSIBLE ROUTES

The least effective design approach to accessibility occurs when the varied needs of users with disabilities are considered after the design concept is already developed and is in use and works well for some people. In such cases, several problems may interfere with form and function, and designers may begin to take out desirable features in public settings (Sommer, 1969), resulting in discrimination on the basis of the users' disabilities. Then, the built environment cannot evolve into better solutions for universal design. The accessible route is a key concept that must be developed at the earliest design stage.

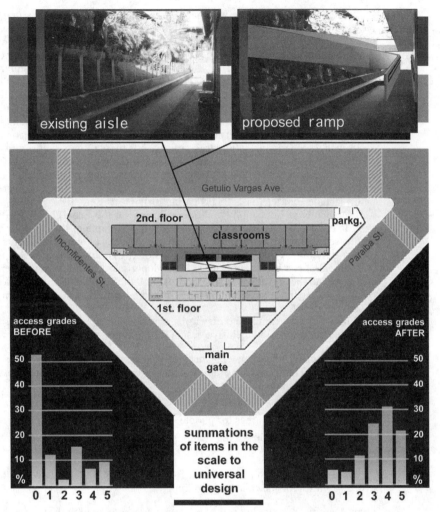

FIGURE 57.4   Different profiles of ratings of the BB School before and after the students' proposal.

A good starting point for assessing a concept for accessible building design is to break down the large size of the building into smaller areas at a more human scale (Silverstein and Jacobson, 1978). This consideration is very important, because it relates to three of the seven Principles of Universal Design (see Chap. 10, "The Principles of Universal Design," by Story):

Principle 3—*Intuitive Use.*   As the size of the building is not so large, it is easier for users to understand how to reach different areas of the building.

Principle 6—*Low Physical Effort.*   The accessible route is not too long, and, therefore, people with walking impairments will not get exhausted when going from one area to another.

Principle 2—*Flexibility in Use.*   Finally, accessible public services rely on respect for customers as individuals by better understanding particular aspects of user needs, resulting in better customized service.

In Belo Horizonte, there are many large buildings and each has unique characteristics. However, those with multiple accessibility characteristics in different areas have higher potential for satisfaction and convenience for users with permanent or temporary disabilities.

The Mercado Central is an old building, which has functioned as a typical downtown mall since late 1940s. The layout of the building is symmetrical and square. There are five entrances, and one of them is adjacent to where visitors can park their cars in the parking area that is on the second floor. At street level, the other four entrances lead to a central circulation area, which features handcrafted goods in shops in a maze of corridors. As tourists visit the building, they may feel lost for some time, because there is no familiar reference or clue to where the exits are. In such situation, potential accessible routes become meaningless.

In such a scenario, a useful design approach is to work with colors and texture in order to provide users a unique reference to each area in the mall plan. The use of four different colors in banners and structural elements, for example, may enhance users' orientation so they can find their location in the building. Also, textures in the floor can certainly help people with visual disabilities to get their bearings using tactile clues.

The Faculty of Philosophy and Social Science at the Federal University of Minas Gerais (FAFICH—UFMG) also presents an interesting example of the need for considering reducing the scale of buildings (Guimarães, 1998a). There are four departments that share the use of one building: (1) the department of history, (2) the department of psychology, (3) the department of communication, and (4) the department of philosophy. There is only one main entrance, and there are secondary ones that link the FAFICH to other university buildings. In the four-story building, circulation takes up more than 1200 m^2 at approximately 10 percent of construction rate.

While there are ramps and elevators, as well as independent entrances, the building complex does not offer good inclusive design for the following reasons:

- The lack of independent entrances for each school results in long corridors with poor wayfinding. That is a structural problem for accessibility that should have been resolved at the early design stage.
- The accessible route is too long.
- There are no resting areas for people who need to walk to distant classrooms.
- The structural modules of the building design make it very difficult for newcomers to find the exit.

Some adaptations to the design of FAFICH have not made it accessible for many people with disabilities for the following reasons:

- For example, reserved parking spaces for cars of people with disabilities are far from the front of the building.
- There are reserved washrooms for wheelchair users who must request a key before using them.
- Identification plates that have both raised letters and Braille can be found next to some but not all classroom doors.
- There are elevators with high, hard-to-reach control panels.

57.6 ACCESSIBLE PARKING FACILITIES

An inclusive society must provide accessible parking facilities wherever people with mobility problems need to go. As the population grows, one may think this assumption is false. It does not really benefit everyone since the high number of vehicles already overloads streets and open areas. However, an automobile is certainly a very important means of personal transportation for people with disabilities. It may be considered a natural extension of the body for legs that do not function properly. Otherwise, people with mobility limitations may not be able to get to where they need to be performing active social roles. Thus, conflicting perspectives arise for a policy that implements accessible parking spaces in the city. Here is an example of the application of the inclusive design model in Belo Horizonte, Brazil.

Belo Horizonte is a hilly city. In the downtown area, there are regular blocks shaped like squares and triangles as street corners meet avenues at 45° angles. In the suburbs, street blocks are irregular, but the overall design of the urban area does not address the problems that result from topographic and geological factors. In some districts, there are streets with very steep inclines. Strong pressure for increased construction results in a very dense concentration of inhabitants who live in hilly areas that should have been preserved as natural parks. Thus, the density and location of many multifamily residential and official buildings for public services does not provide easy access from sidewalks as there are not enough parking spaces in the crowded streets.

The policy for reserved accessible parking spaces in the streets of Belo Horizonte does not seem to work properly. Although Brazilian standards for accessibility contain instructions on how to design such spaces, the steep slope of sidewalks is a strong barrier for pedestrians and the ability to make wheelchair transfers in and out of cars. Also, there is little official effort to create new ramps and curb cuts for reserved spaces. Besides, people with no apparent disabilities always seem to disregard signs that should prevent illegal parking in spaces reserved for people with mobility impairments. In buildings that have indoor parking, there are no extra parking spaces for visitors' cars.

Over 4 years of relative success, an interesting and very flexible program was developed for priority parking for the registered cars of people with disabilities in Belo Horizonte (Guimarães, 1997). In that program, called "Action for Access," positive attitudes toward the needs of people with mobility impairments replaced the practice of strict enforcement of laws and transit codes that regulate parking for everyone. Since there are no special signs for reserved parking, people who carry a badge that certifies mobility impairments may park as close as possible to the entrances of buildings. The fine for parking in prohibited spaces can be waived by arrangements between badge owners who may be drivers or passengers, and parking inspectors. Both consider parking conditions good, if they do not interfere with pedestrians and traffic flows. Furthermore, technical training instructs police officers and other parking inspectors to offer help if people with disabilities require assistance. Evaluations of the "Action for Access" program have demonstrated that sometimes people with disabilities prefer to park in certain places that do not cause any problems. Some people park on wide sidewalks where there is enough space for pedestrians passing by. That way, drivers with disabilities may be able to get assistance from door attendants of the buildings. Others prefer to park in spaces designed for taxicabs, or in spaces for trucks that unload goods according to certain schedules. In such situations, parking inspectors can interfere by solving conflicts and by protecting the rights and needs of people with disabilities.

Currently, changes in city legislation state that new buildings must offer extra parking spaces for visitors with disabilities. However, that seems to be not enough as the number of vehicles is growing quickly. This example on striving for positive attitudes demonstrates that, in order to avoid prejudice, the conceptual framework for inclusive design depends very much on human behavior. It may also be a good example of universal design. As criteria for selecting people who deserve the "Action for Access" badge become more comprehensive, they may include people with temporary disabling conditions, such as those of 7-month pregnant women, or problems of people who are elderly and those with injuries.

57.7 TRANSPORTATION BOARDING AREAS LINK ACCESSIBLE BUILDINGS

Accessible buildings are only fully usable if they have accessible connections to one another and to an integrated system of urban environmental resources. One cannot consider a building accessible when there is no means for users to access transportation. Public light-rail trains, buses, private cars, taxicabs, vans, motor-tricycles, scooters, and motorized wheelchairs are the vehicles usually considered for daily activities. Planes, helicopters, ships, and other means of transportation are also to be considered on the list of the rating scale for inclusive design. However, the unique configurations of each require that designers must adjust guideline specifications in each case.

The main issue in relating transportation to buildings is the provision of accessible boarding areas, such as train platforms, bus stops, drop-off zones, canopies, and reserved parking spaces. These areas must be connected to the accessible route in buildings. In addition, accessibility to vehicles may cause conflicts between pedestrians and automobiles at intercity road services. They are all general concerns that are not part of any specific section on the checklist for inclusive design (Fig. 57.5).

FIGURE 57.5 Buildings' entrances in hilly areas provide access to a car's passenger as well as pedestrians.

Therefore, on the checklist of the Rating Scale for Inclusive Design, alternatives for accessible boarding areas have priority ratings from 1 through 5. For instance, the characteristics of a train platform as well as a drop-off zone may not include level horizontal and vertical gaps between the platform (or drop-zone drainage gratings) and the vehicle floor (access grade 1). There may be no bus service (access grade 1) and no flat floors between drop-off zones and the accessible entrance (access grade 2). Bus stops may be located across the street from the building (access grade 2). At drop-off zones, there may not be enough space for both the driver and the passenger to transfer sideways from a wheelchair (access grade 3). Also, bus stops may not have rain shelters (access grade 2), benches, or drop-off zones (access grade

3). Reserved parking spaces may be far from an accessible entrance to buildings (access grade 2), and the number of reserved spaces may sometimes be less than required (access grade 1).

In listing good solutions, however, the checklist displays a large number of high ratings for these conditions that vary up to access grades 4 and 5.

An easy solution to the design of an accessible building is in the case where the building area is situated on flat terrain. Bus service may be available, and bus stops can be far as there are resting areas on the sidewalk. The bus stop will offer shelter for rainy- and cold-weather conditions. If the bus stop floor is raised up to 1 ft, low-floor buses can approach quite easily. A difficult situation is for the building location to be on a hilly street where there may be no accessible route on sidewalks that connects the building to the streets. Unfortunately, this occurs too often in many Brazilian cities like Belo Horizonte.

It is possible that a building on a hilly street can be made very accessible, since accessible entrances can be provided at the level of driveways that link the building to transportation boarding areas. Driveways to the main lobby and passenger drop-off areas should be provided so that vans and small cars can stop on a flat landing. Preferably, a passenger drop-off area can be provided under the building canopy, and as close as possible to the accessible front entrances.

In public buildings that are a tourist attraction (e.g., museums, monuments, etc.), designers must plan for a wide driveway, so that bus services can be provided in open areas in front of public buildings as seen in Fig. 57.6.

Oscar Niemeyer's Art Museum of Pampulha, Belo Horizonte's Cassino, is an example of the aforementioned requirement. It is located at the upper landing of a large site that provides a nice view of the Pampulha Lake. A wide cul-de-sac is situated in front of the building, and a small canopy protects visitors as they get off tour buses or vans.

FIGURE 57.6 The driveway leads to the front entrance of the Art Museum of Pampulha.

In existing buildings, adaptation is sometimes possible when the driveway to the covered parking also provides enough space for visitors' cars to park near the main entrance. However, that does not occur very often. Generally, designers and their clients prefer to adopt a design solution that emphasizes the security of steps and steel fences as opposed to accessibility at street level. On sloping streets, the main entrance can be located up to one floor higher than the street level. Only the secondary entrance for covered parking inside buildings can be at the lower level. Therefore, the main entrance is usually situated in connection with slippery inclined sidewalks for pedestrians and ordinary visitors. In such cases, car parking close to the secondary entrances is the only way for people with mobility impairments to enter the building.

One interesting proposal for the adaptation of an existing building suggested a new location of the main entrance at street level. This occurred in a residential building that is located in a calm district of Belo Horizonte. That way, both cars and people could share a same gate as seen in Fig. 57.7.

The prisonlike image of the wall for the parking area was to be changed to a more attractive and open configuration. Concerns about prevention of assaults on drivers and visitors were dismissed, due to better visual control over pedestrians. The move of the main entrance was possible because the level difference between the street and the parking area was suitable for the addition of an accessible ramp. Also, the trash room had to be moved to the higher

FIGURE 57.7 Before and after views show changes for easy access to a new, main entrance.

floor in order to provide enough space for the glass walls of the new lobby. Finally, the new configuration included a flat landing after the ramp, and created a passenger drop-off area in front of the new lobby. In the end, the parking entrance offered more attractive access to the building, and the previous lobby at the upper floor of the building was changed into a large room for private parties and entertainment.

Small properties in urban hilly areas of Belo Horizonte usually do not offer enough space for such driveways, canopies, and passenger drop-off zones. Nevertheless, the city administration can promote greater impact for inclusive design by the use of incentives. For example, changing legislation to allow increases in the amount of construction would result in higher revenue to investors who finance accessibility elements like ramps, large bathrooms, and elevators.

57.8   SERVICE LINES IN PUBLIC WAITING AREAS

Bank cashiers, ticket booths, and public offices usually do not provide services upon request. In Brazil, although legislation regulates the minimum time for cashiers to provide assistance, clients who wish to get information only often have to wait in long lines. Through that procedure, according to the Brazilian Constitution, pregnant women, people with disabilities, and the elderly people over age 65 must get priority attention. However, during rush hour, crowds of people may stand in line for hours. When someone comes and has visible disabilities, there is usually not a big problem. Other people who are in line may offer their place. Problems exist when elderly, but able-bodied, people wish to get the same attention, resulting in conflicts and requiring intervention of security staff.

Recently, a nationwide regulation for banks forced the establishment of special cashiers to provide customer service to "special" people. Besides creating a segregated solution, this resulted in a parallel line to those for "ordinary", "normal" people. The parallel line for people with mobility impairments sometimes takes more time to move than ordinary lines, since some of those "special" clients require more attention.

An alternative idea that illustrates an inclusive design model consists of setting up seating for waiting areas in front of accessible low counters, where cashiers call clients in order of their arrival (Guimarães, 1998b). This allows people who need this option to wait comfortably, whether they use a crutch, a cane, a wheelchair, or whether they cannot stand up for a long time. In summary, bank service areas need to be changed into more supportive environments, which is good from the user perspective.

57.9   ACCESSIBILITY AND HISTORICAL PRESERVATION

Preserving architectural sites of historical interest frequently conflicts with practices based on the conceptual framework for inclusive design. Usually, notions about providing minimum accessibility for the population with disabilities prevail over attempts to develop universal design solutions.

A proposal for adaptation of the entry facade of the City Hall in Belo Horizonte illustrates the conflict at the City Council for Historical Preservation (Guimarães, 1999b). The front doors of the building are eight steps above street level. The renovation work consists of installing a vertical lift outside the building, in combination with special arrangements for an accessible parking lot. The controversy focuses on the addition of new extraneous elements to the original building for it to become accessible.

Some preservationists have the position that people with disabilities can get services in other city agencies, and that they do not need to get into the building through the front door. Other specialists recognize the importance of installing the lift adjacent to the main entrance where everyone can see it. They understand the need of people with disabilities to experience equality in the most representative building of the city, the City Hall. This way, the building becomes a cultural symbol that expresses how the community of Belo Horizonte values democracy and addresses the elimination of stigma.

It is important to mention that accessibility to City Hall is currently possible through secondary entrances to the basement, which are located far away on the backside of the building. Other options to provide accessibility to the front entrance are unfeasible, since the structural outer wall, which is made out of large granite stones, does not allow the insertion of a lift into the facade.

The proposal consists of locating the lift inside a protective steel structure that is close to the main staircase wall. The steel structure is linked to a new wall that covers the motor of the lift. The wall treatment consists of horizontal lines of color and texture that resemble the surface of the building, although it uses different materials. A small passage cuts through the wall to the street, and thus, pedestrians can choose whether they prefer to enter the building using the front stairway, or the accessible lift. Finally, two accessible parking spaces were reserved next to the new accessible route as seen in Fig. 57.8.

The idea of adapting important historical buildings for accessibility should demonstrate a symbolic shift. Moving from an emphasis of preserving building elements and environments as they were originally built on inaccessible sites, inclusive design proposals must provide appropriate settings for meaningful activities. It is hoped that in the future, new technology may replace adapted structures, if they permit consistency to restore previous conditions. Nevertheless, adequate design must conform to the setting with architectural harmony, which is essential to eliminate conflicts in distinct historical styles for use of elements of the past and current technology. Discrete architectural solutions for inclusive design reduces a negative impact on the general public understanding of accessibility and on the acceptance of innovative technology that truly benefits everyone.

FIGURE 57.8   Before and after views of City Hall's renovated entrance in Belo Horizonte.

57.10 *LESSONS LEARNED FOR UNIVERSAL DESIGN*

The Rating Scale for Inclusive Design has actually proved to be a strong advocacy tool for creating buildings that are truly accessible for everybody. Instead of a dualistic perspective of considering design solutions for those with disabilities as opposed to an ordinary design approach, the Rating Scale for Inclusive Design enables designers to understand the challenge of achieving higher levels of user satisfaction. However, it may make people believe in the creation of a quality scale for progressive design solutions that can always be used in any situation. This is not the intent nor is it possible.

Universal design may be realized by progressing from one solution that satisfies very few people to other solutions that accommodate a range of different needs. For instance, a ramp with a steep slope may be replaced by another longer ramp with a gradual incline and railings at both sides. In such cases, the Rating Scale for Inclusive Design works well. A strong link between ideas of progression for design solutions depends very much on the designers' understanding of the seven Principles of Universal Design. Also, the conceptual basis for designs, from access paradoxes to adequate design that provides full adaptability, is very important. The idea of an accessible route that connects everything in short distances is very helpful. Transportation systems, sidewalks, drop-off passenger areas, reserved parking spaces, ramps, stairs, elevators, and resting areas—all provide links that can be used to meet different user needs.

On the other hand, universal design can also result in innovative ways of handling a problem, and that cannot be controlled by means of percentage figures, summaries, and so on. For example, allowing wheelchairs and scooters inside trains may be an innovative way of reducing the use of automobiles in large cities. It is just a matter of integrating transportation systems: a personal transportation (i.e., pedestrian walking distance) and public transportation, which covers great distances. This could provide the means for accessibility for everybody, including elderly people, people with mobility problems, and people who prefer to keep their large cars outside the downtown area so that they can drive electric personal vehicles that are easy to park everywhere. In that case, the Rating Scale of Inclusive Design does not work properly for assessing the quality of solutions.

Therefore, the rating scale cannot be thought of as the only source of information regarding the needs of people with different disabilities. Innovative concepts will broaden the spectrum of items contained in the list of the Rating Scale for Inclusive Design by addressing creativity and context-related solutions in different cultures.

57.11 *CONCLUSION*

It is important to emphasize that an inclusive society must have accessible resources for users with permanent or temporary disabilities at the scales of urban planning, architectural design, and product design. If that does not occur, the goals of an inclusive society may fail.

The Rating Scale of Inclusive Design presents a list of environmental conditions that can guide the designer's attention. Each item on the list connects to others that contain similar characteristics, but which are at lower levels of priority. Together, they address the specific configuration of a city, the built environment, and an object's form and function.

In this chapter, the need to foster positive attitudes between service providers and taxpayers was discussed. Temporary and informal use of prohibited areas for general parking as reserved parking for specific needs should receive everyone's respect and total attention. Even hilly areas of the city of Belo Horizonte can change into accessible places where driveways link canopies of accessible entrances to personal transportation. There must be as many entrances as is necessary to connect shorter accessible routes at human scale. Finally, the building stock for accessible dwellings may increase the acceptance rate of adaptable homes as larger bathrooms are considered a public area in apartments.

Thus, city administrators have an important role in advancing an inclusive society, by providing incentives that address builders' interests, direct designers' attention, and emphasize the application of inclusive design models in the public realm.

The roles of policy makers and consumers as well as their interests are part of the criteria used for establishing the list of environmental conditions that best facilitate accessibility. Ultimately, designers and builders will be better positioned to offer inclusive design when they shift from access paradox solutions to adequate settings. The Rating Scale for Inclusive Design that includes more than 400 items is a work in progress. Perhaps it will never be finished, since it represents an ongoing, evolutionary process.

57.12   NOTE

1. The UFMG's graduate students of architecture whose work is cited in this chapter are Daniella Cronenberg, Luciana Mancini, and Sandra Fernandins.

57.13   BIBLIOGRAPHY

Aino, Elizabeth A., et al., *Access for All: An Illustrated Handbook of Barrier-Free Design,* by The Ohio Committee on Employment of The Handicapped & Schooley Cornelius Associates (ed.), Special Press, Columbus, OH, 1978.

Altman, I., *The Environment and Social Behavior: Privacy, Personal Space, Territory, Crowding,* Brooks/Cole Publishing, Monterey, CA, 1975.

Amengual, Clotilde, et al., *Barreras Arquitectonicas e Urbanísticas en Museos,* Asociacion Mutual Sociedad central de Arquitectos (AMSCA) (ed.), Madrid, Spain, 1983.

American National Standard Institute (ANSI), ANSI A 117.1-1986, *American National Standard for Building and Facilities—Providing Accessibility and Usability for Physically Handicapped People,* ANSI, New York, 1986.

Architectural and Transportation Barriers Compliance Board, *Federal Register: American with Disabilities Act (ADA) Accessibility Guidelines for Buildings and Facilities; Transportation Facilities; Amendment to Final Guidelines,* vol. 56, no. 173, Rules and Regulations, Government Printing Office, Washington, DC, 1991.

Aslaksen, Finn, et al., *Universal Design: Planning and Design for All,* The Norwegian State Council on Disability, Oslo, Norway, 1997.

Associação Brasileira de Normas Técnicas (ABNT), *Acessibilidade de Pessoas Portadoras de Deficiência às Edificações, Espaço, Mobiliário e Equipamento Urbanos,* NBR 9050/1994, ABNT, Rio de Janeiro, Brazil, 1994.

Barrier Free Environments, Inc., *The System: Accessible Design and Product Information System,* Barrier Free Environments, Raleigh, NC, 1984.

Beasley, Kim, and Thomas Davies, Jr., *Design for Hospitality: Planning for Accessible Hotels and Motels,* Nichols Publishing, New York, 1988.

EGM Architecten, Gemeenshappelijke Medische Dienst, et al., *Geboden Toegang: Handboek voor het Toegankelijk en Bruikbaar Ontwerpen en Bouwen voor Gehandicapte Mensen,* Stichting Nederlandse Gehandicaptenraad. ill., Utrecht, Netherlands, 1986.

Goldsmith, Selwyn, *Designing for the Disabled,* 2nd ed., McGraw-Hill, New York, 1967.

Guimarães, Marcelo P., *Behavioral Factors in Barrier-Free Environments,* Master's thesis in architecture, State University of New York, Buffalo, NY, 1991a.

———, *Fundamentos do Barrier-Free Design,* edição especial para o Prêmio Nacional de Design, Pesquisa e de Adequação do Mobiliário Urbano à Pessoa Portadora de Deficiência, IAB-MG, Belo Horizonte, Brazil, 1991b.

———, "Análise do Cadastramento do Público Alvo e da Operação nas Ruas," in *Programa Ação Pró-acesso: Credencial de Acessibilidade para Estacionamento Não Regulamentado,* 1a ed., Centro de Vida Independente de Belo Horizonte (CVI-BH), Belo Horizonte, Brazil, 1997.

————, "Acessibilidade: Diretriz Para a Sociedade Inclusiva," in *Anais do Seminário Handicap: As Vantagens de se Contratar um Portador de Deficiência,* Prefeitura Municipal de Belo Horizonte—SMDS, Belo Horizonte, Brazil, 1998a.

————, *Direfrizes para a Prática do Design Inclusivo na UFMG,* Centro de Vida Independente de Belo Horizonte (CVI-BH), Belo Horizonte, Brazil, 1998b.

————, *A Graduação da Acessibilidade Versus A NBR 9050/1994: Uma Análise de Conteúdo,* 3a ed., Centro de Vida Independente de Belo Horizonte (CVI-BH), Belo Horizonte, Brazil, 1999a.

————, *Reforma de Adaptação e Procedimentos de Facilitação para Acessibilidade no Edificio Sde da Prefeitura Municipal de Belo Horizonte,* Centro de Vida Independente de Belo Horizonte (CVI-BH), Belo Horizonte, Brazil, 1999b.

Hall, Edward, *The Hidden Dimension,* Anchor, Garden City, NY, 1966.

Mace, Ronald, *An Illustrated Handbook of The Handicapped Section of The North Carolina State Building Code,* B. Laslett, ed., North Carolina Building Code Council and North Carolina Department of Insurance, Raleigh, NC, 1974.

————, James Bostrom, and Maria Long, *Adaptable Housing,* U.S. DHUD, Barrier Free Environments, Raleigh, NC, 1987.

Proshansky, Harold, William Ittleson, and Leanne Rivlin, "The Use of Behavioral Maps in Environmental Psychology," in *Environmental Psychology: People and Their Physical Setting,* Holt, Rinehart & Winston, New York, 1970.

Ribeiro, Darcy, *O Povo Brasileiro: A Formação e o Sentido de Brasil,* Companhia das Letras, São Paulo, Brazil, 1995.

Silverstein, Murray, and Max Jacobson, "Restructuring The Hidden Program: Toward an Architecture of Social Change," in W. Preiser (ed.), *Facility Programming: Methods and Applications,* Community Development Series. Dowden, Hutchinson & Ross, Stroudsburg, PA, 1978.

Smith, Eleanor, *Entryways: Creating Attractive, Inexpensive No-Step Entrances to Houses,* Concrete Change, Atlanta, GA, 1996.

Sommer, Robert, *Personal Space: The Behavioral Basis of Design,* Prentice-Hall, Englewood Cliffs, NJ, 1969.

Steinfeld, Edward, *Access to the Built Environment: A Review of Literature,* Dept. of Housing and Urban Development, Government Printing Office, Washington, DC, 1979a.

————, "Barrier-Free Design Begins to React to Legislation: Research," *Architectural Record* (March 1979), 1979b.

————, "Designing Entrances and Internal Circulation to Meet Barrier-Free Goals," *Architectural Record,* (July 1979), 1979c.

————, J. Duncan, and P. Cardell, "Towards a Responsive Environment: The Psychological Effects of Inaccessibility," in M. Bednar (ed.), *Barrier-Free Environments,* Community Development Series, vol. 33, Dowden, Hutchinson & Ross, Stroudsburg, PA, 1977.

United Nations, "Disability: Situation, Strategies and Policies," in *United Nations Decade of Disabled Persons, 1983–1992,* V. 86-50069, United Nations Publications, New York, 1986.

van Zuylen, Marjan, et al., *The European Concept of Accessibility,* Central Coordinating Commission for the Promotion of Accessibility (CCPT), Doorn, Netherlands, 1996.

P · A · R · T · 8

CASE STUDIES

CHAPTER 58
CREATING AN ACCESSIBLE PUBLIC REALM

Sandra Manley, M.R.T.P.I.
University of the West of England, United Kingdom

58.1 INTRODUCTION

Celebrating the new millennium in cities throughout the world meant that many people shared an unforgettable and perhaps a once-in-a-lifetime experience. This was not the experience of the extravagant displays of fireworks or the dramatic parades. It was the experience of being able to reclaim the center of the city for people. Walking freely without danger and barriers to movement has become a novel experience. This is unacceptable. This chapter will argue that it is essential for a healthy society and for the freedom of all citizens to extend to everyday life the experience of being able to walk without interruption in city streets and not reserve it for special events and celebrations. It means abandoning the attitude that assumes that the street experience as it currently exists is only disabling for a few people with special needs and therefore not a major issue for the rest of society. It means that the Principles of Universal Design (see Chap. 10, "Principles of Universal Design," by Story), which have provided a clear framework for the design of products and buildings, should be expanded to create a framework for design at the scale of the street, neighborhood, and city. If this universal or inclusive approach to design at the urban scale were adopted more widely, the effect would be streets that would be more liveable and enjoyable for everyone and more inclusive neighborhoods and cities.

58.2 BACKGROUND

Adopting a new attitude to the function of the streets, squares, parks, and other spaces that make up the public realm is implicit in the universal approach to street and neighborhood design. This attitude is one that accepts the universal right of everyone to enjoy using the public realm in safety and security and without impediment to movement. Increasingly, attention is being paid to the need to make buildings more accessible to all users and particularly to people with disabilities. Legislative codes exist in most Western countries to enforce these requirements (Imrie, 1996), but the accessibility of public spaces has received less attention. This is surprising, as interviews with people who have particular knowledge of the disabling nature of the built form of towns and cities, reveal that the inability to gain access to the spaces between buildings is even more of a problem than gaining access to individual buildings (Manley, 1996). This point is borne out by many other users of public space. Ask almost any parent who has tried to make a crosstown journey on foot with a child in a stroller or baby carriage and the response will be an expression of frustration and annoyance—annoyance that the needs of parents and small children seem to be low on the agenda for the design and layout of streets (see Fig. 58.1).

It is argued in this chapter that it is necessary to accept the basic point that the street experience has become disabling for almost everyone and that it is time to change this situation by taking steps to create a more accessible public realm.

FIGURE 58.1 Crossing the street: a hazardous occupation.

The first part of the chapter will explore some of the problems created by the disabling nature of the public spaces of cities and towns. The second part will suggest some ways in which change might be achieved. It will draw on the experience of conducting access audits in a number of locations between 1994 and 2000 and will focus particularly on a comprehensive audit of Bristol City Center undertaken in 1998–1999, which will form the basis of a case study. While recognizing that it is an easier matter to identify the problems of the inaccessible nature of the public realm than to suggest how the problems can be solved, the case study will aim to make positive suggestions to bring about change. The use of public realm audits as techniques to draw attention to the problems in the public realm will be examined. The idea of using a collaborative approach to auditing that helps to make the point that all street users benefit from improved access at the urban scale will also be explored. The overall aim of the chapter is first to draw attention to the need to adopt a universal approach to the design of streets and public spaces. Second, it aims to promote the development of more proactive ways of making the public realm more accessible for everyone.

Disabling Streets—Disabling Society

The disabling nature of the public realm is to a great extent a result of the increased use of the automobile as the primary means of transport. The motor car has brought freedom to travel for many people, but at the tremendous cost of restricting other aspects of freedom. It has been allowed to take over public space to such an extent that safety measures channel the pedestrian into enclosures that restrict free movement for many people and effectively deny access for those who are unable to keep up with the pace of movement required. This often necessitates sprinting across traffic lanes or negotiating high curbs; this is difficult for many—and impossible for people whose mobility is affected by physical impairments or other difficulties (see Fig. 58.2). This does not create a pleasurable public realm experience for anyone and has almost certainly contributed to the increased use of motor vehicles for short journeys and the decline in the number of city and town center residents as the urban environment becomes degraded by traffic-dominated streets. Measures required to direct and control the huge increase in traffic have also been responsible for the introduction of the vast range of traffic-related paraphernalia described as "street furniture" by highway engineers. This "furniture" often acts as a series of unsightly obstacles to movement and free access rather than as complements to the success of public streets and spaces as pleasant places to enjoy.

FIGURE 58.2 Obstructions caused by high curbs.

Furthermore, the layout of streets in many cities and neighborhoods has been planned to facilitate easy movement for the motor vehicle. This means that pedestrian ways often involve more circuitous routes that do not respect the obvious desire of the pedestrian to take the easiest and most direct route to the required destination. For example, the pedestrian may need to take subways under roads that involve longer and more difficult journeys, and in many cases the desired pedestrian line is ignored because the direct route for the highway takes precedence. The lack of permeability created by the proliferation of culs-de-sac in residential areas is a typical example of a situation where the pedestrian's needs are not given high priority by the designers.

The car is not the only culprit in the proliferation of barriers to public access to the street. The rise in crime and perhaps, more important, the rise in the fear of crime means that many people are afraid to use streets, particularly in town centers and after dark (Oc and Tiesdell, 1997). The increasing attractiveness of the sanitized shopping mall, which appears to many people to be a safer and more acceptable place to be than the traditional street, is not unrelated to the fact that the street is becoming a place where people feel less comfortable and safe. The exclusion of "undesirable" people as well as the removal of risks associated with traffic conflict makes the shopping mall appear to be a safer and more pleasing urban experience for many users. This sense of security may well be false, and there are obvious concerns that the screening of "undesirable" people has worrying connotations about the rights of individuals and the moral acceptability of making judgments about an individual's acceptability to the rest of society. Nevertheless, it is evident that if traditional streets and spaces are to compete commercially with the malls, it is necessary to take action to remove from traditional streets some of the barriers that are created by people's perception that danger exists (see Fig. 58.3). Fear of danger is particularly apparent after dark, when many women, and an increasing number of older men, simply decide to stay at home rather than risk encountering antisocial behavior.

FIGURE 58.3 An alienating environment.

The recognition of the fact that the street environment had become disabling to many people is not a new concept. In Goldsmith's seminal work on designing to meet the needs of disabled people (Goldsmith, 1963), it was pointed out that many people with physical disabilities were disadvantaged by barriers in the public realm as well as by problems of gaining access to individual buildings. The point that the inaccessibility of the spaces between buildings can restrict the rights of people with disabilities to participate in all aspects of mainstream community life was well expressed by Goldman (1983):

Accessibility permeates all aspects of a disabled person's civil rights. Without access, rights to be abroad in the land and the full panoply of protections and duties can be rendered meaningless. To a disabled person a six-inch curb may loom as large as the Berlin Wall.

There is a growing recognition that it is necessary to move toward a barrier-free society as part of the acceptance that people with disabilities have a right to good access to all aspects of community life. There is also an acceptance that facilitating these rights makes economic sense by enabling many people with disabilities to become economically active as both workers and consumers and thus making a contribution to the economic life of nations. In spite of this, little progress has been made in relation to the accessibility of the public realm. It is evident that almost 40 years after Goldsmith's work, barriers still remain, even though these barriers may inhibit economic participation. Unlike the Berlin Wall, the obstacles to free movement continue to exist and are even increasing in number in some areas. For example, an audit of key routes in the center of Bristol, a city in the southwest of the United Kingdom, found over 400 obstructions to movement that would be inconvenient or difficult for many people to negotiate—and impossible barriers for people with some types of disability (Fig. 58.4). Furthermore, a new public urban space created for the millennium celebrations includes some features that are unusable, or at best uncomfortable, for some people, and the dominance of fast-moving traffic remains a barrier.

FIGURE 58.4   Human needs or aesthetic considerations.

Barriers, in whatever form, have the effect of excluding some people from participation in mainstream community life. The problems that exist in Bristol's streets and spaces are not untypical. Similar audits undertaken in cities as far flung as Bangkok, Kuala Lumpur, San Francisco, and Sydney found similar results.

It is true to say that in a number of Western countries progress has been made to remove some of the discriminatory practices that have limited the civil rights of disabled people. Many countries, using the Americans with Disabilities Act as a model, have enacted antidiscriminatory legislation (Imrie, 1996) and produced legal codes to ensure that new buildings are made accessible to a wider range of users. Furthermore, progress is being made in countries, such as the United Kingdom, to remove discrimination in employment and in relation to access to goods and services. However, as Lifchez (1987) has pointed out, "an emphasis on technical specifications alone simply transfers disabled people into impersonal objects, wheelchairs within a given turning radius. While specifications are important, they should serve as adjuncts to, not replacements for, an understanding of how disabled people can live independently in a world designed by and for the able-bodied."

Indeed, it is most unlikely that technical specifications or equal rights legislation designed to end discriminatory practices against people with disabilities will succeed unless a proactive approach to the removal of barriers in the public places of towns and cities is undertaken. In consequence, people with disabilities will continue to be disadvantaged by the disabling nature of the urban environment.

However, it is important to realize that disabled people are not the only groups that are discriminated against by the inaccessible nature of the public realm. It is argued in this chapter that the situation has reached a point where everyone is being adversely affected. The consequence for some is a change in behavior. Instead of walking to the local store, school, or community facility, the easiest option is to take the car. Driving, instead of walking, for local journeys has undesirable environmental as well as social consequences and may even affect people's physical health by reducing the opportunity for exercise and mental health by reducing the scope for social interaction. Furthermore, the freedom of children to roam freely is a particular issue. The erosion of the quality of the street environment seriously diminishes the scope for personal development and socialization through exploration and play well out of the direct control of adults. The behavior change for children may be an adult-imposed limitation of children's opportunities so that children only experience the organized play scheme or orchestrated adult-centered social interaction. This takes the place of unrestricted play and more adventurous situations that mold character and develop intelligence as well as enhance enjoyment. If fewer people use the streets because of their inaccessible or undesirable nature, the scope for antisocial behavior and fear for the personal safety of children becomes an even greater issue, and the problems of alienation outlined in Jane Jacobs' seminal work on the city multiply (Jacobs, 1961).

The social and economic consequences of a disabling environment for the whole population require further study to assist in developing an argument to support the idea of channelling a much greater percentage of public expenditure on roads and road transport into the improvement of the pedestrian environment. The environmental effects, for example, of increased car use have received more attention than the social considerations, perhaps because it is possible to quantify some of the undesirable effects of atmospheric pollution by measuring CO_2 emissions and noting changes to the polar ice caps. Social implications are less easy to quantify. On environmental grounds alone there is a strong case for creating a more accessible public realm as a way of encouraging walking and other more environmentally friendly forms of transport as part of a world strategy for more sustainable development. Indeed, although the encouragement of walking is an essential part of the move toward sustainable and healthier lifestyles, developing strategies for the enhancement of the walking environment fails to be seen as a high-priority issue by most Western governments.

There is a need to conduct studies to assess the social implications of a disabling environment in the move to achieve greater awareness of the problem. This might help to convince governments that the cost of a disabling public realm has social and economic as well as environmental consequences—consequences that may add to the difficulties facing governments that wish to secure an urban renaissance and a return to city living. In effect, this means that it is necessary for governments to accept that it is not a question of whether society can afford the removal of barriers, but whether society can afford not to adopt this policy.

58.3 ACCESS FOR ALL: MOVING TOWARD A UNIVERSAL APPROACH

The rather bleak picture of the disabling built environment that has been described implies that there is an urgent need for a more proactive stance to bring about change. To achieve change, it is necessary to recognize that society has a responsibility to remove barriers; it requires a paradigm shift (Hayden, 1981; Goldsmith, 1997). This shift or move up the ladder toward a more inclusive approach will only be achieved if it is recognized that people with disabilities are not the only ones who are disadvantaged by disabling environments. Studies of the work of local planning authorities in England and Wales (Manley, 1996) found that organizations pursuing policies designed to remove barriers were more likely to be effective in achieving their aims if the organization had accepted that providing good access is a mainstream issue. Local authorities that were able to see access as something far more strategic than simply building a few ramps or removing obvious steps had abandoned the idea that improving access is a question of meeting the "special" needs of a small minority of the population. Such an organization might be described as having ascended the ladder toward a universal approach to access (see Table 58.1).

TABLE 58.1 The Ladder Toward an Inclusive Approach

Universally inclusive
  Universal right to an accessible environment at all levels of provision
  Principles of universal/inclusive design accepted
  An holistic approach
  Part of the agenda for social justice
  Collaborative approach including interprofessional and user collaboration
Barrier-free
  Awareness of wider applicability
  Disabling characteristics recognized for a wide group of people
  Awareness of barriers beyond consideration of access to buildings
  Rights legislation in embryo stage
  Equal opportunities policies
Special needs
  Recognize wider definition of disability to include physical and sensory impairments
  Little appreciation of effects of cognitive disorders
  Technical specifications for access to buildings dominate
Design for people who are disabled
  Limited view of disability
  Technical specifications dominate provision
  Designs an "add-on extra"
Unaware
  No provision
  No accessibility codes
  No rights legislation or equal opportunities policies

At the bottom rung of the ladder, authorities that based their work on the medical or charitable model of disability tended to see the provision of facilities for people with disabilities as an optional extra that is dispensed to the disabled population at the whim of the "able-bodied." Implicit in the approach that sees provision for people with disabilities as a "special needs" matter is the belief that society only has responsibility to provide those facilities and benefits that are deemed to be affordable, and that provision is only made for those who deserve these benefits. In terms of planning and designing the public realm, a "special needs" approach is likely to result in very minimal commitment to ensuring that streets and public spaces are accessible. Regrettably, a large percentage of local authorities in England and Wales were found to be in the "special needs" paradigm (Manley, 1996), and findings elsewhere, such as Gleeson's work in New Zealand (Gleeson, 1997) seem to indicate that the situation in England and Wales is by no means untypical (see Fig. 58.5).

FIGURE 58.5   Crossing the street in Kuala Lumpur.

The challenge facing people who wish to see a more inclusive society is to move organizations, whether public or private, into a position that abandons the "special needs" approach to the creation of built environments. It means persuading organizations to accept a much wider definition of who is disabled by the environment than now exists and thus to an agreement that all citizens, regardless of age or ability, have an equal right to full participation in society. The seven Principles of Universal Design, developed by The Center for Universal Design (fully explained in Chap. 10, "The Principles of Universal Design," by Molly Story), provide a basis for a rights-based approach to barrier removal.

The Principles can be applied at the scale of the street—or indeed the city as a whole—as well as to the scale of buildings and products. Indeed, this is essential if equality of opportunity for everyone is to be achieved. People need to be able to travel unhindered from their home to their intended destination. To achieve this, the journey needs to be considered as a series of links—each one of which must be accessible (see Fig. 58.6). In Chap. 24, "Universal Design in Mass Transportation," Steinfeld details the concept of the *travel chain.*

FIGURE 58.6   Darling Harbor Metro, Sydney, Australia.

Creating seamless journeys from people's homes to accessible transport facilities via barrier-free pedestrian routes and on to the final destination needs to be considered strategically in order to create more accessible towns and cities. Table 58.2, based on the principles, demonstrates how they might be embraced by city planning authorities and highway engineers and applied to the spatial scale of the city and street.

The table incorporates an additional principle to add to the original list produced by the Center, which relates to the importance of considering the impact on people of the aesthetic dimension of the environment, or indeed product. This additional principle is included because it is considered that cities and streets will only be truly accessible when it is recognized that it is essential to add to the quality of the human experience by making streets and spaces pleasant or even delightful places to be. The quality of the street environment is made up not just of the functionally effective and accessible street surfaces, but of a wide range of considerations that contribute to its overall quality. These contributors to quality include the architecture of the buildings, the functionality and visual interest of paved surfaces and landscape settings, and the range and vitality of public art and sculpture. Furthermore, perception of what creates a delightful place to be must be taken into consideration in a debate about the definition of a good-quality place. These perceptions include matters such as the impression of safety, the levels of excitement and interest created by the range of activities in the place, and the qualities that make a place locally distinctive, memorable, and genuinely worth experiencing (see Fig. 58.7).

TABLE 58.2   The Principles of Universal Design: Application at the Scale of City and Street

Principle	Definition	Implications for planning the city and street
1 Equitable Use	The design is useful and marketable to any group of users.	Reconsidering the role of neighborhood planning and the development of home zones to facilitate equal access to facilities and liveable streets. Developing a strategic approach to transport policy to prioritize non-motorized transport. Streets accessible to all and development of strategies for barrier removal based on street audit. Development of techniques of social auditing as a means of assessing development proposals.
2 Flexibility in Use	The design accommodates a wide range of individual preferences and abilities.	Adaptability of development proposals judged at the time proposals are scrutinized for planning permission. New streets laid out to facilitate choice for user.
3 Simple and Intuitive Use	Use of the design is easy to understand regardless of the user's experience and abilities.	Legibility assessments of development proposals at the scale of areas, streets, and buildings. Direct routes for pedestrians not cars.
4 Perceptible Information	The design communicates necessary information effectively to the user, regardless of ambient conditions or the user's sensory abilities.	Input to planning process of consultative groups representing wide range of users. Consideration of ways of making the planning process more inclusive.
5 Tolerance for Error	The design minimizes hazards and the adverse consequences of accidental or unintended actions.	Assessment of safety to be given higher priority including road safety, crime, health, and general well-being of population.
6 Low Physical Effort	The design can be used efficiently and comfortably and with a minimum of fatigue.	Priority to pedestrians, cyclists in neighborhood, and street designs. Permeable road networks to minimize abortive journeys.
7 Size and Space for Appropriate Use	Appropriate size and space is provided for approach, reach, manipulation, and use regardless of the user's body size, posture, or mobility.	Attention to need for minimum space standards. Reconsideration of density and relationship to built form.
8 Adding to Human Delight [addition by Manley (2000)]	The environmental outcome is pleasing in itself and adds to the quality of the human experience.	Recognition of the centrality of urban design to the process of planning.

Source: Adapted from the Principles of Universal Design developed by The Center for Universal Design, North Carolina State University, Raleigh, North Carolina, 1997.

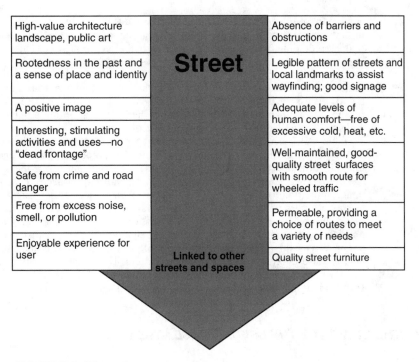

	Street	
High-value architecture landscape, public art		Absence of barriers and obstructions
Rootedness in the past and a sense of place and identity		Legible pattern of streets and local landmarks to assist wayfinding; good signage
A positive image		Adequate levels of human comfort—free of excessive cold, heat, etc.
Interesting, stimulating activities and uses—no "dead frontage"		
Safe from crime and road danger		Well-maintained, good-quality street surfaces with smooth route for wheeled traffic
Free from excess noise, smell, or pollution		Permeable, providing a choice of routes to meet a variety of needs
Enjoyable experience for user	Linked to other streets and spaces	Quality street furniture

FIGURE 58.7   What makes a good street?

It has already been indicated that one reason why the removal of barriers in the public realm is not given high priority is because the problem is seen as one that only affects a minority group. A further problem relates to the lack of evidence available to demonstrate specifically the nature of the problem and prove its existence. This is exacerbated by the fact that many real barriers may be relatively small and may appear insignificant. For example, a high curb may seem to be of minimal consequence to a person who does not use a wheelchair or push a baby stroller. The lack of tactile warnings will be of little interest to a sighted person, and people without young children will not appreciate the fear of road accidents or child abduction. Furthermore, the cumulative effect of a series of apparently insignificant barriers is not always recognized. Even people who regularly take the car because of barriers to easy and free movement that they encounter may not be aware of what is influencing their pattern of behavior. One method of raising awareness of the sheer number of barriers in public spaces is to a carry out systematic street audits to measure the extent to which the public spaces of cities meet the criteria for a good street. The second part of this chapter will explore the use of audits as a way of raising awareness of the importance of creating a high-quality street environment that is accessible to everyone. First, however, we will examine the impediments to the adoption of a universal approach to design in the public realm, since gaining an understanding of the factors that mitigate against the removal of barriers may assist with the development of strategies for barrier removal.

Impediments to the Adoption of a Universal Approach to Street Design

There are many impediments to the adoption of universal design principles for the design of new streets or for the improvement of the existing street environment. Finding ways of overcoming these impediments may be even more difficult than overcoming objections to adopting a universal approach to the design of buildings or products. This is because there is a sense that a street, unlike a building, does not really belong to anyone. This is partly because many streets have evolved over time and were not deliberately planned, but perhaps more significantly, it is through the plethora of bodies and organizations that have some impact on the design and layout of the street. In pursuing an active campaign to remove barriers, it is essential to address the issue of who is responsible for the disabling nature of streets. Exploring the reasons for the problems of disseminating the idea of adopting a universal approach to city and street planning may help to make sense of this issue.

Table 58.3, based on the ideas advanced by Steinfeld (1998), uses Utterback's (1974) theories on the rate of diffusion of an innovative idea to examine the positive and negative indicators that affect the adoption of proactive policies for barrier removal at the scale of city and street. Clearly, it is necessary to adopt a number of alternative approaches to attempt to change public policy in relation to these matters, and it may even be necessary to legislate in many countries to create bodies with comprehensive powers over the street environment. Indeed, it is likely that a multiplicity of campaigns will be necessary to change public opinion—and ultimately the policies of governments. The next part of this chapter will focus on just one possible method that has proved useful in Bristol, Weston-super-Mare, and other towns and cities in the west of England to raise public awareness of these matters. This is the idea of conducting street audits, which will be explored in some depth in the remainder of this chapter. In Chap. 6, "User/Expert Involvement in Universal Design," Ringaert discusses another street audit example in Canada.

58.4   STREET AUDITS: THE BRISTOL CASE STUDY

A street audit is a systematic survey of an area, designed to record all the barriers to access that exist in a street or series of streets. This means recording all the physical obstructions to free movement such as steps, curbs, inappropriately sited street furniture, uneven surfaces, pedestrian-vehicular conflict, and other physical barriers (see Table 58.4). In addition, it means noting the less obvious barriers that limit access to the street—barriers created by the fear of crime or isolation, alienating surroundings, and areas that are simply unpleasant places to be in.

The experience of conducting a range of audits over a 6-year period has led to the conclusion that to be effective an audit should involve direct contact with pedestrians and other street users as part of the process of making judgments about people's perception of the street environment. In effect, the whole process should be a collaborative one. This is regarded as crucial to ensure that the values of the auditors are validated or supplemented by the views of a wide range of users. The audit should also be a collaborative process, in recognition of the fact that there are a wide variety of different types of users. Collaboration also recognizes the fact that a street, unlike a building, is rarely designed as an entity.

The activity of street making has many contributors, ranging from architects who design frontage buildings to highway engineers, planners, urban designers, highway maintenance workers, landscape architects, town center managers, and utility companies, as well as people whose property fronts the street. Involving as many of these contributors to street design as possible in the audit activity will help to give the audit greater meaning and influence and stress that there is a joint responsibility to design and maintain more liveable and more accessible streets. It will also maximize the dissemination of the audit's findings and reduce the likelihood that it will be seen as something that is only relevant to the community of people with disabilities.

One of the advantages of the audit technique is that it can be instigated by local groups who have direct experience of negotiating the problems of the inaccessible nature of streets. In Weston-super-Mare, for example, a locally based group of people with disabilities was the leading player in the commencement of the audit activity. It is worth noting that the changes

TABLE 58.3 Scope for the Dissemination and Adoption of a Universal Approach

Attributes of the innovation	Negative indicators	Positive indicators
Degree of compulsion		
The nature of legal framework for rights. Regulatory codes and rules.	Laws may exist, but may not be enforced. Regulatory codes may inhibit innovative design by being too prescriptive. May encourage "code book" responses rather than a holistic approach.	Compulsion with penalties for non-compliance is likely to lead to acceptance that there is an issue to be addressed. Demand for rights amounts to a worldwide awakening.
Perceived advantages		
Enhanced diffusion of ideas if advantages clear. Social.		Social justice and human rights. A more inclusive society.
Financial.	Perception of costs to business and society.	Reduction of benefit dependency. Economic contribution as consumers. Contribution of skills, knowledge, etc.
Compatibility		
The extent to which the idea conforms with current norms, values, or structures.	May be overwhelmed by other more pressing problems (e.g., health, poverty, disaster, economic situation). May be seen as threatening to traditional built-environment professional groups—questioning the nature of planning and design values.	Acceptance of universal benefits may encourage acceptance. Political unacceptability of denying rights.
Communicability		
The extent to which the idea can be explained and identified.	Lack of awareness of principles among professionals and educators.	Developing a set of clear principles could enhance dissemination. Global cooperation to disseminate ideas widely is already emerging.
Nonpervasiveness		
The greater the number of aspects of the organization or society affected by change, the less likely that change will take place.	The complexity of the issue—interprofessional nature. Many different interest groups to be convinced.	
Reversibility		
Idea is more likely to be accepted if it can be experimented with at low cost of time, money, and commitment and if it is reversible.	Perception of costs of introducing new approaches to design.	
Number of gatekeepers		
The fewer the number of people "keeping the gates," the greater chance of adoption of the idea.	The ideas of universal design involve cooperation across traditional boundaries of professions. Gatekeepers at every level.	

TABLE 58.4   Access Audit Checklist*

Physical (quantitative) indicators	Qualitative indicators
Curb barrier No dropped curb/curb cut Poorly aligned dropped curb/curb cut	Lighting Insufficient after-dark lighting Lighting targeted to road users
Steps barrier No alternative route via ramp Steps poorly constructed Insufficient demarcation of steps Inadequate handrail	Disorientation Paths not following desired lines Lack of clear structure to route—illegible Signage confusion
Slopes or ramps Too steep Adverse camber	Threatening area Fringe area—potential for ambush Potential threats from certain groups.
Surface conditions Poorly maintained—damaged Slippery—at all times or in certain weather conditions	Blank alienating walls/no active frontage Dead ends—no perceived escape route Dangerous corners Overgrown vegetation
Narrow pavement or sidewalk Insufficient width for traffic volume Insufficient width for wheelchairs, etc.	
Danger from vehicles No safe crossing places Crossing places inadequate No tactile warnings	Uncomfortable area Exposed/cold Excessive heat and lack of shade
Lack of safety barrier or other hazard	
Doors, gates, or other boundaries Difficult to close/open Too narrow to negotiate	No activities/stimulation Boring environment No sensory or visual delight Lack of color/interest
Eye-level hazard Overhanging vegetation/building, etc.	
Street furniture Inadequate size, location or design Insufficient or poorly designed (e.g., seating) Poorly sited causing obstruction Excessive use	
Incidence of litter Excessive litter Excessive fouling by dogs Streets not cleared of leaves, snow, or standing surface water	

* This is a shortened version of the checklist used by auditors for the Bristol City and Weston-super-Mare projects. The list was developed in conjunction with representatives of disability groups and consulting engineers W.A. Fairhurst and partners. It was adapted after the Bristol audit to include qualitative indicators. The items in the left-hand column are more easily quantified by auditors than the items in the right-hand column. To validate the auditors' opinions, separate pedestrian perception interviews were carried out to complement the on-site observation.

to the legal systems of many countries to recognize the equal rights of people with disabilities have largely been brought about by a worldwide awakening of these rights that has been orchestrated by the direct campaigning of people with disabilities (Davis, 1993). It seems wise to use similar methods—that is, methods that involve people with disabilities in their conduct. In fact, the process of participating in the audit activity can in itself be a method of empowering people who may otherwise feel that their voices of complaint have been ignored. In summary, an audit can:

- Raise public awareness of the problem of the inaccessible nature of streets
- Draw the attention of governments both centrally and locally to the fact that rights for people with disabilities must be considered at the scale of the street, neighborhood, and city, as well as individual buildings or service provision
- Provide quantifiable evidence that can be compared from area to area and over time to determine whether improvement has taken place
- Cross professional boundaries by addressing the street in a holistic way and drawing the attention of a wide range of professionals whose work impinges on the accessibility of the public realm to access issues
- Draw attention to the scope for improvements to streets that can be carried out on an incremental basis (e.g., minor changes that can be undertaken for minimal cost when business premises are refurbished or street works undertaken)
- Contribute to the production of policies and strategies designed to improve access
- Draw attention to the way in which different groups of people are affected by barriers (e.g., people with different types of disability, women, children, elderly people)
- Provide a basis for the production of a prioritized action plan to remove barriers based on a rolling program of works
- Provide a basis for bids for funding from a variety of sources
- Enable the publication of access maps to indicate accessible routes and premises
- Be of educational value and empower people to take responsibility for making changes

One important point is that undertaking an audit is a positive activity rather than a negative protest. The fact that undertaking an audit produces quantifiable results in the form of data gives credibility to the argument that barriers to access really exist. When faced with concrete evidence, it becomes more difficult for the public or professionals to deny that there is a problem. The audit findings can thus become a central feature of a campaign for radical change, or at the least, incremental improvements. It is possible that results may be questioned in terms of validity in respect of the aspects that involve opinion, such as whether an area feels safe or threatening, but the majority of matters recorded are concrete facts that are difficult to refute. In the Bristol and Weston-super-Mare studies, the audits were carried out in a cooperative way by groups of disabled people, university students, local planners, and consultant highway engineers.

In the case of the Bristol study, the results have been entered into a geographic information system (GIS) so that there is a record available to both local officials in the planning and highway departments and to the public. This means that the many professionals whose work may affect the accessibility of streets can access the information. The information has been used to contribute to the city council's Legible City Initiative. This initiative aims to make the city easier to understand and more livable for all citizens and visitors by not only making the streets more accessible but also improving street signage and reinforcing the identity of its various districts. The intention is to reinforce existing identifiable areas by celebrating the things that make them special to local people, and in less well-defined areas seeking to contribute to the creation of a locally distinctive environment through design interventions. In the Weston-super-Mare audit (Manley, 2000) the intention is to use the information to obtain publicity through exhibitions, seminars, and media coverage, and thus lobby for change. To date, this dissemination has included a seminar attended by auditors, disability groups, social services representatives, local authority planners, and access officers, as well as retailers' trade organizations and central government bodies.

It would be misleading to suggest that the audit activity does not have any disadvantages. One of the problems is that, by its very nature, it is essential to record minute detail as well as the more obvious barriers. These details may seem unimportant and not worth recording to the auditors, and this can lead to inaccuracies. Furthermore, the process of recording the information can be tedious and time-consuming, and hence costly. This can be offset, to some extent, by the use of voluntary labor, although this does devalue to some extent the time and expertise of participants. Even when using voluntary labor there is considerable time input for the effective training of auditors in order to ensure consistency in the audit results. The audit activity may also bring to light conflicts of interest in that people conducting the audit may have very different views about the matters that constitute barriers. For example, the traditional paving of some streets may be seen as an important part of the cultural heritage of the street—or as an obstruction to wheelchair or baby stroller users. Bringing these issues to light may make the audit process more time-consuming, but the result may be a greater understanding among the auditors of the legitimate views of other interest groups.

A further problem in relation to the audit process is that it may raise people's hopes that action will take place and then they will feel cheated if the action is not carried out or if the time scale is too slow. This latter point can be a problem for nonprofessional auditors, who may have little idea how the financial cycle and political decision-making process of public organizations work in practice. This can be mitigated by proper briefing of auditors and the setting of realistic targets for the outcome of the audit. The work undertaken at Bristol, however, had some immediate effects that were heartening for the auditors. For example, a number of obstacles were found to be a threat to public safety, and once published it was evident that the city council felt obliged to take action to remedy the problems for fear of litigation in the event of an accident. A campaign to remove temporary obstructions caused by thoughtless behavior was launched. Other matters that were drawn to the attention of the council were relatively trivial in terms of cost, although the effects for pedestrians were considerable in terms of improving access and safety. For example, it was possible at little or no cost to adjust the "walk" period on crosswalks or pedestrian crossings to give a longer time for pedestrians to cross the street in safety. Furthermore, the owners of business premises identified as having poor access arrangements were inclined to improve the situation without any cost to the public. In the Weston-super-Mare audit, the holding of the seminar with a wide range of interest groups has in itself had positive results. It acted as a focus for reminding local business owners of the need to act to comply with the United Kingdom's legal requirement to make access to goods and services possible by 2004. It has also drawn the attention of the newly formed Disability Rights Commission to the importance of access to the public realm and created a local center for the exchange of ideas and information. Contributors to the seminar are now working cooperatively to find solutions to at least some of the problems identified by the audit. Taking part in the audit has been judged by all contributors to be a very positive experience.

Conducting an Audit

The main aim of the public realm study carried out in Bristol in 1998 was to contribute to the creation of a high-quality pedestrian environment in the city as a means of both encouraging pedestrian journeys and removing barriers that result in social exclusion. The study involved a comprehensive audit of the key pedestrian routes and a pedestrian attitude survey. Auditors surveyed the routes allocated to them and noted every barrier to pedestrian movement on a plan and accompanying survey sheet. A comprehensive list of matters to look out for was provided for each auditor. (For a simplified version, see Table 58.4.) The intention was to identify barriers that limit accessibility for people with disabilities (who are perhaps one of the most vulnerable groups using the public realm) on the basis that developing a strategy that removed the barriers that inhibited people with mobility difficulties would benefit the maximum number of people. The routes selected for detailed scrutiny for the audit had already been identified by the city council's pedestrian survey as the routes with the greatest volume of pedestrian traffic. Clearly, in considering a large built-up area it is impossible and possibly counterproductive to audit every street in the city. Selection of the most important routes is necessary, as this avoids the possibility of collecting an overwhelming quantity of data and enables the action plan to focus on the most heavily used areas. For example, as well as auditing the key city center routes, the study involved audits of the routes to district centers in the city and suburbs, but it did not attempt to survey every street.

An important part of the process of ensuring consistency in the results was the definition of the audit indicators and training the auditors to understand the nature of the indicators. During the 6-year period that audits have been carried out in the city and its surroundings, the indicators have been amended and updated to take into account the views of disabled consultants as well as the views of professionals from a wide variety of built-environment professions. The most recent change was the inclusion of the pedestrian survey and the development of additional items on the checklist to accommodate those barriers to pedestrian use that were based on attitudes to matters such as safety and the experience of using the public realm. The survey method involved interviews with street users that were designed to find out people's attitude to the quality of the street environment. Questions focused on matters that were more difficult to quantify than the more obvious physical obstructions. It was considered to be important to probe opinions to obtain the views of as many people as possible to ensure that people of different sexes, ages, races, and cultural perceptions and abilities were taken into account. The results of this process, during which 596 people were interviewed, were interesting. The results generally supported the auditors' findings about the irritating nature of particular obstacles to free movement and over 37 percent felt that the streets were difficult to negotiate. Safety and security were major issues for most interviewees. Over 50 percent felt that the arrangements for crossing traffic routes were inadequate, and 27 percent felt that street signage was poor and often confusing for pedestrians. Few people were totally satisfied with the quality of the pedestrian experience. Many made it clear that the decision to walk was affected by whether the area was stimulating, pleasant, clean, safe, and comfortable, as well as easily accessible. This was an interesting finding, as it seemed to bear out the importance of considering ways of designing for enjoyment and delight as well as functional efficiency.

The Importance of a Collaborative Approach

It can be seen in Table 58.5, which summarizes the audit process, that the audit process was a collaborative one. People with disabilities, experienced professionals, and students were involved as auditors. The results of the audit were entered into a computer and GIS (see Fig. 58.8) and made available to the city council's planning, transportation, highway maintenance, and leisure departments, as well as to other organizations. Copies of the survey sheets, which were cross-referenced with the plan, could also be accessed from the GIS by clicking on a symbol. This enabled interested parties to obtain details of the nature of the problem identified and the suggested work needed to solve the problem. Many of the problems identified were associated with poor standards of street maintenance and a failure to repair damaged street furniture or remove redundant items. In pedestrianized areas, the number of obstructions of miscellaneous items such as bollards, kiosks, and signage was problematic. In some cases, it appears that when a pedestrian area had been formed there was temptation on the part of designers to try to fill the space created with items of street furniture or public art. This seems to ignore the fact that most of the world's great urban places are notable for their spaciousness, the simplicity of their designs, and their uncluttered appearance. Many of the items of street furniture deemed necessary to fill pedestrianized streets in Bristol simply added to

TABLE 58.5   Bristol Case Study–The Audit Process

Development of access audit guidelines
  Development of pictorial checklist (with R. Guise)
  Experimental use of checklist for street audits—1994
  Literature review
  Adaptation of checklist in consultation with people with disabilities
  Audits in district centers—1995, 1996
  Audit of parks and open spaces—1996, 1997
  Revision of Access Audit Guidelines (with Fairhurst Consulting Engineers and Bristol City Council)
    in the light of experience in use

Selection of routes
  Legible city key routes selected
  Division into acceptable audit lengths taking into account attention span of auditors

Auditor training
  Selection of audit teams where possible with consultants representing variety of different user groups
  Preparation of audit information packs including maps of study area, survey sheets
  Training of auditors in Principles of Universal Design
  Resolving conflicts of interest and securing consistency

On-site audit
  Stage 1: Record of physical barriers
  Stage 2: Pedestrian perception study—593 interviews with random pedestrian users to obtain
    perceptions about the quality of the pedestrian environment. Interviewers stationed along all the
    selected routes.

Data entry onto spreadsheets and GIS

Evaluation of results and recommendations
  Preparation of report of findings and recommendations
  Development of phased action plan for street enhancement
  Feedback to auditors and review of process

Action—some improvements to the quality of the street environment

Dissemination of findings
  Communication of findings to the public, retailers, professionals through the media, exhibitions,
    seminars, etc.

---

street clutter and did little to enhance the quality of the environment. The final report of findings gave details of all the audit, summarized the main problems encountered, and made recommendations for action based on identifying immediate works to be carried out and matters that should form part of a 5-year program of improvements.

58.5   GENERAL LESSONS LEARNED FOR UNIVERSAL DESIGN

The activity of conducting the audits can be seen to have a number of both short-term and long-term benefits and thus implications for universal design. The short-term benefits included:

- Immediate removal of obstructions that were hazards to public safety.
- Results used by highway maintenance to attend to matters such as raised manhole covers, cracked and broken paving, etc.

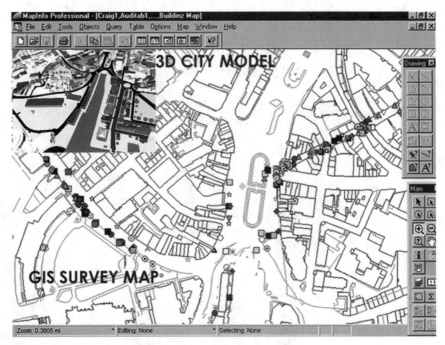

FIGURE 58.8   GIS map of Bristol City Centre: extract from the Bristol City Centre audit.

- Immediate alteration of a series of traffic lights to extend the "walk" period to facilitate crossing by slower pedestrians.
- Production of an access guide for the city center showing the extent of accessible properties and routes.
- Results used by shopmobility scheme (provider of advice and electric scooters for hire to people with disabilities) to justify funding bid for the enhancement and development of their facility.
- Production of an information leaflet for shopkeepers, explaining how to avoid temporary obstructions caused by the display of goods or advertisements.
- Use of the audit information by development control planning staff considering the proposals for the development of several city-center sites. The information enabled more effective negotiations to take place with the developers.
- Launch of a campaign in specified streets to prevent pavement/sidewalk car parking.
- Bids for the funding of new pedestrian crossings for a number of district centers.
- Bid for Heritage Lottery funding for the refurbishment of a Victorian park supported by the audit information.

Long-term benefits included:

- Development of a 5-year program of pedestrian route enhancement.
- Use of the results to feed into traffic-calming schemes.
- Results used to inform the Legible City Initiative (a scheme to improve signage and wayfinding in the city center).

- Establishment of a local center for universal design to act as an information point/training agency for auditors.
- Influence of the audit on central and local government policy.

In addition to these quantifiable benefits that had a tangible outcome in terms of physical alterations (or proposals for physical alterations), there were other less-quantifiable benefits associated with the educational aspects of the audit activity. The effects on students was notable, and it was evident from a follow-up survey of their attitudes to the activity that it had made a dramatic impact on their opinions about access issues. Working with the disabled consultants (Fig. 58.9) had the most marked effect on attitudes by making students conscious of the abilities and skills of the consultants. The fact that the work was "real" rather than simulated was found in practice to be a motivating force for the majority of students. Furthermore, the future architects and planners increased their understanding of the difference between a client and a user of a development scheme. The contact with practitioners also gave the students a sense of responsibility and realization that they had to take the work very seriously. It undoubtedly alerted the future designers to the importance of giving attention to detailed aspects of design, as well as familiarizing the students with the principles of a universal approach to design.

FIGURE 58.9   Conducting an access audit.

Auditors with disabilities who had participated in the study also developed new skills, such as survey and data entry using computer programs. This increased their confidence and sense of their capability to take control of their own lives. Professionals who took part in the audit activities, almost without fail, expressed amazement about the disabling effects of the built environment and appeared to have been influenced considerably by participation in the audit.

The final report on the audit activity concluded that the task had proved worthwhile to all participants. Some of the physical changes were apparent immediately, and it is likely that as the Legible City initiative develops in the future, the implications of the audit will be influential. Planners hope to undertake a second study during 2001–2002 to monitor progress in achieving a more accessible environment in the city and to keep the issue in the public eye.

58.6 CONCLUSIONS

It is evident that undertaking audits is not likely to result in miraculous changes to the built form of towns and cities to the extent that the public realm becomes completely accessible for all users. However, the program does provide a way of raising awareness of the importance of considering access issues as a central part of the design process and has a value as a pedagogic process in itself both for students at the beginning of their training and for midcareer professionals. The challenge for those who wish to disseminate the Principles of Universal Design as widely as possible is to convince university and college lecturers that including the Principles of Universal Design in the curriculum is essential. A great start has been made on this task through the work of the Adaptive Environments Center (Welch, 1995), but much more needs to be done, such as producing learning materials to assist reluctant educators. Success with faculty is also linked to the need to convince the professional bodies and other organizations that have an influence on curriculum design of the importance of universal design. This in turn may only occur if central governments can be convinced of the value of universal design and are prepared to legislate in countries where no effective legal framework for equal rights exists.

The lessons learned from conducting audits may also have the added value of assisting in the development of strategic approaches to street design so that new areas of development are laid out from the beginning in a more accessible way. This means rethinking approaches to city planning and the layout of new areas to incorporate universal urban design strategies that include street design as a priority of new neighborhood planning. To achieve this, it would be necessary for governments to reconsider the way in which new neighborhood planning is undertaken, possibly by rethinking the nature of the process of achieving permission for new development schemes to secure greater scope for public involvement. Valerie Fletcher, in Chap. 60, "A Neighborhood Fit for People: Universal Design on the South Boston Waterfront," of this handbook, provides an insight into how this might be achieved in practice by adopting a collaborative approach that places the needs of people at the center of the development process. At the moment this is far from typical, and many developments take place on the basis that they do not cause demonstrable harm rather than that they create good environments. Perhaps the message for governments and the development industry is that when a new area is developed, before the plan is allowed to proceed it should be essential for the developers to demonstrate how the scheme enhances quality of life and is socially as well as environmentally sustainable.

It has already been explained that conducting audits can contribute to the achievement of some incremental improvements to the accessibility of the public realm, although the task of achieving satisfactory levels of accessibility in existing street environments appears overwhelming. The fact that in many countries the responsibility for streets and public spaces is fragmented and no one body has a sense of ownership of the street exacerbates the problem. It is always some other body or organization that has created a barrier or obstruction. This seems to imply that there is a need, in many areas, to consider how streets are managed and possibly to legislate to create bodies with specific responsibility for the quality of the street environment. The task of tackling the problem of the inaccessible nature of the public realm may appear to be an impossible task and beyond hope. However, it is important to recognize that in the 1970s the battle to improve access to individual buildings seemed to be an overwhelming task. Although the battle is far from over in relation to improving access to buildings, it is evident that great strides have been made; the production of this handbook makes this clear by its many examples of practice throughout the world.

The next major task for universal design is to take up the challenge of the inaccessible nature of the built environment and focus attention on the need to ensure that everyone has access to the streets, spaces, parks, and sidewalks that make up the public realm.

Worldwide concern about the increased use of the motor car and its undesirable environmental effects makes this an opportune moment to try to reclaim the streets for people and encourage walking by creating good-quality street environments that everyone can enjoy. Without this, as Goldman pointed out, "the right to be abroad in the land" becomes meaningless for many people.

58.7 BIBLIOGRAPHY

Davis, K., "On the Movement," in J. Swain, V. Finkelstein, F. French, and M. Oliver (eds.), *Disabling Barriers—Enabling Environments,* Sage Publications in association with the Open University Press, Milton Keynes, London, UK, 1993.

Gleeson, B., "The Regulation of Environmental Accessibility in New Zealand," *International Planning Studies,* 2(3): 367–390, 1997.

Goldman, C., "Architectural Barriers: A Perspective on Progress," *Western New England Law Review,* 5: 465–493, 1983.

Goldsmith, S., *Designing for the Disabled,* RIBA Publications, London, UK, 1963.

———, *Designing for the Disabled—The New Paradigm,* Architectural Press-Butterworth-Heinemann, Oxford, UK, 1997.

Hayden, D., "What Would a Non-Sexist City Be Like? Speculations on Housing, Urban Design and Human Work," C. Stimpson, E. Dixler, M. Nelson, and K. Yatrakis (eds.), *Women and the American City,* University of Chicago Press, Chicago, 1981.

Imrie, R., *Disability and the City: International Perspectives,* Paul Chapman Publishing Ltd., London, UK, 1996.

Jacobs, J., *The Death and Life of Great American Cities: The Failure of Town Planning,* Peregrine Books in association with Jonathan Cape, London, UK, 1961, 1984.

Lifchez, R., *Rethinking Architecture: Design Students and Physically Disabled People,* University of California Press, Berkeley, CA, 1987.

Manley, S., "Walls of Exclusion: The Role of Local Authorities In Creating Barrier-Free Streets," *Landscape and Urban Planning,* 35:137–152, 1996.

Manley, S., "Creating Accessible Environments," in C. Greed and M. Roberts (eds.), *Introducing Urban Design: Interventions and Responses,* 153–167. Addison Wesley Longman Limited, Harlow, UK, 1997.

Oc, T., and S. Tiesdell, *Safer City Centers: Reviving the Public Realm,* Paul Chapman Publishing Ltd., London, UK, 1997.

Steinfeld, E., "Universal Design as Innovation," in J. Reagan and L. Trachtman (eds.), *Proceedings of Designing for the 21st Century, An International Conference on Universal Design of Information, Products and Environments,* The Center for Universal Design, North Carolina State University, Raleigh, NC, 1998, pp. 85–94.

Utterback, J. M., *Diffusion of innovations,* Free Press, New York, 1974.

Welch, P. (ed.), *Strategies for Teaching Universal Design,* Adaptive Environments Center, Boston, and MIG Communications, Berkeley, CA, 1995.

58.8 RESOURCES

A full copy of the access audit checklist that appears in summary in Table 58.6 can be obtained from the author. Contact the University of the West of England, Faculty of the Built Environment, School of Planning and Architecture, Frenchay Campus, Coldharbour Lane, Bristol, BS 16 1QY U.K., or e-mail sandra.manley@uwe.ac.uk. Copies of the Weston-super-Mare access audit report can be obtained from the same address. Readers may also like to compare the audit technique used by the Universal Design Institute for an audit of the city of Winnipeg with the approach adopted in Bristol, United Kingdom. Details of the Winnipeg audit can be obtained from the Universal Design Institute, Faculty of Architecture, 201 Russell Building, University of Manitoba, Winnipeg, Manitoba, R3T 2N2A Canada. Web site: www.arch. umanitoba.ca/uofm/cibfd.

CHAPTER 59

PROMOTING NONHANDICAPPING ENVIRONMENTS IN THE ASIA-PACIFIC REGION

Katsushi Sato
Department of Housing and Architecture
Japan Women's University, Tokyo, Japan

59.1   INTRODUCTION

This chapter reports on work initiated by the United Nations Economic and Social Commission for Asia and the Pacific (UN ESCAP). As part of a series of regional initiatives to translate the agenda for the Asian and Pacific Decade of Disabled Persons 1993–2002 into practical actions and goals, UN ESCAP implemented a project from 1993 to 1997 entitled "Promotion of Non-Handicapping Environments for Disabled and Elderly Persons in the Asia-Pacific Region," which aimed to promote barrier-free built environments in the developing countries of the Asia-Pacific region.

This chapter used information and materials generated in collaboration with San Yuenwah, Social Development Division, UN ESCAP, during the author's tenure at ESCAP as a JICA (Japan International Cooperation Agency) project expert on accessible environments from August 1995 through March 1998.

59.2   PROJECT BACKGROUND

The majority of people with disabilities in the Asia-Pacific region are poor. Their families, on whom they depend, are also poor. Their own opportunities for breaking out of poverty are limited by the physical obstacles that exist in the built environment. Those obstacles often prevent persons with disabilities from attending schools and training programs, visiting government offices, or using other public amenities and services intended for all citizens. With the rapid aging of most societies in the region, there is a growing number of older persons. In view of this fact, one must ensure that all new construction and renovation works incorporate barrier-free features.

Under Phase I of the project, in 1994, UN ESCAP developed a set of guidelines for the promotion of nonhandicapping physical environments for disabled and older persons. The guidelines cover planning and how to make building design accessible, access policy provisions and legislation, and the promotion of public awareness to improve access. The pilot projects on promotion of nonhandicapping environments were followed by Phase I of the project.

At the beginning of the pilot projects, ESCAP prepared informational material to explain objectives and activities of the project. The following concept about universal design was introduced in the informational material:

> A nonhandicapping environment is one which enables people with disabilities to move about freely and safely and to use its facilities and services without undue inconvenience and danger. Ramp and lift access enable people who have difficulty walking to overcome differences in ground levels. At the same time, such access features also reduce the fatigue of nondisabled users who have to carry heavy items. A built environment which meets disabled people's standards for safety, convenience, and usability will also benefit many other user groups. These groups include older persons, expectant women, and parents with babies in prams and toddlers in strollers. People who are physically weak or ill are also beneficiaries. The improved signage and sensitivity to user perspectives that are part of barrier-free design result in built environments that are as easy for first-time visitors as for people with cognitive impairments. A well-designed, barrier-free environment minimizes the overall levels of fatigue and stress among all users. The reduction of handicapping features will greatly benefit diverse social groups whose right to freedom of movement in a safe physical environment has hitherto been neglected. This calls for early action on universal access design. Timely action is essential toward laying a firm foundation for barrier-free Asian and Pacific environments in the twenty-first century.

59.3   OBJECTIVES OF PILOT PROJECTS AND PARTICIPANTS

The objectives of the pilot projects were to support selected ESCAP developing countries in the development and implementation of pilot projects on the elimination of physical barriers in the built environment for people with disabilities and elderly persons, using the previously mentioned technical guidelines developed in Phase I of the project.

Through the pilot projects' activities, models for access promotion were expected to be created in ESCAP regions (i.e., East Asia, Southeast Asia, and South Asia). The models were also expected to generate important lessons for pursuing similar work in other cities in developing ESCAP countries and also pave the way for related works (e.g., improving access to public transportation, legislative measures concerning accessibility, etc.).

Three local authorities from developing countries of the ESCAP region—namely Bangkok, Beijing, and New Delhi—participated through the implementation of pilot projects. These three cities were identified for inclusion in the project in view of their strategic importance (for example, their ability to generate demonstration effects based on a large number of people) and the long-term role that they can play in supporting other cities in their respective countries and subregions to advance the promotion of nonhandicapping environments. Furthermore, in all three cities, self-help organizations of people with disabilities (e.g., Council of Disabled People of Thailand, China Disabled Persons' Federation, and the National Forum of Organizations Working with People with Disabilities/National Federation of the Blind in India) had begun to work on access issues and were expected to assist in following up closely on the implementation of the pilot projects.

59.4   ESCAP'S ACTIVITIES IN THE THREE PILOT PROJECTS

Project Orientation

At the beginning of the pilot projects, ESCAP pursued consultations with the City of Yokohama, Japan, and CITYNET (Regional Network of Local Authorities for Management of

Human Settlements) concerning the co-organization of a workshop in Yokohama. The workshop provided an opportunity for participants from the pilot project cities (Bangkok, Beijing, and New Delhi) to gain a better conceptual and technical understanding of the issues, as well as to observe firsthand examples of barrier-free environments. The participants visited several sites in Yokohama to study accessible features. They also participated in an exercise to simulate diverse types of disability. Furthermore, from the generous sharing of experiences by the city of Yokohama and the other local resource persons, the participants acquired insights into the priority accorded to barrier-free design and the need for urgent action in terms of its broad societal and financial implications.

ESCAP Design Recommendations

For design work, ESCAP proposed illustrated technical design recommendations to assist in the technical design aspects of project implementation (see Fig. 59.1).

59.5   BEIJING PILOT PROJECT

Characteristics of the Beijing Pilot Project Site

The pilot project site, namely Fangzhuang Residential District, is located in the southeastern part of Beijing and is a part of the urban center of Beijing. The area consists mainly of high-rise residential buildings and facilities for finance, trade, commerce, culture, and entertainment. The reasons for the selection of the district were as follows:

1. The district did not have any accessible features because there were no requirements for accessibility when the district was developed.
2. Nearly 4000 people with disabilities and 20,000 older persons live in the district. The project had great significance for the residents.
3. The district has high visibility and value for drawing the attention of people in China.
4. Since the Fangzhuang Property Management Corporation manages the facilities in the Fangzhuang Residential District, it was expected that activities of the project (e.g., public awareness campaign and renovation work) were effectively ensured.

Working Committee Structure and Involvement of Persons with Disabilities and Older People

To carry out the pilot project, strong leadership of the relevant government departments and agencies was needed. Therefore, the Working Committee on the Beijing pilot project was established under the chairmanship of the vice mayor of the Beijing municipal government. This Working Committee, consisting of 22 persons, was a leading body that included representatives of people with disabilities, experts, professionals, managers of local business groups, and government officials, and it had full authority to ensure the implementation of the pilot project. Under the committee, an office was set up to coordinate and organize pilot project activities.

The Working Committee included two persons with disabilities, who played an important role in the decision making. In addition, 12 representatives of people with disabilities and their family members, and over 10 persons with disabilities and elderly persons who live in the pilot project site were invited to participate in every important meeting. They had the chance to discuss renovation plans together with designers. At every stage of construction, their

Building Entrances

Observed obstacles in the pilot project sites

- Staircases / level differences at building entrances.

- Staircases without handrails.

- No guiding/warning signal for visually impaired people.

General solutions

Where there is a staircase at the entrance,
 - A ramp may also be constructed.
 - Warning blocks (tactile surface) may be installed at the top and bottom of the staircase.
 - Handrails may also be installed along the staircase and ramp.

Guiding blocks may be installed to lead visually impaired people to building entrances.

Design recommendations for access improvements

Ramps

 Maximum slope: 1:12
 Maximum rise in single run: 75 cm
 Minimum clear width: 90 cm

Handrails of ramp and staircases

 Height of handrails: 80 to 85 cm above the nosing of the treads / the ramp surface.
 Diameter of rails: 3.8 to 4.5 cm or shape to provide equivalent gripping surface.
 Handrails should extend beyond the top and bottom nosing.

Warning blocks at staircases

 The surface of warning blocks should be raised truncated domes and should provide a visual contrast with adjoining surfaces.
 Warning blocks should be installed 30 cm away from the top and bottom steps.

Striping of treads

 All treads of staircases should be marked by a strip of clearly contrasting color at the nose of each step, to help guide people with visual impairments.

Guiding blocks

 Guiding blocks (line-type blocks) indicate the correct route to follow.
 Guiding blocks should be installed in a sidewalk section of an approach to a building.

FIGURE 59.1 An example of illustrated design recommendations proposed by ESCAP.

views and suggestions were taken into consideration. For instance, audible traffic signals were developed in accordance with suggestions of the Beijing Association of Blind Persons. In the whole process, three technical meetings were held and more than 100 people with disabilities and older persons attended.

Targets of Access Improvements and Mobilization of Resources

After preliminary surveys and discussions, the Working Committee selected the following targets:

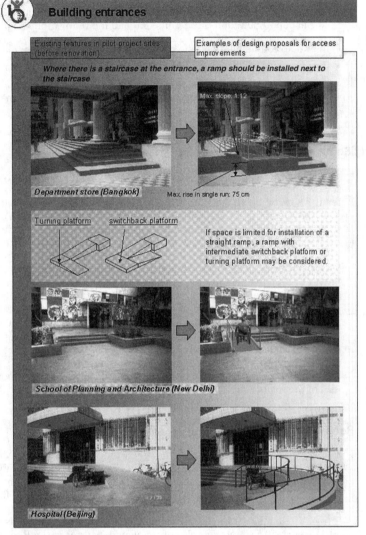

FIGURE 59.1   (*Continued*) An example of illustrated design recommendations proposed by ESCAP.

- *External Environment.* Sidewalks (warning and guiding pavers for persons with visual impairment, curb cuts), audible traffic signals, bus stops (warning and guiding pavers, Braille information), and parking space for motor tricycles used by people with disabilities.
- *Public Facilities.* A primary school, a kindergarten, a community service center, a district office, a police station, a street garden, public toilets, a hospital, a post office, a shopping center, a bank, office buildings, and a grocery store (entrance ramps, toilets, stair handrails etc.).
- *Residential Buildings.* Ten residential buildings (entrance ramps), two housing units (toilet and kitchen).

The Working Committee had to mobilize the funds needed for renovation works from various channels. The following agreement was reached after consultations with the municipal financial department and those units that had the renovation assignment:

- In principle, units that had the responsibility of renovation would bear their own costs.
- If some units had financial difficulties, the municipal government would allocate funds to them.
- Some companies would provide funds. The Urban Construction and Development Group agreed to support the project.

Conducting Seminars and Workshops

The inaugural seminar and a workshop with disabled persons convened in February 1996 to review and finalize a draft action plan during the technical advisory mission of the ESCAP team. The action plan included (1) a public awareness campaign plan, (2) a list of target buildings and facilities, (3) accessibility features to be considered, (4) generation of additional resources, (5) evaluation and monitoring methods, and (6) a schedule for each activity.

The mid-pilot-project review workshop and a workshop involving persons with disabilities were held in April 1997. Twenty-five participants from 6 cities in China, the representatives of all key government agencies concerned with the built environment in Beijing, and representatives of disabled persons and older persons participated in the review workshop. All these participants visited the pilot project site in Beijing to observe examples of barrier-free environments and experienced an on-the-spot simulation exercise on the site.

The final workshop was held in May 1998; in addition to the Chinese participants, there were 15 participants from India, Indonesia, Japan, Malaysia, Singapore, Sri Lanka, and Thailand to exchange and observe the experiences of the Beijing pilot project (see Fig. 59.2) as part of Technical Cooperation among Developing Countries (TCDC) activities. This TCDC program encouraged the promotion of nonhandicapping environments by other local authorities of the region.

Design and Renovation Work

The Working Committee asked some organizations to establish special construction teams for implementing design and renovation work. The Municipal Public Works Management Department formed a construction team that was in charge of installing tactile pavers and curb cuts. The Municipal Construction Committee formed a team that was in charge of renovation of accessibility to buildings, and the Municipal Communications Management Bureau formed a team that was in charge of installing audible traffic signals for blind persons. Figures 59.3 through 59.5 show examples of renovated features in the Beijing pilot project site.

Public Awareness Activities

On the national "Day of Disabled Persons," the Working Committee printed and distributed 500 copies of a promotional brochure on the pilot project among citizens. The groundbreaking ceremony of the pilot project was held in July 1996. Nearly 100 people representing local residents, disabled persons, construction workers, and reporters attended the ceremony. These two major activities made a significant impact on this area.

Effective media involvement was critical to the success of raising public awareness. From July to August 1996, five newspapers reported the access improvement in the Fangzhuang area. Some broadcasting agencies also arranged programs on this project. A reporter from Beijing TV interviewed pedestrians on the streets, and 90 percent of the interviewees knew

FIGURE 59.2   Study visit to the Beijing pilot project site.

FIGURE 59.3   Renovated sidewalks in Beijing.

FIGURE 59.4 Ramp installed at an entrance of a primary school in Beijing.

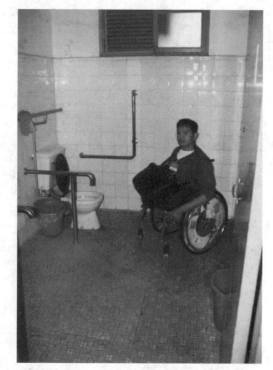

FIGURE 59.5 Accessible toilet provided at a department store in Beijing.

what tactile pavers were used for. In short, the pilot project in Fangzhuang became the focus of everyone's attention. As part of the public awareness campaign, the Working Committee prepared promotion materials for children (e.g., school timetables with color pictures and simple text about barrier-free environments) to enable them to understand what a barrier-free environment is.

ESCAP's Support for the Beijing Pilot Project

During the first technical advisory mission to Beijing, the ESCAP team assisted in the final selection of the pilot project site. The ESCAP team also assisted in the organization of workshops of persons with disabilities to inform them about the Beijing pilot project and to seek their views on its development and implementation. ESCAP collaborated with the Beijing pilot project team in the organization of the final workshop. The final workshop focused on the evaluation of the pilot project and formulation of a citywide plan of action based on the experiences gained through the project. ESCAP arranged for the assistance of a development education specialist to prepare video documentation of the final workshop of the Beijing pilot project, especially the field visits to study examples of barrier-free environments in Beijing.

59.6 NEW DELHI PILOT PROJECT

Characteristics of the New Delhi Pilot Project Site

The selected project site, namely I. P. Estate, is a nodal point in the city, located near public utility offices, recreational and cultural centers, and educational, research, and training institutions. The reasons for the choice of the district were as follows:

1. Since the project area has various facilities, a large number of persons—including disabled and elderly persons—visit the site.
2. The project area has high visibility and publicity value and attracts the attention of people not only from New Delhi but also other cities.
3. Since the project site has important government and private offices, it was recognized that the area was an important place for employment of various disability groups.
4. The project site is adjacent to offices of the print media. They were expected to facilitate wider publicity for the project.
5. The area has offices of the Institute of Town Planners and the Institute of Engineers of India, as well as the School of Planning and Architecture. Therefore, the site has potential to generate greater awareness about barrier-free built environments among professionals.

Working Committee Structure and Involvement of Disabled Persons and Older People

An organizational structure at various levels helped in steering the project and in sorting out issues that were confronted from time to time:

1. *Nodal Agency.* The Ministry of Urban Affairs and Employment, Government of India, was responsible for selection and approval of the project site and coverage of the project and was signatory to the agreement of the pilot project with UN ESCAP.
2. *Working Committee.* The composition of the Working Committee under the chairmanship of Joint Secretary (Urban Development), Government of India, included representatives

from ESCAP, Central Ministries and Departments, Government of National Capital Territory of Delhi, Municipal Corporation of Delhi, Central Public Works Department, Town and Country Planning Organization, and various disability groups and voluntary agencies. The Working Committee was responsible for interdepartmental coordination, review of the progress of the project, and approval of the project plan and proposals. To facilitate the progress of the project and to attend to the day-to-day work, the Working Committee constituted a small Core Committee under the chairmanship of the chief planner of the Town and Country Planning Organization.

3. *Coordination Agency.* The Town and Country Planning Organization of the Ministry of Urban Affairs and Employment was identified as the overall coordination agency to pursue and coordinate the progress of the project with the implementing agencies. The coordination agency was also responsible for providing technical assistance and logistic support in various activities of the project.

Representatives of various disability groups and voluntary agencies were involved as members of the Working Committee. People with diverse disabilities participated in the initial discussions on site selection and the preliminary survey.

Targets of Access Improvements and Mobilization of Resources

The coordination agency (i.e., Town and Country Planning Organization) constituted four survey teams, with architects, engineers, urban planners, and representatives of diverse disability groups. The survey teams were responsible for undertaking surveys of the pilot project area to identify obstacles to accessibility and to suggest areas for improvement. After preliminary surveys, consultation with diverse disabled groups, technical feasibility studies, and taking into account needed time and resources, the Working Committee selected 14 buildings as targets of the pilot project. These included governmental and public offices, a public library, an auditorium, and a school of planning and architecture. Walkways, ramps, curb cuts, parking areas, entries and exits, corridors, toilets, signage and symbols, and other services were targeted for renovation work.

Concerning resources, it was initially hoped that a separate budget allocation for implementation of the pilot project would be available from the government. But later on, due to nonavailability of a specific budget for the project, the Ministry of Urban Affairs and Employment decided that necessary additions and alterations in the buildings selected under the pilot project should be carried out by the respective controlling agencies with their own resources, including the budget for building maintenance. In addition to the targets selected by the working committee, by mobilizing its own resources the Delhi Transport Corporation took up improvement of accessibility conditions of the four local bus stops located in the pilot project area, as part of a comprehensive plan for developing model bus shelters in Delhi.

Conducting Seminars and Workshops

Before the inaugural seminar of the pilot, a series of workshops was organized for the officers of the participating agencies. Through these workshops, simulation exercises about the feelings of disabled persons in inaccessible situations were discussed and demonstrated, as were possible improvements to create accessible conditions. The officers of the participating agencies were sensitized to the promotion of barrier-free built environments. In all, five workshops were organized on the premises of the respective participating agencies, with active collaboration of ESCAP officials.

Following the workshops, in December 1996 an inaugural seminar was organized jointly by UN ESCAP and the Ministry of Urban Affairs and Employment of the Government of India. The seminar emphasized enhancing opportunities for people with diverse disabilities to have greater involvement and participation in the mainstream of development. The seminar called for reorienting the approaches for planning and design of public buildings and townships. The existing built environments should provide opportunities for the full participation of persons with disabilities. It was stressed in the seminar that all public buildings should have at least minimum access for people with disabilities so that they can function effectively. Nongovernmental organizations present in the seminar urged the government to make the necessary amendments and modifications in a policy framework to introduce and promote accessible environments in existing building bylaws.

More than 300 delegates, including policy makers, planners, engineers, architects, user groups, nongovernmental organizations, implementing agencies, and representatives of disabled groups, as well as people from the press and mass media, attended the seminar (see Fig. 59.6).

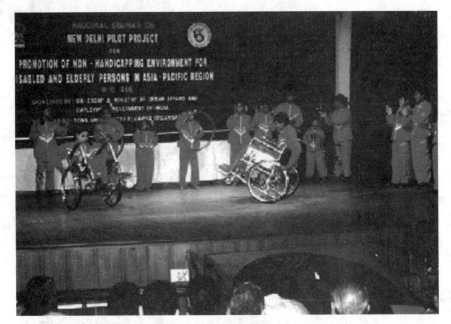

FIGURE 59.6 Inaugural seminar of the New Delhi pilot project.

Design and Renovation Work

The construction drawings of ramps, toilets, walkways, parking spaces, handrails, and other items for renovation and their cost estimates were sent to the respective implementing agencies for realization. All target buildings were made barrier-free for people with disabilities through the addition of parking space for disabled persons, walkways with tactile pavers for blind persons, ramps wherever necessary (see Fig. 59.7), sufficient width of entry, wide-enough walkways, and one toilet either on the ground floor or on another lower floor accessible by elevator.

Curb cuts with appropriate slope along the footpath were constructed for ease of movement by disabled persons. Signage and symbols where renovation work was completed were installed at appropriate locations in a standardized pattern.

FIGURE 59.7   Ramp installed at a public office in New Delhi.

Public Awareness Activities

In the inaugural seminar, survey results and proposals for improving the accessible conditions in various buildings identified in the project were discussed and presented for general public awareness.

The seminar proved to be an important event to create greater awareness about barrier-free built environments for people with disabilities and older persons, and it helped in enlisting the support of the implementing agencies and the public at large in the realization of the project. The event was reported by all the national daily newspapers both in English and the local Hindi language, as well as in the journal of the Institute of Town Planners in India, thereby generating good publicity and awareness about the implementation of the New Delhi Pilot Project.

ESCAP's Support for the New Delhi Pilot Project

The ESCAP team facilitated discussions among diverse groups aimed at the selection of a pilot project site and the development of the pilot project. The team also assisted in the organization of user group participation in a preliminary survey for site selection. Further, assistance was rendered in the planning and conducting of the inaugural seminar and workshops.

59.7   BANGKOK PILOT PROJECT

Favorable Factors for the Bangkok Pilot Project and Characteristics of the Bangkok Pilot Project Site

The Bangkok Metropolitan Administration (BMA) officially approved the Bangkok pilot project, focusing on access needs of people with disabilities in the improvement of a major pedestrian area in Bangkok. There was a good possibility of mobilizing private sector entities on the site in order to set examples of socially responsible business practice by introducing access features on their premises. The Ministry of Labour and Social Welfare and a Committee on Rehabilitation for Disabled Persons were working on the draft of the ministerial regulations concerning accessibility for people with disabilities. The disability movement was actively engaged in promoting public awareness of access issues, and in advocating government and public support for access improvements.

The selected project site is one of the busiest commercial areas of Bangkok. It is famous for business and entertainment activities. It is frequented by business people, college and university students, government officials, and tourists. It is also a popular destination for family outings on weekends and holidays. Therefore, it has high visibility and publicity value for drawing the attention of large numbers of people in Bangkok.

Although it was anticipated that construction of a sky train project would be initiated on the site during the period of the pilot project, representatives of organizations of people with disabilities insisted that the site should be selected even if the pilot project could not be completed within the scheduled time frame, and that the efforts should be continued even after the scheduled completion date.

The selected site covered an area of 2 km^2 and contained a large number of facilities, including 7 shopping centers, 3 five-star hotels, 1 national sports stadium, many entertainment places (1 ice-skating rink, 3 cinemas, 15 minitheaters), 1 library, hundreds of business offices, 1 hospital and many private health care clinics, 10 banks, and 1 temple.

Working Committee Structure and Involvement of Disabled and Older People

The governor of Bangkok appointed a Working Committee to improve pedestrian ways and street furniture to create better-quality environments. Eleven persons were appointed under the chairmanship of the deputy governor for public works. The Working Group comprised private-sector people, representatives of a vocational school and of a university located in the site, and the BMA officials concerned with design and construction. Five meetings and many informal discussions were held by the committee, with active involvement of a senior architect of the BMA.

The National Association of the Blind of Thailand (NABT) assisted in developing the design of tactile pavers by testing them. The NABT and the Association of the Physically Handicapped of Thailand (APHT) cooperated in surveying inaccessible features of the pilot project site.

Targets of Access Improvements and Mobilization of Resources

Twenty students and teachers of vocational schools conducted a survey. The survey focused on (1) width of sidewalks; (2) conditions and locations of sidewalks, traffic lights, bus stops, telephone booths, electricity poles, police shelters, and pedestrian bridges and other facilities which, because of their structure and location, constituted barriers for pedestrians; and (3) owners of major buildings along the sidewalks.

After the survey of the existing conditions of sidewalks on the pilot project site, 11 streets near the National Stadium, totaling approximately 15 km in length, were selected as targets for improvement.

The BMA allocated over 120 million Thai baht, approximately $4.8 million U.S. as of June 1997, for improvement of sidewalks on the pilot project site. In addition, two groups—the Tourism Authority of Thailand and a private-sector organization which manages hotels, restaurants, and other commercial entities—mobilized 80 million Thai baht to complement the BMA's allocation for barrier-free improvements on the pilot project site in a popular downtown area. The concerned parties signed the agreement in April 1998 at a meeting chaired by the deputy prime minister of Thailand.

Conducting Seminars and Workshops

ESCAP organized a training workshop in May 1997 for 41 technical personnel of BMA. Among the participants were technical personnel from the Departments of Public Works in Bangkok and Songkhla, as well as from the Pattaya Municipal Council. Simulation exercises and small-group discussions on nonhandicapping environments were conducted in the workshop.

ESCAP also facilitated the organization of a joint presentation by ESCAP staff, persons with disabilities, and concerned staff of BMA and the Department of Public Welfare to the students of Chulalongkorn University. As a result of the presentation, the students undertook research on the level of awareness of pedestrians on the pilot project site concerning disabled persons, the Thai Rehabilitation Act (1991), and accessibility. The results were presented at a workshop held in October 1997. The workshop highlighted existing disability legislation and the critical need for intensive efforts to reduce obstacles for people with disabilities and to improve awareness about the rights of persons with disabilities.

In 1998, the BMA Architect gave six hours of lectures on barrier-free street furniture to the School of Architecture at Silpakorn University, and three hours of the same to the faculty of the Schools of Architecture and Social Sciences at Chulalongkorn University. Following the lectures at Silpakorn University, the students undertook project work on designing street furniture for Rajdamnern Avenue.

Design and Renovation Works

Design and Production of Tactile Pavers.   Design and production of tactile pavers were difficult, and the cooperation of the private sector was imperative. The Siam Cement Group formed a team to study the design and production of tactile pavers. The cost of developing prototypes was approximately 1 million baht. The Siam Cement Group Company considered the production of tactile pavers as one of its social contribution activities.

After the prototypes were developed, blind persons tested them to find out whether by using the white cane and soles of their shoes they could sense the strips and dots in the pavers.

It was also important to ensure that the strips and dots did not cause any danger or inconvenience to any pedestrians who might step on them, including ladies with high-heeled shoes. The original height of strips and dots on the tactile pavers was 3 mm; however, after several trials by a group of blind persons, the height was increased to 5 mm.

The cost of Braille block was 23 Thai baht per paver (compared to 17 Thai baht for the regular paver). Since the use of the tactile pavers was approximately 10 percent of the total number of pavers, the increase in cost was very small (see Fig. 59.8).

Development of Design Recommendations for Improvement of Sidewalks.   The senior architect and his colleagues developed design recommendations for buildings and environments for all people, including people with disabilities. Most recommendations were excerpts from the ESCAP guidelines on the promotion of physical environments for disabled persons. The Governor's Advisory Committee approved the recommendations, and 2000 copies were printed and distributed to the public works sections of the 50 districts in Bangkok, university libraries, architecture students and faculty members, architects, engineers, building material manufacturers and building contractors, hospitals and nursing home managers, the architects of the Bangkok Mass Transit System, and the Philippine College of Gerontology. Detailed architectural drawings for the renovation of sidewalks were made based on the design recommendations.

FIGURE 59.8   Curb cuts and tactile pavers installed in sidewalks in Bangkok.

59.8   SUPPLEMENTARY ACTIONS ON THE THREE PROJECTS

Beijing Pilot Project

Through the pilot project's activities, the Detailed Rules and Regulations on Implementation of the "Design Codes for the Accessibility of People with Disabilities to Urban Roads and Buildings" was formulated and "Notifications on Strengthening Construction of Accessible Facilities and Sustainability Management" were also issued.

Further, the Municipal Institute of Architectural Design and Research conducted training on access design for all designers in the Institute. The authority of the Institute also incorporated access issues into the required professional examination for promotions.

New Delhi Pilot Project

To promote a nonhandicapping built environment, model building bylaws and guidelines were essential for reference and guidance. Therefore, as part of the exercise on barrier-free built environments, model bylaws to provide facilities for physically disabled persons were formulated. These bylaws contained specifications for various elements of barrier-free environments such as sidewalks, ramps, walkways, entry/exit, toilet size, staircases specifications, landings, and so on. The model bylaws were circulated to all state governments for comments and adoption by the urban local bodies. The Municipal Corporation of Delhi (MCD) accepted the model bylaws, and an official notification was issued by MCD to make the bylaws a mandatory provision for public buildings immediately.

The Ministry of Urban Affairs and Employment also constituted an expert committee to formulate guidelines and space standards for barrier-free built environments for disabled and older persons. The guidelines were prepared with detailed illustrations on various activity areas and public buildings. The guidelines would be circulated to the state government as a guide and reference manual.

TABLE 59.1 Comparative Analysis of Key Characteristics of the Three Pilot Projects

	Bangkok	Beijing	New Delhi
Responsible agency	Bangkok Metropolitan Administration (BMA)	Beijing Municipal Government	Ministry of Urban Affairs and Employment
Site characteristics	Commercial and educational	Residential, commercial and educational	Major government buildings
Committees set up	Working committee for footpath (11 members from BMA departments and university, institute and private company representatives)	Vice mayor (chairperson); 22 representing disabled persons; experts; professional managers of local business; government officials	Working committee (representatives from central and local government, disabled persons, and other NGOs); core committee; survey teams
Focus of renovations	Footpaths (curb ramps, tactile pavers), street furniture and bus stops	23 targets: entrance ramps tactile pavers, curb ramps, crossing sound signals, bus stop posts, public toilets, accessible housing	14 public buildings, and external environments; parking; roads and footpath surrounding them
Funding (resource mobilization)	BMA (120 million baht) and private sector joint contribution (80 million baht) for footpath development	3,213,000 yuan: Each unit was responsible (in principle), supplemented by Beijing Municipal Government according to need	Estimated budget (U.S. $457,875): Each agency was responsible for the renovation of its own premises
Consumer groups' participation	Organization of blind persons—in design of tactile pavers, on-the-spot inspection	Involvement in all main activities in the entire process: needs assessment; planning; on-the-spot inspection	Involvement in the entire process: site surveys (for site selection and technical details); planning; on-the-spot inspection; decision-making in the Working Committee
Nongovernmental/private sector involvement	Tourism Authority of Thailand and group of shopping centers and hotels, tactile block production by cement company	Day care centers; bank; shops	Library and cultural center

Educational institution involvement	Survey on the site for footpath (vocational schools); socio-cultural survey (university students) and presentation of results under the project (at ESCAP)	Participation in the second workshop	Training workshop held for students and faculty of three schools of architecture; renovation by a school of planning and architecture located on the site
Media and publicity coverage	Featured articles in English and local language newspapers	Extensive TV and radio coverage; national day of disabled persons; school schedule for children to publicize access	Coverage of national seminars by English and local language dailies, and the town planner journal
Expansion outside pilot project site	Footpath development in other districts (4,537 curb ramps and 14 km tactile pavers)	24 pavements with 200 km tactile pavers, 20 underpasses and overhead walkways with ramps	Parliament House, Supreme Court, Nirman Bhawan, Shastri Bhawan
Legislative development	Guidelines for BMA; Draft Ministerial regulations are under consideration	Design code already exists. Rules and regulations on implementation of the design code were formulated	Model bylaws were formulated and circulated; guidelines and space standards for barrier-free built environment for disabled and elderly persons were formulated.
Future development and sustainability	1. Installation of accessible street furniture (e.g., access toilets) 2. Focus on the improvement of buildings on the site 3. Issuance and enforcement of ministerial regulations on accessibility by the Thai Government	1. Include accessibility in teaching curriculum 2. Revision of design code based on pilot project experiences by 1999 3. All new buildings to comply with design code 4. Public awareness campaign 5. Training of technical personnel	1. Model bylaws to be incorporated into the building bylaws 2. Barrier-free design as part of the curriculum of all schools of planning and architecture 3. Mass awareness campaign

Bangkok Pilot Project

The Thailand Environmental Institute, a nongovernmental organization on environment promotion, planted trees, installed street furniture, and provided tactile pavers totaling 700 m in length on their project site. This project became complementary to the BMA's demonstration project to install tactile pavers.

There were several associated projects to install tactile pavers and curb cuts on sidewalks in many districts in Bangkok. As of May 1998, BMA had installed 4537 curb cuts and had renovated 13,932 m of sidewalks with tactile pavers.

59.9   KEY LESSONS LEARNT FROM PILOT PROJECTS

The pilot projects were useful examples for demonstrating that it is possible to modify existing urban environments and buildings even given limited fiscal and time constraints. The disadvantage was that they were not totally new environments, which limits the scope of work to that of adaptation and upgrading. They were also in urban settings, and therefore the experiences of the pilot project have limited applications in rural situations. Low-cost and low-technology solutions to access problems should be developed.

Table 59.1 contains a comparative analysis of the key characteristics and experiences of the three pilot projects.

James Harrison and Kenneth Parker, both of the National University of Singapore (who are also authors of chapters 32 and 40 in this book), participated in several seminars and workshops as resource persons and summarized the lessons drawn from the discussions on pilot project experiences and related observations:

General Strategy Lessons
- Improve accessibility in the built environment, as this is feasible.
- Integrate barrier-free features attractively into architecture.
- Include barrier-free features in all new construction, as this is possible and beneficial in the interest of reducing building costs.
- Ensure improvements in legislation by integrating access into building bylaws.
- Ensure strong user participation.

Lessons on Starting Projects
- Identify strong arguments and formulate targets.
- Select approaches based on local conditions, as well as local and national priorities.
- Make special and personal efforts to mobilize senior executives' support.
- Develop action plan(s) to closely monitor progress.

Lessons on Continuing Projects
- Need strong leadership of responsible agency.
- Need political and personal commitment.
- Regularize fund allocation from existing budgets for long-term sustainability.
- Monitor closely.
- Raise awareness of technical personnel at all levels, and in the community.
- Promote research and development.
- Encourage local design and production.
- Exchange experience and ideas through active networking.

End-User Involvement
- Understand universal access and nonhandicapping design.
- Emphasize the functional purpose of access features.
- Share experience, build user group solidarity, and form working alliances.
- Be actively involved.
- Take follow-up action.
- Befriend journalists.
- Monitor and follow up.

59.10   CONCLUSION

Based on the experiences of the pilot projects, ESCAP is now implementing a two-year project called "Development of Guidelines for Trainers of People with Disabilities for the Promotion of Non-Handicapping Environments." Participating countries are targeted to include Fiji, India, Japan, Malaysia, and Thailand. The project includes skills development for the promotion of nonhandicapping environments in six key areas: building confidence and group identity; social mobilization and networking skills; advocacy and negotiation skills; information collection and documentation skills; training skills; and strategy development skills.

In March 2000 the Royal Thai Government, UN ESCAP, and the Japan International Cooperation Agency (JICA) jointly organized a Regional Training of Trainers Course on the Promotion of Non-Handicapping Environments for Persons with Disabilities. Objectives of the training course were: (1) to train architects and urban planners from governmental agencies and nongovernmental organizations, and disabled persons with a record of active advocacy for access promotion in nonhandicapping environments for persons with disabilities, and (2) to foster the building of an access initiative network of technical personnel, persons with disabilities, and personnel from governmental agencies and nongovernmental organizations concerned with access issues in the ESCAP region.

Dissemination of good practice based on experience, analysis of potential problems, effective solutions, and information on implementation can give confidence and enhance skill levels of those involved in similar efforts. Through "training the trainers" schemes and continued advocacy, confidence will also encourage the users to articulate their needs effectively. Use of all the experience and know-how gained in the first rounds of these projects, as well as follow up, will be needed to sustain the effort and monitor progress.

59.11   BIBLIOGRAPHY

Harrison, J. D., and K. J. Parker, *Awareness and Promotion of Non-Handicapping Environments, International Workshop on Universal Design Preconference Proceedings* (workshop held in Yokohama, Japan, November 30–December 4, 1999), Building Research Institute, Japan, pp. 18.1–18.11.

UN ESCAP, *Draft Report of the Final Workshop of the Beijing Pilot Project on the Promotion of Non-Handicapping Environments for Persons with Disabilities and Older Persons,* (workshop held in Beijing, China) May 1998.

———, *Preliminary Background Information on the Promotion of Non-Handicapping Environments for Disabled and Elderly Persons in the Asia-Pacific Region* (Phase II) (unpublished communication to the donor consultation meeting), 1998.

———, *Promotion of Non-Handicapping Environments for Disable Persons* (pilot projects in three cities), 1999.

Working Committee of the Bangkok Pilot Project, *Report of the Bangkok Pilot Project on the Promotion of Non-Handicapping Environments for Persons with Disabilities and Older Persons* (unpublished report presented at the final workshop of the Beijing pilot project), 1998.

Working Committee of the Beijing Pilot Project, *Report of the Beijing Pilot Project on the Promotion of Non-Handicapping Environments for Persons with Disabilities and Older Persons* (unpublished report presented at the final workshop of the Beijing pilot project), 1998.

Working Committee of the New Delhi Pilot Project, *Report of the New Delhi Pilot Project on the Promotion of Non-Handicapping Environments for Persons with Disabilities and Older Persons* (unpublished report presented at the final workshop of the Beijing pilot project), 1998.

CHAPTER 60

A NEIGHBORHOOD FIT FOR PEOPLE: UNIVERSAL DESIGN ON THE SOUTH BOSTON WATERFRONT

Valerie Fletcher, M.T.S.
Adaptive Environments Center, Boston, Massachusetts

60.1 INTRODUCTION

Boston is building its first new neighborhood in 150 years, a once-in-many-lifetimes opportunity to define the city and the citizens through decisions about design. A record-breaking $20 billion public investment in infrastructure improvements has transformed a district of warehouses, parking lots, and a working port into prime development land. Bordered by downtown, an interstate highway, Logan International Airport, a deepwater port, and the residential section of South Boston, the South Boston Waterfront is 1000 acres facing a pristine harbor. This case study details the initiative of a not-for-profit organization, Adaptive Environments Center, to make universal design a defining characteristic of all aspects of the development. The project takes the position that universal design complements Boston's proud claim of being a walkable city, a city at a human scale. The chapter describes the flexible, multifaceted, persistent strategy necessary to make universal design a reality at the urban scale.

60.2 BACKGROUND

This is the story of Adaptive Environments' unfolding strategy to integrate universal design into the development of a large, new urban neighborhood in one of the oldest cities in the United States. Adaptive Environments is a small, not-for-profit educational organization committed to the full social inclusion of all people, regardless of disability or age. The organization educates about the power of design as a tool of social justice and promotes universal design in all its publications and projects. Adaptive Environments' offices are physically located next to the border of the South Boston Waterfront development. The City of Boston began the waterfront development process in early 1998 when the Boston Redevelopment Authority (BRA) issued its first plan for development of the 1000-acre parcel and invited public comment. Adaptive Environments acted quickly to seize the opportunity to integrate universal design as a hallmark of the development.

Its initiative has been characterized by a multipronged strategy of collaboration, advocacy, education, training, and technical assistance. Building a complex network of collaborators was a critical initial priority. Committed allies included the City of Boston (especially its Redevelopment Authority), several state agencies, individuals with disabilities as well as disability and elder organizations, more than a dozen nonprofit organizations with an interest in the South Boston Waterfront development, and the private development community. Although substantial construction in the area closest to the existing downtown will be completed by 2003, the development of all public and private space is anticipated to continue until approximately 2020.

60.3 BOSTON'S CONTEXT

Boston

Boston is a small city of 48 square miles with approximately 590,000 residents. It also is the hub of a metropolitan area of 3.1 million people; the capital of Massachusetts; and the largest city in New England (City of Boston, 1999). Fifty years after a disastrous economic downturn caused the exodus of tens of thousands of residents, Boston has become a vibrant, economically vigorous, and culturally diverse city at the start of the twenty-first century. A bitter experience with disastrous urban renewal of Boston's West End in the 1960s left the city with an abhorrence for growth when that growth depends on destruction of the historic fabric. So, growth by eliminating the old is impossible in Boston. However, the 1 percent vacancy rates for office space, hotel rooms, and housing during the late 1990s created a crisis for space that demanded new solutions.

South Boston Waterfront

The city created a solution on its waterfront close to its financial and retail centers and did not have to destroy anything. How? There are two reasons. The first is the distinct historical pattern of making land from water or marshland by filling it in. Like some other places bordered by water, Boston recreated its geography over the last 300 years by extending the land into the water. The last neighborhood built in the city was Back Bay in the mid-nineteenth century. As is evident from its name, landfill turned a marshy bay into a neighborhood. It evolved to become one of the city's premier residential, retail, and civic spaces. In similar fashion, the South Boston Waterfront landmass substantially evolved in the nineteenth century through the creation of a claw-shaped, mile-long area for piers. The U.S. Navy added to it in the first half of the twentieth century. During the second half of the twentieth century, the area housed a working port, a small number of scattered, mostly water-related businesses, and acres of inexpensive parking lots.

The second reason for existing space to become prime development land is the record-breaking $20 billion public investment made in infrastructure improvements. They began in the 1990s and are expected to be completed by 2005. The expenditures, taken together, transformed the Waterfront from largely forgotten space surrounding a working port into a very attractive development opportunity primarily for private developers who owned or who could acquire land in the district. Boston Harbor, so polluted that it was nationally renowned for its toxicity, was so thoroughly restored with a $4 billion investment that the native marine life returned in force. The State of Massachusetts constructed a third harbor tunnel that links the area to Logan International Airport in minutes. New extensions of highways and public transit that are under construction connect the Waterfront to major interstate roads and destinations. Federal and state funds support the largest current construction project in the world that will build an underground alternative to an elevated roadway. The Central Artery that bifurcated the downtown from the Waterfront will be replaced with 27 acres of land. This gateway to the Waterfront is planned to be a mix of open space and mixed-use development. The state has committed $700 million to construct a new convention center in the South Boston Waterfront that will be able to accommodate 25,000 people at a time. Despite its fortunate location adjacent to downtown (see Fig. 60.1), this long-neglected and nearly inaccessible area became a valuable real estate opportunity only because of the enormous investment of public funds.

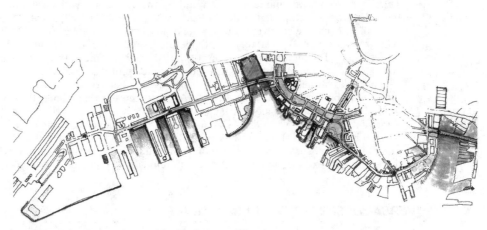

FIGURE 60.1   Illustrated map of the South Boston Waterfront.

Regulatory Environment

Ownership of the 1000 acres of the South Boston Waterfront is divided between a handful of private developers and three public agencies—the City of Boston, the Massachusetts Convention Center Authority, and the Massachusetts Port Authority. The regulatory environment is complex. The City of Boston's Redevelopment Authority has planning and zoning responsibilities. A specific set of Waterfront Design Guidelines has been developed as a detailed, district-specific supplement to the citywide urban design guidelines. The Boston Department of Inspectional Services must issue a building permit for any nonstate construction. Approval of plans is contingent upon compliance with the Massachusetts Building Code. The state also has regulatory authority over the area through Chapter 91, the State Waterways Law. Water, tidelands, and filled land at one time under water are covered. One requirement of the State Waterways Law is that buildings containing other than water-dependent use that are constructed within 100 feet of the shoreline must be "facilities of public accommodation"—an interesting advantage given the requirements for compliance with the Americans with Disabilities Act (ADA). A public accommodation is a private entity that owns, operates, leases, or leases to, a place of public accommodation. The term is broad and includes a wide range of entities such as restaurants, retail stores, parks, hotels, theaters, museums, libraries, and day care centers.

The Mayor of Boston

Mayor Thomas M. Menino and the Boston Redevelopment Authority are not insensitive to the implications of the public investment that has made the South Boston Waterfront a valuable development opportunity. In speeches and public communications, the mayor and Redevelopment Authority leaders consistently stated that *public good must drive private development in this area* (Boston Redevelopment Authority, 1999). They have touted the South Boston Waterfront as a place that would finally give all residents and visitors access to the harbor that has been obscured and difficult to reach for generations.

Legally Mandated Accessibility

There are two sets of minimum access requirements for people with disabilities that apply in the Waterfront development. Title II of the ADA requires that all new buildings and public rights of way that are constructed by a state or local government be accessible. Title III of the ADA requires that all privately owned, operated, or leased places of public accommodation be accessible. The minimum guidelines of the ADA are enforced through voluntary compliance and by complaint. The Massachusetts Architectural Access Board (MAAB) standards also apply to all new development constructed by a state or local government or that are places of public accommodation. The MAAB guidelines are sometimes the same and sometimes different from (and more stringent than) the requirements of the ADA. The most substantive difference between the ADA and the MAAB is the mechanism of enforcement. The MAAB is part of the Massachusetts State Building Code. One cannot obtain a building permit if the design does not comply with the State Building Code. Although a mechanism exists for securing an exception to the MAAB requirements, it is difficult and time-consuming.

60.4   *UNIVERSAL DESIGN AT THE URBAN SCALE*

Principles of Universal Design

Project staff initially struggled with using the seven Principles of Universal Design [(1) Equitable Use, (2) Flexibility in Use, (3) Simple and Intuitive Use, (4) Perceptible Information, (5) Tolerance for Error, (6) Low Physical Effort, and (7) Size and Space for Approach and Use] (Center for Universal Design, 1997) at the urban scale. The utility of the principles for product design and some elements of architecture was well-documented. It was not obvious that the principles could serve as well at the scale and scope of city building. The need to provide a simple set of ideas that could be quickly grasped by diverse audiences argued that the principles were a useful tool. As the detailed recommendations of universal urban design evolved, it became clear that the principles were more relevant than initially perceived. They were used as the organizing structure for the universal design section of the City of Boston's "Municipal Harbor Plan of the City of Boston." However, the challenge remains of how to incorporate some key elements of the urban scale within the current iteration of the principles. The most difficult fit is indoor air quality. It is a key feature of the goal for human-centered design in the district and invaluable as an issue of shared mission with "green" advocates involved in the development. It fits awkwardly within the current seven principles and is a stimulus to thinking about the evolution of the principles.

Adaptive Environments

Adaptive Environments is a nonprofit organization founded in 1978 by Elaine Ostroff and Cora Beth Abel. It was an outgrowth of the Arts and Human Services Project, a multidisciplinary graduate program supported by the Massachusetts Department of Mental Health at the Massachusetts College of Art. The graduate program emphasized the leadership role of artists and designers in creating community-based programs for people with disabilities. Most of the programs and services that the Adaptive Environments Center has offered are educational and exemplify the organization's founding vision of the role of designers together with consumers in addressing social issues. In 1989, the organization initiated the Universal Design Education project, which is now international (Welch, 1995). Adaptive Environments built an international reputation for creating user-friendly, accurate information on design and related civil rights issues. It also sponsors a biennial international conference on universal design— "Designing for the Twenty-First Century"—that brings together leaders from around the world to share ideas and experiences and struggle together to integrate universal design into the common practice of design.

In early 1998, the Adaptive Environments board of directors hired a new executive director and undertook the creation of a new strategic plan. A new tag line was added to Adaptive Environments' name: "Working to Make the World Fit for All People." The organization's mission was refined as: "To promote, facilitate, and advocate for international adoption of policies and designs that enable every individual, regardless of disability or age, to participate fully in all aspects of society." Adaptive Environments began a process of identifying potential projects in which the organization could be a tool for applying universal design in a specific setting. For a project to be considered, it needed to meet two criteria: (1) likelihood of effecting a measurable outcome, and (2) educational value that could benefit similar projects in other places.

The South Boston Waterfront development was ideally suited to meet the criteria. The development parcel was literally at Adaptive Environments' front door. Its offices sit in a 1906 warehouse building bordering the expanse of the South Boston Waterfront. Given the network of relationships with local government entities and other organizations and individuals, it was realistic to start from an assumption that it would be possible to influence the outcome. And the project had educational potential. Making universal design meaningful at the urban scale in the United States required making the case for universal design in an environment characterized by pervasive if minimal legal mandates for access for people with disabilities. Documenting the process and results in the South Boston Waterfront project would offer a case study particularly useful for other cities in the United States.

The board of directors of Adaptive Environments committed to the project in the spring of 1998. *A Neighborhood Fit for People* was chosen as the project name and used in all communications. From its inception, Adaptive Environments understood that a successful outcome depended upon a multipronged, simultaneous, and flexible process. The standard elements, varying in emphasis from time to time, were advocacy, education, collaboration, and technical assistance/consultation.

60.5 *ADVOCACY ROLES*

Boston Redevelopment Authority

The advocacy argument for universal design at the urban scale began with an invitation to exercise creativity in service to a vision. The first targets of advocacy were the planning and design decision makers in the BRA. They prided themselves on their commitment to full compliance with the ADA and MAAB. However, compliance was well short of a vision for seamlessly inclusive urban space.

Prodding them to see opportunity in a commitment to universal design required that they understand the limits of legal compliance. Meeting the requirement of the laws did not invite the quality or scope of creative problem solving envisioned by universal design. The argument aligned well with the city's general goal of public good from private development. Human-centered design would be a corrective to a century-long fashion of form over function. In a place with five schools of architecture, it is a familiar topic. It aligned with Boston's favorite claim of being a city at the human scale, a walkable city. And it offered a corollary to the enviable civic vision that had characterized the building of the beloved Back Bay that was the most recent "new" neighborhood and 150 years old. Those Boston leaders of the mid-nineteenth century established a grid of engaging public spaces and graceful residential streets with a clear priority on civic life. Their most enduring choice was the Boston Public Library, the first public lending library in the United States. Any current public servant with an eye toward a legacy of public benefit could hear the appeal of human-centered design.

But, unlike Back Bay, the South Boston Waterfront is mostly privately owned. It is typical of most urban development in the United States today, where new civic spaces must be carved out of private space (Dixon, 1999). As in any other city, Boston has examples of private development that failed to pay attention to the needs and preferences of users and resulted in sterile environments that are lively only during the workday and empty at night and on weekends. If newly built places are to attract people as destinations, they must meet several requirements. First, they must be legible (Lynch, 1962) and communicate that the space is available for some variety of activities. However, that is not enough. The space must also be comfortable and allow positive connections to others, "connections that create a sense of belonging, of safety, a feeling that personal rights will be protected" (Carr et al., 1992). The case for universal design with Boston officials had to reinforce the importance of making a deliberate commitment to meet user needs and preferences.

The Boston Foundation

The Boston Foundation was another early focus of advocacy. As the city's community foundation, it is the largest philanthropic organization in Boston. Its financial support of the project would not only ensure that a commitment to the project could be sustained; its stature as the premier community foundation would provide credence for the project's goals. An initial planning grant made it possible to pull together an all-day strategy session with important allies and provided the support for the development of educational materials. Participants included state senator Byron Rushing, who has written and spoken often about discrimination by design; Polly Welch, architect on the faculty of the School of Architecture at the University of Oregon, Eugene, and editor of *Strategies for Teaching Universal Design* (Welch, 1995); Robert Shibley, architect, urban designer, and professor at the State University of New York at Buffalo; Donald Stull, architect and leader in Boston's design community; Stephen Spinetto, director of the Commission for Persons with Disabilities for the City of Boston; Lawrence Braman, senior mapping specialist for the City of Boston and an Adaptive Environments board member; Constance Budorow, architect and cochair of the Boston Society of Architects Seaport Focus Team; George Terrien, architect, former president of the Boston Architectural Center, and an Adaptive Environments board member; Allen Crocker, M.D., pediatrician and chairperson of the Adaptive Environments board of directors; Paul Grayson, architect and consultant on universal design/lifespan design; Fred Kent, president of Projects for Public Spaces in New York; and Ana Gomez-Moreno, architect and graduate student at the Massachusetts Institute of Technology Department of Urban Planning and Design. A subsequent implementation grant from the Boston Foundation made it possible to hire a full-time project coordinator. Eric Dietrich, an urban planner and a person with a disability, began in January of 2000. Other philanthropic partners have also been approached for additional financial support.

General

Advocacy, making a verbal case for *A Neighborhood Fit for People,* is an important daily ingredient in all interactions with individuals and organizations. It is important to develop a keen ear for knowing what aspect of the case for universal design a particular person or group can hear and appreciate. In all advocacy conversations, being attuned to the audience is more important than specific terminology. Universal design, design-for-all, inclusive design, lifespan design, human-centered design—gaining a recruit requires that some choice of words resonate. More often than not, getting someone's attention for a minute or two is necessary to open the door to more substantive alliances. Developers and their designers are a critically important audience and the first approach is usually verbal. Some respond to a common-sense demographic argument that universal design speaks to the increasing diversity of users. Others see market opportunity in the way in which design can help to make their projects attractive destinations. Developers of office towers fear the epidemic rates of repetitive stress injuries and sick building syndrome, and can appreciate the long-range advantage of building in solutions from the start that enhance productivity and decrease the need for expensive corrective measures over time. Others know the research about the overwhelming preference by older Americans to age in place, and they see the wisdom of making that possible in new residential construction.

Media

Finding opportunities to advocate for *A Neighborhood Fit for People* in the general media is a continuing goal of the project. Making the language of universal design part of the common parlance requires using any communications route to get to an audience. In 1999, the project was featured in a cover article in *Banker and Tradesman,* a weekly newspaper for the real estate, business, and financial services community under the title: "Can Boston Create a Neighborhood for the Twenty-First Century?" Examples of universal design from popular local places were included. Post Office Square Park in the center of Boston's financial district has many universal features, among them a sculptural fountain with auditory and tactile features and a level cutout to pass into the inner ring of the fountain (see Fig. 60.2). A subsequent interview in *Banker and Tradesman* featured the *Neighborhood Fit for People* project and initiated a series of articles marking the tenth anniversary of the passage of the ADA.

FIGURE 60.2 Leventhal Park/Post Office Square Park in Boston. Sculptural fountain offers auditory and tactile appeal, and the segmented outer ring is designed to permit easy access.

60.6   EDUCATION AND TRAINING

Materials

If advocacy opened the door, it was necessary to follow quickly with education and training about concrete examples of universal design at the urban scale. It was clear early on that good visual materials were needed. A brochure and slide show were developed, with images gathered from dozens of sources in the United States and around the world. The task was far more difficult than had been anticipated. Product designs were the most available and plentiful materials. Too often the images of universal design in the built environment showed examples of legal compliance without sensitivity to all users. For instance, good entrance design and dimensions are depicted in space with blinding glare from glass window walls. The finest examples wove universal design so seamlessly into the whole that it was always necessary to describe the feature. An elegant sloped walkway with a waterfall to the side and textured stone surfaces at Yerba Buena Gardens in San Francisco could be appreciated as a beautiful public space and spoke volumes about human-centered design (see Fig. 60.3). The Boston Seaport Hotel's lobby features counters at two heights, and good acoustics with its lower ceiling and moveable furniture (see Fig. 60.4).

The color brochure was designed to tell the story with few words and lots of images. There was a deliberate choice to use the brochure to educate readers about the ordinariness of variation in human ability. Sally Levine, a multifaceted designer and educator knowledgeable about universal design, worked with Adaptive Environments to create a very engaging and unusual brochure. A short page was sliced in between each pair of full pages and featured a silhouette of a figure on one side and a brief demographic fact on the reverse. Examples included: fewer than 15 percent of people with disabilities were born with them; 27 million Americans report some difficulty in walking; a 60-year-old needs twice as much light as a 40-year-old does to read a menu. No one misses the message of those short pages. There has been a consistently positive response to this colorful, pocket-sized brochure. The highly visual message in its distinctive package has proven essential to promoting the message of universal design.

FIGURE 60.3 Yerba Buena Gardens in San Francisco, California. Detail depicts the sloped walkway designed with elegant textural, auditory, and visual features seamlessly woven into the whole and used by everyone.

Lectures

A Neighborhood Fit for People became one of Adaptive Environments' primary topics for invitational lectures on universal design in 1999. Classes at the Boston Architectural Center, Suffolk University, and Boston University Medical School responded with enthusiasm and provided a source of volunteers. A presentation to all staff at the Massachusetts

FIGURE 60.4 Boston's Seaport Hotel, with careful attention to acoustics, variable-height counters, and flexible seating.

Rehabilitation Commission built new bridges with an important ally. A slide presentation and panel discussion were part of the annual Build Boston conference, the largest regional design/build event sponsored each November by the Boston Society of Architects.

Children

If universal design is to be successfully integrated into our expectations of good design, children must be a prime educational focus. Adaptive Environments found an ideal partner with whom to build a children's educational component to *A Neighborhood Fit for People*. Learning by Design in Massachusetts was born in 1999 as a joint project of the Boston Society of Architects and the Architectural Education Resource Center. Volunteer architects work with children using design to teach math and science. Adaptive Environments suggested building a collaborative multiyear project that incorporated universal design into the design of a box city by Boston children.

The location of the South Boston Waterfront project offered an ideal partner for bringing design-for-all to the attention of children. The Boston Children's Museum is located on the Fort Point Channel in the warehouse district adjacent to the South Boston Waterfront and near Adaptive Environments. Last year the Children's Museum mounted an eight-week summer program called "Camp on the Harbor" that had a total of 200 participants between the ages of 6 and 12. In the summer of 2000, "Camp on the Harbor" joined forces with Learning by Design in Massachusetts and Adaptive Environments to incorporate *A Neighborhood Fit for People* into the educational program. They studied the portion of the Waterfront parcel that is being developed first and designed and built a universally designed box city. A visual record of the project was displayed in the gallery at the Boston Society of Architect's Build Boston event in November 2000.

60.7   COLLABORATION

Adaptive Environments knew that with this ambitious initiative it could not succeed alone. Even with the endorsement and full support of the Boston Redevelopment Authority, not much would happen absent extensive collaboration. There needed to be shared ownership of the goal by a constellation of partners in government, the nonprofit sector, education, and design. One of the methods of sustaining the involvement of collaborators and ensuring a mechanism for evaluation was to develop an Oversight Committee. The group was created as a committee of the Adaptive Environments' board of directors and meets three times annually.

Government

The collaborative commitment of government has been a priority. The Boston Redevelopment Authority, as the body with responsibility for planning, zoning, and the development of design guidelines, was the first necessary ally. The BRA made an initial commitment in the 1999 Seaport Public Realm Plan: "Transportation, open space, access to the harbor, pedestrian facilities and residential, civic and commercial buildings should be usable by all people, to the greatest extent possible, without the need for adaptation or specialized design." The commitment then moved to the next level of specificity in the Harbor Design Guidelines released in the spring of 2000. However, the most critical juncture occurs when each proposed parcel design is reviewed for BRA approval before being allowed to proceed to applying for a building permit.

In addition to the BRA, there are key constituents represented in city government. Director of the Commission on Persons with Disabilities, Stephen Spinetto, was one of the first allies and regularly presents jointly with the Adaptive Environments Center about the project. Commissioner of Elderly Affairs, Joyce Williams Mitchell, endorsed the project early and assigned senior staff to the Oversight Committee, as did Commissioner of Fair Housing, Victoria Williams. Inspectional Services grants building permits based upon compliance with the Massachusetts Building Code. Although their duties are limited to assessing compliance with the code, they agreed to have literature about the project on hand for developers seeking building permits for the South Boston Waterfront.

Adaptive Environments continues to pursue other city agencies such as Neighborhood Development, Parks and Recreation, and Transportation as supporters. However, no appointed public agency official is as critical to the project as are the elected representatives of the Boston City Council and Massachusetts Legislature from South Boston, the residential neighborhood adjacent to the Waterfront. They have wielded impressive power to shape the development to the advantage of South Boston, including commitments of jobs, housing, and subsidies for housing in their neighborhood. Timing and the need to clarify a specific role of support have argued for waiting to approach the elected officials. The overture will be made on the basis of specific parcels nearest the residential neighborhood.

The Massachusetts Architectural Access Board endorsed the project. Its executive director, Deborah Ryan, provides training for staff and volunteers involved in *A Neighborhood Fit for People.*

The Massachusetts Port Authority (MassPort) owns 70 prime acres in the South Boston Waterfront. Its building plans include a park, a hotel, an office tower, and 600 to 800 units of rental housing. The Massachusetts Port Authority's development agenda helps to underwrite the expenses of Logan International Airport. The Authority was an enthusiastic supporter and familiar with the concept of universal design because it had recently incorporated universal design features in Piers Park in East Boston. The Park is on the Boston Harbor opposite the South Boston Waterfront. Universal design features at Piers Park include: scattered seating with umbrellas to break the long walkways to the Harbor; a playground designed to work for all children; accessible public rest rooms for adults and for children; and an exercise area for adults that includes a full set of equipment for people who transfer from wheelchairs. MassPort believed that the universal features substantially enhanced the community's high level of satisfaction with the design.

The Massachusetts Convention Center Authority (MCCA) has broken ground for a $700 million building designed by internationally renowned designer Rafael Vinoly, of Rafael Vinoly Architects of New York, Tokyo, and Buenos Aires, and designer of the Tokyo International Forum. MCCA staff committed to a short-term advisory group of disability advocates who will walk the design team through a point-by-point analysis of the various entry and interior features that should be made universal if they are to work comfortably and efficiently for all users in what will become the largest structure in Boston.

Nonprofit Organizations

There are a variety of nonprofit entities with a strong presence in the development of the South Boston Waterfront. Most have an environmental or harbor-specific mission. They include the Conservation Law Foundation, the Boston Harbor Association, Save the Harbor/Save the Bay, Seaport Alliance for a Neighborhood Development, and the Boston Greenspace Alliance. All of them have been supportive of *A Neighborhood Fit for People*. The issues of indoor air quality as an element of universal design at the urban scale and the human-centered features of open space have been helpful as common ground with the environmental groups.

Adaptive Environments recruited the involvement of several disability-related nonprofits: the Boston Center for Independent Living, the Federation for Children with Special Needs, and Minorities with Disabilities/Vivienne Thomson Center. Each group has a member on the Oversight Committee. The city's largest and most diverse senior center, the Kit Clark Center, also participates on the Oversight Committee and helps to recruit user experts.

Education

To date, three academic institutions have committed to collaboration. The Boston Architectural Center has hosted a lecture series and provided opportunities to present the project to classes, and the director of interior design and the president have been visible and articulate supporters.

Boston University School of Public Health and School of Medicine, through Professor Allan Meyer, provided a variety of opportunities to educate their faculty and students. Boston University also loaned part-time research staff to Adaptive Environments for the project. One BU research assistant, who is a person with low vision, works on the project 12 hours per week analyzing the specific elements of each development parcel to assess universal design features that would work well for people with low vision. A postdoctoral geographer from India with a two-year appointment at Boston University works with the project identifying best practices in indoor and outdoor wayfinding for people with disabilities at the urban scale. An initial literature review indicated a lack of integration of discipline-specific (e.g., psychology, geography, neuroscience, design) theory as well as a dearth of user participation in the design process.

In January of 2000, the Massachusetts Institute of Technology awarded Adaptive Environments an 18-month internship for a student from the Graduate School of Architecture and Planning to work on *A Neighborhood Fit for People*. The student is the project liaison for the 20-acre Fan Pier Project.

Design Community

To date, the initiative's primary ally in the design community in the United States has been the Boston Society of Architects (BSA). Their staff and local and national network provide an ongoing set of invaluable resources. The BSA's Seaport Focus Team has been a vocal supporter. As of spring 2000, the new chair of the BSA Focus Team agreed to serve on the Adaptive Environments Oversight Committee for the project. The annual BSA convention and trade show, "Build Boston," is New England's largest annual design professional event and has included a workshop presenting progress on *A Neighborhood Fit for People* each year.

No large-scale universal design project should miss an opportunity to tap the expertise of the expanding international universal design community. Frances Aragall, president of the European Institute for Design and Disability, generously shared images of the Barcelona waterfront development. Andrew Walker, architect and instructor at the Architecture Association in London, shared insights and suggestions from recent work at the urban scale in the United Kingdom. Sandra Manley, in Chapter 58, "Creating an Accessible Public Realm," offers a compelling argument for the automobile as a culprit in creating a disabling public realm that affects everyone. She also offers the appealing recommendation that the Principles of Universal Design should be expanded to include an eighth principle, "Adding to human delight." There are few examples of large-scale models of successful implementation of universal design at the urban scale, though the pervasive negative human experience of urban design and the specific discrimination toward people with disabilities by urban design are well documented (Imrie, 1996).

60.8   *TECHNICAL ASSISTANCE AND CONSULTATION*

Adaptive Environments decided at the start that universal design at the urban scale in the United States must start with a core commitment to full compliance with legally mandated access for people with disabilities. Legal compliance is the floor upon which to build. It is also the wedge in the door—government entities, developers, and designers are all aware of their obligation to federal and state laws. Assisting entities with responsibilities under the law to meet minimum requirements opens the door to a conversation about universal design.

Technical Assistance

The scope of technical assistance varies. Through a grant from the National Institute on Disability Rehabilitation and Research (NIDRR), Adaptive Environments serves as the New England ADA Technical Assistance Center, one of 10 geographically based centers in the United States. Expert technical assistance specialists answer ADA telephone inquiries 40 hours a week. Up-to-date federally approved materials are available at little or no charge. Adaptive Environments provides training on the ADA tailored to the client's needs. For example, Adaptive Environments annually provides training to real estate brokers in Massachusetts that meet continuing education requirements. Another forum that appeals to large architectural firms is lunchtime seminars designed to address ADA guidance pertinent to their dominant design specialties. These seminars are effective forums for discussion of dilemmas they encounter in meeting their obligations on specific projects, and Adaptive Environments is approved to provide the professional education credits required annually for architects.

In Massachusetts, developers must also meet the requirements of the Massachusetts Architectural Access Board (MAAB). With the assistance of the executive director of the MAAB, Debbie Ryan, Adaptive Environments provides technical assistance for Waterfront developers.

Adaptive Environments makes technical assistance on universal design available through the extensive resources of its research library. Specific lists of universal design features pertinent to each building type and element proposed for the South Boston Waterfront are developed as parcel proposals are submitted. Information on products and suppliers is on hand.

A characteristic technical assistance request has been from firms under contract with public agencies such as the Massachusetts Highway Department and the federally funded Central Artery Project, who are working on sidewalk and street designs and on designs on or abutting the water's edge—floating docks, the Harborwalk, accessible water transportation. The United States Access Board has convened a Pedestrian Rights of Way Committee and a Water Transportation Committee, though the recommendations of the committees and the possible new guidance will not be available for a couple of years. Designers and engineers want information about best practices now. Through communication with the United States Access Board, the Federal Highway Department, and their contractors, the project has been able to offer pertinent and timely information on best practices in pedestrian and trail design.

Consultation

Consultation services are more labor intensive and more likely to involve on-site time and design review for a fee. Beyond in-house capacity, Adaptive Environments has identified universal design experts, primarily designers across the design disciplines, who are able to work with Adaptive Environments to create universal design solutions for particular projects in the South Boston Waterfront.

User/Experts

Elaine Ostroff, founding director at Adaptive Environments, coined the term *user/expert* to describe the diverse consultants that worked with faculty and students in Adaptive Environments' Universal Design Education Project. Ostroff is a leading advocate for the practical value of a role for the user/expert in universal design education and practice (Ostroff, 1997). She emphasized the opportunity to learn from people whose life experience is substantially different from the typical designer's and described the power of learning with user/experts. Examples of user/experts include people with disabilities, older people, individuals with unusual stature, and parents with small children. *A Neighborhood Fit for People* embraced the practicality and opportunity of incorporating user/experts into the project from its inception. In Chap. 6, "User/Expert Involvement in Universal Design," Ringaert describes their participation in Canada.

Adaptive Environments recruited a core group of user/experts and expected to recruit up to a total of 25 people in 2000. Each person was asked to commit to a two-year period that includes voluntary training on the legal requirements for access under state and federal laws, followed by direct engagement in technical assistance and consultation. A user/expert curriculum has been assembled. An extraordinary group of designers, including designers with disabilities and other experts on the legal mandates for access, provide training. Adaptive Environments plans to engage the trained user/experts in direct technical assistance and consultation with designers and developers and compensate user/experts for the service.

60.9   *GENERAL LESSONS LEARNED FOR UNIVERSAL DESIGN*

The challenge of universal design at the urban scale differs from nation to nation. Even if the intended outcomes are similar, the regulatory environment substantially changes the task. In the United States, it is necessary to confront a prevalent perception that laws—504, the ADA, state laws—have resolved the issue. "Just tell me what I need to do" is a common refrain from developers and designers. Getting their attention turned to universal design must first break the orientation toward rules and open them to thinking differently. Time and effort spent on communication—indeed on *marketing* the concept—are critically important. As in any creative process, it is necessary to experiment with the message, avoid rote declarations, and listen to every audience to discern what aspect of the case for universal design they can hear. No project can ignore the value of building consumer demand.

Everyone has had the experience of city places that work very badly for people. Some of them are commonplace across the globe: deplorably poor information about finding your way; faceless facades; pedestrian routes more challenging to negotiate than the rural wild; windy, dark urban caverns bordered by towers. Most people, when invited to consider their favorite examples, can identify city places that work beautifully and may even inspire or delight. When you study those choices, they are inevitably places that satisfy because they've achieved a symbiosis of function and form.

Universal design at the urban scale inevitably must make a case that the decisions made about the design of cities are very long-term and impact legions of users. And it is time, at the start of the twenty-first century, to think about making cities that embrace and meet the demands of a remarkably changed human condition. Life spans, on average, are more than 60 percent longer than in 1900. Thanks to the rapid pace of discovery in medicine and science, it is ordinary to survive illness and injury that would have ended life just a decade or two ago. Not surprisingly, expectations have changed. Age and ability are no longer rigid determinants of life choices but just two more personal characteristics in a shrinking and more diverse world. Cities, defined by density and diversity, are a very good place for design to catch up to the facts of human life. Universal design offers an organized way to think about human-centered design until that day when working well for all people is simply definitive of good design.

60.10 CONCLUSION

With *A Neighborhood Fit for People: Universal Design on the South Boston Waterfront*, Adaptive Environments leaped into uncharted territory. The scope is daunting and the goals are ambitious. Whatever the degree of success that can be achieved in the next several years, a successful outcome long-term will be defined by having built an appreciation and appetite for human-centered design that is not dependent upon the organization's continued direct participation.

To date, the project has encountered a responsive audience in nearly every sector. It is not difficult to make the case for common sense design that has its antennae tuned to the demographics of the market. It is still too soon to judge the outcome, though it is far enough along to know that it would be dangerous to let down the guard on any front. Advocacy, education, collaboration, technical assistance, and consultation will need to be balanced for at least the next several years. An Oversight Committee of 15 members has been established, comprising leaders from the disability community, state and city government, the academic community, the design community, and not-for-profit organizations. The Oversight Committee is a committee of the Adaptive Environments board of directors, and three members sit on the Adaptive Environments board. They will provide an ongoing means for evaluation of the effectiveness of the initiative and report to the Adaptive Environments board of directors three times per year. Discussions are underway with a number of urban planning academic departments to explore the feasibility of supporting a five-year study of the project that would include (though not be limited to) process evaluation, assessment of the effectiveness of user/experts, and post-occupancy assessment.

This project began with a conviction that universal design in the United States must overtly build from an embrace of access under the law. Despite 25 years of legally mandated access in the United States and layers of law and regulation, it remains commonplace to see poor, unthinking design that falls well short of compliance. The willingness to help meet legal compliance is also the wedge in the door for attention to the broader issues of universal design.

In keeping with the perspective that, at least in the United States, legal compliance is the initial responsibility, Adaptive Environments has also not shied away from making the case for universal design by talking about disability and age. Emphasizing the facts of disability deflates the myths. It is important to talk about the mutability of the human condition and to chart the demographics of an aging society. It is also useful to stress the great news of progress at the start of the new century: People are living longer, more active lives and survive injury and illness at rates unimaginable only a few years ago. Now design must catch up. The old design standards were based on a six-foot-tall, 20-year-old male, with perfect vision and a good grip. Universal design is for the rest of us.

60.11 BIBLIOGRAPHY

Boston Redevelopment Authority, *The Seaport Public Realm Plan,* Boston Redevelopment Authority, Boston, 1999.

Carr, Stephen, Mark Francis, Leanne G. Rivlin, and Andrew M. Stone, *Public Space,* Cambridge University Press, Cambridge, MA, 1992.

Center for Universal Design, North Carolina State University at Raleigh, *The Principles of Universal Design,* Center for Universal Design, North Carolina State University, Raleigh, NC, 1997.

"Cities and Suburbs: A Harvard Magazine Roundtable," *Harvard Magazine,* January/February, 2000, p. 111.

City of Boston, *Boston's Indicators of Progress, Change and Sustainability,* City of Boston and the Boston Foundation, Boston, 1999.

City of Portland, *Portland Pedestrian Design Guide,* Office of Transportation, Portland, OR, 1999.

Dixon, John Morris (ed.), *Urban Spaces,* Visual Reference, New York, 1999.

Environmental Design Research Association, *Public and Private Places, Proceedings of the Twenty-Seventh Annual Conference of the Environmental Design Research Association,* EDRA, Edmond, OK, 1996.

Imrie, Rob, *Disability and the City, International Perspectives,* Paul Chapman, LTD., London, UK, 1996.

Kreiger, Alex, and David Cobb with Amy Turner (ed.), *Mapping Boston,* the Muriel G. and Norman B. Leventhal Family Foundation, Cambridge, MA, 1999.

Lynch, Kevin, *Site Planning,* MIT Press, Cambridge, MA, 1962.

Marcus, Clare Cooper, and Carolyn Francis (eds.), *People Places: Design Guidelines for Urban Open Space,* John Wiley & Sons, New York, 1998.

Ostroff, Elaine, "Mining Our Natural Resources: The User as Expert," *Innovation* (quarterly journal of the Industrial Designers Society of America), 16(1): 1997.

Weisman, Leslie Kanes, "Creating Justice, Sustaining Life: The Role of Universal Design in the Twenty-First Century" (unpublished speech presented in Boston on April 10, 1999, at an Adaptive Environments Colloquium, available on the Web at www.adaptenv.org/examples/article2.asp).

Welch, Polly (ed.), *Strategies for Teaching Universal Design,* Adaptive Environments Center, Boston, and MIG Communications, Berkeley, CA, 1995.

CHAPTER 61
ADDING VISION TO UNIVERSAL DESIGN

Cynthia Stuen, D.S.W., and Roxane Offner, M.S.S.W.
Lighthouse International, New York, New York

61.1 INTRODUCTION

As the approach and language regarding design have changed from *barrier-free* to *universal,* there has been a movement to be as inclusive as possible (Center for Universal Design, 1997). Generally, the example of curb cuts in sidewalks has been cited as "good for everyone"—parents with strollers, shoppers with their carts, rollerbladers, as well as people who use wheelchairs—everyone, that is, except people who have vision impairments. For a person who is blind or partially sighted, a curb cut eliminates the cue needed to know where the sidewalk ends and the street begins—a crucial transition. Zebra striping in the crosswalk and color-contrasted curbs, on the other hand, may help people with partial sight as well as provide additional reinforcement for those with sight. For people who are blind, additional information is needed.

This chapter provides an introduction to normal and not-normal vision changes and provides guidelines for designing environments that are user friendly to individuals with impaired vision as well as those who experience normal age-related vision changes. Highlights of a building renovation of Lighthouse International's headquarters will illustrate the Principles of Universal Design with attention to the needs of users with impaired vision.

61.2 PREVALENCE OF VISION IMPAIRMENT

Worldwide, there are an estimated 45 million people who are blind, and an additional 135 million people with low vision (World Health Organization, 1997). The number of people with low vision—vision that cannot be corrected to the normal range by corrective lenses, surgery, or medication—is growing very rapidly and could double by the year 2020. The growth is primarily due to the aging of the global population and the increased prevalence of vision impairment among the older adult population.

In the United States, one in six Americans (13.5 million people) over the age of 45 reports some problem with their vision even while wearing glasses, according to a national survey conducted for Lighthouse International by Louis Harris and Associates. The proportion of people reporting a vision problem increases with age. These facts are from *The Lighthouse National Survey on Vision Loss: The Experience, Attitudes and Knowledge of Middle-Aged and Older Americans,* the first survey to ask randomly selected community-dwelling Americans, over the age of 45, about their visual status. These data reveal very high prevalence rates for the aging population. Among the population age 45 and over, 17 percent (one in six) report a vision problem; among those age 65 and over, 20 percent (one in five) report a problem; at age 75 and over, 26 percent (one in four) report a problem with vision (The Lighthouse, 1995).

Other Lighthouse research studies of older adults residing in nursing home settings have documented that over 40 percent of the residents have impaired vision. Too often, visual needs of nursing home residents are not taken into account in the design or service delivery by the staff in these settings (Horowitz, 1988; Stuen and Fangmeier, 1994).

61.3   AGING AND VISION

Older adults constitute the most vulnerable group for common eye disorders, such as cataract, macular degeneration, glaucoma, and diabetic retinopathy. These disorders result in either overall blur, or central or peripheral visual field loss. Low vision is said to exist when ordinary corrective lenses, medical treatment, and/or surgery are unable to correct a person's sight to the normal range. The goal of low-vision care and other vision rehabilitation services is to maximize the usable sight of a person to perform usual, routine tasks, thus ameliorating functional disability due to impaired vision. Enabling persons with impaired vision to function independently in the mainstream of society is the ultimate goal. However, critical to this goal is an educated and sensitive design community.

Vision impairment spans a continuum, from total blindness at one extreme to partial sight, or low vision, at the other. The type of impairment and its functional impact are important to understand when designing any environment. Diminishment of other senses, particularly hearing, in combination with vision loss makes it even more difficult for the older adult to function independently.

Normal Age-Related Vision Changes

There are normal age-related vision changes that have significant implications for designers to address. The cornea generally remains clear as the eye ages, but it may become slightly thicker and more likely to scatter light. The lens invariably becomes denser, more yellow, and less elastic. These changes account for the loss of accommodation (focusing power). Also, presbyopia, or the loss of accommodation, is the most universal age-related ocular deficiency. The pupil tends to become smaller, permitting less light to be admitted to the eye; hence, older individuals need more time than their youthful counterparts to adjust to changing levels of illumination. Sometime in one's 40s, near vision is "out of focus," and optical correction is required with bifocals, trifocals, or reading glasses. Visual acuity is remarkably well preserved in the elderly population. However, the blue and green ends of the color spectrum become more difficult to distinguish than the red and yellow end.

Generally, older adults have some loss of contrast sensitivity, meaning that they need sharper contrasts and sharper edges in order to discriminate between objects. These difficulties are compounded by increased sensitivity to glare. The thickened cornea or yellowed lens scatters light and interferes with vision. Older adults are more susceptible to discomfort, and sometimes, disability occurs under bright indoor or outdoor conditions.

When lighting is optimal for the older adult, vision is quite good for those without eye disorders. However, many older adults experience difficulty performing visual tasks under adverse lighting or changing illumination levels. For these reasons, designers and architects must be extremely sensitive to selection, placement, and flexibility of lighting.

Age-Related Eye Conditions

The most common age-related eye conditions generally affect the central field of vision, the peripheral (side) vision, or they create an overall blur or patchy field of vision. Overall blur in the visual field is most often caused by cataract, which is a clouding of the normally clear and transparent lens of the eye (see Fig. 61.1). This clouding of the lens reduces the passage of light. Everything looks hazy to a person with a cataract, and there is increased sensitivity to glare. Patchy or splotchy vision is often caused by diabetic retinopathy, and parts of the visual field are obliterated.

The loss of one's central field of vision is often caused by macular degeneration (see Fig. 61.2). This creates difficulty in reading, writing, recognizing faces, or anything that requires detailed visual work. There is no cure for macular degeneration. However, vision rehabilitation can be very effective in enabling people to maintain their independence.

Glaucoma is another of the common age-related eye disorders that results in the loss of peripheral or side vision (see Fig. 61.3). Furthermore, retinitis pigmentosa is another eye disease that results in the loss of peripheral vision. Without peripheral vision, environments with protruding objects can be extremely dangerous.

FIGURE 61.1 Simulation of overall blur.

FIGURE 61.2   Simulation of central field of vision loss.

61.4   *GENERAL PRINCIPLES FOR DESIGN CONSIDERATION*

Until the passage of the Americans with Disabilities Act (ADA) in 1990, federal laws, standards [e.g., the American National Standards Institute (ANSI)], and national and state building codes had focused on the requirements for people who use wheelchairs as the bottom line in designing buildings and facilities. Now, the needs of people with sensory and communication disabilities are being emphasized.

The Americans with Disabilities Act Accessibility Guidelines (ADAAG), issued in 1991, included specifications to ensure access for people with vision impairments to automated teller machines (ATMs), elevators, signs, transportation facilities, and vehicles. Also, to ensure hazard-free access, there are specifications on detectable warnings, handrails, illumination levels, and protruding objects. With the updated version, expected in 2001, many of the current provisions are expanded.

What does this mean in designing a building or facility? Following are some general principles to keep in mind.

- Since vision impairment covers a wide range of visual ability as described earlier, there is frequently no one way to provide access. The keyword is *redundancy,* particularly in signs,

FIGURE 61.3   Simulation of peripheral field of vision loss.

maps, and ATMs/AVMs (automated vending machines). For people who read Braille, providing Braille on signage is essential; high-contrast letters on signs will not help persons without sight. Conversely, people who have partial sight may not be able to read Braille, but tactile letters and strong visual contrast letters on signs can provide the needed information.

* *Consistency* is also key. Room identification signs should always be on the latch side of the door so that a person will know where to look for it. The restrooms on each floor of an office building should be in the same location for each sex.

General Considerations

For anyone entering a building, a railway station, or an airport, understanding the layout and figuring out where to go is essential. For a person with a vision impairment, the task is much more complicated. Information that the sighted person picks up immediately by a casual glance can only be grasped item by item and then assembled into a mental picture by the visually impaired individual.

Large Open Spaces.   These spaces, such as museums and terminals, can be disorienting and audible cues can be masked. Where possible, large spaces can be broken down into smaller areas through a different textured flooring or the placement of furniture out of the path of travel. Main walkways should be consistent in floor texture, color, and resiliency.

Doors.   Where possible, revolving doors should be avoided. Even the large, automatic revolving doors can be hazardous. Sliding power doors are the best solution, with adequate space between two sets of doors to allow for a person with a guide dog to get in and out safely. If the doors are all glass, color-contrasted decals should be placed across the door at 60-in centerline height.

Protruding Objects.   ADAAG addresses protruding objects in the path of travel. They are dangerous. Most people with severe vision impairment do not have a guide dog but rather use a long cane to assist them in moving through the environment. A potentially hazardous object can only be detected if it falls within the sweep of the cane's range. An object cannot protrude more than 4 in into the path of travel if it is placed between 27 and 80 in above the finished floor, since the person using a cane will not detect it. Recessing both water fountains and telephones ensures everyone's safety.

Stairway Overhangs.   These should be enclosed or designed so that a person cannot walk into them. One commonly suggested solution, after the fact, is the placement of planters under the stairway to prevent cross traffic or to prevent an individual from walking into it (see Fig. 61.4).

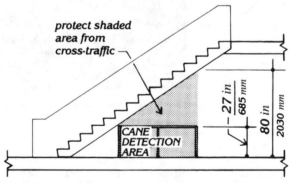

FIGURE 61.4   Overhead hazards.

Stairway Markings.   Consistent color-contrasted markings of stairways, while very important visually for impaired users, are generally appreciated by everyone. In Fig. 61.5, the stair nosings are color contrasted from the treads and large, prominently placed floor numbers, which could also be tactile and have a Braille label, clearly mark the floors. Tactile bumps on the handrail at the top and bottom of the stairway can be used to indicate floor level and/or that the end of the stair flight is approaching.

Color Contrast.   This is a very important technique to employ in all settings to promote a user-friendly environment for people of all ages. There are three perceptual attributes of color: (1) hue, (2) lightness, and (3) saturation. *Color-contrasted* means understanding how to use these attributes effectively. There are changes in perception for the older person and for

FIGURE 61.5   Stairway markings.

someone with partial sight or congenital color deficits; the visual contrast of certain color combinations is reduced. In general, the differences between foreground and background colors should be exaggerated, and using contrasting hues that are adjacent to each other in the color circle should be avoided (Arditi, 1999a).

Floor Patterns.   These can be used to provide wayfinding cues. The main path of travel can be differentiated tactilely and can be color contrasted from the reception or office area. For example, consistent contrasting tile or carpet markers in front of elevators can be helpful to people with partial sight. Busy floor patterns can be confusing for everybody and should be avoided.

Glare.   Both natural light and illumination produce glare and should be avoided. Matte-finished flooring, tinted windows, shades or blinds, light-filtering screens, task lighting—all can be used to eliminate glare. By providing a different texture and color contrast on nosings on steps, stairways can be made safer for all. Railings that are color contrasted as well as continuous provide additional safety cues.

Elevators.   In a crowded elevator, it is often difficult to know which floor the elevator has stopped at. A verbal annunciator, in addition to the beeps, is helpful for almost everyone. Two sets of elevator buttons at different heights within the car provide access for all. The lower set for wheelchair users, short people, and children; the higher set for taller persons and individuals who use the Braille and tactile numbers. A large color-contrasted plate for the elevator call buttons should be considered.

61.5   *THE NEW LIGHTHOUSE BUILDING*

Lighthouse International is a leading resource worldwide on vision impairment and vision rehabilitation. Through its pioneering work in vision rehabilitation services, education, research, and advocacy, Lighthouse International enables people of all ages who are blind or partially sighted to lead independent and productive lives. Founded in 1905 and headquartered in New York, Lighthouse International is a not-for-profit organization.

In 1990, faced with the need to modernize and expand, the Lighthouse decided to move out of its signature building in the heart of Manhattan in New York City for several years, to allow for a complete rebuilding.

While the philosophy of the Lighthouse has always been one of seeking the greatest independence for the person who is visually impaired, the new building had to reflect the principle of inclusion in society. In addition, the people seeking Lighthouse services are now increasingly older people who retain some sight and who have lived their lives as part of their communities and want whatever assistance is available to continue to be part of those communities.

Thus, the building had to include a variety of functions:

- Street-level retail store
- Performing arts and conference center
- Cafeteria and dining area
- Multifunction assembly space
- Child development center
- Technology and employment-training facilities
- Music school
- Library and reader services
- Low-vision clinic
- Teaching facilities
- Vision research laboratories
- Administrative offices

The users of the building include staff, consumers of services, and the general public who come to attend meetings or special events. The wide range of users and their physical abilities, particularly visual abilities, meant that often there was not one solution to a problem that would work for everyone.

Design Process

From the beginning, the project designers (architects Steve Goldberg and Jan Keane from Mitchell/Giurgola Architects) and their consultants (Roger Whitehouse of Whitehouse and Company, wayfinding and graphic design consultants; and H.M. Brandston and Partners, lighting designers) worked together. The goal was to respond to the wide range of abilities of Lighthouse users and to meet their needs by creating a "model of accessibility" for people who have vision, hearing, or mobility impairments.

A unique design process was developed that enabled the design team to use Lighthouse researchers, consumers, and staff for feedback through the establishment of four groups: (1) environmental, (2) orientation and mobility instructors, (3) the Lighthouse Consumer Council, and (4) user groups. From discussions and mock-up testing with these groups, the following general guidelines were developed.

General Guidelines

Orientation and Wayfinding.   To assist everyone in finding their way, the overall plan for each floor is the same with consistent location of basic elements such as elevator lobbies, fire stairs, restrooms, and reception areas. Since techniques for teaching mobility to blind individuals use rectilinear space, no curves were used in circulation areas. To avoid confusion, waiting areas are separate from the path of travel. Carpeting is used to differentiate office space from the halls since it can be detected both visually and tactilely.

Lighting.   To achieve good visibility without glare or dark patches, a rounded ceiling fixture was developed for the hallways. For offices and other spaces, the goal was to have evenness of lighting with no bare lamps exposed. Traditional fluorescent lamps were used, but the surface of the standard open parabolic reflector fixtures was painted soft white and a thin Plexiglass diffuser was placed on top of the louvers.

Color and Contrast.   Since the two most common losses in vision for people with partial sight involve visual acuity and contrast sensitivity, it was essential to develop appropriate design solutions. Brighter lighting is often seen as the method to increase visibility and heighten contrast. Unfortunately, for many people with impaired vision, it produces glare and reduces visibility. At the Lighthouse, choosing appropriate color contrast as a border (e.g., around door frames) enables the user to find the door without additional glare.

Signage.   To meet the need for information for a wide range of visual and other disabilities, signs and maps would have to be tactile as well as visual, and in some locations, audible as well. Based on vision research conducted at the Lighthouse, people with partial sight can more easily differentiate upper- and lowercase letters that are color contrasted than if the letters are all uppercase (Arditi, 1999b). Since only 10 percent of people with impaired vision read Braille, it was important to have tactile letters as well. For tactile letters, thinner, uppercase letters were found to be most discernible.

61.6   DESIGN FEATURES AND SOLUTIONS

The Lighthouse International headquarters building fronts 59th Street in Manhattan, a busy thoroughfare with crosstown buses and a subway entrance at one corner. The Lighthouse entrance has a mat outside the sliding power door that guides the person to the correct place to stand; as the isometric in Fig. 61.6 shows, there is a rail that separates the ingress and egress paths; the distance between the two sets of sliding doors is 23 ft to allow safe passage for people with guide dogs, several people entering at the same time, and for someone using a wheelchair.

A low-pile carpet runner leads the person to the main reception desk, which is on axis with the entrance and is always staffed. Tactile and visual maps are displayed on the desk at a 45° angle. It is more difficult to read Braille and tactile letters if they are flat on a vertical surface (see Fig. 61.7).

The circulation paths are delineated by carpeting that leads both to the elevators, offset from the main lobby, and to conference rooms in the rear. Attractively designed seating is

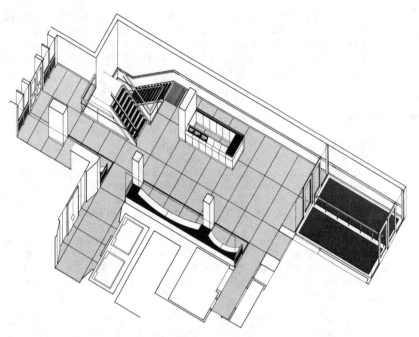

FIGURE 61.6 Floor diagram of Lighthouse lobby.

provided along one wall of the lobby out of the main path of travel, and includes space under the bench for guide dogs.

The main stairway, opposite the elevators, leads to the public areas—an auditorium one flight down, and the cafeteria and multipurpose room one flight up. The nosings on the step edge are highly contrasted with the treads and risers.

In the elevator lobby, oversized color-contrasted wall plates have oversized, illuminated call buttons. The directional signals above the elevator car doors are arrowhead shaped instead of circles, use white light for "up" and red light for "down," and also audibly differentiate between up and down signals to facilitate recognition. The elevator car has audible announcements and two control panels, one lower to meet ADA requirements, the other at standard height, which is a more comfortable height for someone to read Braille or tactile letters.

To facilitate wayfinding, there are many common features on the individual floors. On arrival at a floor, the elevator annunciator not only announces the floor but directs a person to turn left or right so that one will be facing a reception desk, which has a tactile map of the floor. One is now standing in the main path of travel, a north-south axis, with natural light from windows at each end.

All permanent rooms and spaces have signs that incorporate high-contrast upper- and lowercase letters (i.e., black background, white letters on a flat sign) and Braille and uppercase tactile letters on a raised railing. The railing is placed at a 45° angle from the wall, with the entire unit centered at 60 in. This enables the person reading the Braille to do so in a more natural and comfortable position. Since the letters are not meant to be read visually, they are the same color as the plate (in this case, magenta), which is also used to contrast the door frame from the wall and door. In addition, many locations have tiny transmitters on the bottom of the sign that, if one is carrying a special receiver, will identify the location out loud. Where there is no sign, as for example in the elevator lobby, the transmitter is placed separately in the ceiling.

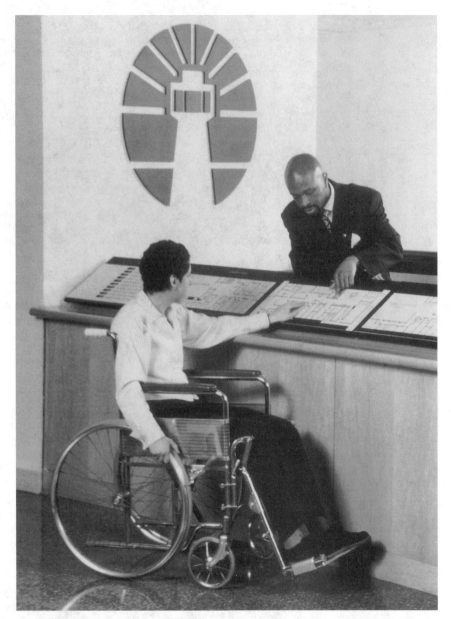

FIGURE 61.7   Reception desk of Lighthouse lobby.

For restrooms, additional symbols were used (see Fig. 61.8). A blue tactile triangle and a blue tactile circle with pictographs were placed above the signs for men and women. As well as consistency in location of these important rooms, the signs provide redundant means of communication.

Strong color contrast is used throughout the building. Warm white, low-reflectance wall color is continued in the cove moldings so that there is a clear contrast with the gray and blue-

FIGURE 61.8 Two samples of restroom signage.

violet floor of medium-brightness color. Magenta is used around door frames except doors that are to utility and maintenance closets. These are painted the same color as the walls and are kept locked.

Since this 170,000-ft^2 building encompasses so many varying activities, each of the 16 floors required a different, thoughtful approach. For example:

- The retail store on the street level was no longer to be an outlet for products "made by the blind," but rather an inviting introduction to products usable by everyone such as large-dial clocks, including specialized products for people with vision impairment. A separate street entrance was needed, as well as an entrance from within the lobby.

- The conference and creative arts center in the lower lobby has seats for 237 people, an assistive listening system for people who are hard of hearing, and a descriptive audiosystem for people with impaired vision. The stage can be accessed by a level passageway alongside the auditorium, or from the front of the auditorium by steps or wheelchair lift.

- In the former building, the children's floor was on an upper floor. Here, it was decided to place the Child Development Center on the third floor so that they would have a short ride in the elevators. Thus, the roof of the cafeteria could be used as 1000-ft^2 outdoor play area. The classrooms are behind the doors, which were designed with large circular windows so that the children could see out (see Fig. 61.9). This accommodates children who are crawling or walking to see or be seen.

- The music school on the fourth floor needed soundproof music studios, rehearsal rooms, and a dance exercise room with a sprung-wood floor.

- The rehabilitation training center on the seventh floor encompasses two kitchens, one for a wheelchair user and one standard, as well as a laundry, bathroom, and bedroom setup to help people who are visually impaired learn to function as independently as possible.

- The print access center on the eighth floor needed space for the library, which has Braille, large-print books and audiocassettes, soundproof rooms for one-on-one reading and soundproof booths for recording books on tape, equipment such as optical scanners that read text aloud, a computer room, and facilities for Brailling materials.

- The low-vision service on the ninth floor required that examining rooms be oversized to allow family members to be present. Small classrooms with large one-way windows for training low-vision specialists had to be included.

61.7 SIX YEARS LATER

Staff were pleased to leave rented quarters and move back to the new Lighthouse. The building maximizes natural light; wayfinding is simplified; staff and consumers find it a comfortable, welcoming place; and the general public, who comes for events that have been planned by outside not-for-profit organizations that rent the facility, have provided positive feedback on its user friendliness.

Some elements have been modified. Doors in the back of the lobby designed to separate the lobby area from conference rooms and the mailroom were not kept closed as originally envisioned. When open, they protruded into the path of travel. As a solution, hinges that allowed the doors to be folded back almost flat when open were installed. A two-story support column in line with the storefront, but in front of the entrance, caused the most controversy. Some people found it useful as a landmark, particularly if they were cane users; others found it hazardous. Although there was a contrasting stone border circling the bottom, it did not provide sufficient contrast. To harmonize with the interior colors and to provide strong contrast, 10-ft 8-in slats of galvanized aluminum painted a rust color were attached to the pillar and now make the site visible—for almost everybody—an attractive landmark.

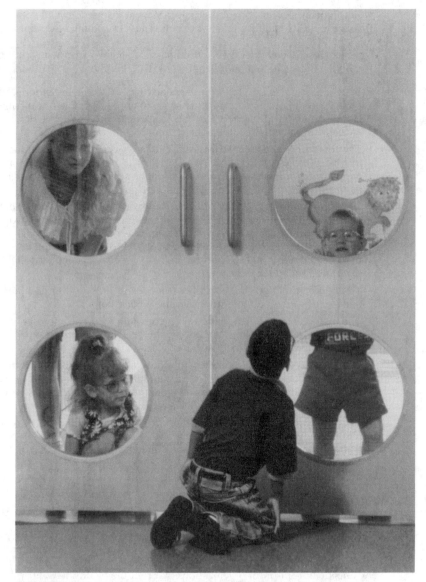

FIGURE 61.9   Doors with cutouts for Lighthouse Child Development Center.

61.8   *GENERAL LESSONS LEARNED FOR UNIVERSAL DESIGN*

Universal design is inclusive to the greatest extent possible; however, there are situations that may require a different approach or approaches to achieve the same end. For people with impaired vision or no usable vision, additional information needs to be provided. Based on the experiences at Lighthouse International, a few highlights are noted here to enhance design for people with impaired vision. These guidelines reinforce the need for redundancy and consistency as guiding principles.

- Use rectilinear space in common areas for ease of wayfinding.
- Invest in good lighting design to avoid glare or dark areas: More is not always better, but building in lighting controls is essential.
- Choose appropriate color contrast to identify space and/or borders such as door frames.
- Signs and maps should have tactile as well as visual information and be audible where appropriate.
- Utilize upper- and lowercase letters that are color contrasted with the background on signage.

61.9 CONCLUSION

The next phase of assessment is a long-awaited external evaluation by users of the Lighthouse space. Lighthouse International will be one of the several sites evaluated by a recently funded Rehabilitation Engineering Research Center at the State University of New York in Buffalo. The 5-year project will focus on universal design in the built environment. Individuals with and without impaired vision will be involved in feedback on the design elements of the building.

Given the phenomenon of global aging and the documented prevalence of vision impairment across the life span, architects and interior design professionals should consider how people with impaired vision will function in the spaces they design. The experience of designing Lighthouse headquarters to meet the needs of a broad range of users illustrates that considering the needs of a special population can benefit everyone . . . and is just good design.

61.10 BIBLIOGRAPHY

Americans with Disabilities Act Accessibility Guidelines for Building and Facilities, Federal Register (July 26, 1991; corrected January 14, 1992), 36 CFR Part 1191, Washington, DC, 1991a.

Americans with Disabilities Act Accessibility Guidelines for Transportation Facilities, Federal Register (September 6, 1991; corrected January 14, 1992); 36 CFR Part 1191, Washington, DC, 1991b.

Arditi, A., *Effective Color Contrast: Designing for People with Partial Sight and Color Deficiencies,* Lighthouse International, New York, 1999a.

———, *Making Text Legible: Designing for People with Partial Sight,* Lighthouse International, New York, 1999b.

Center for Universal Design, *The Principles of Universal Design.* The Center for Universal Design, North Carolina State University, Raleigh, NC, 1997.

Horowitz, A., *The Prevalence and Consequences of Vision Impairment Among Nursing Home Residents,* The Lighthouse, Inc., New York, 1988.

The Lighthouse, Inc., *The Lighthouse National Survey on Vision Loss: The Experience, Attitudes and Knowledge of Middle-Aged and Older Americans,* The Lighthouse, Inc., New York, 1995.

Stuen, C., and R. Fangmeier, *Field Initiated Research to Evaluate Methods for the Identification and Treatment of Visually Impaired Nursing Home Residents,* Final Report, Part I, The Lighthouse, Inc., New York, 1994.

World Health Organization Programme for Prevention of Blindness and Dearness, *Global Initiative for the Elimination of Avoidable Blindness,* World Health Organization, Geneva, Switzerland, 1997.

61.11 RESOURCES

Anyone may contact the Lighthouse Information and Resource Service for educational resources for designers and architects at info@lighthouse.org or 1-800-829-0500.

Arditi, A., R. Cagenello, and B. Jacobs, "Letter Stroke Width, Spacing, and Legibility," in *Vision Science and its Applications,* vol. 1, OSA Technical Digest Series, Optical Society of America, Washington, DC, 1995, pp. 324–327.

———, E. Holmes, P. Reedijk, and R. Whitehouse, "Interactive Tactile Maps: Visual Disability and Accessibility of Building Interiors," *Visual Impairment Research,* 1: 11–21, 1999.

Holmes, E., and A. Arditi, "Wall vs. Path Tactile Maps for Route Planning in Buildings," *Journal of Visual Impairment and Blindness,* 92: 531–534, 1998.

Lighthouse International, *ADA Accessibility Guidelines: Provisions for People with Impaired Vision,* Lighthouse International, New York, 1994. (Compilation of the ADA accessibility guidelines that relate to people with impaired vision: useful for anyone involved in designing or renovating buildings and facilities.)

Stuen, C., and E. E. Faye, *The Aging Eye and Low Vision,* The Lighthouse, Inc., New York, 1994.

CHAPTER 62

A CAPITAL PLANNING APPROACH TO ADA IMPLEMENTATION IN LOCAL PUBLIC SCHOOL DISTRICTS

John P. Petronis, A.I.A., A.I.C.P., and Robert W. Robie, R.A.
Architectural Research Consultants Inc., Albuquerque, New Mexico

62.1 INTRODUCTION

The United States has been gradually moving toward integrating everyone, regardless of physical ability, into all aspects of the built environment. Because of their legal mandate to educate all children within their boundaries, public school districts are challenged with making learning environments supportive for all students, regardless of their learning or physical abilities. Minimizing the negative impacts of the environment on people with disabilities is one of the key goals of universal design. Meeting this challenge in the real world is often burdened with confusion about regulations, threats of litigation, and balancing accessibility funding with other capital needs. This chapter describes a process for public elementary and secondary schools (kindergarten through 12th grade) in the United States to meet their accessibility requirements while balancing the cost of compliance with other capital needs. The process was developed from the project experiences of a private architectural research consulting firm working with school districts in New Mexico, varying in size from large and urban (85,000 students) to small and rural (2000 students).

62.2 BACKGROUND: REGULATORY CONTEXT

In 1973, the federal government passed Sec. 504 of the Rehabilitation Act that prohibited discrimination in programs that received federal funding. Since school districts are a part of state government that receive federal funding, they were and are still subject to this law. Federal guidelines for accessibility in schools were promulgated in 1977, and this law was a major impetus for creating design guidelines and modifications to local building codes to accommodate the needs of people with disabilities.

The Americans with Disabilities Act (ADA) was signed into federal law on July 26, 1990. The ADA is civil rights legislation that requires equal access to programs and facilities in public- and private-sector institutions receiving federal funding. The ADA requirements for state and local governments, Title II, extended the provisions of Sec. 504 of the Rehabilitation Act of 1973 (and subsequent revisions through 1988) regardless of the level of federal funding. Title II of the ADA, effective January 26, 1992, prohibits discrimination on the basis of disability in the services and programs of state and local governments. This includes public schools. The Americans with Disabilities Act Architectural Guidelines (ADAAG), an appendix to Title III of the ADA, contain the specific design requirements and standards of the act. Also known as the ADA Standards for Accessible Design, these standards can be used to plan architectural accessibility in public schools.

The ADA and the ADAAG do not constitute a building code. Public schools are subject to Title II of the ADA that requires program access for all educational services and for all educational associated program space that can be readily modified to achieve access by anyone enrolled or doing business within any school facility. As was noted previously, as recipients of public funds, public schools are also subject to the requirements of Secs. 504 and 502 of the Rehabilitation Act of 1973 and have generally had to comply with the provisions of that law. Since 1977, Sec. 502 of the law initially adopted the American National Standards Institute (ANSI) Specifications for Making Buildings and Facilities Accessible to and Usable by Physically Handicapped People, A117.1-1980. In 1984, the Architectural and Transportation Barriers Compliance Board (ATBCB) adopted the Uniform Federal Accessibility Standards (UFAS) as the standard for accessibility design to be used in all federally funded construction projects. Under the ADA, state and local governments had a choice between two standards to use for design and construction. In addition to the ADA Standards for Accessible Design, state and local governments could also use the UFAS. Many districts chose the design standards described in the UFAS. In addition, many state and local governments have adopted regulations based on federal standards to codify accessibility compliance within their jurisdictions. For example, the state of New Mexico adopted the UFAS and added greater clarification to design issues relating to needs of young students with profound disabilities. New Mexico also reviews construction drawings for most of the school districts in the state to ensure compliance with accessibility regulations.

Although accessibility has been a requirement since 1977 when the regulations for the 1973 Rehabilitation Act were signed, the ADA has had more citizen involvement since it was so highly publicized. Although people with disabilities had the same rights to make complaints under Sec. 504, this was not as well known, nor was there as much expectation or activism in the disability community for equal access. Meeting the access requirements of children, staff, and parents with disabilities presents planning challenges to school districts that ultimately result in making a variety of physical improvements. The vast majority of all funding available to meet the capital needs of a public school district, to build new schools or make major renovations, is ultimately derived from local property taxes. Most school districts have more general capital needs than can be funded from available revenue. Allocating too much to accessibility issues may prevent other important projects from taking place. Yet failure to comply with federal and local accessibility regulations can expose the school district to litigation, accreditation issues, and compliance problems with state building codes and regulations.

62.3 ACCESSIBILITY ISSUES FACING SCHOOL DISTRICTS

Public schools face many issues when seeking to comply with accessibility requirements. Some of the issues are:

- *Status of the existing facility inventory.* School districts generally have a range of facilities to manage. Many school districts had to build new facilities in the 1960s and 1970s to accommodate the baby boom generation. These facilities were constructed prior to the general integration of accessibility requirements into local building codes. While some of these sites and facilities may be relatively easy to modify, most will require extensive and costly changes. Some schools may never be able to cost-effectively comply with all access requirements. For example, a school on a steep slope may require extensive ramping to be accessible to mobility-impaired people.

- *Balancing capital needs.*   School districts must keep investing in their facilities in order to accommodate growing school enrollments, meet changing educational requirements, and renew building systems and sites as they age. There are always more improvements to be accomplished in school facilities than funding can match. Since the ADA is a federal mandate that establishes requirements but provides no money to implement these requirements, accessibility projects must compete for funding with other district capital needs (e.g., new science buildings, roofing, and renovation of older schools). How these brick-and-mortar tradeoffs are balanced to guarantee a sustained quality learning environment is an important capital planning issue. Planning must consider both the mandated regulations and laws, the educational drivers of the school district, and the special needs of the students, teachers, and community. (See Fig. 62.1.)

- *Transition planning.*   School districts are required, if accepting federal funding, to develop a transition plan that identifies how they are going to bring themselves into compliance with federal accessibility laws. The transition plan identifies program and facility deficiencies and how the district will rectify the deficiencies. Districts face liability risks if they do not comply with the law, but this is not the only reason to develop a comprehensive transition plan. The transition plan provides the opportunity for a district to think comprehensively about accessibility requirements and how they integrate with both long-range educational program and facility needs.

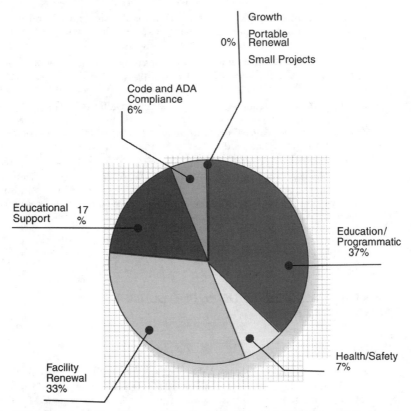

FIGURE 62.1   Addressing accessibility requirements (code and ADA) is a relatively small part of a total capital improvement program of a typical district.

In summary, the questions facing a school district include:

- What architectural barriers face staff, students, and family members with disabilities within the district?
- Are there means to meet the program accessibility intent of the law without physical modifications to a facility (e.g., changing program location, timing, media of presentation, or staffing)?
- How can the district best develop an implementation strategy that balances the capital funding needs between accessibility compliance and the general facility improvement requirements in the district?

62.4   COMPLIANCE PLANNING PROCESS

The authors have assisted a number of school districts in accessibility compliance planning, using a five-step process. The process is developed with a representative committee made up of educators, administrators, public representatives, and design professionals. (See Fig. 62.2.)

Step 1. Organize for planning.   During this step, existing data is collected and assessed, questionnaires distributed, information about the facilities and sites collected, and assessment schedules established. A kickoff meeting is held to introduce all participants, clarify deliverable services, and to establish the goals, schedule, and expectations for the project.

Step 2. Identify compliance target.   During this step, the committee establishes a compliance target based on an assessment of the scope of work and availability of funds. The committee also sets the parameters of how the process is monitored.

Step 3. Assess the sites and facilities to identify compliance needs.   A detailed evaluation of each site and facility is performed during this step. An elementary school may take 1 to 2 days, and a secondary school may require up to 5 days for the evaluation.

Step 4. Identify projects necessary to meet compliance.   This step identifies specific projects and associated costs to rectify any accessibility deficiencies. Often, further discussion with district educators is required to understand the spatial requirements of the educational programs that need to be met by the proposed solution.

Step 5. Develop a strategy to address needs.   This implementation strategy sets the guidelines for decision making relating to each project described. Timing of need, ability to accomplish, and funding sources are discussed and decided on in this step. This strategy realizes the long-range capital planning goals of the district, as well as meets the immediate accessibility needs for students, staff, and involved community members at each school.

A perfect example of the process can be found with the Albuquerque Public School (APS) District in the next section.

62.5   CASE STUDY: ALBUQUERQUE PUBLIC SCHOOLS TRANSITION PLANNING

Planning Context

Albuquerque Public Schools (APS) is a large, 85,000-student urban school district with 124 school sites. In 1989, the district initiated a long-range Facilities Master Planning (FMP) process to ensure that school-based needs were documented, assessed for urgency, prioritized to be resolved in an equitable manner, and presented as a district-wide greatest-needs-first list

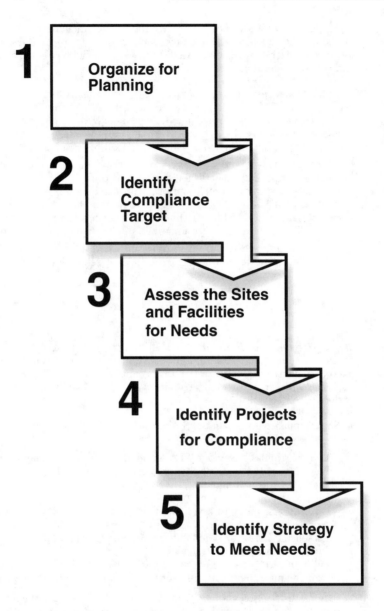

FIGURE 62.2   Compliance planning process diagram.

for voter approval. The plan is updated every 5 years prior to local elections, where district voters are asked to approve local funding initiatives based on the property tax authority of the district. The plan guides about $55 million of annual improvements. For additional discussion regarding the Albuquerque Public Schools Facilities Master Plan, see "Strategic Asset Management: An Expanded Role for Facility Programmers" in *Professional Practice in Facility Programming* (Petronis, 1993).

While the APS FMP is based on a thorough assessment of facilities that considers the site, physical plant, and adequacy and environment for education, it does not provide the detailed information necessary to develop a district-wide transition plan. APS recognized a number of accessibility issues that needed to be resolved, including:

- About 80 percent of the school facilities [approximately 8 million gross square feet (GSF)], were constructed prior to the enactment of the ADA. The existing (FY2000) inventory of school facilities is an average age of 39 years old. Many schools were constructed prior to federal accessibility law, and most with difficult-to-change materials or architectural styles (e.g., narrow, double-loaded corridors with load-bearing masonry construction).

- A large population base with a very broad range of disabilities that needed to be accommodated—the spectrum of disabilities ranges from people with simple mobility problems to those people who self-propel their own wheelchairs, from asthma to extreme chemical sensitivities, as well as from mental illnesses to profound multiple disabilities requiring nursing and specialized hospital-like settings. These settings can require areas for catheterization, tube feeding, or Hoyer lifts for immobile students with profound multiple disabilities.

- A commitment to community-based schools where all students, including students with disabilities, can share the learning experience at all school levels with siblings and peers.

Planning Steps

Step 1—Organize for Planning. The district created an Accessibility Oversight Committee to address compliance issues in a proactive manner. The committee is made up of teachers, parents, and district support staff. District support staff includes design and construction professionals who serve as technical advisors as well as district accessibility advocates. The role of the committee is to:

Set compliance goals. The APS Accessibility Oversight Committee adopted the following goals:

- Use the stipulations under UFAS as guidance for its designers and contractors.
- Identify needed accessibility projects across the district that cannot be otherwise addressed through administrative means. This is to prevent discrimination under Sec. 504 and to comply with the ADA.
- Develop a strategy to select what projects at which sites can be done with available funds.
- Establish criteria for scheduling project implementation to supplement the APS FMP facilities improvement strategy.
- Fulfill the APS district goal to meet school and student needs first. Needs at schools and their campuses have priority over those at administration and support sites.

Identify district accessibility needs. This includes implementing a professional assessment of district facilities as well as reviewing and taking action on petitions by anyone using a school facility with an accessibility issue that cannot be addressed by the site administrator.

Establish a process to identify urgent accessibility issues and their priority.

Monitor and control funding for accessibility compliance work.

Create a transition plan required per federal regulations and give advice and comment to staff, those who petition the committee, and design professionals relating to its implementation. The Accessibility Oversight Committee works cooperatively with district legal and risk management staff in this effort.

Serve as a resource to create design solutions for unique disability issues not addressed by UFAS. For example, committee members will critique design options that may include custom-made furniture, specialized access ways, air conditioning/filtering systems, and bathing/changing room configurations to address unique situations. When possible, committee design recommendations attempt to minimize repetitive motion and back movement injuries by melding UFAS and ergonomic design standards.

Step 2—Identify Compliance Target.   The first task the committee had was to decide whether to bring every facility (124 schools) up to an achievable compliance level or use a strategy that concentrated compliance efforts at a select number of school facilities. In consideration of the high cost of full compliance, conservatively estimated to be $53 million or more, and the necessary long implementation time frame, the committee chose a strategy that targeted work at 56 schools to be upgraded over 10 years. The target strategy was based on the following criteria:

Geographic distribution.   Schools in APS are organized by clusters. A cluster is made up of a high school (grade levels 9th through 12th), a number of middle schools (grade levels 6th through 8th) and elementary schools (grade levels kindergarten through 5th). Each cluster serves a certain geographic area. Generally, three to four elementary schools promote students into a middle school, and two to three middle schools, in turn, promote students into a high school. The cluster was chosen as the basic unit for compliance planning.

Site/facility appropriateness.   Within the cluster, sites and facilities were chosen that were the best candidates for achieving full compliance. These included the newer schools, which were designed and constructed to be ADA compliant or close to ADA compliant, schools built on level terrain, and schools with one level. Also included were existing or planned facilities with educational programs for children with multiple disabilities. Schools not on the list were the oldest schools, those that are the most costly to upgrade to current accessibility standards, and schools with multiple buildings on a sloping site. Since there is only one high school per cluster, all of the high schools were selected regardless of age, site, or building configuration.

This process resulted in a target list of schools that would be modified to be as fully compliant as physically possible. The target list represents about 45 percent of the district's school inventory and consists of 11 (or 100 percent) of the high schools, 14 (or 66 percent) of the middle schools, and 31 (or 40 percent) of the elementary schools. Accessibility issues at schools not on the target list would be addressed on a case-by-case basis through the petitioning process established by committee.

Step 3—Assess the Site and Facilities to Identify Compliance Needs.   Once the target school list was established, a detailed accessibility assessment of each school was conducted to identify specific accessibility projects needed for compliance. This accessibility assessment was designed to complement comprehensive school evaluations done as part of facility master planning for the district.

All target schools were evaluated for compliance with applicable accessibility requirements. The evaluator completed a workbook, comprising six accessibility assessment categories, including a five-part summary section for each site and facility. Figures 62.3 through 62.5 show checklists of the design regulations for that accessibility assessment category. The summary section contains the evaluator's narrative on the overall status of the site and facility, photographs, and floor plans marked to show locations of noncompliant features as well as the status of existing compliant elements.

Step 4—Identify Projects Necessary to Meet Full Compliance.   Based on the field evaluation, capital improvement projects (CIPs) were identified and entered into a computer database. The CIPs are recommendations for correcting deficiencies or for adding accessibil-

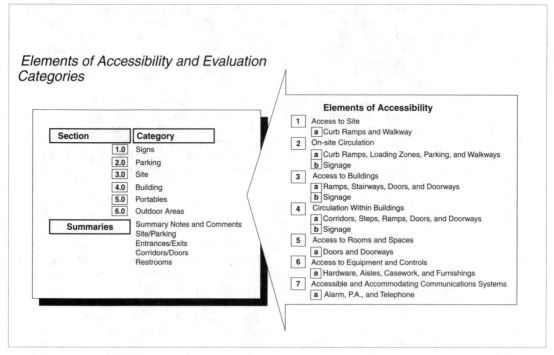

FIGURE 62.3   An outline of the accessibility evaluation forms used by APS to determine capital improvement needs.

ity features to sites and buildings to achieve compliance with ADA requirements. Each CIP is coded in the same manner as the general capital improvement projects of the district's FMP showing different category, type, and priority of work. In addition, accessibility capital improvement projects were also coded according to their level of:

Difficulty level 1.   Work required to meet readily accessible status. This work can be easy to accomplish under normal construction conditions and costs.

Difficulty level 2.   Work required to meet readily accessible status is possible but at a high cost. These projects can be accomplished, but work is more difficult, often requiring special equipment or major structural/HVAC changes.

Difficulty level 3.   Work required to meet readily accessible status is difficult with high costs. These projects are very difficult or unreasonably expensive to accomplish.

This rating system allowed the district to distinguish between those projects that are affordable and those that entail great difficulty and substantial capital cost (see Fig. 62.6). Often, a program change will occur to resolve an accessibility issue that requires a level 3 solution.

The total need of the target schools identified in 1995 exceeded $23.5 million, equivalent to approximately $7.00 per gross square foot (GSF) of evaluated facilities. These accessibility needs are in addition to the $820 million of other capital needs that address the growth, health-safety, educational/programmatic, and facility renewal needs of the district and that are identified in the general facilities master planning process (see Fig. 62.7).

Facility Information

4.0 Building Access

4.1 Are there any power operated or power assisted doors in the building?

Yes   No

4.2 Do any buildings have more than one story?

Yes   No

4.2.1 Is there an elevator between floors?

4.2.2 Is there an *EVAC Chair*?
(Chair to transfer student down stairs in emergency)

4.2.3 Is there an *Area of Refuge*?
(Also referred to as "Safe Haven," a fire safe area for persons with physical impairments to await rescue)

4.3 Do any buildings have more than one level less than one story in height?

Yes   No

4.3.1 Are different floor levels connected by steps?

4.3.2 Are different floor levels connected by ramps?

4.4 Are there wheelchair accessible restrooms on each level / story?

Yes   No

Staff

Student

Unisex

Albuquerque Public Schools
Facilities Access Evaluation

Principal's Questionnaire   6
Form ARC 9417.040
3/23/95

FIGURE 62.4 An example of a self-evaluation form completed by each principal at each school. The evaluator as part of the assessment process reviews this information.

Accessibility Worksheet

4.10 CAFETERIA:

a. Does cafeteria meet requirements?
b. Is the cafeteria on an accessible route?
c. Does the cafeteria have an accessible entrance?
d. Are there any tables that are handicap accessible?
e. Are the tables at least 36" clear between parallel edges?
f. Are tables at least 28" to 34" above the floor?
g. Is the knee space beneath the tables at least 27" high, 30" wide, and 19" deep?
h. Is there a 48" clear floor space approach at the tables?
i. Do food service lines have a minimum clearance of 36" for wheelchair passage?
j. Do self-serve shelves provide for reaching capabilities?
k. Are hot tables protected from touch?

4.11 LIBRARIES:

a. Is the library on an accessible route?
b. Is there proper seating at tables?
c. Are the seating spaces provided with tables at least 27" high for knee space, 30" wide and 19" deep?
d. Is there at least one check-out counter that is 34" above the floor?
e. Is there a 30" minimum between edge of table and accessible path of travel?
f. Are book shelves accessible?

Complete the boxes which apply:

YES   NO   NUMBER OF:

Albuquerque Public Schools
Facilities Accessibility Evaluation

Form ARC 9417.004   4/20/95

13

FIGURE 62.5   An example of a field evaluation form.

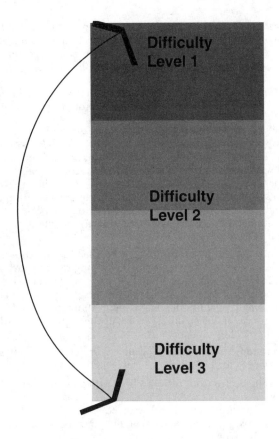

Difficulty Level 1
Difficulty Level 2
Difficulty Level 3
Readily achievable
✓ Not difficult to implement
✓ Relatively inexpensive (low unit cost)
✓ In-house labor (possibly)
✓ Relatively quick (generally no architectural/ engineering required)
More difficult to achieve
✓ Relatively expensive (medium unit cost)
✓ Requires skilled labor
✓ Does not require major alteration/structural modification
✓ Not quick to implement (may require architectural/ engineering)
Difficult to achieve
✓ Relatively expensive (high unit cost)
✓ Requires high-skill labor
✓ Requires major alteration/ structural modification
✓ Not quick to implement (generally requires architectural/ engineering)

FIGURE 62.6 Difficulty levels of access capital improvement projects.

Difficulty level	Description	Summary costs
0	Issues	$0
1	Readily achievable	$13,228,900
2	Achievable—high costs	$ 7,969,600
3	Difficult—high costs	$ 2,124,800

FIGURE 62.7 A summary chart of costs associated with the levels of difficulty.

Step 5—Develop a Strategy to Address Needs. The committee reviewed the compliance needs and adopted a two-pronged strategy that includes a formal capital plan to incrementally make improvements to target schools and an ongoing process where all members of the school community can petition to resolve immediate accessibility issues.

Step 6—Long-Range Capital Planning. To meet the long-range capital needs in target schools, the Accessibility Oversight Committee adopted the following strategy:

Establish a fund to address projects with a difficulty level of 1 and 2. A compliance fund with an annual appropriation of $1.3 million was established to address projects with a difficulty level of 1 or 2. Projects within these difficulty levels are defined as achievable. The budget for this fund was determined by a separate district committee charged with oversight of all capital needs of the district. The intent was to provide a reliable source of money to be used to systematically address needed improvements at target schools. The fund is also used to address other needs at nontarget schools on a case-by-case basis, and to supplement other funded new construction and major renovation projects that foster long-range accessibility goals.

Address level 3 projects when major renovation is scheduled for the school. Level 3 projects (difficult, high costs) will be implemented over the long-term as part of major school renovation projects identified in the FMP. The main driver for this decision is to avoid the potential of making short-term accessibility improvements in a school and then later tearing up the work and doing it again during major renovation activities. It also recognizes that a better long-term accessibility solution can usually be implemented without trying to work within the current building layout constraints.

Address the needs of students as the first priority. Projects at target schools will be addressed in the following priority order:

1. Needs impacting students are to be resolved first, with an emphasis on treating all target schools equally.
2. Needs impacting staff at target schools are resolved next.
3. Needs impacting parents, guardians, and community members are then to be resolved. This category is usually met when students' needs are addressed.

Establish priorities of project type. Within the guidelines already stated, the Accessibility Oversight Committee established the following priorities by project type:

1. *Upgrade or replace areas for nursing.* Schools with large numbers of students with profound disabilities require a larger and specially equipped nursing area. The nursing area includes: a help alarm connected between the nursing office, associated learning areas, and the main office; restrooms designed to allow privacy to students with special requirements (e.g., catheterization); and separate specialized shower/bathing areas.
2. *Provide accessible student/staff restrooms and drinking fountains.* The intent is to provide a distributed group of compliant restrooms and drinking fountains that serve the basic needs of the site as a first priority. The remainder of the restrooms will be addressed after other accessibility needs are met. Modifications generally impact only a part of a restroom such as simple changes to toilet stall doors to allow pregnant teens enough clearance to get in and out. Accessibility upgrades coincide with a district effort to renovate restrooms older than 30 years to address failing piping and fixture renewal. When completed, this work will result in fully accessible restrooms in older schools.
3. *Complete an accessible path from the public access point(s) (parking, drop-off zone, and sidewalk) to the administration office area and classrooms.* This is a local city regulation to allow any person access from the city sidewalk into the school. This is beneficial to the city since school sites also generally serve as voting locations.
4. *Make student educational space accessible, especially one-of-a-kind areas.* Educational spaces include the classroom, library, gymnasiums, and the exterior physical education

areas (e.g., playgrounds and fields). For interior spaces, this may include making door/hardware modifications, providing exterior/interior ramp and path modifications, installing interior signage, and providing accessible furniture and fixed casework. If the program space is too small, the district will attempt to relocate the program to a larger area that allows adequate clearances and space for teaching needed curriculum activities. Exterior spaces may require any number of modifications, including: grading to minimize slopes; and reconstructing play areas to include age- and disability-appropriate equipment; additional shade structures; cushioned play surfaces; and larger hard-surfaced areas. APS recently completed, in joint venture with the city of Albuquerque, a fully accessible playground and park at one of the school centers where 24 children with severe disabilities share classroom space with the neighborhood children.

  5. *Ensure access to assembly and performance spaces, including cafeteria eating and serving areas, gyms, and performing arts centers.*   Currently, the district is minimally meeting this requirement. For example, all performing arts and gym spaces are accessible to the stage and backstage facilities as well as to seating areas. Over time, the goal is to provide opportunities for dispersed seating, multiple exit routes, and additional restroom availability.

  6. *Correct any impediments in fire drill exit routes and install strobe alarm (both visual and horn) systems.*   This work is a health/safety consideration that has priority funding through the district's regular capital program. The Accessibility Oversight Committee will often hear an application for a modification to an existing alarm system. For example, the frequency range of the strobe and horn system at the high school for pregnant teens was modified to reduce stress on in vitro babies and lessen damage to newborn hearing.

  7. *Provide casework modifications.*   Appropriate modifications to nursing, library, science lab, and administration area casework are now done automatically as part of renovation work in the district. However, separate accessibility funding may be used in cases where a renovation is planned too far in the future to be of value to students or staff now in the school.

Step 7—Meeting Immediate Accessibility Needs.   In addition to the formal capital plan, the Accessibility Oversight Committee established a flexible process where all members of the school community (students, parents, staff, and other members of the public) can petition the committee with any accessibility issue that they believe exists. The committee staff investigates on a case-by-case basis all petitions. The investigation considers all potential physical and nonphysical solutions to the issue at hand and makes a final recommendation to the full committee for action. This process recognizes that there is no generic disability and that personal attention is often required to make accommodation. It also serves as an important avenue for public input, as well as a method to implement nonscheduled capital improvements.

The needs of the nonschool community are also heard by a petition process. The district attempts to meet accessibility needs for school-sponsored uses. Nonschool user organizations (e.g., churches, scouting groups, community clubs, and similar educational groups that meet before or after school) have the responsibility to provide assistance and program accommodations as requested by their participants or to move to another facility that meets their needs.

62.6 *LESSONS LEARNED FOR UNIVERSAL DESIGN*

The planning process described in this chapter serves to educate participants on the need for universal design and how the built environment can be a positive force to integrate all members of the school community, and the community in general. Universal design principles promote design solutions that consider the widest range of human needs. Universal design goes beyond the effort to provide special things to special groups that meet the minimum requirements dictated by codes and regulations to a process that seeks to provide optimal solutions for all users.

The authors believe that the planning for accessibility compliance can easily be adapted to a more general universal design orientation. School districts are ripe environments to apply and implement universal design principles. School district facilities accommodate a tremendous diversity of users on a daily basis that range from the very young to the very old, with every variation of ability. Most districts are oriented toward community-based education models that seek participation and inclusion of all groups. Organizational structures currently in place to identify and plan for the needs of special populations that move through the school system can also serve as advocates to promote solutions that meet the widest needs of all users. As universal design solutions are implemented within school districts, they will serve to change the perceptions and expectations of what is good design for the next generation.

62.7   CONCLUSION

Accessibility compliance requires careful planning, with active communication between teachers, their students, and the staff that implement the overall district capital improvements. The authors' experience shows that accessibility planning can be integrated with these other capital planning activities in a systematic and equitable way. By establishing priorities and targeting achievable goals, school districts can transition toward compliance, stay within their capital improvement budget, and thereby make the most effective use of district resources. Through establishment of a systematic planning process, accessibility compliance considerations become a fundamental element in facility programming and evaluation activities of the district and, therefore, are a driving force toward general universal design acceptability and implementation.

62.8   BIBLIOGRAPHY

Petronis, John P., "Strategic Asset Management: An Expanded Role for Facility Programmers," in Wolfgang F. E. Preiser (ed.), *Professional Practice in Facility Programming,* Van Nostrand Reinhold, New York, 1993.

CHAPTER 63
ACCESS IN REBUILDING BEIRUT'S CENTER

Riadh Tappuni, Ph.D. (Arch.)[1]
UN-ESCWA, Beirut, Lebanon

63.1 INTRODUCTION

This chapter describes the attempt to create a barrier-free center for the city of Beirut. The underlying theme was to develop an urban environment that is socially inclusive by being physically accessible to everybody. Applying principles of universal design to an urban-scale project meant having to tackle the issue at all levels and stages of urban planning and design. The resulting outcome, therefore, was not only a more user-friendly city center, but also a set of design guidelines, implementation checklists, troubleshooting tables, and a post-occupancy evaluation process that involves the beneficiaries. The lessons learned from all this relate to policy making, partnership, training, education, awareness and advocacy, and information exchange.

63.2 BACKGROUND

The growing awareness among professionals, decision makers, and legislators of the importance to make our environments barrier-free can be observed on a wide international scale, but its manifestations are fairly diversified. The more developed countries, through the mobilization of their political establishments and their highly developed social structures, have achieved considerable progress in institutionalizing barrier-free issues by completing the process of standard setting, legislation, and implementation monitoring. This process often evolved in harmony with the overall process of development and was enhanced and supported by its mature mechanisms. Yet, in spite of the continuously rising awareness of barrier-free issues, the approach and achievements in this field have varied considerably, and are significantly lacking in many respects.

Attitudes of society toward people with disabilities have shifted considerably from one that confined these people to special institutions or homes, to provision of special accessible facilities and routes. Both attitudes are, to different extents, exclusionary. Universal design, on the other hand, is by definition an inclusive process.

The concept of universal design is a paradigm that should be viewed as an advanced phase of development in the domain of barrier-free planning and design. Perceived within this context, universal design addresses the fundamentals of planning and design, targeting the full spectrum of the population, thus aiming at an inclusive environment for all, including children, elderly people, and people who are disabled. To provide equitable access is to enable important sectors of society to participate in the socioeconomic development of the country.

A similar awareness has evolved in the developing world. But the picture is extremely diverse and the rate of success varied. Fueled by negative social attitudes and stigma, people with disabilities are among the marginalized groups of society, and the extent of marginalization varies with gender. The restrictions to full and fair participation are especially severe for women and girls with disabilities. But this cause at the national levels lags behind a clear international recognition as manifested in the Platform of Action that was adopted by the 1996 Beijing Fourth World Conference on Women, and the moral commitment made by governments toward its implementation (Sará-Serrano, 1999). But at the grassroots and popular levels, lack of awareness and comprehension of the issues at hand have led to the following misconceptions:

- The technical requirements are beyond the technical capabilities of a developing economy.
- A building or a city that provides for accessibility is too costly and economically unjustifiable.
- The contributions of persons with disabilities to development are less than those of others, and their well-being is dependent on the welfare system.
- The population group addressed by this is not large or significant enough to justify the expense or effort for a total or wide application.

Traditionally, Arab governments viewed issues of disability as social issues that were handled by ministries of social affairs or departments of social welfare. Public concern was usually expressed by forming civil society organizations of a charity character, often affiliated to religious factions and financially supported through charity donations. This enforced the image that providing for people who are disabled is a burden to be encumbered by the community.

However, a major shift has recently occurred in developmental thinking toward social issues, and terms like *social affairs* are often being replaced by *social development,* influencing development agendas in general, and those of nongovernmental organizations (NGOs) in particular. Disability NGOs in the Arab countries are among the most active community-based organizations, and they have good advocacy and lobbying capabilities.

63.3   BARRIER-FREE URBAN DESIGN AND PLANNING IN THE DEVELOPING WORLD

Regional specificity in approach results in standards, building techniques, and technologies that are appropriate to the local environment. It is within this context that the African Decade for Persons with Disabilities (2000–2009) emphasizes "implementable accessibility." Moreover, an initiative is being developed in South Africa to establish a Disability Technology Institute that includes a Center for Universal Design. The institute will serve the sub-Saharan region and coordinate efforts in the field (Thompson, 1999). But in order to assess the scope of the problem, it is important to consider the developmental background of cities.

Arab cities have evolved over a long span of time in an incremental manner, resulting in a great mass of building fabric that is sympathetic to the human scale but limiting to human movement. Although the problem of making the existing fabric accessible is far greater and more difficult to tackle than new construction, most of the available literature is geared toward new planning and design. This can be attributed to a bias in the educational paradigms and current urban development legislation.

Driven by their development circumstances, cities with developing economies have approached the issue of accessibility in a variety of ways. Nations of the Middle East suffered long bouts of wars and civil unrest and attach a special significance to disability issues.

But one can observe that many Arab cities are actually full of social and physical barriers for even the healthy adult individual. Incremental and disjointed problem solving and inefficient use of resources often lead to inconsistent and badly distributed development. Many programs and projects, which are potentially well designed and implemented for the benefit of people with disabilities, are not reachable by the beneficiaries and stand as isolated pockets in the sprawling urban fabric. Developing cities are facing great challenges. They have the monumental task of catching up with modern-day development, accommodating for the new technological advances, acting as hubs of national economic activities, and linking to the international global economy. They are also required to preserve their urban and social heritage. All this often occurs with limited resources.

Thus, it is important to think in terms of techniques that are appropriate for the socioeconomic environment. Developing economies are usually overburdened, and solutions that are costly to implement or maintain would pose an unacceptable additional load.

In assessing progress in the field and identifying inadequacies, one needs to look at the major components that are essential for a successful and sustainable national effort (Table 63.1). But seldom do circumstances allow for a coherent convergence of all of these components, especially in light of the pace of progress needed to be pursued, and the instability of the economic and political systems. It is within this context that the experience gained in the reconstruction of Beirut represents a unique case that will be discussed in this chapter.

63.4 THE USER, TECHNOLOGY, AND THE BUILT ENVIRONMENT

A building is a set of solutions that satisfy a number of human requirements. The measure of success of a building is the extent of satisfaction of these solutions to the requirements. The conventional design process defines the physical requirements as those of a healthy adult, thus excluding the needs of significant sectors of the population. Through this approach, children, old people, pregnant mothers, and persons with temporary or permanent disabilities attain various environmentally exacerbated handicaps.

Building users attempt to overcome these handicaps by extending their level of ability through the use of assistive devices, like eyeglasses, crutches, white canes, and wheelchairs. These devices help to create a better interface between the user and the environment. A pair of eyeglasses helps to read signs that are otherwise unclear or too small, and a white cane identifies barriers and protruding elements in a pathway.

Another level of interface is that between transport vehicles and the environment. A person's mobility is limited by the ability to use transport means, which is facilitated or limited by their level of interface with the built environment—a bus and the bus station, an airplane and the terminal, a private car and the parking space. Automobiles have the advantage of being able to penetrate the built environment, getting closer to the user. But modern airports have

TABLE 63.1   The Five Main Factors Essential to Barrier-Free Environments

Political will	Public awareness	Guidelines and legislation	Implementation, monitoring, and testing mechanisms	Training and sensitization
At the level of national decision making	Of the importance of inclusive design	Ensuring universal design and suitability to local practices	Appropriate to local practices and allowing for post-occupancy evaluation	Of professionals in universal design

been more successful in facilitating the move of the individual to the airplane through the use of moving belts, telescopic tubes linking plane to terminal, electric passenger carts, and electrically powered doors—all through a stepless environment. Steinfeld provides a more elaborate review of universal design in transportation in Chap. 24, "Universal Design in Mass Transportation."

63.5 TOWARD A SOCIALLY INCLUSIVE CENTER FOR BEIRUT

After vicious armed hostilities that lasted for more than 15 years, Lebanon witnessed an energetic drive to reconstruct the center of Beirut. This presented a rare opportunity for planners and architects for a distinctive and modern approach to urban development issues. The conflict took its toll on urban settlements throughout the country, but nowhere was the devastation more evident than in the center of the capital, Beirut, where the infrastructure was completely demolished and only the skeletons of most buildings were left standing. With the end of hostilities in the early 1990s, a special company was formed to develop the Beirut Central District (BCD), comprising 1.6 million m^2 in area. The Lebanese Company for the Development and Reconstruction of Beirut Central District (Solidere) devised a scheme with a total of 4.4 million m^2 of built-up space accommodating a wide variety of activities. Businesses at the BCD will employ about 100,000 people. The market complex, replacing the old bustling Beiruti souks that were a commercial Middle Eastern landmark, is still under construction. It is 102,000 square m^2 in built-up area—by far, the largest task undertaken by the company. Hotels are expected to reach a total of 200,000 m^2. The residential part of the development constitutes 42 percent of the total area and will be home to 40,000 residents. What was most significant about the project was that it included the construction of a completely new and comprehensive infrastructure, a task usually undertaken by government institutions. The plan classified buildings into two categories: (1) new construction (new buildings on empty plots or to replace buildings that have been completely destroyed or destined for demolition) and (2) structures that were earmarked for rehabilitation or conservation. This last category includes buildings of cultural and historical value, or that are considered of significance to the collective memory of the Lebanese population.

Regional Support Toward a National Effort

Due to the considerable symbolic significance attached to the geographical location of the project, and due to the prevailing political circumstances, a lively and heated public debate ensued around the proposed plans. As a result, the controversial issue of the social context of the project was considered in need of more emphasis. Within the context of social inclusion, the Economic and Social Commission for Western Asia (ESCWA) was approached in 1994 to provide technical assistance in the field of accessibility for people who are disabled in the Beirut city center. ESCWA's efforts started with a definition of the size of the problem, aiming to place all efforts in a national perspective. Meetings with Lebanese officials and NGO representatives indicated that issues of disability are considered in need of urgent attention. Other than the existence of many NGOs that are generally structured around factional affiliations, and are charity oriented in outlook, work in this field seems to have lacked a clear national framework.

The first step in this effort was, therefore, to suggest a national strategy for accessibility in Lebanon. In view of the circumstances of the country, and in order not to delay development or miss the chances that the current reconstruction efforts provide, the strategy suggested a two-track approach. In the long term, the aim was to achieve a barrier-free environment throughout the country. The immediate objective was to provide accessibility through the ongoing reconstruction of the BCD. It was recognized that efforts within each track should not preclude the other; it was later proven that the two tracks complement or consolidate each other. Most of the tangible efforts that have been achieved so far in Lebanon fall under the second track, namely within the Beirut city center reconstruction effort. This chapter will demonstrate that the barrier-free efforts of Beirut are establishing the foundations for a methodology that will support similar efforts at the national level and possibly in other parts of the developing world.

Setting Appropriate Guidelines: A Design Manual for a Barrier-Free Environment

With the endorsement of the Lebanese National Council for the Disabled, which is the highest formal national institution for people with disabilities, the ESCWA-SOLIDERE working team started by identifying guidelines that are applicable to the environment of Beirut. This entailed looking at internationally published literature as well as standards that have been adopted in a few countries of the region. Providing a barrier-free environment in the BCD required addressing accessibility in buildings that already exist but were originally designed and constructed with little or no concern for people who are disabled. The plans were to make these buildings suitable for contemporary living and business needs. Their rehabilitation provided an important opportunity for making them also responsive to the special needs of groups like children, elderly people, and people who are disabled.

On the other hand, the project involved the construction of new buildings, as well as a new infrastructure. Hence, there was the urgency for compiling guidelines, standards, and specifications that architects, planners, and engineers can use.

The team recognized at an early stage in the endeavor that a clear, systematic methodology has to be followed in order to achieve consistency in the approach and sustainability in the effort. As illustrated in Fig. 63.1, one of the first tasks that were undertaken aimed to identify and determine guidelines that are suitable for application in Lebanon. This involved the study and comparative analysis of well-tested international accessibility guidelines and legislation, and selecting what is suitable for Lebanon.

As a result, a manual entitled *Accessibility for the Disabled: A Design Manual for a Barrier-Free Environment* was drafted, put to use in its prepublication form in 1995, and later published in English in 1998 (Solidere, 1998). In the meantime, Solidere has made it mandatory for professionals to apply these guidelines on all construction work in the BCD. Aspiring to

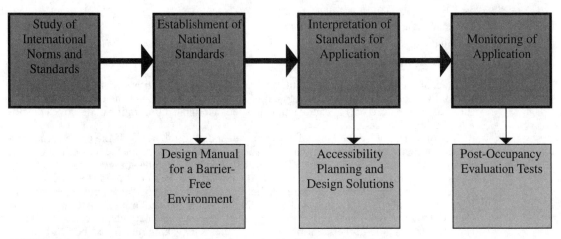

FIGURE 63.1   The development sequence followed toward achieving a barrier-free BCD.

achieve a certain level of quality of construction, Solidere made it conditional for any developer to abide by certain guidelines that replaced the relatively lax or deficient prevailing building regulations. But for pragmatic reasons, there was more flexibility toward rehabilitated buildings. This was reflected in the manual, which included a special section headed "Existing Construction," pertaining to every urban and architectural design consideration. Since public facilities were the direct responsibility of Solidere, the company had a better chance to control the application of the guidelines in streets, pathways, and public gardens, without the pressure of the private property owners.

The Manual as a Design Aid

The manual is designed to achieve the following objectives:

- To be easily usable through the planning and design processes, by taking into consideration the various stages of conceptual planning, architectural design, detailed design, and implementation
- To address the three categories of urban fabric: (1) urban public elements, (2) new buildings, and (3) the rehabilitation of old buildings
- To allow for feedback from the field by facilitating post-occupancy evaluation to test for barriers
- To adopt guidelines that emanate from inclusive anthropometrics, suitable for the Lebanese individual

Since no anthropometric data have previously been tabulated for Lebanon, the team made a comparative analysis of data for Canada, the Netherlands, France, Jordan, and Sweden. It was noticed that the data were given separately for males and females, which could result in exclusionary design decisions. Consequently, a range of anthropometric measurements was identified for Lebanon that can serve the purposes of the manual, which can apply to the adult individual, inclusive of both genders. This decision was made in light of the fact that, for certain cases, the lower limit would satisfy all (e.g., the height of a doorknob or electric switch), and in other cases, the upper limit should be used (e.g., the upper edge of a mirror). To achieve the aforementioned objectives, the primary information of the manual is organized in four sections:

1. Urban design considerations
2. Architectural design considerations
3. Building types
4. Implementation checklists

Each of the first three sections starts with problem identification, where a concise definition of the problem is followed by "Planning Principle," which states the overall objective of the section. The third part of each section, labeled "Design Considerations," describes the relevant technical guidelines.

The organization of the manual allows it to respond to the needs of the professional as a tool for design, checking implementation, and troubleshooting. A special part of the manual is devoted to building types (Sec. III), in which concise guidelines are given for 13 building types. It contains provisions like accessible theater seating, height of food shelves in restaurants, and minimum number of accessible hotel rooms. Once again, these guidelines differentiate between new construction and rehabilitated buildings. The implementation checklists (Sec. IV) are a very important part of the manual. They are problem identification tools that can be used to guide in the identification of problems by asking questions about urban and building design elements, and suggesting remedial measures. Each of the 17 checklists relates to a category of urban or building design elements. The checklist in Table 63.2 covers the design of street furniture.

Moreover, a set of six troubleshooting tables is annexed to the manual. These are classified according to disabilities, list the problems that are likely to be encountered, and suggest measures to overcome them. An example of these is given in Table 63.3.

Interpretation of Guidelines for Application

Accessibility building elements are often designed and implemented as added features, and frequently convey an afterthought attitude. Consequently, professionals think of them as unaesthetic features that are cumbersome to design. To facilitate application and to overcome the inadequate experience in the field, ESCWA published a monograph entitled *Accessibility for the Disabled in the Urban Environment in the ESCWA Region: Planning and Design Solutions* (United Nations, 1997). Being the output of the third phase of the strategy (Fig. 63.1), its aim is to interpret the standards into three-dimensional planning and design solutions. The information in this publication is organized so as to correspond to the structure of the man-

TABLE 63.2 The Implementation Checklist for Street Furniture

3. Street furniture	
Question	Possible solutions
• Does the location of street furniture obstruct the free passage of pedestrians?	• Change the location of street furniture. • Mark the location of street furniture with tactile marking.
Resting facilities • Are resting facilities provided at regular intervals?	• Provide seating facilities at regular intervals between 100 and 200 m.
• Is there an adjoining space for a wheelchair next to benches and public seats? • Are public seats between 0.45 and 0.50 m high? • Are the tops of tables between 0.75 and 0.90 m high? • Are knee spaces at accessible tables at least 0.70 m high, 0.85 m wide, and 0.60 m deep?	• Rearrange the layout of seats to allow an adjoining space of at least 1.20 m. • Modify or replace seats and tables that are too low or too high.
Public telephones • Is there at least one telephone accessible to a wheelchair user? • Is there at least one telephone equipped with hearing aids? • Are the numerals on the telepone raised to allow identification by touch? • Is the coin slot mounted at a maximum height of 1.20 m (1.40 m)? • Are accessible facilities identified?	• Enlarge or adjust one telephone booth. • Install volume controls and induction loops. • Install push buttons with raised numerals. • Reduce the mounting height. • Add signage.
Mailboxes • Are mailbox slots mounted at a maximum height of 1.20 m (1.40 m)?	• Modify the height of the letter slot.
Water fountains • Are water fountain spouts mounted at an approximate height of 0.90 m? • Are controls easy to operate with one closed fist?	• Modify the height of high drinking fountains. • Install a double-tiered fountain. • Replace controls.

TABLE 63.3   Troubleshooting Table for Partially Sighted People

5. The partially sighted	
Problem	Measure
• Identifying obstructions within the path of travel	• Provide bright-colored markings or signals to identify obstructions.
• Orientation	• Provide clearly legible lettering and sufficiently large dimensions for direction signs.
• Crossing roads	• Provide audible traffic signals.
• Maneuvering in elevators and in emergency situations	• Use contrasting color for doors, handrails, tactile signs, etc.
• Locating facilities	• Provide alarm signals.

ual, with design presentation that is easy to read by the nonspecialist. The design solutions also demonstrate to the specialist the possibilities at hand. In addition to urban design, architectural design, and building types, the monograph devotes a small section to some important architectural details and fixtures in which matters like wall fixtures, pay telephones, and thresholds are covered.

The implementation of a barrier-free environment passes through the many stages of application of guidelines to final completion of construction. Ideally, guidelines should be legislated into planning and building regulations before they are interpreted into planning and design concepts and tangible constructions (ESCWA, 1997). The pace of physical reconstruction is often faster than that of institutional building and the establishment of legislation. The two-track approach that was adopted for the BCD at the outset was meant to overcome this problem. But the institutional weakness resulted in inadequate monitoring of the application of guidelines. This fact added to the importance of acquiring feedback from the field (Fig. 63.2).

Sustainability of such efforts can only be achieved when a legislative framework is put in place, linking building standards and specifications to national codes of practice, and rendering them legally binding. The next step, therefore, would be to work closely with the Lebanese government to devise a legislative framework that can be adopted and applied throughout the country.

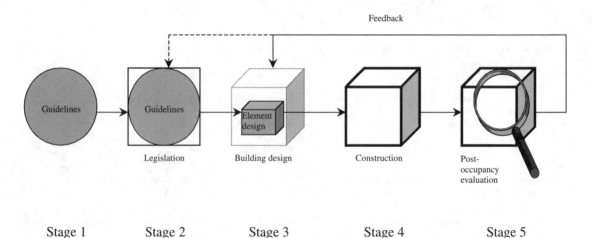

FIGURE 63.2   Stages of application of accessibility guidelines.

Partnership and Roles at the National Level

Like any urban development process, a strategy for national accessibility requires the concerted effort of many partners as elaborated in Table 63.4.

Although participatory planning is a concept that is well tested and discussed, architects often find it difficult to accept client participation in design. The user, be it an elderly person or a wheelchair user, can play an important role in the design and testing of the built environment. Their expertise can be very beneficial when they are involved in a design studio, whether at a professional or educational level. Making studio designs legible to the nonspecialist, however, should facilitate user, or client, participation. Many of the problems that were identified in the postconstruction tests of the BCD could have been avoided had the users been involved during the design stage.

Post-Occupancy Evaluation

Post-occupancy research in architecture assumes its importance from its ability to provide feedback from the field, built on the experiences of the users. It is an essential tool in the monitoring of the extent of success or failure of the process, expressed in terms of the appropriateness of the guidelines and the assessment of the level of accessibility achieved in the end product, the physically constructed environment. The yardstick here is the satisfaction of the targeted population. Although the infrastructure was substantially complete in 1999, and many of its buildings were back in use, the BCD was still far from full occupancy. Moreover, testing for satisfactory accessibility requires fulfilling the technical needs of various disabilities. Consequently, ESCWA designed and conducted two empirical tests. The first test was aimed at the testing of a building, and the second test was of the urban elements and public spaces of the BCD. The objectives of the tests were to:

TABLE 63.4   The Role of Partners in Accessibility

Partners	Functions
Regional and international organizations	International resolutions International technical exchange Transfer of knowledge
Civil society organizations	Mobilizing the disabled population Representing disabled groups and acting as liaison Initiating awareness campaigns Monitoring the application of codes Providing feedback from the field through post-occupancy evaluation
Government bodies and legislators	Formulating and adopting legal instruments Disseminating information on codes and regulations Legal enforcement of instruments Monitoring compliance
The disabled population	Providing feedback on standards and legal instruments Post-occupancy evaluation Identifying areas for further development
Planning and construction professionals	Developing guidelines and regulations Applying guidelines and regulations

- Identify accessibility barriers and suggest remedies
- Test the effectiveness of the manual as a design tool
- Use the checklists and troubleshooting tables of the manual
- Provide feedback from the field
- Check compliance with the manual guidelines

Test Design. Aiming at a good representation of the disability groups, a team was formed comprising wheelchair users, sightless and partially sighted individuals, and persons with limited walking ability. The scope of the test took two considerations into account:

1. *Functional.* By testing the freedom of pedestrian movement of all team members
2. *Technical.* By observing planning and design details and monitoring the performance of building elements

The first test was carried out on the UN House (ESCWA, 1999a), which is the first new building in Beirut's center. It is located at Riad El Solh, in the business center of Beirut, and it accommodates the offices of ESCWA and other UN offices. The UN House is a 14-floor building with 20,000 m^2 of office space, conference rooms, and parking spaces. After studying the floor plans, test paths were selected in order to assess the accessibility of:

- All spaces of public nature (e.g., conference facilities and parking)
- All circulation spaces (e.g., main entrance and floor lobbies)
- Office rooms and furniture
- Doors, windows, and their accessories
- Toilets
- Elevators

To test the planning and design aspects of the BCD, the research identified a representative route that includes pedestrian pathways, sidewalks and street crossings, public gardens, and building entrances. The test team was first briefed of the background of the project and of the objectives of the test before they were asked to move through the route.

Functionally, users can be divided into two subgroups: those who work in the city center, and those who visit. The test related to the needs of the visitors and part of the needs of the other groups, since testing for those who worked in the BCD would have required testing the building interiors.

Test Outcome. Members of the research team accompanied the test team and recorded the feedback. Conducting the test was a very direct and simple method to reveal existing barriers and to discover solutions to removing them. The photographs in Figs. 63.3 through 63.6 were shot by members of the ESCWA research team to illustrate some design failures. As can be seen in Fig. 63.3, a wall-installed ATM was too high for a wheelchair user. A small adjustment could remove the step in Fig. 63.4. The smoother white paving stones in this picture were used decoratively, although they are more user friendly than the gray stones. The difference in color and texture provides an opportunity that could have been used to guide and orient people. The position of the tree and the grating design in Fig. 63.5 do not take into consideration the easy passage of pedestrians, especially wheelchair users. The textural changes in the footpaths allow sightless people to identify their path, but the type of stone and method of construction used for some pedestrian streets are not convenient for people using wheelchairs or crutches or for sightless people. The time interval of the pedestrian crossing green light is too short for the safe street crossing by wheelchair users. The ramps and resting places in the Roman Bath public gardens in Fig. 63.6 were well integrated in the landscape and a joy to use, but they do not meet all the design requirements. The upper ramp is longer than the allowed standard: It should have landing intervals to allow for rests or wheelchair maneuvering, and it should have handrails. These are examples of the practical observations recorded in the field.

Further recommendations were later arrived at by referring to the manual. Recommended solutions were prioritized in three categories: to indicate whether modification (1) is essential, (2) should be made if circumstances and budget allow, or (3) just recommended. This was based on two factors: (1) the importance or severity of the barrier, and (2) the cost or feasibility of the correction. Trimming overhanging vegetation or adding a street sign are much easier and cheaper than changing the street's paving stones. The findings of the test and the recommended solutions were documented and presented to Solidere (ESCWA, 1999b).

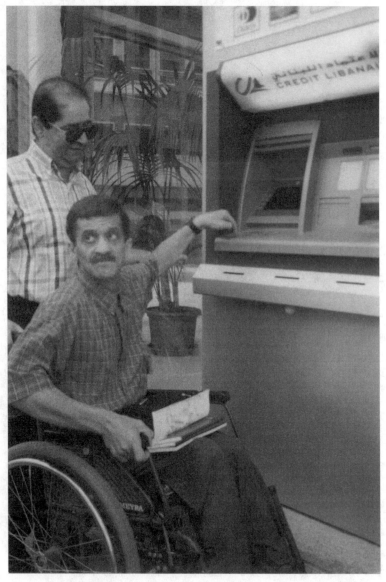

FIGURE 63.3   This ATM could easily have been installed at an accessible height.

FIGURE 63.4   Careful design could have avoided the step and made the pavement more user friendly.

FIGURE 63.5   The tree, grating, and wall configurations hinder the passage of people in general.

FIGURE 63.6   This ramp at the Roman Bath public gardens should have handrails.

Role of Professional Education

The accessibility guidelines resulted in a more user-friendly environment at the UN House. But after more than 2 years of using the building, it has become evident that the building suffers from significant shortcomings in its functional and environmental design. Creating an aesthetically impressive form was the overriding occupation of the building architect. Analysis of findings of the two tests leads to the conclusion that many barriers were created due to a lack of conviction among design professionals of the issues at hand (e.g., a shop window facing a sloped street pavement had its door at the higher end, thus necessitating a step, a barrier that could have been avoided by making the right decision).

A great deal can be achieved through influencing professional education. Developing the curricula of schools of architecture and urban design cannot be successful by itself. The teaching staff have to be convinced and retrained for a new design thinking that is inclusive in its fundamentals. It should be presented as a novel way of thinking that can be explored through creativity, not as a code that should be satisfied. Universal design touches at every aspect of the urban environment and, thus, has to be practiced by urban designers, architects, interior designers, landscape architects, and industrial designers. The Universal Design Education Project, which stimulated the participation of many U.S. universities and covered community involvement, experiential learning, and use of user/expert, is an excellent example of how to influence the academic environment.

Post-Beirut Endeavor

The infrastructure, many new and rehabilitated buildings, and a number of public gardens are now complete. Although many mistakes are still to be corrected, the reconstruction effort made the center a wonderful laboratory for the many issues of universal design. These ranged from major policy-planning decisions at the inception stage, to the detailed implementation, testing, and problem identification. The center, proving to be a good demonstration case for universal design, has become a catalyst in promoting universal design in other parts of the country. A number of municipalities have expressed keen interest in applying accessibility guidelines. In the town of Nabatieh, in southern Lebanon, a project is underway to make public buildings accessible. But the implementation techniques of the BCD are not necessarily appropriate in other localities. The guidelines and the Beirut experience can be used for developing approaches that are suitable to other circumstances. In the absence of binding regulations, the municipality of Nabatieh decided to forfeit building permit fees for designs that apply accessibility guidelines.

It is important to disseminate information about appropriate, low-cost solutions through documentation and exchange of information. Techniques that are implementable in the capital city may not be appropriate for the less wealthy cities or villages, and assistive devices that are designed and produced in an industrialized country may be too costly or unsuitable for the urban or rural environment of the developing world.

As a spin-off from the Beirut project, ESCWA organized in November 1999 two activities: (1) an international technical exchange seminar, and (2) a regional training workshop. Both activities benefited from the experience gained in Beirut.

The seminar was attended by international experts on accessibility from many regions of the world. Experience gained in various parts of the world was discussed.

The workshop was aimed at persons that are concerned with issues of accessibility and are expected to play a role in creating barrier-free environments in their countries. It was regional in scope. It used the manual as a reference and was held in the UN House. It was hoped that the participants would become well equipped to apply the guidelines as well as advocate barrier-free environments in their countries. The training modules were designed so as to enable replication of the workshop in other countries, aiming to disseminate the Beirut knowledge regionally.

Lessons Learned

The scale and breadth of the Beirut experience spans the whole development process, and its value stems from its broad applicability. This makes it possible to draw the following lessons on each component of this process:

At the Policy-Making Level

1. The political will and commitment to the cause by the decision makers at the highest levels is a basic and essential element in overcoming the many obstacles that are likely to occur along the way.
2. Good accessibility guidelines are not sufficient by themselves. Their application should be mandated by legislation, or, until that is established, by the terms and conditions of the project.

On Partnership

3. Interest groups, such as disabled people's NGOs, should be involved at the very start of the project. Their roles can be that of identifying priorities, monitoring implementation, and post-occupancy evaluation. Creating awareness and advocacy throughout the process can ensure a better level of success.
4. It is important to adopt a decision-making process that is inclusive of beneficiary groups and that would expedite the empowerment of the targeted constituency.

Professional Training and Awareness

5. There is a strong need for an awareness campaign among professionals in the construction industry. Planners, architects, interior designers, and engineers should be won over, so as to think in terms of inclusive design.

6. The post-occupancy tests proved that, often, physical barriers are the result of negligence in design or lack of proper consideration. Successful application of the guidelines can best be achieved with professional conviction in the cause.

7. Training should be organized for professionals on accessibility guidelines and their application, sensitizing them to the needs of children, elderly people, and people with disabilities in the built environment.

On Professional Education

8. University curricula should be developed to train the students to take into consideration all sectors of the community. A shift has to be made from the conventional thinking that designing for accessibility is merely prompted by code requirements.

9. Curricula of professional education in the fields of planning, architecture, interior design, landscape design, and industrial design should be developed to train for universal design.

On Advocacy

10. Universal design is a qualitative achievement. An urban environment that is accessible and inclusive is a clear sign of positive social development: It indicates a more civilized state of the society. This argument can be effective in advocacy and awareness-building campaigns.

11. A well-executed and high-profile pilot project can serve as an excellent showcase to advocate universal design thinking at the national and regional levels.

12. A pilot project can be used as a vehicle to build a good knowledge base on universal design for a national effort.

13. Progress and achievements of the project at all its stages should be publicized to stimulate public interest.

On Technical Exchange

14. Identification and dissemination of information on best practices in universal design that are appropriate to developing environments is a good method for consolidating their knowledge base.

15. It is important to promote good practices by identifying and rewarding planning and design ideas that convey the true meaning of inclusion, and where the building would be naturally accessible and the needs of all users are equally respected.

63.6 CONCLUSION

Lebanon is a country that is eager to make the development leap. Going through the process described in this chapter meant instituting criteria and methodologies, thus building capacities for universal planning and design.

Analyzing this experience highlighted the main factors that are essential to achieving a barrier-free environment within the context of developing economies. The Beirut experience was trend-setting for the country, and in many aspects, for the region. Having devised the appropriate planning and design tools at each stage, it involved establishing guidelines, applying them, as well as monitoring and checking the level of accessibility of the built environment.

One of the profound conclusions is that the reluctance or resistance to apply accessibility guidelines is imbedded in the attitude of the design profession where the users' needs are often blurred by the architects' urge for aesthetic creativity. This is nurtured by the popular architectural literature, where buildings are discussed and admired similar to products of fashion.

In environmental design terms, a good building should act as an interface between the user and the prevailing local environment. Electromechanical equipment consists of assistive devices that should be used to complement a building's performance. A good building design is one that minimizes dependence on such artificial means. A similar analogy can be drawn for barrier-free design.

A new way of thinking is required in the planning and design professions. The fulfillment of the needs of the user should be placed at the heart of professional ethics. It is the moral responsibility of the professional to provide a building that is rationally conceptualized, a design that is geared toward the comfort of the user, appropriate to its surroundings, and that functions in harmony with the physical environment.

63.7   NOTE

1. The views expressed in this chapter are those of the author and do not necessarily represent the views of the United Nations.

63.8   BIBLIOGRAPHY

Lebanese Company for the Development and Reconstruction of Beirut City Center (Solidere), *Accessibility for the Disabled: A Design Manual for a Barrier-Free Environment,* Solidere, Beirut, Lebanon, 1998.

Mullick, Abir, *Strategies for Teaching Universal Design,* paper presented to the International Seminar on Environmental Accessibility, United Nations Economic and Social Commission for Western Asia, Beirut, Lebanon, 30 November–3 December 1999.

Sará-Serrano, Maria-Christina, *Getting Women with Disabilities on the Development Agenda of the United Nations,* United Nations Division of Social Policy and Development, New York, 1999; www.visionoffice.com/women/wwdis1.htm.

Thompson, Phillip, *An Overview of Environmental Accessibility in South Africa,* Presentation made at the International Seminar on Environmental Accessibility, United Nations Economic and Social Commission for Western Asia, Beirut, Lebanon, 30 November–3 December 1999.

United Nations Economic and Social Commission for Western Asia (ESCWA), *Accessibility for the Disabled in the Beirut Central District,* a report presented to the Lebanese Company for the Development and Reconstruction of the Beirut Central District (Solidere), UN-ESCWA, Beirut, Lebanon, 1994.

———, *Accessibility for the Disabled in the Urban Environment in the ESCWA Region: Planning and Design Solutions,* United Nations, New York, 1997.

———, *Accessibility of the UN House for Persons with Disability: A Post-Occupancy Assessment,* Unpublished Report, UN-ESCWA, Beirut, Lebanon, 1999a.

———, *Accessibility of the Beirut Central District for Persons with Disability: A Post-Occupancy Assessment,* Unpublished Report, UN-ESCWA, Beirut, Lebanon, 1999b.

63.9   RESOURCES

The following Web sites provide further sources of online information related to this chapter:

The Lebanese Company for the Development and Reconstruction of Beirut City Center (Solidere), Beirut, Lebanon.   www.solidere.com.lb.

The Economic and Social Commission for Western Asia (ESCWA) is a United Nations regional commission. ESCWA member countries include: Bahrain, Egypt, Iraq, Jordan, Kuwait, Lebanon, Oman, Palestine, Qatar, Saudi Arabia, Syria, United Arab Emirates, and Yemen.

www.escwa.org.lb/.www.un.org/esa/socdev/enable/designm/.   Contains an electronic version of the manual *Accessibility for the Disabled: A Design Manual for a Barrier Free Environment.*

CHAPTER 64

THE ED ROBERTS CAMPUS: BUILDING A DREAM[1]

Susan Goltsman, F.A.S.L.A.
Moore Iacofano Goltsman, Berkeley, California

64.1 INTRODUCTION

The Ed Roberts Campus (ERC), when complete, will be a universally designed, mixed-use development that will provide a home for nine disability organizations in Berkeley, California. Currently in design development, the proposed ±170,000 ft^2 building will be the first facility to be located over a Bay Area Rapid Transit (BART) station.

The campus will be the complete demonstration of universal design in terms of site, building, mechanical systems, circulation, furniture, wayfinding, security, technology, transit connections, and relationship to the street and urban fabric. As a transit-oriented development, the facility will emphasize a pedestrian environment that reinforces the use of public transportation, thus increasing accessibility to the facility by individuals of all abilities. Technology will play a large role as well, providing innovative solutions and new opportunities for making communications design universal and accessible worldwide. Through its architecture, its use of technology, and its planning for full inclusion, the Ed Roberts Campus will be a model of what can be achieved with universal design.

64.2 BACKGROUND

The ERC is a partnership formed by nine organizations that share a common history in the independent living movement of people with disabilities. The partner organizations are: Bay Area Outreach and Recreation Program (BORP); Center for Accessible Technology (CforAT); Center for Independent Living (CIL); Computer Technologies Program (CTP); Disability Rights Advocates (DRA); Disability Rights and Education and Defense Fund (DREDF); Through the Looking Glass (TLG); Whirlwind Wheelchairs International (WWI); and World Institute on Disability (WID).

These organizations have joined together to develop the Ed Roberts Campus, a universally designed, transit-oriented campus located at the Ashby Bay Area Rapid Transit (BART) Station in south Berkeley. The campus will house the organizations' offices and services, as well as shared facilities such as a conference center, related retail and commercial services, and a gym/fitness center.

The concept of a campus took form after the death of Ed Roberts, a leader in the disability community. Community leaders decided they could best commemorate Ed's work by supporting the organizations that he helped start and the independent living movement that he championed. They decided to do this by establishing a center dedicated to fostering collaboration and improving the services and opportunities for people with disabilities worldwide.

Community-Based Planning Effort

In a community-based planning effort, the Ed Roberts Campus is working cooperatively with Bay Area cities, the Metropolitan Transportation Commission, BART, AC Transit, disability groups, the major Bay Area universities (including the University of California, Berkeley; San Francisco State University; and the Peralta Community College District), and local neighborhood organizations. The ERC is currently in the process of reaching out to more interest groups to gain their support and to seek their advice and collaboration on the campus development.

Participatory Planning Process

The preliminary planning process for the Ed Roberts Campus took place over a two-year period and involved the active participation of many groups. Each partner organization helped develop and refine their space requirements through interactive meetings, interviews, and discussions. Overall guidelines were then developed with the partners and through meetings with the City of Berkeley, BART, and the community. These guidelines provided the framework for the Design Program, a document detailing information about the site, each of the partner organizations' exact spatial requirements, shared use spaces, and organizational relationships on each of the floors. The Design Program serves as a framework and reference for the design of the campus.

In addition to the Design Program, a series of design charrettes were planned. These daylong work sessions involved assembling national experts in the fields of universal design, technology, and transit-oriented development to brainstorm concepts for the facility. The charrettes resulted in a set of core principles and concepts that should be considered at every level of the ERC design process.

Recruitment and Selection of the Design Team

The recruitment and selection of the design team involved assembling a group of qualified architects and subconsultants with the professional skills necessary to implement the complex project. The team needed expertise that combined knowledge of universal design and experience with disability, a complete understanding of the "transit village" concept, and the technical know-how to build the facility over a continually functioning BART station. Also important was the ability to work with multiple jurisdictions, the community, other consultants, and the ERC partners, as well as the demonstration of innovative community design.

To recruit the design team, the ERC established selection criteria and released a request for proposals (RFP) calling for qualified architects. Fifteen teams submitted proposals, from which five were "short-listed" for an interview. The short-listed teams made presentations before a special ERC committee who evaluated them using a scoring matrix developed from the selection criteria. Three teams were then selected for the final selection process. Each of these design teams put together a tour of their work for the selection committee. After the tour, a second and third round of interviews were conducted. After much discussion and consideration, the Siegel Diamond/Michael Willis and Associates architecture team was selected. During the course of the project, technical consultants will be called in to lend their expertise in the areas of structural engineering, traffic, environmental issues, technology, transit-oriented developments (TODs), universal design, wayfinding, and community planning.

64.3 THE VISION

For people who live in the Bay Area, the Ed Roberts Campus will offer an impressive array of disability-related services and programs in one totally accessible location. For people who live throughout the state and the country, the campus will be a model of integrated and accessible service delivery and a national resource for research, legal analysis, education, training, and model program development. Internationally, the ERC will stand as a beacon of the independent living movement of people with disabilities, providing training, technical assistance, and collaboration opportunities for people with disabilities and disability organizations worldwide.

In addition to the offices of the nine partner organizations, the $\pm 170,000$ ft^2 facility will house a conference facility, a library on the disability movement, a computer/media resource center, a fitness center, a café, a children's play center, a mix of neighborhood-serving retail spaces, and office lease space. The facility and site design will be based on three guiding concepts:

1. Universal design
2. The transit village (transit-oriented development)
3. Intelligent use of technology

Universal Design

Universal design incorporates the general principles of its predecessor, barrier-free design, which emphasized the removal of physical barriers and the creation of specially designed features for people with disabilities. Unlike barrier-free design, however, universal design is not based on the assumption that wheelchair-accessible facilities are also accessible to individuals with other disabilities—for some people, barrier-free features can even be hazardous. Universal design avoids these limitations by incorporating a more comprehensive view of human needs and abilities.

Universal design is not a set of inflexible rules. Its proponents, while recognizing the value of standards such as the ADA Accessibility Guidelines (ADAAG), realize that compliance alone does not guarantee accessibility for all people. Instead, universal design focuses on the complicated interrelationships that exist between the physical environment and the users. Four commonly held goals provide the basis for universal design:

- *Accommodate human movement characteristics.* Universal design addresses three aspects of human movement: *body space, reach range,* and *effort.* "Body space" represents the area immediately surrounding a person and any mobility aid she or he may use—in other words, the space needed to move through an environment. Accessible design requirements for clear space, such as vertical clearance and minimum passage width, address this need for maneuvering space. "Reach range" represents the distance that users can reach to retrieve an object. These ranges are used to determine where items should be placed to be accessible. "Effort" represents the physical exertion required to perform a function such as flipping a switch or ascending a ramp. The required level of effort is determined by the dexterity (i.e., required degree of manipulation), force, and sequence of steps needed to perform the function.
- *Ensure safety.* When facilities are designed to accommodate the way people work and move through their environments, obstructions and hazards are minimized. A well-

designed pathway, for example, provides a smooth and secure path of travel for someone walking, using a wheelchair, or carrying a bulky item.

- *Provide adaptability.*   Facilities must be planned with both present and future needs in mind to accommodate constant changes in population, technology, and building regulations. Every aspect of a facility should be designed for maximum flexibility and use by the broadest spectrum of people.
- *Be cost-effective.*   Affordability and cost-effectiveness are valued in universal design. Expenses are reduced when designs accommodate the easy rearrangement, addition, or removal of structural elements rather than requiring constant retrofitting or renovation. Furthermore, the selection of products based on the general requirements of human movement eliminates the need to purchase costly specialized equipment. Lever-type door handles, for example, are not significantly more expensive than other types of handles, yet they make doors easier to open for all users.

These four goals of universal design will be considered in every aspect of the campus design including siting, circulation, wayfinding, technology, communications, transit connections, and relationship to the street and urban fabric. As the design process continues, each of these features will be articulated through interactive sessions with the partners and the design team. Each part of the design will be evaluated to ensure maximum use by the widest variety of people.

Transit-Oriented Development

A transit-oriented development (TOD) is a mixed-use community within a typical 2000-foot walking distance of a transit stop and core commercial area. The design, configuration, and mix of uses emphasize a pedestrian-oriented environment and reinforce the use of public transportation, without ignoring the role of the automobile. TODs mix residential, retail, office, open space, and public uses within a comfortable walking distance, making it convenient for residents and employees to travel by transit, bicycle, or on foot, as well as by car.

TODs offer an alternative to traditional development patterns by providing housing and employment opportunities for a diverse population, and physical environments that facilitate pedestrian and transit access. Developing a network of TODs throughout an urban region serves to strengthen the overall performance of the regional transit systems.

Proximity to transit is a key factor in determining the suitability of a site for higher density, mixed-use developments. A fundamental purpose of the TOD concept is to create a land use pattern that will ultimately support transit. For TODs to successfully reduce auto travel, they must be located within easy walking distance of, or within convenient feeder bus connections to, dedicated transit lines. Studies by regional transit agencies throughout the country have shown that the greatest pedestrian "capture rate" for public transit occurs when transit stops are within a 10-minute walking distance from home or office, have frequent headways, or are close to a dedicated transit right-of-way.

The location, design, configuration, and mix of uses in a TOD provide an alternative to traditional development by emphasizing a pedestrian-oriented environment and reinforcing the use of public transportation. This linkage between land use and transit is designed to result in an efficient pattern of development that supports the transit system and makes significant progress in reducing sprawl, traffic congestion, and air pollution. The TOD's mixed-use clustering of land uses within a pedestrian-friendly area connected to transit provides for growth with minimum environmental and social costs.

The ERC property is next to the Ashby BART station and is served by the AC Transit system. The ERC is a mixed commercial use facility and is seen as a major part of the emerging transit village around the Ashby BART station in south Berkeley.

As the design evolves, each opportunity to enhance the connection to available transit will be explored; from providing adequate bicycle racks and pedestrian amenities to creating easy, seamless access to the BART station and bus stops. The goal of the ERC will be to have 80 percent of its employees and visitors use means of transportation other than a single-occupant vehicle to arrive at the campus.

Intelligent Use of Technology

Technology is much broader than computers, machinery, and software. Technology in the twenty-first century has to do with how things are created and how they are reused, as well as the tools for producing goods and delivering services. Technology in terms of accessibility can be a device or a building part that helps equalize the function for the users of the facility. Technology as a tool in the ERC enhances the understanding, use, and sustainability of the environment.

The Ed Roberts Campus will seek the help of technology companies to develop the technology vision for the campus. That vision includes making the facility fully accessible to and usable by people with disabilities; creating a work environment that accommodates people's needs; providing services that prepare people for a future in which the tools of technology shape how they will live and work; and developing them in a cost-effective, sustainable manner.

Technology is also an important part of the service program envisioned by the campus. Several of the partners will offer a range of educational programs focused on all phases of employment preparation and disability research and policy. These educational programs not only need to be available to students who can actually come to the campus, but also to students everywhere through distance-learning technology. The same is true of conferences and seminars. The Ed Roberts Campus hopes to hold international meetings of disability leaders and experts, with some attending in person and others participating through video conferencing and other technologies. This is all part of the technology vision.

Another part of the technology vision is the concept of sustainable design. Sustainability requires the understanding of where materials come from, how they perform while in use, and how they can be reused when the needs of the users change and their current life is completed. In addition to choosing appropriate building materials and finishes, sustainability can be achieved through the siting and form of the building; the strategic use of vegetation; the way the building is ventilated, heated and cooled; how everyday trash is handled; how people use and circulate through the facility; and what is purchased and what is stored.

Technology plays a large role in sustainable design. Technological innovations make it possible to create designs that are more sustainable through heating, ventilating, and air-conditioning (HVAC) systems, building materials, construction practices, and more. This broader definition of technology that sustains both people and the environment can be a key to healthier places. The goal of the Ed Roberts Campus is to demonstrate how conscious design and decision making can create an urban setting that works for everyone without diminishing resources.

64.4   THE SITE

The campus will be located above the eastern parking lot of the Ashby BART station in south Berkeley. The property (see Fig. 64.1) fronts Adeline Street and takes up about one-third of the existing parking lot. The remaining two-thirds of the site will remain parking.

The site has been under joint ownership of the City of Berkeley and BART since the agreement in 1965 to put the BART tracks underground in Berkeley. To purchase the site, the ERC has had to negotiate with both the city and BART. In July of 1998 the ERC gained a memorandum of understanding from the city, giving it a three-year option to buy the property. Similarly, the BART board is expected to approve the sale of the site to the ERC.

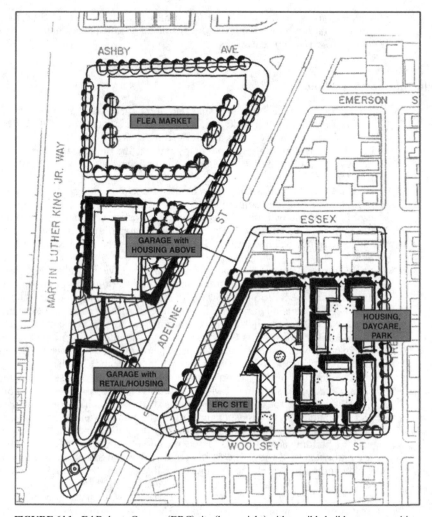

FIGURE 64.1   Ed Roberts Campus (ERC) site (lower right) with possible buildout opportunities.

Neighborhood Context

South Berkeley is a mixed-use neighborhood with a variety of retail, commercial, institutional, and residential development. The immediate neighborhood has a strong residential area with an active community. The community is concerned about the type of development planned for the entire BART property, the traffic it will generate, and the impact it will have on the neighborhood.

Adjacent businesses such as antique stores, a liquor store, and a small café are not particularly neighborhood-serving, and some have a high rate of turnover. On the west side, there are a variety of commercial businesses, a church, and restaurants. The Black Repertory Theater is located on the next block at 3201 Adeline Street. On Saturdays and Sundays, the Berkeley Flea Market takes place on the west side of the BART station.

Overall, the area around the ERC site is not lively with regard to commercial activity. Adeline Street is very wide and not pedestrian friendly. The BART station and parking are sited below street level, and this adds to the feeling that there is not much community life on the streets.

Transit Connections

The neighborhood and site are well-served by public transit. The Ashby BART station connects the site to most of the Bay Area, including the greater San Francisco area as far as Colma, Contra Costa County as far as Pittsburg/Bay Point, and the East Bay as far east as Dublin/Pleasanton and as far south as Fremont. AC Transit buses travel by the site every 15 minutes, providing access to all of Alameda County either directly or by transferring routes.

In addition, Berkeley Electric Shuttle Transit (BEST), a pilot project operated by the City of Berkeley, UC Berkeley, and the Berkeley Gateway Transportation Management Association, operates electric buses for BART commuters who work in West Berkeley.

Pedestrian Plaza and Streetscape

The concept for the pedestrian environment is to create opportunities for a variety of activities and visual connections in a three-dimensional spatial setting to enhance safety and social interaction, and to help create a sense of community.

- *Connections.* Exterior corridors, seating areas, and gathering places will occur at all levels. Balconies or atrium-like spaces and the circulation system of ramps and elevators will allow verbal communication and visual connections from one floor to another, helping to reinforce social interaction and a sense of community.

- *Pedestrian crossings.* At street level, the pedestrian environment interfaces with Adeline Street, a six-lane roadway separated by a grass median. Because traffic in this section moves relatively fast and unimpeded, pedestrian crossing of Adeline Street will be encouraged through the BART station rather than at street level. If future development should bring about significant changes, such as similar development over parking on the west side of Adeline, and elimination of one lane of traffic in each direction, then streetscape design to enhance the crossing of Adeline should be considered.

- *Sidewalks.* Wide sidewalks are recommended to accommodate the increased pedestrian activity, provide ease of access, and facilitate getting on and off buses and shuttles. The paving finish should be enhanced to relate to the entry plaza and reinforce the identity of the project, especially the pedestrian environment.

- *Outdoor seating.* Benches with backs and armrests should be provided near the bus stop and passenger pickup areas. In addition, protected seating areas are recommended at the arcade and at the plaza.

- *Entry plaza.* An entry plaza to the campus should be seen as a welcoming environment that unifies the development and fosters a sense of community. The architectural framework could create a sense of gateway to the entire complex. Arcades at the sidewalk would provide transition from the street and lead to the entry. On the parking lot side, vertical circulation design should reinforce the connection to the BART-level entry. A small café providing carryout and limited food service should be located at the entry plaza, with space for tables provided at the edge. The entry plaza is an appropriate location for artwork, and a small water feature could be provided to mitigate traffic noise and enliven the space.

- *Landscaping.* Vegetation should be designed to soften the acoustics, provide shade, and create color and textural interest to balance the hardscape. Colorful flowers or foliage should be placed near eye level in raised planters. Elevated planters with trailing plants should be considered where trees are not feasible. Strategically placed planters at the upper floors would extend the green character three-dimensionally. Because most of the project will be on an elevated structure, the planting design will need to carefully consider appropriate growing medium.

- *Lighting.*   Pedestrian lighting should be designed to enhance the site and to increase safety. Lighting design should consider illumination levels that are adequate for people who are visually impaired. Consider washing the building façade with light, and installing bollard or wall inset lights that illuminate the ground plane. Bright lights that produce glare near eye level should be avoided.

- *Bicycle parking.*   Limited bicycle parking for the ERC should be provided to one side of the entry on Adeline. The possibility for colocating bike parking with BART bicycle parking could also be explored.

64.5   THE BUILDING

In addition to containing the offices of the nine partner organizations, the ERC building will house a variety of facilities that will be shared by the partner organizations and the community, including a conference center, library, and fitness center (see Figs. 64.2 and 64.3). There will also be space for retail stores that serve the building and neighborhood, as well as lease space for other organizations that are compatible with the ERC.

Building Entries

Because of the characteristics of the site and the needs of the various organizations, the building will probably have more than one entry. Each organization wants to keep its own separate identity. This has implications for the design and layout of the floors. The ERC, which will have its own management and an office in the building, will also need to have its own identity. The ERC will be organizing and running programs in the conference center sometimes separately from any of the partner organizations.

The main entry will need a reception/information area that is staffed during normal business hours. A small security office with video surveillance cameras should be obvious but not central. The feeling in the entry should always remain friendly but secure. To minimize unintentional visitors, the plaza entry to the building will need to be clearly differentiated from the entry to the BART station.

Some of the partner organizations will require special entries. The Center for Independent Living, because of its heavy service focus, clientele, and amount of foot traffic generated may require a direct entrance from the street. Whirlwind Wheelchairs International, which builds wheelchairs and other mobility devices, will need access to a loading area and doors large enough to bring in fabrication materials and tools. Through the Looking Glass will also need access from the parking lot so that parents can drop off children. It will also need direct access to the outdoors for a play area if it has a childcare program.

The third floor requires a prominent entrance that can be used after building hours for access to the fitness center and the conference center, enabling visitors to enter and exit without accessing the rest of the building. On the first floor, the retail stores will require entry from the street.

Building Circulation

The building circulation system is envisioned as the centerpiece of universal design for the building. Without predetermining the elements or form, the circulation system should provide a clear, visible route vertically through the building.

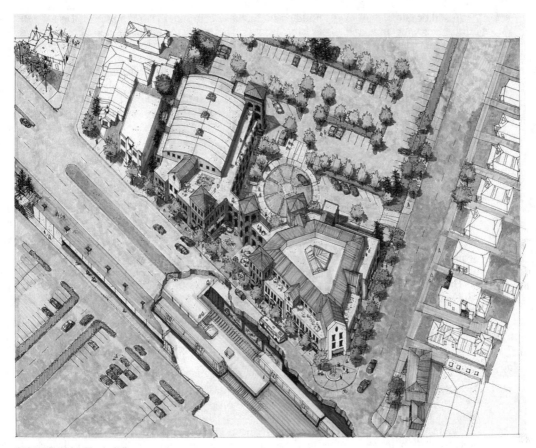

FIGURE 64.2   The building.

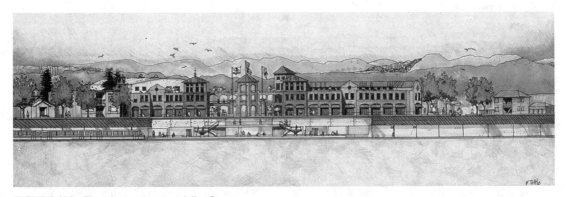

FIGURE 64.3   Elevation/section on Adeline Street.

In addition to elevators, the building should have components of ramps as places for gathering, communicating, and showcasing displays. The ramps should move people vertically up and through the space. Handrails should be communication devices allowing a person who reads Braille to navigate throughout the building by "reading" the handrails.

The wayfinding system should be integrated into the circulation system in a sculptural way. It should be a piece of art that is technologically smart and environmentally clever. The emergency egress from any part of the building should not rely on stairs. The circulation of this building will be its signature and will distinguish it from other architecture.

Retail and Other Commercial Space

About 30,000 ft^2 of commercial space will be available for lease in the entire campus. Space available for retail on the first floor will be about 10,000 ft^2. Some of the ideas for the first-floor retail space would be a café/coffee shop, a copy center, a dry cleaner, a store that would sell products made by the partner organizations or related organizations, and other businesses that serve the neighborhood.

There has been some discussion about helping start businesses that could be run and owned by people with disabilities or neighborhood residents. The type and mix of retail opportunities require further discussion.

The extra office space will be rented to organizations or businesses that are compatible with the ERC mission and have the potential for creating synergistic opportunities. The amount of available office space to lease in the ERC can be adjusted to accommodate the growth/contraction of the partner organizations from year to year.

Technology and Communications

The building and site will reflect the technology vision for the ERC. The vision has yet to be fully articulated, but the design team will develop a facility that can be easily adapted to changing technology.

The technology will also ensure a communications system that connects all of the partner organizations, helps secure the building, and provides outreach capabilities worldwide. The wayfinding system is broader than just building signage. It will be integrated into the technology vision and the communication system for the campus. These functions need to be fully developed in conjunction with the facility design with a clear program of uses.

64.6   CONCLUSION

As of July 2000, the design process was just beginning. The design team had been hired, and the money had been raised to cover the design costs. It is anticipated that as the design process is documented, it will result in a universal design reference book for buildings, workplace, wayfinding, site design and the urban transit/pedestrian interface. As the project evolves, the Center for Universal Design will post information on its Web site, as will the partner organizations. It is anticipated that the Ed Roberts Campus should be completed by 2003.

64.7   NOTE

1. Susan Goltsman would like to thank Julia Ohlemeyer, Joan Leon, Rick Spittler, Gil Kelley, and the Ed Roberts Campus Partnership for their contributions to this chapter.

P · A · R · T · 9

INFORMATION TECHNOLOGY

CHAPTER 65

FUNDAMENTALS AND PRIORITIES FOR DESIGN OF INFORMATION AND TELECOMMUNICATION TECHNOLOGIES[1,2]

Gregg Vanderheiden, Ph.D.
Trace Center, University of Wisconsin, Madison, Wisconsin

65.1 INTRODUCTION

This chapter focuses specifically on the problem of trying to extend the design of standard information and telecommunication products so that these products can be used effectively by individuals with all disabilities. Furthermore, it tries to address the design process pragmatically, in a manner similar to that required of product designers in companies. That is, it tries to answer questions such as, "How does one reach the broadest number of users with the least impact on product design and cost?" "If there is a variety of features that could be added, which would be of highest priority to people with disabilities in general?" "Is there a small set of interface strategies that would provide good cross-disability access, or is it necessary to put multiple strategies or features into my product for each and every disability?"

This chapter attempts to delineate some of the key dimensions of product accessibility/ usability and to provide some initial rationale for prioritizing design options. This chapter discusses a multidimensional prioritization approach, coupled with a vector-based usability evaluation procedure in development at the Trace Center.

65.2 BACKGROUND

Universal design is the process of looking at the design of a product from a general perspective, which would include all users. Much has been written about design for the 90th-percentile users, but much less is available on how to design products so that they will be usable by people with disabilities, particularly severe disabilities. The problem is further complicated when one thinks about the fact that when designing standard products for use by people with disabilities and severe disabilities, one is actually talking about designing a product that is simultaneously usable by individuals with all, or as many as possible, different disabilities.

Universal design has many different specific definitions, all of which deal with the expansion of design beyond focusing on specific target populations and, instead, to keep everyone in mind when designing a product. The author expresses *universal design* as: "The process of designing products so that they are usable by the widest range of people operating in the widest range of situations as is commercially practical" (Vanderheiden, 1997).

This definition highlights that universal design is a process, not an outcome, and that when applied to standard mass-market products, or actually any commercial products, it is subject to the constraints and realities of commerce.

As one might imagine, there are no universal designs or universally usable products. There simply is too great a range of human abilities and too great a range of situations or limitations that individuals may find themselves in. Thus, universal design is more a function of keeping all of the people and all of the situations in mind, and trying to create a product that is as flexible as is commercially practical so that it can accommodate different users and situations. It is important to note that universal design does not refer to designing specifically for people with disabilities. Although disability is one type of limitation or variation in human performance that universal design attempts to address, it is not the only type. In fact, for every type of disability there are situational constraints that present the need for the same requirements. (See, for example, Table 65.1.)

Thus, universal design is not the same as design for disability as is practiced in the design of assistive technology. In fact, accessible design is not the same as design for disability. Accessible design is focused on standard products rather than special products.

This chapter focuses on both *accessible design* (i.e., making a product design more usable by people with disabilities) and *universal design,* or "design for all" (i.e., focusing on keeping

TABLE 65.1   Parallel Between Disability Needs and Situations Everyone Experiences

Requirement	Disability-related need	Situation-related need
Operable without vision	People who are blind	People whose eyes are busy (e.g., driving a car or phone browsing) or who are in darkness
Operable with low vision	People with visual impairment	People using a small display or in a high glare, dimly lit environment
Operable with no hearing	People who are deaf	People in very loud environments or whose ears are busy or are in forced silence (e.g., in a library or meeting)
Operable with limited hearing	People who are hard of hearing	People in noisy environments
Operable with limited manual dexterity	People with a physical disability	People in a space suit or chemical suit or who are in a bouncing vehicle
Operable with limited position or reach	People who use a wheelchair or have limited reach	People who are out of position or have multiple devices to operate
Operable with limited cognition	People with a cognitive disability	People who are distracted, panicked, or under the influence of alcohol
Operable without reading	People with a cognitive, language, or learning disability	People who just have not learned to read this language, people who are visitors, people who left reading glasses behind

all users in mind from the beginning). Clearly, universal design, or design for all, is preferable, but the approach of industrial designers who are facing accessibility regulations is often more like accessible design. Thus, both are addressed.

65.3   *INCREASED FOCUS ON UNIVERSAL DESIGN*

In recent years, there has been an increased focus on universal design in general, and on creating more accessible mass-market products in particular. There are three major driving forces behind the increased focus on universal design and accessible products. One is based on laws resulting from civil rights and societal values. The second is based on market demographic trends. The third one is based on technology trends.

In the United States and other countries, there have been new laws and regulations of late that are increasing the requirements on companies to create products that are usable by people with disabilities. At this time, the most comprehensive of these regulations in the area of information telecommunication technologies are in the United States. These include three major laws (i.e., the Americans with Disabilities Act, Sec. 255 of the Telecommunication Act, and Sec. 508 of the Rehabilitation Act) and a number of smaller bills (e.g., the Hearing Aid Compatibility Act, the Television Captioned Decoder Act, etc.). The Americans with Disabilities Act (ADA) generally requires that all public facilities and services offered to the public must be made accessible to people with disabilities where it does not pose an undue burden. Section 255 of the Telecommunication Act is more specific and requires that anyone producing telecommunication products or services must design those products and services from the beginning to be accessible to people with disabilities wherever it is readily achievable (i.e., can be done "with little effort or expense"). Section 508 of the Rehabilitation Act requires that the government purchase electronic and information technologies that are accessible if it is not an undue burden. These laws were all passed based on the principle that people with disabilities have a right to be able to access and use the new information and telecommunication technologies that are rapidly becoming integral to our education, workplace, and daily living. The first two of these, the ADA and the Telecommunication Act, are examples of what might be called "push" legislation, in that they tell industry that they must make their facilities, under the ADA, and products, under the Telecom Act, accessible to people with disabilities. The third law, Sec. 508 of the Rehabilitation Act, is what might be termed "pull" legislation. Although Sec. 508 pushes the government to create accessible workplaces and to buy accessible products, industry does not specifically have to do anything under Sec. 508. Instead, the preference by the government for accessible products creates a market "pull" on industry, providing industry with an incentive to make accessible products if they want to sell them to the federal government. This combination of push and pull legislation has created a significant mobilization of effort within companies that was not previously there. In particular, it has caused the issue of accessible use by people with disabilities to be brought to the attention of the designers who are creating the companies' mass-market products. This not only has an effect on specific products affected by the regulations, but it also raises awareness of usability issues for these designers in general, which can have an impact on a broader range of products.

The second major force is also "pull" in nature and is the result of aging of the population. Due to the post–World War II baby boom and the changes in family size, the average age of the population in many countries is increasing. This, coupled with the fact that functional limitations increase dramatically as people age (see Fig. 65.1), has resulted in a population with an increasing percentage of individuals with disabilities and severe disabilities. More significantly, there is a growing population of individuals who are older and who have disposable income who will be looking for and purchasing electronic and telecommunication technologies. Figure 65.2 shows that a large percentage of the problems faced by these individuals will be in the physical, visual, and hearing areas. These are all areas where product design can dramatically affect usability of products for older persons.

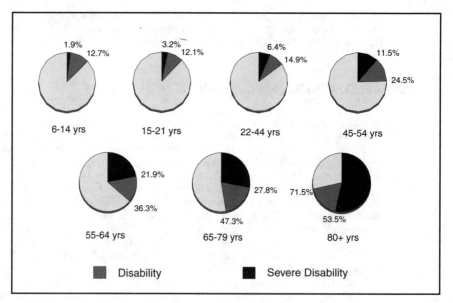

FIGURE 65.1   Disability as a function of age.

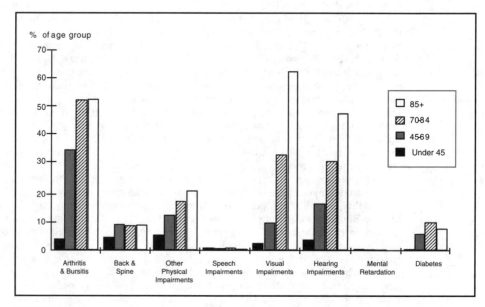

FIGURE 65.2   Prevalence of selected impairments within age groups. Data categories are not exclusive.

The third major influence in the area of universal and accessible design has come about due to the development of ever-more mobile technologies. As telephones and information appliances get smaller so that they can be carried in pockets and purses, the controls and displays are getting smaller. Information that could easily be displayed on large screens cannot be seen on small displays and needs to be reformatted in the same way as it would be reformatted for individuals with low vision who would be viewing it in enlarged form on a standard-sized screen. In addition, people are using appliances while they are driving or otherwise have their eyes occupied, leading them to look for technologies that can be used purely through speech. Finally, the fact that information providers have no idea which type of device, workstation, pocket browser, or phone will be used to access their services is causing them to develop new ways of delivering their information that can adapt to the needs and constraints of individual users and the technologies they are using at the moment. All of these lead to the development of more flexible and adaptable products. As noted earlier, there is a great similarity between the needs of individuals with disabilities and mobile technology users who do not have disabilities. At a recent World Wide Web conference, there was a panel, half of which comprised individuals from the mobile computing field and the other half from the disability access field. As a part of the presentation, a question was posed as to whether there was anything that was beneficial for individuals who had disabilities that was not a benefit to people with mobile computing and vice versa. Although the panel believed that there should be some elements, none could be identified by either panel members or members of the audience that provided benefits for one side and were not of use to the other. The existence of peripherals like Braille printers were identified, but nothing was noted in the way of the overall architecture or structure and operation of the World Wide Web. This convergence of requirements and benefits between disability access and mobile computers is both a very powerful example of the broad benefits of universal design and a strong additional incentive to look at disability access as a part of the continuum of flexible, adaptable design and not as a separate endeavor.

65.4 THE PRACTICAL APPLICATION OF UNIVERSAL DESIGN: THE NEED TO SET PRIORITIES

The multidimensional nature of disabilities, including vision, hearing, physical, and cognitive dimensions, coupled with different degrees of disability and combination of disabilities, results in a very large number of target audiences. Therefore, when looking at designing products so that they are usable across a broad range of users, it is usually necessary to set priorities. In addition, there are multiple strategies that might be implemented for each dimension of disability. A recent compilation of strategies carried out at the University of Wisconsin yielded between 200 and 300 different unique strategies for making products more accessible and usable by people who have disabilities (Schauer, Barnicle, and Vanderheiden, 2000). This number does not include the large number of strategies documented in the general usability literature, which apply to people with all levels of disabilities. Without a means to prioritize among the myriad of access strategies to implement, two behaviors have been observed on the part of designers.

The first behavior is that product designers become overwhelmed with the sheer number of different techniques and strategies. Just the thought of looking at 100 different strategies for making their products more accessible causes many to walk away or to approach feature selection—what they build into their products—in a somewhat random fashion. Turning to usability tests to set priorities is not a solution to the problem of being overwhelmed, since the usability tests themselves quickly generate long lists of problems that, in turn, point back to even longer lists of potential solution strategies.

The second behavior observed in designers is poor prioritization of efforts. Very often, features that were thought of first, or are easiest to implement, are chosen rather than features that are more important to the overall accessibility of the product. The result is a product that has multiple low-priority features, which are helpful but not essential for access, while lacking other key high-priority features that are needed to make the product or key functions of the product accessible. Often, "helpful" features are built into a product where "key" features have been left out. This results in a product that has helpful features for a disability group, but that cannot be used by the members of that same disability group because of missing features. This would be equivalent to having a building where the designers installed tight-nap carpet rather than plush carpet so that it was easier for wheelchair users to get around, but built steps at the front door with no ramp.

To help address the prioritization problem and to better understand how priorities should be set in selecting features for incorporation in products, a number of dimensions for prioritization have been defined. These include:

- Accessibility versus usability
- Independence versus codependence
- Efficiency and urgency
- Ease of implementation

The First Dimension for Prioritization: Accessibility/Usability

In looking at the usability of a product from the perspective of different people, there is a continuous range that runs all the way from people who have no problems at all in using all of the functions of a product, usually a small number of people, to:

- People who have little difficulty with all features
- People who have difficulty with some features
- People who have trouble with most features
- People who are unable to use the product at all

Even the features vary in importance to the overall use of the product. Some features are essential while others are merely convenient. One approach, which has been found useful, is to adopt a three-tier system for evaluating the importance of product features, as follows:

Level 1. Features that, if not implemented, will cause a product to be unusable for certain groups of users or situations

Level 2. Features that, if not implemented, will make the product very difficult to use for some groups of users/situations

Level 3. Features that, if they are implemented, will make the product easier to use but do not make a product usable or unusable, except for individuals who are just on a margin due to other factors and this small amount of usability pushes them over the threshold

The Second Dimension Affecting Prioritization: Independence/Codependence

In addition to the accessibility/usability dimension, there is a second dimension that deals with independence versus codependence. Everyone depends on others for some aspects of their lives. Some individuals know how to repair their own car or television set. However, other individuals do not know how to change printer cartridges, clear paper jams, or reformat their hard drives. In daily life there are some things people need to be able to do independently and some things for which they can depend on others. This independence/codependence can be taken into account to facilitate decisions regarding expenditure of effort.

For example, it is more important that an individual be able to load their work into the input hopper on a copier and operate the controls to get the required number and type of copies than it is for them to be able to change the toner or clear a paper jam. In fact, in many offices only people trained in clearing paper jams are allowed to do so. Loading new reams of blank paper into the copier generally falls somewhere in between. Similarly, it is more important for an individual to be able to launch and operate programs than it is for them to be able to configure their modem settings. This importance stems not from the technical difficulty of the two tasks, but from the fact that one is an activity that is required continuously as a part of a product's daily operation. However, the other task is something that needs to only be done once, or that can be planned for and scheduled when there is someone to assist.

Figure 65.3 shows a rough hierarchy based on the need for independence versus the ability to deal with the product via codependence (i.e., depending on others for infrequent and/or predictable tasks that are outside of one's own abilities). The exact ordering of the items will vary for different types of products and different environments (e.g., the availability of the co-dependent facilitator), but the general ordering can be seen. This hierarchy can be used to set priorities in a resource-constrained or time-constrained product design program. Items at the top are those that it would be difficult to depend on others for help. As one moves down the items, they increase in their ability to be delegated or to depend on assistance from others.

---

1. Functions/features needed for basic use of the product
2. Unpredictable, but typically user-serviceable (by "average" user) maintenance or recovery operations
3. Unpredictable service, maintenance, or recovery, typically corrected by support personnel
4. Predictable or schedulable maintenance that can be delegated to others
5. Unpacking and initial setup
6. Repair

Note: The location and availability (or not) of support personnel (e.g., in a home office) affect this dimension.

FIGURE 65.3   Independence versus codependence hierarchy of needs.

A Third Dimension Affecting Prioritization: Efficiency/Urgency Requirement

A third dimension to prioritization deals with the need for efficiency. If a task is performed only once a day and there is no particular time constraint on its accomplishment (e.g., the person is not trying to disarm an alarm before it goes off), then the relative efficiency of operation is not as critical as a function that must be used continuously throughout the day. For example, if it takes an individual five times longer to operate the on switch on their computer than the average worker, it will not have a major impact on their productivity or effectiveness. In fact, operating the power switch is such a small part of booting a computer that the total time it takes for them to boot up the computer is likely to be only negligibly longer than the time for anyone else to boot up their computer. If it takes an individual five times as long to type characters on their computer, however, and they spend the bulk of their day entering information into their computer, the difference in efficiency could be catastrophic. If it took them five days to get an average day's worth of work done, it would be hard for them to compete in either an educational or work environment. Thus, for the on/off switch, level 1 accessibility may be all that is required. However, for data entry, accessibility level 1 (accessible), 2 (easily usable), and 3 (efficient) may all be critical on an individual's workstation.

A close parallel to efficiency is the urgency issue. If there are situations where a user must do something within a particular time constraint in order to avoid an adverse situation, then, even if it is rarely done, it may be important to strive for level 2 or level 3 usability in order to allow the individual to be able to carry out the activity within the time allowed. The importance that is attached to the efficiency and urgency dimension is the function of at least three factors.

1. The reversibility of the action
2. The severity of the consequence for failure
3. The ability of the person to adjust the time span to meet their increased reaction times

Situations where the result is not reversible or is dire in nature and is also of a type that does not allow for user adjustment or extension (e.g., as in some security-related situations) would create the highest priority for providing not only an accessible but a highly usable interface for the disability group or situation.

A Pseudopriority Dimension: Ease of Implementation

In setting priorities for implementation of features promoting accessibility/usability in products, a factor that is often used to select features is the ease with which they can be implemented in the product. In this context, *ease* may have many different characteristics, including: cost in dollars, cost in time, ease in getting clearance from supervisors, minimized impact on other features, minimized impact on testing, minimal impact on documentation, and so on. Often referred to as "low-hanging fruit," such features are often very tempting when compared with features that are much more difficult to implement. Although it is always good to look at this dimension, there is great danger here as well. Often, this strategy leads one to believe that five low-hanging-fruit features must be better than one that is more difficult to achieve. This can lead to the implementation of multiple level 3 features instead of key level 1 features. People often implement low-hanging-fruit features to help a disability group, but omit key access features for the same disability group, so they are unable to use the product.

Within level 1 accessibility features, however, one will also often find either low-hanging fruit or features that have such mass-market appeal that their costs are offset by their market benefit. As long as ease of implementation does not cause the focus to shift away from the features that are really needed, this can be a good effect. Care must be taken, however, since often companies will think in numbers (e.g., "already five features have been added in this revision"). Thus, getting a number of low-priority, easy-to-implement features included may make it difficult to get the key access features into the product.

65.5   BASIC COMPONENTS OF CROSS-DISABILITY-ACCESSIBLE DESIGN

A number of different approaches have been developed to define the different dimensions of "accessibility across disability" groups as they relate to different product types. The Telecommunication Access Advisory Committee, the Access Board, and the FCC created 23 guidelines for access to telecommunication products. The Electronic and Information Technology Access Advisory Committee devised over 90 guidelines for electronic and information technologies. The Access Board, in its Sec. 508–proposed rulemaking, reduced these 90 guidelines to just under 70. In each of these cases, the guidelines are a combination of general performance-based guidelines and specific design-based guidelines.

To provide a more generic approach, the idea of collapsing the guidelines into their essential components has been explored. The process of collapsing the guidelines takes the form of identifying the key objectives or requirements for providing more flexible, universally usable interfaces, and then identifying key strategies for meeting these objectives. The current working draft is presented in Table 65.2.

This approach for defining the fundamentals for accessibility breaks the requirements for cross-disability-accessible design into five major objectives.

1. Ensuring that all information presented by or through the device can be perceived, even if all sensory channels are not available to the individual

TABLE 65.2   Key Principles and Strategies for Providing Access

Basic access principle	5.0 Why	How—general
Make all information perceivable, including keys and controls—status and labels • Without vision • Without hearing • Without reading, low vision or cognition • Without color perception • Without causing seizure	Information that is presented in a form that is only perceivable with a single sense (e.g., only vision or only hearing is not accessible to people without that sense). Also not accessible by anyone using a mobile technology that does not present all modalities such as a phone—or automobile audio-only browser.	Make all information available either in: • *Presentation-independent form* (e.g., electronic text that can be presented in any sensory form)* • *Sensory-parallel form* where forms suitable for presentation in different sensory modalities are all provided in synchronized form (e.g., a captioned and described movie) Provide a mechanism for presenting information in visual, enlarged visual, auditory, enhanced auditory louder, and, if possible, better signal-to-noise ratio and, where possible, tactile form.

Provide at least one mode for all product features that is operable:
• Without pointing
• Without vision
• Without requirement to respond quickly
• Without fine motor movement
• Without simultaneous action
• Without speech
• Without requiring presence or use of biological parts (touch, fingerprint, iris, etc.)
	Interfaces that are technology or technique specific cannot be operated by individuals who cannot use that technique [e.g., a person who is blind cannot point to a point in an image map; some people cannot use pointers accurately (also not accessible to mobile users who are using voice to users who are using voice to navigate, for example)].	• Make all interfaces controllable via ASCII/UNICODE input and output.
		• Have all text output voiced or compatible with a device that will voice it.
		• Make all input and displayed information nontimed, or allow user to freeze timer or set it to very long time (e.g., 5–10 s to do single action, 2–4 s to stop action)
		• Have at least one mode for achieving each and every function of the product that meets the following conditions:
		  • No simultaneous activations
		  • No twisting motions
		  • No fine motor control required
		  • No biological contact required
		  • No user speech required
		• If biological techniques are used for security, have at least two alternatives and preferably a non biological alternative unless required.

TABLE 65.2 (*Continued*) Key Principles and Strategies for Providing Access

Basic access principle	5.0 Why	How—General
Facilitate navigation • Without sight • Without pointing ability • Without fine motor control • Without prior understanding of the content • Without the ability to hear	Many individuals will not be able to use alternate access techniques if their layout is too difficult to understand. Many individuals will not be able to operate products, such as workstations, with sufficient efficiency to be competitive if navigation is not easy.	• Make overall organization understandable (e.g., provide overview, table of contents, site maps, etc.). • Do not mislead/confuse. Be consistent in use of icons or metaphors. Do not ignore or misuse conventions. • Allow users to jump over blocks of undesired information (e.g., repetitive info), especially if browsing auditorially.
Facilitate understanding of content • Without skill in language used on product • Without good memory • Without background	People with cognitive difficulties may not be able to access and use complex devices or products with language. Many others may find that they are unable to master alternate access techniques if layered on top of complex interfaces or content.	• Use the simplest, easiest-to-understand language as is appropriate for the material/site/situation. • If unexpected languages are used, be sure the language used is identified to allow translation.

Compatible with assistive technologies commonly used by people
• With low vision
• Without vision
• Who are hard of hearing
• Who are deaf
• Without physical reach and manipulation
• Who have cognitive or language disabilities
	In many cases, a person coming to a task includes the assistive technologies they have with them. If they cannot use products directly, it is important that the products be designed to allow them to use the tools they carry with them to access and use the products. This also applies to mobile users, people with glasses, gloves, or other extensions to themselves.	• Support standard points for connection of
		  • Audio amplification devices
		  • Alternate input and output devices or software
		• Do not interfere with use of assistive technologies
		  • Hearing aids
		  • System-based technologies
		• Support use of wheelchairs, walker, etc.

* To meet these requirements, text formatting must be electronically readable and presentable without vision.

2. Ensuring that the device is operable by the users, even if they are operating under constraints

3. Facilitating the ability of individuals to navigate through the information and controls, even if operating with constraints

4. Facilitating their ability to understand the content

If it is not possible to achieve the preceding four objectives directly, then the goal becomes:

5. Making the product compatible with the common tools (e.g., assistive technologies) that users may have with them in order to maximally achieve the aforementioned four objectives

From the wording of the four objectives, it can be seen that the first two tend to be level 1–like objectives, in that they refer to providing the basic ability to perceive and operate a device. Objectives 2 and 3 are more level 2/level 3–oriented in that they facilitate use of the product. Compatibility overlaps all four of the objectives.

Despite the level 1/level 2 appearance of the four basic objectives, elements of level 1, 2, and 3 usability can be seen in all four objectives.

Cognitive Constraints: A Unique Dimension

It is important to note that the cognitive dimension is unique from the other mentioned dimensions. Whereas it is possible to make most products usable to individuals with no vision, or no hearing, and even with essentially no physical ability, there are very few products, if any, that are usable by individuals with no cognitive abilities. This is because it is possible to translate most types of information between sensory modalities, and most types of activities between physical interface techniques, but there is no mechanism for transferring cognitive processing into another domain. While it is true that there are some activities and some types of information that do not constitute good strategies for providing access to individuals with severe or total visual limitation, severe or total hearing limitations, or severe or total physical limitations, the number of devices and activities that are excluded is much smaller than for severe or total cognitive limitations. For this reason, strategies for *enabling access* for people with cognitive disabilities basically look like techniques to *facilitate,* with each technique pushing a few more people over the threshold into the category of individuals who can use a product.

It is also important to note that there are a number of dimensions, such as learning disabilities, reading problems, and such, which are often lumped in with cognitive disabilities when discussing ways to make products accessible. For example, there are strategies that can allow individuals who are completely unable to read to be able to effectively use a very wide variety of products. In this case, the difficulty is not in general cognitive processing or memory but rather in a specific skill, which is reading.

Generic Strategies

Matched with each of the basic objectives is a list of generic strategies for addressing the objective. These are shown alongside the basic access principles in Table 65.2.

Each of these generic strategies would take a different form in different technology families. In some cases, such as the area of Web technologies, some of the strategies would be implemented on the Web content pages, while others would be implemented in the user agents or browsers. The goal of the access principle and strategy consolidation effort represented in Table 65.2 is to try to boil down the overall universe of requirements and strategies to their essence so that it is easier to understand them. In doing so, a certain amount of simplification is necessary. As a result, a condensed set of requirements and strategies needs to be elaborated within each of the different technology families with which it is intended to be applied. Work is continuing in the process of refining this simplified view, as well as providing details on how these might be applied to the latest technology areas. Information on this process, as well as sets of features that can provide cross-disability access on a variety of different products, can be found at the Trace Center Web site at http://trace.wisc.edu

65.6 LESSONS LEARNED

When considering universal design, it is important to distinguish between theory and practice. If one wants to think of the term *universal design* as being a global concept that focuses purely on the topic of making designs that are usable to the widest range of people, then one might want to coin a new term, *commercial universal design,* to represent the practical application of universal design concepts within. The important thing that has been learned is that when universal design is practiced within a commercial context, it must be done in a way that recognizes the constraints and realities of industry and that works within them. Moreover, it must allow or even facilitate the ability of companies to create successful, profitable products.

A second lesson learned is that not all aspects of accessibility have the same importance. There are some operations, such as changing toner cartridges or even replacing batteries, that are more predictable and can be delegated to an assistant or bystander; whereas, there are other aspects of the operation, having full access to the commands, menus, and functions of the device, which are much more important. There are also products or functions of products where efficiency is not only desired but critical for the successful use of the device, or the ability of the person to successfully function in their job. In reviewing all of the options for improving the functionality and usability of a product it is, therefore, important to recognize these differences in priority, to focus on high-priority items, and to ensure access across disabilities.

A third lesson is recognition of the importance of the mass-market customer when considering universal design features or enhancements. Although the term *universal design* is based on the fact that these principles or practices benefit everyone, insufficient data or evidence of the benefit to people without disabilities is usually presented when universal/accessible features are introduced. It is suggested that universal design advocates even consider creating designs that have no disability access features, only mass-market features that happen to also increase or provide access to individuals with disabilities. One such example for a cellular telephone was shown at *Designing for the 21st Century, II: An International Conference on Universal Design* (Vanderheiden, Vanderheiden, and Tobias, 2000).

Finally, it is important to remember that universal design is a *process* and not an outcome. Too often, people refer to "universally designed products" or to products that are "universal designs." These statements are not only false (i.e., they are not universally usable), but they also seek to perpetuate a myth that industry is being asked to create universally usable products when they are asked to practice universal design. In working with industry, this perception that industry is being asked to create universally usable products is the single greatest reason for resistance to the term and for fighting its application within their business practice.

65.7   CONCLUSION

Because of the tremendous variety of forms, degrees, and combinations that a disability can take, and the very large number of strategies that can be found for addressing access, it can seem impractical to try to create standard products that are, indeed, accessible and usable by the regular population and also by the myriad of people with disabilities. However, the process of boiling the access issues down to their essentials, setting priorities for accessing features, and identifying small sets of strategies that work together to provide maximum coverage across disabilities can result in products that are cross-disability accessible without requiring the incorporation of large numbers of features. Like most areas of design and, indeed, most areas of learning, the area of accessible design and extending standard design to include people with disabilities can seem daunting at first. Once the different dimensions are mastered, however, and the fundamental principles and goals are understood, design to include people with functional limitations becomes a natural part of the design process and, in fact, facilitates the design of products that are easier to use by all users.

65.8   NOTES

1. The assistance of Katherine (Kitch) Barnicle, Ph.D., in reviewing and commenting on this paper during its preparation is gratefully acknowledged.

2. This is a publication of the Trace Research & Development Center, which is funded, in part, by the National Institute on Disability and Rehabilitation Research of the Department of Education under grant numbers H133E980008, H133E990006, and H133A60030. The opinions contained in this publication are those of the grantee and do not necessarily reflect those of the Department of Education.

65.9   BIBLIOGRAPHY

Bureau of the Census, series P-70, no. 8, SIPP, Survey, 1984.

LaPlante, M. P., *Data on Disability from the National Health Interview Survey, 1983–85: An InfoUse Report,* National Institute on Disability and Rehabilitation Research, US Department of Education, Washington, DC, 1988.

Schauer, J., K. Barnicle, and G. C. Vanderheiden, "Facilitating the Development of Cross-Disability Accessible Products: The Product Design/Interface Evaluation Toolkit," *Proceedings of the 2000 CSUN Conference on Technology and Disability,* Los Angeles, 2000.

Vanderheiden, G., "Design for People with Functional Limitations Due to Disability, Aging, or Circumstances," in Gavriel Salvendy (ed.), *Handbook of Human Factors and Ergonomics,* John Wiley & Sons, New York, 1997.

Vanderheiden, G., K. Vanderheiden, and J. Tobias, "Universal Design Motivators and Facilitators," in *Proceedings of Designing for the 21st Century, II: An International Conference on Universal Design,* 2000; http://trace.wisc.edu/docs/phones/index.htm.

CHAPTER 66

ACCESS TO THE WORLD WIDE WEB: TECHNICAL AND POLICY PERSPECTIVES

Judy Brewer, M.A.
World Wide Web Consortium: MIT, INRIA, Keio

66.1 INTRODUCTION

The World Wide Web has, in a few short years, become one of the cornerstone technologies of the information society. In many countries the Web already plays a leading role in providing access to information, educational resources, employment opportunities, workplace interactions, government services, and entertainment. To what extent does this new information resource welcome people with disabilities, and what factors will contribute to or hinder its accessibility for people with disabilities?

This chapter examines issues of accessibility and universal design of the Web from several perspectives: barriers to accessibility on the Web, dimensions of universal design on the Web, policies that impact on Web accessibility, and guidelines and implementation issues that affect accessibility of Web sites and Web-based applications.

66.2 ACCESSIBILITY OF THE WEB

Why is accessibility a concern for a medium such as the Web? After all, there are no overt physical barriers such as stairs on the Web. Moreover, for some people with disabilities, access to Web-based information is already a vast improvement over the barriers endemic in existing information resources, such as libraries without ramps, or the unavailability of alternative formats for many texts. However, this new information medium is unfortunately just as prone to accessibility problems as is any other environment.

Accessibility problems on the Web can affect people with a variety of disabilities. For example, users with visual disabilities may miss information that is represented in graphics or video unless it is accompanied by alternative text. They may be unable to access navigational hot spots in image maps, find their way through unlabeled frames, or understand poorly constructed tables. Uncaptioned audio similarly creates barriers for people with hearing impairments who may be unable to hear audio on a Web site. Lack of keyboard support for mouse commands can slow down individuals with physical disabilities. Web sites without consistent navigation options, or with flashing or scrolling text, can impede navigation and comprehension for users with dyslexia, short-term memory problems, or other cognitive or neurological disabilities.

The rapid rise in the role of the Web in society has been a mixed blessing for people with disabilities. When accessible, the Web can provide access to information at far greater speed and volume than was available in the past. But when inaccessible due to the kinds of barriers just described, the Web excludes people with disabilities from accessing critical information and resources.

Given that the Web is already displacing traditional means of accessing information—for instance, local offices where in-person contact is possible, or telephone-based information services where voice contact is possible—it is paramount that industry, government, and the user community work together to ensure accessibility of Web-based services. Otherwise, for many people with disabilities, continued expansion of the Web will mean exclusion from educational and employment opportunities, and from participation in commerce, civic life, and more.

Accessibility and Universal Design of the Web

As with so many other areas of the built environment and information technologies, accessible Web design, or design for people with disabilities, contributes substantially to universal design—design for the needs and capabilities of a broad diversity of Web users.

In the context of the World Wide Web, universal design, or design for all, has been described as design that encompasses accessibility for people with disabilities, as well as issues such as device independence for any Web user, internationalization of the Web, usability, and affordability. In the context of this chapter, *accessibility* refers to access by people with disabilities to Web-based information. *Device independence* refers to comparable access to Web-based information, regardless of the type of device that one is using (e.g., a desktop computer with a mouse, a mobile phone, a personal digital assistant, an information kiosk, a TV, or a land-line-based telephone). *Internationalization* means ensuring that Web technologies can represent Web content equally well across all of the world's languages. *Usability* refers to general human factors issues. *Affordability* refers to issues of economic access to computers and Internet services.

In considering the spread of ideas within the Web community, it is important to note that the Web community is more heavily international than many other sectors of the information technology industry, and one that functions on many levels. It includes those who build Web technologies, those who develop content for it, and those who use its content.

While it is unlikely that consistent definitions of accessibility and universal design will prevail throughout such a diverse and geographically dispersed community, it is remarkable to observe the spread of concepts such as accessibility and device independence that has already occurred in a relatively short time.

Efforts to promote awareness of the need for Web accessibility first emerged in 1995. In 1997, with the help of Mike Paciello from the Yuri Rubinsky Insight Foundation, the World Wide Web Consortium (W3C) took on a project called the Web Accessibility Initiative (WAI). Since 1997, WAI's efforts, coupled with the efforts of many other organizations in the field, have generated a steady stream of information about the benefits of accessibility toward ensuring that the Web is usable by all. During 1999 and 2000, this coincided with the explosive growth of the mobile phone industry. As mobile phones have converged into Web space, a high degree of interest has emerged within the industry around carryover benefits of accessibility for device independence in general, and mobile phone access in particular (see Chap. 65, "Fundamentals and Priorities for Design of Information and Telecommunication Technologies," by Vanderheiden).

An informal comparison by Daniel Dardailler, W3C (unpublished), of the difference between provisions in the Web Content Accessibility Guidelines and provisions that would likely be included in a hypothetical set of guidelines for device independence showed a differential of only 5 percent. While this has not yet been formally substantiated, it indicates a potentially minimal difference between the provisions needed for accessibility and those needed to ensure device independence of Web content.

This perspective is reinforced by growing support for accessibility among those involved in the Wireless Access Protocol Forum (WAP Forum) and other organizations providing access to the Web over mobile phones. For instance, at the Ninth International World Wide Web Conference (WWW9) held in May 2000, almost every presenter from the mobile phone industry spontaneously mentioned during their presentations that W3C/WAI's accessibility guidelines provided a strong foundation for ensuring access to content via mobile phones.

Less is known about the similarities between provisions supporting other dimensions of universal design, such as internationalization and usability, in part because there are not yet guidelines for those other dimensions with which accessibility provisions could be compared. Consideration of other dimensions would very likely broaden that differential. For instance, effective design for internationalization requires not only following accessibility checkpoints (e.g., "Identify the primary natural language of a document," and "Clearly identify changes in the natural language of a document's text and any text equivalents") but also recommendations on the use of a particular character model, such as Universal Character Set (Unicode/ISO 10646), which supports more universal access for the world's languages. Developing appropriate universal design guidelines in the context of the Web would presuppose existence of guidelines-type provisions, comparable to the accessibility guidelines, within each of the areas of device independence, internationalization, and so on.

The Role of the Web Accessibility Initiative at the World Wide Web Consortium

The W3C is an international, vendor-neutral, primarily industry consortium that develops the core technologies—the data formats and transport protocols—used on the Web. It has close to 500 Member organizations, and is hosted at MIT's Laboratory for Computer Science in the United States, at Institut National de Recherche en Informatique et en Automatique (INRIA) in France, and at the Shonan-Fujisawa campus of Keio University in Japan.

WAI is one of four domains, or areas of work, of W3C. W3C's Web Accessibility Initiative addresses accessibility at several levels to identify and promote comprehensive accessibility solutions for the Web. On the most fundamental level, WAI ensures that the core technologies of the Web can support accessibility—for instance, by supporting mechanisms for device-independent navigation and access to accessibility information in HyperText Markup Language (HTML 4.0) (Raggett et al., 1999), Cascading Style Sheets Level 2 (CSS 2) (Bos et al., 1998), Synchronized Multimedia Integration Language (SMIL) 1.0 (Hoschka, 1998), Extensible Markup Language (XML) 1.0, second edition (Bray et al., 2000), and a score of other evolving Web technologies (World Wide Web Consortium, 1994–2000).

WAI also provides guidance on how to use Web technologies in ways that support accessibility, through three WAI guidelines and related documents. In the *Web Content Accessibility Guidelines 1.0 (WCAG 1.0)* (Chisholm et al., 1999), WAI describes how to design Web sites that are accessible, and yet also are advanced from a technical and design standpoint. The *Authoring Tool Accessibility Guidelines 1.0 (ATAG 1.0)* (Treviranus et al., 2000a) addresses developers of the software used to build Web sites by providing guidance on the use of Web standards, on support for accessible authoring practices, and on accessibility of the user interface. The *User Agent Accessibility Guidelines 1.0 (UAAG 1.0)* (Gunderson et al., 2000b) provides guidance for user agents, including browsers, multimedia players, and their inter-operability with assistive technologies, focusing on accessibility of the user interface and on rendering of accessibility information on Web sites.

In addition to development of these guidelines, WAI works on different approaches for evaluation and retrofitting of Web sites and transformation tools for inaccessible sites; develops education and outreach materials; and coordinates with other organizations researching future Web technologies.

66.3   *A POLICY CONTEXT FOR ACCESSIBILITY OF THE WEB*

Over the past several years, guidelines and other resources developed by WAI have emerged as a common reference point throughout many policies relating to Web accessibility around the world. Because the Web is not a single entity but rather a decentralized information space to which people all over the world—businesses, community organizations, governments, and individuals—contribute in various ways, it is extremely unlikely that any one set of laws or policies will apply consistently across all these communities around the world.

Regulation of Rapidly Evolving Technologies

In many areas of our technologically complex society, it is difficult for public policy to keep pace with the rapid evolution of emerging technologies. Technology that may seem a curiosity one year, and a luxury the next, suddenly emerges an essential element of business communications—but often, this happens before policy makers have learned how to use the new technologies, let alone had time to consider meaningful policies that might allay any potentially negative ramifications.

Similarly, the user community cannot always learn about the potential impact of each new technology at a sufficiently rapid pace to articulate and protect their interests. An instance of this, which has received much media attention, is the threat to personal privacy engendered by the growth of the Internet, with many people not even aware of privacy risks that are caused by simple interactions with this new medium.

Typically, the media play a role in educating the public, including people with disabilities, on new issues. But the phenomenon of the "disability digital divide"—the effects of accessibility barriers to new technologies combined with affordability issues that disproportionately impact the disability community—has received little attention from the media. Without a fluency in the technical jargon, an understanding of the technology development and standardization process within industry, or a command of the existing information technology policy context, the disability community has less chance of gaining timely access to key policy discussions that can affect the extent to which accessibility and universal design are considered in new technologies.

Public Policy Trends Around Web Accessibility

Approaches to regulating the accessibility of information technologies generally fall into one of several categories. Governments can establish that individuals with disabilities have a civil right to certain kinds of information; they can require that products or services sold within a country must meet certain criteria for accessibility; or they can require that information technologies procured by certain kinds of organizations must be accessible.

The first approach—establishing that access to certain kinds of information by individuals with disabilities is a civil right—is usually tied to a disability discrimination act, such as the Disability Discrimination Act in Australia or the Americans with Disabilities Act (ADA) in the United States. Recently, as in Portugal, some legislation has directly required Web accessibility. Usually, this approach applies primarily to government Web sites, but in some cases, it can extend beyond those to apply to some kinds of commercial sites. This has been the interpretation in both Australia and the United States. In the United States, a congressional hearing in February 2000 attempted to reverse the U.S. Department of Justice opinion issued in September 1996 (Patrick, 1996) that the ADA applies to the Internet and the Web, but the hearing was unsuccessful in this regard.

A hypothetical example of the second approach might have been Sec. 255 of the Telecommunications Act, which requires accessibility, to the extent readily achievable, of all telecommunications products and services sold in the United States—if the definition of telecommunications products and services had been interpreted differently. As it was, the definition of telecommunications products and services excludes most Internet- and Web-related information and services in the regulations promulgated by the U.S. Federal Communications Commission to enforce Sec. 255. Given that few other countries have legislation similar to Sec. 255, we have yet to see what effect that type of provision might have on accessibility of the Web.

For an example of the third approach, we again turn to the United States. Section 508 of the Rehabilitation Act Amendments of 1998 sets requirements for accessibility of information technology procured by the federal government. It has the dual goal of ensuring that federal employees with disabilities have comparable access as nondisabled employees to the information technology in their workplaces, and that members of the public with disabilities will have access to government information in accessible formats. Among other aspects of federal information technology procurement addressed by Sec. 508 are federal Web sites.

The final rule for Sec. 508, issued in December 2000 (Access Board, 2000), included reuse of some provisions from *WCAG 1.0,* albeit with some rewording, additions, and omissions of certain items. While these regulations apply only to government Web sites in the United States, Web design and development represent a nontrivial part of the estimated $27 billion U.S. government information technology marketplace. Section 508 is, therefore, expected to have a constructive ripple effect on the Web industry in the United States with regard to awareness of accessibility, development of accessibility expertise within the Web design community, and demand for authoring tools that support accessible design.

The period of 1999 through 2000 has been an active one for policy developments related to Web accessibility. WAI maintains a listing of policy reference links providing more detail on the following:

- Canada has established a *Common Look and Feel for the Internet* (Treasury Board of Canada, Secretariat, 2000), which references priority levels 1 and 2 of *WCAG 1.0* as requirements for federal Web sites.

- In Australia, the Equal Opportunity Commission held hearings on Internet accessibility, which resulted in requiring conformance to WAI guidelines for Web sites (Office for Government Online, 2000).

- In Europe, a European Union Ministerial Conference adopted the WAI guidelines as requirements for European Union and member-state Web sites (eEurope 2002, 2000).

- A number of individual European countries passed or started action on Web accessibility requirements for government sites, including France, Ireland, Portugal, Sweden, and the United Kingdom (Brewer, 1998–2000).

- The *United Nations Standard Rules for the Equalization of Opportunities for Persons with Disabilities* provide a normative framework for public policies that includes access to the physical environment as well as access to information. Sara-Serrano and Mathiason (Chap. 11, "United Nations Standards and Rules") highlight this policy, and also highlight implementation strategies developed by several countries in collaboration with the United Nations.

66.4 GUIDELINES FOR ACCESSIBILITY OF WEB CONTENT

From 1997 to 1999, W3C/WAI developed the *Web Content Accessibility Guidelines 1.0* (*WCAG 1.0*) with the help of organizations from around the world interested in Web accessibility. These organizations included representatives of industry, disability communities, accessibility research, and government, participating in or contributing to development of *WCAG 1.0* according to W3C process. Their goal was to develop authoritative guidance on what constitutes an accessible Web site.

This proved to be a complex task. There were already a number of Web accessibility guidelines in existence, but few of these guidelines comprehensively addressed the needs of all disability groups, and few were up to date with the latest Web technologies. It took extensive exploration of issues and consensus building to come up with accessibility solutions that were cross-disability, that were forward-compatible with emerging Web technologies such as XML applications, and that also took into account the prevalence of "legacy" browsers and assistive technologies used by many people with disabilities.

The result of this effort was a document with 14 guidelines—abstract principles of accessible Web site design—and a total of 65 discrete checkpoints correlating with those guidelines. These checkpoints are also available via a checklist that breaks out the checkpoints according to three priority levels, the priority 1 checkpoints being those most crucial for ensuring accessibility of Web sites.

In addition, the working group produced a techniques document as a reference note. *Techniques for Web Content Accessibility Guidelines 1.0* (Chisholm, Vanderheiden, and Jacobs, 2000) includes nonnormative implementation approaches for each of the *WCAG 1.0* checkpoints and provides detail regarding language-specific markup (e.g., for HTML, CSS, SMIL, etc.). Additional implementation support materials of a promotion or training nature have been developed separately through WAI's Education and Outreach Working Group.

This approach of developing complementary resources, including an authoritative guidelines document, prioritized list of checkpoints, an accompanying nonnormative techniques document, and additional promotion and training materials, has also been followed for WAI's other guidelines. As Web technologies evolve, WAI updates each of these resources as necessary to address new technologies.

General Principles for Accessibility of Web Content

The *WCAG 1.0* starts with a discussion of two overriding themes of accessibility on Web pages: (1) ensuring that Web content will "transform gracefully," and (2) "making content understandable and navigable."

In explaining graceful transformation, *WCAG 1.0* states:

Pages that transform gracefully remain accessible despite any of the constraints described in the introduction, including physical, sensory, and cognitive disabilities, work constraints, and technological barriers. Here are some keys to designing pages that transform gracefully:

- Separate structure from presentation (refer to the difference between content, structure, and presentation).
- Provide text (including text equivalents). Text can be rendered in ways that are available to almost all browsing devices and accessible to almost all users.
- Create documents that work even if the user cannot see and/or hear. Provide information that serves the same purpose or function as audio or video in ways suited to alternate sensory channels as well. This does not mean creating a prerecorded audio version of an entire site to make it accessible to users who are blind. Users who are blind can use screen reader technology to render all text information in a page.
- Create documents that do not rely on one type of hardware. Pages should be usable by people without mice, with small screens, low resolution screens, black and white screens, no screens, with only voice or text output, etc.

In explaining the intent of making content understandable and navigable, *WCAG 1.0* states:

Content developers should make content understandable and navigable. This includes not only making the language clear and simple, but also providing understandable mechanisms for navigating within and between pages. Providing navigation tools and orientation information in pages will maximize accessibility and usability. Not all users can make use of visual clues such as image maps, proportional scroll bars, side-by-side frames, or graphics that guide sighted users of graphical desktop browsers. Users also lose contextual information when they can only view a portion of a page, either because they are accessing the page one word at a time (speech synthesis or Braille display), or one section at a time (small display, or a magnified display). Without orientation information, users may not be able to understand very large tables, lists, menus, etc.

Overview of Guidelines for Accessible Web Content

The 14 individual guidelines in *WCAG 1.0* state abstract principles of Web site design:

1. Provide equivalent alternatives to auditory and visual content.
2. Don't rely on color alone.
3. Use markup and style sheets and do so properly.
4. Clarify natural language usage.
5. Create tables that transform gracefully.
6. Ensure that pages featuring new technologies transform gracefully.
7. Ensure user control of time-sensitive content changes.
8. Ensure direct accessibility of embedded user interfaces.
9. Design for device independence.
10. Use interim solutions.
11. Use W3C technologies and guidelines.
12. Provide context and orientation information.
13. Provide clear navigation mechanisms.
14. Ensure that documents are clear and simple.

These guidelines are presented here only as an overview. When using them for Web site design, one should refer to the full guidelines with explanatory text and the specific checkpoints against which implementation can be evaluated.

Implementation of Web Content Accessibility Guidelines

Some of the most common questions about Web accessibility involve the cost of compliance. Perhaps, surprisingly, accessible design in some cases actually presents a cost savings in the development and maintenance of sites; in other cases it entails an effective return on investment given that it enables a deeper market penetration.

In some cases, receptivity to Web accessibility solutions is at first colored by urban legends seemingly spawned by the backlash against accessibility of the built environment still prevalent in some sectors of society. These urban legends include misconceptions such as "the WAI guidelines ban the use of color or graphics on the Web." Quite to the contrary, *WCAG 1.0* encourages the use of color and graphics on accessible Web sites, as these can help with navigation and with differentiation of types of content.

66.5 GUIDELINES FOR ACCESSIBILITY OF WEB-BASED APPLICATIONS

Accessibility of the Web relies on implementation of complementary solutions at different levels. While addressing the accessibility of the Web content itself provides an immediate—and crucial—starting point, it is also necessary to increase the accessibility of the software used on the Web.

There are several kinds of applications (software) relevant to Web accessibility: *user agents* (applications that one uses to access Web content, such as browsers, multimedia players, and assistive technologies used by some people with disabilities), *authoring tools* (applications used to create Web content and Web sites), and various utilities used to evaluate, transform, or retrofit Web sites for accessibility.

Guidelines for Authoring Tools

Web authoring tools hold the greatest promise for increasing the amount of accessible content on the Web. Consider, first, that the Web is an "information space," where tens of millions of people are already publishing information. Then, consider the effort involved in educating every one of those people to understand and use accessibility guidelines whenever they put information on the Web. It quickly becomes apparent that it is far easier to ensure that the software that people use to publish Web content makes it easy to create accessible Web sites than to expect each person to follow accessibility guidelines.

W3C's *Authoring Tool Accessibility Guidelines 1.0 (ATAG 1.0)* explain to the developers of authoring tools how to develop software that facilitates the process of producing accessible content. For instance, the guidelines explain that for a software feature that allows an author to add an image to a Web page, it is important to have a user-configurable prompt available to remind the author to add alternative text to the image. In addition, *ATAG 1.0* explains how to ensure that the authoring tools themselves are accessible to people with disabilities, so that people will be able to publish information to the Web, whether they have a disability or not. For example, mouse-driven commands should also be able to be activated by keyboard strokes, since authors with physical or visual disabilities might not be able to use a mouse.

ATAG 1.0 has seven guidelines, or abstract principles. These are further elaborated by 28 specific checkpoints prioritized in three levels, as are the WCAG checkpoints. The seven guidelines include:

1. Support accessible authoring practices.

2. Generate standard markup.

3. Support the creation of accessible content.

4. Provide ways of checking and correcting inaccessible content.

5. Integrate accessibility solutions into the overall look and feel.

6. Promote accessibility in help and documentation.

7. Ensure that the authoring tool is accessible to authors with disabilities.

As with *WCAG 1.0,* WAI has produced complementary resources for authoring tool accessibility, which are being updated as Web technologies evolve. These include a checklist of prioritized checkpoints, a nonnormative techniques document with implementation detail, and additional promotional materials such as a "frequently asked questions" (FAQs) page.

Guidelines for Browsers, Multimedia Players, and Assistive Technologies

Another area where improved accessibility is needed is in the browsers, multimedia players, and assistive technologies that people use to access the Web. WAI provides guidance for developers of these types of applications through the *User Agent Accessibility Guidelines 1.0 (UAAG 1.0)*. The primary goal of *UAAG 1.0* is to increase accessibility of browsers and multimedia players for people with disabilities, for instance, by explaining the need to render certain types of accessibility information such as long descriptions for complex charts or graphs, or captions for audio files. In addition, *UAAG 1.0* addresses interoperability between mainstream browsers or multimedia players, and various kinds of assistive technologies, such as screen readers, screen magnifiers, or speech recognition software (see Goldberg, Chap. 67, "Universal Design in Film and Media").

There is some synergy between the provisions of *UAAG 1.0* and *WCAG 1.0:* once certain accessibility improvements are implemented within mainstream browsers and multimedia players, related provisions within *WCAG 1.0* will drop priority levels, or will be eliminated entirely. To put this another way, the sooner user agent developers implement *UAAG 1.0,* the fewer *WCAG 1.0* checkpoints Web content developers will need to follow. For example, *WCAG 1.0* includes a priority 2–level provision stating that

> Until user agents allow users to turn off spawned windows, do not cause pop-ups or other windows to appear and do not change the current window without informing the user.

Therefore, once most browsers and multimedia players allow the user to control window pop-ups, Web content developers would not have to restrict their use of pop-up windows.

UAAG 1.0 includes the following 10 guidelines or abstract principles:

1. Support input and output device independence.
2. Ensure user access to all content.
3. Allow the user to configure the user agent not to render some content that may reduce accessibility.
4. Ensure user control of styles.
5. Observe system conventions and standard interfaces.
6. Implement specifications that promote accessibility.
7. Provide navigation mechanisms.
8. Orient the user.
9. Allow configuration and customization.
10. Provide accessible product documentation and help.

These guidelines are further elaborated by specific checkpoints, prioritized in three different levels. The document is accompanied by a checklist of those checkpoints, and by a techniques reference note providing implementation detail for different markup languages.

Implementation of Guidelines for Web Applications

Software development cycles often range from 6 to 18 months. There is, therefore, a very different implementation pattern for accessibility solutions in Web applications compared with Web content where one can implement *WCAG 1.0* directly on a Web site in a short amount of time. Applications developers typically must wait until a guideline is finished before they can implement the majority of its provisions into the feature set of a given product.

The consensus-based W3C process provides opportunities for developers to get an early perspective on features that they may eventually want to implement, by participating in the working group, commenting on drafts, and/or prototyping implementations of accessibility guidelines while they are still in progress. Often, the input of product engineers makes the crucial difference in ensuring the feasibility of implementation of accessibility features, since these features must compete with cascades of other priorities under consideration within product development groups. A number of developers have taken the opportunity to participate in one or more of these matters, thereby getting a heads-up on potential features of either *ATAG 1.0* or *UAAG 1.0;* these and other developers are working on implementations of WAI guidelines as this book goes to press.

66.6   *ACCESSIBILITY OF CORE WEB TECHNOLOGIES*

Much of the work needed to ensure the future accessibility of the Web is at the level of what is called *core Web technologies.* These are the markup languages such as HTML, which was the original common document format on the Web. But instead of one Web markup language as in 1990, 10 years later the W3C "Technical Reports" page shows 20 to 30 different specifications for markup and presentation languages, each of which must be reviewed for potential accessibility issues while in development.

As W3C technologies continually advance through stages of development from initial concept to finished specification, WAI participants review these specifications for potential impact on accessibility. This process starts with reviewing working group charters, requirements documents, early working drafts, then last call drafts, candidate recommendations, and proposed recommendations, and continues until the specifications are issued as final W3C recommendations.

Such reviews have resulted in improvements such as better support for representation of alternative content and easier navigation in HTML 4.0; user override of author style sheets in CSS 2; the capacity for synchronized captions in SMIL 1.0; and accessibility features in Scalable Vector Graphics 1.0 (SVG 1.0). Many of these improvements, which would not have been possible without the close integration of WAI's work within W3C, are detailed in W3C Notes [*Accessibility Features of CSS* (Jacobs and Brewer, 1999), *Accessibility Features of SMIL* (Koivunen and Jacobs, 1999), and *Accessibility Features of SVG* (McCathieNevile and Koivunen, 2000)].

66.7   *TOOLS FOR EVALUATING AND RETROFITTING WEB SITES*

Almost everyone who tries making an accessible Web site wants some way to evaluate that Web site. WAI provides a centralized forum for coordination of development of a variety of evaluation and retrofitting tools through the Evaluation and Repair Tools Working Group, in which many tool developers participate.

One popular evaluation tool is Bobby, developed by the Center for Applied Special Technology (CAST, 1998–2000). This accessibility checker is available in online and downloadable versions, and can provide quick feedback on a number of accessibility problems on Web sites. The WAVE (Kasday, 2000), currently under development by Len Kasday, provides a different type of feedback on Web site accessibility. Lynx-me (Oskoboiny), which is a Lynx emulator (Lynx is a text-only browser), provides yet another type of feedback on Web site accessibility. There are also several retrofitting tools under development, such as A-Prompt (Treviranus et al., 1999–2000) from the University of Toronto's Adaptive Technology Resource Centre, which identifies accessibility problems and then walks the page author through accessibility solutions for those problems. When evaluating or retrofitting a Web site, it can be most effective to use several of these tools in combination.

WAI has fostered dialogue between users and developers regarding desirable feature sets for accessibility checkers and retrofitting tools, and has captured recommended approaches in a W3C Note, *Techniques for Accessibility Evaluation and Repair Tools* (Ridpath and Chisholm, 2000). WAI has found that its most important role in this regard is to develop common resources that can help increase the functionality and the quality of evaluation, repair, and transformation tools.

66.8 EDUCATION AND OUTREACH

Given the enormity of the Web community and the complexity of some aspects of Web accessibility, it is essential to have a variety of materials available to both promote awareness of the need for Web accessibility, and to support training on how to implement Web accessibility. WAI's Education and Outreach Working Group focuses on this area.

Education and outreach materials range from online curricula to videos and hard-copy promotional materials. The *Curriculum for Web Content Accessibility Guidelines* (Letourneau and Freed, 2000), an online and/or downloadable electronic curriculum that explains the guidelines in detail and provides marked-up examples of each checkpoint, is in use in a variety of Web training courses.

Another popular implementation support material started as a challenge to see how many key concepts of accessible Web site development could fit on two sides of a business card. The result of this was the *Quick Tips to Make Accessible Web Sites* (World Wide Web Consortium, 2000) available in a vinyl business card in 11 languages with more on the way. WAI has distributed over 100,000 of these, and the demand continues to increase. These are not complete guidelines, so Web site designers must follow the URL on the card to refer to the complete guidelines, which include information critical to understanding and implementing the *Quick Tips*.

Additional educational or outreach materials include frequently asked question (FAQ) sheets, technical reference notes, links to policy reference pages, and a variety of other implementation support materials for *WCAG 1.0*.

66.9 CONCLUSIONS

In general, much of the Web community has seemed receptive to learning about Web accessibility and appreciative of the economic benefits that universal design brings for the proliferation of new devices to access the Web. In a field that is evolving and expanding as rapidly as the World Wide Web, however, this receptivity does not automatically translate into widespread and comprehensive implementation of accessibility solutions. Therefore, one challenge is to ensure that the availability of accessibility solutions keeps pace with—or, better yet, outstrips—the availability of new technologies for the Web.

Another challenge will be to continue to push awareness of the need for accessibility out to the ever-expanding circle of people participating in the Web community. In some countries, this push for accessibility awareness parallels a broadening societal appreciation of the need for accessibility and universal design; in other countries, the accessibility message may be quite new.

A further challenge is to articulate how the principles of Web accessibility work in specific application areas such as online learning environments and electronic commerce. As the Web becomes more and more essential in our societies, the success or failure of these efforts will determine the extent to which large parts of the population are included in or excluded from social and economic interactions. Continued international collaboration around development of Web accessibility solutions will also be essential toward ensuring accessibility of the Web. As an increasing number of governments adopt Web accessibility requirements, it is important to converge on common solutions and expectations for Web accessibility, in order to avoid the pitfalls of conflicting guidelines and requirements, which can create confusion among Web content developers and narrow the market for evaluation tools, retrofitting tools, and authoring tools that support production of accessible content.

To a far greater extent than with the built environment, it is important to avoid the fragmentation of standards for Web accessibility, since so many Web accessibility solutions rely on complementary approaches on several levels at once. WAI will continue to play an important role in bringing together international collaborations around Web accessibility; these collaborations should reinforce an awareness of the importance of harmonizing standards.

66.10 **BIBLIOGRAPHY**

Access Board, *Electronic and Information Technology Accessibility Standards,* 2000; http://www.access-board.gov/sec508/nprm.htm

Bos, Bert, Håkon Wium Lie, Chris Lilley, and Ian Jacobs, *Cascading Style Sheets, Level 2 Specification (CSS2),* W3C (MIT, INRIA, Keio), 1998; http://www.w3.org/TR/REC-CSS2

Bray, Tim, Jean Paoli, C. M. Sperberg-McQueen, and Eve Maler, *Extensible Markup Language (XML) 1.0,* 2nd ed., 2000; http://www.w3.org/TR/REC-xml

Brewer, Judy, *Policies Relating to Web Accessibility,* W3C (MIT, INRIA, Keio), 1998–2000; http://www.w3.org/WAI/References/Policy

Center for Applied Special Technology (CAST), Bobby, 1998–2000; http://www.cast.org/bobby

Chisholm, Wendy, Gregg Vanderheiden, and Ian Jacobs, *Techniques for Web Content Accessibility Guidelines 1.0,* W3C (MIT, INRIA, Keio), 2000; http://www.w3.org/TR/WCAG10-TECHS

———, ———, and ———, *Web Content Accessibility Guidelines 1.0,* W3C (MIT, INRIA, Keio), 1999; http://www.w3.org/TR/WCAG10

Dürst, Martin J., *WWW Character Model,* W3C (MIT, INRIA, Keio), 1999; http://www.w3.org/Talks/1999/0901-charmod-mjd/Overview

eEurope 2002, *An Information Society for All, Participation for All in the Knowledge-Based Economy,* European Commission, 2000; http://europa.eu.int/comm/information_society/eeurope/actionplan/actline2c_en.htm

Evaluation, Transformation, and Repair Tools for Web Content Accessibility, W3C (MIT, INRIA, Keio), 1998–2000; http://www.w3.org/WAI/ER/existingtools.html

Fact Sheet for Authoring Tool Accessibility Guidelines 1.0, W3C (MIT, INRIA, Keio), 2000; http://www.w3.org/2000/02/ATAG-FAQ

Fact Sheet for Web Content Accessibility Guidelines 1.0, W3C (MIT, INRIA, Keio), 1999; http://www.w3.org/1999/05/WCAG-REC-fact

Gunderson, Jon, and Ian Jacobs, *Techniques for User Agent Accessibility,* W3C (MIT, INRIA, Keio), 2000a; http://www.w3.org/TR/UAAG10-TECHS

———, ———, and Eric Hansen, *User Agent Accessibility Guidelines (UAAG) 1.0* (working draft), W3C (MIT, INRIA, Keio), 2000b; http://www.w3.org/TR/UAAG10

Hoschka, Philipp, *Synchronized Multimedia Integration Language (SMIL),* W3C (MIT, INRIA, Keio), 1998; http://www.w3.org/TR/REC-smil

Jacobs, Ian, and Judy Brewer, *Accessibility Features of CSS,* W3C (MIT, INRIA, Keio), 1999; http://www.w3.org/TR/CSS-access

Kasday, Len, *The WAVE: Pennsylvania's Initiative on Assistive Technology,* Institute on Disabilities/UAP, Temple University, Philadelphia, 2000; http://www.temple.edu/inst_disabilities/piat/wave

Koivunen, Marja-Riitta, and Ian Jacobs, *Accessibility Features of SMIL,* W3C (MIT, INRIA, Keio), 1999; http://www.w3.org/TR/SMIL-access

Letourneau, Chuck, and Geoff Freed, *Curriculum for Web Content Accessibility Guidelines,* W3C (MIT, INRIA, Keio), 2000; http://www.w3.org/WAI/wcag-curric

McCathie-Nevile, Charles, and Marja-Riitta Koivunen, *Accessibility Features of SVG,* W3C (MIT, INRIA, Keio), 2000; http://www.w3.org/TR/SVG-access

Office for Government Online, *Accessibility,* Commonwealth of Australia, 2000; http://www.govonline.gov.au/projects/standards/accessibility.htm

Oskoboiny, Gerald, Lynx-me, http://ugweb.cs.ualberta.ca/~gerald/lynx-me.cgi

Patrick, Deval, Letter to U.S. Senator Tom Harkin, 1996; http://www.usdoj.gov/crt/foia/tal712.txt

Raggett, Dave, Arnaud Le Hors, and Ian Jacobs, *HTML 4.01 Specification,* W3C (MIT, INRIA, Keio), 1999; http://www.w3.org/TR/html401/

Ridpath, Chris, and Wendy Chisholm, *Techniques for Accessibility Evaluation and Repair Tools,* W3C (MIT, INRIA, Keio), 2000; http://www.w3.org/TR/AERT

Technical Reports, W3C (MIT, INRIA, Keio), 1995–2000; www.w3.org/TR

Treasury Board of Canada, Secretariat, *Common Look and Feel for the Internet,* 2000; http://www.cio-dpi.gc.ca/clf-upe/index_e.asp

Treviranus, Jutta, Charles McCathie-Nevile, Ian Jacobs, and Jan Richards, *Authoring Tool Accessibility Guidelines 1.0,* W3C (MIT, INRIA, Keio), 2000a; http://www.w3.org/TR/ATAG10

———, ———, ———, and ———, *Techniques for Authoring Tool Accessibility,* W3C (MIT, INRIA, Keio), 2000b; http://www.w3.org/TR/ATAG10-TECHS

———, et al., *A-Prompt, a Web Accessibility Verifier,* University of Toronto, Adaptive Technology Resource Centre, Toronto, Ontario, Canada, 1999–2000; http://aprompt.snow.utoronto.ca

Web Accessibility Initiative (WAI), W3C (MIT, INRIA, Keio), 1997–2000; http://www.w3.org/WAI

Wireless Access Protocol Forum (WAP Forum), Wireless Access Protocol Forum Ltd., 1999–2000; http://www.wapforum.com

World Wide Web Consortium (W3C), W3C (MIT, INRIA, Keio), 1994–2000; http://www.w3.org

———, *Quick Tips to Make Accessible Web Sites,* W3C (MIT, INRIA, Keio), 2000; http://www.w3.org/WAI/References/QuickTips

World Wide Web 9 Conference, International WWW Conference Committee, 1999; http://www.www9.org

CHAPTER 67
UNIVERSAL DESIGN IN FILM AND MEDIA

Larry Goldberg, B.A.
CPB/WGBH National Center for Accessible Media,
Boston, Massachusetts

67.1 INTRODUCTION

This chapter describes how universal design principles are applied to mass media, in particular film and television of the present and future. Though rarely referred to as "universal design" by the producers of the media or the facilitators of its accessibility for people with disabilities, activities which open up media to all demonstrate universal design concepts in clear and pragmatic ways. From print to silent film to radio, and then to television (both analog and digital) and the varieties of emerging multimedia, when made accessible for people with sensory or physical disabilities, these media become more useful for all users. In short, as in so many other areas, accessible, or inclusive, or universally designed media are simply better media.

67.2 BACKGROUND

In film and other media, perhaps the best examples of universal design are those that are inadvertent or unintended by the creator of the technology or content. More than 100 years ago, the United States experienced the launch of a new medium and technology that foreshadowed the massive impacts of the information technologies of the late twentieth century. That medium was motion pictures, and from their first moments, movies for mass audiences demonstrated how quickly and pervasively a medium can affect society. The movies also had a purely unintended universal design aspect. In their inability to provide sound, "silent pictures" instead included title cards and subtitles, in essence the precursor of closed captions, that proved inclusive of audiences with hearing impairments. Of course, people who were blind or had visual impairments were as underserved as they are today (or were, until a new development in movie-going, discussed later in this chapter, was developed in the early 1990s). And, as is so common in today's phenomenally rapid pace of technology development, the next "upgrade" in movie technology—the "talkies" introduced in 1929—removed the heretofore inclusive feature for one group of users while adding improvements for others.

All of this early demonstration of universal design and its lack was inadvertent and unnoted by the designers and developers of the day.

Until the closing days of the twentieth century, accessible and inclusive means of understanding and enjoying media, particularly for people with sensory impairments, followed the pattern of the transition and evolution of movie technology. That is, all access has been via retrofitted enhancement, more properly called "assistive technology" than "universal design." But as the 1990s came to a close, and as analog paradigms gave way to the all-encompassing digital domain of the new Information Age, built-in solutions and inadvertent inclusive design became more and more possible and even prominent. All forms of media have been affected by their migration from analog to digital, and the opportunities for accessibility have followed rapidly.

67.3   *PRINT*

No medium has demonstrated such accessibility successes and failures as the world of print as it evolved technologically and arrived in the digital age. From the days of quill and ink to movable type and the printing press to electronic typesetting and distribution, print publishing has come ever closer to being inherently and automatically accessible to the very people most closed out by the medium—people who are blind or visually impaired. In each of the developments in print and publishing technology, an opportunity was presented to be more inclusive for individuals who require alternate means of accessing information, whether in large print, via audio output, or on a Braille display.

When publishers of books, newspapers, periodicals, and other print forms began to mark up and typeset their materials on computers, software was developed and used which complicated the transference to alternate presentation formats. Quark Express and the other leading software tools for electronic typesetting added certain proprietary mark-up codes that required tedious reformatting to assure compatibility and readability by the tools in most common use by people who are blind or visually impaired. Nonetheless, such reformatting can and has been done so that, for instance, students who are blind can gain access to the electronic contents of textbooks used in the classroom.

In 1992, the Media Access Research and Development Office of the WGBH Educational Foundation[1] in Boston was funded by the Corporation for Public Broadcasting to begin *An Investigation into Making Daily Newspapers Accessible to Print-Disabled People.* What emerged in March of 1993 was a report of "The Print Access Project," which discussed the recent history and future technological options for providing access to print materials, newspapers in particular, for what has now become an accepted terminology: *print-disabled people.* The implication of this language is that not only do people with visual impairments have difficulty gaining access to print materials, but so do people with dyslexia and other learning disabilities and people with certain mobility and physical impairments.

While representing one of the most comprehensive surveys of the field at that time (astute readers will note that the date—March of 1993—is also notable as the beginning of the growth period of the Internet's World Wide Web), from this point on virtually all of the report's conclusions, and much of the worlds of computing, commerce, and common user access to global information, would require drastic reevaluation. The Mosaic browser was first being widely tested and distributed at that time, and one year later, Marc Andreesen and his colleagues would leave NCSA (the National Center for Supercomputing Applications) to form the Mosaic Communications Corporation (now Netscape).

That 1993 report on print access covered such forms of print access as radio reading services, dial-in talking newspapers, audiotext, and computer-based information services such as Compuserve on-line, videotext and wire services, and the emerging digital newspapers that were experimenting with data delivery to the home via broadcasting and private carriers. Regarding news delivered via on-line computer services, the report noted:

Generally, on-lines are commercial services: their information is available for a fee. These services vary in the amount of newspaper information they offer as well as the format in which it comes. Some services carry whole newspapers. CompuServe, for example, has a database with the full text of 48 major newspapers, but they are always a day or two old. Other services only offer access to "re-packaged" news. Prodigy rewrites news information gathered from Associated Press and other wire services; its subscribers have no direct access to newspapers at all. *(WGBH Print Access Project, 1993, 13)*

As an indication of how the years have changed the landscape of access to print media, CompuServe and Netscape are now owned by America Online, which had only 500,000 subscribers in 1993, compared to today's 28 million in the first quarter of 2001—and Prodigy, which was once the world leader and a pioneer Internet Service Provider (ISP), had only 2.5 million subscribers at the start of 2001. The Yahoo! search engine lists thousands of local and national newspapers online, from the *Shelbyville Daily Union* in Illinois to the *New York Times* (which is updated on-line hourly and available via fax as well as on the Web) and the *Wall Street Journal,* one of the most successful subscription-based online newspapers.

The Print Access Report's "Survey of Future News Options" examined radio-, telephone-, and computer-based print access technologies for the coming years, and, while missing the Web boat, did look over the horizon toward the still-anticipated news appliance of the future:

Here we find real advances possible, not just in pathways for delivery but in major qualitative steps toward true access. The most specific vision for a future news delivery system comes from the Knight-Ridder newspaper chain and proceeds from the assumption that ink-on-paper newspapers are obsolescent. One device that could replace them is called a multimedia panel, a task-specific computer appliance capable of delivering voice, sound, video, and print. Designers envision something the size of a magazine, designed for reading and browsing material from sources like local, national and international newspapers. Information would come in to the panel as digital data, either by wire or fiber optic cable, or over the air as a broadcast or satellite signal. As long as it was connected or tuned in, the panel could remain constantly up to date with news, stock quotes, sports scores or whatever the user wanted, and could, for instance, be set to notify the user of incoming bulletins, news stories containing key words, or designated stock price changes, much as specialized on-line services can now be set up. *(WGBH Print Access Project, 1993, 20)*

What is common to all of the exciting options for delivery and receipt of electronic "print" is that each technology has both inherent challenges and solutions for access for people who are blind or visually impaired. *The New York Times* on the Web (www.nytimes.com) is available free of charge and offers up-to-the-minute international, national, and local news, all formatted using the standard HyperText Markup Language (HTML). HTML is a flexible authoring standard with long-accepted guidelines for providing equal access for people who are blind or visually impaired who use screen-reader-equipped computers. However, the *Times,* like so many other information-bearing Web sites, errs in a few small but critical ways in how they design their Web site, resulting in a frustrating experience for print-disabled people who use personal computers. The accessibility barriers of the *Times* Web site include, among other problems, an inconsistent use of "alt text tags." These are key words and phrases that "hide behind" a Web site's graphics and which are read out loud to blind computer users by their screen-reading software. In addition, the *Times'* overreliance on columns creates a confusing audible display for people who are blind. The guidelines for proper "Web content" authoring are available from the World Wide Web Consortium (www.w3.org/wai) and are discussed in Chap. 66, "Access to the World Wide Web: Technical and Policy Perspectives," by Brewer.

While radio-reading services and dial-up newspapers will be important means for access for a few more years by people who are print-disabled, in feverish development are the "information appliance," "e-book," and "digital ink" projects at Knight-Ridder, MIT's Media Lab, Microsoft, and elsewhere. Already two competing pioneering products, the Rocket eBook (www.rocket-ebook.com) and the SoftBook (www.softbook.com), have been bought by a single corporation, and Microsoft is in intensive development of its own "information appliance." While universal design issues are being introduced to the developers of these new products and services, and customizable user interfaces are readily achievable, the rush to market has resulted in first-generation versions which neglect to include speech output, equal access to navigation elements, and industry-standard protocols which facilitate interface with existing access devices and software.

In the case of these new means of accessing print, the universal design community has shown its awareness of the challenges and has demonstrated its readiness to meet the problems head-on. The Daisy Consortium (www.daisy.org), the Open-eBook Authoring Group, Recordings for the Blind and Dyslexic, the Library of Congress' National Library Service for the Blind and Physically Handicapped (NLS), Productivity Works (makers of pwWebSpeak), and other public and private organizations are deeply involved in awareness-raising and standards development which bode well for the successful application of universal design thinking in the digital print world.

67.4   RADIO

Though for most people radio is an aural-only medium, universal design and access issues have arisen since radio's golden age in the 1920s through the 1940s. Clearly a medium of choice for people who are blind or visually impaired, radio became a means of access to other media for just such users. With the advent of radio reading services in the late 1950s, often on the FM subcarrier of public radio stations, this early voice transmission service became a valued means of access to publications read by volunteers at specific times throughout the day. Newspapers, magazines, TV guide listings, and other publications are still to this day read over these niche broadcast networks for reception by people with specially tuned radio receivers.

But for people who are deaf or hard-of-hearing, radio has been yet another pervasive communications technology enjoyed by the masses but not by them. For an insight into why a hearing-impaired person might want access to radio, one must look past the top 40, classic rock, K-Rock, smooth jazz, oldies, classical, and other music formats and think about talk radio, news, and highly thought of public radio programs such as *Morning Edition* and *All Things Considered*. Could a deaf person gain access to these programs? In what circumstances would a person who couldn't hear want, or need, access? Certainly a daily text transcript of the Howard Stern or Don Imus program would help certain people join in on watercooler conversation about the most recent outrageous and controversial programs of the day.

But more important, and more achievable due to another new and exciting digital broadcast service, is access to essential weather and traffic information for drivers who have trouble hearing. Imagine driving, with no means of communication, on your local freeway at 5:30 P.M. on a normal weekday, trying to decide which of three or four routes home would avoid the traffic snarls ahead. Or think of driving in the Midwest's "Tornado Alley" as the skies darken and having no access to the emergency information being broadcast on every channel on your car radio! These are the situations where a deaf person needs access to radio, and where Radio Broadcast Data Services (RBDS) could play an important role.

RBDS is a technology that enables radio stations to send short text messages to specially equipped radios with small visual displays. The text message may consist of the name of the song being played or the format of the station (i.e., jazz, NPR, talk). About 600 stations throughout the country now are using RBDS, according to the National Association of Broadcasters. The text displays on these high-end radios are normally 8 to 16 characters, which scroll through a small LCD window. But there is nothing to preclude a longer scrolling message about weather or traffic emergencies that could be printed out in a car like the receipts generated in taxicabs or like the tape of a telephone device for the deaf (TTD). If implemented, this digital technology could solve a long-standing problem for deaf and hard-of-hearing drivers.

The coming wave for radio, in the home or on the road, is digital, satellite-delivered radio signals that will provide hundreds of channels of music, talk, and audio services to miniature, receive-only dishes that may be as small as a half-dollar. The opportunities for people who are blind or visually impaired include multiple channels of radio-reading services, perhaps categorized by content and type of publication. With so many channels of high-quality audio services available, there will be room not only for transferring previously analog formats such as radio reading already mentioned, but also for video description, a service that has to date only been available via broadcast and cable TV as part of their Secondary Audio Program subcarriers or on special videotapes.

The challenges for access to this new medium are numerous as well. With so many channels to choose from, how will a user navigate among the hundreds of choices? The electronic program guide is likely to be complex and virtually unusable for persons with visual impairments. Solutions are at hand, via the use of speech synthesizers, which are getting cheaper and smaller in terms of the memory space they utilize. For blind users as well as drivers would can't, or shouldn't, take their eyes off the road, a speech in/speech out interface for digital radio command and control makes the ultimate universal design sense. In Chap. 50, "Universal Design in Automobile Design," Steinfeld and Steinfeld highlight guidelines for communication tools in cars. But there are no guarantees that developers of these new services are taking heed of these barriers and opportunities to serve all potential users.

67.5 TELEVISION

Television was first demonstrated to the public at the 1939 New York World's Fair. Soon after, regular, experimental TV broadcasts began. This milestone was preceded by at least a decade of laboratory experiments and trial and error. But it wasn't until 1972—33 years later—that the first television program was (open) captioned for deaf and hard-of-hearing viewers. Finally in 1980, the public was introduced to closed captioning—four decades after they were introduced to television itself. In 1985, the first public demonstration of video description for blind and visually impaired people was broadcast, and the world's most pervasive medium finally had the means to become universally available and accessible.

Television—Closed Captioning

Closed captioning's early progress was slow but steady in the United States (see "Captioning Milestones" at www.wgbh.org/caption). With significant support from grants awarded competitively by the U.S. Department of Education and with logistical and financial support as well by the four major broadcast networks (ABC, CBS, NBC, and PBS), the early years of closed captioning opened up access for hundreds of thousands of deaf and hard-of-hearing viewers who purchased set-top boxes known as "closed caption decoders." These stand-alone decoders were the antithesis of universal design—they were separately purchased, expensive, and cumbersome pieces of electronic hardware that were difficult to install and that often required troubleshooting for effective operation. In later years, though the size of the hardware began to shrink along with its cost (from an early price tag of $300 to less than $200), the complexities for users grew. As homes began to add VCRs and cable converter boxes in ever greater numbers, a user in the late 1980s and early 1990s was confronted by a rat's nest of cables and connectors behind the TV set. All this to achieve equal access to what a hearing viewer received via audio control at the touch of a button.

The solution for Americans arrived on the heels of the drafting and passage of the Americans with Disabilities Act (ADA) in 1990. And though the eventual outcome is now cited as one of the best examples of universal design in widespread use today (right after the sidewalk curbcut), it did take an act of Congress to achieve. With no mandate for closed captioning in the ADA due to political concerns, Senator Tom Harkin of Iowa and Representative Ed Markey of Massachusetts drafted a bill which came to be known as the "Television Decoder Circuitry Act of 1990." With little fanfare at first, but with growing excitement among deaf and hard-of-hearing consumers and only token opposition by the TV manufacturing industry, the Decoder Act was passed. Recognizing that inclusion of caption decoding capabilities in the tens of millions of TV sets manufactured and sold each year would result in such economies of scale that the cost was almost too small to measure, Congress mandated that

> ... apparatus designed to receive television pictures broadcast simultaneously with sound be equipped with built-in decoder circuitry designed to display closed captioned television transmissions when such apparatus is manufactured in the United States or imported for use in the United States, and its television picture screen is thirteen inches or greater in size.

EFFECTIVE DATE
> ... this Act shall take effect on July 1, 1993. *(TV Decoder Circuitry Act, 1990)*

And with that one short piece of legislation, decades of struggle with the inaccessibility of the world's most popular medium were wiped away for people who are deaf or hard-of-hearing. Or so the drafters of the law hoped at the time. Although the Decoder Act created a huge installed base of potential end-users in rapid fashion, the marketplace leverage of more captioned programs did not occur as widely as had been hoped for or expected. It has been estimated that from 1980, when the first caption decoder box was sold, until 1993, when the first caption-equipped TV set hit the stores, no more than 450,000 set-top decoders were sold. In the first full year of the Decoder Act (1994), some 10 million caption-display-capable TV sets were sold in the United States.

Unfortunately, though the major broadcast and pay-cable networks were arranging to have many of their programs captioned, many local TV stations, syndicators, and basic cable networks still lagged far behind in their captioning efforts throughout the mid-1990s. In 1994, the first drafts of the law which eventually became the Telecommunications Act of 1996 began to circulate and included a little-noticed requirement for closed captioning of all TV programs. As the first major rewrite of the original Communications Act of 1934, the bill was vast and complex, engendering major turf battles between regional telephone companies and long-distance carriers, between broadcasters and cablecasters, and among many other major corporations with billions of dollars at stake in how Congress wrote the bill and how the Federal Communications Commission passed its subsequent regulations. Language requiring accessibility to telecommunications hardware, systems, and services and to video programming resided in the bill as some of the least controversial aspects.

With each new draft, the caption mandates remained. Finally, after some of the most intense lobbying and horse trading Capitol Hill has ever seen, the Act was passed and signed by President Clinton on February 8, 1996. It contained the following words:

SEC. 713. VIDEO PROGRAMMING ACCESSIBILITY
> (b) Accountability Criteria—Within 18 months after such date of enactment, the Commission shall prescribe such regulations as are necessary to implement this section. Such regulations shall ensure that—
> (1) video programming first published or exhibited after the effective date of such regulations is fully accessible through the provision of closed captions, except as provided in subsection (d); and
> (2) video programming providers or owners maximize the accessibility of video programming first published or exhibited prior to the effective date of such regulations through the provision of closed captions, except as provided in subsection (d). (c) Deadlines for Captioning.—Such regulations shall include an appropriate schedule of deadlines for the provision of closed captioning of video programming.

(Telecommunications Act of 1996)

Except for certain narrow exemptions, it was now the law that television would have an irrefutable deadline for near-100 percent accessibility for people who are deaf or hard-of-hearing. After a public rulemaking process by the FCC, the schedule was set:

New programs.   Programs first published or exhibited on or after January 1, 1998. New programs must be captioned over an eight-year period. The FCC set the following guidelines to increase gradually the amount of captioning:

At least 25 percent of new programs by January 1, 2000

At least 50 percent of new programs by January 1, 2002

At least 75 percent of new programs by January 1, 2004

100 percent of new programs by January 1, 2006

Older or pre-rule programs.   Programs published or exhibited before January 1, 1998. At least 30 percent of pre-rule programs aired by nonexempt networks must be captioned by January 1, 2003. Seventy-five percent of older programs must be captioned over a ten-year period (by January 1, 2008).

Captioning was now moving toward becoming a normal and accepted practice in the post-production and budgetary aspects of the creation of TV programs. At this point, it became necessary to begin tackling one lingering access issue for television and one new one: First, access to television for people who are blind or visually impaired still lagged significantly behind that which had been fought for and won by deaf and hard-of-hearing viewers. And second, digital TV had begun to emerge as a major new opportunity and series of challenges for continuing the hard-won gains of activists in the disability community.

International Captioning and Subtitling

While at least five incompatible technologies exist worldwide to broadcast captions or subtitles, there are two dominant systems—line-21 and World System Teletext. Captions and subtitles created using these technologies have a different appearance, use different amounts of bandwidth in the vertical blanking interval, and require different consumer and professional hardware to create, encode, and decode. Line-21 captioning, used in North America, is the most prevalent technology and is employed for hundreds of hours of TV programming weekly, both live and prerecorded.

In the United States, "captions" commonly means text on-screen for deaf and hard-of-hearing viewers, while "subtitles" are generally taken to be words on-screen for speakers of languages other than English. However, in England and elsewhere in the world, TV text services for deaf and hard-of-hearing viewers are often called "subtitling."

The ability to record a program off-the-air or to rent a video with embedded and recoverable closed captions or subtitles is considered to be an essential feature of any captioning or subtitling system. In markets like Asia, Europe, and North America where VCR penetration is high, a home-recordable system is a must. Because the speed at which data are transmitted in the line-21 system is relatively slow (approximately 60 characters per second) and the data are robust (not readily degradable), the line-21 system easily supports home video closed caption display. Teletext subtitles, on the other hand, are delivered using a much faster system (approximately 12 kilobytes per second). This speed allows for the transmission of more data and multiple streams of subtitles, but makes it very difficult for the viewer to record a program off-air with the data intact. It also makes renting or purchasing a closed-subtitled video for home use impossible.

The technology employed in the United States and abroad to make television accessible to deaf and hard-of-hearing viewers shows great promise for building additional bridges among people who speak different languages. In markets like Latin America, dual-language subtitling can make a program simultaneously accessible in both Spanish and Portuguese. In Asia, where a satellite feed often serves a vast audience that speaks several languages and is eager to learn English, a multilingual broadcast can help reach millions of additional viewers. Most important, as international broadcasters like CNN and MTV seek to supply a single, basic program service for markets that speak different languages, multilingual broadcasts offer a practical and logical solution. This practice has already begun in the United States, where CBS broadcasts the newsmagazine, *60 Minutes,* with both English and Spanish captions.

Television—Video Description

Video description (sometimes known as *audio description, descriptive video,* or *narrative description*) is a service that makes television programs accessible to people who are blind or visually impaired. Video description provides narrated descriptions of the key visual elements of visual media without interfering with the audio or dialog of a program or movie. The narration describes visual elements such as actions, settings, body language, and graphics. Video description began at WGBH in Boston, where it is known by its trademark, Descriptive Video Service® or DVS®. DVS provides description for programs on the Public Broadcasting Service (PBS), the Turner Classic Movie (TCM) cable channel, Hollywood movies on video, and other visual media. DVS was launched nationally in 1990 by the WGBH Educational Foundation; home of the world's first caption agency, the Caption Center. Another provider of video description is the Narrative Television Network of Tulsa, Oklahoma (NTN). Initial training of WGBH's describers was provided by audio description pioneer Margaret Pfanstiehl, founder of the Washington Ear, a dial-up newspaper service. Funding for much of the video description available in the United States today has been provided by the U.S. Department of Education.

In order to use DVS, a viewer must have cable service for accessing TCM and eventually other cable networks or live within range of a PBS station that carries DVS. The viewer must also have a stereo TV or a stereo VCR that includes the Second Audio Program (SAP) feature, standard on most newer stereo televisions and video cassette recorders. Inexpensive receivers that convert TV sets to stereo with SAP can also be purchased. Viewers who subscribe to cable should ask the cable company to "pass through" stereo with SAP. NTN provides "open-described" programs which do not require the use of any specialized equipment because the added narration is heard by all viewers. NTN also provides some public domain described programming through its Web site: www.narrativetv.com.

DVS is broadcast free to viewers by 160 participating public television stations reaching 80 percent of U.S. television households. To carry DVS, a station must be equipped to broadcast in stereo with SAP.

While popular home entertainment in the form of VHS videos can be encoded with closed captions and decoded at the user's option, video description cannot be hidden or encoded in the VHS tape format. Therefore, a DVS Home Video service was created by WGBH with Federal funds; these videos have their descriptions recorded in the open and cannot be turned off. The videos are distributed outside the usual home video marketplace. Presently, more than 200 described movies and PBS programs are available for purchase by direct mail; DVS home videos are also available for loan at many public libraries.

Video description, as a newer service introduced during a much more competitive media environment, was not afforded the same legislative and regulatory protections as closed captioning. The Telecommunications Act, while not setting out a mandate for video description as it did for captioning, directed the FCC to:

> commence an inquiry to examine the use of video descriptions on video programming in order to ensure the accessibility of video programming to persons with visual impairments, and report to Congress on its findings. The Commission's report shall assess appropriate methods and schedules for phasing video descriptions into the marketplace, technical and quality standards for video descriptions, a definition of programming for which video descriptions would apply, and other technical and legal issues that the Commission deems appropriate.

Subsequent to a series of inquiries and reports, the FCC published a *Notice of Proposed Rulemaking* (*NPRM*) in November 1999 as a first step toward a report and order to require the addition of video descriptions to a limited number of broadcast and cable-delivered programs. The *NPRM* was followed by a period of comments and replies that resulted in the release, on 7 August 2000, of a report and order (FCC 00-258, MM Docket No. 99-339), which will greatly increase the amount of described video programming available to consumers. Though a handful of petitions for reconsideration were filed by trade associations representing broadcasters and the cable industry, if the FCC's report and order stands as written, the new video description regulations will require, beginning with the calendar quarter 1 April–30 June 2002, 50 hours per quarter of described programming on the top four broadcast networks (ABC, CBS, NBC, and Fox) and the top five cable networks (TNT, TBS, Nickelodeon, USA, and Lifetime).

Video description has also found a home in the new DVD home video format, where the digital capacity affords the possibility of adding description tracks that can be chosen by the viewer. Like closed captioning, video description can be added to DVDs as one of the eight alternate language choices, and users can purchase mainstream DVDs instead of the specialized videos available presently only via mail order or at many public libraries. The trickle of DVDs with video description is just becoming available and should soon grow in number and availability.

The Internet also offers new opportunities for adding description to video delivered via Web sites. These new technologies are discussed later, in the section on Multimedia and the Web.

Digital Television

A new television technology, in development for at least the past 10 years but still considered "emerging," provides a great opportunity to build in accessibility as early as possible during the design and development process. That technology is *digital television*—alternately referred to as *advanced TV* or *high-definition TV*. Work on incorporating and improving the closed captioning and video description capabilities of this new technology began shortly after the 1993 implementation of the TV Decoder Circuitry Act, partly driven by the forward-thinking language inserted in the Act:

> As new video technology is developed, the Commission shall take such action as the Commission determines appropriate to ensure that closed-captioning service continues to be available to consumers.
>
> *(TV Decoder Circuitry Act)*

What is digital television (DTV)?

> The picture television viewers currently receive is based on an analog transmission system that is more than 50 years old. In December 1996 the Federal Communications Commission approved the U.S. standard for a new era of television—digital television. In a digital system, images and sound are captured using the same digital code found in computers—ones and zeroes.
>
> The digital revolution will not only dramatically improve the quality of the television picture and sound, but also makes possible the over-the-air delivery of several simultaneous services to viewers. This is due primarily to the three main benefits of the digital system: high-definition television, multicasting in standard definition television, and data transmission.
>
> *(www.PBS.org)*

All across the country, beginning in the fall of 1999, new digital television (DTV) stations began transmission, promising crystal-clear wide-screen pictures, multichannel sound, and bold new services for entertainment and education. These early entrants to the DTV marketplace commenced operations even as the standards, hardware, and software were still in development. In fact, almost none of the new digital programming aired with closed captions or video descriptions in the early months of launch (and an equally small amount of consumer reception equipment) had the capability to display access features even if the programming had included them.

A working group of companies from the television industry and closed-captioning field has developed a closed captioning standard for the new digital TV system. Referred to as "EIA-708B," the standard outlined an exciting array of new capabilities, including consumer control of character size and shape, an improved "look and feel," and the ability to attach up to 64 different caption services for each program, such as alternate languages and captions edited for beginning readers.

While many in industry could agree on how such a system should work technically, there was not agreement within industry or among consumers as to how many of these new features would be required to be supported in the new digital TV reception equipment. Therefore, in October of 1999, the FCC released a Notice of Proposed Rulemaking regarding closed captioning requirements for digital television receivers. Final comments were received by the FCC in November 1999, and the exact configuration of consumer equipment was pending as of this writing. For more information on digital television and accessibility, see the Web site: www.dtvaccess.org.

Digital television also shows great promise for enhanced services for people who are blind or visually impaired. Better-sounding description tracks and new delivery formats which could reduce production costs are only two of the enhancements that the ongoing development of digital television is likely to lead to throughout the first decade of the twenty-first century. With development continuing even as products and programs are rolled out, improvements for accessibility and universal design will both be built-in and added-on as each new version and generation is released.

67.6   MULTIMEDIA

Multimedia is simply the combination of more than one or two forms of media within a common technical environment. While it could be said that movies or television are multimedia because they combine pictures and sound, multimedia more often refers to computer-based software that gives users a dynamic and interactive environment for education and entertainment. The multiple media often consists of text, audio, video, graphics, and animation. The most common delivery platforms for multimedia have in the past been interactive videodisks and personal computers and more recently computers and CD-ROMs with World Wide Web links or computers controlling direct Web-based media.

Currently, the graphic-rich content of most multimedia software prevents blind people from participating in the dynamic learning environment this technology has brought to the home, classroom, and workplace. Although technology is available to make computers accessible to blind people, no one piece of access technology works with all software or hardware products, and some products have access barriers that cannot be overcome by any existing access technology. The obvious solution is to build access into multimedia software so that only a very few affordable pieces of access technology are required to make the information available. Not only must the graphic environment be made understandable to blind students; the navigation must be usable as well.

Experimental CD-ROM-based projects have demonstrated solutions that employ such technologies as Java, QuickTime, and accessible HyperText Markup Language (HTML) browsers. One such prototype built a new accessible interface and added digitized speech assists to an existing math game for elementary school children—"How the West Was One + Three × Four," by Sunburst Communications.

A more advanced prototype was based on a simulation of photosynthesis, part of a series of biology simulations from LOGAL, Inc. Photosynthesis Explorer is a simulation in which students vary the light, temperature, humidity, and other influences on a plant and observe the resulting amounts of sugar and oxygen produced by the plant. Diagrams show where different chemicals enter and leave a leaf and also present an abstract view of the chemical pathway of photosynthesis.

This prototype was built in Java for several reasons. Java is a programming language designed for Internet-based delivery but it is also useful for other kinds of distribution. Most important, Java is intended for cross-platform use, which means that programs can be written once and then run on different kinds of computers. This is crucial for educational software because some schools use Macintosh computers while others use Windows computers. Without Java software, companies must create two versions of each product. Having a cross-platform language also means that accessibility solutions for Java can meet the needs of users on both platforms.

The real challenges for universal design and access in multimedia are caused by the fact that there is no single, unified platform for creation, distribution, and display. Solutions can make use of the cross-platform benefits of Java, but improvements will come through the raising of industry awareness and the distribution of more general guidelines for authoring accessible software. One example of substantive guidelines in this area was published by NCAM in October 2000 and are entitled *Making Educational Software Accessible: Design Guidelines Including Math and Science Solutions*. This document is available at www.ncam.wgbh.org/cdrom; hard copies are available by e-mailing requests to to <Mary_Watkins@wgbh.org>.

67.7   THE WEB

Due to their reliance on unified standards for interconnection and global usability, the World Wide Web and other Internet technologies can overcome the barriers created by the plethora of multimedia technologies. The basic underlying code of the Web—HTML—and other Web standards such as XML, CSS, SMIL, and others, have been agreed upon by a wide range and large number of Web-based companies and organizations across the globe. While initial elements of these standards erected barriers for computer users who are blind or visually impaired, recent advances and guidelines issuing from the World Wide Web Consortium (W3C) are confronting these barriers head-on. The World Wide Web Consortium is an international industry consortium dedicated to promoting the evolution and ensuring the interoperability of the World Wide Web by developing common protocols. As noted earlier, Web accessibility guidelines are discussed in Chap. 66. In particular the "Web Content Accessibility Guidelines" and "User Agent Accessibility Guidelines" are reliable sources for assuring accessibility of Web pages and Web browsers respectively. Both of these documents are available at www.w3.org/wai.

One of the emerging technologies on the Web involves downloading or streaming audio, video, animation, and graphics. Significant economic activity is buzzing around this high-profile new technological development, but the real impact on consumers won't be felt until significant numbers of Web users have high-speed, large-bandwidth access to the Web. Many analysts are predicting that just such access is coming in the year 2000–2002 time period though cable modem, DSL, and satellite-delivered Internet access. With such large "data pipelines" to the home, classroom, and workplace, provision of radio- and TV-like computing experiences become more likely.

Already on the Web are numerous sites offering audio and video on demand, such as Broadcast.com, QuickTime TV, and radio-tv.com. Content providers like CNN, *Rolling Stone* magazine, National Public Radio, and hundreds of others are transferring their analog media assets onto the Web. With this transference comes the usual and expected accessibility questions: Will a deaf person be able to watch Web videos with captions; will a blind person be able to navigate these media sites and hear the images described; and will Web-based multimedia support the addition of captions and descriptions like their analog predecessors?

At least this last content question can be answered emphatically, *yes*. The major formats for providing media via the Web—QuickTime, Windows Media, and RealMedia—have built-in means for adding or pointing to text and ancillary audio for providing access for people with sensory disabilities. QText, Synchronized Accessible Media Interchange (SAMI), and Synchronized Multimedia Interchange Language (SMIL) are standards for authoring media in this new environment, and the developers of each have recognized the need to build in these sorts of supports.

Apple's QuickTime gives a media producer the ability to add captions for deaf and hard-of-hearing viewers, alternate-language subtitles, alternate-language audio tracks, and video description for people who are blind or visually impaired. These services are embedded in the media file and selectable by the user, who employs a piece of software known as a *player*. In the case of Apple, the player is known as the "QuickTime Player." As of version 4.1 of this standard, QuickTime offered rudimentary support of the SMIL standard.

SMIL is a standard that has emerged from the World Wide Web Consortium. The SMIL standard enables a media producer to create a dynamic Web-based presentation which incorporates and synchronizes such media types as video, audio, text, graphics, and animation. A very popular media player that is based on the SMIL standard is the G2 from RealNetworks. This player has been downloaded millions of times in its free version and can accommodate added captions and descriptions when provided by media producers. As already noted, Apple's QuickTime, also freely available, will now support SMIL. Microsoft has developed its own similar standard, called SAMI (Synchronized Accessible Media Interchange), which provides accessibility supports for the Windows Media Player.

The catch for all of these players and standards is that they rely on the media creators and distributors to add the captions and descriptions. Like analog TV, these services are not automatic and do require a certain degree of effort and expertise by skilled operators. A recently released, free software program called "MAGpie" will make the authoring of captions and descriptions for Web media easier. It is available at www.wgbh.org/ncam.

67.8 THEATRICAL MOTION PICTURES

Finally, at the turn of the century, digital technology has been applied to the century-old, previously inaccessible medium of theatrical movies. In 1992, the Motion Picture Access Project was begun to research and develop ways of making movies in theaters accessible to deaf, hard-of-hearing, blind, and visually impaired people through closed captions and descriptive narration. The project was funded in part by a grant from the National Institute on Disability and Rehabilitation Research (NIDRR), a division of the U.S. Department of Education to the WGBH Educational Foundation.

What resulted was a patented digital technology (dubbed *Rear Window*™) that makes it possible for movie exhibitors to provide closed captions for those who need or desire them without displaying them to the entire audience, and without the need for special prints or separate screenings.

The Rear Window system displays reversed captions on a light-emitting diode (LED) text display which is mounted in the rear of a theater. Deaf and hard-of-hearing patrons use transparent plastic panels attached to their seats to reflect the captions so that they appear superimposed on the movie screen. The reflective panels are portable and adjustable, enabling the caption user to sit anywhere in the theater. The Rear Window system was codeveloped by WGBH and Rufus Butler Seder of Boston, Massachusetts.

A companion system for blind moviegoers has also been developed. DVS Theatrical™ delivers descriptive narration via infrared or FM listening systems, enabling blind and visually impaired moviegoers to hear the descriptive narration on headsets without disturbing other audience members. The descriptions provide narrated information about key visual elements such as actions, settings, and scene changes, making movies more meaningful to people with vision loss.

These technologies have been available in specialty theaters—such as large-format movie theaters and theme parks—for several years. The key to bringing access to movie theaters is the synchronization of captions and descriptions enabled by the digital sound system known as Digital Theater Systems (DTS). DTS provides multichannel digital audio on a CD-ROM. A reader attached to the film projector reads a timecode track printed on the film and signals the DTS player to play the audio in sync with the film. For the Motion Picture Access Project, DTS adapted its technology to include the caption and descriptive narration tracks on a separate CD-ROM, which plays alongside the other discs in the DTS player. In turn the DTS player sends the captions to the LED display and the descriptive narration to the infrared or FM emitter.

67.9 CONCLUSIONS

When one considers universal design in relation to the world's media, it's hard to imagine any information technology being made accessible without the exploitation of digital technology itself. The flexibility offered by computer software and the malleability of digitally based services allows a high degree of adaptation for individual needs and profiles. The coming of the Digital Age has resulted in an exponential growth in the amount of new media and systems introduced annually and a concomitant growth in the vigilance needed to assure accessibility.

In the United States, many of the leaps forward in accessibility of media have been accomplished through the intervention of federal law, regulations, or public-sector funding, or a combination of the three. Though marketplace pressures helped closed captioning become widespread even before the mandates of the Telecommunications Act, in many other circumstances it has been the guiding hand of government that has assured a level playing field for people with disabilities. The extremely competitive and constantly shifting international media environment will probably require regulatory interventions in the coming decades to assure accessibility of merging new media. As loathe as many are to consider government regulation of the Internet, for example, the vast riches available to fast-moving entrepreneurs unfortunately signal that public interest issues and public service uses of new media are often low priorities at best.

In the final analysis, universal design principles must be applied to the *education* of the next generation of engineers, designers, software programmers, and media producers to ensure that at the very moment that each new technology is imagined, the creator's imagination is universal as well.

67.10 NOTE

1. The author was the founder and first director of this office, which subsequently became the CPB/WGBH National Center for Accessible Media (NCAM).

67.11 BIBLIOGRAPHY

CD-ROM Access Project, guidelines and report on prototypes by Madeleine Rothberg and Tom Wlodkowski, www.ncam.wgbh.org/cdrom.

Daisy Consortium: www.daisy.org.

FCC Notice of Proposed Rulemaking regarding Closed Captioning Requirements for Digital Television Receivers—ET Docket No. 99-254.

FCC Report and Order on Television Closed Captioning—47 CFR Part 79, FCC Report and Order MM, Docket No. 95-176, FCC 97-279.

Federal Communications Commission's Disability Rights Office, www.fcc.gov/cib/dro.

National Library Service for the Blind and Physically Handicapped (NLS), www.loc.gov/nls/.

Open eBook Initiative: www.openebook.org.

Print Access Project: An Investigation into Making Daily Newspapers Accessible to Print-Disabled People, Media Access Research and Development Office, WGBH Educational Foundation, Boston, March 1993.

Productivity Works, www.prodworks.com.

Public Broadcasting Service, www.PBS.org.

Telecommunications Act of 1996, Pub. LA. No. 104-104, 110 Stat. 56 (1996).

Television Decoder Circuitry Act—104 STAT.960-961, Public Law 101-431—15 October 1990.

World Wide Web Consortium (W3C) Web Accessibility Initiative (WAI), www.w3.org/wai.

67.12   *RESOURCES*

Apple's QuickTime.   www.apple.com/quicktime/

Caption Center.   www.wgbh.org/caption

CAST.   www.cast.org

CPB/WGBH National Center for Accessible Media (NCAM).   www.wgbh.org/ncam

Descriptive Video Service® (DVS®).   www.wgbh.org/dvs

Microsoft's Windows Media.   www.microsoft.com/windows/windowsmedia/

Motion Picture Access Project.   www.mopix.org

Narrative Television Network.   www.narrativetv.com

RealNetworks' RealPlayer.   www.real.com

Trace Research and Development Center.   www.tracecenter.org

CHAPTER 68

USER-CENTERED DEPLOYMENT AND INTEGRATION OF SMART CARD TECHNOLOGY: THE DISTINCT PROJECT

Jim S. Sandhu, M.Des.R.C.A., F.C.S.D., F.R.S.A., Churchill Fellow, and Alan Leibert
Card Europe Limited, Hertfordshire, United Kingdom

68.1 INTRODUCTION

This chapter describes the DISTINCT Project, which was initiated as a two-year project funded under the Integrated Applications of Digital Sites (IADS) Program of the European Commission. It started in February 1998 and officially ended in January 2000; however, due to its practical applications and continuing deployment strategy, in actuality it is an ongoing project. DISTINCT stands for *Deployment and Integration of Smart Card Technology and Information Networks for Cross-Sector Telematics*. Smart cards and multimedia delivery networks are the key and cover access, identification, payment, and booking of facilities or services encapsulating transport, education, health care libraries, administrations, and more. Universal design is at the heart of the project, which covers seven major demonstration and application sites across Europe.

The vision of the project team was to focus on creating an open, democratic society, accessible to all citizens without discrimination. In such a society older and disabled people are valued as a resource, as active, participating, independent citizens and not a burden. DISTINCT has shown that maximizing access to services through the use of universal design principles can enhance social integration of marginalized groups.

68.2 BACKGROUND

Recent telematic developments are reshaping many basic aspects of the world's social and economic foundations that have developed since the Industrial Revolution. The transition from an industrial society to an information society is accelerating due to convergence of the telecommunication, computer, consumer electronics, and media industries.

Several technologies have now reached the stage where their widespread application is having a major impact on people. The ability to create, access, and transfer information electronically is beginning to remodel our concept of citizenship and the role of the designer in enhancing the process through universal design.

To be competitive, companies and professionals are being forced to rethink their organization and operating structures in order to fit in with the changes initiated by telematics. These changes are closely linked with a new type of social and professional behavior: more individuality and more flexibility. The changes also offer an opportunity to the public sector to recover efficiency by providing its existing services more rapidly, by delivering new services more cheaply, and by outsourcing uneconomic services to specialists. More fundamentally, access to information technology is breaking up the traditional boundaries between the role of the state (local and national) and the market.

The DISTINCT project, which was a two-year project funded by the European Commission, fits in with the type of trends and changes just described. More precisely, it fits into the Commission's Telematics Applications Program: Integrated Applications for Digital Sites, where design-for-all is central to all research and development projects. After the initial funding by the Commission up to January 2000, the project has been self-sustaining through local collaboration funding from universities, municipalities, and other service providers (transport, for example).

68.3 DISTINCT

The project brought together many key developments from previous EU projects through the enhancement and integration of numerous services for the citizen. DISTINCT, which is ongoing through local efforts, uses the platforms of a citizen's "smart card" and multimedia delivery systems to achieve this integration. The applications that are deployed in the five main sites across Europe are wide and varied, with particular special reference to older and disabled people, and rural and urban telematics.

The underlying assumption of DISTINCT is that public and commercial services are often fragmented, with many different professionals involved, and with little cooperation and coordination between services. Users invariably end up carrying a multitude of different identity formats in order to access these services. The DISTINCT smart card has made a valuable contribution in this area by serving as a single conduit to the services. Various multisite evaluation studies indicate not only cost-benefits of this approach, but also greater choice and independence for the citizen (see Web site links at end of chapter).

The $15 million project was coordinated by Newcastle University and focused on seven cities and regions in five countries: Newcastle upon Tyne (United Kingdom); Lapland, Espoo, and Vantaa (Finland); region of Zeeland (the Netherlands); city of Torino (Italy); and the city of Thessaloniki (Greece). The services, which are fully harmonized among the different sites, provide access to "democracy online," tourist and travel information, booking facilities, banking (the Netherlands to date), teleshopping, teleworking, libraries, education, distance learning, special-offer vouchers, information networks, public transport, road tolls, parking facilities, and health care services and support for older and disabled people. One major objective of DISTINCT was to try to integrate and harmonize these services—a key challenge to the application of universal design principles.

Stakeholders and User Needs Assessment

In universal design terms, the fundamental cross-site focus was: What requirements did the project need to address for integration of access to services? In view of the large consortium spread over five countries; the lack of harmonization and varying conceptual frameworks; the range of active stakeholders; and differences in language, political, economic, and social circumstances, this was a difficult question to answer. Over three months of intensive cross-site activity entailing surveys, interactive user forums, and through exchanges with professional and service providers, an extensive list of user requirements was established. These broadly follow universal design principles. Some of the key requirements include:

- Customize services as precisely as possible for the individual or target group, based on profile and preferences.
- Control, secure, and differentiate access to services (e.g., secured identification of user, age-based transport fares, etc.).
- Ensure better quality of access to services in terms of man-machine interface requirements, multilingual access, and information based on physical and cognitive access requirements.
- Provide support for structural relationships such as customer, financial, and social links.
- Ensure direct, maximum communication for the individual.
- Maximize direct distribution in terms of teleshopping.
- Provide remote access to services to the home, in public buildings and on the streets to minimize travel.
- Integrate functions in order to identify entitlement to other services.
- Provide for active control over service provision by users.
- Reduce costs in terms of efficiency gains through using smart cards.
- Facilitate self-service in order to lower labor costs and waiting times.
- Share the same access points.

Universal Design and DISTINCT

Interestingly, the key to DISTINCT access issues is not so much access to terminals, operating instructions, keypads, touchscreens, typefaces and legibility, and/or anthropometrics, as important as these are, but rather the methodology of encoding user requirements on the DISTINCT card. This may appear to remove the design focus to electronic data sets, but that is another new area that universal design has to operate within and therefore cannot really ignore.

68.4   THE AIM OF THE DISTINCT IDENTITY (DID)

The aim of DID was to enable systems to recognize consumers as belonging to a specific closed user group (the DISTINCT group), which would enable special services to be supplied to that group by systems included under agreement between the partners in the project or parties associated with the project by agreement.

In general, users identify themselves with smart cards using the DID. The DID is the essential data to uniquely identify the card, confirm that the card is part of a relationship scheme, identify the issuer, and also provide some essential information about the user—for example, preferred language, special interface requirements, and mobility/physical access needs.

Other more tailored information is optionally available in a standard DISTINCT format, although not all of this data will necessarily be held on the card. This optional DID data is referred to later in this chapter as the *Distinct Profile and Preferences* (DPP) *extension*. In practice some or all of the DPP extension identifying data may be spread across a smart card and a remotely accessed database.

The DID allows the following tasks to be performed:

- Identification of the user.
- Identification of the unique DISTINCT reference number that will identify the DID country, site, and issuer, and through this the person to whom the DID was issued.
- Identification of special needs, language profile, and preferences of the user, to allow the terminal to adapt as appropriate and possible.
- Identification of services available on the card, and correlation with those services available at the terminal.
- Allow provision of "guest" services to the user, based upon services available at the terminal and background agreements between the local DISTINCT/service operator and that of the user's smart card. (For example, allow a visitor with a smart card containing a DID to use local library facilities—perhaps only if he or she has another library application on their card.)

In line with universal design philosophy, the DID operates across all boundaries as well as in local on-line, remote on-line, and off-line environments. The implication is that the DID may be used as a mechanism to link services within a city or region, as well as across regions. Using the DID, a Turin-based smart card holder may inquire from his or her home site about services and facilities operating in Newcastle. The same person may then travel to Newcastle and use those services, as well as have access to their own services operating in Turin—all presented in the Italian language.

Brief Outline of the Architecture of the DID

The architecture of the DID sets down the definition of a set of mandatory and optional data elements that form the DID. The architecture describes and formally defines an Application Program Interface (API), which includes a function set to handle the DID and the structure and layout of the fields within the DID as presented at the interface. Using this concept, the API allows any system implementor to deal with the DID data regardless of whether it is card-based, in a remote database, or split across the two, as well as regardless of the physical configuration of the system and smart card.

The architecture does not specify how the DID is to be implemented at the physical level. This option is entirely left open to the managers of each site, who may or may not wish to implement it onto an existing smart card with limited space availability. Thus the physical definition would allow existing smart card data elements to be mapped into the DID or new elements defined in any way that suits the system.

Site-Specific Data Within the DID

The DID is an identifier that is used to allow systems to recognize that the user has a token that is part of the DISTINCT environment. The DID additionally contains a unique reference number to identify the specific country and site of the DID issuer and the party to whom the DID was issued. Finally, the DID contains some data describing the profile and preferences (DPP) of the person to whom the DID was issued. The DPP extension contains the optional data describing the profile and preferences of the person to whom the DID was issued.

It is central to DISTINCT that there are a number of common applications across sites where interoperability is desirable. The DID plays no part in the application itself, and application files on a smart card or elsewhere will be shared and mutually accessible by agreement between the parties. However, the DPP extension may be used to specify additional profile and preference information that is useful to the interoperating parties but is not included within the application level datasets.

This data is known as *site-specific data*. It may be appended to the DID on a data element by data element basis to create a nonstandard DID. The mechanism and architecture of the DID makes allowance for this option and site implementors may choose to make use of the DID mechanism to create their own nonstandard version of the DID.

This mechanism allows two sites to share additional data through the DID mechanism without upsetting the general concept or the ability for the card to be used at any other site in the normal way. All that will happen is that sites not subscribing to these additional data elements will not see them, and the normal DID will be presented with or without the DPP extension. The architecture makes provision for the extra buffer space that may be required to hold the DID in its extended form.

Guidelines to a Harmonized Approach

It could well be that new members already have a smart card in operation. Once again, the DID architecture allows them to join the fold and for the DID to be mapped onto their card in a unique manner. The degree of interoperability they achieve will then be a function of their ability to export their card-specific software drivers and the ability of others to retrofit these drivers into their systems. This has happened on a very large scale in the Netherlands.

68.5 THE DISTINCT ARCHITECTURE SPECIFICATION

Tables 68.1 to 68.5 summarize the fields of the mandatory part of the DID as specified in the architecture and as expected to be available at the API through various function calls.

All three of the mandatory fields within the DID Profile have to be defined. However, for personal data protection, the user can decide not to enter information into these fields, in which case the standard default conditions apply.

All other fields are optional and are included in the extended DID Profile and Preference structures (see Tables 68.6 to 68.8). For ease of identification, these are logically referred to in this document as the DPP extension.

TABLE 68.1 Summary of Mandatory Fields

Distinct ID	Mandatory	Description
Version	Yes	Identifies specific versions of the DID which will map to specific field structure descriptions. In general will be used to describe different Profile and Preference structure layouts.
Base	Yes	Contains the basic fields of the DID that will always be present and accessible.
Validity	Yes	Contains validity date information.
Certificate	No	Contains a security certificate.
DID profile	Yes	Contains profile information relating to language, information display requirements, and E&D primary mobility codes.
DID preference	No	Contains preference information that may be used to enhance usability as well as enable application and data sharing.

TABLE 68.2 DID Version

DID version	Description
DID version	Identifies the version of the DID. It is to be decided how version numbering will be used in practice. The version number may be unique by consensus or mandate, or it may be unique only when combined with the DID issuer.
DID version data	Identifies the date of the version.

TABLE 68.3 Country and Issuer

DID base	Description
DID country	Identifies the country of the DID issuer.
DID issuer	Identifies the DID issuer. The DID country plus DID issuer fields uniquely identify the DID issuer.
DID card ID	A unique identifier for the card which may or may not map to the card serial supplied by the card manufacturer/issuer.
DID cardholder ID	A unique identifier for the cardholder within DID issuer.

TABLE 68.4 Validity Dates

DID validity	Description
DID creation date	Date of DID creation. This may be mapped to the card issue date, an application creation date, or any other relevant date on the card.
DID start of validity	Start of validity of the DID. This may be mapped to some other start date on the card.
DID end of validity	End of validity of the DID. This may be mapped to some other end date on the card.

TABLE 68.5 Special Requirements

DID profile	Description
DID disability need code	Defines interface requirements (software controlled).
DID physical access code*	Defines accessibility to terminal equipment (hardware construction).
DID language preference	Defines primary language preference.

* The DID physical access code is based on user access requirements and is used in the table to maintain compatibility with the formal DID architecture specification and documentation.

TABLE 68.6 Summary of Profile and Preference Fields

Distinct PP* (DPP) extension	Mandatory	Description
Profile	No	Contains profile information. Some of the fields in the profile are mandatory. These are contained within the main DID and are not listed here.
Preference	No	Contains optional preference fields.

* DISTINCT Profile and Preferences.

TABLE 68.7   Cross-Site Optional Fields

DID profile—DPP extension data	Description
Cross-site common optional fields	
Gender code	Male or female
Date or year of birth	If year only, day and month fields set to zero
Second choice language code	Defines secondary language preference
Name	Name of cardholder
Address	Address of cardholder
Postcode	Post code
Place of Residence	Place name
Telephone number	Related to the cardholder
Fax number	Related to the cardholder
Serious health problems	Coded indicator
Blood group	Standard coding
Immunization emergency category	Standard coding
Social security number	Standard coding
Other optional fields as required	

TABLE 68.8   Additional Need Codes

DID preference—DPP extension data	Description
Cross-site common optional fields	
Pointer to additional need codes	Identifies a structure containing optional disability needs and access codes.
Sport code	Code and meaning to be determined
Cultural/historic code	Code and meaning to be determined
Entertainment code	Code and meaning to be determined
Food and drink code	Code and meaning to be determined
Hobby code	Code and meaning to be determined
Media category code	Code and meaning to be determined
Mobility code	Code and meaning to be determined
Loyalty code	Code and meaning to be determined
Bank account number	Card holder's bank account number
Bank sort code	Bank unique sort code to identify bank
Customer ID	Code and meaning to be determined
Other optional fields as required	

Harmonized Card Specification

In the context of the contactless card, the draft ISO14443 standard allows for two incompatible interface options (Types A and B), which means that where interoperability across these is concerned, contactless smart card readers will have to be able to cope with both standards. Presently, the Mifare interface standard (Type A) is the most widely used, and early contactless card issuers will probably standardize on this, whether a Combi card approach is taken or not. Existing applications within the DISTINCT project make use of both contact and contactless cards.

Smart Card Data Size

The DID minimum size depends upon how data fields are represented on the card. In addition, some fields may be mapped onto existing card fields such as card serial, card validity, and start and end dates.

In the minimum case the DID may be mapped into the card serial, and the DID validity dates mapped into other validity dates on the card, leaving around 30 bytes required for the rest of the DID. In other cases with significant optional DPP extension Profile, Preference, and disability needs data, one could consider the complete overhead to take up to 300 bytes without certificates. It is doubtful that more space will be required if the DID is to usefully act as an aid to interoperability. However, it must be reiterated that the mandatory section of the DID is the only essential part for the smart card, if a smart-card-based DID solution is selected. The DPP extension may be held wholly or partially on the card or in an external database.

Smart Card Standards

The DISTINCT smartcard conforms to ISO 7816-4, which defines a common command set, and ISO 7816-5, which defines the application naming structure. Although contactless cards do not in general follow ISO 7816-3 (which defines the physical and electrical interface) but instead follow draft ISO 14443, they should conform to ISO 7816-4 and ISO 7816-5 (see Appendix at end of chapter for more details).

DID and DPP Extension Field Structure

It is assumed that the mandatory parts of the DID will be openly readable by all parties and are consequently grouped together into a single data area if so desired. However, the architecture will support disparate data as well as missing or inaccessible data.

Table 68.9 is drawn from the architecture and shows a specific DID code structure for cards that have available space. It should be noted that the mandatory profile fields are

TABLE 68.9   Codes for Additional Space

Field	Length	Card field format	Card field length
Version	4 digits	BCD 4 bits/digit (nnn.n)	2 bytes
Version date	Ddmmyyyy	16 bit binary	2 bytes
Country*	2 characters	Character	2 bytes
Issuer[†]	6 digits	BCD 4 bits/digit	3 bytes
Card ID[‡]	8 bytes	Any	8 bytes
Card holder ID	16 characters	Character	16 bytes
Creation date	Ddmmyyyy	16 bit binary	2 bytes
Start of validity	Ddmmyyyy	16 bit binary	2 bytes
End of validity	Ddmmyyyy	16 bit binary	2 bytes
Disability need code	7 digits	BCD 4 bits/digit	4 bytes[§]
User access reqts.	5 digits	BCD 4 bits/digit	3 bytes[§]
Language preference[¶]	2 characters	Character	2 bytes
Pointer to profile	2 byte file pointer	Pointer	2 bytes
Pointer to preference	2 byte file pointer	Pointer	2 bytes
Optional certificate	16 bytes	Binary	16 bytes

   * ISO 3166 2 character alpha country code.
   [†] ISO 7812 IIN or equivalent.
   [‡] Card ID or card number is card-dependent and varies according to card in terms of length and format. 8 bytes is a large enough maximum field. The field may be treated as a unique set of binary bits. Shorter card numbers duplicated in the DID should be left justified and padded with binary zeros.
   [§] To be confirmed.
   [¶] ISO 639.

mapped directly into the base structure. This policy means that where no other Profile and Preference data is present, the Profile and Preference structures and their pointers may be omitted where space is at a premium. If space is available, null pointers can be used. In addition, this layout also permits the essential DID data to be read directly.

Disability Needs Code

Table 68.10 shows the disability needs coding structure. These codes affect the software within the terminal that controls the user interface. Coding is 4-bit BCD. The coding 0 in each of the digit positions in the table indicates that this feature is not a problem. The "R" digit is reserved for future use.

TABLE 68.10 Disability Needs Codes

Field	Position	Meaning
Hearing	YXXXXXX R	1 = mild, 2 = medium, 3 = severe, 4 = totally deaf
Sight	XYXXXXX R	1 = mild, 2 = medium, 3 = severe, 4 = totally blind
Sight color RG	XXYXXXX R	1 = mild, 2 = medium, 3 = severe
Sight color BY	XXXYXXX R	1 = mild, 2 = medium, 3 = severe
Reduced intellectual processing	XXXXYXX R	1 = mild, 2 = medium, 3 = severe
Reduced short memory	XXXXXYX R	1 = mild, 2 = medium, 3 = severe
Dyslexic reading	XXXXXXY R	1 = mild, 2 = medium, 3 = severe

User Access Requirements Code

Table 68.11 shows the user access requirements coding structure. These codes affect the terminal hardware construction and access to the terminal. Coding is 4-bit BCD. As in the Disability Needs Code, the coding 0 in each of the digit positions in the table would indicate that this feature is not a problem or not used. The "R" digit is reserved for future use.

TABLE 68.11 Access Requirement Codes

Field	Position	Meaning
Wheelchair	YXXXX R	1 = manual, 2 = motorised
Upper limbs	XYXXX R	1 = left lacking, 2 = right lacking, 3 = both lacking
Reduced finger mobility	XXYXX R	1 = mild, 2 = medium, 3 = severe
Reduced reach	XXXYX R	1 = mild, 2 = medium, 3 = severe
Tremor	XXXXY R	1 = mild, 2 = medium, 3 = severe

Profile and Preference Extension Codes

It could be that there are a number of common applications across sites where interoperability is desirable and mutually agreed. In these circumstances the DPP (or optional part of the DID) could be used to specify additional profile and preference information that is useful to the interoperating parties but is not included within the application datasets (see Table 68.12).

TABLE 68.12   Profile and Preference Extension Codes

Field	Length	Card field format	Card field length
Gender code	1 character	Character "M," "F"	1 byte
Date or year of birth	Ddmmyyyy	Character	8 bytes
Second choice language code	2 characters	Character	2 bytes
Last name	25 characters	Character	25 bytes
Title	10 characters	Character	10 bytes
Initials	6 characters	Character	6 bytes
First (given) names	20 characters	Character	20 bytes
Street name	24 characters	Character	24 bytes
House number	5 characters	Character	5 bytes*
Addition to house number	4 characters	Character	4 bytes*
Postal code	8 characters	Character	8 bytes[†]
Place name	24 characters	Character	24 bytes
Telephone number	14 characters	Character	14 bytes
Fax number	14 characters	Character	14 bytes
Serious health problems	2 digits	BCD 4 bits/digit	1 byte[‡]
Blood group	2 digits	BCD 4 bits/digit	1 byte[‡]
Immunization emergency category	2 digits	BCD 4 bits/digit	1 byte[‡]
Social security number	14 characters	Character	14 bytes[†]
Optional fields	—	—	—
Optional certificate	16 bytes	Binary	16 bytes

* House number and addition to house number may be taken together where a house is named rather than numbered.
[†] To be correlated with the relevant standard.
[‡] To be confirmed.

DPP Extension Codes

In the case of site-specific data appended to the main DID within the DPP, the issue becomes much more case-by-case sensitive. (See Table 68.13.) The different fields may or may not be returned by agreement between the sites, the application providers, and the card holder. It could be that some DPP data elements are considered sensitive and are access-secured. The likelihood in this event will be that this data will be returned at the home site but not at a remote site.

TABLE 68.13   DPP Extension Codes

Field	Length	Card field format	Card field length
Pointer to optional needs codes	2 byte file pointer	Pointer	2 bytes
Sport code	2 digits	BCD 4 bits/digit	1 byte*
Cultural/historic code	2 digits	BCD 4 bits/digit	1 byte*
Entertainment code	2 digits	BCD 4 bits/digit	1 byte*
Food and drinks code	2 digits	BCD 4 bits/digit	1 byte*
Hobby code	2 digits	BCD 4 bits/digit	1 byte*
Media category code	2 digits	BCD 4 bits/digit	1 byte*
Mobility code	2 digits	BCD 4 bits/digit	1 byte*
Loyalty code	2 digits	BCD 4 bits/digit	1 byte[†]
Bank account number	10 digits	BCD 4 bits/digit	5 bytes[†]
Bank sort code	6 digits	BCD 4 bits/digit	3 bytes
Customer ID	16 characters	Character	16 bytes*
Optional fields	—	—	—
Optional certificate	16 bytes	Binary	16 bytes

* To be confirmed
[†] To be correlated with the relevant standard

Optional Disability Needs Codes

This structure, if present, is pointed to from the DPP Extension Preference structure which must be present if the Optional Disability Needs Codes structure is present. However, it need only contain one field, which is the pointer to this structure. If the DPP Extension Preference structure is present but the Disability Needs Codes structure is not present, the pointer will be null.

Coding is 4-bit BCD. As before, the coding 0 in each of the digit positions in the table indicates that this feature is not a problem or not used. In Table 68.14 a "Y" indicates the 4-bit position of the byte used by a given parameter, while "X" indicates the 4-bit position of the byte not used by the specified parameter. Where 4 bits are not used, they are coded "R" ("reserved for future use"). Where a whole byte is used as a binary field by a parameter, the coding in the table will be "YY."

In this layout the specific card fields do not map exactly in format to those specified by the architecture at the API. This is because the API specification is for programmer convenience, as well as conforming to international standards and conventions where appropriate. At the card level, space is of the essence and inefficient data structures are avoided. The interface module will convert from the API to the card format and vice versa.

The Preference fields are entirely optional, and the DPP Extension Preference structure is laid out similarly to the DPP extension Profile structure. The distinction between Profile and Preference is made to identify which data structure the data will be returned in at the API. In general, the separation is on the basis of Profile data being fixed while Preference data may vary over time. But this distinction is not absolute.

All other data elements defined as potential data elements for the DPP extension are described as site-specific and outside of the formal definition. It will be up to each site/application provider to decide what additional information it wishes to make available through the mechanism of the DPP extension and to add this data to its DPP extension or DID definition on its cards. It will then make this data available via a supplied interface module and data dictionary definition.

DID Mapping on Space-Limited Cards

It will be apparent that the DID duplicates some data that may already exist on the card as card level data elements. They are duplicated in the DID, space permitting, for the sake of

TABLE 68.14   Optional Needs Codes

Field	Position	Meaning
Auditory feedback	Byte 1 YX	1 = standard, 1 = loud
Instructional sessions	Byte 1 XY	1 = standard, 2 = extended
Auto gate opening and close	Byte 2 YX	Minutes (1–15)
Contactless operation in turnstiles	Byte 2 XY	Minutes (1–15)
Readability from wheelchair	Byte 3 YX	1 = standard, 2 = extended
Speech device	Byte 3 XY	1 = loudspeaker, 2 = handset, 3 = both
Pitch	Byte 4 YR	1 = male voice, 2 = female voice
Speech rate	Byte 5 YY	Words/minute (1–255) (1 byte)
Sound amplification	Byte 6 YY	Most significant bit: 0 = positive value, 1 = negative value; 7 least significant bits denote the offset amount in dBA from the terminal's normal sound level (1 byte)
High frequency amplification	Byte 7 YY	As above (1 byte)
Low frequency amplification	Byte 8 YY	As above (1 byte)
Hearing aid communication	Byte 9 YX	1 = magnetic inductive coupling
Braille output	Byte 9 XY	1 = yes
Voice input	Byte 10 YX	1 = yes
Keyboard use	Byte 10 XY	1 = yes
PIN pad use	Byte 11 YX	1 = type 123, 2 = type 789
Button size	Byte 11 XY	1 = normal, 2 = medium, 3 = large, 4 = very large
Click type	Byte 12 YX	1 = click up, 2 = click down
Sensitivity	Byte 12 XY	1 = low, 2 = normal, 3 = moderate, 4 = high
Position of input device X	Byte 13 YY	X co-ordinate (1–256) (1 byte)
Position of input device Y	Byte 14 YY	Y co-ordinate (1–256) (1 byte)
Position of input device Z	Byte 15 YY	Z co-ordinate (1–256) (1 byte)
Time out	Byte 16 YY	Time in decaseconds (1–256) (1 byte)
Reduced concentration	Byte 17 YX	1 = mild, 2 = medium, 3 = severe
Reduced long-term memory	Byte 17 XY	1 = mild, 2 = medium, 3 = severe
Dialogue level	Byte 18 YX	1 = normal, 2 = simple, 3 = very simple
Text complexity	Byte 18 XY	1 = normal, 2 = simple, 3 = very simple
Text density	Byte 19 YX	1 = normal, 2 = low, 3 = very low
Symbols/icons	Byte 19 XY	1 = text only, 2 = text & symbols, 3 = symbols only
Sign language	Byte 20 YX	1 = yes
Text color R	Byte 20 XY	Red coding (0–15)
Text color G	Byte 21 YX	Green coding (0–15)
Text color B	Byte 21 XY	Blue coding (0–15)
Background color R	Byte 22 YX	Red coding (0–15)
Background color G	Byte 22 XY	Green coding (0–15)
Background color B	Byte 23 YR	Blue coding (0–15)
Text size	Byte 24 YY	Millimeters (1–256) (1 byte)
Position of output device X	Byte 25 YY	X co-ordinate (1–256) (1 byte)
Position of output device Y	Byte 26 YY	Y co-ordinate (1–256) (1 byte)
Position of output device Z	Byte 27 YY	Z co-ordinate (1–256) (1 byte)
Reduced field of vision	Byte 28 YX	1 = light, 2 = medium, 3 = severe
Stamina	Byte 28 XY	1 = mild, 2 = medium, 3 = severe
Fainting/dizziness or seizures	Byte 29 YR	1 = mild, 2 = medium, 3 = severe
Optional certificate	16 Bytes	—

completeness and ease of use. However, where space is limited, it will be necessary to map directly to the existing data fields on the card, read the appropriate data through a series of separate Read commands to the card, possibly reformat the data, and then build the formal DID according to the DISTINCT architecture rules before passing it to the DID API.

Site-Specific Extensions to the DID

It is apparent that application data is held within the application file structures on the smart card. In the case of an integrated application that is shared across sites, this data must be made available in order for the application to run successfully.

The DID therefore acts primarily to identify that this is a card participating in the DISTINCT grouping and to provide base information about the issuer. From this a judgment can be made about whether a specific application can be run from this terminal. If so, then the DID can be logged for later accounting purposes, and it can be examined for further information such as language preference and disability information.

The secondary purpose of the DID is to act as a standard communication vehicle whereby an application provider wishing to share an application can pass information that may be required but is not found within the application data itself. This information is passed as site-specific data within the DID as optional extensions to the DID DPP extension. Although the sites within the DISTINCT project have indicated a large amount of possible data that might fit into this category, in practice it is likely that this data will be kept to a minimum in order to avoid access rights conflicts.

Where optional site-specific data is commonly held and used across and within sites, it may be expedient to "standardize" on a specific structure by agreement and to publish this extended DID structure as a directory pertaining to a specific version of DID. This will correspond to the version number in the DID. It is expected that there will be relatively few versions of the DID and that the multiplicity of field level variations that will occur will be dealt with by agreement between sites and the passing of information about field layouts between them.

The architecture supports the concept of unknown data and data-not-present such that fields not available could be left blank, while extra buffer space could be allocated for optional data returned at the API interface, whether or not the caller is aware of the presence of the optional data.

Access Rights, Data Structures, and Security Management

It is intended that site managers will organize their own access rights to data within the DID, based on separate file access rights as previously described. It must be noted that the ability to pass a test or provide a key to open access to data at one's home site may be different from that at a remote site, and due allowance must be made for this when permitting applications to be shared.

The DID makes no provision for the secure passing of data across the API boundary, and the site managers are discouraged from attempting this.

One way that some access rights can be managed and tests passed is from within the interface program code that sits between the application on the card and the DISTINCT API; that is, sitting wholly within the DISTINCT application domain. For example, code may be present to ask a cardholder to enter a PIN, subject of course to an entry device being available. In this case, the application provider wishing to share an application will make the appropriate code and facilities available as part of the distributed application interface.

However, one aim of a harmonized approach is to try to avoid the need for the distribution of code and for a standard interface to be used. Given this, it is hoped that applications can be shared without such access controls being applied. It is hoped that it will be sufficient for the cardholder to signal his or her approval for access to nonsensitive data under their control as being implied by placing their card into the slot. All other more sensitive data will not be provided to remote sites. Of course, this may mean that applications operate in a more restricted manner in remote sites compared to home site operation.

68.6   EVALUATION

Detailed descriptions of user evaluations for the demonstration phase covering all seven sites can be found on the Web sites listed at the end of this chapter. The evaluations focused not only on usability and access issues, but also on whether the card architecture and coding worked in the various test sites. This was critical, given the cultural and linguistic diversity of these sites. It should be stressed that as DISTINCT smart cards continue to be deployed at an increasing pace, ongoing evaluations continue to take place covering new applications.

68.7   WHY DISTINCT AND NOT VISA

The initial rationale for a card like DISTINCT was that it had to be based on local community needs in Europe. This necessarily meant a focus on public sector services and third-party payment as opposed to direct payment, as in the case of the VISA card. The latter entirely operates at the level of a versatile electronic purse. Although its commercial applications are constantly being expanded, its uses are still very much tied with the exchange of monetary value for a service. For organizational, political, and financial reasons, VISA has not ventured into services accessed by, for instance, a Citypass. In the context of European communities this means access to health, career, housing, leisure transport, library, school meals, training, education, and so forth. Up to the development of DISTINCT, most of these were provided by different bureaucracies with each individual user having a range of different means of identification—instead of just one. Moreover, this single-conduit approach ensures that special interface requirements such as large letters, speech output and mobility, and access needs are taken account of—attributes which few commercial cards seem to have.

Another area where the highly commercial VISA card cannot enter is the direct exercise of citizenship. Some member countries of the European Union (especially the U.K. government) are seriously considering extending the use of the DISTINCT card for electronic voting, both at the local and national levels.

68.8   CONCLUSION

The key to the project's importance is, for example, that in the short time it has been operating, 25 million commercial "Chipper" and "Knipchip" cards have been retrofitted with the DISTINCT ID in The Netherlands. In the United Kingdom the Newcastle site is fast enlarging its DISTINCT card base, which presently stands at over 50,000—covering health, education, shopping, career services, transport, and so forth. Card Europe, which is a consortium partner, is currently issuing 250,000 World Point Loyalty multiapplication cards based on DISTINCT. Similar activities are being undertaken in Italy, Finland, and Greece. In fact, the percolation and distribution of the DISTINCT card is exponential and will have a significant impact outside of Europe.

The project has shown that it is possible to utilize and expand universal design principles in areas of increasing importance that were previously considered too remote for the design professions to tackle. DISTINCT is a prime example and proof that it can be done and that it can expand our understanding of universal design principles and the changing role of designers. As design is simply a mirror of its age, designers need to take the fast-expanding and increasingly pervasive information society into account.

The project has also highlighted the consortium's vision of the Information Society as one of an open, democratic society, accessible to all citizens without discrimination in a way that has not been possible hitherto.

In line with universal design the DISTINCT project can best be summed up in the words of the European Commission:

> . . . the challenge is to use the Information Society (IS) to strengthen social cohesion and enhance people's ability to participate fully in every aspect of social and economic life, to make it a tool for the creation of an inclusive society. The IS should be about people and it should be used for people and by people to unlock the power of information, not to create new, or reinforce existing inequalities between the information rich and the information poor.
>
> *(Green Paper, "Living and Working in the Information Society," 1996)*

68.9 APPENDIX

Standards:

Comité Européen de Normalisation
Rue de Stassart 36
B-1050 Brussels, Belgium
Tel: +32 2 519 6811
Fax: +32 2 519 6819.

- *EN 726.* Requirements for IC cards and terminals for telecommunications use.
- *PrEN 1332.* Machine-readable cards, related device interfaces and operations. Part 1 Design principles and symbols for the user interface; Part 2 Dimension and location of tactile identifier for ID I cards; Part 3 Keypads; Part 4 Coding of user requirements for people with special needs.
- *EN 29241.* Part 4 Keyboard requirements; Part II Usability statements.

European Telecommunications Standards Institute
B.P. 52, Route des Lucioles, Sophia-Antipolis, Valbonne
F-06561 Alpes Maritimes, France
Tel: +33 92 94 42 00
Fax: +33 93 65 47 16

- *ETR 029.* Access to telecommunications for people with special needs: recommendations for improving and adapting telecommunication terminals and services for people with impairments.
- *ETR 068.* European standardization situation of telecommunication facilities for people with special needs.
- *DTR/HF-02009: 1996.* Characteristics of telephone keypads.
- *ETS 300 381.* Telephony for hearing-impaired people; inductive coupling of telephones, earphones to hearing aids.

- *ETS 300 488.* Telephony for hearing-impaired people; characteristics of telephone sets that provide additional receiving amplification for the benefit of hearing-impaired.
- *ETS 300 679.* Telephony for the hearing-impaired; electrical coupling of telephone sets to hearing aids.
- *ETS 300 767.* Human factors: telephone prepayment cards tactile identifier.
- *ETR 039: 1992.* Human factors standards for telecommunications applications.
- *ETR 068: 1993.* European standardization situation of telecommunication facilities for people with special needs.
- *TCR-TR 023: 1994.* Assignment of alphabetic letters to digits on push-button dialling keypads.
- *ETR 160: 1995.* Human factors aspects of multimedia telecommunications.
- *ETR 165: 1995.* Recommendation for a tactile identifier on machine-readable cards for telecommunication terminals.
- *DTR/HF-02003:1996.* The implication of human aging for the design of telephone terminals.

International Electrotechnical Commission
3 rue de Varembe, CH-1211
Geneva 20, Switzerland
Tel: +41 22 73 40 150
Fax: +41 22 73 33 843

- *IEC 73.* Colors of pushbuttons and their meanings.
- *IEC 118-4.* Hearing aids: magnetic field strength in audio frequency induction loops or hearing aid purposes.

International Organization for Standardization
1 rue de Varembe, Case postale 56, CH-1211
Geneva 20, Switzerland
Tel: +41 22 749 0111
Fax: +41 22 733 3430

- *ISO 7001: 1991.* Public information symbols.
- *ISO 7816.* Identification cards—Integrated circuit cards with contacts.
- *ISO 9241.* Ergonomic requirements for office work with visual display terminals.
- *ISO/IEC 9995.* Information technology: keyboard layout for text and office systems.
- *ISO/IEC 10536.* Identification cards—contactless integrated circuit cards.

International Telecommunications Union
Place des Nations, CH-1211
Geneva 20, Switzerland
Tel: +41 22 730 5111
Fax: +41 22 733 7256

- *ITU E118.* Automatic international telephone credit cards.
- *ITU E134.* Human factors aspects of public terminals—generic operating procedures.
- *ITU E135.* Human factors aspects of public telecommunications terminals for people with disabilities.
- *ITU E136.* Tactile identifier on prepaid telephone cards.
- *ITU E161.* Arrangement of figures, letters, and symbols on telephones.

68.10 BIBLIOGRAPHY

Gill, J. M., *Smart Cards,* RNIB, 1996.

Gill, J. M., *Access Prohibited? Information for Designers of Public Access Terminals,* RNIB, 1997.

Gill, J. M., *The Use of Electronic Purses by Disabled People: What Are the Needs?* RNIB, 1998.

Roe, P. R. (ed.), *Telecommunications for All,* COST 219, European Commission, 1995.

68.11 RESOURCES

The DISTINCT project has a very extensive range of internal publications targeted at members of the team and the European Commission. Due to commercial reasons these have not been widely publicized, as a number of contractual issues internal to the consortium are still being sorted out. However, a considerable number of these are available on various Web sites. For further detailed information on technical issues, deployment, usability, and user forums, please visit the following Web sites:

www.newcastle.gov.uk/distweb

www.distinct.org.uk

www.tagish.co.uk/distinct/main

www.cenorm.be/iss/workshop/distinctID

www.cenorm.be/iss

www.tag.co.uk/turtle

THE FUTURE OF UNIVERSAL DESIGN

CHAPTER 69

CREATING THE UNIVERSALLY DESIGNED CITY: PROSPECTS FOR THE NEW CENTURY

Leslie Kanes Weisman, M.A.
New Jersey Institute of Technology, Newark, New Jersey

69.1 INTRODUCTION

Balanced at the fulcrum of the millennium, humanity faces a future whose course has been set by the epic events of our recent past. How does the legacy of the twentieth century define the challenges and opportunities that our increasingly global society will face in the coming decades? What role will the growing international movement for universal design play in shaping our future world? This chapter addresses these questions in three ways: first, by discussing universal design as a vehicle for promoting human well-being, environmental wholeness, and the principles of participatory democracy; second, by expanding the concept, meaning, and context of disability to include the effects of substandard housing and infrastructure, unhealthy cities and buildings, and inadequate public transit and safety; and third, by speculating on three broad criteria that will define the universally designed city of the future. The chapter concludes with a discussion about the unique contributions that universal design can make as a part of both historic and contemporary efforts to create and promote healthy, inclusive, sustainable communities.

69.2 BACKGROUND

As people throughout the world entered the year 2000 with celebration and ritual, they simultaneously brought along their own personal recollections of the past. For an entire generation of American baby boomers, whose values and lives were profoundly shaped by the antiwar, civil rights, environmental, and women's movements of the 1960s and 1970s, the twentieth century will be remembered as the century that fought a futile "war to end all wars;" struggled to abolish racism and apartheid, sexism, and environmental degradation; and bore witness to shocking human and ecological holocausts: appalling death camps and refugee camps, Nagasaki, Hiroshima, Viet Nam, Agent Orange, oil spills in the Gulf, Chernobyl, Bhopal, and Love Canal.

But the twentieth century will also be remembered as a century of courage and triumph: the founding of the United Nations; the 1963 historic civil rights march on the U.S. Capitol; Woodstock, flower power, and the love generation; the first Earth Day that placed ecological issues on the public policy table in the United States; the first "Take Back the Night" march in San Francisco in 1978 in which 5000 women took to the streets at nightfall to protest violence against women; and the fall of the Berlin Wall in Germany and white supremacy in South Africa. It is within this milieu of human struggle to create a democratic, equitable, and sustainable existence that universal design is historically situated and discussed in this chapter.

69.3   THE SPATIAL CONTEXT OF DEMOCRACY

Fundamental in any democratic society are the principles of nondiscrimination, equal opportunity, and personal empowerment. When the promise of democracy is compromised or withheld, the result is often social protest. Undoubtedly, civil rights, feminism, and environmentalism will be remembered as three of the most powerful and enduring social protest movements of the late twentieth century.

Although not readily apparent, these three movements are interconnected in complex and profound ways that are embodied in physical and spatial forms that are universally designed. To understand how, it is important to recognize that in the industrialized world of the nineteenth and twentieth centuries, virtually all design and technology were based upon the analysis and disassembling of organic wholes into fragmented parts. Cities and suburbs, workplaces and dwellings, architecture and nature evolved as detached spatial realms, supported by a social worldview that segregates and assigns different status and power to women and men, rich and poor, black and white, young and old, gay and straight, able-bodied and disabled.

When protest movements arise to oppose social inequalities and ecological imbalances like these, activists usually begin by seeking to eliminate discriminatory practices through legal remedies that impose fines and penalties and threaten legal prosecution for noncompliance—for example, civil rights laws like the U.S. Fair Housing Act (see Chap. 35, "Fair Housing: Toward Universal Design in Multifamily Housing," by Steinfeld and Shea) and the Americans with Disabilities Act (see Chap. 12, "U.S. Accessibility Codes and Standards: Challenges for Universal Design," by Salmen). As social movements mature, they begin to look beyond the "letter of the law" to the "spirit of the law," which emphasizes ethics and values, and promulgates systemic changes in attitudes, behaviors, and institutional structures.

In this regard, the overarching, long-term contribution of contemporary social and environmental justice movements is not to be found, as many would assume, in the landmark civil rights and conservation laws that activists fought so hard to achieve, although these laws are certainly of great importance. But rather, it is in the fostering of a transformed global cultural and environmental consciousness and set of priorities. Slowly, people and their governments are beginning to move away from the politics of separation, fragmentation, and competing interests that created ecological breakdown and social injustice in the twentieth century. Slowly, the world is moving toward a politics of greater inclusion, connection, and regeneration that will restore healing and wholeness within the art of living in the twenty-first century. This paradigm shift represents, at the very least, a rightful and timely expression of "universal civility," and at the very most, an essential ethos for sustaining life in the coming decades.

69.4   THE BUILT ENVIRONMENT AS CULTURAL ARTIFACT

As a moral philosophy, universal design holds great promise for the realization of this transformed vision of social life, because advocates and practitioners of universal design recognize that the world is an unjust place and that design has a significant role to play in eliminating human suffering and advancing greater human dignity. Universal design is particularly relevant to the environmental design fields—architecture, planning, and landscape architecture—

because in its making, use, and design, the built environment shapes human experience, identity, and consciousness, and reinforces assumptions about culture and politics. Any serious effort to establish equitable and sustainable communities must involve redefining and restructuring both how people inhabit physical space and how designers teach and practice "place making."

Whereas most environmental design disciplines, particularly architecture, have traditionally been male-centric, object-centered, and technologically focused, universal design situates education, theory, and practice within the wider public debate about human diversity, justice, and equity; links the design process and product to the imperatives of democracy; and seeks to understand and formalize how different people and cultures value their world and act within it.

69.5   CELEBRATING DIVERSITY

Universal design upholds the democratic ideals of social equality and personal empowerment, because universal designers strive to create products and spatial environments that are designed to provide the same level of comfort, accessibility, and assistance to multiple users and multiple publics. Even though advocates of universal design recognize that it is nearly impossible to design all things for all people, the ultimate objective is to be as inclusive as possible. The first Principle of Universal Design, Equitable Use, mandates that no one be segregated, stigmatized, or disadvantaged by design (see Chap. 10, "The Principles of Universal Design," by Story; and Story, 1997). In contrast, architecture has traditionally defined the user, or the public in the case of urban planning, in very narrow terms based on a conception of the user/citizen, which is inherently masculine, and the public, which tends to be made up of middle-class white people living in nuclear families. Therefore, when architects and planners attend to the provision of housing, transportation, and community services, they have tended to design and plan for only a small segment of the population, thereby creating many problems for the ever-increasing numbers of people who do not fit into this assumed definition and life pattern.

Universal design offers a different approach to designing and planning by recognizing both genders and the vast array of different ages, abilities, cultures, and lifestyles that use buildings and actually exist in communities. Universal design challenges designers across the environmental design disciplines to unearth and dispel the narrow assumptions and stereotypes about users and citizens that have misguided these professions for so long. It demands proactive involvement because policies and practices that are gender-, age-, race-, class-, and ability-neutral have historically perpetuated unequal access to housing, public transit, neighborhoods, and workplaces.

69.6   CHALLENGING DUALISTIC THINKING

Universal design reminds us that there is no separation between mind and body, and between people and their environments. Universal design begins with the insight that the environment is not somehow a separate realm that exists beyond the ordinariness of everyday life. All of us—irrespective of gender, race, class, age, size, or ability—develop, grow, and change physically, emotionally, and intellectually throughout our lives. And at any point in our lives, personal self-esteem, identity, and well-being are deeply affected by the ability to function in our physical surroundings with a sense of comfort, independence, and control.

For the burgeoning numbers of older people and people with disabilities, these relationships are especially critical to life quality. In the 75 years from 1950 to 2025, the world's population of people who are 60 years old and over will have increased from 200 million to 1.2

billion, and from 8 to 14 percent of the total global population. In the same 75 years, the very old, those aged 80 and above, will have grown from 13 million to 137 million. In short, between 1950 and 2025, the total world population will have grown by approximately a factor of 3, the elderly by a factor of 6, and the very old by a factor of 10 (United Nations Department of Public Information, 1994).

In the "age of aging," improved survival rates will cause a dramatic increase in the numbers of people with age-related disabilities, such as dementia, mobility, hearing, and vision impairments. Current United Nations estimates are in the range of 500 million, 80 percent in the developing countries (N'Dow, 1995). Because the global aging population is soaring at an unprecedented rate, the demand for universally designed products and living arrangements will accelerate concomitantly in the coming decades.

69.7 RACISM AND ENVIRONMENTAL RISK

Universal design recognizes that there is no separation between human health, environmental health, and social justice. Extensive research has now fully documented that poor people and people of color have borne a disproportionate burden of environmental pollution and degradation. The geography of racism concentrates minority populations in areas where risks from industrialization are often extreme. For example, in the United States, the three super-fund toxic waste sites accounting for more than 40 percent of the nation's total disposal capacity are close to or located in predominantly African-American and Hispanic neighborhoods. Consequently, these populations have higher instances of blood lead levels in children, respiratory disease, and certain cancers than does the population as a whole. Understandably, the primary environmental concern in low-income communities and communities of color is urban public health and safety, in contrast to the conservation of wildlife and the preservation of nature that concerns the white-dominated environmental movement (Seager, 1993).

69.8 RETHINKING DISABILITY

Since the origin of the universal design movement resides in the pioneering actions and achievements of the disability community, it stands to reason that thinking and designing has emphasized ways of providing access and accommodation for people with disabilities, as well as older people who typically experience varying degrees of loss in physical, sensory, or cognitive ability as a normal part of aging. Feminist scholars have shown how knowledge is always of a particular group; how each of us always constructs, sees, and understands the world through our own culture's eyes and thinks within its assumptions (Ahrentzen, 1996). What, then, would happen if disability was defined from the different life experiences of homeless people; city dwellers who are bombarded by pollution and deteriorated infrastructure; women who live with the relentless fear of being assaulted in public places; children who are forced to play in streets with dangerous traffic; and people who are ghettoized by their poverty or race into deplorably substandard public housing projects and neighborhoods located near toxic waste sites?

How would the differing points of view of these groups inform and expand our understanding of spatial discrimination and exclusion? How might these insights help the proponents of universal design to shape plans for urban life that will enable all people to associate freely with others; to assume a responsible role in a free society; and to live long, healthy, productive, and meaningful lives? The following examination of the architectural, planning, environmental, and human consequences of substandard housing and infrastructure, unhealthy cities and buildings, and inadequate public transit and safety will begin to direct us toward some of the speculative answers described later in this chapter.

69.9 SUBSTANDARD HOUSING AND INFRASTRUCTURE

Housing shortages and poor housing conditions are literally life-threatening problems. Overcrowding in slums, shanty towns, refugee camps, and public housing projects worldwide causes transmission of infectious and parasitic diseases, psychological stress, exposure to air and water pollution, accidents, violence, and exposure to extremes in temperature. Substandard housing triggers asthma attacks in children by exposing them to irritants such as smoke, dust mites, molds, rats, cockroaches, and dry heat or no heat (McLead, 1998).

Although lead-based paints have been banned from residential use in the United States since 1978, the U.S. Department of Housing and Urban Development estimates that half of the country's housing still contains some lead-based paint, with some 3.8 million units identified as priority lead hazards because of the presence of deteriorating paint or high levels of lead dust in the air created by renovation work or from friction created from opening and closing windows. According to the Centers for Disease Control, about 1 million children in the United States between the ages of 1 and 5 have damaged brain and neurological function from inhaling or ingesting lead-based paint.

Accidents associated with substandard housing cause some 5000 deaths annually in the United States. Some 54,000 Americans are admitted to the hospital each year from burns and fires alone, caused by tap water temperatures, and home heating burns and fires from wood stoves or kerosene heaters when oil or gas is too expensive. Because the poor who live in cold climates must make the frequent decision whether to eat or heat their homes, malnutrition in children is a common plight among low-income families who are forced to spend large amounts of their income on rent and utility bills, leaving little money left for food (Free, 1998). According to the U.S. National Center for Health Statistics, roughly 60,000 lives are lost each year due to problems associated with cold weather, including fires, carbon monoxide poisoning, pneumonia, influenza, and hypothermia (Colton, 1994).

The health problems related to inadequate housing and infrastructure are by no means unique to North America. Of the 4.4 billion people in developing countries, nearly three-fifths lack access to safe sewers, one-third have no access to clean water, one-quarter do not have adequate housing, and one-fifth have no access to modern health services of any kind. Of the estimated 2.7 million annual deaths from air pollution, 2.2 million are from indoor pollution—including smoke from dung and wood burned as fuel, which is more harmful than tobacco smoke. Eighty percent of the victims are rural poor in developing countries (WEDO, 1998). Substandard housing, unsafe water, and poor sanitation in densely populated cities are responsible for 10 million deaths worldwide each year. Waterborne diseases alone kill 4 million children annually (United Nations Conference on Human Settlements, 1996).

The problem is getting worse because housing cannot keep up with an exploding urban population, which will double from 2.4 billion in 1995 to 5 billion in 2025. Currently, the world's urban population is growing 2.5 times faster than the rural population. In 1996, about 500 million urban dwellers were homeless or lived in inadequate housing (United Nations Conference on Human Settlements, 1996). A disproportionate number are women and children who are among the poorest and most vulnerable of the world's population. According to United Nations statistics, 70 percent of the world's 1.3 billion poorest people are female (WEDO, 1996).

Worldwide poverty among women means that many can afford only limited infrastructural services such as pit latrines, public water hydrants, open drains, and unpaved roads. Lack of adequate sanitation increases women's health risks. Women's universally low wages mean that fewer housing units are affordable and that household income is frequently insufficient to meet eligibility requirements for subsidized housing. In many countries, women's legal standing denies them the right to own land, which means they cannot protect themselves and their children from domestic instability and violence or provide collateral to gain access to credit or capital.

An estimated one-third of the world's households are now headed by women; in parts of Africa and many urban areas, especially in Latin America, the figure is greater than 50 percent, and in the squalid refugee camps of Central America and the disgraceful public housing projects of North America, the figure exceeds 90 percent. Women are segregated everywhere in appallingly overcrowded, hazardous, unsanitary dwellings that lack basic facilities because they are too poor to live anywhere else (Weisman, 1992). Add to these oppressive life circumstances the human suffering and physical damage to dwellings and communities in the late 1990s caused by military conflicts in Kosovo and elsewhere, and by natural disasters like floods in Bangladesh, earthquakes in Turkey, and hurricanes in the Caribbean region and Eastern seaboard of the United States, and it is clear that in global terms, providing access to adequate housing and infrastructure for all people everywhere is among the biggest challenges facing the world in the twenty-first century.

69.10   UNHEALTHY CITIES AND BUILDINGS

The disabilities caused by substandard housing and infrastructure that poor people, people of color, women, and children experience worldwide are the results of gender, race, and class oppression and massive urbanization. At the beginning of the twentieth century, only 1 in 10 people lived in cities. Today, one-half of all humanity lives in urban areas, and by 2030, urban populations will be twice the size of rural populations, most in the developing world (N'Dow, 1995). "As cities grow in absolute size and proportional importance, they play a larger part in shaping the global environment. Urbanization puts enormous stress on natural habitats, and as cities grow, their internal ecology changes: the world's urban centers are increasingly hazardous environments in which to live" (Seager, 1993).

Worldwide, more than 1.1 billion people live in urban areas with unhealthy air. And while there are numerous sources of airborne pollution, in most urban areas motor vehicles have become the single largest source of local air pollution. Today, there are nearly 600 million cars and trucks in the world, and this number is growing by 35 million per year—more than one every second, 100,000 a day. Motor vehicles are responsible for nearly 50 percent of the emissions of smog precursors worldwide.

The main pollutants from automobile exhausts are carbon dioxide, carbon monoxide, nitrogen oxides, hydrocarbons, and lead. Identified health effects include pulmonary diseases like asthma and emphysema, cardiovascular diseases that lead to heart attacks and strokes, lung cancer, and mental retardation in children. Economic and environmental effects include damage to buildings, food crops, and forests.

Another major health and environmental problem related to automobile emissions is global warming. Every gallon of gasoline that is burned releases over 20 pounds of carbon dioxide gas that traps the energy that reaches the earth from the sun. According to experts, this so-called greenhouse effect will cause global temperatures to rise 2° to 6° Fahrenheit over the next century—changing agricultural zones, disease areas, and natural habitats, raising sea levels, and increasing the ferocity of extreme weather events such as hurricanes, droughts, and floods. Reducing the emissions of air pollutants and greenhouse gases is critical to protecting human health and the stability of the global climate we all live in and depend upon (Nichols, 1998).

Blame for much of the environmental damage occurring today, from climate destabilization and the destruction of forests and rivers, to air and water pollution, must be placed squarely at the doorsteps of modern buildings. As much as one-tenth of the global economy is dedicated to constructing, operating, and equipping buildings. In terms of materials, this economic activity uses one-sixth to one-half of the world's wood, minerals, water, and energy (Nichols, 1998). Other studies reveal that close to 70 percent of the sulfur oxides and some 50 percent of the carbon dioxide emissions that are causing global warming are produced through the generation of electricity used to power, heat, and cool our homes and offices, and the production processes of basic building materials such as steel, cement, lime, bricks, and aluminum. The construction industry also contributes to air pollution through emissions of dust, fibers, particles, and various toxic gases (Der-Petrossian, 1999).

Many buildings do harm on the inside as well, by subjecting us to unhealthy air or alienating physical environments, making us both less productive and less healthy than we are capable of being. Sick building syndrome is the result of ventilation systems that subject occupants to stale air for hours on end, or harbor and spread unhealthy molds. Sealed, climate-controlled buildings also trap volatile organic compounds (VOCs), particularly formaldehyde, that can seep out from adhesives and drying agents in furniture, paint, and carpets, often resulting in concentrations hundreds of times higher than those measured outside. According to the U.S. Environmental Protection Agency, sick building syndrome costs American society tens of billions of dollars each year in medical bills, absenteeism, and lost worker productivity (Roodman and Lenssen, 1994). Indoor air quality is inadequate in 30 percent of buildings around the world (Der-Petrossian, 1999).

Sustainable cities, healthy buildings, economic equality, social justice, and environmental protection are inseparable from each other. And they are also inseparable from universal design's mandate to create products, buildings, and communities that promote health and prevent illness.

69.11 INADEQUATE PUBLIC TRANSIT AND SAFETY

The unsustainable patterns of urban transportation each of us must deal with on a day-to-day basis, regardless of income or social status, have become more and more intolerable. Whether in a private car, public bus, or train, riding a bicycle, or simply walking, the time people around the world spend transporting themselves is longer, often up to 4 hours a day, and the psychological stress and health care costs are higher (Williams, 1998). For women, especially those who are older or have low incomes or disabilities, these costs are exacerbated by male-centric planning that ignores gender differences in the use of public transit in North America. A 1989 study by the Toronto Transit Commission showed that the majority of women in the city, some 66 percent, are *transit captives,* defined as people who do not have a driver's license, own a car, or have access to a car. In other words, captive riders are the opposite of *choice riders,* who are usually men who own or have access to a car but who choose to use public transit in order to save money or commuting time or to avoid the stress of traffic jams. This high proportion of women transit captives is based upon surveys of transit users by transit authorities. These surveys do not even measure the percentage of *deep captives*—people who never use the system and, instead, rely on friends and relatives for mobility—a group that is highly likely to include proportionately more women than men. Yet, despite these demographic data, most North American transit authorities, like Toronto, have assumed that a "fare is a fare and a rider is a rider" (Toronto Transit Commission, 1989; Wekerle et al., 1992).

Transportation planners, who are mostly men, typically fail to see and subsequently address the needs of countless wage-earning mothers who take their children to school or drop them off at a day care center en route to their jobs, and pick them up—along with the dry cleaning and groceries—on their way home at night. Neither do they consider the needs of women who depend on public transit to get from the inner city to their jobs as domestic workers in the suburbs, or to their doctors' offices, schools and training programs, social services, and the homes of friends and family members. Neither the fare structure nor the scheduling and routing of buses and trains supports women's travel patterns. Instead, public transit is overwhelmingly designed to reflect a male model of journey-to-work, suburb-to-city, and city-to-city commuting patterns during peak rush hours.

These same transit policies will have an increasingly adverse impact on the mobility of older people. Every 7 seconds, one of the 77 million baby boomers, born in the United States between 1945 and 1964, turns 50. The lives and identities of this generation of Americans are inextricably linked to the freedom and independence that comes with driving and owning automobiles. Research studies project that frailty, especially after the age of 74, and the diminished capacity in general of older people to drive safely, will cause the number of older adults killed annually in automobile accidents in the United States to rise from the 7000 fatalities that occurred in 1999 to 23,000 in 2030. According to transportation experts, "driving for aging boomers will become a major public safety, health and consumer issue in the next [21st] century" (Cobb and Coughlin, 1999).

No matter the age, everyone understands the high cost of losing mobility, and the subsequent loss of independence that leads to isolation, which hastens poor mental and physical health, which results in rapid decline and death. Perhaps the greatest challenge facing the transportation community today is to understand and respond to the changing mobility needs of not only the one in five people who are currently over 65, but the fastest-growing group over 85 years old. Present-day public paratransit, door-to-door service is unreliable, usually requires a 24-hour advanced reservation, and public funding is limited, so services are usually used by disabled people, although older people can qualify, and prioritized for medical and food shopping trips first. Although paratransit fares are generally comparable to regular transit, inflexible and inadequate service and poor communication about the limited services that actually are available leave many older adults who are unable or unwilling to drive stranded at home (Cobb and Coughlin, 1999).

In addition to appropriately designed, affordable transportation systems, safety is especially important in creating accessible public space for vulnerable groups like women, children, and older and disabled people. Women, especially if they are older or disabled, are far more likely to fear crime and sexual assault than men, since they are far more frequently the victims of such violence, and will restrict their own and their children's activities as a result. According to a Canadian study of urban victimization, 56 percent of women are afraid or very afraid of walking alone in their neighborhoods at night compared with 18 percent of men. The same study determined that 27 percent of women in Canada can expect to be sexually assaulted at some point in their lives, and that women are victims in 90 percent of the reported incidents of sexual assault (Toronto Transit Commission, 1989). A different national survey of Canadian women found that in 1993, one out of every two women in Canada had been physically or sexually assaulted at least once, and that the rate of incidence was significantly higher among women with disabilities—67 percent for women with physical disabilities and 98 percent for women with developmental disabilities (La France, 1999).

If the fear of sexual harassment on the streets causes women stress, the very realistic fear of rape and assault keeps women off the streets at night, away from public parks and dangerous parts of town, and unconsciously afraid of half the human race. Women work very hard to avoid sexual assault in public places by using hundreds of avoidance strategies. They vary their walking routes, avoid making eye contact and talking to strangers, refrain from taking evening courses, and take a taxi instead of a bus at night despite the higher costs. They refuse jobs that involve late hours or working night shifts, and they shun recreational activities like jogging, seeing a movie alone, or even visiting friends (Wekerle et al., 1992). Many urban elderly women, especially those with limited education who live alone in apartments, feel so vulnerable to the dangers of public life that they rarely leave their homes at all, especially at night. Fear undermines women's autonomy and well-being by making them anxious, apprehensive, and isolated, and it limits their mobility and ability to participate fully in public life.

City neighborhoods can also be threatening places for children, both physically and socially. Girls in particular are often confined to homes that cannot begin to meet the full range of their social and physical needs. In the deteriorated inner cities of the North and the crowded settlements of the South, children play in dangerous abandoned buildings and among piles of debris and open sewers, and take their chances with heavy traffic. In areas of political tension and violence, they face chronic anxiety, which can have long-term effects on their own social development (Bartlett, 1999). While universal design is not a substitute for systemic remedies to the social forces that isolate women, children, people who are disabled, and elderly people in their homes and place them at risk in the public realm, it can play a significant part in fostering or inhibiting their mobility and independence, and creating or preventing opportunities for assault.

69.12   *FROM ACCESS TO INCLUSION: IMAGINING THE UNIVERSALLY DESIGNED CITY*

While the focus on accessibility and aging that gave rise to the universal design movement in the last decade of the twentieth century has begun to produce significant innovation in communication technologies and product and graphic design—fields in which there are few or no regulations that define *bottom-line design standards*—at the larger scales of buildings and communities the perception of universal design still remains limited to code compliance and wheelchair access. *Compliance minimalism,* as noted by Ostroff in Chap. 1, "Universal Design: The New Paradigm," has frequently meant the addition of ramps, widened doors, and "handicapped" parking, approaches that have rarely produced visually inspiring architecture and public spaces. If the movement for universal design is to effectively generate impacts in the new century that are worthy of its life-sustaining potential, universal design educators and practitioners must go beyond the "letter of the law" to the "spirit of the law"—to shift their focus from redressing human and environmental problems through remedial design to preventing problems through holistic, equitable design.

If we were to define universal design as a values-based framework for design decision making, based on an ethic of inclusiveness and interdependence that values and celebrates human difference and acknowledges humanity's debt to the earth, how might this affect the way we design our neighborhoods and services, provide housing, and construct buildings? Three broad criteria will characterize and identify the universally designed city of the future: (1) establishing housing for all as a human right, (2) supporting human and environmental health, and (3) achieving safety and mobility in the public realm.

69.13   *ESTABLISHING HOUSING FOR ALL AS A HUMAN RIGHT*

The universally designed city will promote and support human diversity and social equity by providing a wide spectrum of housing options that enable people of different incomes, ages, cultures, races, lifestyles, and household types to live within a single neighborhood or district. One-race, one-income-level, or one-family-type housing markets and neighborhoods will be replaced with integrated and balanced living patterns. Amended zoning laws will allow for communal living in single-family neighborhoods so that those with lesser financial resources may pool together and have more choices about where and how they live. The legalization of accessory apartments in single-family houses will allow designers to create new affordable rental apartments in the suburbs in patterns and densities that support household diversity and environmentally responsible land use development by leveraging new housing units out of existing housing stock, instead of building new construction on increasingly scarce open land. The passage of affordable energy bills will guarantee the sustainability of overall shelter, because people will be able to both afford their mortgage or rental payment and their monthly utility bills (Colton, 1994).

Universally designed housing will be dynamic, not static, and changeable, adaptable, and recyclable over time in response to the changing needs of the people who use them. A household established for a family with young children will be able to be modified by the members of the family as they change and grow. Within the dwelling, the relationships and uses of various rooms will be nonspecific, lending themselves equally well to bedrooms, living rooms, workshops, or offices (Weisman, 1992). All homes will have electronic outlets high enough for older people to plug in appliances without bending down, and reinforcement in the lavatory walls near the commode and tub so that grab bars can be easily and inexpensively installed if an occupant needs them at some future time. In the universally designed city, all homes will be visitable by all people because they will feature level thresholds, zero-step entrances, a downstairs lavatory, and wider internal and external doors and corridors (Concrete Change,

1998). Persons with heart conditions or arthritis, or who are temporarily on crutches; those with baby carriages, shopping carts, or bicycles; and children and adults of short or tall stature will benefit equally with persons whose mobility is affected from birth or by accident, illness, or aging in terms of comfort, convenience, and access to friends and family members.

Cities cannot be inclusive without the guarantee of housing rights for all people. In the universally designed city, there will be no gated communities or room for intolerance of any kind. Housing rights will be established and upheld through the full participation of all social and economic groups, supported by local, municipal, and national governments working to increase the stock of affordable, centrally located land that can be accessed by poor and homeless people and other historically marginalized groups. Universally designed cities around the world will loudly and proudly proclaim themselves as "housing rights cities." They will publicly assert their intention to fully comply with the housing rights obligations incumbent on their nation that will be defined in the United Nations Universal Declaration of Human Rights, in which every man, woman, and child will be guaranteed adequate housing, not as a matter of need, or as a "gift," but as an entitlement, a right, and a matter of simple human decency and dignity (Leckie, 1998).

69.14 SUPPORTING HUMAN AND ENVIRONMENTAL HEALTH

The universally designed city will promote and support human and environmental health by reconfiguring suburban sprawl and dependence on automobiles, and by replacing dangerous, substandard housing with adequate, available, and affordable housing and infrastructure. In the universally designed city there will be no environmental racism. The most vulnerable citizens will get the services they need to prevent health problems. A grassroots neighborhood wellness corps will coordinate with other community-based organizations to improve the physical environment to prevent environmentally based health problems like air pollution, water contamination, exposure to lead-based paint and asbestos, toxic dumping, and sick building syndrome. A neighborhood wellness center will provide preventative medical care for all residents. Public buildings and private dwellings will be accessible for people with environmental sensitivities, because they will be built with nontoxic building materials, products, and technologies.

Architects and the construction industry will improve energy efficiency and reduce pollution by specifying local materials that are not dependent on limited, nonrenewable resources, by using local transport, by increasing the use of recycled and agricultural and industrial waste materials in the production of building products, and by incorporating insulation systems that enhance the thermal efficiency of buildings. Physical structures and their sites will be designed to provide flexibility in their uses: A church might also serve as a public auditorium, the deck of a parking garage as a concert space on summer nights. As the needs of the community change, buildings will be adapted accordingly; a warehouse might be converted to a shopping mall or apartments, a single-family house to a collective home. Architecture will be responsive and resistant to the effects of local climate and sustain the economy of local communities by employing the skills and labor of local people.

In the universally designed city, older and lower-income adults will form income-generating neighborhood recycling teams that will transform refuse into cash by collecting, sorting, and selling household and commercial waste and used magazines and newspapers to local recycling centers. Every neighborhood will contain a universal garden park, designed and maintained by a local universal garden society, which will welcome all people and accommodate their different backgrounds, abilities, and experiences. The universal garden will provide connections with nature; healing; a context for horticultural therapy; a break from stress, recreation, gardening, picnics; places for children and adolescents to play in safety; and opportunities for neighbors to meet and work together (Bailey, 1999).

Universal design will transform sprawling suburbs into communities of real and diverse neighborhoods that will conserve the natural environment and be compact, pedestrian friendly, and mixed use. Many activities of daily living will occur within walking distance, allowing independence for those who do not drive. Streets and public places will be comfortable and safe and encourage walking and biking, and enable neighbors to know each other. Public transit, pedestrian, and bicycle systems will maximize access and mobility for people of all ages, abilities, incomes, and lifestyles throughout a region while reducing dependence upon the automobile, thereby reducing pollution from auto emissions and reducing stress from traffic jams and long commutes.

69.15 ACHIEVING SAFETY AND MOBILITY IN THE PUBLIC REALM

The universally designed city will promote and support independent mobility and safe access to public buildings and places. Neighborhoods will be well lit and both feel and be safe. In the long run, fear of crime and violence destroys one of the most important elements necessary for inclusive communities, the assumption that strangers on the street are potential allies, not assailants. Safety audits and walkabouts organized by local governments and police forces will be conducted regularly by diverse user groups—especially women, children, people who are disabled, and older people—to review and improve accessibility, standards of lighting, signage, and security in public parking garages, subways and train systems, bus routes, parks, streets and sidewalks, and buildings.

The universally designed city will provide transportation systems that are safe, accessible, and affordable and that are scheduled and routed to support a wide range of different travel patterns. Because bus stops located in isolated areas are frightening places, transit authorities will permit drivers to do demand stops at night and on routes where there have been sexual assaults. Share-a-fare programs will transport those without transportation to jobs, health services, clinics and social services, schools, and training programs. For people who still cannot use the improved mass transit system, automobile access will be provided by a subsidized paratransit system that will offer demand-response service 24 hours a day, 7 days a week.

The universally designed city will be legible. Wayfinding will be easily accomplished regardless of age, culture, language skills, and physical abilities. Movement throughout public buildings and places and access to intelligent transportation systems will be facilitated by large-type signage and multisensory and multilingual information. All members of our larger human family will feel welcomed in the universally designed city because our "civic home" will provide clean and safe toilet facilities, water for washing and drinking, and places where children can rest without fear of being turned away as an essential and regular feature of collective public life.

69.16 BUILDING ON HISTORIC FOUNDATIONS

The commitment to create safe, healthy, socially inclusive, ecologically sustainable buildings and communities is by no means unique to the proponents of universal design. Though the strategies for achieving these goals vary, they are at least partially found in numerous experiments, including the late-nineteenth- and early-twentieth-century garden cities and new towns that were planned in England; the socialist utopian communities that were established during the same era in Europe and North America; and in the United States at Radburn, located in Fairlawn, New Jersey, designed in 1929 as a comprehensively planned suburban community that would foster neighborly involvement and protect children from the dangers of the increasingly omnipresent automobile (Weisman, 1992). These sentiments and concerns are expressed in the contemporary manifesto of the Congress for the New Urbanism embodied in two built-neighborhood developments in Florida: Seaside, and Celebration Village owned by the Disney Corporation; and in the master planning documents of countless cities all across North America. Many of these aspirations were articulated by the solar and advocacy architects of the 1960s; the feminist architects, planners, and environmental researchers that emerged in the 1970s and 1980s; and are claimed today by a growing number of socially and environmentally responsible architects, interior designers, landscape architects, and planners.

No single scheme, be it realized or imagined, past or present, comprehensively addresses the problems and possibilities described in this chapter; but the one that comes the closest is the radical, visionary blueprint for designing a feminist neighborhood that architect and scholar Dolores Hayden proposed in her pioneering essay, "What Would a Non-Sexist City Be Like? Speculations on Housing, Urban Design, and Human Work" (Hayden, 1980). Inspired by the Radburn model, in which the designers Clarence Stein and Henry Wright flipped the traditional gridiron block inside out so that private houses faced inward toward shared park land, Hayden prioritized "recycling" existing antisocial, automobile-dependent, energy-consumptive suburban neighborhoods designed for male-headed nuclear families, over building new developments. In her plans, she joined together individual backyards to form public commons and gardens; side yards and front lawns became fenced, private spaces; and private garages and sheds were converted to shared public garages with a dial-a-ride van, laundries, day care facilities, community centers, kitchens serving take-out meals and meals-on-wheels to homebound and elderly people, garden plots for growing food, a grocery and food cooperative, a home help office, and rental apartments. These services would be run as businesses for both customers and members of the community and, theoretically, create some 37 jobs for residents.

Hayden envisioned that her project would be implemented by homemakers' organizations made up of 40 households of women and men, thus the acronym HOMES (standing for *Homemakers Organization for a More Egalitarian Society*). Each HOMES group represented the composition of American households in 1980—15 percent single parents and their children, 40 percent two-worker couples and their children, 35 percent one-worker couples and their children, and 10 percent single residents—adding up to a total population of 69 adults and 64 children. Her nonprofit, cooperative plans were designed to relieve employed women from the sole responsibility for domestic life and child rearing (Hayden, 1980).

Twenty years later, the existing for-profit traditional neighborhood development (TND) of Celebration Village, designed largely according to the paradigm of the "new urbanism," manages to avoid the issues of racial and economic diversity and gender equity raised by Dolores Hayden. On the positive side, Celebration Village does include a school, recreation, convenience shopping, and work within a short walk of homes, single-family houses that vary in size and price, rental apartments and small studios for live-in help above the garages, small lots that create higher densities and more shared open space, and curving streets designed to slow traffic and create safety for children. However, "it does so in the context of a rigid, controlled social machine of the type satirized in the movie *The Truman Show*—which may be satisfying to most of Celebration's current inhabitants but can hardly be taken as a model for the long term or on a wide scale in a nation founded on the right to cranky individualism" (Torre, 1999).

In marked contrast to Celebration Village and its corporate developers, architects Kathryn McCamant and Charles Durrett have been successfully promoting *cohousing* in the United States since the late 1980s. Derived from the Danish *bofaelleskab*, which literally means *livingtogetherness*, cohousing communities in both Denmark and the United States are organized, planned, designed, and managed by the residents, in collaboration with architects and technical advisors, to anticipate changes in family structure and to support the lifestyles of couples, nuclear families, singles, single parents, and retirees (Weisman, 1992). Each household has a private residence, but also shares extensive common facilities with the larger group, such as a kitchen and dining hall, children's playrooms, workshops, guestrooms, and laundry facilities. Although individual dwellings are designed to be self-sufficient and each has its own kitchen, the common facilities, particularly common dinners, are an important aspect of community life both for social and practical reasons (McCamant and Durrett, 1988).

In Sweden, the *servicehus,* or family hotel, established in 1907 by the Swedish government to support workers and their families, provides residents with the benefits of cohousing but also offers housecleaning services, child care, maternity and well-baby clinics, medical care for elderly people, errand running for the sick, hobby rooms, youth clubs, plant watering during vacations, and gymnasia (Weisman, 1992). In Scandinavia and the Netherlands, housing arrangements for older people are frequently designed to extend their services and amenities, such as restaurants, bars, swimming pools, and meetings rooms, to people from the surrounding community (Regnier, 1994).

In both Europe and the United States, cohousing developments have been built in cities and suburbs and range in size from 6 to 80 households (McCamant and Durrett, 1988). The first sustainably designed cohousing development in the United States, Winslow Cohousing, consists of 30 households of 80 people, 30 of them children. Like Radburn, Winslow is pedestrian based and children play safely in shared parks and gardens.

The desire to create a sustainable neighborhood that balances architecture and nature and the autonomy of private dwellings with the advantages of community living was also central to the design of Village Homes, built in Davis, California. Architect/developer Michael Corbett laid out the streets primarily in an east-west direction and building lots in a north-south direction, so that most of the windows could be located on the south side of the houses to capture the warmth of the winter sun, while overhangs would keep out summer sun to avoid excessive cooling loads. The streets were made atypically narrow so that they would be easier to shade with trees that would keep the microclimate very comfortable. Corbett broke so many conventional rules—like building in higher densities on smaller lots to preserve more shared green space, including a community garden that produces food, and using natural creek ways and surface drainage, that it took 3 years to gain the approvals and financing needed to build the project. Today, Village Homes is an overwhelming success: Property values have risen dramatically; no one wants to move; and there is a long waiting list to buy into the development (Danitz and Zelov, 1994).

At the urban scale, Jaime Lerner, an accomplished architect and three-time mayor, involved all the citizens of Curitiba in the transformation of their city into the ecological capital of Brazil by reducing the use of private cars and separating garbage. Curitiba's inexpensive transit system has been made so efficient that it has cut down on pollution, traffic congestion, and energy consumption. By increasing the frequency of service, designating exclusive bus lanes, using articulated buses with drop-down boarding ramps instead of stairs, and making boarding and disembarking quick, painless, and universally accessible with elevated boarding "tubes" that serve as glass shelters at bus stops, the buses have become an enjoyable and much more affordable alternative to cars. Both Steinfeld (in Chap. 24, "Universal Design in Mass Transportation") and Guimarães (in Chap. 57, "Universal Design Evaluation in Brazil: Development of Rating Scales") discuss Curitiba's advanced system.

City planners from around the world have gone to Curitiba for advice on how to relieve their own traffic problems and to study Curitiba's "Garbage That Is Not Garbage" program. In 1989, Curitiba held a referendum for a voluntary program of recycling that would involve the entire population of 1.5 million, especially the poorer communities, after they had emphatically rejected a mandatory recycling program. To provide an incentive for people that they could understand, Mayor Lerner announced that the city would pay for bags of recyclable trash with bags of food or transit tokens. Organized through existing neighborhood associations, within months, 120,000 people from 55 communities were participating in the incentive programs. While the city collects 23 tons of recyclables a day, "private entrepreneurs" are collecting over 100 tons (Danitz and Zelov, 1994).

69.17 *MOVING THE DESIGN AGENDA FORWARD IN THE TWENTY-FIRST CENTURY*

These and countless unnamed examples of innovative historic and contemporary urban/suburban design affirm the fact that remapping the geography of our established social order, while certainly possible, is incredibly complex and slow. The ideas for transforming the public realm set forth in this chapter are not just utopian fantasies; some already exist, while others have yet to be actualized but may prove to be survival imperatives. Who could have imagined at the turn of the last century looking out of the window of an aircraft flying across the United States at night? "Streetlights have spread like galaxies across this continent, reaching far out into farmland and desert, claiming rural spaces that once seemed infinite. We are spending our nation's capital in sprawl, with potentially dire consequences for our children and ourselves" (Ivy, 1999). Who among us can imagine with certainty what will be 100 years from now? One thing is certain, we will not create inclusive spatial patterns of living and working until society values those aspects of human experience that have been devalued through the marginalization of women, children, poor, disabled, and older people, gay people, and people of color. Because our cities and buildings embody our common humanity, because they are a record of deeds done by those who have had the power to build, they will not be easily changed until the society that produced them is changed. A democratic and sustainable vision for living can only be realized if it is shared by the majority of people and chosen as a preferred way of life.

So what does the movement for universal design bring to historic and ongoing efforts to create equitable, sustainable environments? Universal design can have a unique and powerfully positive impact in two ways. First, universal design can elevate design and its value in the public consciousness. By improving the person-to-product or person-to-environment interface, universal design enhances comfort, health, safety, and human performance. The tangible results in a workplace, for example, might include a reduction in human error, which impacts risk management, increases in productivity, reductions in sick days and job turnover, and decreased insurance claims (see Chap. 45, "Office and Workplace Design," by Mueller). By promoting the benefits of universal designing to companies, their employees, and the general public, we are also promoting the importance of good design and the value that designers bring to their clients and to society.

Proponents of universal design are particularly well positioned to convince the public and those who make decisions about what actually gets built—the developers, city managers and mayors, real estate industry, corporations, and financial institutions—that design expertise should be used as a tool to create rather than respond to public policy, legal regulations, and building codes. Further, unhealthy and ugly buildings cost society greatly, both economically and in terms of our social existence, which is increasingly characterized by so much violence that gated communities; security devices in our cars, homes, and workplaces; and restrictions on our ability to walk in public parks and on city streets have become a way of life.

Second, universal design education and practice has the potential to address social, cultural, and ecological sustainability as inseparable from tectonics and technologies, and to foster the practice of high ethical standards. While not the rule in landscape architecture and urban planning, traditional architectural praxis has often assumed that form making is an autonomous act that is divorced from the notion of social responsibility and civic engagement. Universal designers resolve the inherently false dichotomy between these two opposing ideologies—architecture as a formal art versus architecture as a social art—by embracing creative work that is equally dedicated to both the making of formally beautiful products, buildings, and communities, and the making of a better, more just society through addressing human and environmental problems and concerns. Universal design fosters ethical behavior among design professionals because it fundamentally recognizes that when we design, we affect the lives of others by our decisions, and that we are therefore responsible and accountable to the people who will be affected by what we do.

69.18 CONCLUSION

Ending the spread of placeless sprawl, the loss of wilderness and rural landscapes, and the increasing separation by race and income; restoring deteriorated and unhealthy cities, the fights to end environmental racism and gender discrimination, and working to create an environment in which the talents of children, and disabled and older people can find full expression and their needs be met, are all interrelated community building challenges. While physical design solutions alone will not solve the social and economic problems of the twenty-first century, community stability, economic vitality, and environmental health cannot be sustained without a coherent and supportive physical framework (Congress for the New Urbanism, undated).

If we are to design a society in which all people and all living things matter, in this new century we will have to move beyond the politics of human and environmental exploitation that defined the twentieth century. Universal design provides us with important guidelines for doing so by recognizing the interdependence among all of humanity, the natural world, and the products of human design, including the built and planned environment, and by teaching us to think, to act, and to design out of that recognition and understanding.

69.19 BIBLIOGRAPHY

Ahrentzen, Sherry, "The F Word in Architecture: Feminist Analyses in/of/for Architecture," in Thomas A. Dutton and Lian Hurst Mann (eds.), *Reconstructing Architecture: Critical Discourses and Social Practices,* University of Minnesota Press, Minneapolis, MN, 1996, pp. 86–86.

Bailey, Ann, "A Universal Garden for Vancouver," *Accessible BC* (Summer 1999): 1, 1999.

Bartlett, Sheridan, "How Urban Design Can Support Children's Rights," *Habitat Debate* 5(2): 18, 1999.

Cobb, Roger W., and Joseph F. Coughlin, "Keeping Baby Boomer Mobile," *Transition,* 4(Summer 1999): 7–9, 32–33, 1999.

Colton, Roger D., "Energy Policy Hurts the Poor," *Shelterforce,* 76(July/August 1994): 9, 1994.

Concrete Change, *An International Effort to Make All Homes Visitable,* 1998; Web site: http://concretechange.home.mindspring.com.

Congress for the New Urbanism, *History and Membership,* brochure, undated.

Danitz, Brian, and Chris Zelov, "Ecological Design, Inventing the Future," San Luis Video Publishing, Los Osos, CA, 1994.

Der-Petrossian, Baris, "Conflicts Between the Construction Industry and the Environment," *Habitat Debate,* 5(2): 8, 1999.

Free, Karen, "Millions Are Worried Sick About Housing," *Habitat World* (August/September 1998): 7–8, 1998.

Hayden, Dolores, "What Would a Non-Sexist City Be Like? Speculations on Housing, Urban Design, and Human Work," *Signs: Journal of Women in Culture and Society,* Supplement, 5(3): S170–S187, 1980.

———, *Redesigning the American Dream: The Future of Housing, Work, and Family Life,* Norton, New York, 1984.

Ivy, Robert, "Standing Up to Sprawl," editorial, *Architectural Record,* 188(1): 15, 1999.

La France, Cathy, "Cowichan Valley Safer Futures Project," *Accessible BC,* (Summer 1999): 4, 1999.

Leckie, Scott, "Cities Cannot Be Inclusive Without Housing Rights," *Habitat Debate,* 4(4): 22–23, 1998.

McCamant, Kathryn, and Charles Durrett, *Cohousing, A Contemporary Approach to Housing Ourselves,* Habitat Press, Berkeley, CA, 1988, p. 10.

McLead, Milana, "Remedy for Ill Health: Decent Shelter," *Habitat World* (August/September 1998): 2, 1998.

N'Dow, Wally, "Address by Dr. Wally N'Dow, Secretary-General of the United Nations Conference on Human Settlements," delivered at the International Council for Caring Communities Conference, 10 January 1995, United Nations, New York, 1995, p. 2.

Nichols, Matthew, "Driving Ourselves to Death," *Habitat Debate*, 4(2): 6–7, 1998.

Regnier, Victor A., *Assisted Living Housing for the Elderly, Design Innovations from the United States and Europe,* Van Nostrand Reinhold, New York, 1994, pp. 62–63.

Roodman, David Malin, and Nicholas Lenssen, "Our Buildings, Ourselves," *World Watch* (November/December 1994): 22–24, 1994.

Seager, Joni, *Earth Follies,* Routledge, New York, 1993, pp. 183–184.

Story, Molly Follette, "Is it Universal?" *Innovation* (Spring 1997): 30, 1997.

Toronto Transit Commission, *Moving Forward: Making Transit Safer for Women,* Toronto Transit Commission, Toronto, Ontario, Canada, 1989, p. 19.

Torre, Susana, "Expanding the Urban Design Agenda," in Joan Rothschild (ed.), *Design and Feminism, Re-Visioning Spaces, Places, and Everyday Things,* Rutgers University Press, New Brunswick, NJ, 1999, pp. 37–38.

United Nations Conference on Human Settlements, *Global Report on Human Settlements Reveal: 500 Million Homeless or Poorly Housed in Cities Worldwide,* paper published by the United Nations Conference on Human Settlements, Istanbul, Turkey, 3–14 June 1996, p. 1.

United Nations Department of Public Information, *A New Age for Old Age,* paper published by the United Nations Department of Public Information, New York, 1994.

Weisman, Leslie Kanes, *Discrimination by Design: A Feminist Critique of the Man-Made Environment,* University of Illinois Press, Urbana, IL, 1992, pp. 138, 148–154, 155, 177.

Wekerle, Gerda R., and Carolyn Whitzman, *Safe Cities: Guidelines for Planning, Design, and Management,* Van Nostrand Reinhold, New York, 1995.

———, and City of Toronto Planning and Development Staff, *A Working Guide for Planning and Designing Safer Urban Environments,* Safe City Committee of the City of Toronto Planning and Development Department, Toronto, Ontario, Canada, 1992, p. 2.

Williams, Brian, "The Missing Link: Towards Sustainable Urban Transport," *Habitat Debate,* 4(2): 1, 1998.

Women's Environment & Development Organization (WEDO), "Overview: Habitat II Women in the Cities," *News and Views,* 9(1–2) (June-July 1996): 1, 1996.

———, "WEDO Factsheet," *News and Views* (November 1998): 11, 1998.

INDEX

FIGURE CREDITS

The publisher and the editors of this book have made every attempt to acquire permissions to reproduce all artwork that appears in this book. If a specific figure is not credited, it is assumed to be the sole property of the contributing author. In the event that proper credit is not given at this time, it will be included in a future edition of this book.

Chapter 4
Figure 4.2c *Source: Nick Hayes.*
Figure 4.4 *Source: Susan Hewer.*
Figure 4.5a,b *Source: Susan Hewer.*
Figure 4.5c *Susanna Berger-Steele.*
Figure 4.10 *Source: Alastair Macdonald.*
Figures 4.11 through 4.13 *Source: IDEO.*
Figure 4.14 *Source: Jan Teunen.*

Chapter 9
Figures 9.2 and 9.4 through 9.8 *Source: Graphics courtesy of M. A. Borger Design.*

Chapter 10
Figure 10.2 *Source:* Images of Universal Design Excellence, *John C. Gustavsen, Designer.*
Figure 10.3 *Source: Robb Hornson, ASLA, Little & Little, Landscape Architecture.*

Chapter 15
Figures 15.1 and 15.5 *Source: ITEA, 1999.*
Figure 15.2 *Source: Galli, 1995.*
Figure 15.3 *Source: CNR ICITE, 1991.*
Figure 15.4 *Source: Ministero Affari Esteri et al., 1992.*

Chapter 16
Figures 16.3 *Source: M. Boreskie Architect, Inc.*

Chapter 18
Figure 18.1 *Source: U.S. Department of Justice: U.S. v. Ellerbe Becket, Inc. (1998).*

Chapter 22
Figures 22.7 and 22.8 *Source: Evergreen Retirement Community.*
Figure 22.9 *Source: Center for Universal Design.*

Chapter 27
Figure 27.1 *Source: From Le Corbusier, 1995.*
Figure 27.10 *Architects: Grosbois and Sautet.*
Figure 27.11 *Architects: Chemetov and Huidobro.*

Chapter 28
Figure 28.1 *Architect: Rem Koolhaas.*
Figure 28.4 *Architect: Piet Blom.*
Figure 28.13 *Architect: Wiel Arets.*

Chapter 32
Figures 32.1 through 32.4 *Source: Original illustrations courtesy of Ellen P. S. Sasiang.*

Chapter 33
Figure 33.1 *Source: Concept drawing by E. Siré.*

Chapter 36
Figure 36.9 *Source: ©Generations Architecture Planning Research.*

Chapter 38
Figure 38.1 *Source: Reproduced with kind permission from T. Redway at A. V. Jennings Ltd., South Australia.*

Chapter 39
Figures 39.1 through 39.5 and 39.7 *Source: Photos courtesy of HEWI, Inc., Germany.*

Chapter 41
Figure 41.5 *Source: GE Appliances.*
Figure 41.6 and 41.7 *Source: Rhode Island School of Design.*

Chapter 43
Figures 43.1, 43.4, and 43.5 *Source: Courtesy of Brian Rose.*
Figures 43.2 and 43.3 *Source: Courtesy of Goody, Clancy & Associates.*
Figures 43.6 and 43.8 *Source: Courtesy of Kalman McKinnel and Wood.*
Figure 43.7 *Source: Courtesy of Steve Rosenthal.*
Figures 43.9 through 43.11 *Source: Courtesy of HOK.*

Chapter 45
Figure 45.1 *Source: Courtesy of the Job Development Lab.*
Figure 45.2 and 45.3 *Source: The Computer/Electronic Accommodations Program (CAP), U.S. Department of Defense, 1998 (public domain).*
Figure 45.4 *Source: Access to the Environment: A Review of Literature, HUD, 1997 (public domain).*

Figure 45.5 *Source: From the:* Workplace Workbook 2.0.
Figure 45.6 *Source: Herman Miller, Inc.*

Chapter 46
Figure 46.2 *Source: Dull, Olson, and Weekes, 1999.*
Figure 46.3 *Source: Wilmsen Endicott and Unthank, AIA, 1970.*
Figure 46.6 *Source: Cameron McCarthy Gilbert and Royston Hanamoto Alley and Abey, 1998.*
Figure 46.8 *Source: SRG Partnership, 2000.*
Figure 46.9 *Source: YGH Architects, 1999.*
Figure 46.10 *Source: Dull, Olson, and Weekes, 1999.*
Figure 46.11 *Source: TBG Architects, 1999.*

Chapter 50
Figure 50.1 *Source: Ford Motor Company.*
Figure 50.2 *Source: Lear Corporation (Web site).*
Figure 50.3 *Source: Cadillac Division, General Motors Corporation (Web site).*
Figure 50.4 *Source: The Hertz Corporation (Web site).*
Figure 50.5 *Source: Siemens (Web site).*

Chapter 51
Figure 51.2 *Source: Lifchez, 1987.*
Figure 51.5 *Source: Adaptive Environments.*
Figure 51.6 *Source: Jones.*
Figure 51.7 *Source: Student Project, University of Oregon.*
Figure 51.9 *Source: Adaptive Environments.*
Figures 51.11 and 51.12 *Source: Jones.*
Figure 51.13 *Source: Welch, 1995.*

Chapter 56
Figure 56.1 *Source: Cooper-Hewitt, National Design Museum, S.I. Photo, Andrew Garn.*

Chapter 60
Figure 60.1 *Source: Courtesy of the Boston Redevelopment Authority.*
Figure 60.4 *Source: Courtesy of the Seaport Hotel, Boston.*

Chapter 65
Figure 65.1 *Source: U.S. Census Bureau Report on Americans with Disabilities: 1994–1995 (August 1997).*
Figure 65.2 *Source: Based on data from LaPlante, 1998.*